VOLUME 1

# AMERICAN BEETLES

## Archostemata, Myxophaga, Adephaga, Polyphaga: Staphyliniformia

VOLUME 1

# AMERICAN BEETLES

## Archostemata, Myxophaga, Adephaga, Polyphaga: Staphyliniformia

The late Ross H. Arnett, Jr.
and
Michael C. Thomas

CRC Press
Taylor & Francis Group
Boca Raton London New York

CRC Press is an imprint of the
Taylor & Francis Group, an **informa** business

**Cover photographs courtesy of Michael C. Thomas**

CRC Press
Taylor & Francis Group
6000 Broken Sound Parkway NW, Suite 300
Boca Raton, FL 33487-2742

First issued in hardback 2017

CRC Press is an imprint of Taylor & Francis Group, an Informa business

**Library of Congress Cataloging-in-Publication Data**

American beetles / edited by Ross H. Arnett and Michael C. Thomas.
p. cm.
Contents: v. 1. Archostemata, Myxophaga, Adephaga, Polyphaga: Staphyliniformia.
Includes bibliographical references (p.).
ISBN 0-8493-1925-0 (alk. paper : v. 1)
1. Beetles — North America. I. Arnett, Ross H. II. Thomas, M. C. (Michael Charles), 1948–
QL581 .A43 2000
595.76′097 — dc21 00-050809
CIP


Library of Congress Card Number 00-050809

ISBN 13: 978-1-138-40709-1 (hbk)
ISBN 13: 978-0-8493-1925-9 (pbk)

To Ross H. Arnett, Jr.
1919-1999

# Preface

It has been nearly 40 years since Ross H. Arnett, Jr. published the first fascicle of *The Beetles of the United States: A Manual for Identification.* It quickly became an indispensable tool for professional and amateur coleopterists, general entomologists, and naturalists. Although there were four additional printings it has long been out of print and difficult to obtain. It was prepared to replace Bradley's *A Manual of the Genera of Beetles of America, North of Mexico*, which itself was some 30 years out of date in 1960. *American Beetles* is, in turn, designed to replace *The Beetles of the United States*. It is hoped that it will prove to be as useful as its predecessor.

Ironically, much of the preface to the original edition applies today as well as it did 40 years ago:

> Many genera have since been described and reported within the area concerned, and many families have been revised. Extensive changes have been made in the family classification of the beetles of the United States during this period.
>
> The aim of this series of fascicles is to provide a tool for the identification of adult beetles of the United States to family and genus with the aid of illustrations, keys, descriptions, and references to sources for keys and descriptions of the species of this area. All of the genera known to inhabit this area are included in the keys and lists of genera which follow.

The design and format of this work follow closely that of the original edition, but the way it was put together was quite different. Its predecessor was very much the work of one man, Ross H. Arnett, Jr. With a few exceptions (George Ball wrote the carabid treatment for both the 1960 edition and for this one), Dr. Arnett wrote the family treatments of *The Beetles of the United States*. Many specialists reviewed those chapters, but they were almost entirely Dr. Arnett's work.

When Dr. Arnett announced plans to prepare a work to replace *The Beetles of the United States*, coleopterists literally lined up to volunteer their time and expertise in preparing the family treatments. Ultimately, more than 60 coleopterists participated in the preparation of *American Beetles*. This has truly been a community project.

Due to the size of the ensuing work, *American Beetles* is being printed in two volumes. Volume 1 includes the introductory material, and family treatments for the Archostemata, Myxophaga, Adephaga, and Polyphaga: Staphyliniformia. The remainder of the Polyphaga and the keys to families will appear in Volume 2.

Sadly, although Dr. Arnett initiated this project and was instrumental in its planning, he did not live to see its fruition. He became seriously ill in late 1998 and died on July 16, 1999 at the age of 80. We hope he would be pleased with the outcome.

Michael C. Thomas, Ph.D.
Gainesville, Florida
September 15, 2000

# Acknowledgments

Originally, Ross Arnett was to have authored many of the family treatments, especially for those families with no specialists available. His death in 1999 left many families without an author. Several volunteers stepped forward, but Dan Young of the University of Wisconsin took responsibility for more than his fair share and got several of his enthusiastic graduate students involved in the project also. The members of the Editorial Board, listed in the Introduction, provided guidance, advice, and constructive criticism, but J. Howard Frank of the University of Florida has been outstanding in his unwavering demands for scholarship and proper English. John Sulzycki of CRC Press has been more than helpful throughout some trying times.

Many of the excellent habitus drawings beginning the family treatments were done by Eileen R. Van Tassell of Michigan State University for *The Beetles of the United States*.

Authors of the family treatments often have acknowledgments in their respective chapters throughout the body of the text.

Ross Arnett's widow, Mary, was always his support staff throughout his long and productive career. Since Ross' death, she has helped by providing free and gracious access to Ross' files, and by her steady encouragement and quiet conviction that we would indeed be able to finish this, Ross Arnett's last big project.

And I would like to acknowledge my wife, Sheila, for her patience and forbearance during the long and sometimes difficult path that led to this volume.

# Contributors to Volume 1 of American Beetles


## Authors

**Arnett, Ross H., Jr., Ph. D.**
Senior Editor, Deceased
*5. Rhysodidae*

**Ashe, James S., Ph. D.**
Snow Entomological Museum
University of Kansas
Lawrence, Kansas
*22. Staphylinidae, Aleocharinae*

**Ball, George E., Ph.D.**
Department of Entomology
University of Alberta
Edmonton, Alberta
*6. Carabidae; 9. Trachypachidae*

**Bousquet, Yves. Ph.D.**
Eastern Cereal and Oilseed Research
CentreAgriculture and Agri-Food Canada
Ottawa, Ontario
*6. Carabidae*

**Caterino, Michael S., Ph. D.**
Department of Entomology
The Natural History Museum
London, England
*15. Histeridae.*

**Chandler, Donald S., Ph. D.**
Department of Zoology
University of New Hampshire
Durham, New Hampshire
*22. Staphylinidae: Pselaphinae*

**Hall, W. Eugene, Ph. D.**
Systematics Research Collections
University of Nebraska
Lincoln, Nebraska
*3. Microsporidae; 4. Hydroscaphidae; 17. Ptiliidae.*

**Ivie, Michael A., Ph. D.**
Department of Entomology
Montana State University
Bozeman, Montana
*5. Rhysodidae*

**Kovarik, Peter W., Ph. D.**
Department of Entomology
Florida A & M University
Tallahassee, Florida
*15. Histeridae.*

**Larson, D.J., Ph. D.**
Department of Biology
Memorial University of Newfoundland
St. John's, Newfoundland
*12. Dytiscidae*

**Newton, Alfred F., Ph. D.**
Field Museum of Natural History
Chicago, Illinois
*14. Sphaeritidae; 22. Staphylinidae*

**O'Keefe, Sean, Ph. D.**
Department of Entomology
Texas A & M University
College Station, Texas
*20. Scydmaenidae*

**Peck, Stewart B., Ph. D.**
Department of Biology
Carleton University
Ottawa, Ontario
*18. Agyrtidae; 19. Leiodidae; 21. Silphidae*

**Perkins, Philip D., Ph. D.**
Department of Entomology
Museum of Comparative Zoology
Harvard University
Cambridge, Massachusetts
*16. Hydraenidae*

**Philips, T. Keith, Ph. D.**
Department of Zoology and Entomology
University of Pretoria
Pretoria, South Africa
*2. Micromalthidae; 11. Amphizoidae*

**Roughley, R. E., Ph. D.**
Department of Entomology
University of Manitoba
Winnipeg, Manitoba
*7. Gyrinidae; 8. Haliplidae; 10. Noteridae; 12. Dytiscidae*

**Thayer, Margaret K., Ph. D.**
Field Museum of Natural History
Chicago, Illinois
*22. Staphylinidae*

**Thomas, Michael C., Ph. D.**
Second Editor
Florida State Collection of Arthropods
Florida Department of Agriculture & Consumer Services
Gainesville, Florida

**Van Tassell, Eileen R., Ph. D.**
Department of Entomology
Michigan State University
East Lansing, Michigan
*13. Hydrophilidae*

**Xie, Weiping, Ph.D.**
Urban Entomology Department
University of Toronto
Toronto, Ontario
*11. Amphizoidae*

**Young, Daniel K., Ph. D.**
Department of Entomology
University of Wisconsin
Madison, Wisconsin
*1. Cupedidae; 2. Micromalthidae*


## Editorial Board

# Credits

Family 1

From: Hatch, M.H. The beetles of the Pacific Northwest. Part I: Introduction and Adephaga. University of Washington Publications in Biology, 16:1-340 (1953). With permission for Figure 1.2.

Family 8

From Hilsenhoff, W.L. and Brigham, W.U. 1978. The Great Lakes Entomologist, 11(1):11-22. With permission for Figures 2.8, 3.8, 4.8.

Family 9

From: Lindroth, C.H. 1961. The ground beetles (Carabidae excl. Cicindelinae) of Canada and Alaska, Part 2. Opuscula Entomologica, Supplementum XX. With permission for Figure 1.9.

Family 10

From Beutel, R.G. and Roughley, R.E., Canadian Journal of Zoology, 65(8), 1908, 1968. With permission for Figures 3.10. 4.10, 5.10.

Family 12

From: Larson, D.J. et al. 2000. Predaceous diving beetles (Coleoptera: Dytiscidae) of the Nearctic Region, with emphasis on the fauna of Canada and Alaska. Monographs in Biodiversity, NRC Press, Ottawa, Ont. (in press). With permission for Figures 2.12, 3.12, 4.12, 5.12, 6.12, 7.12, 8.12, 9.12, 10.12, 11.12, 12.12,13.12, 14.12, 15.12, 16.12, 17.12, 18.12, 19.12, 20.12, 21.12, 22.12, 23.12, 24.12, 25.12, 26.12, 27.12, 28.1, 29.12, 32.12, 33.12, 34.12, 35.12, 36.12, 37.12, 38.12, 39.12, 40.12, 41.12, 42.12, 43.12, 44.12, 45.12, 46.12, 47.12.

From Leech, H.B. Key to the Nearctic genera of water beetles of the tribe Agabini, with some generic synonymy (Coleoptera: Dytiscidae). Annals of the Entomological Society of America, 32: 355-362. 1942. With permission for Figures 30.12, 31.12.

Family 13

From Van Tassell, E.R. 1963. A new *Berosus* from Arizona, with a key to the Arizona species (Coleoptera: Hydrophilidae). Coleopterists Bulletin, 17 (1): 1-5. With permission for Figure 13.16.

From Van Tassell, E.R. 1965. An audiospectrographic study of stridulation as an isolating mechanism in the genus *Berosus* (Coleoptera: Hydrophilidae). Annals of the Entomological Society of America, 58(4): 407-413. With permission for Figures 17.31, 18.13.

Family 17

From: Besuchet, C. 1971. Ptiliidae. In: H. Freude, K.W. Harde, G.A. Lohse (eds.) Die Käfer Mitteleuropus, 3. Krefeld, Groecke und Evers. pp. 311-334. With permission for Figures 4.17, 6.17, 9.17, 10.17, 11.17, 12.17, 13.17, 14.17, 20.17, 22.17, 23.17.

From: Dybas, H.S. 1990. Ptiliidae, pp. 1093-1112. In: D. Dindal (ed.), Soil Biology Guide. Reprinted by permission of John Wiley & Sons, Inc., New York, for Figures 5.17, 15.17, 16.17, 18.17, 19.17, 26.17, 27.17, 28.17, 29.17, 30.17, 31.17, 32.17, 33.17, 34.17, 35.17, 36.17, 38.17, 39.17, 40.17, 41.17, 42.17, 43.17, 44.17.

From: Seevers, C.H. and Dybas, H.S. 1943. A synopsis of the Liumulodidae (Coleoptera): A new family proposed for myrmecophiles of the subfamilies Limulodinae (Ptiliidae) and Cephaloplectinae (Staphylinidae). Annals of the Entomological Society of America, 36: 546-586. With permission for Figures 24.17, 45.17, 47.17.

From: Wilson, E.O. et al. 1954. The beetle genus Paralimulodes Bruch in North America, with notes on morphology and behaviour (Coleoptera: Limulodidae). Psyche: 154-161. With permission for Figures 25.17, 48.17.

Family 18

From: Peck, S.B. 1990. Insecta: Coleoptera: Silphidae and the associated families Agyrtidae and Leiodidae. pp. 1113-1136. In: Dindal, D., (ed.), Soil Biology Guide. Reprinted by permission of John Wiley & Sons, Inc., New York, for Figures 1.18 to 4.18.

Family 19

From: Peck, S.B. 1990. Insecta: Coleoptera: Silphidae and the associated families Agyrtidae and Leiodidae. pp. 1113-1136. In: Dindal, D., (ed.), Soil Biology Guide. Reprinted by permission of John Wiley & Sons, Inc., New York, for Figures 1.19, 2.19, 3.19, 6.19 to 20.19.

Family 20

From O'Keefe S.T. 1998. Kenntnis der Scydmaeniden der Welt: The contributions of Herbert Franz to scydmaenid taxonomy 1952-1997 Coleoptera: Scydmaenidae). Koleoperologische Rundschau 68: 119-136. With permission for Figure 1.20.

From O'Keefe S.T. 1997. Revision of the genus *Chevrolatia* Jacquelin Du Val (Coleoptera: Scydmaenidae) for North America. Transactions of the American Entomological Society 123(3): 163-185. With permission for Figure 9.20.

From: O'Keefe, S.T. 1997. Review of the Nearctic genus *Ceramphis* (Coleoptera: Scydmaenidae). Entomological News 108(5): 335-344. With permission for Figure 13.20.

Family 21

From: Peck, S.B. 1990. Insecta: Coleoptera: Silphidae and the associated families Agyrtidae and Leiodidae. pp. 1113-1136. In: Dindal, D., (ed.), Soil Biology Guide. Reprinted by permission of John Wiley & Sons, Inc., New York, for Figures 1.21, 2.21, 3.21, 5.21, 9.21, 10.21.

Family 22

From: Besuchet, C. 1974. 24. Familie: Pselaphidae, pp. 305-362. In: Freude, H., Harde, K.W. and Lohse, G.A. (eds.), Die Käfer Mitteleuropas, Vol. 5, Staphylinidae II (Hypocyphtinae und Aleocharinae), Pselaphidae. Goecke and Evers, Krefeld. With permission for Figures 11.22, 12.22, 13.22, 14.22, 15.22.

From: Campbell, J.M. 1978. A revision of the North American Omaliinae (Coleoptera: Staphylinidae). 1. The genera *Haida* Keen, *Pseudohaida* Hatch, and *Eudectoides* new genus. 2. The tribe Coryphiini. Memoirs of the Entomological Society of Canada, 106:1-87. With permission for Figures 75.22, 77.22.

From: Campbell, J.M. 1991. A revision of the genera *Mycetoporus* Mannerheim and *Ischnosoma* Stephens (Coleoptera: Staphylinidae: Tachyporinae) of North and Central America. Memoirs of the Entomological Society of Canada, 156: 1-169. With permission for Figure 149.22.

From: Campbell, J.M. 1993. A revision of the genera *Bryoporus* kraatz and *Bryophacis* Reitter and two new related genera from America North of Mexico (Coleoptera: Staphylinidae: Tachyporinae). Memoirs of the Entomological Society of Canada, 166:1-85. With permission for Figure 150.22.

From: Freude, Harde, Lohse (eds.) Die Käfer Mitteleuropas, vol. 5, Goecke & Evers, 1974. With permission for Figure 13.22.

From: Hansen, M. 1997. Phylogeny and classification of the staphyliniformia beetle families (Coleoptera). Biologiske Skrifter, Det Kongelige Danske Videnskabernes Selskb, 48:13-39. With permission for Figure 7.22.

From: Hatch, M.H. 1957. The beetles of the Pacific Northwest. Part II. Staphylinformia, University of Washington Publications in Biology 16: ix + 394 pp. With permission for Figure 36.22.

From: Herman, L.H., Jr. 1970. Phylogeny and reclassification of the genera of the rovebeetle subfamily Oxytelinae of the world (Coleopters, Staphylinidae). Bulletin of the American Museum of Natural History, 142: 343-454. With permission for Figures 331.22, 332.22, 333.22, 334.22, 335.22, 337.22.

Modified from Lohse, G.A. 1964. 23. Familie: Staphylinidae. In: Freude, H., Harde, K.W. and Lohse, G.A. (eds.), Die Käfer Mitteleuropas, Vol. 4, Staphylinidae I (Micropeplinae bis Tachyporinae). 264 pp. Goecke and Evers, Krefeld. With permission for Figures 5.22, 6.22, 9.22, 10.22, 16.22, 17.22, 18.22, 27.22, 28.22, 29.22, 30.22, 31.22, 33.22, 34.22, 35.22, 38.22, 39.22, 40.22.

From: Lohse, G.A. 1974. 23. Familie: Staphylinidae. pp. 1-304. In: Freude, H., Harde, K.W. and Lohse, G.A. (eds.), Die Käfer Mitteleuropas, Vol. 5, Staphylinidae II (Hypocyphtinae und Aleochrinae), Pselaphidae, 381 pp. Goecke and Evers, Krefeld. With permission for Figures 19.22, 20.22, 22.22, 23.22.

From: Moore, I. 1964. A new key to the subfamilies of the Nearctic Staphylinidae and notes on their classification, Coleoptorists Bulletin, 18: 83-91. With permission for Figures 25.22, 63.22, 64.22.

From: Moore, I. 1976. *Giulinium campbelli*, a new genus and species of marine beetle from California (Coleoptera: Staphylinidae). PanPacific Entomologist, 52: 56-59. With permission for Figure 74.22.

From: Moore, I. and Legner, E.F. 1979. An illustrated guide to the genera of the Staphylinidae of America North of Mexico exclusive of the Aleocharinae (Coleoptera). University of California Division of Agricultural Sciences Priced Publication, 4093: 1-332. With permission for Figure 26.22.

From: Newton, A.F., J.R. 1990. Insecta: Coleoptera: Staphylinidae adults and larvae, pp, 1137–1174. In: Dindal, D.L. (ed.), Soil Biology Guide. Reprinted by permission of John Wiley & Sons, New York, for Figures 8.22, 32.22, 41.22, 44.22, 43.22, 47.22, 48.22, 49.22, 52.22, 53.22, 54.22, 61.22, 62.22, 65.22, 66.22, 68.22, 69.22, 71.22, 72.22, 73.22, 92.22, 146.22, 147.22, 148.22, 336.22, 339.22, 347.22, 346.22, 348.22, 349.22, 354.22, 355.22, 356.22.

From: Newton, A.F., Jr. 1990. Myrmelibia, a new genus of myrmecophile from Australia, with a generic review of Australian Osoriinae (Coleoptera: Staphylinidae). Invertegrate Taxonomy, 4: 81-94. With permission for Figures 327.22, 329.22, 330.22.

From: Smetana, A. 1982. Revision of the subfamily Xantholininae of America north of Mexico (Coleoptera: Staphylinidae). Memoirs of the Entomological Society of Canada, 120: iv + 389 pp. With permission for Figure 37.22.

From: Van Dyke, E.C. 1934. The North American species of Trigonurus Muls. Et Rey (Coleoptera Staphylinidae). Bulletin of the Brooklyn Entomological Society, 29: 177-183. With permission for Figure 24.22.

# Table of Contents

# Introduction

The beetles comprise the largest and most diverse order of life on Earth. Beetles occupy almost every conceivable habitat except for the purely marine. The immense number of species and great diversity have resulted in an equally immense literature that can be daunting even to the professional systematist. Relatively few works have attempted to treat the entire Order Coleoptera, and then usually only for a restricted geographical area or at the family level only. One of those few was Ross H. Arnett, Jr.'s *The Beetles of the United States (A Manual for Identification)*, published in fascicles from 1960 to 1963 and then reprinted in 1963, 1968, 1971, and 1973. It has been out of print for many years.

This publication is a direct descendant of that volume and attempts to build on that work and to bring it up to date with current thoughts on the classification of the Coleoptera and recent phylogenetic analyses.

The area covered by this publication is North America, minus the Nearctic area of Mexico. The classification adopted here is that of Lawrence and Newton (1995) with the exception that Bruchidae are treated here as a full family rather than as a subfamily of the Chrysomelidae. This first volume covers the suborders Archostemata, Myxophaga, and Adephaga, plus the series Staphyliniformia of the suborder Polyphaga. The second volume includes the family keys and covers the remainder of the Polyphaga. Much of the following introductory material is adopted from Arnett (1960-1963).

## Family treatments

The family chapters have a consistent format and are arranged as follows:

Family name, author, date
Common name
Synonymy
Diagnosis
Description (shape, size, color, vestiture, head, antennae, mouthparts, eyes, thorax, legs, elytra, wings, abdomen, male and female genitalia, larvae)
Habits and habitats
Status of the classification
Distribution and number of species
Key to the genera of the United States
Classification of the genera of the United States
Bibliography

The family names are those used by Lawrence and Newton (1995). The common name has been selected by the individual authors, or based on common usage, or because they were used in Arnett (1960-1963). The common name is to be applied to the family and not to individual species that are already named in the common name list published by the Entomological Society of America. In this sense, then, it is believed that they are appropriate, and may be useful in certain works, i.e., works of a popular nature, or occasionally in titles of reports, and the like, directed to the general public.

The general features particularly useful for the spot identification of a member of a family are given as a diagnosis to the recognition of a group. This will help in determining if a specimen has been keyed to the proper family. There then follows a partial description of the family as a whole. This is the most useful section of the book because it attempts to define a family,

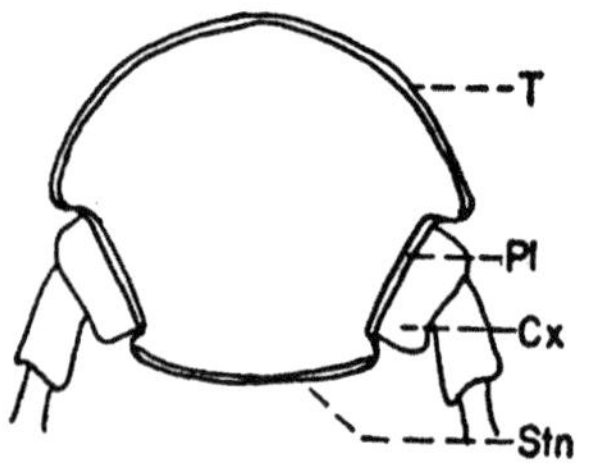


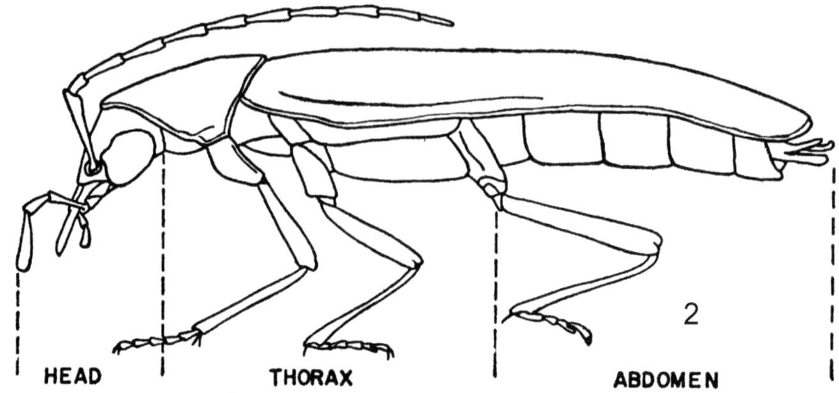


FIGURES 1-2. Fig. 1, cross-section of hypothetical arthropod segment; Fig. 2, *Oxacis taeniata* (LeConte), lateral view, showing tagmata (Oedemeridae).

not simply to distinguish a family. At the same time, it is recognized that these definitions are often incomplete. It is hoped that these descriptions will provide the basis for further, more complete and comprehensive descriptions. Meanwhile their usefulness will be a further aid in checking identifications to family. They will provide the user with some idea of the range and variety of structures in a family. Caution must be exercised, however. The assumption that a particular structure is not found in a family should be made only after a careful evaluation of all of the features discussed in the description.

Ecological information about beetles still is distressingly limited. A large volume of the literature, plus the field experience of the authors, have been condensed into this brief section of the family characterization. Many specialists have a wealth of information on the habits and habitats of beetles which they have not made available in published form. They should be urged to publish these data piecemeal if necessary so that a permanent record is available to future students. Also many of the published statements should be reexamined for accuracy, and erroneous notions be eliminated from the literature wherever possible.

The status of the classification is given as an aid for the evaluation of the work accomplished to date on the family and an indication of research yet to be done. Obviously, revisions that are now underway, but as yet unpublished, will outdate these comments. Where it is known that studies are underway, this is indicated. Also, a brief history of the group is included, and suggestions are sometimes given for possible further studies. When determinations are being made, these comments should be kept in mind.

Most families have a worldwide distribution, but often groups are more concentrated in certain areas. This is indicated, along with estimates on the size of the family for the world and the approximate number of species found in the United States.

The keys to the genera have been compiled by the individual authors, some are modified from those used by Arnett (1960-1963); many are original. In all cases, the keys have been brought up to date by the inclusion of new genera and genera new to the United States.

The classification of the genera proposed here is not to be construed as final in any sense.The number of species occurring in the United States is indicated for each genus, and a brief summary of the range of the genus is given. This will be an aid in the identification of specimens. Those genera with only one species have included also the name of the species. It is possible to make some specific determinations with this manual because of this, but before the determination is accepted, a description of the species should be consulted.

A major change from Arnett (1960-1963) is the addition of a detailed bibliography with each family. The completeness of the bibliographies varies from family to family, but at a minimum they provide an entry into the literature on the family. It is sincerely hoped that there are no serious omissions.

Keys for the identification of organisms, at best, are only guides. As a practical method, the use of external anatomical characters in keys has served us admirably as a means of identification of specimens. Some of the more obvious anatomical features of the various classification groups have been singled out and organized into what we call "keys." These keys are intended to be a short cut to the literature on a particular group by giving us a classification name. The keys on the following pages are meant for this purpose alone and do not reflect the natural classification of the group. It is not possible, under present circumstances, to write "perfect" keys. Not enough is known about a given group, especially at the family level. Mistakes will be found in these keys, as well as any other key. If this were not so, there would be no need for this book. We would be using the "perfect" one already available.

For those who have not had previous experience in using keys for identification, the following brief explanation should suffice. A key is an elimination device. The numbered sections are couplets, or sometimes triplets.

## The external anatomy of adult beetles

For practical reasons, the foundation upon which classification is based is anatomy, even though as many biological features as possible are also employed. Each group of animals has its

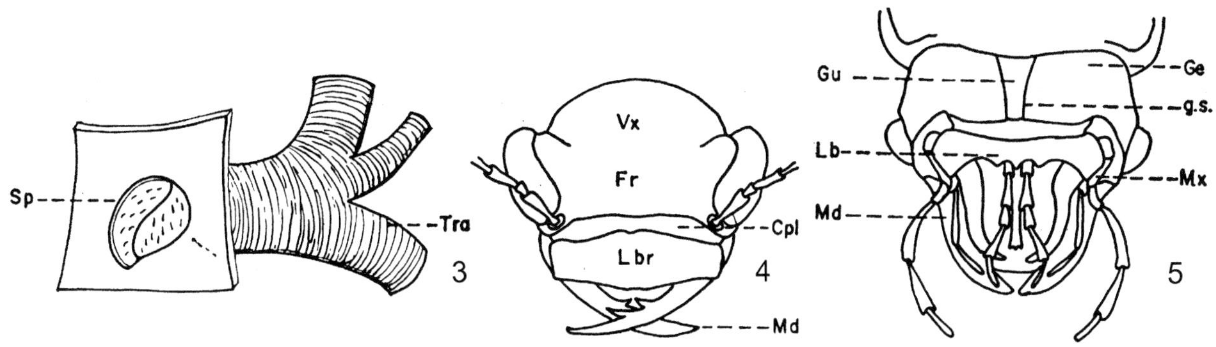


FIGURES 3-5. Fig. 3, section of body wall showing spiracle and tracheae, diagrammatic; Fig. 4, head of *Cicindela* sp. (Carabidae), anterior view; Fig. 5, head of Carabidae, ventral view.

own set of terms peculiar to the anatomy of the group, as well as a number of terms that are common to many groups. It is necessary to have a good working background in beetle anatomy in order to understand the classification and identification descriptions of the beetles discussed in the following sections.

It must be borne in mind that even though the homologies of insect parts have been worked out fairly well, the names used in various groups for various parts have been handed down to us before these homologies were recognized. Therefore, it is frequently more convenient to use the terms as they are found in the early literature rather than to attempt to bring all morphological terms into uniformity. Also, the names for these parts are, even today, likely to change as our knowledge increases with further morphological study.

The following discourse on the external anatomy of beetles is not intended to be anything other than a brief introduction to "traditional" beetle structure, mainly those features generally used in identification keys.

Like all arthropods, the body of a beetle is segmented with a hard exoskeleton. This integument is composed of sclerites or plates separated by a suture, a thin, flexible groove in the exoskeleton. The fundamental pattern of a segment is four sclerites arranged in a ring, the dorsal *tergum* (Fig. 1, T), lateral *pleura* (Fig. 1, Pl), and a ventral *sternum* (Fig. 1, Stn). Various segmented appendages are found on some of the segments located on the pleuron, e.g., the *coxa* (Fig. 1, Cx). This plan is the same from segment to segment and from group to group, but varies in detail.

Beetles have six legs in the adult stage, and the body is divided into three parts: the head, the thorax, and the abdomen (Fig. 2). They breath by means of *tracheae* (Fig. 3, Tra), a series of ramifying tubes which run throughout the body much the same as arteries do in vertebrate animals. The entrance to these tubes is a valve-like arrangement under muscular control, the *spiracle* (Fig. 3, Sp). By means of this respiratory system there is an exchange of the gasses, oxygen and carbon dioxide, with the cells and the external atmosphere. In addition, beetles have one pair of antennae, usually two pairs of wings, and the external organs of copulation near the posterior or caudal end of the body. These characteristics they share with almost all other insects. The following anatomical details will serve to characterize the order Coleoptera and serve to separate this order from the other insects.

## Head, thorax, and abdomen

### The head and its appendages

The anterior portion of the body is the head. It varies greatly in form, and is joined to the prothorax by a membrane. It is always sunken into the prothorax which covers it to a varying degree. Usually the hind portion is but slightly narrowed; sometimes the part behind the eyes is suddenly narrowed and constricted, forming a neck, or gradually narrowed and much prolonged. One large group commonly known as the snout beetles (Curculionidae) have the head elongate in front of the eyes. This is not definitive for this group; however, for some of the snout beetles do not have snouts, and almost every other major group throughout the order have one or two genera which have the head prolonged into a snout.

The top surface of the head is called the *vertex* (Fig. 4, Vx). In other insects the head may be composed of several sclerites limited by distinct sutures. In some beetles (e.g., Hydrophilidae) there may be secondary folding which gives the appearance of sutures, but these are not true sutures and are not homologous with those of other insects. Below the vertex is the *frons* (Fig. 4, Fr) and attached to the frons is the *clypeus* (Fig. 4, Clp) to which is attached the *labrum* (Fig. 4, Lbr), the dorsal movable, flap-like covering of the mouthparts. Usually the clypeus is not separated from the frons by a distinct suture, but often there is a line which may be called the epistomal sulcus.

On the under surface of the head in many species there are two grooves, or sutures, which interrupt the *genae* (Fig. 5, Ge). These are the *gular sutures* (Fig. 5, g.s.) and the area between these is the *gula* (Fig. 5, Gu). This sclerite is interposed between the labium and the *foramen magnum* in beetles. Sometimes the two sutures are joined on the median line so that the gula is no longer evident. This characteristic is used in identifying certain groups and is especially characteristic of the snout beetles mentioned above. The head bears the antennae, the eyes, and the mouthparts.

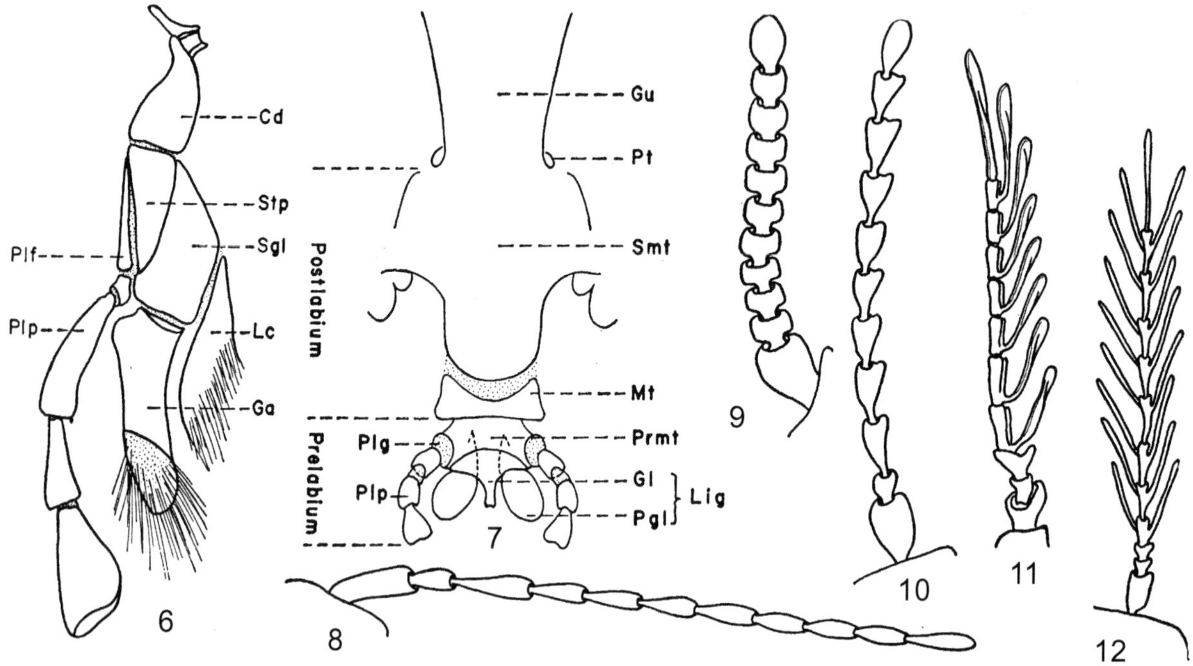


FIGURES 6-12. Fig. 6, maxilla of *Sisenes championi* Horn (Oedemeridae); Fig. 7, labium of *Sisenes championi* Horn (Oedemeridae), ventral view; Fig. 8, filiform antenna, *Harpalinus* sp. (Carabidae); Fig. 9, moniliform antenna of *Clinidium sculptiltis* Newm. (Rhysodidae); Fig. 10, serrate antenna of *Ctenicera* sp. (Elateridae); Fig. 11, pectinate antenna of *Calochromus* sp. (Lycidae); Fig. 12, bipectinate antenna of *Pityobius* sp. (Elateridae).

**Eyes.** The eyes show comparatively little variation throughout the order. They differ in size and shape, usually round or oval, sometimes kidney-shaped, or sausage-shaped, and rarely absent. In a few groups, the eyes are divided so that they form four eyes. This is characteristic of one family, the Gyrinidae, and some genera of several other families. Ocelli are rarely present in the adult (see Staphylinidae and Dermestidae).

**Mouthparts.** The mouthparts are similar to those of the grasshopper, but the various parts show considerable variation throughout the order and have been used by early workers as a basis for the classification of beetles. The position of the parts is referred to morphologically, the assumption being that the head is segmented from anterior to posterior, even when it is hypognathus.

Anterior to the mouth opening is a small, more or less transverse, piece: the labrum mentioned above. It is variable in form, and is in nearly all of the families distinctly visible. It may, however, be solidly united with the clypeus, or retracted beneath it. In the snout beetles, except for some groups, it is entirely absent.

Immediately posterior to the labrum are the *mandibles* (Fig. 4, Md). They also vary much in shape and size but are usually curved, often toothed on the inner side, and, in the males of certain Lucanidae, they are long and branched like the antlers of a deer. In some, they are partly membranous, and in one group, the beaver parasites, they are atrophied. The motion of the mandibles is from side to side. They are used to grasp and crush the food, often as defense organs, and rarely as clasping organs.

Posterior to the mandibles is a second pair of horizontally moving pieces, the *maxillae* (Fig. 5, Mx). They are of complex structure and are of importance in the classification of certain groups. The basal portion of the maxilla is composed of four pieces: the first articulates with the side of the head near the labium, the *cardo* (Fig. 6, Cd); the second is the *stipes* (Fig. 6, Stp) and articulates, usually, at a more or less acute angle with the first; the third, which lies on the inner side of the stipes is the *subgalea* (Fig. 6, Sgl); and the fourth is the *palpifer* (Fig. 6, Plf) which bears the maxillary palpus. The subgalea bears the *lacinia* (Fig. 6, Lc) which is modified in various ways, sometimes comb-like, sometimes sharp and mandible-like. The *galea* (Fig. 6, Ga), if distinct, is often palpus-like, usually of two segments. The *maxillary palpi* (Fig. 6, Plp) are usually composed of four segments, rarely three-segmented, and in at least one case (the genus *Aleochara*, family Staphylinidae) they are five-segmented.

The development and shape of the maxillae of beetles, as of other insects, depend very largely upon the nature of the food, as those organs serve not only to seize and hold the food in the mouth, but also as accessory jaws, aiding the mandibles in the rendering of the food into a form more suitable for swallowing. Their palpi are not only organs of touch, but in many cases act as hands in prehending and carrying morsels of food to the mouth.

The posterior portion of the mouth, the *labium* (Fig. 5, Lb) has at its base, the *submentum* (Fig. 7, Smt) which is attached to the gula. Distal to the submenturn is another piece, the *mentum* (Fig. 7, Mt). These two sclerites together comprise the *postlabium* (Fig. 7) or, as it is sometimes called, the *postmentum*. The apical portion

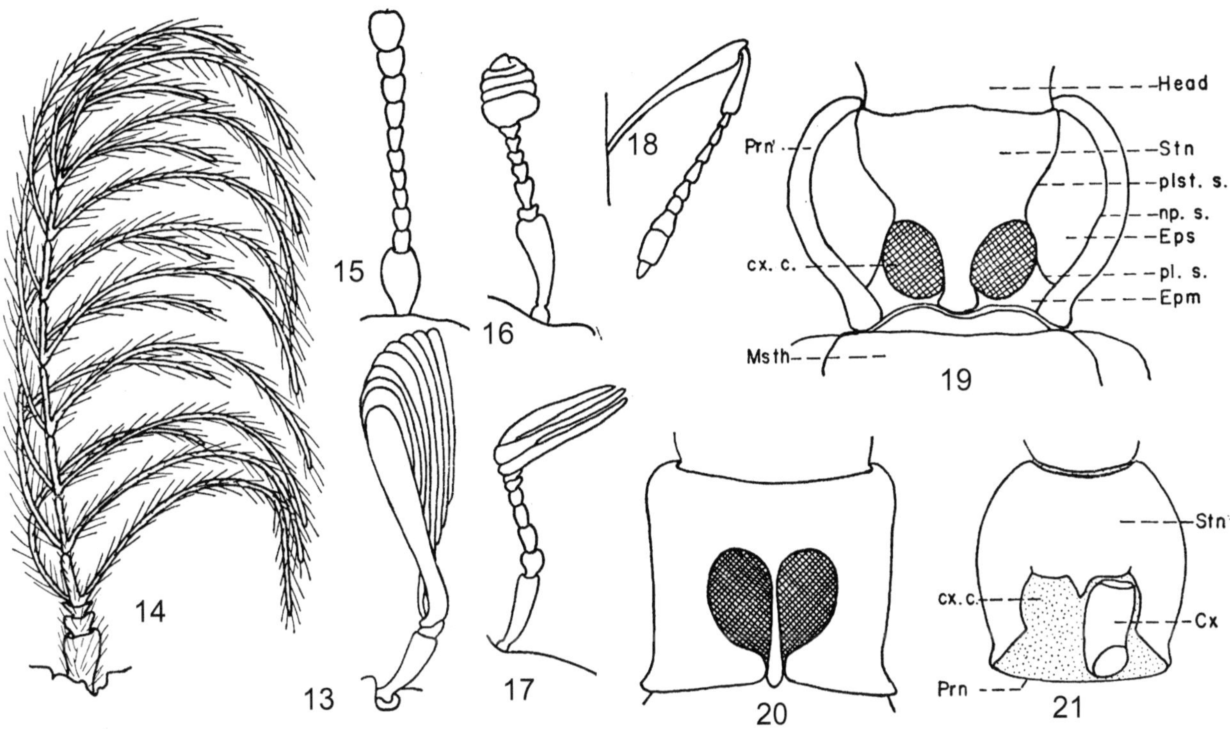


FIGURES 13-21. Fig. 13, flabellate antenna of *Sandalus* sp. (Rhipiceridae); Fig. 14, plumose antenna of *Phengodes* sp. (Phengodidae); Fig. 15, clavate antenna of *Ostoma* sp. (Trogossitidae); Fig. 16, capitate antenna of *Nicrophorus* sp. (Silphidae); Fig. 17, lamellate antenna of *Phyllophaga* sp. (Scarabaeidae); Fig. 18, geniculate antenna of *Rhyssematus* sp. (Curculionidae); Fig. 19, prothorax, ventral view, of *Patrobus* sp. (Carabidae); Fig. 20, prothorax, ventral view, of *Statira* sp. (Tenebrionidae); Fig. 21, prothorax, ventral view, of *Dendroides* sp. (Pyrochroidae).

of the labium is, in the primitive forms, divided into three pieces, a basal portion, the *prementum* (Fig. 7, Prmt), a central portion composed of two lobes, the *glossae* (Fig. 7, Gl), and two side-pieces, the *paraglossae* (Fig. 7, Pgl). The glossae and paraglossae together comprise the *ligula* (Fig. 7, Lig). The *labial palpi* (Fig. 7, Plp) are borne on lobes of the prementum, the palpigers (Fig. 7, Plg). These are similar to the maxillary palpi and are usually three-segmented, but may be reduced to two segments, or less; there are rarely four segments. The entire portion of the labium distal to the mentum is called the *prelabium* (Fig. 7) but this term is not often used in the Coleoptera. There frequently is no suture separating the submentum from the gula, but the two regions may be separated for descriptive purposes by locating the *posterior tentorial pits* (Fig. 7, pt). A line drawn between these pits separates the two regions.

**Antennae.** The antennae of beetles are borne at the side of the head, usually between the eyes and the base of the mandibles, but the position is variable, depending on the group. The size and shape of the antennae are exceedingly diverse, almost every type of antennae found in the insect class is represented in this single order. The number of segments also varies depending on the species. The usual number is 11, and in rare cases 12 segments may be present by the division of the apical segment into two apparent segments which are rarely freely articulated. In a few cases, the number of segments may be thirty or more and vary somewhat with individuals.

The principal forms of the antennae are defined below (figs. 8 to 17). These terms are fairly precise and are extensively used as key characters. The types of antennae are often characteristic of families. Some of the often used terms are defined as follows:

1. **Filiform** (Fig. 8): the simplest form, thread-like, the segments are cylindrical and uniform in shape, or nearly so.
2. **Moniliform** (Fig. 9): bead-like, the segments are nearly uniform in size and rounded, thus resembling a string of beads.
3. **Serrate** (Fig. 10): saw-like, the segments are triangular and compressed.
4. **Pectinate** (Fig. 11): comb-like, the segments are short with the front angles of each much prolonged.
5. **Bipectinate** (Fig. 12): each segment has a comb-like tooth on each side.
6. **Flabellate** (Fig. 13): fan-like, the prolongations of each segment are long and fold together like a fan.
7. **Plumose** (Fig. 14): feather-like, the prolongations are long, slender, and flexible.
8. **Clavate** (Fig. 15): the outer segments are gradually enlarged, forming a distinct club.
9. **Capitate** (Fig. 16): some of the apical segments separated into a round or oval part, distinct from the basal segments.
10. **Lamellate** (Fig. 17): the outer segments are flattened plates which are capable of close approximation, forming a special type of club.

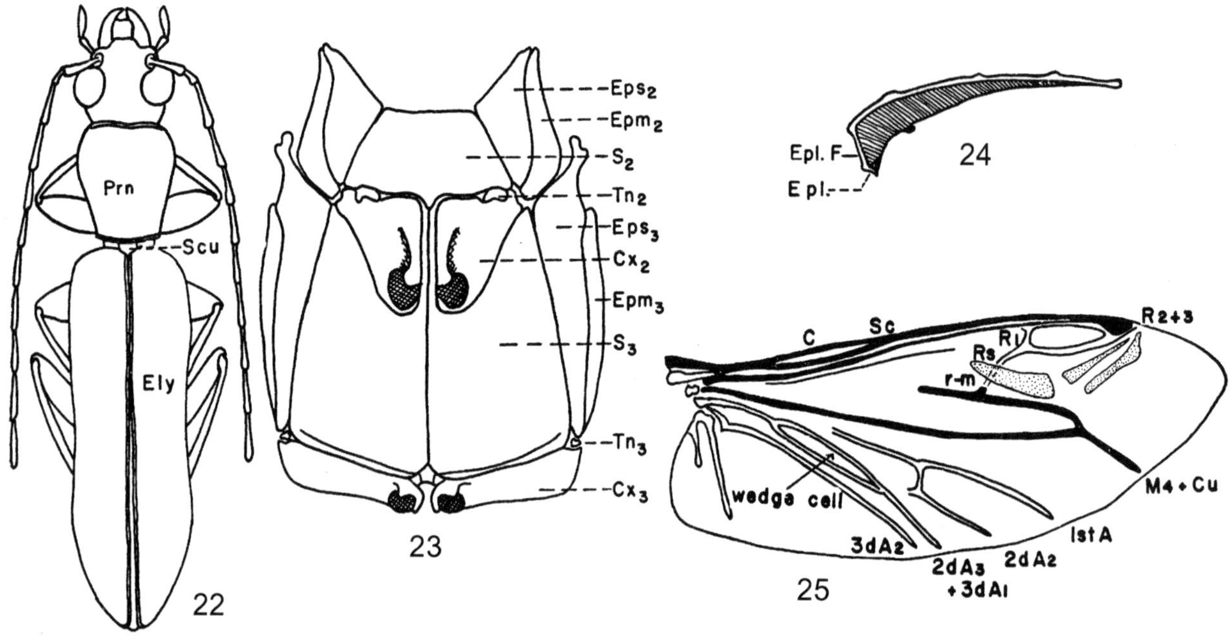


FIGURES 22-25. Fig. 22, *Oxacis xerensis* Arn. (Oedemeridae), dorsal view; Fig. 23, pterosternum of *Sisenes championi* Horn (Oedemeridae); Fig. 24, cross-section view of elytra looking toward the anterior articulation, diagrammatic; Fig. 25, *Sisenes championi* Horn (Oedemeridae), wing.

The antennae are said to be *geniculate* (Fig. 18) or elbowed when the second segment is attached to the first in such a way as to make an obtuse angle, the segments after the second following the same line as the latter. In this form the first or basal segment is usually much longer, and is called the scape. When the geniculate form is at the same time capitate, the segments intermediate between the scape and the club form the *funicle* (Fig. 16).

The antennae of beetles are supposed to be primarily organs of smell, but they also bear organs of touch. In a number of species they are put to other uses. Certain Cerambycidae use the antennae as balancing poles, much the same as a tightwire walker. Many aquatic species use the antennae for gathering in air and passing it under the body for storage while submerged. The males of some of the blister beetles use the antennae as clasping organs. Some species have been observed using their antennae as defense organs, flailing their opponents unmercifully, or others use them in the same manner during courtship procedures.

## The thorax and its appendages

The thorax, the middle portion of the beetle body, is divided into three segments, the first of which is always distinct, the other two, the pterothorax, are more or less firmly united. The prothorax, or first thoracic segment, has as appendages only the front legs. The head is attached to the front part over which it may project to varying degrees, sometimes hiding the head entirely from view from above. Located between the head and the thorax, in the neck region are, in most species, two pairs of articulating sclerites, the cervical sclerites, the exact nature of which remain obscure. The prothorax is often freely movable, or it may be firmly united with the other two thoracic segments. The dorsal surface of the segment is called the *pronotum* (Fig. 22, Prn). Its shape and modifications are very frequently used as specific, generic, and even family characters. Beneath are attached the front or fore legs. The method and position of attachment of these legs serve as useful characters in separating genera and families. Often there are grooves present on the underside of the prothorax for the reception of the antennae, legs, and rarely, the mouthparts. The lateral part of the lower surface of the prothorax is called the *propleura* (Fig. 19); the central portion, the *prosternum* (Fig. 19, Stn). The pleural region is separated from the tergum by the *notopleural suture* (Fig. 19, np. s.) and the sternum by the *pleurosternal suture* (Fig. 19, plst. s.). The presence or absence of lobes, ridges, etc. on the intercoxal piece of the prosternum (also called the *spinasternum*) is useful in beetle classification. When the propleurae have a *pleural suture* (Fig. 19, pl. s.) dividing them into two sclerites, the anterior sclerite is called the prosternal *episternum* (Fig. 19, Eps). The posterior sclerite is the prosternal *epimeron* (Fig. 19, Epm). More frequently, all of these sclerites are solidly fused to the pronotum; in which case, the lower surface is simply referred to as the prosternum. Members of the suborder Adephaga have a distinct suture separating the protergal region from the pleural region of the prothorax, the *notopleural suture* (Fig. 19, np. s.).

The cavities in which the anterior legs are inserted are called anterior coxal cavities. An extremely useful, but difficult character is to be found in the relationship of the sclerites which surround the coxal cavities. If they are enclosed behind by the junction of the prosternum and the epimeron, or by the meeting of the epimera, they are referred to as "anterior coxal cavities closed behind" (Fig. 20). If there is no sclerite exposed behind the coxal cavities, but rather, the cavities open directly on the next thoracic segment, they are referred to as "anterior coxal cavities open be-

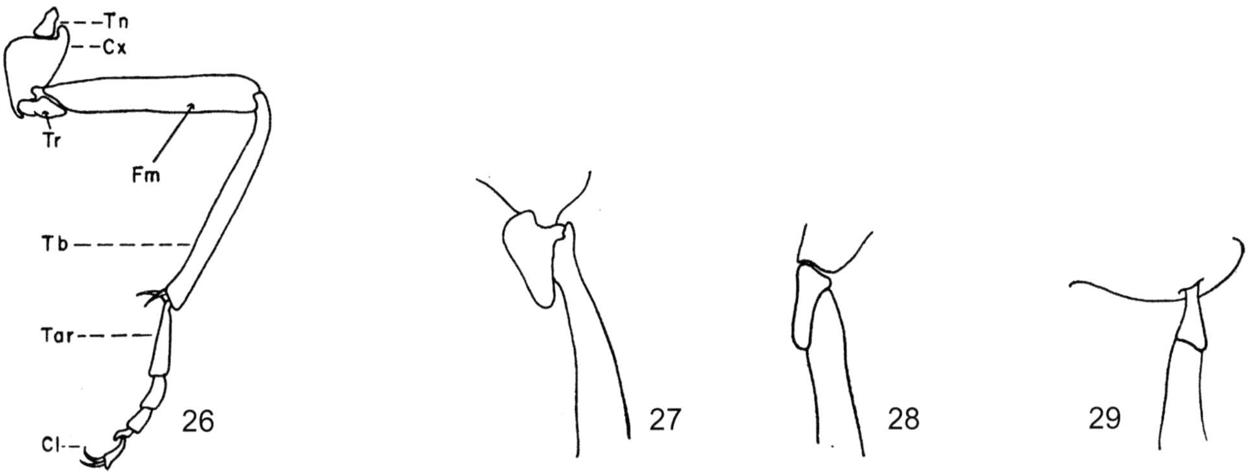


FIGURES 26-29. Fig. 26, *Sisenes championi* Horn (Oedemeridae), middle leg; Fig. 27, *Nebria* sp. (Carabidae), hind leg, part; Fig. 28, *Chauliognathus pennsylvanicus* (DeGeer) (Cantharidae), middle leg, part; Fig. 29, *Lyctus* sp. (Bostrichidae), hind leg, part.

hind" (Fig. 21). These two conditions serve to distinguish families, subfamilies, and tribes.

The middle thoracic segment, or mesothorax, bears the middle pair of legs and the anterior pair of wings, which in beetles are hardened, or leathery and are called *elytra* (singular *elytron*) (Fig. 22, Ely). The dorsal surface is hidden by the elytra when they are in repose, except for a small shield-like portion, appropriately called the *scutellum* (Fig. 22). The scutellum, the shape of which is very useful in distinguishing some genera or groups of genera, is always located immediately behind the pronotum, between the elytra. It varies considerably in size and shape and is rarely entirely hidden. The lower surface of the *mesothorax*. (Fig. 23) has essentially the same parts as the prothorax, which are distinguished by adding the prefix meso- to the terms *episternum* (Fig. 23, $Eps_2$), *epimera* (Fig. 23, $Epm_2$) and *sternum* (Fig. 23, $S_2$). The mesothorax is firmly united to the last thoracic segment, sometimes without distinct sutures separating the two segments. The characters offered by this thoracic segment are rarely used in the classification of the groups. The location of the middle *coxae* (Fig. 23, $Cx_2$) is the best criterion for determining the posterior limit of the mesothorax. The mesocoxal cavities are always closed behind by the metasternum. There can sometimes be located a small *trochantin* (Fig. 23, $Tn_2$) on the middle coxae.

The hind thoracic segment, the *metathorax*, bears the flight wings, which are usually present, and the hind pair of legs. The most important characters are to be found in the manner of insertion of the hind legs, and in the shape, position, and modifications of the posterior portion of the *metepimeron* (Fig. 23, $Epm_3$). The exposed portion of this sclerite is usually a small plate lying at the lateral extremities of the *metacoxae* (Fig. 23, $Cx_3$), the remaining portion is hidden by the elytra. The *metepisternum* (Fig. 23, $Eps_3$) is a long sclerite lying lateral to the long and broad *metasternum* (Fig. 23, $S_3$). The hind coxae are usually transverse. A small sclerite lateral to the coxae and articulating with the metepisternum, the *trochantin* (Fig. 23, $Tn_3$) is sometimes evident. These sclerites are modified by the general shortening of the metathorax in wingless forms. A transverse sclerite is sometimes formed by the presence of a sulcus anterior to the coxae. This is also referred to as the antecoxal piece (see especially Carabidae).

**Wings**. The anterior *wings* (elytra, Fig. 22, Ely) of Coleoptera are greatly modified as protective covers for the hind wings. Usually they are hard or horny, sometimes soft, but always opaque. They are held side by side over the abdomen, and the morphological hind edges generally join forming a median line along the abdomen. When the elytra are in repose this junction point along the hind edge is termed the suture. The morphological front edge is called the lateral margin, and the inflexed portion of the elytra, that is, the portion of the elytra along the lateral margin which is bent down, is called the *epipleural fold* (Fig. 24, Epl. F). Bordering the inner edge of this is a piece of varying width, sometimes extending from the base to the apex, called the *epipleura* (Fig. 24, Epl). The elytra usually cover most of the abdomen and often extend slightly beyond the apex. In one large group, in particular, the Staphylinidae, and in many other widely separated groups, the elytra are shortened, exposing several of the abdominal segments. In a few groups the elytra are small, fleshy flaps (Rhipiphoridae and Stylopidae). Striae and intervals, as well as costae are present on many elytra and are useful characters for identification.

The posterior wings, or flying wings are membranous with supporting structures, called *veins* (Fig. 25), crossing their surface. These veins are so constructed as to allow the wings to be folded under the elytra when they are not in use. Both the arrangement of the veins and the folding patterns of the wings are uniform for a given taxon. This is best described by Forbes (1922 and 1926). However, the characters furnished by the flying wings are difficult to use, and are seldom employed in the characterization of species or genera. They are rarely used in descriptions. The flying wings are lacking in many desert and cave species, in which case, the metathorax is short, and the elytra closely united or connate.

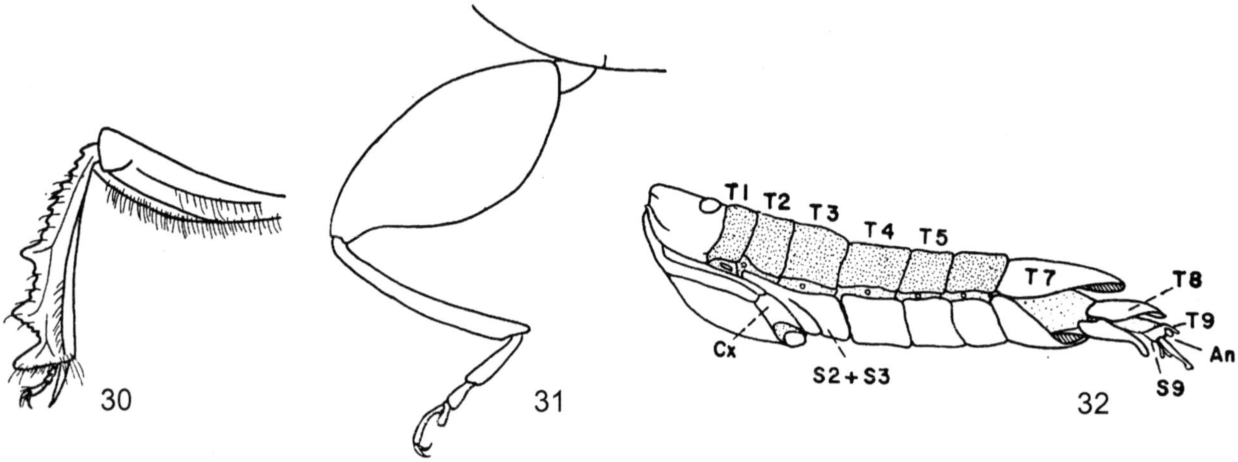


FIGURES 30-32. Fig. 30, foreleg of *Canthon* sp. (Scarabaeidae); Fig. 31, hind leg of *Disonycha* sp. (Chrysomelidae); Fig. 32, male abdomen of *Sessinia decolor* (Fairmaire) (Oedemeridae) (redrawn from Arnett, 1949).

**Legs.** The fundamental structure of insect legs consists of six parts, the trochantin, coxa, trochanter, femur, tibia, and tarsus. This is fairly uniform throughout the class. The *trochantin* (Fig. 26, Tn) is a small sclerite at the base of the coxa, usually partly hidden, or entirely hidden by the base of the coxa when the leg is in normal position. The *coxae* (Fig. 26, Cx) vary considerably in shape. They may be round, oval, elongate, conical, or rarely cylindrical. Inserted between the coxa and the femur is a small piece called the *trochanter* (Fig. 26, Tr). Its point of attachment is a useful character in classification. There are four types of trochanters in beetles. In the Caraboid type (Fig. 27) the trochanter of the hind leg lies entirely to one side of the femur. In the "normal" type, the trochanter is triangular (Fig. 28) and the femur is attached near the base. In the heteromerous type (Fig. 26, Tr) the femur is attached along the side of the trochanter, and in the elongate type (Fig. 29) the femur is attached to the truncate apex of the trochanter. The *femur* (Fig. 26, Fm) is a large piece, sometimes swollen apically or otherwise modified depending on the species and group. As in grasshoppers those beetles that are good jumpers have a greatly enlarged femur. The *tibia* (Fig. 26, Tb) is more slender, often adorned with spines, particularly at the apex. The presence, absence, or size of these spines is useful in determining groups or species. The *tarsi* (Fig. 26, Tar) are the most useful classification structures of the legs. They are always segmented, and end in *claws* (Fig. 26, Cl). The number of tarsal segments, or tarsomeres, varies from one to five, depending on the group, or sometimes, the sex. The early classifications were dependent on the number of tarsomeres, and family groups were named according to their number. It is now believed that the tarsal formula is not really basic to the classification. The number of tarsomeres is always constant for a particular genus or higher group. The number of tarsomeres for the different pairs of legs may also vary. These numbers are expressed as tarsal formulae in this manner. If the front tarsi have five segments, the middle tarsi five segments, and the hind tarsi four segments, the formula is written in this form: 5-5-4. Or further reductions, or arrangements thus: 5-5-5, 4-4-4, 5-4-5, etc. This is extremely useful and is employed in all family keys. The tarsomeres vary greatly in shape, some may be lobed beneath, some may have spongy or brushlike pads beneath, or the length of each tarsomere in relation to the length of another tarsomere may be quite different.

The claws are attached to the apex of the last tarsomere and are paired. The shape of an individual claw may differ on different legs, or different groups. Some are toothed, cleft, serrate, comb-like, bifid, or with ventral pads. However, the usual condition is simple, i.e., unmodified, without the forementioned characteristics. These differences are always used for the separation of species, genera, and whenever possible, even families because such characteristics are not variable with individuals. When the claws are capable of being drawn back upon the last tarsal segment, they are chelate, a condition often found in the family Scarabaeidae. Sometimes there may be a small pad, or a hair, between the claws, but this is not often a useful descriptive character.

In some descriptions the legs are given various names according to their form. Legs fitted for walking are termed ambulatorial (Fig. 2); digging, fossorial (Fig. 30); swimming, natatorial (Fig. 47) (this term is rarely used); jumping, saltatorial (Fig. 31). A few others similarly designate the particular function, but are rarely used.

### Abdomen

The portion of the body behind the metathorax, the *abdomen* (Fig. 32), is a series of rings, or partial rings, each ring with a pair of *spiracles* (Fig. 3, Sp), the openings to the *tracheae* (Fig. 3, Tra). There are no abdominal appendages, but some species may have articulating sternal lobes (see Mordellidae).

Each segment of the abdomen shows the four fundamental sclerites: the tergum, the pleura, and the sternum (Fig. 1). The tergum is usually soft and more or less membranous except for one or two (sometimes more) apical tergites. The pleura are very small and are usually more or less hidden. Each pleuron has a

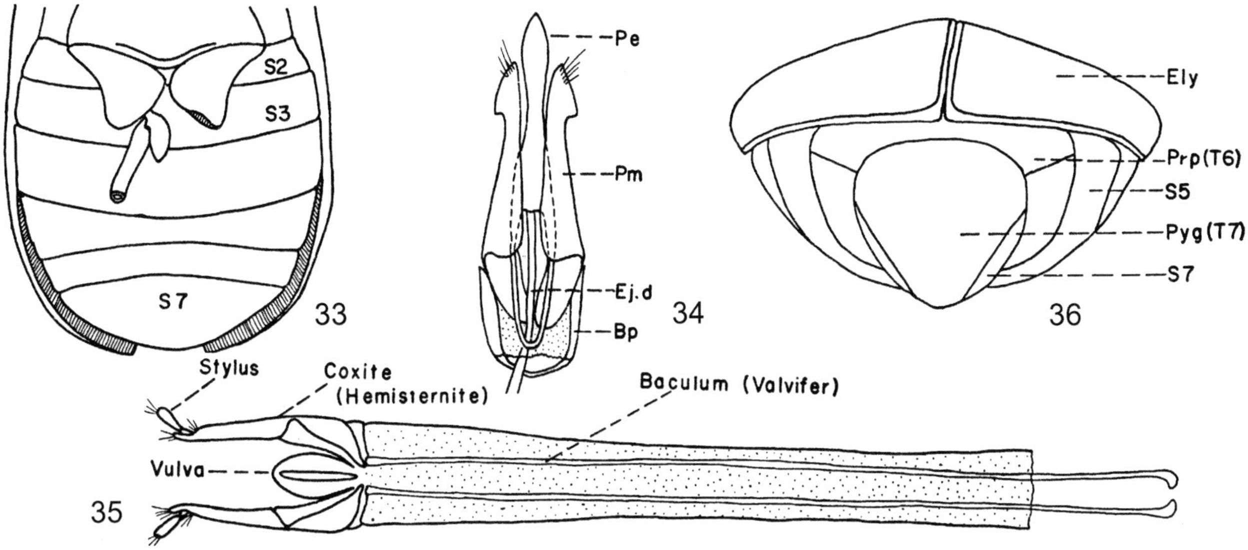


FIGURES 33-36. Fig. 33, ventral view of abdomen of *Nebria* sp. (Carabidae); Fig. 34, dorsal view of male genitalia of *Melanotus* sp. (Elateridae); Fig. 35, ventral view of ninth segment (female genitalia) of *Melanotus* sp. (Elateridae); Fig. 36, posterior view of abdomen of *Saprinus* sp. (Histeridae).

spiracle opening on its surface. Often the pleura are fused with the sternum so that the spiracle appears to be on the sternum. The sternum is a more or less heavily scleritized plate and is the most visible portion of the abdomen, the tergum usually is hidden by the wings and elytra except when the beetle is in flight position.

The usual number of abdominal segments is nine. This may be demonstrated by counting the pairs of spiracles, and counting the terga. The first spiracle is large, the ninth is usually very obscure or absent. There are often only five abdominal sterna visible. The first abdominal sternum is absent or fused with the metasternum of the thorax. The second and third sterna are usually fused and appear as one. In the suborder Adephaga, the second sternum is visible as two triangular pieces lying at the side and in front of the third *sternum* (Fig. 33). The eighth sternum is usually visible, but partly drawn into the seventh segment. The ninth abdominal segment is usually invisible, being covered by the eighth abdominal segment. Often both the eighth and ninth segments are modified in various ways and function as a part of the copulatory organs. The last visible abdominal tergum is referred toas the *pygidium* (tergum 7) (Fig. 36, Pyg), and the next to the last is the *propygidium* (Fig. 36, Prp) (tergum 6).

The genitalia (*aedeagus*) of the male consists of two parts: the intromittent organ, the *penis* (or median lobe), and the *tegmen*, composed of a *pars basalis* and one pair of *parameres* (or lateral lobes) and rarely the *tegminite*. The parameres may be fused to form a single median structure. Four main types of male genitalia (Tuxen, 1956) have been described, of which only one, the trilobate type (Fig. 34), is shown here. There are many variations of these types. The genitalia of the female is a tube composed of the ninth segment, and is poorly known (Fig. 35). These parts have been variously named; the system used by Tanner (1928) differs considerably from that of Lindroth (1957). (See Tuxen, 1956, pp. 69 to 76, and Lindroth, 1957, for further details and a standardized system of nomenclature.)

The genitalia project from the membrane which connects the ninth sternite below the anus, the terminal opening of the alimentary tract, and the tenth sternite. The tenth sternite is rarely visible as a scleritized structure in insects. The tenth tergite has been demonstrated only in a few groups (Arnett, mss.). It is believed, however, that in the primitive insect form there were ten or more abdominal segments. This is borne out by the form of the abdomen in many primitive existing insects. In many groups the male genitalia are extremely useful for separating species and higher categories. The female genitalia have been studied at the family level, but only a few papers have been published using the female genitalia as the basis for the separation of species (e.g., Arnett, 1944).

# Classification

## Order Coleoptera Linnaeus 1758

(**Note:** *Families surrounded by brackets ({ }) are not known to occur in the area covered by this book. They are listed here but not further treated.*)

Suborder **ARCHOSTEMATA** Kolbe, 1908
- Family 1. **CUPEDIDAE** Laporte 1836; The reticulated beetles
  - {Ommatidae Sharp and Muir 1912}
  - {Crowsonellidae Iablokoff-Khnzorian 1983}
- Family 2. **MICROMALTHIDAE** Barber 1913, The telephone-pole beetles

Suborder **MYXOPHAGA** Crowson 1955
  - {Lepiceridae Hinton 1936}
- Family 3. **MICROSPORIDAE** Crotch 1873, The minute bog beetles
- Family 4. **HYDROSCAPHIDAE** LeConte 1874, The skiff beetles
  - {Torridincolidae Steffan 1964}

Suborder **ADEPHAGA** Schellenberg 1806
- Family 5. **RHYSODIDAE** Laporte 1840, The wrinkled bark beetles
- Family 6. **CARABIDAE** Latreille 1802, The ground beetles
- Family 7. **GYRINIDAE** Latreille 1810, The whirligig beetles
- Family 8 **HALIPLIDAE** Aubé 1836, The crawling water beetles
- Family 9. **TRACHYPACHIDAE** C. G. Thomson 1857, The false ground beetles
- Family 10. **NOTERIDAE** C. G. Thomson 1860, The burrowing water beetles
- Family 11. **AMPHIZOIDAE** LeConte 1853, The trout-stream beetles
  - {Hygrobiidae Regimbart 1878}
- Family 12. **DYTISCIDAE** Leach 1815, The predacious diving beetles

Suborder **POLYPHAGA** Emery 1886
- Series **STAPHYLINIFORMIA** Lameere 1900
  - Superfamily **HYDROPHILOIDEA** Latreille 1802
    - Family 13. **HYDROPHILIDAE** Latreille 1802, The water scavenger beetles
    - Family 14. **SPHAERITIDAE** Shuckard 1839, The false clown beetles
      - {Synteliidae Lewis 1882}
    - Family 15. **HISTERIDAE** Gyllenhal, The clown beetles
  - Superfamily **STAPHYLINOIDEA** Latreille 1802
    - Family 16. **HYDRAENIDAE** Mulsant 1844, The minute moss beetles
    - Family 17. **PTILIIDAE** Erichson 1845, The feather-winged beetles
    - Family 18. **AGYRTIDAE** C. G. Thomson 1859, The primitive carrion beetles
    - Family 19. **LEIODIDAE** Fleming 1821, The round fungus beetles
    - Family 20. **SCYDMAENIDAE** Leach 1815, The antlike stone beetles
    - Family 21. **SILPHIDAE** Latreille 1807, The carrion beetles
    - Family 22. **STAPHYLINIDAE** Latreille 1802, The rove beetles
- Series **SCARABAEIFORMIA** Crowson 1960
  - Superfamily **SCARABAEOIDEA** Latreille 1802
    - Family 23. **LUCANIDAE** Latreille 1804, The stag beetles
    - Family 24. **DIPHYLLOSTOMATIDAE** Holloway 1972, The diphyllostomatid beetles
    - Family 25. **PASSALIDAE** Leach 1815, The bess beetles
    - Family 26. **GLARESIDAE** Semenovrlan-Shanskii and Medvedev 1932; The glaresid beetles
    - Family 27. **TROGIDAE** MacLeay 1819, The skin beetles
    - Family 28. **PLEOCOMIDAE** LeConte 1861, The rain beetles
    - Family 29. **GEOTRUPIDAE** Latreille 1802, The earth-boring dung beetles
      - {Belohinidae Paulian 1959}
    - Family 30. **OCHODAEIDAE** Mulsant and Rey 1871, Theochodaeid scarab beetles
    - Family 31. **HYBOSORIDAE** Erichson 1847, The hybosorid scarab beetles

Family 32. **CERATOCANTHIDAE** Cartwright and Gordon 1971, The ceratocanthid scarab beetles
Family 33. **GLAPHYRIDAE** MacLeay 1819, The glaphyrid scarab beetles
Family 34. **SCARABAEIDAE** Latreille 1802, The scarab beetles
Series **ELATERIFORMIA** Crowson 1960
[Podabrocephalidae]
[Rhinorhipidae]
Superfamily **SCIRTOIDEA** Fleming 1821
[Decliniidae Nikitsky *et al.* 1994]
Family 35. **EUCINETIDAE** Lacordaire 1857, The plate-thigh beetles
Family 36. **CLAMBIDAE** Fischer 1821, The minute beetles
Family 37. **SCIRTIDAE** Fleming 1821, The marsh beetles
Superfamily **DASCILLOIDEA** Guérin-Méneville 1843
Family 38. **DASCILLIDAE** Guérin-Méneville 1843
Family 39. **RHIPICERIDAE** Latreille 1834, The cicada parasite beetles
Superfamily **BUPRESTOIDEA** Leach 1815
Family 40. **SCHIZOPODIDAE** LeConte 1861, The schizopodid beetles
Family 41. **BUPRESTIDAE** Leach 1815, The metallic wood-boring beetles
Superfamily **BYRRHOIDEA** Latreille 1804
Family 42 **BYRRHIDAE** Latreille 1804, The pill beetles
Family 43. **ELMIDAE** Curtis 1830, The riffle beetles
Family 44. **DRYOPIDAE** Billberg 1820, The long-toes beetles
Family 45 **LUTROCHIDAE** Kasap and Crowson 1975, The robust marsh-loving beetles
Family 46. **LIMNICHIDAE** Erichson 1846, The minute marsh-loving beetles
Family 47. **HETEROCERIDAE** MacLeay 1825 The variegated mod-loving beetles
Family 48. **PSEPHENIDAE** Lacordaire 1854, The water penny beetles
[Cneoglossidae Champion 1897]
Family 49. **PTILODACTYLIDAE** Laporte 1836, The toe-winged beetles
Family 50. **CHELONARIIDAE** Blanchard 1845, The turtle beetles
Family 51. **EULICHADIDAE** Crowson 1973, The eulichadid beetles
Family 52. **CALLIRHIPIDAE** Emden 1924, The cedar beetles
Superfamily **ELATEROIDEA** Leach 1815
Family 53. **ARTEMATOPODIDAE** Lacordaire 1857, The soft-bodied plant beetles
Family 54. **BRACHYPSECTRIDAE** Leconte and Horn 1883, The Texas beetles
Family 55. **CEROPHYTIDAE** Latreille 1834, The rare click beetles
Family 56. **EUCNEMIDAE** Eschscholtz 1829, The false click beetles
Family 57. **THROSCIDAE** Laporte 1840, The false metallic wood-boring beetles
Family 58. **ELATERIDAE** Leach 1815, The click beetles
[Plastoceridae Crowson 1972]
[Drilidae Blanchard 1845]
[Omalisidae Lacordaire 1857]
Family 59. **LYCIDAE** Laporte 1836, The net-winged beetles
Family 60. **TELEGEUSIDAE** Leng 1920, The long-lipped beetles
Family 61. **PHENGODIDAE** LeConte 1861, The glowworm beetles
Family 62. **LAMPYRIDAE** Latreille 1817, The firefly beetles
Family 63. **OMETHIDAE** LeConte 1861, The false firefly beetles
Family 64. **CANTHARIDAE** Imhoff 1856, The soldier beetles
Series **BOSTRICHIFORMIA** Forbes 1926
Family 65. **JACOBSONIIDAE** Heller 1926, The Jacobson's beetles
Superfamily **DERODONTOIDEA** LeConte 1861
Family 66. **DERODONTIDAE** LeConte 1861, The tooth-necked fungus beetles
Superfamily **BOSTRICHOIDEA** Latreille 1802
Family 67. **NOSODENDRIDAE** Erichson 1846, The wounded-tree beetles
Family 68. **DERMESTIDAE** Latreille 1804, The skin and larder beetles
Family 69. **BOSTRICHIDAE** Latreillew 1802, The horned powder-post beetles
Family 70. **ANOBIIDAE** Fleming 1821, The death-watch beetles

Series **CUCUJIFORMIA** Lameere 1938
Superfamily **LYMEXYLOIDEA** Fleming 1821
Family 71. **LYMEXYLIDAE** Fleming 1821, The ship-timber beetles
Superfamily **CLEROIDEA** Latreille 1802
[Phloiophilidae Kiesenwetter 1863]
Family 72. **TROGOSSITIDAE** Latreille 1802, The bark-gnawing beetles
[Chaetosomatidae Crowson 1952]
Family 73. **CLERIDAE** Latreille 1802, The checkered beetles
[Acanthocnemidae Crowson 1964]
[Phycosecidae Crowson 1952]
[Prionoceridae Lacordaire 1857]
Family 74. **MELYRIDAE** Leach 1815, The soft-winged flower beetles
Superfamily **CUCUJOIDEA** Latreille 1802
[Protocucujidae Crowson 1954]
Family 75. **SPHINDIDAE** Jacquelin du Val 1860, The dry-fungus beetles
Family 76. **BRACHYPTERIDAE** Erichson 1845, The short-winged flower beetles
Family 77. **NITIDULIDAE** Latreille 1802, The sap-feeding beetles
Family 78. **SMICRIPIDAE** Horn 1879, The palmetto beetles
Family 79. **MONOTOMIDAE** Laporte 1840, The root-eating beetles
[Boganiidae Sen Gupta and Crowson 1966]
[Helotidae Reitter 1876]
[Phloeostichidae Reitter 1911]
Family 80. **SILVANIDAE** Kirby 1837, The silvanid flat bark beetles
Family 81. **PASSANDRIDAE** Erichson 1845, The parasitic flat bark beetles
Family 82. **CUCUJIDAE** Latreille 1802, The flat bark beetles
Family 83. **LAEMOPHLOEIDAE** Ganglbauer 1899, The lined flat bark beetles
[Propalticidae Crowson 1952]
Family 84. **PHALACRIDAE** Leach 1815, The shining flower beetles
[Hobartiidae Sen Gupta and Crowson 1966]
[Cavognathidae Sen Gupta and Crowson 1966]
Family 85. **CRYPTOPHAGIDAE** Kirby 1837, The silken fungus beetles
[Lamingtoniidae Sen Gupta and Crowson 1969]
Family 86. **LANGURIIDAE** Crotch 1873, The lizard beetles
Family 87. **EROTYLIDAE** Latreille 1802, The pleasing fungus beetles
Family 88. **BYTURIDAE** Jacquelin du Val 1858, The fruitworm beetles
Family 89. **BIPHYLLIDAE** LeConte 1861, The false skin beetles
Family 90. **BOTHRIDERIDAE** Erichson 1845, The dry bark beetles
Family 91. **CERYLONIDAE** Billberg 1820, The minute bark beetles
[Alexiidae Imhoff 1856]
[Discolomatidae Horn 1878]
Family 92. **ENDOMYCHIDAE** Leach 1815, The handsome fungus beetles
Family 93. **COCCINELLIDAE** Latreille 1807, The ladybird beetles
Family 94. **CORYLOPHIDAE** LeConte 1852, The minute fungus beetles
Family 95. **LATRIDIIDAE** Erichson 1842, The minute brown scavenger beetles
Superfamily **TENEBRIONOIDEA** Latreille 1802
Family 96. **MYCETOPHAGIDAE** Leach 1815, The hairy fungus beetles
Family 97. **ARCHEOCRYPTICIDAE** Kaszab 1964, The archeocryptic beetles
[Pterogeniidae Crowson 1953]
Family 98. **CIIDAE** Leach 1819, The minute tree-fungus beetles
Family 99. **TETRATOMIDAE** Billberg 1820, The polypore fungus beetles
Family 100. **MELANDRYIDAE** Leach 1815, The false darkling beetles
Family 101. **MORDELLIDAE** Latreille 1802, The tumbling flower beetles
Family 102. **RHIPIPHORIDAE** Gemminger and Harold 1870, The wedge-shaped beetles
Family 103. **COLYDIIDAE** Erichson 1842, The cylindrical bark beetles
Family 104. **MONOMMATIDAE** Blanchard 1845, The opossum beetles

Family 105. **ZOPHERIDAE** Solier 1834, The ironclad beetles
[Ulodidae Pascoe 1869]
[Perimylopidae St. George 1939]
[Chalcodryidae Watt 1974]
[Trachelostenidae Lacordaire 1859]
Family 106. **TENEBRIONIDAE** Latreille 1802, The darkling beetles
Family 107. **PROSTOMIDAE** C. G. Thomson 1859, The jugular-horned beetles
Family 108. **SYNCHROIDAE** Lacordaire 1859, The synchroa beetles
Family 109. **OEDEMERIDAE** Latreille 1810, The pollen-feeding beetles
Family 110. **STENOTRACHELIDAE** C. G. Thomson 1859, The false long-horned beetles
Family 111. **MELOIDAE** Gyllenhal 1810, The blister beetles
Family 112. **MYCTERIDAE** Blanchard 1845, The palm and flower beetles
Family 113. **BORIDAE** C. G. Thomson 1859, The conifer bark beetles
[Trictenotomidae Blanchard 1845]
Family 114. **PYTHIDAE** Solier 1834, The dead log bark beetles
Family 115. **PYROCHROIDAE** Latreille 1807, The fire-colored beetles
Family 116. **SALPINGIDAE** Leach 1815, The narrow-waisted bark beetles
Family 117. **ANTHICIDAE** Latreille 1819, The antlike flower beetles
Family 118. **ADERIDAE** Winkler 1927, The antlike leaf beetles
Family 119. **SCRAPTIIDAE** Mulsant 1856, The false flower beetles
Superfamily **CHRYOMELOIDEA** Latreille 1802
Family 120. **CERAMBYCIDAE** Latreille 1802, The long-horned beetles
Family 121. **BRUCHIDAE** Latreille 1802, the pea and bean weevils
Family 122. **MEGALOPODIDAE** Latreille 1802
Family 123. **ORSODACNIDAE** C. G. Thomson 1859
Family 124. **CHRYSOMELIDAE** Latreille 1802, The leaf beetles
Superfamily **CURCULIONOIDEA** Latreille 1802
Family 125. **NEMONYCHIDAE** Bedel 1882, The pine-flower snout beetles
Family 126. **ANTHRIBIDAE** Billberg 1820, The fungus weevils
Family 127. **BELIDAE** Schönherr 1826, The primitive weevils
Family 128. **ATTELABIDAE** Billberg 1820, The tooth-nosed snout beetles
Family 129. **BRENTIDAE** Billberg 1820, The straight-snouted weevils
[Caridae Thompson 1992]
Family 130. **ITHYCERIDAE** Schönherr 1823, The New York weevils
Family 131. **CURCULIONIDAE** Latreille 1802, The snout beetles and true weevils

## Bibliography of General Works

ARNETT, R. H., JR. 1946. The order Coleoptera, parts 1 and 2. Systema Naturae, nos. 3 & 4, pp. 1-32.

ARNETT, R. H., JR. 1958. A list of beetle families. Coleopterists Bulletin, 12: 65-72

ARNETT, R. H., JR. 1960. Coleoptera. McGraw-Hill Encyclopedia of Science and Technology, vol. 3, pp. 275-288.

ARNETT, R. H., JR. 1960. The Beetles of the United States (A manual for identification). The Catholic University of America Press, Washington, D.C. xi + 1112 pp.

BALDUF, W. V. 1935. Bionomics of Entomophagous Coleoptera, New York, 220 pp.

BALL, G. E. 1985. (Editor). Taxonomy, phylogeny and zoogeography of beetles and ants: a volume dedicated to the memory of Philip Jackson Darlington, Jr. (1904-1983). Dr. W. Junk, Publishers, Dordrecht, XIV + 514 pp.

BLACKWELDER, R. E., 1939. Catalogue of the Coleoptera of America, north of Mexico. Fourth Supplement (1933-1938), 146 pp.

BLACKWELDER, R. E. and R. M., 1948. Catalogue of the Coleoptera of America, north of Mexico. Fifth Supplement (1939-1947), 87 pp.

BLACKWELDER, R. E. 1957. Checklist of the Coleopterous insects of Mexico, Central America, the West Indies, and South America. United States National Museum Bulletin 185, 1492 pp. (issued in separate parts, 1944-1957.)

BLATCHLEY, W. S. 1910. An illustrated descriptive catalogue of the Coleoptera or beetles (exclusive of the Rhynchophora) known to occur in Indiana - with bibliography and descriptions of new species. The Nature Publishing Co., Indianapolis. 1386 pp.

BLATCHLEY, W. S. and LENG, C. W. 1916. Rhynchophora or Weevils of North Eastern America. Nature Publishing Co., Indianapolis, 682 pp.

BOUSQUET, Y. (Ed.) 1991. Checklist of the Beetles of Canada and Alaska. Publication 1861/E, Research Branch, Agriculture Canada, Ottawa, Ontario. vi + 430 pp.

BØVING, A. G. and CRAIGHEAD, F. C. 1931. An illustrated synopsis of the principal larval forms of the order Coleoptera. Entomologica Americana, II (new series), 351 pp., 125 pls.

BRADLEY, J. C. 1930. Manual of genera of beetles of America, north of Mexico. Ithaca, N. Y., 360 pp.

BRADLEY, J. C.1947. The classification of Coleoptera. Coleopterists Bulletin, 1: 75-84.

BRIMLEY, C. S. 1938. The insects of North Carolina, being a list of the insects of North Carolina and their close relatives. Raleigh, 560pp.

BRUES, T. C., MELANDER, A. L. and CARPENTER, F. M. 1954. Classification of insects. Bulletin of the Museum of Comparative Zoology, vol. 108, 917 pp., 1197 illus.

CARPENTER, G. H. 1928. The biology of insects. London, 473 pp.

CARPENTER, F. M. 1992. Treatise on Invertebrate Paleontology. Part R, Arthropoda 4, Vol. 3, 4: Superclass Hexapoda. Geological Society of America and University of Kansas, Boulder, Colorado and Lawrence, Kansas, xxi + 655 p.

CASEY, T. L. 1890-96. Coleopterological notices. 7 vols. Annals of the New York Academy of Science (various volumes and pp.)

CASEY, T. L. 1910-24. Memoirs on Coleoptera. 11 vols. Lancaster, Pa. (Published and distributed by the author.)

CHAMBERLIN, W. J. 1939. Bark and Timber beetles of North America. Oregon State College Cooperative Association, 513 pp.

CHU, H. F. 1949. How to know the immature insects. Wm. C. Brown Co., Dubuque, Iowa, 234 pp., 601 figs.

CLAUSEN, C. P. 1940. Entomophagous insects. McGraw-Hill, New York, 688 pp.

COSTA, C., VANIN, S.A., and CASARI-CHEN, S.A. 1988. Larves de Coleoptera do Brasil. Mus. Zool., Universidade de São Paulo, São Paulo. 282 pp, 165 plates.

CRAIGHEAD, F. C. 1950. Insect enemies of eastern forests. United States Department of Agriculture, Miscellaneous Publication no. 657, 679 pp. (Coleoptera, pp. 155-343).

CROWSON, R. A. 1955. The natural classification of the families of Coleoptera. Nathaniel Lloyd and Co., Ltd., London, 187 pp. (first publ. in parts, 1950-1954 in Ent. Mon. Mag., vols. 86-90.)

CROWSON, R. A. 1960. The phylogeny of Coleoptera. Annual Review of Entomology 5:111-134.

CROWSON. R. A. 1975. The evolutionary history of Coleoptera, as documented by fossil and comparative evidence, p. 47-90. *In* Atti del X Congresso Nazionale Italiano di Entomologia. Tip. Coppini, Firenza.

CROWSON, R. A. 1981. The biology of the Coleoptera. Academic Press, London, xii + 802 pp.

DIBB, J. R. 1948. Field book of beetles. A. Brown & Sons, London, 197 pp.

DICE, L. R. 1943. The biotic provinces of North America. University of Michigan Press, Ann Arbor, 78 pp.

DILLON, E. and L. 1960. A manual of common beetles of eastern North America. Row, Peterson and Co., Evanston, 111. 768 pp.

DOWNIE, N.M. and ARNETT, R.H., JR. 1996. The Beetles of Northeastern North America. Volume 1. Introduction, Suborders Archostemata, Adephaga, and Polyphaga thru [sic!] Superfamily Cantharoidea. The Sandhill Crane Press, Publisher,. Gainesville, Florida. xiv + 880 pp.

EDMONDSON, W. T. (ed.) 1959. Fresh-water Biology. Wiley, 1248 pp.

EDWARDS, J. G. 1949. Coleoptera or beetles east of the great plains. Publ. by author, 186 pp., illus.

EDWARDS, J. G. 1950. A bibliographical supplement to "Coleoptera or beetles east of the great plains" applying particularly to western United States. Publ. by author, 183-212 pp.

ERICHSON, W. F., *et al.* 1845-63. Naturgeschichte der Insecten Deutschland. Coleoptera, 4 vols., Berlin.

ERWIN, T. L., BALL, G. E., WHITEHEAD, D. R., and HALPERN, A. L., Eds. 1979. Carabid beetles: their evolution, natural history, and classification. Dr W. Junk bv Publishers, The Hague, The Netherlands. X + 644 pp.

FABRICIUS, J. C. 1775. Systema entomologiae ... 832 pp.

FABRICIUS, J. C. 1776. Genera Insectorum . . . 310 pp.

FABRICIUS, J. C. 1781. Species Insectorum, vol. 1, 552 pp., vol. 2, 494 pp.

FABRICIUS, J. C. 1792-94. Entomologia Systematica, vol. 1, 332 pp., and 538 pp., vol. 2, 519 pp., vol. 3, 487 pp. and 349 pp., vol. 4, 472 pp.

FORBES, W. T. M. 1922. The wing-venation of the Coleoptera. Annals of the Entomological Society of America, 15: 328-352, 7 pls.

FORBES, W. T. M. 1926. The wing folding patterns of the Coleoptera. 34: 42-68; 91-139, 12 pls. 1928.

GAHAN, C. J. 1911. On some recent attempts to classify the Coleoptera in accordance with their phylogeny. Entomologist, 44: 121-125.

GEMMINGER, M. and HAROLD, B. 1868-76. Catalogus Coleopterorum. Monaco, 12 vols., 3986 pp.

HATCH, M. H. 1926. Palaeocoleopterology. Bulletin of the Brooklyn Entomological Society, 21: 137-144.

HATCH, M. H. 1926a. Tillyard on Permian Coleoptera. Bulletin of the Brooklyn Entomological Society, 21: 193.

HATCH, M. H. 1927. A systematic index to the keys for the determination of the Nearctic Coleoptera, Journal of the New York Entomological Society, 35: 279-306.

HATCH, M. H. 1928. A geographical index of the catalogues and local lists of Nearctic Coleoptera. Journal of the New York Entomological Society, 36: 335-354.

HATCH, M. H. 1929. A supplement to the indices to the keys to and local lists of Nearctic Coleoptera. Journal of the New York Entomological Society, 37: 135-143,

HATCH, M. H. 1941. A second supplement to the indices to the keys to and local lists of Nearctic Coleoptera. Journal of the New York Entomological Society, 49: 21-42.

HATCH, M. H. 1953. The beetles of the Pacific Northwest. Part 1: Introduction and Adephaga. University of Washington Publication in Biology, 16: 1-340.

HATCH, M. H. 1957. The Beetles of the Pacific Northwest. Part 11: Staphyliniformia. University of Washington Publication in Biology, 16: 1-384.

HATCH, M.H. 1961. The Beetles of the Pacific Northwest. Part III: Pselaphidae and Diversicornia I. University of Washington Press, Seattle, 503 pp.

HATCH, M.H. 1965. The Beetles of the Pacific Northwest. Part IV: Macrodactyles, Palpicornes, and Heteromera. University of Washington Press, Seattle. viii + 268 pp.

HATCH, M. H. 1971. The beetles of the Pacific Northwest. Part V: Rhipiceroidea, Sternoxi, Phytophaga, Rhynchophora, and Lamellicornia. University of Washington Publications in Biology 16: 662 pp.

HINTON, H. E. 1945. A monograph of the beetles associated with stored products, 443 pp. British Museum, London.

HORION, A. 1949. Käferkunde fur Naturfreunde, Vittorio Klostermann. Frankfurt am Main, 292 pp., 21 pls.

HORN, W. and KAHLE, I. 1935-37. Uber entomologische Sammlungen. Entomologen und EntomoMuseologie. Ent. Beihefte, vols. 2 to 4, pp. 1-536, 38 pls. (contains photographs of determination labels.)

HORN, W. and SCHENKLING, S. 1928-29. Index litteraturae entomologicae. Berlin, 1426 pp.

JACQUES, H. E. 1947. How to know the insects. Wm. C. Brown Co., Dubuque, Iowa, 206 pp., 411 figs.

JACQUES, H. E. 1951. How to know the beetles. Wm. C. Brown Co., Dubuque, Iowa, 372 pp.

JEANNEL, R. (in PIERRE-P., G.). 1949. Les insectes, classification et phylogéne, les insectes fossiles, évolution et géonémie. Traité de zoologie, Masson, Paris, 9: 1-1117. Coleoptera, 771-1077.

JEANNEL, R. and PAULIAN, R. 1944. Morphologie abdominale des Coléoptéres et systématique de l'ordre. Revue Française d'Entomologie, 11: 65-110.

JUNK, W. (publisher) and SCHENKLING, S. (editor) 1910-1940. Coleopterorum Catalogus, 170 pars., 31 vols. (issued in parts, plus supplements, see families sections).

KEEN, F. P. 1952. Insect enemies of western forests. United States Dept. Agric. Misc. Publ. no. 273, 280 pp.

KIRK, V. M. 1969. A list of beetles of South Carolina. Part I - Northern Coastal Plain. South Carolina Agric. Exp. Sta., Tech. Bull. 1033:1-124.

KIRK, V. M. 1970. A list of the beetles of South Carolina. Part 2 - Mountain, Piedmont, and Southern Coastal Plain. South Carolina Agric. Exp. Sta., Tech. Bull. 1038:1-117.

KUKALOVA-PECK, J. and LAWRENCE, J. 1993. Evolution of the hind wing in Coleoptera. Canadian Entomologist, 125: 181-258.

LACORDAIRE, J. T. and CHAPUIS, F. 1854-1876. Histoire naturelle des insectes. Genera des Coléoptéres, 12 vols., Paris.

LAMEERE, A. 1933. Précis de Zoologie. Paris, vol. 5, pp. 273-395.

LAMEERE, A. 1938. Evolution des Coléoptéras. Bull. Ann. Soc. Ent. Belgique, 78: 355-362.

LATREILLE, P. A. 1810. Considerations générales sur l'ordre naturel des animaux composant les classes des Crustacés, des Arachnides et des Insectes . . . Paris, 444 pp.

LAWRENCE, J. F. 1982. Coleoptera, p. 482-553. *In* Synopsis and Classification of Living Organisms. Volume 2. (Ed. S. B. Parker), McGraw-Hill Publ., New York.

LAWRENCE, J. F. (Co-ordinator). Chapter 34. Coleoptera. pp. 144-658. *in* F. Stehr, Editor. Immature Insects. Volume 2. Kendall/Hunt Publishing Co., Dubuque, Iowa.

LAWRENCE, J. F., HASTINGS, A. M., DALLWITZ, M. J., PAINE, T. A. and ZURCHER, E. J. 1999a. Beetle larvae of the World: Descriptions, illustrations, and information retrieval for families and subfamilies. CD-ROM, Version 1.1 for MS-Windows. CSIRO Publishing, Melbourne.

LAWRENCE, J. F., HASTINGS, A. M., DALLWITZ, M. J., PAINE, T. A. and ZURCHER, E. J. 1999b. Beetles of the World: Descriptions, illustrations, and information retrieval for families and subfamilies. CD-ROM, Version 1.0 for MS-Windows. CSIRO Publishing, Melbourne.

LAWRENCE, J. F. and NEWTON, A. F., JR. 1995. Families and subfamilies of Coleoptera (with selected genera, notes, references and data on family-group names), p. 779-1006. *In* Biology, Phylogeny, and Classification of Coleoptera: Papers Celebrating the 80th Birthday of Roy A. Crowson. (Eds. J. Pakaluk and S. A. Slipinski), Muzeum i Institut Zoologii, Polska Academia Nauk, Warsaw.

LECONTE, J. L. 1861-2, 1873. Classification of the Coleoptera of North America. Smithsonian Miscellaneous Collections, 348 pp.

LECONTE, J. L. and HORN, G. H. 1876. The Rhynchophora of America North of Mexico. Proceedings of the American Philosophical Society, 15: 1-455.

LECONTE, J. L. and HORN, G. H. 1883. Classification of the Coleoptera of North America. Smithsonian Miscellaneous Collections, 507, xxxviii + 567 pp.

LEECH, H.B. and CHANDLER, H. P. 1956. Aquatic Coleoptera. pp. 293-371 in Usinger, R. L. Editor. Aquatic insects of California with keys to North American genera and California species. University of California Press, Berkeley and Los Angeles. ix + 508 pp.

LENG, C. W., 1920. Catalogue of the Coleoptera of America, north of Mexico. Mount Vernon, N. Y., 470 pp.

LENG, C. W. and MUTCHLER, A. J., 1927. Catalogue of the Coleoptera of America, north of Mexico. Supplement (1919-1924), 78 pp.

LENG, C. W. and MUTCHLER, A. J., 1933. Catalogue of the Coleoptera of America, north of Mexico. Second and Third Supplements (1925-1932), 112 pp.

LEONARD, D. L. 1928. A list of the insects of New York with a list of the spiders and certain other allied groups. Cornell University Agricultural Experiment Station, Memoir 101:1-1121 (1926).

LINDROTH, C. H. 1957. The principal terms used for male and female genitalia in Coleoptera. Opuscula Entomologica, 22: 241-256.

LINNAEUS, C. 1758. Systema Naturae, 10th ed., 824 pp.

LØDING, H. P. 1945. Catalogue of the beetles of Alabama. Geological Survey of Alabama, Monograph 11:1-172.

LUCAS, R. 1920. Catalogus alphabeticus generum et subgenerurn Coleopterorum. Berlin, 696 pp.

MAYR, E., LINSLEY, E. G. and USINGER, R. L. 1953. Methods and principles of systematic zoology. McGraw-Hill, 328 pp.

MEIXNER, JOSE (in Kukenthal, W. and Krumbach, T.) 1935. Achte dberordnung der Pterygogenea: Coleopteroidea, in Handbuch der Zoologie, 4(2): Insecta 2: 1037-1382.

MICHNER, C. D. and M. H. 1951. American Social Insects, D. Van Nostrand, New York, 267 pp.

NEAVE, S. A. 1939-40. Nomenclator zoologicus, 1758-1935, Zoological Society, London, 4 vols.

NEAVE, S. A. 1950. Nomenclator zoologicus (1936-1945), Zoological Society, vol. 5, 308 pp.

NEEDHAM, J. G., *et al.* 1928. Leaf-mining insects. Baltimore, 351 pp.

PAULIAN, R. 1943. Les Coléoptéres; formes - Moeurs - Role. Bibliothèque Scientifique. Payot. Paris. 396 pp., 164 figs., 14 pls.

PAULIAN, R. 1944. La vie des scarabées. Galliurard, Paris, 235 pp.

PECK, S.B., and M.C. THOMAS. 1998. A distributional checklist of the beetles (Coleoptera) of Florida. Arthropods of Florida and Neighboring Land Areas, 16: i-viii + 1-180.

PENNAK, R. W. 1953. Fresh-water invertebrates of the United States. Ronald Press, New York, 769 pp.

PETERSON, A., 1951. Larvae of insects. An introduction to Nearctic species. Part II. Coleoptera, Diptera, Neuroptera, Siphonaptera, Mecoptera, Trichoptera. Columbus, Ohio, v + 1-416 pp.

PETERSON, A. 1955. A manual of entomological technique. Publ. by author, Columbus, Ohio, 367 pp.

PEYERIMHOFF, P. DE 1933. Les larves des Coléoptéres d'apres A. G. Bøving et F. C. Craighead et les grands criteriums de l'ordre. Ann. de Soc. Ent. France, 102: 77-106.

POOLE, R. W. and P. GENTILI, 1996. Nomina Insecta Nearctica: A Check List of the Insects of North America, Volume 1: Coleoptera, Strepsiptera. Entomological Information Services, Rockville, MD. 827 pp.

REITTER, E. 1908-16. Fauna Germanica. Die Käfer des Deutschen Reiches, Stuttgart. 5 vols.

ROCKCASTLE, V. N. 1957. Beetles. Cornell Rural School Leaflet, 50(4): 1-32.

ROEDER, K. D. (ed.) 1953. Insect physiology. Wiley, New York, 1100 pp.

SEEVERS, C. H. 1838. Termitophilous Coleoptera in the United States. Ann. Ent. Soc. America, 31: 422-441.

SHARP, D., CHAMPION, G. C., *et al.* 1887-1909. Coleoptera. Biologia Central i-Americana, 7 vols.

SHARP, D. 1909. Insects, part II, pp. 184-303 (Coleoptera). Cambridge Natural History, Macmillan, New York.

SHARP, D. and MUIR, F. 1912. The comparative anatomy of the male genital tube in Coleoptera. Transactions of the Entomological Society of London, 1912, 477-642 pp., 78 pls.

SNODGRASS, R. E. 1935. Principles of Insect Morphology. New York, 667 pp.

STEHR, F. (Ed.) 1991. Immature Insects. Volume 2. Kendall/Hunt Publishing Co., Dubuque, Iowa. xvi + 975 pp.

STICKNEY, F. S. 1923. The head-capsule of Coleoptera. Illinois Biological Monographs, 8(l): 1-51, 26 pls.

TANNER, V. M. 1927. A preliminary study of the genitalia of female Coleoptera. Transactions of the American Entomological Society, 53: 5-50.

THEODORIDES, J. 1950. The parasitological, medical and veterinary importance of Coleoptera. Acta Tropica, 7: 48-60 (contains extensive bibliography).

THEODORIDES, J. 1952. Les Coléoptéres fossiles. Ann. Soc. Ent. France, 121: 23-48 (extensive bibliography).

TILLYARD, R. J. 1924. The evolution of the class Insecta. American Journal of Science, 23: 529-539.

DE LA TORRE-BUENO, J. R. 1937. A glossary of entomology. Brooklyn Entomological Society, 336 pp.

TUXEN, S. L. (ed.) 1956. Taxonomists' glossary of genitalia in insects, Cophenhagen. 284 pp.

USINGER, R. L., *et al.* 1948. Biology of aquatic and littoral insects. University of California Press, 244 pp.

USINGER, R. L., *et al.* 1956. Aquatic insects of California with keys to North American genera and California species. Univ. California Press, 508 pp.

WHEELER, W. M. 1913. Ants. Columbia Univ. Press, 663 pp.

WHEELER, W. M. 1923. The Social Insects. Harcourt Brace and Co., 378 pp.

WILLIAMS, I. W. 1938. The comparative morphology of the mouthparts of the order Coleoptera treated from the standpoint of phylogeny. Journal of the New York Entomological Society, 46: 245-289, 11 pls.

WILSON, C. B. 1923. Water beetles in relation to pondfish culture, with life histories of those found in fishponds at Fairport, Iowa. Bulletin of the Bureau of Fisheries, 39: 231-345.

WADE, J. S. 1935. A contribution to a bibliography of the described immature stages of North American Coleoptera, United States Department of Agriculture, E-358, 114 pp.

WYTSMAN, P. (ed.) 1902-date. Genera Insectorum. Bruxelles. (various vols. on Coleoptera.)

YOUNG, F. N. 1954. The water beetles of Florida. University of Florida Press, 238 pp.

Suborder ARCHOSTEMATA Kolbe, 1908

# 1. CUPEDIDAE Laporte, 1836

by Daniel K. Young

**Family common name**: The reticulated beetles

**Family synonyms**: Cupesidae Alluaud, 1900.

Extant cupedids are but relictual members of a once more diverse lineage dating from the Triassic (Carpenter 1992). Adults can sometimes be found around and beneath bark of dead logs, the general habitat of larvae, or in the evening at light. The western Priacma serrata (LeConte) can frequently be attracted in large numbers to bleach containing sodium hypochlorite (Edwards 1951; Atkins 1957). They are easily distinguished by their flattened, parallel-sided body, rows of closely placed, square elytral "window punctures," and vestiture of broad scale-like setae.

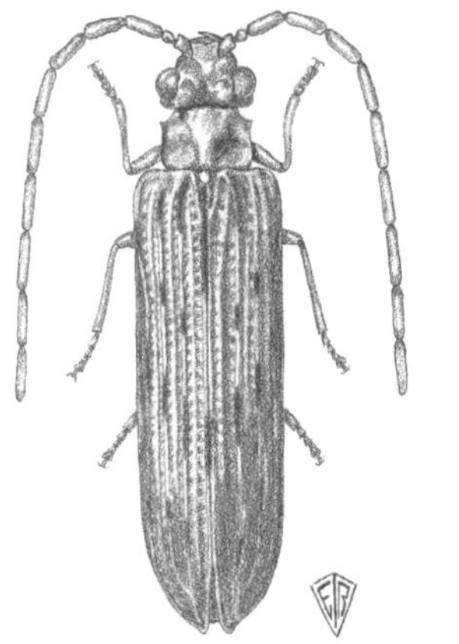

FIGURE 1.1. *Tenomerga cinereus* (Say).

**Description**: Body elongate, length 5-25 mm, more or less parallel-sided, moderately to strongly flattened, the dorsal surface irregularly sculptured; color brownish or black and gray; vestiture consisting of broad scale-like setae nearly covering body.

Head usually short, subquadrate with dorsal protuberances; antennae 11-segmented, filiform but thick, longer than combined head and prothorax, situated dorsally, with insertions elevated; mandibles short, blunt, with a single apical tooth; palpi small, apical segment of maxillary palpi bearing exposed, digitiform sensillae [sensillae enclosed in deep pit in Micromalthidae and Ommatidae (Australia, South America) and lacking in Crowsoniellidae (Italy)]; eyes usually large, protruding, entire, finely faceted.

Thorax: Pronotum irregularly sculptured, with more or less distinct lateral margins and acute subapical angles; notopleural suture always distinct; prosternum with distinct protarsal grooves (absent in *Priacma*); legs slender, prothoracic coxae separated by prosternum which is extended behind and fits into small, mesoventral cavity; tarsal formula 5-5-5, tarsomeres tomentose beneath, penultimate tarsomere expanded, apically emarginate and/or with ventral lobe; elytra broader than pronotum, elongate, strongly costate, with rows of quadrate "window punctures," parallel-sided; elytral epipleura narrow, lacking cells.

Abdomen with five visible, distinctly overlapping sterna; male genitalia usually with bifurcate V-shaped sclerite between tergite nine and aedeagus, parameres usually with supplementary subapical lobes and long, attenuate ventromarginal spines; female genitalia with styli on the coxites and with valvifers mostly membranous. Three thoracic and eight abdominal ganglia present in *Priacma serrata*; four Malpighian tubules in *Priacma serrata* (as in Adephaga).

Larvae of known Cupedidae [*Distocupes* (Australia), *Priacma serrata* (first instar only), and *Tenomerga*] lack stemmata, have a single stemma on either side of the head capsule, or possess a few poorly developed eye spots; possess a single, simple median endocarina; have six-segmented legs; and possess a median sclerotized process on the ninth abdominal tergum.

**Habits and habitats**. Larvae are wood-borers, living in fairly firm, but fungal-infested wood through which they navigate with the aid of their asperate thoracic and abdominal ampullae. Sometimes they have been found in structural timbers. *Priacma serrata* males have been attracted in great numbers to chlorine bleaches. Adults of *Tenomerga* are occasionally attracted to lights at night.

**Status of the classification**. Cupedid classification has had a colorful history, starting with Fabricius who described the first species as a chrysomelid, presumably because of the hispine-like elytral punctures. Latreille (1810) realized it was not a leaf-beetle and placed it in the lampyroid complex because of the 5-5-5 tarsal formula, but later suggested it might belong near Rhysodidae. Lacordaire did not agree, and placed the group near Ptinidae (= Anobiidae) because of the similarly inserted antennae. Ganglbauer (1903) placed them in Adephaga. Kolbe (1908) recognized the distinctiveness of these beetles and erected Archostemata, including it in his division Symphytogastra of the suborder Heterophaga. Sharp and Muir (1912) placed Cupedidae in Byrrhoidea. In 1920, Leng concluded that they should be in a superfamily by themselves, but placed it near Lymexylidae. Forbes (1926) placed Cupedidae together with Micromalthidae in Archostemata as a separate suborder, based on the spiral rolling of the distal metathoracic wings during folding. Bøving and Craighead (1931) supported the suborder treatment based upon larval structures. Since the 1930s, Cupedidae has generally been placed as one of the most primitive beetle families, in the suborder Archostemata. Ommatidae, based on the Australiam *Omma*

FIGURE 2.1. *Priacma serrata* (LeConte) (after Hatch 1953).

was first proposed as a family distinct from Cupedidae on the basis of male genitalia (Sharp and Muir 1912), but most subsequent works continued to include it in Cupedidae until 1976, when Crowson supported elevation of his Ommatini (*Omma* Newman) to Ommatidae and combined his Tetraphalerini (*Tetraphalerus* Waterhouse, South America) with the Italian *Crowsoniella* to form the family Tetraphaleridae. Lawrence and Newton (1995) noted the putitive maxillary palpal synapomorphy (digitiform sensillae enclosed in a deep pit) in uniting *Omma* and *Tetraphalerus* in Ommatidae, but excluding *Crowsoniella* (as Crowsoniellidae). In the most recent classification (Lawrence 1999), Cupedidae has three subfamilies: Mesocupedinae (Lower Triassic to Lower Cretaceous), Priacminae (Upper Jurassic to Recent) and Cupedinae (Oligocene to Recent).

**Distribution.** Cupedidae exhibits an ancient, Pangean distribution with species known from all continents and most of the larger continental islands. The extant world fauna consists of approximately 30 species. Four genera, each with a single species, are known from the United States and Canada.

Only nine genera are known: *Adinolepis* Neboiss has four Australian species; *Ascioplaga* Neboiss is represented by two species endemic to New Caledonia; *Cupes capitatus* Fabricius (monotypic) is indigenous to eastern North America; *Distocupes varians* (Lea) is known only from eastern Australia; *Paracupes brasiliensis* Kolbe occurs in Brasil and Ecuador; the monotypic *Priacma serrata* (LeConte) is known from montane forests of western North America; *Prolixocupes* Neboiss consists of *P. laterillei* (Solier) from the mountains of Argentina and Chile, while *P. lobiceps* (LeConte) is known from California and the southern regions of Arizona; *Rhipsideigma* Neboiss is represented by one East African and three Malagesian species. The most diverse, extant genus is *Tenomerga* Neboiss, with 10 widely distributed species.

## Key to the Nearctic Genera and Species

1. Prosternum without distinct tarsal grooves; antennae short, scarcely half as long as body (Fig. 1.2) ................................ *Priacma serrata* (LeConte)

— Prosternum with distinct tarsal grooves; antennae longer, usually distinctly longer than half length of body (Fig. 1.1) ................................................ 2

2 (1). All four head tubercles obtuse, posterior pair distinctly larger than anterior pair; antennae slightly serrate; tarsal grooves of prosternum separated anteriorly by nearly entire width of prosternum ........................ *Prolixocupes lobiceps* (LeConte)

— One or both pairs of head tubercles conical, of subequal size; antennae filiform; tarsal grooves of prosternum separated anteriorly by one or two mesal carinae or low ridges .............................. 3

3(2). Prosternal tarsal grooves separated anteriorly by a pair of low ridges near the meson ...................... .................................. *Cupes capitatus* Fabricius

— Prosternal tarsal grooves separated anteriorly by a distinct, single mesal carina .............................. ................................ *Tenomerga cinereus* (Say)

## Classification of the Nearctic Genera

### Cupedidae

*Cupes* Fabricius, 1801, one species, *C. capitatus* Fabricius, Eastern United States and Canada.

*Priacma* LeConte, 1874, one species, *P. serrata* (LeConte), mountains of Western United States and Canada (BC, WA, OR, ID, MT, CA).

*Prolixocupes* Neboiss, 1959, one species, *P. lobiceps* (LeConte), southern Arizona and California.

*Tenomerga* Neboiss, 1984, one species, *T. cinereus* (Say), Eastern United States and Canada.

## Bibliography

ATKINS, M. D. 1957. An interesting attractant for *Priacma serrata* (Lec.), (Cupesidae: Coleoptera). The Canadian Entomologist, 89: 214-219.

ATKINS, M. D. 1958a. Observations on the flight, wing movements and wing structure of male *Priacma serrata* (Lec.) (Coleoptera: Cupedidae). The Canadian Entomologist, 90: 339-347.

ATKINS, M. D. 1958b. On the phylogeny and biogeography of the family Cupedidae (Coleoptera). The Canadian Entomologist, 90: 532-537.

ATKINS, M. D. 1963. The Cupedidae of the World. The Canadian Entomologist, 95: 140-162.

ATKINS, M. D. 1979. A catalog of the Coleoptera of America North of Mexico. Family Cupesidae. United State Department of Agriculture, Agriculture Handbook, 529-1: 1-5.

BARBER, G. W. and ELLIS, W. O. 1920. The Cupesidae of North America north of Mexico. Journal of the New York Entomological Society, 28: 197-208.

BØVING, A. G. and CRAIGHEAD, F. C. 1931. An illustrated synopsis of the principal larval forms of the order Coleoptera. Entomologica Americana (N.S.) 11(1930): 1-351.

CARPENTER, F. M. 1992. Treatise on Invertebrate Paleontology. Part R, Arthropoda 4, Vol. 3, 4: Superclass Hexapoda. Geological Society of America and University of Kansas, Boulder, Colorado and Lawrence, Kansas, xxi + 655 p.

CROWSON. R. A. 1975. The evolutionary history of Coleoptera, as documented by fossil and comparative evidence, p. 47-90. *In* Atti del X Congresso Nazionale Italiano di Entomologia. Tip. Coppini, Firenza.

EDWARDS, J. G. 1951. Beetles attracted to soap in Montana. Coleopterists Bulletin, 5: 42-43.

EDWARDS, J. G. 1953a. Peculiar clasping mechanisms of males of *Priacma serrata* (Lec.). Coleopterists Bulletin, 7: 17-20.

EDWARDS, J. G. 1953b. The morphology of the male terminalia of beetles belonging to the genus *Priacma* (Cupesidae). Bulletin of the Institute Royal Science Nat. Belgique, 29(28): 1-8.

FORBES, W. T. M. 1922. The wing-venation of Coleoptera. Annals of the Entomological Society of America, 15: 328-345, pls. 29-35.

FORBES, W. T. M. 1926. The wing folding patterns of the Coleoptera. Journal of the New York Entomological Society, 34: 42-115, pls. 7-18.

HATCH, M. H. 1953. The beetles of the Pacific Northwest. Part I: Introduction and Adephaga. University of Washington Publications in Biology, 16: 1-340.

LAWRENCE, J. F. 1982. Coleoptera, p. 482-553. *In* Synopsis and Classification of Living Organisms. Volume 2. (Ed. S. B. Parker), McGraw-Hill Publ., New York.

LAWRENCE, J. F. 1991. Cupedidae, p. 298-300. *In* Immature Insects. Volume 2. (Ed. F. W. Stehr), Kendall-Hunt Publ., Dubuque, Iowa.

LAWRENCE, J. F. 1999. The Australian Ommatidae (Coleoptera: Archostemata): new species, larva, and discussion of relationships. Invertebrate Taxonomy, 13: 369-390.

LAWRENCE, J. F. and NEWTON, A. F., JR. 1995. Families and subfamilies of Coleoptera (with selected genera, notes, references and data on family-group names), p. 779-1006. *In* Biology, Phylogeny, and Classification of Coleoptera: Papers Celebrating the 80th Birthday of Roy A. Crowson. (Eds. J. Pakaluk and S. A. Slipinski), Muzeum i Institut Zoologii, Polska Academia Nauk, Warsaw.

SHARP, D. and MUIR, F. A. G. 1912. The comparative anatomy of the male genital tube in Coleoptera. Transactions of the Entomological Society of London, (1912): 477-642, pls. 42-78.

TANNER, V. M., 1927. A preliminary study of the genitalia of female Coleoptera. Transactions of the American Entomological Society, 53: 5-50.

# 2. MICROMALTHIDAE Barber, 1913

by T. Keith Philips and Daniel K. Young

**Common name:** The telephone-pole beetles

These very small beetles are easily recognized by their cantharid-like appearance. Unlike cantharids, they have large heads, and moniliform antennae. They are dark brown to black in color with yellowish legs and antennae.

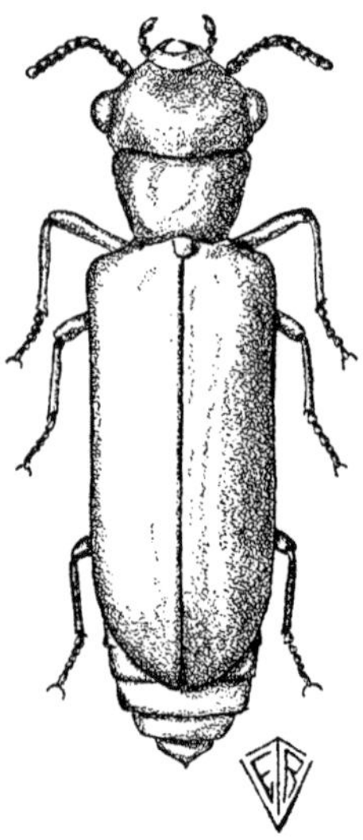

FIGURE 1.2. *Micromalthus debilis* LeConte.

**Description**: Elongate, somewhat depressed, and relatively soft bodied, 1.5 to 2.5 mm in length; dark brown to piceous, with pale yellow legs and antennae: vestiture sparse, limited to scattered long, erect, pale setae on lateral margin of elytra and pronotum, front, antennae, legs, and apex of abdomen.

Head (Stickney 1923) large and prognathous, broader than thorax, with large round eyes protruding from head; antennae short with eleven antennomeres, moniliform, first two antennomeres large, third small, 4 to 11 gradually increasing in diameter with 11th nearly as large as 2nd; mandibles large, curved toward each other with apices acutely three-toothed; clypeus and labium small, labial and maxillary palpi large, apical maxillary palpomere with digitiform appendages located in a large cavity.

Thorax with pronotum small, flattened above, oval, slightly narrower than the head, without margins or sutures, also lacking prosternal process and both notopleural and sternopleural sutures; legs with 5-5-5 tarsi; trochanters interstitial; hind coxae cylindrical, separate, metatrochantins exposed; elytra with prominent shoulders, flattened above, sides parallel, apices rounded, shortened, exposing about five abdominal tergites; wings with apical half spirally rolled, venation reduced, oblongum cell absent (Forbes 1926).

Abdomen with six overlapping ventrites, with the second sternite exposed, the male has ventrites 3-5 each with large median setose cavities; male genitalia distinctive and similar to other Archostemata in shape of pregenital ring, lack of phallobase, and parameres from near base with two long slender spines (Paterson 1938).

Larvae: [Micromalthidae exhibits hypermetamorphic development; the following description is based largely on the actively feeding, cerambycoid instars, unless specifically noted otherwise] vestiture sparse but stout on most segments; color white. Head enlarged, larger than prothorax, occasionally with a single stemma on each side. Antennae three-segmented, with an elongate third segment and well-developed sensorium. Clypeus and free labrum present; mandibles robust, longer than wide; apices bidentate, mola transversely ridged, protheca distinct, elongate, narrow, and sclerotized. Maxilla short, with short, transverse cardo, stipes distinct with three-segmented palpi; lobe-like setiferous articulated galea, and small fixed lacinia. Mentum with a large ligular sclerome, and one-segmented palpi, with an elongate appendage at their base. Thorax three-segmented, transverse, dorsal and ventral ampullae beginning on the mesothorax; legs when present (caraboid 1st instar) four-segmented and long, with a very long and narrow tarsus and two pretarsal claws, each with a long basal seta. Abdomen nine-segmented, anal region surrounded by opposing, sclerotized, distally toothed dorsal and ventral acuminations, with an additional pair of large, fleshy lobes bilaterally. Spiracles annular, small, on mesothorax and abdominal segments one to eight (Bøving and Craighead 1931, Lawrence 1991, Peterson 1951).

**Habits and habitats.** The larvae are wood borers, feeding in moist, decaying oak or chestnut logs in the red-rotten or yellowish-brown stage of decay, and are reported as causing damage to buildings and poles. They may be common, but are seldom recognized and individual collecting events appear rare. The complicated life cycle involves paedogenesis and several types of parthenogenesis. The larvae exhibit several stages and shapes in their development, including caraboid, cerambycoid, and curculionoid larval types. The caraboid larva is very mobile and molts into a cerambycoid larva. Perhaps the most unusual feature is that the cerambycoid larva can either develop into an adult female, or a paedogenetic female-producing larva which gives birth to caraboid larvae. The cerambycoid larva can also develop into a male producer. These individuals produce eggs which hatch as a curculionoid larvae and later develop into haploid adult males (Barber 1913a, b; Pringle 1938; Scott 1936, 1938, 1941).

**Status of the classification.** This family, represented by a single described species, has been subject to much discussion and controversy. This species has even recently been considered part of the Polyphaga and included in the Lymexylidae (Lymexyloidea), the Telegusidae (Cantharoidea), or as a separate family in the Cantharoidea (or Elateroidea *sensu* Lawrence and Newton [1995]). But based on overwhelming evidence from larval, wing, and male genitalic characters, the family is now placed in the suborder Archostemata (Lawrence 1999). Although the family contains only one described species (Schenkling 1915), what may be a second species is known from first instars discovered in Hong Kong. Fossils (Cretaceous through to perhaps Miocene in age) representing additional species are known from amber collected in Mexico, Lebanon, the Baltic, and the Dominican Republic (Rozen 1971, Ivie and Miller, *in litt.*).

**Distribution.** Native to the eastern United States and perhaps Belize, the single species has been widely distributed by commerce, and its range now includes British Columbia, New Mexico and the overseas localities of Brazil, Cuba, Hong Kong, Hawaii, and South Africa (Borror *et al.* 1986, Downie and Arnett 1996, Marshall and Thornton 1963, Philips 2000, Scholtz and Holm 1985). Although there is no reason why it may not be present in Europe, this record is in error as Silvestri (1941) only thought the species should be found there due to the importation of timber from the New World.

## Classification of the Nearctic Genus

Micromalthidae

*Micromalthus* LeConte, 1879, 1 sp., *M. debilis* LeConte, 1878, eastern United States, Belize.

## Bibliography


BARBER, H. S. 1913a. The remarkable life history of a new family (Micromalthidae) of beetles. Proceedings of the Entomological Society of Washington, 15: 31-38.

BARBER, H. S. 1913b. Observations on the life history of *Micromalthus debilus* Lec. (Coleoptera). Proceedings of the Biological Society of Washington, 26: 185-190.

BORROR, D. J., TRIPLEHORN, C.A., and JOHNSON, N.F. 1986. An introduction to the study of insects, Sixth Edition, Saunders College Publishing, Philadelphia, 875 pp.

BOVING, A. G. and CRAIGHEAD, F. C. 1931. An illustrated synopsis of the principle larval forms of the order of Coleoptera. Brooklyn Entomological Society, Brooklyn, NY, 351 pp.

DOWNIE N. M. and ARNETT, R. H. 1996. The beetles of Northeastern North America, Vol. 1. The Sandhill Crane Press, Gainesville, Florida, 880 pp.

FORBES. W. T. M. 1926. The wing folding patterns of the Coleoptera. Journal of the New York Entomological Society, 34:42-68, 91-138.

LAWRENCE, J. F. 1991. Ommatidae, Cupedidae, Micromalthidae. *In:* Stehr, F. (ed.) Immature Insects, Vol. 2. Kendall-Hunt Publishing Co., Dubuque, Iowa, [xvi] + 975 pp.

LAWRENCE, J. F. 1999. The Australian Ommatidae (Coleoptera: Archostemata): new species, larva and discussion of relationships. Invertebrate Taxonomy, 13: 369-390.

LAWRENCE, J. F. and NEWTON, A. F., Jr. 1995. Families and subfamilies of Coleoptera (with selected genera, notes, references, and data on family-group names). pp. 779-1006. *In*: Pakaluk, J. and Slipinski, S. A. (eds.), Biology, Phylogeny, and Classification of the Coleoptera: Papers celebrating the 80th birthday of Roy A. Crowson. Muzeum i Instytut Zoologii PAN, Warzawa.

MARSHALL, A. T. and THORNTON, I. W. B. 1963. *Micromalthus* (Coleoptera: Micromalthidae) in Hong Kong. Pacific Insects, 5: 715-720.

PATERSON, N. F. 1938. Notes on the external morphology of South African specimens of *Micromalthus* (Coleoptera). Transactions of the Royal Entomological Society of London, 87: 287-290.

PETERSON, A. 1951. Larvae of Insects, Part II. Edwards Brothers, Inc., Ann Arbor, MI, 416 pp.

PHILIPS, T. K. 2000. A record of *Micromalthus debilus* (Coleoptera: Micromalthidae) from Central America and a discussion of its distribution. Florida Entomologist, 83: (in press).

PRINGLE, J. A. 1938. A contribution to the knowledge of *Micromalthus debilus* LeC. (Coleoptera). Transactions of the Royal Entomological Society of London, 87: 271-286.

ROSEN, J. G. 1971. *Micromalthus debilus* LeConte from amber of Chiapas, Mexico (Coleoptera: Micromalthidae). University of California Publications in Entomology, 63: 75-76.

SCHENKLING, S. 1915. Micromalthidae. Coleopterorum Catalogus, 10 (64): 14.

SCHOLTZ, C. H. and HOLM, E. 1985. Coleoptera. pp. 188-280. *In*: Scholtz, C. H. and Holm, E. (eds.), Insects of Southern Africa, Butterworth Publishers (Pty) Ltd., Durban.

SCOTT, A. C. 1936. Haploidy and aberrant spermatogenesis in a coleopteran, *Micromalthus debilus* LeConte. Journal of Morphology, 59: 485-515.

SCOTT, A. C. 1938. Paedogenesis in the Coleoptera. Z. Morphol. Ökol. Tiere, 3: 633-653.

SCOTT, A. C. 1941. Reversal of sex production in *Micromalthus*. Biological Bulletin, 81: 220-231.

SILVESTRI, F. 1941. Distribuzione geografica del *Micromalthus debilus* LeConte (Coleoptera: Micromalthidae). Bolletinno della Societa Entomologica Italiana, 73: 1-2.

STICKNEY, F. S., 1923. The head capsule of Coleoptera. Illinois Biological Monographs, 8(1): 1-105.

## Suborder MYXOPHAGA Crowson, 1956

# 3. MICROSPORIDAE Crotch, 1873

by W. Eugene Hall

**Family common name:** The minute bog beetles

**Family synonym:** Sphaeriidae Erichson, 1845

The small, obscure beetles of this family resemble the Orthoperidae, but may be distinguished from other families by the comparatively large and prominent head, capitate antennae, design of meso- and metasternum, large posterior coxal plates and the unequal length of the visible abdominal sternites.

FIGURE 1.3. *Microsporus* sp.

**Description:** Form greatly convex, subhemispherical, color dark brown or black, body dorsal surface smooth, glabrous, glossy, occasional markings, head prominent, antennae yellowish with long setae on terminal segments. Length: 0.5 - 1.2 mm.

Head large, somewhat produced; antennae inserted near eyes, extended well beyond margin of head, 11 antennomeres, segment I large, segment II nearly as large, segment III long and slender, as long as segments I and II combined, segments IV - VI nearly equal in length, but combined shorter than segment III, segments VII and VIII short, segments IX - VI forming a conical club, segment XI with long setae and sensorial discs on apices; labrum nearly as long as wide; mentum widest at base, narrowed anteriorly; mandibles bifid, curved, prostheca present on left mandible; maxillary palp 4-segmented, segment I small, segment II long, slender, segment III large, suboval, segment IV slender, dilated at apicex; galea and lacinia forming a single mala; eyes prominent, interfacetal setae absent.

Prothorax expanded posteriorly, widest at hind angles; prosternum short, procoxae moderately separated, coxal cavities open, procoxae globular, prosternal process present between procoxae, narrow; notopleural suture present. Mesosternum reduced in size, fused to larger metasternum, forming a large plate, mesocoxae widely separated. Metacoxae nearly contiguous, coxal plates large, covering hind femur and first visible abdominal sternite, trochanters of legs short, femora robust, widely grooved on under side with edge of grooves acutely angulated near middle; tibiae broadly dilated, widest near middle, outer margin deeply excised near apex and armed with short spines in the upper angle of excision, tarsi 3-3-3, first long, subcylindrical, second much smaller than first, third elongate, claws moderate, simple, unequal.

Elytra complete, covering all abdominal segments, with distinct epipleural fold. Wing membrane large, wide, fringed with long setae along margin of membrane, possessing a distinct closed cell close to the posterior margin near middle.

Abdomen with three visible sternites, first large and long, second narrow, third wide and long; adominal tergite VIII extended anteriorly, medially narrowing segments VI - VII. Male and female genitalia undescribed.

**Larva:** Description based on synopsis by Beutel *et al.* (1999) and Britton (1966): Body small, flattened, 0.65 mm in length, head large, posteroventral area of head retracted into prothorax, head broader than long, deflexed, 4 large stemmata present lateral to antennae, labrum articulated, frontoclypeal, frontal and coronal sutures absent, clypeal area smooth, labrum well developed, covering mandibles and maxillary mala, antennae short but well developed, directed posteriorly, segment I long and broad, segments II and III fused, sensorial appendage elongate, narrow, mandibles small, molar well developed, apical teeth and subapical prostheca reduced or absent, maxilla deeply inserted into articulatory area, cardo short, palpifer well developed, 2-segmented, proximal palpomere possessing sensory structures, galea and lacinia fused and forming sclerotized mala, mentum and submentum fused, prementum unsclerotized, ligula large and divided, hypopharynx covered by labrum. Thorax broader than head, rounded laterally, approximately half the length of abdomen, pronotum 2.5 times broader than long with lateral extensions, legs short, 5-segmented, widely separated, coxae large, trochanter indistinct, femur and tibiotarsus short, single claw, mesothorax broad, shorter than pronotum, sternum broad, unsclerotized, metathorax broad, shorter than mesonotum. Abdomen nearly as broad as thorax, narrowed posteriorly, lancet-shaped setae on tergites I - V, segments I - VIII possessing lateral projections and balloon-shaped spiracular gills, sternum unsclerotized, segment IX with posterolateral projections, urogomphi absent, gills and lancet-shaped setae absent, sternum well developed, segment X inserted on ventral side of segment and possessing 3 membranous lobes, each with a pair of hooks.

**Habits and habitats.** These beetles occur in mud, under stones or on algae along the edge of streams and rivers, among the roots of plants, in mosses associated with bogs, or inhabiting moist leaf litter further away from bodies of water (Löbl, 1995). Adult microsporids store air beneath their elytra, but lack a dense mat of setae forming a plastron as in *Hydroscapha* (Beutel 1998/1999; Britton 1966; Hinton 1967; Messner and Joost 1984). Females produce a large, single egg at a time (Britton 1966). Adults and larvae feed on algae.

**Status of the classification.** This small group of beetles has long been recognized as a distinct family, but in the past has been placed in various suborders and superfamilies. While it is generally agreed upon that Microsporidae currently belong in the suborder Myxophaga, controversy remains regarding the family-group name (Jach, 1999).

LeConte and Horn (1883) placed Microsporidae (=Sphaeriidae) between Hydroscaphidae and Scaphidiidae, and stated rather prematurely: "The relations between this family and Trichopterygidae (=Ptiliidae) are so obvious as to require no further elucidation." Matthews (1899) placed the microsporids near the hydroscaphids. Kolbe (1901) and Ganglbauer (1903) both placed Microsporidae within Staphylinoidea. Stickney (1923) and Williams (1938), using Leng's catalog (1920), include microsporids within Staphylinoidea, and briefly comment on shared characters between microsporids and ptiliids. On the basis of hind wing venation, Forbes (1926) compared "the spring-like structure which closes the wing" to adephagans, but conceded that the body structures lacked adephagan attributes, yet placed microsporids in the adephagan superfamily Hydradephaga. Bøving and Craighead (1931) and Jeannel and Paulian (1944) placed microsporids among the Staphylinoidea. Crowson (1955) erected a new suborder, Myxophaga, partially based on the presence of notopleural sutures. Myxophaga currently contain the families Microsporidae, Hydroscaphidae, Lepiceridae and Torridincolidae. Hinton (1967) proposed the type of respiratory system in larval myxophagans, including the presence of spiracular gills, validates the suborder. Reichardt (1973) presents evidence suggesting Myxophaga represent a primitive group of beetles possessing characters intermediate between Adephaga and Polyphaga. Based on characters of the hind wing and other morphological and ecological features, Kukalova-Peck and Lawrence (1993) recognize Myxophaga as a sister group to Adephaga. Recent coleopteran classifications (Lawrence, 1982; Lawrence and Newton 1995, 1982) retain Microsporidae within Myxophaga. Phylogenetic analyses of adult and larval morphological characters (Beutel 1998/1999; Beutel *et al.* 1999) support the monophyly of Myxophaga.

**Distribution.** Microsporidae consist of 19 species that in Africa (including Madagascar), Asia, Australia, Europe and North and Central America (Endrödy-Younga 1997).

## Classification of the Nearctic Species

*Microsporus* Kolenati 1846

*Sphaerius* Waltl 1838

Three nominal species are known from Texas, Arizona, California, and Washington. The genus is in need of revision.

## Bibliography

BEUTEL, R. G., MADDISON, D. R. and HAAS, A. 1999. Phylogenetic analysis of Myxophaga (Coleoptera) using larval characters. Systematic Entomology, 24: 171 - 192.

BEUTEL, R. G. 1998/1999. Phylogenetic analysis of Myxophaga (Coleoptera) with a redescription of *Lepicerus horni* (Lepiceridae). Zoologischer Anzeiger: 291 - 308.

BØVING, A. G. and CRAIGHEAD, F. C. 1931. An illustrated synopsis of the principal larval forms of the order Coleoptera. Entomologia Americana (New Series), 11: 1 - 351.

BRITTON, E. B. 1966. On the larva of *Sphaerius* and the systematic position of the Sphaeriidae (Coleoptera). Australian Journal of Zoology, 14: 1193 - 1198.

CROTCH, G. R. 1873. On the arrangement of the families of Coleoptera. Proceedings of the American Philosophical Society, 13: 75 - 87.

CROWSON, R. A. 1955. The natural classification of the families of Coleoptera. N. Lloyd, London, 187 pp.

CSIKI, E. 1910. Coleopterorum Catalogus, Volume 8 (18). Family Sphaeriidae. W. Junk, Berlin. pp. 34 - 35.

ENDRÖDY-YOUNGA, S. 1997. Microsporidae (Coleoptera: Myxophaga), a new family for the African continent. Annals of the Transvaal Museum, 36 (23): 309 - 311.

ERICHSON, W. F. 1845. Naturgeschichte der Insecten Deutschlands, 3: 1 - 320.

FORBES, W. T. M. 1926. The wing folding patterns of the Coleoptera. Journal of the New York Entomological Society, 34: 61, fig. 21.

GANGLBAUER, L. 1903. Systematisch-koleopterologische Studien. Münchener Koleopterologische Zeitschrift, 1: 271 - 319.

HINTON, H. E. 1967. On the spiracles of the larvae of the suborder Myxophaga (Coleoptera). Australian Journal of Zoology, 15: 955 - 959.

JACH, M. A. 1999. *Sphaerius* Waltl, 1838 and Sphaeriusidae Erichson, 1845 (Insecta, Coleoptera): proposed conservation by the partial revocation of Opinion 1331. Bulletin of Zoological Nomenclature, 56 (2): 117 - 120.

JEANNEL, R. and PAULIAN, R. 1944. Morphologie abdominale des coléoptères et systématique de l'ordre. Revue Française d'Entomologie, 11: 65 - 110.

KOLBE, H. J. 1901. Vergleichend-morphologische Untersuchungen an Coleopteren nebst Grundlagen zu einem System und zur Systematik derselben. Archiv für Naturgeschichte, 67 (Beiheft): 89 - 150, Tafel II - III.

KOLENATI, F. A. 1846. Insecta Caucasi. Coleoptera, Dermaptera, Lepidoptera, Neuroptera, Mutillidae, Aphaniptera, Anoplura. Meletemata Entomologica, 5: 169 pp.

KUKALOVA-PECK, J. and LAWRENCE, J. F. 1993. Evolution of the hind wing in Coleoptera. The Canadian Entomologist, 125: 181 - 258.

LAWRENCE, J. L. 1982. Coleoptera. pp. 482 - 553. *In*: Synopsis and Classification of Living Organisms, Vol. 2 (ed. S. B. Parker), McGraw-Hill, New York.

LAWRENCE, J. L. and NEWTON, A. F. 1995. Families and subfamilies of Coleoptera (with selected genera, notes, references and data on family-group names). pp. 779 - 1006 + 48. *In*: Biology, Phylogeny, and Classification of the Coleoptera: Papers Celebrating the 80th Birthday of Roy A. Crowson (ed. by J. Pakaluk and S. A. Slipinski). Muzeum i Instytut Zoologii PAN, Warsaw.

LAWRENCE, J. L. and NEWTON, A. F. 1982. Evolution and classification of beetles. Annual Review of Ecology and Systematics, 13: 261 - 290.

LECONTE, J. L. and HORN, G. H. 1883. Classification of the Coleoptera of North America. Smithsonian Miscellaneous Collections, 26 (507): xxxviii + 567 pp.

LENG, C. W. 1920. Catalogue of the Coleoptera of America, North of Mexico. John D. Sherman, Jr., Mount Vernon, New York, X + 470 pp.

LÖBL, I. 1995. New species of terrestrial *Microsporus* from Himalaya (Coleoptera: Microsporidae). Entomologische Blätter, 91 (3): 129 - 138.

MATTHEWS, A. 1899. A monograph of the coleopterous families Corylophidae and Sphaeriidae: pp. 209-215, pl. 8. O. E. Janson and Son, London.

MESSNER, B. and W. JOOST. 1984. Die Plastronatmung von *Hydroscapha granulum* Imagines (Coleoptera, Hydroscaphidae). Zool. Jb. Anat. Ontog. Tiere, 112(3): 269 - 278.

REICHARDT, H. 1973. A critical study of the suborder Myxophaga, with a taxonomic revision of the Brazilian Torridincolidae and Hydroscaphidae (Coleoptera). Arquivos de Zoologia, 24 (2): 73 - 162.

STICKNEY, F. S. 1923. The head capsule of Coleoptera. Illinois Biological Monographs, 8 (1): 1 - 51, 26 pls.

WALTL, J. 1838. Beiträge zur nähern naturhistorischen Kenntniss des Unterdonaukreises in Bayern. Isis von Oken: 250 - 273.

WILLIAMS, I. W. 1938. The comparative morphology of the mouthparts of the order Coleoptera treated from the standpoint of phylogeny. Journal of the New York Entomological Society: 256, fig. 20.

# 4. HYDROSCAPHIDAE LeConte, 1874

by W. Eugene Hall

**Family common name**: The skiff beetles

These small beetles somewhat resemble tachyporine staphylinids, but the presence of a distinct notopleural suture and aquatic habits immediately separates the two groups. The hind femora are partially covered by metacoxal plates.

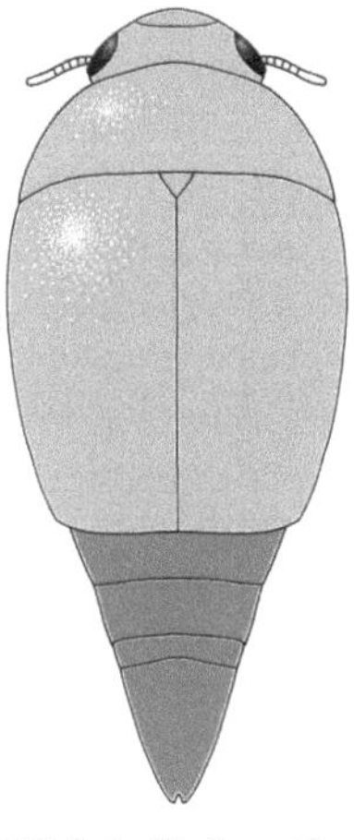

FIGURE 1.4. *Hydroscapha natans* LeConte.

**Description**: Body fusiform, elongate, somewhat depressed; length 1.0 - 2.0 mm; color tan to brown, margins darkened; pubescence fine, recumbent, sparse on body, longer and somewhat denser on appendages.

Head moderate, prognathous, anteriorly arcuate with pronounced frontal ridge; surface with moderately shallow punctures. Antennae sub-clavate, cylindrical, extended beyond anterolateral margins of head, possessing 9 antennomeres, segment IX enlarged, nearly as long as segments V - VIII combined, antennae inserted beneath the frontal ridge, between eyes and base of mandibles; labrum well-developed, fused to clypeus, apically arcuate. Mandibles small, apices blunt, prostheca present on left mandible. Maxillary palpi prominent, possessing four palpomeres, apical one small, no longer than wide, apex acute; penultimate enlarged and three times as long as apical. Labial palpi possessing three palpomeres, small and obscure; gular sutures present, widely separated then converging anteriorly. Eyes small, oval, widely separated, not protuberant, interfacetal setae absent.

Pronotum broader than head, broadened posteriorly and widest at hind angles, nearly as wide as elytra; anterior margin excavated to receive head; lateral margins arcuate, sides deeply explanate, posterior margin arcuate. Prosternum short, spinasternum depressed, obscure. Procoxal cavities open posteriorly, procoxae conical, prominent. Mesosternum short, mesocoxal cavities widely separated, mesocoxae small, ovate. Metasternum long, metacoxal cavities widely separated, metacoxae transverse, posteriorly enlarged and forming a coxal plate partially covering hind femora. Trochanters small, triangular. Femora and tibiae normal. Tarsi 3-3-3; claws moderate.

Elytra truncate, exposing three or four abdominal tergites. Wings fringed with long setae along margin; venation reduced, oblong cell absent. Epipleural fold prominent basally, narrowing apically.

Abdomen narrowed posteriorly, possessing a dense mat of recumbent setae on tergite III, forming a plastron beneath elytra; six visible sternites, sternite VI of female tapered at posterior margin, male with segment VI possessing two widely separated, acute teeth along posterior margin; genitalia as in Reichardt and Hinton (1976).

The following synopsis of larval characters is based on Beutel and Haas (1998) and Beutel *et al.*, (1999) for *Hydroscapha natans* LeConte 1874, which occurs in North America. Body small, flattened anteriorly, 1.1 - 1.3 mm in length, narrowed toward posterior apex. Head subprognathous, broad compared to body size, posteriorly retracted into prothorax, stemmata in two rows of three and two, labrum extended and fused with clypeus, clypeolabrum nearly covering mandibles, frontal suture distinct and lyriform, antennae short, two segmented, segment I broad and short, segment II slender, sensorial appendage present, mandibles short and flattened, mola well-developed, subapical pseudomola present, apical and subapical teeth present, maxilla maxillary articulatory area and membrane well-developed, cardo moderate in size, stipes elongated, galea and lacinia fused, forming hook-like mala, palpifer absent, two palpomeres present, mentum and submentum fused, prementum semimembranous, ligula broad, hypopharynx separated from dorsal wall of prementum by a distinct fold. Thorax slightly broader than head, longer than half the length of abdomen, more than twice as broad as abdominal segments V - IX, prothorax 2 times broader than long and possessing lateral extension, balloon-like spiracular gill present on posterolateral margin, sternum weakly sclerotized, legs nearly adjacent, 5-segmented, coxae large and conical, trochanter inconspicuous, femur long, tibiotarsus short, single claw elongate, mesonotum as broad but shorter than pronotum, spiracular gill absent, metanotum similar to mesonotum. Abdomen narrowed posteriorly, tergite I narrower and shorter than thoracic terga, lateral extensions present, lancet-shaped setae present posteriorly, spiracular gills present posterolaterally, sternum moderately broad, segment II narrower than segment I, spiracular gills absent, lateral extensions indistinct, lancet-shaped setae present posteriorly, sternum as broad as tergum, segments III - VII narrowed posteriorly, lateral extensions and spiracular gills absent, lancet-shapes setae present posteriorly, sternum broad, segment VIII longer than preceding segments, balloon-shaped spiracular gills present posterolaterally, lancet-shaped setae absent posteriorly, segment IX narrower than segment VIII, posterior margin with row of long setae, urogomphi absent, segment X composed of a flat, opercular plate possessing numerous spines, thin hooks attached dorsally.

**Habits and habitats.** Skiff beetles are abundant in streams, on filamentous algae growing on rocks, especially in the marginal shallows. Adults and larvae are non-predacious, feeding on algae. Hydroscaphids can tolerate a wide range of water temperatures ranging from icy streams to hot springs. Reproduction occurs with one large egg being produced at a time and the eggs are then laid on mats of algae. The developing larvae of *Hydroscapha* are entirely aquatic (Beutel 1998/1999; Beutel and Haas 1998; Bøving 1914). Adults carry a supply of air under the elytra via recumbent setae that act as a plastron on tergite III, assisting in respiration (Beutel 1998/1999; Messner and Joost 1984).

**Status of the classification.** Upon describing the type species *Hydroscapha natans*, LeConte (1874) proposed placing the new genus within its own family. Matthews (1884: 114-116) believed the group deserved tribal ranking within Ptiliidae (=Trichopterygidae) based on morphological characters, including the form of the hindwing. Later, Matthews (1900: 13-16) determined Hydroscaphidae deserved family status representing a "connecting link" between Staphylinidae and Ptiliidae. Lameere (1900), Kolbe (1901) and Ganglbauer (1903) placed Hydroscaphidae among families of the Staphyliniformia. Based on larval characters, Bøving (1914) placed the hydroscaphids as a subfamily within Hydrophilidae. Stickney (1923), using Leng's (1920) catalog, included the hydroscaphids as a subfamily of hydrophilids. Forbes (1926) included Hydroscaphidae in the suborder Adephaga within the superfamily Hydradephaga. Bøving and Craighead (1931) and Jeannel and Paulian (1944) placed Hydroscaphidae as a family within the suborder Polyphaga, superfamily Staphylinoidea. D'Orchymont (1945) retained hydroscaphids within Staphyliniformia. Crowson (1955) included Hydroscaphidae within the new suborder Myxophaga, which currently includes the families Hydroscaphidae, Lepiceridae, Microsporidae and Torridincolidae. Reichardt (1973) suggested that Myxophagans possess primitive characteristics intermediate between Adephaga and Polyphaga. Reichardt (1974) discusses the close relationship of Hydroscaphidae and Torridincolidae. Reichardt and Hinton (1976) reviewed Hydroscaphidae from the New World. Based on characters of the hindwing, Kukalova-Peck and Lawrence (1993) recognize Myxophaga as a sister group to Adephaga. Recent coleopteran classifications (Lawrence 1982; Lawrence and Newton 1995, 1982) retain Hydroscaphidae within Myxophaga. Phylogenetic analyses based on larval and adult morphological characters (Beutel 1998/1999; Beutel and Haas 1998; Beutel *et al.*, 1999) support the monophyly of Myxophaga and inclusion of Hydroscaphidae.

**Distribution.** Hydroscaphidae contain three genera: *Hydroscapha* LeConte 1874, *Scaphydra* Reichardt 1973 and *Yara* Reichardt and Hinton 1976. The genus *Hydroscapha* contains 11 species and occurs in North America, Mexico, Eurasia, North Africa and Madagascar (Arce-Perez *et al.*, 1996; Arce-Perez and Novelo-Gutierrez 1990; Lawrence and Newton 1995; Löbl 1994; Reichardt 1973; Reichardt and Hinton 1976). The remaining genera of Hydroscaphidae that occur outside the United States include *Scaphydra* (Brazil) and *Yara* (Brazil, Panama).

## Classification of the Nearctic Species

*Hydroscapha* LeConte, 1874

One species, *H. natans* LeConte, 1874, has been described in the United States from Arizona, southern California, and southern Nevada.

## Bibliography


ARCE-PEREZ, R. and NOVELO-GUTIERREZ, R. 1990. Contribución al conocimiento de los coleópteros acuáticos del Rio Amacuzac, Morelos, México. Folia Entomologica Mexicana 78:29-47.

ARCE-PEREZ, R., NOVELO-GUTIERREZ, R. and GOMEZ-ANAYA, J. A. 1996. Nuevo registro estatal de *Hydroscapha natans* LeConte, 1874 (Coleoptera: Myxophaga) para Mexico. Folia Entomologica Mexicana, 98: 67 - 68.

BEUTEL, R. G. 1998/1999. Phylogenetic analysis of Myxophaga (Coleoptera) with a description of *Lepicerus horni* (Lepiceridae). Zoologischer Anzeiger: 291 - 308.

BEUTEL, R. G. and HAAS, A. 1998. Larval head morphology of *Hydroscapha natans* (Coleoptera, Myxophaga) with reference to miniaturization and the systematic position of Hydroscaphidae. Zoomorphology, 118: 103 - 116.

BEUTEL, R. G., MADDISON, D. R. and HAAS, A. 1999. Phylogenetic analysis of Myxophaga (Coleoptera) using larval characters. Systematic Entomology, 24: 171 - 192.

BØVING, A.G. 1914. Notes on the larva of Hydroscapha and some other aquatic larvae from Arizona. Proceedings of the Entomological Society of Washington, 16(4): 169 - 174, figs. 1 -17.

BØVING, A. G. and CRAIGHEAD, F. C. 1931. An illustrated synopsis of the principal larval forms of the order Coleoptera. Entomologia Americana (new Series), 11: 1 - 351.

CROWSON, R. A. 1955. The natural classification of the families of Coleoptera. N. Lloyd, London, 187 pp.

CSIKI, E. 1911. Coleopterorum Catalogus, 8(32): 3 - 4. Family Hydroscaphidae. W. Junk, Berlin.

FORBES, W. T. M. 1926. The wing folding patterns of the Coleoptera. Journal of the New York Entomological Society, 34:61, fig. 22.

GANGLBAUER, L. 1903. Systematisch-koleopterlogische Studien. Münchener Koleopterologische Zeitschrift, 1: 271 - 319.

JEANNEL, R. and PAULIAN, R. 1944. Morphologie abdominale des Coléoptères et systématique de l'ordre. Revue Française d'Entomologie, 11: 65 - 110.

KOLBE, H. J. 1901. Vergleichend-morphologische Untersuchungen an Coleopteren nebst Grundlagen zu einem System und zur Systematik derselben. Archiv für Naturgeschichte, 67 (Beiheft): 89 - 150, Tafel II - III.

KUKALOVA-PECK, J. and LAWRENCE, J. F. 1993. Evolution of the hind wing in Coleoptera. The Canadian Entomologist, 125: 181 - 258.

LAMEERE, A. 1900. Notes pour la classification des coléoptères. Annales de la Société Entomologique de Belgique, 44: 355 - 377.

LAWRENCE, J. L. 1982. Coleoptera. Synopsis and Classification of Living Organisms, Vol. 2 (ed. S. B. Parker), pp. 482 - 553. McGraw-Hill, New York.

LAWRENCE, J. L. and NEWTON, A. F. 1982. Evolution and classification of beetles. Annual Review of Ecology and Systematics, 13: 261 - 290.

LAWRENCE, J. L. and NEWTON, A. F. 1995. Families and subfamilies of Coleoptera (with selected genera, notes, references and data on family-group names. Biology, Phylogeny, and Classification of the Coleoptera: Papers Celebrating the 80th Birthday of Roy A. Crowson (ed. by J. Pakaluk and S. A. Slipinski), pp. 779 - 1006, +48. Muzeum i Instytut Zoologii PAN, Warszawa.

LECONTE, J. L. 1874. Descriptions of new Coleoptera chiefly from the Pacific slope of North America. Transactions of the American Entomological Society, 5: 43 - 72.

LENG, C. W. 1920. Catalogue of the Coleoptera of America, North of Mexico. John D. Sherman, Jr., Mount Vernon, New York, X + 470 pp.

LÖBL, I. 1994. Les espèces asiatiques du genre *Hydroscapha* LeConte (Coleoptera: Hydroscaphidae). Archs. Sci. Genève, 47(1): 15 - 34.

MATTHEWS, A. 1884. Synopsis of North American Trichopterygidae. Transactions of the American Entomological Society, 11: 113 - 156.

MATTHEWS, A. 1900. Trichopterygia Illustrata et descripta. Supplement. O. E. Janson and Son, London, 112 pp., 7 pls.

MESSNER, B. and JOOST, W. 1984. Die Plastronatmung von *Hydroscapha granulum* imagines (Coleoptera, Hydroscaphidae). Zool. Jb. Anat. Ontog. Tiere, 112(3): 269 - 278.

D'ORCHYMONT, A. 1945. Notes sur le genre *Hydroscapha* LeConte (Coleoptera, Polyphaga, Staphyliniformia). Bulletin du Musée Royal d'Histoire Naturelle de Belgique, 21(25): 1 - 16.

REICHARDT, H. 1973. A critical study of the suborder Myxophaga, with a taxonomic revision of the Brazilian Torridincolidae and Hydroscaphidae (Coleoptera). Arquivos de Zoologia, 24(2): 73 - 162.

REICHARDT, H. 1974. Relationships between Hydroscaphidae and Torridincolidae, based on larvae and pupae, with the description of the immature stages of *Scaphydra angra* (Coleoptera, Myxophaga). Revista Brasileira de Entomologia, 18(4): 117 - 122.

REICHARDT, H. and HINTON, H.E. 1976. On the New World beetles of the family Hydroscaphidae. Papéis Avulsos de Zoologia, 30(1): 1 - 24.

STICKNEY, F. S. 1923. The head capsule of Coleoptera. Illinois Biological Monographs, 8(1): 1 - 51, 26 pls.

# 5. RHYSODIDAE Laporte, 1840

by Ross H. Arnett, Jr. and Michael A. Ivie

**Family Common Name**: The wrinkled bark beetles

**Family synonym**: Rhyssodidae Jacquelin du Val, 1857

The members of this small family closely resemble in a superficial way, some of the species of the family Colydiidae and Brentidae. However, they are really highly specialized members of the suborder Adephaga, and recent workers have included them in this suborder. The notopleural suture and divided first visible sternite will clearly distinguish this group from any of the similar looking Polyphaga, and the moniliform antennae combined with the distinctively grooved prothorax and head will separate them from other members of the Adephaga.

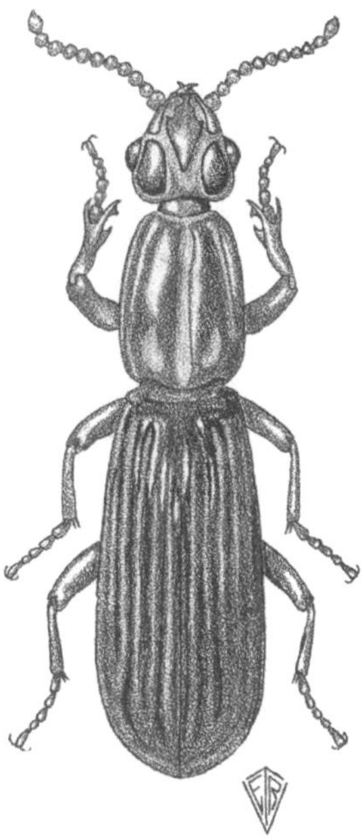

FIGURE 1.5. *Clinidium sculptile* Newman.

**Description**: Shape cylindrical, elongate; size 5-8 mm; color testaceous to dark, immaculate; vestiture absent except for sparse, moderate hairs on antennae and abdomen.

Head moderate, grooved dorsally, constricted behind into a neck which forms a ball and socket joint with the prothorax; antennae short, 11-segmented, each segment as broad as long, bead-like; mouthparts unique, labrum triangular, bi- to quadrisetose; mandibles nonfunctional, lacking cutting edges, each with a scrobal seta and hollowed external tooth; maxillae, except tip of palps, hidden by mentum, galea and lacinia styliform; labial palpi reduced, hidden by mentum; mentum fused laterally to the head capsule, extending anteriorly beyond other mouthparts to form cutting edge; labial palpi reduced, hidden by mentum when the jaws are closed but emerge when the jaws are open; maxillae and their palpi also completely hidden when jaws are shut, except for tip of palpus; galea and lacinia of maxillae converted into stylets; eyes narrow, obscure, facets obscure, or round, with more or less distinct facets.

Prothorax with notopleural sutures; elongate with sides arcuate, narrowed in front and behind, dorsal surface with deep longitudinal grooves; legs slender, tarsi 5-5-5, claws simple; elytra covering abdomen, rounded behind, sides parallel, surface with deep longitudinal grooves; wings when present with oblong area absent, as in Cicindellidae, m-cu straight.

Abdomen with five visible sternites, the first visible sternite divided into 3 pieces, one on each side of the coxae, and one between the coxae; sternites in *Clinidium* with pollinose sulci, the arrangment of which is important in species identification; male genitalia a modified trilobed type, the parameres short and concave, closely resembling certain Carabidae.

The grub-like larvae are of caraboid type with labial palpi. latent, prementum and ligula fused into an unpaired anteriorly bilobed piece, retracted ventral mouthparts, one claw, urogomphi absent.

**Habits and habitats**. Both adults and larvae are found in moist rotten logs infested with slime molds (Myxomycetes), which are their presumed food (Bell 1991, Lawrence and Britton 1994). Both hard woods and conifers are commonly inhabited. Feeding is done in a unique manner (Bell 1994). The mandibles cannot be used to bite, and the anterior margin of the mentum is used as a cutting edge. The mentum and mandibles enclose the rest of the mouthparts when the mandibles are closed, but opening the mandibles exposes/extends the maxillary stylets and labial palps.

**Status of the classification**. The hard body and unusual wrinkled appearance led early authors to place this family near the Colydiidae and Cucujidae, at that time both extremely diverse groups themselves. Modern workers have recognized the clearly adephagan membership of this group from the notopleural sutures, divided first abdomenal segment, fused hind coxae, and unambiguous characters of the wing and larva. The current controversy is whether to continue to recognize the rhysodids as a family, or as a member of the Carabidae. Bell and Bell (1962), Bell (1970), and Forsyth (1970) provided strong evidence for placement in the Carabidae based on characters of the tactile setae, antenna cleaner, and pygidial defense glands. Beutel (1990) and Bell (1998) went even further, showing support for a specific relationship with the carabid tribe Scaritini. However, consideration of larval characters leaves a less clear picture, leading Beutel (1992, 1993, 1995) to reverse himself, and follow larval workers Grandi (1972), Burakowski (1975), and Costa *et al.* (1988) in considering rhysodids as a full family, placing them as the sister-group of the rest of the Carabidae. The final answer is still out on the question of the phylogenetic placement of this family, but we are following Lawrence and Newton (1995). The family has been relatively recently revised for the world by Bell and Bell (1978, 1979,1982,1985), and cataloged for North America by Bell (1985).

**Distribution**. Approximately 170 species in 20 genera are described nearly worldwide, with the richest concentration in the Southeast Asia-New Guinea area. There are 8 species in 2 genera in North America north of Mexico.

### Key to the Nearctic Genera

1. Sutural margin of elytra depressed below level of rest of elytra at base, striae strongly sulcate and not or indistinctly punctate; pronotum with lateral grooves of disc restricted to basal half .............. *Clinidium*

— Sutural margin of elytra level with rest of elytra at base; striae indistinctly sulcate, strongly punctate; pronotum with lateral grooves of disc long, reaching nearly to apical margin ........... *Omoglymmius*

### Classification of the Nearctic Genera

Rhysodidae

*Omoglymmius* Ganglbauer 1892, 2 species, eastern and western North America.

*Rhysodes* Dalman 1823 [not North American]

*Clinidium* Kirby, 1835 6 species, eastern United States and Pacific coastal states and provinces

### Bibliography

ARROW, G. J. 1942. The beetle family Rhysodidae, with some new species and a key to those at present known. Proceedings of the Royal Entomological Society of London (Series B), Taxonomy, 11: 171-183.

BELL, R. T. 1970. The Rhysodini of North America, Central America, and the West Indies (Coleoptera: Carabidae or Rhysodidae). Miscellaneous Publications of the Entomological Society of America, 6(6): 289-324.

BELL, R. T. 1975. *Omoglymmius* Ganglbauer, a separate genus (Coleoptera: Carabidae or Rhysodidae). Coleopterists Bulletin, 29(4): 351-352.

BELL, R. T. 1985. A catalog of the Coleoptera of America North of Mexico. Family: Rhysodidae. United States Department of Agriculture, Agriculture Handbook No. 529-4. pp. i-x + 1-4.

BELL, R. T. 1991. Rhysodidae (Adephaga). Pages 304-305 in Immature Insects Volume 2 (F. W. Stehr, ed.). Kendall/Hunt Publishing Company, Dubuque, Iowa.

BELL, R. T. 1994. Beetles that cannot bite: functional morphology of the head of adult Rhysodines (Coleoptera: Carabidae or Rhysodidae). The Canadian Entomologist 126:667-672.

BELL, R. T. 1998. Where do the Rhysodini (Coleoptera) belong? Pages 261-272 in Phylogeny and Classification of Caraboidea. XX I.C.E. (1996, Firenze, Italy) (G. E. Ball, A. Casale, and A. Vigna Taglianti, eds.). Museo Regionale di Scienze Naturali, Torino.

BELL, R. T. and BELL, J. R. 1962. The taxonomic position of the Rhysodidae (Coleoptera). Coleopterists Bulletin, 16(4): 96-106

BELL, R. T. and BELL, J. R. 1975. Two new taxa of *Clinidium* (Coleoptera: Rhysodidae or Carabidae) from the eastern U.S., with a revised key to U.S. *Clinidium*. Coleopterists Bulletin, 29(2): 65-68.

BELL, R. T. and BELL, J. R. 1978. Rhysodini of the world. Part I. A new classification of the tribe, and a synopsis of *Omoglymmius*, subgenus *Nitiglymmius*, new subgenus (Coleoptera: Carabdidae or Rhysodidae). Quaestiones Entomologicae, 14: 43-88.

BELL, R. T. and BELL, J. R. 1979. Rhysodini of the World Part II. Revisions of the smaller genera (Coleoptera: Carabidae or Rhysodidae). Quaestiones Entomologicae 15: 377-446.

BELL, R. T. and BELL, J. R. 1982. Rhysodini of the World Part III. Revision of *Omoglymmius* Ganglbauer (Coleoptera: Carabidae or Rhysodidae) and substitutions for preoccupied generic names. Quaestiones Entomologicae 18: 127-259.

BELL, R. T. and BELL, J. R. 1985. Rhysodini of the World Part IV. Revisions of *Rhyzodiastes* and *Clinidium*, with new species in other genera (Coleoptera: Carabidae or Rhysodidae). Quaestiones Entomologicae 21(1): 1-172.

BEUTEL, R. G. 1990. Metathoracic features of *Omoglymmius hamatus* and their significance for classification of Rhysodini (Coleoptera: Adephaga). Entomologica Generalis 15: 185-201.

BEUTEL, R. G. 1992. Larval head structures of *Omoglymmius hamatus* and their implications for the relationships of Rhysodidae (Coleoptera: Adephaga). Entomologica Scandinavica 23: 169-184.

BEUTEL, R. G. 1993. Phylogenetic analysis of Adephaga (Coleoptera) based on characters of the larval head. Systematic Entomology 18: 27-147.

BEUTEL, R. G. 1995. The Adephaga (Coleoptera): phylogeny and evolutionary history, pp. 173-217. *in* J. Pakaluk and S. A. Slipinski (eds.) Biology, Phylogeny, and Classification of Coleoptera: Papers Celebrating the 80th Birthday of Roy A. Crowson. Museum i Instytut Zoologii, PAN, Warszawa.

BURAKOWSKI, B. 1975. Description of larva and pupa of *Rhysodes sulcatus* (F) (Coleoptera, Rhysodidae) and notes on the bionomy of this species. Annales Zoologici 32: 271-287.

COSTA, C., VANIN, S. A., and CASARI-CHEN, S. A. 1988. Larvas de Coleoptera do Brazil. Universidade de São Paulo (Museo de Zoologica, São Paulo, 447 pp.

FORSYTH, D. J. 1972. The stucture of the pygidial defense glands of Carabidae (Coleoptera). Transactions of the zoological society of London 32: 249-309.

GRANDI, G. 1972. Comparative morphology and ethology of insects with a specialized diet, *Rhysodes germarii* Ganglb. Bolletino dell'Instituto di Entomoligia dell'Universitá di Bolologna 30: 31-47.

LAWRENCE, J. F. and BRITTON, E. B. 1994. Australian Beetles. Melbourne University Press, Melbourne. 192 pp

LAWRENCE, J. F. and NEWTON, A. F., JR. 1995. Families and subfamilies of Coleoptera (with selected genera, notes, references and data on family-group names), p. 779-1006. *In* Biology, Phylogeny, and Classification of Coleoptera: Papers Celebrating the 80th Birthday of Roy A. Crowson. (Eds. J. Pakaluk and S. A. Slipinski), Muzeum i Institut Zoologii, Polska Academia Nauk, Warsaw.

# 6. CARABIDAE Latreille, 1810

by George E. Ball and Yves Bousquet

**Family common name**: The ground beetles

**Family synonyms**: Agridae Kirby 1837; Anchomenidae Laporte de Castelnau 1834; Anthiidae Hope 1838; Apotomidae Jacquelin duVal 1857; Bembidiidae Westwood 1838; Brachinidae Bonelli 1810; Broscidae Hope 1837; Callistidae Jeannel 1941; Calophaenidae Jeannel 1942; Chlaeniidae Westwood 1838, Cnemacanthidae Lacordaire 1854; Ctenodactylidae Castelnau 1834; Cyclosomidae Laporte de Castelnau 1846; Cymbionotidae Jeannel 1941; Dryptidae Jeannel 1941; Elaphridae Stephens 1827; Feronidae Laporte de Castelnau 1834, Gehringiidae Darlington 1933; Glyptidae Horn 1881; Harpalidae MacLeay 1833, Hiletidae Lacordaire; 1854; Lebiidae Bonelli 1810; Licinidae Bonelli 1810; Loroceridae Bonelli; 1810; Masoreidae Chaudoir 1876; Melanodidae Jeannel 1943; Metriidae LeConte 1861; Migadopidae Chaudoir 1961; Nebriidae Castelnau 1834; Odacanthidae Laporte de Castelnau 1834; Omophronidae Latreille 1810; Orthogoniidae Chaudoir 1871; Ozaenidae Hope 1839; Panagaeidae Bonelli 1910; Patrobidae Kirby 1937; Paussidae Latreille 1806; Peleciidae Horn 1891; Pentagonicidae Bates 1873; Pericalidae Hope 1838; Perigonidae Horn 1881; Pseudomorphidae Horn 1881; Psydridae LeConte 1861; Pterostichidae Erichson 1837; Scaritidae Bonelli 1810; Siagonidae Bonelli 1810; Thyreopteridae Chaudoir 1868; Trechidae Bonelli 1810; Zuphiidae Jeannel 1941.

Most adults of this family present a somber appearance, that is, a uniformly dark color, although some are bicolored or tricolored dorsally, and marked variously. Range in size is substantial. The margined pronotum, large head and mandibles, and striate elytra help characterize this family. These features vary considerably, however, throughout the family.

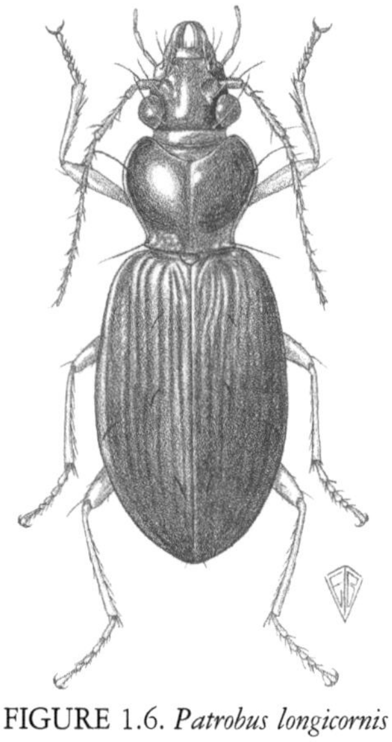

FIGURE 1.6. *Patrobus longicornis* (Say).

**Prologue**. We offer the following comments and instructions preceding the details of the treatment of this family. *The Beetles of the United States* (Arnett 1960), the volume on which the present one is based, was published toward the beginning of an era of intense active investigation of carabid beetles by North American workers. Within that decade, Wallis (1961) published his treatment of Canadian tiger beetles. A monumental treatment of the Carabidae (Cicindelinae excluded) of Canada and Alaska appeared in six parts (Lindroth 1961-1969b). This latter work and the magnetic influence of its talented, friendly, inspirational and cooperative author seemed to set the stage for all of the work by North American carabid specialists who followed, to the present. Most of the publications have been in the form of monographic treatments of tribes or genera, with very limited publication of isolated descriptions. Many of these monographs, referred to in appropriate places in the following taxonomic treatment, addressed broader issues, principally phylogeny and geographical history of taxa. Many of these publications treated larvae as well as adults. Most emphasized morphological features as the basis for taxonomic decisions, but with the passage of time, electrophoretic and molecular features were used. Illustrations have been a prominent feature of the monographs, in the form of line drawings, photographs, and SEM photographs.

Much progress has been made with the study of larvae. Building on the substantial contribution of van Emden (1942), Thompson (1979) provided a treatment of larvae of North American Carabidae, featuring a key to tribes. Bousquet and Goulet (1994) proposed a system for designation of primary setae and pores, a system that has received general approval and use in the years following publication of this contribution. Luff (1993) treated the larvae of Fennoscandian Carabidae. Arndt (1993 and 1998) and Arndt and Putchkov (1996/97) treated carabid larvae from a phylogenetic perspective. Zetto Brandmayr *et al.* (1998) offered an excellent analysis of the relationship among structural features, functional features, and way of life of carabid larvae, with appropriate conclusions about evolutionary aspects.

**Acknowledgments.** The authors are pleased to express their gratitude to the Editor, Michael C. Thomas for his encouragement and good humor as he waited patiently to receive the manuscript on which this chapter is based. We acknowledge also the encouragement received previously from Ross H. Arnett, Jr., to participate in the revision of *Beetles of the United States*. Our feelings of appreciation are matched by our disappointment that he did not survive to see the publication of *American Beetles*, which is surely destined to be another major contribution to development of coleopterology in North America. For his thorough review of the bibliography of this chapter and advice about related matters, the senior author extends his thanks to his longtime friend and entomological associate, Andrew P. Nimmo (Edmonton, Alberta).

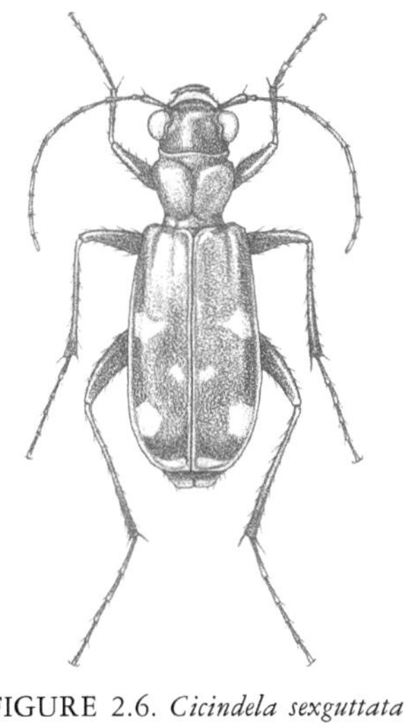

FIGURE 2.6. *Cicindela sexguttata* Fabricius.

Much of the taxonomic work is summarized or at least alluded to, in various symposium volumes, especially: Erwin *et al.* (1979); Ball (1985a); Liebherr (1988a); Noonan *et al.* (1992); Ball *et al.* (1998); and Zamotajlov and Sciaky (1999). Nomenclatural aspects are summarized in two catalogues for Nearctic North America: Bousquet and Larochelle (1993) and Freitag (1999). Madge (1989) provided a useful catalogue of suprageneric names for Carabidae. Lorenz (1998) produced in two volumes a splendid, virtually exhaustive treatment of names of the genus- and species-groups, as well as a useful arrangement of taxa. A checklist of the Carabidae of the Western Hemisphere, prepared by T. L. Erwin, is available on the Internet: <http://nmnhwww.si.edu/gopher-menus/ChecklistoftheWesternHemisphereCaraboideaColeoptera.html>.

Devoted exclusively to research (principally taxonomic) on cicindelines is the journal *Cicindela*, which, beginning in 1968, is published four times a year, by Ronald L. Huber.

Because carabid adults seem to be abundant in especially Quaternary age deposits, and because the fragments of many species can be identified at least to genus, they have been studied extensively by geologists interested in the Quaternary Period. Most of the fossil carabids represent extant species, and thus are useful in tracking faunal movements associated with waxing and waning of glacial ice. See Elias (1994) for a general summary of Quaternary palaeoentolology, and Matthews (1979), Matthews and Telka (1997) for more detailed considerations of carabids. Also, see Ball and Currie (1997) for an example of the use of carabid fossils in interpreting distribution patterns of the extant fauna.

These efforts outlined above by neontologists and palaeontologists have modified appreciably knowledge and understanding of the Nearctic carabid fauna. Further, as noted, the tiger beetles have been included, as a carabid subfamily, and the genus *Trachypachus* Motschulsky has been assigned to a separate family. So, the present chapter differs substantially from that which appeared in 1960. Although these changes reflect substantial progress, much remains to be learned, as noted below.

**Terms for structural features**. Most of the words used here to designate details of structures are found in general textbooks of entomology, or are used by coleopterists, generally. Other words, used to designate particular structures or parts thereof are not in general use, though they have been used in the carabid literature. We provide information about these latter words here.

MICROSCULPTURE. A "sculpticell" is the space on the surface of the cuticle enclosed by adjacent microlines of the integumental system of microsculpture (Allen and Ball 1980: 485-486). Microsculpture of the elytra varies from mesh pattern isodiametric, with sculpticells shingled, to transverse, with sculpticells flat, to linear, with single sculpticells extended and very long. In present usage, length of a sculpticell refers to its longer dimension, with width referring to the shorter dimension.

FIXED SETAE. These are the long, evidently tactile setae, quite precisely located, that are characteristic of carabids. They are: dorsal labral (six in most taxa, but more or less numerous); supraorbital (one or two pair) on the frons, near eyes; suborbital (one pair), ventrally on the head capsule, near the eyes; submental; mental; palpiferal; palpigeral; pronotal (on lateral margins); elytral, parascutellar, discal (on or near interval 3), and umbilicate (lateral series). Fixed setae on the legs are coxal, trochanteral, femoral, and tarsal.

BODY PARTS. The term "segment" is restricted for use to those body parts that reflect embryonic somites; thus, somite-like portions of the abdomen are referred to as segments. Abdominal segments are designated by Roman numerals corresponding to their respective somites. The first complete sternum is III, and the last one normally exposed is VII. In brachinines, additional segments are exposed.

Portions of appendages (articles) are designated by the suffix "-mere", the prefix depending upon the appendage in question: antenno-, palpo-, tarso-, etc. Individual articles are designated by a number, following the appropriate name. Hence, rather than referring to articles by word in terms of their position relative to the terminal one, the articles are numbered from proximity to insertion of the basal-most article. The antennal scape is antennomere 1; the pedicel antennomere 2, and so on to antennomere 11. The terminal maxillary papal article is maxillary palpomere 4 and the terminal labial palpal article is labial palpomere 3.

**Geographical terms**. North America refers to the continent, which extends from the Canadian Arctic Islands southward through Panama. Middle America refers to Mexico plus the republics of Central America, collectively. The West Indies refers to the Greater and Lesser Antilles and the Bahamian archipelago.

**Biogeographical terms**. Following Frank and McCoy (1990), we use "indigenous" (synonyms autochthonous, native) for a taxon that achieved, in an evolutionary sense, its current taxonomic status in the area in which it is living. If an indigenous taxon occurs nowhere else, it is referred to as "precinctive" (from Latin, meaning to gird or encircle). A taxon that achieved its taxonomic status elsewhere than in a given area where it occurs now is "adventive" in that area. An adventive taxon is either "introduced" if moved to a given area by man, either accidentally or on purpose, or it is "immigrant" if it was not introduced.

To designate the broader distribution patterns, we use the standard names for zoogeographical regions: the primarily temperate Nearctic and Palaearctic Regions; and the primarily tropical (or Southern Hemisphere) Australian, Oriental, Afrotropical, and Neotropical Regions. A taxon represented in both Palaearctic and Nearctic Regions is referred to as Holarctic. The term "Megagaea", proposed by Darlington (1957: 424-425), refers to a "realm" including the Afrotropical (=Ethiopian), Oriental, Palaearctic, and Nearctic Regions. The southern counterparts are the Neogaean Realm (for the Neotropical Region), and Notogaean Realm (the Australian Region).

As appropriate, "Holarctica" is used as a noun to designate the Holarctic Region. In the Western Hemisphere, the boundary between the Neotropical and Nearctic Regions is placed arbitrarily at the Mexican border, with the latter region including only the continental United States and Canada. The term "Nearctic North America", used to achieve some verbal variety in the text, means the same as Nearctic Region. "Pantropical" designates taxa whose ranges extend through the primarily tropical zoogeographical regions (Australian + Oriental + Afrotropical + Neotropical).

**Keys.** With very few exceptions, we have confined the characters and character states used for identification to those that are easily seen, or at least do not require dissection. Thus, it may seem that many of the recognized taxa are based on flimsy evidence. In fact, most of the genera recognized are securely established on the basis of many distinctive character states, many of them derived from the normally concealed male and female reproductive organs.

**Taxonomic treatments.** For subfamilies, we provide detailed characterization including structural features (external, male and female [ovipositor and genital tract] genitalia) and defensive gland secretions. For tribes, genera, and subgenera, the information presented is confined to statements about numbers (based principally on Lorenz (1998)) of included taxa, geographical distribution, and ecological information, if available. We offer also for some taxa our appraisal of current knowledge, and suggestions for further work. Concerning this latter point, knowledge of almost all Nearctic taxa could be improved markedly through life history studies, with concomitant descriptions of life stages other than adult, as well as investigation of cytological aspects and defensive secretions of adults. Molecular studies, the current, virtually all-consuming focus of academic biology, may be expected to provide appreciable enlightenment about phylogenetic relationships of taxa at all levels. Hopefully, this emphasis will not sweep away efforts to gain knowledge of the topics noted above.

**Species identification.** This publication focuses on identification and knowledge of diversity and geographical distribution of supraspecific taxa. As appropriate, we cite references to recent monographs and revisions of species, with omission of much of the older literature referred to by Ball (1960a). We note here regional treatments which are not cited consistently for individual taxa, but which are valuable for identifying species: Ciegler (2000)-southeastern species; Downie and Arnett (1996)- northeastern species; Hatch (1953)- species of the Pacific Northwest; Kavanaugh (1992)- species of the Queen Charlotte Islands; Lindroth (1961-1969b)- species of Canada and Alaska; Larochelle (1976) and Gariepy *et al.* (1977)- species of Quebec.

The treatment of the Carabidae (and Cicindelidae) of the Pacific Northwest, by M. H. Hatch, is out of date and with many errors. Nonetheless, it provides a useful regional synopsis, enhanced by illustrations which are helpful in making identifications of genera and species.

C. H. Lindroth's taxonomic treatment of the northern Nearctic carabid fauna (minus cicindelines) is exemplary for regional studies. Its treatment of taxa is noted for accuracy (based on study of types, as well as extensive field experience), brevity, clarity, and perceptiveness, as well as excellent illustrations of structural features and of adult habitus of most of the genera treated.

Like Lindroth's publications, Kavanaugh's (1992) is noteworthy for general high quality, including illustrations.

Structured much like Blatchley's *Coleoptera of Indiana* (1910) and intended as a replacement for that excellent but dated volume, N. Downie's and R. H. Arnett's treatment of the northeastern carabids provides a useful though rather superficial means for species identification.

**Extra-limital regional publications.** We note here some relatively recent publications treating the carabids of regions geographically more or less proximal to the Nearctic Region. For the Neotropical Region, Reichardt (1977) treated the genera. Erwin and Sims (1984) summarized knowledge of the West Indian carabid genera, including a list of the species. Erwin (1982 and 1991a) presented the first two parts of what is to be a species-level treatment of the carabid fauna of Central America (Guatemala south to Panamá). Additionally, Erwin (1991b) provided a general classification and useful information about characteristic ways of life of many Neotropical indigenous genera, some of which are represented by species in the Nearctic Region. Costa *et al.* (1988) provided a treatment of the beetle larvae (including Carabidae) of Brazil that is useful for other parts of the world, too. Ball and Shpeley (2000) summarized information about biodiversity of Carabidae in Mexico.

For Eastern Palaearctic faunal assemblages, the seminal studies by A. Habu (1967, 1973, and 1978) provide insight into the Japanese carabid fauna, and the basis for useful comparisons with the Nearctic fauna. Similarly, O. L. Kryzhanovskij's (1983) treatment of the carabid supraspecific taxa of the U.S.S.R. is useful especially for consideration of Holarctic distribution. An appreciation of the Western Palaearctic carabid fauna, the source of many introduced species, may be obtained by consulting Lindroth's (1974 and 1985/86) regional treatments, Freude *et al.* (1976), and Hurka (1996).

**Description**: Adults. Shape flat to oval, pronotum with lateral margins rounded, nearly straight, or distinctly sinuate, in most taxa prothorax narrower than elytra; length 0.7 to 66 mm (Erwin 1991b: 28); color black, immaculate, to variously bi- or tri-colored, with elytra spotted with orange or yellow; adults of some species with dorsal surface metallic green or purple, or a combination of these, and also metallic colors with orange prothorax; adults of other species brownish or testaceous, and marked or not with black. Surface (dorsal, ventral or both surfaces) with or without distinct pelage. Setation sparse, adults of most species with a definite number of precisely located (fixed) tactile setae (absent from Paussini).

Head narrower than prothorax, prominent, prognathous. Eyes various from large and semi-globose to small and flat, to absent. Antennae inserted between eyes and base of mandibles under a frontal ridge, or anteriad and laterad bases of mandibles and eyes; filiform to markedly transverse, with two to eleven antennomeres. Labrum of most taxa distinctly transverse, or as long or longer than wide; with apical margin subtruncate to markedly concave; epipharynx various. Mandibles prominent, acute at apices, occlusal margins variously toothed. Maxillae with occlusal margin of lacinia more or less densely trichiate, apically with more or less prominent tooth; galea of most taxa palpiform, with one or two articles; palpus of four articles, of various size relative to each other, palpomere 4 narrow to very broad (securiform) apically. Labium with submentum and mentum separated by a suture, or suture absent; mentum deeply emarginate; ligula with glossal sclerite more or less prominent, sclerotized, on each side laterally with a pair of more or less distinct paraglossae, or paraglossae not evident; palpus of 3 articles, palpomere 3 with apex narrow to very broad apically.

Thorax. Prothorax of most adults narrower than paired elytra together, with lateral margins inflexed and a distinct submarginal suture between pronotum and proepisternum; pronotum with lateral margins of many taxa more or less reflexed. Metepimera posteriorly distinctly lobate or not, if not lobate, then metepisterna in contact with anterior margin of abdominal sternum II.

Elytra margined laterally or not, humeri broadly rounded, broad, to narrow and sloped; lateral margins curved; posteriorly sinuate or not, apical margin narrowly rounded to broadly truncate; dorsal surface striate, each stria linear, continuous, or each represented by row of punctures, or striae absent, surface smooth. Tactile setae (fixed in position) sparse, variously situated, but precisely located.

Metathoracic wings typical of Adephaga, venation not distinctive; folding patterns fundamental for Adephaga, with some distinctive features; many species with brachypterous or apterous adults.

Legs cursorial, gressorial, or fossorial, slender to wide, with femora and tibiae flattened; front and middle coxae globular, hind dilated internally, not extended to lateral margins of body; tarsi each with five tarsomeres, tarsomere 5 terminated by pair of claws; claws with inner margins smooth, serrate, or pectinate.

Abdomen with six pregenital sterna (II-VII), seven or eight in Tribe Brachinini, sterna II-IV connate, sternum II (first visible sternum) interrupted by hind coxae, remnants visible only at sides.

Male genitalia trilobed, median lobe variously developed, parameres varied and closely connected with dorsal margin of median foramen by prominent condyle; basal piece entirely membranous or absent; internal sac (endophallus) various, more or less armored with microtrichia and spines.

Ovipositor with sternum VIII divided through middle by longitudinal membranous strip; gonocoxa IX variously developed, of one or two articles; remnant of sternum X present in most species.

Chromosomes: 2n= 8 (Graphipterini) to 69 (ditomine Harpalini), with or without multiple sex chromosomes (Serrano and Galián 1998).

Larvae. Campodeiform, subcylindrical, or slightly flattened, distinctly segmented, and completely or partially sclerotized on the dorsum and to a lesser degree on the venter; some (Cychrini) distinctly depressed and almost onisciform. Color yellow, brown or reddish to black; prominent and definitely arranged setae on sclerotized areas. Head prominent, exserted, prognathous to hyperprognathous, pigmented or not; epicranial suture present in most taxa, delimiting the fused frons, clypeus and labrum; cephalic margin of frons serrate or deeply dentate; a prominent and frequently bifurcate nasale present or not. Antennae prominent; three or four (four in most taxa) articled. Most taxa eyed, with six distinct ocelli on each side of head, some taxa eyeless. Mouthparts: mandibles cultriform or not, three or more times as long as wide at base, with or without distinct tooth-like retinaculum along middle and with or without a membranous or setose or spine-like basal structure; maxillae prominent, cardo small, stipes moderately prominent, palpus of four or five (four in most taxa) articles, galea 2-segmented in most taxa, lacinia absent or developed as a small, rounded or acuminate tubercule in most taxa, with prominent apical or lateral seta; labial palpus prominent, palpomeres two (most taxa) or four. Thorax large. Legs each of six (most taxa) or fewer articles, with or without prominent spines; claws moveable, single or paired. Respiratory system: spiracles annular or elliptical, one pair on each of mesothorax and abdominal segments one to eight. Abdomen ten-segmented with urogomphi variously developed on segment IX; segment X may function as an anal proleg. The larvae of many North American species have been described.

**Habits and habitats.** Overall, carabids range widely, from cavernicolous environments to the surface, and to arboreal situations. Ground inhabitants (geophiles, or terricoles) are: hygrophiles, occupying riparian zones, sunlit marshes, and dark swamp forests; mesophiles, living in damp or wet forest or meadows, but independently of permanent surface water; and xerophiles, living in dry forests to grassland and desert situations. In the tropics, especially, many carabids are arboreal; they are associated with vegetation, standing or fallen, living on tree trunks, branches, and some even on leaves, where they hunt for prey. Many species, principally geophiles, occupy anthropogenic habitats, where, from time to time, they become very abundant.

According to Brandmayr and Zetto Brandmayr (1980), most carabids are polyphagous olfactory-tactile predators or scavengers, seeking and eating dead and dying arthropods, or are specialized seeking active prey, such as molluscs, millipedes, or ants. Some are day-active optical predators, such as tiger beetles, notiophilines and loricerines, using primarily eyesight to cap-

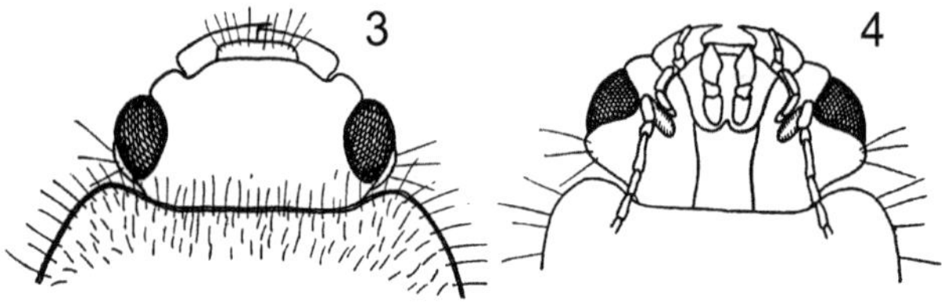


FIGURES 3.6-4.6. Fig. 3.6 *Pseudomorpha* sp., head, dorsal view; Fig. 4.6, *Pseudomorpha* sp., head, ventral view.

ture active prey, including collembolans and other small soil-inhabiting arthropods. Many other carabids are phytophagous, eating especially seeds which have fallen to the ground, or which the beetles obtained *in situ*, on the plants. Some species are myrmecophiles and some are parasitoids, the carabid larva consuming its host over a period of days. For details, see under the taxa discussed below. Most species are nocturnal, with adults pursuing their activities on foot, or in flight. On some nights, flying adults come to lights in great numbers. Many species are active during the winter as adults. Pupation is in the ground. The life cycle is one year long in most species for which the life history is known, adults living two or three, even four years. For general coverage of food and feeding, see Larochelle (1990) and Den Boer *et al.* (1986).

Females of several genera (*Chlaenius, Calleida, Brachinus, Galerita, Pterostichus, Craspedonotus, Carabus*, and *Calosoma*) lay their eggs in especially constructed egg cells made of mud, twigs, leaves, etc. The larvae of carabids have three instars in most taxa. Overall, though, very little is known about the way of life of the North American ground beetles. Here is a fertile field for investigation.

**Status of the classification**. The composition of most of the tribes is fairly well settled, even though the taxonomic status of a few of them is in doubt. However, the grouping of the tribes into taxa of higher rank is in a state of flux, with virtually every classification of carabid beetles published during the past 40 years differing more or less substantially from any other one (Ball *et al.* 1998). Rather than pondering which one to use, we have adopted the one proposed by general coleopterists for general purposes (Lawrence and Newton 1995). We have, however, added two subfamilies (Nebriinae and Promecognathinae) on the basis of current publications by specialists.

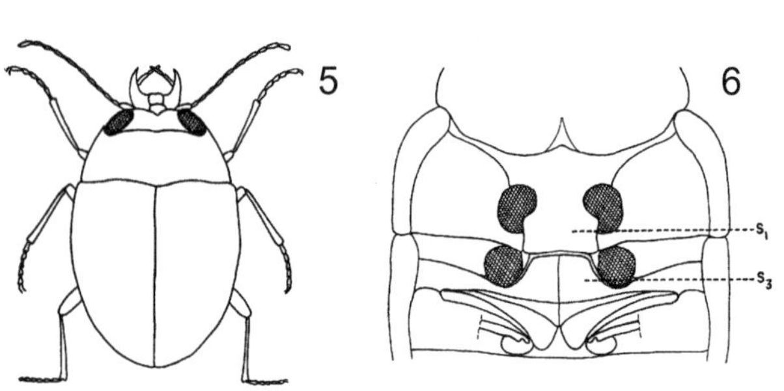


FIGURES 5.6-6.6. Fig. 5.6, *Omophron* sp., habitus, dorsal view; Fig. 6.6, *Omophron* sp., ventral view of thorax.

**Distribution**. Carabids are found in all areas throughout the world, the large genera occurring in all areas as well, between 78°56' N lat. and 55° S lat. Approximately 40,000 species are known (Erwin 1991b: 28). Altitudinally, the range of the family extends from sea level (and below) to 5300 m. in the Himalaya (Mani 1968).

In this work, 2635 species and subspecies inhabiting Nearctic North America are grouped into 189 genera, which in turn comprise 47 tribes and 14 subfamilies. In the future, as revisionary studies are conducted, the number of species may remain about as is. Although many species recognized presently will be found not to warrant recognition, species previously undescribed may be expected to be discovered, especially in groups such as cavernicoles, and the small tachyine and anilline Bembidiini.

## I. Key to the Tribes of Nearctic Carabidae

1. Head beneath with a short, deep, antennal groove on each side between eye and mouthparts (Fig. 4.6); antennal scape not visible from above (Fig. 3.6); labium without suture between submentum and mentum (Fig. 4.6) (*Pseudomorpha*) .................... .................................................. Pseudomorphini

— Head beneath without short grooves for antennae; scape of antenna visible from above; labium with or without suture between submentum and mentum ................................................................ 2

2(1). Scutellum of mesothorax not visible (Fig. 5.6); mesosternum completely covered by scoop-shaped intercoxal process of prosternum (Fig. 6.6); body oval, not pedunculate. (*Omophron*) .................... ..................................................... Omophronini

— Scutellum of mesothorax normally exposed, but difficult to see if body is pedunculate; prosternum not scoop-shaped, mesosternum not completely covered by it .................................................... 3

3(2). Abdomen with seven or eight sterna normally exposed; mandible with a setigerous puncture in scrobe; head with single supraorbital setigerous puncture over each eye; elytron truncate at apex; head and prothorax testaceous, elytra blue or brownish. (*Brachinus*) .......................... Brachinini

— Abdomen with six sterna normally visible. ............ 4

4(3). Elytron with lateral portion of disc as well as epipleuron bent under abdomen (Fig. 8.6); palpus with palpomere 4 very small, nearly subulate; mandible with a setigerous puncture in scrobe; head with single supraorbital setigerous puncture over each eye; hind coxae separated from one another in mid line of body; total length 2.0 mm or less (Fig. 7.6) (*Gehringia*) .......................... Gehringiini

— Elytron with only epipleuron bent under abdomen .. 5

5(4). Elytron with prominent outward-projected ridge (flange of Coanda) in apical one-fifth of lateral margin (Fig. 9.6); anterior tibia with both spurs apical (Fig. 10.6), emarginate or not; antennomeres

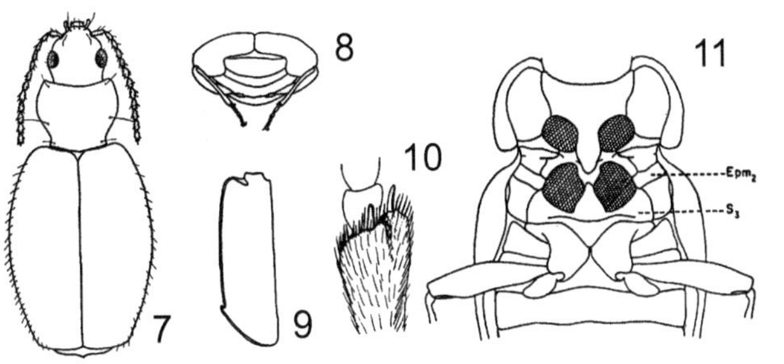


FIGURES 7.6-11.6. Fig. 7.6, *Gehringia olympica* Darlington, habitus, dorsal view; Fig. 8.6, *Gehringia olympica* Darlington, abdomen and elytra, posterior view; Fig. 9.6, *Pachyteles* sp., left elytron, dorsal view; Fig. 10.6, *Pachyteles* sp., front tibia; Fig. 11.6, *Carabus* sp., thorax, ventral view.

thickened, more or less moniliform; anterior coxal cavities closed; mesepimeron disjunct (i.e., extended inward to outer edge of middle coxa); metepimeron distinct; hind coxae separated (Key II) ........................................................ Ozaenini

— Elytron without flange of Coanda; anterior tibia with spurs various in position; antennomeres various; metepimeron distinct or not; hind coxae various . 6

6(5). Middle coxal cavities disjunct (not entirely enclosed by sterna, Fig. 11.6 ........................................... 7

— Middle coxal cavities conjunct (entirely enclosed by sterna, Fig. 12.6), mesepimera not extended to middle coxal cavities ...................................... 21

7(6). Antennae with insertions on frons, between bases of mandibles (Cicindelinae) .................................... 8

— Antenna with insertion in line with and posteriad adjacent mandibular base .................................... 10

8(7). Anterolateral pronotal angles not more prominent than prosternal margin (Fig. 16.6); anterior pronotal sulcus continuous with anterior prosternal sulcus (*Cicindela*) ........................................... Cicindelini

— Anterolateral angles of pronotum more prominent than anterior margins of prosternum (Fig. 15.6); anterior pronotal sulcus separated from anterior prosternal sulcus and also from prosterno-episternal sulcus .......................................................... 9

9(8). Posterior coxae separated; eyes small; body dorsally dark (color uniformly black), dull (Key X) .... Omini

— Posterior coxae contiguous; eyes large, prominent; body dorsally colored brilliantly (*Megacephala*) ... .................................................. (Megacephalini)

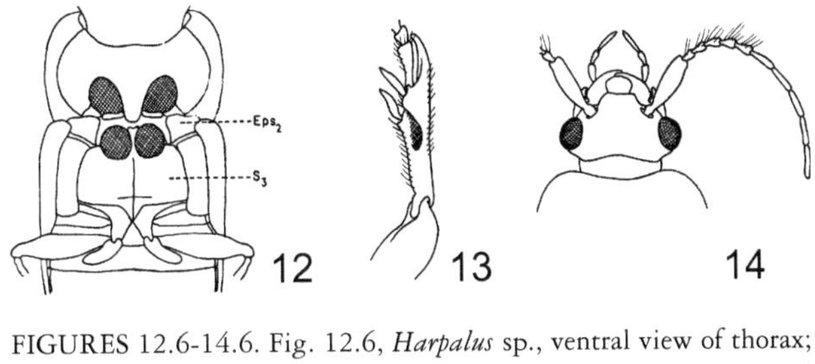


FIGURES 12.6-14.6. Fig. 12.6, *Harpalus* sp., ventral view of thorax; Fig. 13.6, *Pasimachus* sp., anterior tibia; Fig. 14.6, *Loricera* sp., head, dorsal view.

10(7) Anterior coxal cavities open posteriorly (Fig. 11.6) . ...................................................................... 11

— Anterior coxal cavities closed posteriorly (Fig. 12.6) ...................................................................... 16

11(10). Eyes absent; humeral margin of elytron serrate (*Horologion*, Key XXI) (in part) ............. Bembidiini

— Eyes present, distinctly developed ...................... 12

12(11). Mandible without setigerous puncture in scrobe 13

— Mandible with setigerous puncture in scrobe...... 14

13(12). Posterior coxae contiguous in mid-line of body (Key V) .......................................................... Carabini

— Posterior coxae separated in mid-line of body (Key VIII) ....................................................... Cychrini

14(12). Head with two supraorbital setigerous punctures over each eye; frons deflexed, mandibles directed vertically; dorsal surface of elytron with ocellate foveae (*Opisthius*)........................................... Opisthiini

— Head with one supraorbital setigerous puncture over each eye, frons flat, mandibles directed horizontally; elytron without ocellate foveae ............... 15

15(14). Elytron with parascutellar stria extended about 4/5 length of elytron. (*Pelophila*)............... Pelophilini

— Elytron with parascutellar stria short, extended less than 1/4 length of elytron. (Key IV) ......... Nebriini

16(10). Both spurs of front tibia apical; head with one supraorbital setigerous puncture over each eye; hind coxae separated in mid-line of body (*Metrius*) ..... .............................................................. Metriini

— Only one spur of front tibia apical, other preapical, above apical one, near deep emargination (antenna cleaner, Fig. 13.6) .......................................... 17

17(16). Mandible with setigerous puncture in scrobe; head with two supraorbital setigerous punctures over each eye; hind coxae contiguous (Key XI) ........... .............................................................. Elaphrini

— Mandible without setigerous puncture in scrobe; head with one or two supraorbital setigerous punctures over each eye; hind coxae contiguous or not .. 18

18(17). Hind coxae separated in mid-line of body; head with two supraorbital setigerous punctures over each eye; body pedunculate, base of pronotum and elytra remote (*Promecognathus*) ......... Promecognathini

— Hind coxae contiguous in mid-line of body ......... 19

19(18). Antennomere 1 with insertion not concealed under distinct frontal plate or ridge, antennomeres 2-6 with long, projected setae (Fig. 14.6); body not pedunculate (*Loricera*).............................. Loricerini

— Insertion of antennal scape concealed under distinct frontal plate; antennomeres 2-6 without long, projected setae; body pedunculate (Fig. 17.6) ...... 20

20(19). Tarsus with unguitractor plate prominent, setiform; size relatively small, overall length 10 mm. or less (Key XIV)................................................. Clivinini

— Tarsus with unguitractor plate small, not prominent, hardly evident; overall length 15 mm or more (Key XIII) ....................................................... Scaritini

21(6). Maxillary palpomere 4 inserted obliquely in palpomere 3 (Fig. 18.6); integument coarsely punctate, with

dense covering of yellow setae, long on head and pronotum; elytra black and red, black color in form of fascia toward middle (Key XLIX) ........................................................................ Panagaeini
— Maxillary palpomere 4 inserted normally in palpomere 3 .................................................................. 22

22(21). Elytron with broad transverse sulcus in basal one-third, and with several patches of white scales; dorsal surface setose, setae not in form of dense pile – of two colors and sizes – longer black setae, these almost as long as antennomere 1, and shorter yellow setae, about length of antennomere 2; labial palpomere 3 oval in shape, apex sharply pointed; antennomeres 2–11 densely setose; body form ant-like (*Ega*, Key LVIII) (in part) ........................................................................ Lachnophorini
— Elytron without broad transverse sulcus in basal third; combination of characters not as above .......... 23

23(22). Head with one supraorbital setigerous puncture over each eye ........................................................ 24
— Head with two supraorbital setigerous punctures over each eye ........................................................ 33

24(23). Eyes absent or minute; mandible with setigerous puncture in scrobe; integument testaceous (*Neaphaenops*, Key XX) (in part) ............... Trechini
— Mandible without setigerous puncture in scrobe; eyes normal ........................................................ 25

25(24). Apex of elytron truncate (Fig. 19.6); antennae entirely hirsute; body covered with short setae; testaceous (Key LXVI) ............................................. Zuphiini
— Apex of elytron not truncate (Fig. 20.6); antennae entirely hirsute or not; body covered with short setae or not; color various ................................. 26

26(25). Labial palpomere 3 minute, subulate, and inserted within cavity at tip of palpomere 2; antennae entirely hirsute (*Micratopus*) (in part) ...... Bembidiini
— Labial palpomere 3 of average size, not subulate, subequal in length to palpomere 2 ................. 27

27(26). Antennomeres 5-11 covered with setae; antennomeres 1-4 with ring of setae at apex, otherwise glabrous; metepimeron absent or very narrow, posterior margin obliquely truncate; mandible with or without seta in scrobe .................................................. 28
— Antennomeres 3-11 or 4-11 covered with setae and/or pubescence; antennomeres 1-2 or 1-3 with ring of setae at apex only, otherwise glabrous; metepimeron large, posterior margin broadly rounded, or subtruncate; mandible without a seta in scrobe ........................................................ 29

28(27). Body pedunculate, convex; surface of elytra smooth or normally striate, interval 2 not unusually wide; anterior coxal cavities closed; eyes of normal size; middle of frons without series of longitudinally directed ridges, with or without one or two ridges close to eyes; with or without setigerous puncture in scrobe of mandible (Key XVIII) ............ Broscini
— Body flattened, not pedunculate; surface of elytron striate, interval 2 about as wide as next 2–4 intervals together (Fig. 21.6); eyes unusually large; anterior coxal cavities open; at least middle of frons with several longitudinally directed ridges; mandible with seta in scrobe (*Notiophilus*) ........................................................................ Notiophilini

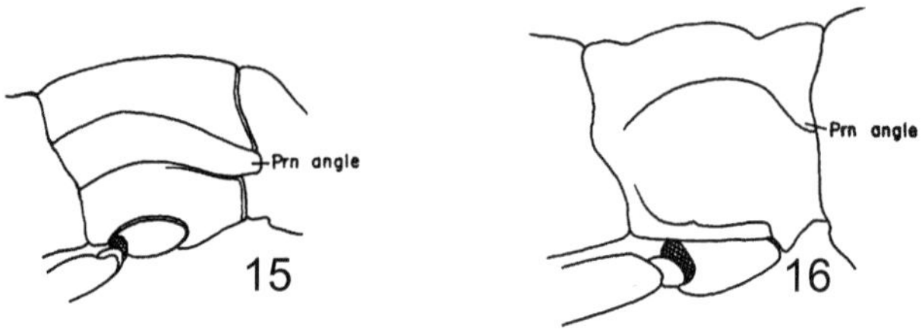

FIGURES 15.6-16.6. Fig. 15.6, *Omus* sp., prothorax, lateral view; Fig. 16.6, *Cicindela* sp., prothorax, lateral view.

29(27). Elytral margin without internal plica toward apex; antennomeres 1–2 with only ring of long setae toward apex, antennomere 3 with or without dense covering of setae in apical one-third, in some taxa, antennomere 3 like 2 (Key XXXIV) .......... Harpalini
— Elytral margin with internal plica toward apex (Fig. 22.6); antennomeres 2–3 with ring of long setae toward apex only, antennomere 1 with single seta (most taxa); without pubescence characteristic of antennomeres 4–11 ........................................ 30

30(29). Labial palpomere 2 with anterior margin plurisetose (*Amara hyperborea* Dejean) ...................... Zabrini
— Labial palpomere 2 bisetose or glabrous ............ 31

31(30). Surface of elytra and pronotum very finely and densely punctate, covered with short setae (Key L) (*Chlaenius*) ........................................... Chlaeniini
— Pronotum and elytra not densely punctate, not covered with setae ............................................... 32

32(31). Elytron with interval 9 a narrow, broken carina; stria 8 in form of deep groove in apical half (Fig. 23.6) (Key LI) ..................................................... Oodini
— Interval 9 normal throughout its length; stria 8 not in form of broad groove toward apex (*Gastrellarius honestus* Say, Key XXVII) .................... Pterostichini

33(23). Maxillary palpomere 4 small, about one-third the length of palpomere 3, subulate (Fig. 24.6) (Key XXI) (in part) ........................................ Bembidiini

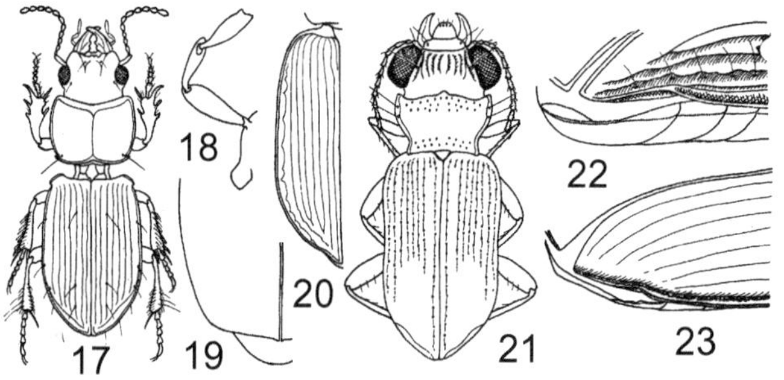

FIGURES 17.6-23.6. Fig. 17.6, *Scarites subterraneus* Fabricius, habitus, dorsal view; Fig. 18.6, *Panagaeus* sp., maxillary palpus; Fig. 19.6, *Galerita* sp., elytra. apical portion, dorsal view; Fig. 20.6, *Agonum* sp., left elytron, dorsal view; Fig. 21.6, *Notiophilus aeneus* Herbst, habitus, dorsal view; Fig. 22.6, *Pterostichus diligendus* Chaudoir, right elytron, apical portion, lateral view; Fig. 23.6, *Oodes americanus* Dejean, right elytron apical region, lateral view.

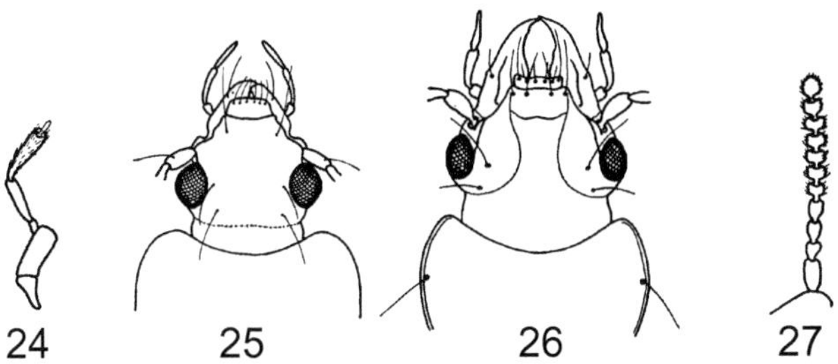


FIGURES 24.6-27.6. Fig. 24.6, *Bembidion* sp., maxillary palpus; Fig. 25.6, *Patrobus* sp., head, dorsal view; Fig. 26.6, *Trechus* sp., head, dorsal view, Fig. 27.6, *Morion* sp., antenna.

— Maxillary palpomere 4 not subulate, about subequal in length to palpomere 3 .................................. 34

34(33). Mandible with setigerous puncture in scrobe (Fig. 26.6) ................................................................ 35

— Mandible without setigerous puncture in scrobe .... ........................................................................ 38

35(34). Dorsal surface of head with transverse sulcus behind eyes (Fig. 25.6) ................................................ 36

— Dorsal surface of head without transverse sulcus behind eyes ......................................................... 37

36(35). Antennomeres 5-10 short and broad, submoniliform (Key XIX) .................................................. Psydrini

— Antennomeres 5-10 longer than wide, filiform (Key XXIV) ..................................................... Patrobini

37(35). Head with frontal grooves broad, shallow, not extended to plane of posterior margin of eyes; dorsal surface of hind tarsomeres each with single pair of setae (Key XXIII) ................................. Pogonini

— Head with frontal grooves narrow, deep, curved posteriorly behind plane of posterior margin of eyes (Fig. 26.6), or eyes lacking; dorsal surface of hind tarsomeres with numerous setae (Key XX) ........... ................................................................ Trechini

38(34). Antennomeres 5-10 submoniliform, short, slightly compressed (Fig. 27.6); margin of pronotum with about seven setigerous punctures; eighth stria appearing as zig-zag groove, with numerous, evenly spaced ocellate punctures (Fig. 28.6); body subpedunculate, legs flattened (*Morion*) ............. ........................................................... Morionini

— Antennomeres 5–10 slender, distinctly filiform; if submoniliform, then pronotum with only single pair of setae on each side and/or other characters not as above ......................................................... 39

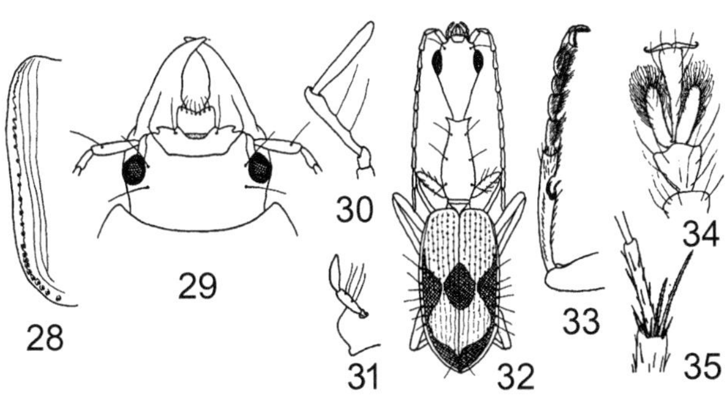


FIGURES 28.6-35.6. Fig. 28.6, *Morion* sp., left elytron, lateral portion, with striae, dorsal view; Fig. 29.6, *Dicaelus dilatatus* Say, head, dorsal view; Fig. 30.6, *Pterostichus diligendus* Chaudoir, labial palpus; Fig. 31.6, *Amara* sp., labial palpus; Fig. 32.6, *Colliuris pensylvanica* (Linnaeus), habitus, dorsal view; Fig. 33.6, *Agra* sp., tibia and tarsus; Fig. 34.6 *Leptotrachelus* sp., tarsomeres 2-5, dorsal view; Fig. 35.6, *Tetragonoderus* sp., tibia, apical portion, with spurs and tarsomere 1.

39(38). Clypeus with dorsal surface concave, deflected, meeting labrum at obtuse angle; anterior margin of clypeus concave; basal margin of labrum normally exposed, apical margin of labrum more or less deeply incised, resulting lobes more or less asymmetrical (Fig. 29.6); elytron without internal plica (Key XLVI) .................................................. Licinini

— Clypeus plane, not deflected ventrally, labrum and clypeus meeting in same plane; anterior margin of clypeus either straight or only slightly emarginate ....................................................................... 40

40(39). Elytron with internal plica near apical portion of lateral margin (Fig. 22.6) .................................... 41

— Elytron without internal plica near apical portion of lateral margin ................................................ 44

41(40). Labial palpomere 2 plurisetose (Fig. 31.6); elytron without discal setigerous puncture (*Amara*, Key XXXIII) (in part) ......................................... Zabrini

— Labial palpomere 2 bisetose (Fig. 30.6) OR, if plurisetose, then elytron with one or more discal setigerous punctures ...................................... 42

42(41). Male with front tarsomeres 1-3 with apical margin diagonal AND elytron without parascutellar stria (Key XXVI) ........................................... Loxandrini

— Male with front tarsomeres 1-3 with apical margin truncate or shallowly notched; elytron with or without parascutellar stria ...................................... 43

43(42). Maxillary palpomeres 3 and 4 with numerous long setae; labial palpomere 2 bisetose (*Pseudamara*) (in part) ..................................................... Zabrini

— Maxillary palpomeres 3 and 4 glabrous or 3 with few short setae near apex; labial palpomere 2 bisetose or plurisetose (Key XXVII) .................. Pterostichini

44(40). Pronotum slender, distinctly longer than wide, at apex no wider than posterior portion of head (Fig. 32.6) ....................................................................... 45

— Pronotum not distinctly longer than wide AND/OR wider at apex than posterior portion of head .. 49

45(44). Labial palpomere 3 triangular, maxillary palpomere 4 cylindrical; antennomeres 1-2 with between 10 and 15 long setae, each; tarsomere 4 bilobed; tarsal claws pectinate (Fig. 33.6) (*Agra*) (in part) .......... ................................................................ Lebiini

— Labial palpomere 3 and maxillary palpomere 4 similar, cylindrical or slightly oval or triangular, OR labial palpomere 3 short; tarsomere 4 deeply bilobed or normal; tarsal claws not pectinate ............. 46

46(45). Tarsomere 4 deeply cleft, resulting lobes more than half length of tarsomere 5 (Fig. 34.6); elytra entire, covering abdomen completely (*Leptotrachelus*)... ..................................................... Ctenodactylini

— Tarsomere 4 not deeply cleft .............................. 47

47(46). All antennomeres setose; labial palpomere 2 plurisetose; maxillary palpomere 4 and labial palpomere 3 both triangular; elytron truncate at apex; body covered with setae (*Galerita*, Key LXVII) ............................................................ Galeritini

— Antennomeres 1-3 with few setae; labial palpomere 2 bisetose; maxillary palpomere 4 and labial palpomere 3 not triangular; body glabrous ..... 48

48(47). Elytra entire; color of integument uniformly testaceous; eyes of average size, or very markedly reduced (*Rhadine*, Key LIII) (in part) ........... Platynini

— Elytron with apical margin obliquely subtruncate; color various, but not testaceous; eyes normal (*Colliuris*, Key LIX) ............................ Odacanthini

49(44). Hind tibia with inner spur much longer than outer one, one-half or more length of hind tarsomere 1; elytron with apical margin sinuate, subtruncate............ 50

— Tibial spurs of hind leg subequal, less than one-half length of hind tarsomere 1 .............................. 51

50(49). Tibial spurs serrulate (Fig. 35.6); tarsal claws not pectinate; head not constricted in form of neck (*Tetragonoderus*, Key LX) .................. Cyclosomini

— Tibial spurs not serrulate; tarsal claws pectinate; head constricted in form of neck (*Nemotarsus*, Key LXI) (in part) ..................................................... Lebiini

51(49). Elytron with apical margin truncate ..................... 52

— Elytron with apical margin not truncate, broadly rounded or sinuate ......................................... 57

52(51). Head markedly constricted behind eyes in form of neck; width one-half or less than that of width of head across eyes (if slightly more, then tarsal claws pectinate) ........................................................ 53

— Head not markedly constricted behind eyes in form of narrow neck; tarsal claws pectinate or not .. 55

53(52). Dorsal integument covered with setae; labial palpomere 2 plurisetose, labial palpomere 3 triangular, oval or fusiform; antennomeres 1-3 sparsely or densely setose; tarsal claws not pectinate (*Zuphium*, Key LXVI) (in part) .................... Zuphiini

— Dorsal integument glabrous, except for usual fixed setae; labial palpomere 2 bisetose, labial palpomere 3 fusiform; antennomeres 1-3 glabrous except for single long seta on scape and ring of setae toward apex of 2 and 3, tarsal claws pectinate or not . 54

54(53). Tarsal claws pectinate; lateral margins of pronotum rounded (Fig. 36.6) or markedly sinuate between anterior and posterior setae (*Lebia* in part, and *Onota*, Key LXI) ......................................... Lebiini

— Tarsal claws not pectinate; sides of pronotum angular (Fig. 37.6) (*Pentagonica*) ............ Pentagonicini

55(52). Tarsal claws pectinate (Key LXI)-(in part) ....... Lebiini

— Tarsal claws simple ........................................... 56

56(55). Legs flattened, anterior tibia broad (Fig. 38.6) its width at apex subequal to length of scape of antenna; antennomere 4 slightly trianguloid, antennomeres 5-11 more or less flattened (Fig. 39.6), broad surfaces of each with depressed triangular area at base, this smoother than adjacent areas (Fig. 40.6); mentum with tooth; integument covered with short setae;

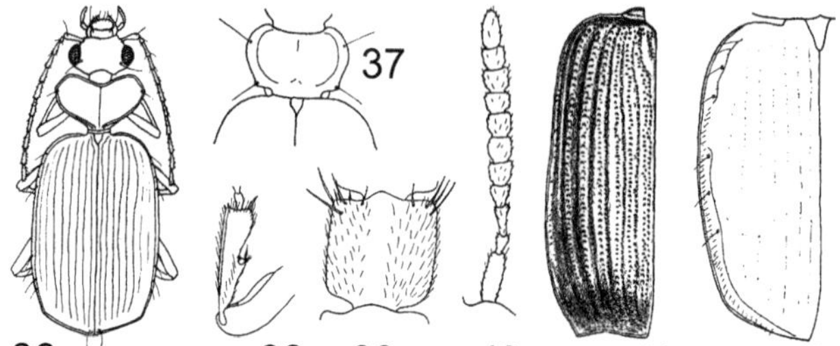


FIGURES 36.6-42.6. Fig. 36.6, *Lebia* sp., pronotum, dorsal view; Fig. 37.6, *Pentagonica flavipes* LeConte, habitus, dorsal view; Fig. 38.6, *Helluomorphoides* sp., anterior tibia; Fig. 39.6, *Helluomorphoides texanus* LeConte, antenna; Fig. 40.6, *Helluomorphoides texanus* LeConte, antennomere 7; Fig. 41.6, *Eucheila (Inna) boyeri* (Solier), left elytron, dorsal view; Fig. 42.6, *Perigona nigriceps* Dejean, left elytron, dorsal view.

elytral striae (Fig. 41.6) about as wide as intervals, coarsely punctate, punctures bi- or tri-seriate or confused (*Helluomorphoides*) ....................... Helluonini

— Legs not flattened, anterior tibia slender, its width at apex evidently less than length of antennomere 1; antennomere 4 not trianguloid, antennomeres 5-11 not flattened, without depressed triangular areas; elytral striae various, narrow, or as wide as intervals (Fig. 41.6); integument glabrous or setose; mentum with or without a tooth (Key LXI)........... ................................................................. Lebiini

57(51). Elytron with stria 8 toward apex deeply impressed and obliquely extended nearly to suture (Fig. 42.6), apical margin subarcuate; hind trochanter nearly one-half length of hind femur (*Perigona* ) ........... ............................................................ Perigonini

— Stria 8 only slightly more deeply impressed than the others, and toward apex not nearly attaining suture ................................................................. 58

58(57). Dorsal integument glabrous, except for usual fixed setae .............................................................. 59

— Dorsal integument with more or less dense covering of setae ........................................................... 61

59(58). Tarsal claws not pectinate (Key LIII) (in part) .......... .............................................................. Platynini

— Tarsal claws pectinate ......................................... 60

60(59). Apical portion of elytron slightly sinuate, subtruncate; pronotum with broad punctate lateral margins of equal width throughout, OR pronotum about 1.75 times wider than long and elytra metallic green (Key LXI) (in part) .............................................. Lebiini

— Apical portion of elytron not subtruncate; pronotum about 0.95 times wider than long, lateral margins impunctate, narrower anteriorly than posteriorly; elytra black or piceous, not metallic green (Key LIII) (in part) .................................................. Platynini

61(58). Elytra with only odd-numbered intervals set with coarse setigerous punctures, otherwise glabrous (Key LIII) (in part) ................................... Platynini

— Elytra with all intervals bearing setae .................. 62

62(61). Elytra with striae more deeply impressed in anterior half than posterior half; AND/OR anterior half of striae coarsely punctate, posterior half finely punc-

tate or impunctate; setae of integument erect and at least some about as long as antennal scape (Key LVIII) (in part) .................................. Lachnophorini
— Elytra with striae equally impressed throughout, finely punctate throughout or impunctate; setae of body shorter, decumbent, and dense ........................ 63

63(62). Hind tarsus with tarsomeres 1-3 each with pair of grooves on dorsal surface (*Agonum*, subgenus *Stictanchus*, in part, Key LVI) .................... Platynini
— Dorsal surfaces of hind tarsomeres not grooved .... 64

64(63). Labial palpomere 2 plurisetose; tarsal claws pectinate (Key LXI) (in part) ....................................... Lebiini
— Labial palpomere bisetose; tarsal claws not pectinate ........................................................................ 65

65(64). Setae of dorsal surface of elytra very short, about length of antennal pubescence (*Agonum*, in part, Key LVI) ............................................................. Platynini
— Setae of dorsal surface longer than antennal pubescence ........................................................................ 66

66(65). Hind corners of pronotum rounded; dorsal surface concolorous (*Atranus* ) ............................ Platynini
— Hind corners of pronotum angulate; dorsal surface bicolored, head and pronotum black, elytra brownish or testaceous ................................................... 67

67(66). Intervals of elytra broad and flat, densely and finely punctate (*Anchonoderus*, Key LVIII) ....................... ....................................................... Lachnophorini
— Intervals of elytra convex, each with at most two rows of punctures; hind angles of pronotum distinctly margined (*Oxypselaphus*) .............................. Platynini

## Keys To The Genera And Subgenera Of Nearctic Carabidae

II. Key To The Nearctic Genera Of Ozaenini

1. Tibia very wide and compressed, dorsal surface in form of knife-like edge; anterior tibia with shallow emargination; each femur with deep groove on underside .................................................................. *Physea*
— Tibia normal, more or less round in cross section; femora on underside not or only shallowly grooved ........................................................................ 2

2(1). Anterior tibia shallowly emarginate, without prominent tooth at base of emargination; head, pronotum and elytra without normal long setae, glabrous or with vestiture of short thick setae; mandibles with dorsal surfaces more or less densely punctate; antennomere 11 shining, slightly swollen and wider than preceding, narrowed abruptly distally, apical portion wedge-shaped, thickly granulate and mat .............................................................. *Ozaena*
— Anterior tibia deeply emarginate, with prominent tooth at base of emargination; head, pronotum and elytra with normal long setae; mandibles with dorsal surfaces impunctate; antennomere 11 dull, of about same width as preceding, narrowed gradually apically, granulate mat area extended along each margin almost to base ................ *Pachyteles*

III. Key to Nearctic Subgenera of *Pachyteles*

1. Middle coxae in contact with one another, not separated by intercoxal extensions of meso- and meta-sterna ................................................ *Pachyteles*
— Middle coxae separated by intercoxal extensions of meso- and meta-sterna ..................... *Goniotropis*

IV. Key to Nearctic Genera of Nebriini

1. Outer margin of maxilla with 5-6 processes, each with stout seta; gula with transverse, curved row of long, stout setae each inserted in process like those of maxilla; lateroventral edge of mandible produced laterally, in form of broad lamella; labial palpomere 2 bisetose, about 1.25 times longer than palpomere 3 ............................................................ *Leistus*
— Outer margin of maxilla smooth, without processes, setose or not; gula glabrous; submentum with or without transverse row of slender setae, these not arising from prominent processes; lower outer edge of mandible not produced laterally in form of broad lamella; labial palpomere 2 bi- or pluri-setose, subequal in length to, or shorter than palpomere 3 ......................................................................... 2

2. Pronotum distinctly cordate, basal sinuation of lateral margin moderately to very deep; hind tarsus with tarsomere 4, ventrolaterally extended as short to moderately long lobe, and with apicoventral setae longer than medial setae and displaced toward apex of lobe; labial prementum with paraglossae indistinct, fused with lateral portion of glossal sclerite ............................................................ *Nebria*
— Pronotum semiovoid, basal sinuation of lateral margin absent or very short and shallow; hind tarsomere 4 truncate ventrally, with medial and lateral apicoventral setae equal in length and similar in position; prementum with paraglossae distinct, each a short apicolateral lobe on glossal sclerite .... *Nippononebria*

V. Key to the Nearctic Genera of Carabini

1. Antenna with antennomere 2 globose, antennomere 3 long, more than twice length of 2, and with dorsal margin of both articles distinctly carinate; mandibles with surfaces transversely corrugate (Key VI) ........................................................... *Calosoma*
— Antenna with antennomere 2 not longer than antennomere 3, both ecarinate, or antennomere 3 with carina at base, only; surfaces of mandibles smooth (Key VII) .................................... *Carabus*

VI. Key to the Nearctic Subgenera of *Calosoma* (After Gidaspow 1959)

1. Metepisternum short, lateral margin as long as basal (anterior) margin ............................................... 2
— Metepisternum long, lateral margin longer than anterior margin .................................................. 3

2(1). Antenna with antennomeres 5-11 uniformly setose (pubescent) ...................................... *Blaptosoma*
— Antennomeres 5-11 each with medial area glabrous, setae concentrated dorsally and ventrally .......... ............................................... *Microcallisthenes*

3(1). Pronotum with basolateral and mediolateral setae each side; tibiae, especially front pair with fine dense punctation (in addition to normal grooves and rows of spines); male with hind trochanter markedly arcuate and pointed at apex .... *Castrida*

— Pronotum without basolateral setae; mediolateral setae present or absent; tibiae (beside usual grooves and spines), smooth or with few fine punctures; hind trochanter of male at most slightly arcuate, apex rounded .................................................. 4

4(3). Maxillary palpomere 4 of same length and hardly wider than palpomere 3; tooth of mentum small, blunt; ventral surface metallic green or with distinct green or bluish luster ............................... 5

— Maxillary palpomere 4 notably wider, of same length as or shorter than palpomere 3; tooth of mentum much longer and pointed in most taxa; ventral surface with or without very faint metallic luster.... 6

5(4). Femora black or dark brown, without metallic green, violet or blue luster; elytra either brilliant green, often with tint of red on sides, but not margins, and foveae small, OR dark, almost black, with large coppery or green foveae ....................... *Calosoma*

— Femora reddish brown or dark brown, with metallic green, blue or violet luster; elytra brilliant green or dark green; foveae inconspicuous or very small .......................................................... *Calodrepa*

6(4). Pronotum with angulate or markedly arcuate sides and small, mostly pointed, hind angles; dorsal surface black ........................................................ 7

— Pronotum with slightly arcuate sides, flattened or not at base; hind angles rounded; elytra bronze, green, brown, or black .............................................. 10

7(6). Elytra elongate, with striae deep, and convex scaly intervals; pronotum narrower, about 1.5 times as wide as long, sides markedly angulate ............... ........................................................ *Carabosoma*

— Elytra wider, striae mostly obliterated or rather fine toward apex; intervals flat, scaly at basal part only; pronotum wider, nearly twice, or more than twice as wide as long, with lateral margins markedly or moderately angulate ......................................... 8

8(7). Hind trochanter without setae; head either with sparse, large punctures and pronotum with rounded side angles, OR head more finely and densely punctate and pronotum with lateral margins markedly angulate ............................................ *Camegonia*

— Hind trochanter of most specimens with a seta; head finely punctate, densely so in many specimens; pronotum with lateral margins more rounded ... 9

9(8). Pronotum with angulate or markedly arcuate lateral margins and very small pointed hind angles; mentum with tooth small and pointed ......... *Camedula*

— Pronotum with lateral margins more rounded, and rounded (though small), hind angles; tooth of mentum either blunt or long and pointed. (in part) .... ...................................................... *Chrysostigma*

10(6). Head with sparse, large punctures; antennomeres 5-11 uniformly pubescent; metepisternum with large, sparse punctures ................................ *Callitropa*

— Head finely and densely punctate; antennomeres 5-6 with elongate glabrous areas; metepisternum with punctation various (in part) ............................ 11

11(10). Elytron with intervals 4, 8, and 12 catenate (i.e., with elongate tubercles between foveae); head large, wider than single elytron; pronotum with basolateral angles broadly rounded, distinctly projected posteriad median portion of basal margin ..................................................... *Tapinosthenes*

— Elytron with intervals 4, 8 and 12 uniform, not catenate; head normal in size, not wider than single elytron; pronotum with basolateral angles narrowly rounded, smaller, only slightly projected posteriad median potion of basal margin (in part) ............. ...................................................... *Chrysostigma*

VII. Key to the Nearctic Subgenera of *Carabus*

1. Labial palpomere 2 plurisetose; proepisternum with few punctures, elytral intervals convex, broken by very coarse punctures, surface rugose; disc of elytra purple, lateral margins red and green ................ ......................................................... *Megodontus*

— Labial palpomere 2 bisetose; disc of elytra bronze, black, or green .................................................. 2

2(1). Elytron with three broad, smooth, prominent costae, interspaces rugulose; legs, mouthparts and antennomeres 1–4 reddish-orange. . *Autocarabus*

— Elytron not as above, if with prominent costae, some of these broken ................................................ 3

3(2). Anterior tibia with prominent pointed apophysis at external apical angle (fig 43.6); elytron with three, prominent, convex, catenate intervals; between adjacent catenate intervals three convex but less prominent intervals, broken by cross striations .. ....................................................... *Hemicarabus*

— Anterior tibia without prominent pointed apophysis at external apical angle; elytral sculpture not as described above ................................................ 4

4(3). Femora and tibiae red with black tips; each elytron with three rows of large, shallow foveae, intervals numerous, convex, evenly spaced, and of about same size ..................................... *Aulonocarabus*

— Legs concolorous, black; sculpture of elytra various ........................................................................ 5

5(4). Elytron with three catenate intervals, these either more prominent than remaining intervals, or, if not, then discal striae narrow, finely punctate, and uninterrupted in basal half .................................. 6

— Elytron with two or three series of large foveae intervals numerous and narrow AND/OR indistinct and interrupted by shallow, crenulate, transverse lines ......................................................................... 7

6(5). Proepisternum impunctate; elytron with humeral margin smooth, not serrate .. *Carabus (sensu stricto).*

— At least anterior half of proepisternum punctate; elytron with humeral margin serrate (Fig. 44.6)... ................................................. *Homoeocarabus*

7(5). Dorsal surface concolorous, black, OR with head and pronotum black, elytra brown ......... *Tomocarabus*

— Dorsal surface bronze or black, lateral margins of pronotum and elytra purple or metallic green ... 8

8(7). Hind angles of pronotum with lobes at least 1.5 times longer than antennomere 2, disc of pronotum and

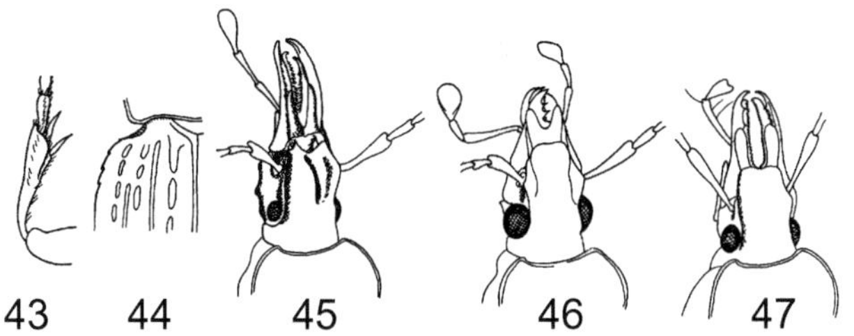


FIGURES 43.6-47.6. Fig. 43.6, *Carabus* sp., anterior tibia; Fig. 44.6, *Carabus maeander* Fischer von Waldheim, left elytron, basal portion, dorsal view; Fig. 45.6, *Scaphinotus (Neocychrus) angulatus* Harris, head, dorso-lateral view; Fig. 46.6, *Scaphinotus (Brennus)* sp., head, dorso-lateral view; Fig. 47.6, *Scaphinotus (Steniridia) andrewsii* Harris, head, dorso-lateral view.

elytra dull bronze, or dull metallic green ............................................................ *Archicarabus*

— Hind angles of pronotum with lobes shorter than antennal pedicel; disc of pronotum and elytra black, not bronze .................................... *Tanaocarabus*

VIII. Key to the Nearctic Genera of Cychrini

1. Suture between proepisternum and proepimeron present, at least in medial half of pleuron; base of labral emargination with or without two setae; labial palpomere 2 with more than two setae (Key IX) ........................................................ *Scaphinotus*

— Suture between proepisternum and proepimeron absent, (but area representing proepimeron densely punctate or not); base of labral emargination bisetose; labial palpomere 2 with two or more setae .................................................................... 2

2(1). Labial palpomere 2 bisetose; range, eastern North America .......................................... *Sphaeroderus*

— Labial palpomere 2 plurisetose; range, northwestern North America .......................................... *Cychrus*

IX. Key to the Subgenera of *Scaphinotus*
(From Gidaspow 1973: 55-56)

1. Antenna with antennomeres 5-11 pubescent; lateral margin of pronotum asetose or with no more than two setae; hind angles various .......................... 2

— Antenna with antennomeres 3-11 or 4-11 pubescent; lateral margin of pronotum with setation various-if with two or more setae, form of hind angles and punctation of pronotal disk various; if setae absent or only one middle seta, then hind angles of pronotum pointed, produced, and disc coarsely punctate ............................................................ 8

2(1). Pronotum broader toward base, hind angles pointed, produced beyond base ........................................................................... *Scaphinotus* (*sensu stricto*)

— Pronotum narrowed toward base, hind angles not produced posteriad base, or only slightly so, pointed or not ................................................................ 3

3(2). Head laterally with gena either wide laterally, as prominent process, or notched (as seen underneath and in front of eyes, figs. 45.6, 46.6); elytron with epipleuron smooth, or sparsely, finely punctate .................. 4

— Head with gena simple, not expanded (Fig. 47.6); elytron with epipleuron densely punctate, punctures coarse or fine ........................................... 7

4(3). Head laterally with gena moderately or markedly widened, in form of undivided plate (Fig. 45.6); head dorso-anteriorly either with crest or deeply transversely wrinkled; pronotum with lateral margins angulate or long and narrow; male with front tarsomeres only slightly dilated, similar to front tarsomeres of females ...................... *Neocychrus*

— Head laterally with gena notched (Fig. 46.6), dorsally with or without crest; pronotum about as long as wide, lateral margins angulate or not; male with front tarsomeres moderately to markedly dilated ......................................................................... 5

5(4). Elytron with striae 18-20, partly irregular, shallowly or only moderately deeply impressed, surface either not shiny and velvety, or if more shiny, then color of pronotum and elytral margin violet red. Head with supraorbital seta over each eye, hind coxa with front seta .................... *Stenocantharus*

— Elytron with striae various in number, distinctly impressed, color of pronotum various, but not violet red. If striae shallow and elytral punctures scattered, then either pronotum distinctly wider than long, or supraorbital setae and front seta of hind coxa absent; legs normal .................................. 6

6(5). Elytral intervals 4 and 8 each with shallow foveae. If striae irregular and foveae inconspicuous then pronotum with lateral margins bisetose and front tarsomeres of male markedly dilated (in part) ..... ............................................ *Pseudonomaretus*

— Elytral intervals 4 and 8 not foveate; striae either regular or merged (impossible to count); pronotum with lateral margins unisetose (middle seta each side, only); and front tarsomeres of male moderately dilated ................................................... *Brennus*

7(3). Pronotum with lateral margins wide, markedly reflexed ........................ *Scaphinotus* (*sensu stricto*)

— Pronotum with lateral margins narrow ..... *Steniridia*

8(1). Labrum at base with four setae. Labial palpomere 2 plurisetose (more than two setae); elytral striae at least 10 ............................................................ 9

— Labrum at base with two setae. Labial palpomere 2 bisetose; elytral striae 10 or fewer ...... *Maronetus*

9(8). Elytron with epipleuron densely, coarsely punctate; dorsal surface with purple or metallic luster ... 10

— Elytral epipleuron smooth or with few shallow punctures; dorsal surface without metallic luster (in part) ................................................ *Pseudonomaretus*

10(9). Pronotum with hind angles pointed, produced, lateral margins asetose or with middle seta only .... ................................ *Scaphinotus* (*sensu stricto*)

— Pronotum with hind angles not extended posteriorly, lateral margins bisetose or plurisetose .............................................................................. *Nomaretus*

X. Key to the Nearctic Genera of Cicindelinae

1. Prothorax in lateral aspect with anterior pronotal angles more prominent than anterior margin of prosternum (Fig. 15.6); anterior pronotal sulcus

separated from anterior prosternal sulcus and also from prosterno-episternal sulcus ....................... 2
— Anterior pronotal angles not more prominent than prosternal margin (Fig. 16.6), anterior pronotal sulcus continuous with anterior prosternal sulcus ... ............................................................ *Cicindela*

2(1). Hind coxae separated; eyes small; body dorsally dark (color uniformly black), dull .............................. 3
— Hind coxae contiguous with one another; eyes large, prominent; body dorsally colored brilliantly ........ ....................................................... *Megacephala*

3. Mandible with scrobe with several long setae. Inflexed part of elytron coarsely punctate. Male sternum VII with posterior margin entire ............ *Amblycheila*
— Mandibular scrobe without setae. Inflexed part of elytron impunctate. Male sternum VII with posterior margin deeply emarginate .................. *Omus*

XI. Key to the Nearctic Genera of Elaphrini
(After Goulet 1983)

1. Pronotum asetose or with single seta on each lateral margin near hind angle. Elytron with striae absent or at most suggested near base; middle coxa with numerous setae. (Key XII) ...................... *Elaphrus*
— Pronotum with two setae on each lateral margin, one near hind angle and one near middle. Elytral striae distinctly developed on disc; middle coxa with one or two setae ...................................................... 2

2(1). Pronotum with lateral portion explanate. Head dorsally with frontal furrows sharply impressed and eight-shaped. Mentum with two pairs of setae. Clypeus with sublateral impressions ........ *Blethisa*
— Pronotum with lateral portion not explanate. Frontal furrows shallowly impressed and linear. Mentum with one pair of setae. Clypeus without impressions ............................................................ *Diacheila*

XII Key to Nearctic Subgenera of *Elaphrus*
(After Goulet 1983)

1. Pronotum with fringe of setae along posterior margin extended to posterior angles; front tarsomere 4 of male without adhesive vestiture ventrally .. 2
— Pronotum with fringe of setae along posterior margin not extended to posterior angles; front tarsomere 4 of male with adhesive vestiture ventrally ................................................................ 3

2(1). Clypeus with four to six setigerous punctures; disc of prosternum with setae; front and middle trochanters with three setigerous punctures ........... ....................................... *Elaphrus (sensu stricto)*
— Clypeus with two setigerous punctures; disc of prosternum without setae; front trochanter with two setigerous punctures, middle trochanter with one or two setigerous punctures ..... *Elaphroterus*

3(1). Maxillary palpomere 3 0.5 X as long as palpomere 4; disc of prosternum and lateral portion of metasternum with setae ................ *Arctelaphrus*
— Maxillary palpomere 3 0.3 X as long as palpomere 4; disc of prosternum and lateral portion of metasternum without setae ........... *Neoelaphrus*

XIII. Key to the Nearctic Genera of Scaritini

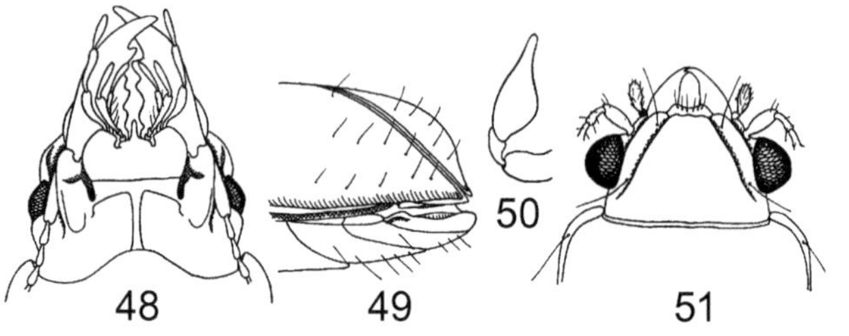


FIGURES 48.6-51.6. Fig. 48.6, *Scarites substriatus* Fabricius, head, ventral view; Fig. 49.6, *Semiardistomis viridis* (Say), abdomen and elytron, left lateral view; Fig. 50.6, *Oxydrepanus* sp., maxillary palpus; Fig. 51.6, *Bembidion (Trepanedoris) anguliferum* LeConte, head, dorsal view.

1. Antennomere 1 with long seta in apical half; pronotum with hind angles rectangular or nearly so; rectangular submentum and paragenae not separated by fissure ............................................... *Pasimachus*
— Antennomere 1 without long seta; pronotum with rounded hind angles, each with small tooth; submentum and paragenae separated by distinct fissure (Fig. 48.6) .................................... *Scarites*

XIV. Key to the Nearctic Genera of Clivinini

1. Elytron with umbilicate series with less than eight punctures, these in two distinct groups (subhumeral group [absent from some species] and subapical group); marginal groove of elytron not extended to sutural angle, at least interval 2 extended to apical margin ................................................... 2
— Elytral umbilicate series with more than 15 continuous punctures; marginal groove of elytron extended almost to sutural angle, interval 2 separated from apical margin .................................................... 4

2(1). Pronotum and elytra without microsculpture. (Key XVIII) (in part) .................................. *Dyschiriodes*
— Pronotum and elytra with microsculpture mesh pattern .................................................................. 3

3(2). ...............................................................................
Elytron without subapical setigerous punctures; base of elytron somewhat margined inside shoulder; wings markedly reduced ............... *Akephorus*
— Elytron with one or two subapical setigerous punctures; base of elytron unmargined inside shoulder; wings fully developed (KEY XVII) (in part) ........... ....................................................... *Dyschiriodes*

4(1). Abdominal sternum VII with dorsolateral projection on each side fitted into subapical elytral plica (Fig. 49.6); marginal elytral groove united at base with stria 3 or 4 ....................................................... 5
— Abdominal sternum VII without dorsolateral projections, elytral epipleuron without plica; elytron with marginal groove not directly united with stria 3 or 4 at base ........................................................... 7

5(4). Metasternum without transverse row of setigerous punctures anteriad hind coxae; abdomen with groove between sterna III and IV distinct at middle; marginal elytral groove united with stria 4 at base ....................................................... *Aspidoglossa*

— Metasternum with transverse row of setigerous punctures anteriad hind coxae; abdomen with groove between sterna III and IV obsolete at middle; elytron with marginal groove united with stria 3 at base . 6

6(5). Front tibia with proximal spur longer than apical spur; lateral margins of pronotum with several long setae; elytron with striae coarsely punctate, each puncture bearing seta; antennomere 2 shorter than antennomere 1 ............................ *Semiardistomis*

— Front tibia with proximal spur shorter than apical spur; lateral margins of pronotum with 2 long setae on each side; elytron with striae impunctate; antennomere 2 about as long as antennomere 1 ... ............................................................ *Ardistomis*

7(4). Frons with series of longitudinal carinae between eyes ......................................................................... 8

— Frons without series of longitudinal carinae ......... 9

8(7). Length of body less than 3.0 mm; antennomere 2 plurisetose; pygidium without series of fine paramedian longitudinal striae ................. *Halocoryza*

— Length of body more than 3.0 mm; antennomere 2 at most bisetose; pygidium with series of fine paramedian longitudinal striae (Key XV) .................... ........................................................ *Schizogenius*

9(7). Maxillary palpomere 4 (Fig. 50.6) subulate; length of body less than 3.5 mm ................... *Oxydrepanus*

— Maxillary palpomere 4 not subulate, subconical, tapered gradually toward apex; length of body more than 3.5 mm; (Key XVI) .............................. *Clivina*

XV. Key to Nearctic Subgenera of *Schizogenius*

1. Marginal groove of elytron expanded near apex, with one or more deep subapical pits ..... *Genioschizus*

— Marginal groove of elytron narrowed near apex, without deep subapical pits ....................................... ................................ *Schizogenius* (*sensu stricto*)

XVI. Key to Nearctic Subgenera (incl. one species group) of *Clivina*

1. Lateral bead of pronotum rounded behind posterior setigerous puncture; elytron with two setigerous punctures in interval/stria 3; medial setae on sternum VII approximate, distance between them much less than between medial and lateral setae ......... ......................................................... *Paraclivina*

— Lateral bead of pronotum convergent behind posterior setigerous puncture, in form of distinct angle near base; elytron with four or five setigerous punctures in interval/stria 3; medial setae on sternum VII distant, distance between them equal to or longer than between medial and lateral setae ... 2

2(1). Middle tibia with small, rounded, preapical protuberance; seta on protuberance apical ...................... ............................................... *americana* group

— Middle tibia with long, acuminate, preapical protuberance; seta on protuberance lateral .............. 3

3(2). Proepisternum with narrow, sculptured band extended more or less parallel to lateral margin; front femur with posteroapical, toothlike process; distance between medial setae on sternum VII longer than between medial and lateral setae; elytron with five setigerous punctures in interval 3 ....................... ......................................................... *Semiclivina*

— Proepisternum without sculptured band; front femur without process or with posteroapical, rounded process; distance between medial setae on sternum VII subequal to distance between medial and lateral setae; elytron with four or five setigerous punctures in interval 3 .............. *Clivina* (*sensu stricto*)

XVII. Key to Nearctic Subgenera of *Dyschiriodes* Jeannel

1. Elytron with coppery transverse spots ................... ..................................................... *Antidyschirius*

— Elytron without coppery spots .............................. 2

2(1). Parameres of male genitalia each with apical seta .. ....................................................... *Eudyschirius*

— Parameres without apical seta .............................. 3

3(2). Elytra dark with apex yellowish; anterior margin of clypeus with two short projections at middle ...... .................................................... *Paradyschirius*

— Elytra uniformly dark or pale with diffuse cloud in middle; anterior margin of clypeus straight or convex at middle ........... *Dyschiriodes* (*sensu stricto*)

XVIII. Key to the Nearctic Genera of Broscini
(From Davidson and Ball 1998)

1. Pronotum with single pair of setae (basal pair absent); pronotum markedly constricted posteriorly, with deep transverse groove; labial mentum with pair of paramedian pits ............................... *Miscodera*

— Pronotum with two pairs of marginal setae; pronotum not markedly constricted posteriorly; mentum with or without paramedian pits .............................. 2

2(1). Head with frons and vertex more or less grooved and rugose; mentum without paramedian pits .......... ............................................................... *Zacotus*

— Head with dorsal surface smooth, except frontal impressions and sparse punctation; mentum with or without paramedian pits .................................. 3

3(2). Tarsomeres 1–4 with dorsal surfaces smooth, glabrous; elytron laterally with eight or nine setae in umbilical series; mentum without paramedian pits; middle femur with row of about six setae ventrad, on anterior surface; size larger, overall body length about 20 mm; Palaearctic, introduced in North America, known only from coastal localities in eastern Canada ............................................. *Broscus*

— Tarsomeres 1–4 with dorsal surfaces rugulose and sparsely setose; elytron with three or four setae in umbilical series; mentum with paramedian pits; middle femur with row of not more than three setae ventrad on anterior face; size smaller, overall body length about 10 mm; western North America .......................................................... *Broscodera*

XIX. Key to the Nearctic Genera of Psydrini

1. Antenna with antennomeres 3–11 pubescent. Elytron without discal setigerous punctures ......... *Nomius*

— Antenna with antennomeres 1–11 pubescent. Elytral interval 3 with one or two discal setigerous punctures ........................................................ *Psydrus*

XX. Key to the Nearctic Genera of Trechini
(Modified from Barr 1972)

1. Eyes present, of average size; color of integument rufotestaceous, dark brown to black ................ 2
— Eyes absent or vestigial; color of integument rufotestaceous; cave inhabitants ...................... 4

2(1). Color black or piceous; integument dorsally glabrous, shining; geographical range most of Canada, Alaska, much of conterminous United States, excluding southern Great Plains and Atlantic and Gulf coastal plains .......................................... *Trechus*
— Color more or less testaceous; integument dorsally densely pubescent ............................................ 3

3(2). Dorsal surface darker, with prominent blue-black fascia across elytra; pronotum cordiform; eyes very convex; elytron with striae absent anteriad apex; apical recurrent groove oblique to suture; geographical range, eastern Canada and northeastern United States ........................................... *Blemus*
— Dorsal surface paler, elytral fascia nebulous or absent; pronotum transverse-subquadrate; eyes slightly convex; elytron with at least striae 1-3 extended to apical margin; apical recurrent groove subparallel in relation to suture, conspicuously joined to stria 3; geographical range, northwestern Oregon .................................... *Trechoblemus*

4 (1). Labium with mentum separated from submentum by distinct suture; submentum with row of four long (prebasilar) setae; posteriolateral margins of pronotum with two or three prominent teeth; geographical range, Missouri, caves in vicinity of St. Louis and Ste. Genevieve ................. *Xenotrechus*
— Mentum and submentum fused; submentum with transverse row of six to 12 setae; posteriolateral margins of pronotum without teeth; geographical range east of Mississippi River .......................... 5

5(4). Head with one pair of supraorbital setae; form elongate, subconvex; maxillary palpomere 4 shorter than 3; length 6-7 mm; geographical range: Kentucky, western Pennyroyal plateau ...................... ...................................................... *Neaphaenops*
— Head with two or more pairs of supraorbital setae; maxillary palpomeres 3 and 4 subequal ............ 6

6(5). Front tibia with external face glabrous; body length 6-8 mm; geographical range. Kentucky (Estil, Jackson, Rockcastle, Pulaski, Clinton, Wayne, and McCreary counties) and Tennessee (Fentress County) .......................................... *Darlingtonea*
— Front tibia with external face rather densely pubescent; body length 4.5-7 mm; geographical range various ............................................................. 7

7(6). Antenna as long as body, extended to elytral apex in backfolded position; body length 5.5-7 mm; head and mandibles very large, all appendages elongate and slender; elytron with prehumeral (basal) margin sharply oblique; geographical range: Jackson County, Kentucky to Van Buren County, Tennessee ........................................................... *Nelsonites*
— Antenna in backfolded position not extended to elytral apex; body length 4.5-6 mm ................... 8

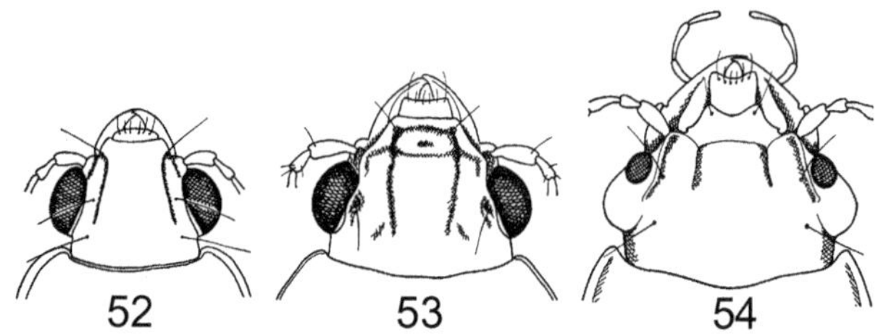

FIGURES 52.6-54.6. Fig. 52.6, *Bembidion* (s. str.) *quadrimaculatum* (Linnaeus), head dorsal view; Fig. 53.6, *Bembidion (Eupetedromus) incrematum* LeConte, head, dorsal view; Fig. 54.6, *Pterostichus (Hypherpes) morionoides* Chaudoir, head, dorsal view.

8(7). Right mandible with retinaculum with five tubercles, anterior two and posterior three teeth separated by a deep emargination; geographical range: Kentucky (Jackson, Rockcastle, Pulaski, northern Wayne and McCreary counties) ................ *Ameroduvalius*
— Right mandible of most specimens with retinaculum with two to four tubercles, without conspicuous gap between teeth; geographical range: various ........ ............................................ *Pseudanophthalmus*

XXI. Key to the Nearctic Genera of Bembidiini

1. Maxillary palpomere 4 long, nearly as long as palpomere 3; palpomere 3 not pubescent .......... .......................................................... *Horologion*
— Maxillary palpomere 4 rudimentary, much shorter and narrower than palpomere 3; palpomere 3 pubescent .................................................................. 2

2(1). Eyes absent (*Anillina*) ........................................... 3
— Eyes present, reduced or not ................................. 7

3(2). Head with deep frontal sulci; frons without lateral carinae (California)................................ *Anillaspis*
— Head without or with very shallow frontal impressions; frons with lateral carinae .................................. 4

4(3). Left paramere of aedeagus large, ovoid, without long apical setae; geographical range: eastern United States ................................................................ 5
— Left paramere of aedeagus narrower, styliform, with 2 rather long apical setae; geographical range: Texas, Arizona, California, Idaho ....................... 6

5(4). Abdominal sternum VII with apical margin distinctly toothed in male; right paramere absent ............. ...........................................................*Serranillus*
— Abdominal sternum VII with apical margin not toothed in male; right paramere present ............. *Anillinus*

6(4). Elytral apex entire, not lobed; long apical seta of umbilical series (8th in series) located inside marginal gutter .................................................... *Anillodes*
— Elytral apex shortened, somewhat lobed; long apical seta of umbilical series (8th in the series) displaced outside marginal gutter .................. *Micranillodes*

7(2). Elytron with sutural stria recurved at apex; front tibia with apex oblique ........................................... 8
— Elytron with sutural stria not recurved at apex; front tibia with apex truncate or rounded ............... 16

8(7). Mentum with pair of paramedian large, deep, circular foveae ........................................................ 9
— Mentum without deep foveae, at most with shallow impressions .................................................. 13

9(8). Recurrent groove of elytron elongate, prolonged anteriorly beyond posterior discal seta, then curved posteriorly in form of a hook .......................... 10
— Recurrent groove of elytron not prolonged anteriorly beyond posterior discal seta ............................ 11

10(9). Elytral stria 8 bent medially just posteriad setae Eo5 and Eo6 (see Erwin 1974 for chaetotaxy system); elytra pronouncedly iridescent in most species ... ......................................................... *Paratachys*
— Elytral stria 8 not bent medially near setae Eo5 and Eo6; elytra not iridescent in most species ........... .............................................................. *Tachys*

11(9). Elytron with recurrent groove elongate and very close and parallel to lateral margin ............. *Porotachys*
— Recurrent groove variously developed but not very close and parallel to lateral margin ................. 12

12(11). Elytral striae distinctly impressed, punctate ........... ........................................................ *Pericompsus*
— Elytral striae absent or very shallow .......... *Polyderis*

13(8). Frons with one supraorbital setigerous puncture over each eye; elytra more or less truncate; labrum bilobed, covering mandibles ................. *Micratopus*
— Frons with two supraorbital setigerous punctures over each eye; elytra entire, not truncate; labrum truncate, not covering mandibles .......................... 14

14(13). Elytron with recurrent groove nearly parallel to lateral margin; body flattened, elytra with microsculpture ................................................ 15
— Recurrent groove divergent from lateral margin; body very convex, elytra without microsculpture ......... ......................................................... *Elaphropus*

15(14). Pronotum with lateral margins broadly and evenly reflexed; elytron with recurrent groove broken at apex, and with apical 2/5 pale, testaceous .......... ......................................................... *Mioptachys*
— Pronotum with lateral margins very narrowly reflexed; elytron with recurrent groove uninterrupted at apex, entirely dark, piceous to black ....... *Tachyta*

16(7). Maxillary palpomere 4 somewhat conical, about 0.33 X as long as palpomere 3; internal sac of median lobe with large spine ............................ *Phrypeus*
— Maxillary palpomere 4 somewhat cylindrical or almost so, about 0.1- 0.25 X as long as palpomere 3; internal sac of median lobe without spine .......... 17

17(16). Elytra without striae .............................. *Asaphidion*
— Elytra with at least one (sutural) stria ...... *Bembidion*

XXII. Key to North American Subgenera (including some species groups) of *Bembidion* (Modified from Lindroth 1963)

1. Elytron with discal punctures on interval 3 each surrounded by a dull, more densely microsculptured, field .................................................... *Bracteon*
— Elytron with interval 3 equally microsculptured .... 2

2(1). Eyes small, longitudinal diameter less than or equal to length of antennomere 1 .............................. 3
— Eyes larger, longitudinal diameter much greater than length of antennomere 1 .................................. 4

3(2). Elytron with discal setigerous punctures confluent with stria 3 ........................................ *Lymnaeum*
— Elytron with discal setigerous punctures on interval 3 .......................................................... *Amerizus*

4(2). Elytron with lateral bead prolonged inside shoulder, either angulately or arcuately, with forward directed concavity ....................................................... 5
— Elytron with lateral bead terminated at shoulder or evenly prolonged upon base ........................... 12

5(4). Elytral intervals each with row of setigerous punctures .................................................... *Hydrium*
— Elytral intervals asetose except for two discal setigerous punctures on interval 3 ......................... 6

6(5). Elytral stria 8 remote from lateral bead at middle, at least half as much as from stria 7 ..................... 7
— Elytral stria 8 closer to lateral bead ..................... 8

7(6). Elytron with discal setigerous punctures in interval 3 large, foveate; submentum with more than eight long setae .................................... *Ochthedromus*
— Elytron with discal setigerous punctures in interval 3 smaller, not foveate; submentum with six or fewer long setae ........................................... *Odontium*

8(6). Elytron with only sutural stria complete to apex, stria 2 more or less obliterated toward apex ............ 9
— Elytron with at least sutural and stria 2 complete to apex ............................................................... 10

9(8). Pronotum wide, with lateral margins hardly sinuate in posterior half; lateral margin without small setae near front angle; mentum tooth shallowly emarginate at apex ................................ *Eurytrachelus*
— Pronotum narrower, with lateral margins markedly sinuate in posterior half; lateral margin with group of small setae at or near front angle; mentum tooth entire ................................................... *Metallina*

10(8). Pronotum without microsculpture ................... *Phyla*
— Pronotum with microsculpture ............................ 11

11(10). Legs very pale, yellowish; elytra pronouncedly iridescent; microsculpture linear, with microlines dense ....................................................... *Trechonepha*
— Legs darker, rufous or somewhat infuscated; elytra at most moderately iridescent, microsculpture coarser, with microlines less dense .. *Plataphodes*

12(4). Frons with coarse, confluent punctures; frontal furrows obsolete; mentum with pair of large pits, tooth very long ............................................... *Actedium*
— Frons smooth or with fine, non-confluent punctures; frontal furrows evident; mentum without large pits in most species, length of tooth various .......... 13

13(12). Pronotum with hind angles rounded (geographical range southern New Mexico and Arizona) .......... ........................................................ *Cyclolopha*
— Pronotum with hind angles about right .............. 14

14(13). Elytron with one to three clearly evident striae, others at most suggested .................................... 15
— Elytron with at least five evident striae, abbreviated or not apically ................................................ 17

15(14). Elytron without spots .................. *Lionepha* (in part)
— Elytron with two pale spots ................................ 16

16(15). Elytron with sutural stria abbreviated anteriorly; elytral striae minutely punctate; metasternal process not margined ............................. *Liocosmius*
— Elytron with sutural stria evident anteriorly; elytral striae coarsely punctate; metasternal process margined ......................................... *Wickhami* Group

17(14). Frontal furrows parallel or almost so, not prolonged upon clypeus .................................................. 18
— Frontal furrows more or less convergent, at least anteriorly, and prolonged upon clypeus ............. 38

18(17). Elytral intervals each with row of setigerous punctures ............................................................... 19
— Elytral intervals glabrous (except for the discal setigerous punctures on interval 3) ....................... 20

19(18). Head and pronotum without additional setae; elytron with lateral striae obsolete posteriad level of posterior discal seta ................ *Hydriomicrus* (in part)
— Head and pronotum with several scattered additional setae; elytron with striae evident almost to apex ............................ *Constricticollis* Group (in part)

20(18). Elytron with discal punctures, at least anterior one, confluent with stria 3 ...................................... 21
— Elytron with discal punctures located in interval 3 .. ...................................................................... 33

21(20). Abdominal sterna IV-VI each with several setae .. 22
— Abdominal sterna IV-VI each with one (exceptionally two) pair of setae ........................................... 23

22(21). Abdominal sterna IV-VI with setae more or less aligned in transverse fringe; pronotum with distinct posteriolateral carina; sternum VII without or with few additional small setae ...... *Blepharoplataphus*
— Abdominal sterna IV-VI with setae irregularly scattered; pronotum with rudimentary posteriolateral carina; sternum VII with numerous additional small setae .......................................... *Trichoplataphus*

23(21). Metasternal process (between middle coxae) entirely unmargined .................................................... 24
— Metasternal process margined at least laterally .. 27

24(23). Elytron with three discal setigerous punctures........ ............................................................... *Cillenus*
— Elytron with two discal setigerous punctures ...... 25

25(24). Elytron with striae, except sutural one, obsolete before apex (in part) .......................... *Hydriomicrus*
— Elytron with at least stria 2 evident to apex ........ 26

26(25). Length of body 4.7 mm or more, usually about 6.0 mm; pronotum without posteriiolateral carina; elytron with striae evident to apex ..................... .................................................. *Pseudoperyphus*
— Length of body 5.2 mm or less, usually about 4.0 mm; pronotum at least with rudiment of posteriolateral carina OR elytron with lateral striae abbreviated at apex .......................................... *Hirmoplataphus*

27(23). Elytral striae complete, lateral striae finer or not toward apex ...................................................... 28
— Elytral striae obliterated, or at least not quite continuous, before apex ...................................... 30

28(27). Elytra dark, without fasciae; elytron with microsculpture transverse, linear or isodiametric ........................................................... *Plataphus*
— Elytra pale with dark fascia(e) (expanded or not posteriorly to near apex); elytron with microsculpture isodiametric, in part absent from males of some species ........................................................... 29

29(28). Pronotum with distinct posteriolateral carina; elytron with striae conspicuously punctate, apex rounded at sutural angle ................................. *Peryphodes*
— Pronotum without posteriolateral carina; elytron with striae inconspicuously punctate, with apex angulate at sutural angle ............................... *Leuchydrium*

30(27). Elytron with preapical setigerous puncture free, microsculpture mesh pattern isodiametric (in part) ............................................................. *Lionepha*
— Elytron with preapical setigerous puncture in a groove (prolongation of stria 5 or 7), microsculpture mesh pattern isodiametric or transverse ................. 31

31(30). Margin of metasternal process distinctly posteriad apex; pronotum with posteriolateral carina rudimentary or absent (geographical range: Alaska and northwestern Canada) ..................................... 32
— Margin of metasternal process close to apex; pronotum with posteriolateral carina distinct or not (geographical range, all over continent) ............. ............................................................. *Peryphus*

32(31). Length of body less than 4.5 mm; elytron with microsculpture mesh pattern linear .... *Lenae* Group
— Length of body greater than 5.0 mm; elytron with microsculpture mesh pattern more or less isodiametric, or microlines indistinct, pattern less evident ..................................... *Bembidionetolitzkya*

33(20). Elytron with three discal setigerous punctures........ .................................................... *Scudderi* Group
— Elytron with two discal setigerous punctures ...... 34

34(33). Metasternal process not margined; anterior supraorbital setigerous puncture surrounded by an elevated field more shiny than adjacent surface. Head with frontal furrows (Fig. 53.6) extended posteriad plane of anterior pair of supraorbital setae ......... ..................................................... *Eupetedromus*
— Metasternal process entirely margined; anterior supraorbital setigerous puncture without surrounding elevated field or with elevated field as dull as adjacent surface. Head with frontal furrows various ................................................................. 35

35(34). Prosternum with intercoxal process, mesosternum and apex of metasternum (near hind coxae) with few scattered setae; elytral intervals each with row

of small punctures (visible only under high magnification) (in part) ............... *Constricticollis* Group
— Prosternal intercoxal process, mesosternum and apex of metasternum without setae; elytral intervals without micropunctures .................................. 36

36(35). Elytron with striae more or less obsolete or interrupted before apex ................ *Contractum* Group
— Elytral striae distinctly engraved to apex ........... 37

37(36). Pronotum without microsculpture on disc, or microlines very shallow....... *Obtusangulum* Group
— Pronotum with evident microsculpture sculpticells even on disc ........................................ *Notaphus*

38(17). Frontal furrows double for their entire length......... ......................................................... *Diplocampa*
— Frontal furrows simple at least anteriad anterior supraorbital setigerous puncture ........................ 39

39(38). Head each side with two frontal furrows posteriad anterior supraorbital setigerous puncture ........... ........................................................ *Furcacampa*
— Head each side with one frontal furrow posteriad anterior supraorbital setigerous puncture........... 40

40(39). Pronotum with basal margin with short, deep sinuation inside hind angles; head with dorsal surface as in Fig. 52.6........ *Bembidion* (*sensu stricto*)
— Basal margin of pronotum without sinuation inside hind angles ...................................................... 41

41(40). Frontal furrows (Fig. 51.6) very deep and markedly convergent, almost joined on front margin of clypeus............................................ *Trepanedoris*
— Frontal furrows almost parallel on frons, moderately convergent on clypeus .................................... 42

42(41). Frontal furrows shallow and single on clypeus.... 43
— Frontal furrows deep and double on clypeus ...... 44

43(42). Dorsal surface without microsculpture; elytra unspotted ...................................................... *Emphanes*
— Dorsal surface with microlines, at least on sides of head and base of pronotum; most individuals with pale spots on each elytron......... *Versicolor* Group

44(42). Length of body more than 4.0 mm; elytral striae evident to apex ............................. *Oberthueri* Group
— Length of body less than 3.5 mm; lateral elytral striae obliterated or interrupted before apex ................ .......................................................... *Semicampa*

XXIII. Key to the North American Genera of Pogonini
(After Bousquet and Laplante 1997)

1. Elytron without striae; basal ridge absent; metasternum not margined anteromedially; metepisternum subquadrate; hind coxae narrowly separated from each other.................................. *Thalassotrechus*
— Elytron distinctly striate; basal ridge present, broken (or not) in form of a prominent carina laterally; metasternum margined anteromedially; metepisternum elongate; hind coxae contiguous ......................... 2

2(1). Elytron with basal ridge entire and regular; submentum with three pairs of setae; body mostly black with metallic luster ........................ *Pogonus*
— Elytron with basal ridge entire or broken, in form of an oblique carina laterally; submentum with two pairs of setae; body yellowish to reddish brown, without metallic luster ..................... *Diplochaetus*

XXIV. Key to the Nearctic Genera of Patrobini
(Modified from Darlington 1938)

1. Elytron with numerous setigerous punctures on interval 3, and basally on intervals 1 and 5 ........... ..................................................... *Platypatrobus*
— Elytron with three or four setigerous punctures on stria 3 or on outer edge of interval 3 ................ 2

2(1). Total length less than 6.0 mm; genae longer than eyes; palpus with apical palpomere subconical ... ........................................................... *Platidiolus*
— Total length more than 8.0 mm; genae shorter than eyes; palpus with apical palpomere subcylindrical, subtruncate....................................................... 3

3(2) Head dorsally with occipital constriction shallow; posteriolateral impressions of pronotum shallow, median longitudinal impression shallow; epimeron of mesothorax broadly triangular; body relatively depressed ................................................ *Diplous*
— Head with occipital constriction deep; pronotum with posteriolateral impressions and median longitudinal impression deep; epimeron of mesothorax narrow; body more convex (Key XXV) ......... *Patrobus*

XXV. Key to the Nearctic Subgenera of *Patrobus*

1. Pronotum with anterior transverse impression very deep, practically impunctate; middle of prosternum anteriorly without distinct, coarse punctures ...... ....................................................... *Neopatrobus*
— Pronotum with anterior transverse impression with coarse punctures; prosternum at middle anteriorly with or without coarse punctation ....................... ...................................... *Patrobus* (*sensu stricto*)

XXVI. Key to the Nearctic Genera of Loxandrini

1. Pronotum markedly constricted at base, with posterior lateral setigerous punctures located distinctly anteriad basal margin; lateral grooves terminated clearly anteriad base; antennomeres 7-11 white . ............................................................ *Oxycrepis*
— Pronotum not or only slightly constricted at base, with posterior lateral setigerous punctures on basal margin; lateral grooves extended to base; antennomeres 7-11 rufotestaceous ..... *Loxandrus*

XXVII. Key to Nearctic Genera of Pterostichini

1. Elytral interval 3 without setigerous punctures ..... 2
— Elytral interval 3 with one to seven setigerous punctures ................................................................. 7

2(1). Antennomere 1 at least as long as antennomeres 2 and 3 combined; maxillary palpomere 3 with row of five or six distinct apical setae; hind trochanter with seta (Key XXVIII) ................................ *Stomis*

— Antennomere 1 shorter than antennomeres 2 and 3 combined; maxillary palpomere 3 without or with two or three barely distinct apical setae; hind trochanter with or without seta .............................. 3

3(2). Glossal sclerite with two or three pairs of setae on anterior margin; labial palpomere 3 conspicuously widened toward apex; elytra with metallic luster ......... ..................................................................... *Myas*

— Glossal sclerite with one pair of setae on anterior margin; labial palpomere 3 not widened toward apex; elytra with or without metallic luster ................ 4

4(3). Metepisternum with lateral margin distinctly longer than anterior margin; wings more or less fully developed ............................................................ 5

— Metepisternum with lateral margin subequal or shorter than anterior margin; wings markedly reduced ................................................................ 6

5(4). Elytron without parascutellar stria absent, parascutellar seta present; mesh pattern of elytral microsculpture transverse; labium with paramedian mental pits distinct .................. *Ophryogaster*

— Elytron with parascutellar stria, parascutellar seta absent; mesh pattern of elytral microsculpture isodiametric; labium with paramedian mental pits indistinct .............................................. *Stereocerus*

6(4). Middle trochanter without seta; elytron with interval 7 carinate anteriorly and posteriorly ........... *Abax*

— Middle trochanter with seta; elytron with interval 7 not carinate (Key XXXI) (in part) ........ *Pterostichus*

7(1). Hind trochanter without seta ............................... 8

— Hind trochanter with seta .................................. 12

8(7). Middle tibia near apex with ctenidium comprised of apical transverse row of rather stout setae; mesepisternum without punctures; labial palpomere 2 without or with one to three rather long apical setae nearly as long as median setae (Key XXIX) ..................................... *Cyclotrachelus*

— Ctenidium near apex of middle tibia comprised of subapical more or less semicircular row of slender setae; mesepisternum with or without at least a few punctures; labial palpomere 2 with one to three small (much smaller than median ones) apical setae or without setae .......................................... 9

9(8). Tarsus with claws finely pectinate; eyes voluminous, hemispherical ........................................... *Abaris*

— Tarsal claws smooth; eyes normal, more or less rounded ......................................................... 10

10(9). Elytral interval 3 with one setigerous puncture (Key XXXI) (in part) ................................... *Pterostichus*

— Elytral interval 3 with two to six setigerous punctures ......................................................................... 11

11(10). Hind tarsus with carina more or less distinct on tarsomere 1, only; male abdominal sternum VII emarginate medioapically ................ *Lophoglossus*

— Hind tarsus with carina distinct on tarsomeres 1-2 or tarsomeres 1-3; male abdominal sternum VII rounded, not emarginate (Key XXXI) (in part) ....... ........................................................ *Pterostichus*

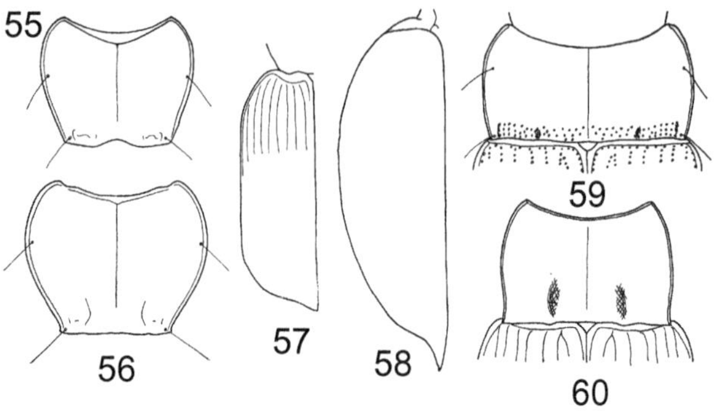


FIGURES 55.6-60.6. Fig. 55.6, *Cyclotrachelus (Evarthrus) sigillatus* Say, pronotum, dorsal view; Fig. 56.6, *Cyclotrachelus (Evarthrus) furtivus* LeConte, pronotum, dorsal view; Fig. 57.6, *Agonum cupripenne* Say, left elytron, dorsal view, showing striae basally, only; Fig. 58.6, *Rhadine* sp., elytron, dorsal view; Fig. 59.6, *Amara (Bradytus) apricaria* (Paykull), pronotum, dorsal view; Fig. 60.6, *Amara (Celia) californica* Dejean, pronotum, dorsal view.

12(7). Submentum without lateral setae; antennomeres 2 and 3 (also 1 in some specimens) carinate medially or ecarinate (Key XXX) .............................. *Poecilus*

— Submentum with lateral setae; antennomeres 1-3 not carinate ........................................................... 13

13(12). Metepisternum with lateral margin distinctly longer than anterior margin; wings fully developed or reduced ............................................................. 14

— Metepisternum with lateral margin subequal or shorter than anterior margin; wings markedly reduced .............................................................. 15

14(13). Elytral interval 3 with two setigerous punctures in posterior half ........................................... *Piesmus*

— Elytral interval 3 with three setigerous punctures (or two, asymmetrically located) (Key XXXI) (in part) . ......................................................... *Pterostichus*

15(13). Elytral interval 3 with one setigerous puncture ....... ....................................................... *Gastrellarius*

— Elytral interval 3 with two or three setigerous punctures (Key XXXI) (in part) ................... *Pterostichus*

XXVIII. Key to Nearctic Subgenera of *Stomis*

1. Pronotum without posteriolateral setae; glossal sclerite with two pairs of setae on anterior margin; head nearly as wide as pronotum ................ *Neostomis*

— Pronotum with posteriolateral setae; glossal sclerite with one pair of setae on anterior margin; head narrower than pronotum .... *Stomis (sensu stricto)*

XXIX. Key to Subgenera of *Cyclotrachelus*

1. Pronotum with posteriolateral impression single, linear or punctiform ... *Cyclotrachelus (sensu stricto)*

— Pronotum with posteriolateral impression double, basinlike in most specimens (Figs. 55.6 and 56.6) ............................................................ *Evarthrus*

XXX. Key to Subgenera of *Poecilus*

1. Antennomeres 2 and 3 (or 1, 2, and 3) carinate medially .................................. *Poecilus (sensu stricto)*
— Antennomeres 1-3 not carinate ...................... *Derus*

XXXI. Key to Subgenera of *Pterostichus*

1. Elytral interval 3 without setigerous punctures ..... 2
— Elytral interval 3 with one to seven setigerous punctures .................................................................. 4

2(1). Taxa distributed in or west of the Rocky Mountains (in part) ............................................... *Hypherpes*
— Taxa distributed east of the Mississippi River ....... 3

3(2). Hind trochanter without seta; elytral microsculpture mesh pattern isodiametric ............. *Cylindrocharis*
— Hind trochanter with seta; elytral microsculpture mesh pattern markedly transverse, in part linear (in part) ............................................................ *Hypherpes*

4(1). Hind trochanter without seta ................................ 5
— Hind trochanter with seta .................................... 14

5(4). Elytral interval 5 with five or six setigerous punctures .................................................. *Orsonjohnsonus*
— Elytral interval 5 without setigerous punctures ..... 6

6(5). Tarsomere 5 with two rows of setae ventrally ....... 7
— Tarsomere 5 without setae ventrally ................... 10

7(6). Pronotum with posteriolateral impressions punctate; front tibia with two clip setae (in part) .. *Abacidus*
— Pronotum with posteriolateral impressions smooth; front tibia with three clip setae .......................... 8

8(7). Middle femur posteriorly with five to eleven setae; elytral plica slightly developed ........ *Metallophilus*
— Middle femur posteriorly with two setae; elytral plica clearly developed .............................................. 9

9(8). Pronotum with hind angles denticulate; metepisternum with lateral margin longer than anterior margin ............................. *Morphnosoma*
— Pronotum with hind angles not denticulate; metepisternum with lateral margin subequal to anterior margin (in part) ...................... *Euferonia*

10(6). Elytral interval 3 with one setigerous puncture ....... ........................................................ *Monoferonia*
— Elytral interval 3 with two to six setigerous punctures ....................................................................... 11

11(10). Metepisternum with lateral margin distinctly longer than anterior margin; hind wings fully developed; labium with paramedian mental pits indistinct .... ...................................................... *Bothriopterus*
— Metepisternum with lateral margin subequal to anterior margin; hind wings very short; labium with paramedian mental pits distinct ..................... 12

12(11). Middle femur with three to five (or two, usually on one side) setae along the posterior half of the ventral side; elytron with parascutellar stria between suture and stria 1(in part) ..................... *Abacidus*
— Middle femur with two setae along posterior half of ventral side; parascutellar stria, if present, between striae 1 and 2 .................................................. 13

13(12). Abdominal sterna II-VI (most specimens also VII) with rather coarse punctures laterally; front tibia with two clip setae; elytron without parascutellar stria (or parascutellar stria present but short); length of body 7-12 mm ................................ *Gastrosticta*
— Abdominal sterna II-IV with small punctures laterally, sterna V-VII smooth; front tibia with three clip setae; elytron with parascutellar stria long or short; length of body 13-19 mm (in part) ........ *Euferonia*

14(4). Metepisternum with lateral margin distinctly longer than anterior margin; hind wings fully developed or not ............................................................ 15
— Metepisternum with lateral margin subequal or shorter than anterior margin; hind wings very short ...................................................................... 19

15(14). Pronotum with posteriolateral angles rounded; labium with mental pits indistinct ...................... *Argutor*
— Pronotum with posteriolateral angles about right or slightly obtuse; labium with mental pits distinct . ....................................................................... 16

16(15). Length of body less than 8 mm; pronotum with posteriolateral impressions linear; tarsomere 5 with or without two rows of setae ventrally .... *Phonias*
— Length of body more than 8 mm; pronotum with posteriolateral impressions rounded or basinlike; tarsomere 5 without rows of setae ventrally ... 17

17(16). Hind tarsus with tarsomeres 1-3 each with distinct carina ........................................... *Pseudomaseus*
— Hind tarsus with tarsomeres 1-2 each with distinct carina ............................................................ 18

18(17). Pronotum markedly constricted basally; gonocoxite 2 without ensiform setae; length, 10-13 mm ....... ............................................................ *Lamenius*
— Pronotum less constricted basally; gonocoxite2 with ensiform setae; length, 11.5-17 mm ....... *Melanius*

19(14). Tarsomere 5 with two rows of setae ventrally..... 20
— Tarsomere 5 without rows of setae ventrally ...... 21

20(19). Hind coxa with medial seta; pronotum with posteriolateral angles sharp or rounded; length of body 4.0-9.5 mm ................................. *Cryobius*
— Hind coxa without medial seta; pronotum with posteriolateral angles rounded; length of body 10-12 mm (in part) ................................. *Eosteropus*

21(19). Elytron with sculpture of disc more or less irregular, some striae wavy or interrupted or intervals unequal in width and convexity; hind tarsus with tarsomere 1 with carina more or less distinct, tarsomere 2 at base carinate or not ... *Lenapterus*
— Elytral sculpture regular, striae straight and intervals equal in width and convexity; hind tarsus with tarsomeres 1 and 2 distinctly carinate ............ 22

22(21). Elytral interval 3 with four setigerous punctures; elytral microsculpture isodiametric; length of body more than 14 mm (in part) ................. *Eosteropus*
— Elytral interval 3 with two or three setigerous punctures; elytral microsculpture transverse or linear; length of body less than 12 mm ...................... 23

23(22). Species distributed west of the Rocky Mountains (Oregon, Washington, Idaho); male genitalia with me-

dian lobe with or without slightly sclerotized band ventrally .................................... *Pseudoferonina*
— Species distributed east of the Mississippi River (southern Appalachian Mountains); male genitalia with median lobe without slightly sclerotized band ventrally ........................................................ 24

24(23). Elytron with lateral striae distinctly impressed; male with abdominal sternum VII with or without small, more or less distinct protuberance; male genitalia with right paramere elongate ................ *Feronina*
— Elytron with lateral striae indistinctly impressed, suggested; male with abdominal sternum VII with prominent protuberance; male genitalia with right paramere short ................................. *Paraferonia*

XXXII. Key to the Nearctic Genera of Zabrini

1. Labial palpomere 2 bisetose; maxillary palpomeres 3 and 4 with numerous long setae; head with frontal impressions transverse, shallow; elytron with plica small, not very evident ..................... *Pseudamara*
— Labial palpomere 2 plurisetose; maxillary palpomeres 3 and 4 glabrous, or palpomere 3 with few short setae near apex; head with frontal impressions more or less linear, moderately deep; elytron with plica moderately prominent (Key XXXIII) .... *Amara*

XXXIII. Key to the Nearctic Subgenera of *Amara*
(After Lindroth 1968, with modifications)

1. Front tibia with apical spur trifid .................... *Zezea*
— Front tibia with apical spur simple, not forked...... 2

2(1). Prosternum with intercoxal process not margined at apex ................................................. *Curtonotus*
— Prosternum with intercoxal process margined at apex ........................................................................ 3

3(2). Prosternum with intercoxal process apically plurisetose ........................................................ 4
— Prosternum with intercoxal process apically asetose, or bi- or quadri-setose ...................................... 5

4(3). Pronotum with posteriolateral impression delimited laterally by a pronounced carina; geographical range transcontinental ........................... *Percosia*
— Pronotum with posteriolateral carina suggested, only; geographical range, Texas only ........ *Neopercosia*

5(3). Pronotum basally more or less constricted (Fig. 59.6), sinuate or not anteriad posteriolateral angles, or very narrow, with dense and pronounced punctation over entire basal area; dorsal surface dark (rufopiceous to black ), dull or shiny, not metallic ............................................................ *Bradytus*
— Pronotum basally broad, lateral margins slightly curved, straight or slightly sinuate (Fig. 60.6) anteriad posteriolateral angles; dorsal surface dark, dull, to metallic ................................................. 6

6(5). Elytron with stria 7 preapically with four to five punctures; geographical range, California ................. ........................................ *Insignis* Species Group
— Elytron with stria 7 preapically without punctures, or with number of punctures less than four ............ ......... *Amara* (*sensu stricto*), *Celia*, and *Paracelia*

XXXIV. Key to the Nearctic Genera of Harpalini

1. Head each side, anteriad eye and dorsad antennal insertion, with prominent angulate projection; ventrad projection with preocular sulcus ................. ...................................................... *Cratacanthus*
— Head without angulate projections and without preocular sulci .................................................. 2

2(1). Front tibia with outer apical portion expanded in form of broadly rounded plate; apical spur broad, about half length of preapical spur; hind tibia with both spurs spatulate; integument pallid ........ *Geopinus*
— Front tibia with outer apical portion extended as prominent spine, or not thus modified; apical and preapical spurs similar in length; hind tibia with spurs more or less slender, lanceolate, not spatulate .................................................................... 3

3(2). Front tibia with outer apical portion extended laterally as prominent spine, subequal in length to apical spur, latter slightly shorter but much broader than preapical spur; pronotum with lateral margins plurisetose, setae long; elytron with at least odd-numbered intervals each with row of large setigerous punctures; even-numbered intervals with or without setae ....................................... *Euryderus*
— Front tibia lateroapically not extended as prominent lateral spine; pronotum each side unisetose or plurisetose; elytral intervals setose or not ........ 4

4(3). Pronotum with more than two pairs of lateral setae; elytron with intervals (all or odd-numbered, only) with row of large setigerous punctures, each puncture with seta longer than antennomere 1 ......... 5
— Pronotum with one or two pairs of lateral setae, only; elytron with or without series of large setigerous punctures in intervals ....................................... 6

5(4). Head with vertex with only single pair of (supraorbital) setae, otherwise glabrous; pronotum with disc asetose ........................................... *Hartonymus*
— Head with vertex with transverse row of about six setae, in addition to single pair of supraorbital setae; pronotum with disc sparsely setose ............ ............................................................... *Piosoma*

6(4). Pronotum with two pairs of lateral setae (one pair toward middle, one pair near posteriolateral angles); eyes with fine, short setae; dorsal surface more or less setose (Key XXXIX) ... *Dicheirotrichus*
— Pronotum with single pair of lateral setae; eyes glabrous ................................................................ 7

7(6). Elytron with striae 2, 5 and 7 or 2 and 5 with row of fine setigerous punctures, setae short .............. 8
— Elytral striae 2, 5, and 7 without row of setigerous punctures ......................................................... 14

8(7). Prothorax elongate in front of coxae, distance from anterior margin of prosternum to anterior rim of front coxal cavity twice distance from latter point to tip of intercoxal process; elytron with striae 2 and 5 with few very fine setigerous punctures .... ..................................................... *Stenomorphus*
— Prothorax of normal proportions, distance from anterior margin of prosternum to anterior rim of front coxa subequal to distance from latter point to tip of intercoxal process ......................................... 9

9(8). Elytron with intervals rather densely, evenly punctate ........................................................................ 10
— Elytron with at least discal intervals (1-7) smooth, not punctate ............................................................ 11

10(9). Dorsal surface of pronotum laterally and elytra not setose, with iridescent luster, microsculpture mesh pattern linear; head, including clypeus, and lateral portions of pronotum with surface smooth, except fine punctures; hind tarsus long, tarsomere 1 distinctly longer than inner tibial spur .. *Athrostictus*
— Dorsal surface of pronotum laterally and elytra setose, shining, faintly aeneous, but not iridescent, microsculpture mesh pattern on elytra isodiametric to slightly transverse, pronotum with most of surface smooth, without microlines; head, including clypeus and pronotum laterally rugulose-punctate and setose; hind tarsus short, tarsomere 1 and inner tibial spur subequal in length (Key XLII) (in part) ............................................... *Selenophorus*

11(9). Clypeus with anterior margin deeply, broadly emarginate and base of labrum visible in emargination ..................................................... *Amblygnathus*
— Clypeus with anterior margin not deeply, broadly emarginate, approximately straight or shallowly curved ............................................................ 12

12(11). Front tibia with six or more spines on outer margin, near apex; pronotum (Fig. 61.6) with posteriolateral angles broadly rounded, and disc more or less markedly convex; male with middle tibia bowed or not; female with abdominal sternum VII with apical margin medially thickened or plate-like (Key XLIII) (in part) ............................................. *Discoderus*
— Front tibia with five or fewer spines on outer margin, near apex; pronotum with posteriolateral angles and convexity of disc various; male with middle tibia not bowed; female with abdominal sternum VII with apical margin medially thickened or not . ........................................................................ 13

13(12). Pronotum cordate in form, with posteriolateral angles broadly rounded; elytron without parascutellar stria; female with abdominal sternum VII with apical margin medially thickened (Key XLIII) (in part) .......................................................... *Discoderus*
— Pronotum not cordate; elytron with parascutellar stria; female with abdominal sternum VII with apical margin medially not thickened (Key XLII) (in part)...... ....................................................... *Selenophorus*

14(7). Labial palpomere 2 plurisetose .......................... 15
— Labial palpomere 2 bisetose or trisetose ........... 24

15(14). Labium with mentum and submentum fused completely ............................................................ 16
— Labial mentum separated from submentum partially or completely by suture .................................. 19

16(15). Dorsal surface, including all of elytra, densely setose ........................................................................ 17
— Dorsal surface mostly glabrous, elytra setose at most along lateral margins, discal intervals glabrous .. ........................................................................ 18

17(16). Front tibia with apical spur trifid; geographical range: North America west of the Sierra Nevada- Baja California to southern British Columbia ....... *Dicheirus*
— Front tibia with apical spur lanceolate; range: eastern North America to the Rocky Mountains (Key XXXV) *Amphasia*

18(16). Labium with glossal sclerite markedly expanded preapically, lateral margins sinuate; geographical range: eastern North America ............. *Xestonotus*
— Labium with glossal sclerite not expanded, parallel sided, or expanded apically; geographical range: continent-wide (Key XXXVI) ............. *Anisodactylus*

19(15). Hind tarsus with tarsomere 1 elongate, 1.3 or more times length of antennomere 1 (antennal scape). ........................................................................ 20
— Hind tarsus with tarsomere 1 shorter, less than 1.3 times length of antennomere 1 ........................ 21

20(19). Elytron with microsculpture mesh pattern isodiametric or nearly so, surface dull to shiny, but not iridescent; male with front tarsomeres 1-3 ventrally each with dense pad of adhesive vestiture .......... .......................................... *(Anisotarsus) Notiobia*
— Elytron with microsculpture mesh pattern markedly transverse or linear, surface iridescent; male with front tarsomeres 1-3 ventrally each with biseriate adhesive vestiture (in part) ............. *Aztecarpalus*

21(19). Hind tarsus with dorsal surfaces of tarsomeres more or less setose .................................................. 22
— Hind tarsus with dorsal surfaces of tarsomeres glabrous ............................................................ 23

22(21). Head with frons and temples impunctate, smooth (without vestiture of short setae) (Key XLV) (in part) ............................................................. *Harpalus*
— Head with frons and temples punctate, with vestiture of short setae ........................................ *Ophonus*

23(21). Elytron with microsculpture mesh pattern markedly transverse or linear, surface iridescent, not setose, at base without parascutellar setigerous puncture or parascutellar stria; geographical range: eastern North America (in part) .................... *Aztecarpalus*
— Elytron with microsculpture mesh pattern isodiametric or nearly so, or microlines absent, surface smooth and shiny, but not iridescent, setose or not, at base with parascutellar setigerous puncture and parascutellar stria; geographical range: continent-wide (Key XLV) (in part) .......................... *Harpalus*

24(14). Head with frontal impressions extended posteriolaterally on each side to supraorbital setigerous punctures ............................ *Pogonodaptus*
— Head with frontal impressions linear or broad, if linear, then extended laterally toward eyes, but distinctly anteriad supraorbital setigerous punctures. ........................................................................ 25

25(24). Total length 15.0 mm or more; integument concolorous; range: southern Arizona and New Mexico (Key XLI) .................................. *Polpochila*
— Total length 10.0 mm or less, but if slightly more, then pronotum and elytra bicolored ................ 26

26(25). Mentum with distinct tooth ................................ 27
— Mentum without tooth ......................................... 29

27(26). Pronotum elongate, longer than maximum width, lateral margins distinctly sinuate posteriorly; man-

dibles elongate, projected markedly anteriorly .... ............................................................ *Amerinus*
— Pronotum transverse, wider than long, lateral margins not or only slightly sinuate; mandibles short (i.e., normal), not projected markedly anteriorly . ........................................................................ 28

28(27). Pronotum with base completely but thinly beaded AND posteriolateral angles rounded; male with tarsomeres 1-4 of front and middle tarsi expanded, ventral surfaces of these with dense covering of adhesive vestiture; geographical range: southern New Mexico and Arizona .................................... .................................... *(Pelmatellus) Pelmatellus*
— Pronotum with base beaded only laterally, OR, if completely beaded, posteriolateral angles almost rectangular; male with tarsomeres 1-4 of front and in some taxa, middle tarsus, only slightly broadened, and with or without two rows of adhesive vestiture ventrally; geographical range: most of Nearctic Region, including southern New Mexico and Arizona (Key XXXVIII) .................. *Bradycellus*

29(26). Elytron with stria 8 with posterior series of umbilicate punctures not divided into two groups of four punctures (Fig. 62.6) (Key XL) ............. *Acupalpus*
— Elytron with stria 8 with posterior series of umbilicate punctures divided into two groups of four punctures each, groups separated by distance equal to half or more length of one group (Fig. 63.6) (Key XXXVII) ............................................. *Stenolophus*

XXXV. Key to the Nearctic Subgenera of *Amphasia*

1. Body with dorsal surface bicolored, head and pronotum rufo-testaceous, elytra darker; legs pale (testaceous); elytra with punctation uniform, surface iridescent, microsculpture mesh pattern transverse or linear ............... *Amphasia (sensu stricto)*
— Body with dorsal surface concolorous, dark piceous to black; legs with tibiae and tarsi pale, femora black; elytron with intervals 3, 5, and 7 each with row of larger punctures, surface dull, microsculpture mesh pattern about isodiametric .................................................. *Pseudamphasia*

XXXVI. Key to the Nearctic Subgenera of *Anisodactylus*

1. Front tibia with apical spur trifid .......................... 2
— Front tibia with apical spur lanceolate, or enlarged with margin more or less angulate ................... 3

2(1). Following with vestiture of short setae: pronotum laterally and at base, and prosternum medially; clypeus laterally with two or three pairs of setae (in part) .................... *Anisodactylus (sensu stricto)*
— Pronotum and prosternum glabrous; clypeus laterally with single pair of setae ....... *Gynandrotarsus*

3(1). Mandibles elongate, projected anteriorly, with dorsal surfaces strigulose ....................... *Spongopus*
— Mandibles of normal length, with dorsal surfaces smooth ............................................................ 4

4(3). Front tibia with apical spur slender, lanceolate, not widened basally, margin not angulate .............. 5
— Front tibia with apical spur widened basally, margin angulate .......................................................... 6

5(4). Body form broad, Amara-like; pronotum widest near base, lateral depressions each side not extended to anteriolateral angles; legs with femora infuscated .................................................... *Aplocentroides*
— Body form narrow, not Amara-like; pronotum widest slightly anteriad middle, lateral depressions each side extended to anteriolateral angles; femora rufous, like tibiae and tarsi .......... *Pseudaplocentrus*

6(4). Hind tarsus with tarsomere 1 approximately three times length of antennomere 1 (in part) ............. ................................ *Anisodactylus (sensu stricto)*
— Hind tarsus with tarsomere 1 short less than three times length of antennomere 1 ........... *Anadaptus*

XXXVII. Key to the Nearctic Subgenera of *Stenolophus*

1. Hind tarsus with tarsomere 5 with one pair of setae ventrolaterally; prosternum medially with or without vestiture of short setae; size small, total length less than 5.5 mm ............................. *Agonoleptus*
— Hind tarsus with tarsomere 5 glabrous ventrally; size various, total length less or more than 5.5 mm . 2

2(1). Legs with tarsi short, hind tarsus with tarsomere 1 not or only slightly longer than tarsomere 2; tarsomere 1 externo-laterally smooth, not longitudinally carinate; front tibia in apical half with five or six setae along outer margin ......... *Agonoderus*
— Legs with tarsi longer, hind tarsus with tarsomere 1 distinctly longer than tarsomere 2, and with or without longitudinal carina externo-laterally; front tibia in apical half with two or three spines ................ ................................. *Stenolophus (sensu stricto)*

XXXVIII. Key to the Nearctic Subgenera of *Bradycellus*

1(2). Elytron with only sutural stria impressed, others shallow, indistinct, or completely effaced ..... *Liocellus*
— Elytron with all striae equally impressed .............. 2

2(1). Antenna with antennomere 3 glabrous (except normal few apical setae) ......................... *Triliarthrus*
— Antenna with antennomere 3 with fine depressed pubescence at least in apical half .......................... 3

3(2). Elytron without parascutellar or discal setigerous punctures .......................................... *Catharellus*
— Elytron with parascutellar and single dorsal setigerous puncture in interval 3 ................................. 4

4(3). Antenna with antennomeres 1 and 2 contrasted in color: 1, rufous, and 2, black; head laterally with frontal impression (clypeo-ocular line) not extended to margin of eye ............................... *Lipalocellus*
— Antenna with antennomeres 1 and 2 both rufous; head laterally with frontal impression extended to margin of eye ......................................................... 5

5(4). Prosternum medially and abdominal sterna with vestiture of short setae ....................................... .................................. *Bradycellus (sensu stricto)*
— Prosternum and abdominal sterna without vestiture of short setae .................................... *Stenocellus*

XXXIX. Key to the Nearctic Subgenera of *Dicheirotrichus*

1. Disc of elytron with large setigerous puncture in interval 3; microsculpture lines evident on margins

of head, posteriolateral angles of pronotum, and at least laterally on elytra; elytron with vestiture of very short depressed setae ................ *Trichocellus*

— Elytral disc without setigerous puncture in interval 3; dorsal surface without microsculptural lines; elytron with vestiture of erect setae .... *Oreoxenus*

XL. Key to the Nearctic Subgenera of *Acupalpus*

1. Elytron with interval 3 with row of three or more setigerous punctures; interval 5 with or without setigerous punctures .................................... *Philodes*

— Elytron with interval 3 with one setigerous punctures; interval 5 impunctate ........................................ 2

2(1). Pronotum with posteriolateral angles angulate ....... ........................................................... *Anthracus*

— Pronotum with posteriolateral angles rounded ..... 3

3(2). Dorsal surface with microlines dense, mesh pattern linear, elytra with surface iridescent (in part) ...... .................................... *Acupalpus* (*sensu stricto*).

— Dorsal surface mostly smooth, without microlines; surface of elytra shiny, not iridescent ................ 4

4(3). Elytron with striae punctulate anteriorly, bicolored, principally black but base and sutural area abruptly pale; prosternum and abdominal sterna with vestiture of short setae (in part) .......................... .................................... *Acupalpus* (*sensu stricto*).

— Elytron with striae impunctate, concolorous or indistinctly bicolored; prosternum and abdominal sterna without vestiture of short setae, glabrous ........... .......................................................... *Tachistodes*

XLI. Key to the Nearctic Subgenera of *Polpochila*

1. Head large; gena (ventrad eye) wider than width of antennomere 1; mandibles dorsally with surface strigules extended nearly to base ........................ ................................................. *Phymatocephalus*

— Head normal, gena narrower than antennomere 1; mandibles dorsally with strigules confined to apical one-third ................ *Polpochila* (*sensu stricto*)

XLII. Key to the Nearctic Subgenera of *Selenophorus*

1. Prosternum with intercoxal process narrow, toward apex curved slightly dorsally, sharply margined laterally .......................................... *Celiamorphus*

— Intercoxal process of prosternum short and broad, immarginate laterally, toward apex markedly curved dorsally ......... *Selenophorus* (*sensu stricto*)

XLIII. Key to the Nearctic Subgenera of *Discoderus*

1. Front tibia with six or more broad spines on outer margin near apex; pronotum transverse in form, with posteriolateral angles broadly rounded ....... ................................... *Discoderus* (*sensu stricto*)

— Front tibia with five or fewer broad spines on outer margin near apex; pronotum cordate in form, with posteriolateral angles more narrowly rounded .... .............................................................. *Selenalius*

XLIV. Key to the Nearctic Subgenera of *Trichotichnus*

1. Pronotum posteriorly with surface moderately densely punctate, especially posteriolateral impressions; elytron with parascutellar setigerous puncture; labium with paraglossae narrow, each paraglossa separated from glossal sclerite by a wide notch .. ................................ *Trichotichnus* (*sensu stricto*)

— Pronotum posteriorly with surface impunctate, or punctures confined to posteriolateral impressions; elytron with or without parascutellar setigerous puncture; labium with paraglossae broad, each paraglossa separated from glossal sclerite by a narrow notch ......................................... *Iridessus*

XLV. Key to the Nearctic Subgenera of *Harpalus*

1. Elytron with short row of large punctures toward apex, in intervals 3, 5, and 7; pronotum cordate ........ 2

— Elytron impunctate, or with single puncture on disc, in interval 3, or with single discal puncture and few punctures apically in interval 7; pronotum various in form ............................................................. 3

2(1). Pronotum constricted slightly toward base, posteriolateral angles rectangular ........... *Opadius*

— Pronotum markedly constricted posteriorly, posteriolateral angles obtuse ................ *Glanodes*

3(1) Pronotum markedly cordate, lateral margins posteriorly markedly sinuate, without distinct lateral bead; elytron without setigerous puncture in interval 3 .................................................. *Harpalobrachys*

— Pronotum not markedly cordate, lateral margins evenly rounded or slightly sinuate, OR if cordate, then lateral bead distinct at least posteriorly; elytron with or without setigerous puncture in interval 3 ............................................................. 4

4(3). Elytron basad basal ridge with densely punctate, finely setose field near scutellum, and without discal setigerous puncture in interval 3 .......................... 5

— Elytron basally smooth and glabrous toward scutellum ................................................................... 6

5(4). Dorsal surface black; pronotum with lateral bead coarse through most of length but narrowed abruptly anteriad posterior margin; elytron with preapical sinuation shallow; size large, total length 17.5 mm or more ............................. *Megapangus*

— Dorsal surface rufous to rufo-piceous, not black; pronotum with lateral bead uniform throughout length; elytron with preapical sinuation deep; size smaller, total length not more than 18 mm ......... ......................................................... *Plectralidus*

6(4). Elytron with at least intervals 7 and 8 more or less punctate and setose, setae evident or small and easily overlooked ............................................... 7

— Elytron with intervals 1-8 smooth and glabrous .... 8

7(6). Dorsal surface more or less metallic green or aeneous; elytron with humerus rounded; tarsomeres with dorsal surfaces glabrous .................................... ...................................... *Harpalus (sensu stricto)*

— Dorsal surface not metallic-colored; elytron with humerus various, and tarsomeres dorsally glabrous or setose (in part) ....................... *Pseudoophonus*

8(6). Elytron with apical margin subtruncate to truncate, dorsal surface metallic green, aeneous, blue, or black ............................................ *Harpalomerus*

— Elytron with apical margin more or less sinuate, but not subtruncate; dorsal surface various in color ........ 9

9(8). Antenna distinctly bicolored: antennomere 1 or antennomeres 1 and 2 testaceous, 2-11 or 3-11 infuscated to black; abdominal sterna IV-VI with accessory setae arranged in an irregular transverse row laterad paired ambulatory setae .................. ........................................................ *Harpalobius*
— Antenna concolorous, uniformly testaceous or infuscated, or infuscated basally and gradually paler apically; abdominal sterna IV-VI with or without accessory setae ............................................... 10

10(9). Elytron without setigerous puncture in interval 3 .. 11
— Elytron with setigerous puncture in interval 3........ 12

11(10). Pronotum each side with punctate deplanation next to bead on lateral margin; deplanation evident but narrowed to anteriolateral angles; humerus without tooth (in part) ....................... *Pseudoophonus*
— Pronotum virtually without lateral deplanation; humerus with prominent tooth (in part) ..... *Pharalus*

12(10). Elytra metallic green (in part) .................... *Pharalus*
— Elytra piceous or black ....................................... 13

13(12). Abdominal sterna with only one pair of ambulatory setae (in part) ............. *Euharpalops*, and *Cautus, Obnixus, Atrichatus, Opacipennis* and *Herbivagus* Species Group
— Abdominal sterna IV-VI each with long (accessory) setae in addition to pair of ambulatory setae ..... 14

14(13). Accessory setae on abdominal sterna IV-VI few; dorsal surface black (in part) ................ *Euharpalops*
— Accessory setae on abdominal sterna IV-VI numerous, in row each side of paired ambulatory setae; dorsal surface rufo-piceous to rufous (in part) .... ............................................................... *Pharalus*

XLVI. Key to the Nearctic Genera of Licinini
(After Ball 1959)

1. Dorsal surface of one mandible with broad, deep transverse notch (Key XLVIII) .................. *Badister*
— Neither mandible as above ................................... 2

2(1). Episterna of metathorax markedly transverse, approximately rectangular in shape, outer margin little, if at all, longer than anterior margin, about as wide posteriorly as anteriorly; tarsus with tarsomere 5 with row of setae on each ventro-lateral margin (Key XLVII) ........................... *Dicaelus*
— Episterna of metathorax elongate, outer margin at least 1.25 times longer than anterior margin, distinctly narrower posteriorly than anteriorly; tarsus with tarsomere 5 ventrally without setae............ .......................................................... *Diplocheila*

XLVII. Key to the Subgenera of *Dicaelus*
(After Ball 1959)

1. Labium with palpomere 2 plurisetose .................... ....................................... *Dicaelus* (*sensu strictu*)
— Labial palpomere 2 bisetose ................................ 2

2(1). Elytron with striae moderately deep, intervals convex ................................................. *Paradicaelus*
— Elytral striae absent, or indicated only by rows of shallow punctures, intervals flat .............. *Liodicaelus*

XLVIII. Key to the Nearctic Subgenera of *Badister*
(After Ball 1959)

1. Left mandible with deep notch in dorsal surface, right mandible normal ....................................... *Baudia*
— Right mandible with deep notch in dorsal surface, left mandible normal ............................................... 2

2(1) Tarsus with tarsomere 5 with row of setae on each ventro-lateral margin; isodiametric meshes, surface not markedly iridescent ... *Badister* (*sensu strictu*)
— Tarsus with tarsomere 5 without setae ventrally; pronotum with microsculpture mesh pattern transverse or linear, surface markedly iridescent ........ ......................................................... *Trimorphus*

XLIX. Key to the Nearctic Genera of Panagaeini

1. Anterior margin of clypeus with shallow notch, on either side with short tooth; mandible with dorsal surface extended laterally, ventro-lateral margin not evident from dorsal aspect ............... *Micrixys*
— Anterior margin of clypeus straight, without teeth; mandible with ventro-lateral margin visible from dorsal aspect ...................................... *Panagaeus*

L. Key to the Nearctic Subgenera of *Chlaenius*
(Modified from Bell 1960)

1. Pronotum with four or more long setae on each side, in row parallel to, and somewhat distant from, margin ............................................ *Eurydactylus*
— Pronotum with single large seta at or near each posteriolateral angle.......................................... 2

2(1). Mandibles elongate, terebral blades at least twice as long as eye; scrobe much less than half as long as mandible .......................................................... 3
— Mandibles short, curved, terebral blade no more than slightly longer than eye; scrobe extended more than half length of mandible ..................................... 4

3(2). Labrum truncate; mentum with prominent, median tooth; mandibles slender, abruptly decurved ...... ............................................ *Pseudanomoglossus*
— Labrum at least shallowly emarginate; median tooth of mentum very small or absent; mandibles flat, not appreciably decurved ............. *Anomoglossus*

4(2). Antennomere 3 elongate, longer than antennomeres 1+2, and distinctly longer than antennomere 4; middle tibia of male with dense brush of setae on outer surface, near apex .............................. .................................... *Chlaenius* (*sensu stricto*)
— Antennomere 3 not elongate, equal to antennomere 4, shorter than antennomeres 1+2; middle tibia of male without dense brush of setae ...... *Agostenus*

LI. Key to the Nearctic Genera of Oodini
(After Bousquet 1996)

1. Clypeus with pair of setae .................................... 2
— Clypeus without setae .......................................... 6

2(1). Mesepisternum with conspicuous apodemal pit; labrum with four setae, two median setae connate ........ *Oodinus*
— Mesepisternum without apodemal pit; labrum with 6 more or less equidistant setae ........ 3

3(2). Eye smaller, vertical diameter less than length of antennomere 1; antenna shorter, ratio length/width of antennomere 9 less than 1.6 ........ 4
— Eye larger, vertical diameter greater than length of antennomere 1; antenna longer, ratio length/width of antennomere 9 more than 2 ........ 5

4(3). Labial palpomere 2 with two apical setae; anterior margin of labrum straight; eye not prominent; pronotum with slight oblique impression laterally; parascutellar seta absent ........ *Evolenes*
— Labial palpomere 2 without setae; anterior margin of labrum emarginate medially; eye prominent; pronotum with deep oblique impression laterally; parascutellar seta present ........ *Dercylinus*

5(3). Tarsomeres 1-4 of middle and hind legs with dense cluster of yellowish setae on ventral surface ........ *Lachnocrepis*
— Tarsomeres 1-4 of middle and hind legs without dense cluster of yellowish setae on ventral surface ........ *Oodes*

6(1). Labrum with six setae, four median ones close but not connate; stipes with anterior seta ........ *Anatrichis*
— Labrum with four to six setae, 2 median ones connate; stipes without anterior seta (Key LII) ........ *Stenocrepis*

LII. Key to the Nearctic Subgenera of *Stenocrepis*

1. Lateral margins of metasternum, metepisterna, and abdominal sterna laterally with distinct iridescence; microsculpture meshes indiscernible except in depressions ........ *Stenocrepis* (*sensu stricto*)
— Lateral margins of metasternum, metepisterna, and abdominal sterna laterally with microsculpture mesh pattern isodiametric to slightly transverse, surface not iridescent ........ *Stenous*

LIII. Key to the Nearctic Genera of Platynini
(Modified from Liebherr 1986)

1. Mentum without tooth ........ *Olisthopus*
— Mentum with median tooth ........ 2

2(1). Middle and hind tarsomeres setose dorsally, tarsomeres without lateral sulcus (Key LIV) ........ *Laemostenus*
— Middle and hind tarsomeres glabrous dorsally and/or with lateral sulcus ........ 3

3(2). Tarsus with claws pectinate ........ 4
— Tarsal claws smooth ........ 5

4(3). Pronotum with posteriolateral angles angulate, or posterior setigerous puncture of pronotum confluent with side margin; male genitalia with right paramere styloid; ovipositor stylomere 2 with nematiform setae near apex (Key LV) ........ *Calathus*
— Pronotum with posteriolateral angles rounded; posterior setigerous puncture remote from side margin; right paramere violin-shaped; stylomere 2 without nematiform setae ........ *Synuchus*

5(3). Male genital parameres with apical setae ........ 6
— Parameres without apical setae ........ 7

6(5). Dorsal surface flat; elytra with lateral margins parallel to one another; hind coxa with two setae, medial one absent; wings fully developed ... *Sericoda*
— Dorsal surface convex; elytra broadly ovoid; hind coxa with three setae, medial one present; wings markedly reduced ........ *Elliptoleus*

7(5). Tarsomeres of hind tarsus each with distinct median furrow dorsally ........ 8
— Hind tarsomeres each without, or with indistinct median furrow dorsally ........ 12

8(7). Head constricted posteriad eyes, with evident transverse impression also on dorsum of neck ........ 9
— Head without constriction posteriad eyes ........ 10

9(8). Antennomere 3 subequal in length to antennomere 4, glabrous except for apical ring of setae; metepisternum elongate; hind wings fully developed ........ *Paranchus*
— Antennomere 3 at least 1.15 X longer than antennomere 4, covered or not with sparse, fine pubescence; metepisternum subquadrate; elytron markedly constricted at base, humerus broadly rounded (Fig. 58.6); hind wings markedly reduced (in part) ........ *Rhadine*

10(8). Antennomere 3 pubescent on apical half; elytron with humerus not markedly narrowed, angulate (Fig. 57.6) (Key LVI) (in part) ........ *Agonum*
— Antennomere 3 glabrous except for apical ring of setae ........ 11

11(10). Dorsal surface of body bright metallic green to aeneous; tibiae testaceous (Key LVI) (in part) ........ *Agonum*
— Dorsal surface of body dark, slightly metallic or not; legs concolorous with body ........ *Anchomenus*

12(7). Male genitalia with internal sac of median lobe with spines or fields of stout spicules ........ 13
— Internal sac of median lobe without spines or fields of stout spicules ........ 16

13(12). Head constricted posteriad eyes, with evident transverse impression also on dorsum of neck ........ 14
— Head without constriction posteriad eyes ........ 15

14(13). Dorsal surface bicolored, elytra metallic green; ovipositor stylomere 1 with fringe of setae ........ *Platynus, Metacolpodes*
(No easily used characters separate *Platynus* adults from those of *Metacolpodes buchannani*, the sole species of this genus in North America. The species was introduced in the Pacific Northwest, adults of which are distinguished from most North American *Platynus* adults by the metallic purple to green luster of the dorsal surface of the elytra.)
— Dorsal surface concolorous, piceous to rufous; stylomere 1 glabrous or with only a few setae (in part) ........ *Rhadine*

15(13). Elytron with more than three discal setae; apex of antennomere 2 with four or more setae ............. ........................................................... *Tanystoma*
— Elytron with 3 discal setae; apex of antennomere 2 with less than four setae (Key LVI) (in part) ........ ............................................................... *Agonum*

16(12). Head constricted posteriad eyes, with evident transverse impression also on dorsum of neck ....... 19
— Head without constriction posteriad eyes ........... 17

17(16). Antennomeres 8-11 abruptly pale, yellow to white 18
— Antennomeres without such color contrast (Key LVI) (in part) .................................................. *Agonum*

18(17). Pronotum without latero-posterior seta; legs ferrugineus ....................................... *Tetraleucus*
— Pronotum with latero-posterior seta; femora dark .. ............................................................... *Agonum*

19(16). Body without vestiture of fine short setae (Key LVII) (in part) .................................................. *Platynus*
— Body with vestiture of fine short pubescence ...... 20

20(19). Labial mentum with tooth bifid apically ...... *Atranus*
— Mentum with tooth rounded apically .... *Oxypselaphus*

LIV. Key to the Nearctic Subgenera of *Laemostenus*

1. Hind tibia with dense patch of short setae medially on apical half ................................... *Pristonychus*
— Hind tibia without patch of setae on apical half ...... ................................ *Laemostenus (sensu stricto)*

LV. Key to the Nearctic Subgenera of *Calathus*

1. Pronotum with posteriolateral setigerous puncture confluent with side margin; prosternum with apex not margined .................................... *Procalathus*
— Pronotum with posteriolateral setigerous puncture remote from side margin; prosternum with apex margined ........................................................... 2

2(1). Elytral interval 3 with more than six setigerous punctures, interval 5 with setigerous punctures; elytron with humerus toothed; pronotum with posteriolateral impressions coarsely punctate ..... ...................................... *Calathus (sensu stricto)*
— Elytral interval 3 with six setigerous punctures or less, interval 5 without setigerous punctures; elytron with humerus not toothed; pronotum with posterolateral impressions impunctate, or with fine, sparse punctures ............................ *Neocalathus*

LVI. Key to the Nearctic Subgenera of *Agonum*
(Modified from Liebherr 1986)

1. Antennomeres 1, 2, or 3 with pubescence ........... 2
— Antennomeres 1-3 glabrous except for apical setae ........................................................................ 6

2(1). Elytra piceous to black with basal and subapical orange spots on each side, pronotum orange; pronotum without posteriolateral setae ............. ....................................................... *Deratanchus*
— Elytra and pronotum not colored as above; pronotum with or without posteriolateral setae ................ 3

3(2). Dorsal surface with fine pubescence (in part)......... ........................................................ *Stictanchus*
— Dorsal surface without pubescence ...................... 4

4(3). Elytron with coarse setigerous punctures in the odd-numbered intervals (in part) ............................... ....................................... *Agonum (sensu stricto)*
— Elytron with setigerous punctures in interval 3 only ........................................................................ 5

5(4). Middle coxal ridge with one seta............. *Europhilus*
— Middle coxal ridge with two or more setae (in part) ........................................................ *Stictanchus*

6(1). Apex of antennomere 1 with one long seta and three or more smaller setae; middle coxal ridge with two or more setae (in part) ....................... *Stictanchus*
— Apex of antennomere 1 with one long seta and fewer than three smaller setae; if three small setae present, then mesocoxal ridge with one seta only ........................................................................ 7

7(6). Tarsomere 5 ventrally glabrous or with as many as four small setae ................................................ 8
— Tarsomere 5 with six or more setae underneath ... 13

8(7). Elytron with three foveate discal setigerous punctures (in part) ........................... *Agonum (sensu stricto)*
— Elytron with three to six non-foveate discal setigerous punctures ................................................... 9

9(8). Pronotum narrow, markedly cordate, with protruded, somewhat acute, hind angles (in part) ................ ...................................................... *Stereagonum*
— Pronotum broader, less constricted at base, with obtuse or rounded hind angles ........................... 10

10(9). Pronotum very broad, lateral depression evident to front angles (in part)....................... *Stereagonum*
— Pronotum narrower, lateral depression more or less evident in anterior half .................................. 11

11(10). Elytron with transverse impression about middle and with longitudinal impression toward apex of stria 5 (in part) ......................................... *Micragonum*
— Elytron not impressed ....................................... 12

12(11). Pronotum shallowly punctate around posteriolateral impressions; legs testaceous (in part) ................ ........................................................ *Micragonum*
— Pronotum with entire base distinctly punctate; femora dark, piceous to black (in part) ........ *Stereagonum*

13(7). Pronotum with postriolateral impressions nearly obsolete or not; body slightly sclerotized, ferrugineous; length 5.0-7.8 mm.................. *Platynomicrus*
— Pronotum with posteriolateral impressions evident; body, at least head and pronotum, thickly sclerotized and darkly pigmented; length 5.4-10+ mm ........ 14

14(13). Pronotum with very small, deep, punctiform posteriolateral impressions; pronotum almost circular in form, hind angles completely rounded ...... 15
— Pronotum with broader, less defined posteriolateral impressions; pronotum of various shapes, hind angles prominent or not ................................. 16

15(14). Elytra with shallow impression at middle, near second discal puncture; pronotum slightly wider than

head; body with dorsal surface slightly metallic (in part) ............................................ *Micragonum*
— Elytra without impression; pronotum much broader than head; body shiny but not metallic dorsally .. ............................................................ *Olisares*

16(14). Pronotum narrow (width/length less than 1.30), hind angles obsolete (in part) .................. *Micragonum*
— Pronotum wider (width/length usually greater than 1.30), hind angles slight to distinct (in part)........ ....................................... *Agonum* (*sensu stricto*)

LVII. Key to Nearctic Subgenera of *Platynus*
(Modified from Liebherr and Will 1996)

1. Hind femur anteriorly without setae on apical half .. 2
— Hind femur anteriorly with 2-20 longer or very short setae on apical half ............................................ 5

2(1). Hind tarsomere 5 setose ventrally (each of 6-14 seta subequal to longer than thickness of tarsomere) . 3
— Hind tarsomere 5 apparently glabrous ventrally (setae visible in some species but only at high magnification) (in part) ............. *Platynus* (*sensu stricto*)

3(2). Size larger, length of body more than 9.0 mm; body color, including legs, dark piceous to black (in part) ....................................... *Platynus* (*sensu stricto*)
— Body smaller, length less than 9.0 mm; color, including legs, ferrugineous to rufopiceous ................ 4

4(3). Pronotum with lateral margins rounded, posteriolateral angles obsolete; neck without evident constriction ........................... *Glyptolenopsis*
— Pronotum with lateral margins convergent or sinuate in posterior half, posteriolateral angles evident; neck with evident constriction ....... *Microplatynus*

5(1). Hind tarsomere 5 with four to 14 setae ventrally.. 6
— Hind tarsomere 5 apparently without setae ventrally (setae in some species visible but only at high magnification) ........................................................... 7

6(5). Pronotum with posteriolateral angles evident, about right; hind tarsomere 5 with four to six setae ventrally (in part) .................. *Platynus* (*sensu stricto*)
— Pronotum with posteriolateral angles obsolete, rounded; hind tarsomere 5 with nine to 14 setae ventrally (in part) .................................... *Batenus*

7(5). Frons with two rufous spots. (in part) ......... *Batenus*
— Frons without rufous spots .................................. 8

8(7). Mandibles elongate, sickle shaped, retinacular teeth evident anteriad labrum; hind tarsus without sulci dorsally (in part) ..................................... *Batenus*
— Mandibles proportionally shorter, less sickle shaped, retinacular teeth covered by labrum; hind tarsus with lateral and medial sulci dorsally (in part) ..... ....................................... *Platynus* (*sensu stricto*)

LVIII. Key to the Nearctic Genera of Lachnophorini

1. Dorsal surface smooth and glabrous, except normal fixed setae; elytron with microsculpture mesh pattern linear, surface markedly iridescent; antenna with antennomeres 6-11 white, contrasted markedly with more or less piceous antennomeres 1-5 ............................................................ *Eucaerus*
— Dorsal surface more or less setose; elytron with microsculpture mesh pattern isodiametric to slightly transverse, not linear, surface shiny to dull, not iridescent; antenna with antennomeres concolorous, testaceous to rufopiceous, or bicolored, with antennomeres 1-4 testaceous and 5-11 piceous, or tricolored, with antennomeres 1-4 testaceous, 5-7 white, and 8-11 piceous .................. 2

2(1). Elytron with broad, transverse sulcus in basal one-third, and with several patches of white scales; dorsal surface with setae of two colors and sizes: black setae, these almost as long as antennomere 1 and yellow ones, about length of antennomere 2, labial palpomere 3 oval in shape, apex sharply pointed; body form antlike *(Ega)* .............. *Calybe*
— Elytron without broad transverse sulcus in basal one-third; setae of elytra either short or long, all testaceous in color; palpi various .............................. 3

3(2). Elytron with striae with small setigerous punctures throughout; discal setae relatively short, length less than length of antennomere 2; elytron bicolored, with posterior 1/4 black, anterior 3/4 rufous...... ..................................................... *Anchonoderus*
— Elytron with strial punctures of two sizes, anteriorly larger, posteriorly smaller; dorsal surface uniformly black, or with pronotum coppery-green and elytra variegated piceous and testaceous .................... 4

4(3). Dorsal surface with integument uniformly black; setae of dorsum erect, sparse, some as long as antennomere 1; labial palpomere 3 fusiform ....... ......................................................... *Euphorticus*
— Dorsal surface of head and pronotum dull coppery green, elytra variegated piceous and testaceous; dorsal surface densely setose, setae much shorter than antennomere 2; labial palpomere 3 oval, apex sharply pointed ............................. *Lachnophorus*

LIX. Key to the Nearctic Subgenera of *Colliuris*

1. Head and prothorax black throughout; elytra uniformly dark, OR red except for three black spots in form of interrupted or constricted band, and apices black ............................................... *Cosnania*
— Posterior portion of head, and anterior and posterior portions of prothorax red; elytra red except for an unconstricted black transverse band, and apices black ................................................ *Calocolliuris*

LX. Key to the Nearctic Subgenera of *Tetragonoderus*

1. Tarsal claws more or less serrate; elytra in great part piceous, AND intercoxal process of prosternum margined at tip; OR, if intercoxal process immarginate, then pronotum piceous ................ ........................... *Tetragonoderus* (*sensu strictu*)
— Tarsal claws not serrate; elytra in great part testaceous, AND intercoxal process of prosternum margined at tip; OR, if latter immarginate, then pronotum and elytra entirely testaceous ............ ......................................................... *Peronoscelis*

LXI. Key to the Nearctic Genera of Lebiini

1. Hind tibia with inner spur almost as long as tarsomere 1 of hind tarsus; tarsal claws pectinate; head mark-

edly constricted posteriorly in form of narrow neck; labial palpomere 3 slender ............... *Nemotarsus*
— Hind tibia with inner spur less than half length of hind tarsomere 1; other features various .......... 2

2(1). Head capsule ventrally with pair of suborbital setae, each seta about as long as supraorbital seta ..... 3
— Head capsule ventrally without suborbital setae ... 8

3(2). Dorsal surface with integument with vestiture of short setae; pronotum on each lateral margin with three or more long setae .......................... *Somotrichus*
— Integument glabrous except normal fixed setae; pronotum with not more than two setae on each lateral margin .................................................... 4

4(3). Labrum transverse, distinctly wider than long; labial mentum with margin of median portion broadly convex, but not distinctly toothed ..... *Euproctinus*
— Labrum distinctly longer than width at base, or quadrate, or slightly wider than long, but not transverse; mentum various ................................................ 5

5(4). Labium with mentum edentate, median portion broadly concave; palpiger with long seta ........... .......................................................... *Coptodera*
— Mentum distinctly toothed, or at least margin medially broadly bisinuate; palpiger asetose ........... 6

6(5). Head and pronotum with dorsal surface densely, coarsely punctate; elytron with striae wide, about as wide as markedly convex intervals; each stria with two or more rows of punctures throughout length *(Inna)* .......................................... *Eucheila*
— Head with dorsal surface smooth or irregularly longitudinally rugulose, pronotum smooth; elytron with striae wider or narrower, but not coarsely punctate, intervals slightly convex, not markedly so .......... 7

7(6). Elytron with striae shallowly impressed, about as wide as intervals; abdominal sterna asetose *(Oenaphelox)* .......................................................... *Phloeoxena*
— Elytron with striae narrow, much narrower than intervals; abdominal sterna with vestiture of short setae .................................................... *Mochtherus*

8(2). Labium without transverse suture between mentum and submentum ................................................ 9
— Labium with mental-submental suture ............... 12

9(8). Labial palpus with palpomere 3 securiform, apical margin much wider than apical margin of maxillary palpomere 4 .............................. *Axinopalpus*
— Labial palpomere 3 not securiform, apical margin similar in width to apical margin of maxillary palpomere 4 ...................................................................... 10

10(9). Pronotum with posterior (basal ) margin lobate medially ................................................. *Microlestes*
— Pronotum with posterior margin truncate, not lobate ...................................................................... 11

11(10). Body slender, head across eyes as wide as maximum width of pronotum; elytron with humerus markedly sloped, narrowed; dorsal surface with integument bicolored—pronotum and elytra testaceous, head piceous ............................................. *Philorhizus*

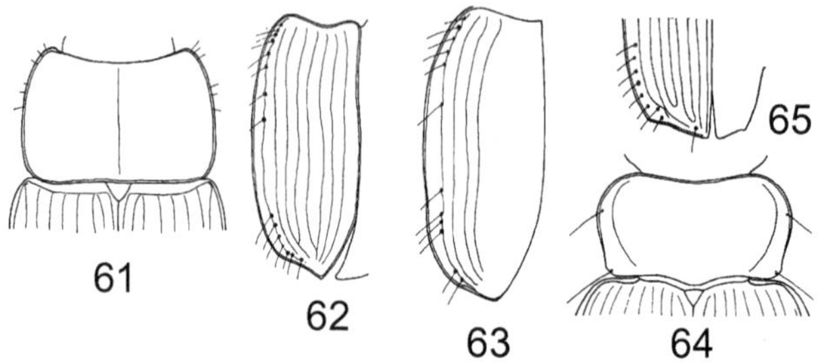


FIGURES 61.6-65.6. Fig. 61.6, *Discoderus parallelus* Horn, pronotum, dorsal view; Fig. 62.6, *Acupalpus hydropicus* LeConte, left elytron, dorsal view; Fig. 63.6, *Stenolophus lecontei* Chaudoir, left elytron, dorsal view; Fig. 64.6, *Lebia grandis* Hentz, left elytron, apical portion, dorsal view; Fig. 65.6 *Plochionus amandus* Newman, pronotum, dorsal view.

— Body broader, head across eyes narrower than maximum width of pronotum; elytron with humerus quadrate, lateral margin not markedly sloped mediad; dorsal surface concolorous–rufous to piceous–or with bicolored elytra .......................... *Dromius*

12(8). Elytron laterally at outer apical angle with three terminal umbilical punctures in form of a triangle (Fig. 64.6); tarsal claws pectinate ........................... 13
— Elytron with umbilicasl punctures in curved row, not in form of triangle; tarsal claws pectinate or not . ......................................................................... 14

13(12). Elytron posteriorly with sutural interval distinctly flat ("pinched") (Key LXIII) .................................. *Lebia*
— Elytron posteriorly with sutural interval not flattened ................................................................ *Apenes*

14(12). Size small, total length of body less than 6 mm; dorsal surface concolorous, black; mandible not conspicuously widened near base, scrobe narrow, ventro-lateral margin basad not conspicuously rounded .......................................................... 15
— Size larger, total length more than 6 mm; dorsal surface with color of integument various, bicolored or not; mandible conspicuously widened near base, scrobe wide, shallow, ventro-lateral margin basad conspicuously rounded ................................... 16

15(14). Pronotum with posterior (basal) margin truncate, not lobate medially ...................................... *Apristus*
— Pronotum with posterior margin lobate medially..... ............................................................. *Syntomus*

16(14). Head capsule markedly narrowed and prolonged posteriad eyes; prothorax more or less tubular, narrowed anteriorly, without lateral flanges, at least mediad ........................................................ *Agra*
— Head capsule not markedly prolonged posteriad eyes; prothorax not tubular, pronotum with lateral flanges .................................................... 17

17(16). Labial mentum with tooth ................................. 18
— Mentum without tooth ........................................ 22

18(17). Labial prementum with glossal sclerite apically quadrisetose; pronotum (Fig. 65.6) with posterior (basal) margin slightly lobed medially (Key LXIV). .......................................................... *Plochionus*
— Glossal sclerite apically bisetose; pronotum with posterior margin straight, not lobed .................... 19

19(18). Tarsus with tarsomeres dorsomedially longitudinally sulcate (*sensu stricto*, Key LXV) ............... *Calleida*
— Tarsus with tarsomeres dorsally smooth, not longitudinally sulcate ............................................. 20

20(19). Tarsal claws smooth, not pectinate; integument dorsally bicolored, with head and pronotum rufotestaceous, elytra metallic green or blue ...... ........................................................ *Tecnophilus*
— Tarsal claws pectinate; elytra dorsally metallic colored or not .................................................... 21

21(20). Abdominal sternum VII near posterior margin with at least six pairs of moderately long setae in female; male with rather dense cluster of setae each side, brush-like ....................................... *Infernophilus*
— Abdominal sternum VII with not more than two pairs of setae each side in males and females (Key LXII) ............................................................ *Cymindis*

22(17). Tarsus with tarsomeres dorsally smooth; prothorax elongate, pronotum longer than wide, and narrower than head .................................... *Cylindronotum*
— Tarsomeres dorsally sulcate ............................... 23

23(22). Elytron with discal intervals smooth ............... *Onota*
— Elytron with discal intervals tuberculate .. *Hyboptera*

LXII. Key to the Nearctic Subgenera of *Cymindis*

1. Elytral intervals impunctate, glabrous; tarsomeres with dorsal surfaces glabrous, except at apices (in part) .................................................. *Pinacodera*
— Elytral intervals punctate, punctures bearing short setae; tarsi with dorsal surfaces setose ................ 2

2(1). Tarsi with each tarsomere dorsally with six or fewer setae; geographical range: southern California and Arizona only (in part) ......................... *Pinacodera*
— Tarsi with each tarsomere dorsally with more than six setae; geographical range: Nearctic North America, including southern California and Arizona ...................................... *Cymindis* (*sensu strictu*)

LXIII. Key to the Nearctic Subgenera of *Lebia*
(After Madge 1967)

1. Front leg with tibia with single (apical) spur, preapical spur absent .......................... *Lebia* (*sensu stricto*)
— Front tibia with both spurs ................................... 2

2(1). Dorsal surface of head capsule with frons and pronotum with many coarse setigerous punctures; elytron with disc metallic, with basal 1/3 pale..... ............................................................ *Lamprias*
— Frons and pronotum without setigerous punctures; elytron with disc either metallic color or entirely pale ................................................................... 3

3(2). Elytron with surface of disc entirely pale; proepisternum with surface longitudinally rugulose ....................................................... *Polycheloma*
— Elytron with discal integument entirely metallic; proepisternum with surface smooth ...... *Loxopeza*

LXIV. Key to the Nearctic Subgenera of *Plochionus*

1. Tarsus with tarsomere 4 bilobed, the length of lobes greater than half length of fourth .......... *Menidius*
— Tarsomere 4 emarginate, not bilobed ..................... .................................... *Plochionus* (*sensu strictu*)

LXV. Key to the Nearctic Subgenera of *Calleida*

1. Tarsus with tarsomere 4 bilobed ................. *Calleida*
— Tarsomere 4 at most emarginate ........... *Philophuga*

LXVI. Key to the Nearctic Genera of Zuphiini

1. Head markedly constricted posteriad eyes, in form of neck; width across eyes more than twice width of neck ................................................... *Zuphium*
— Head not markedly constricted posteriad eyes; width of head across eyes much less than twice width of area behind eyes ............................................... 2

2. Pronotum truncate at base, antenna with antennomere 3 shorter than 4 ......... *Pseudaptinus*
— Pronotum subpedunculate at base, antennomeres 2-4 of nearly equal length .......................... *Thalpius*

LXVII. Key to the Nearctic Subgenera of *Galerita*
(After Reichardt 1967)

1. Elytron with dorsal surface striate-punctate, intervals broad, flat or slightly convex ........... *Progaleritina*
— Elytron with dorsal surface carinate; two rows of carinulae between each adjacent pair of carinae . ......................................... *Galerita* (*sensu stricto*)

### A Classification of the Nearctic Genera
(After Lawrence and Newton 1995, with modifications)

I. Subfamily Paussinae
 1. Tribe Metriini
 2. Tribe Ozaenini
II. Subfamily Gehringiinae
 3. Tribe Gehringiini
III. Subfamily Omophroninae
 4. Tribe Omophronini
IV. Subfamily Nebriinae
 5. Tribe Pelophilini
 6. Tribe Nebriini
 7. Tribe Opisthiini
 8. Tribe Notiophilini
V. Subfamily Carabinae
 9. Tribe Carabini
 10. Tribe Cychrini
VI. Subfamily Cicindelinae
 11. Tribe Omini
 12. Tribe Megacephalini
 13. Tribe Cicindelini
VII. Subfamily Loricerinae
 14. Tribe Loricerini
VIII. Subfamily Elaphrinae
 15. Tribe Elaphrini
IX. Subfamily Promecognathinae
 16. Tribe Promecognathini
X. Subfamily Scaritinae
 17. Tribe Clivinini

18. Tribe Scaritini
XI. Subfamily Trechinae
19. Tribe Broscini
20. Tribe Psydrini
21. Tribe Trechini
22. Tribe Bembidiini
23. Tribe Micratopini
24. Tribe Pogonini
25. Tribe Patrobini
XII. Subfamily Harpalinae
26. Tribe Morionini
27. Tribe Loxandrini
28. Tribe Pterostichini
29. Tribe Zabrini
30. Tribe Harpalini
31. Tribe Licinini
32. Tribe Panagaeini
33. Tribe Chlaeniini
34. Tribe Oodini
35. Tribe Pentagonicini
36. Tribe Platynini
37. Tribe Perigonini
38. Tribe Lachnophorini
39. Tribe Ctenodactylini
40. Tribe Odacanthini
41. Tribe Cyclosomini
42. Tribe Lebiini
43. Tribe Zuphiini
44. Tribe Dryptini
45. Tribe Helluonini
XIII. Subfamily Pseudomorphinae
46. Tribe Pseudomorphini
XIV. Subfamily Brachininae
47. Tribe Brachinini

## I. Subfamily Paussinae

The diagnostic features of adults of this subfamily are the following: size moderate (total length ca. 6-20 mm); head capsule ventrally without antennal grooves; antennal insertions laterad, each between base of mandible and anterior margin of eye; antenna filiform to clavate, number of antennomeres 11 to 2 (lower numbers result of fusion of flagellar articles); antennomeres 2-4 not nodose, 2-6 without unusually long setae; labrum with 12 setae; mandible with scrobe unisetose or asetose, occlusal margin with one terebral teeth, retinacular tooth about size of terebral tooth, not enlarged and molariform; terminal palpomeres various, from about cylindrical to distinctly securiform; pterothorax with mesonotal scutellum dorsally visible, not concealed beneath elytra; front coxal cavities closed, with lateral arm of prosternum overlapped by tip of adjacent proepimeron; middle coxal cavities disjunct-separated, and hind coxal cavities lobate (disjunct)-confluent (Bell 1967: 105); elytra indistinctly striate to smooth dorsally, extended to apex of abdominal tergum VIII, and apical margins subsinuate; with or without a prominent ridge or fold in apical one-fifth (flange of Coanda); metathoracic wings large to variously reduced; front tibia isochaete, antenna cleaner, if present, independent of spurs, sulcate or emarginate; abdomen with six (II-VII) pregenital sterna normally exposed, sterna III and IV separated by suture; male genitalia with parameres asymmetric in form (left shorter and broader, right styliform, left paramere with one or few setae apically, right with more or less dense brush of setae preapically and apically (Ball and McCleve 1990: 65, Figs. 82-85C, D); ovipositor with gonocoxae not rotated 90 ° in repose, with or without ramus, gonocoxa single, not divided, without accessory lobe; reproductive tract with spermatheca 2, only (Liebherr and Will 1998: 126, Figs. 13-15). Secretion of pygidial glands quinones, delivered hot; discharged by crepitation (Moore 1979: 194), or by oozing (Eisner, *et al.*, 2000).

Larvae with anal plates formed by fusion of tergites of abdominal segment VIII with epipleurites of abdominal segment IX; cerci either antler-like, each with several branches, or plate-like (Vigna Taglianti *et al.* 1998; see also Arndt 1998: 180-182).

Important features not seen in preserved specimens are: larvae hold the body in a curved position, with the posterior part of the abdomen (the anal plate) held over the head; and defensive secretion of adults consists of quinones (Eisner *et al.* 1977; Moore 1979: 198).

This subfamily of 724 species in 34 genera is represented in Nearctic North America by the Tribes Metriini and Ozaenini. The nominotypical tribe, Paussini (including Protopaussini), is restricted largely to the tropics of the southern hemisphere, ranging in the Neotropical Region north to southern México.

Various taxa of Paussini are myrmecophilous, as adults and larvae. Because of similarities in striking modifications of larval structure, we assume that all members of this tribe are in some way myrmecophilous, though this conjecture remains to be established as fact.

Relationships of the subfamily. Some authors take the position that the Paussinae is the adelphotaxon of the remaining groups of Carabidae (for example, Beutel 1992; and Liebherr and Will 1998). Others (for example, Erwin 1985: 446) place the paussines as adelphotaxon of the Brachininae and together these taxa form the adelphotaxon of an assemblage identified as the Loxomeriformes, a major middle-grade taxon of carabids. We favor the first alternative, based in part on the numerous plesiotypic features of the Paussinae, including the ovipositor with non-segmented gonocoxae, and its absence of rotation in the infolded position (see also Ball and McCleve 1990), and in part on the features of the brachinines, indicating their relationships with the higher carabids. Thus, we agree with Beutel (1992) that paussines and brachinines are not closely related.

Relationships of the tribes. Because of overall similarity, we include the Protopaussini in the Paussini. Based primarily on the study of adult structure by Darlington (1950) and larval features by Bousquet (1986), Ball and McCleve (1990) adopted the following system: Metriini + (Ozaenini + (Protopaussini + Paussini). In contrast, Vigna Taglianti *et al.* (1998), based on numerical cladistic analysis of larval features, proposed the following system: (Metriini + Ozaenini) + Paussini. Protopaussines were not included for want of larval features. Considering the morphoclines in adult features as well as the one in details of the urogomphi, we believe that some ozaenine group is the adelphotaxon of the Paussini, rather than the arrangement postulated by Vigna Taglianti *et al.* (1998).

### Metriini

This monogeneric tribe is precinctive in western Nearctic North America. Adults lack the elytral flange of Coanda that is characteristic of all other paussines, and on this basis, it is postulated to be the most primitive group of this subfamily. As shown by the work of Eisner *et al.* (2000), the delivery mechanism of defensive secretions by adult *Metrius contractus* Eschscholtz is the most primitive in the subfamily, thus supporting the postulated basal position for the Metriini.

*Metrius* Eschscholtz 1829
This genus includes two allopatric western North American species. The polytypic *M. contractus* Eschscholtz, with three subspecies, is more coastal, ranging from California northward to southern British Columbia. The monotypic *M. explodens* Bousquet and Goulet 1990 is known only from Idaho. References. Bousquet 1986: 373-388. Bousquet and Goulet 1990: 13-18.

### Ozaenini

This Gondwanan tribe of 130 species arranged in 14 genera represents in structural features a primitive group of paussines, based on relatively unmodified mouthparts and antennae. Pantropical, it is represented geographically in the Oriental Region, and in the Afrotropical, Australian, and Neotropical Regions of the Southern Hemisphere. In the Western Hemisphere, the group ranges north to the southwestern reaches of the Nearctic Region.

Little is known of ozaenine way of life. Representatives of the Oriental genera *Eustra* Schmidt-Goebel and *Itamus* Schmidt-Goebel and *Pseudozaena lutea* Hope 1842 have been taken in rotten wood, and *Eustra japonica* Bates has been found under stones partially embedded in soil, in the shelter of shady thickets (Andrewes 1929: 165, 168, and 170). Larvae of *Pachyteles* have been found under bark, and those of *P. mexicanus* (Chaudoir 1848) have been collected on the surface of bat guano, in a cave (Vigna Taglianti *et al.* 1998: 275). Species of *Physea* evidently are myrmecophilous. References. Ball and McCleve 1990: 30-116 (rev. N. A. spp.). Bänninger 1927: 177-216 (rev. tribe).

*Physea* Brullé 1834
Six species of this indigenous Neotropical genus are known, ranging collectively from Brazil northward to southeastern Texas. The one species which reaches Texas is *P. hirta* LeConte 1853. Larvae of the South American species *Physea setosa* Chaudoir 1868 were collected in nests of the leaf-cutter ant genus *Atta*, and adults of *Physea hirta* LeConte 1853, were collected in midden heaps of the same ant genus.

*Pachyteles* Perty 1830
The 69 species of this precinctive Western Hemisphere genus are grouped in three subgenera, two of which (*Pachyteles* (s. str.) and *Goniotropis* Gray 1832) are represented in the southwestern United States by three species. One species of the former subgenus (*P. gyllenhalii* Dejean 1825) and two species of the latter (*P. parca* LeConte 1844, and *P. kuntzeni* Bänninger 1927) reach southern Arizona in southwestern United States. Adults have been found under bark of cottonwood trees (*Populus*) and at UV light, at night.

*Ozaena* Olivier 1812
This genus is also restricted to the Neotropical Region and the southern reaches of the Nearctic Region. One species is recorded from the United States: *O. lemoulti* Bänninger 1932; specimens were collected at Nogales, and in the Pajarito Mountains, Arizona. Reference. Ball and Shpeley 1990 (rev. spp.)

## Subfamily II. Gehringiinae

Adult characters of this distinctive group are the following: size very small (total length 1.6-1.7 mm); head capsule ventrally without antennal grooves; antennal insertions laterad, each between base of mandible and anterior margin of eye; antennomeres filiform, 2-6 without long setae; labrum with six setae; mandible with scrobe unisetose, occlusal margin with one terebral tooth, retinacular tooth about size of terebral tooth, not enlarged and molariform; terminal palpomeres subulate; pterothorax with scutellum visible dorsally, not covered by elytra; front coxal cavities open, middle coxal cavities disjunct-confluent, hind coxal cavities separated from one another and incomplete (Bell 1967: 106); elytra smooth, not striate, relatively short, not extended to apex of abdominal tergum VIII, and apices subtruncate; metathoracic wings large, with marginal fringe (Lindroth 1961: 4), front tibia isochaete (both spurs apical), antenna cleaner emarginate (Grade C, Hlavac 1971: 57); abdomen with six (II-VII) pregenital sterna normally exposed, sterna III and IV partially fused; male genitalia with parameres slender, about equal in size, and setose apically; ovipositor with gonocoxae rotated 90° in repose, without ramus, gonocoxa with two gonocoxites, gonocoxite 2 without accessory lobe; and reproductive tract with spermathecae 1 and 2 (Liebherr and Will 1998: 125, Fig. 10). Nature of secretions from pygidial glands unknown.

The combination of character states of this taxon is enigmatic (Lawrence and Newton 1995: 811; Maddison 1985: 35), with some states indicating a near-basal position in an evolutionary sequence, and other states indicating a median position (Bell 1964). In the absence of agreement about where to place it, this taxon is best ranked as a subfamily. This high rank will ensure that this group is not submerged in some other higher taxon, and its remarkable combination of features thus forgotten. Only one species is included, assigned to a monospecific genus and a monogeneric tribe.

Gehringiini

*Gehringia* Darlington 1933

The single precinctive Nearctic species of *Gehringia, G. olympica* Darlington 1933, ranges from the Pacific Northwest (Oregon to British Columbia) eastward to western Montana and Alberta, and northward to the Northwest Territories. The beetles live on bars of very fine and clean gravel along cold mountain streams. Because of their small size and slow gait, adults are difficult to find.

## Subfamily III. Omophroninae

The diagnostic adult features of this monotribal taxon are the following: size moderate (total length 5-10 mm); head capsule ventrally without antennal grooves; antennal insertions laterad, each between base of mandible and anterior margin of eye; antennae filiform, antennomeres 2-4 not nodose, 2-6 without especially long setae; labrum with six setae; mandible with scrobe unisetose, occlusal margin with one terebral tooth, retinacular tooth about size of terebral tooth, not enlarged and molariform; terminal palpomeres about cylindrical; pterothorax with mesonotal scutellum dorsally concealed beneath elytra; front coxal cavities closed, with projection of prosternum inserted into cavity in proepimeron; middle coxal cavities disjunct-separate, hind coxal cavities conjunct-confluent (Bell 1967: 106); elytra striate dorsally, extended to apex of abdominal sternum VIII, and apical margins subsinuate; metathoracic wings large; front tibia anisochaete, antenna cleaner sulcate, grade B (Hlavac 1971: 57); abdomen with six (II-VII) pregenital sterna normally exposed, sterna III and IV separated by suture; male genitalia with parameres slender, about equal in size, and setose apically; ovipositor with gonocoxae rotated 90° in repose, with ramus, gonocoxa single, without accessory lobe; and reproductive tract with spermathecae 1 and 2, or only with 2 (Liebherr and Will 1998: 126, Figs. 13-15). Secretions of pygidial glands higher saturated acids, discharge by oozing (Moore 1979: 194).

Omophronini

Relationships of this tribe have been investigated from a phylogenetic perspective by Bils (1976), Nichols (1985a), Beutel and Haas (1996), Kavanaugh (1998), and Liebherr and Will (1998) (see these last-named authors for a review of the topic), with somewhat divergent results. All authors agree that the group represents a primitive (basal grade) carabid lineage, but disagree about its adelphotaxon. So, the arbitrary solution adopted is to place the group in its own subfamily, pending settlement of the points of disagreement.

*Omophron* Latreille 1802
 *Homophron* Semenov 1922
 *Paromophron* Semenov 1922
 *Istor* Semenov 1922
 *Prosecon* Semenov 1922

This genus, with a total of 60 species arranged in two subgenera, is represented in all of the major zoogeographical regions of the world, except the Australian. Most of the species are Oriental or Afrotropical. Of the 17 Western Hemisphere species, 11 are confined to the Nearctic Region; five species are in the northern Neotropical Region, the range extending south to Costa Rica (Reichardt 1977: 379). In the Nearctic Region, the group is widespread, from the Mexican border and Gulf Coast north to southern Canada.

The beetles are convex, almost hemispherical in outline. Most are variegated, with shades of brown and green predominating. Both larvae and adults live in damp sand on the margins of lakes, rivers, and streams, and are collected by pouring water over the ground which they inhabit. This forces the beetles out of their burrows, and although they run swiftly, they do not attempt to fly. They are nocturnal, predacious, and gregarious. They probably pass the winter as adults. Adults of several species fly to light sources on warm summer nights. References. Benschoter and Cook 1956: 411-429. (rev. N.A. spp.). Landry and Bousquet 1984: 1557-1569 (descr. larvae). Nichols 1981 (classification of species).

## Subfamily IV. Nebriinae

The diagnostic features of adults of this subfamily are the following: size various (total length 4-17 mm); head capsule ventrally without antennal grooves; antennal insertions laterad, each between base of mandible and anterior margin of eye; antenna filiform, antennomeres 2-4 not nodose, 2-6 without unusually long setae; labrum with six setae; mandible with scrobe unisetose, occlusal margin with one terebral tooth, retinacular tooth about size of terebral tooth, not enlarged and molariform; terminal palpomeres about cylindrical; pterothorax with mesonotal scutellum dorsally visible, not concealed beneath elytra; front coxal cavities open, middle coxal cavities disjunct-confluent, hind coxal cavities conjunct-confluent (Bell 1967: 105); elytra striate dorsally, extended to apex of abdominal sternum VIII, and apical margins subsinuate; metathoracic wings large to variously reduced; front tibia anisochaete; antenna cleaner sulcate (Grades A and B, Hlavac 1971: 57; abdomen with six (II-VII) pregenital sterna normally exposed, sterna III and IV separated by suture; male genitalia with parameres asymmetric in form but similar in length, asetose; ovipositor with gonocoxae rotated 90° in repose, with or without ramus, gonocoxa single or with two gonocoxites, gonocoxite 2 without accessory lobe; and reproductive tract with spermatheca 2 only (Liebherr and Will 1998: 126, Figs. 13-15). Secretions of pygidial glands higher saturated or unsaturated acids; discharge by oozing (Moore 1979: 194).

The Nebriinae includes five tribes, of which four are Laurasian; one, the monogeneric Notiokasiini Kavanaugh and Nègre 1983, is precinctive in the Neotropical Region.

## Pelophilini

This Holarctic tribe includes only *Pelophila* Mannerheim.

*Pelophila* Dejean 1821
A resident of the arctic and subarctic regions, the bispecific *Pelophila* in North America is represented by both species. They are transcontinental in the north, are hygrophilous, and live in marshes and on the borders of slow-moving streams. The species *P. borealis* Paykull 1790 is circumpolar; *P. rudis* LeConte 1863 is precinctive in the Nearctic Region. References. Kavanaugh 1996: 31-37 (phylogenetic relationships).

## Nebriini

This tribe, of five genera with 510 species, is Holarctic, and is also represented in the northern portion of the Oriental Region. The Nearctic assemblage includes 60 species in three genera.

*Nippononebria* Uéno 1955
A Holarctic genus of eight species, *Nippononebria* comprises two subgenera: the Palaearctic (Japanese) *Nippononebria (sensu stricto)*, and the Nearctic *Vancouveria* Kavanaugh 1995. Confined to the west, *Vancouveria* includes three species with a range extending in coastal areas from northern California to southwestern British Columbia, inland to the Sierra Nevada in the south and in the north to south-central British Columbia. References. Kavanaugh 1995: 153-160. (taxonomic treatment of subgenera; illus. of *N. (V.) virescens* (Horn 1870); notes about spp. distr.).

*Nebria* Latreille 1806
*Nebriola* Daniel 1903
*Paranebria* Jeannel 1937
*Neonebria* Hatch 1939
This genus ranges throughout the cooler portions of the Holarctic Region and in the northern, mountainous portion of the Oriental Region. In the Nearctic Region, represented by 53 species in four (or five) subgenera, *Nebria* ranges from the Arctic tundra (one species) southward in the Rocky Mountains to northern New Mexico and Arizona (White Mountains). Most of the species inhabit the mountains of the west; a few are transcontinental in the boreal forest, and a few species are restricted to the Appalachian region of the east. Adults of most of the species rest and hunt for prey under stones and rubble along the margins of cool or cold streams; a few inhabit wet places in forested areas, and rest under leaf litter; high montane species hunt for prey on snow fields, at night.

subgenus *Nebria (s. str.)*
This predominantly Palaearctic subgenus, with 38 species, is represented in the Nearctic Region by the introduced West Palaearctic *N. brevicollis* (Fabricius 1792). It is known only from St. Pierre and Miquelon, and Quebec, and may not be established (Lindroth 1961: 78).

subgenus *Boreonebria* Jeannel 1937
*gyllenhali* species group
Also predominantly Palaearctic, *Boreonebria* includes 40 species, of which nine are Nearctic. The geographical range in the Nearctic Region is transcontinental, predominantly in the north, but extending southward in the east to Virginia and Tennessee, in the Appalachians, and in the west, to northern California.

subgenus *Reductonebria* Shilenkov 1975
*gregaria* species group
Including 29 species, this subgenus is predominantly Nearctic, with 23 precinctive species. The geographical range of the Nearctic species is predominantly western, with only two species in the east: *N. pallipes* Say, ranging southward in the Appalachian Mountains to Georgia, and westward to Minnesota; and *N. suturalis* LeConte 1850, with isolated montane populations in the northeast. The latter species is also represented by populations isolated in the Rocky Mountains of Colorado, Wyoming and western Alberta.

subgenus *Catonebria* Shilenkov 1975
*metallica* species group
*gebleri* species group
Represented by 25 species, *Catonebria* is Holarctic, with 20 Nearctic precinctive species and five Palaearctic precinctive species. Principally western montane, the Nearctic members of this subgenus range collectively from the southern Yukon Territory south to New Mexico (at high altitude) and east not farther than Wyoming.

*Leistus* Frölich 1799
subgenus *Neoleistus* Erwin 1970
This genus, with 139 species in six subgenera, is predominantly Palaearctic- northern Oriental, with only one Nearctic precinctive subgenus (*Neoleistus* Erwin; but see Shilenkov 1999) of three species, restricted to the west (California north to southern coastal Alaska, and east to western Alberta). These species live in wet, shady places, usually under leaves in forests, near stream margins. The introduced Palaearctic *L. (s. str.) ferrugineus* (Linnaeus 1758) lives in Newfoundland. Reference. Erwin 1970: 111-119 (rev. N. Amer. precinctive spp).

## Opisthiini

This tribe, related to the Nebriini, is represented by only two genera: *Paropisthius* Casey 1920, known from southwestern China, northern India, Bhutan, Nepal and Taiwan; and the precinctive Nearctic *Opisthius* Kirby 1837. References. Bousquet and Smetana 1991:104-114. (larvae). Bousquet and Smetana 1996: 215-232 (rev. spp.).

*Opisthius* Kirby 1837
The single species, *O. richardsoni* Kirby 1837, is a resident of the boreal forest, and ranges from southern Alaska to northern California, and in Canada, eastward to Hudson's Bay. It lives along stream margins, on sandy-clay soil. Adults appear very much like large specimens of *Elaphrus*, and fly on warm days, if disturbed. References. Lindroth 1960: 30-42 (larva).

Notiophilini

This tribe comprises the single genus *Notiophilus*, and is represented in the Palaearctic, Oriental, Nearctic, and Neotropical Regions. For a taxonomic conspectus of the adjacent Central American Notiophilini, see Erwin (1991a).

*Notiophilus* Duméril 1806
Fifty-one species comprise this genus, of which 15 are in the Nearctic Region. Two additional species occupy Neotropical Middle America, ranging in the mountains from northern México to Costa Rica. Most species live in the mountains or in the boreal forest, but two species are on the Arctic tundra, and one species ranges in lowland habitats southward to southern Alabama, in eastern United States.

Most species are in conifer forests, on the ground, among the fallen needles and are found by disturbing the forest litter. Adults run quickly when disturbed, and their small size and smooth integument render them difficult to capture. Both larvae and adults prey on collembolans, the adults being diurnal and hunt using visual cues (Bauer 1979 and 1981).

The species *N. aquaticus* (Linnaeus 1758), is circumpolar. Two species were apparently accidentally introduced from western Europe: *N. biguttatus* (Fabricius 1779), known in North America from British Columbia in the west and from the Maritime provinces (and St. Pierre and Miquelon), and Maine and New Hampshire in the east; and *N. palustris* (Duftschmid 1812), known from Nova Scotia and Prince Edward Island. Although Lindroth (1961: 90-101) provided an excellent treatment of the Canadian-Alaskan taxa, knowledge of the genus in the Western Hemisphere would benefit from a general revision.

Subfamily V. Carabinae

The diagnostic adult features of this monotribal taxon are the following: size moderate to large (total length ca. 10-50 mm); head capsule ventrally without antennal grooves; antennal insertions laterad, each between base of mandible and anterior margin of eye; antenna filiform, antennomeres 2-4 not nodose, 2-6 without unusually long setae, with 11 antennomeres; labrum with six setae; mandible with scrobe asetose; terminal palpomeres various, from about cylindrical to distinctly securiform; pterothorax with mesonotal scutellum dorsally evident, not concealed beneath elytra; front coxal cavities open, middle coxal cavities disjunct-confluent, hind coxal cavities conjunct-confluent (Bell 1967: 105); elytra striate to smooth dorsally, extended to apex of abdominal sternum VIII, and apical margins subsinuate; metathoracic wings various, long (macropterous) to short (brachypterous); front tibia anisochaete; antenna cleaner sulcate or emarginate, grade A or B (Hlavac 1971: 57); abdomen with six (II-VII) pregenital sterna normally exposed, sterna III and IV separated by suture; male genitalia with parameres slender, about equal in size, asetose apically; ovipositor with gonocoxae rotated 90° in repose, with ramus, gonocoxa of two gonocoxites, gonocoxite 2 without accessory lobe; and reproductive tract with spermatheca 1, without spermatheca 2 (Liebherr and Will 1998: 128, Fig. 22). Secretions of pygidial glands saturated acids and aldehydes, or unsaturated acids; discharge by spraying (Moore 1979: 194).

Carabini

In a restricted sense, this tribe, which is worldwide in geographical distribution, includes ground beetle species with large adults of predatory habits, many being molluscivorous. Adults of many species are brightly colored, and some are of unusual form. The genera are arrayed in two subtribes, the bigeneric Calosomatina and the monogeneric Carabina (Prüser and Mossakowski 1998: 312). References. Lapouge 1929 192: 153 pp.; 1930, ibid. 192A: 155-291 (phylogeny of Carabini); 1931 192B: 293-580; 1932 192C: 581-747; 1953 192E, plates 2-10.

*Calosoma* Weber, 1801
Taken in the broad sense (Lindroth 1961: 53) this genus includes all of the species assigned to the subtribe Calosomatina. Prüser and Mossakowski (1998: 300) excluded only the monospecific *Aplothorax* Waterhouse 1843 from *Calosoma* (Basilewsky 1985: 265-266 placed this genus, which is precinctive on the South Atlantic island of St. Helena, in a separate subtribe). Gidaspow (1959: 231), in her treatment of the Nearctic calosomatines, recognized *Callisthenes* Fischer von Waldheim 1821 as a genus separate from *Calosoma*. We follow Lindroth, including all of the calosomatines in Holarctica, at least, in *Calosoma*.

Between genus and species, only one other rank (subgenus) was used by Gidaspow and by Lindroth (species group). Jeannel (1940) in contrast, who recognized 20 genera of calosomatines, classified the species using four ranks: [unspecified], phyletic series, genus, and subgenus. The latter, more complex, system, with its potential to show phylogenetic relationships more clearly, is the one that we would prefer to use, though with lower ranks. Unfortunately, some of the features used by Jeannel are more varied than he indicated, and so his system cannot be used without a re-analysis of said features. Nonetheless, because of Jeannel's work, such a re-analysis could be conducted relatively easily, though not without facing some interesting challenges. For the present, though, we use the system of Gidaspow, as modified by Lindroth, but with the sequence of groupings proposed by Jeannel.

The 170 species are arranged in 22 subgenera. Of these, the 41 species that live in the Nearctic Region are arrayed in 10 subgenera. Most of the species occur in the south and southwest. Adults and/or larvae of some of the species are climbers and eat caterpillars and pupae of tree-inhabiting moths, sawflies and other insects, eliminating substantial numbers of these defoliators. Adults of a number of species are attracted to light sources at night, on some evenings in great numbers. In the southwest, it is not uncommon to see at light several individuals consuming the crushed but still kicking remains of one of their less fortunate fellows. When disturbed, these beetles give off a distinctive, fetid odor. References. Burgess and Collins 1917 (way of life of North American spp. of *Calosoma* ). Gidaspow 1959 (rev. N.A. Mex. spp.). Jeannel 1940 (rev. of *Calosoma* [*sensu lato*] for the world). Erwin 1991a (rev. Centr. Amer. spp.).

subgenus *Calodrepa* Motschulsky 1865
*C. sycophanta* species group (in part), Lindroth 1961

This subgenus, precinctive in the Western Hemisphere, has four species in Nearctic North America. All have bright green elytra, the lateral margins red or golden green or not. The distribution is as follows: the West Indian *C. splendidum* Dejean 1831- Florida and Georgia, only; *C. wilcoxi* LeConte 1848 - southern Ontario southward to Texas and westward to California; *C. scrutator* (Fabricius 1775) throughout United States and southeastern Canada; and the Mexican *C. aurocinctum* Chaudoir 1850 in southern Texas. Adults of *C. scrutator* and *C. wilcoxi* climb trees in search of prey, whereas the larvae of these taxa remain on the ground.

subgenus *Calosoma (sensu stricto)*
*Callipara* Motschulsky 1865
*Syncalosoma* Breuning 1927
*C. sycophanta* species group (in part), Lindroth 1961

Two species represent this Holarctic subgenus in North America: the Palaearctic indigenous *C. sycophanta* (Linnaeus 1758), with metallic green elytra and dark blue pronotum; and the black Nearctic precinctive *C. frigidum* Kirby 1837. The former species was introduced into Massachusetts and other New England states as a potential predator of the gypsy moth. (For details, see Burgess and Collins 1917.) Attempts were also made to establish colonies in California, Washington, and British Columbia, but with no success, for the only area where the species has persisted is the New England states. Both larvae and adults climb trees, and this habit accounts to a considerable extent for the value of this beetle in reducing populations of the gypsy moth. Adults may live as long as four years, some individuals hibernating over one complete season.

*Calosoma frigidum* is a transcontinental species of the southern portions of the boreal forest. It ranges southward to Nebraska and to New Jersey on the east coast. Adults climb trees, but the larvae remain on the ground.

subgenus *Castrida* Motschulsky 1865
*Callistriga* Motschulsky 1865
*Calamata* Motschulsky 1865
*Acampalita* Lapouge 1929
*Catastriga* Lapouge 1929
*C. sayi* species group Bousquet and Larochelle 1993

One Nearctic species, *C. sayi* Dejean 1826, is one of 10 included in this precinctive Western Hemisphere subgenus. This species ranges from the Greater Antillean West Indies and Guatemala (Gidaspow 1963: 301) northward to Iowa and southern New York. Adults of *C. sayi*, capable of climbing, hunt caterpillars in low shrubbery and herbage.

subgenus *Chrysostigma* Kirby 1837
*Lyperostenia* Lapouge 1929
*C. calidum* species group Lindroth 1961

This precinctive Western Hemisphere subgenus ranges from Brazil to southern Canada. In Nearctic North America, it is represented by 10 species, ranging collectively from the highlands of Panamá northward to southern Canada. The fiery hunter, *Calosoma calidum* (Fabricius 1775) is incapable of climbing, and thus both larvae and adults confine their hunting activities to the ground.

subgenus *Tapinosthenes* Kolbe 1895
*C. cancellatum* species group Lindroth 1961

This monospecific precinctive Nearctic subgenus includes *C. cancellatum* Eschscholtz 1833, a species of western North America that ranges from California northward to southeastern British Columbia and eastward to Arizona and Montana. Included in *Chrysostigma* by Jeannel (1940: 106) and Gidaspow (1959: 262), we accept Lindroth's (1961: 52) opinion that this species is sufficiently distinctive to be isolated in a subgenus of its own.

subgenus *Microcallisthenes* Apfelbeck 1918
*Isotenia* Lapouge 1929
*Callistenia* Lapouge 1929
*C. moniliatum* species group Lindroth 1961

This Holarctic subgenus seems to be related to two Palaearctic subgenera: *Callisthenes* Fischer von Waldheim 1821, and *Teratexis* Semenov and Znojko 1933 (Gidaspow 1959: 301). The 14 Nearctic species comprising this subgenus live in the western part of the United States and southwestern Canada and are not found in the east or in Mexico. These species are ground-dwelling xerophiles, and range from prairie habitat at low altitudes (*C. luxatum* Say 1823) to dry woodland at higher elevations (*C. moniliatum* LeConte 1852). Adults of *Microcallisthenes* are found most commonly in spring and early summer.

subgenus *Camedula* Motschulsky 1865
*Acamegonia* Lapouge 1924
*C. peregrinator* species group (in part) Bousquet and Larochelle 1993

Of the four species which represent this precinctive Western Hemisphere subgenus, three occur in Nearctic North America. They are confined to the southwest, ranging eastward to Texas and northward to Colorado. Adults of one species, *C. peregrinator* Guérin-Méneville 1844, climb trees.

subgenus *Camegonia* Lapouge 1924
*C. prominens* species group Lindroth 1961

This Western Hemisphere precinctive subgenus is represented in Nearctic North America by three species all of which are confined to the southern plains area (Nebraska and southward) and to the southwest. Neither adults nor larvae of *C. marginale* Casey 1897 (one of the species of *Camegonia*) climb vegetation in search of prey.

subgenus *Carabosoma* Géhin 1885

*C. angulatum* species group Bousquet and Larochelle 1993

This Western Hemisphere precinctive subgenus contains a single species, *C. angulatum* Chevrolat 1834, which ranges from South America northward to southwestern United States, from California to Texas.

subgenus *Callitropa* Motschulsky 1865

*Paratropa* Lapouge 1929

*C. externum* species group Lindroth 1961

Three species comprise this Nearctic precinctive subgenus, all of which occur in Nearctic North America: *C. externum* (Say 1823) is eastern and ranges northward to southern Ontario and Massachusetts; two are southwestern, and one of these (*C. macrum* LeConte 1853) ranges eastward to Louisiana. Adults of *C. externum* climb trees in search of prey.

subgenus *Blaptosoma* Géhin 1876

*Aulacopterum* Géhin 1885

*C. haydeni* species group Bousquet and Larochelle 1993

This Western Hemisphere precinctive subgenus comprises seven species. Of these, *C. haydeni* Horn 1870, ranges northward to Colorado, and westward to Arizona.

*Carabus* Linnaeus 1758

The Holarctic genus *Carabus* includes 715 species (Deuve 1994), arrayed in 86 subgenera, which, in turn, comprise five "fundamental groups" (l.c., p. 60). We have elected Deuve's arrangement here, though not without some reservation. The Nearctic carabid fauna includes only 14 species, three of which were introduced through the agency of man. The species are arranged in nine subgenera, and collectively represent only three of Deuve's fundamental groups: the Digitulati, including subgenus *Carabus (sensu stricto)*; the Archicarabomorphi, including *Archicarabus* Seidlitz 1887; and the Lobifera, including *Autocarabus* Seidlitz 1887, *Hemicarabus* Géhin 1876, *Homoeocarabus* Reitter 1896, *Tomocarabus* Reitter 1896, *Aulonocarabus* 1896, *Tanaocarabus* Reitter *1*896, and *Megodontus* Solier 1848.

Courtship patterns have been observed for some species. Egg laying is in cavities in the soil, made by the ovipositor. References. Deuve 1994. Van Dyke 1944b: 87-137 (rev. N.A. species).

subgenus *Carabus (sensu stricto)*

*Lichnocarabus* Reitter 1896

*Neocarabus* Hatch 1949 (not Bengtsson 1927; not Lapouge 1931)

*granulatus* species group, Lindroth 1961

The subgenus *Carabus* is Holarctic in distribution, including 46 species. Three species are in the Nearctic Region. Two species, *C. goryi* Dejean and *C. vinctus* Walker 1801, are precinctive, living in the east, and ranging from northern Florida to southern Ontario. They are sympatric over much of their range. Both species inhabit wet woodlands. *C. granulatus* Linnaeus 1758 is an introduced Palaearctic species, which is essentially transcontinental in Canada, occurring also in the state of Washington. A synanthrope, living near water in open country or open woodland, this species is represented by two subspecies: *C. g. granulatus*, which ranges from the west coast eastward to Quebec, and *C. g. hibernicus* Lindroth, with a range in the east, including Québec, the Maritime provinces, and the islands of St. Pierre and Miquelon (Lindroth 1961a: 36-37).

subgenus *Archicarabus* Seidlitz 1887

*granulatus* species group (in part), Lindroth 1961

*nemoralis* species group, Bousquet and Larochelle 1993

This subgenus includes 10 western Palaearctic species, with a collective geographical range from Western Europe to Asia Minor. A single synanthropic species, *C. nemoralis* Müller 1764, represents *Archicarabus* in North America. An introduced species, *C. nemoralis* is transcontinental in Canada, ranging south to California in the west, and to Virginia in the east. This species is common about human habitations, and in the spring of the year, dead, squashed specimens are frequently seen on sidewalks. In the coastal areas of British Columbia *C. nemoralis* is an important predator on slugs (Glendenning 1952: 5).

subgenus *Autocarabus* Seidlitz 1887

*auratus* species group, Bousquet and Larochelle 1993

With four species, this subgenus is Palaearctic, ranging from Siberia to western Europe. Two species were introduced to North America as potential predators of the gypsy moth. For further details, see Smith (1959: 7-10). Of these two species, *C. cancellatus* Illiger 1798 did not become established (Bousquet and Larochelle 1993: 77); *C. auratus* Linnaeus 1758 is known presently in North America from the New England states only (Bell 1957: 254).

subgenus *Hemicarabus* Géhin 1876

*serratus* species group, Lindroth 1961

This Holarctic subgenus, with four species in eastern Asia and Siberia, includes the single precinctive Nearctic species *C. serratus* Say 1823, which is a resident of the southern portions of the boreal forest and the northern plains. Transamerican in southern Canada and northern United States, this species ranges southward in the Rocky Mountains to New Mexico; southward in the east to North Carolina, in the mountains, and to Kansas in the plains. Most specimens are found in the vicinity of lakes and streams.

subgenus *Homoeocarabus* Reitter 1896

*Paracarabus* Lapouge (in part) 1931

*granulatus* species group (in part), Lindroth 1961

This subgenus contains the single Holarctic species, *C. maeander* Fischer von Waldheim 1822. It is a cool-temperate species, living mainly in the northern prairie regions of North America, in the vicinity of sloughs and small lakes. Its range also extends northward into the southern reaches of the boreal forest and eastward to Newfoundland and Maine. Specimens are taken by treading along the edges of *Carex* marshes.

subgenus *Tomocarabus* Reitter 1896
*Cryocarabus* Lapouge 1930
*Eucarabus* Géhin 1876
*Neocarabus* Lapouge (in part) 1930
*granulatus* species group (in part), Lindroth 1961
*taedatus* species group (in part), Lindroth 1961
Twenty species are included in this Holarctic subgenus, many of which are east Asian. Two species are precinctive Nearctic: *C. taedatus* Fabricius 1787, a widespread species of the boreal forest and Aleutian Islands and the Arctic-subarctic *C. chamissonis* Fischer von Waldheim 1822.

subgenus *Aulonocarabus* Reitter 1896
*Diocarabus* Reitter 1896
*taedatus* species group (in part), Lindroth 1961
*truncaticollis* species group, Bousquet and Larochelle 1993
This subgenus includes 13 species, 12 of which are east Asian-east Siberian in geographical distribution. One species, *C. truncaticollis* Eschscholtz 1833, is Holarctic, lives on dry tundra, and in North America is known only from the west: some of the islands of the Bering Sea, and on the mainland, ranging from Alaska eastward to the Anderson River, Northwest Territories.

subgenus *Tanaocarabus* Reitter 1896
*sylvosus* species group, Bousquet and Larochelle 1993
This Nearctic precinctive subgenus includes four species, one of which, *C. hendrichsi* Bolivar, Rotger and Coronado 1967 is known only from the Sierra Madre Oriental, in northeastern Mexico. Its close relative, *C. forreri* Bates 1882, ranges from Durango, in the Mexican Sierra Madre Occidental, northward to the mountains of southern Arizona. Two species, *C. sylvosus* Say 1823 and *C. finitimus* Haldeman 1852, are eastern in the United States; *C. sylvosus*, with most of its range east of the Mississippi River from Florida to Québec, and westward to Minnesota and Texas; and *C. finitimus*, with its range in eastern Texas and Oklahoma. All four of these species are inhabitants of forests.

subgenus *Megodontus* Solier 1848
*vietinghoffi* species group, Lindroth 1961
A widespread group in the Palaearctic Region, *Megodontus* includes the Holarctic species *C. vietinghoffi* Adams 1812, which lives in Arctic and subarctic North America from western Alaska eastward to Bathurst Inlet, Northwest Territories, and southward in the Mackenzie Basin to Fort Norman, N.W.T.

Cychrini

This tribe includes four genera: two are Nearctic precinctive, one is Holarctic, and one (*Cychropsis* Boileau 1901) is confined to the mountains of southern China, Tibet, and the Himalaya in Sikkim (Deuve 1997: 184-185).

The Cychrini are molluscivorous, and the head and mouthparts of the adult are particularly adapted for entering the aperture of a snail shell. The beetles inhabit cool, damp woods and ravines, where, especially at night, they run on the ground, or climb about on logs and tree trunks. References. Lapouge 1932: 580-747 (rev. tribe).

*Sphaeroderus* Dejean 1826
This group comprises 10 taxa grouped in six species, whose aggregate range extends from northern Newfoundland to the northern portion of Georgia, and westward to the eastern portion of the Mississippi Basin (Iowa, Minnesota, and Manitoba) and eastern Saskatchewan. This genus requires study, both to establish clear species limits and phylogenetic relationships.

*Cychrus* Fabricius 1794
This cool-temperature Holarctic genus includes two very similar, allopatric, Nearctic species, which range from southern British Columbia southward to Oregon and southeastern Idaho, and farther east to western Wyoming. Reference. Gidaspow 1973: 89-98.

*Scaphinotus* Dejean 1826
Based on a broad concept of this genus, Van Dyke (1936 and 1944a) included in *Scaphinotus* all of the subgenera treated below. His arrangement was accepted by Lindroth (1961a) and Gidaspow (1968). Bell (1959: 11-13) treated *Irichroa* Newman 1838, as consubgeneric with *Scaphinotus* (*sensu stricto*), as did Ball (1966a). Barr (1969: 75) accepted most of this, at least implicitly, but he treated *Maronetus* as generically distinct.

subgenus *Nomaretus* LeConte 1853
Five species are included here. They inhabit the Mississippi Basin in the midwest and eastern Texas; northward, the range extends to the north shore of the Great Lakes region, and eastward to the northeastern states and the adjacent portion of Canada. References. Allen and Carlton 1988: 132. Gidaspow 1973: 78-88.

subgenus *Pseudonomaretus* Roeschke 1907
This group includes four species which collectively occupy the mountains of eastern Oregon, eastern Washington, Idaho, western Montana, southeastern British Columbia, and western Alberta, on the eastern slopes of the Rocky Mountains. Reference. Gidaspow 1973: 73-78.

subgenus *Maronetus* Casey 1914
The seven species (one with two subspecies) of this subgenus are restricted to the mountains of eastern United States, from

southern Pennsylvania to northern Georgia. References. Schwarz 1895: 269-273 (rev. *Nomaretus*; following are *Maronetus*: *S. imperfectus* (Horn 1860); *S. hubbardi* (Schwarz 1895); *S. incompletus* (Schwarz 1895); and *S. debilis* (LeConte 1853)).

subgenus *Scaphinotus* (*sensu stricto*)
*Irichroa* Newman 1838
*Megaliridia* Casey 1920

This group comprises 25 taxa arrayed in nine species, distributed as follows: two, mountains of northern Mexico; 10, mountains of New Mexico and southern Arizona; seven, Appalachian mountain system and adjacent lowlands, northward to southern Québec; one, *S. elevatus* (Fabricius 1787), with six subspecies, which occupy lowland as well as montane habitats, ranging from the east coast westward to the eastern slopes of the Rocky Mountains, and northward in the Mississippi Basin to southern Manitoba. Except *S. elevatus neomexicanus* Van Dyke 1924, the other taxa occurring in New Mexico, Arizona and the Sierra Madre Occidental, south to Durango, México, are all probably conspecific. References. Ball 1966a, 92: 687-722 (rev. *S. petersi* species group). Bell 1959, 11-13 (new sp., and notes on classification).Van Dyke 1938: 93-133 (rev. spp.).

subgenus *Steniridia* Casey 1924

This subgenus comprises 21 taxa in seven species, which are confined to the Appalachian Mountain system, and the adjacent lowlands. References. Barr 1969: 74-75. Valentine 1935: 341-375 (rev. spp.), 1936, 223- 234 (subspecies of *S. andrewsii*).

subgenus *Stenocantharus* Gistel 1857
*Pemphus* Motschulsky 1866

The three species of this subgenus are restricted to the mountain forests of the Pacific Northwest, and range collectively from northern California to southeastern coastal Alaska, and inland to southeastern British Columbia (Creston). The most widespread species of the three is *S. angusticollis* (Fischer von Waldheim 1823), adults of which frequent rotten logs, and emit a vile-smelling fluid from the pygidial glands, when disturbed. Reference. Gidaspow 1973: 63-73.

subgenus *Brennus* Motschulsky 1865

The 15 species (one with two subspecies) of this subgenus live in the mountain forests and adjacent lowlands of the Pacific Coast, from southern California northward to the Aleutian Islands (where one species, *S. marginatus* (Fischer von Waldheim 1822), lives on the tundra); and inland to central Alberta. References. Gidaspow 1968 (rev. spp.).

subgenus *Neocychrus* Roeschke 1907

This subgenus comprises three species which live in montane forests from northern California to southwestern British Columbia. Reference.Gidaspow 1973: 57-63 (rev. spp.).

## Subfamily VI. Cicindelinae (Latreille 1804)

Collyridae Hope 1838
Ctenostomidae Castelnau 1835
Mantichoridae Castelnau 1835
Megacephalinae Castelnau 1835

**Common name**: The tiger beetles

The diagnostic adult features of this subfamily are the following: size moderate to large (total length 7-70 mm); head capsule ventrally without antennal grooves; antennal insertions mediad, above and closer together than bases of mandibles; antenna filiform, antennomeres 2-4 not nodose, 2-6 without unusually long setae, with 11 antennomeres; labrum with six to twelve setae; mandible with scrobe multisetose or asetose, occlusal margin with two to four terebral teeth, retinacular tooth very large, molariform; terminal palpomeres about cylindrical; pterothorax with mesonotal scutellum dorsally visible, not concealed beneath elytra; front coxal cavities open, middle and hind coxal cavities disjunct-confluent (Bell 1967: 105); elytra smooth, not striate dorsally, extended to apex of abdominal tergum VIII, and apical margins subsinuate; metathoracic wings large to variously reduced (Ward 1979); front tibia anisochaete, antenna cleaner sulcate, type A (Hlavac 1971: 57); abdomen with six (II-VII) pregenital sterna normally exposed, sterna III and IV separated by suture; male genitalia with parameres but similar to one another in form (slender) and in length, asetose, joined medio-dorsally to narrow jugum; ovipositor with gonocoxae rotated in retracted position, with or without ramus, gonocoxa with two gonocoxites, and gonocoxite 2 with accessory lobe, not articulated or articulated; reproductive tract with spermatheca 2 only (Liebherr and Will 1998: 128-129, Figs. 23-24); secretions of pygidial glands aromatic aldehydes, discharge by oozing (Moore 1979: 194). Distinctive features of larvae are large prothorax with pronotum semicircular; abdominal tergum V with two or three pairs of sclerotized, hook-like projections; and abdominal segment IX without urogomphi.

Although many authors have treated the tiger beetles as a family separate from the Carabidae, many workers accept a close relationship of this group to the carabid tribe Carabini. It is more a matter of personal opinion whether the group should or shouldn't be recognized as a separate family; there are no set rules for the determination of family rank other than applying cladistic analysis to seek monophyletic relationships. Consistent treatment of the families of beetles requires that these beetles be placed with the rest of the ground beetles, with the group ranked as a subfamily.

Recent studies summarized by Vogler and Barraclough (1998) indicate that the widely used classification (W. Horn 1908-1915) of tiger beetles is not a satisfactory representation of relationships as inferred from studies of molecular and larval features. A formal replacement system based on this evidence has not been proposed. For present purposes, and emerging from our interpretation of the analysis of Vogler and

Barraclough (1998: 255, Fig. 1, and 256, Fig. 2), we recognize four tribes: Omini, Manticorini, Megacephalini, and Cicindelini.

This subfamily, with more than 2100 species arranged conservatively in 61 genera and four tribes, is worldwide in distribution. About half the total number of species belong to the genus *Cicindela (sensu latissime)* Linnaeus 1758.

**Habits and habitats.** For an incisive and perceptive review of this topic, see Pearson (1988 and 1999). Tiger beetles are predacious as adults and larvae. Further details are provided below, for the adults of each tribe that is represented in the Nearctic Region. The larvae live in a tunnel they have dug into the earth. They hold themselves in the vertical part of the tunnel with the aid of the hooks on the dorsal side of the fifth abdominal segment. When suitable prey is near, they seize the organism, usually some small insect, and pull it into the lair. The abdominal hooks also serve to hold the larvae in the tunnel if the prey is too large for them.

**The Nearctic tiger beetles.** The 111 species are included in three tribes and four genera: Omini (two genera), Megacephalini (one genus), and Cicindelini (one genus).

**Status of the taxa.** Most genera are readily identifiable, provided that one adopts a broad definition of the genus *Cicindela*. At the species and subspecies level, much remains to be done before taxonomic stability is achieved. Range maps are available for each species, as well as a list of regional distributional and taxonomic publications useful for identification of tiger beetles (primarily adults) (Pearson *et al.* 1997: 33-84). See also Hamilton (1925) for a taxonomic treatment of larvae.

### Omini

This tribe, known from the Western Hemisphere only, contains 12 species in three genera. The range is discontinuous, with *Omus* Eschscholtz 1829 known only from far western United States and Canada, *Amblycheila* Say 1830, more centrally located in the United States and northern México, and *Picnochile* Motschulsky 1856, in the southern temperate part of the Neotropical Region, far removed from the Nearctic genera.

The adult beetles are black, relatively small-eyed, and brachypterous, and hence flightless.

*Amblycheila* Say 1830
This indigenous Nearctic genus includes seven species which range from South Dakota to San Luis Potosi, in Mexico, west to California. Five of these species inhabit Nearctic North America. Adults are crepuscular or nocturnal, and rest in holes made by themselves or by other animals. Reference. Sumlin 1991: 1-16 (rev. Nearctic spp.).

*Omus* Eschscholtz 1829
*Leptomus* Casey 1914
*Megomus* Casey 1914
Five species belong to this Nearctic genus. They are restricted to the Pacific Coast from southern British Columbia south to California. As for the preceding group, adults are crepuscular or nocturnal. Adults are found during the day under cover in fields and open forests. The genus is taxonomically difficult and agreement about the actual number of species has not been achieved. Reference. Leffler 1985: 37-50 (rev. spp.).

### Megacephalini

*Megacephala* Latreille 1802
This genus is represented by some 98 species inhabiting the Western Hemisphere and the Australian, Afrotropical and southern part of the Palaearctic Regions. Only three species occur in the Nearctic Region. They range over most of the United States except the northernmost states. Adults are crepuscular or nocturnal in habits, and many are attracted to light sources. Reference. Knisley and Schultz 1997: 96-98 (S.E. spp.).

### Cicindelini

This tribe (which includes the collyridines, or collyrines) of slightly more than 2000 species in 54 genera is represented in all zoogeographical regions. Adults of most species are brilliantly colored, very active, predators. The large eyes, long setose legs, brilliant colors, characteristic shape (Fig. 2.6), high speed, and nervous activity are characteristic of this group.

Adults of many species run very rapidly over the ground, particularly along sandy banks of streams, along roadways and paths and other such sunny places. Adults of such species fly readily if disturbed and in spite of their brilliant metallic colors, they are difficult to see because of their extreme rapidity, both running and flying.

Some exotic species of *Cicindela* live in termite nests. The tropical collyridines are arboreal, adults searching out their prey among the branches of trees and bushes, larvae living in burrows in twigs and tree trunks. Adults of the Neotropical *Oxycheila* and *Cheiloxya* live along water courses, and enter the flowing water to capture prey or to avoid predators (Pearson 1999: 58-59). They are nocturnal. Unlike most ground beetles, a net is usually necessary to collect cicindeline adults, although adults of desert inhabiting species under some circumstances can be collected by hand in abundance.

*Cicindela* Linnaeus 1758
This genus, with slightly more than 1000 species in 85 subgenera (many ranked by other authors as genera, see Lorenz 1998: 39-60), ranges through all major zoogeographical regions of the world. In the Nearctic Region, the genus contains 98 species and many subspecies which as a whole, range all over the boreal and temperate regions of Canada and United States. The Nearctic species are arrayed in 10 subgenera. A key to these subgenera is not included here because such key is non-existent and identifications rely mostly on character states of the internal sac of the male genitalia (see Rivalier 1954). Such characters are not practical for routine identifications.

Adults of most species are brillantly colored; most are diurnal and live in sandy habitats, many close to bodies of water. They fly rapidly when disturbed.

Knisley and Schultz (1997: 13-88) provide a wealth of information about development and way of life of the species of this genus. References. Knisley and Schultz 1997: 98-137 (S.E. species, U.S.A.). Leonard and Bell 1999 (N.E. species, U.S.A.). Leng 1902: 93-186 (rev. spp.). Wallis 1961: 11-68 (Can. spp.). Willis 1968: 303-317 (key N.A. spp.). See also Pearson *et al.* (1997: 33-84).

## Subfamily VII. Loricerinae

The diagnostic adult features of this subfamily are the following: size moderate (total length ca. 6-9 mm); head capsule ventrally without antennal grooves; antennal insertions laterad, each between base of mandible and anterior margin of eye; antennomeres 2-4 nodose, 2-6 each with several long setae directed antero-ventrally; labrum with six setae; mandible without seta in scrobe; terminal palpomeres about cylindrical; pterothorax with mesonotal scutellum dorsally visible, not concealed beneath elytra; front coxal cavities closed by insertion of tip of proepimeron each side in cavity of lateral projection of prosternum, middle and hind coxal cavities disjunct-confluent (Bell 1967: 105); elytra striate dorsally, extended to apex of abdominal sternum VIII, and apical margins subsinuate; metathoracic wings large to variously reduced; front tibia anisochaete, antenna cleaner sulcate, grade B (Hlavac 1971: 57); abdomen with six (II-VII) pregenital sterna normally exposed, sterna III and IV separated by suture; male genitalia with parameres similar in form, moderately broad, and moderately long, asetose; ovipositor with gonocoxae rotated 90° in repose, without ramus, gonocoxa with two articles, gonocoxite 2 without accessory lobe; and reproductive tract with spermatheca 2 only (Liebherr and Will 1998: 169); secretions of pygidial glands higher saturated acids, discharge by oozing (Moore 1979: 194).

The subfamily Loricerinae is monotribal.

### Loricerini

This tribe contains only the Holarctic-Oriental genus *Loricera* Latreille 1802.

*Loricera* Latreille 1802
*Lorocera* Bedel 1878

The 13 species of *Loricera* are arranged in three subgenera: the monospecific *Elliptosoma* Wollaston 1854, from Madeira; the monospecific *Plesioloricera* Sciaky and Facchini 1999; and *Loricera* (s. str.), which occupies the rest of the range of the genus. Three species represent this subgenus in the Nearctic Region, and two more are known from the mountains of Middle America, from northern México south to Guatemala. One species, *L. pilicornis* Fabricius 1775, is circumpolar and transamerican, including the Aleutian Islands, but apparently does not live on the tundra of the Arctic; two species (*L. decempunctata* and *L. foveata* LeConte 1851) are restricted to the west. Of these, *L. decempunctata* Eschscholtz 1829 ranges northward to southwestern Alaska and Kodiak Island, and eastward to the eastern slopes of the Rocky Mountains, in Alberta. Reference. Ball and Erwin 1969, 47: 877-907 (key to spp.; descr. of larva). Erwin 1991a (Cent. Amer. spp.)

## Subfamily VIII. Elaphrinae

The diagnostic adult features of this subfamily are the following: size moderate to large (total length ca. 6-18 mm); head capsule ventrally without antennal grooves; antennal insertions laterad, each between base of mandible and anterior margin of eye; antenna filiform, antennomeres 2-4 not nodose, 2-6 without unusually long setae, antennomeres 11; labrum with six setae; mandible with seta in scrobe; terminal palpomeres about cylindrical; pterothorax with mesonotal scutellum dorsally visible, not concealed beneath elytra; front coxal cavities closed by insertion of tip of proepimeron each side in cavity of lateral projection of prosternum, middle and hind coxal cavities disjunct-confluent (Bell 1967: 105); elytra extended to apex of abdominal sternum VIII, with apical margins subsinuate, striate or not dorsally; metathoracic wings large; front tibia anisochaete, antenna cleaner sulcate, grade B (Hlavac 1971: 57); abdomen with six (II-VII) pregenital sterna normally exposed, sterna III and IV separated by suture; male genitalia with parameres similar in form, and both moderately long, setose preapically and apically; ovipositor with gonocoxae rotated 90° in repose, without ramus, gonocoxa with two articles, gonocoxite 2 without accessory lobe; and reproductive tract without spermatheca 1, with spermatheca 2 (Liebherr and Will 1998: 169); secretions of pygidial glands higher saturated acids, discharge by oozing (Moore 1979: 194).

The Elaphrinae is monotribal.

### Elaphrini

This tribe, with 50 species arranged in three genera, is Holarctic. Stridulation by adults is accomplished by rubbing a fine file, located on the medial margin of the elytral epipleuron, against a row of fine tubercles on the elytral tergum VII (Andrewes 1929; Lindroth 1954).

*Diacheila* Motschulsky 1844

This genus includes three species; in distribution, one is Palaearctic; and two are Holarctic. In the Nearctic Region, the geographical range extends on the Arctic tundra from Alaska to Labrador and south into the northern forest zone. Adults live at the margins of small lakes or ponds, or on dry tundra.

*Blethisa* Bonelli 1810

This Holarctic genus includes eight species of which six inhabit the Nearctic Region. The geographical range extends from the Arctic tundra and boreal Canada, southward to Oregon in the west, and eastward to the north central and northeastern states.

Two species, *B. multipunctata aurata* Fischer von Waldheim 1828 and *B. catenaria* Brown 1944, are Holarctic. Adults and larvae are markedly hygrophilous and live in marshes and bogs. References. Lindroth 1954: 10-20 (rev. spp.). Goulet and Smetana 1983: 552-555 (key to spp.).

*Elaphrus* Fabricius 1775
This distinctive genus includes 39 species distributed over the Palaearctic and Nearctic Regions. Nineteen species inhabit the Nearctic Region, where they live in arctic, boreal, and temperate habitats. The species are arranged in four subgenera, all represented in North America. The species are hygrophilous, either riparian or marsh-inhabiting; adults run with agility in the sunshine. References. Goulet 1983: 238-340 (rev. spp.).

subgenus *Arctelaphrus* Semenov 1926
The sole species in this subgenus, *E. lapponicus* Gyllenhal 1810, is represented by two subspecies. One of them is confined to Alaska and the other is Holarctic in distribution, being found in the Nearctic Region in subarctic areas from Alaska to Labrador.

subgenus *Neoelaphrus* Hatch 1951
*Elaphrus* Semenov (not Fabricius 1775)
This subgenus contains 15 species in the Palaearctic and Nearctic Regions. Six species inhabit North America where they are found over most of the United States and Canada, as far north as the tree line.

subgenus *Elaphrus s. str.*
This subgenus is represented by 18 species of which 11 occur in the Nearctic Region. They range in North America over most of United States and Canada, including the arctic regions. Two species are Holarctic.

subgenus *Elaphroterus* Semenov 1895
Five species belong to this Holarctic subgenus. Two species are in the Nearctic Region; one is precinctive in western North America and the other is Holarctic, ranging in North America from Alaska to the Mackenzie Delta in the Northwest Territories.

## Subfamily IX. Promecognathinae

The diagnostic adult features of this monotribal taxon are the following: size moderate (total length ca. 8-15 mm); head capsule ventrally without antennal grooves; antennal insertions laterad, each between base of mandible and anterior margin of eye; antenna filiform, antennomeres 2-4 not nodose, 2-6 without unusually long setae, antennomeres 11; labrum with four setae; mandibles protruded, with scrobes asetose, left mandible with occlusal margin with one terebral (or supraterebral) tooth located about midway between base of terebra and apical incisor, retinacular tooth about size of terebral tooth, not enlarged and molariform; terminal palpomeres with apices slightly expanded, truncate, or maxillary palpus with terminal palpomere distinctly securiform; body pedunculate, with mesothorax and extreme base of elytra markedly constricted; pterothorax with mesonotal scutellum dorsally evident, not concealed beneath elytra; front coxal cavities closed by projection of proepimeron into cavity each side of lateral projection of prosternum, middle coxal cavities disjunct-confluent, hind coxal cavities conjunct-confluent (Bell 1967: 106); elytra smooth dorsally, extended to apex of abdominal tergum VIII, and apical margins subsinuate; metathoracic wings short (brachypterous); front tibia anisochaete, antenna cleaner sulcate, grade B (Hlavac 1971: 57); abdomen with six (II-VII) pregenital sterna normally exposed, sterna III and IV separated by suture; male genitalia with parameres asymmetric, left much broader than right, about equal in length, setose apically; ovipositor with gonocoxae rotated 90° in repose, with ramus, gonocoxa of two gonocoxites, gonocoxite 2 without accessory lobe; and reproductive tract with spermatheca 1 without spermatheca 2 (Liebherr and Will 1998: 127, Fig. 19); secretions of pygidial glands not studied.

### Promecognathini

Nine species included in six genera and two subtribes, comprise this tribe. The Afrotropical subtribe Palaeoaxinidiina includes only the monospecific *Palaeoaxinidium*, based on a single Cretaceous age fossil (McKay 1991). The subtribe Promecognathina includes the extant taxa, which are arranged in two genus-groups: the precinctive Afrotropical *Axinidium* genus group, with four genera, which are known from the Union of South Africa only (Basilewsky 1963); and the *Promecognathus* genus group, with the single genus, *Promecognathus* Chaudoir 1846, which is precinctive in the Nearctic Region.

*Promecognathus* Chaudoir 1846
This genus contains two species *P. laevissimus* (Dejean 1829), and *P. crassus* LeConte 1868, both of which live in the forests of the west coast, their collective range extending from Nevada and California northward to southwestern British Columbia (including Vancouver Island). Labonte (1983: 681-682) discovered that members of this genus are specialized predators on millipedes. Reference. Bousquet and Smetana 1986, 11: 25-31 (descr. of larva).

## Subfamily X. Scaritinae

The diagnostic adult features of this taxon are the following: size small to large (total length ca. 2-30 mm); head capsule ventrally without antennal grooves, laterally with or without antennal grooves; antennal insertions laterad, each between base of mandible and anterior margin of eye; antenna filiform or antennomeres 4-11 quadrate, antennomeres 2-4 not nodose, 2-6 without unusually long setae, antennomeres 11; labrum with six setae; mandible with scrobe unisetose, occlusal margin with one terebral tooth, retinacular tooth about size of terebral tooth, not enlarged and molariform; terminal palpomeres various, securiform or not; body pedunculate, with mesothorax and extreme base of elytra markedly constricted; pterothorax with mesonotal scutellum dorsally evident, not concealed beneath elytra; front coxal cavities closed by projection of proepimeron into cavity each side of lateral projection of prosternum, middle and hind coxal cavities disjunct-confluent (Bell 1967: 105); elytra extended to apex of abdominal tergum VIII, apical margins subsinuate; dorsal surface striate or smooth; metathoracic wings long (macropterous) or short (brachypterous); front tibia anisochaete, antenna cleaner sulcate, grade B (Hlavac 1971: 57); abdomen with six (II-VII) pregenital sterna normally exposed, sterna III and IV separated by suture; male genitalia with parameres similar in form and size, setose or asetose apically; ovipositor with gonocoxae rotated 90° in repose, with or without ramus, gonocoxa of one or two gonocoxites, without accessory lobe; reproductive tract with spermatheca 1, without spermatheca 2 (Liebherr and Will 1998: 128, Fig. 22); secretions of pygidial glands higher saturated or unsaturated acids (Scaritini), or quinones or aliphatic ketones (Clivinini) (Moore 1979: 194).

This subfamily includes two tribes: Clivinini and Scaritini. Reference. Nichols 1988 (rev. West Indian taxa).

### Clivinini

With about 70 genera, this tribe constitutes a large group represented in all the major zoogeographical regions of the world. More than 110 species currently placed in 17 supraspecific taxa inhabit the Nearctic Region.

*Oxydrepanus* Putzeys 1866
This genus of small species is precinctive in the Western Hemisphere and includes eight described species ranging from northern South America and the West Indies through México to Florida, in the Nearctic Region. One species only, *O. rufus* (Putzeys 1846), is in the Nearctic Region, in southern Florida. The habitat requirement of this species is inadequately known though some specimens have been found in floating vegetation in pools and in hammock forest litter.

*Halocoryza* Alluaud 1919
This genus of small beetles is closely related to the following one. It contains three species with a geographical range extending from islands of the western part of the Indian Ocean, to the southern part of the Red Sea, and western Africa in the Old World, and the West Indies, Mexico, and southeastern United States in the Western Hemisphere. The sole species present in the Nearctic Region, *H. arenaria* (Darlington 1939), lives in southern Florida. The species has been recorded also from the West Indies, Yucatan Peninsula, and West Africa. The species are halobiontic, with adults being found on sea beaches, above or below the tide line. References. Bruneau de Miré 1979: 328. Whitehead 1966. 1969.

*Schizogenius* Putzeys 1846
Seventy-five species in three subgenera comprise this genus. Indigenous in the Western Hemisphere, *Schizogenius* ranges widely in both the Neotropical and Nearctic Regions. Baehr (1983) described *S. freyi* from the Fiji Islands in the Central Pacific Ocean based upon a single specimen. Most species are hygrophilous, living along rivers or streams on barren gravel or on sand. The 24 species inhabiting Nearctic North America belong to two subgenera. Although the northern Neotropical-Nearctic species have been studied carefully (Whitehead 1972), the South American species present a taxonomic challenge yet to be met.

subgenus *Genioschizus* Whitehead 1972
Ten species belong to this subgenus of which one subspecies, *S. crenulatus crenulatus* LeConte 1852, occurs in Nearctic North America, in southern California, Arizona, and Mexico.

subgenus *Schizogenius s. str.*
Fifty-seven species belong to the nominotypical subgenus. Twenty-three of them are Nearctic, ranging collectively throughout the United States, and as far north as southern Canada.

*Semiardistomis* Kult 1950
This is a precinctive Western Hemisphere genus of 12 species, with most of them living in the tropics. Two species are in eastern Nearctic North America, *S. puncticollis* (Dejean 1831) and *S. viridis* (Say 1823), whose aggregate range extends from Texas north to southern New York. Both species are hygrophilous. This group was previously combined with the genus *Ardistomis* Putzeys 1846.

*Ardistomis* Putzeys 1846
Like the preceding, this is a precinctive Western Hemisphere genus, with most of its species living in the tropics. The group is represented by 54 species, of which only three live in Nearctic North America, in eastern United States. Little is known about the habitat requirement of these species but some information suggests that they live in swamps, and are thus hygrophilous. The group is in need of revision.

*Aspidoglossa* Putzeys 1846
This is another precinctive Western Hemisphere genus. It includes about 26 species of which one only enters United States: *A. subangulata* (Chaudoir 1843), which ranges from southeastern Texas northward to Washington, D.C., and in the Mississippi Basin, to southern Michigan. This species is hygrophilous, adults being found among sparse vegetation growing on wet clay near the margins of ponds and streams. Adults fly to light sources on warm nights during the summer.

*Clivina* Latreille 1802
A very diverse, worldwide genus, with 389 species arranged in nine subgenera, and more than 100 species in the Western Hemisphere. These species occupy temperate, subtropical, and tropi-

cal regions. The 27 Nearctic species are placed in four taxa of subgeneric rank. The genus is in need of revision.

subgenus *Semiclivina* Kult 1947
This Western Hemisphere precinctive subgenus contains two species in North America: *C. dentipes* Dejean 1825 which occurs over most of the United States and in southern Ontario, and *C. vespertina* Putzeys 1866, which seems to be introduced from South America and is now established in the Gulf States. The species are hygrophilous, with adults being found in swamps and along bodies of water. References. Nichols 1985b: 380.

subgenus *Clivina s. str.*
*Reichardtula* Whitehead 1977 (in Reichardt 1977)
This widely distributed subgenus of 352 species is represented in the Nearctic Region by ten species ranging as far north as southern Canada. Two of these species are European introductions that have become established on both east and west coasts: *C. collaris* (Herbst 1784) and *C. fossor* (Linnaeus 1758). The species are hygrophilous or mesophilous, adults being found in open fields and along river banks. The species *C. impressefrons* LeConte 1844 is a minor pest in eastern United States, as a result of feeding on corn seed after planting. Reference. Bousquet 1997: 347-348 (key to N.A. spp.).

*americana* species group
In North America, five species belong to this group. They occur principally in southern United States with one species, *C. americana* Dejean 1831, ranging as far north as southeastern Canada. The species are hygrophilous; adults are found mainly along margins of fresh water.

subgenus *Paraclivina* Kult 1947
This subgenus of 11 species includes ten species in Nearctic North America. These species are distributed all over United States except the states along the Pacific, with one species, *C. bipustulata* (Fabricius 1801), ranging as far north as southern Ontario in Canada. The species are hygrophilous; adults are found mainly along margins of standing fresh water.

*Akephorus* LeConte 1852
This precinctive Western Hemisphere genus is represented by two species, *A. marinus* LeConte 1852 and *A. obesus* (LeConte 1866), which inhabit the Pacific coast, ranging from southern Vancouver Island south to California. The genus is halobiontic, with adults found on sandy seashores, resting by day under logs and other debris on sea beaches. Adults are brachypterous. Reference. Fedorenko 1996: 69-70.

*Dyschiriodes* Jeannel 1941
This genus is represented in all of the major zoogeographical regions but is more diverse both in term of lineages and species in the Northern Hemisphere. About 55 valid species inhabit Nearctic North America, with a collective range extending from the arctic tundra to the subtropical lands of southern Florida. The species are hygrophilous, inhabiting burrows in sand and clay soil, usually near water; some species live in saline places. Adults and larvae of this genus are frequently found in association with adults and larvae of the heterocerid genus *Heterocerus* and, especially, the staphylinid genus *Bledius*, upon which they feed. Members of *Dyschiriodes* were previously included in the genus *Dyschirius* Bonelli 1810 until Fedorenko (1996) recognized two distinct lineages within *Dyschirius*. The species of *Dyschiriodes* were arrayed by Federenko (1996) in five subgenera of which four inhabit Nearctic North America. References. Fedorenko 1996. Bousquet 1988: 366-369 (key to spp.).

subgenus *Antidyschirius* Fedorenko 1996
This Holarctic group includes two species: the Nearctic precinctive *D. laevifasciatus* (Horn 1878) from western North America and the Palaearctic precinctive *D. matisi* (Lafer 1989) from eastern Russia. They are hygrophilous and riparian.

subgenus *Eudyschirius* Fedorenko 1996
About 60 species belong to this subgenus, which inhabits all of the major zoogeographical regions except the Australian. In Nearctic North America, the group contains about ten species. They are hygrophilous and either riparian or live near the margins of saline pools. This subgenus includes the *tridentatus*, *ferrugineus*, and *brevispinus* groups of Bousquet (1988).

subgenus *Dyschiriodes s. str.*
More than 100 species are included in this subgenus. As for the preceding group, this one also occupies all major zoogeographical regions except the Australian. It is represented in North America by about 35 species. Many species are riparian and some are confined to saline places. This subgenus includes the *integer*, *pilosus*, *filiformis*, *exochus*, *politus*, and *sellatus* groups of Bousquet (1988).

subgenus *Paradyschirius* Fedorenko 1996
This subgenus, known from the Palaearctic, Oriental, Afrotropical, Nearctic, and the northern part of the Neotropical Regions, includes about 20 species. The species are hygrophilous and riparian, adults being found along exposed, sandy river banks. This subgenus corresponds to the *analis* group of Bousquet (1988).

## Scaritini

The tribe Scaritini forms a large element, represented in all of the major zoogeographical regions of the world. The group is predominantly tropical and of the 40+ genera presently recognized, only two are present in Nearctic North America. As a rule, their members are active predators and have a fossorial way of life. They are variously hygrophilous, mesophilous, and xerophilous.

*Pasimachus* Bonelli 1813
This genus of 32 species arrayed in two subgenera (*Pasimachus s. str.* and *Emydopterus* Lacordaire 1854) is precinctive in North

and Middle America. The eleven species which occur in the Nearctic Region are members of the nominotypical subgenus. The group is predominantly southern but the range of one species, *P. elongatus* LeConte 1846, extends northward into the southern part of the Canadian prairie provinces. The species inhabit dry, open woodlands, or prairies with sparse vegetation. All species are brachypterous, with elytra coalesced along the suture. In spite of the excellent analysis by Bänninger (1950; 1956), challenging taxonomic problems remain to be resolved. Reference. Bänninger 1950: 481-511 (rev. spp.).

*Scarites* Fabricius 1775
This genus of large burrowing beetles includes 233 species arranged in eight subgenera. All of the major zoogeographical regions, except the Australian, have representative species of *Scarites*. Seven species, all members of subgenus *Scarites*, live in the Nearctic Region. They are mainly southern in distribution, but one reaches southern Ontario. These insects inhabit burrows in damp clay soil, usually in the vicinity of lakes or marshes. Adults of one species, *S. subterraneus* Fabricius 1775, are encountered frequently in seed beds, where the individuals damage young seedlings. The group, though small in our area, requires further thorough study to determine the validity of the species currently recognized. References. Bänninger 1938: 149-152. Nichols 1986: 257-264.

## Subfamily XI. Trechinae

The diagnostic adult features of this taxon are the following: size very small to moderate (total length ca. 0.7-10 mm); head capsule ventrally without antennal grooves; antennal insertions laterad, each between base of mandible and anterior margin of eye; antenna filiform, antennomeres 11, antennomeres 2-4 not nodose, 2-6 without unusually long setae; labrum with six setae; mandible with scrobe unisetose, occlusal margin with one terebral tooth, retinacular tooth about size of terebral tooth, not enlarged and molariform; terminal palpomeres various, subulate or not; body pedunculate (with mesothorax and extreme base of elytra markedly constricted) or not; pterothorax with mesonotal scutellum dorsally evident, not concealed beneath elytra; front coxal cavities closed, front coxal cavities closed by projection of proepimeron into cavity each side of lateral projection of prosternum, middle coxal cavities conjunct-confluent, hind coxal cavities lobate (disjunct)-confluent (Bell 1967: 105); elytra extended to apex of abdominal tergum VIII, apical margins subsinuate, dorsally striate (more or less) or smooth; metathoracic wings long (macropterous) or short (brachypterous); front tibia emarginate-anisochaete; abdomen with six (II-VII) pregenital sterna normally exposed, sterna III and IV separated by suture; male genitalia with parameres similar in form and size, setose apically; ovipositor with gonocoxae rotated 90° in repose, with or without ramus, gonocoxa of two gonocoxites, gonocoxite 2 without accessory lobe; reproductive tract with spermatheca 1, without spermatheca 2 (Liebherr and Will 1998: 130, Figs. 30-31; 131, Figs. 32-33, and 35); secretions of pygidial glands: various hydrocarbons, higher saturated or unsaturated acids, aliphatic ketones, saturated esters, and formic acid (Moore 1979: 194).

This subfamily, the equivalent of the "Stylifera" of Jeannel (1941: 79, 80-81) includes 11 tribes: Melaenini, Broscini, Apotomini, Trechini, Zolini, Bembidiini, Psydrini, Melisoderini, Amblytelini, Pogonini, and Patrobini. Of these, only the broscines, trechines, bembidiines, psydrines, pogonines, and patrobines are represented in the Nearctic Region.

### Broscini

This tribe, with 264 species arranged in 31 genera and four subtribes, is represented in all of the major zoogeographical regions, but only in the temperate portions thereof. Four species included in as many genera inhabit the Nearctic Region. All belong in the subtribe Broscina.

*Broscus* Panzer 1813
This wide-ranging Palaearctic genus of some 22 species is represented in the Nearctic Region by *B. cephalotes cephalotes* (Linnaeus 1758). Probably introduced fairly recently, this species lives on sandy sea beaches on the coasts of Nova Scotia and Prince Edward Island (Larochelle and Larivière 1989: 69-73).

*Zacotus* LeConte 1869
This monospecific precinctive Nearctic genus includes only the species *Z. matthewsii* LeConte 1869. This mesophilous species inhabits the coniferous forests of the Pacific Northwest, ranging from northern California to extreme southwestern coastal Alaska, and eastward from Vancouver Island and the Queen Charlotte Islands to the Bitter Root Mountains in southwestern Montana. Adults are active principally in the fall and early spring.

*Miscodera* Eschscholtz 1830
This Holarctic monospecific genus, containing *M. arctica* (Paykull 1800), in the Nearctic Region is transcontinental in the boreal forest, where it lives in dry coniferous forest and on glacial moraines.

*Broscodera* Lindroth 1961
This is a Holarctic genus, with two subgenera: the bispecific *Sinobrosculus* Deuve 1990, known only from Nepal and the Tibetan Plateau, at high altitude (Deuve 1998: 228); and the Nearctic monospecific nominate subgenus *Broscodera*, containing *B. insignis* (Mannerheim 1852), which is precinctive in the west, ranging in the coastal states and provinces from the Queen Charlotte Islands, and on the mainland of southeastern Alaska south to Oregon and east to Wyoming. This species is widespread altitudinally, from near sea level to above treeline on mountain peaks. Adults occur in greatest numbers at the margins of small streams. By day, the beetles rest deep in loose rocky substrate, especially where it is very moist. Adults are found also away from streams, especially above treeline, under

stones on barren slopes, and at the edges of snowfields (Kavanaugh 1992: 59).

### Psydrini

This tribe as presently restricted (Baehr 1998: 360) contains three genera, two of which occur in Holarctica, with one (the monospecific *Laccocenus* Sloane 1890) confined to the Australian Region. Lindroth (1961a: 173) used the name Nomiini for this group, but Psydrini has priority (Madge 1989: 465 and 467). Reference. Baehr 1998: 359-367.

*Psydrus* LeConte 1846
*Monillipatrobus* Hatch 1933
This precinctive Nearctic genus includes only the species *P. piceus* LeConte 1846, which is nearly transcontinental in the north (British Columbia to Quebec), ranging southward in the western mountains to northern California, and Arizona and New Mexico. Larvae and adults live under bark of conifers References. Schwarz 1884. McCleve 1975: 176.

*Nomius* Castelnau 1835
*Haplochile* LeConte 1848
This is a Holarctic and central African-Madagascan genus (Basilewsky 1954 and 1967). Represented in Holarctica by *Nomius pygmaeus* (Dejean 1831), the range of this species in the Nearctic Region is transcontinental in the north, extending southward in the Rocky Mountains to northern New Mexico, and southward along both east and west coasts. *N. pygmaeus* is known also from the Neotropical Region, in the highlands of Chiapas, Mexico (Reichardt 1977: 394). If disturbed, adults emit from the pygidial glands an odor that is markedly offensive, slightly nauseous, and reminds one of overripe cheese. These beetles fly at night, are attracted to light, and their invasions of human habitations have on occasion earned them considerable enmity and the unpleasant, though not inappropriate, epithet "stink beetle" (Barrows 1897 and Hatch 1931). Nothing is known of the way of life of this species.

### Trechini

This is a markedly diverse tribe, with more than 2500 species, arranged in 172 genera, worldwide in distribution, but, like the two preceding groups, most of the species are located in temperate regions, both north and south of the equator. On the small side (adults of most species are between 2 and 7 mm long), and monotonous in color (uniformly black to rufotestaceous) and form (rather slender, about average for carabids), trechines are markedly divergent in features not readily observed, particularly details of the mouthparts and male genitalia. Most are brachypterous in the adult stage, but some groups have macropterous adults that fly readily if disturbed. Small eyes or lack of eyes are common features among trechines.

Trechini are divergent ecologically, from gravel-inhabiting, readily flying tropical riparian hygrophiles living at low altitude, through temperate-subarctic forest-inhabiting brachypterous mesophiles and endoges, to temperate montane blind troglobites. No trechines are known to be arboreal, however. In Nearctic North America, trechines are particularly abundant as troglobitic cavernicoles, concentrated in the Applachian Mountains, where 81% of the continental fauna resides.

Although a rich source for study of evolutionary problems, few workers have studied Trechini during the 20th Century. The primary worker on this group in the Nearctic Region during the past five decades is Thomas C. Barr (Nashville, Tennessee), whose masterful synthesis (1985a) traces speciation and dispersal patterns of trechines in the Appalachians. That publication is based on extensive surveys, both on the surface and underground in the extensive systems of caverns and small caves that formed by erosion in the Palaeozoic limestone formations that are a prominent feature of the region. Carl Lindroth treated the much less diverse epigean forest-inhabiting Trechini (principally genus *Trechus*) in northern North America.

Nine genera are Nearctic, grouped by Barr (1985a: 351) in four series, as follows: *Trechus* series (*Trechus* Clairville 1806); *Trechoblemus* series (*Trechoblemus* Ganglbauer 1892; *Blemus* Dejean 1821 [=*Lasiotrechus* Ganglbauer; 1892]); *Pseudanophthalmus* Jeannel 1920; *Neaphaenops* Jeannel 1920; and *Nelsonites* Casale and Laneyrie 1982); *Darlingtonea* series (*Darlingtonea* Valentine 1952; and *Ameroduvalius* Valentine 1952); and *Aphaenops* series (*Xenotrechus* Barr and Krekeler 1967).

Each series is probably a monophyletic assemblage of early Tertiary age. References. Barr 1985a: 350-407. Casale and Laneyrie 1982 (catalogue and worldwide taxonomic treatment). Jeannel 1926: 221-550.1927: 1-592.1928: 1-808. 1930: 59-122 (detailed taxonomic treatment of the world fauna). 1931: 403-499 (Nearctic species). 1949b: 37-104.

*Trechus* Clairville 1806
*Microtrechus* Jeannel 1927
Numerous species of this genus have been described, distributed throughout the Palaearctic and Nearctic Regions, the Mediterranean Basin, the Himalaya, and the Philippine Islands. Forty-eight species are known from Nearctic North America, arrayed in two subgenera: *Trechus* (*s. str.*), which is wide ranging, and *Microtrechus* Jeannel, known only from the Appalachian Mountains, in North Carolina and Tennessee. Most species live in montane forests of the Appalachians, with many fewer species in the mountain ranges of the west. The beetles live in deep litter in forested areas, but occupy the Aleutian tundra in the west, and open fields in Newfoundland. The species *T. apicalis* Motschulsky 1845 is transcontinental in the boreal forest. Three species were introduced, probably accidentally, from the Old World: *T. obtusus* Erichson 1837, reported from central California northward in British Columbia to the Queen Charlotte Islands (Kavanaugh and Erwin 1985; Kavanaugh 1992: 61); *T. rubens* (Fabricius 1801), reported from Newfoundland, Quebec, and St. Pierre and Miquelon, southward to Vermont; and *T. quadristriatus* (Schrank 1781), reported originally from the Great Lakes region (Bousquet *et al.* 1984: 215-220). Thanks to the work

of Lindroth (1961) and Barr (1962), the northern species of *Trechus* (s. str.) and of the subgenus *Microtrechus* are fairly well understood. However, little is known about the species that live in the southern part of the Rocky Mountains and the outliers of this mountain range, in New Mexico and Arizona. A thorough study of the Nearctic species of *Trechus* is required. References. Barr 1962: 65-92 (rev. *Microtrechus*). 1979a: 29-75 (rev. Appalachian spp.).

*Trechoblemus* Ganglbauer 1892
This Holarctic genus includes two species: the Palaearctic *T. micros* (Herbst 1784), and *T. westcotti* Barr 1972, which is known only from Oregon. Reference. Barr 1972: 140-149.

*Blemus* Dejean 1821
*Lasiotrechus* Ganglbauer 1892
This genus contains one species, *B. discus* (Fabricius 1792), known in North America from the northeast (Nova Scotia to eastern Ontario and northeastern United States), westward to Ohio and Michigan. The species is wide-ranging in the Old World, and was probably introduced by accident into eastern Canada, from whence it has spread.

*Pseudanophthalmus* Jeannel 1920
*Aphanotrechus* Barr 1960
*Anophthalmus* Sturm 1844, not North American.
This genus, which is restricted to the mountains and caves of unglaciated eastern United States, includes 138 described species. Barr (1985a: 351) reported an additional 103 species that were known to him, but undescribed. References. Barr 1959: 5-30 (descr. of new spp., and refs. to earlier literature). 1960a: 65-70. 1960c: 307-320. 1980: 85-96. 1981: 37-94.

*Neaphaenops* Jeannel 1920
The single species of this genus, *N. tellkampfi* (Erichson 1844), with its four subspecies, is found in the eastern part of the western Pennyroyal Plateau of Kentucky from the Ohio River southward to the Tennessee border. Reference. Barr 1979b.

*Nelsonites* Casale and Laneyrie 1982
*Nelsonites* Valentine 1952 (invalid proposal)
This genus contains two cave-inhabiting species, from the western margin of the Cumberland Plateau in southern Kentucky and northern Tennessee. References. Bousquet and Larochelle 1993: 118 (nomenclature). Casale and Laneyrie 1982: 111. Valentine 1952: 18 (descr.).

*Darlingtonea* Valentine 1952
This genus comprises one ditypic cave-inhabiting species (*D. kentuckensis* Valentine 1952) from the Cumberland Plateau of southeastern Kentucky and adjacent portions of Tennessee. Reference. Valentine 1952: 22 (descr.).

*Ameroduvalius* Valentine 1952
This genus comprises one cave-inhabiting species (*A. jeanneli* Valentine 1952) from the Cumberland Plateau of southeastern Kentucky. Reference. Valentine 1952: 24 (descr.).

*Xenotrechus* Barr and Krekeler 1967
This ditypic genus is known only from caves in Missouri. Reference. Barr and Krekeler 1967: 1322.

### Bembidiini

This tribe, with about 3000 species, is represented in all the zoogeographical regions of the world. In Nearctic North America, it includes about 360 species of which more than 250 belong to the worldwide genus *Bembidion* Latreille 1802. For a taxonomic conspectus of the adjacent Central American bembidiine fauna, see Erwin (1982). Most bembidiines, as adults, are surface-dwelling hygrophiles, though some are xerophilous. The anillines are endogeous, living deep in leaf litter in mesophytic forests, or in the upper layers of the soil (Jeannel 1937, and 1963a and b). Reference. Casey 1918: 1-223 (rev. spp.).

*Asaphidion* des Gozis 1886
This genus of 39 species is represented in the Nearctic Region, the Palaearctic Region, and the northern part of the Oriental Region. Three species live in North America, of which two are restricted to the northwest, as far north as the tundra of the Arctic Coast. Both live on moist to dry sandy-clay soil, where adults are found running on the surface, or resting in the shade of bushes, along the banks of rivers and streams. One species, *A. flavipes* (Linnaeus 1761), has been accidentally introduced from Europe into Long Island, New York and is now found in several northeastern states. References. Lindroth 1963: 203-206 (rev. N.A. spp.).

*Bembidion* Latreille 1802
This is a very diverse genus, worldwide in distribution, but better represented in the temperate and subarctic or subantarctic zones of both hemispheres. Some European authors recognize a number of genera instead of one, but until worldwide studies reveal distinct major lineages we prefer to retain a broad generic concept. About 250 species are currently recorded from Nearctic North America. They range from the Arctic coast south to the subtropical lands of Florida. Most species are markedly hygrophilous, living close to water, adults being found particularly on bare sandy-clay soil along lakes and rivers. Some species are independent of water and live in open, often cultivated, fields. Others are confined to alkaline places. Adults of most species are active in daylight. The wings are fully developed in adults of most species and some are good fliers and fly rapidly when disturbed. The Nearctic species currently are arranged in 42 subgenera and species groups.

subgenus *Bracteon* Bedel 1879
*Chrysobracteon* Netolitzky 1914
*Parabracteon* Notman 1929
*Foveobracteon* Netolitzky 1939

This Holarctic subgenus is represented by 17 species. The 11 Nearctic species range over most of the temperate and subarctic regions of the continent. Of these, two are Holarctic in distribution. They are hygrophilous, living on barren, sandy or clayish substrate along rivers and lakes, where adults are active, especially on warm, sunny days. Individuals fly readily when disturbed. Reference. Maddison 1993: 143-299 (rev. spp.)

subgenus *Odontium* LeConte 1848
This Holarctic subgenus includes 21 species of which nine inhabit North America. They range throughout United States and southern Canada. The species are hygrophilous and riparian, with adults being found along the margins of rivers and lakes, usually on bare sandy-clay soil. As in the preceding group, individuals fly readily when disturbed.

subgenus *Pseudoperyphus* Hatch 1950
*Bracteomimus* Lindroth 1955
Four species belong to this precinctive Nearctic group. Their aggregate range extends over most of the United States, mainly east of the Rocky Mountains, and southern Canada. The species are hygrophilous and riparian, adults being found on gravel or coarse sand along rivers and brooks, and fly readily when disturbed.

subgenus *Ochthedromus* LeConte 1848
This precinctive Nearctic subgenus includes two species. Their range covers most of the United States and the southern part of Canada. The species are hygrophilous and riparian, adults being found near water on bare, clay or sandy-clay soil, and at the margins of saline lakes and ponds. Adults fly readily when disturbed.

subgenus *Eurytrachelus* Motschulsky 1850
*Eudromus* Kirby 1837
*Platytrachelus* Motschulsky 1844
*Pogonidium* Ganglbauer 1892
This Holarctic subgenus is represented in Nearctic North America by three species. They are mainly northern in distribution and range from Alaska to Quebec, south to New York in the east and to Colorado and California in the west. The species are hygrophilous and riparian or xerophilous. Adults of the hygrophilous-riparian *B. interventor* Lindroth 1963, and *B. obliquulum* LeConte 1859 hunt along the banks of rivers, on bare soil; adults of the more xerophilous *B. nitidum* Kirby 1837 are active hunters in open fields on sandy soil.

subgenus *Hydrium* LeConte 1848
A single species, *B. levigatum* Say 1823, belongs to this precinctive Nearctic subgenus. Its geographical range extends all over United States east of the Rocky Mountains. The species is hygrophilous and riparian, adults hunting on bare soil along streams.

subgenus *Metallina* Motschulsky 1850
This is a Holarctic subgenus, represented in Nearctic North America by three species, one of which, the precinctive Nearctic *B. dyschirinum* LeConte 1861, ranges over western North America as far south as California and Colorado. The other two species were accidentally introduced into North America from Europe: *B. lampros* (Herbst 1784), on both east and west coasts, and *B. properans* (Stephens 1827) on the east coast alone. The three species are wing-dimorphic and live in open places, principally on sandy ground, where adults hunt actively.

subgenus *Phyla* Motschulsky 1844
This indigenous Palaearctic group is represented in northeastern North America by one introduced species, *B. obtusum* Audinet-Serville 1821. The species is mesophilous, inhabiting open, often cultivated, ground. The species is wing-dimorphic.

*wickhami* species group
The sole species included in this precinctive Nearctic group, *B. wickhami* Hayward 1897, ranges from Nevada and California, north to Oregon. The species is hygrophilous and riparian, adults hunting along streams. Reference. Erwin and Kavanaugh 1981: 36-40.

subgenus *Lionepha* Casey 1918
The geographical range of this precinctive Nearctic group is restricted to western North America. The subgenus contains nine species. Little is known about their habitat requirement. Adults of some species have been found along cool or cold streams and rills, often among gravel, or in drift debris on river banks. This subgenus corresponds to the *erasum* and *disjunctum* groups of Lindroth (1963). Reference. Erwin and Kavanaugh 1981: 40-64 (rev. spp.)

subgenus *Lymnaeum* Stephens 1828
*Lymneops* Casey 1918
This Holarctic subgenus is represented in Nearctic North America by two species, of which *B. laticeps* LeConte, 1858 is known only from California. The species *B. nigropiceum* (Marsham, 1802) is European, was introduced into Massachusetts, and probably is not established in the Nearctic Region (Erwin and Kavanaugh, 1980: 241-242). The species *B. utahense* Van Dyke, 1925, assigned originally to *Lymneops*, is a member of *Amerizus* Chaudoir, 1868 (Van Dyke, 1949: 56).

subgenus *Actedium* Motschulsky 1864
This Holarctic subgenus includes a single precinctive Nearctic species, *B. lachnophoroides* Darlington 1926. The species is known only from southern and central Alberta and Saskatchewan. It is hygrophilous and riparian, living on coarse sand along the margins of rivers. As the specific epithet suggests, adults resemble markedly adults of some species of *Lachnophorus*.

subgenus *Trechonepha* Casey 1918
Two Nearctic species belong to this subgenus. They live in the western part of the continent. Little is known about their habitat requirement but one of them, *B. iridescens* (LeConte 1852),

seems to prefer mosses and leaves in damp places, usually near trickles and small creeks.

subgenus *Plataphodes* Ganglbauer 1892
*kuprianovi* species group, Lindroth 1963
*incertum* species group, Lindroth 1963
This Holarctic subgenus has 18 species in the Nearctic Region, of which only one, *B. occultator* Notman 1919, occurs in the east. The other species are mainly in the northwestern part of the continent. They are hygrophilous, living chiefly on the banks of running waters. The group is taxonomically very difficult and the male genitalia offer the only reliable characters to distinguish among some of the species.

subgenus *Plataphus* Motschulsky 1864
*Melomalus* Casey 1918
*Micromelomalus* Casey 1918
*Trachelonepha* Casey 1918
*simplex* species group, Lindroth 1963
*planiusculum* species group, Lindroth 1963
This is a Holarctic subgenus, more or less confined to the northern and mountainous regions. The group includes 19 Nearctic species and is more diverse in the west. The species are hygrophilous and riparian, with adults hunting on gravel banks of running waters. As for the preceding, this group is taxonomically very difficult.

subgenus *Blepharoplataphus* Netolitzky 1920
This Holarctic subgenus is represented by two Palaearctic and one Holarctic species, *B. hastii* C.R. Sahlberg 1827. This species is transcontinental in the Nearctic Region, ranging from Alaska to Labrador. The species live on gravel along the margins of rivers, lakes, and pools, where adults hunt actively.

subgenus *Trichoplataphus* Netolitzky 1914
This Holarctic subgenus is restricted to Asia and eastern Nearctic North America where it includes four species. The species are hygrophilous and riparian, with adults hunting along banks of flowing water.

subgenus *Hirmoplataphus* Netolitzky 1942
This Holarctic subgenus includes three Palaearctic and nine Nearctic species. The aggregate range of the North American species extends from central Alaska to Newfoundland south to Alabama in the east and Arizona in the west. The species are hygrophilous and riparian, adults being found in gravel or on sand, along the margins of streams, creeks, and rivers.

subgenus *Hydriomicrus* Casey 1918
Five species belong to this precinctive Nearctic subgenus. Three of these occur in California and Oregon, and the remaining two are eastern in distribution and range from Newfoundland southward to Alabama. The eastern species live in sphagnum bogs or on the banks of flowing waters.

subgenus *Bembidionetolitzkya* Strand 1929
*Daniela* Netolitzky 1910 [preoccup.]
This Holarctic subgenus is represented in North America by a single Holarctic species, *B. mckinleyi* Fall 1926, with two subspecies. The taxa are restricted to Alaska and western Canada. This species is hygrophilous and riparian, with adults hunting among rocks and debris deposited by the flowing water.

*lenae* species group
Of the two species that belong to this group, one is confined to Pacific Asia and the other one, *B. lenae* Csiki 1928, is Holarctic, ranging from eastern Siberia to western Northwest Territories. This species is hygrophilous and riparian, with adults hunting among rocks and gravel, along river banks.

subgenus *Peryphus* Dejean 1821
*Peryphanes* Jeannel 1941
*nebraskense* species group, Lindroth 1963
*striola* species group, Lindroth 1963
*grapii* species group, Lindroth 1963
*bimaculatum* species group, Lindroth 1963
*tetracolum* species group, Lindroth 1963
*transversale* species group, Lindroth 1963
*scopulinum* species group, Lindroth 1963
This subgenus is Holarctic in distribution and comprises about 30 species in the Nearctic Region. Their aggregate range extends from Alaska to Newfoundland south to California and Arizona in the west, to Georgia in the east. Four of these species have been introduced accidentally from Europe and six species are Holarctic in distribution. The species live in a variety of habitats, some being confined to the vicinity of water, often on barren gravel or coarse sand, others being mostly xerophilous and occurring in cultivated fields, gravel pits, etc.

subgenus *Liocosmius* Casey 1918
This precinctive Nearctic group is represented by three species. They range in western North America as far north as southwestern British Columbia. The species are hygrophilous and riparian, adults being found on sandy river banks and on gravelly sides of small brooks.

subgenus *Peryphodes* Casey 1918
Two species belong to this precinctive Nearctic subgenus. One, *B. ephippigerum* (LeConte 1852), occurs along the Pacific coast, the other one, *B. salinarium* Casey 1918, is mainly in the prairie provinces and adjacent states. The species are hygrophilous and lacustrine, with adults being found in saline places, such as sandy seashore or at the edges of lakes and ponds. Both species are wing-dimorphic.

*constricticollis* species group
Like the preceding taxon, this group is precinctive Nearctic, and includes two species. They live in western North America from the Canadian Prairies south to Arizona and New Mexico. Very little is known about their habitat requirement except for

the fact that one species (*B. nudipenne* Lindroth 1963) appears xerophilous and is not confined to the vicinity of water.

subgenus *Eupetedromus* Netolitzky 1911
Holarctic, this subgenus includes four species in the Nearctic Region. They range from Alaska to Newfoundland south to California and Arizona in the west and to Virginia in the east. The species are hygrophilous and lacustrine, adults being found on wet mud, among dead vegetation, usually along the margins of ponds or lakes.

subgenus *Leuchydrium* Casey 1918
The species *B. tigrinum* LeConte 1879 is the only representative of this precinctive Nearctic subgenus. The species, a halobiont, lives on the seacoast of western North America from southern California northward to southern Vancouver Island, British Columbia.

subgenus *Cillenus* Samouelle 1819
This group includes one species widely distributed in western Europe and Morocco, a few others in eastern Asia (including Japan), and one species, *B. palosverdes* Kavanaugh and Erwin 1992, known only from the Palos Verdes Peninsula in California. Adults are found in the rocky intertidal zone.

*scudderi* species group
This precinctive Nearctic group contains four species which are found only west of the Mississippi River although none of them reaches the Pacific coast. The species are hygrophilous, principally lacustrine, on alkaline soil, most adults being found at the borders of lakes and ponds.

*obtusangulum* species group
This group, which is closely related to the preceding, contains three species, all of them restricted to North America west of the Mississippi River. Lacustrine, the species live on alkaline soil.

subgenus *Notaphus* Dejean 1821
This subgenus has representatives in the Nearctic and Palaearctic Regions, in the montane region of tropical Africa, in Central America, in temperate regions of South America, and in Australia. It contains about 30 species in Nearctic North America. The species are hygrophilous and riparian or lacustrine, adults being found at the edges of standing or running waters in gravel clay or sand substrate, among vegetation or on bare soil; some species are apparently halobionts. This group is taxonomically difficult and a revision of the species would be useful.

*contractum* species group
This precinctive Nearctic group is represented by seven species which range east of the Rocky Mountains, from Northwest Territories south to Texas. The group is also represented in Neotropical México. The species are hygrophilous and halophile, adults being found at the border of saline lakes and ponds or on sea beaches.

*oberthueri* species group
This precinctive Nearctic group includes a single species, *B. oberthueri* Hayward 1901. Hygrophilous and riparian, it occurs east of the Rocky Mountains, as far south as Illinois. Adults are found mainly on the banks of flowing water.

subgenus *Furcacampa* Netolitzky 1931
Of the two species that belong to this precinctive Western Hemisphere subgenus, one is precinctive in Panama and *B. affine* Say 1823 ranges over eastern Nearctic North America, as far north as southern Ontario and Québec. The species is hygrophilous and lacustrine, with adults being found at the margins of marshy pools or on sandy beaches of lakes.

*versicolor* species group
Nine species belong to this indigenous Nearctic group. Collectively, they range over most of the temperate and boreal regions of the continent. The species is hygrophilous and primarily lacustrine, with adults being found among gravel or on sand, clay, or peat substrate.

subgenus *Emphanes* Motschulsky 1850
*Omala* Motschulsky 1844 [preoccup.]
This Holarctic subgenus includes two species in North America: *B. vile* (LeConte 1852) is confined to the Pacific coast and *B. diligens* Casey 1918, lives in the interior plains. Both are hygrophilous and lacustrine, and seem to be restricted to saline habitats.

subgenus *Bembidion s. str.*
This Holarctic subgenus contains eight Nearctic species. The range of this group covers most of the United States and Canada, south of the Arctic tundra. The species inhabit dry or moderately moist sandy areas.

subgenus *Cyclolopha* Casey 1918
Eight species belong to this precinctive Western Hemisphere subgenus. They range from Guatemala northward to Arizona and New Mexico. Only two species, *B. poculare* Bates 1884 and *B. sphaeroderum* Bates 1882, enter the Nearctic Region. The habitat requirements of these species are known inadequately, but they seem to be xerophilous, the adults being found in or in the vicinity of mountain meadows. References. Perrault 1982: 97-110 (rev. spp.).

subgenus *Semicampa* Netolitzky 1910
This Holarctic subgenus is represented in Nearctic North America by seven species. Their aggregate range extends from central Alaska to Newfoundland, south to California in the west, and New Jersey in the east. The species are hygrophilous, with adults being found among moss and leaves, usually near pools and small lakes or along rivers. One species, *B. roosvelti* Pic 1902, seems to be halobiontic. Members of this group are brachypterous or wing dimorphic with a low frequency of macropterous individuals.

subgenus *Diplocampa* Bedel 1896
This is a Holarctic subgenus represented in North America by one circumpolar species, *B. transparens* (Gebler 1829). The species ranges from Alaska to Newfoundland, south to Oregon in the west and to Indiana in the east. The species is hygrophilous, with adults being found in wet places, at the edges of standing waters (pools, marshes, lakes) where the vegetation is rich. The adult stage is wing-dimorphic.

subgenus *Trepanedoris* Netolitzky 1918
This Holarctic group includes 13 Nearctic species. They are hygrophilous, with adults found in wet places at the margin of standing waters, such as swamps, or in dry places on river banks. The adult stage of at least two species is wing-dimorphic, with adults of the remaining 11 species being macropterous.

subgenus *Amerizus* Chaudoir 1868
This distinctive subgenus is precinctive in the Nearctic Region, including five species. Four are in western North America, only with *B. utahense* Van Dyke, 1925 occurring in the Rocky Mountains of Utah. In contrast, *B. wingatei* Bland 1864, ranges over the eastern part of the continent. The species are mesophilous, with adults usually found under logs, stones, or in leaf litter in coniferous and mixed forests, or along seeps and small streams. The wings are fully developed in *B. spectabile* Mannerheim 1852; adults of the other species are brachypterous (Lindroth 1963: 403). Reference. Van Dyke, 1949: 56.

*Phrypeus* Casey 1924
This monospecific precinctive Nearctic genus contains only *P. rickseckeri* (Hayward 1897), restricted to the Pacific Northwest from California to southern British Columbia. The species is hygrophilous, adults living on the banks of large rivers or on lake shores, on sand and gravel substrate.

*Mioptachys* Bates 1882
*Tachymenis* Motschulsky 1862 [preoccup.]
This precinctive Western Hemisphere genus includes 13 species of which one only, *M. flavicauda* (Say 1823), occurs in Nearctic North America. It is known from both eastern and western United States and the adjacent portions of Canada. This species is mesophilous and is associated with forest vegetation, with adults being found beneath the bark of logs.

*Tachyta* Kirby 1837
Members of this group are found in boreal, temperate, and tropical areas in all major biogeographical regions. Four species inhabit the Nearctic Region. Their aggregate range includes most of the continent from Alaska to Nova Scotia, south to Texas. The group is associated with forest vegetation, with adults of all species being found beneath the bark of logs. References. Erwin 1975: 1-68 (rev. spp.).

*Elaphropus* Motschulsky 1839
*Tachyura* Motschulsky 1862
*Barytachys* Chaudoir 1868
This genus, which is represented in all major zoogeographical regions, includes 340 species arranged in five subgenera. In the Nearctic Region, this group is represented by about 30 species, all of which are included in the wide-ranging subgenus *Tachyura* Motschulsky 1862. Their aggregate range includes all of the United States and southern Canada. One species, *E. parvulus* (Dejean 1831), has been accidentally introduced along the west coast. The species are hygrophilous and primarily lacustrine, adults are found on bare sand, clay, or gravel. Some species appear to be myrmecophilous. The group is taxonomically difficult and in need of a revision.

*Micratopus* Casey 1914
Five species are presently assigned to this indigenous Neotropical genus, of which one, *M. aenescens* (LeConte 1848), is in the Nearctic Region. This species ranges from Trinidad and Panama in the Neotropical Region north to the Gulf Coastal Plain and Interior Low Plateaus of southeastern United States. The species are mesophilous, inhabiting swampy areas in hardwood forest. Adults are best collected by sifting forest floor litter. References. Barr 1971a: 32-37 (rev. N.A. spp.).

*Pericompsus* LeConte 1852
*Tachysops* Casey 1918
*Tachysalia* Casey 1918
The 69 species of this Western Hemisphere -Australian genus are arrayed in three subgenera: the precinctive Australian *Upocompsus* Erwin 1974, with nine species; the precinctive Neotropical *Eidocompsus* Erwin 1974, with 13 species; and the indigenous Neotropical *Pericompsus* (*s. str.*), with 47 species. Three species inhabit the Nearctic Region, two of which (*P. laetulus* LeConte 1852 and *P. sellatus* LeConte 1852) are restricted to southwestern United States. The species *P. ephippiatus* (Say 1830), ranges from Texas eastward and northward to the New England States. This species is hygrophilous and riparian, with adults being found in open sandy places, along streams. Little is known about the habitat requirements of the other two species. Reference. Erwin 1974: 1-95 (rev. spp.).

*Porotachys* Netolitzky 1914
This Palaearctic-Oriental genus that includes five species is represented in the Nearctic Region by *P. bisulcatus* (Nicolai 1822), which was accidentally introduced both on the west coast, in Washington State, and on the east coast, where it is now known from several provinces and states in the northeast, west to North Dakota. Probably a mesophilous species, adults are found in sawdust and under loose pieces of bark and bark on logs, notably in the vicinity of sawmills.

*Polyderis* Motschulsky 1862
*Microtachys* Casey 1918
This genus of 40 species is known from the southern parts of Holarctica, and from the tropical Oriental, Afrotropical, and Neotropical Regions. One species, *P. antigua* Erwin 1971, is represented by a single Tertiary Age amber fossil from the Mexican state of Chiapas. In the Nearctic Region, *Polyderis* is repre-

sented by three species. Their aggregate range includes most of the United States and southeastern Canada. Little is known about their habitat requirements. Larochelle (1975) collected adults of *P. laevis* (Say 1823) on sandy hills, while Blatchley (1910) found his material among leaves at the border of marshes.

*Tachys* Dejean 1821
*Isotachys* Casey 1918
More than 40 species are included in this genus, with a collective range that includes all of the zoogeographical regions. Eleven species are Nearctic. Most of them live in southern United States. A few species reach the Canadian Prairies and southwestern Canada. Apparently all the species are halobionts or halophiles. A revision of the group is needed.

*Paratachys* Casey 1918
With 143 species, *Paratachys* is represented in most zoogeographical regions, including the southern parts of Holarctica. Twenty species are Nearctic. They range over most of the temperate region of the continent but are more diverse in the east and southeast. The species are hygrophilous, living principally in more or less shaded swamps. Adults are found mainly on clay or sandy-clay soil. The group is in need of a taxonomic revision.

*Micranillodes* Jeannel 1963
The sole species of this seemingly precinctive Nearctic genus, *M. depressus* Jeannel 1963, is known from a single female collected in Travis county, Texas.

*Anillodes* Jeannel 1963
Five Nearctic species are assigned to this seemingly precinctive Nearctic genus. They occur in California and Texas and each is known from only a few specimens. Reference. Jeannel 1963a: 54-57 (rev. spp.).

*Anillinus* Casey 1918
*Troglanillus* Jeannel 1963
This precinctive Nearctic group is represented by 11 species, all of which are restricted to eastern United States and particularly the Appalachian region. Adults are found in leaf litter or under large, deeply embedded stones on mountain slopes or in ravines in forested areas. Adults of at least one species, *A. valentinei* (Jeannel 1963), is regularly found in caves and may be a troglobite. The group is in need of a taxonomic revision. Reference. Jeannel 1963b: 148-152 (key to spp.).

*Serranillus* Barr 1995
One species, *S. jeanneli* Barr 1995, belongs to this precinctive Nearctic genus, which is known from localities in North Carolina and Georgia. Reference. Barr 1995: 246-248 (descr. sp.).

*Anillaspis* Casey 1918
This precinctive Nearctic genus includes two California species, each of them being known from a single female specimen. References. Jeannel 1963a: 77-80 (rev. spp.).

*Horologion* Valentine 1932
Placed originally in the Trechini by its describer, this genus was subsequently moved to the Psydrini by Van Emden (1936) and finally to the Bembidiini, subtribe Anillina, by Erwin (1982: 459). The group includes a single species, *H. speokoites* Valentine 1932, known only from one specimen collected from a cave in West Virginia. Reference. Valentine 1932 (desc. sp.).

### Pogonini

This tribe is represented in all zoogeographical regions of the world but is more diverse both in lineages and species in the Palaearctic Region. The group includes about 70 species of which six inhabit North America. The species are halobiontic: some are restricted to littoral places, others to inland salt lakes, ponds, or pans. Reference. Bousquet and Laplante 1997: 699-731 (rev. Western Hemisphere spp.).

*Thalassotrechus* Van Dyke 1918
*Anatrechus* Casey 1918
This indigenous Nearctic genus includes a single species, *T. barbarae* (Horn 1892), which is distributed along the coast of California and Baja California. Adults of this species exhibit geographical variation in color (Evans 1977). Adults and larvae live in rock crevices in the upper intertidal zone, their activities endogenously controlled by circadian and circatidal rhythms (Evans 1976 ).

*Diplochaetus* Chaudoir 1871
This indigenous Nearctic genus includes four species. All of them are represented in Nearctic North America with one, *D. rutilus* (Chevrolat 1863), ranging from New Jersey and southern United States south in the Neotropical Region to Colombia and Venezuela, including the West Indies. *Diplochaetus planatus* (Horn 1876) was assigned mistakenly to *Pogonistes* Chaudoir 1871. Adults live on saline flats along the margins of saline lakes.

*Pogonus* Dejean 1821
This genus is represented in all zoogeographical regions, except the Neotropical Region, and includes about 45 species. One species only, *P. texanus* Chaudoir 1868, inhabits Nearctic North America; it lives along the Gulf of Mexico from southwestern Louisiana to southeastern Texas. A revision of the genus is required if relationships of *P. texanus* are to be revealed.

### Patrobini

This group of 182 species in 15 genera is represented in the Oriental, Palaearctic, and Nearctic Regions. Twelve species included in four genera occur in the Nearctic Region. Reference. Darlington 1938: 135-183.

*Platidiolus* Chaudoir 1878
*Patroboidea* Van Dyke 1925

This Holarctic genus includes two species: one, precinctive Palaearctic, in eastern Siberia, and one, *P. vandykei* Kurnakov 1960, precinctive Nearctic, ranging in the northwest from Washington to southern Alaska, and eastward to the eastern slopes of the Rocky Mountains, in Alberta. The species is hygrophilous and riparian. Adults are small-eyed and pale, living in coarse gravel and rubble, along the margins of cold, swift streams.

*Platypatrobus* Darlington 1938
This precinctive Nearctic genus is monospecific, containing only *P. lacustris* Darlington 1938, ranging through the boreal forest from Alberta and the Northwest Territories, eastward to Maine and Vermont, and southward in the east to Pennsylvania. This species is hygrophilous, living in association with beavers in their houses. Adults are fully winged, and evidently fly at night, but the flight season probably is brief.

*Patrobus* Dejean 1821
This Holarctic genus of 15 species is represented in the Nearctic Region by six species, arrayed in two subgenera.

subgenus *Neopatrobus* Darlington 1938
This subgenus, precinctive in the Nearctic Region, includes a single species, *P. longicornis* (Say 1823), whose range extends from northern Alaska and British Columbia across the north in the boreal forest to Newfoundland, and in the east, southward to northern Alabama. The species is hygrophilous, probably principally riparian, with adults living in wet, muddy places along the margins of streams and lakes.

subgenus *Patrobus* (*sensu stricto*)
*Geopatrobus* Darlington 1938
This Holarctic group is represented in the Nearctic Region by five species, three of which, *P. fossifrons* (Eschscholtz 1823), *P. septentrionis* Dejean 1828, and *P. foveocollis* (Eschscholtz 1823), are Holarctic. The collective Nearctic range of this assemblage extends eastward from the Aleutian and Pribilof Islands to Labrador and Newfoundland, south to Colorado, the Great Lakes and northern Maine. The species *P. foveocollis* is mesophilous, living in damp rotten logs in forests, or under cover on the ground in alpine meadows. The other Nearctic species are hygrophilous, with adults being found at the margins of cold bogs and damp meadows.

*Diplous* Motschulsky 1850
*Platidius* Chaudoir 1871
Twenty-one species arranged in two subgenera represent this Holarctic genus. The Palaearctic species are included in subgenus *Diplous* (s. str.), whereas the four Nearctic species are included in the subgenus *Platidius* Chaudoir 1871. Of the four precinctive Nearctic species, three are northwestern in geographical distribution, one (*D. rugicollis* (Randall 1838)) being northeastern (Nova Scotia and New Brunswick, southward to Vermont and Massachusetts). The species are hygrophilous and riparian, adults living along the margins of cold, swift-flowing streams, where they rest under stones and in gravel.

## Subfamily XII. Harpalinae

The diagnostic adult features of this taxon are the following: size small to large (total length ca. 2.0-40 mm); head capsule ventrally without antennal grooves; antennal insertions laterad, each between base of mandible and anterior margin of eye; antenna filiform or moniliform, or clavate, antennomeres 2-4 not nodose, 2-6 without unusually long setae, antennomeres 11; labrum with six setae; mandibles with scrobes asetose, occlusal margin with or without one terebral tooth, retinacular tooth about size of terebral tooth, not enlarged and molariform; terminal palpomeres various, subulate or not; body pedunculate (with mesothorax and extreme base of elytra markedly constricted) or not; pterothorax with mesonotal scutellum dorsally evident, not concealed beneath elytra; front coxal cavities closed by projection of proepimeron into cavity each side of lateral projection of prosternum, middle coxal cavities conjunct-confluent, hind coxal cavities lobate (disjunct)-confluent (Bell 1967: 105); elytra extended or not to apex of abdominal tergum VIII, and apical margins subsinuate to truncate; dorsal surface striate (more or less) or smooth dorsally; metathoracic wings long (macropterous) or short (brachypterous); front tibia emarginate-anisochaete; abdomen with six (II-VII) pregenital sterna normally exposed, sterna III and IV separated or not by suture; male genitalia with parameres markedly different in form, and in many taxa, in size, left paramere enlarged, conchoid, right paramere more slender, as long as to much shorter than left paramere, parameres of most taxa asetose apically; ovipositor with gonocoxae rotated 90° in repose, without ramus, gonocoxa of two gonocoxites (or gonocoxa undivided either through loss of gonocoxite 2 or through fusion of 1 and 2), without accessory lobe; reproductive tract with spermatheca 1, without spermatheca 2 (Liebherr and Will 1998: 132-136, Figs. 38-52 and 54-55); secretions of pygidial glands markedly varied: most taxa with hydrocarbons, formic acid and unsaturated acids; some taxa with aliphatic ketones, saturated esters, higher saturated or unsaturated acids, or aromatic aldehydes, but not quinones; discharge by spraying (most taxa), or oozing (few taxa) (Moore 1979: 194-195).

This subfamily, the equivalent of the "Conchifera" of Jeannel (1941: 79, 80-81) includes 36 tribes, according to the classification of Kryzhanovskij (1976), or 34 tribes, according to the classification of Erwin (1991b). Recent studies (for example, Deuve 1986 and 1988) confirm the probable monophyly of the Harpalinae as defined here, though Deuve includes also the pseudomorphines. Kryzhanovskij (1976) introduced the rank of supertribe (indicated by the suffix "itae") to distinguish probable monophyletic assemblages of tribes. However, differences among different authors about composition of the supertribes are sufficient that we advocate use of the latter only informally. Our ideas about this topic are reflected in the following classification of the Nearctic tribes of Harpalinae:

| | |
|---|---|
| Pterostichitae: | Morionini; Loxandrini; Pterostichini; Zabrini |
| Harpalitae: | Harpalini |
| Chlaeniitae: | Licinini; Panagaeini; Chlaeniini; Oodini |
| Pentagonicitae: | Pentagonicini; |
| Platynitae: | Platynini |
| Ctenodactylitae: | Ctenodactylini |
| Lebiitae: | Perigonini, Lachnophorini, Odacanthini; Cyclosomini, Lebiini |
| Dryptitae: | Zuphiini; Galeritini |
| Anthiitae: | Helluonini |

Morionini

This pantropical tribe includes nine genera, of which only *Morion* Latreille 1810 is in the Nearctic Region. The position of this tribe is uncertain. The structure of the male genitalia seems to indicate that it should be placed near the Pterostichini, whereas a number of larval characters are distinctly scaritine. Van Emden (1953: 54) attributed the scaritine affinities to convergent evolution, and we incline toward this explanation. Lindroth (1969b: XXIII) accepted the opposite interpretation of the facts.

*Morion* Latreille 1810
*Morio auctorum*
This tropical-warm temperate genus with 41 species is represented in the following zoogeographic regions: Australian, Oriental, Palaearctic (eastern, north to the Japanese archipelago, and southern, north and west to Asia Minor), Afrotropical, Neotropical, and Nearctic. The two Nearctic species are the eastern *M. monilicornis* (Latreille 1806), ranging from the Gulf Coast northward to New York and South Dakota, and the southwestern *M. aridus* Allen 1968, which ranges from Baja California northward to southern Arizona. Evidently arboreal, both larvae and adults of *M. monilicornis* live under the bark of pine logs (Schwarz 1884), whereas adults of *M. aridus* have been collected in the stems of saguaro-type cacti. Reference. Allen 1968: 141-163 ( rev. W. Hemisph. spp.).

Loxandrini

This tribe, which comprises 11 genera, occupies the Western Hemisphere, Australia, New Guinea, and Celebes. Two genera are represented in the southern reaches of the Nearctic Region, primarily the southeastern United States. This group has been considered as part of the Pterostichini for a long time.

*Oxycrepis* Reiche 1843
subgenus *Stolonis* Motschulsky 1865
*Prostolonis* Mateu 1976
This indigenous Neotropical genus of 17 species in two subgenera is represented in the Nearctic Region by the single species, *O. (Stolonis) intercepta* (Chaudoir 1873), which ranges from southern Mexico to southern Texas and Arizona. An hygrophilous species, adults were collected in damp leaf litter near dry water courses of previously flowing streams. References. Allen and Ball 1980: 530-532 (descr. sp.).

*Loxandrus* LeConte 1852
This genus, with 225 species is represented in the Oriental Region (Indo-Australian Archipelago), the Australian Region, and the Western Hemisphere Neotropical and Nearctic Regions. Sixty or so species, inhabit the Nearctic Region, most of which have been reported from the Gulf Coastal states. Only a few species extend as far north as southern Ontario. Most of the species are markedly hygrophilous, living in swamp forests, and around the margins of sunlit ponds, and *Carex* and *Typha* marshes. Some, however, are mesophilous, living in open places, such as meadows, but not in the vicinity of standing surface water. Adults of most species are macropterous, flying at night, many coming to light sources. References. Allen 1972: 1-184 (rev. N.A. spp.). Will and Liebherr 1997: 230-238 (descr. of two species).

Pterostichini

This very diverse group is represented in all major regions of the world but is more diverse in the temperate areas, particularly of the Palaearctic Region. In Nearctic North America, the group includes about 250 species. They range all over the continent including the arctic tundra. The Platynini and Loxandrini, included in this group by several authors, are excluded from the Pterostichini in the present work. Reference. Bousquet 1999: 1-292 (rev. N.A. genera and subgenera).

*Abaris* Dejean 1831
This indigenous Neotropical genus includes 27 species which range from Brazil northward to Arizona and southeastern California. Only one species, *A. splendidula* (LeConte 1863), occurs in Nearctic North America. The species are mesophilous, adults being found in leaf litter in various types of tropical and subtropical forests, from lowland to lower mountain slopes. Reference. Bousquet: 1984: 384-388 (descr. N.A. sp.). Will 2000: 142-250 (thesis; spp. rev.).

*Ophryogaster* Chaudoir 1878
Five species are included in this indigenous Neotropical genus: two live in southwestern United States and Mexico and three in South America. Only *O. flohri* Bates 1882 (referred to as *Pterostichus arizonicus* Schaeffer 1910 in most North American publications) occurs in the Nearctic Region, in southern Arizona. Evidently mesophilous or hygrophilous, this species occupies riparian forest, adults being found in leaf litter, and at fruit falls (specifically palm fruit). The wings are fully developed in adults of *O. flohri* and many specimens have been collected at lights in Arizona. Reference. Darlington 1936: 150-152.

*Poecilus* Bonelli 1810

*Feronia* Latreille 1817

This genus is represented by about 130 species inhabiting the Palaearctic and Nearctic Regions, including the northern part of Mexico and Africa north of the Sahara Desert; 13 of these species live in Nearctic North America. Of the seven subgenera presently recognized, two are represented in the Nearctic Region.

subgenus *Poecilus s. str.*

*Sogines* Stephens 1828

*Americobius* Lutshnik 1915

*Leconteus* Lutshnik 1915

*Parapoecilus* Jeannel 1942

This Holarctic subgenus includes about 70 species of which 12 live in Nearctic North America. Adults are found in open places such as vacant or cultivated fields, meadows, ditches, sides of roads, and along edges of forests. The wings are fully developed in adults of most species.

subgenus *Derus* Motschulsky 1844

*Derulus* Tschitschérine 1896

This group includes one Nearctic and 25 Palaearctic species of which all but one occur in Asia. The Nearctic species, *P. nearcticus* Lindroth 1966, is presently known only from some islands of the Anderson River Delta in the Northwest Territories on tundra, but fossils of that species have been unearthed from the Yukon Territory, Alaska, and northeastern Siberia. Very little is known about way of life of these species.

*Lophoglossus* LeConte 1852

This North American genus comprises six species inhabiting eastern United States with one species, *L. scrutator* (LeConte 1848), ranging as far as southern Canada in the north. Hygrophilous and lacustrine or swamp-inhabiting, adults of these species live mainly in and around marshes and swamps, and in wet places near lakes. References. LeConte 1873: 316-317. Casey 1913: 143-146. Will 1999: 259-276 (key to spp.).

*Piesmus* LeConte 1848

The single species of this precinctive Nearctic genus, *P. submarginatus* (Say 1823), occurs in southeastern United States, north to North Carolina and Tennessee; Bates (1882) reports it from Mexico. Very little is known about way of life of *P. submarginatus*, other than that the species is a forest-inhabiting mesophile, with adults and larvae living under the bark of dead pine trees (J.V. Capogreco, personal communication). Reference. LeConte 1873: 302.

*Gastrellarius* Casey 1918

This precinctive Nearctic genus includes three species confined to the temperate regions east of the Mississippi River. The species are forest-inhabiting mesophiles. *Gastrellarius honestus* (Say 1823) ranges from southern Canada to North Carolina; adults being found almost exclusively under the bark of logs or fallen trees. The two other species are precinctive in the southern part of the Appalachian Mountains and live apparently in leaf litter. Adults are brachypterous. Reference. - Darlington 1931: 155-156.

*Stomis* Clairville 1806

This Holarctic genus is represented by about 29 species. The species of *Stomis* described by Say (1830) from Mexico is probably not a pterostichine. Two subgenera are presently recognized.

subgenus *Neostomis* Bousquet 1983

The sole species of this taxon, *S. termitiformis* (Van Dyke 1925), is known from a few adults collected in western Oregon and Vancouver Island, British Columbia (Bousquet 1999b). The species is a mesophile; adults have been caught in a variety of habitats including dense forests, sandy soil under bushes near seashore beaches, sand dunes, and open forests. Adults are brachypterous. Reference. Hacker 1968: 41-42.

subgenus *Stomis* s. str.

This indigenous Palaearctic subgenus contains 28 species with a collective range extending over most of the temperate part of the Palaearctic Region. One European species, *S. pumicatus* (Panzer 1795), was accidentally introduced into Nearctic North America and is known from a few specimens collected in Nova Scotia and in Prince Edward Island. The previous record from Quebec is very likely based on a mislabelled specimen. In Europe, most adults are found in moist habitats along rivers and creeks, under rocks and debris on open land and in deciduous forests, around buildings and particularly in gardens, or in nests of rodents. Reference. Lindroth 1966: 442-443.

*Stereocerus* Kirby 1837

*Boreobia* Tschitschérine 1896

This Holarctic genus includes two Holarctic species: *S. haematopus* (Dejean 1831) lives in Lapland, western Siberia, Manchuria, and in North America from Alaska eastward to Newfoundland and southward to the mountains of New England and Wyoming. The second species, *S. rubripes* (Motschulsky 1860), is known from Siberia and Alaska. Mesophilic, adults of both species are found under rocks or in moss in dry places in tundra and alpine areas; those of *S. haematopus* also occur in the northern coniferous region.

*Myas* Stürm 1826

*Trigonognatha* Motschulsky 1857

*Neomyas* Allen 1980

This Holarctic genus includes about 20 species inhabiting the temperate areas of the Palaearctic and Nearctic Regions and the mountains of Taiwan. Of these species, three are precinctive in Nearctic North America east of the Mississippi River, 15 occupy eastern Asia, and one is precinctive in southeastern Europe. The North American species are forest-inhabiting mesophiles, adults being found in leaf litter and under rotten logs in deciduous and mixed forests. Adults of many species are brachypterous, with elytra fused along the suture. Reference. Allen 1980: 1-29.

*Pterostichus* Bonelli 1810

This large genus is represented in Holarctica, as well as in the northern parts of the Neotropical and Oriental Regions. The Nearctic fauna comprises about 180 species distributed among 21 subgenera.

subgenus *Argutor* Dejean 1821

*Lagarus* Chaudoir 1838

*Pseudargutor* Casey 1918

*Pseudolagarus* Lutshnik 1922

Until recently, this subgenus was known under the name *Lagarus* Chaudoir. However, the name *Argutor* applies to this subgenus since the type species of the first valid type species designation of *Argutor* is the same as that of *Lagarus* (see Silfverberg 1983).

This group is Holarctic in distribution and includes nine species. Only one of these, *P. commutabilis* (Motschulsky 1866) (until recently reported in the literature under the name *P. leconteianus* Lutshnik 1922), occurs in North America where it is transcontinental in temperate regions. Hygrophilous, most adults are found on sandy wet places, often near bodies of water, such as temporary ponds and marshes.

subgenus *Phonias* des Gozis 1886

*Micromaseus* Casey 1918

*Americomaseus* Csiki 1930

Four species of this Holarctic subgenus are known from the Nearctic Region, one of which, *P. strenuus* (Panzer 1797), was accidentally introduced from Europe on both the Atlantic and Pacific coasts. About 25 species of *Phonias* have been described from the Palaearctic Region. The species are hygrophilous, with adults being found in moist to wet places in open or forested habitats; on occasion, they are found in great numbers in moors, mires, eurytopic swamps, wet meadows, bogs, and beaver houses. This subgenus was previously known under the name *Argutor*.

subgenus *Bothriopterus* Chaudoir 1835

*Dysidius* Chaudoir 1838

*Parargutor* Casey 1918

This group of 20 species is distributed mainly in Holarctica, with two species living in the Himalaya. Six of these species are precinctive in the Nearctic Region, with *P. tropicalis* Bates 1882 precinctive in the mountains of Mexico, *P. adstrictus* Eschscholtz 1823, circumboreal, and the remaining occur in the Palaearctic Region. The species are mesophilic or xerophilic, living mainly along forest edges and in nearby fields, where adults are found under rocks and various kinds of debris, as well as in woods and forests in leaf litter and under rocks and logs.

subgenus *Melanius* Bonelli 1810

*Omaseus* Dejean 1821

*Lyperus* Chaudoir 1838

*Metamelanius* Tschitschérine 1900

This Holarctic group includes eight species of which three are precinctive in the Nearctic Region, east of the Rocky Mountains, three are in China and Japan, and two are found over most of Europe and into the northern parts of Africa and Central Asia. The species are hygrophilous, with adults being found along river banks and lakes under rocks and various kinds of debris, in swamps, and marshes, and occasionally in or near bogs and fens. The Nearctic species, *P. castor* Goulet and Bousquet 1983, lives exclusively in inhabited or recently deserted beaver houses. Adults are macropterous.

subgenus *Pseudomaseus* Chaudoir 1838

In addition to the two very similar precinctive Nearctic species, the group contains about 15 species in the Palaearctic Region. The species are hygrophilous, adults being found in damp places, such as swamps, marshes, moors, mires, fens, bogs, occasionally beaver houses, and also along river banks and lakes, resting under rocks and various kinds of debris on soil rich in humus. Adults are macropterous. Reference. Bousquet and Pilon 1983: 389-395.

subgenus *Monoferonia* Casey 1918

This precinctive Nearctic group includes four species that range in the temperate regions east of the Mississippi River. The range of *P. diligendus* Chaudoir 1868 extends from southern Québec south to North Carolina. Mesophilous, this species lives in deciduous and mixed forests, in damp places, such as swamps and along small creeks and streams, where adults rest under rocks and logs. The remaining species occur in the mountains of southern Appalachians, with adults resting in leaf litter, and under moss and logs. Adults are brachypterous. Reference. Darlington 1931: 157-163.

subgenus *Feronina* Casey 1918

The precinctive Nearctic subgenus includes two North American species, with *P. palmi* Schaeffer 1910, at high altitude in the southern part of the Appalachians. An undescribed species occurs in northeastern Kentucky, eastern West Virginia, and western Virginia. The species are probably mesophilous and forest-inhabiting. Adults are brachypterous. Reference. Ball 1965: 106-107.

subgenus *Paraferonia* Casey 1918

The sole species of this precinctive Nearctic subgenus, *P. lubricus* LeConte 1852, is in the southern part of the Appalachians. Mesophilous and forest-inhabiting, adults of this species rest under rocks and litter along streams, springs or seeps. Adults are brachypterous. Reference. Ball 1965: 106.

subgenus *Pseudoferonina* Ball 1965

*Melvilleus* Ball 1965

This precinctive Nearctic subgenus including seven species is confined to the Pacific Northwest ranging from Idaho and Oregon, north to Washington. The species are mesophilous forest inhabitants, with adults resting in leaf litter or beneath other forms of cover. Adults are brachypterous. Reference. Bousquet 1985: 253-260.

subgenus *Gastrosticta* Casey 1918
This precinctive Nearctic group of 10 species is restricted to the temperate regions of United States east of the Rocky Mountains. They are mesophilous forest inhabitants, with adults resting in leaf litter or under other forms of forest cover. Adults are brachypterous. Reference. Bousquet 1992: 510-519.

subgenus *Morphnosoma* Lutshnik 1915
*Omaseidius* Jeannel 1942
This indigenous Palaearctic group is represented by three species whose range extends from Western Europe south to the Caucasus Mountains and east to Siberia. One of these, *P. melanarius* (Illiger 1798), has been accidentally introduced in North America on both coasts and has spread rapidly. Pronouncedly synanthropic, the species occupies meadows, cultivated fields, road sides, ditches, land around human habitations, and forests. The species is wing-dimorphic.

subgenus *Euferonia* Casey 1918
This precinctive Nearctic subgenus includes six species which occur east of the Rocky Mountains. Mesophilous, they live in forested places as well as in open fields and meadows, adults resting in leaf litter and under logs and rocks. Adults are brachypterous.

subgenus *Lenapterus* Berlov 1996
This Holarctic subgenus includes nine species. Of these, four are precinctive Nearctic, including *P. punctatissimus* (Randall 1838) adults of which rest mainly inside small rotten logs and in leaf litter in the boreal forest from Yukon to Newfoundland. Three species are Holarctic in distribution and live in the tundra.

This group was known under the name *Lyperopherus* Motschulsky. However, the type species of this taxon, designated by Lindroth (1966), belongs to another group. Reference. Budarin 1976: 32-38.

subgenus *Metallophilus* Chaudoir 1838
*Lyperopherus* Motschulsky 1844
This northern Holarctic group includes seven species of which one, *P. sublaevis* Sahlberg 1880, is itself Holarctic, occurring in the subarctic and arctic regions of Siberia and northwestern North America; the other species are precinctive in the mountainous regions of Siberia and northern Mongolia. The species is a tundra-inhabiting hygrophile. Adults are brachypterous.

subgenus *Abacidus* LeConte 1863
*Peristethus* LeConte 1863
The five species of this precinctive Nearctic subgenus are restricted to the temperate regions of eastern United States and southern Ontario. Mesophilous, the species occupy open places, such as cornfields, vacant areas, and ravines, as well as forests, adults resting under the cover available in such habitats. Adults are brachypterous.

subgenus *Orsonjohnsonus* Hatch 1933
This precinctive Nearctic subgenus contains only *P. johnsoni* Ulke 1889, which is found in Washington and Oregon. Adults rest under rocks or among gravel along forest creeks and in the vicinity of water falls. Adults are brachypterous. Reference. Hatch 1953: 114. Lindroth 1966: 473.

subgenus *Lamenius* Bousquet 1999
This precinctive Nearctic subgenus includes only *P. caudicalis* Say 1823, which ranges from Newfoundland to British Columbia south to Virginia in the east and Colorado in the west. The species is hygrophilous, with adults resting under rocks and various kinds of debris along rivers, lakes and swamps.

subgenus *Eosteropus* Tschitschérine 1902
*Refonia* Casey 1918
This Holarctic group includes 23 species distributed throughout the Northern Hemisphere. Three species live in North America, of which *P. circulosus* Lindroth 1966, is northern, ranging from central Alaska to the central Yukon Territory (Ball and Currie 1997: 451). The species is hygrophilous, with adults inhabiting bogs and small marshes, resting near the surfaces thereof, in moss, or under other types of cover. The mesophilous forest-inhabiting species *P. moestus* (Say 1823) and *P. superciliosus* (Say 1823) occur far to the south and east, in the Appalachian region as far north as Pennsylvania, the adults resting in leaf litter and under rocks, logs and moss in deciduous and coniferous forests. Adults of these species are brachypterous.

This subgenus has been known under the name *Steropus* Dejean 1821 but its type species, *P. madidus* Fabricius 1775 (see Silfverberg 1983), belongs to another subgenus, exclusively Palaearctic.

subgenus *Cylindrocharis* Casey 1918
This precinctive Nearctic group of three species occupies the Appalachian region. Mesophilous and forest-inhabiting, the species live in dense, mature forests, where adults rest inside rotten logs or under rocks, logs, and bark of fallen trees. Adults are brachypterous. Reference. Barr 1971b: 1-14.

subgenus *Hypherpes* Chaudoir 1838
*Haplocoelus* Chaudoir 1838
*Brachystilus* Chaudoir 1838
*Holciophorus* LeConte 1852
*Gonoderus* Motschulsky 1859
*Hammatomerus* Chaudoir 1868
*Leptoferonia* Casey 1918
*Pheryphes* Casey 1920
*Anilloferonia* Van Dyke 1926
As presently conceived, this precinctive Nearctic group is the most diverse subgenus of *Pterostichus* in North America. This group comprises about one hundred valid species. All but two species occur along the West Coast and in the Rocky Mountains from Alaska to the mountains of New Mexico and Ari-

zona. Characteristic of adults of *P. morionoides* Chaudoir 1868 is a very large head, with bulged temples and small eyes (Fig. 54.6). The two eastern species, *P. adoxus* (Say 1823) and *P. tristis* (Dejean 1828), range from southeastern Canada southward to North Carolina and Tennessee. Most species are mesophilous forest inhabitants, where adults rest under rocks, logs, and the bark of fallen trees, or in leaf litter and moss. Some species are hygrophilous, and occur near creeks and streams. Adults are brachypterous. Reference. Casey 1913: 99-126.

subgenus *Cryobius* Chaudoir 1838
This Holarctic group includes approximately 110 species. The Nearctic Region has 28 species of which 21 are precinctive; collectively, the species range all across the boreal and arctic regions, south along the Rocky Mountains to Colorado, and in the east, along the northern Appalachians to the New England states. The species are mesophilous, inhabiting forests, riparian situations, and arctic or alpine tundra, the adults resting in moss, under rocks and various kinds of debris. Those living in the boreal forests are collected usually in leaf litter and under logs, often near rivers. Adults are brachypterous. Reference. Ball 1966b: 1-166.

*Cyclotrachelus* Chaudoir 1838
This indigenous Nearctic genus includes 43 species living in the temperate regions of North America east of the Rocky Mountains, north to southern Ontario and south to the state of Durango in México. All species are brachypterous. Two subgenera are recognized. Reference. Freitag 1969: 88-211 (rev. spp.).

subgenus *Cyclotrachelus s. str.*
*Fortax* Motschulsky 1866
*Ferestria* Leng 1915
This subgenus comprises 18 species ranging from Pennsylvania to Illinois and south to eastern Texas and Florida. They are mesophilous forest inhabitants, with adults resting in leaf litter or under logs and under bark on logs.

subgenus *Evarthrus* LeConte 1852
*Anaferonia* Casey 1918
*Megasteropus* Casey 1918
*Eumolops* Casey 1918
*Evarthrinus* Casey 1918
*Evarthrops* Casey 1920
This group includes 25 species which range all over eastern United States west to South Dakota and eastern Arizona. The species *C. sodalis* (LeConte 1848) ranges northward into extreme southern Ontario and *C. substriatus* (LeConte 1848), reaches the state of Durango, in Mexico. The mesophilous species inhabit forests, adults resting in leaf litter or under other forms of cover. The xerophilous species occupy open places, such as corn fields and pastures, the adults resting under rocks and various kinds of debris.

*Abax* Bonelli 1810
Members of this indigenous Palaearctic genus of 17 species range over most of Europe and Anatolia but are better represented in the Alps than anywhere else. One species, *A. parallelepipedus* (Piller and Mitterpacher 1783), has been accidentally introduced along the Atlantic coast of Canada. The species are primarily mesophilous forest inhabitants, but they occur also in alpine prairies, and in open places. Adults are brachypterous and the elytra are fused.

Zabrini

Known previously as Tribe Amarini in most North American publications treating Carabidae, Zabrini is the older and therefore correct name (Madge 1989: 468; cf. 460). Primarily Holarctic, this tribe is represented also in the northern portions of the Oriental, Ethiopian, and Neotropical Regions. Zabrines are not known from the Australian Region nor from South America. The numerous species of this tribe are included in three genera: the monospecific precinctive Nearctic *Pseudamara* Lindroth 1968; the Holarctic *Amara* Bonelli 1810, whose 517 species are arrayed in 46 subgenera; and the precinctive Palaearctic *Zabrus* Clairville 1806, with 100 species, in 11 subgenera.

The Zabrini consists primarily of xerophilous-mesophilous geophile species, occupying grassland, pastures, and areas that have been opened to sunlight, or otherwise disturbed, by man. Some species are riparian, but they live in dry, sandy areas, on flood plains.

Closely related to the Pterostichini, larval features seem to provide the best diagnostic features for the Zabrini. However, if phylogenetic analysis indicates a position for the zabrine lineage within, rather than at the base of the pterostichine radiation, then the two groups should be combined.

The arrangement of the accounts and ranking of supraspecific groups below are based partly on Hieke (1995). References. Casey 1918: 224-318. Hieke 1995: 1-163.

*Amara* Bonelli 1810
*Lirus* Zimmermann 1832
Like the Tribe Zabrini, this genus of 520 species arrayed in 45 subgenera is primarily Holarctic, with numbers declining in the tropical regions of Megagaea. In Nearctic North America, *Amara* is represented by 102 species distributed among six subgenera and one species group. Twenty-two species are shared between the Nearctic and Palaearctic Regions: 11 are naturally Holarctic, and 11 were introduced. The range of this genus extends throughout the Nearctic Region, though the species are relatively few in the Arctic zone, and in the humid southeast. Adults of most species are macropterous.

subgenus *Zezea* Csiki 1929
*Triaena* LeConte 1848, not Hübner 1818
*pallipes* species group Lindroth 1968
This Holarctic subgenus includes in Nearctic North America eight precinctive species whose collective range is transcontinental at middle latitudes.

subgenus *Amara (sensu stricto)*
*Celia* Zimmermann 1832
*Isopleurus* Kirby 1837
*quenseli* species group Lindroth 1968
*aurata* species group Lindroth 1968
*lunicollis* species group Lindroth 1968
This Holarctic subgenus (or complex of subgenera), with 168 species, includes in Nearctic North America 58 species, of which seven are Holarctic and seven were introduced from the Palaearctic Region. As treated here, this complex includes three species groups recognized by Lindroth (1968). Hieke (1995) recognized *Celia* and *Paracelia* as distinct from *Amara (sensu* Lindroth). The collective range of the group extends from the tundra in the Northwest Territories and Newfoundland southward to the Gulf Coast in the east, and to California and Arizona in the southwest.

Identification of many of the species is accomplished with considerable difficulty. Thanks to the careful work of Lindroth (1968: 692-735), adults of most of the more northern species may be identified, but either close relationship among sympatric species, or local, more or less extensive, hybridization may make identification of some specimens difficult or impossible. A study based on molecular methods might be required to solve some species pairs or complexes.

subgenus *Percosia* Zimmermann 1832
*obesa* species group Lindroth 1968
This Holarctic subgenus of six species includes in Nearctic North America one precinctive species, *A. obesa* Say 1823, whose geographical range extends throughout the middle latitudes of the continent, from southern Canada south to Arizona and Georgia. This species inhabits situations with sandy soil and sparse vegetation. A second species, referred to this group by Lindroth (1968: 691) was removed to the following subgenus, *Neopercosia* Hieke.

subgenus *Neopercosia* Hieke 1978
*obesa* species group (in part) Lindroth 1968
This is a monospecific group including only the precinctive Nearctic North American species *A. fortis* LeConte 1880. It is known only from eastern Texas. Adults are brachypterous. Reference. Hieke 1978: 289.

subgenus *Bradytus* Stephens 1827
*Leiocnemis* Zimmermann 1832
*apricaria* species group Lindroth 1968
Holarctic in distribution, this subgenus of 45 species includes 13 species in Nearctic North America, one of which is Holarctic naturally, and two were introduced from the Palaearctic Region. The collective ranges of the included species extend throughout Nearctic North America. We continue to include here *A. colvillensis* Lindroth 1968, which was assigned to the subgenus *Reductocelia* Lafer 1989 by Hieke (1999).

The *insignis* species group
Recognized by Lindroth (1968:691-692), the two precinctive Nearctic species included here were placed in *Bradytus* by Hieke (1995: 40). Both species are markedly western, with *A. insularis* Horn 1875 known only from San Clemente Island in the Channel Islands, off the west coast of southern California, and *A. insignis* Dejean 1831 ranges from southern California to Oregon, occurring also in the Channel Islands and Guadalupe Island.

subgenus *Curtonotus* Stephens 1828
*Cyrtonotus* (emendation)
*aulica* species group Lindroth 1968
This Holarctic subgenus of 84 species includes 19 species in Nearctic North America, five of which are naturally Holarctic, with one having been introduced from the Palaearctic Region. Predominantly northern, in North America, this subgenus ranges from tundra southward to Georgia in the east and to California in the west. Adults are rather long and slender, *Pterostichus*-like in appearance, in contrast to the broad form generally characteristic of *Amara*. The species occupy open, rather dry places, grassy fields on clay soil; some live on riparian flood plains. References. Hieke 1994: 310.

*Pseudamara* Lindroth 1968
*Leironotus auct.* (not Ganglbauer 1892)
*Disamara* Lindroth 1976
*Harpaloamara* Hieke 1995
This precinctive Nearctic genus is represented in eastern North America by the single species *A. arenaria* LeConte 1848, with its geographical range extending from New Brunswick westward to Illinois, and southward to West Virginia. The adult stage is wing-dimorphic.

## Harpalini

This is one of the more diverse tribes of carabid beetles, and its members occupy all of the major zoogeographical regions. Like the pterostichines, harpalines range throughout the whole of North America, save for the Arctic Coast. Noonan (1976), based on study of the world fauna, proposed the classification of Harpalini adopted here. The subtribes, genus-groups and genera in the Nearctic Region are as follows.

Subtribe Pelmatellina
*Pelmatellus*
Subtribe Anisodactylina
Genus-group Notiobii
*Notiobia*
Genus-group Anisodactyli
*Xestonotus, Anisodactylus, Geopinus, Amphasia,* and *Dicheirus*
Subtribe Stenolophina
Genus-group Stenolophi
*Stenolophus, Bradycellus, Amerinus, Dicheirotrichus*, and *Acupalpus*
Genus-group Polpochili

*Polpochila,* and *Pogonodaptus*

Subtribe Harpalina

Genus-group Selenophori

*Trichotichnus, Selenophorus, Discoderus,* and *Stenomorphus*

Genus-group Trichotichni

*Aztecarpalus* and *Trichotichnus*

Genus-group Harpali

*Piosoma, Euryderus, Ophonus, Cratacanthus, Harpalus,* and *Hartonymus.*

Genus-group Dapti

*Cratacanthus*

The Pelmatellina constitute a small group, probably related to the Anisodactylina, the species of which occupy the Australian, Neotropical and southern Nearctic Regions (Noonan 1976: 6).

Anisodactylines are worldwide, with Notiobii exhibiting a distribution pattern consistent with time of origin prior to separation of South America and Africa, hence Gondwanan, and time of origin of Anisodactyli following the breakup of Gondwana (Noonan 1973: 476-479, figs. 246-252).

The Stenolophi include many "microharpalines" (an informal designation for an assemblage with adults of most of the species being less than 7 mm in length); most of the species being warm temperate or tropical in distribution. The far north, however, has at least one group of species. The Polpochili (Cratocarina auct.) are restricted to the Western Hemisphere, ranging from southern South America to southern Texas and Arizona (Noonan 1976: 15-16). The group comprises wet-adapted forms, with the northern taxa occurring in desert areas, but living there in relict wet meadows.

Trichotichni and Selenophori of the subtribe Harpalina are too closely related to the Harpali to be placed in a separate subtribe. They occupy all zoogeographical regions, but are especially abundant in the tropical and temperate areas (Noonan 1976: 29-30). Trichotichnines, which are forest-adapted, in the Nearctic Region are confined to eastern North America. Both subgenera of Nearctic *Trichotichnus* are represented by numerous species in eastern Asia. In North America, selenophorines reach only extreme southeastern and southwestern Canada. The group is ecologically diverse, with some genera primarily dry-adapted, while others are inhabitants of marshes.

The genus-group Harpali of the subtribe Harpalina is restricted to the Holarctic, Afrotropical, and Oriental Regions (Megagaea of Darlington 1957). This group, as a whole, is adapted to dry conditions, most of the species occurring either in desert areas, or in dry situations in mesic areas. Some taxa exhibit fossorial adaptations. From an ecological perspective, the Palaearctic Ditomi probably are the most interesting, for at least some of the species exhibit subsocial behavior, with adults supplying food in the form of umbelliferous seeds for the developing larvae (Brandmayr and Zetto Brandmayr 1974 and 1979: 36).

The Nearctic harpaline fauna shares genera with the Palaearctic Region or has there putative close relatives. Various elements are shared with the Neotropical Region and, as well, some genera are indigenous, especially among the Harpalina. At the species level, the ranges of most of those representing Neotropical elements extend into (or emanate from) the Neotropical Region. Palaearctic associations are more varied: some species are themselves Holarctic; others have close Palaearctic relatives; and others have more distant Palaearctic relationships.

Thus the distribution pattern of the Harpalini is quite complex, and it suggests that the tribe has had a long and complex history. On the whole, the Nearctic Harpalini are understood inadequately, though during the last 40 or so years, much progress has been made. Some of the genera with larger adults (i.e., *Notiobia*, *Dicheirus*, *Anisodactylus*, and *Harpalus*) have been the subjects of intensive study. Among the groups of smaller harpalines, *Pelmatellus* has been reviewed, in part. But the Stenolophi and Selenophori are in need of study to define the species, to determine the affinities of the species groups, and to determine the limits of the genera. We have yet to have a review of the Nearctic harpalines (a symphony of "harps") as comprehensive as that produced for the Harpalini of the Afrotropical Region (Basilewsky 1950-51).

### Subtribe Pelmatellina

This subtribe comprises seven genera collectively with a Gondwanan distribution pattern, including Australia, New Zealand, Andean South America, and north through Middle America to southwestern United States.

*Pelmatellus* Bates 1882

This indigenous Neotropical genus includes 17 species arranged in three subgenera. The subgenus *Pelmatellus* s. str. includes 10 species, ranging in the mountains through Middle America and into southwestern United States, where it is represented by two species, known from southern Arizona and New Mexico. Both species range widely in Mexico. References. Goulet 1974: 80-102. Reichardt 1977: 422.

### Subtribe Anisodactylina

This subtribe is worldwide in distribution, but the 30 genera are included in two zoogeographically distinctive assemblages: the Notiobii are principally in the Southern Hemisphere, exhibiting a Gondwanan-type pattern, while the Anisodactyli are abundant in the Afrotropical Region but have a good representation in Holarctica (Noonan 1973: 403-406). Both complexes are represented in Nearctic North America, where as one might expect, anisodactyloids are the more diverse. In the Nearctic Region six genera are represented: one notiobioid (*Notiobia* Perty 1830); and five anisodactyloids (*Xestonotus* LeConte 1853; *Anisodactylus* Dejean 1821; *Geopinus* LeConte 1848; *Amphasia* Newman 1838; and *Dicheirus* Mannerheim 1843). Reference. Noonan 1973: 267-480 (rev. genera, worldwide. 1976: 8-15. Bousquet and Tchang 1992: 751-775 (larvae, E. N.A.: key to genera).

*Notiobia* Perty 1830
This genus includes 90 species arranged in three subgenera: the Australian-Neotropical-Nearctic *Anisotarsus* Chaudoir 1837; the Neotropical *Notiobia* (*s. str.*), and the Afrotropical *Diatypus* Murray 1858 (Noonan 1973: 293-294).

subgenus *Anisotarsus* Chaudoir 1837
*Diapheromerus* Chaudoir 1843
*Eurytrichus* LeConte 1848
*Stilbolidus* Casey 1914
This subgenus, with one species, *N. cupripennis* (Germar 1824) (Desender and Baert 1996: 345) on remote Easter Island, is Australian-Western Hemisphere in distribution, including 53 species, of which eight occur in Nearctic North America, or are confined thereto. They range from the Mexican border and Gulf Coast north to southern Canada in the east. The species are mesophilous or xerophilous, living in open deciduous woods, meadows and disturbed areas. Adults are found in such areas, resting during the day under available cover of leaf litter, stones, or logs. Adults of most species are macropterous, are nocturnal, and many fly to ultra violet light sources. Adults of at least one species, *N. terminata* (Say 1823), eat seeds of fireweed (*Epilobium*). References. Noonan 1973: 295-321 (rev. N.A. and M. A. spp).

*Amphasia* Newman 1838
Two species, each in its own subgenus, represent this precinctive North American group; both are in northeastern United States and the adjacent portions of Canada. Xerophilous, the species occupy open deciduous forests and fields, where adults are found resting under available cover during the day, on dry or damp clay soil.

subgenus *Pseudamphasia* Casey 1914
*sericea* species group Lindroth 1968
The species *A. sericea* (Harris 1828) is the sole representative of this subgenus, and occurs only in eastern and midwestern United States and the adjacent portions of Canada, from southern Québec and Ontario southward to South Carolina and westward to Missouri and Manitoba. Adults eat seeds, which they obtain from flowering heads of herbaceous plants; they fly at night, and many are seen at light sources, such as U-V light traps, especially in the southeast.

subgenus *Amphasia* (*sensu stricto.*)
*interstitialis* species group Lindroth 1968
This subgenus includes only *A. interstitialis* (Say 1823), whose geographical range is about the same as that of *A. sericea* (above). Habits are similar to those of *A. sericea*, but adults are less frequent at light, at night.

*Xestonotus* LeConte 1853
Ranked by some recent authors as a subgenus of *Anisodactylus* (*s. lat.*), Noonan (1973: 347-348) stated that the distinction between the two groups indicated sufficient divergence to support generic rank. This precinctive Nearctic group is monospecific, containing only *A. lugubris* (Dejean 1829), which lives at mid-latitudes in the east, ranging from North Carolina to Nova Scotia, and west to Iowa and South Dakota.

*Anisodactylus* Dejean 1829
This genus includes 53 species, arrayed in 10 subgenera. It is Holarctic in distribution (marginal in the Oriental Region), and is represented in Nearctic North America by 34 species in six subgenera. Its range extends from the Mexican border and Gulf Coast northward to central Canada.

subgenus *Aplocentroides*, **new name**
*Aplocentrus* LeConte 1848 (not Rafinesque-Schmalz 1819)
*A. amaroides* species group Lindroth 1968
*A. caenus* species group (in part) Lindroth 1968
Two species comprise this precinctive Nearctic group, one (*A. caenus* (Say 1823)) from eastern United States and the Mississippi Basin, and one (*A. amaroides* LeConte 1851) from the Pacific Northwest.

subgenus *Pseudaplocentrus* Noonan 1973
*A. caenus* species group (in part) Lindroth 1968
Monospecific, this precinctive Nearctic subgenus is based on *A. laetus* Dejean 1829, which is eastern in geographical distribution, ranging from the Gulf Coast northward to southern Ontario, and westward to Kansas.

subgenus *Spongopus* LeConte 1848
Bousquet and Tchang (1992: 774), on the basis of larval features, suggest that this monospecific, Nearctic precinctive subgenus might better be ranked as a separate genus. The one species, *A. verticalis* (LeConte 1848), lives in eastern United States and southern Canada, westward to the Mississippi Basin. It is mesophilous, adults being found in damp open woods or fields, with dense herbaceous growth.

subgenus *Anisodactylus* (*sensu stricto*)
*Cephalogyna* Casey 1918
This subgenus contains 22 species arranged in five species groups. It is Holarctic and marginally Oriental in distribution. Fourteen species are in Nearctic North America, one of which, *A. californicus* Dejean 1829, occurs also in northern and central Mexico. On the whole, the group is continent-wide, ranging northward to central Canada in the west. One species, *A. binotatus* (Fabricius 1787), is a Palaearctic introduction in the west, now recorded from southwestern British Columbia, western Washington, and northwestern Oregon. The species are mesophilous or hygrophilous, inhabiting pastures and disturbed areas, usually on substrate of damp clay soil rich in organic matter. References. Noonan 1996: 1-210 ( rev. world spp.).

subgenus *Anadaptus* Casey 1914
This precinctive Nearctic group of eight species is transcontinental at middle latitudes, ranging from southern Canada to at least northern Mexico (Baja California). The eastern species are

hygrophilous, living in the vicinity of standing water (sunlit *Carex* marshes, etc.), on clay soil.

subgenus *Gynandrotarsus* La Ferté-Sénectère 1841
*Triplectrus* LeConte 1848
This is an indigenous Nearctic subgenus, the nine species of which are found in eastern United States, including the Mississippi Basin, the adjacent parts of Canada, the southwest, as far as Arizona, and Mexico south to the Sierra Transvolcanica. The species are xerophilous, living in areas of sandy soil, with sparse vegetation composed of grasses and herbs. Adults fly at night, and some are taken at light sources. Reference. Noonan 1973: 354-373 (rev. spp).

*Geopinus* LeConte 1848
Bousquet and Tchang (1992: 774), on the basis of larval features, suggest that this monospecific Nearctic genus might be the adelphotaxon of *Gynandrotarsus*, and thus more appropriately ranked as a subgenus of *Anisodactylus*. The single species, *G. incrassatus* (Dejean 1829), ranges from Massachusetts and Georgia in United States westward to Nevada, and in southern Canada, from Quebec westward to Alberta. The habitat appears to be the sandy banks of rivers; adults fly at night, and many are seen at light sources, such as U-V light traps.

*Dicheirus* Mannerheim 1843
The five species representing this indigenous Nearctic genus are restricted to the West Coast, ranging from Baja California (Mexico) northward to southern British Columbia. References. Noonan 1975: 1-15 (species rev.).

Subtribe Stenolophina

Bradycellina
Acupalpina
This subtribe is worldwide, including numerous species arranged in 35 genera. In the Nearctic Region, 106 species are included in seven genera. The group is continent-wide, ranging from Arctic tundra southward to the Mexican border and Gulf Coast. References. Casey 1913: 218-229. Noonan 1976: 15-28.

*Stenolophus* Dejean 1821
This genus is represented in all of the major zoogeographical regions of the world, but most of the species are in Holarctica. The 75 species are arranged in four subgenera. The 32 Nearctic species, all of which are indigenous, are arranged in three subgenera. The range of the group is transcontinental, from southern Canada southward to the Gulf Coast and Mexican border. The species occupy open areas, wet or dry, depending upon the group to which they belong.

subgenus *Stenolophus* (*sensu stricto*)
Fifty-seven species are included in this Holarctic subgenus. Of these, 20 are indigenous in the Nearctic Region. Their collective range is transcontinental, extending from southern Canada in the north, southward to the Gulf Coast and Mexican border. The species are hygrophilous, living in damp meadows and near the border of ponds and lakes. Adults fly at night, and many are seen at light sources.

subgenus *Agonoderus* Dejean 1829
This is an indigenous Nearctic subgenus of seven species, restricted to the temperate portion of the continent, from southern Canada southward to the Gulf Coast and Mexican border. The group as a whole ranges over all of United States, as well as southern Canada. The species live in grassy areas, agricultural fields, and on the borders of eutrophic ponds. Adults fly at night, and many are seen at light sources. The species *S. lecontei* (Chaudoir 1868) is the seed-corn beetle of the official list of common names.

subgenus *Agonoleptus* Casey 1914
*conjunctus* species group Lindroth 1968
This Nearctic indigenous subgenus includes five species with a transcontinental range extending from southern Canada southward to the Gulf Coast and Mexican border. The species live in open places, in dry situations, with sandy substrate. Reference. Bousquet 1990: 203-204.

*Bradycellus* Erichson 1837
The 128 species of this virtually worldwide genus of small beetles are arranged in 10 subgenera. Fifty-one species are in Nearctic North America, arranged in five subgenera. The geographical range of the group is transcontinental, from southern Canada, south to the Gulf Coast and Mexican border. The species occupy a variety of habitats; adults fly at night.

subgenus *Bradycellus* (s. str.)
*Tetraplatypus* Tschitschérine 1897
*harpalinus* species group Lindroth 1968
This subgenus, whose aggregate range includes the Palaearctic Region and part of the Western Hemisphere, is represented in North America by two species, known from the West Coast. These are *B. fenderi* Hatch 1951, and *B. harpalinus* (Audinet-Serville 1821). The latter species was apparently introduced by accident at Vancouver, British Columbia, from the western part of the Palaearctic Region. It occupies disturbed habitats, with Europeanized vegetation. Reference. Hatch 1953: 186. 1955: 10. Lindroth 1968: 883-884.

subgenus *Liocellus* Motschulsky 1864
*Glycerius* Casey 1884
*nitidus* species group Lindroth 1968
Eight species comprise this precinctive Nearctic subgenus. Their aggregate range, primarily western extends from central Mexico (Sierra Transvolcanica) to southeastern British Columbia. Xerophilous, the species live in dry open situations, such as grassland and pastures, at high altitude farther south.

subgenus *Catharellus* Casey 1914

*lecontei* species group, Lindroth 1968

This precinctive Nearctic subgenus includes only *Bradycellus lecontei* Csiki 1932, a transcontinental species with a range extending northward to Dawson, Yukon Territory, and southward to Colorado in the west and Illinois in the east. This species is hygrophilous, living in wet open situations, near margins of small ponds.

subgenus *Stenocellus* Casey 1914

*rupestris* species group, Lindroth 1968

Thirty-five species comprise this indigenous North American subgenus, of which 33 occupy Nearctic North America. The group as a whole is transcontinental, ranging from southern Canada southward to the Gulf Coast and Mexican border. Principally hygrophilous, the species live in the vicinity of standing surface water. Nocturnal flight activity results in numerous individuals being attracted to light, especially in the southern states. The species are inadequately understood, and the group needs revision.

subgenus *Lipalocellus*, **new name**

*Liocellus* Tschitschérine 1901 (not Motschulsky 1864)

*nigrinus* species group, Lindroth 1968

Precinctive in the Nearctic Region, this subgenus includes two species (*B. nigrinus* (Dejean 1829), and *B. semipubescens* Lindroth 1968) ranging from southern Alaska southward to northern California, eastward to Newfoundland, south in the Rocky Mountains to New Mexico and Arizona, and south to North Carolina, on the east coast. Hygrophilous, both species occupy open areas, on moist ground, such as the vicinity of marshes.

subgenus *Triliarthrus* Casey 1914

*badiipennis* species group, Lindroth 1968

This precinctive Nearctic subgenus is represented by six species whose aggregate range extends throughout United States and Canada to about 57°N. Lat. The species are hygrophilous, living in open areas, in the vicinity of *Carex* marshes, etc. Adults have been swept from herbaceous plants, which are evidently climbed by the beetles to obtain seeds from the flowering heads.

*Amerinus* Casey 1914

This precinctive eastern Nearctic genus includes only *A. linearis* LeConte 1863, which ranges from the Gulf Coast north to southern Ontario and Massachusetts, and west to Kansas. Specimens have been collected in sandy soil, near Lake Michigan.

*Dicheirotrichus* Jacquelin du Val 1857

The 47 species of this Holarctic genus are arranged in four subgenera. Two species, each representing a different subgenus, are themselves Holarctic. Previous workers on the Nearctic carabid fauna have included these species in the genus *Trichocellus*, that name now used for a subgenus (for an explanation, see Kryzhanovskij *et al.* 1995: 135, footnote 269).

subgenus *Oreoxenus* Tschitschérine 1899

This Holarctic monospecific subgenus has only the boreo-montane species *D. mannerheimi* (R. F. Sahlberg 1844), which ranges in the Nearctic Region from Alaska eastward to Quebec, and southward at high altitude in the Rocky Mountains to Colorado.

subgenus *Trichocellus* Ganglbauer 1892

*Bradycellus* Westwood 1838, not Erichson 1837

This subgenus includes 31 species, one of which, *D. cognatus* (Gyllenhal 1827), is Holarctic, occurring in the Nearctic Region, where it is transcontinental, boreo-montane, ranging from Alaska (Attu, in the Aleutian Islands), east to Newfoundland south to Pennsylvania, South Dakota, and California, and in the mountains, south to New Mexico and Arizona. The species is found on tundra and southward in forests, around the margins of marshes and in leaf litter on drier ground.

*Acupalpus* Latreille 1829

This genus, with 130 species arranged in nine subgenera, is represented in all of the major zoogeographical regions of the world. Four subgenera with 17 species (one introduced) are known from the Nearctic Region, where the group is predominantly eastern, ranging from the Gulf Coast north to Newfoundland and Manitoba. The species are hygrophilous, inhabiting pond margins, in dense vegetation (*Carex*, etc.).

subgenus *Philodes* LeConte 1861

*Aepus* LeConte 1848, *nec* Samouelle 1819

*Goniolophus* Casey 1914

Five species comprise this indigenous eastern Nearctic subgenus. The geographical range extends from the Gulf Coast northward to southern Quebec and Ontario in the east, and South Dakota in the west. Reference. Casey 1914: 260-264.

subgenus *Acupalpus* (*sensu stricto*)

This worldwide group of 77 species is represented in the Nearctic Region by six species. This subgenus is transcontinental at middle and southern latitudes. One species, *A. meridianus* (Linnaeus 1761), is a Palaearctic species which was accidentally introduced on the west coast and has become almost transcontinental (Pollock 1991: 705-706). Unlike the other Nearctic species of *Acupalpus*, *A. meridianus* lives in mesic or dry sites, in disturbed habitats.

subgenus *Tachistodes* Casey 1914

*pauperculus* species group, Lindroth 1968

Four species are included in this precinctive Nearctic group, all of which are eastern, ranging from the Gulf Coast north to southeastern Canada. Several range as far west as Kansas and Texas. Reference. Casey 1914: 286-289.

subgenus *Anthracus* Motschulsky 1850

*tener* species group, Bousquet and Larochelle 1993: 229

This subgenus, with 31 species, ranging throughout the Palaearctic and Oriental Regions, includes two western Nearctic species: *A. tener* (LeConte 1857), described from California, and *A. punctulatus* Hatch 1953, described from northwestern Oregon.

*Polpochila* Solier 1849
This indigenous Neotropical genus includes 21 species in two subgenera, ranging from Chile, in southern South America northward to southern Arizona in the Nearctic region. Three species are in the Nearctic Region, each of them represented farther south, in Mexico. Although adults of these taxa fly to light sources at night in desert areas, they are really hygrophilous. Reference. Nègre 1963: 216-239.

subgenus *Phymatocephalus* Schaum 1864
*capitata* species group, Bousquet and Larochelle 1993
The range of this indigenous northern Neotropical group is confined to northern Mexico and southern Arizona. Of the three species, two (*P. erro* (LeConte 1854) and *P. capitata* (Chaudoir 1852)) are represented in the Nearctic Region.

subgenus *Polpochila* (s. str.)
*Melanotus* Dejean 1831 (not Eschscholtz 1829)
*Cratocara* LeConte 1863 (replacement name for *Melanotus* Dejean)
*rotundicollis* species group, Bousquet and Larochelle 1993
Eighteen species are included in this subgenus, most of them occurring in South America. One, *P. rotundicollis* Bates 1882, enters the Nearctic Region, in southern Arizona.

*Pogonodaptus* Horn 1881
This indigenous Neotropical genus includes three species: *P. mexicanus* (Bates 1878), which ranges from Central America to southern Texas; *P. rostratus* Darlington 1935, which is known from the island of Hispaniola only; and *P. impressiceps* Casey 1914 from Panama. The species live at the margins of standing water, in marshes and swamps.

Subtribe Harpalina

With 60 genera and about 1200 species, this subtribe is represented in all zoogeographical regions. These taxa are arranged in eight genus groups by Noonan (1976: 29-59). To these we add the Trichotichni, which includes some genera assigned by Noonan to the Selenophori. Four genus-groups are represented in the Nearctic Region: Selenophori, tropically based, with extensive temperate representation, especially in the Nearctic Region; Trichotichni, principally Old World tropical and warm temperate, and in only the northern portion of the Neotropical Region; Harpali, which is concentrated in Holarctica; and Dapti, with two genera that are widely separated geographically. Kataev (1996: 159) postulated that the Harpali are the most recent of these assemblages to have evolved, and they probably did so in the ancient Mediterranean area, in Palaeocene time.

The Nearctic assemblage of Harpalina comprises 11 genera and 138 species. Reference. Noonan 1985: 1-92 (rev. gen. Selenophori and Trichotichni).

*Stenomorphus* Dejean 1831
*Agaosoma* Ménétriés 1843
This indigenous Neotropical genus ranges from northern South America to the Great Plains in southern United States, and east to Georgia, on the Gulf Coast, in the Nearctic Region. Three species range into southern United States: *S. californicus* Ménétriés 1843 is widespread in Mexico, and occurs from western Georgia to California in United States; *S. convexior* Notman 1922 is known from southern Arizona and western México; *S. sinaloae* Darlington 1936 is known from southern Arizona and New Mexico. Adults are nocturnal, and specimens of all three species have been taken at light. Reference. Ball *et al.* 1991: 933-988 (rev. spp.).

*Amblygnathus* Dejean 1829
An indigenous Neotropical genus of 24 species in three species groups, ranging from the Atlantic Forest of Brazil and Amazon Basin north to New Jersey and Pennsylvania in northeastern United States, this genus is represented in the Nearctic Region by four species, all members of the *iripennis* species group. The species *A. iripennis* (Say 1823) is precinctive in the Nearctic Region, ranging in the east from Florida to New Jersey and southeastern Pennsylvania. Reference. Ball and Maddison 1987: 189-307 (rev. spp.).

*Athrostictus* Bates 1878
This indigenous Neotropical genus includes 16 species, ranging from Argentina northward to southeastern Texas. The range of only one species, *A. punctatulus* (Putzeys 1878) (not *punctulatus* auctorum; not *punctulatus* Dejean 1829) extends into the Nearctic Region, as noted above. The natural habitat seems to be open dry forest, but dry, disturbed areas, such as pastures and gardens around houses, are also occupied. Reference. Bousquet and Larochelle 1993: 330.

*Selenophorus* Dejean 1829
This indigenous Neotropical genus includes 186 species in three subgenera or species groups, ranging from the pampas of Argentina northward in the Nearctic Region to southern Canada. The Nearctic fauna includes 39 species, some of which are hygrophilous, living in wet forest leaf litter or at the margins of sunlit swamps. Others live in dry places (meadows, or disturbed habitats which support sparse vegetation, or in bare sand, but close to vegetation). Adults of some species are diurnal, but most are nocturnally active; some fly to light. The group is in need of revision. References. Casey 1914: 134-157. Horn 1882a: 8.

subgenus *Selenophorus (sensu stricto)*
*Gynandropus* Dejean 1831
*Hemisopalus* Casey 1914
The range of this subgenus with 171 species, is co-extensive with the range of its genus, as noted above. In the Nearctic

Region, *Selenophorus* (*s. str.*) is represented by 29 species, some of which are hygrophilous forest inhabitants, while others are xerophilous, living in relatively dry open areas, including the southwestern deserts.

subgenus *Celiamorphus* Casey 1914
This subgenus is represented by 14 species, with a range that extends from the Amazon Basin in South America northward to southern Ontario in the eastern part of the Nearctic Region. Nine species of *Celiamorphus* inhabit the Nearctic Region, living in the east and southwest. They are xerophilous, and are associated with sandy soil.

*breviusculus* species group, Bousquet and Larochelle 1993
This precinctive Nearctic group includes only the one markedly distinctive species, *S. breviusculus* Horn 1880, whose range is confined to Texas and Oklahoma. Reference. Bousquet and Larochelle 1993: 242.

*Discoderus* LeConte 1853
Twenty-six species represent this indigenous Neotropical genus, whose range extends from Chiapas, in southern Mexico northward to southern Canada. Basically xerophilous, the species occupy dry oak pine forest to grassland and desert. Adults are nocturnal, and in the American Southwest in particular, they are an important element in the carabid fauna that assembles at light on warm nights. This genus is in need of revision. Reference. Casey 1914 : 157-165.

subgenus *Discoderus* (*s. str.*)
Twenty species of this precinctive Western Hemisphere genus are found in the United States. All except *D. parallelus* (Haldeman 1843) are southwestern. The latter species ranges widely, from the Gulf coastal states (except Florida) and the Mexican border northward to southern Canada.

subgenus *Selenalius* Casey 1914
This subgenus is represented by two species which are known from Texas (*D. parilis* Casey 1914) and Texas to Arizona, and northwestern Mexico (*D. cordicollis* Horn 1891).

*Trichotichnus* Morawitz 1863
This Holarctic-Afrotropical-Oriental-Australian genus with 166 species arranged in five subgenera is represented on the North American continent by four species included in two subgenera, which are restricted to eastern United States and southeastern Canada. The species are residents of deciduous forest, adults being found therein in leaf litter or under other types of cover.

Lindroth (1968: 812-813) discussed the position of *Episcopellus* Casey at some length, treating it as a separate genus. Noonan (1985: 67) included *Episcopellus* in subgenus *Trichotichnus*. As noted below, Kataev included *Episcopellus* in *Trichotichnus*, combining it with subgenus *Iridessus* Bates.

subgenus *Trichotichnus* (*s. str.*)
*Pteropalus* Casey 1914
The range of this subgenus, with 110 species, is Holarctic-Oriental, extending eastward to New Guinea from the Indo-Australian archipelago. The two Nearctic species, *T. dichrous* (Dejean 1829) and *T. vulpeculus* (Say 1823), range from Georgia northward to Vermont, and westward to the Mississippi Basin.

subgenus *Iridessus* Bates 1883
*Episcopellus* Casey 1914
*orientalis* species group, Habu 1973
This subgenus, as indicated by Kataev (*in litteris*, 9 December 1996) includes six species: four in the eastern Palaearctic Region (eastern China and Japan); and two in the eastern Nearctic Region (*T. autumnalis* (Say 1823), and *T. fulgens* (Csiki 1932)), with a geographical range extending from the Gulf Coast northward to Vermont and westward to eastern Texas. Previous recent authors (Lindroth 1968: 811, and Noonan 1991: 136) included *T. fulgens* in the genus *Harpalus*. References. Habu 1973: 237-248.

*Aztecarpalus* Ball 1970
*Selenophorus* (in part) *auct.*
This indigenous northern Neotropical genus includes nine species, one of which, *A. schaefferi* Ball 1970, occurs in southeastern Texas. This is the only species in the genus with macropterous adults, and the only one that occupies dry lowland forest. The other species are in montane subtropical or temperate forest, in eastern Mexico, where they live in damp leaf litter. References. Ball 1970: 97-1213. 1976a: 61-72.

*Piosoma* LeConte 1848
This indigenous Nearctic monospecific genus includes *Piosoma setosum* LeConte 1848, whose geographical range extends from southwestern Canada (southeastern British Columbia and Alberta) southward to northern Mexico and eastward to Arkansas. This species occupies grassland habitats. Adults are nocturnal, and some fly to light sources.

*Euryderus* LeConte 1846
*Nothopus* LeConte 1852
One species comprises this indigenous Nearctic genus: *E. grossus* (Say 1830), whose geographical range extends from western New York westward to the coast of Washington and from southwestern Canada southward to Arizona. This species, adults of which eat plant seeds, frequents sandy areas. Adults are nocturnal, and some fly to light sources.

*Ophonus* Dejean 1821
*Harpalus* (in part) *auct.*
*rufibarbis* species group, Lindroth 1968
This group was excluded from the genus *Harpalus* Latreille on the basis of structural features of adults (Kataev 1996: 158-159) and of larvae (Brandmayr *et al.* 1980). This indigenous Palaearctic genus of 76 species is represented in the Nearctic Region by two introduced species: *O. puncticeps* Stephens 1828

(Nova Scotia-Pennsylvania and west to Ohio, and *O. rufibarbis* (Fabricius 1792) eastern Quebec, only). Adults are nocturnal, and some fly to light sources.

*Harpalus* Latreille 1802
This genus of more than 450 species is principally Holarctic, but has precinctive species in the Oriental and Afrotropical Regions. The species are arranged in about 50 subgenera or groups of subgeneric rank. In the Nearctic Region, *Harpalus* is represented by 61 species, arranged in 19 subgenera or groups equivalent thereto. As noted above, the species *H. fulgens* Csiki 1932 was removed to *Trichotichnus*, and subgenus *Ophonus* was ranked as a genus.

Noonan (1991), extending Lindroth's (1968: 749-812) excellent taxonomic treatment of the more northern fauna, achieved the following. At the species level, he studied variation exhaustively, and on that basis, he managed to combine putative species, thus synonymizing many names, especially those proposed by Casey (1914). At the supraspecific level, of the groups studied (the subgenera *Pseudoophonus* and *Glanodes* were excluded, and the introduced species were in the key, only), he recognized two major assemblages of subgenera and species groups, referred to as "stocks": the *fraternus* stock, including subgenera *Euharpalops, Cordoharpalus, Pharalus,* and the *nigritarsis* species group; and the *caliginosus* stock, including the subgenera *Megapangus, Plectralidus,* and *desertus* species group. These stocks were based on phylogenetic analysis, and together, they included 21 species. Most of the remaining species were grouped in the text as "taxa not placed in cladograms" (l.c., p. 135), this assemblage of convenience including nine groups: *fulgens, cautus, obnixus, atrichatus, opacipennis, amputatus, somnulentus, leiroides, and rewolinskii.* Each of these groups was regarded as monophyletic, but without close Nearctic relatives.

We believe that the two stocks and the other groups recognized are monophyletic or natural (except the *rewolinskii* species group[see below]). However, many of them are not useful in identification because of the discordant distribution of character states, or the need to rely on genitalic features. Therefore, the key deals only with those groups with relatively easily perceived external diagnostic features. For identification of members of the 'difficult' groups, it is best to proceed from species identification, which is achieved by consulting at once the basic works (Lindroth and/or Noonan). The notes that follow provide a review of divergence and diversity of the genus *Harpalus* in the Nearctic Region.

subgenus *Pseudoophonus* Motschulsky 1844
*Pardileus auctorum, nec* Gozis 1882
*Pseudophonus auctorum*
This Holarctic subgenus, with 41 species, is represented in the Nearctic Region by 12 species, whose aggregate range is transamerican, from southern British Columbia southward to the Mexican Plateau. The species inhabit meadows and pastures, and some species are common in disturbed areas. Adults are nocturnal, and many fly to light sources.

The one introduced Palaearctic species, *H. rufipes* (De Geer 1774), lives in the east, ranging from Newfoundland, and Saint-Pierre and Miquelon, westward to Quebec and southward to Connecticut (Zhang *et al.* 1994). Adults eat seeds of various herbaceous plants, and the species is a recognized pest of strawberries. Nocturnal flight is common, and many adults come to light sources. Larvae live in burrows, in which they store seeds of foxtail (Gramineae, *Setaria* spp., introductions from the Palaearctic Region), pressing them into the burrow walls (Kirk 1972: 1426-1428). Kirk (1973) described the life history of *H. pensylvanicus* (DeGeer 1774).

Although substantial effort has been expended in attempting to understand the Nearctic species of *Pseudoophonus*, much taxonomic work remains to be done. References. Ball and Anderson 1962: 1-94.

subgenus *Harpalus (sensu stricto)*
*Epiharpalus* Reitter 1900
*Harpalophonus auctorum, nec* Ganglbauer 1892
This naturally Palaearctic precinctive subgenus of numerous species is represented in the Nearctic Region by the following two introduced species. *H. affinis* (Schrank 1781) has a bicentric distribution pattern: in the east, from Newfoundland to Florida and west to Kansas; in the west, from Oregon and British Columbia, east to Alberta. In contrast, *H. rubripes* (Duftschmid 1812) is known only from the eastern United States: Rhode Island, Connecticut, and New Hampshire. Both species are synanthropic, living in association with man, in gardens and in agricultural fields, pastures and meadows. According to Lindroth (1968: 768) the food of this species consists both of seeds and other insects.

subgenus *Euharpalops* Casey 1914
*Haploharpalus* Schauberger 1926
*Lasioharpalus* Hatch 1953, in part (not Reitter 1900)
*Amblystus* Hatch 1953 (not Motschulsky 1864)
*fraternus* species group, Lindroth 1968
*fulvilabris* species group (in part), Lindroth 1968
*spadiceus* species group, Lindroth 1968
*ventralis* species group, Lindroth 1968
*viduus* species group, Noonan 1991
*laticeps* species group, Noonan 1991
*fraternus* stock, Noonan 1991
*Euharpalus*, Hatch 1953 (*lapsus calami* for *Euharpalops* Casey)
A Holarctic assemblage, we do not know how many species are included, overall. In the Nearctic Region, this subgenus includes 11 species, arranged in seven groups and subgroups, a reflection of the marked divergence in habitus and size that the assemblage exhibits. This subgenus is transcontinental in the north (central Canada and northern tier of states), extending south in the Rocky Mountains and the Mexican Sierra Occidental to the Transvolcanic Sierra, and south in the Appalachian Mountains to northern Georgia. The species range in habitat from alpine and subalpine meadows and deciduous forest in leaf litter to dry prairie. Some species seem to be associated with burned areas (Holliday 1991: 1375).

Adults of most species are capable of flight, though relatively few appear at light sources. Reference. Noonan 1976: 35-36.1991: 21.

subgenus *Opadius* Casey 1914
*Cordoharpalus* Hatch 1949
*cordifer* species group, Lindroth 1968
*fraternus* stock (in part), Noonan 1991

Two species represent this precinctive Nearctic subgenus: *H. cordifer* Notman 1919, which is Pacific Northwestern ranging from Oregon north to southern Alaska; and *H. tadorcus* Ball 1972, which is southwestern, ranging from Colorado south to New Mexico and Arizona. The two species differ in wing development: *H. cordifer* is functionally wing dimorphic, whereas adults of *H. tadorcus* are brachypterous, only. The two species are markedly divergent ecologically, with *H. cordifer* inhabiting forested areas, in leaf litter, and *H. tadorcus* occurring in grassland. Reference. Noonan 1991: 92-99.

*nigritarsis* species group, Noonan 1991
*Pheuginus* Hatch 1953 (in part), not Motschulsky 1844
*herbivagus* species group (in part), Lindroth 1991
*fulvilabris* species group (in part), Lindroth 1968
*fraternus* stock (in part), Noonan 1991

Two species represent this Holarctic species group in the Nearctic Region. The Holarctic *H. nigritarsis* C. R. Sahlberg 1827 is northern transcontinental, ranging southward in the Rocky Mountains to New Mexico and Arizona, and in the northern Applachian Mountains in the east to New Hampshire. This species lives in montane open places, such as meadows. It is also in low altitude localities in eastern Canada. The Nearctic precinctive species, *H. megacephalus* LeConte 1848, with a more restricted distribution, ranges only from the Lake Superior area northward to near Hudson Bay and westward to central Alberta. Probably it is forest-inhabiting, but no data about its requirements are available. Reference. Noonan 1991: 99-110.

subgenus *Pharalus* Casey 1914
*fraternus* stock (in part), Noonan 1991

This indigenous Nearctic subgenus includes three species, arranged in two markedly distinctive subgroups: the *indianus* subgroup, with two eastern species, *H. indianus* Csiki 1932, from the Mississippi Basin, south to Louisiana, and *H. gravis* LeConte 1858, a southeastern species ranging south to northeastern Mexico; and the western *desertus* subgroup, with a single markedly varied species *H. desertus* LeConte 1859 that ranges north in the Great Plains to southern Canada, and southwestward to Arizona.

With some doubt, we include here also *H. cyrtonotoides* Notman 1919, based on a single specimen collected in Colorado. Noonan (1991: 216), in spite of an intense search, was unable to locate the type specimen of this species, and from the description was not able to place it in a group, classifying it as *incertae sedis*. To us, the description seems to fit *H. desertus* LeConte, with reference especially to the form of pronotum. This species is widespread in Colorado (and elsewhere, in the west). Reference. Noonan 1991: 110-127.

subgenus *Megapangus* Casey 1914
*caliginosus* stock (in part, Noonan 1991

This indigenous Nearctic subgenus includes two species with large black adults: *H. caliginosus* (Fabricius 1775), which ranges from coast to coast, and from southern Canada into northwestern Mexico; and *H. katiae* Battoni 1985, with a range extending from northeastern Gulf Coastal Mexico to northern Florida and North Carolina in the east, west to southern Arizona, and north in the Great Plains to Colorado and Nebraska. These species live in open areas, many adults being found in agricultural fields. Adults are nocturnal, and many flying individuals are attracted to light sources. Adults of *H. caliginosus* have been observed eating seeds of ragweed (*Ambrosia* sp.). References. Will 1997: 43-51 (sp. rev.).

subgenus *Plectralidus* Casey 1914
*caliginosus* stock (in part), Noonan 1991

Two species are included in this precinctive Nearctic subgenus: *H. eraticus* Say 1823, which lives in the Mississippi Basin and eastward; and *H. retractus* LeConte 1863, which is farther west, in Colorado, New Mexico, Utah, and Arizona. The species occupy sandy areas, are nocturnal, and some adults are taken at light sources. Larvae of *H. eraticus* live in burrows, in which, like those of *H. pensylvanicus*, they store seeds of foxtail (Kirk 1972: 1426-1428). Reference. Noonan 1991: 131-135.

*cautus* species group, Noonan 1991
*Pheuginus* Hatch 1953, in part (not Motschulsky 1844)
*opacipennis* species group (in part), Lindroth 1968
*innocuus* species group, Lindroth 1968

Six species represent this Holarctic species group in the Nearctic Region, one of which, *H. vittatus* Gebler 1833 is itself Holarctic, being represented in Alaska by the subspecies *H. v. alaskensis* Lindroth 1968. The group is predominantly western and northern, but two species are nearly transcontinental, and *H. balli* Noonan 1991 seems to be confined to coastal localities in northeastern United States. The species live in grassy areas both in the lowland prairies and in the mountains. Reference. Noonan 1991: 138-159.

*obnixus* species group, Noonan 1991
*cautus* species group (in part), Lindroth 1968

The overall distribution of this species group is unknown. Like the *cautus* group, probably it is Holarctic. Two Nearctic species are included, one of which (*H. obnixus* Casey 1924) is confined to the west, and one (*H. plenalis* Casey 1914) is nearly transcontinental, but with a gap through the central part of the Great Plains. The species inhabit xeric grasslands and woodlands. Reference. Noonan 1991: 159-166.

*atrichatus* species group, Noonan 1991
*Pheuginus* Hatch 1953 (in part, not Motschulsky 1844)
*herbivagus* species group (in part), Lindroth 1968

The overall distribution of this presently monospecific species group is unknown. Like the *cautus* group, probably it is Holarctic. The included species is *H. atrichatus* Hatch 1949, which is known only from the Pacific Northwest (Idaho, Oregon, Washington, and British Columbia). Habitat data are not available. Reference. Noonan 1991: 166-168

*opacipennis* species group, Noonan 1991
*cautus* species group (in part), Lindroth 1968
The overall distribution of this presently monospecific species group is unknown. Like the *cautus* group, probably it is Holarctic. The included species is *H. opacipennis* (Haldeman 1843), which is transcontinental at northern and mid-latitudes: Alaska to northern California and east to eastern Quebec and western Pennsylvania; south in the Rocky Mountains to central Arizona, at high altitude. The species lives in open dry country, with sparse vegetation, on a substrate of sand and gravel. Reference. Noonan 1991: 169-173.

subgenus *Harpalomerus* Casey 1914
*Lasioharpalus* Hatch 1953, in part (not Reitter 1900)
*amputatus* species group, Lindroth 1968
One Holarctic species, *H. amputatus* Say 1830, represents this Holarctic species group in the Nearctic Region. The nominotypical subspecies, *H. a. amputatus*, is Nearctic, with the other three subspecies, previously treated as members of a separate species, *H. obtusus* (Gebler 1833), being Palaearctic (Kataev 1997). Two males, collected in southern Alberta, exhibit genitalia of the *rewolinskii* type (see below). The range of *H. a. amputatus* is western, extending southward from Alaska to the Sierra Transvolcanica, in Mexico, and eastward to the western part of the Mississippi Basin. Adults of this species are encountered commonly in grassland and in disturbed habitats, including pastures grazed by cattle. Reference. Noonan 1991: 173-180.

*herbivagus* species group, Lindroth 1968
*Pheuginus* Hatch 1953, in part (not Motschulsky 1844)
*Lasioharpalus* Hatch 1953, in part (not Reitter 1900)
*somnulentus* species group, Noonan 1991
*rewolinskii* species group, Noonan 1991
This species group is Holarctic. Three precinctive Nearctic species and one that is Holarctic (*H. solitaris* Dejean 1829) are included, which collectively are wide ranging, from Alaska and northern Canada southward to Georgia in the southeast, and California in the southwest. Noonan (1991: 180) stated that this species group is "a purely phenetic assemblage of species", based on plesiotypic spines and spine patterns of the internal sac of the male genitalia.

Among the species is *H. somnulentus* Dejean 1829, which is remarkably varied in structural features, including armature of the internal sac of the male genitalia. We believe that *H. rewolinskii* Noonan 1991 is conspecific with *H. somnulentus*. The type material of the former is *somnulentus*-like, the three males differing from males of *H. somnulentus* in armature of the internal sac: *rewolinskii* with two very large spines (Noonan 1991: 269, Fig. 194); and *somnulentus* with clusters of small spines (l.c., 266-269, Figs. 164-192). The difference between the two types of internal sac is striking, but Kataev (1996) reported a similar situation in five Palaearctic species: two of *Harpalus*; and three of *Ophonus*. Further, as noted above, two males of *H. amputatus* exhibit two large spines in contrast to the normal multispined condition of males of that species. Kataev (1996: 163) postulated that the two-spined condition is plesiotypic for *Harpalus*, and that its appearance in males belonging to multispined species is an atavistic feature. Our conclusion, then, is that *H. rewolinskii* is not a valid species, and accordingly combine it with *H. somnulentus* (**NEW SYNONYMY**).

The species inhabit grassy areas, xeric or mesic, including disturbed ones, such as pastures. Reference. Noonan 1991: 180-211, 214-216.

subgenus *Glanodes* Casey 1914
*obliquus* species group, Bousquet and Larochelle 1993
Seven species, restricted to the southwestern United States (western Texas to Arizona and north to Utah), represent this precinctive North American group. Adults are brachypterous. The species live in grassland, including disturbed areas, such as pastures. Reference. Ball 1972: 179-204 (rev. spp.).

subgenus *Harpalobrachys* Tschitschérine 1898
*leiroides* species group, Noonan 1991
Ranked previously as a genus, Noonan (1991: 211) emphasized the similarities of *Harpalobrachys* to *Harpalus*, and combined the two, ranking *Harpalobrachys* as a subgenus. Monospecific, *Harpalobrachys* includes only *H. leiroides* Motschulsky 1844. This species ranges from eastern Siberia and Alaska eastward to Fort Smith, Northwest Territories, and southward to central Alberta. The species lives in sandy areas near rivers and lakes.

subgenus *Harpalobius* Reitter 1900
*Lasioharpalus* Hatch 1953, in part (not Reitter 1900)
*Harpalellus* Lindroth 1968
*fuscipalpis* species group, Kataev 1989
Lindroth (1968: 815) proposed the genus *Harpalellus* to include the single species *Harpalus basilaris* Kirby. Kataev (1989: 219) established the synonymy *H. fuscipalpis* Sturm 1818 = *H. basilaris* Kirby 1837, transferring thereby *Harpalellus* to *Harpalobius*, and establishing the synonymy of these two names. The subgenus *Harpalobius*, then, includes four species, three of which are eastern Palaearctic, and one, *H. fuscipalpis*, being Holarctic. In the Nearctic Region, this species is nearly transcontinental northern, ranging from Alaska eastward to Québec, southward in the east to New Hampshire, and in the Rocky Mountains, southward to New Mexico and Arizona. The beetles live in open environments, such as gravel pits and dry grassland, with sparse vegetation. References. Kataev 1989: 219. Lindroth 1968: 813-817.

*Hartonymus* Casey 1914
This precinctive Nearctic genus includes two species: *H. hoodi* Casey 1914, ranging in the Mississippi Basin from Wisconsin south to Oklahoma; and *H. alternatus* (LeConte 1863), from Oklahoma and Texas west to New Mexico. Both species live on sand, *H. hoodi* in riparian situations, and *H. alternatus* on dunes, remote from water. Reference. Ball 1976b: 417-430.

*Cratacanthus* Dejean 1829
This precinctive North American genus is represented by one species, *C. dubius* (Palisot de Beauvois 1811), in eastern United States, ranging from the Gulf Coast as far north as New Jersey, and westward to Arizona. The right mandible of the male has a prominent dorsal ridge, and is markedly curved, whereas the mandibles of females are normal. The species is wing dimorphic; adults are nocturnal and many macropterous ones fly to light sources. Open, dry fields, with clay soil, seem to be the preferred habitat. The closest relative of *Cratacanthus* appears to be the Palaearctic *Daptus* Fischer von Waldheim 1824.

Licinini

Comprised of 232 species in 23 genera and four subtribes, this tribe has representatives in all of the major zoogeographical regions, but South America has only a single precinctive monospecific genus, and one species of a widespread genus. Because of the restricted distributions of many of the genera, each region has a distinctive licinine fauna. The group is widespread in Nearctic North America , but most of the species occur southward, and none have been taken north of about 57°N. Lat. The Licinini, Panagaeini, Oodini, and Chlaeniini seem to form a natural group (Laferté Sénectère 1851), based on similarities in structure of the male front tarsi, genitalia, and more especially, on similarities in structure of the larvae. This group probably shares a common ancestry with the Pterostichini. The front coxal cavities of the adults of these three tribes are supposed to be biperforate, but actually, this condition is realized only in the Chlaeniini. Licinines are geophilous: some are swamp-inhabiting, but most live away from standing water. A number of taxa are known to be molluscivorous, and this is thought to be a general characteristic of the tribe (Ball 1992: 344). With the distinctive mandibles of the adults, probably the members of each subtribe deal in a different manner with the problem of invading snail shells to gain access to the soft tissue.

Two of the four subtribes of Licinini are represented in the Nearctic Region. References. Ball 1959: 1-250 (rev. N. A. spp.). 1992: 325-380 (subtribal classfn.).

Subtribe Dicaelina

Rhembidae Gistel 1856
Submerini Lafer 1989
This group is basically Megagaean, with 45 species arranged in two genera, both of which are represented in the Nearctic Region.

Adults, with their thick mandibles, attack snails simply by biting through the shell.

*Diplocheila* Brullé 1834
*Rembus* Dejean 1826 (*nec* Germar 1824)
*Eccoptogenius* Chaudoir 1852
*Symphyus* Nietner 1858
*Submera* Habu 1956

This genus, with 28 species arranged in three subgenera, is represented in the Oriental and Afrotropical Regions, along the southern and eastern edges of the Palaearctic Region, the central and southern portions of the Nearctic Region, and the northern part of the Neotropical Region. In Nearctic North America, *Diplocheila* is represented by nine species, all members of the indigenous *D. striatopunctata* species group, of the subgenus *Isorembus*. The range of this subgenus is coextensive with that of *Diplocheila* (*s. lat.*). Eight of the Nearctic species are found east of the Mississippi River, and none are known from north of about 57°N. Lat. Eight species are hygrophilous, with adults being found either at the margins of marshes and lakes, or else over the water, in beds of floating plant debris and mosses. One species, *D. obtusa* (LeConte 1848) prefers drier areas, with adults found under cover along river banks, and in pastures. El Moursy and Ball (1959) reported observing adults of *D. oregona* (Hatch 1951) eating snails.

*Dicaelus* Bonelli 1813
This is an indigenous Nearctic genus, whose 17 species are arrayed in three subgenera. The inclusive range of the group extends from the Mexican Transvolcanic Sierra northward, east of the Sierra Nevada Mountain Range to southern Canada.

subgenus *Paradicaelus* Ball 1959
*furvus* species group, Lindroth 1969
Eight species represent this precinctive Nearctic subgenus. They live in eastern North America, westward to the Rocky Mountains. All appear to be woodland species. Adults of several species are known to eat snails.

subgenus *Liodicaelus* Casey 1913
*laevipennis* species group, Lindroth 1989
This subgenus of five species is represented in southwestern Nearctic North America by three species, one of which ranges as far north as southeastern Alberta. The species inhabit damp montane meadows, as well as moist localities in the prairies and desert.

subgenus *Dicaelus* (*sensu stricto*)
*purpuratus* species group, Lindroth 1969
Four southeastern species comprise this indigenous Nearctic group. They inhabit forests in the east, and grassland in the west. One species, *D. purpuratus*, Bonelli 1813, is known to eat snails.

Subtribe Licinina

The 99 species of this worldwide subtribe are arranged in 10 genera. Only one genus, *Badister* Clairville 1806, is known from the Nearctic Region. One other genus, the Palaearctic-indigenous *Licinus* Latreille 1802, was represented in eastern Massachusetts by a single introduced species, *L. silphoides* (Rossi 1790). Specimens have not been seen recently, so it seems that this species did not become established (Bousquet and Larochelle

1993: 212). Thus, it seems best to drop the name from the list of Nearctic carabids.

Both adults and larvae of the Licinina are molluscivores. Adults have markedly modified mandibles (Ball 1992: 332-333). They are used rather like a can opener to enter a snail's shell (Brandmayr and Zetto Brandmayr 1986: 174-175, Figs. 2-3).

*Badister* Clairville 1806

The 48 species of this genus are arranged in three subgenera. The group as a whole is found throughout the warmer portions of Holarctica, the Oriental, Afrotropical and Neotropical Regions. Closely related genera occur in the Oriental Region, and one more distantly related genus lives in the Australian Region. Because of the similarities in the markedly characteristic structure of the mandibles exhibited by adults of this genus and of *Licinus*, which is known to be molluscivorous, we assume that *Badister* adults not only eat snails, but use their mandibles to do so in the same way as adult *Licinus* do. This assumption should be tested.

Most species are hygrophilous, adults living among the emergent vegetation in eutrophic marshes, or else in damp woodland, in dense litter.

subgenus *Badister* (*sensu stricto*)

*Amblychus* Gyllenhal 1810

*pulchellus* species group, Lindroth 1969

*notatus* species group, Lindroth 1969

This Oriental-Holarctic-Neotropical group of 21 species (the South American one is undescribed) is represented in Nearctic North America by nine species, one of which ranges southward in México, south of the Transvolcanic Sierra. The Nearctic distribution is transcontinental, from the Mexican border and Gulf coastal states northward to southern Canada. Lindroth (1969: 957) placed *B. notatus* Haldeman 1843 in a group of its own, based on strikingly distinctive features of adults, including brachyptery.

subgenus *Trimorphus* Stephens 1828

*transversus* species group, Lindroth 1969

Lindroth (1969a: 962) declined to include the precinctive Nearctic species, *B. transversus* Casey 1920, in *Trimorphus*. Pending a revision of the entire genus, we prefer to leave *B. transversus* in this subgenus, for lack of a better place to put it. This subgenus is Holarctic, with two Palaearctic species, and the Nearctic *B. transversus*, whose range is confined to northeastern United States and adjacent parts of Canada.

subgenus *Baudia* Ragusa 1884

*reflexus* species group Lindroth 1969

This subgenus, with 24 species, is worldwide in distribution. In Nearctic North America, *Baudia* is represented by five species, one of which, *B. reflexus* LeConte 1880, is in Mexico and the West Indies, also. Reference. Ball 1959: 219-230.

Panagaeini

Tefflini Basilewsky

This tribe, with 268 species in 21 genera and three subtribes is represented in all of the major zoogeographical regions. The Western Hemisphere has five genera, two of which (*Brachygnathus* Perty 1830 and *Geobius* Dejean 1831) are confined in the Neotropical Region to South America; one (*Coptia* Brullé 1835) is in both Middle and South America, and two (*Panagaeus* Latreille 1804 and *Micrixys* LeConte 1854) are shared between Neotropical Middle America and Nearctic North America. These last two genera are members of the subtribe Panagaeina. Bousquet and Larochelle (1993: 34) suggested a close relationship between panagaeines and some chlaeniines, with adults of both groups spraying phenols from the pygidial glands. They discussed also relationships of Panagaeini to other genera, such as *Brachygnathus* and *Bascanus* Peringuey 1896, taxa included by Lorenz (1998: 299), each in its own monogeneric subtribe of Panagaeini.

Subtribe Panagaeina

The 251 species of this subtribe are distributed among 19 genera including the following two, in the Nearctic Region.

*Panagaeus* Latreille 1804

Twelve species distributed between two subgenera comprise this Holarctic genus. The subgenera exhibit a vicariant relationship, with *Panagaeus* (s. str.) being Palaearctic-northern Oriental, and *Hologaeus* Ogueta 1966, being precinctive in the Western Hemisphere. Three species are in the Nearctic Region, living in eastern and southwestern United States. The species *P. sallei* Chaudoir 1861 ranges from Oaxaca, in Neotropical México northward to southern Texas and Arizona, in the Nearctic Region. Two species, *P. crucigerus* Say 1823, and *P. fasciatus* Say 1823 are primarily southern and central eastern, though the latter species ranges westward to Arizona, and northward in the Great Plains to Kansas. Two species, *P. fasciatus* and *P. sallei* inhabit dry forest; adults are found in leaf litter. Adults of *P. crucigerus* live on wet ground. Seen infrequently in daylight, adults are taken most frequently at night, at light sources. Reference. Ogueta 1966: 1-13 (key to spp. of *Hologaeus*).

*Micrixys* LeConte 1854

*Eugnathus* LeConte 1853 [preoccup.]

Three species, one of which is undescribed, comprise this Neotropical indigenous genus. All three are in Mexico. The only species known from the Nearctic Region is *M. distincta* (Haldeman 1852), which ranges from the Mexican state of Guanajuato on the Mexican Plateau north to Kansas, in the Nearctic Region. This species lives in dry open grassland, adults resting during the day under stones, or at the bases of semi-desert plants. Specimens of *M. mexicana* Van Dyke 1927, and of the undescribed species were collected on the western slopes of the Sierra Madre Occidental, to the east of Mazatlan, Sinaloa, Mexico. Both collections were made in tropical deciduous forest. The peculiarly modified mouthparts of adults suggest that the species are mollucivorous. A synopsis treating all three species and their relationships would be useful.

Chlaeniini

Callistini
This large and varied group of conspicuous ground beetles includes 985 species, arranged in 18 genera and two subtribes (see also Basilewsky and Grundmann 1955, who recognized for the same assemblage two subfamilies and 10 tribes). Worldwide in distribution, chlaeniines occur on many oceanic islands, as well as on the continents. The group is most diverse in the Afrotropical and Oriental Regions, both in numbers of species and in numbers of lineages.

Subtribe Chlaeniina

This assemblage includes 907 species in 15 genera. Its range is co-extensive with that of the tribe.

*Chlaenius* Bonelli 1810
This large, taxonomically complex, varied genus of moderately sized to large, predominantly brightly colored, foul-smelling, pubescent beetles, comprising 864 species in 58 subgenera, is worldwide in distribution, with a range co-extensive with that of the tribe and subtribe. Diversity declines northward, and markedly so in Australia, for which continent only nine species are known (Moore *et al.* 1987: 256-258). Bousquet (1987a: 165) and Bousquet and Larochelle (1993: 34-35) recorded that two kinds of defensive secretions are produced by the pygidial glands, and that such differences might be useful in classification. Adults of one group of species discharge phenols, the other, quinones. Panageines also discharge phenols, so there may be a close relationship with the phenol-producing chlaeniines. Regardless, the nature of chemicals produced is known presently for too few chlaeniines.

The 51 Nearctic species were arranged by Bell (1960) in five subgenera. On the basis of primarily phenetic considerations, and following in part Basilewsky and Grundmann (1955), Lindroth (1969a: 969-995), Bousquet and Larochelle (1993: 203-208), and Davidson (1980) adopted arrangements differing from that of Bell. Because Bell's revision treats the Nearctic fauna in a consistent and detailed manner, we continue to use his arrangement.

The Nearctic chlaeniine species are either precinctive or shared with the Neotropical Region. The species inhabit mainly the warm-temperate areas of the continent; few are known from the northern two-thirds of Canada, or from Alaska. The species are varied ecologically, as noted below. Thanks to the excellent revisionary work by Bell (1960), identification of Nearctic *Chlaenius* adults is relatively easy, but much is to be learned about the other life stages, especially larvae, and about life history. Many of the species are found without much difficulty, and individuals are reasonably large, and thus easy to work with. Oviposition by females of a few species has been observed; it is rather complex, and offers opportunity for further rewarding study. In brief, this genus is a fine one for research in comparative ecology and behavior. Reference. Bell 1960, 97-166 (rev. N. A. spp.).

subgenus *Pseudanomoglossus* Bell 1960
*maxillosus* species group, Bousquet and Larochelle 1993
This precinctive Nearctic group includes only *Chlaenius maxillosus* Horn 1876, which is known from the Gulf states of Alabama, Georgia and Florida, and from the Bahama Islands. Its way of life is unknown, with the peculiarly modified mandibles of adults suggesting some special food item, or a unique way of eating.

subgenus *Eurydactylus* LaFerté Sénectère 1851
*Glyptoderus* Laferté Sénectère 1851
This indigenous Neotropical subgenus is represented by two species in the Nearctic Region. East of the Rocky Mountains, *C. tomentosus* (Say 1823) ranges from northern México to southern Canada (Alberta to Quebec); west of the Rockies, that species and *C. pimalicus* Casey 1914 occur only in Arizona and New Mexico. They are not hygrophilous and seem to prefer dry, usually open, situations. The species *C. tomentosus* lives in dry, grassy meadows, where adults are found at the bases of growing plants. Adults are nocturnal, and some fly to light sources. The structure of the mandibles indicates that the species of *Eurydactylus* have feeding habits similar to those of *Chlaenius* (*sensu stricto*).

subgenus *Anomoglossus* Chaudoir 1856
Three species comprise this precinctive Nearctic subgenus. Their aggregate range covers most of United States east of the Great Plains and the adjacent southern portions of Canada. The species are evidently not pronouncedly hygrophilous, for they occur in leaf litter in deciduous forests, not necessarily near water. The form and size of the mandibles indicate that they prey on large, soft-bodied organisms.

subgenus *Chlaenius* (*sensu stricto*)
*Hemichlaenius* Lutschnik 1933 [preoccup.]
*Callistometus* Grundmann 1956
*solitarius* species group, Bell 1960
*ruficauda* species group, Bell 1960
*cursor* species group, Bell 1960
*aestivus* species group, Bell 1960
*sericeus* species group, Bell 1960
*laticollis* species group, Bell 1960
*Lithochlaenius* Kryzhanovskij 1976 (replacement name for *Hemichlaenius* Lutschnik)
Representatives of this subgenus of about 170 species occur throughout the world. The aggregate range of the 25 species living in Nearctic North America extends throughout this area, but none are known from localities north of 57°N. Lat. Most of the species are clearly hygrophilous, with the *C. solitarius* species group (*Lithochlaenius* Kryzhanovskij 1976) and *C. ruficauda* species group (*Callistometus* Grundmann 1956) being primarily

riparian. Although adults of most species of *Chlaenius* have fully developed wings, and use them in dispersal, those of three species of the *C. aestivus* species group (*C. platyderus* Chaudoir 1856, *C. augustus* Newman 1838, and *C. viduus* Horn 1871) are brachypterous, and *C. aestivus* Say 1823 is wing-dimorphic, with some adults macropterous and others brachypterous (Bell 1960: 119-120).

Adults eat largely dead or injured insects, for they show little inclination to attack uninjured specimens of suitable size. (Observations based on captive *C. sericeus* (Forster 1771), and *C. cordicollis* Kirby 1837.) The gregariousness of many species is perhaps adaptive, enabling them to dismember larger prey than each could manage individually. (If several specimens are confined together, and are offered a food item, they pull it apart in a tug-of-war.)

Females of *C. aestivus* and of *C. sericeus* lay their eggs in mudballs, which are stuck to vegetation (King 1919). See notes for subgenus *Agostenus*, below.

subgenus *Agostenus* Motschulsky 1850
*Pelasmus* Motschulsky 1850
*Brachylobus* Chaudoir 1876
*Chlaeniellus* Reitter 1908
*Agostenus lithophilus* species group, Bell 1960
*Agostenus purpuricollis* species group, Bell 1960
*Agostenus nemoralis* species group, Bell 1960
*Agostenus glaucus* species group, Bell 1960
*Agostenus pennsylvanicus* species group, Bell 1960
*Agostenus variabilipes* species group, Bell 1960
*Agostenus vafer* species group, Bell 1960

This geographically widespread subgenus includes 102 species, of which 25 are precinctive in the Western Hemisphere (Nearctic-precinctive) or indigenous, and shared with the northern part of the Neotropical Region. The Nearctic species collectively occupy the whole of the region, excluding the northern tundra. Most of the species are clearly hygrophilous, and some are semi-aquatic, occurring in marshes and sphagnum bogs in the north. The species *C. purpuricollis* Randall 1838, however, is seemingly dry-adapted, inhabiting, for example, short grass prairie, and not in the vicinity of surface water.

The mouthparts and, presumably, feeding habits, show no consistent differences from *Chlaenius* (*sensu stricto*). Females of *C. tricolor* Dejean 1826 lay their eggs in mud cells, which are stuck to vegetation (King 1919). Females of *C. impunctifrons* Say 1823 attach their eggs in capped cells to tips of *Typha* leaves, a meter or so above the surface of the water in which the plants are growing (Claassen 1919).

### Oodini

This clearly defined tribe is represented in the temperate and tropical areas of all major zoogeographical regions of the world. About 270 species have been described to date in about 30 genera. This tribe has been combined with the Chlaeniini by several authors. However, the components of the pygidial secretion suggest that the two groups are probably not adelphotaxa (Bousquet 1987a). The Nearctic-Central American Neotropical and Palaearctic western European faunas are taxonomically well-known but much remains to be done for the others regions. Most species, including all of those in the Nearctic Region are pronouncedly hygrophilous, inhabiting the margins principally of standing water, though some species are riparian. Reference. Bousquet 1996: 443-537 (rev. N.A. spp.)

*Dercylinus* Chaudoir 1883
Only one species, *D. impressus* (LeConte 1853), belongs to this genus. It is known from a few specimens collected in Louisiana and Oklahoma. It was also reported from Georgia by Fattig (1949: 45) but this record requires confirmation.

*Evolenes* LeConte 1853
This genus, which is closely related to the preceding one, includes a single species, *E. exarata* (Dejean 1831). The species is restricted to the Coastal Plain where it is known from Washington, D.C. to Alabama.

*Anatrichis* LeConte 1853
This small genus is represented in Australia, New Guinea, India, and North and Central America. Many species previously assigned to *Anatrichis* have been transferred recently to the genus *Nanodiodes* Bousquet 1996 (= *Nanodes* Habu 1956). In Nearctic North America, the genus is represented by two species, *A. minuta* (Dejean 1831) and *A. oblonga* Horn 1891, whose combined range extends from Massachusetts to Honduras.

*Oodinus* Motschulsky 1864
This genus includes about 10 species which live in the Western Hemisphere and the Malay Archipelago. It is represented in Nearctic North America by two species, *O. pseudopiceus* Bousquet 1996 which occurs in Florida and the West Indies, and *O. alutaceus* (Bates 1882) which ranges from southern Texas south to Costa Rica. Very little is known about their habitat requirement but adults of *O. alutaceus* have been collected in swamps and marshes.

*Lachnocrepis* LeConte 1853
Three species belong to this genus; two live in the Far East and one, *L. parallela* (Say 1830), occurs in eastern North America from southern Canada south to Florida and Louisiana. Adults are most commonly found by treading vegetation in *Typha* marshes.

*Oodes* Bonelli 1810
This genus comprises about 55 species and has representatives in all of the major zoogeographical regions. In the Nearctic Region, it is represented by four species which are restricted to the eastern part of the continent, as far north as southern Canada. These species are inhabitants of marshes and swamps, and may be collected by treading vegetation in these areas.

*Stenocrepis* Chaudoir 1857
This indigenous Neotropical genus includes about 30 species. Of these, six are found in North America, as far north as south-

ern Ontario. Adults are usually collected in swamps, marshes, and under debris along shores. The species are grouped into two subgenera.

subgenus *Stenocrepis s. str.*

This group is represented by about seven Neotropical species of which one, *S. insulana* (Jacquelin du Val 1857), reaches southern Texas.

subgenus *Stenous* Chaudoir 1857

Of the twenty species included in this subgenus, five are known to occur in Nearctic North America. One of these, *S. elegans* (LeConte 1851), lives in northern Mexico and southwestern United States, and the others range from southern Texas northward to southern Ontario. Some of these species extend into the Neotropical Region.

Pentagonicini

Scopodini, Csiki 1932

This tribe was combined with the Odacanthini by Liebherr (1988b: 17-18), based specifically on similarities between the two groups, but generally on cladistic analysis of genitalic and other features. However, the defensive (pygidial gland) secretions (Moore 1979: 201) of the two groups are markedly dissimilar: pentagonicines, saturated acids, which are produced by Harpalinae of median grade derivation; and odacanthines, formic acid, a compound with very strong repellent properties that is characteristic of the more derived harpaline tribes. So, it seems possible that the Pentagonicini belong elsewhere than with the highly derived groups, where they have been placed previously. Accepting the opinion of Bousquet and Larochelle (1993: 37) that the evidence is indecisive for including the Pentagonicini with the Odacanthini or other highly derived Harpalinae, we continue to recognize the former group as a distinct tribe, which we place for the present near the callistomorph tribes (Licinini, Panagaeini, Chlaeniini), as proposed by Jeannel (1949a: 767). We include in the Pentagonicini the genera *Aeolodermus* Andrewes 1929 and *Homethes* Newman 1842, based on the postulate by Liebherr (1991b: 319, Fig. 10) of a close relationship between these genera and *Pentagonica* Schmidt-Goebel and *Scopodes* Erichson 1842.

The tribe Pentagonicini with 162 species in five genera occupies all of the major land masses, with three of the genera restricted to the Australian Region.

*Pentagonica* Schmidt-Goebel 1846

*Rhombodera* Reiche 1842 [preoccup.]

*Didetus* LeConte 1853

Representatives of this genus live in all of the major faunal regions. In the Palaearctic Region, however, *Pentagonica* is known only from the islands of Japan and Taiwan. Five species occur in the southern part of the Nearctic Region, concentrated in states bordering northern Mexico and the Gulf Coast. One of those species, *P. picticornis* Bates 1883, ranges east of the Appalachian Mountains northward to New Hampshire and southern Ontario; in the Mississippi Basin, northward to Illinois; and westward, to Arizona. Although the Nearctic species are easily identified (but *P. felix* Bell 1987 must be incorporated into Reichardt's (1968) key to species), the Neotropical species require much additional work. References. Bell 1987: 373-376 (descr. *P. felix* Bell). Reichardt 1968: 145-147 (key to spp.).

Platynini

This large and complex tribe, with a worldwide distribution, is more diverse both in term of lineages and species in the tropics than in the temperate regions. In Nearctic North America, it includes about 175 species which are arrayed in 16 genera. The group has been included by many authors within the Pterostichini. However, the pygidial gland secretions suggest that the Platynini are probably more closely related to the Truncatipennes complex.

*Calathus* Bonelli 1810

This genus of 177 species arranged in ten subgenera is represented throughout Holarctica, and in the mountains of the northern portion of both the Oriental and Neotropical Regions. We include in the group *Acalathus* Semenov 1889, and *Procalathus* Jedlicka 1953. Eleven species, distributed among three subgenera, occur in Nearctic North America. Of the known species, 64 (or slightly more than one third of the total) have not been assigned to subgenus. Clearly, a worldwide revision of this genus is called for. Reference. Ball and Negre 1972: 413-533 (rev. spp.).

subgenus *Calathus s. str.*

This indigenous Palaearctic subgenus of 34 species is represented in North America by the introduced *C. fuscipes* (Goeze 1777), known on this continent only from British Columbia and Washington, where it is clearly synanthropic.

subgenus *Neocalathus* Ball and Nègre 1972

This Holarctic subgenus of 36 species includes nine species in Nearctic North America. Their aggregate range covers most of the boreal and temperate regions. One species, *C. melanocephalus* (Linnaeus 1758), is a European introduction, caught in New York but probably not established on this continent. The species are mesophilous, inhabiting open sandy grounds or forests, where adults are found resting under cover of leaf litter, or stones or logs.

subgenus *Procalathus* Jedlicka 1953

This Holarctic subgenus includes two species: the northern Asian *C. fallax* (Semenov 1889) and the northern Nearctic *C. advena* (LeConte 1848). The range of the latter species is transcontinental though with a gap in the center. The eastern segment extends from Newfoundland to the Lake Superior region and southward to the mountains of New Hampshire and New York. The western portion extends from the Aleutian Islands to the foothills of the Rocky Mountains south to the mountains of New Mexico and Arizona. A mesophilous species, adults are most commonly found resting among leaves and debris in forests.

*Laemostenus* Bonelli 1810
This genus is represented by more than 150 species occurring in the Palaearctic Region, central Asia and the Himalaya. Two introduced and synanthropic species live in the Nearctic Region. Each belongs to a different subgenus. Reference. Casale 1988: 448-892 (rev. spp.).

subgenus *Laemostenus s. str.*
The sole species of this subgenus present in Nearctic North America, *L. complanatus* (Dejean 1828), ranges along the west coast from southern British Columbia to California. The species has been dispersed by trade to several continents and islands.

subgenus *Pristonychus* Dejean 1828
The single Nearctic species of this subgenus, *L. terricola* (Herbst 1784), is found in British Columbia and in eastern Canada, from Newfoundland to Quebec. Adults are usually found in dark cellars, stables and so forth.

*Synuchus* Gyllenhal 1810
This genus of 76 species, abundantly represented in eastern Asia, includes only three species in the Western Hemisphere: *S. semirufus* Casey 1913, in northwestern México; and two species farther north, in the Nearctic region. The species *S. dubius* (LeConte 1854) is restricted to the mountains in southwestern United States and *S. impunctatus* (Say 1823), is transcontinental in Canadian latitudes, ranging southward to Georgia in the east. A synopsis treating all three species and their relationships would be useful. Reference. Lindroth 1956: 485-537 (rev. spp.).

*Olisthopus* Dejean 1828
This Holarctic genus, of 23 species, includes seven species in Nearctic North America, which range throughout the eastern part of the continent from southeastern Canada south to Georgia, Louisiana, and Texas. Rather hygrophilous, the species live in deciduous or mixed forests, where adults rest by day in leaf litter, on damp soil. Reference. Horn 1882b: 63. Lindroth 1966: 552-555 ( rev. Can. sp.).

*Tetraleucus* Casey 1920
*Agonum picticorne* species group, Lindroth 1966
This precinctive Nearctic genus includes a single species, *T. picticornis* (Newman 1844), broadly distributed across the eastern United States from Vermont to Florida, west to Mississippi and Iowa. Hygrophilus, *T. picticornis* lives in swamp forests or along wooded lake shore, where adults are found resting under available cover. Reference. Liebherr 1991a: 30-33 (rev. sp.).

*Anchomenus* Bonelli 1810
*Pseudanchus* Casey 1920
*Agonum errans* species group (in part), Lindroth 1966
*Agonum funebre* species group, Lindroth 1966
This Holarctic genus is represented by six species in the Palaearctic Region and by three species in the Nearctic Region, including *A. quadratus* (LeConte 1854) that was included by Lindroth (1966: 617) in the *Agonum errans* species group. An additional species is found in Baja California. All three Nearctic species live in western North America, from southern British Columbia and Alberta south to Baja California. Adults of most species are found along gravelly or stony shores of rivers, lakes, and streams. Reference. Liebherr 1991a: 33-57 (rev. spp.).

*Sericoda* Kirby 1837
*Agonodromius* Reitter 1908
*Agonum quadripunctatum* species group, Lindroth 1966
This genus contains seven species. Two of them occur in the Old World, two are Holarctic and the remaining three are precinctive in the Western Hemisphere, with one species being confined to Cuba. Of the four species living in Nearctic North America, two are transcontinental, and two are confined to the western half of the continent. Adults of Nearctic species are attracted to burning wood and have been observed running actively on hot ashes in the vicinity of forest fires. Reference. Liebherr 1991a: 60-90 (rev. spp.).

*Elliptoleus* Bates 1882
This indigenous northern Neotropical genus is represented by 11 species of which all but one are precinctive in Mexico. One species found in Mexico, *E. acutesculptus* Bates 1882, reaches Texas and Arizona. According to Liebherr (1991), this species occupies open pine-grassland savannah habitats, where adults are found under available cover. Reference. Liebherr 1991a: 90-113 (rev. spp.).

*Atranus* LeConte 1848
This genus includes two species: *A. ruficollis* (Gautier des Cottes 1857) in Europe and Caucasus, and *A. pubescens* (Dejean 1828), in eastern North America, from the temperate Quebec and Ontario south to Texas. The Nearctic species is mesophilous, adults having been found in caves in the southern Appalachians, along wooded margins of small and slow streams, among deep litter in forests and, in the northern part of its distribution, in beaver houses. Reference. Lindroth 1966: 648; 1968: 649 (descr. N.A. sp.).

*Rhadine* LeConte 1848
*Comstockia* Van Dyke 1918
*Agonum larvale* species group, Lindroth 1966
This precinctive Western Hemisphere genus ranges from central Mexico north to the Canadian prairies, but is more diverse in term of species in southwestern United States, particularly Arizona, New Mexico, and Texas. About 40 species live in Nearctic North America. Adults are most commonly found under rotting logs and rocks in moist situations, in the burrows of mammals, and in caves. One troglobitic species, *R. subterranea* (Van Dyke 1918), eats mainly eggs of cave crickets (*Ceuthophilus* spp.) (Mitchell 1971). References. Barr 1960b: 45-65. 1974: 1-30. Barr and Lawrence 1960: 137-145.

*Tanystoma* Motschulsky 1845
*Tanystola* Motschulsky 1850
*Leucagonum* Casey 1920

This precinctive Nearctic genus comprises five species living along the Pacific Coast of North America from Oregon to Baja California. Adults are found mainly in open grassland areas. *Tanystoma maculicolle* (Dejean 1828) is listed in the common name index as the "tule beetle". References. Liebherr 1985: 1182-1211 (rev. spp.); Liebherr 1989: 178 (key to spp.).

*Paranchus* Lindroth 1974
*Agonum albipes* species group, Lindroth 1966

Proposed as a subgenus of *Agonum*, Liebherr (1986: 177) changed the rank of *Paranchus* to that of genus, based on a cladistic analysis of Holarctic Platynini. A single species, *P. albipes* (Fabricius 1796), belongs to this indigenous Palaearctic genus. Its range extends over Western Europe, south to northern Africa, and east to Asia Minor. The species was accidentally introduced along the Atlantic coast of Canada and northern United States. Hygrophilous, *P. albipes* frequents lake shores, the seashore, and river banks, though adults are more frequent in the former two habitats. Reference. Lindroth 1966: 630 (descr. sp.). 1974: 81.

*Oxypselaphus* Chaudoir 1843
*Anchus* LeConte 1854
*Agonum puncticeps* species group, Lindroth 1966

This Holarctic genus is represented by four species: three are precinctive in the western part of the Palaearctic Region; the Nearctic species, *O. pusillus* (LeConte 1854), is transcontinental in southern Canada, ranging south to South Carolina in the east and to Oregon in the west. This species is more or less hygrophilous, living near rivers and lakes, where adults are found resting among dead leaves and debris.

*Agonum* Bonelli 1810

Of the 269 species of this Megagaean genus, 214 are distributed among 27 subgenera, with 55 species being unclassified. In the Western Hemisphere, 74 species arrayed in eight subgenera range collectively from the Canadian and Alaskan arctic to the mountain forests of Chiapas and Nicaragua. All but two of these species inhabit the Nearctic Region. Most of the species are hygrophilous, living in the vicinity of standing or flowing water, but some are forest-inhabiting mesophiles, and one is a pronounced xerophile. Reference. Liebherr 1994: 8-16 (key to Western Hemisphere spp.).

subgenus *Europhilus* Chaudoir 1859
*sordens* species group, Lindroth 1966

This is a predominantly northern Holarctic subgenus with 30 species. In Nearctic North America, *Europhilus* is represented by 15 species of which three are Holarctic. Adults of most species are found most commonly in marshes, sphagnum bogs, and in the vicinity of lakes, pools, and ponds where the vegetation is rich. One species, *A. retractum* LeConte 1848, is not associated with water bodies and lives in forest litter.

subgenus *Platynomicrus* Casey 1920
*bicolor* species group, Lindroth 1966

This is also a predominantly northern Holarctic group of nine species. In Nearctic North America, this subgenus includes four species of which two are Holarctic. Adults are found on the ground among the aquatic vegetation of eutrophic marshes, or are on the plants, like those of *A. nigriceps* LeConte 1848, or they live along pond margins and river banks, often at some distance from water.

subgenus *Agonum s. str.*
*Melanagonum* Casey 1920
*Paragonum* Casey 1920
*Taphranchus* Casey 1920
*Punctagonum* Gray 1937
*cupripenne* species group, Lindroth 1966
*octopunctatum* species group, Lindroth 1966
*melanarium* species group, Lindroth 1966
*excavatum* species group, Lindroth 1966
*marginatus* species group, Liebherr 1986
*collare* species group, Liebherr 1986

This Holarctic subgenus contains 80 species, of which about 30 are Nearctic. The species *A. muelleri* (Herbst 1784), was accidentally introduced from Europe on both east and west coasts. The group is ecologically divergent, extremes being represented by *A. belleri* (Hatch 1933), which inhabits sphagnum bogs, and *A. placidum* (Say 1823), an inhabitant of dry pastures and woods. The group as a whole ranges throughout North America.

subgenus *Micragonum* Casey 1920
*Circinalidia* Casey 1920
*bicolor* species group (in part), Lindroth 1966
*nutans* species group, Lindroth 1966
*punctiforme* species group (in part), Lindroth 1966

Six species belong to this precinctive Nearctic subgenus. Five are distributed in eastern North America, from southern Canada south to Texas; *A. limbatum* Motschulsky 1845, ranges along the Pacific coast, from southern British Columbia south to California.

subgenus *Olisares* Motschulsky 1865
*Circinalia* Casey 1920
*punctiforme* species group (in part), Lindroth 1966

This precinctive Nearctic subgenus includes five species which inhabit eastern and southwestern United States and southeastern Canada. The species are hygrophilous or mesophilous, with adult hygrophiles occurring in swamps and marshes, while the mesophiles are found most commonly in open fields.

subgenus *Stereagonum* Casey 1920
*errans* species group (in part), Lindroth 1966
*ferreum* species group, Lindroth 1966

This precinctive Nearctic subgenus is represented by three species. Two are restricted to eastern North America, with *A. errans* (Say 1823) more widely distributed, its range extending to British Columbia and the western states. The species are hygrophilous or mesophilous, living in damp areas in forests, or near standing water.

subgenus *Stictanchus* Casey 1920
*extensicolle* species group, Lindroth 1966
This Nearctic group includes seven species. Their aggregate range covers southern Canada, all United States, and Mexico. The species are hygrophilous, adults being found at the margins of standing or flowing water. Reference. Liebherr 1986: 1-198 (rev. spp.).

subgenus *Deratanchus* Casey 1920
*quadrimaculatum* species group, Lindroth 1966
A single species, *A. quadrimaculatum* (Horn 1885), is included in this precinctive Nearctic subgenus. It is restricted to the southeastern United States. It is evidently hygrophilous, with adults being found in marshes.

*Platynus* Bonelli 1810
*Agonum* (in part) Lindroth 1966
This large genus is represented in the Holarctic, Oriental, and Neotropical Regions. The 22 species present in Nearctic North America are arrayed in four subgenera although five of these species are considered as incertae sedis and are not included in the key to subgenera of *Platynus*. Reference. Liebherr and Will 1996: 303-307 (key to N.A. spp.).

subgenus *Platynus s.str.*
*Agonum decentis* species group (in part), Lindroth 1966
This Holarctic subgenus is represented in North America by nine species, their aggregate range including most of the boreal and temperate regions of the continent. The species are mesophilous, with adults being found in deciduous and mixed forests, under stones, logs, and bark, on stony places along rivers, lakes, or streams.

subgenus *Microplatynus* Barr 1982
This precinctive Nearctic subgenus includes two species distributed in southern California and New Mexico. Little is known about the habitat requirement of these species except for the fact that the specimens of the type series of *P. pecki* Barr 1982 were collected beneath boulders in spruce forest.

subgenus *Batenus* Motschulsky 1865
*Platynidius* Casey 1920
*Paranchomenus* Casey 1920
*Agonum decentis* species group (in part), Lindroth 1966
*Agonum hypolithos* species group, Lindroth 1966
This Holarctic subgenus is represented in Nearctic North America by five species. One, *P. mannerheimii* (Dejean 1828), is Holarctic and transcontinental in Canada. The range of the other species covers eastern United States and southeastern Canada. Adults are hygrophilous and found most commonly along river banks or in forest swamps.

subgenus *Glyptolenopsis* Perrault 1991
This indigenous Neotropical subgenus contains 36 species. The aggregate range of the species extends from northern South America, through Middle America to southeastern Arizona, in the Nearctic Region, where *Glyptolenopsis* is represented by *P. ovatulus* (Bates 1884). Adults are most commonly collected by searching in leaf litter in mountain forests. Reference. Liebherr 1992: 1-115 (rev. spp.).

*Metacolpodes* Jeannel 1948
This is a widely distributed Oriental-eastern, tropical-subtropical Palaearctic genus of 22 species. In the Nearctic Region, it is represented by a single species, *M. buchannani* Hope 1831, which was accidentally introduced from southeastern Asia to Oregon and Idaho. This species was included previously in the genus *Colpodes* Macleay 1825, a group redefined by Liebherr (1998) to include only three species from the island of Java. Reference. Hatch 1953: 132-133.

### Ctenodactylini

This Neotropical indigenous tribe is closely related to the Hexagoniini of the Afrotropical and Oriental Regions. The larva of the genus *Leptotrachelus* possesses a small inner lobe on its maxilla, a characteristic of the pterostichine assemblage of tribes, but not of the more derived groups (see below, and van Emden 1942: 7). This feature serves to emphasize the annectant position of the Ctenodactylini, a position which was recognized by Bates (1883: 158) on the basis of form of the elytral apices.

The genera included in this tribe by most previous authors are associated with monocots, with adults climbing on the stems of grasses, Carices, and *Typha*. As one might expect, the tarsi of adults are equipped ventrally with adhesive setae. These features are shared with adults of the Neotropical genus *Calophaena* Klug 1821, which are associated closely with leaves of *Heliconia*, a member of the monocot family Musaceae. According to N. E. Stork (personal communication) the tarsi of adult *Calophaena* exhibit the adhesive setae characteristic of the ctenodactylines, but in a more modified, much elaborated form. Based on these considerations and on overall similarity in structural features, in spite of the opinions of other authors, we accept Erwin's (1991b: 44) assignment of this genus to the Ctenodactylini, and include also the similar *Calophaenoidea* Liebke 1930.

As understood here, this tribe, with 115 species in 18 genera is indigenous to the Neotropical Region. In the Nearctic Region, the Ctenodactylini is represented by the single genus *Leptotrachelus*.

*Leptotrachelus* Latreille 1829
*Spheracra* Say 1830
This indigenous Neotropical genus of 35 species is represented in the Nearctic Region by four species (one undescribed), found primarily in the east, with a combined range extending from the Gulf Coast north to southern New York in the east and to Illinois and Iowa in the Mississippi Basin. The species live in *Typha* marshes, adults climbing on the cattail leaves and stems.

In the tropics, many adults of this genus come to light sources, at night. No single publication treats the described Nearctic species. A synopsis that provided such treatment would be useful. Reference. Motschulsky 1864: 218.

Perigonini

Trechicinae Bates 1873

This pantropical tribe includes 116 species, arranged in four genera, two of which are restricted to the Neotropical Region. One genus, *Ripogenites* Basilewsky 1954, is Afrotropical. The range of *Perigona* Castelnau 1835, is co-extensive with the range of the Perigonini.

*Perigona* Castelnau 1835

*Nestra* Motschulsky 1851

*Spathinus* Nietner 1858

The 104 species of this genus are arranged in 11 subgenera. The genus is represented in all of the zoogeographical regions (but only in the southern part of Holarctica), with 14 species in the Neotropical Region. In the Nearctic Region, *Perigona* is represented by two species in the subgenus *Trechicus* LeConte 1853.

The species *P. nigriceps* (Dejean 1831) was introduced by accident to both the eastern (Florida to Maine (Nelson 1991: 284-285), and west to Arkansas and Illinois) and western (Oregon and California) coasts of the United States (Lindroth 1968: 652) and into Europe as well. It appears to be native to the lands bordering the Indian Ocean (Jeannel 1942: 581). This species lives in decaying vegetation, or beneath the surface of the ground, in the vicinity of roots and peanuts (Jeannel 1948: 737). Adults fly to light sources at night.

The species *P. pallipennis* (LeConte 1853) is Nearctic-precinctive, ranging on the Gulf Coast from Florida to Mississippi, and north in the Mississippi Basin to Iowa in the midwest and to Maine and southern Ontario in the east. Adults have been collected in leaf litter near pond margins (Bousquet 1987b: 126).

Although the two species of *Perigona* noted above seem to be geophilous, many (most?) other species of this genus are arboreal, with adults and larvae living under bark of fallen tree trunks, and pursuing a predatory existence (Erwin 1991b: 44). This genus requires study and revision (Erwin, l.c.).

Lachnophorini

The limits of this tribe and its composition in the Western Hemisphere have been discussed and at least partially clarified by Ball and Hilchie (1983: 101-108) and by Liebherr (1983: 257, 258, and 1988b). We agree with most of Liebherr's conclusions about relationships except those concerning the Neotropical *Calophaena* Klug 1821. Erwin (1991b: 44) assigned this genus to the tribe Ctenodactylini, and we accept this decision (see above).

This tribe includes two genus groups. Adults of the *Eucaerus* (or eucaerine) complex are glabrous dorsally, and the microsculpture mesh pattern of at least the elytra is transverse-linear. This complex includes four genera: *Eucaerus* LeConte 1853; *Aporesthus* Bates 1871; *Amphitasus* Bates 1871; and *Asklepia* Liebke 1938. Adults of the *Lachnophorus* genus group are setose dorsally, and have elytra with approximately isodiametric microsculpture. This group includes four Western Hemisphere precinctives: *Lachnophorus* Dejean 1831; *Stigmaphorus* Motschulsky 1861 (according to Liebherr 1988: 34, incorrectly combined with *Lachnophorus*); *Euphorticus* Horn 1881; and *Calybe* Laporte de Castelnau 1834; and the Afrotropical-Oriental genus *Selina* Motschulsky 1858. All of these genera require detailed study and revision of the species. Reference. Horn 1881b: 144-145 and 152-153.

*Eucaerus* LeConte 1853

subgenus *Lachnaces* Bates 1872

*E. varicornis* species group, Ball and Hilchie 1983

*E. hilaris* species group, Ball and Hilchie 1983

Combination of the taxa noted above was proposed by Ball and Hilchie (1983: 107). This is an indigenous Neotropical genus including 12 species in two subgenera, 11 of which are restricted to the Neotropical Region. The species *E. (s. str.) varicornis* LeConte 1853 is found in the southeastern states which border the Gulf of Mexico, from Texas to Florida. The species of this genus are hygrophilous, inhabiting the margins of sunlit marshes and dark swamp forests, adults being found in leaf litter.

*Anchonoderus* Reiche 1843

*Axylosius* Liebke 1936

This indigenous Neotropical genus of 26 species is represented in the Nearctic Region by three species, restricted in geographical distribution to the southwestern United States: Texas to Arizona only. Hygrophilous and riparian, the species live along slow-flowing water courses, where adults are found running on sand and mud, or resting under rocks. References. G. H. Horn 1894: 360. Schaeffer 1910: 395. Schaupp 1879: 85

*Euphorticus* G. H. Horn 1881

This indigenous Neotropical genus of four species ranges from the Amazon Basin of South America to the southern part of the Nearctic Region, where it is represented by two species. *Euphorticus occidentalis* G. H. Horn 1891 is known from California only, but probably is in northwestern Mexico, also. The species *E. pubescens* (Dejean 1831) is wide-ranging in the Neotropics. In the Nearctic Region, *E. pubescens* is confined to the east, ranging on the Gulf Coast from Louisiana to Florida, and north in the Mississippi Basin to Ohio, and east of the Appalachian Mountains, on the Atlantic coastal plain to North Carolina. This latter species lives on the ground close to the margins of small ponds, with adults running in the sunlight.

*Lachnophorus* Dejean 1831

*Aretaonus* Liebke 1936

This indigenous Neotropical genus of 40 species ranges from the Amazon Basin north to the southern part of the Nearctic Region, where it is represented by one species, *L. elegantulus*

Mannerheim 1843, which ranges from West Texas to California and north to Oregon, west of the Sierra Nevada. Adults of this species, which resemble adults of *Bembidion* in form, size, and behavior, live in gravel along the margins of streams, where they run during the day.

*Calybe* Castelnau 1834
The 25 species of this indigenous Neotropical genus are arranged in two subgenera, *Calybe s. str.*, and *Ega*. Only *Ega* is represented in the Nearctic Region.

subgenus *Ega* Castelnau 1834
Seventeen species belong to this subgenus. All are in the Neotropical Region, with two ranging into the southern parts of the Nearctic Region. Primarily southeastern, *C. sallei* (Chevrolat 1839), ranges from Arizona to Florida, and north on the Atlantic coastal plain to South Carolina. The more western species, *C. laetula* (LeConte 1851), is known from southern California and Arizona north to Colorado. These small insects live on wet mud, along the margins of slow-flowing streams, or around ponds with relatively bare banks. Adults of both species are ant-like, in movement as well as in appearance. References. Horn 1881b: 152-153. Liebherr 1983 (descr. of larva). Schaupp 1879: 85 (key to species).

Odacanthini

Casnoniae LeConte 1851
Colliurini Bedel 1910
Lasiocerini Jeannel 1948
Liebherr (1988b: 17-18 and 1991b), on the basis of cladistic principles, postulated an adelphotaxon relationship between the Lachnophorini and Odacanthini, and proposed inclusion of the tribe Pentagonicini with the Odacanthini (Liebherr 1988b: 17-18), as defined by Jeannel (1948: 745-758). For the reasons noted above (see Pentagonicini), we continue to recognize the pentagonicines as representing a distinct tribe that is related more to the callistomorph complex than to the lebiomorphs.

The Tribe Odacanthini occurs in all of the major zoogeographical regions, but is especially well represented in the tropics. The 295 species of this tribe are arranged in 30 genera.

*Colliuris* De Geer 1774
*Casnonia* Latreille 1822
This large genus, composed of small, brightly colored beetles with tubular, more or less slender pronota, is indigenous in the Western Hemisphere. Of the 81 species, 77 are distributed among 19 subgenera, with four being unclassified. This arrangement, according to Erwin (1991b: 46) "is totally confusing", and the group is in need of extensive revision.

The ranges of only two subgenera, with four species, extend into Nearctic North America. References. Liebke 1930, 658-660. Schaupp 1879: 85.

subgenus *Cosnania* Dejean 1821
*Odacanthella* Liebke 1930
*pensylvanica* species group, Bousquet and Larochelle 1993
This indigenous Neotropical subgenus, with 14 primarily Neotropical species, is represented in Nearctic North America by three species: all are southwestern or eastern. The most widespread species, *C. pensylvanica* (Linnaeus 1767), ranges northward to southeastern Canada. It lives on wet ground around marshes. The southwestern *C. lengi* (Schaeffer 1910) lives in dry open oak forest, among grass clumps that grow among the oaks. The habitat of *C. lioptera* (Bates 1891), also southwestern, has not been specified, but adults of that species and of *C. pensylvanica* are commonly attracted to light sources at night. A key to these species is not available.

subgenus *Calocolliuris* Liebke 1938
*Calocollius* Leng and Mutchler 1933 (*lapsus calami*)
*Calocolliuris* Csiki 1932 (*nomen nudum*)
*ludoviciana* species group, Bousquet and Larochelle 1993
This indigenous Neotropical subgenus, with five species, four of which are precinctive in the Neotropical Region, is represented in Nearctic North America by the single species *C. ludoviciana* (Sallé 1849). It ranges from southern Mexico to eastern United States, and as far north as southern New York.

Cyclosomini

Masoreini (in part)
In spite of previous statements (Ball 1960a: 156, and 1983: 519-528) to the contrary, we do not accept a broad definition of the lebiomorph group, referred to as "Masoreini", and characterized principally by especially long tibial (masoreimorph) spurs of adults. Although that feature seems to be apotypic, it also seems to be simple enough that it could have evolved a number of times. Considering the marked differences in other features among the groups with masoreimorph spurs, it seems unlikely that the complex as a whole is monophyletic. So, we advocate recognition as separate tribes the following, with their included genera:

Somoplatini (proposed by Basilewsky 1987-- *Somoplatus* Dejean 1829, *Somoplatodes* Basilewsky 1987, *Lophidius* Dejean 1831, and *Paralophidius* Basilewsky 1987);

Graphipterini (*Graphipterus* Latreille 1802, *Piezia* Brullé 1834, *Trichopiezia* Nègre 1955, *Corsyra* Dejean 1825, and *Discoptera* Semenov 1889);

Masoreini (*Masoreus* Dejean 1831, *Atlantomasoreus* Mateu 1984, *Aephnidius* W. S. MacLeay 1825, *Microus* Chaudoir 1876, *Caphora* Schmidt-Goebel 1846, *Macracanthus* Chaudoir 1846, *Odontomasoreus* Darlington 1968, and *Leuropus* Andrewes 1947);

Sarothrocrepidini (*Sarothrocrepis* Chaudoir 1850); and

Cyclosomini (*Mnuphorus* Chaudoir 1873, *Cyclosomus* Latreille 1829, *Cyclicus* Jeannel 1949a, and *Tetragonoderus* Dejean 1829).

The genus *Nemotarsus* LeConte 1853 was transferred previously to the Lebiini (Ball 1960a: 157), though this transfer was not accepted by other authors (see, for example, Basilewsky 1984: 527, and Lorenz 1998: 429).

The tribe Cyclosomini with 118 species in four genera is primarily tropical, with most species in the Oriental, Afrotropical, and Neotropical Regions. One genus, *Mnuphorus*, is principally Palaearctic, but near the northern border of the Oriental Region, and *Tetragonoderus* enters the Nearctic Region.

*Tetragonoderus* Dejean 1829
This predominantly tropical genus with 74 species in two subgenera is widespread in the Oriental, Afrotropical, and Neotropical Regions, on the southern fringe of the Palaearctic Region (Bedel 1906: 230-231), and more widespread in the southern part of the Nearctic Region. We recognize presently two subgenera, the Eastern Hemisphere tropical-Neotropical *Tetragonoderus s. str.*, and the Neotropical indigenous *Peronoscelis* Chaudoir 1876. The species live in riparian situations, on sand, the adults concealed in leaf litter during the day, and hunting at night, either in their daytime resting places, or on open sand-- rather like tiger beetles. If disturbed during the day, adults are quick to run and fly (Erwin 1991b: 45). At night, many fly to light sources. The species in the Western Hemisphere require revision. Reference. Horn 1881a: 39.

subgenus *Tetragonoderus* (*sensu stricto*)
*Crossonychus* Chaudoir 1848
*intersectus* species group, Bousquet and Larochelle 1993
*pallidus* species group, Bousquet and Larochelle 1993
This subgenus includes 60 species, of which 23 are in the Western Hemisphere, and primarily in the Neotropical Region. Two wide-ranging species enter the Nearctic Region: *T. intersectus* (Germar 1824), in the southeast, northward to Kentucky; and *T. fasciatus* (Haldeman 1843), transcontinental in the south (California to Florida), northward in the east to New York and southern Ontario, through the Mississippi Basin to Iowa and Illinois; *T. pallidus* G. H. Horn 1868 is found in the Sonoran desert, in Arizona, California, and northern Mexico. Adults of this species are very distinctive (as indicated implicitly by Bousquet and Larochelle 1993: 267) in color (dorsal surface uniformly rufous or rufotestaceous) and, unlike other species of *Tetragonoderus*, it lives away from sites with surface water.

subgenus *Peronoscelis* Chaudoir 1876
*Personocellus* Leng and Mutchler 1933 (*lapsus calami*)
*latipennis* species group, Bousquet and Larochelle 1993
This indigenous Neotropical group of 14 species is represented in the southwestern United States by two allopatric and closely related species: *T. mexicanus* Chaudoir 1876, and *T. latipennis* LeConte 1874. In turn, adults of these taxa are very similar to those of the South American species, *T. undatus* Dejean 1829.

## Lebiini

This taxon is the equivalent of the supertribe Lebiitae, one of three specified by Basilewsky (1984: 528) in his magisterial treatment of the African-Madagascan members of the Lebiinae. In his arrangement of the lebiine genera, however, Basilewsky placed undue value on the development of the mental-submental suture of the labium, and as a result, reached some unfortunate conclusions (Casale 1998: 388), details of which are of no concern here.

The 220 genera of this tribe are arrayed tentatively in 16 subtribes. To the 12 recorded by Ball *et al.* (1995; cf. Casale 1998: 410, Fig. 90), we add the Western Hemisphere Nemotarsina and the Eastern Hemisphere Demetriina and Peliocypina, and remove from the Calleidina the indigenous Neotropical genus *Agra* Fabricius, to form the monogeneric subtribe Agrina (Casale 1998: 411, Fig. 91). The 24 genera (two introduced) in the Nearctic Region are arranged in nine subtribes, as follows: Pericalina, Apenina, Cymindidina, Dromiina, Lebiina, Nemotarsina, Metallicina, Calleidina, and Agrina.

This tribe is represented in all of the major zoogeographical regions of the world. Some of the genera are wide-ranging, while many others are restricted to only a single region. The lebiine fauna of the tropics is especially diverse, while genera are fewer in the warm temperate regions, and fewer still in cool temperate zones. Many of the lebiine genera are arboreal, and the adults of many species are brightly colored, and are active during the day. Some groups inhabit very dry, or desert, areas, whereas others live on plants which grow in marshes. A few lebiines, such as the African *Oecornis* Britton 1949, and the Madagascan *Paulianites nidicola* Jeannel 1949, inhabit bird nests.

### Subtribe Pericalina

This pantropical group includes 73 genera, worldwide, with 10 indigenous genera in the Neotropical Region and two of these in the Nearctic Region. The latter has also two introduced pericaline genera. Pericalines occur in the Australian Region, the eastern edge of the Palaearctic Region, in the Japanese Archipelago; and in the Nearctic Region, from the Mexican border in Arizona east to the Gulf Coast and north to New Jersey and Pennsylvania in the east, and to Kansas in the west. Adults are primarily arboreal, living on fallen (logs) and standing tree trunks, and branches. References. Ball 1975: 143-242. Ball *et al.* 1995. Basilewsky 1984: 538-242.

*Phloeoxena* Chaudoir 1869
This indigenous Neotropical genus of 32 species in five subgenera is represented in the Nearctic Region by a single species, *P. (Oenaphelox) signata* (Dejean 1825), which inhabits the Gulf States from Louisiana eastward, and northward to Virginia. Adults are found under bark, in hardwood forests. References. Ball 1975: 143-242. Shpeley and Ball (in press).

*Mochtherus* Schmidt-Goebel 1846
This indigenous Oriental genus of seven species is represented in the Nearctic Region by a single species, *M. tetraspilotus* (W. S. MacLeay 1825). Evidently, this species was introduced recently in Florida. Specimens have been collected in several localities, so *M. tetraspilotus* seems to be established. Like other pericalines, this species is arboreal, with adults being found on trunks and branches of standing

or fallen trees. Reference. Habu 1967:104-105 (and Fig. 2, Plate XIV- habitus illus.) (Habu ranks *Mochtherus* as a subgenus, including it in the genus *Dolichoctis* Schmidt-Goebel 1846).

*Coptodera* Dejean 1825
The 103 species of this genus are arranged in four subgenera. Pantropical, the range of this genus is co-extensive with that of the subtribe Pericalina. The nominotypical subgenus, with 44 species, is indigenous in the Neotropical Region. The five species in the Nearctic Region are arranged in three species groups (Shpeley and Ball 1993). The Nearctic range of these species is co-extensive with the Nearctic range of the Pericalina. Four of the Nearctic species have extensive ranges in the Neotropical Region; one species, the southeastern *C. aerata* Dejean 1825, is Nearctic precinctive. Primarily arboreal, adults live on tree trunks, standing or fallen, resting under bark by day. Adults of some species come to fruit falls in the tropics. Reference. Shpeley and Ball 1993 (rev. spp.).

*Eucheila* Dejean 1831
subgenus *Inna* Putzeys 1863
subgenus *Hansus* Ball and Shpeley 1983
subgenus *Bordoniella* Mateu 1989
subgenus *Pseudoinna* Mateu 1989
This genus includes 24 species arranged in five subgenera. The group is Neotropical-indigenous, with one species, *E. (Inna) boyeri* (Solier 1835) ranging northward from the Amazon Basin to Brownsville, Texas, in the Nearctic Region. The genus is arboreal, with adults inhabiting tree crowns, bromeliads, and clumps of dry leaves in the tropical undercanopy. Adults are nocturnal, and are found on tree trunks at night (Erwin 1991b: 48). References. Ball and Shpeley 1983: 775-777. Shpeley and Ball (in press).

*Somotrichus* Seidlitz 1887
Probably Afrotropical in origin (Ball 1975: 169-170), this genus contains two species, one of which is known only from Madagascar, while the second one, *S. unifasciatus* (Dejean 1831), is worldwide in distribution, having been transported by human commerce. The latter has been reported from Seattle, Washington, where five specimens were intercepted in a shipment of Brazil nuts from Brazil. According to Hatch (1953, 153) the species has probably not become established in that area. In the east, *S. unifasciatus* is known from Florida and Texas. References. See above, and the following. Ball 1975: 170 (photograph). Basilewsky 1984: 552-553. Jeannel 1949a. - 917 (habitus drawing).

Subtribe Apenina

Apenini Basilewsky 1984
This subtribe includes 116 species in three genera. Nearly pantropical, the range of the Apenina extends in the Eastern Hemisphere from New Guinea in the Australian Region to the Afrotropical Region, and northward in the eastern Palaearctic Region to the Japanese Archipelago, and in the west, to southern Europe. In the Western Hemisphere, one genus, *Apenes* LeConte 1851 is represented in the Neotropical and Nearctic Regions. References. Ball and Hilchie 1983: 1225-126. Basilewsky 1984: 552.

*Apenes* LeConte 1851
This genus includes 75 species arranged in two subgenera. Only the Neotropical-indigenous *Apenes* (*s. str.*) enters the Nearctic Region, where it is represented by six species, two of which are precinctive in the east. The Nearctic range of *Apenes* (*s. str.*) extends from the Mexican border and Gulf Coast north to Rhode Island, southern Ontario, and Kansas. The species are geophilous, ranging from xerophilous to mesophilous. *Apenes sinuata* (Say 1823), was reported by Blatchley (1910: 154) to occur in open woodland about the bases of trees and stumps. Adults of that species and *A. lucidula* (Dejean 1831) come to light sources at night. References. Blatchley 1910: 154. Horn 1882c: 156-158.

Subtribe Cymindidina

Cymindina *auctorum*
This subtribe includes 305 species arrayed in 11 genera. The group is wide-ranging, occurring in the Oriental, Afrotropical, Holarctic, and Neotropical Regions. Only one genus, *Cymindis* Latreille 1806, is known from the Western Hemisphere.

*Cymindis* Latreille 1806
This genus includes 207 species arrayed in four subgenera, its overall range including the Oriental, northern Afrotropical, Holarctic, and northern Neotropical Regions. In the Western Hemisphere, *Cymindis* is represented by 22 species in two subgenera, with the geographical range extending from the tundra in the north (Alaska to Newfoundland) southward to Costa Rica, in the Neotropical Region. See below for additional remarks about the subgenera.

Subgenus *Cymindis s. str.*
This Holarctic subgenus, which is represented on the northern fringes of the Afrotropical and Oriental Regions, includes 111 species, with 15 species in the Nearctic Region of which one (*C. vaporariorum* (Linnaeus 1758)) is Holarctic (Shpeley and Ball 1999). The geographical range in the Western Hemisphere extends in the north from the tundra transcontinentally, southward to northwestern Mexico, in the Neotropical Region. The members of this subgenus are xerophilous, most of the species living in open, dry grassland. Lindroth's (1969: 1070-1086) excellent treatment of the northern species provides a basis for revision of all of the Nearctic species.

Subgenus *Pinacodera* Schaum 1859
This precinctive Western Hemisphere subgenus includes 19 species. The range extends in the Neotropical Region from Costa Rica to northern México and in the Nearctic Region from southern California in the west eastward to Florida, and northward to Colorado in the west and Massachusetts and southern Canada in the east. Adults are arboreal, hunting for prey on tree trunks

at night, and resting by day under bark, or in bromeliads (in the tropics and subtropics) Reference. Horn 1882c: 146-149.

Subtribe Dromiina

Lionychina
Pseudotrechina
Metadromiina
Peliocypidina
Cymindidina (in part)

This subtribe, worldwide in distribution, includes 704 species in 46 genera. It seems to be especially diverse and divergent in the Afrotropical Region. The genera form two complexes: the *Lionychus* genus-group and the *Dromius* genus-group.

The *Dromius* group includes Basilewsky's (1984) Dromiini, Pseudotrechini, Metadromiini, and Peliocypini, characterized by fusion of the mentum and submentum (i.e., a mental-submental suture is absent).

The Nearctic genera are grouped as follows:

*Lionychus* genus-group (*Apristus* Chaudoir 1846, and *Syntomus* Hope 1838); and *Dromius* genus-group (*Dromius* Bonelli 1810, *Philorhizus* Hope 1838, *Microlestes* Schmidt-Goebel 1846, and *Axinopalpus* LeConte 1849).

*Apristus* Chaudoir 1846

A Megagaean indigenous genus, this assemblage of 61 species enters only the northern part of the Neotropical Region (Middle America, south to Guatemala). Thirteen species are Nearctic, the combined range principally western, extending from British Columbia to California in the west and in the east from New Brunswick south to Texas. The species are geophilous and xerophilic, living in sandy soil among rocks, close to or remote from the margins of streams. Adults run swiftly and fly readily, and are difficult to catch, especially on warm days. The genus requires study. Lindroth's (1969: 1043-1047) careful work with the northern species provides a good basis for a needed general revision of the Western Hemisphere taxa. Reference. Casey 1920, 272-276.

*Syntomus* Hope 1838
*Metabletus* Schmidt-Goebel 1846

This genus of 53 species is Megagaean, occupying all of the major zoogeographical regions except the Australian and Neotropical. The sole Nearctic representative of *Syntomus*, *S. americanus* (Dejean 1831), is transcontinental, ranging from Newfoundland to British Columbia, and southward to the Mexican border. This species is xerophilous, adults occurring on open spots exposed to sun, and with sparse, low vegetation, on soil of gravel or sand, in some places with a mixture of peat. The beetles are diurnal.

*Dromius* Bonelli 1810
subgenus *Dinodromius* Casey 1920
*piceus* species group, Lindroth 1969

This widespread genus of 101 species, arranged in three subgenera and one unranked complex, occupies all of the major zoogeographical regions except the Australian. In the north, however, it does not occur in the arctic regions. The subgenus *Dinodromius,* proposed for the Nearctic species, is regarded as consubgeneric with *Dromius s. str.*

Two species of *Dromius* are in the Nearctic Region: one indigenous, and one introduced. The indigenous *D. piceus* Dejean 1831 is wide-ranging, from the northern Neotropical Region northward through the United States to southern Canada, from Nova Scotia to British Columbia. The introduced species, *D. fenestratus* (Fabricius 1794), is known from Newfoundland (Larson 1998: 126) and Nova Scotia. The species are basically arboreal, adults being found under bark, but also on the ground, under debris. Adults are nocturnal, some flying to light sources.

*Philorhizus* Hope 1838
*Dromius, auct.*

This Afrotropical-Holarctic genus includes 49 species. In the Nearctic Region, it is represented by one precinctive species, *P. atriceps* (LeConte 1880), ranging on the Gulf Coast from Louisiana eastward to Florida, and northward to Massachusetts. Adults (and probably larvae) inhabit bunches of grass (Schwarz 1884) in wetland habitats. Reference. Bousquet and Larochelle 1993: 271.

*Microlestes* Schmidt-Goebel 1846
*Blechrus* Motschulsky 1847

This primarily Megagaean genus (not represented in the Australian Region or in South America) includes 129 species. Seven species inhabit the Nearctic Region, ranging from southern Canada (British Columbia to Newfoundland) to the Mexican border and Gulf Coast, and southward into Middle America. The species are basically xerophilous, living in open areas with sparse vegetation. Adults are diurnal. Lindroth's (1969: 1043-1055) careful work with the northern species provides a good basis for a needed general revision of the Western Hemisphere taxa.

*Axinopalpus* LeConte 1849
*Variopalpis* Solier 1849

This indigenous Neotropical genus includes 19 species, the aggregate range extending from Chile, in trans-Andean South America, northward through Middle America to southern Canada, in the Nearctic Region. Eleven species are Nearctic. Blatchley (1910: 151) states that *A. biplagiatus* (Dejean 1825) is found under the bark of logs in damp localities, but adults have been found also on the ground in dry, open places. A revision of the species is required. Reference. Horn 1882c: 135-136.

Subtribe Lebiina

This taxon is worldwide and is at least moderately diverse in all zoogeographical regions. The 766 species are arrayed in 25 genera. The Nearctic fauna consists of 49 species arrayed in two genera. References. Basilewsky 1984: 543-548.

*Lebia* Latreille 1802
*Loxopeza* Chaudoir 1870
*Metabola* Chaudoir 1871
*Dianchomena* Chaudoir 1871
*Aphelogenia* Chaudoir 1871
This worldwide genus, with a range co-extensive with that of its subtribe, includes 715 species in 17 subgenera. In the Nearctic Region, 48 species are included in four subgenera. The Nearctic range extends throughout the United States northward to the southern part of the Yukon Territory and the Northwest Territories in the west, and to northern Quebec, in the east. Most of the species are planticolous or arboreal, living in association with standing vegetation, from herbs to shrubs and forest trees. Adults of some species are diurnal, climbing about on plants in daylight; others are nocturnal, some flying to light sources. Some are geophiles, living on the ground in dry, open places. A few species are parasitoids on chrysomelids. The early stage larvae are campodeiform, while the later stages exhibit the structural degeneration which is characteristic of a parasitic mode of life. Pupation takes place near or within the remains of the host (Capogreco 1989). It is suspected that a parasitoid mode of life is common to the group as a whole. Adults of two Nearctic species mimic their host adults (Lindroth 1971; Balsbaugh 1967). Determining the way of life of the species of this genus remains a major challenge for those interested in carabid beetles. More generally, although the adults of the Nearctic species are easily identified, thanks to the perceptive analysis of Madge (1967), the Neotropical components of *Lebia* are hardly known, and many species await description and systematic analysis. Reference. Madge 1967: 139-142 (rev. Nearctic spp.).

subgenus *Loxopeza* Chaudoir 1871
*grandis* species group, Lindroth 1969
This indigenous Neotropical subgenus includes 25 species, eight living in the Nearctic Region, ranging from the Mexican border and the Gulf Coast northward to South Dakota in the west and Quebec in the east.

subgenus *Polycheloma* Madge 1967
*lecontei* species group, Bousquet and Larochelle 1993
This monospecific subgenus is probably a Middle American indigene, with a range extending into the southern part of the Nearctic Region. The sole member, *L. lecontei* Madge 1967, is known from Texas and Arizona, only. We have seen specimens of undescribed species from Mexico that probably belong in *Polycheloma*, but the confirmatory studies have yet to be conducted. Reference. Bousquet and Larochelle 1993: 275.

subgenus *Lamprias* Bonelli 1810
*divisa* species group, Lindroth 1969
This Holarctic subgenus includes 15 species. The Nearctic Region, however, has only one species, *L. divisa* LeConte 1850, which lives in the Great Plains area, ranging from Kansas and Colorado northward to southern Canada, and in the east, to Indiana.

subgenus *Lebia* (*sensu stricto*)
*Metabola* Chaudoir 1871
*Aphelogenia* Chaudoir 1871
*Dianchomena* Chaudoir 1871
*Poecilothais* Chaudoir 1871
*pulchella* species group (in part), Lindroth 1969
*viridis* species group (in part), Lindroth 1969
*analis* species group (in part), Lindroth 1969
*vittata* species group (in part), Lindroth 1969
*bivittata* species group (in part), Lindroth 1969
*vittata* species group, Bousquet and Larochelle 1993
This subgenus is markedly diverse, with a total of 506 species and a worldwide geographical range. In the Nearctic Region live 38 indigenous and precinctive species, with their collective range co-extensive with that of the genus. The group is quite heterogeneous, as shown by the numerous species groups in which Lindroth (1969: 1015-1039) arranged the species.

*Hyboptera* Chaudoir 1872
This indigenous Neotropical genus contains four species, with a combined range extending from the Amazon Basin in South America northward to southeastern Texas, in the Nearctic Region. Adults are arboreal, living on tree trunks in tropical forests. Probably *Hyboptera* should be removed from the subtribe Lebiina and placed in a separate subtribe (presently uncharacterized) to include also at least the Neotropical precinctive genera *Aspasiola* Chaudoir 1877 and *Cryptobatis* Eschscholtz 1829 (T. L. Erwin, personal communication). One species, *H. dilutior* Oberthür 1884, is known from Brownsville, Texas. (We are indebted to E. G. Riley [Texas A&M University] and M. K. Thayer [Field Museum of Natural History] for drawing our attention to Texas records of this species). References. Erwin 1991b: 46, and 74, Fig. 60 (habitus illustration). Reichardt 1973: 47-55 (rev. spp.).

Subtribe Nemotarsina

This is a monogeneric indigenous Neotropical taxon, including only the genus *Nemotarsus* LeConte 1853. A group of uncertain relationships, it was first recognized as distinctive by Bates (1883: 173) who proposed the subfamily Nemotarsinae for its reception. Jeannel (1949a: 860) placed this group as a tribe in the subfamily Masoreitae. Ball (1960a:158) suggested that *Nemotarsus* was related to the Afrotropical-southern Palaearctic lebiine genus *Pseudomasoreus* Desbrochers 1904. Lindroth (1969a: 1014) accepted *Nemotarsus* as a lebiine, including it in a subtribe of its own, followed by Reichardt (1977: 444) and Ball and Hilchie (1983: 112). Basilewsky (1984: 527), on the other hand, agreed with Jeannel that *Nemotarsus* was a masoreimorph, placing it in the subtribe Cyclosomina, on the basis of the shared size and form of the tibial spurs of the hind legs. Ball *et al.* (1995), in their analysis of relationships of the lebiine subtribes, neglected to include the Nemotarsina—an unfortunate oversight. Here, we continue to postulate that *Nemotarsus* is a lebiine, and the features shared with the Cyclosomini are examples of evolutionary convergence. Nonetheless, we recognize that this decision is arbitrary, and that the position of this group remains to be determined.

*Nemotarsus* LeConte 1853
*Nematotarsus* Gemminger and Harold 1868
This indigenous Neotropical genus of nine species is represented in the eastern part of the Nearctic Region by one species, *N. elegans* LeConte 1853, with a range extending from the Gulf Coast north to Pennsylvania and west to Iowa. Evidently, adults are nocturnal, being taken at light sources which are reached by flight. Erwin (1991b: 48) records adults of Amazonian species as living in the canopy and subcanopy of tropical forests.

Subtribe Metallicina

Represented by 71 species in four genera, this subtribe is in the tropics of the Eastern and Western Hemispheres, ranging northward in Eastern Asia to the Palaearctic Japanese Archipelago, and in the Western Hemisphere to Nearctic southern United States. References. Ball *et al.* 1995: 297. Basilewsky 1984: 542-543. Casale 1998: 411. Shpeley 1986: 261-349.

*Euproctinus* Leng and Mutchler 1927
*Euproctus* Solier 1849, not Gene 1839
*Andrewesella* Csiki 1932
The only Western Hemisphere member of the Metallicina, the indigenous Neotropical *Euproctinus* is represented by 16 species in two subgenera. The nominotypical subgenus *Euproctinus* is represented by a single Argentinian-Chilean species, *E. fasciatus* (Solier 1849). The subgenus *Neoeuproctus* Shpeley 1986, with 15 species, is represented in the Nearctic Region by three species: *E. abjectus* (Bates 1883) from California and Texas; *E. balli* Shpeley 1986 from Arizona; and *E. trivittatus* (LeConte 1878) from peninsular Florida.

Erwin (1991b: 48) recorded adults of this genus from forest canopy, and from clumps of dry leaves suspended above the ground. In Mexico, adults have been taken from *Acacia* bushes in dry forest and in pastures. Reference. Shpeley 1986: 261-349. (rev. spp.).

Subtribe Calleidina

This worldwide group of primarily tropical carabids includes 629 species arranged in 41 genera. The Western Hemisphere component includes 298 species in 29 genera; of these, 23 species in five genera are in the Nearctic Region. References. Ball and Hilchie 1983: 173-196. Basilewsky 1984: 548. Casale 1998: 381-428 (phyl. analysis). Habu 1967: 121-168. Larson 1969: 381-428.

*Calleida* Latreille and Dejean 1824
This genus, with its 277 species in four subgenera, is primarily tropical, the range including the Neotropical, Afrotropical, and Oriental Regions, with northern extensions in the eastern Palaearctic Region, to the Japanese Archipelago, and in the Nearctic Region to southern Canada (Casale 1998: 415, Fig. 95). In the Western Hemisphere, *Calleida* is represented by two subgenera with 180 species. Most of the species are arboreal, at least as adults.

subgenus *Calleida s. str.*
*Lecalida* Casey 1920
*decora* species group, Lindroth 1969
With 176 species, *Calleida* (*s. str.*) is confined to the Western Hemisphere; most species are Neotropical. Of the total, 11 species enter or are precinctive in the Nearctic Region. The species are vegetation-associated, adults being found primarily on trees, or on herbaceous vegetation. Reference. Horn 1882a: 138-142. Zhou and Goyer 1993: 234-237 (descr. larva).

subgenus *Philophuga* Motschulsky 1859
*Glycia auctorum*, not Chaudoir 1842
*viridis* species group, Lindroth 1969
Four species comprise this indigenous Nearctic subgenus. One is precinctive in Mexico; two occur in Mexico and the southern United States, from Arizona eastward to Alabama, and northward to Colorado; and one, *C. viridis* (Dejean 1831), is precinctive in the Nearctic Region, ranging widely in the west, from California east to Kansas and north to southern Canada (British Columbia to Manitoba). This species is geophilous-planticolous, adults living in dry fields, on clay soil, and resting under stones by day, though they climb on vegetation when active (Larson 1969: 65). Adults of the Mexican precinctive species *C. brachinoides* Bates 1883 evidently are arboreal, adults having been collected in bromeliads growing on trees in dry pine-oak forest. Reference. Larson 1969: 25-26 (larvae); 29-43 (adults).

*Tecnophilus* Chaudoir 1877
*Philotecnus* LeConte 1961, not Mannerheim 1837
This precinctive Nearctic genus contains two species: *T. pilatei* Chaudoir 1877, confined to the Gulf Coast of Texas, and *T. croceicollis* (Ménétriés 1843), ranging from California and southern Texas northward to Oregon and southern Alberta. As noted by Larson (1969: 63) adults of *Tecnophilus* are principally terrestrial, though females climb low plants, on the stems of which they attach their eggs, which are packed in mud balls. Reference. Larson 1969: 26-28 (larvae); 44-66 (adults).

*Infernophilus* Larson 1969
This monospecific taxon contains the species *I. castaneus* (Horn 1882), ranging from southern California and Nevada south to northwestern Mexico. The species is desert-inhabiting (Sonoran and Mojave). Adults live on the ground and on low vegetation. Reference. Larson 1969: 43-44.

*Onota* Chaudoir 1872
An indigenous Neotropical genus, *Onota* includes 10 species, of which one, *O. floridana* Horn 1881, is known only from Florida. Adults of this species have been collected only from the leaves of *Sabal palmetto*. References. Erwin 1991b: 48. Horn 1882c: 159-160.

*Plochionus* Dejean 1825
This indigenous Neotropical genus includes 16 species arranged in two subgenera: *Plochionus* (*s. str.*), and *Menidius* Chaudoir. To

place the Nearctic species in proper context, a general analysis of the species of this genus is required. References. Erwin 1991b: 48. Reichardt 1977: 442.

subgenus *Plochionus s. str.*

*Plocionus* Agassiz (invalid emendation)

This subgenus includes three species: one from Brazil, one from New Caledonia, and one, *P. pallens* (Fabricius 1775), is cosmopolitan. In the Nearctic Region, *P. pallens* is known from Florida, Maryland, and Pennsylvania in the east, and from California in the west.

As presently constituted, the subgenus *Plochionus* must be polyphyletic. At least the New Caledonian species, *P. niger* Fauvel 1903, probably belongs elsewhere, unless it is conspecific with *P. pallens*.

subgenus *Menidius* Chaudoir 1872

This Neotropical indigenous subgenus includes 13 species, three of which are in the Nearctic Region. The species *P. timidus* Haldeman 1843 is widespread in the southern part of the Nearctic Region, but is not known from Canada. This species is arboreal, adults and larvae eating the larvae of the arctiid moth species *Hyphantria cunea* Drury 1773, the Fall Webworm (Duffey 1891: 189). Two species, *P. amandus* Newman 1840 and *P. bicolor* Notman 1919, are confined to the southeast—Florida north to Georgia and west to Alabama. They live in the Bahamas and Greater Antilles (Erwin and Sims 1984: 446), also. The relationships of these latter two species are not well understood. References. Horn 1882c: 145-149. Notman 1919: 234. Zhou and Goyer 1993: 237-242 (descr. larva).

*Cylindronotum* Putzeys 1845

*Micragra* Chaudoir 1872

An indigenous Neotropical genus of seven species, one species, *C. aeneum* Putzeys 1845 ranges northward to southeastern Texas. The species of this genus are arboreal, living in the canopy in tropical rain forest, and in shorter trees in more northern, drier areas. References. Bates 1883: 199-200. Erwin 1991b: 47. Reichardt 1977: 441.

Subtribe Agrina

A monogeneric, Neotropical indigenous group of 520 species, this subtribe contains some of the most spectacular carabids in terms of elegant form and brilliant color. Previously, the agrine assemblage was treated as a separate tribe (or subfamily), or included in the Calleidina. As shown by Casale (1998: 411, Fig. 91) agrines might be closest to the Oriental-Australian subtribe Physoderina. Thus, it is best not to include the former group in the Calleidina.

*Agra* Fabricius 1801

This genus exhibits its greatest diversity in Neotropical South America, diminishing rapidly northward, with not more than two species (both undescribed; T.L. Erwin, personal communication) reaching the northern end of the Neotropical Region, in southeastern Texas. Little is known about the way of life of the species of *Agra*, except that adults are arboreal, many living in the high canopy of tropical evergreen forest; some individuals are attracted to light sources, which are reached by flight, at night. The foremost living authority, who has published extensively on this group, is T. L. Erwin (Smithsonian Institution). Reference. Erwin 1991b: 47 (habitus illustration).

Zuphiini

This tribe, with 274 species in 20 genera and two subtribes, has representatives in all of the major zoogeographical regions. Most of the genera are restricted to the tropics of either the Eastern or Western Hemisphere. The three genera in the Nearctic Region are members of the subtribe Zuphiina. In the northern hemisphere, the Zuphiini occur only in warm temperate areas. Knowledge of the taxa of the Western Hemisphere is very limited, and taxonomic study is required.

Virtually nothing is known about way of life of Zuphiina (Erwin 1991b: 42). Most of the adults known of the taxa reported below were collected at light sources, at night.

*Zuphium* Latreille 1806

*Zophium* Gistel 1839 [unjust. emendation]

*Zoyphium* Agassiz 1847 [unjust. emendation]

This genus, with 72 species, is represented in all of the major zoogeographical regions of the world. It is primarily tropical, and is not represented in the more northern portions of Holarctica. Six species are Nearctic. The collective range is transcontinental from California to Florida, extending northward through the Mississippi Basin to Michigan and southern Ontario (Point Pelee), and on the eastern coastal plain, to Maryland. References. Hatch 1953: 150. Mateu 1981: 1-12 (rev. spp.).

*Pseudaptinus* Castelnau 1834

*Diaphorus* Dejean 1831 [preoccup.]

*Tiphys* Gistel 1848 [preoccup.]

This indigenous Neotropical genus includes 18 species, three of which are in the Nearctic Region, collectively with a transcontinental range in the south; *P. lecontei* (Dejean 1831), is known from the Gulf States northward on the eastern coastal plain to North Carolina; and *P. tenuicollis* (LeConte 1851), ranges from Texas westward to California and northward to southwestern Oregon. References. LeConte 1879: 62. Liebke 1934: 372-388.

*Thalpius* LeConte 1851

*Enaphorus* LeConte 1851

*Zuphiosoma* Laporte de Castelnau 1867

This indigenous Neotropical genus includes 34 species, eight of which are in the Nearctic Region, collectively with a transcontinental range in the south, from California to Florida, and northward on the Atlantic coastal plain to the area of the District of Columbia. References. See above, and Darlington 1934: 125-126, and 128 for West Indian spp. that are in Florida, also.

## Galeritini

Galeritinini Jeannel 1949a
Galeritulini Jedlicka 1963
Closely related to the Zuphiini, this tribe of 136 species in five genera and two subtribes is primarily pantropical, with extensions into Holarctica. Lorenz (1998: 481) does not agree that the monogeneric Afrotropical-East Asian subtribe Planetina has galeritine relationships; rather, he places the group as a subtribe of Zuphiini. The three genera (*Ancystroglossus* Chaudoir 1863, *Trichognathus* Latreille 1829, and *Galerita* Fabricius 1801) inhabiting the Western Hemisphere belong to the subtribe Galeritina. Of these, only members of the genus *Galerita* occur in the Nearctic Region. References. Ball 1985b: 276-321 (evolution of genera). Reichardt 1967: 1-76 (rev. spp.).

*Galerita* Fabricius 1801, not Gouan 1770
*Galeritula* Strand 1936
Widespread, *Galerita* with 88 species in two subgenera (*Progaleritina* Jeannel 1949a and *Galerita* [*s. str.*]) is represented in all of the major zoogeographical regions except the Australian. It is not found far north, however, and is restricted to the eastern part of the Palaearctic Region, including Formosa and Japan. More than half of the known species are in the Neotropical Region. Eight species are in the Nearctic Region, the collective range being transcontinental in the south, from California to Florida, and northward through the Mississippi Basin to southern Manitoba in the west, and to Massachusetts to the east of the Appalachian Mountains. As reported by Reichardt (1967: 9) larvae of several species representing both subgenera have been described. The larvae exhibit a striking frontal projection similar to that of the nebriite genera *Leistus* and *Notiophilus* (Thompson 1979: 282, Fig. 82; cf. 240, Fig. 35, and 242, Fig. 38). Little is known about life history of *Galerita,* with one detail being outstanding: ovipositing females of *G. (Progaleritina) bicolor* (Drury 1773) encapsulate their eggs in mud cells, which they attach to the undersides of smooth leaves (King 1919: 383). The distinctive larva and oviposition habit suggest an unusual way of life, and the basis for a thorough, comparative investigation of the way of life of the members of this genus.

subgenus *Progaleritina* Jeannel 1949
With seven species, this subgenus is precinctive in the Western Hemisphere, its range extending through Neotropical Middle America to the cool temperate parts of the Nearctic Region. All seven species are in the Nearctic Region, with two (*G. bicolor* and *G. janus* (Fabricius 1792)) precinctive in the east (Gulf Coast northward in the Mississippi Basin and to the east of the Appalachians, as noted above). The geographical range of only one species (*G. lecontei* Dejean 1831) extends as far west as California. The ranges of five species include both the Neotropical and Nearctic Regions. Most adults in collections were taken at light sources, at night. Most specimens collected during the day or away from light sources at night were found on the ground in rather dry, open forest, or in open areas adjacent to such woodland. Reference. Ball and Nimmo 1983: 295-356 (rev. spp.).

subgenus *Galerita* (s. str.)
*Diabena* Fairmaire 1901
*Galeritula* Strand 1936
*Galeritina* Jeannel 1949
*Galericeps* Jeannel 1949
*Galeritella* Jeannel 1949
*Galeritiola* Jeannel 1949
This subgenus, with 81 species, is Neotropical-Afrotropical-Oriental-Eastern Palaearctic in distribution. The Neotropical-Nearctic species are included in the *G. americana* complex and the Afrotropical-Oriental-Palaearctic species in the *G. perrieri* complex. In turn, the species in each complex are arranged in two species groups. The Western Hemisphere *G. americana* complex comprises 54 species, 53 of which are precinctive in the Neotropical Region, with one, *G. aequinoctialis* Chaudoir 1852 being Neotropical-indigenous, because it ranges into the southernmost part of the Nearctic Region (southeastern Texas). The total range of the *G. americana* complex extends from northern Argentina in South America to southeastern Texas. Most of the species are forest inhabitants in both tropical lowlands and highlands. One species, *G. ruficollis* Dejean 1825, is a resident of more open habitats, such as meadows and stream margins, adjacent to rather open woodland.

## Helluonini

The Helluonini are included in the Anthiitae, a supertribe, with the Eastern Hemisphere tropical Anthiini and Helluodini. Helluonines occur in most zoogeographical regions, but the group is marginal in Holarctica. The 163 species are included in 25 genera and two subtribes, the Helluonina and Omphrina. Only the latter group is represented in the Western Hemisphere. Reference. Reichardt 1974: 211-302 (class. and rev. Western Hemisphere spp.)

### Subtribe Omphrina

Helluomorphina Reichardt 1974
With 129 species in 13 genera, omphrines are represented in the Western Hemisphere and in the Eastern Hemisphere Afrotropical, Oriental, and eastern Palaearctic Regions. Five genera are in the Western Hemisphere, all in the Neotropical Region, and one, *Helluomorphoides* Ball 1951, in the Nearctic Region, as well.

*Helluomorphoides* Ball 1951
*Helluomorpha auctorum*, not Castelnau 1840
A Neotropical indigenous genus with 23 species, the range of *Helluomorphoides* extends from Argentina through Middle America to northeastern and central United States. Eight species are in the Nearctic Region: three (*H. praeustus* (Dejean 1825), *H. clairvillei* (Dejean 1831), and *H. nigripennis* (Dejean 1831)) are

precinctive in the southeast; and the ranges of five (*H. ferrugineus* (LeConte 1853), *H. texanus* (LeConte 1853), *H. mexicanus* (Chaudoir 1872), *H. latitarsis* (Casey 1913), *H. papago* (Casey 1913)) are principally northern Mexican (Neotropical), extending varying distances into the Nearctic Region. The Nearctic species inhabit open areas, such as pastures, as well as forest. Most adults in collections were taken at light sources to which they evidently flew, at night. Topoff (1969) observed predation on army ants by adults of *H. latitarsis* and *H. ferrugineus*. The taxonomic range of this habit remains to be determined: Is it characteristic of just those two species, of the genus *Helluomorphoides*, or of the entire tribe? The taxonomic work of Reichardt (1974) and Ball (1956) provide a satisfactory basis for a more detailed study of the Western Hemisphere species of this genus, many of which remain to be described, especially in Middle America. Reference. Ball 1956: 67-91 (rev. N. A. spp.).

## Subfamily XI. Pseudomorphinae

The diagnostic adult features of this subfamily are the following: size moderate (total length ca. 3-17 mm); head capsule ventrally with antennal grooves; antennal insertions laterad, each between base of mandible and anterior margin of eye; antenna filiform, number of antennomeres 11; antennomeres 2-4 not nodose, 2-6 without unusually long setae; labrum with two or four setae; mandible asetose, [tooth pattern not studied]; submentum and mentum fused, not separated by suture; terminal palpomeres subsecuriform, apical margins obliquely truncate; pterothorax with mesonotal scutellum dorsally visible, not concealed beneath elytra; front coxal cavities closed, with projection of proepimeron inserted into lateral cavity of prosternum; middle coxal cavities conjunct-confluent, and hind coxal cavities lobate (disjunct)-confluent (Bell 1967: 105); elytra smooth dorsally, not extended to apex of abdominal tergum VIII, apical margins truncate; without flange of Coanda posteriolaterally; metathoracic wings large; front tibia anisochaete, antenna cleaner emarginate, type C (Hlavac 1971: 57); abdomen with six (II-VII) pregenital sterna normally exposed, sterna III and IV separated by suture; male genitalia with parameres nearly symmetric to markedly asymmetric in form, without or with few setae apically; ovipositor with gonocoxae rotated 90° in repose, without ramus, gonocoxa with two gonocoxites, gonocoxite 2 without accessory lobe; reproductive tract with spermatheca 2 only (Liebherr and Will 1998: 136, Fig. 53); secretions of pygidial glands hydrocarbons and formic acid; discharge by spraying (Moore 1979: 195).

This subfamily is monotribal.

### Pseudomorphini

Included in this group are 311 species in six genera. Of these, four are confined to the Australian Region; one (*Cryptocephalomorpha* Ritsema 1874) is in the Afrotropical (South Africa), Oriental, and Australian Regions; and the genus *Pseudomorpha* Kirby 1825 occurs in the Western Hemisphere Nearctic and Neotropical Regions, and in the Australian Region. Thanks to the outstanding publications of Baehr (1992 1994, and 1997), the important Australian pseudomorphine fauna has been the subject of recent study, and a firm basis is available for studies of pseudomorphines living in other zoogeographical regions.

Scattered observations suggest very interesting features of the development and way of life of pseudomorphines. Many Australian pseudomorphines have been found under the bark of *Eucalyptus* trees, and anecdotal evidence and some direct observations indicate that pseudomorphines are myrmecophilous (Baehr 1994: 15, and 1997: 455-456). Liebherr and Kavanaugh (1985) presented evidence for ovoviviparity in *Pseudomorpha*, and Baehr (1994: 14) referred to "unpublished" observations suggesting this mode of development in other pseudomorphine genera.

*Pseudomorpha* Kirby 1825
*Heteromorpha* Kirby 1825
*Axinophorus* Dejean and Boisduval 1829
*Drepanus* Dejean 1831

Thirty-two species in three subgenera comprise this genus, distributed as follows: Australia, *Austropseudomorpha* Baehr 1997, three species; South America, *Notopseudomorpha* Baehr 1997, two species; and Mexico and United States, *Pseudomorpha* (*s. str.*), 27 species, of which two are known only from Mexico. Two of the Nearctic species occur in the states which border the Gulf Coast; the remainder are confined to the southwest, including California and Utah.

Little is known about the way of life of the species of this genus. Most adults have been taken at light sources, and a few specimens have been collected in montane dry oak forest in Texas and Arizona, under stones on gravelly soil. One adult was observed entering an ant nest.

Presently, it is virtually impossible to identify adults of subgenus *Pseudomorpha*. A thorough revision is required, which will have to be based on more material than is available currently in collections. References. Notman 1925: 1-34.Van Dyke 1943: 30.

## Subfamily XII. Brachininae

The diagnostic adult features of this subfamily are the following: size small to large (total length ca. 2-30 mm); head capsule ventrally without antennal grooves; antennal insertions laterad, each between base of mandible and anterior margin of eye; antenna filiform, number of antennomeres 11; antennomeres 2-4 not nodose, 2-6 without unusually long setae; labrum with six to eight setae; mandible with scrobe unisetose or plurisetose, occlusal margin with one terebral tooth, retinacular teeth about size of terebral tooth, not enlarged and molariform; terminal palpomeres about cylindrical not securiform; pterothorax with mesonotal scutellum dorsally visible, not concealed beneath elytra; front coxal cavities closed, with projection of proepimeron inserted each side into cavity of prosternum; middle coxal cavities conjunct-confluent, and hind coxal cavities lobate (disjunct)-confluent (Bell 1967: 105); elytra striate (intervals and striae broad) to smooth dorsally, extended to apex of abdominal tergum VI, and apical margins truncate; without flange of Coanda; metathoracic wings large to variously reduced; front tibia anisochaete, antenna cleaner emarginate, type C (Hlavac 1971: 57); abdomen with seven (II-VIII) or eight (II-IX) sterna normally exposed, sterna III and IV separated by suture; male genitalia with parameres glabrous, reduced, right minute, left larger, and in adults of most species baldric-like in position, across basal part of median lobe (this type of arrangement of the left paramere was designated "balteate" by Jeannel 1941: 79); ovipositor with gonocoxae rotated 90° in repose, without ramus, gonocoxa with two gonocoxites, gonocoxite 2 without accessory lobe; reproductive tract with spermatheca 1 only (Liebherr and Will 1998: 1132, Figs. 36-37); secretion of pygidial glands quinones, delivered hot; discharge by crepitation (Moore 1979: 194).

### Brachinini

This tribe of moderate size contains 649 species in 20 genera, and is represented in all of the major zoogeographical regions. Most of the genera, however, are restricted to the tropics. One genus, *Pheropsophus* Solier 1833, is precinctive in the Neotropical Region, with its adelphotaxon, *Stenaptinus* Maindron, 1906, in the tropics of the Eastern Hemisphere (Erwin, 1971). The brachinine fauna of Africa seems to be especially rich and divergent, and even a totally blind species, *Brachynillus varendorffi* Reitter 1904, is known from that area (Basilewsky 1959: 329). Some of the species are known to be ectoparasitoid (larvae of *Brachinus*, on pupae of water beetles, and larvae of *Pheropsophus* on eggs of mole crickets [Erwin 1967,1976 and 1979; Habu and Sadanaga 1965; Frank 1994: 470]).

*Brachinus* Weber 1801
This genus comprises 304 species, of which 208 are distributed among eight subgenera, with 96 species not thus assigned. The group as a whole is worldwide in distribution, but only a single species has been recorded from the Australian Region, and that one from New Guinea. Only the subgenus *Neobrachinus* Erwin 1970 is represented in the Nearctic Region.

subgenus *Neobrachinus* Erwin 1970
This subgenus of 83 species (82 precinctive in the Western Hemisphere, one relict species in Sikkim [Erwin 1970: 48]) is represented in the Nearctic Region by 48 species, whose aggregate range includes all of the United States north to about the latitude of 50° in the west and 45° to the east of Lake Superior. The beetles live in moist situations, close to the margins of lakes and streams. These beetles have been reared from the pupal cells of the whirligig beetle, *Dineutes* spp. The female lays her eggs in mud cells formed on stems, twigs, or stones. The larvae have reduced legs and show parasitic adaptations. Thanks to the work of Erwin (1970), adults of the Nearctic species and of those inhabiting Mexico may be determined to species. The more southern Neotropical species continue to present a problem. Reference. Erwin 1970: 47-165 (rev. spp.).

### Bibliography


ALLEN, R.T. 1968. A synopsis of the tribe Morionini in the Western Hemisphere with descriptions of two new species. Caribbean Journal of Science, 8: 141-163.

ALLEN, R.T. 1972. A revision of the genus *Loxandrus* LeConte (Coleoptera: Carabidae) in North America. Entomologica Americana (New Series), 46: 1-184.

ALLEN, R.T. 1980. A review of the subtribe Myadi: description of a new genus and species, phylogenetic relationships, and biogeography (Coleoptera: Carabidae: Pterostichini). Coleopterists Bulletin, 34: 1-29.

ALLEN, R.T. and Ball, G.E.. 1980. Synopsis of Mexican taxa of the *Loxandrus* series (Coleoptera: Carabidae: Pterostichini). Transactions of the American Entomological Society, 105 (1979): 481-576.

ALLEN, R.T. and CARLTON, C.E. 1988. Two new *Scaphinotus* from Arkansas with notes on other Arkansas species (Coleoptera: Carabidae: Cychrini). Journal of the New York Entomological Society, 96: 129-139.

ANDREWES, H.E. 1929. Coleoptera. Carabidae. Vol. I. - Carabinae. The fauna of British India, including Ceylon and Burma. Taylor and Francis, London. xviii + 431 pp. (+ 9 pls.).

ARNDT, E. 1993. Phylogenetische Untersuchungen larvalmorphologischer Merkemale der Carabidae (Insecta: Coleoptera). Stuttgarter Beiträge zur Naturkunde (Serie A), Nr. 488: 1-56.

ARNDT, E. 1998. Phylogenetic investigation of Carabidae (Coleoptera) using larval characters, pp. 171-190. *In*, Phylogeny and classification of Caraboidea (Coleoptera: Adephaga) (Ball, G.E., Casale, A., and Vigna Taglianti, A., Eds.). Atti, Museo Regionale di Scienze Naturali, Torino. 543 pp.

ARNDT, E. and PUTCHKOV, A.V. 1997. Phylogenetic investigation of Cicindelidae (Insecta: Coleoptera) using larval morphological characters. Zoologischer Anzeiger, 1996/97: 231-241.

ARNETT, R.H., JR. 1946. Systema Naturae, no. 2, 7 pp. (key to gen. world).

ARNETT, R.H., JR. 1960. The Beetles of the United States (A manual for identification). The Catholic University of America Press, Washington, D.C. xi + 1112 pp.

BAEHR, M. 1983. *Schizogenius freyi*, sp. nov., die erste *Schizogenius* Art ausserhalb Amerikas. Entomologischen Arbeiten aus dem Museum G. Frey, 31/32: 91-95.

BAEHR, M. 1992. Revision of the Pseudomorphinae of the Australian Region. 1. The previous genera *Sphallomorpha* Westwood and *Silphomorpha* Westwood. Spixiana, Supplement 18, München. 440 pp.

BAEHR, M. 1994. Phylogenetic relations and biogeography of the genera of Pseudomorphinae (Coleoptera:

Carabidae), pp. 11-17. *In*, Carabid beetles: ecology and evolution (Desender, K., Dufrene, M., Loreau, M., Luff, M. L., and Maelfait, J.P., Eds.). Kluwer Academic Publishers, Dordrecht/Boston/London. xii + 474 pp.

BAEHR, M. 1997. Revision of the Pseudomorphinae of the Australian Region. 2. The genera *Pseudomorpha* Kirby, *Adelotopus* Hope, *Cainogenion* Notman, *Paussotropus* Waterhouse and *Cryptocephalomorpha* Ritsema. Spixiana, Supplement 23, München. 508 pp.

BAEHR, M. 1998. A preliminary survey of the classification of the Psydrinae (Coleoptera: Carabidae), pp. 359-368. *In,* Phylogeny and classification of Caraboidea (Coleoptera: Adephaga) (Ball, G.E., Casale, A., and Vigna Taglianti, A. , Eds.). Atti, Museo Regionale di Scienze Naturali, Torino, Italy. 543 pp.

BALL, G. E. 1956. A revision of the North American species of the genus *Helluomorphoides* Ball 1951 (Coleoptera, Carabidae, Helluonini). Proceedings of the Entomological Society of Washington, 58: 67-91.

BALL, G. E. 1959. A taxonomic study of the North American Licinini with notes on the Old World species of the genus *Diplocheila* Brullé (Coleoptera). Memoirs of the American Entomological Society. Philadelphia, Pa., Memoir 16, 258 pp.

BALL, G. E. 1960a. Carabidae, Fascicle 4, pp. 55-210. *In*, The beetles of the United States (A manual for identification) (R.H. Arnett, Jr., Ed.). The Catholic University of America Press, Washington, D. C. xi + 1112 pp.

BALL, G. E. 1960b. A review of the taxonomy of the genus *Euryderus* LeConte 1848, with notes on the North American Dapti (of authors) (Carabidae: Harpalini). Coleopterists Bulletin, 14: 44-64.

BALL, G. E. 1965. Two new subgenera of *Pterostichus* Bonelli from western United States, with notes on characteristics and relationships of the subgenera *Paraferonia* Casey and *Feronina* Casey (Coleoptera: Carabidae). Coleopterists Bulletin, 19: 104-112.

BALL, G. E. 1966a. The taxonomy of the subgenus *Scaphinotus* Dejean, with particular reference to the subspecies of *Scaphinotus petersi* Roeschke (Coleoptera: Carabidae: Cychrini). Transactions of the American Entomological Society, 92: 687-722.

BALL, G. E. 1966b. A revision of the North American species of the subgenus *Cryobius* Chaudoir (*Pterostichus*, Carabidae, Coleoptera). Opuscula Entomologica Supplementum, No. 28, 166 pp.

BALL, G. E. 1970. The species of the Mexican genus *Aztecarpalus*, new genus (Coleoptera: Carabidae: Harpalini). Coleopterists Bulletin, 24: 97-123.

BALL, G. E. 1972. Classification of the species of the *Harpalus* subgenus *Glanodes* Casey (Carabidae, Coleoptera). Coleopterists Bulletin, 26: 179-204.

BALL, G. E. 1975. Pericaline Lebiini: notes on classification, a synopsis of the New World genera, and a revision of the genus *Phloeoxena* Chaudoir (Coleoptera: Carabidae). Quaestiones Entomologicae, 11: 143-242.

BALL, G. E. 1976a. *Aztecarpalus* Ball: new species from Oaxaca, Mexico, re-classification, and a reconstructed phylogeny of the *hebescens* Group (Coleoptera: Carabidae: Harpalini). Coleopterists Bulletin, 30: 61-72.

BALL, G. E. 1976b. Notes about the species and relationships of *Hartonymus* Casey (Coleoptera: Carabidae: Harpalini). Proceedings of the Entomological Society of Washington, 78: 417-430.

BALL, G. E. 1983. Evaluation of the Baron Maximilien de Chaudoir's contribution to classification of cymindine Lebiini and Masoreimorphi (Coleoptera: Carabidae). Coleopterists Bulletin, 36 (1982): 61-72.

BALL, G. E. 1985a. (Editor). Taxonomy, phylogeny and zoogeography of beetles and ants: a volume dedicated to the memory of Philip Jackson Darlington, Jr. (1904-1983). Dr. W. Junk, Publishers, Dordrecht/ Boston/ Lancaster XIV + 514 pp.

BALL, G. E. 1985b. Reconstructed phylogeny and geographical history of genera in the tribe Galeritini, pp. 276-321. *In,* Taxonomy, phylogeny and zoogeography of beetles and ants: a volume dedicated to the memory of Philip Jackson Darlington, Jr. (1904-1983) (Ball, G.E., Ed.). Dr. W. Junk, Publishers, Dordrecht/ Boston/ Lancaster XIV + 514 pp.

BALL, G. E. 1992. The tribe Licinini (Coleoptera: Carabidae): a review of the genus-groups and of the species of selected genera. Journal of the New York Entomological Society, 100: 325-380.

BALL, G.E. and ANDERSON, J.N. 1962. The taxonomy and speciation of *Pseudophonus*. The Catholic University of America Press, Washington, D. C. XI + 94 pp.

BALL, G.E., CASALE, A., and VIGNA TAGLIANTI, A. (Eds.). 1998. Phylogeny and classification of Caraboidea (Coleoptera: Adephaga). Atti, Museo Regionale di Scienze Naturali, Torino, Italy. 543 pp.

BALL, G.E. and CURRIE, D.C. 1997. Ground beetles (Coleoptera: Trachypachidae and Carabidae) of the Yukon: geographical distribution, ecological aspects, and origin of the extant fauna, pp. 445-489. *In,* Insects of the Yukon (Danks, H.V., and Downes, J.A., Eds.). Biological Survey of Canada (Terrestrial Arthropods). Ottawa. x + 1034 pp.

BALL, G.E. and ERWIN, T.L. 1969. A taxonomic synopsis of the Tribe Loricerini. Canadian Journal of Zoology, 47: 877-907.

BALL, G.E. and HILCHIE, G.J. 1983. Cymindine Lebiini of authors: redefinition and reclassification of genera (Coleoptera: Carabidae). Quaestiones Entomologicae, 19: 93-216.

BALL, G.E., KAVANAUGH, D.H., and MOORE, B.P. 1995. *Sugimotoa parallela* Habu (Coleoptera: Carabidae: Lebiini): redescription, geographical distribution, and relationships based on cladistic analysis of adult structural features, pp. 275-311. *In,* Beetles and Nature (Watanabe, Y., Sato, M., and Owada, M., Eds.). Special Bulletin of the Japanese Society of Coleopterology, Tokyo, No. 4, ix + 510 pp.

BALL, G.E. and MADDISON, D.R. 1987. Classification and evolutionary aspects of the species of the New World genus *Amblygnathus* Dejean, with description of *Platymetopsis* new genus, and notes about selected species of *Selenophorus* Dejean (Coleoptera: Carabidae: Harpalini). Transactions of the American Entomological Society, 113: 189-307.

BALL, G.E. and MCCLEVE, S. 1990. The Middle American genera of the tribe Ozaenini with notes about the species in southwestern United States and selected species from México. Quaestiones Entomologicae, 26: 30-116.

BALL, G.E. and NEGRE, J. 1972. The taxonomy of the Nearctic species of the genus *Calathus* Bonelli (Coleoptera: Carabidae: Agonini). Transactions of the American Entomological Society, 98: 413-533.

BALL, G.E. and NIMMO, A.P. 1983. Synopsis of the species of subgenus *Progaleritina* Jeannel, including reconstructed phylogeny and geographical history (Coleoptera: Carabidae: *Galerita* Fabricius). Transactions of the American Entomological Society, 109(4): 295-356.

BALL, G.E. and SHPELEY, D. 1983. The species of eucheiloid Pericalina: classification and evolutionary considerations. Canadian Entomologist, 115: 743-806.

BALL, G.E. and SHPELEY, D. 1990. Synopsis of the Neotropical genus *Ozaena* Olivier: classification and reconstructed evolutionary history (Coleoptera: Carabidae: Ozaenini). Canadian Entomologist, 122: 779-815.

BALL, G.E. and SHPELEY, D. 2000. [CHAPTER] 19. Carabidae (Coleoptera), pp. 363-399. *In,* Biodiversidad, taxonomía y biogeografía de México: Hacia una syntesis de su conociemiento, Volumen II (Llorente Bousquets, J., and González Soriano, E., Eds.). Universidad Nacional Autónoma de México, México.XVI + 676 PP.

BALL, G.E., SHPELEY, D. and CURRIE, D.C. 1991. The New World genus *Stenomorphus* Dejean (Coleoptera: Carabidae: Harpalini): classification, allometry, and evolutionary considerations. Canadian Entomologist, 123: 933-988.

BALSBAUGH, E.U. 1967. Possible mimicry between certain Carabidae and Chrysomelidae. Coleopterists Bulletin, 21: 139-140.

BÄNNINGER, M. 1927. Die Ozaenini. Deutsche Entomologische Zeitschrift (1927): 177-216.

BÄNNINGER, M. 1930. Die Gattung *Pelophila* Dej. (Col. Carab.). Notulae Entomologicae, 10: 95-102.

BÄNNINGER, M. 1938. Monographie der subtribus Scaritina (Col. Carab.). II. Deutsche Entomologische Zeitschrift (1938), part 1: 41-182, 4 figs. and 4 tables.

BÄNNINGER, M. 1950. The subtribe Pasimachina (Coleoptera, Carabidae, Scaritini). Revista de Entomología, 21(3): 481-511.

BÄNNINGER, M. 1956. Über Carabinae, Ergänzungen und Berichtigungen (IV). Entomologischen Arbeiten aus dem Museum G. Frey, 7(1): 398-411.

BARR, T. C., JR. 1959. New cave beetles (Carabidae, Trechini) from Tennessee and Kentucky. Journal of the Tennessee Academy of Science, 34(1): 5-30, 9 figs.

BARR, T. C., JR. 1960a. A new genus of cave beetles (Carabidae: Trechinae) from southwestern Virginia with a key to the genera of Trechinae of North America north of Mexico. Coleopterists Bulletin, 14(3): 65-70.

BARR, T. C., JR. 1960b. The cavernicolous beetles of the subgenus *Rhadine*, genus *Agonum* (Coleoptera: Carabidae). American Midland Naturalist, 64: 45-65.

BARR, T. C., JR. 1960c. A synopsis of the cave beetles of the genus *Pseudanophthalmus* of the Mitchell Plain in southern Indiana. American Midland Naturalist, 63: 307-320.

BARR, T. C., JR. 1962. The genus *Trechus* (Coleoptera: Carabidae: Trechini) in the southern Appalachians. Coleopterists Bulletin, 16: 65-92.

BARR, T. C., JR. 1969. Evolution of the (Coleoptera) Carabidae in the southern Appalachians, pp. 67-92. *In*, The Distributional history of the Southern Appalachians, Part I: Invertebrates (Holt, P. C., Hoffman, R. L., and Hart, C. W., Eds.). Research Division Monograph 1, Virginia Polytechnic Institute, Blacksburg, Virginia.

BARR, T. C., JR. 1971a. *Micratopus* Casey in the United States (Coleoptera: Carabidae: Bembidiinae). Psyche, 78: 32-37.

BARR, T. C., JR. 1971b. The North American *Pterostichus* of the subgenus *Cylindrocharis* Casey (Coleoptera, Carabidae). American Museum Novitates, No. 2445. 14 pp.

BARR, T. C., JR. 1972. *Trechoblemus* in North America, with a key to North American genera of Trechinae. Psyche, 78 (1971): 140-149.

BARR, T. C., JR. 1974. Revision of *Rhadine* LeConte (Coleoptera, Carabidae) I. The *subterranea* group. American Museum Novitates, No. 2539. 30 pp.

BARR, T. C., JR. 1979a. Revision of Appalachian *Trechus* (Coleoptera: Carabidae). Brimleyana, 2: 29-75.

BARR, T. C., JR. 1979b. The taxonomy, distribution, and affinities of *Neaphaenops*, with notes on associated species of *Pseudanophthalmus* (Coleoptera, Carabidae). American Museum Novitates, No. 2682, 20 pp.

BARR, T. C., JR. 1980. New species groups of *Pseudanophthalmus* from the central basin of Tennessee (Coleoptera: Carabidae: Trechinae). Brimleyana, 3: 85-86.

BARR, T. C., JR. 1981. *Pseudanophthalmus* from Appalachian caves (Coleoptera: Carabidae): The *engelhardti* complex. Brimleyana, 5: 37-94.

BARR, T. C., JR. 1985a. Pattern and process in speciation of trechine beetles in eastern North America (Coleoptera: Carabidae: Trechini). pp. 350-407. *In*, Taxonomy, phylogeny and zoogeography of beetles and ants (Ball, G.E.,

Ed.). Dr. W. Junk, Publishers, Dordrecht/Boston/ Lancaster. XIII + 514 pp.

BARR, T. C., JR. 1985b. New trechine beetles from the Appalachian region (Coleoptera: Carabidae). Brimleyana, 11: 119-132.

BARR, T. C., JR. 1995. Notes on some anillines (Coleoptera, Carabidae, Bembidiinae) from southeastern United States, with descriptions of a new genus and two new species, pp. 239-248. *In*, Beetles and Nature (Watanabe, Y., Sato, M., and Owada, M., Eds.). Special Bulletin of the Japanese Society of Coleopterology, Tokyo, No. 4, ix + 510 pp.

BARR, T.C. and KREKELER, C.H. 1967. *Xenotrechus*, a new genus of cave trechines from Missouri. Annals of the Entomological Society of America, 60: 1322-1375.

BARR, T. C., JR. and LAWRENCE, J.F. 1960. New cavernicolous species of *Agonum (Rhadine)* from Texas (Coleoptera: Carabidae). Wasmann Journal of Biology 18(1): 137-145.

BARROWS, W. B. 1897. Notes on the malodorous carabid, *Nomius pygmaeus* Dej. United States Department of Agriculture Entomology Bulletin (New Series), 9: 49-53.

BASILEWSKY, P. 1950. Révision générale des Harpalinae d'Afrique et de Madagascar. Premiere partie. Annales du Musée du Congo Belge, Tervuren, Sciences Zoologiques, 6: 1-283, pls. 1-9.

BASILEWSKY, P. 1951. Révision générale des Harpalinae d'Afrique et de Madagascar. Deuxième partie. Annales du Musée du Congo Belge, Tervuren, Sciences Zoologiques, 9: 1-333.

BASILEWSKY, P. 1954. Description d'un carabique nouveau du Ruanda-Urundi, représentant d'une sous-famille inédite pour l'Afrique noire. Annales du Musée du Congo Belge, Tervuren, in 4°. Miscellanea Zoologica, 1: 301-303.

BASILEWSKY, P. 1959. Revision des Crepidogastrini (Coleoptera, Carabidae, Brachyninae). Rev. Ent. Mozambique, 2(1): 229-352, 39 figs.

BASILEWSKY, P. 1963. Révision des Promecognathinae d'Afrique (Coleoptera; Carabidae). Annals of the Transvaal Museum, 24: 305-314.

BASILEWSKY, P. 1967. IV. Description de deux carabides de Madagascar, représentants de groupes encore inconnus dans la région malgache. Bulletin de la Société Entomologique de France, 72: 248-252.

BASILEWSKY, P. 1984. Essai d'une classification supragénérique naturelle des Carabides Lébiens d'Afrique et de Madagascar. Revue de Zoologie Africaine, 98(3): 525-559.

BASILEWSKY, P. 1985. The South Atlantic Island of Saint Helena and the origin of the beetle fauna, pp. 257-275. *In*, Taxonomy, Phylogeny and Zoogeography of Beetles and Ants (Ball, G. E., Ed.). Dr. W. Junk Publishers, Dordrecht/Boston/Lancaster. XIII + 514 pp.

BASILEWSKY, P. 1987. Sur quelques genres de Masoreini africains (Coleoptera Carabidae). Revue de Zoologie Africaine, 100: 237-258.

BASILEWSKY P. and GRUNDMANN, E. 1955. Contributions à l'étude systématique des Chlaéniens. Bulletin et annales de la Societé Entomologique de Belgique, 91: 199-206.

BATES, H. W. 1882. Insecta. Coleoptera. Vol. I. Part 1. *In*, Biologia Centrali-Americana (Godman, F.D. and Salvin O., Eds.). Taylor and Francis, London. pp. 41-152 (plates iii-v).

BATES, H. W. 1883. Insecta, Coleoptera, Carabidae, Cicindelidae, supplement, Vol. 1, part 1, *In*, Biologia Centrali-Americana (Godman, F.D. and Salvin O., Eds.). Taylor and Francis, London. pp. 153-256, plates vi-xii.

BATES, H. W. 1884. Insecta, Coleoptera, Cicindelidae suppl., Carabidae suppl., Vol. 1, Part 1. *In*, Biologia Centrali-Americana (Godman, F.D. and Salvin O., Eds.). Taylor and Francis, London. pp. 257-299, plate xiii.

BAUER, T. 1979. The behavioural strategy used by imago and larva of *Notiophilus biguttatus* (Coleoptera, Carabidae) in locating Collembola, pp. 133-142. *In*, On the evolution of behaviour in carabid beetles (Den Boer, P. J., Thiele, H. V., and Wiker, F., Eds.).

BAUER, T. 1981. Prey capture and structure of the visual space of an insect that hunts by sight on the litter layer (*Notiophilus biguttatus* F., Carabidae, Coleoptera). Behavior, Ecology and Sociobiology, 8: 91-97.

BAUER, T. 1982a. Predation by a carabid beetle specialized for catching Collembola. Pedobiologia, 24(3): 169-179.

BAUER, T. 1982b. Prey-capture in a ground-beetle larva. Animal Behaviour, 30: 203-208.

BEDEL, L. 1906. Catalogue raisonné des coléoptères du Nord de l'Afrique (Maroc, Algérie, Tunisie, et Tripolitaine) avec notes sur la faune des Îsles Canarie et de Madère. pp. 201-264. Paris.

BELL, R. T. 1957. *Carabus auratus* L. (Coleoptera: Carabidae) in North America. Proceedings of the Entomological Society of Washington, 59: 254.

BELL, R. T. 1959. A new species of *Scaphinotus* Dejean, intermediate between *Scaphinotus s. str.*, and *Irichroa* Newman (Coleoptera, Carabidae). Proceedings of the Entomological Society of Washington, 61: 11-13, 8 figs.

BELL, R. T. 1960. A revision of the genus *Chlaenius* Bonelli (Coleoptera: Carabidae) in North America. Miscellaneous Publications of the Entomological Society of America, 1: 97-166.

BELL, R. T. 1964. Does *Gehringia* belong to the Isochaeta? Coleopterists Bulletin 18: 59-61.

BELL, R. T. 1967. Coxal cavities and the classification of the Adephaga (Coleoptera). Annals of the Entomological Society of America, 60: 101-107.

BELL, R. T. 1983. What is *Trachypachus*? (Coleoptera: Trachypachidae), pp. 590-596. *In*, The Baron Maximilien de Chaudoir (1816-1881): a symposium to honour the memory of a great Coleopterist during the centennial of his death. (Whitehead, D.R., Ed.). Coleopterists Bulletin, 36(1982): 459-609.

BELL, R. T. 1987. A new species of *Pentagonica* Schmidt-Goebel (Coleoptera: Carabidae) from the southwestern United States. Coleopterists Bulletin, 41: 373-376.

BENSCHOTER, C. A. and COOK, E.F. 1956. A revision of the genus *Omophron* (Carabidae, Coleoptera) of North America north of Mexico. Annals of the Entomological Society of America, 49(5): 411-429.

BEUTEL, R.G. 1992. Study on the systematic position of Metriini based on characters of the larval head (Coleoptera: Carabidae). Systematic Entomology, 17: 207-218.

BEUTEL, R.G. 1998. Trachypachidae and the phylogeny of Adephaga (Coleoptera), pp. 81-105. *In,* Phylogeny and classification of Caraboidea (Coleoptera: Adephaga) (Ball, G.E., Casale, A., and Vigna Taglianti, A., Eds.). Atti, Museo Regionale di Scienze Naturali, Torino, Italy. 543 pp.

BEUTEL, R.G. and HAAS, A. 1996. Phylogenetic analysis of larval and adult characters of Adephaga (Coleoptera) using cladistic computer programs. Entomologica Scandinavica, 27: 197-205.

BILS, W. 1976. Die Abdomenende weiblicher, terrestrisch lebender Adephaga (Coleoptera), und seine Bedeutung für die Phylogenie. Zoomorphologie, 84: 118-193.

BLATCHLEY, W. S. 1910. An illustrated descriptive catalogue of the Coleoptera or beetles (exclusive of the Rhynchophora) known to occur in Indiana - with bibliography and descriptions of new species. The Nature Publishing Co., Indianapolis. 1386 pp.

BOUSQUET, Y. 1984. Redescription of *Abaris splendidula* (LeConte) new combination (Coleoptera: Carabidae: Pterostichini). Coleopterists Bulletin, 38: 384-389.

BOUSQUET, Y. 1985. The subgenus *Pseudoferonina* Ball (Coleoptera: Carabidae: *Pterostichus*): description of three new species with a key to all known species. The Pan-Pacific Entomologist, 61: 253-260.

BOUSQUET, Y. 1986. Description of first instar larva of *Metrius contractus* Eschscholtz (Coleoptera: Carabidae) with remarks about phylogenetic relationships and ranking of the genus *Metrius* Eschscholtz. Canadian Entomologist, 118: 373-388.

BOUSQUET, Y. 1987a. Notes about the relationships of the Callistini (=Chlaeniini) (Coleoptera: Carabidae). Coleopterists Bulletin, 41: 165-166.

BOUSQUET, Y. 1987b. The carabid fauna of Canada and Alaska: range extensions, additions, and descriptions of two new species of *Dyschirius*. Coleopterists Bulletin, 41: 111-135.

BOUSQUET, Y. 1988. *Dyschirius* of America north of Mexico: descriptions of new species with keys to species groups and species (Coleoptera: Carabidae). Canadian Entomologist, 120: 361-387.

BOUSQUET, Y. 1990. On the taxonomic position of *Agonoleptus parviceps* Casey (Coleoptera: Carabidae). Coleopterists Bulletin, 44: 203-204.

BOUSQUET, Y. 1992. Descriptions of new or poorly known species of *Gastrosticta* Casey 1918 and *Paraferonina* Ball 1965 (Coleoptera: Carabidae: *Pterostichus* Bonelli 1810). Journal of the New York Entomological Society, 100: 510-521.

BOUSQUET, Y. 1996. Taxonomic revision of Nearctic, Mexican, and West Indian Oodini (Coleoptera: Carabidae). Canadian Entomologist, 128: 443-537.

BOUSQUET, Y. 1997. Description of a new species of *Clivina* Latreille from southeastern United States with a key to North American species of the *fossor* group (Coleoptera: Carabidae: Clivinini). Coleopterists Bulletin, 51: 343-349.

BOUSQUET, Y. 1999a. Supraspecific classification of the Nearctic Pterostichini (Coleoptera: Carabidae). Fabreries, Supplement 9, 292 pp.

BOUSQUET, Y. 1999b. Presence of *Stomis termitiformis* (Van Dyke) in Canada (Coleoptera: Carabidae). Fabreries, 24: 81-82.

BOUSQUET, Y. and GOULET, H. 1984. Notation of primary setae and pores on larvae of Carabidae (Coleoptera: Adephaga). Canadian Journal of Zoology, 62: 573-588.

BOUSQUET, Y. and GOULET, H. 1990. Description of a new species of *Metrius* (Coleoptera: Carabidae: Paussini) from Idaho with comments on the taxonomic status of the other taxa of the genus. The Pan-Pacific Entomologist, 66: 13-18.

BOUSQUET, Y. and LAPLANTE, S. 1997. Taxonomic review of the New World Pogonini (Coleoptera: Carabidae). Canadian Entomologist, 129: 699-731.

BOUSQUET, Y. and LAROCHELLE, A. 1993. Catalogue of the Geadephaga (Col. Trachypachidae, Rhysodidae, Carabidae, incl. Cicindelini) of America north of Mexico. Entomological Society of Canada, Memoir No. 167. 395 pp.

BOUSQUET, Y. and PILON, J.G. 1983. Redescription of *Pterostichus (Pseudomaseus) tenuis* (Casey), a valid species (Coleoptera: Carabidae). Coleopterists Bulletin, 37: 389-396.

BOUSQUET, Y. and SMETANA, A. 1986. A description of the first instar larva of *Promecognathus* Chaudoir (Coleoptera: Carabidae). Systematic Entomology, 11: 25-31.

BOUSQUET, Y. and SMETANA, A. 1991. The tribe Opisthiini: description of the larvae, notes on habitat, and brief discussion on its relationships. Journal of the New York Entomological Society, 99: 104-114.

BOUSQUET, Y. and SMETANA, A. 1996. A review of the tribe Opisthiini (Coleoptera: Carabidae). Nouvelle Revue d'Entomologie (N.S.), 12 (1995): 215-232.

BOUSQUET, Y., SMETANA, A., and MADDISON, D.R. 1984. *Trechus quadristriatus*, a Palaearctic species introduced into North America (Coleoptera: Carabidae). Canadian Entomologist, 116: 215-220.

BOUSQUET, Y. and TCHANG, J.P. 1992. Anisodactyline larvae (Coleoptera: Carabidae: Harpalini): descriptions of genus-group taxa of eastern Canada and phylogenetic remarks. Canadian Entomologist, 124: 751-783.

BRANDMAYR, P. and BRANDMAYR ZETTO, T. 1974. Sulle cure parentali e su altri aspetti della biologia di *Carterus (Sabienus) calydonius* Rossi, con alcune considerazioni sui fenomeni di cura della prole sino ad oggi ricontrati in carabidi (Coleoptera, Carabidae). Redia, 55: 1243-175.

BRANDMAYR, P. and ZETTO BRANDMAYR, T. 1979. The evolution of parental care phenomena in pterostichine ground beetles with special reference to the genera *Abax* and *Molops* (Coleoptera, Carabidae), pp. 35-49. *In*, On the evolution of behaviour in carabid beetles (Boer, P.J. den, Thiele, H.V., and Wiker, F., Eds.). Miscellaneous 18. Agricultural University, Wageningen, The Netherlands. 222 pp.

BRANDMAYR, P. and ZETTO BRANDMAYR, T. 1980. "Life forms" in imaginal Carabidae (Coleoptera): a morphofunctional and behavioural synthesis. Monitore Zoologia Italiana (N.S.), 14: 97-99.

BRANDMAYR, P. and ZETTO BRANDMAYR, T. 1986. Food and feeding behaviour of some *Licinus* species (Coleoptera Carabidae Licinini). Monitore Zoologia Italiana (N.S.), 20: 171-181.

BRANDMAYR, P., FERRERO, E., and ZETTO BRANDMAYR, T. 1980. Larval versus imaginal taxonomy and the systematic status of the ground beetle taxa *Harpalus* and *Ophonus* (Coleoptera: Carabidae: Harpalini). Entomologia Generalis, 6: 335-353.

BRITTON, E.B. 1940. The insect fauna of a nest of the silvery-cheeked hornbill, including the description of *Oecornis nidicola* g. et sp. n. (Col., Carabidae) from Tanganyika. Entomologists' Monthly Magazine, 76: 108-112, 3 figs.

BROWN, W. J. 1932. New species of Coleoptera III. Canadian Entomologist, 64: 3-12.

BROWN, W. J. 1949. On the American species of *Lyperopherus* Mots. (Coleoptera: Carabidae). Canadian Entomologist, 81: 231-232.

BRUNEAU DE MIRÉ, P. 1979. Trans-Atlantic dispersal: several examples of colonization of the Gulf of Biafra by Middle American stock of Carabidae, pp. 327-330. *In*, Carabid beetles: their evolution, natural history, and classification (Erwin, T. L., Ball, G. E., Whitehead, D. R., and Halpern, A.L., Eds.). Dr. W. Junk bv Publishers, The Hague, The Netherlands. x+644 pp.

BUDARIN, A.M. 1976. Review of ground-beetles of the subgenus *Lyperopherus* Motsch. of the genus *Pterostichus* Bon. (Coleoptera, Carabidae) [in Russian]. Trudy Zoologicheskkaya, 67: 32-38.

BURGESS, A. F. and COLLINS, C.W. 1917. The genus *Calosoma*, including studies of seasonal histories, habits and economic importance of American species north of Mexico and several introduced species. United States Department of Agriculture Bulletin, No. 417, 124 pp.

CAPOGRECO, J.V. 1989. Immature *Lebia viridis* Say (Coleoptera Carabidae): bionomics, descriptions, and comparisons to other *Lebia* species. Coleopterists Bulletin, 43: 183-194.

CASALE, A. 1988. Revisione degli Sphodrina (Coleoptera, Carabidae, Sphodrini). Monografie V. Museo Regionale di Scienze Naturali - Torino. 1024 pp.

CASALE, A. 1998. Phylogeny and biogeography of Calleidina (Coleoptera: Carabidae: Lebiini), pp. 381-428. *In*, Phylogeny and Classification of Caraboidea (Coleoptera: Carabidae: Lebiini) (Ball, G.E., Casale, A., and Vigna Taglianti, A., Eds.). Atti, Museo Regionale di Scienze Naturali, Torino. 543 pp.

CASALE, A. and LANEYRIE, R. 1982. Trechodinae et Trechinae du Monde. Memoires de Biospeologie, 9: 1-226.

CASEY, T. L. 1913. Studies in the Cicindelidae and Carabidae of America. Memoirs on the Coleoptera, vol. IV, pp. 1-192. The New Era Printing Company, Lancaster, Pa. pp.1-388.

CASEY, T. L. 1914. Memoirs on the Coleoptera, vol. V. The New Era Printing Company, Lancaster, Pa. pp. 1-378.

CASEY, T. L. 1918. Memoirs on the Coleoptera, vol. VIII. The New Era Printing Company, Lancaster, Pa. pp. 1-416.

CASEY, T. L. 1920. Some observations on the Carabidae, including a new subfamily, pp. 25-299. Memoirs on the Coleoptera, vol. IX. The New Era Printing Company, Lancaster, Pa. pp. 1-516.

CASEY, T. L. 1924. Additions to the known Coleoptera of North America. Memoirs on the Coleoptera, vol. XI. The New Era Printing Company, Lancaster, Pa. pp. 1-347.

CIEGLER, J.C. 2000. Ground beetles and wrinkled bark beetles of South Carolina (Coleoptera: Geadephaga: Carabidae and Rhysodidae). Biota of South Carolina. Vol. 1. Clemson University, Clemson, South Carolina. 149 pp.

CLAASSEN, P.W. 1919. Life history and biological notes on *Chlaenius impunctifrons* Say (Coleoptera: Carabidae). Annals of the Entomological Society of America, 12: 95-101.

COSTA, C., VANIN, S.A., and CASARI-CHEN, S.A. 1988. Larves de Coleoptera do Brasil. Mus. Zool., Universidade de São Paulo, São Paulo. 282 pp, 165 plates .

CSIKI, E. 1927-1933. Carabidae. *In*, Coleopterorum. Catalogus auspiciis et auxilio W. Junk editus a S. Schenkling, Berlin and s'Gravenhage.

CSIKI, E. 1927. Carabinae I, 1, 91: 1-313.

CSIKI, E. 1927. Carabinae II, 1, 92: 317-621.

CSIKI, E. 1928. Mormolycinae, 97: 3-4. Harpalinae, I, 97: 5-226.

CSIKI, E. 1928. Harpalinae II, 98: 227-345.

CSIKI, E. 1929. Harpalinae III, 11, 104: 347-527.

CSIKI, E. 1930. Harpalinae IV, II, 112: 529-737.

CSIKI, E. 1931. Harpalinae V, III, 115: 739-1022.

CSIKI, E. 1932. Harpalinae VI, III, 121: 1023-1278.

CSIKI, E. 1932. Harpalinae VII, III, 124: 1279-1598.

CSIKI, E. 1933. Harpalinae VIII, III, 126: 1599-1933.

CSIKI, E. 1933. Carabinae: Corrigenda et Addenda, I, 127: 623-648.

DARLINGTON, P. J., JR. 1931. On some Carabidae, including new species, from the mountains of North Carolina and Tennessee. Psyche, 38(4): 145-164.

DARLINGTON, P. J., JR. 1934. New West Indian Carabidae, with a list of the Cuban species. Psyche, 41: 66-131.

DARLINGTON, P. J., JR. 1936. The species of *Stenomorphus* (Coleoptera: Carabidae), with data on heterogony in *S. californicus* (Men.). The Pan-Pacific Entomologist, 12(1): 33-44.

DARLINGTON, P. J., JR. 1938. The American Patrobini (Coleoptera, Carabidae). Entomologica Americana 18 (New Series) 4: 135-183.

DARLINGTON, P. J., JR. 1950. Paussid beetles. Transactions of the American Entomological Society, 76: 47-142.

DARLINGTON, P. J., JR. 1957. Zoogeography: The Geographical Distribution of Animals. Wiley, New York; Chapman and Hall, London. xi + 675 pp.

DAVIDSON, R.L. 1980. A taxonomic revision of the genus *Chlaenius* Bonelli (Coleoptera: Carabidae) in Mexico and Central America, with species revisions of the subgenera *Callistometus* Grundmann, *Agostenus* Motschulsky, and *Chlaenius* (*sensu stricto*). MSc thesis, The University of Vermont. ix + 194 pp.

DAVIDSON, R.L. and BALL, G.E. 1998. The tribe Broscini in Mexico: *Rawlinsius papillatus*, new genus and new species (Insecta: Coleoptera: Carabidae), with notes on natural history and evolution. Annals of the Carnegie Museum, 67: 349-378.

DEN BOER, P.J., GRÜN, L., and SZYSZKO, J. 1986. Feeding behaviour and accessibility of food for carabid beetles. Report of the Fifth Meeting of European Carabidologists held at the Field Station Staraya Brda Polska of the Institute of Forest Protection and Ecology of Warsaw Agricultural University - SGGW-AR September 13-15 1982. Warsaw Agricultural University Press, Warsaw. 167 pp.

DESENDER, K. and BAERT, L. 1996. Easter Island revisited: Carabid beetles. Coleopterists Bulletin, 50: 343-356.

DEUVE, T. 1986. Hypothèse sur le caractère monophylétique des Harpalidae sensu novo (= Conchifera Jeannel + Pseudomorphinae Lacordaire). Nouvelle Revue d' Entomologie, n.s., 3: 320.

DEUVE, T. 1988. Etude phylogénétique des Coléoptères Adephaga: redéfinition de la famille des Harpalidae, sensu novo, et position systématique des Pseudomorphinae et Brachinidae. Bulletin de la Société Entomologique de France, 92: 161-182.

DEUVE, T. 1994. Une classification de genre *Carabus*. Bibliothèque Entomologique, Vol. 5. Sciences Nat., Vennette, France. 296 pp.

DEUVE, T. 1997. Catalogue des Carabini et Cychrini de Chine. Memoires de la Société Entomologique de France, No 1. Société Entomologique de France, Paris. 236 pp.

DEUVE, T. 1998. Note sur le genre *Broscodera* Lindroth répandu depuis l'Himalaya jusqu'à l'Amerique du Nord (Coleoptera, Broscidae). Nouvelle Revue d'Entomologie 14(1997): 227-228.

DOWNIE, N.M. and ARNETT, R.H., JR. 1996. The Beetles of Northeastern North America. Volume 1. Introduction, Suborders Archostemata, Adephaga, and Polyphaga thru [sic!] Superfamily Cantharoidea. The Sandhill Crane Press, Publisher. Gainesville, Florida. xiv + 880 pp.

DUFFEY, J.C. 1891. Transformations of a carabid (*Plochionus timidus*), etc. Transactions of the Academy of Sciences, St. Louis, 5: 533-542.

EISNER, T., ANESHANSLEY, D.J., EISNER, M., ATTYGALLE, A.B., ALSOP, D.W., and MEINWALD, J. 2000. Spray mechanism of the most primitive bombardier beetle (*Metrius contractus*). The Journal of Experimental Biology, 203: 1265-1275.

EISNER, T., JONES, T.H., ANESHANSLEY, D.J., TSCHINKEL, W.R., SILBERGLIED, R.E., and MEINWALD, J. 1977. Chemistry of defensive secretions of bombardier beetles (Brachinini, Metriini, Ozaenini, Paussini). Journal of Insect Physiology, 23: 1383-1386.

ELIAS, S.A. 1994. Quaternary insects and their environments. Smithsonian Institution Press, Washington/ London. xiii + 284 pp.

EL MOURSY, A. A. and BALL, G.E. 1959. A study of the diagnostic characters of two North American species of the genus *Diplocheila* Brullé, with notes on their ecology. Coleopterists Bulletin, 13: 47-57, 6 figs.

EMDEN, F. I. VAN 1942. A key to the genera of larval Carabidae. Transactions of the Royal Entomological Society of London, 92: 1-99, 100 figs.

EMDEN, F. I. VAN 1953. The larva of *Morion* and its systematic position (Coleoptera: Carabidae). Proceedings of the Hawaiian Entomological Society, 15(1): 51-54.

ERWIN, T.L. 1967. Bombardier beetles of North America: Part II. Biology and behavior of *Brachinus pallidus* Erwin in California. Coleopterists Bulletin, 21: 41-55.

ERWIN, T.L. 1970. A reclassification of bombardier beetles and a taxonomic revision of the North and Middle American species (Carabidae: Brachinida). Quaestiones Entomologicae, 6: 4-216.

ERWIN, T.L. 1971. Notes and corrections to a reclassification of bombardier beetles (Carabidae, Brachinida). Quaestiones Entomologicae, 7: 281

ERWIN, T.L. 1974. Studies of the subtribe Tachyina (Coleoptera: Carabidae: Bembidiini), Part II: A revision of the New World-Australian genus *Pericompsus* LeConte. Smithsonian Contributions to Zoology, 162: 1-96.

ERWIN, T.L. 1975. Studies of the subtribe Tachyina (Coleoptera: Carabidae: Bembidiini), Part III: Systematics, phylogeny, and zoogeography of the genus *Tachyta* Kirby. Smithsonian Contributions to Zoology, 208: 1-68.

ERWIN, T.L. 1979. 3.13. A review of the natural history and evolution of ecto-parasitoid relationships in carabid beetles. pp. 479-484. *In*, Carabid beetles: their evolution, natural history, and classification. (Erwin, T. L., Ball, G. E., Whitehead, D. R., and Halpern, A. L., Eds.). Dr. W. Junk bv Publishers, The Hague, The Netherlands. X + 644 pp.

ERWIN, T.L. 1981. A synopsis of the immature stages of Pseudomorphini (Coleoptera: Carabidae) with notes on tribal affinities and behavior in relation to life with ants. Coleopterists Bulletin, 35: 53-68.

ERWIN, T.L. 1982. Small terrestrial ground-beetles of Central America (Carabidae: Bembidiina and Anillina). Proceedings of the California Academy of Sciences, 42: 455-496.

ERWIN, T.L. 1985. The taxon pulse: a general pattern of lineage radiation and extinction among carabid beetles, pp. 437-493. *In,* Taxonomy, Phylogeny and Zoogeography of beetles and Ants (Ball, G.E., Ed.). Dr. W. Junk Publishers, Dordrecht/Boston/Lancaster. XIII + 514 pp.

ERWIN, T.L. 1991a. The ground-beetles of Central America (Carabidae), Part II: Notiophilini, Loricerini, and Carabini. Smithsonian Contributions to Zoology, 501: 1-30.

ERWIN, T.L. 1991b. Natural history of the carabid beetles at the BIOLAT Biological Station, Rio Manu, Pakitza, Peru. Revista Peruana de Entomologia, 33: 1-85.

ERWIN, T.L., BALL, G.E., WHITEHEAD, D.R., and HALPERN, A.L. (Editors). 1979. Carabid beetles: their evolution, natural history, and classification. Dr. W. Junk bv Publishers, The Hague, The Netherlands. x+644 pp.

ERWIN, T.L. and D.H. KAVANAUGH. 1980. On the identity of *Bembidion puritanum* Hayward (Coleoptera: Carabidae: Bembidiini). Coleopterists Bulletin, 34: 241-242.

ERWIN, T.L. and KAVANAUGH, D.H. 1981. Systematics and zoogeography of *Bembidion* Latreille: I. The *carlhi* and *erasum* groups of western North America (Coleoptera: Carabidae, Bembidiini). Entomologica Scandinavica (Supplementum), 15: 33-72.

ERWIN, T.L. and SIMS, L.L. 1984. Carabid beetles of the West Indies (Insecta: Coleoptera): a synopsis of the genera and checklist of tribes of Caraboidea, and of the West Indian species. Quaestiones Entomologicae, 20: 350-466.

EVANS, W.G. 1976. Circadian and circatidal locomotory rhythms in the intertidal beetle *Thalassotrechus barbarae* (Horn): Carabidae. Journal of Experimental and Marine Biology and Ecology, 22: 79-90.

EVANS, W.G. 1977. Geographic variation, distribution, and taxonomic status of the intertidal insect *Thalassotrechus barbarae* (Horn) (Coleoptera: Carabidae). Quaestiones Entomologicae, 13: 83-90.

FATTIG, P.W. 1949. The Carabidae or Ground Beetles of Georgia. Emory University Museum Bulletin, 7: 1-62.

FEDORENKO, D.N. 1996. Reclassification of world Dyschiriini, with a revision of the Palearctic fauna (Coleoptera, Carabidae). Pensoft, Sofia. 224 pp.

FRANK, J.H. 1994. Chapter 42. Inoculative biological control of male crickets, pp. 467-475. *In*, Integrated Pest Management for Turf and Ornamentals. (Leslie, A., Ed.). CRC Press, Inc.

FRANK, J.H. and MCCOY, E.D. 1990. Endemics and epidemics of shibboleths and other things causing chaos. The Florida Entomologist, 73(1): 109.

FREITAG, R. 1969. A revision of the species of the genus *Evarthrus* LeConte (Coleoptera: Carabidae). Quaestiones Entomologicae 5: 88-211.

FREITAG, R. 1999. Catalogue of the Tiger Beetles of Canada and the United States. NRC Research Press, Ottawa. vii + 195 pp.

FREUDE, H., HARDE, K.W., and LOHSE, G.A. 1976. [Chapter] 1. Familie: Carabidae (Laufkäfer), pp. 7-302. *In*, Die Käfer Mitteleuropas Band 2 Adephaga 1. (Freude, H., Harde, K. W., and Lohse, G. A., Eds.). Goecke and Evers, Krefeld. 302 pp.

GARIEPY, C.M., LAROCHELLE, A., and BOUSQUET, Y. 1977. Guide photographique des Carabidae du Québec. Cordulia, Supplement 2. Rigaud, Québec. 134 pp.

GIDASPOW, T. 1959. North American caterpillar hunters of the genera *Calosoma* and *Callisthenes* (Coleoptera, Carabidae). Bulletin of the American Museum of Natural History, 116(3): 225-344.

GIDASPOW, T. 1963. The genus *Calosoma* in Central America, the Antilles, and South America (Coleoptera: Carabidae). Bulletin of the American Museum of Natural History, 124: 279-313.

GIDASPOW, T. 1968. A revision of the ground beetles belonging to *Scaphinotus*, subgenus *Brennus* (Coleoptera: Carabidae). Bulletin of the American Museum of Natural History, 140: 137-192.

GIDASPOW, T. 1973. Revision of ground beetles of American genus *Cychrus* and four subgenera of genus *Scaphinotus* (Coleoptera: Carabidae). Bulletin of the American Museum of Natural History,152: 53-102.

GLENDENNING, R. 1952. Slug control in Canada. Canada Department of Agriculture, Processed Publications Series, No. 85, 5 pp.

GOULET, H. 1974. Classification of the North and Middle American species of the genus *Pelmatellus* Bates (Coleoptera: Carabidae: Harpalini). Quaestiones Entomologicae, 10: 80-102.

GOULET, H. 1983. The genera of Holarctic Elaphrini and species of *Elaphrus* Fabricius (Coleoptera: Carabidae): classification, phylogeny and zoogeography. Quaestiones Entomologicae, 19: 219-481.

GOULET, H. and SMETANA, A. 1983. A new species of *Blethisa* Bonelli from Alaska, with proposed phylogeny, biogeography and key to known species (Coleoptera: Carabidae). Canadian Entomologist, 115: 551-558.

HABU, A. 1967. Carabidae Truncatipennes group (Insecta: Coleoptera). Fauna Japonica, Biogeographical Society of Japan, Tokyo. xiv + 338 pp., 27 plates.

HABU, A. 1973. Carabidae: Harpalini (Insecta: Coleoptera). Fauna Japonica, Biogeographical Society of Japan, Tokyo. Keigaw Publishing Co., Tokyo. xiii + 430 pp., 24 plates.

HABU, A. 1978. Carabidae Platynini. Fauna Japonica, Biogeographical Society of Japan, Tokyo. Yugaku-Sho Limited, Tokyo, Japan. viii + 447 pp., 37 plates.

HABU, A. and SADANAKA, K. 1965. Illustrations for identification of larvae of the Carabidae found in cultivated fields and paddy fields. Bulletin of the National Institute of Agricultural Science (Japan) Ser. C, 3(19): 169-177.

HACKER, H.A. 1968. The species of the subgenus *Leptoferonia* Casey (Coleoptera: Carabidae: *Pterostichus*). Proceedings of the United States National Museum, 124 (3649): 1-61.

HAMILTON, C. C. 1925. Studies on the morphology, taxonomy and ecology of Holarctic tiger beetles (Coleoptera: Cicindelidae). Proceedings of the United States National Museum, 65: 1-87, 12 pls.

HATCH, M. H. 1931. Notes on another pest-the malodorous ground beetle (*Nomius pygmaeus* Dej.). Monthly News Letter, Puget Sound Academy of Science, 3(9): 2.

HATCH, M. H. 1953. The beetles of the Pacific Northwest. Part 1: Introduction and Adephaga, vii + 340 pp., 37 pls., 2 text-figs. University of Washington Press, Seattle.

HATCH, M.H. 1955. *Bradycellus harpalinus* Serv. in North America. Coleopterists Bulletin, 9: 10.

HIEKE, F. 1978. Revision der *Amara*-Untergattung *Percosia* Zimm. und Bemerkungen zu anderen *Amara*-Arten (Col.; Carabidae). Deutsche Entomologische Zeitschrift (N.F.), 25: 215-326.

HIEKE, F. 1994. Sieben neue asiatische Arten und weitere neue Synonyme aus der Gattung *Amara* Bon. (Coleoptera, Carabidae). Deutsche Entomologische Zeitschrift (N.F.), 41(2): 299-350.

HIEKE, F. 1995. Namenverzeichnis der Gattung *Amara* Bonelli 1810. Coleoptera Carabidae. Coleoptera Schwanfelder Coleopterologische Mitteilungen, 2: 1-163.

HIEKE, F. 1999. The *Amara* of the subgenus *Reductocelia* Lafer 1989 (Coleoptera Carabidae Zabrini), pp. 333-362. *In,* Advances in Carabidology (A. Zamotajlov and R. Sciaky, Eds.). MUISO Publishers, Krasnodar, Russia. 473 pp.

HLAVAC, T. 1971. Differentiation of the carabid antenna cleaner. Psyche, 78: 51-66.

HOLLIDAY, N.J. 1991. Species responses of carabid beetles (Coleoptera: Carabidae) during post-fire regeneration of boreal forest. Canadian Entomologist, 123: 1369-1389.

HORN, G.H. 1881a. Synoptic tables of Coleoptera. *Tetragonoderus* Dej., *Nemotarsus* Lec., *Dromius* Bon., *Axinopalpus* Lec., *Metabletus* Schmidt, *Apenes* Lec., *Pinacodera* Schaum. Bulletin of the Brooklyn Entomological Society, 4(7and8): 39-40.

HORN, G.H. 1881b. On the genera of Carabidae with special reference to the fauna of Boreal America. Transactions of the American Entomological Society, 9: 91-196, pls. 1-10.

HORN, G.H. 1882a. *Selenophorus* Dej. Synoptic Table. Bulletin of the Brooklyn Entomological Society, 5: 8.

HORN, G.H. 1882b. *Olisthopus* Dej. Synoptic Table. Bulletin of the Brooklyn Entomological Society, 5: 63.

HORN, G.H. 1882c. Synopsis of the species of the tribe Lebiini. Transactions of the American Entomological Society, 10: 126-163.

HORN, G.H. 1892. A study of *Amara*, s.g. *Celia*. Transactions of the American Entomological Society, 10: 19: 19-40.

HORN, G.H. 1894. The Coleoptera of Baja California. Proceedings of the California Academy of Sciences, series 2, 4: 302-449, illus.

HORN, W. 1908-1915. Genera Insectorum, Coleoptera, Adephaga, fam. Carabidae, subfam. Cicindelinae. V. Verteneuil, Bruxelles (1908-1915).

HORN, W. 1908. Fascicle 82A, pp. 1-104 pls. 1-5.

HORN, W. 1910. Fascicle 82B, pp. 105-208, pls. 6-15.

HORN, W. 1915. Fascicle 82C, pp. 209-486, pls. 16-23

HURKA, K. 1996. Carabidae of the Czech and Slovak Republics. Kabourek, Zlín. 565 pp.

JEANNEL, R. 1926-1931. Monographie des Trechinae. Morphologie comparée et distribution géographique d'un groupe de Coléoptères

JEANNEL, R. 1926. Première livraison. L'Abeille, 32: 221-550.

JEANNEL, R. 1927. Deuxième livraison. L'Abeille, 33: 1-592.

JEANNEL, R. 1928. Troisième livraison. L'Abeille, 35: 1-808.

JEANNEL, R. 1930. Quatrième livraison. L'Abeille, 34: 59-122.

JEANNEL, R. 1931. Revision des Trechinae de l'Amérique du Nord. Biospeologica, vol. 61. Arch. Zool. Exp. et Gener., 71: 403-499.

JEANNEL, R. 1937. Les Bembidiides endogés. Revue Francaise d'Entomologie, 4: 1-23.

JEANNEL, R. 1940. Les calosomes. Memoires du Muséum National d'Histoire Naturelle, n.s., 13(1): 1-233, 202 text-figs., 8 pls.

JEANNEL, R. 1941. Coléoptères carabiques, première partie. Faune de France, 39: 1 -571, figs. 1-213. Paris.

JEANNEL, R. 1942. Coléoptères carabiques, deuxième partie. Faune de France, 40: 573-1173, figs. 214-368. Paris.

JEANNEL, R. 1948. Coléoptères carabiques de la Région Malgache (deuxième partie). Faune de l'Empire francais, vol. X, 765 pp., 364 figs.

JEANNEL, R. 1949a. Coléoptères carabiques de la Région Malgache (troisième partie). Faune de l'Empire francais, vol. XI, 767-1146, figs. 365-548.

JEANNEL, R. 1949b. Les coléoptères cavernicole de la région des Appalaches: étude systematique. Notes Biospeologiques. 4. Publications du Muséum National d'Histoire Naturelle de Paris, No. 12: 37-104.

JEANNEL, R. 1963a. Monographie des "Anillini", bembidiides endogés [Coleoptera Trechidae]. Mémoires du Muséum National d'Histoire Naturelle (Série A) Zoologie, 28: 33-204.

JEANNEL, R. 1963b. Supplément à la monographie des Anillini. Sur quelques espèces nouvelles de l'Amérique du Nord. Revue Française d'Entomologie, 30: 145-152.

KATAEV, B.M. 1989. New data on carabid beetles of the genera *Pangus* and *Harpalus* (Coleoptera, Carabidae) of Mongolia with revision of some palaearctic species groups. *Nasekomii Mongolii*, Leningrad. Nauka, 10: 188-277.

KATAEV, B.M. 1996. On similar anomalies in the structure of the male genitalia of some species of ground beetles of the genus *Harpalus* Latr. and *Ophonus* Dejean (Coleoptera, Carabidae) and their significance for reconstructing the phylogeny of the Harpali genus groups. Entomological Review, 75: 151-166 [See pp. 163 and 166].

KATAEV, B.M. 1997. Ground-beetles of the genus *Harpalus* Latreille 1802. Steenstrupia, 23: 123-160.

KAVANAUGH, D.H. 1992. Carabid beetles (Insecta: Coleoptera: Carabidae) of the Queen Charlotte Islands, British Columbia. Memoir 16, California Academy of Sciences, San Francisco. viii + 113 pp.

KAVANAUGH, D.H. 1995. The genus *Nippon准ebria* in the Nearctic region, with description of a new subgenus. Entomological News, 106: 153-160.

KAVANAUGH, D.H. 1996. Phylogenetic relationships of genus *Pelophila* Dejean to other basal grade Carabidae (Coleoptera). Annales Zoologicae Fennici, 33: 17-21.

KAVANAUGH, D.H. 1998. Investigations on phylogenetic relationships among some basal grade Carabidae (Coleoptera): a report upon work in progress, pp. 329-342. *In*, Phylogeny and classification of Carabidae (Coleoptera: Adephaga). (Ball, G.E., Casale, A., and Vigna-Taglianti, A., Eds.). Atti Museo Regionale di Scienze, Torino. 543 pp.

KAVANAUGH, D.H. and ERWIN, T.L. 1985. *Trechus obtusus* Erichson (Coleoptera: Carabidae), a European ground beetle on the Pacific coast of North America: its distribution, introduction and spread. The Pan-Pacific Entomologist, 61: 170-179.

KING, J.L. 1919. Notes on the biology of the carabid genera *Brachinus*, *Galerita*, and *Chlaenius*. Annals of the Entomological Society of America, 12: 382-390.

KIRK, V.M. 1972. Seed-caching by larvae of two ground beetles, *Harpalus pensylvanicus* and *H. erraticus*. Annals of the Entomological Society of America, 5: 1426-1428.

KIRK, V.M. 1973. Biology of a ground beetle, *Harpalus pensylvanicus*. Annals of the Entomological Society of America, 66: 513-518.

KNISLEY, C.B. and SCHULTZ, T.D. 1997. The biology of tiger beetles and a guide to the species of the South Atlantic States. Virginia Museum of Natural History Special Publication No. 5. Martinsville, Virginia. viii + 210 pp.

KRYZHANOVSKIJ, O.L. 1976. Revised classification of the family Carabidae. Entomological Review, 1: 80-91.

KRYZHANOVSKIJ, O.L. 1983. Fauna of the USSR. Coleoptera. Vol. 1, No. 2. New Series, No. 128. Beetles of the suborder Adephaga: families Rhysodidae, Trachypachidae, family Carabidae [introductory part]. Science Press, Leningrad Division, pp. 1-230 [341 in original]. [English translation by Multilingual Service Division, Translation Bureau, Agriculture Canada.]

KRYZHANOVSKIJ, O.L., BELUSOV, I.A., KABAK, I.I., KATAEV, B.M., MAKAROV, K.V., and SHILENKOV, V.G. 1995. A checklist of the ground-beetles of Russia and adjacent lands (Insecta, Coleoptera, Carabidae). Pensoft Publishers, Sofia and Moscow. 271 pp.

LABONTE, J.R. 1983. Feeding behavior in the carabid beetle *Promecognathus*. Bulletin of the Oregon Entomological Society, No. 86, pp. 681-682. [Abstract]

LAFERTÉ-SÉNECTÈRE, F. T. 1851. Revision de la tribu des Patellimanes de Dejean. Annales de la Société Entomologique de France, 2[me] serie, 9: 209-294.

LANDRY, J.F. and BOUSQUET, Y. 1984. The genus *Omophron* Latreille (Coleoptera: Carabidae): redescription of the larval stages and phylogenetic considerations. Canadian Entomologist, 116: 1557-1569.

LAPOUGE, G. V. DE 1929-1953. Genera Insectorum, dirigés par P. Wytsman. Coleoptera. Bruxelles. Carabidae: Carabinae.

LAPOUGE, G. V. DE 1929. I, fascicule 192A, pp. 1-153, 1 pl.

LAPOUGE, G. V. DE 1930. II, fascicule 192A, pp. 155-291.

LAPOUGE, G. V. DE 1931. III, fascicule 192B, pp. 293-580.

LAPOUGE, G. V. DE 1932. IV, fascicule 192C, pp. 581-747.

LAPOUGE, G. V. DE 1953. Fascicule 192E, pls. 2-10.

LAROCHELLE, A. 1975. Les Carabidae du Québec et du Labrador. Département de Biologie du Collège Bourget, Bulletin 1. 255 pp.

LAROCHELLE, A. 1976. Manuel d'identification des Carabidae du Québec. Cordulia, Supplement 1. Rigaud, Québec. 127 pp. + additions et corrections.

LAROCHELLE, A. 1990. The food of carabid beetles (Coleoptera: Carabidae, including Cicindelinae). Fabreries, Supplement 5. Québec, Canada. 132 pp.

LAROCHELLE, A. and LARIVIÈRE, M.C. 1989. First records of *Broscus cephalotes* (Linnaeus) (Coleoptera: Carabidae: Broscini) for North America. Coleopterists Bulletin, 43(1): 69-73.

LARSON, D. J. 1969. A revision of the genera *Philophuga* Motschoulsky and *Technophilus* Chaudoir, with notes on the North American Callidina (Coleoptera: Carabidae). Quaestiones Entomologicae, 5(1): 15-84.

LARSON, D. J. 1998. *Dromius fenestratus* (Fabricius) possibly established in North America. Coleopterists Bulletin, 52: 126.

LAWRENCE, J.F. and NEWTON, A.F., JR. 1995. Families and subfamilies of Coleoptera (with selected genera, notes, references and data on family-group names), pp. 779-1006. *In*, Biology, phylogeny, and classification of Coleoptera: Papers celebrating the 80th birthday of Roy A. Crowson. (Pakaluk, J., and Slipinski, S.A., Eds.). Museum i Instytut Zoologii PAN, Warsaw.

LECONTE, J. L. 1848. A descriptive catalogue of the geodephagous Coleoptera inhabiting the United States east of the Rocky Mountains. Annals of the Lycaeum of Natural History of New York, 4: 173-474.

LECONTE, J. L. 1854. Synopsis of the species of *Platynus* and allied genera inhabiting the United States. Proceedings of the Academy of Natural Sciences of Philadelphia, 7: 35-59.

LECONTE, J. L. 1857. Synopsis of the species of *Clivina* and allied genera inhabiting the United States. Proceedings of the Academy of Natural Sciences of Philadelphia, 9: 75-83.

LECONTE, J. L. 1862. Notes on the species of *Brachinus* inhabiting the United States. Proceedings of the Academy of Natural Sciences of Philadelphia, 14: 523-525.

LECONTE, J. L. 1863. List of the Coleoptera of North America. Prepared for the Smithsonian Institution. Smithsonian Miscellaneous Collections No. 140, pp. 1-49. (Part I, continued).

LECONTE, J. L. 1873. The Pterostichi of the United States. Proceedings of the Academy of Natural Sciences of Philadelphia 1873, pp. 302-320.

LECONTE, J. L. 1879. Synoptic tables, *Panagaeus* Latr., *Micrixys* Lec., *Morio* Latr., *Helluomorpha* Lap., *Galerita* Fab., *Zuphium* Latr., *Diaphorus* Dej. Bulletin of the Brooklyn Entomological Society, 2: 59-62.

LECONTE, J. L. and HORN, G. H. 1883. Classification of the Coleoptera of North America. Smithsonian Miscellaneous Collections, 507, xxxviii + 567 pp.

LEFFLER, S.R. 1985. *Omus submetallicus* G. Horn: historical perspective, systematic position, type locality, and habitat. Cicindela, 17: 37-50.

LENG, C. W. 1902. Revision of Cicindelidae of boreal America. Transactions of the American Entomological Society, 28: 93-186.

LEONARD, J.G. and BELL, R.T. 1999. Northeastern Tiger Beetles: A Field Guide to Tiger Beetles of New England and Eastern Canada. CRC Press, Boca Raton. xii + 176 pp.

LIEBHERR, J.K. 1983. Larval description of *Calybe* (*Ega*) *sallei* Chevrolat with a preliminary assessment of lachnophorine affinities. Coleopterists Bulletin, 37: 254-260.

LIEBHERR, J.K. 1985. Revision of the Platynine carabid genus *Tanystoma* Motschulsky (Coleoptera). Journal of the New York Entomological Society, 93: 1182-1211.

LIEBHERR, J.K. 1986. Cladistic analysis of North American Platynini and revision of the *Agonum extensicolle* species group (Coleoptera: Carabidae). University of California Publications in Entomology, 106: 1-198.

LIEBHERR, J.K. 1988a. [Ed.]. Zoogeography of Caribbean insects. Cornell University Press, Ithaca. ix + 285 pp.

LIEBHERR, J.K. 1988b. Redefinition of the supertribe Odacanthitae and revision of the West Indian Lachnophorini (Coleoptera: Carabidae). Quaestiones Entomologicae, 24: 1-42.

LIEBHERR, J.K. 1989. *Tanystoma diabolica* new species (Coleoptera: Carabidae: Platynini) from Baja California and its biogeographic significance. Journal of the New York Entomological Society, 97: 173-186.

LIEBHERR, J.K. 1991a. Phylogeny and revision of the *Anchomenus* clade: the genera *Tetraleucus*, *Anchomenus*, *Sericoda*, and *Elliptoleus* (Coleoptera: Carabidae: Platynini). Bulletin of the American Museum of Natural History, 202: 1-163.

LIEBHERR, J.K. 1991b. A new tribal placement for the Australasian genera *Homethes* and *Aeolodermus* (Coleoptera: Carabidae: Odacanthini). The Pan-Pacific Entomologist, 66: 312-321.

LIEBHERR, J.K. 1992. Phylogeny and revision of the *Platynus degallieri* species group (Coleoptera: Carabidae: Platynini). Bulletin of the American Museum of Natural History, 214: 1-115.

LIEBHERR, J.K. 1994. Identification of New World *Agonum*, review of the Mexican fauna, and description of *Incagonum*, new genus, from South America (Coleoptera: Carabidae: Platynini). Journal of the New York Entomological Society, 102: 1-55.

LIEBHERR, J.K. 1998. On *Rembus* (*Colpodes*) *brunneus* MacLeay (Coleoptera: Carabidae: Platynini): redesignation and relationships. The Journal of Natural History, 32: 987-1000.

LIEBHERR, J.K. and KAVANAUGH, D.H. 1985. Ovoviviparity in carabid beetles of the genus *Pseudomorpha* (Insecta: Coleoptera). The Journal of Natural History, 19: 1079-1086.

LIEBHERR, J.K. and WILL, K.W. 1996. New North American *Platynus* Bonelli (Coleoptera: Carabidae), a key to species north of Mexico, and notes on species from the southwestern United States. Coleopterists Bulletin, 50: 301-320.

LIEBHERR, J.K. and WILL, K.W. 1998. Inferring phylogenetic relationships within the Carabidae (Insecta, Coleoptera) from characters of the female reproductive tract, pp. 107-170. *In,* Phylogeny and classification of Caraboidea (Coleoptera: Adephaga). (Ball, G.E., Casale, A., and Vigna Taglianti, A., Eds.). Atti Museo Regionale di Scienze, Torino. 543 pp.

LIEBKE, M. 1930. Revision der americanischen Arten der Unterfamilie Colliurinae (Col. Carab.). Mitteilungen aus dem Zoologischen Museum in Berlin, 15: 647-726.

LIEBKE, M. 1934. Die Arten der Gattung *Pseudaptinus* Castelnau (Col. Carabidae). Revista de Entomologi*a*, 4: 372-388.

LINDROTH, C. H. 1954. A revision of *Diachila* Motsch. and *Blethisa* Bon., with remarks on *Elaphrus* larvae. Kungl. Fysiografiska Sallskepets Handlingar N.F., 65: 1-28.

LINDROTH, C. H. 1956. A revision of the genus *Synuchus* Gyllenhal (Coleoptera: Carabidae) in the widest sense, with notes on *Pristosia* Motschulsky (*Eucalathus* Bates) and *Calathus* Bonelli. Transactions of the Royal Entomological Society of London, 108(11): 485-576, 39 text-figs.

LINDROTH, C. H. 1960. The larvae of *Trachypachus* Mtsch., *Gehringia* Darl., and *Opisthius* Kby. (Coleoptera: Carabidae). Opuscula Entomologica, 25: 30-42.

LINDROTH, C. H. 1961-1969. The ground-beetles (Carabidae, excl. Cicindelinae) of Canada and Alaska. Opuscula Entomologica.

LINDROTH, C. H. 1961. Part 2. Supplementum No. 20. pp. 1-200.

LINDROTH, C. H. 1963. Part 3. Supplementum No. 24. pp. 201-408.

LINDROTH, C. H. 1966. Part 4. Supplementum No. 29. pp. 409-648.

LINDROTH, C. H. 1968. Part 5. Supplementum No. 33. pp. 649-944.

LINDROTH, C. H. 1969a. Part 6. Supplementum No. 34. pp. 945-1192.

LINDROTH, C. H. 1969b. Part 1. Supplementum No. 35. pp. I-XLVIII.

LINDROTH, C. H. 1971. Disappearance as a protective factor. A supposed case of Bates'ian mimicry among beetles (Carabidae and Chrysomelidae). Entomologica Scandinavica, 2: 41-48.

LINDROTH, C. H. 1974. Coleoptera Family Carabidae. Handbooks for the identification of British insects. Royal Entomological Society of London, London. 148 pp.

LINDROTH, C. H. 1985 and 1986. The Carabidae (Coleoptera) of Fennoscandia and Denmark. Fauna Entomologica Scandinavica Vol. 15. (E.J. Brill, Ed.). Scandinavian Science Press Ltd, Leiden, Copenhagen.

LORENZ, W. 1998. Systematic list of extant ground beetles of the world (Insecta Coleoptera 'Geodephaga': Trachypachidae and Carabidae and Paussinae, Cicindelinae, Rhysodinae). Published by author, Tutzing, Germany. 502 pp.

LUFF, M.L. 1993. The Carabidae (Coleoptera) larvae of Fennoscandia and Denmark. Fauna Entomologica Scandinavica, 27: 1-186.

MADDISON, D.R. 1985. The discovery of *Gehringia olympica* Darlington (Coleoptera; Carabidae), pp. 35-37. *In,* Taxonomy, phylogeny, and zoogeography of beetles and ants (Ball, G.E., Ed.). Dr. W. Junk Publisher, Dordrecht. xiii + 514 pp.

MADDISON, D.R. 1993. Systematics of the Holarctic beetle subgenus *Bracteon* and related *Bembidion* (Coleoptera: Carabidae). Bulletin of the Museum of Comparative Zoology, 153: 143-299.

MADGE, R. B. 1967. A revision of the genus *Lebia* Latreille in America north of Mexico. Quaestiones Entomologicae, 3: 139-242.

MADGE, R. B. 1989. A catalogue of the family-group names in the Geodephaga, 1758-1985 (Coleoptera: Carabidae *s. lat.*). Entomologica Scandinavica 19: 459-474.

MANI, M.S. 1968. Ecology and biogeography of high altitude insects. Dr. W. Junk n. v. Publishers, The Hague. XIV + 527 pp.

MATEU, J. 1981. Revision de los *Zuphium* Latreille del continente Americano (Coleoptera: Carabidae) 1ª Nota. Folia Entomológica Mexicana, 47: 111-128.

MATTHEWS, J.V. 1979. [Chapter] 2.44. Late Tertiary carabid fossils from Alaska and the Canadian Archipelago, pp. 425-445. *In,* Carabid beetles: their evolution, natural history, and classification. (Erwin, T. L., Ball, G.E., Whitehead, D. R., and Halpern, A.L., Eds.). Proceedings of the First International Symposium of Carabidology, Smithsonian Institution, Washington, D.C., August 21, 23, and 25 1976. Dr. W. Junk b.v., Publishers, The Hague, The Netherlands. x + 635 pp.

MATTHEWS, J.V. and TELKA, A. 1997. Insect fossils from the Yukon, pp. 911-962. *In,* Insects of the Yukon (Danks, H. V. and Downes, J. A., Eds.). Biological Survey of Canada (Terrestrial Arthropods). Ottawa. x + 1034 pp.

MCCLEVE, S. 1975. *Psydrus piceus* LeConte from Arizona. Coleopterists Bulletin, 29(3): 176.

MCKAY, I.J. 1991. Cretaceous Promecognathinae (Coleoptera: Carabidae): a new genus, phylogenetic reconstruction and zoogeography. Biological Journal of the Linnean Society, 44: 1-12.

MITCHELL, R.W. 1971. Food and feeding habits of the troglobitic carabid beetle *Rhadine subterranea*. International Journal of Speleology, 3: 249-270.

MOORE, I. 1956. Notes on some intertidal Coleoptera with descriptions of the earlystages (Carabidae, Staphylinidae,

Malachiidae). Transactions of the San Diego Society of Natural History, 12(11): 207-230, pls 14-17.

MOORE, B.P. 1979. 2.24. Chemical defense and its bearing on phylogeny, pp. 193-203. *In,*Carabid beetles: their evolution, natural history, and classification. (Erwin, T. L., Ball, G.E., Whitehead, D. R., and Halpern, A.L., Eds.). Proceedings of the First International Symposium of Carabidology, Smithsonian Institution, Washington, D.C., August 21, 23, and 25 1976. Dr. W. Junk b.v., Publishers, The Hague, The Netherlands. x+635 pp.

MOORE, B.P., WEIR, T.A., and PYKE, J.E. 1987. Rhysodidae and Carabidae, pp. 20-320. in Vol. 4, Coleoptera: Archostemata, Myxophaga and Adephaga, viii + 444 pp. *In,* Zoological Catalogue of Australia (D.W. Walton, Executive Ed.). Australian Government Publishing Service, Canberra.

MOTSCHULSKY, T. V. VON. 1864. Énumération des nouvelles espèces de coléoptères rapportés de ses voyages. Bulletin de la Société Imperiale de Naturalistes de Moscou, 37: 171-240, 297-355.

NÈGRE, J. 1963. Revision du genre *Polpochila* Solier. Revue Francaise d'Entomologie, 30(4): 216-239, 35 fig.

NELSON, R.E. 1991. First records of *Perigona pallipennis* (LeC) and *Perigona nigriceps* (Dej.) from Maine: easternmost records for the genus in North America. Coleopterists Bulletin, 45: 284-285.

NICHOLS, S.W. 1981. The internal female genitalia of the genus *Omophron* (Coleoptera: Carabidae) with special reference to the taxonomic and phylogenetic usefulness of the basal conduit. MSc thesis, The Ohio State University, Columbus. v+20 pp, 11 plates.

NICHOLS, S.W. 1985a. *Omophron* and the origin of Hydradephaga (Insecta: Coleoptera: Adephaga). Proceedings of the Academy of Natural Sciences of Philadelphia, 137: 182-201.

NICHOLS, S.W. 1985b. *Clivina (Semiclivina) vespertina* Putzeys, a probable introduction to the United States from South America (Coleoptera: Carabidae: Scaritini). Coleopterists Bulletin, 39: 380.

NICHOLS, S.W. 1986. Two new flightless species of *Scarites s. str.* inhabiting Florida and the West Indies. Proceedings of the Entomological Society of Washington, 88: 257-264.

NICHOLS, S.W. 1988. Systematics and biogeography of West Indian Scaritinae (Coleoptera: Carabidae). Doctoral Dissertation, Cornell University, Ithaca, New York. xiii + 393 pp.

NOONAN, G.R. 1973. The anisodactylines (Insecta: Coleoptera: Carabidae: Harpalini): classification, evolution and zoogeography. Quaestiones Entomologicae, 9(4): 266-480.

NOONAN, G.R. 1975. Bionomics, evolution, and zoogeography of members of the genus *Dicheirus* (Coleoptera: Carabidae). The Pan-Pacific Entomologist, 51(1): 1-15.

NOONAN, G.R. 1976. Synopsis of the supra-specific taxa of the tribe Harpalini (Coleoptera: Carabidae). Quaestiones Entomologicae, 12(1): 3-87.

NOONAN, G.R. 1985. Classification and names of the Selenophori group (Coleoptera: Carabidae: Harpalini) and of nine genera and subgenera placed incertae sedis within Harpalina. Milwaukee Public Museum, Contributions in Biology and Geology, No. 64, 92 pp.

NOONAN, G.R. 1991. Classification, cladistics, and natural history of native North American *Harpalus* Latreille (Insecta: Coleoptera: Carabidae: Harpalini) excluding subgenera *Glanodes* and *Pseudoophonus*. The Thomas Say Foundation Monographs, 13, Entomology of America, Lanham, Maryland. viii + 319 pp.

NOONAN, G.R. 1996. Classification, cladistics, and natural history of species of the subgenus *Anisodactylus* Dejean (Insecta: Coleoptera: Carabidae: Harpalini: *Anisodactylus*). Milwaukee Public Museum, Contributions in Biology and Geology, No. 89, iii + 210 pp.

NOONAN, G.R., BALL, G.E., and STORK, N.E. (Eds.). 1992. The biogeography of ground beetles of mountains and islands. Intercept, Andover. vii + 256 pp.

NOTMAN, H. 1919. Records and new species of Carabidae. Journal of the New York Entomological Society, 27(2-3): 225-237.

NOTMAN, H. 1925. A review of the beetle family Pseudomorphidae and a suggestion for a rearrangement of the Adephaga, with descriptions of a new genus and new species. Proceedings of the United States National Museum, No. 2586, 67(14): 1-34.

OGUETA, E. 1966. Las especies americanas de *Panagaeus* Latreille 1804 (Coleoptera, Carabidae). Physis, 26(71): 1-13.

PEARSON, D.L. 1988. Biology of tiger beetles. Annual Review of Entomology, 33: 123-147.

PEARSON, D.L. 1999. Natural history of selected genera of tiger beetles (Coleoptera: Cicindelidae) of the World. Cicindela, 31: 53-71.

PEARSON, D.L., BARRACLOUGH, T.G., and VOGLER, A.P. 1997. Distributional maps for North American species of tiger beetles (Coleoptera: Cicindelidae). Cicindela, 29: 33-84.

PERRAULT, G.G. 1982. Etudes sur la tribu des Bembidiini *s. str.* (Coleoptera, Carabidae). II. Révision du sous-genre *Cyclolopha* (Casey). Entomologica Basiliensia, 7: 89-126.

POLLOCK, D. A. 1991. Notes on the distribution of, and range extension for, *Acupalpus meridianus* (Linné) (Coleoptera: Carabidae). Canadian Entomologist, 123: 705-706.

PRÜSER, F. and MOSSAKOWSKI, D. 1998. Conflicts in phylogenetic relationships and dispersal history of the supertribe Carabitae (Coleoptera: Carabidae), pp. 297-328. *In,* Phylogeny and classification of Caraboidea (Coleoptera: Adephaga). (Ball, G.E., Casale, A., and Vigna

Taglianti, A., Eds.). Atti Museo Regionale di Scienze, Torino. 543 pp.

REICHARDT, H. 1967. A monographic revision of the American Galeritini (Coleoptera; Carabidae). Arquives de Zoologia, 25(1-2): 1-176.

REICHARDT, H. 1968. Revisionary notes on the American Pentagonicini (Coleoptera, Carabidae). Papéies Avulsos de Zoologia, 15: 143-160.

REICHARDT, H. 1973. A review of *Hyboptera* Chaudoir (Coleoptera Carabidae). Revista Brasiliera de Entomologia, 17: 47-55.

REICHARDT, H. 1974. Monograph of the Neotropical Helluonini, with notes and discussions on Old World forms (Coleoptera Carabidae). Studia Entomologica, 17: 211-302.

REICHARDT, H. 1977. A synopsis of the genera of Neotropical Carabidae (Insecta: Coleoptera). Quaestiones Entomologicae, 13(4): 346-493.

REITTER, E. 1908. Fauna Germanica. Die Käfer des Deutschen Reiches, Vol. I, 248 pp., 66 text-figs., 40 colored pls. Stuttgart.

RIVALIER, E. 1954. Démembrement du genre *Cicindela* Linné. II. Faune américaine. Revue Francaise d'Entomologie, 21: 249-268.

ROUGHLEY, R. E. 1981. Trachypachidae and Hydradephaga (Coleoptera), a monophyletic unit? Pan-Pacific Entomologist, 57: 273-285.

SAY, T. L. 1830. Descriptions of new species of North American insects and observations on some already described. The Disseminator, 1(1): [3]; 1(3): [3]; 1(4): [3]; 1(5): [3]; 1(6): [3]; 1(7): [3].

SCHAEFFER, C. 1910. Additions to the Carabidae of North America with notes on species already known. Museum of the Brooklyn Institute of Arts and Sciences, Science Bulletin, 1 (17): 391-405.

SCHAUPP, F. G. 1879. Synoptic tables of Coleoptera. *Ega* Lap. Bulletin of the Brooklyn Entomological Society, 2: 85.

SCHAUPP, F. G. 1882. Synoptic tables of Coleoptera. *Lachnophorus* Sturm. Bulletin of the Brooklyn Entomological Society, 5: 63.

SCHWARZ, E. A. 1878. The Coleoptera of Florida. Proceedings of the American Philosophical Society, XVII: 353-472.

SCHWARZ, E. A. 1884. Carabidae confined to single plants. Bulletin of the Brooklyn Entomological Society, 6: 135-136.

SCHWARZ, E. A. 1895. Notes on *Nomaretus* with descriptions of two new species. Proceedings of the Entomological Society of Washington, 3(4): 269-273.

SCIAKY, R. and FACCHINI, S. 1999. A review of the Chinese *Loricera*, with description of a new subgenus and three new species (Coleoptera Carabidae Loricerinae), pp. 95-108. *In,* Advances in Carabidology (Zamotajlov, A. and Sciaky, R., Eds.). MUISO Publishers, Krasnodar, Russia. 473 pp.

SERRANO, J. and GALIÁN, J. 1998. A review of karyotypic evolution and phylogeny of carabid beetles (Coleoptera), pp. 191-228. *In*, Phylogeny and classification of Caraboidea (Coleoptera: Adephaga). (Ball, G.E., Casale, A., and Vigna Taglianti, A., Eds.). Atti Museo Regionale di Scienze, Torino. 543 pp.

SHILENKOV, V. 1999. Ground-beetles of the genus *Leistus* Froehlich of the Caucasus (Coleoptera Carabidae Nebriini). *In,* Advances in Carabidology (Zamotajlov, A. and Sciaky, R., Eds.). MUISO Publishers, Krasnodar, Russia. 473 pp.

SHPELEY, D. 1986. Genera of the subtribe Metallicina and classification, reconstructed phylogeny and geographical history of the species of *Euproctinus* Leng and Mutchler (Coleoptera: Carabidae: Lebiini). Quaestiones Entomologicae, 22: 261-349.

SHPELEY, D. and BALL, G.E. 1993. Classification, reconstructed phylogeny and geographical history of the New World species of *Coptodera* Dejean (Coleoptera: Carabidae: Lebiini). Proceedings of the Entomological Society of Ontario, 124: 3-182.

SHPELEY, D. and BALL, G.E. (In press). A taxonomic review of the subtribe Pericalina (Coleoptera: Carabidae: Lebiini) in the western Hemisphere, with descriptions of new species and notes about classification and zoogeography. Insecta Mundi.

SILFERBERG, H. 1983. The coleopteran genera of Dejean 1821. 1. Carabidae. Annales Entomologici Fennici, 49: 115-116.

SUMLIN, W.D., III. 1991. Studies on the Mexican Cicindelidae II: two new species from Coahuila and Nuevo Leon (Coleoptera). Cicindelidae: Bulletin of Worldwide Research, 1: 1-6.

THOMPSON, R.G. 1979. Larvae of North American Carabidae with a key to the tribes, pp. 209-291. *In*, Carabid beetles: their evolution, natural history, and classification. (Erwin, T. L., Ball, G. E., Whitehead, D. R., and Halpern, A. L., Eds.). Dr. W. Junk b. v. Publishers, The Hague, The Netherlands. X + 644 pp.

TOPOFF, H.R. 1969. A unique association between carabid beetles of the genus *Helluomorphoides* and colonies of the army ant *Neivamyrmex nigrescens*. Psyche, 76(4): 375-381.

VALENTINE, J. M. 1932. *Horologion*, a new genus of cave beetles (Fam. Carabidae). Annals of the Entomological Society of America, 25(1): 1-11, 12 figs.

VALENTINE, J. M. 1935. Speciation in *Steniridia* a group of cychrine beetles. Journal of the Elisha Mitchell Scientific Society, 51: 341-375, pls. 65-73 (1 map).

VALENTINE, J. M. 1936. Raciation in *Steniridia andrewsii* Harris, a supplement to speciation in *Steniridia*. Journal of the Elisha Mitchell Science Society, 52: 223-234, pl. 17.

VALENTINE, J. M. 1952. New genera of anophthalmid beetles from Cumberland caves (Carabidae: Trechini). Alabama Geological Survey Museum Papers, No. 34, 41 pp.

VAN DYKE, E. C. 1938. A review of the subgenus *Scaphinotus*, subgenus *Scaphinotus* Dejean (Coleoptera, Carabidae). Entomologica Americana, 18 (New Series): 93-133.

VAN DYKE, E. C. 1943. New species and subspecies of North American Carabidae. The Pan-Pacific Entomologist, 19: 17-30.

VAN DYKE, E. C. 1944a. A review of the subgenera *Stenocantharis* Gistel and *Neocychrus* Roeschke of the genus *Scaphinotus* Dejean. Entomologica Americana, 24 (New Series): 1-19, 14 figs.

VAN DYKE, E. C. 1944b. A review of the North American species of the genus *Carabus* Linnaeus. Entomologica Americana, 87-137.

VAN DYKE, E.C. 1949. Notes on *Bembidion*. The Pan-Pacific Entomologist, 25: 56.

VAURIE, P. 1955. A review of the North American genus *Amblycheila* (Coleoptera, Cicindelidae). American Museum Novitates, No. 1724, 26 pp.

VIGNA TAGLIANTI, A., SANTARELLI, F., DI GIULIO, A., and OLIVERIO, M. 1998. Phylogenetic implications of larval morphology in the tribe Ozaenini, pp. 273-246. *In,* Phylogeny and classification of Caraboidea (Coleoptera: Adephaga). (Ball, G.E., Casale, A., and Vigna Taglianti, A., Eds). Atti Museo Regionale di Scienze, Torino. 543 pp.

VOGLER, A. P. and BARRACLOUGH, T. G. 1998. Reconstructing shifts in diversification rate during the radiation of tiger beetles, pp. 251-260. *In,* Phylogeny and classification of Caraboidea (Coleoptera: Adephaga). (Ball, G.E., Casale, A., and Vigna Taglianti, A., Eds.). Atti Museo Regionale di Scienze, Torino. 543 pp.

WALLIS, J.B. 1961. The Cicindelidae of Canada. University of Toronto Press, Toronto. xii + 74 pp.

WARD, R.D. 1979. Metathoracic wing structures as phylogenetic indicators in the Adephaga (Coleoptera), pp. 181-191. *In*, Carabid beetles, their evolution, natural history, and classification. (Erwin, T. L., Ball, G. E., Whitehead, D. R., and Halpern, A. L., Eds.). Dr. W. Junk Publishers, The Hague. x + 635 pp.

WHITEHEAD, D.R. 1966. A review of *Halocoryza* Alluaud, with notes on its relationships to *Schizogenius* Putzeys (Coleoptera, Carabidae). Psyche, 73: 217-228.

WHITEHEAD, D.R. 1969. Variation and distribution of the intertidal beetle *Halocoryza arenaria* (Darlington), in Mexico and the United States (Coleoptera: Carabidae). Journal of the New York Entomological Society, 77: 36-39.

WHITEHEAD, D.R. 1972. Classification, phylogeny, and zoogeography of *Schizogenius* Putzeys (Coleoptera: Carabidae: Scaritini). Quaestiones Entomologicae, 8: 131-348.

WICKHAM, H. F. 1897. Coleoptera of the Lower Rio Grande. Bulletin of the Laboratory of Natural History of the State University of Iowa, 4: 96-115.

WILL, K.W. 1997. Review of the species of the subgenus *Megapangus* Casey (Coleoptera; Harpalini; *Harpalus* Latreille). Coleopterists Bulletin, 51: 43-51.

WILL, K.W. 1999. Systematics and zoogeography of the genus *Lophoglossus* Leconte (Coleoptera, Carabidae, Pterostichini), pp. 259-276. *In,* Advances in Carabidology (Zamotajlov, A., and Sciaky, R., Eds.). MUISO Publishers, Krasnodar, Russia. 473 pp.

WILL, K.W. 2000. Systematics and zoogeography of the abaryform genera (Coleoptera: Carabidae: Pterostichini), and a phylogenetic hypothesis for pterostichine genera. PhD dissertation, Cornell University, Ithaca, New York. xviii + 289 pp.

WILL, K.W. and LIEBHERR, J.K. 1997. New and little known species of *Loxandrus* LeConte 1852 (Coleoptera: Carabidae) from North and South America. Studies on Neotropical Fauna and Environment, 32: 230-238.

WILLIS, H.L. 1968. Artificial key to the species of *Cicindela* of North America north of Mexico (Coleoptera: Cicindelidae). Journal of the Kansas Entomological Society, 41: 303-317.

ZETTO BRANDMAYR, T., GIGLIO, A., MARANO, I., and BRANDMAYR, P. 1998. Morphofunctional and ecological features in carabid (Coleoptera) larvae, pp. 449-490. *In,* Phylogeny and classification of Caraboidea (Coleoptera: Adephaga). (Ball, G.E., Casale, A., and Vigna Taglianti, A., Eds.). Atti Museo Regionale di Scienze, Torino. 543 pp.

ZHANG, J., DRUMMOND, F., and LIEBMAN, M. 1994. Spread of *Harpalus rufipes* DeGeer (Coleoptera: Carabidae) in eastern Canada and the United States. Entomology (Trends in Agricultural Science), 2: 67-71.

ZHOU, J. and GOYER, R.A. 1993. Descriptions of the immature stages of *Calleida viridipennis* (Say) and *Plochionus timidus* Haldeman. Coleopterists Bulletin, 47: 233-242.

# 7. GYRINIDAE Latreille, 1810

by R. E. Roughley

**Family common name**: The whirligig beetles

The ovate to broadly ovate, depressed body form, flattened and greatly modified swimming legs and remarkable divided eyes serve to distinguish adults of this family from the other water beetles. Their habit of swimming in circles when alarmed has earned them their common name.

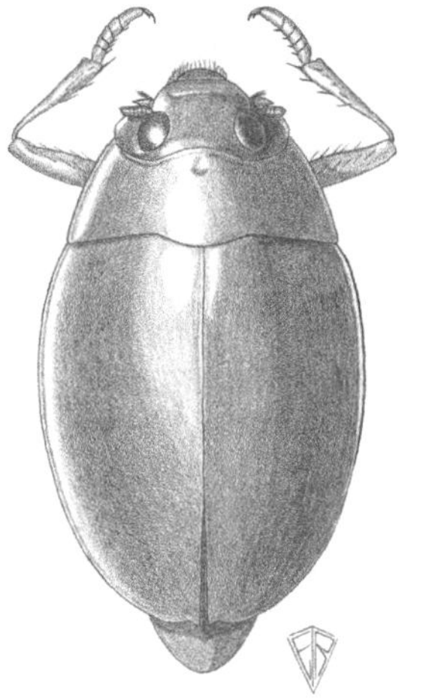

FIGURE 1.7. *Dineutus nigrior* Roberts.

**Description**: Overall structure of gyrinids is described in Hatch (1927) and, Larsén (1954, 1966) and detailed descriptions of *Spanglerogyrus albiventris* were presented by Beutel (1989a, 1989b, 1990) and Burmeister (1990). Body ovate to broadly ovate, depressed or strongly convex; size 2.5-15 mm; dorsal color entirely black or dark metallic green (Gyrininae) or with pronotum white laterally (Spanglerogyrinae), some extralimital taxa with yellow lateral margins, ventral color white (Spanglerogyrinae) or entirely red, brown or black with various contrasting markings (Gyrininae); setation generally reduced except around mouthparts, antennae, and in some species the lateral margin of the body with very short dense setae, legs with natatory setae on tibiae and tarsi; labrum and antennomere 2 (pedicel) with long, fine, erect setae in contact with the water line.

Head prognathous (Stickney 1923) and strongly retracted into prothorax, relatively short, broad and depressed (Gyrininae) or relatively short and more convex (Spanglerogyrinae); surface smooth except for tactile setae on labrum, clypeus and frons. Eyes divided, so placed that upper pair remain above the water line and lower pair below the water line as the beetle swims; eye divided by either a narrow (Spanglerogyrinae) or a broader (Gyrininae) bridge of cuticle. Antennae with eight to eleven antennomeres, antennomere 1 (scape) hemispherical, antennomere 2 (pedicel) enlarged, with scoop-shaped lobe and tactile setae, antennomeres 3-11 short, compact, forming an incrassate club which is about 1.5 times longer than antennomeres 1+2, and not as broad as 2; club without pubescence; antennae inserted between upper and lower eye near base of the mandibles; antennal insertion exposed and visible from above. Labrum distinct and transverse, anterior margin arcuate with fringe of setae as long as the labrum (1 row of setae present in Spanglerogyrinae, three rows in Gyrininae). Clypeus distinct, frontoclypeal suture distinct, clypeus about two times (Spanglerogyrinae) to three or more times (Gyrininae) as wide as long. Mandibles short, relatively blunt, with three apical teeth; lacinia arcuate with sharp apex, inner surface with stout setae; galea two-segmented in Spanglerogyrinae, one-segmented in Gyrinini, absent in Orectochilini and Enhydrini. Maxillary palpus with four palpomeres (Williams 1938). Median gular apodeme present. Head internally with large air sacs.

Pronotum somewhat broader than the head, broadened behind to width of elytra at base; anterior and posterior border sinuate (sinuation more developed in Spanglerogyrinae), lateral borders evenly arcuate, with short tactile setae along lateral margin in Spanglerogyrinae and Orectochilini, dorsal surface smooth or with pubescent margins, posterior margin with even series of setae (Spanglerogyrinae) or a specialized proprioreceptive organ (Gyrininae). Ventral portion of prothorax with notopleural and pleurosternal sutures present and distinct; mediolaterally with impressed area continued onto mesoventrite (=mesosternum, of authors) for reception of fore legs in Gyrininae. Prosternal process very short and not reaching mesoventrite, with socket of procoxal articulation present only in Spanglerogyrinae. Procoxal cavities open behind externally but closed internally. Anterior legs adapted for grasping prey on water surface. Procoxae transverse (Gyrininae), to round with ventral condyle (Spanglerogyrinae), trochantin at least partially exposed. Protrochanter large. Profemur and protibia elongate, protarsi 5-segmented, somewhat twisted, protarsomeres subequal in length, males with protarsomeres 1 to 5 with elongate, broad pad of articulo-setae (Stork 1980) used as an adhesive pad for grasping the female; two short claws present on protarsomere 5.

Mesoventrite unusually large and with discrimen (=median mesosternal suture, of authors) present only in Gyrininae. Mesocoxae strongly transverse and somewhat triangular, limited laterally by epimeron. Mesotrochanter large. Mesofemur short and broad (Gyrininae) to more elongate (Spanglerogyrinae, see Beutel 1990, Fig. 12), mesotibia short and broad (Gyrininae) to greatly elongated distal to articulation of mesotarsi, mesotarsi 5-segmented, much broader than long and modified for swim-

**Acknowledgment**. It has been a privilege, over the years, to associate with an excellent German morphologist, Dr. Rolf G. Beutel (Friedrich Schiller Universität, Jena). I thank him for his help, knowledge and the spirited and comprehensive review of this chapter. Any errors, misinterpretations or omissions are my own.

ming (Gyrininae) or elongate with mesotarsomere 1 longer than mesotarsomeres 2-5 (Spanglerogyrinae); two short claws present on mesotarsomere 5.

Metaventrite (=metasternum, of authors) moderately large in Spanglerogyrinae and Enhydrini, strongly narrowed posteriorly by metacoxal expansion in Gyrinini and Orectochilini; with a transverse suture only in Spanglerogyrinae. Metacoxae triangular, extended laterally to epipleuron, elongate on midline and acutely narrowed laterally (Spanglerogyrinae), to trapezoidal and fairly broad laterally but not expanded anterolaterally (Enhydrini), to parallelogram-shaped and markedly expanded anterolaterally (Gyrinini and Orectochilini) (Beutel 1990). Metatrochanter large. Metafemur short and broad (Gyrininae) to more elongate (Spanglerogyrinae, see Beutel 1990, Figs. 13, 14); metatibia short and broad (Gyrininae) to greatly elongated distal to articulation of metatarsi, metatarsi five-segmented with metatarsomere 1 much broader than long and modified for swimming (Gyrininae) or elongate with metatarsomere 1 longer than metatarsomeres 2-5 (Spanglerogyrinae); two short claws present on metatarsomere 5.

Scutellum concealed (*Gyretes* and *Dineutus)* or exposed (remaining genera). Elytra with sides evenly arcuate, base as broad as base of pronotum, widest in basal one-quarter, apices subacute, variously modified, surface smooth or striate-punctate; with (Spanglerogyrinae and Orectochilini) or without (Gyrinini and Enhydrini) broad lateral field of pubescence; epipleuron broad, nearly reaching apex. Wing venation typically adephagan with oblongum cell; folding pattern as in Amphizoidae and part of the Dytiscidae (Kukalová-Peck and Lawrence 1993); wings rolled apically in Spanglerogyrinae (see Beutel, 1990, Figs. 9, 10).

Abdomen with six apparent sterna, surface smooth, visible sternum 1 (=true sterna I-III) divided by hind coxae (as in other Adephaga), apparent apical sternum made up of fused gonocoxasterna in Gyrininae (Burmeister 1976) which remain separate in Spanglerogyrinae (Burmeister 1990). Male genitalia of the tri-lobed type; penis simple and somewhat flattened in Gyrininae, less flattened in Spanglerogyrinae (Beutel and Roughley 1988); parameres large, somewhat expanded apically; pars basalis (terga IX) large, partially surrounding base of parameres. Male genitalia without torsion (90° rotation, at rest, 180° during copulation) characteristic of other Adephaga (Sharp and Muir 1912, Beutel and Roughley 1988). Female genitalia with genital appendages VIII membranous, terga IX forming a more narrowed and somewhat elongated support for the enlarged, distal gonocoxa (Burmeister 1990, Hilsenhoff, 1990b). The term gonocoxae is equivalent to the coxites of Tanner (1927).

Larvae (Bøving and Craighead 1931; Petersen 1951; Beutel and Roughley 1988, 1995; and Spangler 1991) elongate, somewhat flattened, more or less uniform in width; size 10-30 mm (larvae of *Spanglerogyrus* not known); color white or tinted with yellowish-brown; head and prothorax may be heavily pigmented. Head prognathous and exserted, narrow, elliptical, depressed, distinct neck region present or absent; frontoclypeal suture absent, anterior margin truncate or toothed (=nasal teeth); antennae four-segmented, antennomeres extremely slender in Enhydrini and Orectochilini; mandible without mola, slender, curved, acute, with inner, more or less closed liquid channel; maxilla slender and elongate, longer than antenna in some species; stipes quadrangular with both galea and lacinia present, lacinia long, hook-like and probably mobile, cardo large, with deep maxillary groove, freely articulated with stipes; maxillary palpus two-segmented (palpus appearing three-segmented, prementum completely divided longitudinally and appearing to be an extra palpomere); labium slender and elongate, without ligula extended anteriorly between the three-segmented palpus; six pairs of conspicuous stemmata. Gula represented by a narrow strip between gular sutures, posterior tentorial pits shifted anteriorly, narrowly separated. Prothorax narrower than rest of thorax with two large tergal sclerotizations; tergum of prothorax may be distinctly pigmented and sclerotized; meso- and metathoracic terga semi-membranous; legs six-segmented with apical segment (=pretarsus) consisting of a pair of claws. Abdomen with 10 segments, segments 9 and 10 narrower than first 8; paired elongate lateral tracheal gills on segments 1-8, segment 9 with two pairs of gills, segment 10 without gills, with 4 sclerotized posteriorly projecting, curved hooks. Larval structure is discussed in relation to identification by Spangler (1991) and in relation to phylogenetic analysis by Beutel and Roughley (1995).

**Habits and habitats.** A summary of the life history of gyrinids was provided by Hatch (1925) and Omer-Cooper (1934), Balduf (1935), Ochs (1969) and Sanderson (1982). Nevertheless, much is yet to be done in the study of their habits. Some gyrinids cling to roots of undercut stream banks, whereas most members of this family glide or skate on the surface of the water. Swimming may be aided by use of a surfactant secreted from the pygidial glands (Vulinec 1987). Adults cluster together and often swim rapidly in circles when alarmed. Also, when disturbed they will dive and scatter widely. Some species inhabit standing water, others moving water. Congregations of newly emerged adults are often seen in the latter part of the summer and in the fall. The adult is a scavenger and its food consists of dead and dying insects floating on the water surface. Larvae are predacious, feeding on other aquatic insect larvae and nymphs. Eggs are laid on the surface of submerged parts of either dead or living plants. Pupation takes place in a pupal case which is constructed out of sand and debris either on land near shore, or attached to plants several cm above the water surface.

**Status of the classification.** Because of the remarkable adaptations for aquatic life of this group, the status of the family has varied from that of a suborder or a superfamily, to simply another family of the Adephaga. Beutel and Roughley (1988) placed gyrinids as the most primitive family of Adephaga; they represent an independent invasion of the water from other water beetles. While detailed analysis is complete for only some few genera, the relationships seem rather conclusive for a sister-group relationship between the monophyletic Spanglerogyrinae and Gyrininae. Within Gyrininae, Gyrinini is the sister-group to Enhydrini + Orectochilini. The nomenclature of genera is reviewed by Balfour-Browne (1945). The genera are all easily separated and distinctive; the species within a genus are very similar and often they are difficult to separate. Most of the species are best separated on the basis of the structure of the male genitalia or the form of the gonocoxae (Hilsenhoff 1990b). All of the species in North America north of Mexico have been characterized recently and reliable identification can be achieved by nonspecialists. Excellent regional works are available for Quebec (Morissette 1979), North and South Carolina (Sanderson 1982), Florida (Epler 1996), Minnesota and environs (Ferkinhoff and Gundersen 1983), Wisconsin (Hilsenhoff 1990a), North Dakota (Gordon and Post 1965), the Pacific Northwest (Hatch 1953) and California (Leech and Chandler 1956).

**Distribution.** The family is worldwide in distribution. The most recent world catalogue is Ahlwarth (1910) with 423 species; today there are 12 genera and ca. 700 species recognized worldwide. There are 41 species of *Gyrinus*, 11 species of *Dineutus*, 3 species of *Gyretes* and one species of *Spanglerogyrus* recorded from North America north of Mexico for a total of 56 species. Arce-Pérez and Roughley (1999) listed 19 species of gyrinids for Mexico; Roughley (1991) provided a list of the species of Canada and Alaska.

### Key to the Nearctic Genera

1. Dorsal and ventral portion of eye separated by narrow septum; meso- and metatibia extended well past insertion of tarsi; venter white; size smaller, < 3.0 mm (Spanglerogyrinae) ................................................... *Spanglerogyrus*
— Dorsal and ventral portion of eye broadly separated; meso- and metatibia not extended well past insertion of tarsi; venter dark; size larger, > 3 mm (Gyrininae) ........................................................ 2

2(1). Scutellum exposed; elytra each with eleven striae of punctures; size small, 4-7 mm in length (Gyrinini) ................................................ *Gyrinus*
— Scutellum concealed; elytra smooth or with nine vague or moderately impressed striae; size various, 3-15 mm ..................................................... 3

3 (2). Elytra and pronotum laterally with broadly pubescent margins; last abdominal segment elongate, conical, ventrally with median longitudinal row of erect setae; size smaller, 3-5 mm (Orectochilini) ............................................................... *Gyretes*
— Elytra and pronotum laterally glabrous; last abdominal segment as wide as long rounded, ventrally without median longitudinal row of setae; size larger, 9-15 mm (Enhydrini) .................. *Dineutus*

### Classification of the Nearctic Genera

Gyrinidae

Spanglerogyrinae

**Diagnosis.** Many of the diagnostic features of Spanglerogyrinae are listed above in the description. They are the only Nearctic gyrinids with a white venter and large white markings laterally on the pronotum. The ventral surface is without an impressed area for reception of the anterior legs at rest. The form of the meso- and metatibia which are elongated well past the point of articulation of the meso- and metatarsi is distinctive.

*Spanglerogyrus* Folkerts, 1979.
A single species, *S. albiventris* Folkerts, was described from Alabama. There are unpublished reports of this species from the District of Columbia and Tennessee. The habitat of adults is deep undercut stream banks with trailing roots of terrestrial vegetation (Folkerts 1979, Steiner and Anderson 1981) and this habitat is poorly collected, suggesting that this genus may be more generally distributed.

Gyrininae

**Diagnosis.** Many of the diagnostic features of Gyrininae are listed above in the description. No Nearctic adults have a white venter nor white markings on the pronotum. Ventral surface with an impressed area for reception of the anterior legs at rest. The short, very wide and paddle-like form of the meso- and metafemora, and meso- and metatibia are distinctive.

Enhydrini

**Diagnosis.** Scutellum concealed. Last visible sternite without a median, longitudinal row of erect setae. Pronotum laterally and elytra laterally glabrous, without a broad field of tomentum-like pubescence.

*Dineutus* MacLeay 1825, 11 spp., generally distributed
    *Dineutes* Aubé 1838
    *Cyclous* Kirby 1837
    subgenus *Cyclinus* Kirby 1837
Members of *Dineutus* are large, surface inhabiting gyrinids ranging in length from 9.0 to 15.5 mm. Specimens are larger than those of *Gyrinus*. They are found on the surface of ponds, streams, lakes and rivers. Those on lakes can form large rafts of immense numbers. Keys to species are found in Roberts (1895), Hatch (1930), and Wood (1968) as well as many of the regional studies listed above.

Gyrinini

**Diagnosis.** Scutellum visible. Last visible sternite without a median, longitudinal row of erect setae. Pronotum laterally and elytra laterally glabrous, without a broad field of tomentum-like pubescence.

*Gyrinus* Geoffroy 1762, 41 spp., widely distributed
    subgenus *Gyrinulus* Zaitzev 1907
    subgenus *Oreogyrinus* Ochs 1935
Members of *Gyrinus* are smaller, surface inhabiting gyrinids ranging in length from 3.0 to 8.0 mm. Specimens are smaller than those of *Dineutus*. They are found on the surface of ponds, streams, lakes and rivers and along the margins of undercut

stream banks. Many species form large aggregations of one or more species. Keys to species are found in Oygur and Wolfe (1991) and Atton (1990).

Orectochilini

**Diagnosis.** Scutellum concealed. Last visible sternite with a median, longitudinal row of erect setae. Pronotum laterally and elytra laterally with a broad field of tomentum-like pubescence.

*Gyretes* Brullé 1834.
Adults are 3 to 5 mm in length; they are not surface inhabiting species and are usually found under overhanging banks of small, cool streams and rivers. When captured within a net they will use the abdominal apex as a spring and flip around within the net. There are three species reported from California, New Mexico, Texas, Alabama, Louisiana, Mississippi, Illinois, Indiana, Missouri, Tennessee, and Florida. Keys to the three species are provided in Walls (1974) and Ochs (1929).

Bibliography

ACORN, J.H. and BALL, G.E. 1991. The mandibles of some adult ground beetles: structure, function, and the evolution of herbivory (Coleoptera: Carabidae). Canadian Journal of Zoology, 69(3): 638-650.

AHLWARTH, K., 1910. Gyrinidae. Coleopterorum Catalogus, 4(21) 1-42.

ARCE-PÉREZ, R. and R. E. ROUGHLEY. 1999. Lista anotada y claves para los Hydradephaga (Coleoptera: Adephaga: Dytiscidae, Noteridae, Haliplidae, Gyrinidae) de México. Dugesiana, 6: 69-104.

ATTON, F.M. 1990. *Gyrinus (Gyrinulus) cavatus* sp. nov. from North America described and compared with *Gyrinus (Gyrinulus) minutus* Fabricius (Coleoptera: Gyrinidae). The Canadian Entomologist, 122: 651-657.

BALDUF, W. V. 1935. The bionomics of entomophagous Coleoptera. John S. Swift, N.Y. 220 pp. [Reprinted 1969, E.W. Classey, Ltd., Middlesex, England. i + 220 pp].

BALFOUR-BROWNE, J. 1945. The genera of the Gyrinoidea and their genotypes. The Annals and Magazine of Natural History, (Series 11), 12: 103-111.

BEUTEL, R. G. 1989a. The head of *Spanglerogyrus albiventris* Folkerts (Coleoptera: Gyrinidae). Contribution towards clarification of the phylogeny of Gyrinidae and Adephaga. Zoologische Jahrbücher für Anatomie, 118: 431-461.

BEUTEL, R. G. 1989b. The prothorax of *Spanglerogyrus albiventris* Folkerts, 1979 (Coleoptera, Gyrinidae). Contribution towards clarification of the phylogeny of Gyrinidae. Entomologica Basiliensia, 13: 151-173.

BEUTEL, R. G. 1990. Phylogenetic analysis of the family Gyrinidae (Coleoptera) based on meso- and metathoracic characters. Quaestiones Entomologicae, 26: 163-191.

BEUTEL, R. G. and ROUGHLEY, R. E. 1988. On the systematic position of the family Gyrinidae (Coleoptera: Adephaga). Zeitschrift für Zoologische Systematik und Evolutionsforschung, 26: 380-400.

BEUTEL, R. G. and ROUGHLEY, R. E. 1995. Phylogenetic analysis of Gyrinidae based on characters of the larval head (Coleoptera: Adephaga). Entomologica Scandinavica, 24: 459-68.

BØVING, A. G. and CRAIGHEAD, F. C. 1931. An illustrated synopsis of the principal larval forms of the order Coleoptera. Brooklyn Entomological Society, Brooklyn, N.Y. viii + 351 pp.

BURMEISTER, E.-G. 1976. Der Ovipositor der Hydradephaga (Coleoptera) und seine phylogenetische Bedeutung unter besonderer Berucksichtigung der Dytiscidae. Zoomorphologie, 85: 165-257.

BURMEISTER, E-G. 1990. The female genital structures of *Spanglerogyrus albiventris* Folkerts, 1979. A contribution to the systematic position of the Gyrinidae. Spixiana 13: 253-265.

EPLER, J. H. 1996. Identification manual for the water beetles of Florida (Coleoptera: Dryopidae, Dytiscidae, Elmidae, Gyrinidae, Haliplidae, Hydraenidae, Hydrophilidae, Noteridae, Psephenidae, Ptilodactylidae, Scirtidae). Bureau of Water Resource Protection, Florida Department of Environmental Protection, Tallahassee.

FERKINHOFF, W. D. and GUNDERSEN, R. W. 1983. A key to the whirligig beetles of Minnesota and adjacent States and Canadian Provinces. Scientific Publications of the Science Museum of Minnesota (Saint Paul), (n.s.) 5 (3): 1-53.

FOLKERTS, G. W. 1979. *Spanglerogyrus albiventris*, a primitive new genus and species of Gyrinidae (Coleoptera) from Alabama. Coleopterists Bulletin, 33: 1-8.

FORBES, W. T. M. 1923. The wing-venation of the Coleoptera. Annals of the Entomological Society of America, 15 (1922): 328-345 + plates XXIX-XXXV.

FORBES, W. T. M. 1926. The wing-venation of the Coleoptera. Annals of the Entomological Society of America, 15: 328-345 + plates XXIX-XXXV.

GORDON, R.D. and POST, R. L. 1965. North Dakota Water Beetles. North Dakota Insects, Publication No. 5. Department of Entomology, Agricultural Experiment Station, North Dakota State University. 53 pp.

HATCH, M. H. 1925. An outline of the ecology of Gyrinidae. Bulletin of the Brooklyn Entomological Society, 20: 101-114.

HATCH, M.H. 1927. The morphology of Gyrinidae. Papers of the Michigan Academy of Science, Arts and Letters, 6: 311-350 + plates xx-xxiv.

HATCH, M. H. 1930. Records and new species of Coleoptera from Oklahoma and western Arkansas, with subsidiary studies. Publications of the University of Oklahoma Biological Survey, 2: 15-26.

HATCH, M. H. 1953. The beetles of the Pacific Northwest, Part 1. Introduction and Adephaga. University of Washington Publications in Biology, 16: 1-340.

HILSENHOFF, W. L. 1990a. Gyrinidae of Wisconsin, with a key to adults of both sexes and notes on distribution and habitat. Great Lakes Entomologist, 23: 77-91.

HILSENHOFF, W.L. 1990b. Use of gonocoxae and the sternal apex to identify adult females of North American *Gyrinus* Geoffroy (Coleoptera: Gyrinidae). Quaestiones Entomologicae, 26: 193-97.

KUKALOVA-PECK, J. and LAWRENCE, J. 1993. Evolution of the hind wing in Coleoptera. Canadian Entomologist, 125: 181-258.

LARSÉN, O. 1954. Die Flugorgane der Gyriniden (Col.). Opuscula Entomologica, 19: 5-17.

LARSÉN, O. 1966. On the morphology and function of the locomotor organs of the Gyrinidae and other Coleoptera. Opuscula Entomologica Supplementum, 30: 1-242.

LEECH, H.B. and CHANDLER, H. P. 1956. Aquatic Coleoptera. pp. 293-371 *in* Usinger, R. L. Editor. Aquatic insects of California with keys to North American genera and California species. University of California Press, Berkeley and Los Angeles. ix + 508 pp.

MORRISSETTE, R. 1979. Les coléoptères Gyrinidae du Québec. Cordulia, Supplément, 8: 1-43.

OCHS, G., 1929. Bestimmungstabelle der Gyrinidengattung *Gyretes* Brullé, nebst Neubeschreibungen und kritischen Bemerkungen. Koleopterologische Rundschau, 15: 62-93.

OCHS, G. 1969. Zur Ethökologie der Taumelkäfer (Col., Gyrinoidea). Archiv für Hydrobiologie, Supplementum, 35: 373-410.

OMER-COOPER, J. 1934. Notes on the Gyrinidae. Archiwum Hydrobiologji i Rybactwa (Suwa»ki), 8: 1-26.

OYGUR, S. and WOLFE, G.W. 1991. Classification, distribution and phylogeny of North American (north of Mexico) species of *Gyrinus* Müller (Coleoptera: Gyrinidae). Bulletin of the American Museum of Natural History, 207: 1-97.

PETERSON, A. 1951. Larvae of Insects. An introduction to Nearctic species. Part II. Coleoptera, Diptera, Neuroptera, Siphonaptera, Mecoptera, Trichoptera. Privately printed. Columbus, Ohio. v + 416 pp.

ROBERTS, C. H., 1895. The species of *Dineutes* of America north of Mexico. Transactions of the American Entomological Society, 22: 279-288 + plates V-VI.

ROUGHLEY, R. E. 1991. Family Gyrinidae. pp. 72-73 *in* Bousquet, Y. Editor. Checklist of the Beetles of Canada and Alaska. Publication 1861/E, Research Branch, Agriculture Canada, Ottawa, Ontario. vi + 430 pp.

SANDERSON, M. W. 1982. Gyrinidae. pp. 10.29-10.38 *in* Brigham, W.U. Chapter 10. Aquatic Coleoptera. pp. 10.1-10.136 *in* Brigham, A. R., W. U. Brigham, and A. Gnilka. Editors. Aquatic Insects and Oligochaetes of North and South Carolina. Midwest Aquatic Enterprises, Mahomet, Illinois. xi + 837 pp.

SHARP, D. and MUIR, F., 1912. The comparative anatomy of the male genital tube in Coleoptera. Transactions of the Entomological Society of London, 1912: 477-642 + plates 42-78.

SPANGLER, P.J. 1991. Gyrinidae (Adephaga). pp. 319-320, *in* Lawrence, J.F. Co-ordinator. Chapter 34. Coleoptera. Pp. 144-658 *in* F. Stehr. Editor. Immature Insects. Volume 2. Kendall/Hunt Publishing Co., Dubuque, Iowa. xvi + 975 pp.

STEINER, W. E. Jr. and ANDERSON, J. J. 1981. Notes on the natural history of *Spanglerogyrus albiventris* Folkerts, with a new distribution record (Coleoptera: Gyrinidae). Pan-Pacific Entomologist, 57: 124-132.

STICKNEY, F. S. 1923. The head capsule of Coleoptera. Illinois Biological Monographs, 8: 1-104.

STORK, N.E. 1980. A scanning electron microscope study of tarsal adhesive setae in the Coleoptera. Zoological Journal of the Linnean Society, 68: 173-306.

TANNER, V. M. 1927. A preliminary study of the genitalia of female Coleoptera. Transactions of the American Entomological Society, 53: 5-50.

VULINEC, K. 1987. Swimming in whirligig beetles (Coleoptera: Gyrinidae): A possible role of the pygidial gland secretion. Coleopterists Bulletin, 41: 151-153.

WALLS, J.G. 1974. Distribution and recognition of United States whirligig beetles of the genus *Gyretes* (Coleoptera: Gyrinidae). Studies in Arthropoda, 1: 1-10.

WILLIAMS, I. W. 1938. The comparative morphology of the mouthparts of the order Coleoptera treated from the standpoint of phylogeny. Journal of the New York Entomological Society, 46: 245-289.

WOOD, F. E. 1968. The taxonomic status of *Dineutus serrulatus* and *Dineutus analis* in North America (Gyrinidae: Coleoptera). Proceedings of the United States National Museum, Vol. 124 (#3646): 1-9.

# 8. HALIPLIDAE Aubé, 1836

by R.E. Roughley

**Family common name**: The crawling water beetles

The principal features for recognition of these beetles are the extremely large hind coxal plates which cover most of the abdominal ventrites, tarsal formula of 5-5-5, yellowish ground color of adults and the body form adapted for aquatic habits.

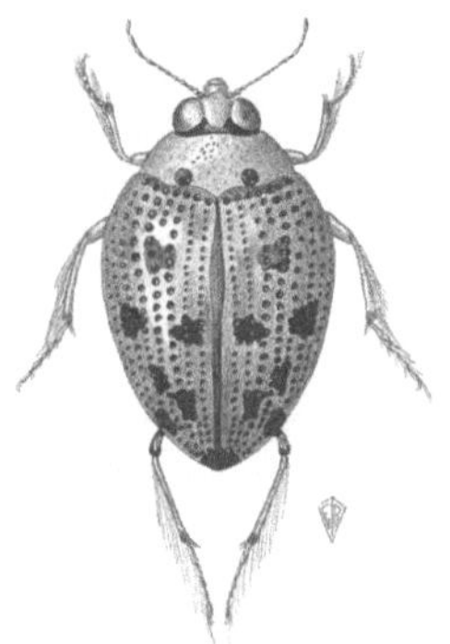

FIGURE 1.8. *Peltodytes edentulus* (LeConte).

**Description.** Shape oval or broadly ovate to somewhat fusiform (*Brychius, Apteraliplus*); length 1.5 to 5 mm; color yellowish to very light brownish with 10+ rows of large punctures, darkened on most species; devoid of obvious setae except long swimming hairs on tibiae and tarsi.

Head (Stickney 1923) prognathous, small and somewhat elongate, about two-thirds (*Apteraliplus* and *Brychius*) to about half (*Peltodytes* and *Haliplus*) the width of pronotum at base, somewhat retracted into prothorax, dorsally convex, surface punctate but without setae except for long setae on labrum. Eyes more or less bulging (in dorsal view), outline not emarginate. Antennae with 11 glabrous antennomeres, 1st (scape) and 2nd antennomeres broad and short, remainder filiform, antenna shorter than pronotum, antennomeres 3 to 11 of increasing length and slightly longer than wide, nearly uniform in width, inserted between the anterior portion of the eyes well above the bases of the mandibles. Mouthparts (Williams 1938) with labrum transverse and narrow with long setae (shorter in *Peltodytes*) on the anterior margin, covering most of the mandibles. Clypeus fused to frons, frontoclypeal suture not visible. Mandibles small and robust, somewhat asymmetrical, with an apical tooth, retinaculum rounded, with a mesal comb or brush of setae. Maxilla with lacinia with spines along inner margin and an elongate two-segmented galea, maxillary palpus four-segmented with a basal palpiger. Labium with small anterolateral lobes, ligula present and small but extending between the bases of the labial palps, labial palpus short, moderately stout, three segmented with a basal palpiger. Median gular apodeme absent.

Pronotum widest at base, narrower to much narrower in front, excavated for head, without pubescence, coarsely punctate, lateral margins straight or convex, posterior margin sinuate with medial V-shaped posterior extension, most species with short, posterolateral carina or plica. Ventral portion of prothorax with notopleural and pleurosternal sutures. Prosternum at base elevated to form a kind of ventral head rest, in repose, prosternal process long and forming a complex junction with the mesoventrite (=mesosternum, of authors), bordered laterally by raised lines or carinae (absent in *Apteraliplus*). Procoxal cavities round, open behind externally but closed internally. Procoxae globular to subglobular, trochantin not visible. Protrochanter short, globular. Profemur thickened medially, with a row of broadly spaced, stiff, elongate setae on anteroventral surface. Protibia slender with rows of stiff, elongate setae on anterior, dorsal and posterior surfaces, and row of long, natatory setae on dorsal surface. Protarsi slender and elongate, protarsomere 1 longer than protarsomeres 2, 3 and 4 which are longer than wide, protarsomere 5 longest, with row of long, natatory setae on dorsal surface of protarsomeres 1 to 4; males with protarsomeres 1 to 3 perceptibly broader, apex of protarsomere 1 and ventral surface of protarsomeres 2 and 3 with some short, unmodified setae; two rather elongate claws (very long in *Brychius*) present on tarsomere 5 with inner margins minutely serrate to pectinate.

Mesoventrite short, mostly concealed by prosternal process, mesocoxal cavity round, limited laterally by mesepimeron. Mesocoxae subglobular with limited range of motion. Mesotrochanter short and subglobular. Mesofemur widened at middle; with poorly defined row of broadly spaced, stiff, elongate setae on anteroventral surface. Mesotibia slender and elongate with row of stiff, elongate setae on anterior, dorsal and posterior surfaces, with row of long, natatory setae on dorsal surface. Mesotarsi slender and elongate, mesotarsomere 1 longer than mesotarsomeres 2, 3 and 4 which are longer than wide, mesotarsomere 5 longest, with row of long, natatory setae on dorsal margin of mesotarsomeres 1 to 4; males with mesotarsomeres 1 to 3 perceptibly broadened, apex of mesotarsomere 1, and ventral surface of mesotarsomeres 2 and 3 with some short, unmodified setae, two rather elongate claws (very long in *Brychius*) on mesotarsomere 5 with inner margins serrate in some species.

Metaventrite (=metasternum, of authors) (Beutel and Belkaceme 1986) large and extended almost to epipleuron but limited laterally by elongate, narrow metepimeron, with a short discrimen (=median metasternal suture, of authors) and without (*Apteraliplus*, Chandler 1943) or with (remaining genera) a complete, posterior, transverse suture; anteromedially with broad

**Acknowledgment.** I thank the late Hugh B. Leech for his legendary patience, copious informative letters, helpful loans of literature and specimens and his overall perseverance with a struggling coleopterist in the early part of my career. I can only hope, Hugh, that you would be proud of the outcome.

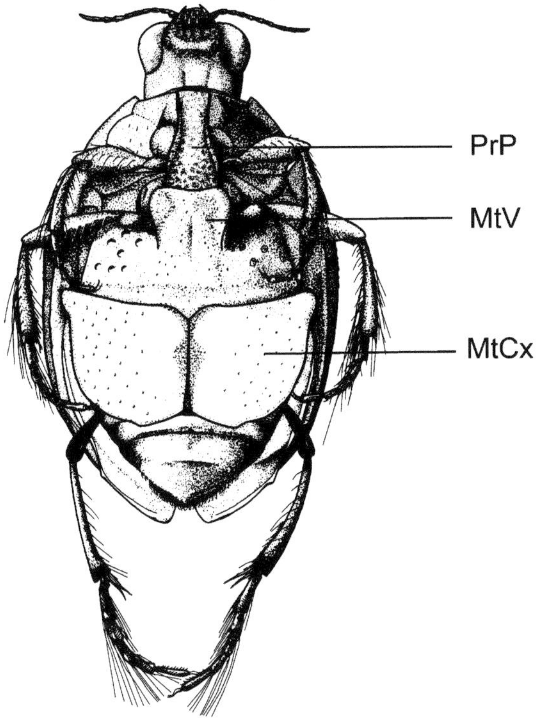


FIGURE 2.8. *Peltodytes* sp., ventral view.

extension (=anterior metasternal process, of authors), partially covering inner margins of mesocoxae and forming a junction with the prosternal process. Metacoxae (Belkaceme 1986) greatly enlarged and extending laterally to epipleuron, greatly elongate on midline and exposing only one (*Peltodytes*), or three (remaining genera) abdominal sternites. Metatrochanter small and subtriangular. Metafemur thin and elongate, mostly covered by metacoxal plates, with a short row of stiff, elongate setae only on exposed dorsal surface. Metatibia long and slender, with numerous rows of stiff setae on anterior, posterior and ventral surfaces, and row of long natatory setae on dorsal surface (=setiferous striole); metatibial spurs short, with posterior margin serrate (*Haliplus* and *Apteraliplus*), or not (*Brychius* and *Peltodytes*). Metatarsi slender and elongate, with row of long natatory setae on dorsal surface of metatarsomeres 1 to 4; metatarsomere 1 much longer than metatarsomeres 2, 3, and 4 which are longer than wide, protarsomere 5 longest; two claws (very long in *Brychius*) present on apex of metatarsomere 5, with or without serration on ventral margin.

Scutellum concealed. Elytra with sides evenly arcuate or tapering, base as broad as base of pronotum, widest in basal quarter, apices rounded, lateral margin at apex and humeral margin serrate on many species, surface with well-impressed microreticulation on most species, striate-punctate with 10+ rows of punctures; sutural stria present in apical half of elytron (*Peltodytes* and *Brychius*) or absent (*Haliplus* and *Apteraliplus*); epipleuron very broad basally, extended almost to the apex (*Peltodytes* and *Brychius*) or ending at level of anterior margin of last visible sternite (*Haliplus* and *Apteraliplus*); epipleuron with a distinctive notch to receive the metacoxae, notch connecting air-storage space beneath metacoxal plates to that beneath elytra. Wing venation typically adephagan with oblongum cell (Forbes 1923, Fig. 19; Kukalovà-Peck and Lawrence 1993, Fig. 16); folding pattern as in Hygrobiidae and part of Dytiscidae (Forbes 1926, Fig. 16), at rest, wings rolled apically.

Abdomen with six apparent sternites (true sternites 2-7), surface generally with large, shallow punctures, sternites 2 to 4 fused and without sutures, metacoxal plates covering all but sternites 5 to 7 (most genera) or exposing only apical sternite (*Peltodytes*); apical sternite (=sternite 7) drawn out medioposteriorly into a point, with median longitudinal groove or incision in apical portion only in *Brychius*.

Male genitalia (Sharp and Muir 1912) caraboid, penis asymmetrical, laterally compressed and apex broad and finger-like or pointed, with a sperm channel on the convex surface covered by a membranous hood; parameres asymmetrical, left paramere shorter and broadly triangular, right paramere more elongate and narrow, with (*Haliplus*, subgenus *Liaphlus* only and *Apteraliplus*) or without (*Haliplus*, remaining subgenera, *Peltodytes* and *Brychius*) stylus-like extension (=digitus of Brigham and Sanderson 1972). Male genitalia with torsion (90° rotation at rest, 180° during copulation) characteristic of most of Adephaga (Beutel and Roughley 1988).

Female genitalia (Tanner 1927, Burmeister 1976) with genital appendages VIII separate, triangular and apically pointed (except *Peltodytes*), withdrawn into the abdomen and lying above sternite 7; gonocoxasterna with well-developed anterior apodeme, which are short (*Peltodytes* and *Brychius*) or more elongate (*Haliplus* and *Apteraliplus*); terga IX forming a narrowed and somewhat elongated support for the distal, pencil-shaped gonocoxae which have apical, sensory setae.

Larvae (Beutel 1986, Holmen 1987, Peterson 1951, Seeger 1971a, Spangler 1991) elongate, slender, cylindrical to subcylindrical; size 5-12 mm; color whitish to yellowish-brown or reddish-brown, surface not darkly pigmented but heavily sclerotized and with pronounced microsculpture, with surfaces of head, thorax and abdomen with numerous tubercle-like microtracheal gills (*Brychius*, *Apteraliplus* and *Haliplus*) or with greatly elongated tracheal gills (*Peltodytes*). Head prognathous to semi-prognathous and exserted, rounded laterally with various prominent setiferous tubercles; distinct neck region absent; frontoclypeal suture absent, anterior margin concave; antennae short to very short (*Brychius*), four-segmented, with apical antennomere very reduced and originating beside sensory appendage; labrum fused to head capsule; mandible short, without mola, with a distinct apical tooth enclosing the external opening of the feeding channel; maxilla short and broad, cardo small and more or less fused to stipes, lacinia and galea reduced or absent; maxillary palpus short, two-segmented. Labium with postmentum small, somewhat quadrangular, prementum small, without ligula, labial palpus short, two-segmented; six pairs of conspicuous stemmata. Gula longer than wide, represented by a broad strip between gular sutures, posterior tentorial pits shifted forward and clearly separated. Pro-, meso- and metathoracic tergites not darkened but well sclerotized; in *Peltodytes* with three pairs of elongate tracheal gills on prothorax, two pairs on meso- and metathorax, other genera with much shorter microtracheal gills. Legs six-segmented with apical segment (=pretarsus) consisting of a single claw; natatory setae absent; femur (*Brychius*) or tibia (remaining genera) modified ventroapically to form a grasping apparatus opposing the claw, leg unmodified in some larvae of *Haliplus* (Holmen 1987). Abdomen 10-segmented, with two pairs of elongate gills on tergites 1 to 8 and one pair on each of tergites 9 and 10 (*Peltodytes*) or with numerous microtracheal gills (remaining genera). Apical abdominal segment (=segment 10) short (*Peltodytes*) or greatly elongate (remaining genera) which has an elongate apical bifurcation (reduced in *Peltodytes*).

**Habits and habitats.** Haliplids are an aquatic family of beetles. Swimming is effected by alternate movement of the legs, and is therefore feeble and is assisted by the natatory setae on the tibiae and tarsi of all legs; crawling is a more normal form of locomotion. Some species are known to fly. These beetles rise to the air-water interface at intervals to renew their air supply which is in the form of a physical gill. The air is stored in the subelytral cavity where it is in contact with the abdominal spiracles. The size of the physical gill is enlarged by using a cavity beneath the metacoxal plates which is in connection with the subelytral space by means of the notch on the epipleuron (Hickman 1931a). Haliplids are found at the edge of small ponds, lakes, or quiet streams (*Peltodytes* and *Haliplus*), on mineral substrate of larger rivers and wave-washed lake shores (*Brychius*) or in temporary, vernal pools (*Apteraliplus*). They are often seen on mats of filamentous algae or *Chara* and similar vegetation. The life cycle and habits can be generalized from Hickman (1930, 1931b), Holmen (1987) and Seeger (1971a, 1971b, and 1971c). Adults are known to eat insect eggs, algae and polyps of Hydrozoa. Larvae are almost exclusively algophagous. Adults are active the year round when weather permits and egg laying takes place from spring to early summer and perhaps again in the fall. Oviposition in *Apteraliplus* and *Brychius* is not known; adult females of *Peltodytes* deposit eggs onto the surface of aquatic plants, those of *Haliplus* chew a cavity in algae or aquatic vascular plants into which eggs are deposited. There are three larval instars and larvae often overwinter. Pupation takes place on land within a pupal chamber constructed by the last larval stage. The pupal stage can last up to two weeks, probably depending on temperature at the pupation site.

**Status of the classification.** The distinctiveness of this group is unquestioned. Haliplidae probably arose from terrestial forms independent of other aquatic beetles (Beutel 1995, 1997; Beutel and Roughley 1988) but for an alternative view of the phylogenetic position of Haliplidae see Kavanaugh (1986). The phylogeny of the five extant genera is provided in Beutel and Ruhnau (1990) and Beutel (1995, 1997). In the classification, *Peltodytes* is the sister-group to the remaining taxa, *Brychius* is the sister-group to *Haliplus* which includes the South African genus *Algophilus* and the Nearctic *Apteraliplus*. The two latter genera are flightless and the genera appear to have been based on derived character states based on features associated with flightlessness. *Apteraliplus* is retained as a distinct genus in this treatment because a formal synonymy with *Haliplus* has not been published and it would be best to wait until a comprehensive revision of *Haliplus* is completed. The genera *Haliplus, Peltodytes* and *Brychius* have not been revised recently; all of them are in need of comprehensive, continent-wide taxonomic revisions. There are numerous faunistic treatments which should allow most species to be be identified. Excellent regional works are available for Virginia (Matta 1976), North and South Carolina (Brigham 1982), Florida (Epler 1996), Minnesota (Gundersen and Otremba 1988), Wisconsin (Hilsenhoff and Brigham 1978), North Dakota (Gordon and Post 1965), the Pacific Northwest (Hatch 1953) and California (Leech and Chandler 1956). A key to the larvae of the Nearctic genera is published in Leech and Chandler (1956).

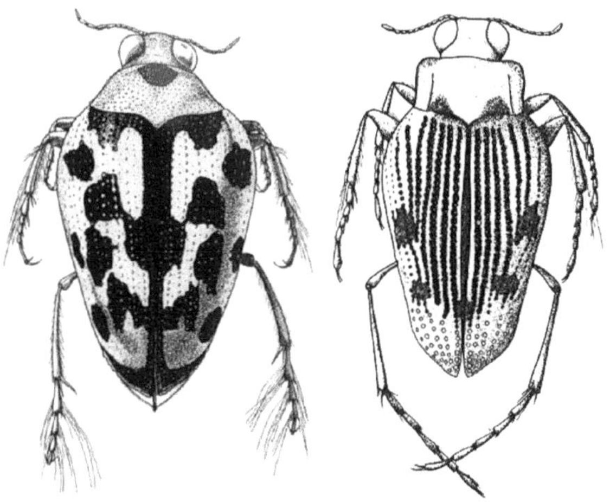

FIGURES 3.8 - 4.8. Fig. 3.8 *Haliplus triopsis*, dorsal view; Fig. 4.8, *Brychius* sp., dorsal view.

**Distribution.** The family is almost worldwide in distribution. The species are more common at temperate latitudes and increasingly rare toward the tropics. The most recent world catalogue is Zimmermann (1920) with 103 species; today there are about 200 known species worldwide with about 67 species known from North America (Spangler 1991). Arce-Pérez and Roughley (1999) listed 9 species of haliplids for México; Roughley (1991a) provided a list of the 37 species of Canada and Alaska.

### Key to the Nearctic Genera

1. Pronotum with lateral margins parallel-sided medially and widened at the base, body shape fusiform; last abdominal ventrite with distinct, median longitudinal groove; epipleuron broad, extending almost to apex of elytra ............................ *Brychius*

— Pronotum with lateral margins rounded, convergent anteriorly and widest at base; body shape various; last abdominal ventrite without median longitudinal groove; epipleuron broad or evenly narrowed, not reaching apex, ending near base of last ventral abdominal segment .................................. 2

2(1). Pronotum with two large black spots on posterior margin (Fig. 3.8); metacoxal plates margined and longer completely exposing only the last visible sternite ........................................................ *Peltodytes*

— Pronotum without large dark spots on posterior margin, some species with small black punctures on posterior margin; metacoxal plates not margined and shorter exposing the last three sternites .... 3

3(2). Pronotum depressed medially between lateral, basal impressions or plicae; last abdominal ventrite at apex densely and coarsely punctate, contrasting with almost impunctate remainder of visble sternites; prosternal process with apex somewhat rounded; prosternum, in cross section, more evenly rounded without raised, prosternal lines or carinae; flight wings reduced ................ *Apteraliplus*

— Pronotum not depressed medially between lateral, basal impressions or plicae; last abdominal segment not as densely and coarsely punctate; prosternal process with apex truncate; prosternum, in cross section, with raised prosternal lines or carinae, angularly separating lateral portions of prosternum ............................................ *Haliplus*

## Classification of the Nearctic Genera

### Haliplidae

*Haliplus* Latreille, 1802
*Cnemidotus* Illiger, 1802, *nec* Erichson, 1832
*Hoplitus* Clairville 1806, *nec* Neumayr, 1875
*Haliplidius* Guignot, 1928
subgen. *Haliplinus* Guignot 1939
*Hoplites* Kinel 1929 *nec* Neumayr,1875
subgen. *Liaphlus* Guignot 1928
subgen. *Neohaliplus* Netolitzky 1911
*Protohaliplus* Scholz 1929
subgen. *Paraliaphlus* Guignot 1930
subgen. *Phalilus* Guignot 1935

Specimens range from 1.75 to 5 mm; they are found throughout North America and in a wide range of habitats. They are most easily characterized by their oval to ovate shape and the absence of two large spots on the posterior margin of the pronotum. Species appear to most common in non-acidic, lentic habitats with algae. The last revision of the Nearctic members of the genus *Haliplus* was by Wallis (1933). Since then, there have been numerous additional species described, synonymies proposed. Wallis (1933) treated 41 of the approximately 56 species of *Haliplus* now known for North and Central America. Holmen (1987), in his discussion of Fennoscandian species, provides comments on some Holarctic species and provides an excellent example of a format for a Nearctic revision.

All six known subgenera of *Haliplus* are listed above but even the subgeneric names are in dispute (see Holmen 1987); some of these subgenera may not occur in North America but because most Nearctic species were not assigned to subgenera when they were first described all available subgenera are listed. Most of our species belong to *Haliplinus, Liaphlus*, *Paraliaphlus* and *Haliplus* (*s. str.*).

*Peltodytes* Régimbart, 1872
*Cnemidotus* Erichson, 1832
subgen. *Neopeltodytes* Satô, 1963

The members of *Peltodytes* all have two large, black spots basolaterally on the pronotum and the metacoxae are expanded leaving only the last visible sternite uncovered. There are about 23 species known for North and Central America and they are generally distributed throughout much of the United States and Canada. They tend to occur in many of the same habitats as the species of *Haliplus.* Early work on North American species was published by Matheson (1912) and Roberts (1913) but the most recent revision is Zimmermann (1924).

*Apteraliplus* Chandler, 1943

A single species, *A. parvulus* (Roberts, 1913), is presently assigned to this genus. A synonym is *Haliplus wallisi* Hatch 1944 (see Hatch 1953 and Zack 1991). This species is small (1.5 to 2.5 mm), flightless, humpbacked and more or less fusiform in shape. It is known from temporary, vernal pools and ponds from California, Oregon, and Washington (Kitayama 1981, Zack 1991). As mentioned above, this genus is almost certainly invalid and it appears to be a member of the subgenus *Liaphlus* of *Haliplus* (Beutel and Ruhnau 1990).

*Brychius* Thomson, 1859

The members of this genus are easily recognized by their distinctive body shape with the bell-shaped pronotum. Adults range from 3.5 to 4.5 mm. Presently there are three described western species (*B. albertanus* Carr, *B. hornii* Crotch and *B. pacificus* Carr, key to western species in Carr 1928) which range collectively from Saskatchewan across to British Columbia south to California, and one eastern species (*B. hungerfordi* Spangler 1954) known from Michigan and Ontario (Roughley 1991b). Collections at hand suggest that there are further undescribed species. These species inhabit rivers, streams and creeks with clear water and mineral substrates; they can also be collected along wave-washed lake shores.

## Bibliography

ARCE-PÉREZ, R. and R. E. ROUGHLEY. 1999. Lista anotada y claves para los Hydradephaga (Coleoptera: Adephaga: Dytiscidae, Noteridae, Haliplidae, Gyrinidae) de México. Dugesiana, 6: 69-104.

BELKACEME, T. 1986. Skelet und Muskulatur der Hinterhüfte von *Haliplus lineatocollis* Mrsh. (Haliplidae, Coleoptera). Stuttgarter Beiträge zur Naturkunde, Serie A (Biologie), 393: 1-12.

BEUTEL, R. G. 1986. Skelet und Muskulatur des Kopfes der Larve von *Haliplus lineaticollis* Mrsh. (Coleoptera:Haliplidae).

Stuttgarter Beiträge zur Naturkunde, Serie A (Biologie), 390: 1-15.

BEUTEL, R. G. 1995. The Adephaga (Coleoptera): phylogeny and evolutionary history. pp. 173-217. *In*: Pakaluk, J. and Slipinski, S. A., editors. Biology, phylogeny, and classification of Coleoptera. Papers celebrating the 80th birthday of Roy A. Crowson. Muzeum i Instytut Zoologii Pan, Warszawa.

BEUTEL, R. G. 1997. Über Phylogenese und Evolution der Coleoptera (Insecta), inbesondere der Adephaga. Abhandlungen des Naturwissenschaftlichen Vereins in Hamburg (N.F.), 31: 1-64.

BEUTEL, R. G. and. BELKACEME, T. 1986. Comparative studies on the metathorax of Hydradephaga and Trachypachidae (Coleoptera). Entomologica Basiliensia, 11: 221-229.

BEUTEL, R. G. and ROUGHLEY, R. E. 1988. On the systematic position of the family Gyrinidae (Coleoptera: Adephaga). Zeitschrift für Zoologische Systematik und Evolutionsforschung, 26: 380-400.

BEUTEL, R. G. and RUHNAU, S. 1990. Phylogenetic analysis of the genera of Haliplidae (Coleoptera) based on characters of adults. Aquatic Insects, 12: 1-17.

BRIGHAM, W. U. 1982. Haliplidae. pp. 10.38-10.47. *In*: Brigham, A. R., Brigham, W. U. and Gnilka, A. Editors. Aquatic Insects and Oligochaetes of North and South Carolina. Midwest Aquatic Enterprises, Mahomet, Illinois.

BRIGHAM, W. U. and SANDERSON, M. W. 1972. A new species of *Haliplus* from Illinois and South Dakota (Coleoptera: Haliplidae). Transactions of the Illinois State Academy of Science, 65: 17-22.

BURMEISTER, E.-G. 1976. Der Ovipositor der Hydradephaga (Coleoptera) und seine phylogenetische Bedeutung unter besonderer Berucksichtigung der Dytiscidae. Zoomorphologie, 85: 165-257.

CARR, F. S. 1928. New species of the genus *Brychius* (Coleoptera). The Canadian Entomologist, 60: 23-26.

CHANDLER, H. P. 1943. A new genus of Haliplidae (Coleoptera) from California. The Pan-Pacific Entomologist, 19: 154-158.

EPLER, J. H. 1996. Identification manual for the water beetles of Florida (Coleoptera: Dryopidae, Dytiscidae, Elmidae, Gyrinidae, Haliplidae, Hydraenidae, Hydrophilidae, Noteridae, Psephenidae, Ptilodactylidae, Scirtidae). Tallahassee, Florida: Bureau of Water Resource Protection, Florida Department of Environmental Protection.

FORBES, W. T. M. 1923. The wing-venation of the Coleoptera. Annals of the Entomological Society of America, 15 (1922): 328-345 + plates XXIX-XXXV.

FORBES, W. T. M, 1926. The wing-venation of the Coleoptera. Annals of the Entomological Society of America, 15: 328-345 + plates XXIX-XXXV.

GORDON, R.D. and POST, R. L. 1965. North Dakota Water Beetles. *North Dakota Insects*, Publication No. 5. Department of Entomology, Agricultural Experiment Station, North Dakota State University. 53 pp.

GUNDERSEN, R. W. and OTREMBA, C. 1988. Haliplidae of Minnesota. Scientific Publications of the Science Museum of Minnesota, 6: 1-43.

HATCH, M. H. 1944. Two new adephagid water beetles from the Pacific Northwest. Bulletin of the Brooklyn Entomological Society, 39: 45-47.

HATCH, M. H. 1953. The beetles of the Pacific Northwest, Part l. Introduction and Adephaga. University of Washington Publications in Biology, 16: fp + 1-340.

HICKMAN, J. R. 1930. Life-histories of Michigan Haliplidae (Coleoptera). Papers of the Michigan Academy of Science, Arts and Letters, 11: 399-424 + plates xlvii-lv.

HICKMAN, J. R. 1931a. Respiration of the Haliplidae (Coleoptera). Papers of the Michigan Academy of Science, Arts and Letters, 13 (1930): 277-289.

HICKMAN, J. R. 1931b. Contribution to the biology of the Haliplidae (Coleoptera). Annals of the Entomological Society of America, 24: 129-142.

HILSENHOFF, W. L. and BRIGHAM, W. U. 1978. Crawling water beetles of Wisconsin (Coleoptera: Haliplidae). The Great Lakes Entomologist, 11: 11-22.

HOLMEN, M. 1987. The aquatic Adephaga (Coleoptera) of Fennoscandia and Denmark. I. Gyrinidae, Haliplidae, Hygrobiidae and Noteridae. Fauna Entomologica Scandinavica, 20: 1-168.

KAVANAUGH, D. H. 1986. A systematic review of amphizoid beetles (Amphizoidae: Coleoptera) and their phylogenetic relationships to other Adephaga. Proceedings of the California Academy of Sciences, 44: 67-109.

KITAYAMA, C.Y. 1981. *Apteraliplus parvulus* Roberts (Coleoptera: Haliplidae), an obscure and possibly endangered species. Atala, 7 (1979): 12-14.

KUKALOVÀ-PECK, J. and LAWRENCE, J. 1993. Evolution of the hind wing in Coleoptera. The Canadian Entomologist, 125: 181-258.

LEECH, H.B. and CHANDLER, H. P. 1956. Aquatic Coleoptera. pp. 293-371 in Usinger, R. L. Editor. Aquatic insects of California with keys to North American genera and California species. University of California Press, Berkeley and Los Angeles. ix + 508 pp.

MATHESON, R. 1912. The Haliplidae of North America, north of Mexico. Journal of the New York Entomological Society, 20: 156-192, plates x-xv.

MATTA, J.F. 1976. The Insects of Virginia. No. 10. The Haliplidae of Virginia (Coleoptera: Adephaga). Research Division Bulletin 109. Virginia Polytechnic Institute and State University, Blacksburg, VA 24061. vi + 26 pp.

PETERSON, A. 1951. Larvae of Insects. An introduction to Nearctic species. Part II. Coleoptera, Diptera, Neuroptera, Siphonaptera, Mecoptera, Trichoptera. Privately printed. Columbus, Ohio. v + 416 pp.

ROBERTS, C.H. 1913. Critical notes on the species of Haliplidae of America north of Mexico with descriptions of new species. Journal of the New York Entomological Society, 21:91-123.

ROUGHLEY, R.E. 1991a. Family Haliplidae. pp. 60-61. *In*: Bousquet, Y. Editor. Checklist of the Beetles of Canada and Alaska. Publication 1861/E, Research Branch, Agriculture Canada, Ottawa, Ontario.

ROUGHLEY, R. E. 1991b. *Brychius hungerfordi* Spangler (Coleoptera: Haliplidae), the first record from Canada with notes about habitat. The Coleopterists Bulletin, 45:295-296.

SEEGER, W. 1971a. Morphologie, Bionomie und Ethologie von Halipliden, unter besonder Berücksichtigung functionsmorphologischer Gesichtspunkte (Haliplidae; Coleoptera). Archiv für Hydrobiologie, 68: 400-435.

SEEGER, W. 1971b. Autökologische Laboruntersuchungen an Halipliden mit zoogeographischen Anmerkungen (Haliplidae; Coleoptera). Archiv für Hydrobiologie, 68: 528-574.

SEEGER, W. 1971c. Die Biotopwahl bei Halipliden, zugleich ein Beitrag zum Problem der synoptischen (sympatrischen s. str.) Arten (Haliplidae: Coleoptera). Archiv für Hydrobiologie, 69: 175-199.

SHARP, D. and MUIR, F., 1912. The comparative anatomy of the male genital tube in Coleoptera. Transactions of the Entomological Society of London, 1912: 477-642 + plates 42-78.

SPANGLER, P. J. 1954. A new species of water beetles from Michigan (Coleoptera, Haliplidae). Entomological News, 65: 113-117.

SPANGLER, P.J. 1991. Haliplidae (Adephaga). pp. 311-312. *In*: Lawrence, J.F. Co-ordinator. Chapter 34. Coleoptera. pp. 144-658 in F. Stehr. Editor. Immature Insects. Volume 2. Kendall/Hunt Publishing Co., Dubuque, Iowa.

STICKNEY, F. S. 1923. The head capsule of Coleoptera. Illinois Biological Monographs, 8: 1-104.

TANNER, V. M. 1927. A preliminary study of the genitalia of female Coleoptera. Transactions of the American Entomological Society, 53: 5-50.

WALLIS, J. B.. 1933. Revision of the North American species, (north of Mexico), of the genus *Haliplus,* Latreille. Transactions of the Royal Canadian Institute, 19: 1-76.

WILLIAMS, I. W. 1938. The comparative morphology of the mouthparts of the order Coleoptera treated from the standpoint of phylogeny. Journal of the New York Entomological Society, 46: 245-289.

ZACK, R.S. 1991. *Apteraliplus parvulus* (Roberts) (Coleoptera: Haliplidae) in the Pacific Northwest. Proceedings of the Entomological Society of Washington, 93: 865-866.

ZIMMERMANN, A. 1920. Haliplidae, pp. 297-321. *In*: Schenkling, S. (ed.), Coleopterorum Catalogus. Pars 71. Dytiscidae, Haliplidae, Hygrobiidae, Amphizoidae. W. Junk, Berlin.

ZIMMERMANN, A. 1924. Die Halipliden der Welt. Entomologische Blätter, 20: 1-16, 65-80, 129-144, 193-213.

# 9. TRACHYPACHIDAE C.G. Thomson, 1857

by George E. Ball

**Family common name**: The false ground beetles

Nearctic adult trachypachids present a somber appearance, that is, uniformly dark color. Range in size is slight, adults being toward the smaller size for beetles. The adults look like small carabids, but are distinguished from members of that family by the smooth, non-pubescent flagellar articles of the antennae, and by the large hind coxae that are extended to the lateral margins of the body. For explanation of morphological and zoogeographical terms that might be unfamiliar, see the prefatory remarks in Family 6, Carabidae.

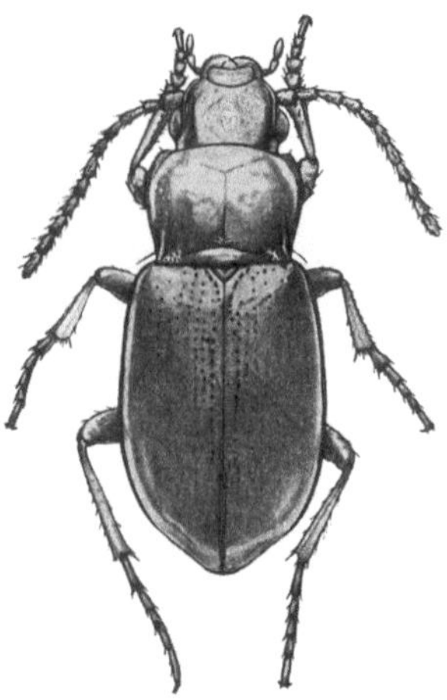

FIGURE 1.9 *Trachypachus zetterstedtii* (Gyllenhal) (From Lindroth 1961).

**Description**: Adults. Shape rather oval in dorsal aspect, terete in cross section, pronotum with lateral margins rounded, prothorax narrower than elytra; length ca. 4-7 mm; color black, immaculate, with distinct metallic (bronze or coppery) luster, or bright green; elytra uniformly glossy or with dull areas alternating with glossy areas. Surface without distinct pelage. Setation sparse, with a definite number of precisely located (fixed) tactile setae.

Head narrower than prothorax, prominent, prognathous, dorsally with two pairs of supraorbital setae. Eyes moderately prominent. Antennae inserted between eyes and base of mandibles under a frontal ridge, filiform, with eleven antennomeres. Labrum distinctly transverse, with apical margin subtruncate, preapically with row of 12 setae; epipharynx not studied. Mandibles prominent, apices (incisor teeth) rather blunt, laterally with (*Systolosoma*) or without (*Trachypachus*) scrobal setae, occlusal margins toothed; retinaculum large, prominent, in occlusal aspect transverse, with two ridges (inferior and superior) directed posteriorly (Acorn and Ball 1991: 640, Figs. 1A-D). Maxillae with occlusal margin of lacinia more or less densely trichiate, apically with prominent tooth; galea palpiform, with two articles; palpus of four articles, palpomeres of similar shape, rather short and broad, palpomere 4 with apical margin subtruncate. Labium with submentum and mentum separated by a suture; mentum anteriorly bi-emarginate, medially with tooth; ligula with glossal sclerite prominent, sclerotized, on each side laterally with an apically membranous paraglossa; palpus of 3 articles, palpomere 3 with apical margin subtruncate.

Thorax. Prothorax narrower than paired elytra together, with lateral margins inflexed and a distinct submarginal suture between proepipleuron and propleuron; pronotum with lateral margins narrowly reflexed; lateral setae three or more pairs. Prosternum with intercoxal process projected prominently, posteriorly. Front coxal cavities open, confluent, bridged; middle coxal cavities disjunct, confluent; hind coxal cavities incomplete, confluent (Bell 1967: 105); metepimera posteriorly not lobate, evident only laterally. Mesosternum with hexagonal groove posteriorly deeply excavated (Beutel 1994: 165).

Elytra margined laterally and basally, humeri broadly rounded, or rectangular; posteriorly not sinuate, apical margin narrowly rounded; dorsal surface striate, each stria linear, each represented by row of punctures, or striae absent laterally, surface smooth. Tactile setae (fixed in position) absent.

Metathoracic wings typical of Adephaga, venation not distinctive, but subcubital binding patch present in *Trachypachus* (Hammond 1979: 171) but absent from *Systolosoma* (Beutel 1994: 161); folding pattern typical for Adephaga.

Legs gressorial, fairly short and thick; front and middle coxae globular, hind coxae dilated internally, and extended each side to lateral margins of body; front tibia isochaetous, with antenna cleaner sulcate, Grade B (Hlavac 1971: 57); tarsi each with five tarsomeres, tarsomere 5 terminated by pair of claws; claws with inner margins smooth; males with front tarsomeres 1 and 2 and middle tarsomere 1 expanded, each ventrally with pad of adhesive squamo-setae (Stork 1980).

Abdomen with six pregenital sterna (II-VII), sternum II (first visible sternum) interrupted by hind coxae, remnants visible only at sides. Sterna IV-VI each with one pair of ambulatory setae; sternum VII of males with single pair of setae near posterior margin; females with two pairs of setae.

Male genitalia trilobed, median lobe of moderate size, parameres rather long, symmetrical, apical portion of each sparsely setose; basal piece entirely membranous or absent; internal sac (endophallus) short, slightly developed (Kavanaugh 1986: 93).

Female genitalia: ovipositor (Deuve 1993: 123, Figs. 165-166) with sternum little differentiated; gonocoxa IX of one article, without ensiform setae, but with subapical setose organ, including nematiform setae; capable of antero-posterior motion only, gonocoxae not folded dorsoventrally in repose; spermatheca 1 present, spermatheca 2 absent (Liebherr and Will 1998: 124, Fig. 4, and 169.

Larvae (based primarily on description of *T. gibbsii* LeConte 1861 [Thompson 1979: 230-231]). Campodeiform, slightly flattened, distinctly segmented, and completely or partially sclerotized on dorsum and to a lesser degree on venter. Head prominent, exserted, prognathous, with pair of egg bursters (Arndt 1993); epicranial suture delimiting fused frons, clypeus and labrum; cephalic margin of frons serrate or deeply dentate; nasale prominent, with eight equally spaced teeth of equal size. Antennae prominent, with four articles; antennomere III without lateral appendage, but with ventral sensorial field (Arndt and Beutel 1995: 445). Ocelli six on each side of head. Mouthparts: mandibles cultriform, three or more times as long as wide at base, with distinct tooth-like retinaculum on occlusal surface, without a membranous or setose or spine-like basal structure (penicillus); maxillae prominent, cardo small, stipes moderately prominent, palpus of three articles, palpomere III with additional setae (Arndt and Beutel, l.c.), galea with two articles, lacinia absent; labium without prementum, labial palpus prominent, palpomeres two. Thorax large. Legs short, each of six articles, with or without prominent spines; claws moveable, paired. Abdomen ten-segmented with urogomphi much shorter than segment X, stout, reflexed apically, fixed (not articulated); tergite IX with long setae (Arndt and Beutel, l.c.),

**Habits and habitats.** Trachypachids probably are olfactory-tactile predators or scavengers, as are most carabids (Brandmayr and Zetto Brandmayr 1980). They are day-active. Trachypachids are mesophilous to xerophilous geophiles, living in coniferous or deciduous forests, the adults resting in, or active in, leaf litter. Specimens of *Trachypachus holmbergi* have been collected in synanthropic situations also, such as urban flower gardens.

**Status of the classification.** The most interesting feature of this small family is its uncertain phylogenetic relationships (Bousquet and Larochelle 1993: 19). This topic is not reviewed in detail here, but instead references are provided for use of interested readers. Ball (1960: 95) and Lindroth (1961: 1) included trachypachids in the Carabidae, as a monotribal subfamily, although Crowson (1955: 6-7) had concluded that this group should be ranked as a family. Bell (1964: 61) proposed that trachypachids be included with the paussines and metriines in a subfamily Paussinae, equivalent to the Isochaeta of Jeannel (1941: 78). Bell (1966: 111-112), Hammond (1979: 171-174), Roughley (1981), Evans (1983: 604), Nichols (1985), Ponomarenko (1992: 60), Beutel (1995), and Beutel and Haas (1996) placed the trachypachids in a separate family, but either with the amphizoids and dytiscids, or between these water beetle groups and the Carabidae. Erwin (1985: 446, Fig. 3) and Kavanaugh (1986: 81-98) ranked the trachypachids as a family, but related to the Carabidae as adelphotaxon. Bell (1983) presented five hypotheses about relationships of trachypachids, but stated that the available evidence in favor of any one of them was equivocal. Fifteen years later, Beutel (1998: 101), after reviewing the accumulated evidence, and the phylogenetic analyses based on that evidence, concluded that "the position of Trachypachidae is not fully clarified," though he did recognize the group as a monophyletic, and ranked it as a family. For the present, then, trachypachids are placed most appropriately between the Carabidae and the hydradephagan dytiscoids.

Ponomarenko (1992: 60) included the extant trachypachids as a subfamily with the Triassic-Lower Cretaceous Eodromeinae, a group of 20 species in seven genera, all described by him. However, Beutel (1998: 83) wrote: "The affinities between eodromeine genera and Trachypachidae are unclear because of no apparent synapomorphic character states." For this reason, the eodromeines are ranked here as a family and thus removed from the Trachypachidae, this latter group then including only presently extant taxa.

The six species of Trachypachididae are included in two genera: *Trachypachus* Motschulsky, 1844; and *Systolosoma* Solier, 1849.

**Distribution.** This family is amphitropical, with *Trachypachus* in north temperate-subarctic Holarctic and *Systolosoma* far to the south, in Chile, in the temperate reaches of the Neotropical Region (Reichardt 1977: 376; Erwin 1985: 459, Fig. 9).

## Classification

Trachypachidae

*Trachypachus* Motschulsky 1844

Holarctic, *Trachypachus* includes four species: *T. zetterstedtii* (Gyllenhal 1827), precinctive in the Palaearctic Region (coniferous region, Norway to eastern Siberia [Lindroth 1985: 42]); and *T. holmbergi* Mannerheim 1853, *T. gibbsii* and *T. slevini* Van Dyke 1925, precinctive in the western part of the Nearctic Region, with *T. holmbergi* ranging from California northward to southern Alaska, and eastward as far as Saskatchewan and Montana. Reference. Lindroth 1961: 1-4.

## Bibliography


ACORN, J.H. and BALL, G.E. 1991. The mandibles of some adult ground beetles: structure, function and the evolution of herbivory. Canadian Journal of Zoology, 69: 638-650.

ARNDT, E. 1993. Phylogenetische Untersuchungen larvalmorphologischer Merkmale der Carabidae (Insecta: Coleoptera). Stuttgarter Beiträge zur Naturkunde Serie A (Biologie), No. 488: 1-56.

ARNDT, E. and BEUTEL, R.G. 1995. Larval morphology of *Systolosoma* Solier and *Trachypachus* Motschulsky (Coleoptera: Trachypachidae) and phylogenetic considerations. Entomologica Scandinavica, 26: 439-446.

BALL, G.E. 1960a. Carabidae, Fascicle 4, pp. 55-210. *In*, The beetles of the United States (A manual for identification) (Arnett, R.H., Jr.). The Catholic University of America Press, Washington, D. C. xi + 1112 pp.

BELL, R.T. 1964. Does *Gehringia* belong to the Isochaeta? Coleopterists Bulletin, 18: 59-61.

BELL, R.T. 1966. *Trachypachus* and the origin of the Hydradephaga. Coleopterists Bulletin, 20: 107-112.

BELL, R.T. 1967. Coxal cavities and the classification of the Adephaga (Coleoptera). Annals of the Entomological Society of America, 60: 101-107.

BELL, R.T. 1983. What is *Trachypachus?* (Coleoptera: Trachypachidae), pp. 590-596. *In*, The Baron Maximilien de Chaudoir (1816-1881): a symposium to honor the memory of a great Coleopterist during the centennial of his death. (D.R. Whitehead, Ed.). The Coleopterists Bulletin, 36 (1982): 459-609.

BEUTEL, R.G. 1994. Study on the systematic position of *Systolosoma breve* Solier (Adephaga: Trachypachidae) based on characters of the thorax. Studies on Neotropical Fauna and Environment, 29: 161-167.

BEUTEL, R.G. 1995. The Adephaga (Coleoptera): phylogeny and evolutionary history, pp. 173-217. *In*, Biology, Phylogeny, and Classification of Coleoptera: Papers Celebrating the 80th Birthday of Roy A. Crowson (Pakaluk, J. and Slipinski, S. A., Eds.). Muzeum i Instytut Zoologii PAN, Warsaw. Two vols., XII + 1092 pp.

BEUTEL, R.G. 1998. Trachypachidae and the phylogeny of Adephaga (Coleoptera), pp. 81-105. *In*, Phylogeny and classification of Caraboidea (Coleoptera: Adephaga) (G.E. Ball, Casale, A. and Vigna Taglianti, A., Eds.). Atti, Museo Regionale di Scienze Naturali, Torino, Italy. 543 pp.

BEUTEL, R.G. and HAAS, A. 1996. Phylogenetic analysis of larval and adult characters of Adephaga (Coleoptera) using cladistic computer programs. Entomologica Scandinavica, 27: 197-205.

BOUSQUET, Y. and LAROCHELLE, A. 1993. Catalogue of the Geadephaga (Col. Trachypachidae, Rhysodidae, Carabidae, incl. Cicindelini) of America north of Mexico. Entomological Society of Canada, Memoir No. 167. 395 pp.

BRANDMAYR, P. and ZETTO BRANDMAYR, T. 1980. "Life forms" in imaginal Carabidae (Coleoptera): a morphofunctional and behavioural synthesis. Monitore Zoologia Italiana (N.S.), 14: 97-99.

CROWSON, R.A. 1955. The natural classification of the families of Coleoptera. Nathaniel Lloyd and Co., Ltd., London. 187 pp.

DEUVE, T. 1993. L'abdomen et les genitalia des femelles de Coléoptères Adephaga. Mémoires du Muséum National d'Histoire Naturelle, 155: 1-184.

ERWIN, T.L. 1985. The taxon pulse: a general pattern of lineage radiation and extinction among carabid beetles, pp. 437-493. *In*, Taxonomy, Phylogeny and Zoogeography of beetles and Ants (Ball, G. E., Ed.). Dr. W. Junk Publishers, Dordrecht. XIII + 514 pp.

EVANS, M.E.G. 1983. Early evolution of the Adephaga-some locomotor speculations, pp. 597-607. *In*, The Baron Maximilien de Chaudoir (1816-1881): a symposium to honor the memory of a great Coleopterist during the centennial of his death. (Whitehead, D.R., Ed.). Coleopterists Bulletin, 36 (1982): 459-609.

HAMMOND, P.M. 1969. Wing-folding mechanisms of beetles, with special reference to adephagan phylogeny, pp. 113-180. *In*, Carabid beetles: their evolution, natural history, and classification (T. L. Erwin, Ball, G. E., Whitehead, D. R. and Halpern, A. L., Eds). Dr W. Junk bv Publishers, The Hague, The Netherlands. X + 644 pp.

HLAVAC, T. 1971. Differentiation of the carabid antenna cleaner. Psyche, 78: 51-66.

JEANNEL, R. 1941. Coléoptères carabiques, première partie. Faune de France, 39: 1 -571. Paris.

KAVANAUGH, D.H. 1986. A systematic review of amphizoid beetles (Amphizoidae: Coleoptera) and their phylogenetic relationships to other Adephaga. Proceedings of the California Academy of Sciences, 44: 67-109.

LIEBHERR, J.K. and WILL, K.W. 1998. Inferring phylogenetic relationships within the Carabidae (Insecta, Coleoptera) from characters of the female reproductive tract, pp. 107-170. *In*, Phylogeny and classification of Caraboidea (Coleoptera: Adephaga). (G.E. Ball, Casale, A. and Vigna Taglianti, A., Eds.). Atti Museo Regionale di Scienze, Torino. 543 pp.

LINDROTH, C.H. 1961. The ground-beetles (Carabidae, excl. Cicindelinae) of Canada and Alaska. Part 2. Opuscula Entomologica Supplementum No. 20. pp. 1-200.

LINDROTH, C.H. 1985. The Carabidae (Coleoptera) of Fennoscandia and Denmark. Fauna Entomologica Scandinavica, 15 (1): 1-225. E.J. Brill/ Scandinavian Science Press Ltd., Leiden.

NICHOLS, S.W. 1985. *Omophron* and the origin of Hydradephaga (Insecta: Coleoptera: Adephaga). Proceedings of the Academy of Natural Sciences of Philadelphia, 137: 182-201.

PONOMARENKO, A.G. 1992. Descriptions of new taxa, suborder Adephaga, pp. 5-130. *In*, Mesozoic Coleoptera (Arnol'di, L. V., Zherikhin, V. V., Nikritin, L. M. and Ponomarenko, A. G.) (Vandenberg, N. J., Scientific Editor). Amerind Publishing Co., New Delhi. viii + 285 pp. [English translation of "Mezozoiskie Zhestkokrylye"; published originally in 1977, by Akademiya Nauk SSSR].

REICHARDT, H. 1977. A synopsis of the genera of Neotropical Carabidae (Insecta: Coleoptera). Quaestiones Entomologicae, 13: 346-493.

ROUGHLEY, R.E. 1981. Trachypachidae and Hydradephaga (Coleoptera): a monophyletic unit? Pan-Pacific Entomologist, 57: 273-285.

STORK, N.E. 1980. A scanning electron microscope study of tarsal adhesive setae in Coleoptera. Zoological Journal of the Linnean Society, 68: 173-306.

THOMPSON, R.G. 1979. [Chapter] 2.26. Larvae of North American Carabidae with a key to tribes, pp. 209-291. *In*, Carabid beetles: their evolution, natural history, and classification (T. L. Erwin, Ball, G. E., Whitehead, D. R. and Halpern, A. L., Eds). Dr W. Junk bv Publishers, The Hague, The Netherlands. X + 644 pp.

# 10. NOTERIDAE C. G. Thomson, 1857

by R. E. Roughley

**Family common name**: The burrowing water beetle family

Members of this family are similar to the Dytiscidae, and burrowing water beetles are not easily distinguished from the predaceous water beetles. The main features for recognition of Noteridae are lack of a visible scutellum, relatively small to moderate size (1 to 5.8 mm) with protibia fitting into an excavation on the ventral margin of the profemur, large hind coxal plates, presence of a large sternal platform called the noterid platform (Fig. 5.10); streamlined body form which is flat ventrally and convex to very convex dorsally and a tarsal formula which is clearly 5-5-5.

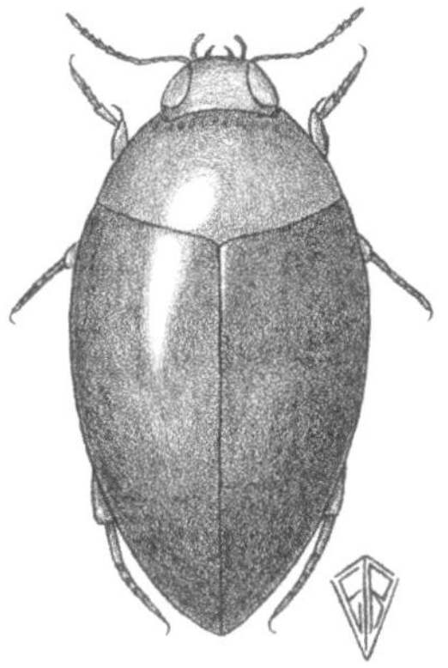

FIGURE 1.10. *Hydrocanthus iricolor* Say.

**Description.** See Belkaceme (1991) and Beutel and Roughley (1987) for detailed description of structure of noterids. The following is abstracted for Nearctic genera only. Shape oval, broadly ovate to fusiform; length 1 to 5 mm; color light brownish to piceous to dark reddish brown, with or without distinct elytral punctation; devoid of obvious setae except long swimming hairs on tibiae and tarsi.

Head prognathous, short, somewhat retracted into prothorax, dorsally convex, surface punctate but without setae except for row of short setae on labrum. Eyes forming a smooth curve from frons to pronotum (in dorsal view), outline not emarginate. Antennae with 11, glabrous antennomeres; 1st antennomere (scape) spherical and appearing two-segmented, 2nd antennomere (pedicel) elongate, remainder filiform; antennomeres 3 to 11 of increasing length and slightly longer than wide, often of varying shape, males of some groups with widened antennomeres, inserted between the anterior portion of the eyes and the bases of the mandibles. Mouthparts with labrum transverse and somewhat narrowed with row of short setae on the anterior margin, covering most of the mandibles. Clypeus fused to frons, frontoclypeal suture not visible. Mandibles short and robust, asymmetrical, with a subapical and two apical teeth, retinaculum rounded, with a mesal comb or brush of setae. Maxilla with lacinia with spines along inner margin and an elongate two-segmented galea, maxillary palpus four-segmented with a basal palpiger, apical palpomere enlarged (Fig. 5.10) (except fusiform in *Notomicrus*, Fig. 4.10). Labium with broad anterolateral lobes, ligula present and small but extending between the bases of the labial palps, labial palpus short, moderately stout, three segmented with a basal palpiger, apical palpomere enlarged. Median gular apodeme present (except absent in *Notomicrus*).

Pronotum widest at base, narrower in front, excavated for head, without pubescence, coarsely punctate to impunctate, lateral margins convex, posterior margin sinuate with median, V-shaped posterior extension. Ventral portion of prothorax with notopleural and pleurosternal sutures. Prosternum flat to somewhat swollen near base (most genera) or with base elevated to form a kind of ventral head rest, in repose, (*Suphis*); prosternal process elongate and rounded apically (*Notomicrus*, Fig. 4.10, *Pronoterus*, *Mesonoterus*) or broad to very broad apically and forming a complex junction (*Suphis*, *Suphisellus* and *Hydrocanthus*, Fig. 5.10) with the mesoventrite (=mesosternum, of authors), laterally with or without distinct, raised lines. Procoxal cavities elongate oval to oval (*Notomicrus, Pronoterus, Mesonoterus*) or round (*Suphis, Suphisellus* and *Hydrocanthus*), open behind externally but closed internally. Procoxae elongate oval to globular, trochantin not visible. Protrochanter triangular and joined to basoventral portion of femur. Profemur highly modified; slender and elongate (*Notomicrus*, Fig. 2.10, *Pronoterus* and *Mesonoterus*) to short and thick (*Suphis, Suphisellus* and *Hydrocanthus*), ventral margin excavated for protibia, excavation delimited on anterior and posterior margin by distinct ridges, with a row of distinctly spaced, stiff, elongate setae on anterodorsal and anteroventral surfaces, dorsal surface with sparse natatory setae; apicoventral margin with a distinct notch with short, dense setae used as an antennal cleaner. Protibia slender and elongate (*Notomicrus*, Fig. 2.10, and *Mesonoterus*) to broadly triangular (*Pronoterus*), to short and thick (*Suphis, Suphisellus* and *Hydrocanthus*), with rows of short, stiff setae angularly across anterior surface, and sparse row of elongate spines (*Notomicrus*) or row of short, uniform, dense spines (remaining genera) along ventral and/or apical margins, row of long, natatory setae on dorsal surface (very sparse or absent in *Notomicrus*); protibial spurs not differentiated (*Notomicrus*) or with (most genera) spur greatly enlarged and hook-like, used for burrowing through substrate. Protarsi unmodified (*Notomicrus*) with protarsomere 1 very short and

**Acknowledgments.** I thank the late Frank N. Young of Bloomington, Indiana, for his enthusiasm, knowledge, and the gift of important specimens. They remain appreciated and extremely useful. The research of Torsten Belkaceme (Tübingen) stands as an example of the reciprocal illumination of morphology and systematics.

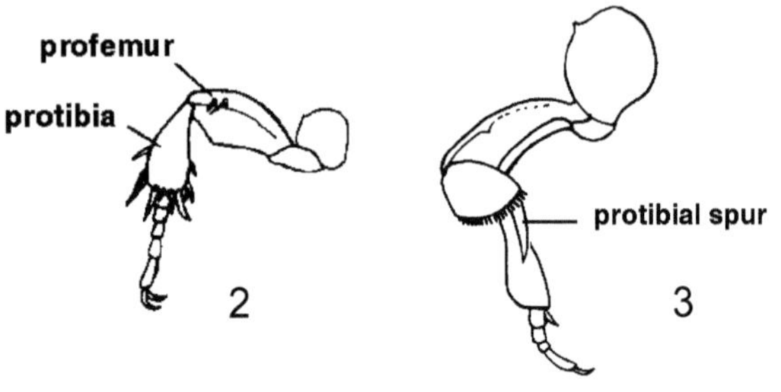


FIGURES 2.10 - 3.10. Fig. 2.10, *Notomicrus* sp., fore leg; Fig. 3.10, *Hydrocanthus* sp., fore leg.

protarsomere 5 longest, or modified (remaining genera) short and thick, with protarsomere 1 enlarged and incrassate, longer than protarsomeres 2 to 4 combined, protarsomeres 2, 3 and 4 about as wide as long, protarsomere 5 subequal in length to protarsomere 1 or much shorter, with sparse row of long, natatory setae on dorsal surface of basal protarsomeres; males with protarsomeres 1 to 3 perceptibly to distinctly broader, apex of protarsomere 1 and ventral surface of protarsomeres 2 and 3 with some short, unmodified setae or with enlarged articulo-setae; two rather short claws present on tarsomere 5.

Mesoventrite short, mostly concealed by prosternal process; mesocoxal cavity round, limited laterally by mesepimeron. Mesocoxae round with limited range of motion. Mesotrochanter short and triangular and fused to basoventral portion of femur. Mesofemur elongate and narrow (*Notomicrus, Pronoterus, Mesonoterus*) to short and very wide (remaining genera); with poorly to well-defined row of broadly spaced, stiff, elongate spines on anteroventral surface. Mesotibia slender and elongate (*Notomicrus, Mesonoterus*) to shorter and widened apically with sparse row of stiff, elongate spines on anterior and anteroventral surfaces and row of short, dense spines on apical margin, with row of long, natatory setae on dorsal surface. Mesotarsi slender and elongate, mesotarsomere 1 longer than mesotarsomeres 2, 3 and 4 which are longer than wide; mesotarsomere 5 longest, with row of long, natatory setae on dorsal margin of basal mesotarsomeres; males with mesotarsomeres 1 to 3 perceptibly to distinctly broadened, apex of mesotarsomere 1, and ventral surface of mesotarsomeres 2 and 3 with some short, unmodified setae or with enlarged articulo-setae; two rather short claws on mesotarsomere 5.

Metaventrite (=metasternum, of authors) (Beutel and Belkaceme 1986) large and extended toward epipleuron but limited laterally by metepimeron, with discrimen (=median metasternal suture, of authors); with transverse suture absent; without (*Notomicrus*, Fig. 4.10) or with (remaining genera, Fig. 5.10) suture separating metacoxae and metaventrite (Beutel and Roughley 1988); anteromedially with broad extension (=anterior metasternal process, of authors), partially covering inner margins of mesocoxae (except *Notomicrus*) and forming a junction with the prosternal process. Metasternum medially forming an extension of the noterid platform (except *Notomicrus*). Metacoxae greatly enlarged and extending laterally to epipleuron and elongate on midline; differentiated into inner and outer lamina which are on different planes, inner lamina depressed ventrally and forming a platform (=noterid platform) with base of hind legs moving between inner and outer lamina. Metatrochanter small and joined to basoventral portion of metafemur. Metafemur thin and elongate with a sparse row of short, spines on ventral margin *(Notomicrus)* to shorter and broader (remaining genera) with a few apicoventral elongate setae (*Mesonoterus, Pronoterus, Suphis*) or an apicoventral comb of setae (*Suphisellus*, and *Hydrocanthus*). Metatibia long and slender (most genera) to short and wide (*Hydrocanthus*), with irregular rows of stiff setae on anterior, posterior and ventral surfaces, row of short, dense spines on apical margin, and row of long natatory setae on dorsal surface and ventral surfaces; metatibial spurs short. Metatarsi slender and elongate, with row of long natatory setae on dorsal surface of basal metatarsomeres; metatarsomere 1 longest, metatarsomeres 2, 3, and 4 which are about as long as wide, protarsomere 5; two short claws present on apex of metatarsomere 5.

Scutellum concealed. Elytra with sides evenly arcuate or tapering, base as broad as base of pronotum, widest in basal quarter, apices rounded to acuminate, surface with well-impressed microreticulation on most species, without discrete rows of punctures (most genera) or glabrous with only three rows of punctures (*Hydrocanthus)*; sutural stria absent; epipleuron very broad basally, abruptly narrowed at about level of hind margin of metacoxa. Wing venation typically adephagan with oblongum cell (Kukalovà-Peck and Lawrence 1993, Fig. 19); folding pattern as in Hygrobiidae and part of Dytiscidae.

Abdomen with five or six apparent sternites (true sternites 2-7), surface punctate or smooth, visible sternite 1 (=true sternite 2) divided by metacoxa and obscured by apex of lateral margin of noterid platform, visible sternites 2 and 3 fused and without sutures, or with poorly defined suture visible only laterally; apical sternite (=true sternite 7) rounded, extended distally medially or pointed.

Male genitalia caraboid, penis symmetrical, laterally compressed and apex broad and finger-like or pointed, with a sperm channel on the convex surface; parameres asymmetrical. Male genitalia with torsion (90° rotation at rest, 180° during copulation) characteristic of most of Adephaga (Beutel and Roughley 1988).

Female genitalia (Burmeister 1976) with genital appendages VIII separate, triangular and apically pointed, withdrawn into the abdomen and lying above sternite 7; gonocoxasterna triangular without anterior apodeme; terga IX forming a narrowed and somewhat elongated support for the distal, pencil-shaped gonocoxae which have apical, sensory setae.

Larvae (Bøving and Craighead 1931, Holmen 1987, Lawrence *et al.* 1993, Peterson 1951, Spangler 1991) elongate, fusiform and cylindrical to parallel-sided to moderately to very broad at midlength; known mature larvae range in size from 2 to 4.5 mm; color yellowish-brown or reddish-brown, surface without pronounced microsculpture. Head prognathous and exserted but retracted into base of pronotum, without tempo-

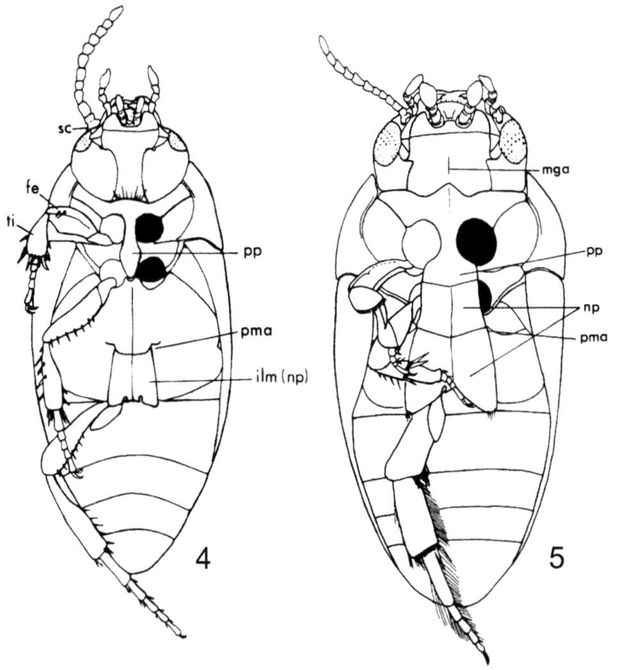


FIGURES 4.10-5.10. Fig. 4.10, *Notomicrus* sp., ventral view, Fig. 5.10, *Hydrocanthus* sp., ventral view.

ral spines or setae; distinct neck region absent; frontoclypeal suture absent, anterior margin concave; antennae short to very short, four-segmented, with apical antennomere long, without sensory appendage; labrum fused to head capsule; mandible short, without mola, with or without retinacular teeth, inner margin smooth, serrate or toothed, without feeding channel or channel present only basally; maxilla short and broad, cardo small, stipes long and broad, lacinia absent, galea two-segmented and articulated with palpifer; maxillary palpus short, three-segmented. Labium with prementum small, somewhat rectangular, without ligula, labial palpus short, three-segmented; six pairs of conspicuous stemmata. Gula wider than long, represented by a short, broad strip between gular sutures, posterior tentorial pits clearly separated and shifted posteriorly almost to posterior edge of head. Pro-, meso- and metathoracic tergites well sclerotized; prothoracic tergite about as long as to shorter than meso- and metathoracic terga combined. Legs six-segmented with apical segment (=pretarsus) consisting of a pair of claws; natatory setae absent; overall form short and robust, modified for burrowing. Abdomen with eight visible segments, apical abdominal segment (=segment 8) elongate and more or less acuminate apically to obtain air from aquatic plants; urogomphi one-segmented and very short.

**Habits and habitats**. Both the larvae and the adults are aquatic and are commonly collected in water bodies with aquatic vascular plants. Pupation takes place under water in an airtight pupal cell prepared by the larva. Larvae and adults burrow through the substrate on the bottom of ponds and marshes as well as more temporary, lentic habitats. Some larvae are known to use the last abdominal segment to obtain air from aquatic plants. Larvae and adults are primarily carnivorous, feeding on chironomid larvae, insect eggs, etc., although they may scavenge also on dead insects. The life cycle is not known for any Nearctic species.

**Status of the classification**. The Noteridae were formerly classified as a subfamily of Dytiscidae but they are a distinct family (Belkaceme 1991; Beutel 1995, 1997; Beutel and Roughley 1987, 1988; Crowson 1955; Young 1954). It is a monophyletic unit quite separate from Dytiscidae; the family belongs to the superfamily Dytiscoidea where it is the sister group of the remaining families: Noteridae + (Amphizoidae + (Hygrobiidae +Dytiscidae)). The Dytiscoidea are an independent invasion of the water from other aquatic Adephaga (Beutel 1995, 1997; Beutel and Roughley 1988) but for an alternative view of the phylogentic position of Noteridae see Kavanaugh (1986). Crowson (1955) emphasized the differences between larval noterids and larval dytiscids to suggest family status whereas other authors (Belkaceme 1991, Young 1954) have emphasized adult features.

Pederzani (1995) included 12 genera in his key to genera of Noteridae of the world (as Noterinae of Dytiscidae). Belkaceme (1991) and Beutel and Roughley (1987) have placed Phreatodytidae (Uéno 1957) as a junior synonym of Noteridae with the single included genus *Phreatodytes* Uéno (1957) as a basal member of Noteridae. The majority of conventional tribes (e.g., Pederzani 1995, Zimmermann 1920) are paraphyletic within Noteridae and are not used here. Spangler and Folkerts (1973) provided a key to three genera of Neartic Noteridae (larvae of *Notomicrus*, *Pronoterus*, and *Mesonoterus* are undescribed).

Young (1978, 1979, 1985) has provided keys to most of the New World species of *Notomicrus*, *Suphisellus* and *Hydrocanthus*, respectively. The principally Neotropical genera, *Pronoterus* and *Mesonoterus*, each with one species extending into the Nearctic region remain in need of revision. Excellent regional works are available for North and South Carolina (Brigham 1982), Florida (Epler 1996), Wisconsin (Hilsenhoff 1992), and California (Leech and Chandler 1956).

**Distribution**. The family is worldwide in distribution. The species are more common in tropical latitudes and increasingly less common toward the temperate zone. The most recent world catalogue is Zimmermann (1920 as a subfamily of Dytiscidae) with 143 species; today there are 12 genera and about 230 known species worldwide with about six genera and 14 species known from the United States and Canada (Spangler 1991, Epler 1996). Arce-Pérez and Roughley (1999) listed 16 species of noterids for Mexico; Roughley (1991) provided a list of the two species of Canada and Alaska.

### Key to the Nearctic Genera

1. Protibia at inner apex with conspicuous, enlarged spur (Fig. 3.10); larger specimens, total length > 1.9 mm ........................................................................ 2

— Protibia at inner apex without conspicuous, enlarged spur, anterior margin with series of distinct spines (Fig. 2.10); smaller specimens, total body length 1.0 to 1.6 mm .................................. *Notomicrus*

2(1). Body form very high and broad, almost hemispherical; prosternal process raised and margined and continued basally almost to head, forming a head rest, in repose; total body length 3.0 to 3.5 mm ................................................................ *Suphis*

— Body form convex and rounded but not hemispherical; prosternal process not raised and not extended basally ........................................................... 3

3(2). Protibia at ventral (inner) apex with spur large, curved and conspicuous and dorsal (outer) apex without spine; protibia with dorsal angle rounded, apex rounded and with apical and dorsal rows of short, uniform spines contiguous (Fig. 3.10) .............. 4

— Protibia at ventral (inner) apex with spur less developed and less conspicuous and dorsal (outer) apex with distinct spine; protibia with apex truncate, dorsal angle acute, and with only apical row of short distinct spines; total body length = 2.7 to 3.0 mm ....................................................... *Mesonoterus*

4 (3). Prosternal process broadly rounded apically; metafemur on ventral (posterior) margin at apex with some few, isolated elongate setae; total body length = 2.3 to 2.6 mm ...................... *Pronoterus*

— Prosternal process very broad apically, forming a junction with the noterid platform (Fig. 5.10); metafemur on ventral (posterior) margin at apex with brush or comb of elongate setae .............. 5

5(4). Prosternal process at apex at least twice as wide as width at procoxae, not broader than long; last maxillary palpomere shallowly emarginate at apex; pronotum with lateral bead narrow and of uniform width, with additional, fine submarginal stria originating inside posterolateral hind angle and disappearing before middle; total body length 1.9 to 3.5 mm ................................................... *Suphisellus*

— Prosternal process at apex very broad, two and one-half to three times as wide as width at procoxae, broader than long; last maxillary palpomere obliquely truncate at apex; pronotum with lateral bead very broad anteriorly and narrowed posteriorly and without submarginal stria; total body length 3.0 to 5.8 mm ..................... *Hydrocanthus*

## Classification of the Nearctic Genera

### Noteridae

*Notomicrus* Sharp 1882

Members of this genus are best recognized by their very small size, elongate shape and reddish yellow dorsal coloration. They are most often confused with members of Bidessini (Dytiscidae, Hydroporinae) but adult specimens of *Notomicrus* have the protarsus clearly five-segmented. There are eight species in the New World (Young 1978) of which two species occur in the southeastern United States: *N. nanulus* (LeConte) (southern Georgia, northern and central Florida and Louisiana) and *N. sharpi* Balfour-Browne (southern Florida). They are sometimes quite common at light. They are most abundant in debris, mats of plant roots, or floating aquatic vascular plants. They are rarely collected with a conventional aquatic net, presumably due to their small size. The most effective method is to place portions of substrate from their preferred habitat onto a mesh screen over a pan or into a Berlese funnel.

*Suphis* Aubé 1836

*Colpius* LeConte 1861

The placement of *Colpius* LeConte as a junior synonym of *Suphis* Aubé was proposed by Spangler and Folkerts (1973). There is only one Nearctic species, *S. inflatus* (LeConte, 1863) which is known from South Carolina, Georgia, Florida, Alabama and Louisiana in the United States and from Cuba. It prefers acidic, relatively permanent water and has been taken in sinkhole ponds, lakes and marshes (Young 1954). Adults have a total length of 3.0 to 3.5 mm and they are distinctive due to their highly hemispherical body shape and the form of the apical portion of the prosternum which is depressed ventrally to form a kind of head rest. The shape of the body of the larva which is broadly hemispherical is quite diagnostic (Spangler and Folkerts 1973).

*Pronoterus* Sharp 1882

One Nearctic species, *P. semipunctatus* (LeConte), is assigned to this genus. It is known from Michigan and Florida but the species is probably overlooked in many collections. Epler (1996) and Young (1954) characterized adults of this species as 2.3 to 2.6 mm in length, body not very convex dorsally, outline elongate oval and with the posterior margin more or less rounded, with elytra reddish brown and with several rows of coarse, sparse punctures; males with one, median antennomere slightly widened. It is found in lentic situations such as ponds filled with aquatic vascular plants.

*Mesonoterus* Sharp 1882

One Nearctic species, *M. addendus* (Blatchley), is assigned to this genus; it was formerly assigned to *Pronoterus* (e.g., Young 1954) but Belkaceme (1991) reassigned it to *Mesonoterus*. It is known only from the central and southern parts of the Florida Peninsula and Cuba. Epler (1996) and Young (1954) characterized this species as 2.7 to 3.0 mm in length, body relatively convex above, outline more elongate oval and with the posterior margin somewhat attenuate, with the elytra dark reddish brown to piceous and numerous rows of fine punctures; males with four, median antennomeres widened. It is found in floating, emergent aquatic plants such as water hyacinth in ponds and canals.

*Suphisellus* Crotch 1873

Young (1979) has provided a key to the species of this genus for North, Central and northern South America including the six species found in the United States and Canada. The broad apex of the prosternal process, conspicuous protibial spur, presence of a comb of spines at the ventro-apical margin of the metafemur and with the apical maxillary palpomere shallowly but distinctly emarginate at the tip will serve to separate members of this genus. In addition, adults have the lateral bead of the pronotum fine and of uniform width throughout its length, as well they have a very fine, impressed line originating at the posterior, lateral margin of the pronotum; this line diverges medially from the pronotal bead and before midlength of the pronotum it is not well impressed but continued anteriorly as a series of punctures. Adults and larvae are commonly collected among decaying vegetation and mats of floating vegetation and algae; specimens may be collected from this substrate by sweeping the habitat vigorously with an aquatic net and placing the contents bag onto a wide-meshed screen over a shallow pan. They are common at light.

*Hydrocanthus* Say 1823.

subgenus *Guignocanthus* Young 1985

Young (1985) provided a key to subgenera of the three subgenera of the world and the New World species. There are six species in the United States and Canada with one species assigned to the subgenus *Guignocanthus* and the remaining five in *Hydrocanthus* (*s. str.*). The broad apex of the prosternal process, conspicuous protibial spur, presence of a comb of spines at the ventro-apical margin of the metafemur and with apical maxillary palpomere angularly truncate or shallowly emarginate at the tip emarginate will serve to separate members of this genus. In addition, adults have the lateral bead of the pronotum much broader, particularly wide anteriorly and narrowed posteriorly. Young (1954) considered the form of the pronotal bead in adults of *Suphisellus* (see above) and *Hydrocanthus* to be different states of the same character, however, in all specimens examined they appear distinct and while the two states are subtly different they work very well for separation of these two genera. The habitat and collecting methods for *Hydrocanthus* are the same as for *Suphisellus* (see above).

## Bibliography


ARCE-PÉREZ, R. and ROUGHLEY, R. E. 1999. Lista anotada y claves para los Hydradephaga (Coleoptera: Adephaga: Dytiscidae, Noteridae, Haliplidae, Gyrinidae) de México. Dugesiana, 6: 69-104.

BELKACEME, T. 1991. Skelet und Muskulatur des Kopfes und Thorax von *Noterus laevis* Sturm. Ein Beitrag zur Morphologie und Phylogenie der Noteridae (Coleoptera: Adephaga). Stuttgarter Beiträge zur Naturkunde, Serie A (Biologie), 462:1-94.

BEUTEL, R. G. 1995. The Adephaga (Coleoptera): phylogeny and evolutionary history. pp. 173-217 *In,* Pakaluk, J. and S. A. Slipinski. Editors. Biology, phylogeny, and classification of Coleoptera. Papers celebrating the 80th birthday of Roy A. Crowson. Muzeum i Instytut Zoologii Pan, Warszawa. X + 1092 pp.

BEUTEL, R. G. 1997. Über Phylogenese und Evolution der Coleoptera (Insecta), inbesondere der Adephaga. Abhandlungen des Naturwissenschaftlichen Vereins in Hamburg (N.F.), 31:1-64.

BEUTEL, R. G. and. BELKACEME, T. 1986. Comparative studies on the metathorax of Hydradephaga and Trachypachidae (Coleoptera). Entomologica Basiliensia, 11: 221-229.

BEUTEL, R. G. and ROUGHLEY, R. E. 1987. On the systematic position of the genus *Notomicrus* Sharp (Hydradephaga, Coleoptera). Canadian Journal of Zoology 65: 1898-1905.

BEUTEL, R. G. and ROUGHLEY, R. E. 1988. On the systematic position of the family Gyrinidae (Coleoptera: Adephaga). Zeitschrift für Zoologische Systematik und Evolutionsforschung, 26: 380-400.

BØVING, A. G. and CRAIGHEAD, F. C. 1931. An illustrated synopsis of the principal larval forms of the order Coleoptera. Entomologia Americana (New Series), 11: 1 - 351.

BRIGHAM, W. U. 1982. Noteridae. pp. 10.44-10.47 in Chapter 10. Aquatic Coleoptera. pp. 10.1- 10.136. *In,* Brigham, A. R., W. U. Brigham, and A. Gnilka. Editors. Aquatic Insects and Oligochaetes of North and South Carolina. Midwest Aquatic Enterprises, Mahomet, Illinois. xi + 837 pp.

BURMEISTER, E.-G. 1976. Der Ovipositor der Hydradephaga (Coleoptera) und seine phylogenetische Bedeutung unter besonderer Berucksichtigung der Dytiscidae. Zoomorphologie, 85: 165-257.

CROWSON, R. A. 1955. The natural classification of the families of Coleoptera. N. Lloyd, London, 187 pp.

EPLER, J. H. 1996. Identification manual for the water beetles of Florida (Coleoptera: Dryopidae, Dytiscidae, Elmidae, Gyrinidae, Haliplidae, Hydraenidae, Hydrophilidae, Noteridae, Psephenidae, Ptilodactylidae, Scirtidae). Tallahassee, Florida: Bureau of Water Resource Protection, Florida Department of Environmental Protection.

HILSENHOFF, W. L. 1992. Dytiscidae and Noteridae of Wisconsin (Coleoptera). I. Introduction, key to genera of adults, and distribution, habitat, life cycle, and identification of species of Agabetinae, Laccophilinae and Noteridae. The Great Lakes Entomologist 25:57-69.

HOLMEN, M. 1987. The aquatic Adephaga (Coleoptera) of Fennoscandia and Denmark. I. Gyrinidae, Haliplidae, Hygrobiidae and Noteridae. Fauna Entomologica Scandinavica, 20: 1-168.

KAVANAUGH, D. H. 1986. A systematic review of amphizoid beetles (Amphizoidae: Coleoptera) and their phylogenetic relationships to other Adephaga. Proceedings of the California Academy of Sciences, 44: 67-109.

KUKALOVÀ-PECK, J. and LAWRENCE, J. 1993. Evolution of the hind wing in Coleoptera. The Canadian Entomologist, 125: 181-258.

LAWRENCE, J., HASTINGS, A., DALLWITZ, M., and PAINE, T. 1993. Beetle larvae of the world. Interactive identification and information retrieval for families and sub-families. CSIRO, Melbourne, Victoria CD ROM, Version 1.0 for DOS.

LEECH, H.B. and CHANDLER, H. P. 1956. Aquatic Coleoptera. pp. 293-371. *In* Usinger, R. L. Editor. Aquatic insects of California with keys to North American genera and California species. University of California Press, Berkeley and Los Angeles. ix + 508 pp.

PEDERZANI, F. 1995. Keys to the identification of the genera and subgenera of adult Dytiscidae (*sensu lato*) of the world (Coleoptera Dytiscidae). Atti dell'Accademia Roveretana degli Agiati 224(1994):5-83.

PETERSON, A., 1951. Larvae of Insects. An introduction to Nearctic species. Part II. Coleoptera, Diptera, Neuroptera, Siphonaptera, Mecoptera, Trichoptera. Privately printed. Columbus, Ohio. v + 416 pp.

ROUGHLEY, R.E. 1991. Family Noteridae. pp. 61-62. *in* Bousquet, Y. Editor. Checklist of the Beetles of Canada and Alaska. Publication 1861/E, Research Branch, Agriculture Canada, Ottawa, Ontario. vi + 430 pp.

SPANGLER, P.J. 1991. Noteridae (Adephaga). pp. 314-315. *in* Lawrence, J. F. Co-ordinator. Chapter 34. Coleoptera. Pp. 144-658 in F. Stehr. Editor. Immature Insects. Volume 2. Kendall/Hunt Publishing Co., Dubuque, Iowa. xvi + 975 pp.

SPANGLER, P. J. and FOLKERTS, G.W. 1973. Reassignment of *Colpius inflatus* and a description of its larva (Coleoptera: Noteridae). Proceedings of the Biological Society of Washington, 86: 501-509.

YOUNG, F. N. 1954. The water beetles of Florida. University of Florida Press, Gainesville ix + 238 pp.

YOUNG, F. N. 1978. The New World species of the water-beetle genus *Notomicrus* (Noteridae). Systematic Entomology, 3: 285-293.

YOUNG, F. N. 1979. Water beetles of the genus *Suphisellus* Crotch in the Americas north of Colombia (Coleoptera: Noteridae). The Southwestern Naturalist, 24: 409-429.

YOUNG, F. N. 1985. A key to the American species of *Hydrocanthus* Say, with descriptions of new taxa (Coleoptera: Noteridae). Proceedings of the Academy of Natural Sciences of Philadelphia, 137: 90-98.

ZIMMERMANN, A. 1920. I. Subfam. Noterinae, pp. 2-15. *in* Schenkling, S. N. (ed.), Coleopterorum Catalogus. Pars 71. Dytiscidae, Haliplidae, Hygrobiidae, Amphizoidae. W. Junk, Berlin. 326 pp.

# 11. AMPHIZOIDAE LeConte, 1853

by T. Keith Philips and Weiping Xie

**Common name:** Trout-stream beetles

Although a small family and rare in collections, this single genus has received considerable attention because of its distinctive morphology and the possibility that it represents an intermediate evolutionary stage between the terrestrial and aquatic Adephaga. Although they are probably not the first group leading to other lineages with a truly aquatic lifestyle, they are one of the more basal relict families in the dytiscoid lineage. Adults resemble carabids of the genus *Metrius*, but have the middle coxal cavities similar to the higher dytiscids. Larvae most closely approach the form of *Silpha* in overall appearance. Both larvae and adults are semi-aquatic and usually live in cold, fast flowing streams.

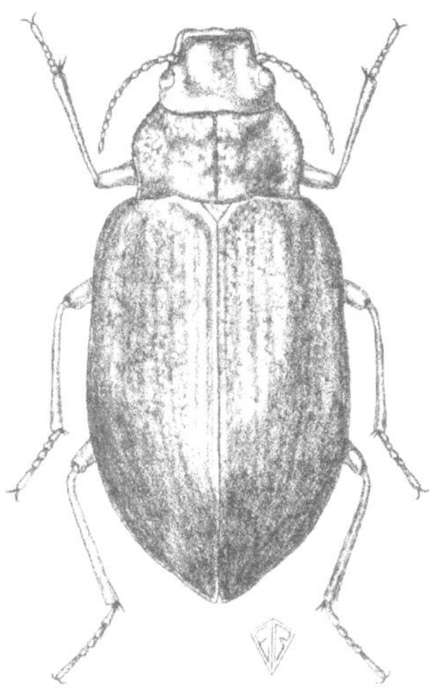

FIGURE 1.11. *Amphizoa lecontei* Matthews.

**Description**: Tenebrionoid in shape, elongate oval, body somewhat convex dorsally; length: 11-16 mm; dull black to piceus; vestiture absent except for silky fringes of long setae in a groove along the outer edge of the middle tibiae and traces of such a fringe of setae on the front and hind tibiae.

Head quadrate, sides rounded, about two-thirds as wide as pronotum; antennae with eleven antennomeres, filiform, stout, each segment slightly longer than wide, expanded somewhat apically, inserted in groove between the eyes and the base of the mandibles; mandibles nearly hidden by the labrum and labium, stout, blunt at apices; maxillary palpi with three palpomeres, galea 1-segmented; labium forming a broad bilobed plate, mental suture absent, labial palpi with three palpomeres, short, inserted near apex of labium; eyes small, ovate, on the sides of the head.

Thorax with pronotum narrower than the elytra, sides bluntly margined and slightly crenulated, narrower in front, sinuate laterally, notopleural suture prominent; prosternum moderately long in front of coxae, with a broad, flat intercoxal process which extends posteriorly and abuts on the metasternal process; mesocoxae widely separated, cavities open laterally and meeting both mesepimeron and metepisternum; metasternum short, metacoxae meet elytra laterally; legs slender, without long setae; tarsi 5-5-5, stout, each tarsomere simple, last tarsomere as long as the preceding three combined; claws simple, arising from central cavity at apex of tarsus; elytra broad, constricted anteriorly, tapering behind, slightly margined laterally, striae very weak, usually rugose; well-developed wings with venation typically adephageous, but closer to Dytiscidae than Carabidae, oblong cell and wedge cell present, cubitus visible only as a trace beyond the oblong cell; metafurca very small and flight musculature degenerated.

Abdomen with six ventrites, first ventrite cleft by hind coxae; first three ventrites connate; last ventrite deeply cleft and forming two triangular sclerites; antecoxal piece obscure and incompletely separated from the metasternum by an abbreviated suture. Male genitalia of the trilobed type, with well-developed basal piece and symmetrical parameres, penis slender, parameres moderately stout, pars basalis detached from the parameres and forms a small T-shaped piece laying in the membrane connecting the parameres and the ninth sternum; genitalia in repose rotated 90° to left, and prior to copulation must extend and rotate an additional 90° into a completely inverted position.

Larvae strongly sclerotized above, pale beneath, fusiform; thorax and abdomen strongly flattened, with dorsal plates broadly expanded as thin lateral lobes; mandible with an open furrow and no internal duct; antennomere IV almost completely reduced and sensory appendage absent; gular sutures are confluent; legs ambulatory, each with two claws and two setae protruding beneath at middle; gills absent, but the eighth pair of abdominal spiracles is very large, protruding dorsally, and located close together on the tergite; eight visible abdominal segments, pair of prominent, fleshy, pointed one-segmented urogomphi extend beyond and beneath eighth abdominal segment.

**Habits and habitats.** The larvae occur with the adults usually in swift to moderate flowing mountain streams, with cool to icy water. They are often found at high altitudes, averaging around 2000 meters for both *A. insolens* LeConte and *A. lecontei* Matthews, although the range of the latter has been recorded from 480 to 2930 meters. In contrast, *A. striata* Van Dyke, lives in relatively warm, slow moving streams and rivers at average altitudes of only about 200 meters. Of interest is one of the Chinese species which may be associated with seepage areas and pools adjacent to cliffs in a coniferous mountain forest.

Adults have been observed briefly surfacing and then crawling into the water carrying a bubble of air beneath and surrounding the elytral apices. Although living under water, larvae stay near the surface and must expose at least the eighth abdominal segment to the air in order to obtain oxygen. They are not capable of being submerged for very long and are usually found clinging to driftwood, pine needles, logs or other debris in frothy eddies and backwaters of mountain streams. They can also be found on the stream margins and undercut banks, clinging to exposed roots. Larvae have also been reported to be gregarious and will remain motionless if disturbed. The larvae and adults do not swim efficiently in the water. Instead, they passively drift with the current. Amphizoids are bivoltine and usually both first instar larvae and adults overwinter.

Both larvae and adults are predacious, but they also scavenge on various types of dead insects including both aquatic and terrestrial forms such as Ephemeroptera, Plecoptera, Trichoptera, Diptera and Hymenoptera. Larvae usually cling head-down in the water on floating debris and quickly enter the water to capture insect prey. They then return to the surface to feed while holding the food item high out of the water and over their heads. Larvae have three instars. When full grown, they pupate underground in sand or mud some distance from the stream bank and hence the newly emerged adults are sometimes thickly coated with mud. If agitated, the adults exude a viscous yellowish fluid from the anus with an odor similar to decaying wood. This defensive system probably first evolved as protection against various anurans. Eggs are typically laid under water in cracks in submerged wood, debris and other similar locations.

**Status of the classification.** This small group was first extensively revised by Edwards who concluded that it is intermediate between Carabidae and Dytiscidae, and more similar to the European Hygrobiidae than to any other family. Kavanaugh, with access to vastly more material, designated lectotypes and improved our understanding of the taxonomy and distribution. Phylogenetic studies by Beutel indicate that the Amphizoidae are a lineage which shares a common ancestor with the Hygobiidae + Dytiscidae. Based on larval studies by Xie, Amphizoidae + Hygrobiidae is the sister group of Dytiscidae. All three families share an ancestor with the more basal Noteridae and together all four compose the Dytiscoidea.

**Distribution.** There is only one genus with six species, three of which live in the Pacific Coast and Rocky Mountain region from Alaska to southern California and east to Arizona, Colorado and Montana. The other three species are known from southwestern and northeastern China.

## Key to Species of North American *Amphizoa* LeConte

1. Elytron with blunt but distinct carina on fifth interval; Rocky Mountains from the Yukon Territory south to Arizona and east to Colorado, Montana and Alberta .......................... *A. lecontei* Matthews

— Elytron without carina on fifth interval .................. 2

2. Size large (standardized body length [SBL] over 13.00 mm); front tarsi with well-developed groove on posterior surface, grooves bearing a fringe of long, natatory setae; Vancouver Island south to Oregon and east as far as Yakima Co., Washington ......... ............................................. *A. striata* Van Dyke

— Size medium (SBL under 13.00 mm); front tarsi without developed grooves, natatory setae absent; Alaska to southern California and east to western Montana and Wyoming .......... *A. insolens* LeConte

## Classification of the Nearctic Genus

*Amphizoa* LeConte 1854, 3 spp., distributed as above.
*Dysmathes* Mannerheim 1853

## Bibliography

BEUTEL, R. G. 1988. Studies of the metathorax of the trout-stream beetle, *Amphizoa lecontei* Matthews (Coleoptera: Amphizoidae): contribution towards clarification of the systematic position of Amphizoidae. International Journal of Insect Morphology and Embryology 17(1): 63-81, 12 figures.

BEUTEL, R. G. 1991. Internal and external structures of the head of 3rd instar larvae of *Amphizoa lecontei* Matthews (Coleoptera: Amphizoidae). A contribution towards the clarification of the systematic position of Amphizoidae. Stuttgarter Beitrage zur Naturkunde Serie A (Biologie) Nr. 469. 1-24, 9 figures.

BEUTEL, R. G. 1993. Phylogenetic analysis of Adephaga (Coleoptera) based on characters of the larval head. Systematic Entomology, 18: 127-147, 31 figures, 1 table.

BEUTEL, R. G. 1995. The Adephaga (Coleoptera); phylogeny and evolutionary history. 173-217. *In*, Pakaluk, J. and Slipinski, S. A. Biology, phylogeny, and classification of Coleoptera. Papers celebrating the 80th birthday of Roy A. Crowson. Vol 1. Muzeum i Instytut Zoologii PAN, Warszawa.

BLACKWELDER, R. E. and R. H. ARNETT, JR. 1974. The water beetles, rove beetles and related groups. (Red Version). 1-165. *In*, Blackwelder, R. E. and R. H. Arnett, Jr. (Associates of the NABF Project). Checklist of the beetles of Canada, United States, Mexico, Central America and the West Indies. (Red Version). Part 1. North America Beetle Fauna Project. The Biological Research Institute of America, Inc., New York.

BONNELL, D. E., AND J. BRUZAS. 1938. A method of collecting *Amphizoa*. Pan-Pacific Entomologist. 14(3): 112.

BOUSQUET, Y. 1991. Checklist of beetles of Canada and Alaska. Research Branch, Agriculture Canada, Publication 1861/E. [vi] + 430 pp.

BØVING, A. G. and F. C. CRAIGHEAD. 1931. An illustrated synopsis of the principal larval forms of the Order Coleoptera. Brooklyn Entomological Society, Brooklyn, New York. [viii] + 351 pp.

BURMEISTER, E. G. 1990. On the systematic position of Amphizoidae, emphasizing features of the female genital organs (Insecta: Coleoptera: Adephaga). Quaestiones Entomologicae, 26: 245-272, 13 figures.

CROWSON, R. A. 1967. The natural classification of the families of Coleoptera. E.W. Classey Ltd., 214 pp + 213 figures.

CROWSON, R. A. 1981. The Biology of the Coleoptera. Academic Press, London. [i-xii] + 802 pp.

DARLINGTON, P. J., JR. (1929)1930. Notes on the habits of *Amphizoa*. Psyche, 36(4): 383-385.

EDWARDS, J. G. (1950) 1951. Amphizoidae (Coleoptera) of the World. Wasmann Journal of Biology, 8(3): 303-332, 4 plates.

EDWARDS, J. G. 1953. The real source of *Amphizoa* secretions. The Coleopterists Bulletin, 7(1): 4.

EDWARDS, J. G. 1954. Observations on the biology of Amphizoidae. The Coleopterists Bulletin, 8(1): 19-24.

FORSYTH, D. J. 1970. The structure of the defence glands of the Cicindelidae, Amphizoidae, and Hygrobiidae (Insecta: Coleoptera). Journal of Zoology, London, 160: 51-69, 19 figures and 2 plates.

HATCH, M. H. 1953. The beetles of the Pacific Northwest. University of Washington Publications in Biology. Part I: Introduction and Adephaga, 16: 1-340. University of Washington Press, Seattle.

HUBBARD, H. G. 1892a. Notes on the larvae of *Amphizoa*. Insect Life, 5: 19-22, 4 figures.

HUBBARD, H. G. 1892b. Description of the larva of *Amphizoa lecontei*. Proceedings of the Entomological Society of Washington, 2(3): 341-346, 1 plate.

HUBBARD, H. G. 1894. Rapports Nat. Phylog. Fam. Col. p. 111, 114, fig. 10a, 10b.

JI, L., and JAECH, M. A. 1995. Amphizoidae (Coleoptera). Water Beetles of China, 1: 103-108, 4 figures.

KAVANAUGH, D. H. 1984. A catalog of the Coleoptera of America north of Mexico. Family: Amphizoidae. Agricultural Handbook of the U.S. Department of Agriculture.

KAVANAUGH, D. H. 1986. A systematic review of Amphizoid beetles (Amphizoidae: Coleoptera) and their phylogenetic relationships to other Adephaga. Proceedings of the California Academy of Sciences, 44(6): 67-109, 28 figures, 2 tables.

KAVANAUGH, D. H. 1991. Amphizoidae (Adephaga). p.312-314, 1 figure. *In*, Stehr, F.W. (ed.), Immature Insects, Volume 2. Kendall/Hunt Publishing Company. [xvi] + 975 pp.

LEECH, H. B., and H. P. CHANDLER. 1968. Aquatic Coleoptera. p. 293-371. *In*, Usinger, R. L. (ed), Aquatic Insects of California, with keys to North American Genera and California species. University of California Press, Berkeley and Los Angeles. [ix] + 508 pp.

LENG, C. W. 1920. Catalogue of the Coleoptera of America, North of Mexico. The Cosmos Press, Cambridge, Massachusetts, U.S.A. 470 pp.

MASON, W. 1949. Amphizoid collecting. The Coleopterists Bulletin, 3(2): 21.

PETERSON, A. 1960. Larvae of insects - an introduction to Nearctic species. Part II. Coleoptera, Diptera, Neuroptera, Siphonaptera, Mecoptera, Trichoptera. Fourth edition. Columbus, Ohio, v and 416 pp., illus.

ROUGHLEY, R. E., WEIPING XIE, and YU PEIYU. 1998. Description of *Amphizoa smetanai* n.sp. and supplementary description of *Amphizoa davidi* Lucas (Coleoptera: Amphizoidae). Water beetles of China, 2: 123.

TANNER, V. M. 1927. A preliminary study of the genitalia of female Coleoptera. The Transactions of the American Entomological Society, 53:5-50, 15 plates.

WILSON, J. W. 1928. The male genital tube of the Amphizoidae. Psyche, 35:98-99, 1 figure.

YU PEIYU and N. E. STORK. 1991. New evidence on the phylogeny and biogeography of the Amphizoidae: discovery of a new species from China (Coleoptera). Systematic Entomology, 16:253-256, 3 figures.

YU PEIYU, XIE WEIPING, and LIN FENG-QIN. 1993. Bionomics and morphology of the larvae of *Amphizoa sinica* Yu et Stork (Coleoptera: Amphizoidae). Scientific Treatise on Systematic and Evolutionary Zoology, 2:107-114, 5 figures, 2 plates.

YU PEIYU, R. E. ROUGHLEY, WEIPING XIE, and LIN FENGQIN. Pupal bionomics of *Amphizoa sinica* Yu et Stork (Coleoptera: Amphizoidae) (unpublished).

ZAITSEV, F. A. 1972. Fauna of the U.S.S.R. Coleoptera Vol.IV, Amphizoidae, Hygrobiidae, Haliplidae, Dytiscidae, Gyrinidae. translated from Russian. Published for the Smithsonian Institution and the National Science Foundation, Washington, D.C. by the Israel Program for Scientific Translations. [iii] + 401 pp.

# 12. DYTISCIDAE Leach, 1815

by R. E. Roughley and D.J. Larson

**Family common name**: The predaceous diving beetle family

These beetles are well-adapted to aquatic life. The streamlined shape and flattened, paddle-like hind legs give them a characteristic appearance. The divided first visible abdominal sternum and short palpi distinguish them from the Hydrophilidae which they resemble, the lack of a prosternal platform will distinguish them from most noterids and the single pair of eyes separate them from Gyrinidae.

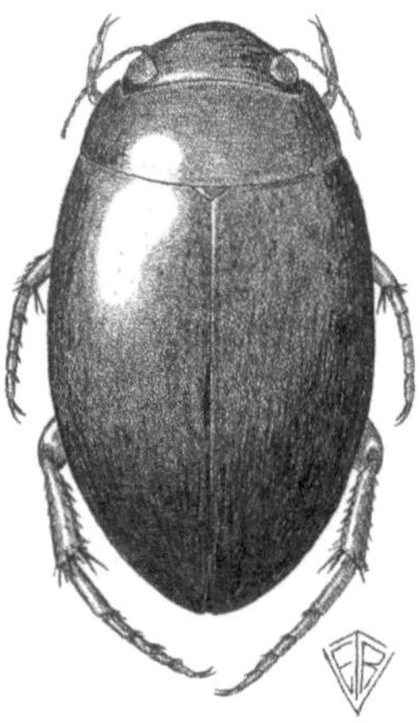

FIGURE 1.12. *Agabus seriatus* (Say) (Agabini, Colymbetinae).

**Description.** See Sharp (1882), Balfour-Browne (1940, 1950), Franciscolo (1979a), and Nilsson and Holmen (1995) for a detailed description of structure of dytiscids. Pederzani (1995) and Larson *et al.* (2000) discuss structural features in relation to identification. The following treatment of structure, keys, diagnoses and characterization of genera is abstracted extensively from Larson *et al.* (2000) for Nearctic genera only. Shape oval, broadly ovate to quite elongate oval; length 1.6 to 40 mm; color various, most species reddish-brown to black, with or without yellow markings, some species yellowish with black markings, the generalized pattern of elytral punctation is a subsutural (ElSS, Fig. 2.12) stria or series of punctures and three rows of serial punctures (ElSP, Fig. 2.12); devoid of obvious setae except long swimming hairs on tibiae and tarsi.

Head prognathous, short, somewhat retracted into prothorax, dorsally convex, surface punctate but without setae except for row of short setae on labrum medially. Eyes, in dorsal aspect, forming a smooth curve from frons to pronotum or somewhat bulging, in frontal view with outline emarginate (Fig. 12.12) or entire (Fig. 11.12). Antennae (Ant, Fig. 3.12) with 11, glabrous antennomeres; 1$^{st}$ antennomere (scape) spherical and appearing two-segmented, 2$^{nd}$ antennomere (pedicel) elongate, remainder filiform; antennomeres 3 to 11 of increasing length and longer than wide, often of various shapes, males of some groups with widened antennomeres inserted between the anterior portion of the eyes and the bases of the mandibles. Mouthparts with labrum transverse and narrow with row of short setae on the anterior margin at middle, covering most of the mandibles. Clypeus (Cl, Fig. 2.12) fused to frons, frontoclypeal suture absent or visible only laterally (Figs. 2.12, 12.12). Mandibles short and robust, asymmetrical, with a subapical and two apical teeth, retinaculum pointed, with a mesal comb or brush of setae. Maxilla with lacinia with spines along inner margin and an elongate two-segmented galea, maxillary palpus (MxPlp, Fig. 3.12) four-segmented with a basal palpiger, apical palpomere elongate and/or enlarged in some groups. Labium with broad anterolateral lobes, ligula (Lig, Fig. 3.12) present and small but extending between the bases of the labial palps, labial palpus (LbPlp, Fig. 3.12) short, moderately stout, three-segmented with a basal palpiger, apical palpomere elongate and/or enlarged in some taxa. Median gular apodeme absent.

Pronotum (PnD, Fig. 2.12) widest at base, narrower in front, excavated for head, without pubescence, coarsely punctate to impunctate, lateral margins convex or sinuate, lateral curvature continuous with outer margin of elytra (El, Fig. 2.12) or posterior angle (PnPL, Fig. 2.12) displaced beyond base of elytra or inward from humeral margin and lateral outline discontinuous in dorsal aspect; posterior margin sinuate with medial V-shaped posterior extension covering scutellum (Sc, Fig. 2.12) in some taxa. Ventral portion of prothorax with notopleural and pleurosternal sutures. Prosternum flat to somewhat swollen near base; prosternal process (PStP, Fig. 3.12,) elongate and rounded to pointed apically and ending of front of mesocoxae (Cx2, Fig. 3.12) (Vatellini) or extended between mesocoxae and contacting metaventrite (MtV, Fig. 3.12, = metasternum, of authors). Procoxal cavities elongate oval to oval or round, open behind externally but closed internally. Procoxae (Cx1, Fig. 3.12) elongate oval to globular, trochantin not visible. Protrochanter triangular and fused to basoventral portion of femur. Profemur (Fe1, Fig. 3.12) simple to highly modified; elongate and robust to short and thick, ventral margin rounded (most groups) or excavated for reception of protibia (males of Dytiscinae), distribution of spines and setae various; apicoventral margin with a distinct notch with short, dense setae used as an antennal cleaner. Protibia (Tb1, Fig. 3.12) slender and elongate (most groups) to short and thick (males of Dytiscinae), with distribution of spines and setae various, protibial spurs not differentiated. Protarsi (Ta1, Fig. 3.12) clearly five- (Figs. 5.12, 6.12, 8.12) or appearing four-segmented (Fig. 7.12), unmodified in shape in females, or modified in males of most groups with protarsomeres 1 to 3 enlarged and widened (Figs. 5.12, 6.12) and with various numbers of enlarged articulo-setae on ventral

**Acknowledgments.** We owe an enormous debt to the late Hugh Bosdin Leech (1910-1990) and Frank Norman Young (1915- ) for their help, teaching and encouragement.

surface, protarsomere 5 longest, distribution of spines and setae various but many groups with sparse row of long, natatory setae on dorsal surface of basal protarsomeres; two short to rather long claws present on tarsomere 5.

Mesoventrite (MsV, Fig. 3.12, = mesosternum, of authors) short, mostly concealed by prosternal process; mesocoxal cavity round, limited laterally by mesepimeron. Mesocoxae (Cx 2, Fig. 3.12) round with limited range of motion. Mesotrochanter (Tr2, Fig. 3.12) short and triangular and fused to basoventral portion of femur. Mesofemur (Fe2, Fig. 3.12) elongate and narrow in most groups to short and very wide; with poorly to well-defined row of broadly spaced, stiff, elongate spines on anteroventral surface. Mesotibia (Tb2, Fig. 3.12) slender and elongate, in most groups, to shorter and widened apically with sparse row of stiff, elongate spines on anterior and anteroventral surfaces and row of short, dense spines on apical margin, with row of long, natatory setae on dorsal surface. Mesotarsi (Ta2, Fig. 3.12) apparently four-segmented or clearly five-segmented, slender and elongate; mesotarsomere 5 longest, with row of long, natatory setae on dorsal margin of basal mesotarsomeres; males with mesotarsomeres 1 to 3 perceptibly to distinctly broadened, ventral surface of mesotarsomeres 1 to 3 with some short, unmodified setae or with enlarged articulo-setae; two short to very long claws on mesotarsomere 5.

Metaventrite (MtV, Fig. 3.12, =metasternum, of authors) (Beutel and Belkaceme 1986) large and extended toward epipleuron but limited laterally by metepimeron, with discrimen (=median metasternal suture, of authors) present at least apically and without a posterior, transverse suture; anteromedially with broad extension (=anterior metasternal process, of authors), between and partially covering inner margins of mesocoxae (Cx2, Fig. 3.12) and contacting the prosternal process, in most groups. Metacoxae greatly enlarged and extending laterally to epipleuron; outer lamina forming a large metacoxal plate (MtCxPl, Fig. 3.12); inner lamina or intralinear area (MtCxi, Fig. 3.12) small and elongate, limited laterally by metacoxal lines (MtCxL, Fig. 3.12) with base of hind legs, of most groups, articulating beneath variously shaped lobes of metacoxal process (MtCxP, Fig. 3.12). Metatrochanter (Tr3, Fig. 3.12) small to large and fused to basoventral portion of metafemur. Metafemur (Fe3, Fig. 3.12) thin and elongate to shorter and more robust with distribution of setae and spines various but generally reduced. Metatibia (Tb3, Fig. 3.12) long and slender to short and wide, with most groups with irregular rows of stiff setae on anterior, posterior and ventral surfaces, row of short, dense spines on apical margin, and row of long natatory setae on dorsal surface and ventral surfaces; metatibial spurs (Tb3Sp Fig. 3.12) of various shapes. Metatarsi (Ta3, Fig. 3.12) slender and elongate to short and robust, with long natatory setae on dorsal surface of basal metatarsomeres; metatarsomere 1 or 5 longest; one or two short claws present on apex of metatarsomere 5.

Scutellum (Sc, Fig. 2.12) exposed or concealed. Elytra (El, Fig. 2.12) subparallel or with sides evenly arcuate or tapering, continuous or discontinuous with base of pronotum, in dorsal aspect, widest in basal quarter, in most groups; apices rounded

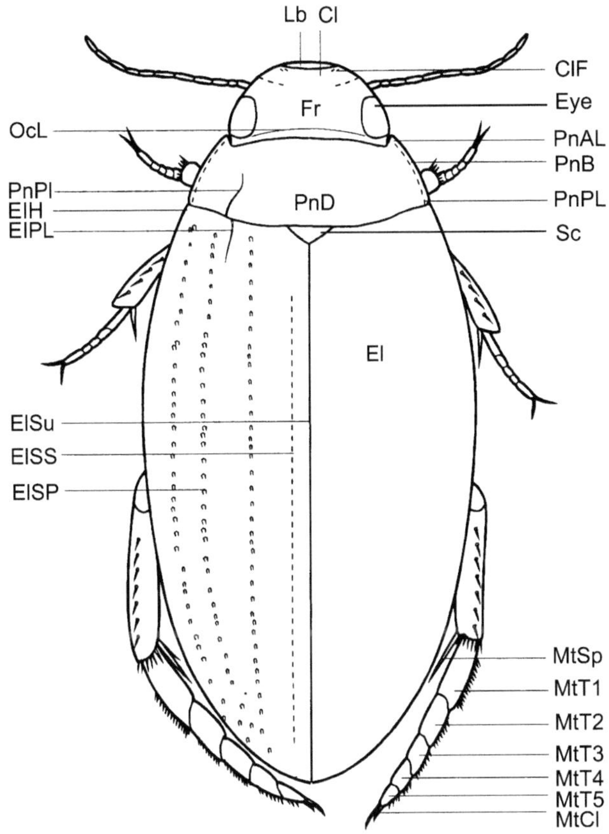


FIGURE 2.12 Generalized dytiscid beetle, dorsal aspect, showing principal structural features. Cl - clypeus; ClF - clypeal fovea; El - elytron; ElH - elytral humerus; ElPl - elytral plica; ElSP - elytral serial punctures; ElSS - elytral subsutural stria; ElSu - elytral suture; Fr - frons; Lb - labrum; MtCl- metatarsal claw; MtSp - metatibial spur; MtT1...5 - metatarsomeres 1 - 5; OcL - occipital line; PnAL - pronotal anterolateral angle; PnB - pronotal lateral bead; PnD - pronotal disc; PnPL - pronotal posterolateral angle; PnPl - pronotal plica; Sc - scutellum. This and all subsequent figures are from Larson *et al.* 2000 unless otherwise mentioned.

to acuminate, with or without apical spine-like process or subapical lateral tooth, surface with well impressed microreticulation on most species, with three discrete rows of serial punctures (ElSP, Fig. 2.12) distinctive, in most groups, or confused, disrupted or absent; subsutural stria (ElSS, Fig. 2.12) present or absent; epipleuron (Epl, Fig. 3.12) broader basally and narrowed posteriorly, with carina (EplC, Fig. 3.12) crossing epipleuron in some groups. Wing venation typically adephagan with oblongum cell (Kukalovà-Peck and Lawrence 1993); folding pattern complex and various (Forbes 1926, Goodliffe 1939, Hammond 1979).

Abdomen with six visible sternites (S1-6, Fig. 3.12, =true sternites 2-7), surface punctate, sculptured or smooth, visible sternite 1 (=true sternite 2) divided by metacoxa as is characteristic for Adephaga, visible sternites 2 and 3 separate or fused and without sutures, or with poorly defined suture visible only

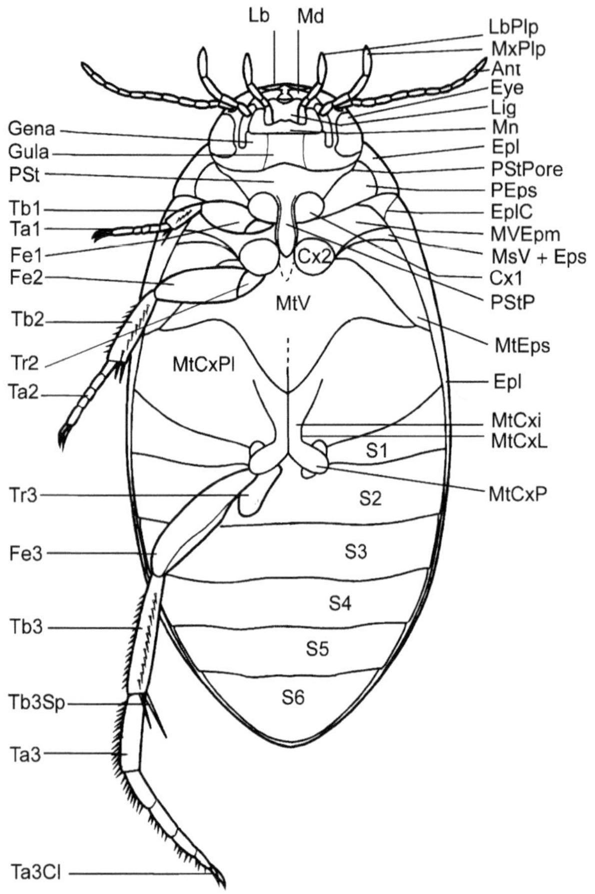


FIGURE 3.12 Generalized dytiscid beetle, ventral aspect, showing principal structural features. Ant - antenna; Cx1, Cx2 - pro- and mesocoxa; EplC - epipleural carina; Epl - epipleuron; Fe1, Fe2, Fe3 - pro-, meso-, and metafemur; Lb - labrum; LbPlp - labial palp; Lig - ligula; Md - mandible; Mn - mentum; MVEpm - mesepimeron; MsV+Eps - mesoventrite plus mesepisternum; MtCxi - metacoxa, interlinear space; MtCxL - metacoxal line; MtCxP - metacoxal process; MtCxPl - metacoxal plate; MtV= metaventrite (=metasternum of authors); MtVEps - episternum of metaventrite; Tb3Sp - metatibial spur; Ta1...3 - pro-, meso- and metatarsus; MxPlp - maxillary palp; PEps - proepisternum; PSt - prosternum; PStP - prosternal process; PStPore - prosternal pore; S1...S6 - visible abdominal sterna 1 to 6; Ta1, Ta2, Ta3 - pro-, meso-, and metatarsus; Ta3Cl - metatarsal claw; Tb1, Tb2, Tb3 - pro-, meso-, and metatibia; Tr1, Tr2, Tr3 - pro-, meso-, and metatrochanter.

laterally; apical sternite (=true sternite 7) rounded, extended distally medially or pointed or with various modifications.

Male genitalia caraboid, penis symmetrical, laterally compressed and apex broad and finger-like or pointed, with a sperm channel on the convex surface; parameres asymmetrical. Male genitalia with torsion (90° rotation at rest, 180° during copulation) characteristic of most of Adephaga (Beutel and Roughley 1988).

Female genitalia complex and phylogenetically informative (Bøving 1913, Balfour-Browne 1950, Burmeister 1976 and Wolfe 1985) with many competing systems of structural terms and differing uses of the same term. The main, generalized structures among dytiscids (following Burmeister 1976) are the basal, somewhat triangular to quadrate gonocoxosternum (=sternum VIII, of authors), a triangular to elongate, transverse or linear tergum IX (=valvifer, paraproct of authors), paired or fused apical gonocoxae (=genital valves, coxites of authors), located dorsal to the paired to fused genital appendages VIII (=vulval sclerite, of authors).

Larvae (Peterson 1951, Spangler 1991, Lawrence *et al.* 1993, Nilsson and Hüolmen 1995, Larson *et al.* 2000) campodeiform, elongate, fusiform and moderately to very broad at prothorax or at midlength; known mature larvae range in size from 1.5 to 70.0 mm; color white, yellowish-brown or reddish-brown to black variously marked with yellow or black, surface without pronounced microsculpture. Head prognathous and exserted but retracted into base of pronotum, with or without temporal spines or setae; distinct neck region present or absent; frontoclypeal suture present, limited laterally by arms of epicranial suture, anterior margin convex to concave or greatly extended anteriorly as a nasale (Hydroporinae); antennae short to very long, four-segmented (some taxa with accessory pseudo-segmentation), with apical antennomere long, with sensory appendage; labrum fused to head capsule; mandible short, without mola, with or without retinacular teeth, and inner margin serrate or toothed and without feeding channel or channel present only basally (some groups) or mandible elongate, narrow and without teeth, and with enclosed feeding channel; maxilla slender, cardo fused with stipes or small, stipes variously shaped, lacinia absent, galea absent or two-segmented (some taxa with accessory pseudo-segmentation) and articulated with palpifer; maxillary palpus short, two- or three-segmented. Labium small, somewhat rectangular, without ligula, labial palpus short, two-segmented; six pairs of conspicuous stemmata (reduced or absent in subterranean taxa). Gula absent or narrow, represented by a narrow, elongate strip between gular sutures, posterior tentorial pits present and located at about midlength. Pro-, meso- and metathoracic tergites well sclerotized; prothoracic tergite larger than meso- and metathoracic terga. Legs six-segmented with apical segment (=pretarsus) consisting of a pair of claws; natatory setae present in most groups on femur, tibia and tarsus; overall form elongate and slender, modified for swimming. Abdomen with eight visible segments, apical abdominal segment (=segment 8) elongate with urogomphi terminal or tergite extended above articulation of urogomphi; urogomphi one- or two-segmented. Our knowledge of the classification of larvae of Nearctic species of Dytiscidae was reviewed by Alarie in Larson *et al.* (2000). This is an excellent review and demonstrates the contribution of larval studies to overall classification. Interested readers should consult it for detailed descriptions and the state of knowledge which are not treated further here.

**Habits and habitats.** Both the larvae and the adults of all of our species are aquatic and good swimmers. Adults use the hind legs extensively in the swimming process and they are used simultaneously. The legs are modified to varying degrees for

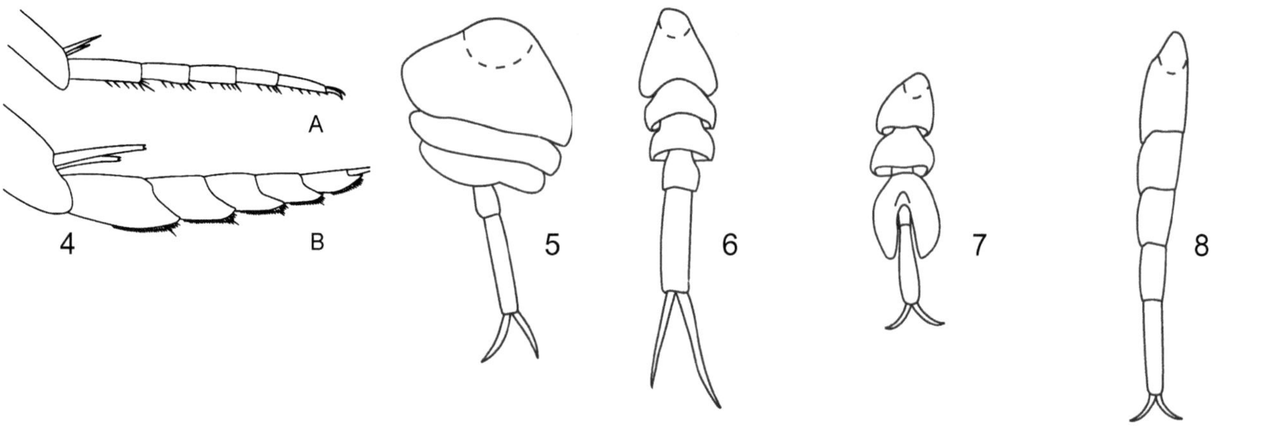


FIGURES 4.12 - 8.12. Fig. 4.12. Lateral view of apex of metatibia and metatarsus. A. *Hydroporus* sp. with tibial spurs simple apically, metatarsomeres not lobed apically and two tarsal claws. B. *Laccophilus* sp. with tibial spurs bifid apically, metatarsomeres lobed apically and one tarsal claw; Fig. 5.12 Dorsal view of protarsus of male *Graphoderus* sp. (Dytiscinae); Fig. 6.12 Dorsal view of protarsus of male *Rhantus* sp. (Colymbetinae); Fig. 7.12 Dorsal view of protarsus of male *Hydroporus* (Hydroporinae); Fig. 8.12 Dorsal view of protarsus of male *Laccophilus* sp. (Laccophilinae).

swimming. The adults of virtually all species can fly and thus migrate from habitat to habitat, sometimes in very large numbers; they are clumsy on land. Many species are found in small ponds, along margins of lakes and streams, others occur in a wide variety of habitats from hot springs to cold mountain streams. Most species overwinter as adults either in water or on land, whereas some taxa overwinter as larvae in water and a very few are known to overwinter as eggs. Copulation and egg laying take place during the warm months. Most if not all species produce thoracic and pygidial substances, used as wetting and defensive agents. Most species are scavengers or carnivorous as adults, feeding on dead or live dragonfly and damselfly nymphs and other small aquatic animals including tadpoles and small fish. The larvae are voracious and will prey on virtually anything that they can overpower, including their siblings. The adults are able to stay submerged for a long period of time making use of a physical gill employing air trapped beneath the elytra. They come to the surface periodically to renew this air, hanging head downward at the surface while replenishing the physical gill. Eggs are deposited upon the submerged surface or in the tissues of aquatic plants in some species; others deposit eggs loosely on debris near the shore. Pupation takes place on land in a cell made of moist soil or plant material prepared by the prepupal larva with most species eclosing in late summer to early fall. For further discussion and references see Larson *et al.* (2000).

**Status of the classification.** The Dytiscidae have long been a family of Adephaga but at times has included other groups of beetles now considered to be separate familes (e.g., Noteridae). Beutel (1995, 1997) and Beutel and Roughley (1987, 1988) have discussed the overall classification of Adephaga with emphasis on the aquatic Adephaga. Dytiscidae belong to the superfamily Dytiscoidea where it is the sister group of the remaining families: Noteridae + (Amphizoidae + (Hygrobiidae +Dytiscidae)). The Dytiscoidea are an independent invasion of the water from other aquatic Adephaga (Beutel 1995, 1997, Beutel and Roughley 1988) but for an alternative view of the phylogentic position of Noteridae see Kavanaugh (1986).

Historically there were four major subfamilies of Dytiscidae (Laccophilinae, Hydroporinae, Colymbetinae and Dytiscinae); however, with detailed study of the structure of adults and larvae and application of phylogenetic methods we have seen the elevation of Copelatinae, Agabetinae, Hydrotrupinae which were all former tribes within Colymbetinae. Within the family, subfamily and tribal classification has been somewhat problematic. For instance, Burmeister (1976) in his excellent study of the female ovipositor was able to provide many synapomorphies of individual subfamilies but few characters which unite any two subfamilies. The limits of the higher taxa will continue to be debated and much work needs to be done in this area. It does seem fair, however, to suggest that many of our species are fairly well-known taxonomically and that we are at an appropriate stage to test the classification with life history studies, ecological associations, and hypotheses about community structure and evolution with respect to habitat.

The nomenclature of higher taxon names (subfamilies, tribes, genera and subgenera) has been examined in detail by Nilsson *et al.* (1989) and Nilsson and Roughley (1997a). There are many, excellent regional works for larger regions such as North America (genera only, White and Brigham 1996), Canada and the northern U.S. (Larson *et al.* 2000), Pacific Northwest (Hatch 1953), the northeastern United States (Downie and Arnett 1996), or for individual provinces and states, such as Alberta (Larson 1975), Manitoba (Wallis 1973), Newfoundland (Larson and Colbo 1983) and Yukon Territory (Larson 1997b), Alabama (Folkerts 1978), California (Leech and Chandler 1956, for which the nomenclature should be correlated with the checklist of Challet and Brett 1998), Florida (Young 1954, Epler 1996), Indiana (White *et al.* 1985), Kansas (Huggins *et al.* 1976, Slater 1979, Coler and Slater 1982), Maine (Malcolm 1971, Boobar *et al.* 1996), Nevada (LaRivers 1951, Anderson 1985), North and South Carolina (Brigham 1982), North Dakota (Gordon and

Post 1965), Utah (Anderson 1962, 1985), Virginia (Michael and Matta 1977) and Wisconsin (Hilsenhoff 1992, 1993a, b, c, 1994, 1995). Arce-Pérez and Roughley (1999) listed 27 genera and 179 species species of dytiscids for México; Larson and Roughley (1991) provided a list of the 262 species known from Canada and Alaska.

**Distribution.** The family is worldwide in distribution. The species are common in tropical latitudes and increasingly common toward the temperate zone. The most recent world catalogue is Zimmermann (1920) and it is very out of date; today there are at least 160 genera encompassing more than 4,000 species (Pederzani 1995, Nilsson and Roughley 1997b) with about 52 genera and more than 500 species (Larson *et al.* 2000).

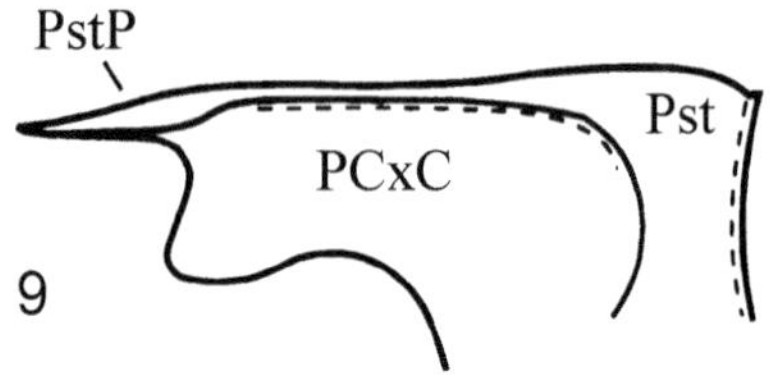


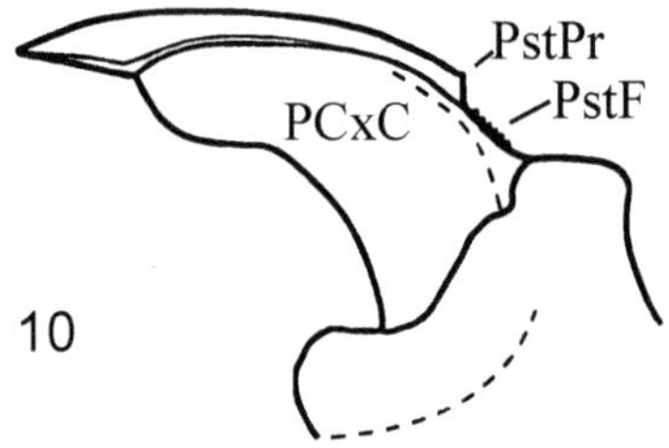


FIGURES 9.12 - 10.12. Prosternum and prosternal process in inverted, lateral view with procoxa and mesoventrite removed. Fig. 9.12 *Laccophilus* sp. (Laccophilinae); Fig. 10.12. *Hydroporus* sp. (Hydroporinae). PCxC - procoxal cavity, Pst - Prosternum, PstP - prosternal process, PstF - prosternal file, PstPr - prosternal prominence.

### Key to the Nearctic Genera
(Modified from Larson *et al.*, 2000)

1. Scutellum not visible ............................................ 2
— Scutellum (Sc, Fig. 2.12) visible ............................ 4

2(1). Prosternum apically and prosternal process on same plane in lateral aspect (Fig. 6.12A); body oval to ovate and elytron with apex broadly rounded; pro- and mesotarsi clearly 5-segmented (Fig. 8.12); metatarsomeres with apical margin strongly lobed (Fig. 4.12B) (Laccophilinae) ............................... 3
— Prosternum apically and prosternal process not in same plane in lateral aspect (Fig. 10.12B), prosternum descending between procoxae to prosternal process which is on a lower plane than midline of prosternum; body elongate and subparallel, elytron with apex produced as an acute spine (Fig. 36.12); pro- and mesotarsi distinctly pentamerous but with tarsomere 3 lobed apically (Fig. 7.12C); metatarsomeres not lobed apically (Fig. 4.12B) ........................................................ *Celina*

3(2). Metatibia with apical spurs emarginate at tip (Fig. 4.12B); total length 3.0 mm; widespread ........... .......................................................... *Laccophilus*
— Metatibia with apical spurs acute at tip (Fig. 4.12A); total length < 2. 5 mm; Florida ........... *Laccodytes*

4(1'). Prosternum in lateral aspect declivous anterior to procoxae, descendant between procoxae to prosternal process which is on a lower plane than midline of prosternum (Fig. 10.12); pro- and mesotarsi pseudotetramerous (Fig. 7.12), tarsomere 3 bilobed and hiding small tarsomere 4 (Hydroporinae) ................................................. 5
— Prosternum and its process on more or less same plane in lateral aspect (Fig. 9.12); pro- and mesotarsi pentamerous (Fig. 5.12, 6.12, 8.12) ....................... 32

5(4). Metepisternum not reaching middle coxal cavity; prosternal process short, apex ending in front of contiguous mesocoxae (Vatellini) ...................... 6
— Metepisternum (Eps, Fig.3.12) contacting lateral margin of middle coxal cavity; most specimens with prosternal process contacting metasternum medially and separating middle coxal cavities (Fig. 3.12) .......................................................................... 7

6(5) Pronotum not much narrower than elytra at base (width of pronotum about 0.9 subhumeral width of elytra), eyes large but not prominent, head and pronotum yellowish with elytra predominantly dark brown; size smaller, length to 3.9 to 4.1 mm ...... ........................................................ *Derovatellus*
— Pronotum much narrower than elytra at base (width of pronotum at base about 0.5 subhumeral width of elytra), eyes large and prominent; head, pronotum and elytra black; size larger, length 5.5 to 6.5 mm ...................................... *Macrovatellus*

7(5'). Body short and broad, length/width less than 1.55 (Fig. 37.12); metacoxal process with lateral lobe covering base of trochanter, apical margin with a deep sublateral emargination making hind margin of prosternal processes conjointly tripartite (Fig. 27.12); elytron with apex shortly produced (Hydrovatini) (Fig. 37.12) .................. *Hydrovatus*
— Body more elongate, or if short and broad, combination of characters not as above ......................... 8

8(7'). Male paramere jointed; pronotum with a sublateral plica (PnPl, Fig. 2.12, Fig. 39.12) on each side which contacts basal margin and extends anteriorly onto disc; many specimens with a basal plica on elytron (ElPL, Fig. 2.12) adjacent to pronotal plica; size minute to small, length = 1.6 to 2.5 mm (Bidessini) ......................................................................... 9
— Paramere not jointed, consisting of a single piece; pronotum without a sublateral plicae which contacts basal margin; or if present, body subglobose, length/width less than 1.60 mm; elytron without a basal plica; size various .................................. 17

9(8). Head dorsally with a transverse occipital line behind eye (OcL, Fig. 2.12) .......................................... 11
— Head dorsally lacking occipital line ..................... 10

10(9'). Metacoxal lines absent; eyes rudimentary ............. ...................................................... *Comaldessus*

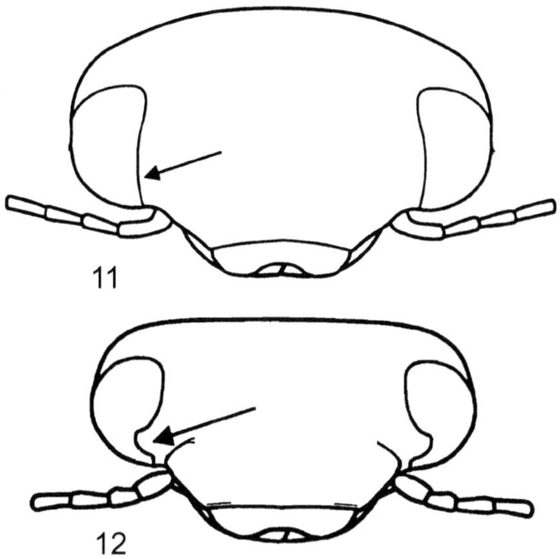


FIGURE 11.12 - 12.12. Anterior view of head showing shape of anterior margin of eye. Fig. 11.12. without emargination, *Hydaticus* sp., (Dytiscinae); Fig. 12.12. with emargination, *Agabus* sp., (Colymbetinae).

— Metacoxal lines (MtCxL, Fig. 3.12) present; eyes normally developed ....................................... *Uvarus*

11(9). Epipleuron with a basal excavation limited laterally by an oblique humeral carina (EplC, Fig. 3.12); elytron lacking a basal plica (PnPl, Fig. 2.12); clypeus thickened anteriorly, with two small tubercles but without a distinct rim ....................... *Brachyvatus*

— Epipleuron without an oblique humeral carina; elytron with a basal plica, but plica may be small and indistinct on some specimens; clypeus various, thickened or not ...................................................... 12

12(11'). Pronotum with basal plicae connected by an irregular transverse groove; elytron with a longitudinal carina extending posteriorly from level of basolateral pronotal plica; clypeus with anterior margin thickened and with two upturned tubercles; body broadly oval, coarsely punctate ................. ...................................................... *Anodocheilus*

— Pronotum with area between basal plicae flat, lacking a transverse groove; elytron without a longitudinal discal carina; clypeus with anterior margin modified or not; body shape and punctation various ................................................................. 13

13(12'). Metasternum medially with a pair of lines continuous with metacoxal lines posteriorly then running anteriorly and converging between mesocoxae, the area bounded by metasternal and metacoxal lines flattened or concave (in male); pro- and mesotarsi distinctly 5-segmented .................. *Bidessonotus*

— Metasternum without such lines; pro- and mesotarsi apparently 4-segmented, tarsomere 4 small and concealed between lobes of tarsomere 3 ....... 14

14(13'). Head with transverse occipital line (OcL, Fig. 2.12) separated from posterior margin of eye; genal line broadly separated from ventral margin of eye; clypeus with anterior margin prominently extended, in lateral aspect angulate between dorsal and ventral surfaces; body relatively elongate, pronotum cordate with greatest width anterior to middle and hind angle produced, acute and slightly sinuate; length = 2.6 mm; color brown, paler laterally ...... ......................................................... *Crinodessus*

— Head with transverse occipital line contiguous with posterior margin of eye; other characters various; widespread ..................................................... 15

15(14'). Clypeus with anterior margin more or less raised or thickened, many specimens with a well-defined marginal bead but some females with bead reduced to a weakly impressed line at middle; elytron of most specimens brightly patterned with a basically transverse pattern of pale spots or dark irregular fascia ......................................... *Neoclypeodytes*

— Clypeus with anterior margin not or only feebly thickened; if slightly thickened, area not delimited by an impressed line; elytron color various but if with strongly transverse pattern, surface densely punctate and setose ............................................ 16

16(15'). Sternum 6 broad, apical margin only slightly more pointed than semicircular; elytron pattern faint to strong, consisting of longitudinal dark vittae; metacoxal plate and epipleuron smooth, punctation obsolete or very fine and sparse; aedeagus in lateral aspect with apex produced ventrally at right angles into a spine- or beak-like process ............ ....................................................... *Neobidessus*

— Sternum 6 narrower, somewhat triangular in shape with apex narrowly rounded; elytron color various, unicolorous, with transverse vittae or broken irregular longitudinal lines and blotches (Fig. 39.12); metacoxal plate and epipleuron with punctation various but on most specimens punctures deep and conspicuous; aedeagus with apex simple, without a spine or beak-like process ................ *Liodessus*

17(8'). Metacoxal process not lobed or produced laterally over base of trochanter (Fig. 29.12); apex of process appressed to sternum 2 (Hyphydrini) ....... 18

— Metacoxal process with apicolateral angle extended over at least base of metatrochanter (Figs. 18-25.12); apex of metacoxal processes on distinctly lower plane than sternum 2, at least laterally .. 19

18 (17). Size minute to small, length = 1.1 to 3.0 mm; labial palpus with apical palpomere bifid apically; pronotum with hind margin sinuate laterally (Fig. 38.12), posterolateral angle acutely produced; prosternal process with apex sharply pointed medially ............................................ *Desmopachria*

— Larger, length = 4.0 to 5.0 mm; labial palpus with apical palpomere not distinctly bifid apically; pronotum with hind margin straight or slightly convex laterally, posterolateral angle right angled to slightly acute, not produced; prosternal process with apex subtruncate or very broadly rounded .. ........................................................... *Pachydrus*

19(17'). Metafemur (Fe3, Fig. 3.12) extending basally to level of metacoxal lobe (Fig. 18.12) (Laccornini) .......... ............................................................ *Laccornis*

13

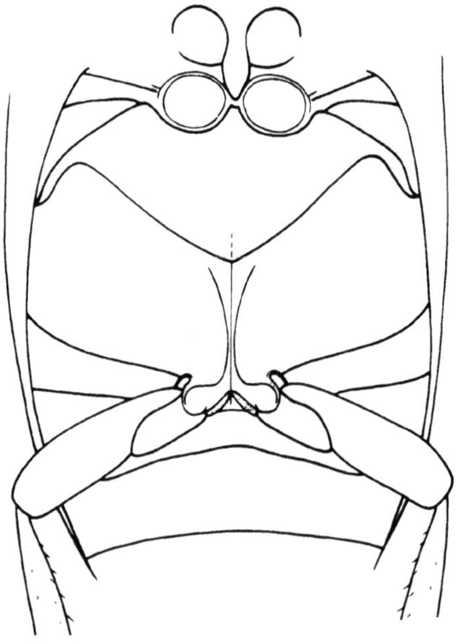

14

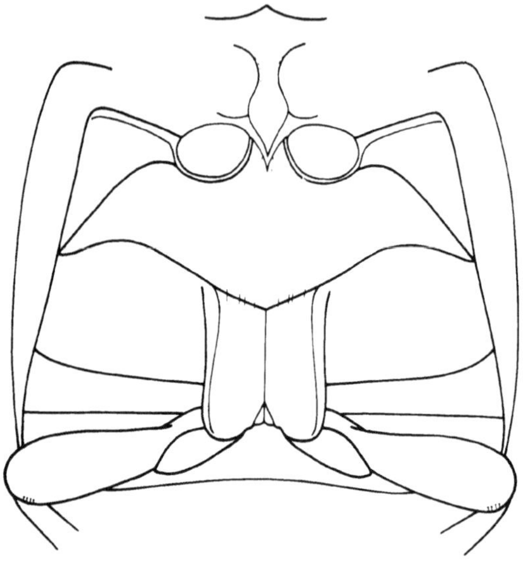

15

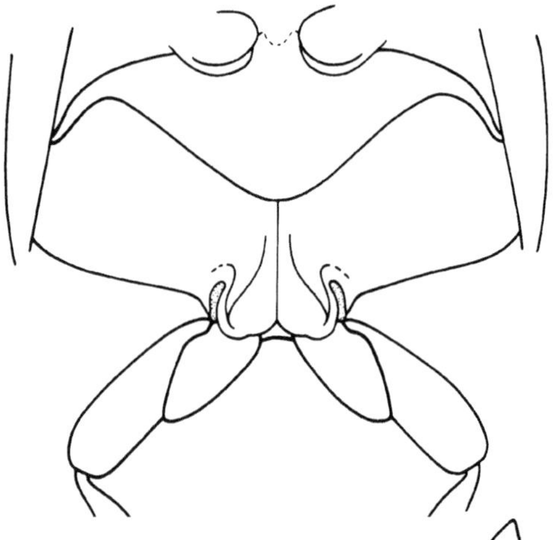

16

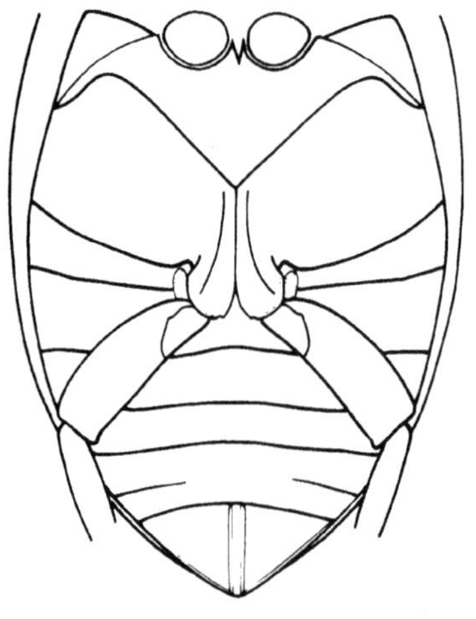

17

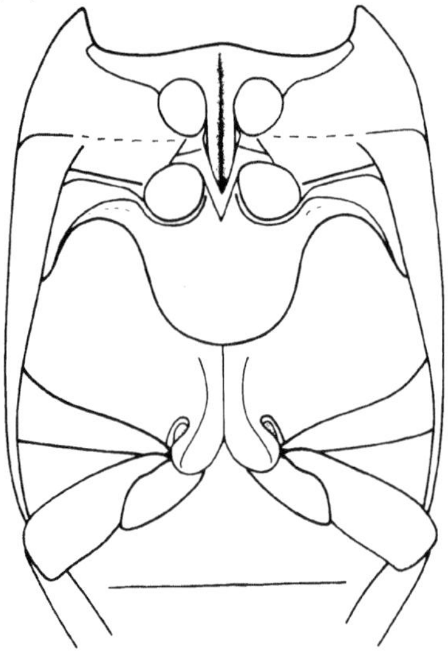

FIGURES 13.12 - 17.12. Ventral view of thorax and basal abdominal sterna. Fig. 13.12. Copelatinae; Fig. 14.12. *Agabinus* sp. (Colymbetinae, Agabini); Fig. 15.12. Hydrotrupinae; Fig. 16.12. Agabetinae; Fig. 17.12. *Matus* sp., (Colymbetinae, Matini).

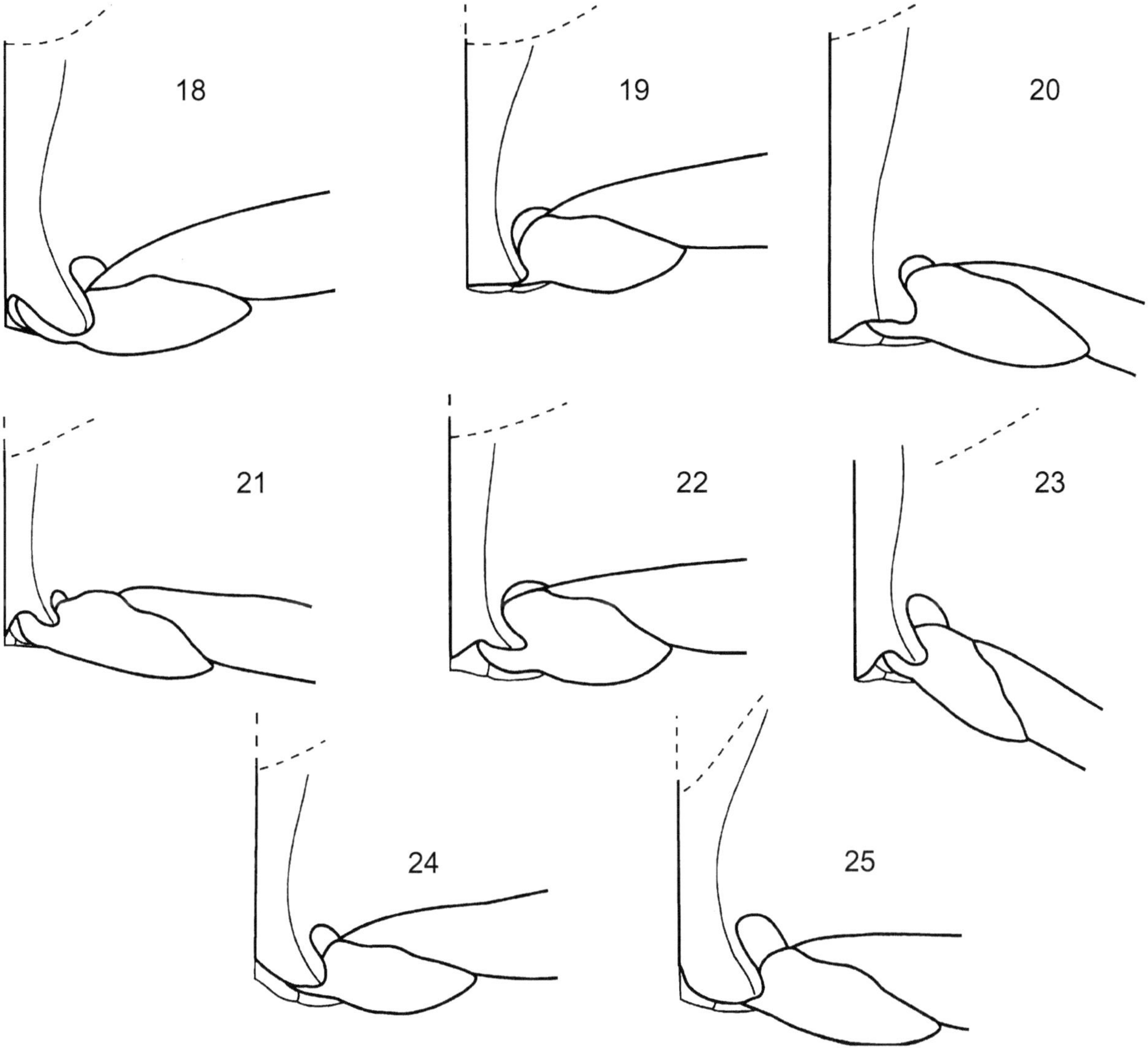


FIGURES 18.12 - 25.12. Left metacoxal process and leg base, ventral aspect. Fig. 18.12. *Laccornis* sp. (Hydroporinae, Laccornini); Fig. 19.12 *Hydroporus* sp. (Hydroporinae, Hydroporini); Fig. 20.12 *Neoporus* sp. (Hydroporinae, Hydroporini); Fig. 21.12 *Sanfilippodytes* sp.(Hydroporinae, Hydroporini); Fig. 22.12 *Hydrocolus* sp. (Hydroporinaei, Hydroporini); Fig. 23.12 *Heterosternuta* sp. (Hydorporinae, Hydroporini); Fig. 24.12 *Oreodytes* sp. (Hydorporinae, Hydroporini); Fig. 25.12 *Stictotarsus* sp. (Hydorporinae, Hydroporini).

— Metafemur basally separated (often only narrowly) from metacoxal lobes (Figs. 19-25.12) by metatrochanter (Hydroporini) .......................... 20

20(19'). Eyes lacking, represented by a vertical stria on each side of head; depigmented, subterranean forms . ........................................................................ 21

— Head with eyes well developed and conspicuous 22

21(20). Length = 3.4 to 3.7 mm; maxillary palpomeres 3 and 4 each with a fan-like brush of elongate setae; metacoxal processes conjointly with hind margin truncate; hind leg with natatorial setae ............... ........................................................ *Haideoporus*

— Length = 1.8 to 2.1 mm; maxillary palpomeres without setal brushes; metacoxal processes conjointly with hind margin produced medially; hind leg without natatorial setae ............................ *Stygoporus*

22(20'). Epipleuron with a transverse carina (EplC, Fig. 3.12) at humeral angle (carina may be weak or obsolete on some specimens of *H. novemlineatus*); body lacking conspicuous vestiture ...................... *Hygrotus*

— Epipleuron lacking a carina on humeral angle; vestiture various, body glabrous to distinctly setose ................................................................. 23

23(22'). Metacoxal processes conjointly truncate or excised laterally but medial line extending as far posteriorly as hind margin of lateral lobes (Fig. 9.12B-F) ...................................................................... 24

— Metacoxal processes conjointly with hind margin incised along medial line so that medial line does not extend as far posteriorly as level of hind margin of lateral lobes (Figs. 24.12, 25.12) ........... 30

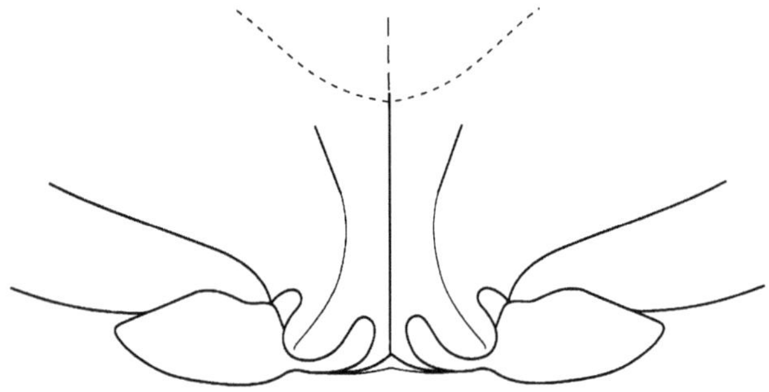

26

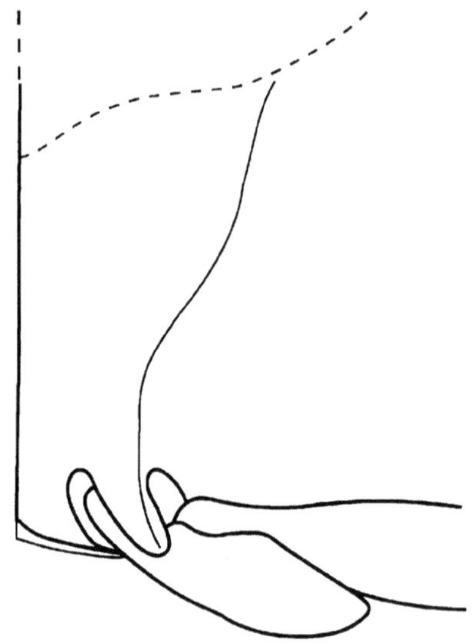

27

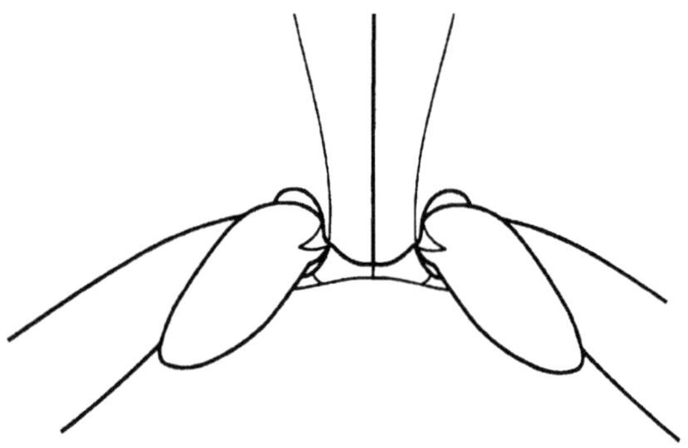

28

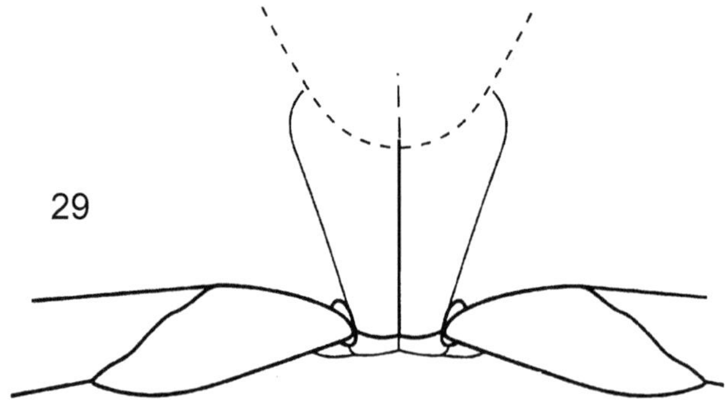

29

FIGURES 26.12 - 29.12. Metacoxal processes, ventral aspect. Fig. 26.12. *Celina* sp. (Hydroporinae, Methlini); Fig. 27.12. *Hydrovatus* sp. (Hydroporinae, Hydrovatini); Fig. 28.12. *Liodessus* sp. (Hydroporinae, Bidessini); Fig. 29.12. *Desmopachria* sp. (Hydroporinae, Hyphydrini).

24(23'). Metacoxal processes conjointly with hind margin straight or angularly projecting at middle (Figs. 19.12, 20.12) .................................................. 25

— Metacoxal processes with hind margin sinuate, incised on each side of midline (Fig. 9.12D-F) .... 26

25(24). Metacoxal processes conjointly with hind margin straight or slightly angulate medially (Fig. 19.12); ventral surface of most specimens mainly piceous to black, elytra of most specimens piceous to black but some specimens with variously developed pale maculation ........................................ *Hydroporus*

— Metacoxal processes distinctly angulate medially (Fig. 20.12C); ventral surface of body of most specimens yellow to reddish, elytra of most specimens distinctly maculate ....................... *Neoporus* (in part)

26(24'). Metatrochanter very large (Fig. 21.12), ratio length of metafemur/ length of metatrochanter = 1.9 to 2.2 .............................................. *Sanfilippodytes*

— Metatrochanter smaller (Fig. 22.12, 23.12), ratio of length of metafemur/ metatrochanter = 2.3 to 2.8 ........................................................................ 27

27(26'). Elytra maculate, most specimens with elytra dark with subbasal, postmedial and subapical transverse spots (either dark or pale areas may be greatly expanded on some specimens); male protibia without a basoventral emargination ....................... 28

— Elytra brown to piceous, without distinct maculations; male protibia with a large basoventral emargination ........................................................... *Hydrocolus*

28(27). Male with antennomere 4 (sometimes also 5) enlarged; protarsomere 1 with a basoventral cupule; prosternal process without a basal protuberance ........................................................... *Lioporeus*

— Male antennomere 4 not modified; male protarsomere 1 without a basoventral cupule; prosternal process with a distinct basal protuberance ................... 29

29(28'). Aedeagus with apex bifid in dorsal aspect; elytron pattern consisting basically of transverse dark maculations not longitudinally interconnected by narrow dark longitudinal projections; male anterior protarsal claw variously modified but not distinctly shorter than posterior........ *Heterosternuta*

— Aedeagus with apex entire; elytron with transverse dark fasciae longitudinally interconnected by 3 or 4 narrow longitudinal extensions; male anterior protarsal claw abbreviated and about one half length of posterior .................. *Neoporus* (in part)

30(23'). Ventral surface of body very densely, confluently punctate; metafemur with ventral surface densely

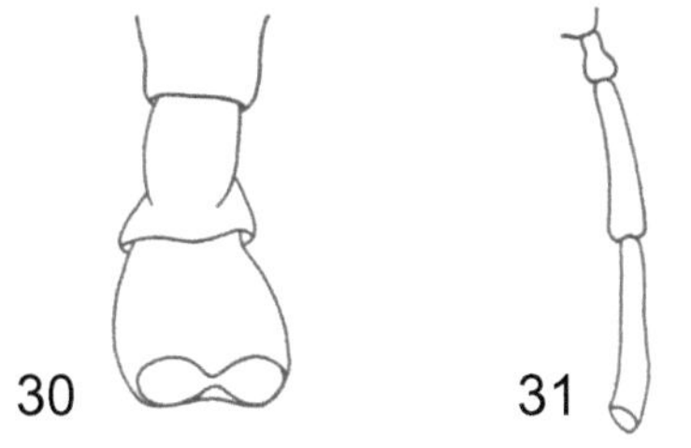


FIGURES 30.12 - 31.12. Labial palpus. Fig. 30.12 *Hydrotrupes* sp. (Hydrotrupinae); Fig. 31.12 *Agabus* sp. (Agabini, Colymbetinae).

punctate; pronotum without sublateral longitudinal grooves or impressions ............................ 31

— Ventral surface of body variously punctate but punctures not dense and confluent; metafemur with a single longitudinal row of setae on ventral surface; pronotum with distinct sublateral longitudinal grooves or lines .................................. *Oreodytes*

31(30'). Elytron with a subapical, lateral tooth .. *Nebrioporus*

— Elytron without a subapical, lateral tooth ............... ........................................................ *Stictotarsus*

32 (4'). Eye with anterior margin rounded (Fig. 11.12); male with protarsomeres 1 to 3 greatly widened to form a large, round (Fig. 5.12) or oval palette (Dytiscinae) ....................................................................... 33

— Eye with anterior margin notched above base of antenna (Fig. 12.12); male with protarsomeres 1 to 3 slightly to moderately widened to form an elongate to elongate-oval palette (Figs. 6.12, 8.12) 40

33 (32). Metatarsomeres 1 to 4 with dorsoapical margin bare; size large, length = 18 to 40 mm .................... 34

— Metatarsomere 1 to 4 with dorsoapical margin bearing a row of flat, yellow setae, smaller, length less than 18 mm .................................................... 36

34(33). Metatibia very short and broad (length/width less than 2.0), its length about 0.5 length of metafemur; metatibial spurs (MtSp, Fig. 2.12) large, ventral spur dilated and much broader than dorsal spur; prosternum in lateral aspect abruptly elevated, its anteroventral angle more or less right-angled (Cybistrini) ..................................................... 35

— Metatibia slender (length/width about 3.0), its length about 0.66 to 0.75 length of metafemur; ventral metatibia spur shorter than dorsal, spurs evenly narrowed to apex; prosternum not abruptly elevated anteriorly, in lateral aspect its anteroventral margin broadly obtuse (Dytiscini) .......... *Dytiscus*

35 (34). Male metatarsus with two subequal claws; female metatarsus with a long outer claw and a rudimentary inner claw; size smaller, total length less than 25 mm; elytron without lateral yellow stripe although lateral margin may be diffusely reddish .. ........................................................... *Megadytes*

— Metatarsus of male with one claw; female metatarsus usually with only one claw; larger, total length greater than 25 mm; elytron with a lateral yellow stripe .................................................... *Cybister*

36(33'). Prosternal process with apex sharply pointed; pronotum with lateral bead (PnB, Fig. 2.12); elytron with lateral margin in apical half bearing short, stout spines (Eretini) .................................. *Eretes*

— Prosternal process rounded apically; pronotum without lateral bead; elytron without spines along lateral margin .................................................... 37

37(36'). Metaventrite (MtV, Fig. 3.12) with anterolateral margin with metepisternum (MtEps, Fig. 3.12) straight; metatibial spurs each with apex acute (Hydaticini) ........................................................... *Hydaticus*

— Metaventrite with anterolateral margin with metepisternum arcuate; metatibial spurs each with apex finely emarginate (as in Fig. 4.12B) (Aciliini) ....................................................................... 38

38(37'). Elytron densely and coarsely punctate; male protarsal palette ventrally with 3 large, round and two dense clumps of smaller suckers; female elytron with or without longitudinal sulci, sulci when present bearing elongate setae .................................... *Acilius*

— Elytron densely punctate but punctures fine and inconspicuous; male protarsal palette with 15 to 30 adhesive suckers, without dense clumps of adhesive hairs; female elytron without sulci ............ 39

39(38'). Mesofemur with hind margin bearing a series of stiff, erect setae which are longer than width of femur at their base; mesotarsomeres 1 to 3 apicoventrally with a pair of elongate stout setae which are subequal to length of following tarsomere; many females with short, deep strioles on basal parts of elytron .......................................... *Thermonectus*

— Mesofemur with setae of hind margin shorter than width of femur at their base; mesotarsomeres with apical setae shorter than length of following tarsomere; female elytron without coarse strioles or at most with slightly deepened and elongated punctures ....................................... *Graphoderus*

40(32'). Metacoxal lines strongly convergent and almost contacting medial line anterior to metacoxal lobes then diverging anteriorly (Fig. 13.12); elytron of many specimens with narrow longitudinal striae (Fig. 32.12); metatibia without group of setae on posteroapical angle (Copelatinae) .................. 41

— Metacoxal lines various but not strongly convergent anterior to metacoxal lobes (Fig. 17.12); elytron without longitudinal striae; metafemur with or without setae on posteroapical angle .................... 42

41(40). Total length > 3.5 mm; pronotum laterally with distinct margin or pronotal bead; venter with, at least, basal abdominal sternites with short curved impressed lines or strioles; elytra with impressed longitudinal lines or striae (Fig. 32.12) ...... *Copelatus*

— Total length < 3.0 mm; pronotum laterally without or with very narrow margin or pronotal bead; venter with basal abdominal sternites smooth and polished; elytra smooth, without impressed lines..... ................................................ *Agaporomorphus*

42(40'). Labial palp very short, apical palpomere subquadrate (Fig. 30.12), (Hydrotrupinae)............ *Hydrotrupes*

— Labial palp elongate, apical palpomere much longer than wide (Fig. 31.12) ..................................... 43

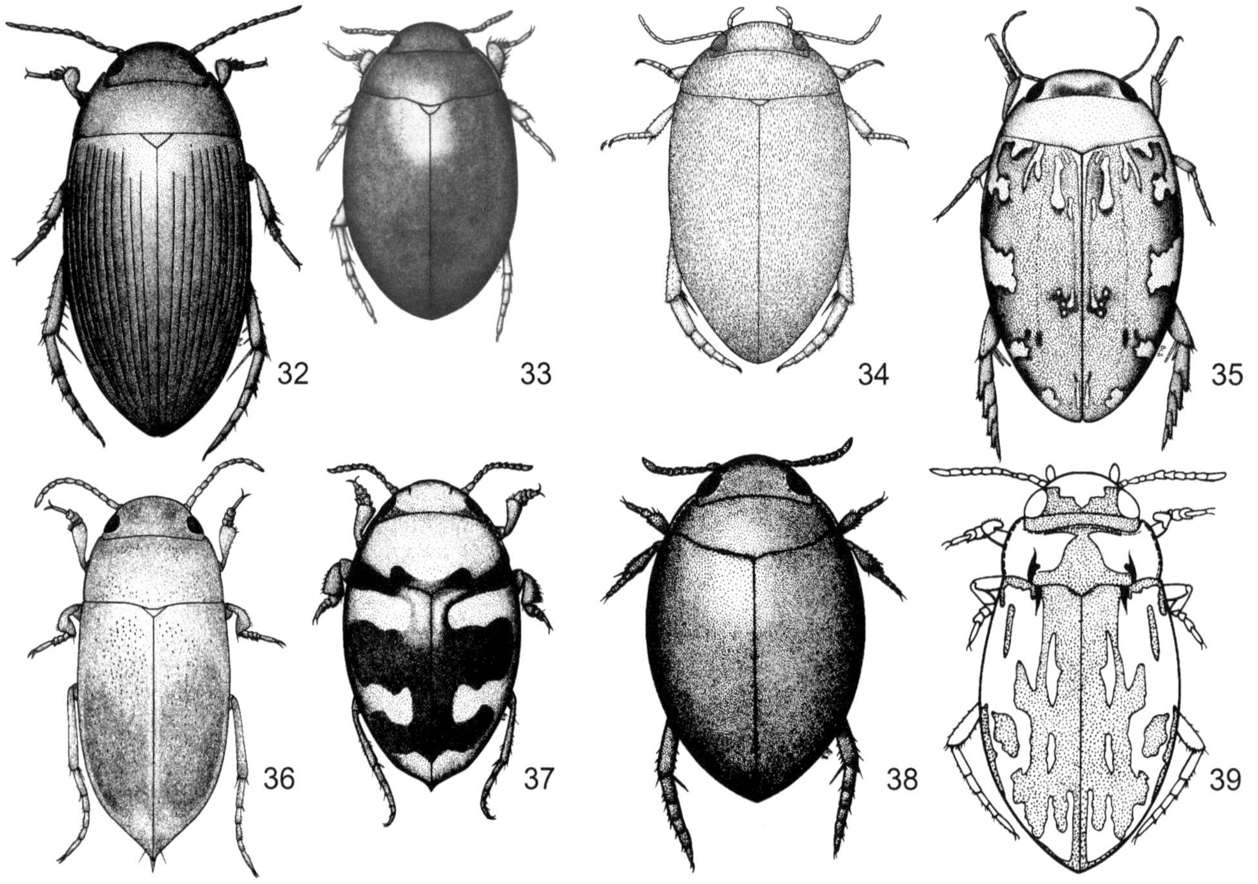


FIGURES 32.12 - 39.12. Habitus. Fig. 32.12 *Copelatus glyphicus* (Say), (Copelatinae); Figure 33.12 *Hydrotrupes palpalis* Sharp, (Hydrotrupinae); Fig. 34.12 *Agabetes acuductus* (Harris), (Agabetinae); Fig. 35.12 *Laccophilus maculosus* Say, (Laccophilinae); Fig. 36.12 *Celina hubelli* Young, (Hydroporinae, Methlini); Fig. 37.12 *Hydrovatus pustulatus* (Melsheimer), (Hydroporinae, Hydrovatini); Fig. 38.12 *Desmopachria convexa* (Aubé), (Hydroporinae, Hyphydrini); Fig. 39.12 *Liodessus affinis* (Say), (Hydroporinae, Bidessini).

43(42'). Pronotum without a lateral bead; abdominal sternum 6 with a pair of sharply impressed longitudinal lines (Fig. 16.12) (Agabetinae) ........................ *Agabetes*
— Pronotum with a lateral bead; sternum 6 various but without a pair of impressed longitudinal lines (Colymbetinae) ................................................ 44

44(43'). Prosternum and its process with a deep, medial, longitudinal groove (Fig. 17.12); clypeus with anterior margin broadly and distinctly emarginate medially (Matini) ............................................ *Matus*
— Prosternum and its process flat or longitudinally convex medially, without a longitudinal groove; clypeus with anterior margin more or less straight or at most faintly concave ....................................... 45

45(44'). Maxillary and labial palpi with apical palpomere asymmetrically bifid apically; prosternum in lateral aspect with anterior margin almost vertical, extending below level of apex of procoxae then angled at almost a right angle and level with midline of prosternal process (Coptotomini) .... *Coptotomus*
— Maxillary and labial palpi with apical palpomeres entire; prosternum in lateral aspect not or very slightly descendant at anterior margin ........................ 46

46(45'). Metafemur ventrally with a linear series of setae on posteroapical angle; or if setae present but irregular in distribution, metatarsal claws equal (Agabini) ........................................................................ 47
— Metafemur ventrally lacking setae on posteroapical angle or if present, setae distributed irregularly and forming a distinct row; metatarsal claws unequal (Colymbetini) ................................................. 50

47(46). Metacoxal lines parallel (Fig. 14.12), more or less parallel to apex of metacoxal lobes ............ *Agabinus*
— Metacoxal lines narrowest anterior to metacoxal processes, diverging posteriorly onto metacoxal processes and anteriorly toward metasternum ..... 48

48(47'). Labial palpus with penultimate palpomere triangular in cross-section, with three well defined longitudinal ridges .......................................... *Carrhydrus*
— Labial palpus more or less cylindrical, without longitudinal ridges .................................................. 49

49(48'). Metatarsal claws subequal in length, or if slightly dissimilar, claws short and not more than 0.25 length of metatarsomere 5; female sternum 6 with apex evenly rounded or truncate medially; gonocoxae

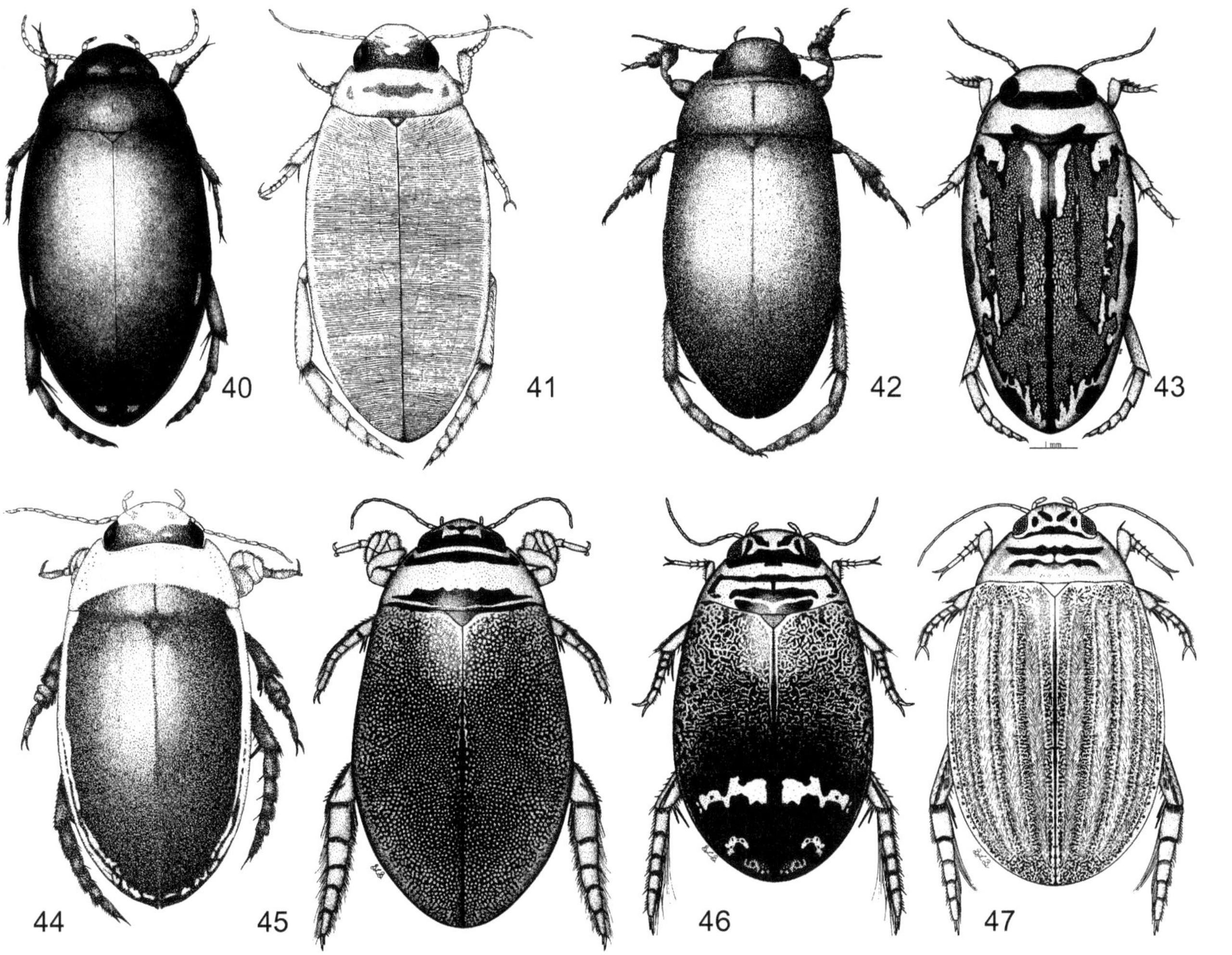


FIGURES 40.12 - 47.12. Habitus. Fig. 40.12 *Ilybius angustior* (Gyllenhal) (Colymbetinae, Agabini); Fig. 41.12 *Colymbetes sculptilis* (Harris) (Colymbetinae, Colymbetini); Fig. 42.12 *Neoscutopterus hornii* (Crotch) (Colymbetinae, Colymbetini); Fig. 43.12 *Coptotomus longulus* LeConte (Colymbetinae, Coptotomini); Fig. 44.12 *Hydaticus aruspex* Clark (Dytiscinae, Hydaticini); Fig. 45.12 *Graphoderus perplexus* Sharp (Dytiscinae, Aciliini); Fig. 46.12 *Acilius mediatus* (Say) (Dytiscinae, Aciliini); Fig. 47.12 *Acilius athabascae* Larson (Dytiscinae, Aciliini).

setose, not strongly sclerotized and without ventral teeth ................................................ *Agabus*

— Metatarsal claws unequal in length; female sternum 6 emarginate medially; gonocoxae strongly sclerotized, ventroapical margin toothed ....... *Ilybius*

50(46'). Elytron with numerous transverse, subparallel grooves ............................................ *Colymbetes*

— Elytron with sculpture various, generally of meshes of various sizes but without strong transverse grooves ......................................................... 51

51(50'). Pronotum without a lateral bead; large, length = 13.5 to 17 mm; elytron with large, irregular polygonal meshes ...................................... *Neoscutopterus*

— Pronotum with lateral bead although bead may be weakly defined on some specimens; size and color various but if large and black, elytron with small, irregular meshes or lines of sculpture partly effaced and densely punctate ..................................... 52

52 (51'). Prosternal process flat; pronotum widely margined laterally ................................................ *Hoperius*

— Prosternal process convex to more or less longitudinally carinate medially; pronotum with width of lateral bead various, usually narrow to moderate .... ............................................................... *Rhantus*

## Classification of Nearctic Genera

### Hydrotrupinae

**Diagnosis.** Eyes emarginate on anterior margin; pronotum with lateral bead; scutellum visible; pro- and mesotarsi distinctly five-segmented; prosternum and prosternal process on same plane (not declivous); metacoxal structure as in Fig. 15.12; metafemur with a distinct group of spine-like setae on apical posterior corner; metatarsus with two apical claws.

*Hydrotrupes* Sharp 1882 (Fig. 33.12).
Presently there is only one species assigned to this genus, *H. palpalis* Sharp, known from isolated localities along the Pacific

Coast from California north to Oregon. Adults and larvae can be found in hygropetric rock seeps in small pools at the head of the seep, and among algae in the flowing water of the seep itself. Adults closely resemble those of Agabini (subf. Colymbetinae) and most authors have placed this genus as an agabine. Beutel (1994) based on features of the larval head (e.g., lack of a closed mandibular feeding channel) assigned *Hydrotrupes* to a basal position within Dytiscidae; Alarie *et al.* (1998) described the larva and assigned it to Agabini based on larval chaetotaxy. Therefore the phylogenetic position of *Hydrotrupes* remains controversial.

## Copelatinae

**Diagnosis.** Eyes emarginate on anterior margin; pronotum with very narrow lateral bead; scutellum visible; pro- and mesotarsi distinctly five-segmented; prosternum and prosternal process on same plane (not declivous); metafemur without distinct group of spine-like setae on apical, posterior corner; metatarsus with two claws.

Burmeister (1976) has elevated this group, formerly a tribe within Colymbetinae, to subfamily rank based on features of the female ovipositor. Other studies based primarily on larval stages (DeMarzo 1976, Ruhnau 1986, Ruhnau and Brancucci 1984) confirm subfamily status. Guéorguiev (1968) and Pederzani (1995) provide keys to the known four genera of Copelatinae as defined by Ruhnau and Brancucci (1984): *Agaporomorphus* Zimmermann, *Aglymbus* Sharp, *Copelatus* Erichson and *Lacconectus* Motshulsky. Two of these genera occur in North America.

*Copelatus* Erichson 1832 (Fig. 32.12)
*Liopterus* Dejean 1833
*Exocelina* Broun 1886
*Pelocatus* Zaitzev 1908

Eight species are known from Canada and the United States and a key to species is presented in Young (1963a); 365 species were catalogued for the world fauna (Guéorguiev 1968). The species tend to colonize seasonally temporary pools but are common in marshes and swamps (Spangler 1962, Hilsenhoff 1993b). All of our species have impressed lines, termed elytral striae, on the elytra; some extralimital species are without elytral striae.

*Agaporomorphus* Zimmermann, 1921

Only one Nearctic species, *A. dodgei* Young 1989, known from Florida is assigned to this genus. Guéorguiev (1968) provides a catalogue and a key to the remaining four species which are found in Brazil.

## Agabetinae

**Diagnosis.** Eyes emarginate on anterior margin; pronotum without lateral bead; scutellum visible; pro- and mesotarsi distinctly five-segmented; prosternum and prosternal process on same plane (not declivous); metafemur without distinct group of spine-like setae on apical, posterior corner; metatarsus with two apical claws.

Burmeister (1990a) elevated Agabetini, formerly as a tribe of Colymbetinae, to subfamily status and placed it near the Laccophilinae. This placement is based primarily on the form of the female ovipositor (Burmeister 1976, 1990a) and requires further study (Nilsson 1989a). There is only one genus, *Agabetes*, and in North America there is one species with another species known from Europe (Nilsson 1989a)

*Agabetes* Crotch, 1873 (Fig. 34.12)

One species, *A. acuductus* (Harris), known from Ontario and Quebec east to Massachusetts and west to Wisconsin and Indiana and south to Texas and Louisiana. Adults of reddish-brown color (head and lateral margins reddish), with dense dorsal sculpture of numerous, short strioles and the last abdominal segment of males with two medial parallel grooves. It is typically a species of woodland ponds and marshes (Young 1954, Hilsenhoff 1992, Epler 1996).

## Laccophilinae

**Diagnosis.** Eyes emarginate on anterior margin; pronotum without lateral bead; scutellum not visible; pro- and mesotarsi distinctly five-segmented; prosternum and prosternal process on same plane (not declivous); metafemur without distinct group of spine-like setae on apical, posterior corner; metatarsus with only one apical claw.

*Laccophilus* Leach 1817 (Fig. 35.12)

Zimmerman (1970) has provided a revision of the species for Canada, United States and México. North of México there are 14 species. Members of *Laccophilus* have some of the more interesting color patterns found among dytiscids, particularly in the southwestern U. S. Zimmerman (1970) used the subspecies category more liberally than most dytiscid workers and many of the subspecies are geographically separated, color forms. Most species occur in relatively open pools and ponds in forested and grassland areas.

*Laccodytes* Régimbart 1895

The only species of this genus extending into the United States is *L. pumilio* (LeConte) known from Florida (Young 1954, Epler 1996). It is from 1.9 to 2.1 mm and looks like a small version of the more widespread *Laccophilus*. It commonly occurs in heavily shaded hammock ponds and pools.

## Hydroporinae

**Diagnosis.** Eyes emarginate on anterior margin; scutellum not visible (except *Celina*); pronotum with lateral bead, in most groups; pro- and mesotarsi pseudotetramerous, tarsomere 3 enlarged laterally, and partially enclosing small tarsomere 4; prosternum and prosternal process not on same plane (declivous); metafemur without distinct group of spine-like setae on apical, posterior corner; metatarsus with two claws.

Laccornini

**Diagnosis.** The relatively large size (for hydroporines) of 4.2 to 7.2 mm, reddish to very dark brown dorsal color, with the lateral margins of the pronotum and elytra continuous in dorsal aspect will serve to sort out most candidate specimens in collections. The most diagnostic feature is that the metacoxal process is lobed apically and extended to overlay the base of the metatrochanter with the base of metafemur extended on the dorsal margin past the metatrochanter and contacting or only very narrowly separated from metacoxal lobe (Fig. 18.12). Adult males exhibit an interesting and complex array of sexual dimorphism in the form of the middle antennomeres, protarsal claws, last visible sternite and elongate setae on the posterior margin of the meso- and metafemora. Adult females have genitalia with a tergite IX (=valvifer) large and distinct, gonocoxa moderately sclerotized, not fused, rounded apically, lateral and apical margins with dense setae. This latter combination of features is unique among members of Hydroporinae as in most hydropines tergite IX is reduced to a very small sclerite (Burmeister 1976).

Wolfe (1985, 1989) and Wolfe and Roughley (1990) have discussed the phylogenetic position of Laccornini and the single included genus, *Laccornis*. It is the most primitive group of the subfamily Hydroporinae. This phylogenetic position has been confirmed through the study of larvae (Alarie 1989).

*Laccornis* Gozis 1914

*Agaporus* Zimmermann 1919

This is the only genus presently assigned to the tribe Laccornini. It is Holarctic in distribution with 10 species of which nine occur in the Nearctic region (Wolfe and Roughley 1990). Most species are associated with organic debris and moss along the margins of small pools in deciduous and boreal forest or in peatland pools.

Vatellini

**Diagnosis.** Metepisternum not reaching the mesocoxal cavities, excluded from forming the coxal wall by the mesepimeron. Mesocoxal cavities contiguous with the prosternal process short and ending in front of contiguous mesocoxae. Our genera with the sutures separating the abdominal sternites shallow and effaced on midline.

The placement of Vatellini has caused confusion. Sharp (1882) used the form of the metepisternum and its role in the formation of the mesocoxal wall as one of his principal divisions of Dytiscidae. His Dytisci Fragmentati (Laccophilinae and Vatellini) have the metepisternum displaced from the mesocoxa by the mesepimeron whereas the Dytisci Complicati have the metepisternum contacting and forming part of the rim of the mesocoxal cavity. However, Wolfe (1985) placed Vatellini as relatively advanced members of Hydroporinae based on his phylogenetic analysis of 17 characters of the subfamily.

There are four genera (*Vatellus* Aubé-Neotropical, *Macrovatellus* Sharp-Neotropical, *Mesovatellus* Trémouilles-Neotropical and *Derovatellus* Sharp-circum-tropical) assigned to Vatellini (Pederzani 1995, Nilsson and Roughley1997a).

*Derovatellus* Sharp 1882

In North America north of Mexico there is only one taxon, *D. lentus floridanus* Fall, which is known from Florida. It is characterized by the features in the key above. Adults are commonly collected at light and among leaf debris in shaded pools.

*Macrovatellus* Sharp 1882

*Platydessus* Guignot 1955

Members of this genus are not recorded from north of México where a single species, *M. mexicanus* Sharp is known from as far north as Baja California (Leech 1948). It is characterized by the features in the key above. Adults are collected from organic debris on the margins of ponds and marshes.

Methlini

**Diagnosis.** The size (Nearctic specimens with total body length of 2.8 to 6.8 mm), yellow to reddish-brown color, elongate subparallel body with apical spines on the elytra, apical tergite and sternite will distinguish specimens of this tribe. The scutellum is visible and the pro- and mesotarsi are clearly five-segmented; however, protarsomere 3 is laterally lobed apically and encompasses the base of protarsomere 4.

This tribe contains two genera. The genus *Celina* is found in the Nearctic and Neotropical region whereas *Methles* Sharp is found in the Palearctic, Afrotropical and Oriental regions. The two genera differ in several characters and Wolfe (1985) suggested the tribe may be polyphyletic. They share in common the spine-like apices of the elytra, tergum 8 and visible sternum 6 and the laterally emarginate metacoxal processes which are interpreted as synapomorphies and may indicate a phylogenetic connection with the tribe Hydrovatini (Wolfe 1989).

Incorporation of Methlini into the Hydroporinae subfamily greatly increases the structural diversity within the subfamily. It does belong to Hydroporinae based on the definition of the subfamily provided by Wolfe (1985), although some authors (Franciscolo 1979a, Pederzani 1995, Trémouilles 1995) have elevated the group to subfamily status as Methlinae which would make the Hydroporinae paraphyletic.

*Celina* Aubé 1837 (Fig. 36.12)

*Hydroporomorpha* Babington 1841

There are nine Nearctic species (Young 1979). The body form and color of adults are highly characteristic and the form of the metacoxa is illustrated in Fig. 26.12. They are relatively small, 2.8 to 6.8 mm long, yellow to reddish-brown in color with a narrow, elongate shape with the pronotum rounded laterally and slightly discontinuous with the elytra, and with the apical portions of the elytra and last abdominal segment drawn out into a common posterior point. Larvae and adults are taken in ponds and marshes with thick organic substrate and aquatic vascular plants (Young 1979) when the substrate is well disturbed. The curiously sharpened posterior of adults may allow them to use air from plant roots and stems when burrowing in sediments (Hilsenhoff 1994).

### Hydrovatini

**Diagnosis**. The body form is characteristic; they are of very small to moderately small size (length = 1.8 to 6.2 mm), metacoxa as in Fig. 27.12, body shape broad and subglobose with the elytra and abdominal apices forming a common acute apex.

Hydrovatini includes two genera, *Queda* Sharp (generally of larger size, 2.5 to 6.2 mm, Neotropical - three species, Biström 1990), and *Hydrovatus* Motschulsky (generally smaller size, 1.6 to 5.3 mm, principally circum-tropical - 202 species worldwide, Biström 1997). Members of Hydrovatini share with Methlini the acutely produced elytral and abdominal apices (possibly adapted for obtaining air from aquatic vascular plants), a very broad head and similarly shaped metacoxal processes (possibly adapted for burrowing). Wolfe (1985, 1989) and Biström (1997) considered the posteriorly acuminate body a synapomorphy of the Methlini + Hydrovatini. Although members of these tribes are of very different body shape, the similar character states appear to reflect a similar life style of burrowing in the bottom mud and silt among rooted aquatic and emergent plants.

*Hydrovatus* Motschulsky 1853 (Fig. 37.12)
*Oynoptilus* Schaum 1868
*Hydatonychus* Kolbe 1883
*Vathydrus* Guignot 1954

Adult specimens are recognized by small size (Nearctic species, 1.8 to 4.0 mm), subglobose shape, margined clypeus, hidden scutellum, apically produced elytral apices, broad prosternal process broadly separating the mesocoxae, and the tripartite shape of apex of prosternal process. This genus contains eight species in North America north of México. Young (1956, 1963b) presented keys and descriptions for some of the North American species and Biström (1997) revised the genus on a world basis.

### Hyphydrini

**Diagnosis**. Members of Hyphydrini are very similar to Hydrovatini in terms of minute to moderate size (1.1 to 5.0 mm) and of a broad, subglobose body form. The clypeus is produced anteriorly as a rim extended over the base of the labrum. However they lack the acute elytral and abdominal apex and the form of the metacoxae (Fig. 29.12) is different in that medially it is depressed and fused to sternite 2 apically seemingly leaving the articulation of the hind legs more prominent.

Biström *et al.* (1997) split this tribe into two tribes, Hyphydrini Sharp and Pachydrini Biström *et al.* They assigned the genus *Desmopachria* to Hyphydrini and *Pachydrus* to Pachydrini. Alarie *et al.* (1997), based on seven larval synapomorphies, found evidence for classification as a single tribe. The North American members of these genera share two characters, the reduced anterior (dorsal) metatarsal claw and the base of the metatrochanter totally exposed and not covered by the lateral lobes of the metacoxal process, which separate them from all other hydroporines in the fauna and suggest a close relationship. Therefore, Larson *et al.* (2000) have maintained the more traditional classification of placing both genera in the same tribe. The two western hemisphere genera, *Desmopachria* and *Pachydrus*, are most diverse in the tropics and subtropics.

*Desmopachria* Babington 1841 (Fig. 38.12)

The small size (1.1 to 3.0 mm), short, broadly oval body, with margined clypeus, labial palpus with terminal palpomere bifid apically and form of the metacoxal processes (Fig. 29.12) will serve to separate adults of this genus.

They are common in a wide variety of habitats, primarily in standing water of various kinds and among organic substrate, moss and other debris; in the desert southwest some species are found in streams and gravelly pond margins. This is a diverse and complex genus in Neotropical and Nearctic regions of the New World. Young (1951) gave a key to species of North America but he has subsequently (1980, 1981a, 1981b, 1986, 1990b) revised many of the species and species groups.

*Pachydrus* Sharp 1882

The larger size (4.0 to 5.0 mm) and the labial palpus with apical palpomere not distinctly bifid apically will separate the one North American species, *P. princeps* (Blatchley), known from the southeastern United States. They are collected in lentic situations among leaf debris and other organic material as well as at light.

### Bidessini

**Diagnosis**. Size minute to small, length = 1.0 to 4.0 mm; the body shape is various, subglobose to elongate. The principal feature defining members of Bidessini is the form of the parameres of the male genitalia which is jointed and divided into two or three distinct segments.

This is a diverse group in warm temperate to tropical regions of the world. Despite their small size, bidessines exhibit considerable structural diversity. Biström (1988) recognized 30 genera for the world fauna and he restricted bidessines to those hydroporines which possess jointed or articulated parameres in the male. Use of this feature as a synapomorphy clearly distinguishes the bidessines but the relationships of this tribe to others is obscure. Young (1967a) placed the taxonomy of the American Bidessini on a sound footing by defining the genera and (Young 1969) compiled a checklist of taxa. Nine genera are recognized from the Nearctic region.

*Anodocheilus* Babington 1841
*Anodontochilus* Régimbart 1895

Young (1974) revised this genus and they are distinguished by the common presence of an irregular transverse groove connecting the basolateral plicae of the pronotum, elevated carinae on the elytra, the clypeus with the anterior margin thickened and with two upturned tubercles, the body form is broadly oval and coarsely punctate. There are two species found in North America north of Mexico: *A. exiguus* (Aubé), 1.5 to 1.7 mm, widespread throughout the southeastern United States to eastern Texas and *A. francescae* Young, 1.7 to 2.0 mm, known from Texas. They are most commonly collected in sandy or silty margins of streams and lakes or at light.

*Brachyvatus* Zimmermann 1919
There is only one North American species, *B. apicatus* (Clark), known from permanent ponds in the Peninsular region of Florida but not yet taken on the Florida Keys. It is quite small (1.6 -1.7 mm), and reddish-brown in colour.

*Bidessonotus* Régimbart 1895
The members of this genus are unique among bidessines in having the fore and middle tarsi distinctly 5-segmented, the metasternite medially with lines extending anteriorly and continuous with the more posterior metacoxal lines which converge anteriorly between the mesocoxae. In males the area bounded by metasternal and metacoxal lines is flattened or concave which seems to form a mechanism for grasping females during copulation. The species range in length from 1.7 to 2.3 mm and have an elongate shape, with the pronotum and elytra slightly discontinuous in outline.

The genus contains 29 described species, the majority of which occur in Central and South America. Four species are known from Canada and the United States. J. Balfour-Browne (1947) revised the genus and Young (1990a) reviewed the genus and described additional species. The species are very similar to one another but can be readily identified by the peculiar shapes of the aedeagus. Young (1990a) stated that these beetles occur in both small pools and very slow streams where there is dense vegetation and an accumulation of organic debris. Specimens are often taken at light.

*Comaldessus* Spangler and Barr 1995
The absence of metacoxal lines, the rudimentary eyes and subterranean habits distinguish the only member of the genus, *C. stygius* Spangler and Barr 1995, known only from Comal Springs, Comal Co., Texas.

*Crinodessus* Miller 1997
This is a relatively elongate (2.6 mm) bidessine with a cordate pronotum. There is only one species, *C. amyae* Miller 1997, known from clear desert streams in Arizona, New Mexico and Texas (Miller 1999).

*Liodessus* Guignot 1939 (Fig. 39.12)
Small to medium length bidessines (1.5 to 2.3 mm) with an elongate shape, metacoxal processes as in Fig. 28.12, and some species with the pronotum more rounded laterally than elytra so that the outline, in dorsal aspect, is discontinuous. Elytron with the color pattern various, unicolorous, with blotches and elongate stripes or with transverse markings. The metacoxal plates are finely to coarsely punctate. These beetles occur in lentic habitats or in very shallow water at the margin of streams. They are usually associated with dense vegetation or filamentous algae.

As presently defined, the genus occurs in North and South America, Africa and Australia. The genus is probably polyphyletic. Young (1969) listed 23 New World species. Larson and Roughley (1990) reviewed the North American species but they recognized that certain common and widespread species were probably species complexes. Miller (1998) studied one of these complexes and showed that it contained four similar species separable only on the shape of the aedeagus. There are now nine species known from North America north of México.

*Neobidessus* Young 1967
Adult members of *Neobidessus* are relatively small (1.8 to 2.2 mm), elytral pattern when present consisting of longitudinal dark vittae, elytron with a discal stria (=accessory stria) of punctures sunk down into a longitudinal groove or sulcus, metacoxae smooth and at most finely punctate, with the last visible sternite broad and semicircular in shape.

There are eight species in Central and North America (Young 1977) with only two species, *N. pullus* (LeConte) ( widespread in the southeastern United States) and *N. pulloides* Young (Texas, Young 1977, 1981c) reaching the United States. Adults prefer standing water of various kinds from temporary ponds, newly formed pools in quarries and rockpits, drainage ditches, canals, roadside pools to swamps and pools in intermittent or permanent streams. As with most bidessines they are abundant at light.

*Neoclypeodytes* Young 1967
The body form is elongate oval with the lateral margins of the pronotum slightly to distinctly discontinuous with those of the elytra in dorsal aspect. They range from 1.8 to 2.4 mm in length. Color various but many specimens with the elytron dark with strongly contrasting pale markings. The clypeus has the anterior margin thickened and raised and limited behind by a distinctly impressed line, at least at middle. The metacoxal plates are moderately to strongly punctate.

This genus is distinctive when only males are examined but it may be closely related to *Liodessus* if these secondary, sexual features are ignored. Some female specimens which have no or little modification of the clypeus will run to *Liodessus* in the key above. These beetles usually occur on sand or gravel substrates, often at the margin of small streams or stream-side pools. This genus is represented by about 20 described species in western North America and Central America.

*Uvarus* Guignot 1939
The body shape of these smallish bidessines (1.5 to 1.9 mm) is various, from broadly oval to elongate-oval and the punctation is various, with the metacoxal plate ranging from smooth to strongly punctate. The North American species can be separated into two distinctive groups, referred to as the *granarius*- and *lacustris*-groups (Larson *et al.* 2000). There is much taxonomic confusion within the

genus as it is presently defined and it is known from North and Central America (unconfirmed in South America), Africa, Asia and Australia. Biström (1988) revised the African species but concluded that the characteristics on which the genus is based are plesiomorphic and that the genus may be paraphyletic. Young (1969) listed 16 species of *Uvarus* from the Americas, of which nine are recognized for the United States (Larson *et al.* 2000). However, the status of several species is uncertain and the genus requires revision, preferably on a world basis.

## Hydroporini

**Diagnosis.** Minute to moderate sized hydroporines, total body length = 1.8 to 7.1 mm; body shape various, subglobose to elongate and parallel-sided; color and sculpture various; body glabrous to densely setose; clypeus is simple on most specimens but with an anterior bead or rim extended anteriorly over the base of the labrum in some species; pronotum with lateral bead and lacking the basolateral plicae characteristic of bidessines but some specimens with a longitudinal sublateral crease on disc; scutellum hidden or apex visible behind base of pronotum; elytron lacking basal plica and the elytral apex is not produced into a spine but some specimens with lateral margin with an acute subapical tooth; prosternal process declivous, slightly to strongly descendent from plane of midline of prosternum; metacoxal process (Figs. 19-24.12) with a well-developed lateral lobe which covers base of metatrochanter; pro- and mesotarsi pseudotetramerous, tarsomere 4 small and hidden between lobes of tarsomere 3; metatarsus with two equal to subequal claws.

The classification of this species-rich and structurally diverse group of beetles is problematic and they may not represent a monophyletic tribe. No single synapomorphy defines it and almost all tribal characters vary within the tribe. There can be a tendency to erect separate tribes from Hydroporini for unique well-supported groups (e.g., Nilsson and Holmen 1995 placed *Hygrotus* in a separate tribe, Hygrotini Portevin). The same kinds of systematic problems exist at the genus level where such genera as *Hygrotus* and *Hydroporus sensu lato* are structurally and phylogenetically diverse and need to be assembled into monophyletic sets. The creation of these within *Hydroporus* in particular is difficult because of the large number of included species, the structural variation exhibited among species and the observation that many genera are quite widespread. Despite sharing many elements of a common Holarctic fauna, North American workers used a conservative, generic definition due to lack of information or difficulty of applying Palearctic genus concepts. In certain examples where appropriate research has demonstrated the monophyly and need for a separate genus these are accepted.

*Haideoporus* Young and Longley 1976

The combination of size (total body length 3.4 to 3.7 mm), form of the metacoxal process with the apical margin truncate, elongate setae on the maxillary palpomeres and posterolateral margin of the pronotum, reduced eyes, and subterranean habits will distinguish the only member of the genus, *H. texanus* Young and Longley (1976), known only from the Edwards Aquifer system in Texas.

*Heterosternuta* Strand 1935

*Heterosternus* Zimmermann 1919 (preoccupied)

*Heterostethus* Falkenström 1938 (preoccupied)

This is the *Hydroporus pulcher*-group of Fall (1923). Specimens are of small to medium size (total body length = 2.7 to 4. 6 mm). The body shape is narrow and elongate oval, relatively subparallel at midlength in dorsal aspect and they are somewhat depressed in lateral aspect. Most specimens have the dorsal surface maculate with the pronotum black or yellow to reddish-brown and the elytron brown to black with variously developed or fused basal, medial and subapical transverse yellow bands. The clypeus is not margined and the pronotum is without a sublateral longitudinal crease or line on each side. The elytral epipleuron near the base is without a carina near the humeral angle. Metacoxal processes (Fig. 23.12F) conjointly with hind margin sinuate, emarginate on each side of medial line and apex produced along medial line; lateral lobe covering base of metatrochanter, metafemur not touching metacoxal lobe. Metatrochanter relatively large. Male protibia without a basoventral emargination. Aedeagus slender with the apex deeply notched in dorsal aspect.

The genus was revised by Matta and Wolfe (1981), who discussed 13 species. This is a monophyletic group recognized by a uniform body shape, generally bright colors and apically bifid median lobe of the aedeagus. Alarie (1991, 1992) using larval features listed five synapomorphies of *Heterosternuta* + *Neoporus*. The species of *Heterosternuta* are found in the eastern half of North America and they are principally Appalachian. They prefer interstitial, gravelly substrate along the margins of temporary or permanent streams (sometimes under undercut stream banks or in leaf pack of pools), springs, and rocky margins of pools and lakes.

*Hydrocolus* Larson *et al.* 2000

This is the *Hydroporus oblitus*-group of Fall (1923). Specimens are of small to moderate size (total body length = 2.6 to 4.7 mm). The body shape is elongate oval to subparallel with the lateral outline continuous to slightly discontinuous in dorsal aspect and distinctly flattened to somewhat depressed dorsoventrally in lateral aspect. The color of the dorsal surface is yellowish- or reddish-brown to rufopiceous and the elytra is not bicolored; the ventral surface is mainly black. The antenna and clypeus are not modified; the pronotum is without sublateral longitudinal creases and the epipleuron is without a carina near the humeral angle. The metacoxal processes (Fig. 22.12) with hind margin broadly emarginate each side of medial line, apex produced along medial line; lateral lobe covering base of metatrochanter, metafemur not touching the metacoxal lobe. Metatrochanter of moderate length. Male protibia with a basoventral emargination. Median lobe of aedeagus with the apex simple (not notched) or complexly flanged in *Hydrocolus deflatus* (Fall).

Recognition of members of *Hydrocolus* is based on the presence of a strongly notched protibia of males, the apex of the scutellum visible at high magnification, smaller metatrochanters and the form of the medial apex of the metacoxae. Larson *et al.* (2000) described and keyed 10 Nearctic species and tentatively assigned one northern Palaearctic species, *H. picicornis* (Sahlberg) to *Hydrocolus*. Adults are most common in springs, seeps and bogs with occasional occurrence in pools in Cypress swamps and moss along the edge of waterfalls.

*Hydroporus* Clairville 1806
*Hydrocoptus* Motschulsky 1853
*Hydatoporus* Gistel 1856
*Sternoporus* Guignot 1945
*Hydroporinus* Guignot 1945
*Hydroporidius* Guignot 1949

This is the *Hydroporus niger-tenebrosus* group of Fall (1923). Specimens are of small to moderate size (total body length = 2.6 to 7.1 mm). The body shape is broad and more or less elongate-oval and continuous or discontinuous in dorsal aspect and they are moderately convex in lateral aspect. Most specimens have the dorsal surface reddish to black, some specimens with elytra more or less distinctly maculate with yellow and the ventral surface of the metaventrite and abdominal sternites very dark. Males with antenna not modified. The clypeus is not modified and the pronotum is without a sublateral longitudinal crease or line on each side. The elytral epipleuron is without a carina near the humeral angle Metacoxal processes (Fig. 19.12) conjointly with hind margin truncate or slightly produced medially, but not strongly bisinuate (emarginate each side of midline) or incised along midline; lateral lobe covering base of metatrochanter; metafemur not touching metacoxal lobe. Metatrochanter of moderate length. Male protibia of most species without a basoventral emargination. Median lobe of aedeagus slender with the apex entire (not notched).

The genus is revised by Larson *et al.* (2000) who treated 55 species assigning most species to species-groups (cf. Nilsson and Holmen 1995). Nilsson (1989b) synonymized several genus-group names proposed for European species with *Hydroporus*. It is not certain that the genus is monophyletic and species-level differences are subtle and difficult without access to a good reference collection. The concept of the genus as *Hydroporus (s. str.)* is now more similar to that of the European definition of the genus and should avoid many of the problems that have plagued the generic placement of many species and the overall taxonomy of the group in the past. Adults are diverse and abundant in many shallow aquatic habitats, especially near the water margin amongst emergent vegetation.

*Hygrotus* Stephens 1828
*Coelambus* Thomson 1860

Size and shape of members of *Hygrotus* are quite various: (total body length = 2.1 to 5.6 mm); body shape from subglobose to elongate-oval with the lateral outline continuous or not. The color pattern of many species consists of having the elytra distinctly maculate with linear vittae or more diffuse blotches. The clypeus with the anterior margin unmodified or anterior bead or rim extended anteriorly over the base of the labrum in some species. The pronotum is without a sublateral longitudinal impressed line or crease on each side. The elytral epipleuron with a fine, transverse carina at humeral angle. The metacoxal processes with rounded lateral lobes covering the bases of the trochanters and the hind margin of processes conjointly truncate or sinuate and somewhat emarginate on each side of the middle.

The concept of the genus used here is broader than that of most European workers who generally regard *Coelambus* and *Hygrotus* as distinct genera (but see Nilsson and Holmen 1995). When sorting hydroporines in collections, the majority of specimens may be separated by their shiny cuticle, striking elytral markings and then confirmed by examining the base of the epipleuron for the carina. Nilsson and Holmen (1995) assigned the Holarctic genus *Hygrotus* and the Old World genera *Heroceras* Guignot, *Herophydrus* Sharp, *Hyphoporus* Sharp and *Pseudhydrovatus* Peschet to the tribe Hygrotini. This tribe may well be valid but we have not accepted it because the remainder of genera are not yet well-known and an overall phylogeny of Hydroporini *s. lat.* has not been completed. The Nearctic species were revised by Anderson (1971, 1976, 1983) and Young and Wolfe (1984) and total about 43 species. They generally prefer exposed lentic habitats but certain other species prefer mossy pools or saline lake margins.

*Lioporeus* Guignot 1950
*Falloporus* Wolfe and Matta 1981

Specimens are of medium size (total body length = 3.8 to 4.3 mm). The body shape is narrow, elongate oval and tapering posteriorly, in dorsal aspect, and they are somewhat depressed, in lateral aspect. The dorsal color is yellow to reddish with variously developed darker markings. Males with one or two median antennomeres expanded. The clypeus is not margined and the pronotum is without a sublateral longitudinal crease or line on each side. The elytral epipleuron near the base is without a carina near the humeral angle. Metacoxal processes (as in Fig. 23.12) conjointly with hind margin sinuate, emarginate on each side of medial line and apex produced along medial line; lateral lobe covering base of metatrochanter, metafemur not touching metacoxal lobe. Metatrochanter of moderate size. Male protibia without a basovental emargination. Aedeagus slender with the apex entire (not notched).

The two species in this exclusively Nearctic genus were revised by Wolfe and Matta (1981). They range over much of the United States east of the Great Plains. Most specimens are collected from undercut stream banks in small, clear, shaded streams.

*Nebrioporus* Régimbart 1906
*Potamonectes* Zimmermann 1921
*Hydroporus*, in part, of authors
*Deronectes*, in part, of authors

Most recent authors have referred to these species as a member of *Deronectes* Sharp. Specimens are of moderate size (total body length = 4.4 to 5.6 mm). The body shape is elongate and the lateral outline is strongly discontinuous, in dorsal aspect, and they are moderately to quite convex in lateral outline. The color of the dorsal surface is predominantly yellow with darker markings in longitudinal vittae; the ventral color is reddish-yellow to black. The antenna and clypeus are not modified; the pronotum is without sublateral longitudinal creases and the epipleuron is without a carina near the humeral angle. Elytron with impressed longitudinal striae and lateral margin with a small, acute subapical tooth. Metacoxal processes (as in Fig. 25.12) conjointly with hind margin medially emarginate and with the apex incised; lateral lobe covering base of metatrochanter, metafemur not touching metacoxal lobe. Metatrochanter relatively large. Male protibia without a basoventral emargination. Median lobe of aedeagus slender and with apex entire (not notched).

The genus *Nebrioporus*, as defined by Nilsson and Angus (1992), is found predominantly in the Palaearctic and Ethiopian regions but is represented in the New World by only three species, one of which is Holarctic. The Nearctic species have been variously assigned to *Hydroporus*, *Deronectes* or *Potamonectes*. Keys to species are presented in Shirt and Angus (1992) and Larson *et al.* (2000). Adults are collected on mineral substrates along the margins of small streams, rivers and lake margins.

*Neoporus* Guignot 1931
*Circinoporus* Guignot 1945

This is the *Hydroporus undulatus*-group of Fall (1923). Specimens are of small to medium large size (total body length = 2.2 to 6.4 mm). The body shape of most species is elongate oval and parallel-sided to oval and broadly rounded laterally. In terms of color, most specimens are brightly colored with the head, pronotum and ventral surface of the body yellow to pale reddish, with the elytron black to dark brown variously marked with transverse or longitudinal yellow bands; other species are predominantly dark brown to black. Male with median antennomeres unmodified. The clypeus is not modified in most species but some species with clypeus swollen and somewhat protuberant and the pronotum is without a longitudinal crease or line on each side. The elytral epipleuron near the base is without a carina near the humeral angle. Metacoxal processes (Fig. 20.12) conjointly with hind margin angulate medially and not evidently emarginate each side of midline; lateral lobe covering base of metatrochanter, metafemur not touching metacoxal lobe. Metatrochanter of moderate size. Male protibia without a basoventral emargination. Median lobe of aedeagus slender and apex entire (not notched).

*Neoporus* is known only from the Nearctic region and is comprised of 40+ species distributed over much of North America but the majority of species are found in the Appalachian region, east of the Great Plains (Wolfe 1984). Identification of species is difficult and requires reference to Fall (1923), and Young (1940, 1967b, 1978, 1984) and Wolfe (1984). About the only common feature of the habitat of species of *Neoporus* is that the majority of species live in protected, depositional areas of lotic habitats although some species occur along undercut stream banks of streams and rivers, lake shores, reservoirs, beaver ponds, springs and peatland streams.

*Oreodytes* Seidlitz 1887
*Neonectes* Balfour-Browne 1944
*Deuteronectes* Guignot 1945
*Nectoporus* Guignot 1950

Specimens are of small to moderate size (total body length = 2.8 to 5.4 mm) The body shape varies from short and broadly oval to elongate and parallel-sided with the lateral outline continuous or discontinuous in dorsal aspect and they are moderately to highly convex in lateral aspect. Most specimens have the dorsal color yellowish with dark blotches on the head and disc of pronotum variously maculate, and elytra with longitudinal dark vittae and the ventral surface mainly black; *Oreodytes quadrimaculatus* dorsally is dark with two yellow spots on each elytron. Antennae and clypeus not modified. Pronotum with a longitudinal sublateral longitudinal impression or crease on each side of disc. Elytron without carina near humeral angle, some specimens with a subapical tooth. Metacoxal processes (Fig. 24.12) conjointly with hind margin incised along midline (except produced along medial line in *O. quadrimaculatus*) lateral lobe covering base of metatochanter, metafemur not touching metacoxal lobe. Male protibia without or with a shallow basoventral emargination. Median lobe of aedeagus with apex simple and entire (not notched).

The genus *Oreodytes* is Holarctic in distribution and seems to form a good, monophyletic unit which may or may not include the problematic *O. quadrimaculatus* (Horn) which differs from other members of the genus somewhat in habitat preferences, color and structural features. Zimmerman (1985), Larson (1990) and Alarie (1993) have revised the Nearctic species and a key to all species is presented in Larson *et al.* (2000). The adults are either lotic or occur along the exposed shorelines of cold lakes where they prefer eroded, gravelly or rocky substrates.

*Sanfilippodytes* Franciscolo 1979b

This is the *Hydroporus vilis*-group of Fall (1923). Specimens are relatively small (total body length = 2.1 to 4.3 mm). The body shape is evenly elongate to oval in dorsal aspect and they are moderately convex to distinctly flattened, in lateral aspect. The color is pale brown to reddish-black and many species have the pronotal disc darker than elytra whereas as few adults have the elytra bicolored. The eyes are relatively small and the pronotum is without a sublateral longitudinal crease or line on each side. The elytral epipleuron near the base is without a carina near the humeral angle. Metacoxal processes (Fig. 21.12) conjointly with hind margin sinuate, emarginate on each side of medial line and apex produced along medial line; lateral lobe covering base of metatrochanter, metafemur not touching metacoxal lobe. Metatrochanter (Fig. 21.12) enlarged and elongate in most species. Male protibia without a basovental emargination.

The members of *Sanfilippodytes* are exclusively Nearctic and range collectively from Alaska to Mexico. The type species, *Sanfilippodytes sbdornii* Franciscolo, was described from a cave in

Mexico and Franciscolo (1979b, 1983) has maintained that this is a troglodytic species. However, it seems appropriate to use this name for all members of the genus as the type species has most of the synapomorphies of the genus. The taxonomy of species of *Sanfilippodytes* is extremely difficult. Rochette (1983a, 1983b, 1983c, 1985, 1986) and Larson *et al.* (2000) provide a provisional key to 26 described species. The species are best characterized by very subtle differences in the form of the median lobe of the aedeagus which in most but not all species is notched apically. Specimens are collected from caves, among gravel and stones along margins of springs, creeks and rivers or cold springs and small creeks, mosses along the margins of springs and seeps and along the margins of alpine and subalpine lakes.

*Stictotarsus* Zimmermann 1919
*Trichonectes* Guignot 1941
*Hydroporus*, in part, of authors
*Deronectes*, in part, of authors
*Potamonectes*, in part, of authors
Most recent authors have referred to these species as members of *Deronectes* Sharp. Specimens are of small to moderate size (total body length = 3.3 to 6.4 mm). The body shape varies from short and broadly ovate to elongate and the lateral outline continuous to strongly discontinuous, in dorsal aspect, and they are moderately to quite convex, in lateral outline. Most specimens have the dorsal surface predominantly yellow with darker markings in the form of blotches or longitudinal vittae; some species are predominantly black. The antenna and clypeus are not modified; the pronotum is without sublateral longitudinal creases and the epipleuron is without a carina near the humeral angle. Metacoxal processes (Fig. 25.12) conjointly with hind margin medially emarginate and with the apex incised; lateral lobe covering base of metatrochanter, metafemur not touching metacoxal lobe. Metatrochanter relatively large. Male protibia without a basoventral emargination. Median lobe of aedeagus slender and with apex entire or shallowly emarginate.

The North American members of this Holarctic genus have been assigned to various genera but Nilsson and Angus (1992) redefined *Stictotarsus* and assigned many Nearctic species to the genus. The majority of our 23 species occur in the southwestern United States. Pederzani (1995), on the other hand, restricted the definition of *Stictotarsus* to two Palaearctic species and moved the Nearctic species to *Potamonectes*, subgenus *Trichonectes* Guignot. Keys to Nearctic species are available in Leech and Chandler (19 56), Zimmerman and A. Smith (1975), Zimmerman (1982) and Larson *et al.* (2000). Adult beetles are collected in lentic and lotic habitats where they prefer eroded substrates with silt and a lack of aquatic plants with some species known to occur in saline waters.

*Stygoporus* Larson and LaBonte 1994
The combination of size (total body length 1.8 to 2.1 mm), form of the metacoxal process with the apical margin angularly extended medially, lack of elongate setae on the maxillary palpomere, lack of eyes, lack of natatorial setae on the hind legs will distinguish the only member of the genus, *S. oregonensis* Larson and LaBonte 1994, known only from shallow wells in Oregon. They suggested that this species might be the sister group to *Sanfilippodytes* Franciscolo.

Colymbetinae

**Diagnosis.** Medium to large in size (total body length 5.0 to 20.0 mm); pronotum with lateral bead present or absent; eyes emarginate on anterior margin; scutellum visible; pro- and mesotarsi distinctly five-segmented; prosternum and prosternal process on same plane (not declivous); metafemur with or without distinct group of spine-like setae on apical, posterior corner; metatarsus with two apical claws.

Historically the definition of this subfamily has used phylogenetically primitive features. In recent years the subfamilies Copelatinae, Agabetinae and Hydrotrupinae have been removed from Colymbetinae and are now separate subfamilies. Much of the remaining assemblage is more uniform but it probably does not form a monophyletic group.

Agabini

**Diagnosis.** Medium sized dytiscids (total body length = 4.9 to 14.0 mm) of various shapes and colors. The palpi with the apex of the apical palpomere entire. Pronotum with lateral bead well developed. Metacoxa with rounded lobes, lines diverging on lobes, converging anterior to lobes then diverging anteriorly. Hind legs slender to broad; metafemur with setae of posterior apicoventral angle arranged in a short row on most specimens; metatibia without bifid setae on posterior (upper) face.

The genera and species of Agabini are very common and structurally diverse on a world basis and in the Holarctic region. They are quite common in the vegetation-rich margins of small, often temporary, ponds although many species occur only in flowing water. Most of the species are well characterized but the higher classification is unstable and much further phylogenetic analysis is required.

*Agabus* Leach 1817 (Fig. 1.12)
*Necticus* Hope 1839 (preoccupied)
*Acatodes* Thomson 1859
*Eriglenus* Thomson 1859
*Gaurodytes* Thomson 1859
*Ilybiosoma* Crotch 1873
*Arctodytes* Thomson 1874
*Metronectes* Sharp 1882
*Dichodytes* Thomson 1886
*Heteronychus* Seidlitz 1887
*Scytodytes* Seidlitz 1887
*Xanthodytes* Seidlitz 1887
*Apator* Semenov 1899
*Allonychus* Zaitzev 1905
*Asternus* Guignot 1931

*Gabinectes* Guignot 1931
*Agabinectes* Guignot 1932
*Colymbinectes* Falkenström 1936
*Nebriogabus* Guignot 1936
*Parasternus* Guignot, 1936
*Ranagabus* Balfour-Browne 1939
*Dichonectes* Guignot 1945
*Allogabus* Guignot 1951
*Neonecticus* Guignot 1951
*Mesogabus* Guéorguiev 1969
*Neoplatynectes* Vazirani 1970

Size moderate (total body length = 5.0 to 14.0 mm). The body shape is various but most species are generally ovate to elongate oval with the outline continuous or discontinuous in dorsal aspect; in lateral aspect, with ventral margin more or less flat and dorsal surface convex. Most specimens have a pair of contrasting frontal spots on the frons between the eyes, the dorsal surface is uniformly black with some species yellow, pale brown to yellow and few species with elytra black with posterolateral yellow spots or yellowish with longitudinal black vittae. Head with antenna of most specimens filiform, but submoniliform, serrate or clavate on some specimens. Pronotum with lateral bead present. Metatarsomeres 1 to 4 without external apical angles lobed and metatarsal claws of most species equal.

This diverse genus is primarily Holarctic with only a few species entering northern Central America and Africa. Over 100 species occur in North America and the genus is probably poly- or paraphyletic. Larson (1989) assigned the described North American species to species groups and Larson (1989, 1991, 1994, 1996, 1997a), Larson and Wolfe (1998), and Larson *et al.* (2000) provide keys to all Nearctic species. In terms of habitat use, the majority of species occur in shallow ponds with extensive vegetation but they occur in almost every conceivable aquatic habitat within a given area.

*Carrhydrus* Fall 1923

There is only one species presently assigned to this genus, *Carrhydrus crassipes* Fall. It is moderately large (total body length = 10.5 to 13.5 mm) The body shape is oval and with a continuous outline, in dorsal aspect; in lateral aspect, the venter is flat but dorsally it is strongly convex and the body shape is sexually dimorphic. The color of the head and the lateral margins of the pronotum and elytra is reddish darkening to black medially. The labial palpus has the second last palpomere expanded and triangular in cross section with faces more or less concave and the dorsal surface with a sharp ridge which lies below a series of longitudinal ridges on the submentum (stridulatory organ?, Larson and Pritchard 1974). Male specimens with various secondary sexual features including more robust legs and a concave metaventrite (see Larson 1975) as well as widened median antennomeres.

Larson (1975) and Larson *et al.* (2000) have discussed the phylogenetic relationship and similarities among *C. crassipes* and the *Agabus clavicornis* group. The former very likely is a member of the genus *Agabus*. Adults are collected from permanent ponds in parkland and boreal forest regions from Ontario to the Alberta foothills north to the Yukon Territory.

*Ilybius* Erichson 1832 (Fig. 40.12)
*Hyobius* Gistel, 1856
*Ilyobius* Gemminger and Harold 1868
*Idiolybius* Gozis 1886
*Agabidius* Seidlitz 1887
*Ilybidius* Guignot 1948

**Diagnosis.** Medium sized specimens (total body length = 7.1 to 11.1 mm). The body shape is elongate oval narrowing anteriorly and posteriorly, in dorsal aspect, and with the venter flat and the dorsum highly convex, in lateral aspect. Most specimens are piceous to black with lateral margins paler, many specimens have a metallic sheen; elytron with a postmedial, sublateral and a subapical spot. The palpi with the apex of the apical palpomere entire. Pronotum with lateral bead well developed. Metacoxa with rounded lobes, lines diverging on lobes, converging anterior to lobes then diverging anteriorly. Metatarsomeres 1 to 4 with external apical angles lobes and metatarsal claws unequal, dorsal claw shorter than ventral claw.

This Holarctic genus contains about 30 species, of which five species are Holarctic with 14 species occurring in the Nearctic region. Keys to species are provided by Larson (1987). Many species lay eggs in the summer which hatch with the larvae overwintering and then pupating the following spring. This semivoltine life cycle is more common in areas with more continental climates whereas more equitable climates often have species with a univoltine life cycle in which adults are the overwintering stage. Most specimens are collected from lentic habitats, primarily among emergent plants along the edge of permanent or semipermanent ponds.

*Agabinus* Crotch 1873

Members of the genus *Agabinus* are among the smallest of Nearctic colymbetines with a total body ength = 4.9 to 6.7 mm. The body shape is broadly oval and with a continuous outline in dorsal aspect, and in lateral aspect the venter is flat but the dorsal surface is depressed compared to other agabines.The palpi with the apex of the apical palpomere entire. Pronotum with lateral bead well developed. Metacoxae (Fig. 14.12) with lobes rounded only on posterior margin, metacoxal lines well impressed and subparallel onto metaventrite. Antenna with median antennomeres not modified. Metafemur narrowed basally.

This genus contains only two species restricted to western North America from British Columbia south to California. Adults are collected among stones and gravel along the margins of streams. Keys to species are found in Leech (1941) and Larson *et al.* (2000).

Coptotomini

**Diagnosis.** Medium sized colymbetines (total body length = 5.7 to 8.6 mm) with the body shape elongate-oval and narrowed, somewhat pointed posteriorly with a continuous outline in dorsal aspect; in lateral aspect with the ventral margin depressed medially and the dorsal margin convex. Specimens are yellow to pale red in color with the pronotum and elytra variously marked with black spots, blotches and vittae. Antennomeres slender. Palpi each with apex of apical palpomere emarginate. Pronotum with lateral bead well developed. Metacoxal lobes rounded, lines diverging anteriorly and more or less parallel with margin of metaventrite. Metafemur without posterior, apical row of setae; metatarsomeres 1 to 4 with the apical angles lobed; and the metatarsal claws are subequal.

The Coptotomini contains only the single genus *Coptotomus* Say which is exclusively Nearctic. However, assignment of the tribe to Colymbetinae is problematic as the members exhibit a suite of unique features making the tribe unique but most of the characters that they share with other colymbetines are assignable to the ground plan of dytiscids.

*Coptotomus* Say 1834 (Fig. 43.12)
There are about 10 species in the genus within North America and the collective range includes most of Canada, the continental U.S. and northern Mexico. A key to eastern species was provided by Hilsenhoff (1980) but the western species remain untreated. Specimens are most commonly collected from the deeper sections of permanent ponds with emergent vegetation, lakes and slowly flowing streams.

### Matini

**Diagnosis.** Relatively small to moderately sized specimens (total body length = 5.4 to 9.5 mm) with the body shape elongate-oval with the outline continuous in dorsal aspect, and the profile, in lateral aspect, somewhat flattened. Palpi and antennae unmodified. Pronotum with lateral bead well developed. A distinctive feature of the members of Matini is a distinct groove extending from the prosternum along the length of the prosternal process (Fig. 17.12). Metacoxa (Fig. 17.12) with lobes large and rounded apically, lines diverging on lobes, converging anterior to lobes and then diverging anteriorly. Metacoxal lines diverging both anteriorly and posteriorly; metacoxal processes large and apically rounded. Hind legs robust; metafemur without comb of setae on apicoventral angle; metatibia without bifid setae on posterior (upper face); metatarsi broad with dorsal and ventral margins somewhat lobed; metatarsal claws unequal.

This tribe contains three genera: *Matus* in North America, and *Allomatus* Mouchamps and *Batrachomatus* Clark in Australia.

*Matus* Aubé 1836
Small to moderately sized colymbetines (total body length = 5.4 to 9.5 mm). The dorsal color pale reddish to dark brown. Young (1953) provided a key to adults of the four Nearctic species. Specimens of *Matus* are lentic and occur on soft substrates among dense vegetation and debris.

### Colymbetini

**Diagnosis.** Medium to larger specimens (total body length = 8.5 to 20.0 mm) of various colors and shapes but predominantly ovate to elongate and lateral profile more or less depressed. Palpi and antennae unmodified. Pronotum with lateral bead well-defined or absent. Metacoxa with rounded lobes, lines diverging on lobes, converging anterior to lobes and then diverging anteriorly. Hind legs slender to moderate; metafemur with setae on the posterior apicoventral angle present but not arranged in a linear comb; metatibia with the dorsal face bearing a row of bifid setae; metatarsomeres 1 to 4 lobed apically; metatarsal claws unequal.

The Nearctic genera assigned to this genus are *Rhantus* Dejean (with a wordlwide distribution), *Colymbetes* Clairville (Holarctic), *Hoperius* Fall (Nearctic) and *Neoscutopterus* Balfour-Browne. Specimens are most commonly collected in areas of dense vegetation at the margins of both temporary and permanent ponds but some species are found in streams and rivers.

*Colymbetes* Clairville 1806 (Fig. 41.12)
*Cymatopterus* Dejean 1833

Medium to large in size (total body length = 11.0 to 20.0 mm) with the body shape elongate oval and with the lateral outline more or less continuous in dorsal aspect; in lateral aspect somewhat depressed. The color is various, dorsally ranging from yellow to dark brown and some specimens with dark spots; ventrally the color is predominantly black with the appendages yellow to black. Pronotum with lateral bead absent. Male with protibia not emarginate basally.

On a worldwide basis the genus contains over 20 species, of which three are Holarctic (Nilsson and Holmen 1995). Zimmerman (1981) recognized seven North American species whereas Larson *et al.* (2000) elevated two subspecies to species rank and recognized nine Nearctic species occurring throughout much of northern North America. Geographical variation in structure and color is complex and species limits are hard to define. Specimens are most commonly collected from permanent (or at least long-lasting) ponds, lakes, bogs, beaver ponds and fens often among emergent vegetation in shaded areas. However, a number of species are quite common in sun-warmed grassland ponds, reservoirs and lakes.

*Hoperius* Fall 1927
Size moderately large (total body length = 12.0 to 15.0 mm). The body shape is broadly oval with the outline continuous in dorsal aspect; in lateral aspect, with the ventral margin flat and the dorsal surface depressed. The dorsal color is dark brown in the middle with broad paler margins and reddish-brown ventrally. Pronotum with broad lateral bead present. Male with protibia not emarginate basally.

There is only one species, *H. planatus* Fall, known from the eastern United States west to Texas and north to about Maryland. It is probably a member of the woodland pond community (Spangler 1973).

*Neoscutopterus* Balfour-Browne 1943 (Fig. 42.12)
*Meladema*, of authors, not Nearctic
*Scutopterus*, of authors, not Nearctic
*Pseudoscutopterus* Hatch 1953

Size moderately large (total body length = 13.5 to 16.7 mm) with coarse, large meshes of sculpture. The body shape is robust and elongate with the pronotum and base of the elytra rounded differently and the outline appearing slightly discontinuous in dorsal aspect; in lateral aspect, with ventral margin more or less flat and dorsal surface in a somewhat low arc. Pronotum with lateral bead absent. Male with protibia emarginate basally.

There are only two species of *Neoscutopterus* known and keys to separate them are included in Larson (1975) and Larson *et al.* (2000). Both species are northern and transcontinental. Specimens with the median antennomeres infuscate and the palpi darkened at the apex and the venter black and without reddish spots are assignable to *N. angustus* (LeConte) whereas those with the entire antenna and palpi reddish-yellow and the abdomen with reddish spots are assignable to *N. hornii* (Crotch). These species occur in cold water of *Carex* and sphagnum fens, bogs, boggy lake margins and beaver ponds.

*Rhantus* Dejean 1833
subgenus *Nartus* Zaitzev 1907
*Rantogiton* Gozis 1910
*Ilybiomorphus* Porta 1923

Medium to large in size (total body length = 8.5 to 16.0 mm) with the body shape elongate oval and with the lateral outline continuous in dorsal aspect; in lateral aspect somewhat depressed. Pronotum with lateral bead well developed or not. Male with protibia not emarginate basally.

This large genus is represented in all major zoogeographic regions and on many oceanic islands. The genus is well-established but many species are variable in color pattern, which has led to the description of many species that are simple synonyms and have long confused the species-level nomenclature. Within the Holarctic region there are two subgenera. The distinguishing features separating members of *Rhantus (Nartus)* are dorsal and ventral color predominantly black and the pronotum with the posterior margin laterally strongly sinuate and with the labrum shallowly emarginate medially. There is only one Nearctic species in this subgenus: *Rhantus (Nartus) sinuatus* (LeConte). Members of *Rhantus* (*s. str.*) are worldwide in distribution and they are varied in dorsal and ventral color but most are yellowish with distinct dark markings dorsally on the head and/or pronotum and the elytra irrorate and the labrum is more deeply emarginate medially. There are nine Nearctic species of *Rhantus* (*s. str.*). Keys to species, descriptions and discussion of problems with nomenclature are presented in Zimmerman and R. Smith (1975) and Balke (1990, 1993). In terms of habitat use, the majority of specimens are collected from shallow ponds with extensive vegetation but they occur in almost every conceivable aquatic habitat within a given area.

## Dytiscinae

**Diagnosis**. Eyes not emarginate on anterior margin; pronotum of most species without lateral bead; scutellum visible; pro- and mesotarsi five-segemented (in males protarsus with basal three tarsomeres expanded laterally and forming a large, round to oval palette equipped with a varied number of ventral suckers and mesotarsomeres 1 to 3 widened and forming an elongate pad with numerous small suckers in most groups); prosternum and prosternal process on same plane (not declivous); metafemur without distinct group of spine-like setae on apical, posterior corner; metatarsus with one or two claws which are equal or unequal in length.

There are five tribes of Dytiscinae known from North America. In general the species limits are well-known and the overall classification is stable. They are generally the large, more quickly swimming active predators although they are well-known to scavenge. Adults are most common in ponds and lakes where they prefer deeper, more open water.

### Dytiscini

**Diagnosis**. Nearctic species of large size (total body length = 22 to 40 mm) with an elongate oval body shape with a continuous outline, in dorsal aspect, and a lateral profile which is flat ventrally and shallowly domed dorsally with the greatest height in anterior third of the body. The color of the dorsal surface is dark brown to black with the head, pronotum and elytra with yellow markings, ventrally the color ranges from yellow or patterned yellow and black or black. Male protarsus with palette round with two large and many smaller suckers. Pronotum with lateral bead absent. The metaventrite laterally (=metasternal wing, of authors) triangular in shape and with anterior, lateral margin arcuate. Hind legs robust; metatibial spurs similar in shape; metartarsomeres 1 to 4 lobed apically and metarsal claws subequal.

This tribe contains two genera: the Holarctic genus, *Dytiscus* Linné, and the Australian genus, *Hyderodes* Hope. The latter is comprised of two species in southern Australia. Therefore, these two genera are in very widely separated, somewhat temperate regions of the world.

*Dytiscus* Linné 1758
*Leionotus* Kirby 1837
*Macrodytes* Thomson 1859

This is the only genus in the tribe in the Northern Hemisphere. It is Holarctic. This Holarctic genus is comprised of 26 species with 12 species known from the Nearctic region and 16 species occurring in the Palaearctic region; two species are Holarctic (Roughley 1990). Adult females only of some species have the elytra closely yet deeply sulcate. The combination of very large size and the presence of broad yellow bands laterally on the pronotum and elytra separate specimens of *Dytiscus* from all other North American genera except *Cybister* and *Megadytes* (Cybistrini). These latter two genera are characterized by a broader, more wedge-shaped habitus, larger metacoxae and broader hind legs. As well, the species of Cybistrini are generally more southerly in distribution than most species of *Dytiscus*.

Hydaticini

**Diagnosis.** Nearctic species of moderate size (total body length = 10.9 to 15.4 mm) with an elongate oval body shape and a continuous outline, in dorsal aspect, and a lateral profile which is flat ventrally and convex dorsally. The color of the dorsal surface is dark brown to black and variously ornamented with yellow markings. Male protarsus with pallette round with moderately large, uniform-sized suckers. Pronotum with lateral bead absent. The metaventrite laterally (=metasternal wing, of authors) narrowly triangular with anterior, lateral margin straight. Hind legs robust, metatibial spurs similar in shape; metatarsomere 1 to 4 are not lobed apically and the metatarsal claws are unequal in length.

*Hydaticus* Leach 1817 (Fig. 44.12)
This is the only genus in the tribe in North America and worldwide there is only the additional genus *Prodaticus* Sharp. There are five Nearctic species of *Hydaticus* Roughley and Pengelly (1981) but the genus is much more diverse in the tropics with over 200 species known worldwide. The North American species are lentic, inhabiting shallow, vegetation-rich margins of usually rather eutrophic ponds and small lakes. They tend to occur in shallower and more densely vegetated water than do other members of Dytiscinae.

Aciliini

**Diagnosis.** Nearctic species are of moderate to large size (total body length = 7.5 to 17 mm with a broadly oval body shape which is widest behind the middle and the outline is continuous, in dorsal aspect; in lateral aspect, they appear rather flattened. The color of the dorsal surface is yellow to reddish with darker markings and the ventral surface is yellow to black. Male protarsus with palette round with three large and a variable number of smaller suckers or with areas of much smaller setae. Pronotum with lateral bead absent. The metaventrite laterally (=metasternal wing, of authors) very narrowly triangular and with anterior, lateral margin arcuate. Hind legs robust with the metatibia short and broad; metatibial spurs finely emarginate apically and similar in shape; metatarsomeres 1 to 4 each with apical margin lobed or emarginate; metatarsal claws unequal.

Worldwide there are seven genera assigned to this tribe and three of these occur in North America. They are among the most dramatically and strikingly colored dytiscids which generally is not a group known for excessive color markings. Adults are strong swimmers and frequently venture into the open water of ponds and lake margins but they can also be taken close to shore. Larvae of some genera are pelagic hunters within these habitats.

*Acilius* Leach 1817 (Figs. 46.12, 47.12)
Nearctic species of moderate size (total body length = 10.2 to 17.0 mm). The body is broadly ovate with the point of maximum width behind midlength. The dorsal surface is yellow with dark markings, in particular the pronotum is yellow with an anterior and a posterior black transverse band and the ventral surface is yellow to black. Males with protarsal palette with three large and two areas of very small suckers or setae. Females of most species with elytron bearing four broad, longitudinal grooves which are setose.

This is a genus of Holarctic distribution which is not particularly rich in species. No species is shared between the Nearctic and Palearctic regions. The North American species range northward from about the middle of the United States. Hilsenhoff (1975) revised the species found in eastern North American. Larson *et al.* (2000) recognized six Nearctic species but acknowledged that the genus needs revision. Most species prefer woodland ponds or forest pools, beaver ponds, roadside ditches with extensive vegetation and the boggy margins of lakes. Very often these habitats are without extensive bottom vegetation and the substrate is composed of peat or decaying leaves.

*Graphoderus* Dejean 1833 (Fig. 45.12)
*Graphothorax* Motschulsky 1853
*Prosciastes* Gistel 1856
*Derographus* Portevin 1929
Nearctic species of moderate size (total body length = 10.4 to 15.0 mm). The body is broadly ovate with the point of maximum width behind midlength. The dorsal surface is yellow with dark markings, head and pronotum with or without dark markings, elytra with small, elongate, irregular markings laterally which in middle are more mesh-like and ventral surface is yellow to reddish. Males with protarsal palette with two or three large and 13 to 38 smaller suckers. Adult females with elytron without deep, longitudinal grooves.

The genus is Holarctic but no species are shared between the Palaearctic and Nearctic regions. There are 5 species of *Graphoderus* known from the Nearctic region and the species are treated by Wallis (1939), Tracy and Hilsenhoff (1982) and Larson *et al.* (2000). As adults the species are lentic and prefer open, sun-warmed ponds, peatland ponds and bog-ringed lakes. Certain species were observed feeding on dead and dying insects trapped on the surface film of ponds and lakes (Larson *et. al.* 2000).

*Thermonectus* Dejean 1833
Nearctic species of medium to moderate size (total body length = 7.5 to 13.5 mm). The body shape is ovate. The dorsal surface is of various colors: black with pale markings to yellow with dark markings and the ventral surface is yellow to near black. Males with protarsal palette with three large and a variable number of smaller suckers. Females with elytron without deep longitudinal grooves.

This genus is exclusively Nearctic and Neotropical. There are six species in North America north of Mexico and a key to these species is presented in Larson *et al.* (2000). Two, very pretty species occurring in the desert southwest of the U.S. exhibit disruptive coloration; they are primarily lotic occurring in pools in intermittent desert streams. Adults of the remaining species are lentic and use temporary to permanent ponds (Hilsenhoff 1993a).

### Eretini

**Diagnosis.** Nearctic specimens of moderate size (10.2 to 16.8 mm) with an elongate oval body shape with a discontinuous outline, in dorsal aspect, and a lateral profile with the dorsal margin flat and the ventral margin shallowly rounded. The color of the dorsal surface is yellow to pale brown with various dark markings and elytron with dark speckles, ventrally the color is yellow to pale brown. Male protarsus with palette round with two large and many much smaller suckers. Pronotum with lateral bead present but narrow. The metaventrite laterally (=metasternal wing, of authors) narrowly triangular in shape and with anterior, lateral margin arcuate. Hind legs slender; metatibial spurs similar in shape; metatarsomeres 1 to 4 not lobed apically and metatarsal claws unequal.

This tribe contains only one genus, *Eretes* Laporte, which is found worldwide including the Neotropical and southern Nearctic regions.

*Eretes* Laporte 1833
    *Eunectes* Erichson 1832, preoccupied.
This is the only genus in the tribe and there is only one species in North, Central and South America. Previously it was thought that there were only two species assigned to the genus (*E. australis* (Erichson) - Australia) and *E. sticticus* Linné - known from all other zoogeographic regions). However, as reported in Larson *et al.* (2000), Nearctic and Neotropical specimens should be assigned to *E. occidentalis* (Erichson). This species prefers temporary pools and ponds in dry regions (Young 1954) where it colonizes temporary water bodies and completes it larval stages in nine to 10 days (Kingsley 1985).

### Cybistrini

**Diagnosis.** Nearctic specimens of large size (total body length = 20 to 34 mm) with an elongate oval body shape, widest behind the middle and a continuous outline, in dorsal aspect, and a lateral profile which is flat to very shallowly rounded ventrally and shallowly domed dorsally with the greatest height in the middle of the body. The color of the dorsal surface is reddish-brown, green or black with the head, pronotum and elytra with or without yellow lateral margins; ventrally the color is yellow, brown or black. Male protarsus elliptical with many small oval suckers. Pronotum with lateral bead absent. The metaventrite laterally (=metasternal wing, of authors) narrow and strap-like in shape with the anterior, lateral margin arcuate. Hind legs very wide; metatibial spurs dissimilar in shape with the outer spur broader at the base; metatarsomeres 1 to 4 not lobed apically; in Nearctic and Neotropical species the metarsus with one (males, females of some species) or two very unequal claws (females of some species).

Pederzani (1995) lists six genera for the world fauna. There is much controversy, however, among subgeneric names and generic versus subgeneric rank within cybisterines. In North, Central and South America we have only two genera: *Cybister* Curtis and *Megadytes* Sharp. The species in these genera prefer deeper, more permanent ponds and lakes and are generally more common in the deeper parts of these habitats. They are rarely collected with a traditional aquatic net but are common at lights.

*Cybister* Curtis 1827
    *Trogus* Leach 1817
    *Cybisteter* Bedel 1881
    *Alocomerus* Brinck 1945
    *Megadytoides* Brinck 1945
    *Meganectes* Brinck 1945
    *Nealocomerus* Brinck 1945
    *Gschwendtnerhydrus* Brinck 1945
    *Trochalus* Dejean 1833 (preoccupied)
Nearctic specimens are of large size (total body length = 26 to 34 mm). Male with metarsomere 5 with one claw, female metatarsus 5 with two unequal claws; body dorsally black with a greenish hue and pronotum and elytron with a yellow stripe along the lateral margins.

The subgeneric names that have been proposed in *Cybister* are problematic in terms of nomenclature (Nilsson *et al.* 1989) and in terms of their taxonomic limits and assignment of species (Pederzani 1995). The Nearctic species probably are best assigned to *Cybister* (*s. str.*) (Larson *et al.* 2000). A key to North American species is given by Larson *et al.* (2000) although species limits and subspecies definition and limits are poorly understood and the genus could use revision.

*Megadytes* Sharp 1882
Nearctic specimens are of large size (total body length = 20 to 24 mm). Male metatarsus with two subequal claws; female metatarsus with a long outer claw and a rudimentary inner claw; pronotum with, but elytron without, lateral yellow stripe, elytron with lateral margin paler.

Only one species occurs in North America, *Megadytes fraternus* Sharp, which occurs in southern Florida. Many other species are known from the Neotropical region.

## Bibliography


ALARIE, Y. 1989. The larvae of *Laccornis* Des Gozis 1914 (Coleoptera: Adephaga: Dytiscidae) with description of *L. latens* (Fall, 1937), and redescription of the mature larva of *L. conoideus* (LeConte, 1850). Coleopterists Bulletin, 43: 365 - 378.

ALARIE, Y. 1991. Description of larvae of 17 Nearctic species of *Hydroporus* Clairville (Coleoptera: Dytiscidae:

Hydroporinae) with an analysis of their phylogenetic relationships. The Canadian Entomologist, 123: 627 - 704.

ALARIE, Y. 1992. Description of the larval stages of *Hydroporus* (*Heterosternuta*) *cocheconis* Fall 1917 (Coleoptera: Dytiscidae: Hydroporinae) with a key to the known larvae of *Heterosternuta* Strand. The Canadian Entomologist, 124: 827-840.

ALARIE, Y. 1993. A systematic review of the North American species of the *Oeodytes alaskanus* clade (Coleoptera: Dytiscidae: Hydroporinae). The Canadian Entomologist, 125: 847 - 867.

ALARIE, Y., SPANGLER, P. J. , and PERKINS, P. D. 1998. Study of the larvae of *Hydrotrupes palpalis* Sharp (Coleoptera: Adephaga: Dytiscidae) with implications for the phylogeny of the Colymbetinae. Coleopterists Bulletin, 52: 313 - 332.

ALARIE, Y., WANG. L.-J., NILSSON, A. N., and SPANGLER, P. J. 1997. Larval morphology of four genera of the tribe Hyphydrini Sharp (Coleoptera: Dytiscidae: Hydroporinae) with an analysis of their phylogenetic relationships. Annals of the Entomological Society of America, 90: 709 - 735.

ANDERSON, R. D. 1962. The Dytiscidae (Coleoptera) of Utah: keys, original citation, types and Utah distribution. Great Basin Naturalist, 22: 54 - 75.

ANDERSON, R. D. 1971. A revision of the Nearctic representatives of *Hygrotus* (Coleoptera: Dytiscidae). Annals of the Entomological Society of America, 64: 503 - 512.

ANDERSON, R. D. 1976. A revision of the Nearctic species of *Hygrotus* groups II and III (Coleoptera: Dytiscidae). Annals of the Entomological Society of America, 69: 577 - 584.

ANDERSON, R. D. 1983. A revision of the Nearctic species of *Hygrotus* groups IV, V, and VI (Coleoptera: Dytiscidae). Annals of the Entomological Society of America, 76: 173 - 196.

ANDERSON, R. D. 1985. Proposed faunal affinities of the Great Basin Dytiscidae (Coleoptera). Proceedings of the Academy of Natural Sciences of Philadelphia, 137: 12 - 21.

ARCE-PÉREZ, R. and ROUGHLEY, R. E. 1999. Lista anotada y claves para los Hydradephaga (Coleoptera: Adephaga: Dytiscidae, Noteridae, Haliplidae, Gyrinidae) de México. Dugesiana, 6: 69-104.

BALFOUR-BROWNE, F. 1940. British water beetles. Volume I. The Ray Society, London, no. 127. 375 pp.

BALFOUR-BROWNE, F. 1950. British water beetles. Volume II. The Ray Society, London, no. 134. 394 pp.

BALFOUR-BROWNE, J. 1947. A revision of the genus *Bidessonotus* Régimbart (Coleoptera: Dytiscidae). Transactions of the Royal Entomological Society of London, 98: 425-448.

BALKE, M. 1990. Die Gattung *Rhantus* Dejean. IV. Taxonomie und Faunistik verschiedener paläarktischer und nearktischer Spezies. Bulletin de la Société Entomologique Suisse, 63: 195-208.

BALKE, M. 1993. Neotropische Wasserkäfer der Gattung *Rhantus* Dejean IV. Liste und Notizen über die "großen" Arten (Insecta: Coleoptera: Dytiscidae). Reichenbachia Staatliches Museum für Tierkunde Dresden, 30: 21 - 32.

BEUTEL, R. G. 1994. On the systematic position of *Hydrotrupes palpalis* Sharp (Coleoptera: Dytiscidae). Aquatic Insects, 16: 157-164.

BEUTEL, R. G. 1995. The Adephaga (Coleoptera): phylogeny and evolutionary history. pp. 173-217, *In* Pakaluk, J. and Slipinski, S. A. Editors. Biology, phylogeny, and classification of Coleoptera. Papers celebrating the 80th birthday of Roy A. Crowson. Muzeum i Instytut Zoologii PAN, Warszawa. X + 1092 pp.

BEUTEL, R. G. 1997. Über Phylogenese und Evolution der Coleoptera (Insecta), inbesondere der Adephaga. Abhandlungen des Naturwissenschaftlichen Vereins in Hamburg (N.F.), 31:1-64.

BEUTEL, R. G. and BELKACEME, T. 1986. Comparative studies on the metathorax of Hydradephaga and Trachypachidae (Coleoptera). Entomologica Basiliensia, 11: 221-229.

BEUTEL, R. G. and ROUGHLEY, R. E. 1987. On the systematic position of the genus *Notomicrus* Sharp (Hydradephaga, Coleoptera). Canadian Journal of Zoology, 65: 1898-1905.

BEUTEL, R. G. and ROUGHLEY, R. E.. 1988. On the systematic position of the family Gyrinidae. Zeitschrift für Zoologische Systematik und Evolutionsforschung, 26: 380-400.

BISTRÖM, O. 1988. Generic review of the Bidessini (Coleoptera, Dytiscidae). Acta Zoologica Fennica, 184: 1-41.

BISTRÖM, O. 1990. Revision of the genus *Queda* Sharp (Coleoptera: Dytiscidae). Quaestiones Entomologicae, 26: 211 - 220.

BISTRÖM, O. 1997. Taxonomic revision of the genus *Hydrovatus* Motschulsky. Entomologica Basiliensia, 19 (1996): 57 - 584.

BISTRÖM, O., NILSSON, A. N., and WEWALKA, G. 1997. A systematic review of the tribes Hyphydrini Sharp and Pachydrini n. trib. (Coleoptera, Dytiscidae). Entomologica Fennica, 8: 57 - 82.

BOOBAR, L. R., GIBBS, K. E., LONGCORE, J. R., and PERILLO, A. M. 1996. New records of predaceous diving beetles (Coleoptera: Dytiscidae) in Maine. Entomological News, 107: 267 - 271.

BØVING, A. G. 1913. Studies relating to the anatomy, the biological adaptations and the mechanism of ovipositor in the various genera of Dytiscidae. Internationale Revue der gesamten Hydrobiologie, Biologische Supplemente 5 (1912): 1-28 + Tafeln I-VI.

BRIGHAM, W. U. 1982. Chapter 10. Aquatic Coleoptera, 10.14 - 10.136. *In* Brigham, A. R., Brigham, W.U. and Gnilka, A. (eds.). Aquatic insects and oligochaetes of North and South Carolina. Midwest Aquatic Enterprises, Mahomet, Illinois.

BURMEISTER, E.-G. 1976. Der Ovipositor der Hydradephaga (Coleoptera) und seine phylogenetische Bedeutung unter besonderer Berucksichtigungder Dytiscidae. Zoomorphologie, 85: 165-257.

BURMEISTER, E.-G. 1990. The systematic position of the genus *Agabetes* Crotch within Dytiscidae (Coleoptera: Adephaga). Quaestiones Entomologicae, 26: 221-238.

CHALLET, G. L. and BRETT, R. 1998. Distribution of the Dytiscidae (Coleoptera) of California by county. Coleopterists Bulletin, 52: 43 - 54.

CHAMBERLAIN, K. F. 1947. Notes on the ecology of *Hydroporus rufiplanulus* Fall (Coleoptera, Dytiscidae). Journal of the New York Entomological Society, 55 (l, March): 57 - 58.

COLER, B. G. and SLATER, A. 1982. Additions and corrections to the list of Dytiscidae of Kansas. Technical Publication of the State Biological Survey of Kansas, 12: 43 - 48.

DE MARZO, L. 1976. Studi sulle larve dei Coleotteri Ditiscidi. IV. Morfologia dei tre stadi arvali di *Copelatus haemorroidalis* F. Entomologica, Bari 12: 89 - 106.

DOWNIE, N. M. and ARNETT, R. H. 1996. The beetles of northeastern North America. Volume 1. Sandhill Crane Press, Gainesville. 880 pp.

EPLER, J. H. 1996. Identification manual for the water beetles of Florida. Bureau of Water Resource Protection, Florida Department of Environmental Protection, Tallahassee.

FALL, H. C. 1923. A revision of the North American species of *Hydroporus* and *Agaporus*. J.D. Sherman, Jr., Mount Vernon, N.Y. 129 pp.

FOLKERTS, G. W. 1978. A preliminary checklist of the Hydadephaga (Coleoptera) of Alabama. Coleopterists Bulletin, 32: 345 - 347.

FORBES, W. T. M, 1926. The wing folding patterns of the Coleoptera. Journal of the New York Entomological Society, 34: 42-139.

FRANCISCOLO, M. E. 1979a. Coleoptera: Haliplidae, Hygrobiidae, Gyrinidae, Dytiscidae. Fauna d'Italia, 14: vi + 804 pp.

FRANCISCOLO, M. E. 1979b. On a new Dytiscidae from a Mexican cave – a preliminary description. Fragmenta Entomologica Roma, 15: 233-241.

FRANCISCOLO, M. E. 1983. Adaptation in hypogean Hydradephaga, with new notes on *Sanfilippodytes* (Dytiscidae and Phreatodytidae). pp. 5-20. *In* Satô, M. Editor. Special issue concerning the aquatic Coleoptera presented at the workshop of the XVI International Congress of Entomology in Kyoto, Japan in 1980. Tokyo Tsûhan Service-sha, Tokyo, Japan. 41 pp.

GOODLIFFE, F. D. 1939. The taxonomic value of wing venation in the larger Dytiscidae (Coleoptera). Transactions of the Society for British Entomology, 6:23-38 + plates I-IV.

GORDON, R. D. and POST, R. L. 1965. North Dakota water beetles. North Dakota Insects, 5: 1-53.

GUÉORGUIEV, V. B. 1968. Essai de classification des Coléoptères Dytiscidae. 1. Tribus Copelatini (Colymbetinae). Izvestia na Zooligicheskia Institut i Musei, Bulgarska Akademia na Naukite, 28: 5-45.

HAMMOND, P. M. 1979. Wing-folding mechanisms of beetles, with special reference to investigations of adephagan phylogeny (Coleoptera). pp. 113-180 *In* Erwin, T. L., Ball, G. E., Whitehead, D. R. and Halpern, A. L. Editors. Carabid beetles: their evolution, natural history, and classification. Dr. W. Junk bv Publishers, The Hague. 635 pp.

HATCH, M. H. 1953. The beetles of the Pacific Northwest. Part 1: Introduction and Adephaga. University of Washington Publications in Biology, 16. fp + 340 pp.

HILSENHOFF, W. L. 1975. Notes on Nearctic *Acilius* (Dytiscidae), with the description of a new species. Annals of the Entomological Society of America, 68: 271-274.

HILSENHOFF, W. L. 1980. *Coptotomus* (Coleoptera: Dytiscidae) in eastern North America with descriptions of two new species. Transactions of the American Entomological Society, 105: 461 - 471.

HILSENHOFF, W. L. 1992. Dytiscidae and Noteridae of Wisconsin (Coleoptera). I. Introduction, key to genera of adults, and distribution, habitat, life cycle, and identification of species of Agabetinae, Laccophilinae and Noteridae. The Great Lakes Entomologist, 25: 57-69.

HILSENHOFF, W. L. 1993a. Dytiscidae and Noteridae of Wisconsin (Coleoptera). II. Distribution, habitat, life cycle, and identification of species of Dytiscinae. The Great Lakes Entomologist, 26: 35-53.

HILSENHOFF, W. L. 1993b. Dytiscidae and Noteridae of Wisconsin (Coleoptera). III. Distribution, habitat, life cycle, and identification of species of Colymbetinae, except Agabini. The Great Lakes Entomologist, 26: 121-136.

HILSENHOFF, W. L. 1993c. Dytiscidae and Noteridae of Wisconsin (Coleoptera). IV. Distribution, habitat, life cycle, and identification of species of Agabini (Colymbetinae). The Great Lakes Entomologist, 26: 173-197.

HILSENHOFF, W. L. 1994. Dytiscidae and Noteridae of Wisconsin (Coleoptera). V. Distribution, habitat, life cycle, and identification of species of Hydroporinae, except *Hydroporus* Clairville *sensu lato*. The Great Lakes Entomologist, 26: 275-295.

HILSENHOFF, W. L. 1995. Dytiscidae and Noteridae of Wisconsin (Coleoptera). VI. Distribution, habitat, life cycle, and identification of species of *Hydroporus* Clairville *sensu lato* (Hydroporinae). The Great Lakes Entomologist, 28: 1-23.

HUGGINS, D. G., LIECHTI, P. M., and ROUBIK, D. W. 1976. Species accounts for certain aquatic macroinvertebrates from Kansas (Odonata, Hemiptera, Coleoptera and Sphaeriidae). Technical Publication of the State Biological Survey of Kansas, 1: 13 - 77.

KAVANAUGH, D. H. 1986. A systematic review of amphizoid beetles (Amphizoidae: Coleoptera) and their phylogenetic relationships to other Adephaga. Proceedings of the California Academy of Sciences, 44: 67-109.

KINGSLEY, K. J. 1985. *Eretes sticticus* (L.) (Coleoptera: Dytiscidae): life history observations and an account of a remarkable event of synchronous emigration from a temporary desert pond. Coleopterists Bulletin, 39: 7-10

KUKALOVÀ-PECK, J. and LAWRENCE, J. 1993. Evolution of the hind wing in Coleoptera. The Canadian Entomologist, 125: 181-258.

LARIVERS, I. 1951. Nevada Dytiscidae (Coleoptera). American Midland Naturalist, 45: 392 - 406.

LARSON, D. J. 1975. The predaceous water beetles (Coleoptera: Dytiscidae) of Alberta: systematics, natural history and distribution. Quaestiones Entomologicae, 11: 245 - 498.

LARSON, D. J. 1987. Revision of North American species of *Ilybius* Erichson (Coleoptera: Dytiscidae), with systematic notes on Palaearctic species. Journal of the New York Entomological Society, 95: 341-413.

LARSON, D. J. 1989. Revision of North American *Agabus* Leach (Coleoptera: Dytiscidae): introduction, key to species groups, and classification of the *ambiguus-*, *tristis-*, and *arcticus*-groups. The Canadian Entomologist, 121: 861 - 919.

LARSON, D. J. 1990. *Oreodytes obesus* (LeConte) and *O. sanmarkii* (C. R. Sahlberg) (Coleoptera: Dytiscidae) in North America. Coleopterists Bulletin, 44: 295-303.

LARSON, D. J. 1991. Revision of North American *Agabus* Leach (Coleoptera: Dytiscidae): *elongatus-*, *zetterstedti-*, and *confinis*-groups. The Canadian Entomologist, 123: 1239-1317.

LARSON, D. J. 1994. Revision of North American *Agabus* Leach (Coleoptera: Dytiscidae): *lutosus-*, *obsoletus-*, and *fuscipennis*-groups. The Canadian Entomologist, 126: 135-181.

LARSON, D. J. 1996. Revision of North American *Agabus* Leach (Coleoptera: Dytiscidae): the *opacus*-group. The Canadian Entomologist, 128: 613-665.

LARSON, D. J. 1997a. Revision of North American *Agabus* Leach (Coleoptera: Dytiscidae): the *seriatus*-group. The Canadian Entomologist, 129: 105-149.

LARSON, D. J. 1997b. Dytiscid water beetles (Coleoptera: Dytiscidae) of the Yukon. pp. 491-522. *In* H. V. Danks and J. A. Downes (Eds.), Insects of the Yukon. Biological Survey of Canada (Terrestrial Arthropods). Ottawa. 1034 pp.

LARSON, D.J., ALARIE, Y., and ROUGHLEY, R.E. 2000. Predaceous diving beetles (Coleoptera: Dytiscidae) of the Nearctic Region, with emphasis on the fauna of Canada and Alaska. Monographs in Biodiversity, NRC Press, Ottawa, Ont. (In press).

LARSON, D. J. and LaBONTE, J.R. 1994. *Stygoporus oregonensis*, a new genus and species of subterranean water beetle (Coleoptera: Dytiscidae: Hydroporini) from the United States. Coleopterists Bulletin, 48: 371-379.

LARSON, D. J. and COLBO, M. H. 1983. The aquatic insects: biogeographic considerations. pp. 593-677. *In* South, G. R. (Ed.). Biogeography and Ecology of the Island of Newfoundland. Dr. W. Junk Publishers, The Hague.

LARSON, D. J. and ROUGHLEY, R. E. 1990. A review of the species of *Liodessus* Guignot of North America north of Mexico with the description of a new species (Coleoptera: Dytiscidae). Journal of the New York Entomological Society, 98: 233-245.

LARSON, D. J. and ROUGHLEY, R. E. 1991. Family Dytiscidae. pp. 93-106. *In* Bousquet, Y. (Ed.). 1991. Checklist of beetles of Canada and Alaska. Agriculture Canada Publication 1861/E. 430 pp.

LARSON, D. J. and WOLFE, R. W. 1998. Revision of North American *Agabus* (Coleoptera: Dytiscidae): the *semivittatus*-group. The Canadian Entomologist, 130: 27-54.

LAWRENCE, J., HASTINGS, A., DALLWITZ, M., and PAINE, T. 1993. Beetle larvae of the world. Interactive identification and information retrieval for families and sub-families. CSIRO, Melbourne, Victoria CD ROM, Version 1.0 for DOS.

LEECH, H.B. 1941. Note on the species of *Agabinus* (Coleoptera: Dytiscidae). The Canadian Entomologist, 73: 53.

LEECH, H.B. 1942. Key to the Nearctic genera of water beetles of the tribe Agabini, with some generic synonymy (Coleoptera: Dytiscidae). Annals of the Entomological Society of America, 35: 355-362.

LEECH, H. B. 1948. Contributions toward a knowledge of the insect fauna of lower California, No. 11. Coleoptera: Haliplidae, Dytiscidae, Gyrinidae, Hydrophilidae, Limnebiidae. Proceedings of the California Academy of Sciences, 24 (11): 375-484 + plates 20-21.

LEECH, H. B. and CHANDLER, H. P. 1956. Aquatic Coleoptera. pp. 293-371. *In* R. L. Usinger. Editor. Aquatic insects of California with keys to North American genera and California species. University of California Press. Berkeley and Los Angeles. ix + 508 pp.

MALCOLM, S. E. 1971. The water beetles of Maine: including the families Gyrinidae, Haliplidae, Dytiscidae, Noteridae and Hydrophilidae. Technical Bulletin 48. Life Sciences and Agriculture Experiment Station, University of Maine at Orono. 49 pp.

MATTA, J. F. and WOLFE, G. W. 1981. A revision of the subgenus *Heterosternuta* Strand of *Hydroporus* Clairville (Coleoptera: Dytiscidae). The Pan-Pacific Entomologist, 57: 176-219.

MICHAEL, A. G. and MATTA, J. F. 1977. The insects of Virginia: No. 12. The Dytiscidae of Virginia (Coleoptera: Adephaga) (Subfamilies: Laccophilinae, Colymbetinae, Dytiscinae, Hydaticinae and Cybistrinae). Virginia Polytechnic Institute and State University, Research Division Bulletin, 124. 53 pp.

MILLER, K. B. 1997. *Crinodessus amyae*, a new Nearctic genus and species of predaceous diving beetle (Coleoptera: Dytiscidae: Hydroporinae: Bidessini) from Texas, U.S.A. Proceedings of the Entomological Society of Washington, 99: 483-486.

MILLER, K. B. 1998. Revision of the Nearctic *Liodessus affinis* (Say 1823) species group (Coleoptera: Dytiscidae, Hydroporinae, Bidessini). Entomologica Scandinavica, 29: 281-314.

MILLER, K. B. 1999. New distribution records for *Crinodessus amyae* Miller (Coleoptera: Dytiscidae: Hydroporinae: Bidessini) with comments on variation and relationships. Coleopterists Bulletin, 53:40-41.

NILSSON, A. N. 1989a. On the genus *Agabetes* Crotch (Coleoptera, Dytiscidae), with a new species from Iran. Annales Entomologici Fennici, 55: 35-40.

NILSSON A. N. 1989b. Larvae of northern European *Hydroporus* (Coleoptera: Dytiscidae). Systematic Entomology, 14: 99-115.

NILSSON, A. N. and ANGUS, R. B. 1992. A reclassification of the *Deronectes*-group of genera (Coleoptera: Dytiscidae) based on a phylogenetic study. Entomologica Scandinavica, 23: 275-288.

NILSSON, A. N. and HOLMEN, M. 1995. The aquatic Adephaga (Coleoptera) of Fennoscandia and Denmark. II. Dytiscidae. Fauna Entomologica Scandinavica, 32: 192 pp.

NILSSON, A. N. and ROUGHLEY, R. E. 1997a. The genus- and family-group names of the Dytiscidae - additions and corrections. Beiträge zur Entomologie, 47:1-6.

NILSSON, A. N. and ROUGHLEY, R. E. 1997b. A classification of the family Dytiscidae (Coleoptera). Latissimus, 8: 1-4.

NILSSON, A. N. and ROUGHLEY, R. E. and BRANCUCCI, M. 1989. A review of the genus- and family-group names of the family Dytiscidae Leach (Coleoptera). Entomologica Scandinavica, 20: 287-316.

PEDERZANI, F. 1995. Keys to the identification of the genera and subgenera of adult Dytiscidae (*sensu lato*) of the world (Coleoptera Dytiscidae). Atti Dell'Accademia Roveretana Degi Agiati, a. 244, ser. VII, vol. IV, B: 5-83.

PETERSON, A., 1951. Larvae of Insects. An introduction to Nearctic species. Part II. Coleoptera, Diptera, Neuroptera, Siphonaptera, Mecoptera, Trichoptera. Privately printed. Columbus, Ohio. v + 416 pp.

ROCHETTE, R. A. 1983a. A new species of water beetle from California belonging to the *Hydroporus vilis* group (Coleoptera: Dytiscidae) with comments on the male genitalia of this group. Coleopterists Bulletin, 37: 148-152.

ROCHETTE, R. A. 1983b. A preliminary checklist of the *Hydroporus vilis*-group with a key to the species groups of the genus *Hydroporus* (Coleoptera: Dytiscidae). Coleopterists Bulletin, 37: 153-158.

ROCHETTE, R. A. 1983c. *Hydroporus adelardi*, a new dytiscid of the *vilis* group from California (Coleoptera: Dytiscidae). Proceedings of the Entomological Society of Washington, 85: 734-736.

ROCHETTE, R. A. 1985. A multivariate morphometric analysis between *Hydroporus veronicae* and *H. terminalis* (Coleoptera: Dytiscidae). The Southwestern Naturalist, 30: 445-448.

ROCHETTE, R. A. 1986. *Hydroporus williami*, a new dytiscid of the *vilis* group from the western United States. The Southwestern Naturalist, 31: 341-344.

ROUGHLEY, R. E. 1990. A systematic revision of species of *Dytiscus* Linnaeus (Coleoptera: Dytiscidae). Part 1. Classification based on adult stage. Quaestiones Entomologicae, 26: 383-557.

ROUGHLEY, R. E. and LARSON, D. J. 1991. Aquatic Coleoptera of springs in Canada. Memoirs of the Entomological Society of Canada, 155: 125-140.

ROUGHLEY, R. E. and PENGELLY, D. H. 1981. Classification, phylogeny, and zoogeography of *Hydaticus* Leach (Coleoptera: Dytiscidae) of North America. Quaestiones Entomologicae, 17 (1981): 249-309.

RUHNAU, S. 1986. Phylogenetic relations within the Hydradephaga (Coleoptera) using larval and pupal characters. Entomologica Basiliensia, 11: 231-271.

RUHNAU, S. and BRANCUCCI, M. 1984. Studies on the genus *Lancetes*. 2. Analysis of its phylogenetic position using preimaginal characters (Coleoptera, Dytiscidae). Entomologica Basiliensia, 9: 80-107.

SHARP, D. 1882. On aquatic carnivorous Coleoptera or Dytiscidae. The Scientific Transactions of the Royal Dublin Society, 2 (New Series), Part II: 179-1003 + plates vii-xviii.

SHIRT, D. B. and ANGUS, R. B. 1992. A revision of the Nearctic water beetles related to *Potamonectes depressus* (Fabricius) (Coleoptera: Dytiscidae). Coleopterists Bulletin, 46: 109-141.

SLATER, A. 1979. Additions to and corrections of the list of aquatic beetles of the families Dytiscidae, Gyrinidae, and Haliplidae from Kansas. Technical Publication of the Sate Biological Survey of Kansas, 8: 31-37.

SPANGLER, P. J. 1962. Natural history of Plummers Island, Maryland. XIV. Biological notes and description of the larva and pupa of *Copelatus glyphicus* (Say) (Coleoptera: Dytistidae). Proceedings of the Biological Society of Washington, 75: 19-24.

SPANGLER, P. J. 1973. A description of the larva of *Celina angustata* Aubé (Coleoptera: Dytiscidae). Journal of Washington Academy of Sciences, 4: 165-168.

SPANGLER, P.J. 1991. Dytiscidae (Adephaga). Pp. 315-319, *In* Lawrence, J.F. Co-ordinator. Chapter 34. Coleoptera. pp. 144-658. *In* F. Stehr. Editor. Immature Insects. Volume 2. Kendall/Hunt Publishing Co., Dubuque, Iowa. xvi + 975 pp.

SPANGLER, P. J. and BARR, C. B. 1995. A new genus and species of stygobiontic dytiscid beetle, *Comaldessus stygius* (Coleoptera: Dytiscidae: Bidessinae) from Comal Springs, Texas. Insecta Mundi, 9: 301-308.

TRACY, B. H. and HILSENHOFF, W. L. 1982. The female of *Graphoderus manitobensis* with notes on identification of female *Graphoderus* (Coleoptera: Dytiscidae). The Great Lakes Entomologist, 15: 163-164.

TRÉMOUILLES, E.R. 1995. Insecta. Coleoptera. Dytiscidae: Methlinae-Hydroporinae. Fauna de agua dulce de la República Argentina, 37 (1): 1-82.

WALLIS, J. B. 1939. The genus *Graphoderus* Aubé in North America (North of Mexico). (Coleoptera). The Canadian Entomologist, 71: 128-130.

WHITE, C. E., YOUNG, F. N., and DOWNIE, N. M. 1985. A checklist of the aquatic Coleoptera of Indiana. Indiana Academy of Science, 94: 357-369.

WHITE, D. S. and BRIGHAM, W. U. 1996. Aquatic Coleoptera, pp. 399-473. *In* Merrit, R. W. and K. W. Cummins (eds.) An introduction to the Aquatic Insects of North America, Kendall/Hunt, Dubuque, Iowa, 862 pp.

WOLFE, G. W. 1984. A revision of the *vittatipennis* species group of *Hydroporus* Clairville, subgenus *Neoporus* Guignot (Coleoptera: Dytiscidae). Transactions of the American Entomological Society, 110: 389-433.

WOLFE, G. W. 1985. A phylogenetic analysis of pleisiotypic hydroporine lineages with an emphasis on *Laccornis* Des Gozis (Coleoptera: Dytiscidae). Proceedings of the Academy of Natural Sciences of Philadelphia, 137: 132-155.

WOLFE, G. W. 1989. A phylogenetic investigation of *Hydrovatus*, Methlini and other plesiotypic hydroporines (Coleoptera: Dytiscidae). Psyche, 95 (1988): 327-344.

WOLFE, G. W. and MATTA, J. F. 1981. Notes on nomenclature and classification of *Hydroporus* subgenera with the description of a new genus of Hydroporini (Coleoptera: Dytiscidae). The Pan- Pacific Entomologist, 57: 149-175.

WOLFE, G. W. and ROUGHLEY, R.E. 1990. A taxonomic, phylogenetic, and zoogeographic analysis of *Laccornis* Gozis (Coleoptera: Dytiscidae) with the description of Laccornini, a new tribe of Hydroporinae. Quaestiones Entomologicae, 26: 273-354.

YOUNG, F. N. 1940. Description of a new dytiscid from Florida (Coleoptera: Dytiscidae). The Florida Entomologist, 23: 28.

YOUNG, F. N. 1951. A new water beetle from Florida, with a key to the species of *Desmopachria* of the United States and Canada (Coleoptera; Dytiscidae). Bulletin of the Brooklyn Entomological Society, 46: 107-112.

YOUNG, F. N. 1953. Two new species of *Matus*, with a key to the known species and subspecies of the group (Coleoptera: Dytiscidae). Annals of the Entomological Society of America, 46: 49-55.

YOUNG, F. N. 1954. The water beetles of Florida. University of Florida Studies in Biology, Science Series 5 (1): ix + 1-238.

YOUNG, F. N. 1956. A preliminary key to the species of *Hydrovatus* of the eastern United States (Coleoptera: Dytiscidae). Coleopterists Bulletin, 10: 53-54.

YOUNG, F. N. 1963a. The Nearctic species of *Copelatus* Erichson (Coleoptera: Dytiscidae). Quarterly Journal of the Florida Academy of Sciences, 26: 56-77.

YOUNG, F. N. 1963b. Two new North American species of *Hydrovatus*, with notes on other species (Coleoptera: Dytiscidae). Psyche, 70: 184-192.

YOUNG, F. N. 1967a. A key to the genera of American bidessine water beetles with descriptions of three new genera (Coleoptera: Dytiscidae, Hydroporinae). Coleopterists Bulletin, 21: 75-84.

YOUNG, F. N. 1967b. The *Hydroporus blanchardi-tigrinus* complex (Coleoptera: Dytiscidae). The Florida Entomologist, 50: 63-69.

YOUNG, F. N. 1969. A checklist of the American Bidessini (Coleoptera: Dytiscidae- Hydroporinae). Smithsonian Contributions to Zoology, 33: 1-5.

YOUNG, F. N. 1974. Review of the predaceous water beetles of genus *Anodocheilus* (Coleoptera: Dytiscidae: Hydroporinae). Occasional Papers of the Museum of Zoology, University of Michigan, No. 670. 28 pp.

YOUNG, F.N. 1977. Predaceous water beetles of the genus *Neobidessus* Young in the Americas north of Columbia (Coleoptera: Dytiscidae, Hydroporinae). Occasional Papers of the Museum of Zoology, University of Michigan 681. 24 pp.

YOUNG, F. N. 1978. A new predaceous water beetle from the eastern United States (Coleoptera: Dytiscidae). Coleopterists Bulletin, 32: 189-192.

YOUNG, F. N. 1979. A key to the Nearctic species of *Celina* with descriptions of new species (Coleoptera: Dytiscidae). Journal of the Kansas Entomological Society, 52: 820-830.

YOUNG, F. N. 1980. Predaceous water beetles of the genus *Desmopachria*: the subgenera with descriptions of new taxa (Coleoptera: Dytiscidae). Revista de Biologia Tropical, 28: 305-321.

YOUNG, F. N. 1981a. Predaceous water beetles of the genus *Desmopachria* Babington: the *leechi-glabricula* group (Coleoptera: Dytiscidae). The Pan-Pacific Entomologist, 57: 57-64.

YOUNG, F. N. 1981b. Predaceous water beetles of the genus *Desmopachria*: the *convexa-grana* group (Coleoptera: Dytiscidae). Occasional Papers of the Florida State Collection of Arthropods, 2: 1-11.

YOUNG, F. N. 1981c. Predaceous water beetles of the genus *Neobidessus* Young from South America (Coleoptera: Dytiscidae). Coleopterists Bulletin, 35: 317-340.

YOUNG, F. N. 1984. Two new species of *Hydroporus* (*Neoporus*) from the southeastern United States (Coleoptera: Dytiscidae). Coleopterists Bulletin, 38: 185-189.

YOUNG, F. N. 1986. Predaceous water beetles of the genus *Desmopachria* Babington: The *nitida* group (Coleoptera: Dytiscidae). Coleopterists Bulletin, 40: 269-271.

YOUNG, F. N. 1989. A new species of *Agaporomorphus* from Florida, a new Nearctic record for the genus (Coleoptera: Dytiscidae). Florida Entomologist, 72: 263-264.

YOUNG, F. N. 1990a. A review of classification of the water beetles of the New World genus *Bidessonotus* Régimbart (Coleoptera: Dytiscidae: Hydroporinae: Bidessini). Quaestiones Entomologicae, 26: 355-381.

YOUNG, F. N. 1990b. Predaceous water beetles of the genus *Desmopachria* Babington: the subgenus *Pachriostrix* Guignot (Coleoptera: Dytiscidae). Coleopterists Bulletin, 44: 224-228.

YOUNG, F. N. and LONGLEY, G. 1976. A new subterranean aquatic beetle from Texas (Coleoptera: Dytiscidae-Hydroporinae). Annals of the Entomological Society of America, 69: 287-292.

YOUNG, F. N. and WOLFE, G. W. 1984. *Hygrotus berneri*, a new water beetle from Florida (Coleoptera: Dytiscidae). Journal of the Kansas Entomological Society, 57: 130-133.

ZIMMERMAN, J. R. 1970. A taxonomic revision of the aquatic beetle genus *Laccophilus* (Dytiscidae) of North America. Memoirs of the American Entomological Society, 26: 1-275.

ZIMMERMAN, J. R. 1981. A revision of the *Colymbetes* of North America (Dytiscidae). Coleopterists Bulletin, 35: 1-52.

ZIMMERMAN, J. R. 1982. The *Deronectes* of the southwestern United States, Mexico and Guatemala (Coleoptera: Dytiscidae). Coleopterists Bulletin, 36: 412-438.

ZIMMERMAN, J. R. 1985. A revision of the genus *Oreodytes* in North America (Coleoptera: Dytiscidae). Proceedings of the Academy of Natural Sciences of Philadelphia, 137: 99-127.

ZIMMERMAN, J. R. and SMITH, A. H. Smith. 1975. A survey of the *Deronectes* (Coleoptera: Dytiscidae) of Canada, the United States and northern Mexico. Transactions of the American Entomological Society, 101: 651-722.

ZIMMERMAN, J. R. and SMITH, R. L. 1975. The genus *Rhantus* (Coleoptera: Dytiscidae) in North America. Part I. General account of the species. Transactions of the American Entomological Society, 101: 33-123.

ZIMMERMANN, A. 1920. pp. 15-296. *In* Schenkling, S. (ed.) Coleopterorum Catalogus. Pars 71. Dytiscidae, Haliplidae, Hygrobiidae, Amphizoidae. W. Junk, Berlin. 326 pp.

Suborder POLYPHAGA Emery, 1886

Series STAPHYLINIFORMIA Lameere, 1900

Superfamily HYDROPHILOIDEA Latreille, 1812

Palpicornia Latreille, 1817

# 13. HYDROPHILIDAE Latreille, 1802

by Eileen R. Van Tassell

**Family common name**: The water scavenger beetles

**Family synonyms**: Helopheridae, error; Helophoridae Leach 1817; Hydrochidae Thomson 1860; Spercheidae Erichson 1837; Sphaeridiidae Latreille 1804; Georyssidae Laporte 1840; Epimetopidae, Zaitzev, 1908

Most members of this family are aquatic and are adapted for aquatic life. There are two major groupings, the aquatic and the terrestrial scavengers. The terrestrial members belong to a single subfamily, the Sphaeridiinae. It should be noted that some members of the Sphaeridiinae are semi-aquatic and form somewhat of a transition between the two groups. Aquatic hydrophilids may usually be recognized by the long maxillary palpi which are longer than the antennae and may be mistaken for antennae. The antennae are short, clavate, inserted in front of the eyes, 6-9 segmented, usually with a 3-segmented pubescent club. Terrestrial hydrophilids usually have maxillary palpi equal in length or shorter than the antennae, with a very fine, straight, or T-shaped, rather than V-shaped, epicranial suture, sometimes visible only at the sides of the head.

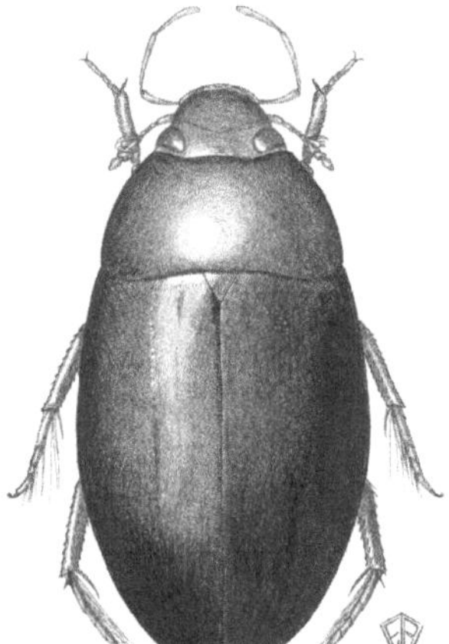

FIGURE 1.13. *Hydrobiomorpha casta* (Say).

**Description**: Adults. Shape usually broadly oval, dorsal surface convex, ventral surface flattened or concave, usually streamlined, dull; size 1-40 mm; color black, black with brownish markings, or brown with black, sometimes metallic markings, rarely yellow or cream with dark markings; pubescence scant on dorsum, moderate on legs, dense on sternum, antennal club and around mouthparts.

Head slightly to strongly deflexed; surface smooth, punctate or rugose, with a Y-shaped epicranial suture separating the vertex from the frons, sometimes only visible at sides of head (some Sphaeridiinae). Antennae clavate, with 6-10 antennomeres, first long (scape), second shorter (pedicel), 3-6 small, narrower than scape, apical 3 antennomeres enlarged, forming a loose, moderately compact or variously modified club, usually with the segment preceding the club forming a cupule; inserted close to eyes, under frontal ridge, often lying in a groove or excavation along lower margin of eye. Clypeus large, usually fused with frons but separated from labrum; labrum small, anterior margin evenly arcuate or emarginate, often fringed with long setae; frons sometimes with setiferous punctures near medial margins of eyes (systematic punctures); mandibles moderate in size, mostly hidden by labrum, apices blunt to acute and bifid; sometimes with small, mobile teeth, rarely small with membranous tips; maxillary palpi often exceed-

**Acknowledgments**. First of course, thanks to the late Ross H. Arnett, Jr., a great mentor and teacher from whom I learned the not-so-gentle art of verbal scientific combat. Fortunately, neither distance nor age diminished the enjoyment of a good argument for either of us, including many issues concerning this new edition. Ross fought very hard for what he felt was good science and always found a way to admit it when he thought he had been wrong. His integrity, generosity and wry sense of humor will be missed by many if not all of the writers of these chapters and perhaps most of all by his former students. I hope he would be proud of this, his final contribution to coleopterology. Michael Hansen gave invaluable assistance in the honing of a workable key. Alfred Newton provided many helpful corrections and comments on the manuscript. Conversations with Miguel Archangelsky helped clarify several points concerning Hydrophilid classification. Any errors still remaining are entirely my own.

ing length of antennae, with 4 palpomeres,1 usually minute, 2-4 very elongate, narrow; labial palpi with 3 palpomeres, first very small, 2-3 subequal, slender; paraglossae well developed. Eyes lateral, equally visible above and below, moderate to large in size, oval, emarginate, sometimes divided by a canthus; gula variable, posterior tentorial pits usually small.

Pronotum broader than head, usually much wider than long; anterior border broadly emarginate, lateral borders usually evenly arcuate, slightly to moderately divergent or constricted posteriorly, posterior border evenly arcuate to slightly sinuate with a very narrow transverse ridge just beneath it; all borders very narrowly margined, surface usually smooth, sometimes with bumps, depressions or grooves, finely to deeply punctate, often with a short line of setiferous punctures near outer anterior edges and /or just past middle (systematic punctures, Figs. 19.13-21.13); notopleural suture absent; prosternum short to moderate, with an intercoxal process extending posteriorly, lateral margins with a narrow glabrous border, procoxal cavities usually open behind or rarely closed (*Hydrochus*). Scutellum triangular, of variable size, usually distinct (except in Georissinae). Mesosternum large, depressed, somewhat keeled, plate-like or with a hood-like process (with a highly variable central keel, anterior projection or flattened area); mesepisternum reaching coxal cavities; mesocoxal cavities open behind. Metasternum large, variable, may be keeled; metepisternum reaching coxal cavities; metacoxal cavities very large.

Legs with trochantin obscure on fore legs, large, broad, freely articulated on mid-legs, obscure or fused on hind legs. Coxae of fore legs usually large, globular, subconical; mid- and hind-coxae broad, usually flattened, with medial part large, prominent, triangular; trochanters small and obscure on fore legs, fused with coxae in Georissinae, moderately developed on mid- and hind legs, rarely enlarged on hind legs. Femora and tibiae usually normal in shape and size, somewhat flattened, tibiae spinose or with heavy spurs, sometimes with a longitudinal line of long, silky hairs (*Berosus*), rarely expanded at apex.

Tarsal formula 5-5-5 or 5-4-4 (*Cymbiodyta*), rarely 4-5-5 in males (*Berosus* ) or highly modified in males (*Sphaeridium*); fore tarsi with uniform subequal tarsomeres, except basal first and second segments swollen in males (*Berosus*), second segment swollen in males (*Hemiosus*), or second and third often swollen in males (*Laccobius*), fifth elongate; mid and hind tarsi with first tarsomere very small to absent, the rest elongate, flattened, with a fringe of setae variable in length; tarsal claws usually simple but sometimes with complex modifications in the males.

Elytra broader at base than pronotum, widest near middle, sometimes with rounded or oval plectral areas on lateromedial ventral surface, apices acute or subacute, sometimes modified; surface smooth, punctate-striate or rugose; epipleural fold broad to narrow, variously separated into an inner pubescent, true epipleuron, and an outer, glabrous, pseudepipleuron.

Hind wings primitively Polyphagan, with a distinct mediocubital loop and lacking an oblongum cell. Basal cell and wedge cell usually well defined, jugal lobe usually present; apical venation complex, often faint, with traces of the radial cell beyond the fold. The folding pattern is as for the superfamily and suborder.

Abdomen with 5 (rarely 6) visible sterna with usually well-defined laterosternites, sometimes with an oval or elongated stridulatory mechanism or file on the first visible laterosternite; surface smooth or rough, densely or finely punctate; sutures between segments distinct, segments movable.

Male genitalia trilobed, penis and parameres well developed, basal piece small to large, symmetrical or with asymmetrical base, position usually with opening of median lobe ventral, or aedeagus facing to side and able to rotate ventrally (*Berosus*), with parameres usually deeply positioned in basal piece. Female genitalia with the proctiger, paraproct, valvifer, coxite, stylus, and tenth sternum present. Construction of the egg cases or cocoons has been described for several genera (Laabs 1939), and the structure and histology of female accessory glands have been described and illustrated for *Hydrophilus olivaceus* Fabricius (Gundevia and Rammurty 1984).

Larvae. Campodeiform, elongate, cylindrical, spindle-shaped or conical; size 4-60 mm; color gray or yellowish-brown, sometimes with a pattern of dark setae; head deeply pigmented and frequently spotted; thorax, especially the tergum, deeply pigmented. Head prominent, exserted, or slightly retracted, usually prognathous, with mouthparts projecting dorsally. Antennae 3-4 segmented, dorsal-laterally inserted near mandibles. Labrum and clypeus fused with frons often producing a distinct nasale which may be asymmetrical. Mandibles sometimes asymmetrical, conspicuous, apices simple or bifid, with one or more teeth on the mesal margin, without prostheca or molar area. Maxilla with palpifer palpiform, large, with minute galea and no lacinia. Labium with modified ligula and 2-segmented palpi; gula normal, reduced, or absent. Ocelli 5-6 pairs, sometimes less. Thorax with three distinct segments; legs 5-segmented with tarsungulin, rarely legless or with segments fused (Georissidae). Abdomen with 10, 9, or usually 8 visible segments, rarely (some Berosini) with long lateral gill-like appendages on some segments; surface may be rugose or coriaceous, usually with numerous folds, sometimes with fine pubescence rarely forming a coarse pattern. Spiracles biforous on segments 1-8, absent in Berosini instars one and two. Urogomphi paired, 2 or 3 segmented on those forms with 10 abdominal segments (second segment vestigial in *Georissus*) or short, 1-2 segmented on those forms with 8 visible segments; the latter also have a terminal breathing cavity (except *Berosus*).

**Habits and habitats**. Several workers have discussed the shortcomings of this family's common name, the water scavengers. With rare exceptions, the immatures are predatory, not scavenging, and the adults may be vegetarians, omnivores, occasionally predators, or scavengers. Perhaps a better name would be the Omnivorous water beetles.

Most species are aquatic but many (mostly Sphaeridiinae) live in fresh dung of mammals, in soil rich in humus, or in moist decaying leaves. The aquatic species are found in stagnant pools, littoral areas of lakes, in the shallow quiet water of streams containing an abundance of vegetation, in springs and margins of streams, or rarely in brackish or strongly saline (potash) water.

Adults of some species are predaceous on snails and other small invertebrates, others feed on infusoria, spores, or decaying vegetation. In the laboratory, adults have been maintained on a combination of algae, fish food flakes, decaying leaves (*Berosus* sp.) and/or dog food pellets (*Tropisternus* sp.); small fish, tadpoles (*Hydrophilus triangularis*) or rarely (*Dibolocelus ovatus*), on snails.

The adults swim by alternate movements of the hind legs unlike the dytiscids. Frequently, workers in this group have referred to setae on the legs of both adults and larvae as "swimming hairs," but in fact there is no evidence that these hairs aid in swimming. Some excellent swimmers do not have them, and some relatively poor swimmers do. Species of *Berosus* which have well-developed tibial setae are frequently seen sweeping these hairs through their air bubble, such that increased aeration seems just as likely a function, especially as many of these species inhabit standing water where oxygen supplies would be lower. Alternatively, a "secretion grooming system" such as that described by Perkins for the Hydraenidae (Perkins 1997) is also a possible function for these setae, especially since many hydrophilids have similar systems of pores and setae, easily seen with the scanning electron microscope (Oliva 1992 ).

In some species the antennae are carried beneath the head when the beetle is in water, but often extend outward in air. When submerged, the antennae are used to break the surface tension of the water, bringing fresh air to the plastron bubble. In most of the aquatic species this extends from the posterior of the head over the entire ventral surface of the abdomen, thus bringing air to the abdominal spiracles. See Hrbacek (1950) for observations on the mechanism of air exchange in various genera.

Many species, mostly in the Hydrophilini, are able to produce sounds when handled or attacked, using an elytral-abdominal mechanism. Masters (1979) has demonstrated in laboratory experiments that these sounds are effective deterrents to predation or at least delay predation.

Some have been observed making a more rhythmical type of sound in a pre-mating context (Maillard 1972, Pirisinu 1985, Ryker 1972, Scheloske 1975, Van Tassell 1965). Maillard and Sellier (1970) have constructed a tentative classification of the stridulatory apparatus in Hydrophilids and have illustrated scanning electron micrographs of rows of raised ridges on *Paracymus* and *Enochrus* suggesting that these may be primitive stridulatory structures. However, species of these taxa have not been heard to stridulate in any context (pers. obs., Hansen, pers. communication), although many attempts have been made by the author to record them. Miller (1874) also describes posssible structures on *Chaetarthria*, but with the comment that these do not look like typical stridulatory structures and that this group has never been heard making any sounds. These are possibly involved with the wing-folding mechanism.

Adults are often attracted to lights at night in great numbers. Adults are occasionally attracted to car tops, as has been reported for the Dytiscidae (Duncan 1927) and to blue plastic tarps spread on the ground to dry (pers. obs.). From above, automobiles, like water surfaces, are very conspicuous and may serve as an attractant to migrating adults.

Mating takes place for aquatic species in the water, following a brief courtship, and lasts from one to five minutes. The eggs are laid singly with a silken covering, or in protective silk cases holding 2-100 or more eggs; these are fastened to plants, substrate, or may rarely float free. In three genera, the egg case is attached to the abdomen of the female; in both *Epimetopus* (Rocha, 1967), and *Spercheus emarginatus* (Bøving and Henriksen, 1938) the egg case is held in place against the abdomen with the hind legs, but is attached by two fine silk threads to the metafemora in *Helochares* (Richmond, 1920). The abdomen of those females producing egg cases have spinnerets on the caudal end near the anus; spinning takes place in the water, either with the abdomen slightly projecting out of the water, or completely submerged. The function of the egg case has been the subject of considerable interest and debate, with most workers considering it to be of respiratory function, providing some air for the eggs. Several authors, including Crowson (1981), have considered the case, especially the "mast" to be a protection from drowning of the eggs, a possibility denied by Hinton (1981). In fact, some have successfully reared the larvae in distilled water outside the case (Van Tassell, unpublished). It has been pointed out (Vlasblom and Wolverkamp 1955) that distilled water may have much more oxygen available than stagnant water, such that the case may yet be an advantage, especially when many eggs are packed together into a single case. In their experiments, they ligated the "funnel" or "mast" of the egg case of *H. piceus* and found that none of the larvae developed. Females of some genera (*Berosus* sp.) draw out part of their air bubble onto the substrate with their hind legs and spin the silk into the extended bubble, thereby incorporating air into the case. A bubble of air also emerges with the first larva to hatch (pers. obs.). Perhaps there are a number of strategies which have evolved in different groups for various oxygen concentrations in the water of their diverse habitats, but this question has not yet been resolved.

With rare exceptions, the larvae are predaceous, voracious, and, except for *Dibolocelus ovatus*, in confined spaces, cannibalistic. In captivity, larvae will eat almost any fleshy material from insect larvae to hamburger. Small pieces of mealworm are acceptable when agitated in front of the mouthparts of many larvae (pers. obs.). Several workers have had success rearing larval stages to adulthood using fruit fly or housefly maggots, brine shrimp, other insect larvae and hamburger (Archangelsky 1992a, 1992b, 1994; Young 1958).

Larvae have three instars of variable length, the first two averaging about one week each and the last longer, depending mostly on the food supply, but also on temperature. When sufficient fat-bodies have accumulated, the larva enters the soil at the edge of its aquatic habitat. In *Berosus*, this stage is indicated by the obscuring of the internal structures such that the larva appears opaque. Species of some genera crawl out of the water at this stage then dig down until the right soil is encountered, but some burrow directly into the bank horizontally. Pupation usually takes place in a mud cell near the shore.

The mud cell is constructed by the larva and it pupates from 5-20 days or even weeks later, depending on circumstances. If the larva is not fully ready to pupate, it may wander in the soil or form an earthen cell to await more favorable circumstances (Richmond 1920). Pupae are usually white, but may be various shades of pale green as well. The head, thorax and abdomen have dorsal setae upon which the pupa rests in the cell. Most also have two cerci at the posterior of the abdomen; in *Berosus* and possibly other genera, the cerci assist the pupa in rotating in the pupal cell (pers. obs.). About a day before eclosion, the eyes, apices of the mandibles and tarsal claws become progressively darker.

The newly emerged adult stays in the pupal cell until it is fully hardened, usually about 5 days. Adults in drier parts of the country may become encased in mud which has hardened and the adult is not freed until rain softens the mud cell. Wilson (1923) accidentally discovered three adult *Berosus* in some hard mud which he discarded. The clump broke apart revealing the three live adults which had been encased for a full month.

Archangelsky (1996) has compiled a bibliography of larval literature for the family. See Wilson (1923), Bøving and Henriksen (1938) and Richmond (1920) for descriptions of larvae and life histories for several genera and Bertrand (1972, 1977) for descriptions of larvae and pupae of the world. Recent keys to larval genera may be found in White and Brigham (1996). A partial key to the egg cases as well as a discussion of the physics of the plastron was published by Hinton (1981).

**Economic importance.** Larger Hydrophilids such as *Hydrophilus triangularis* have been reported to be pests in fish hatcheries (Wilson, 1923) and in England, larvae of *Helophorus* were found to damage wheat (Petherbridge and Thomas 1935) turnips, rutabagas and rape (a European forage crop). In land where the crucifer crops have been planted in previous years in England, Cos lettuce was also attacked. These larvae will also bore into the galls made by the weevil *Ceuthorhynchus pleurostigma* to attack and eat the gall larva (Petherbridge, 1928). Perhaps the usually predacious *Helophorus* acquired a taste for vegetables in the process. *Tropisternus lateralis nimbatus* was found by Wooldridge and Wooldridge (1972) to harbor pathogenic intestinal bacteria that could be transmitted from pond to pond during dispersal flights. Milliger *et al.* (1971) found 8 species in 4 genera of Hydrophilids, as well as numerous Dytiscids and other aquatic beetles to be carrying an average of 3.3-10.8 genera of algae and protozoans per beetle. This could be either positive in the colonization of new habitats or negative, if pathogens are present.

On the positive side, *Hydrophilus triangularis* has also been reported to prey on mosquito larvae (Nelson, 1977). In the laboratory, *H. triangularis* was found able to eat 194 mosquito larvae of *Culex quinquefasciatus* in a 24-hour period. Both this species and a Dytiscid, *Dytiscus marginalis*, were more effective feeding on mosquito larvae than on pupae, largely because of the greater activity of mosquito pupae. Successful predation was also found to differ at different prey densities. Overall, an 80.2% success rate in predation was reported. Niesen and Nielsen, (1953, p. 146) found large numbers of first larvae, then adults, mostly of *Tropisternus lateralis nimbatus* (Say) (100 of 130 specimens identified) and *Berosus infuscatus* LeConte feeding on *Aedes taeniorhynchus* in roadtrack pools in Florida. "They congregated in a few of the small open pools at a time, and when they had exterminated all the mosquito larvae there, they moved to the next pool." Since adult Hydrophilids are primarily vegetarian, the adult predators were probably the 25 dytiscids which were also present. James (1965) used radioactive tracers to determine predation in rock pools in Ontario, Canada. Although several Dytiscids were highly successful predators of *Aedes atropalpus* (Coq), *Enochrus* sp. adults and *Berosus* sp. larvae (of unspecified numbers and species) were found negative after one 24-hour trial. However, this is not a favored habitat of *Berosus* sp., and may not indicate their potential for control of other species of mosquitoes in other environments. Notestine (1971) also found *Enochrus* and *Berosus*, both as larvae and adults in a mosquito-rich waterfowl management area in Utah but concluded, based on James' study, that they were not predators of mosquitoes. However, she did recommend that further studies be undertaken to determine how best to take advantage of the control potential of invertebrate predators. The findings of Eyre *et al.* on the preferences of some water beetle assemblages for sites of short duration and others for long duration give additional support to the need for further study. See Bay (1974) for a review of literature on predation in other families and orders.

**Status of the classification.** This family has received much-needed attention in the past decade, both higher classification worldwide and regionally. Recently, some workers, especially Hansen (1991) using primarily adult characters, Beutel (1994) using characters of the head capsule of larvae and adults, and Archangelsky (1998) employing pre-imaginal stage evidence, have supported the elevation of 4 of the subfamilies which occur in the Nearctic region to family status (Helophoridae, Epimetopidae, Georissidae and Hydrochidae). However, others such as Lawrence and Newton (1995), Newton and Thayer (1992), Spangler (1991) and Smetana(1988), have retained them as subfamilies of the Hydrophilidae. Authors also disagree on the status of the berosines, some (Smetana 1988, Spangler 1991) retaining them as a subfamily, others, (Archangelsky,1998, Hansen, 1991, Lawrence and Newton 1995, Newton and Thayer 1992) treating them as a tribe within the subfamily Hydrophilinae. More study is needed to assess all the supra-generic taxa, but again, Lawrence and Newton's treatment, modified by Newton and Thayer is utilized here, retaining all as subfamilies, including the recently placed Georissinae, and treating the berosines as a tribe. See Hansen (1999) for a more detailed review of this history. It should be noted that Lawrence and Newton relied on Newton and Thayer for their consideration of the systematic position of the Hydrophiloid taxa, but in an addendum to the latter paper, the authors discuss the fact that they did not receive Hansen's work in time to incorporate all of his changes to the classification, but did include a updated classification (modified from the Knisch 1924 catalog), which I have followed here. In the event that the family status of the five subfamilies listed is more widely accepted by North American workers in the future, the subfamily descriptions are more detailed. Newton and Thayer (1992) have pub-

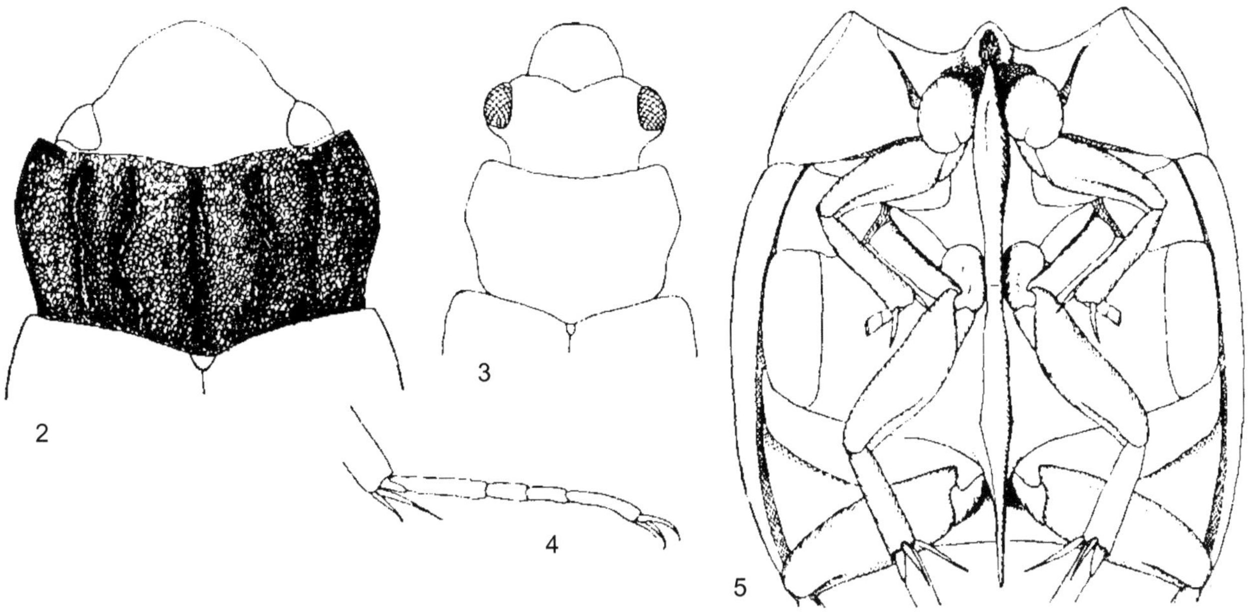


FIGURES 2.13-5.13. Fig. 2.13, *Helophorus* sp., head and pronotum, dorsal view; Fig. 3.13, *Hydrochus* sp., head and pronotum, dorsal view; Fig. 4.13, *Berosus* sp., hind tarsus; Fig. 5.13 *Hydrophilus triangularis* Say, sternal region showing median keel.

lished a worldwide classification of the entire series Staphyliniformia, with families listed alphabetically and sub-taxa above the generic level listed also alphabetically. This work summarizes changes of family-group names in the series required by the International Code of Zoological Nomenclature (1985) and presents an invaluable discussion of the history and usage of many commonly accepted family names and includes correct type genera. See Archangelsky (1998) for a discussion of the influence of the larval spiracular atrium on the relationships of the clades, and for a comparison of his and Hansen's data sets.

See Williams (1838), for a study of the distribution of glandular structures on the mouthparts of several genera. Chromosomal studies have been reported by Angus (1989, 1996) and Shaarawi and Angus (1992). Neither sets of characters are sufficiently complete yet to provide evidence for classification.

The family Hydraenidae has been frequently united with the Hydrophilidae in the past, but has many characters which separate it from the rest of the Hydrophilidae and it is therefore recognized here as a distinct family, but arguably allied to the Hydrophilidae (see Perkins, Chapter 22, this volume for a discussion of Hydraenid relationships).

Regional keys for parts of the Nearctic region are also more available recently for Hydrophilids (Brigham 1982; Epler 1996; Leech and Chandler 1956; Hilsenhoff 1995; Malcolm 1971; Matta 1974a, 1974b, 1974c; Merritt and Cummins 1996; Testa and Lago 1994) replacing or supplementing some of the older works (Hatch 1965; Leech 1948, 1950; Young 1954). Larval studies are not as well-represented in the literature, but several works include them (Merritt and Cummins 1996), or focus on them (Spangler 1991). The species of many groups are well-known, those of other groups are greatly in need of study. Distribution of the family is worldwide; there are about 2475 known species (Hansen 1999); of these, 258 species in 35 genera occur in North America north of Mexico. Larvae of 33 of the 34 genera in North America have been described, including a description of the larva of probably *Hemiosus* sp. White and Brigham (1996) have provided keys to the larvae of 21 of the genera. A world catalog of species is now available for the family (Hansen 1999).

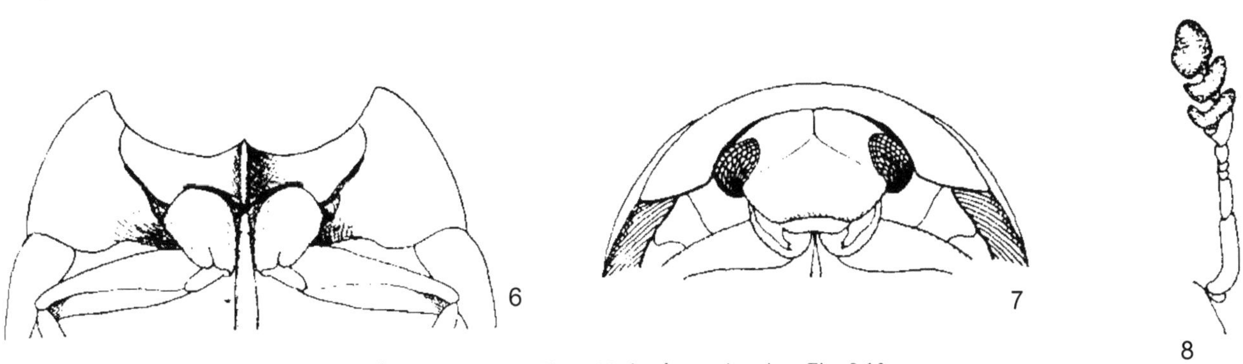


FIGURES 6.13-8.13. *Hydrochara* sp., Fig. 6.13, prosternum; Fig. 7.13, head, anterior view; Fig. 8.13, antenna.

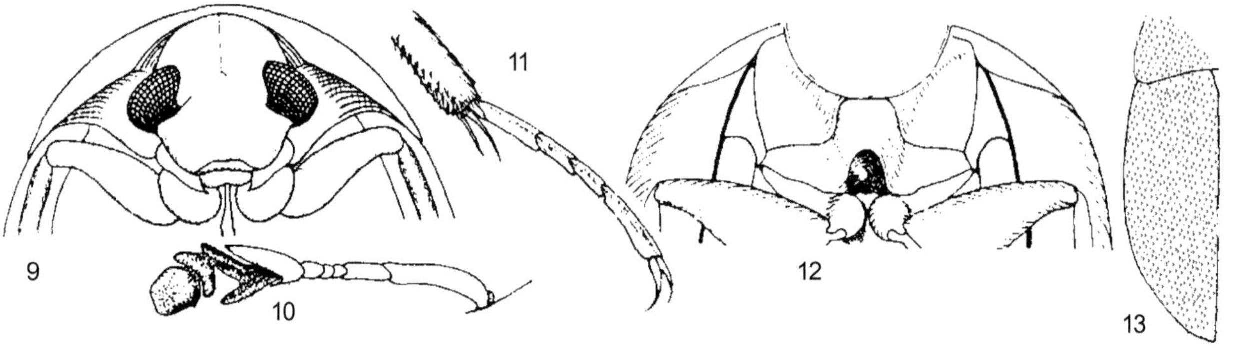


FIGURES 9.13-13.13. Fig. 9.13, *Hydrobiomorpha casta* (Say), head, anterior view; Fig. 10.13, same, antenna; Fig. 11.13 *Hydrobius fuscipes* (Linnaeus), hind tarsus; Fig. 12.13, *Anacaena limbata* (Fabricius), prosternum, showing median spine; Fig. 13.13, *Enochrus ochraceus* (Melsheimer), elytra, dorsal view.

In his 1908 catalog, Zaitzev (1910) notes that in the family Hydrophilidae, in the Gemminger and Harold list of 1868, there were 30 genera and 537 species; in 1908, in his catalog, 114 genera and 1336 species. Now, nearly a century later, there are 161 genera and 2803 species worldwide listed in the Hansen catalog (Hansen 1999). Writing good, working keys has become a much more difficult task. Fortunately, the Nearctic region has seen a far more modest increase in genera, such that the task is easier. Thirty-three genera were keyed in the first edition for the northeastern U.S., now 36 are included in this edition. New are *Hemiosus*, a Western U. S. genus; *Georissus*, treated as a separate family in the first edition; and the new genus *Agna*, described by Smetana in 1978, with a single U.S. species and one from Mexico. *Pemelus* Horn, has been placed in synonymy with *Oosternum*, and *Dibolocelus* has been reduced to a subgenus of *Hydrophilus*. Finally, the genus *Neohydrophilus* d'Orchymont 1911 has been placed in synonymy with *Hydrobiomorpha* Blackburn by Mouchamps (1959). However, there has been a considerable amount of work, including new tribes and genera in other regions of the world. Several genera have been re-defined as tribes and genera are studied in more detail, especially in the Sphaeridiinae. This subfamily still needs work at the world level to make the taxa more sharply defined. See Hansen (1991, pp. 9-11) for a complete historical review of the classification of the family. Fossil record: 21 fossil species in 15 extant genera have been described (Scudder 1878, 1890), mostly from the Green River Station in Wyoming. See Hansen 1999, Appendix, pp. 318-324 for a complete list.

## Key to the Nearctic Subfamilies, Tribes, and Genera

[Modified from Hansen 1991. Key does not include subgenera, and is accurate only for the Nearctic taxa. A relatively new character, called "systematic punctures" (Figs. 19.13-21.13) has been utilized to organize taxa at the world level (Hansen 1991). This feature is also characteristic of many Berosini, but these have long hair-like setae fringing the middle and hind tibiae. Systematic punctures are absent in all of the Anacaenini and in the

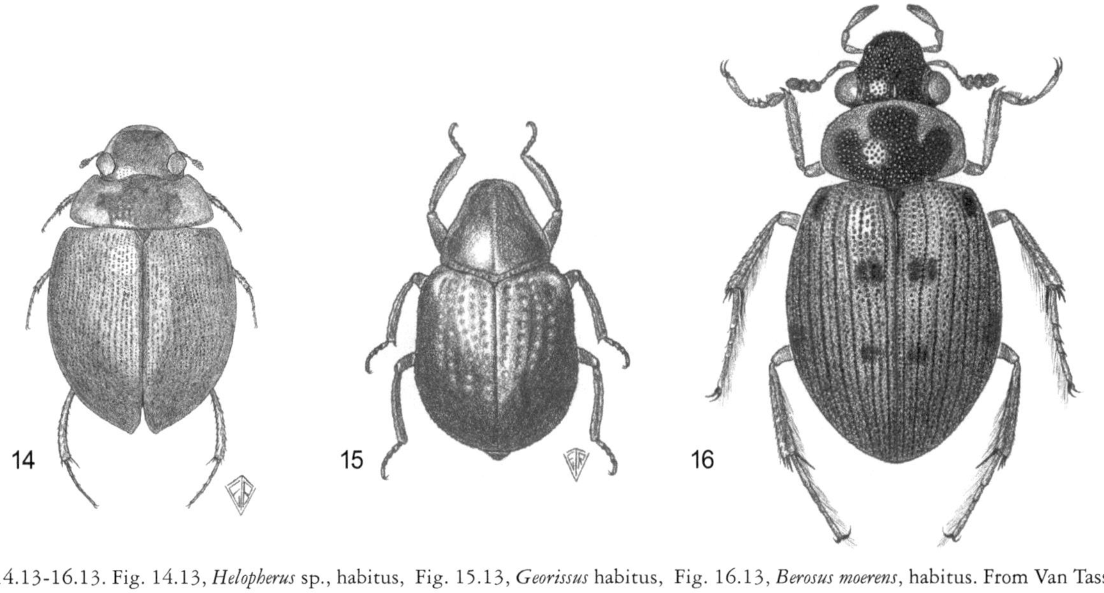


FIGURES 14.13-16.13. Fig. 14.13, *Helopherus* sp., habitus, Fig. 15.13, *Georissus* habitus, Fig. 16.13, *Berosus moerens*, habitus. From Van Tassell, 1963.

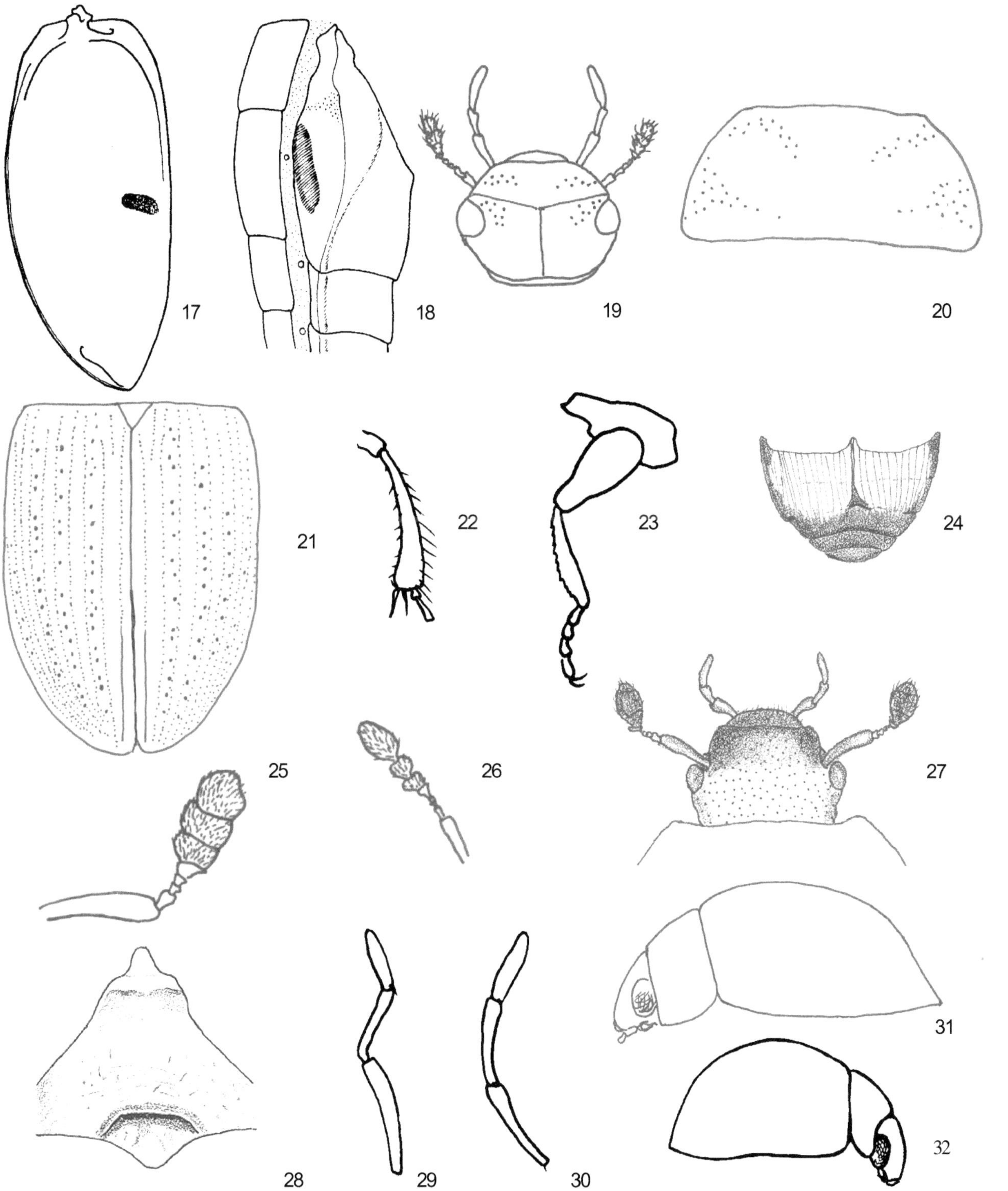


FIGURES 17.13-32.13. Figs. 17.13-18.13, *Berosus* sp., both from Van Tassell 1965; Fig. 17.13, underside of elytron, showing position of plectron; Fig. 18.13, abdomen, lateral view, showing position of stridulatory file on laterosternite; Figs 19.13-21.13, generalized hydrophilid, showing position of systematic punctures, [redrawn from Hansen 1991;] Fig. 19.13, head, dorsal view; Fig. 20.13, pronotum; Fig. 21.13, elytra; Fig. 22.13, *Laccobius* sp., tibia; Fig. 23.13, *Georissus* sp., fused procoxa and trochanter, [redrawn from Hansen 1991;] Fig. 24.13, *Chaetarthria* sp., abdomen; Fig. 25.13, *Sphaeridium* sp., antenna; Fig. 26.13, *Phaenonotum exstriatum*, antenna; Fig. 27.13, *Omicrus palmarum*, head. [Redrawn from Smetana 1975;] Fig. 28.13, *Cymbiodyta* sp., mesosternal transverse crest; Fig. 29.13, *Cymbiodyta* sp., maxillary palp; Fig. 30.13, *Enochrus* sp., maxillary palp; Fig. 31.13, *Enochrus* sp., lateral view; Fig. 32.13, *Derallus altus*, lateral view.

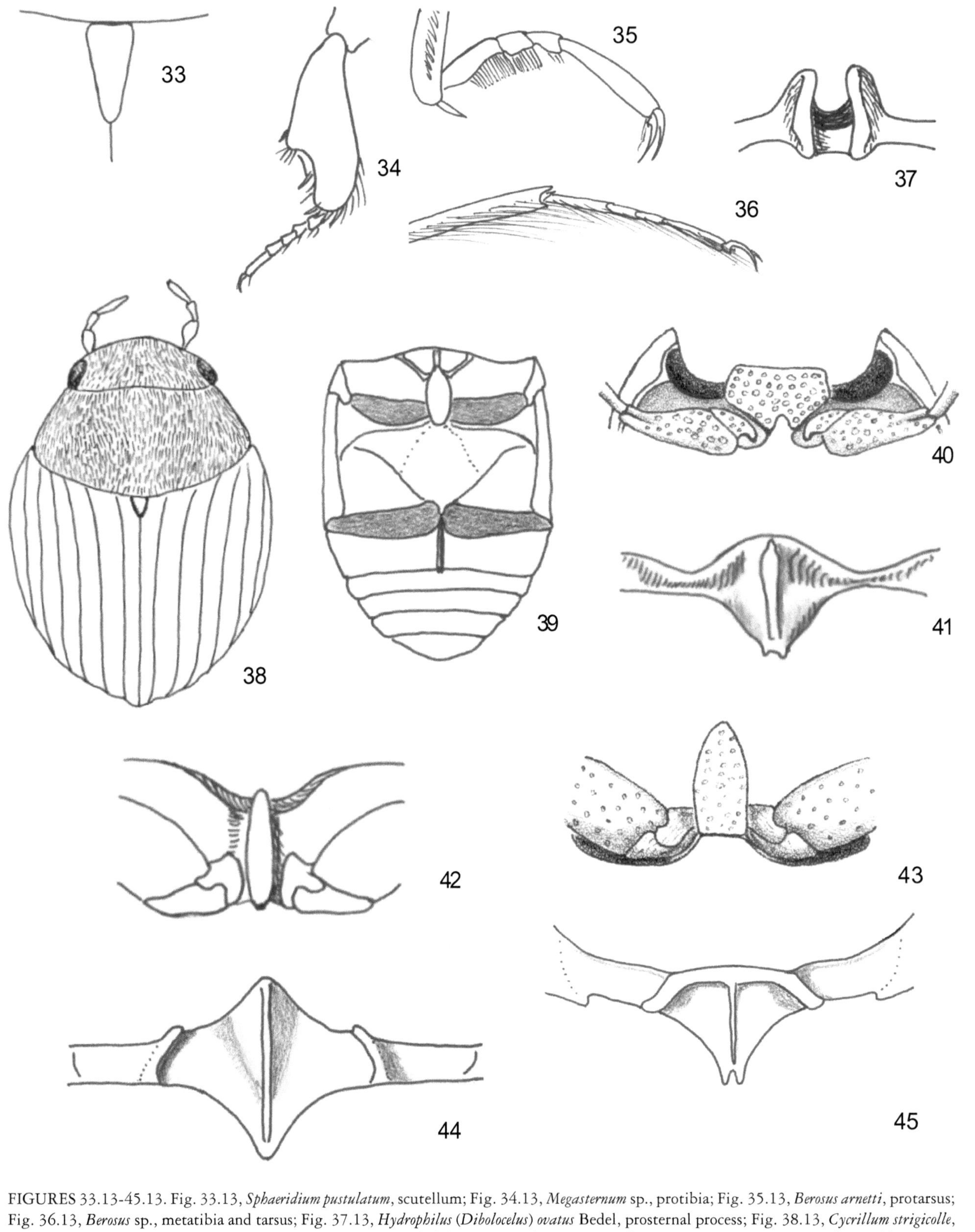

FIGURES 33.13-45.13. Fig. 33.13, *Sphaeridium pustulatum*, scutellum; Fig. 34.13, *Megasternum* sp., protibia; Fig. 35.13, *Berosus arnetti*, protarsus; Fig. 36.13, *Berosus* sp., metatibia and tarsus; Fig. 37.13, *Hydrophilus* (*Dibolocelus*) *ovatus* Bedel, prosternal process; Fig. 38.13, *Cycrillum strigicolle*, dorsal view; Fig. 39.13, *Oosternum* sp., meso-, metathorax, and abdomen; Fig. 40.13, *Cryptopleurum* sp., mesosternum; Fig. 41.13, *Tectosternum naviculare*, prosternum; Fig. 42.13, *Cercyon marinus*, mesosternum; Fig. 43.13, *Pelosoma* sp., mesosternum; Fig. 44.13, *Agna capillata* (LeConte), mesosternum; Fig. 45.13, *Deltostethus columbiensis*, mesosternum.

Sperchopsinae except for the genus *Ametor*, which has a rugose elytral integument.]

1. Eyes nearly or completely divided by a median canthus into dorsal and ventral parts; first abdominal sternum very short (Epimetopinae) .................... ........................................................ *Epimetopus*
— Eyes rounded, not divided; first abdominal sternum well developed ................................................ 2

2(1). Pronotum with seven longitudinal grooves (one subartic species has a single median groove) (Fig. 2.13) (Helophorinae) ........................ *Helophorus*
— Pronotum without such grooves .......................... 3

3(2). Pronotum with a shelf-like projection which extends forward over the head; metacoxae broadly separated; front coxae and trochanters fused to form large plates that conceal the small prosternum (Georissinae) ....................................... *Georissus*
— Pronotum without a forward projecting shelf; metacoxae narrowly separated; front coxae and trochanters separate; prosternum usually well developed ...................................................... 4

4(3). Pronotum distinctly narrower at base than elytra, surface with shallow, bilaterally symmetrical elevations and depressions; scutellum very small; eyes protuberant (Fig. 3.13) (Hydrochinae) ....... ......................................................... *Hydrochus*
— Pronotum nearly as wide at base as basal one-fourth of elytra, but if narrower, scutellum is elongate triangular, surface usually smooth; eyes prominent or not ...................................................... 5

5(4). Antennae shorter than or subequal to maxillary palpi; basal tarsomere of middle and hind tarsi very short, shorter than second and often minute (Fig. 4.13) (except *Helocombus* and *Cymbiodyta*, in which the basal segment has disappeared); aquatic species (Hydrophilinae) ....................... 6
— Antennae usually longer than maxillary palpi; basal tarsomere of middle and hind tarsi elongate, longer than second; usually terrestrial or semiaquatic species (Sphaeridiinae) ..................... 25

Subfamily Hydrophilinae

6(5). First two visible abdominal sterna with a common excavation on each side covered by a fringe of long, dense hairs arising from the first abdominal sternum (Fig. 24.13) (Chaetarthriini) .................. ...................................................... *Chaetarthria*
— Abdomen without such excavations ................... 7

7(6). Head strongly deflexed (Fig. 32.13); antennae usually with seven antennomeres, scutellum a long triangle; middle and hind tibiae fringed with long, fine setae (Fig. 16.13, 36.13) (Berosini) .......... 12
— Head not or only moderately deflexed (Fig. 31.13); antennae usually with nine antennomeres; scutellum not or barely longer than its basal width; tibiae without fringe of long, fine setae ....................... 8

8(7). Hind tibiae arcuate (Fig. 22.13); hind trochanters large, about one-third as long as femora, their apices distinct from femora (Laccobiini) ........... .......................................................... *Laccobius*
— Hind tibiae straight, not arcuate; hind trochanters smaller, more closely applied to femora .......... 9

9(8). Head, and/or pronotum and/or elytra with systematic punctures on at least one part (Fig. 19.13-21.13) (Tribe Hydrophilini) ............................. 16
— Head, pronotum and usually elytra without systematic punctures ................................................ 10

10(9). Elytra striate, species 4-9.5 mm (tribe Sperchopsini) ......................................................................... 11
— Elytra not striate, at most with punctures subserially arranged, impunctate or confusedly punctate; species 3 mm or less in length (tribe Anacaenini) ........................................................................ 14

Tribe Sperchopsini

11(10). Form strongly convex, almost hemispherical in profile; clypeus more deeply emarginate, median part nearly truncate, elytra with second interval keeled near apex, intervals smooth color reddish-brown ............................................. *Sperchopsis*
— Form oval; clypeus evenly, shallowly, emarginate, second elytral interval not keeled at apex, similar to adjacent intervals, intervals rugose-punctate, with systematic punctures on alternate intervals; color usually black or dark brown ...................... *Ametor*

Tribe Berosini

12(7). Eyes very protuberant; front tibiae slender, linear; labrum prominent, color testaceous to brown, usually with darker spots which may have metallic reflections, usually larger than 2 mm ............. 13
— Eyes not protuberant; front tibiae triangular, gradually widening from base to apex, labrum very short, not prominent, color black, without markings, size 2 mm or less .......................... *Derallus*

13(12). Ventral surface with dense, velvety pubescence obscuring integument; mesosternum with a rhomboidal median crest which is excavated anteriorly; males with protarsi 5-5-5, second segment slightly expanded .............................. *Hemiosus*
— Ventral surface pubescent, but not velvety, not obscuring integument; mesosternal crest usually blade-like or only narrowly expanded anteriorly; males with protarsi 4-5-5, basal tarsomere and usually second clearly expanded (Fig. 35.13) .. ............................................................... *Berosus*

Tribe Anacaenini

14(10). Prosternum longitudinally carinate at middle; metafemora lacking hydrofuge pubescence or if present, only at extreme base ......... *Paracymus*
— Prosternum not carinate but may be slightly convex medially; metafemora usually with more extensive hydrofuge pubescence .................... 15

15(14). Mesosternum simple, not carinate, or with a small transverse protuberance anterior to middle coxae ............................................................... *Crenitis*
— Mesosternum with a prominent angularly elevated or dentiform protuberance anterior to middle coxae (Fig. 12.13) .............................. *Anacaena*

Tribe Hydrophilini

16(9). Meso- and metasternum with a united median keel which usually extends between hind coxae as a stout spine (Fig. 5.13); large species, usually over 10 mm (subtribe Hydrophilina) ........................ 17

— Sternal region without such a keel; smaller species, usually less than 10 mm ................................ 20

17(16). Prosternum sulcate to receive anterior part of mesosternal keel (Fig. 5.13, 37.13); metasternal keel projecting beyond hind trochanters as a spine ............................................................ 18

— Prosternum carinate, not sulcate (Fig. 6.13); metasternal keel not or hardly reaching beyond bases of hind trochanters ............................ 19

18(17). Large species, 30 to 45 mm in length; last maxillary palpomere shorter than preceding ... *Hydrophilus*

— Moderately large species, 6 to 15 mm in length; last maxillary palpomere equal to or longer than penultimate .................................... *Tropisternus*

19(17). Front margin of clypeus simply truncate or arcuate (Fig. 7.13); antennal club with first two segments more similar (Fig. 8.13) .................... *Hydrochara*

— Front margin of clypeus arcuate and emarginate at middle to expose a preclypeus (Fig. 9.13); antennae perfoliate, club with first segment very different from second (Fig. 10.13) ..... *Hydrobiomorpha*

20(16). Tarsomeres two to five of middle and hind tarsi with a fringe of long, fine setae which arise from a series of punctures or a narrow groove along upper inner edge of tarsi (Fig.11.13); elytra with distinct striae (subtribe Hydrobiina) .......................... *Hydrobius*

— Tarsomeres two to five of middle and hind tarsi without such a fringe of long setae; Elytra with or without rows of punctures (subtribe Acidocerina) ....................................................................... 21

21(20). Labrum concealed beneath clypeus which extends outward for a distance equal to about the width of an eye; antennae 8-segmented; elytra with broadly explanate margins .................. *Helobata*

— Labrum fully exposed, clypeus not extending laterally in front of eyes; antennae 9-segmented; elytral margins not broadly explanate ........... 22

22(21). Mesosternum at most feebly protuberant, without a longitudinal lamina, or mesosternum with a pyramidal elevation; elytra with longitudinal rows of punctures ...................................................... 23

— Mesosternum with a variously projecting longitudinal or transverse lamina; elytra usually confusedly punctate (Fig. 13.13) ...................................... 24

23(22). All tarsi 5-segmented, basal segment small, mesosternum without a longitudinal mesosternal crest ......................................................... *Helochares*

— Tarsal formula 5-4-4, mesosternum with a pyramidal crest .............................................. *Helocombus*

24(22). Middle and hind tarsi 4-segmented; maxillary palpi evenly, outwardly curved (Fig. 29.13); mesosternum with a transverse crest (Fig. 28.13) ........... ........................................................ *Cymbiodyta*

— Middle and hind tarsi 5-segmented, basal segment usually minute; maxillary palpi with second segment bent toward mid-line, last segment forward or outward facing (Fig. 29.13) mesosternum with a median, longitudinal lamina, often with a small anterior tooth ...................................... *Enochrus*

Subfamily Sphaeridiinae

25 (5). Head abruptly narrowed or excised in front of eyes, exposing antennal bases, or if weakly excised, then first abdominal sternum much longer than second ........................................................... 26

— Head not abruptly narrowed or excised in front of eyes, antennal bases concealed under lateral shelf; first abdominal sternum about as long as second, or slightly shorter ............................ 27

26(25). Clypeus strongly deflexed and somewhat convex in profile, with angulate, strongly deflexed extensions in front of antennae; labrum somewhat large, not retracted under clypeus; epipleura and pseudepipleura well developed (Omicrini) ........ ............................................................ *Omicrus*

— Clypeus not deflexed, usually weakly convex in profile, usually without anterolateral extensions; labrum retracted under clypeus, usually not or only very narrowly visible in front of clypeus; antennal club always compact; epipleura usually not visible next to abdomen (Megasternini) .. 29

27(25). Antennal club compact (Fig. 25.13); labrum not retracted under clypeus, fully visible, well sclerotized and pigmented; first segment of middle and hind tarsi much longer than second; tibiae with very long spines on outer face; eyes deeply emarginate anteriorly; scutellum much longer than wide (Fig. 33.13) (Sphaeridiini) ...... *Sphaeridium*

— Antennal club loosely segmented (Fig. 26.13) or if compact, then labrum is mostly retracted under clypeus and is soft and pale; scutellum barely longer than wide (Coelostomatini) ................ 28

28(27). First abdominal sternum longitudinally carinate at middle; elytral sutural striae distinct ................. ................................................ *Dactylosternum*

— First abdominal sternum not carinate at middle; elytral striae absent ...................... *Phaenonotum*

29(26). Prosternum with very large and deep antennal grooves that reach the lateral prothoracic margin ....................................................................... 30

— Prosternum with antennal grooves not reaching lateral prothoracic margin, sometimes indistinct .. 33

30(29). Femoral lines of metasternum reduced to anterolateral vestiges, protibia with apex steeply narrowed (Fig. 34.13) .................................... *Megasternum*

— Femoral lines complete, protibia with apex smoothly and evenly narrowed ..................................... 31

31(30). Elytra without striae, only rows of fine punctures; epipleura present in anterior half; prosternum steeply tectiform (Fig. 41.13) ....... *Tectosternum*

— Elytra with well-marked striae (Fig. 38.13); epipleura present at extreme base only, prosternum not tectiform, but steeply raised in middle to form a flat pentagonal or quadrilateral tablet ........... 32

32(31). Mesosternal tablet about as wide as long, distinctly margined; mesocoxae moderately widely separated; prosternum shallowly carinate medially .. .......................................................... *Cycrillum*

— Mesosternal tablet somewhat wider than long (Fig. 40.13), mesocoxae very broadly separated; prosternum not carinate medially...................... .................................................. *Cryptopleurum*

33(29). Anterolateral corners of metasternum demarcated from the remainder of metasternum by a transverse, more or less arcuate, ridge that corresponds to the arc described by the tip of hind femora (Fig. 39.13) ........................... *Oosternum*

— Metasternum without such a ridge, having only a fine line running close to the posterior margin of the mesocoxae ............................................... 34

34(33). Mesosternal tablet elongate-oval to almost linear, narrowed both anteriorly and posteriorly, rounded or pointed posteriorly, never truncate, seldom subparallel posteriorly, not closely abutted to anteromedial part of raised mid-metasternum (Fig. 42.13) (except in *Cercyon* subgen. *Paracercyon*, where it is received in an anterior excision of metasternum), mesosternum rarely narrowly laminate medially without well defined tablet.......... ............................................................ *Cercyon*

— Mesosternal tablet pentagonal, parallel sided or nearly so between mesocoxae, truncate posteriorly and closely abutted to anteromedial part of raised mid-metasternum .................................. 35

35(34). Mesosternal tablet about as long as wide ......... 36

— Mesosternal tablet two to three times longer than wide (Fig. 43.13).................................. *Pelosoma*

36(35). Prosternum tectiform and strongly carinate medially, somewhat protruding anteromedially, demarcated from antennal grooves by a pair of moderately strong ridges (Fig. 44.13); antennal grooves hardly defined laterally ............................. *Agna*

— Middle portion of prosternum flat and finely carinate medially, not protruding anteromedially, demarcated from antennal grooves by a pair of very strong ridges (Fig. 45.13); antennal grooves well defined laterally by an arcuate ridge ................ ...................................................... *Deltostethus*

## Classification of the Nearctic Genera

### Hydrophilidae

Recent studies of this family by Hansen (1991) using primarily adult characters, Beutel (1994) using characters of the head capsule of larvae and adults, and Archangelsky (1998) employing total evidence, have supported the elevation of 5 of the subfamilies to family status (Helophoridae, Epimetopidae, Georissidae, Hydrochidae and Spercheidae). Following Lawrence and Newton, Newton and Thayer, Spangler and Smetana, they are here retained as subfamilies. Authors also disagree on the status of the Berosines, with some (Smetana 1988, Spangler 1991) retaining them as a subfamily, while others (Archangelsky 1998, Hansen 1991, Lawrence and Newton 1995, Newton and Thayer 1992) treat them as a tribe within the subfamily Hydrophilinae. More study is needed to assess all the supra-generic taxa, thus the more conservative treatment is utilized here, retaining all as subfamilies, including the recently placed Georissinae. See Hansen (1999), for a more detailed review of this history. It should be noted that Lawrence and Newton relied on Newton and Thayer for their consideration of the systematic position of the hydrophiloid taxa, but in an addendum to the latter paper, the authors discuss the fact that they did not see Hansen's work in time to make a full assessment of its merits.

## Summary of the Subfamilies, Tribes and Genera of the United States

Subfamily Helophorinae
- *Helopherus*

Subfamily Epimetopinae
- *Epimetopus*

Subfamily Georissinae
- *Georissus*

Subfamily Hydrochinae
- *Hydrochus*

Subfamily Hydrophilinae
- Tribe Sperchopsini
  - *Ametor*
  - *Sperchopsis*
- Tribe Berosini
  - *Hemiosus*
  - *Berosus*
  - *Derallus*
- Tribe Chaetarthriini
  - *Chaetarthria*
- Tribe Anacaenini
  - *Paracymus*
  - *Crinitis*
  - *Anacaena*
- Tribe Laccobiini
  - *Laccobius*
- Tribe Hydrophilini
  - Subtribe Acidocerina
    - *Helochares*
    - *Enochrus*
    - *Helobata*
    - *Cymbiodyta*
    - *Helocombus*
  - Subtribe Hydrobiina
    - *Hydrobius*
  - Subtribe Hydrophilina
    - *Hydrochara*
    - *Hydrobiomorpha*
    - *Tropisternus*
    - *Hydrophilus*

Subfamily Sphaeridiinae
- Tribe Coelostomatini
  - *Phaenonotum*

*Dactylosternum*
Tribe Omicrini
*Omicrus*
Tribe Megasternini
*Cercyon*
*Deltostethus*
*Agna*
*Pelosoma*
*Oosternum*
*Tectosternum*
*Megasternum*
*Cryptopleurum*
*Cycrillum*
Tribe Sphaeridiini
*Sphaeridium*

## Helophorinae

**Diagnosis**: These are somewhat elongated and slightly convex beetles with prominent eyes, and seven prominent longitudinal grooves on the pronotum (only a single species, *H. arcticus*, has secondarily lost this feature). In the past, the number of grooves was described as five, omitting the ones at the margins of the pronotum. Size 2.6-3.8 mm. Labrum not emarginate, eyes prominent, antennae 8-9 segmented with a loose, three-segmented club; prothorax with distinct notopleural sutures; elytra striate-punctate; abdomen with five visible sterna; first tarsomere much shorter than second or subequal (terrestrial species). Male with somewhat simple, symmetrical aedeagus.

Larva: Galea present as a small setiferous appendage; lacinia absent; nasale symmetrical; epistomal lobes well developed, symmetrical; first antennal segment longer than second; ocelli well separated; gular sutures confluent; legs well developed; urogomphi very long, three-segmented; abdominal segments somewhat soft, each with some small sclerites, without lateral outgrowths; eighth segment without tergal shield, not concealing ninth; stigmatic atrium absent.

Adults are aquatic, primarily in fresh water, occasionally in brackish water, but a few species in the Palearctic subgenus *Empleurus* sp. are terrestrial and have vegetarian larvae, a condition unique among hydrophilids. Larvae apneustic, with well-developed urogomphi and abdominal segments 1-8 with functional spiracles. Larvae occur among soil and vegetation near but not in the water. One species is known to be parthenogenetic, based on laboratory rearing studies and may be an important colonizer (Angus 1970).

*Helophorus* Fabricius 1775, Palearctic and Nearctic regions, a few species in the Ethiopian region. There are 183 spp. worldwide in a single genus, of which 43 spp. are Nearctic, with 36 in the U.S., including Alaska, 14 in contiguous U.S., with a questionable record from Arizona. (Key to subgenera and spp., Smetana 1988)
*Elophorus* Fabricius 1775. Incorrect original spelling
*Elophorus* Paykull 1798. Incorrect original spelling
*Helophorus* Illiger 1801. Emendation of *Elophorus* Fabricius

## Subfamily Epimetopinae

**Diagnosis**: This small group is easily recognized by the divided eyes and the projecting pronotum which covers most of the deflexed head. The posterior width of the pronotum is much narrower than the elytra. Length 1.2-3.5 mm. Aedeagus of male symmetrical.

Larva: Head with nasale symmetrical, epistomal lobes well developed; ocelli closely aggregated; galea and lacinia absent; antennae with first segment shorter than second; legs well developed; urogomphi 3-segmented, very long; abdominal segments barely sclerotized, without lateral projections; 8th segment without tergal shield; posterior margin of 8th tergum with 4 projecting lobes, not hiding ninth segment; stigmatic atrium absent.

Aquatic. These are the rarest and most unusual members of the family, especially with respect to their dorsal sculpturing. Little is known about their biology or habitats. Rocha (1967) illustrated the ventral surface of the female, showing the egg sac being carried in the same manner as in *Sphercheus* (not Nearctic) and *Helochares*. In two egg sacs he collected still attached to females, there were hatched larvae inside each sac. This subfamily includes 3 genera and 27 species worldwide. Only one, *Epimetopus*, is Nearctic.

*Epimetopus* Lacordaire 1854, 4 spp., Texas and Arizona.
*Ceratoderus* Mulsant 1851, not Westwood 1841
*Sepidulum* LeConte 1874

## Subfamily Hydrochinae

**Diagnosis**: This subfamily is easy to identify using the narrow shape, bulging eyes, and pronotum with shallow depressions and a narrow base; dorsal surface often faintly metallic; ventral surface densely pubescent and the antennal club 7-segmented. There are no systematic punctures. Size 1.5-5.5 mm. Larvae are elongate and cylindrical. The mouthparts have a large retinaculum on the mandible.

*Hydrochus* Leach 1817. There are 164 species in all major regions, of which 26 species are Nearctic. Smetana (1985) has revised the Nearctic species and notes (Smetana 1988) that the species are very difficult to separate without dissection of the aedeagus. (See Hansen (1999) for a discussion of his well-reasoned basis for placing the new genera of Makhan (1984) in synonymy.) (Keys to spp.: Downie and Arnett 1996; Young 1954; Usinger 1956).
*Hydrochus* Provancher 1877
*Hydrochous* Bedel 1881
*Kiransus* Makhan 1994
*Deepakius* Makhan 1998
*Amrishius* Makhan 1998
*Rishwanius* Makhan 1998

## Subfamily Hydrophilinae

**Diagnosis**: Diverse; shape oblong-oval to broadly oval, outline not or only slightly interrupted between pronotum and elytra; pronotum usually widest at base, head rarely strongly deflexed; head and pronotum simply punctate. In general, with systematic punctures on elytra and often also on head and pronotum; maxillary palpi usually two thirds as long as width of head or longer; with 8 or 9 antennomeres; scutellum about as long as wide; sternum variable, often with median carinae; first visible abdominal sternum shorter, often much shorter , than second, fifth sternum often with an apical emargination bordered by dense yellow setae; laterosternite with or without strigil (absent in all Acidocerina); epipleura and pseudepipleura usually well defined, sometimes strongly oblique.

Hind wing with r-m crossvein arising at anterior wing margin from near distal half of pigmented area, wedge cell about as long as basal cell, jugal lobe usually, but not always, demarcated with a sharp excision at the posterior wing margin.

Larva: Galea present as a small setiferous tubercle, rarely indicated only by its seta (some Megasternini); lacinia absent; nasale symmetrical or asymmetrical; epistomal lobes present usually symmetrical, but sometimes (*Berosus*, *Laccobius*) with the left one larger; first antennal segment usually longer than second; ocelli well-separated, rarely closely aggregated; gular sutures confluent; legs usually well developed, rarely reduced or absent; urogomphi present (except *Berosus*), but very short, 2-segmented; abdominal segments somewhat soft and more or less coriaceous, sometimes with more or less long lateral outgrowths, first to seventh segments without distinct sclerites; 8th segment with well-defined tergal shield, and concealing the 9th from above, forming a stigmatic atrium (except *Berosus*). There are 2,336 spp. in the subfamily worldwide, with 188 Nearctic species in 6 tribes and 20 genera.

### Tribe Sperchopsini

**Diagnosis**: Recognizable by the serrate margins of the elytra, absence of pronotal systematic punctures and by the elytra with 10 well-defined punctate striae. Tarsi have randomly arranged fine, dense but not extensive, pubescence. There are 5 genera worldwide, 2 Nearctic.

*Sperchopsis* LeConte, 1862, 1 sp., *S. tesselata* (Ziegler), Canada, eastern and southern U.S. This is a monotypic genus, easy to recognize by the key characters.

*Ametor* Semenov, 1900, 2 spp., *A. latus* (Horn 1873) from California and *A. scabrosus* (Horn 1873) from Canada, Alaska, western contiguous U.S., Central Asia and the Himalayan region. (Key to species, Smetana 1988.)

### Tribe Berosini

**Diagnosis**: Under 12 mm in length; usually brown or testaceous with darker markings on pronotum and/or elytra or completely black (*Derallus*); head strongly deflexed, usually black with metallic reflections, sometimes brown; antennae 7- or 8-segmented; middle and hind tibiae with a fringe of long hairs; elytra punctate-striate; ventral sruface pubescent, sometimes densely so (*Hemiosus*); mesosternum usually with a median longitudinal crest which may be narrow (*Berosus*), rhomboid and excavated (*Hemiosus*) or abbreviated in length (*Derallus*). Fifth sternum with a pronounced apical subrectangular excision flanked by a pair of more or less strong projections; laterosternite 3 (concealed under the elytron) in all genera with an obliquely ribbed area (stridulatory file).

Hind wing with r-m crossvein rising from base or (*Hemiosus*) at least basal half of the pigmented area at anterior wing margin; wedge cell only about half as long as basal cell, jugal lobe demarcated by a more or less sharp excision at posterior wing margin.

Males often with anterior tarsi more or less dilated basally (less so in *Hemiosus*). Aedeagus dorsoventrally oriented as in most Hydrophilids, or (*Berosus*) oriented to one side. Basal piece slightly asymmetrical (*Hemiosus*).

Larva: Nasale symmetrical or asymmetrical; epistomal lobes asymmetrical with the left one being larger (*Berosus*), or symmetrical; ocelli well separated; legs well developed; abdomen with or (*Berosus*) without a stigmatic atrium and urogomphi; abdominal segments with somewhat long or (Berosus) very long lateral outgrowths. There are 5 genera of which 3 are Nearctic; 344 species worldwide, 25 Nearctic.

*Hemiosus* Sharp, 1882, 22 spp., primarily Neotropical; only one species is found in the U.S., *H. exilis* (LeConte), collected in Arizona, near the Gila River, and in Mexico. This species has not been collected in the U.S. for many years despite searches by collectors and may be endangered. The minute first protarsal tarsomere of the male has caused this small species to be transferred from *Hemiosus* to *Berosus* and back once more to *Hemiosus*. (Key to potentially Nearctic species, Van Tassell 1964.)

*Berosus* Leach, 1817, 17 spp., generally distributed. (Keys to species: Leech 1948; Testa and Lago 1994; Young 1954, some eastern spp.; Usinger 1956, Calif. spp.; Van Tassell 1966, Nearctic spp.)

*Derallus* Sharp, 1882, 1 sp., *D. altus* (LeConte, 1855), New Jersey south to Florida and west to eastern Texas in the U.S., Cuba, Dominican Republic, Puerto Rico and Central and South America. (Van Tassell 1966; Matta 1974; Testa and Lago 1994; Oliva 1981, 1995.)

### Tribe Chaetarthriini

**Diagnosis**: The small size, 1.3-2.5 mm, strongly convex form and gelatinous masses on the first abdominal segment covered with long setae will readily distinguish these beetles from any others. The elytra do not have striae or serial punctures except for a sutural stria in the posterior half; 8 antennomeres, with the very long scape and pedicel somewhat globular and slightly wider than the scape, club compact. Only the genus *Chaetarthria* occurs in the Nearctic region. Hind wing as in the subfamily.

Larva: Nasale symmetrical; epistomal lobes symmetrical; ocelli closely aggregated; legs very reduced without claw segment; abdominal segments without long lateral outgrowths. There are 77 species in 5 genera worldwide; one Nearctic genus, *Chaetarthria*, with 14 species. (Key to species: Miller 1974 (revision of Nearctic species), Usinger 1956, Testa and Lago 1994, Smetana 1988.)

*Chaetarthria* Stephens 1833, 4 spp., generally distributed
*Chaetarthrias*, Chenu 1851, incorrect spelling
*Chaethartria*, Thompson 1859, incorrect spelling
*Cyllidium* Erichson 1837

Tribe Anacaenini

**Diagnosis**: These genera lack systematic punctures and do not have an apical emargination on the fifth sternum. The pseudepimeron is narrow and not demarcated from the epipleuron by a distinct ridge. No stridulatory file is present on abdominal laterosternites. Hind wings are as for the subfamily.

Males sometimes (*Paracymus*, some *Anacaena*) with claw segment of anterior tarsi slightly dilated.

Larva: Nasale asymmetrical; epistomal lobes symmetrical; ocelli well separated; legs well developed; abdominal segments without long lateral outgrowths.

*Paracymus* Thompson 1867, 15 Nearctic spp., generally distributed. (Key to spp.: Arnett 1994, northeastern U.S. spp.; Testa and Lago 1994, 6 Mississippi spp.; Young 1954, Florida spp.; Leech 1956, some western U.S. spp.; Smetana 1988, Canadian spp.; Wooldridge 1966, revision of Nearctic species. Wooldridge (1975) has suggested that worldwide the species fall into 6 distinct groups and based on their distribution are probably monophyletic. There are two Nearctic species groups, based on the shape of the aedeagus.)

*Crenitis* Bedel 1881, 12 Nearctic spp., New Hampshire, Pennsylvania, Indiana, and Pacific Coast. (Key to spp.: Leech 1956, Smetana 1988, Canadian species.)

*Anacaena* Thomson 1859, 4 Nearctic spp., Newfoundland, Indiana, Lake Superior, Florida, New Mexico, and California.
*Cryniphilus* Motschulsky 1845, incorrect original spelling
*Crenophilus* Agassiz 1859, unjustified emendation
*Crenitulus* Winters 1926
*Metacymus* Sharp 1882

Tribe Laccobiini Bertrand 1954

*Oocyclini* Hansen 1991

**Diagnosis**: The pale testaceous or brown color, elytra with up to twenty rows of darker punctures but no striae and the arcuate metatibiae will distinguish the Nearctic members of this genus. A few rows of systematic punctures may be present but are seldom evident. A stridulatory file is present in a similar position as in the Berosini. Hind wings are as for the subfamily. Males with second and sometimes third tarsomeres expanded; aedeagus variable, but parameres often with a dorsal apical projection.

Both Cheary (1971) and Miller (1965) state the males to have 4 protarsomeres, but Van Tassell (1966) found five in the Nearctic species and Hansen (pers. com.) knows of no other *Laccobius* with a male protarsus with only 4 tarsomeres. The first tarsomere is very small.

Larva: Nasale asymmetrical; epistomal lobes asymmetrical, the left one being larger than the right; ocelli well separated; legs well developed; abdominal segments without long lateral outgrowths. There are 9 genera and 195 species worldwide, one Nearctic genus, *Laccobius* Erichson, 1837, with 24 Nearctic species. (Key to species: Smetana 1988, Gentili 1985, Testa and Lago 1994, Cheary 1971, Miller 1965.)

Tribe Hydrophilini Latreille, 1802

**Diagnosis**: This is a very diverse tribe, difficult to diagnose except for the characters in the key. These are usually large to medium sized, oval to elliptical in shape, convex with smooth elytra generally without striae. Antennae 8- or 9- segmented. Frons with a distinct group of systematic punctures near inner margin of eyes. Frons with a distinct group of systematic punctures near inner margin of eyes.

Description: Body without power of rolling up, head not strongly deflexed toward ventral face. Labrum fully visible in front of clypeus (except *Helobata*), lateral and anterior margins normally without fringes of setae, anterior margin more or less shallowly emarginate medially, sometimes truncate or subtruncate. Pronotum almost always with detectable systematic punctures on lateral portions, otherwise normally evenly punctate or impunctate. Prosternum somewhat short to very well-developed, variably raised medially, prosternal process often somewhat short, often with fine median carina and more or less pronounced posteromedian process or highly raised to form a narrow keel. Raised middle portion of metasternum usually not projecting anteriorly between mesocoxae, normally terminating posteriorly in a median, more or less well-marked bulge above ventral condyles; this bulge sometimes produced into a shorter or longer horizontal tooth or acute spine (very strongly developed in some species of *Hydrophilus*). Abdomen with 5 visible sterna; first usually distinctly shorter than second, posterior margin of 5th ventrite normally with a small, more or less well defined and sometimes somewhat shallow apical emargination fringed with dense, stiff yellowish setae; emargination sometimes completely absent

(*Hydrophilus*, *Enochrus*, some *Cymbiodyta*. Stridulatory structures are reported in *Paracymus* and *Enochrus* (Maillard and Sellier, 1870), but species in these genera are not known to produce sound (Hansen, pers. com.). Epipleura and pseudepipleura normally somewhat well defined from each other, more or less strongly oblique, especially the latter; the pseudepipleuron normally narrow or even very narrow, but in *Helobata* quite wide, in general narrower than true epipleuron. Lateral margins of elytra not serrate of denticulate. Hind wing with r-m crossvein arising from distal, etc., but sometimes (*Hydrophilus*) not defined by such excision.

Males sometimes with claws (especially on anterior tarsi) more strongly curved and toothed than in females; rarely also with claw segment of anterior tarsi dilated.

Larva: nasale symmetrical or asymmetrical, epistomal lobes symmetrical; ocelli well separated; legs well developed; abdomen without or with (*Berosus*) long lateral outgrowths.

Subtribe Acidocerina

**Diagnosis**: Meso-and metasternum not fused to form a common keel, but with a variably raised process or ridge which may be longitudinal or horizontal. Fifth visible abdominal sternum usually with an apical median emargination or notch. Pronotum and elytra usually with distinct systematic punctures. Maxillary palpi usually as long as width of head, with apical segment often shorter than or subequal to penultimate. Middle and hind tarsi without a fringe of hairs on dorsal surface.

*Helochares* Mulsant 1844, 3 sp., *H. maculicollis* Mulsant from Indiana, Illinois, Florida, and Louisiana, *H. sallaei* Sharp from Mexico and Florida and *H. normatus* from south and central United States.

*Helobata* Bergroth 1888, 1 sp., *H. larvalis* (Horn 1873) from Florida, Louisiana, Mississippi, North Carolina, South Carolina, and Texas.
*Helopeltis* Horn 1873, not Signoret 1858
*Helopeltina* Cockerell 1906

*Enochrus* Thomson 1859, 25 spp., generally distributed. Most recently revised for the Nearctic region by Gundersen (1978), including a key to the Nearctic species. He notes that adults feed in shallow water, on algae or decaying vegetation. Larvae eat large protozoans, insects and very small crustaceans. Members of this genus cannot truly swim, but crawl to the surface using submerged vegetation. Some species have been collected in hot springs in Oregon at water temperatures as high as 52.2° centigrade (Malkin, 1958). Several species may be collected from the same locality and many are attracted to lights at night (Keys to spp.: Downey and Arnett (1996), northeastern U.S. spp.; Leech (1956), western spp.; Gundersen (1977) Nearctic species).

*Philhydriis* Brulle 1835, not Brooks 1828, unjustified emendation.
*Philydrus* Solier 1834, not Duftschmidt 1805
*Methydrus* Rey 1885, now a subgenus
*Agraphilhydrus* Kuwert 1888
*Lumetus* Zaitzev 1908, now a subgenus

*Cymbiodyta* Bedel 1880, 23 Nearctic spp., generally distributed. (Keys to spp.: Smetana 1974.)
*Hydrocombus* Sharp 1882

*Helocombus* Horn 1890, 1 sp., *H. bifidus* (LeConte 1865), eastern United States.

Subtribe Hydrobiina

**Diagnosis**: Head, pronotum and elytra always with detectable systematic punctures. Mesosternum variable, but usually with a median keel or process not joined to metasternum. Metasternum with a posterior median bulge or short tooth above ventral condyles. Middle and hind tarsi with a fringe of yellowish-brown or golden hairs on dorsal surface. Apical segment of maxillary palpi definitely longer than penultimate. Stridulatory files are reported in the Palearctic *Limnoxenus* and some species are known to produce sound (Hansen, pers. com.).

*Hydrobius* Leach 1815, 3 spp., *H. melaenus* (Germar) occurs in eastern United States and eastern Canada,. *H. fuscipes* (Linnaeus) is Holarctic and also occurs in northern Canada and Alaska (Smetana 1988) and *H. melaenum* Germar is found from Maine to Texas. (Key to spp.: Downey and Arnett 1996.)

Hydrophilina

**Diagnosis**: Easily distinguished by the ventral keel formed from the fusion of the median parts fo the meso- and metasternum; metasternum with keel extended posteriorly as a horizontal spine of variable length. Head and pronotum with very distinct systematic punctures. Elytra usually with distinct systematic punctures (few in some *Tropisternus* spp.). Middle and posterior tarsi with a fringe of golden-brown setae on dorsal surface. Maxillary palpi with apical segment of variable length with respect to penultimate segment.

*Hydrochara* Berthold 1827, 10 Nearctic species, mostly northeastern U.S. and Canada. (Key to species: Smetana 1980.)

*Hydrobiomorpha* Blackburn 1888. One Nearctic species, *H. casta* (Say 1835).
*Neohydrophilus* d'Orchymont 1911

*Tropisternus* Solier 1834, 14 spp., generally distributed (Keys to spp.: Smetana 1988, Canadian species; Young 1954, 4 spp.; Usinger 1956, most western spp.; Downey and Arnett 1996, northeastern U.S. spp.).
*Cyphostethus* d'Orchymont 1919, not Fieber 1861
*Strepitornus* Hansen 1989

*Hydrophilus* Geoffroy 1762. 3 spp. (See Testa and Lago, 1994 for a discussion of *Hydrophilus atra*, a third species which is probably not Nearctic, and Hansen (1999) for notes on the authorship of the genus.)

*Hydrous* Linnaeus 1775

*Dibolocelus*, now a subgenus (Hansen 1991)

Subfamily Sphaeridiinae

**Diagnosis**: These are the semi-aquatic or terrestrial Hydrophilids and are generally small, oval beetles often found in dung of mammals. They can be recognized by the lack of systematic punctures on any dorsal surface and by the antennae longer than the maxillary palpi. The antennal club is usually compact, the labrum paler than the clypeus and usually soft, labial palpi often small, third segment never with a long seta on the outer side near the apex.

Description: Body with outline usually not interrupted between pronotum and elytra, pronotum usually widest at base. Head not strongly deflexed, labrum often pale soft, mostly retracted under clypeus, lateral margins usually fringed with setae. Head either not narrowed in front of eyes or somewhat abruptly narrowed in front of eyes; antennomeres usually 9 (8 in *Sphaeridium*), scape somewhat short to very long, pedicel not thicker than scape, cupule distinctly differentiated but somewhat small, club loosely segmented (*Phaenonotum*) or compact. Clypeus with variable shape, flat to convex, deflexed in lateral view; clypeus and frons without systematic punctures, frontoclypeal suture fine to very fine, sometimes clearly visible only at the sides (many Megasternini and Omicrini) or even undetectable, seldom impressed or (a few Megasternini) with secondary transverse (and straight instead of V-shaped) "frontoclypeal" groove. Mandibles very reduced only in the Omicrini and apex usually simply pointed and acute. Maxillary palpi with second segment usually swollen and thicker than the following segments.

Pronotum usually evenly convex, only very rarely with distinct impressions or humps (a few Megasternini), without systematic punctures, usually evenly punctate. Prosternum often carinate, sometimes with well-defined antennal grooves on the anterior portion of the hypomeron, prosternal pointed or notched apically, sometimes broadly concave at apex. Procoxal cavities open or (Sphaeridiini, Megasternini) very narrowly closed posteriorly. Scutellum hardly longer than wide, only in Sphaeridiini quite elongate; mesosternum fused to mesepisterna or if not, strongly narrowed anteriorly; variably raised medially or posteromedially, seldom almost flat. Metasternum sometimes with femoral lines; without posterior spine or tubercle above ventral condyles. Abdomen with 5 visible sterna, first about as long as second or much longer, carinate medially or not, posterior margin of 5th sternum evenly rounded (rarely with small apical emargination). Laterosternites without stridulatory files. Femora often less than 3x as long as wide; tibiae almost cylindrical to strongly and broadly flattened; middle and posterior tibiae with very fine to moderately long, spines on external face, tarsi 5-segmented, often with fine, sometimes quite dense hairlike setae beneath, first segment of hind tarsi of variable length. Elytra often with distinct striae or rows of serial punctures, rarely without striae or serial punctures; rows of systematic punctures and scutellary stria always completely absent.

Hind wing radius not distinct proximal to r-m crossvein, sometimes bifurcate basally; r-m crossvein rising from base or at least basal half of the pigmented area at anterior wing margin; media distinct for some distance proximal to r-m crossvein, or (Megasternini and some Omicrini) not or hardly distinct; cubital spur rising from apex of m-cu loop (except Megasternini); basal cell normally distinct, but sometimes (e.g., Megasternini) small; wedge cell distinct or not; jugal lobe present or not. Aedeagus in general somewhat simple sometimes with asymmetrical basal piece.

Larva: Nasale symmetrical or asymmetrical; ocelli well separated or closely aggregated; legs well developed or very reduced, sometimes absent; abdominal segments without long lateral outgrowths. There are 9 tribes, 101 genera, and 759 species worldwide; 3 Nearctic tribes with 12 genera and 122 species.

Tribe Coelostomatini Heyden 1891

**Description**: Size 1.5-7.5 mm; labrum retracted, body oval to broadly oval, often somewhat parallel sided, moderately to strongly convex; not broad and rounded.

Labrum pale and soft, retracted under clypeus, barely visible from above; anterior margin truncate or nearly truncate and usually fringed with setae; eyes usually weakly emarginate anteriorly; galea with setae curved apically and arranged in well-defined rows; labial palpi about as long as mentum or slightly shorter; antennae 8- or 9-segmented, about 2/3 to 3/4 as long as width of head; prosternum without distinct antennal grooves; prosternal process without apical notch; procoxal cavities open behind; mesosternum usually fused to mesepisterna, strongly and abruptly raised posteromedially to form a variably shaped process, usually without cavities for reception of procoxae (except for a few *Dactylosternum*), but usually with anteromedian pitlike groove; raised middle part of metasternum often noticeably projecting anteriorly between mesocoxae, contiguous with mesosternal process; first segment of metatarsus distinctly longer than second; gula forming a transverse triangle posteriorly, only narrow or confluent anteriorly; antennae 9-segmented; tibiae strongly flattened; epipleura and pseudepipleura usually well developed, somewhat oblique, sometimes nearly horizontal; elytral margins not serrate; hind wing with radius not distinct proximal to r-m crossvein, sometimes bifurcate basally; r-m crossvein rising from base, or at least basal half of the pigmented area at anterior wing margin; media distinct for some distance proximal to r-m crossvein or (Megasternini and some Omicrini) not or hardly distinct; cubital spur rising from apex of m-cu loop (except Megasternini); basal cell normally distinct but sometimes small (Megasternini); wedge cell distinct or not; jugal lobe present or not.

Male without "sucking-disc shaped" appendage on maxilla. Aedeagus somewhat simple, sometimes with asymmetrical basal piece.

Larva with nasale various, ocelli well separated, legs well developed. There are 18 genera worldwide, with 204 species; 2 Nearctic genera with 5 species.

*Dactylosternum* Wollaston 1854, 3 spp., North Carolina, Florida, Arizona, and California. ( See Hansen 1991 for commentary on the synonymy.)

*Dactilosternum* Kuwert 1890, incorrect subsequent spelling

*Phaenonotum* Sharp 1882, 2 spp., eastern United States.

Tribe Omicrini Smetana 1975

**Diagnosis**: Size 1-2 mm. Members of this tribe can be recognized by the abruptly narrowed head with exposed antennal bases; deflexed clypeus, large, exposed labrum and well-developed epipleura and pseudepipleura. Wing venation is as for the subfamily except that the basal cell is not detectable in *Omicrus*. There are 89 spp. worldwide in 15 genera, only one of which occurs in the Nearctic region.

*Omicrus* Sharp 1879, 2 Nearctic species: *O. intermedius* Smetana 1975 and *O. palmarum* (Schwarz 1878), both from Florida and the Neotropics.

*Phaenotypus* Horn 1890

Megasternini Mulsant 1844

**Diagnosis**: Members of the tribe may be recognized by the narrowed or excised head in front of the antennae, exposing their bases from above, the soft labrum usually retracted under the clypeus and epipleura not visible next to abdomen. Hypomeron with distinct antennal grooves usually with an arcuate ridge at sides. Mesosternum with a raised median tablet meeting metasternum. Metasternum shining, with a well-defined pentagonal or hexagonal raised median area. Tibiae flattened, sometimes with outer apex excised. Elytra striate or with 9 or 10 rows of punctures.

Males with a "sucking-disc shaped" appendage on the maxilla; aedeagus usually with basal piece asymmetrical, ending in a sideways curve or hook.

Larva with symmetrical nasale and closely aggregated ocelli. The legs are reduced or absent.

There are 48 genera and 274 species worldwide, of which there are 8 Nearctic genera and 57 species.

*Cercyon* Leach 1817, 205 species worldwide, generally distributed, with 39 Nearctic spp., many introduced from the Palearctic region. (Keys to spp.: Smetana 1988; Downey and Arnett 1965, northeastern species; Leech 1956, some California spp.)

*Deltostethus* Sharp 1882, 4 species, 3 of which are Neotropical; 1 is Nearctic, *D. columbiensis* (Hatch)

*Ictinosternum* Smetana 1978

*Agna* Smetana 1978, 1 Nearctic species, *A. capillata* (LeConte)

*Pelosoma* Mulsant 1844, 2 Nearctic spp. (Diagnosis and key: Smetana 1984.)

*Merosoma* Balfour-Brown 1939

*Oosternum* Sharp 1882, 2 spp., eastern United States.

*Crypteuna* Motschulsky 1863 (suppressed, see Hansen 1991: 304)

*Pemelus* Horn 1890, 8 species, mostly Neotropical; 4 Nearctic spp., mostly from eastern and southern U.S.

*Tectosternum* Balfour-Browne 1958, 3 spp., of which 1 is Nearctic, *T. naviculare* (Zimmerman), from Eastern Canada and United States.

*Genyon* Smetana 1978

*Megasternum* Mulsant 1844, 7 spp., 3 Nearctic spp., Alaska, California, Oregon, Washington, and Louisiana. (See Hansen 1999 for history of *M. concinnum* (Marsham), new combination.)

*Cycrillum* Knisch 1921, 1 sp., *C. strigicolle* (Sharp 1882), from the Neotropical region and in the U.S. from southern Florida

*Cryptopleurum* Mulsant 1844, 21 spp., generally distributed, 5 of which are Nearctic. (Key to species: Smetana 1978.)

Sphaeridiini

**Diagnosis**: Body broadly oval, moderately convex; antennae with basal segment very elongate, First segment of middle and hind tarsi much longer than second; tibiae with very long spines on outer face. Eyes deeply emarginate anteriorly; scutellum much longer than wide; mesosternum fused with mesepisterna; antennae 8-segmented, antennal club compact, labrum fully visible and fully sclerotized and pigmented; front coxae with short, stout, backward-projecting spines; tibiae with long stout spines, front tarsi sexually dimorphic; first tarsomere of hind tarsus very elongate, much longer than second tarsomere. Male often with slightly asymmetrical basal piece.

*Sphaeridium* Fabricius 1775, 42 spp. worldwide, 4 Nearctic spp., generally distributed, introduced from Europe. Species of this terrestrial genus are predators as adults and are fast runners and strong fliers. All live in decaying organic material, especially in fresh mammal droppings. (See Hansen 1999, pp. 312, 314 for notes on possible confusion of species and synonymies.)

*Sphaeridiolinus* Minozzi 1921

*Spaeridiolinus* Menozzi 1921, error in spelling

## Bibliography


AGARWAL, U. 1960. Chromosomes of *Berosus indicus* Mots. (Coleoptera, Hydrophilidae). Current Science, Bangalor, 29:404-405.

AIKEN, R.B. 1985. Sound production by aquatic insects. Biological Review, 65: 163-211.

ANGUS, R.B. 1970. *Helophorus orientalis* (Coleoptera: Hydrophilidae), a parthenogenetic water beetle from Siberia and North America, and a British Pleistocene fossil. The Canadian Entomologist, 102: 129-143.

ANGUS, R.B. 1971. Revisional notes on *Helophorus* F. (Col., Hydrophilidae), 2. The complex round *H. flavipes* F. Entomologist's Monthly Magazine, 106: 129-148.

ANGUS, R.B. 1973. The habitats, life histories and immature stages of *Helophorus* F. (Coleoptera, Hydrophilidae). Transactions of the Royal Entomological Society of London, 125: 1-26.

ANGUS, R.B. 1989. Towards an atlas of *Helophorus* chromosomes. Balfour-Browne Club Newsletter, 44: 13-22.

ANGUS, R.B. 1996. A re-evaluation of the *Helophorus flavipes* group of species (Coleoptera, Hydrophiloidea), based on chromosomal analysis, larvae and biology. Nouvelle Revue d'Entomologie, (N.S.) 13(2): 111-122.

ARCHANGELSKY, M. and DURAND, M.E. 1992. Description of the immature stages and biology of *Phaenonotum exstriatum* (Say 1835) (Coleoptera: Hydrophilidae: Sphaeridiinae). The Coleopterists Bulletin, 46(3): 209-215.

ARCHANGELSKY, M. 1994a. Description of the preimaginal stages of *Dactylosternum cacti* (Coleoptera: Hydrophilidae, Sphaeridiinae). Entomologica Scandinavica, 25: 121-128.

ARCHANGELSKY, M. 1994b. Description of the immature stages of three Nearctic species of the genus *Berosus* Leach (Coleoptera: Hydrophilidae). Internationale Revue der Gesamten Hydrobioligie, 79: 357-373.

ARCHANGELSKY, M. 1994c. Description of the preimaginal stages and biology of *Phaenonotum (Hydroglobus) puncticolle* Bruch (Coleoptera: Hydrophilidae). Aquatic Insects, 16(1): 55-63.

ARCHANGELSKY, M. 1996. A bibliographic compilation on the immature stages of Hydrophiloidea (Insecta, Coleoptera). Entomologica Basiliensia, 19: 653-673.

ARCHANGELSKY, M. 1998. Phylogeny of Hydrophiloidea (Coleoptera: Staphyliniformia) using characters from adult and preimaginal stages. Systematic Entomology, 23: 9-24.

BALDUF, W. V. 1935. Bionomics of entomophagous Coleoptera, New York, 220 pp.

BAY, E.C. 1974. Predator-prey relationships among aquatic insects. Annual Review of Entomology 19:441-453.

BERTRAND, H. P. I. 1972. Larves et nymphes des coléoptères aquatiques du globe. Paris. 804 pp., 561 figs.

BERTRAND, H. P. I. 1977. Larves et nymphes des coléoptères aquatiques du globe. Errata et addenda. Abbeville, 19 pp.

BEUTEL, R.G. 1994. Phylogenetic analysis of Hydrophiloidea based on characters of the head of adults and larvae (Coleoptera: Staphyliniformia). Koleopterologische Rundschau, 64: 103-131.

BEUTEL, R.G. and MOLENDA, R. 1997. Comparative morphology of selected larvae of Staphylinoidea (Coleoptera, Polyphaga) with phylogenetic implications. Zoologischer Anzeiger, 236: 37-67.

BLACKWELDER, R. E. 1931. The Sphaeridiini of the Pacific Coast. Pan Pacific Enomologist, 8: 19-32.

BLATCHLEY, W. S. 1910. An illustrated descriptive catalogue of the Coleoptera or beetles (exclusive of the Rhynchophora) known to occur in Indiana. The Nature Publishing Company of Indianapolis, 1386 pp.

BØVING, A. G. and CRAIGHEAD, F. C. 1931. An illustrated synopsis of the principal larval forms of the order of the Coleoptera. Entomologica Americana, 11 (new series) 351 pp., 125 pls.

BØVING, A. G. and HENRIKSEN, K. L. 1938. The developmental stages of the Danish Hydrophilidae. (Insecta, Coleoptera). Videnskabelige Meddeleser fra Dansk naturhistorisk Forening i KØbenhavn, 102: 25-161.

BRIGHAM, W. V. 1982. Hydrophilidae, pp. 10.75-10.95, in Brigham, A.R., Brigham, W.V. and Gnilka, A. (eds). Aquatic insects and Oligochaeles of North and South Carolina. Midwest Aquatic Enterprises, Mahomet, Illinois.

BROWN, W. J. 1940. Some new and poorly known species of Coleoptera. Canadian Entomologist, 72: 182-187.

BUHK, F. 1910. Stridulationsapparat bei *Spercheus emarginatus* Schall. Zeitschrift für Wissenschäftliche Insektenbiologie, 6: 342-345.

CHEARY, B. S. 1971. The biology, ecology and systematics of the genus *Laccobius* (*Laccobius*) (Coleoptera: Hydrophilidae) of the new world. Riverside 178 pp.

CROWSON, R. A. 1981. The biology of Coleoptera. xii + 802 pp. Academic Press, London, & Co., London

DAJOS, R. 1997. Description de *Cercyon arizonicus* n. sp. et inventaire des Sphaeridiinae (Coléoptères, Hydrophilidae) d'une localité du sud-est de l'Arizona (Êtais-Unis). Bulletin mensuel de la Société Linnéenne de Lyon 66: 273-276.

DOWNIE, N. M. and ARNETT, R. H. Jr. 1996. The beetles of northeastern North America. Vol. I, pp. 275-310, 316.

DUNCAN, D.K. 1927. An unusual condition found in collecting water beetles in Arizona. Bulletin of the Brooklyn Entomological Society, 22: 143.

EDWARDS, J. G. 1949. Coleoptera or beetles east of the Great Plains. Published by the author, 185 pp.

EDWARDS, J. G. 1950. A bibliographical supplement to "Coleoptera or Beetles East of the Great Plains applying particularly to western United States." Published by the author, pp. 183-212.

EMDEN, F.I. van. 1956. The *Georyssus* larva a Hydrophilid. Proceedings of the Royal Entomological Society of London, (Series A) 31: 20-24.

ENGELKE, M. J. 1980. Aestivation of a water scavenger beetle in southwestern Wyoming. (Coleoptera: Hydrophilidae) Coleopterists Bulletin, 34: 176.

EPLER, J. H. 1996. Identification manual for the water beetles of Florida. Bureau of Water Resource Protection, Florida Department of Environmental Protection, Tassahassee.

EYRE, M.D., CARR, R., MCBLANE, R.P. and FOSTER, G.N. 1992. The effects of varying site-water duration on the distribution of water beetle assemblages, adults and larvae (Coleoptera: Haliplidae Dytiscidae, Hydrophilidae).

FORBES, W. T. M. 1922. The wing venation of the Coleoptera. Annals of the Entomogical Society of America, 15: 328-352, 7 plates.

FRANKENBERG, G. VON. 1940. Ein brutpflegender und musizierender Wasserkäfer (Spercheus emarginatus). Senckenbergische Natürforschende Gesellschaft, 70(1): 79-89.

GENTILI, E. 1979. I *Laccobius* della regione orientale (Coleoptera, Hydrophilidae). Annuario Osservatorio di Fisica terrestre e Museo Antonio Stoppani del Seminario Arcivescoville di Milano, (N. S.) 1: 27-50.

GENTILI, E. 1981b. The genera *Laccobius* and *Nothydrus* (Coleoptera: Hydrophilidae) in Australia and New Zealand. Record of the South Australian Museum, 18(7): 143-154.

GENTILI, E. 1985. I *Laccobius* americani- I. I *Laccobius* del Canada (Coleoptera: Hydrophilidae). Annuario Osservatorio di Fisica terrestre e Museo Antonio Stoppani del seminario Arcivescoville di Milano (new series) 6, (1983): 31-45.

GENTILI, E. 1986a. I *Laccobius* americani - II. Il genere *Laccobius* a sud del Canada (Coleoptera, Hydrophilidae). Annuario Osservatorio di Fiscia terrestre e Museo Antonio Stoppani del Seminario Arcivescoville di Milano, (new series) 7 (1984): 31-40.

GENTILI, E. 1986b. I *Laccobius* americani - III. Il genere *Laccobius* a sud del Canada (Coleoptera, Hydrophildae). Annuario Osservatorio di Fiscia terrestre e Museo Antonio Stoppani del Seminario Arcivescoville di Milano, (new series) 8 (1985): 31-52.

GORDON, R. D. North Dakota water beetles. North Dakota State University, 1(5):1-53.

GUNDERSON, R. W. 1978. Neartic *Enochrus* biology, keys, descriptions and distribution. (Coleoptera: Hydrophilidae) 54 pp. (Privately published.)

GUNDEVIA, H. S. and RAMAMURTY, P. S. 1984. The female accessory reproductive glands and cocoon in *Hydrophilus olivaceus* Fabr. (Polyphaga-Coleoptera). Zeitshrift für Mikroskopische Anatomishe Forschuung (Leipzig), 98: 293-301.

HANSEN, M. 1991. The Hydrophiloid beetles, phylogeny, classification and a revision of the genera (Coleoptera, Hydrophiloidea). Biologiske Skrifter, Det Kongelige Danske Videnskabernes Selskab, 40: 1-368.

HANSEN, M. 1995. Evolution and classification of the Hydrophiloidea a systematic review. (pp. 321-353), in Pakaluk, J. and S. A. Slipinski (eds.). Biology, phylogeny, and classification of Coleoptera: Papers Celebrating the 80th Birthday of Roy A. Crowson. Vol. 1, xii + 558 pp. Muzeum i Instytut Zoologii PAN, Warsaw.

HANSEN, M. 1999. Hydrophiloidea (*s. str.*) (Coleoptera). *In*: World Catalogue of Insects, 2: 1-416.

HARDY, A. R., CHEARY, B. S. and MALCOLM, S. E. 1981. A clarification of the authorship of some names in *Laccobius* Erichson (Coleoptera: Hydrophilidae). Pan Pacific Entomologist, 57: 303-305.

HILSENHOFF, W. L., 1995. Aquatic Hydrophilidae and Hydraenidae of Wisconsin (Coleoptera). I. Introduction, key to genera of adults and distribution, habitat, life cycle and identification of species of *Helophorus* Fabricius, *Hydrochus* Leach, and *Berosus* Leach (Hydrophilidae), and Hydraenidae. The Great Lakes Entomologist, 28(1): 25-53.

HINTON, H. E. 1981. Biology of insect eggs. Pergamon Press Inc., New York, Volumes 1-3.

HORN, G. H. 1873. Revision of the genera and species of the tribe Hydrobiini. Proceedings of the American Philosophical Society, 13: 118-137.

HORN, G. H. 1890b. A revision of Sphaeridiini inhabiting Boreal America. Transactions of the American Entomological Society, 17: 279-314.

HRBACEK, J. 1950. On the morphology and function of the antennae of the Central European Hydrophilidae (Coleoptera). Transactions of the Royal Entomological Society of London, 101(7): 239-256.

JASPER, S. K. and R.C. VOGTSBERGER. 1996. First Texas records of five genera of aquatic beetles (Coleoptera: Noteridae, Dytiscidae, Hydrophilidae) with habitat notes. Entomological News, 107: 49-60.

KNISCH, A. 1922. Hydropliliden-Studien (OP. 10) Archiv für Natürgeschichte, 88. A(5): 87-126.

KNISCH, A. 1924. Hydrophilidae. In Junk, W. and Schenkling, S. (ed.): Coleopterorum Catalogus, Vol. 14, part 79, 306 pp. W. Junk, Berlin.

LAABS, A. 1939. Brutfürsorge und brutpflege einiger Hydrophiliden mit berücksichtigung des Spinnapparates, seines Äusseren Baues und seiner Tätigkeit. Zeitschrift für Morphologie und Ökologie der Tiere, 36: 123-178.

LAGO, P. K. and TESTA, S., III. 1989. The aquatic and semi-aquatic Hemiptera and Coleoptera of Point Clear Island, Hancock Country, Mississippi. J. Mississippi Academy of Science, 34: 33-38.

LARIVER, S. I. 1954. Nevada Hydrophilidae (Coleoptera). The American Midland Naturalist, Volume 52 (1): 164-174.

LAWRENCE, J. F. and NEWTON, A. F., Jr. 1982. Evolution and classification of beetles. Annual Review of Ecology and Systematics, 13: 261-290.

LAWRENCE, J. F. and NEWTON, A. F., Jr. 1995. Families and subfamilies of Coleoptera(with selected genera, notes, references and data on family-group names. *In* Pakaluk, J. and Slipinski, S. A., eds., Biology, Phylogeny and Classification of Coleoptera; Papers celebrating the 80th birthday of Roy A. Crowson, pp. 779-1006. Muzeum i Instytut Zoologii PAN, Warszawa.

LEACH, W. E. 1815. Entomology. *In* Brewster, D., Edinburgh Encyclopedia, Vol. 9, 9: 57-172. Balfour, Edinburgh.

LECONTE, J. L. 1855. Synopsis of the Hydrophilidae of the United States. Proceedings of the Academy of Natural Sciences of Philadelphia, 7: 356-375.

LECONTE, J. L. 1861. Classification of the Coleoptera of North America. I. Smithsonian Miscellaneous Collection: xxv + 286 pp. (pp. 282-286 were published in 1862).

LECONTE, J. L. 1874. Descriptions of the new Coleoptera chiefly from the Pacific Slope of North America. Transactions of the American Entomological Society, 5: 43-72.

LEECH, H. B. 1948a. Some Nearctic species of palpicorn water beetles, new and old (Coleoptera: Hyrdophilidae). Wasmann Collector, 7(2): 33-46.

LEECH, H. B. 1948b. Contributions toward a knowledge of the insect fauna of lower California. Number 11. Coleoptera: Haliplidae, Dytiscidae, Gyrinidae, Hydrophilidae, Limnibiidae. Proceedings of the California Academy of Sciences, (4. ser.) 24: 375-483, 2 plates.

LEECH, H. B. and CHANDLER, H. P. 1956. (Aquatic Coleoptera, part) pp. 293-371, *in* Usinger, R.L. Aquatic Insects of California, Berkeley. 508 pp.

MAILLARD, Y. P. 1968. Mise en évidence d'un territoire ovarien modifié en glande de la soie, chez les Coléoptères Hydrophilidae. Données anatomiques comparatives. Comptes Rendus Academie des Sciences (Paris), 266: 500-502

MAILLARD, Y. P. 1970. Étude comparée de la construction du cocon de ponte chez *Hydrophilus piceus* L. et *Hydrochara caraboides* L. (Insecte Coléoptères Hydrophilidae). Séance, Jan. 27, pp. 71-84.

MAILLARD, Y. P. and SELLIER, R. 1970. La pars stridens des Hydrophilidae (Insectes Coléoptères); étude au microscope électronique à balayage. Comptes Rendus Academie des Sciences (Paris), 270: 2969-2972.

MAILLARD, Y. P. 1972. Structure fine de la surface de la filiere et de la soie du cocon de ponte des Hydrophilidae (Ins. Coléoptères). Comptes Rendus Academie des Sciences (Paris), 275: 75-78.

MALCOLM, S. E. 1971. The water beetles of Maine: including the families Gyrinidae, Haliplidae, Dytiscidae, Noteridae and Hydrophilidae. University of Maine Technology Bulletin No. 48, 49 pp.

MALKIN, B. 1958. On some water beetles from Oregon hot springs. Coleopterists Bulletin, 12: 34-35.

MASTERS, W. M. 1979. Insect disturbance stridulation: its defensive role. Behavioral Ecology and Sociobiology, 5: 187-200.

MATTA, J. F. 1973. A checklist of the aquatic Hydrophilidae of Virginia. Virginia Journal of Science, 24: 87-88.

MATTA, J. F. 1974 a. The Insects of Virginia, Number 8. The Aquatic Hydrophilidae of Virginia (Coleoptera: Polyphaga). Research Division Bulletin 94, Virginia Polytechnic Institute, State University of Blacksburg, 44 pp.

MATTA, J. F. 1974b. The aquatic Coleoptera of Dismal Swamp. Virginia Journal of Science, 24: 199-205.

MCCORCKLE, D. V., 1965. (Subfamily Elophorinae). pp. 23-38, *in* Hatch, M.H. (Ed.) The Beetles of the Pacific Northwest. Part IV. University of Washington Publications in Biology, volume 16, 268 pp.

MCCOY, C. J. 1969. Diet of bullfrogs (*Rana catesbiana*) in central Oklahoma farm ponds. Proceedings of the Oklahoma Academy of Science, 48: 44-45.

MERRITT, R. W. and CUMMINS, K. 1996. An introduction to the aquatic insects of North America, 3rd edition, Kendall/Hunt Publishing Company, Dubuque, Iowa, 862 pp.

MILLER, D. C. 1963. *Paracymus tarsalis*, a new species from the northwestern United States (Coleoptera: Hydroplilidea). Coleopterists Bulletin, 18: 69-78.

MILLER, D.C. 1974. Revision of the New World Chaetarthria (Coleoptera: Hydrophilidae). Entomologica Americana, 49: 1-123.

MILLIGER, L. E., STEWART, K. W. and SILVEY, J. K. G. 1971. The passive dispersal of viable algae, protozoans and fungi by aquatic and terrestrial Coleoptera. Annals of the Entomological Society of America, 64: 36-45.

MOUCHAMPS, R. 1959. Remarques concernant les genres *Hydrobiomorpha* Blackburn et *Neohydrophilus* Orchymont (Coléoptères Hydrophilides). Bulletin et Annales de la Société Royale d'Entomologie de Belgique, 95: 295-335.

NELSON, F. R. S. 1977. Predation on mosquito larvae by beetle larvae, *Hydrophilus triangularis* and *Dytiscus marginalis*. Mosquito News, 37(4): 628-630.

NOTESTINE, M. 1971. Population densities of known invertebrate predators of mosquito larvae in Utah marshlands. Mosquito News, 31:331-334.

OLIVA, A. 1981. El genero *Derallus* Sharp en la Argentina (Coleoptera, Hydrophilidae). Revista de la Sociedad Entomologica Argentina, 40: 285-296.

OLIVA, A. 1992. Cuticular microstructure in some genera of Hydrophilidae (Coleoptera) and their phylogenetic significance. Bulletin du l'Institut Royal des Sciences Naturelles de Belgigue, 62: 35-56.

OLIVA, A. 1995 Redescription and lectotype designation of *Derallus altus* (LeConte, 1855) (Coleoptera; Hydrophilidae). Physis (Buenos Aires, B, 50(118-119)(1992): 45-46.

D'ORCHYMONT, A. 1942a. Revision des *Laccobius* americains (Coleoptera, Hydrophilidae, Hydrobiini). Bulletin du l'Institut Royal des Sciences Naturelles de Belgigue, 18: 1-18.

D'ORCHYMONT, A. 1942b. Contribution a l'etude la tribu Hydrobiini Bedel, specialement de sa sous-tribu Hydrobiae (Palpicornia-Hydrophilidae). Mémoires Museum Royal d'Histoire Naturelles de Belgique, (ser. 2) 24: 1-68.

PERKINS, P. D., and SPANGLER, P. J. 1981. A description of the larvae of *Helocombus bifidus* (Coleoptera: Hydrophilidae). Pan-Pacific Entomologist, 57: 52-56.

PETERSON, A. 1951. Larvae of Insects, 2 volumes, published by author, Columbus Ohio, Part II, Coleoptera, pp. 2-218, 388-392.

PETHERBRIDGE, F. R. 1928. The turnip mud beetles (*Helophorus rugosus* Ol. and *Helophorus porculus* Bedel. Annals of Applied Biology, 15(4): 659-679.

PETHERBRIDGE, F. R. and THOMAS, I. 1935. Damage to wheat by *Helophorus nubilus* F. Annals of Applied Biology, 23(3): 640-649.

PIRISINU, Q., SPINELLI, G. and BICCHERAI, M. C. 1988. Stridulatory apparatus in the Italian species of the genus *Laccobius* Erichson (Coleoptera: Hydrophilidae). International Journal of Insect Morphology and Embryology, 17: 95-101.

PIRISINU, Q., SPINELLI, G. and PAPARO, G. 1985. Ulteriori informazioni sui suoni emessi da coletteri palpicorni: *Hydrophilus caraboides* (L.). Atti. XIV Congresso Nazionale Italiano di Entomologia, Palermo, Erice, Bagheria.

RICHMOND, E. A. 1920. Studies on the biology of the aquatic Hydrophilidae. Bulletin of the American Museum of Natural History, 1-94, pl. I-XVI.

ROCHA, A. A. 1967. Biology and first instar larva of *Epimetopus trogoides* (Coleoptera, Hydrophilidae). Papâeis Avulsos Zoologia São Paulo, 20: 175-189.

ROCHA, A. A.. 1969. Sobre o genero *Epimetopus* (Col., Hydrophilidae). Papâeis Avulsos Zoologia São Paulo, 22: 175-189.

RODIONOV, Z.S. 1928. The swamp beetle (*Helophorus micans* Fald.) as a pest of cereals in floodlands. Zashch. Rast Vredit. 4: 951-954. (In Russian)

RYKER, L. C. 1972. Acoustic behvior of four sympatric species of water scavenger beetles (Coleoptera, Hydrophilidae, *Tropisternus*). Occasional Papers of the University of Michigan Museum of Zoology, 666: 163-168.

SCHELOSKE, H. W. 1975. Mating and sound emission in the beetle *Laccobius minutus* (L.) (Coleoptera, Hydrophilidae). Verhandlüngen. Deutsche Zoologische Gesellschaft, 67: 329-34.

SHAARAWI, F. A. and ANGUS, R. B. 1990. A chromosomal investigation of five European species of *Anacaena* Thomson (Coleoptera: Hydrophilidae). Entomologica Scandinavica, 21: 415-426.

SHAARAWI, F. A. and ANGUS, R. B. 1991. A chromosomal analysis of some European species of the genus *Berosus* Leach (Coleoptera: Hydrophilidae). Koleopterologische Rundschau, 61: 105-110.

SHAARAWI, F. A. and ANGUS, R. B. 1992. Chromosomal analysis of the genera *Georissus Latreille*, *Spercheus*, Illiger, and *Hydrochus* Leach (Coleoptera: Hydrophilidae). Koleopterologische Rundschau, 62: 127-135.

SHARP, D. and MUIR, F. 1912. A comparative anatomy of the male genital tube in Coleoptera. Transactions of the Entomological Society of London, 1912: 477-542, 78 plates.

SMETANA, A. 1974. Revision of the genus *Cymbiodyta* Bedel (Coleoptera: Hydrophilidae). Memoirs of the Entomological Society of Canada, 93: 1-113.

SMETANA, A. 1975. Revision of a New World genera of the tribe Omicrini trib. nov. of the Hydrophilid subfamily Sphaeridiini (Coleoptera). Studies on the Neotropical Fauna, 10: 153-182.

SMETANA, A. 1978. Revision of the subfamily Sphaeridiinae of America north of Mexico (Coleoptera: Hydrophilidae). Memoirs of the Entomological Society of Canada, No. 105: 1-292.

SMETANA, A. 1979. Revision of the subfamily Sphaeridiinae of America north of Mexico (Coleoptera: Hydrophilidae). Supplementum 1. The Canadian Entomologist, 111: 959-966.

SMETANA, A. 1980. Revision of the genus *Hydrochara* Berth. (Coleoptera: Hydrophilidae). Memoirs of the Entomological Society of Canada, No. 111: 1-100.

SMETANA, A. 1983. Geographical distribution of the water beetle genus *Hydrochara* (Coleoptera, Hydrophilidae). Special Issue Concerning the aquatic Coleoptera presented by the workshop of the XVI International Congress of Entomology on Kyoto, Japan in 1980, pp. 27-33.

SMETANA, A. 1984. Revision of the subfamily Sphaeridiinae of America north of Mexico (Coleoptera: Hydrophilidae). Supplementum 2. The Canadian Entomologist, 116: 555-566.

SMETANA, A. 1985a. Revision of the subfamily Helophorinae of the Nearctic Region. (Coleoptera: Hydrophilidae). Memoirs of the Entomological Society of Canada, 131: 1-154.

SMETANA, A. 1985b. Synonymical notes on Hydrophilidae. Coleopterists Bulletin, 39: 328.

SMETANA, A. 1987. Replacement name for *Helophorus frater* Smetana 1985 (Coleoptera: Hydrophilidae). Coleopterists Bulletin, 41: 262.

SMETANA, A. 1988. Review of the family Hydrophilidae of Canada and Alaska (Coleoptera). Memoirs of the Entomological Society of Canada, No. 142, 316 pp.

SMITH, S. G. 1953. Chromosomes of Coleoptera. Heredity, 7: 31.

SPANGLER, P. J. 1961. Notes on the biology and distribution of *Sperchopis tessellatus* (Ziegler) (Coleoptera: Hydrophilidae). Coleopterists Bulletin, 15: 105-112.

SPANGLER, P. J. 1982. Hydrophilidae, pp. 355-363. *In*: Aquatic biota of Mexico, Central America and the West Indies, S.H. Hurlbert and A. Villalobos-Figueroa, eds. San Diego State University, San Diego, California.

SPANGLER, P. J. 1991. Hydrophilidae (Hydrophiloidea) (Including Helophoridae, Hydrochidae, Sphaeridiidae, Spercheidae), pp. 355-359, *In* Stehr, F. W., Immature Insects, vol. 2.

TANNER, V. M. 1927. A preliminary study of the genitalia of female Coleoptera. Transactions of the American Entomology Society, 53: 5-50.

TESTA, S. III, and Lago, P. K. 1994. Aquatic Hydrophilidae (Coleoptera) of Mississippi. Mississippi Agriculture and Forestry Experiment Station Technical Bulletin No. 193, Mississippi Entomological Museum, Publication No. 5, 71 pp.

VAN TASSELL, E. R., 1963 A new *Berosus* from Arizona, with a key to the Arizona species (Coleoptera: Hydrophilidae). Coleopterists Bulletin, 17(1): 1-5.

VAN TASSELL, E. R. 1964. A note on *Hemiosus exilis* Leconte (Coleoptera: Hydrophilidae). Coleopterists Bulletin, 18: 53-56.

VAN TASSELL, E. R. 1965. An audiospectrographic study of stridulation as an isolating mechanism in the genus *Berosus* (Coleoptera: Hydrophilidae). Annals of the Entomogical Society of America, 58(4): 407-413.

VAN TASSELL, E. R. 1966. Taxonomy and biology of the subfamily Berosinae of North and Central America and the West Indies. The Catholic University of America. Unpublished Ph.D. thesis.

VLASBLOM, A. G. and WOLVERKAMP, H. P. 1955. On the function of the "funnel" on the nest of the waterbeetle *Hydrous piceus* L. Physiologia Comparata et Oecologia, 4: 240-246.

WHITE, D. S. and BRIGHAM, W. U. 1996. Aquatic Coleoptera, Hydrophiloidea, pp.427-434. *In* Merritt, R. W. and Cummins, K.W. An Introduction to the Aquatic Insects of North America, 3rd ed., Kendall/Hunt Publishing Company, Dubuque, Iowa, 862 pp.

WILLIAMS, I. W. 1938. The comparative morphology of the mouthparts of the Coleoptera treated from the standpoint of phylogeny. Journal of the New York Entomological Society, 46: 245-289, 11 plates.

WILSON, C. B. 1923. Water beetles in relation to pondfish culture, with life histories of those found in fishponds at Fairport, Iowa. Bulletin of the Bureau of Fisheries, 39: 231-345.

WINTERS, F. 1926. Notes on the Hydrobiini (Coleoptera: Hydrophilidae) of Boreal America. Pan Pacific Entomologist, 3:49-58.

WINTERS, F. 1944. Sphaeridini inhabiting Boreal America. Bulletin of the Brooklyn Entomological Society, 39: 94-95.

WOOLDRIDGE, D. P. 1966. Notes of Nearctic *Paracymus* with descriptions of new species (Coleoptera: Hydrophilidae). Journal of the Kansas Entomological Society, 39: 712-725.

WOOLDRIDGE, D. P. 1967. The Aquatic Hydrophilidae of Illinois. Transactions of the Illinois State Academy of Science, 60:422-431.

WOOLDRIDGE, D.P. AND WOOLDRIDGE, C.R. 1972. Bacteria from the digestive tract of the phytophagous aquatic beetle *Tropisternus lateralis nimbatus* (Coleoptera: Hydrophilidae). Scientific Notes, 1(4):533-534.

YOUNG, F. N. 1954. Water Beetles of Florida. Gainesville: University of Florida Studies, Biological Science Series, 5(1): x+238 pp.

YOUNG, F. N. 1960. Notes on the care and rearing of *Tropisternus* in the laboratory (Coleoptera: Hydrophilidae). Ecology, 39: 166-167.

ZAITZEV, P. 1910. Georyssidae. *In* Junk, W. and Schenkling, S. (ed.): Coleopterorum Catalogus, Volume 14, part 17, pp. 49-52. Berlin.

# 14. SPHAERITIDAE Shuckard, 1839

by Alfred F. Newton

**Common name:** The false clown beetles

The four species in one genus that form this small family resemble histerid beetles, but are much less compact: the head and legs are less retractile, the antennae are only weakly geniculate, the elytra are truncate but less snug-fitting and expose only one abdominal tergite, and the base of the abdomen ventrally is not closely molded to the thorax. Sphaeritids also differ from most histerids in having completely separated gular sutures, contiguous projecting procoxae, exposed pro- and mesotrochantins, contiguous metacoxae, and bisetose tarsal empodia, and having lost abdominal spiracles only on segments VII and VIII. Some nitidulids (e.g., *Oxycnemus*) superficially resemble sphaeritids but have low, well-separated procoxae, among other differences. Sphaeritid larvae differ from those of the otherwise very similar histerid larvae in having 4-segmented urogomphi, a strong basal molar lobe on the mandibles, and a free mentum and maxillary cardines.

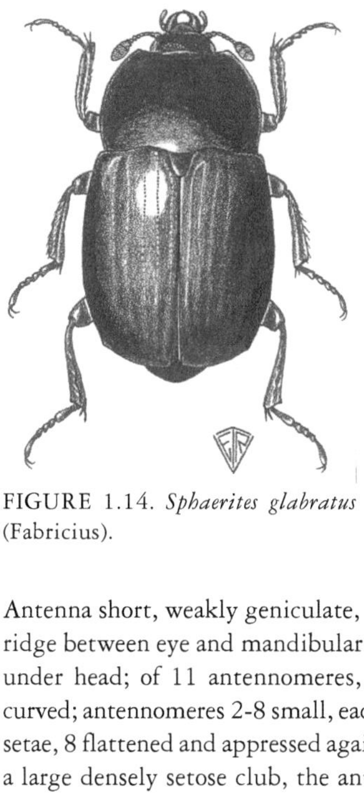

FIGURE 1.14. *Sphaerites glabratus* (Fabricius).

**Description (adults):** Length about 4.5-7 mm; body broadly oval, convex; color black (elytra partly red in one Asian species), moderately shining with slight blue-green metallic reflection; vestiture generally absent except on appendages.

Head less than half as wide as pronotum, deflexed but not retractile, without differentiated neck, surface punctate; eyes large, oval. Epistomal suture present, broadly curved or more or less angulate and with short medial stem (i.e., Y-shaped). Antenna short, weakly geniculate, inserted under slight frontal ridge between eye and mandibular base, antenna in repose bent under head; of 11 antennomeres, scape long and moderately curved; antennomeres 2-8 small, each with or without a few large setae, 8 flattened and appressed against 9; 9-11 together forming a large densely setose club, the antennomeres distinct but immovably associated, 11 with a pair of internal vesicles. Labrum short, strongly transverse with straight anterior edge and rounded anterior angles, with 2 adjacent very long setae near each side; underside and epipharynx with dense setae or microtrichia. Mandibles elongate, strongly projecting, with contiguous molar areas at base, each mandible with a large medially curved acute apex, 1-2 strong subapical teeth, and a setose prostheca that is more than 2/3 as long as the mandible. Maxilla with densely setose galea and lacinia, the latter at apex with a pair of strong non-articulated medially curved teeth, the palpus of 4 palpomeres, the first very small, the remaining segments stout, the apical palpomere longest, cylindrical or slightly depressed. Labium with densely setose hypopharynx and bilobed ligula, palp of 3 palpomeres, the apical palpomere longest and stoutest. Mentum sexually dimorphic: trapezoidal, flat and well sclerotized in female, transversely oval, slightly concave and translucent in male. Gular sutures complete, well separated throughout. Tentorium: see Stickney (1923).

Prothorax freely movable, not tightly co-adapted to rest of thorax. Pronotum transverse, broadly emarginate anteriorly, sides evenly convex and broader posteriorly, base sinuate, sides strongly and anterior border weakly margined, surface punctate especially laterally; hypomeron broad, explanate, with long acute postcoxal process that does not close the procoxal cavities; notosternal suture complete; prosternum short, with small fingerlike intercoxal process that fits into groove in coxae but does not separate the apices of the coxae. Scutellum moderately large, subtriangular; mesosternum short, with elevated intercoxal plate; mesepisternal and mesepimeral sutures complete. Elytra convex, long but abruptly truncate, exposing only pygidium (abdominal tergum VII); striae represented by 9 unimpressed rows of punctures which become more or less irregular or obsolescent near base and apex; epipleural fold present, complete. Metathoracic wings present, with large anal lobe and relatively complete venation including distinct radial cell and large RP-$MP_{1+2}$ loop (former "M-Cu" loop); see Kukalová-Peck and Lawrence (1993: fig. 41) for details and modern venational terminology, but note that $MP_{1+2}$ is mislabeled $MA_{1+2}$ there. Metasternum long, metepisternal and metepimeral sutures complete; metendosternite with long basal stalk and pairs of long furcal and short anterior arms. Legs with pro- and mesotrochantins exposed; anterior coxae transverse, contiguous, prominent; middle coxae small, globular, well separated by elevated mesosternal plate plus flat metasternal process; hind coxae transverse, subcontiguous; trochanters small; femora swollen; tibiae relatively robust with small spines along outer edge but without teeth; tarsae 5-5-5, tarsomeres

**Acknowledgments:** I thank M. S. Caterino, P. W. Kovarik and M. K. Thayer for helpful comments on the manuscript, and M. A. Ivie for additional distribution records.

**Color Figure 1**
FAMILY: Buprestidae
SPECIES: *Trachykele b. blondeli* Marseul
PHOTOGRAPHER: R. L. Westcott

**Color Figure 2**
FAMILY: Curculionidae
SPECIES: *Artipus floridanus* Horn
PHOTOGRAPHER: M. C. Thomas

**Color Figure 3**
FAMILY: Chrysomelidae
SPECIES: *Anomoea* sp.
PHOTOGRAPHER: M. C. Thomas

**Color Figure 4**
FAMILY: Chrysomelidae
SPECIES: *Griburius larvatus* (Newman)
PHOTOGRAPHER: M. C. Thomas

**Color Figure 5**
FAMILY: Scarabaeidae
SPECIES: *Stephanucha thoracica* Casey
PHOTOGRAPHER: P. E. Skelley

**Color Figure 6**
FAMILY: Scarabaeidae
SPECIES: *Cyclocephala* sp.
PHOTOGRAPHER: M. C. Thomas

**Color Figure 7**
FAMILY: Chrysomelidae
SPECIES: *Bassareus croceipennis* (LeConte)
PHOTOGRAPHER: M. C. Thomas

**Color Figure 8**
FAMILY: Chrysomelidae
SPECIES: *Disonycha l. leptolineata* Blatchley
PHOTOGRAPHER: M. C. Thomas

**Color Figure 9**
FAMILY: Buprestidae
SPECIES: *Chalcophora virginiensis* (Drury)
PHOTOGRAPHER: M. C. Thomas

**Color Figure 10**
FAMILY: Geotrupidae
SPECIES: *Bolbocerosoma hamatum* Brown
PHOTOGRAPHER: M. C. Thomas

**Color Figure 11**
FAMILY: Meloidae
SPECIES: *Nemognatha* sp.
PHOTOGRAPHER: M. C. Thomas

**Color Figure 12**
FAMILY: Scarabaeidae
SPECIES: *Trigonopeltastes delta* (Forster)
PHOTOGRAPHER: M. C. Thomas

**Color Figure 13**
FAMILY: Chrysomelidae
SPECIES: *Agasicles hygrophila* Selman and Vogt
PHOTOGRAPHER: M. C. Thomas

**Color Figure 14**
FAMILY: Curculionidae
SPECIES: *Rhynchophorus cruentatus* (Fabricius)
PHOTOGRAPHER: M. C. Thomas

**Color Figure 15**
FAMILY: Elateridae
SPECIES: *Alaus oculatus* (Fabricius)
PHOTOGRAPHER: M. C. Thomas

**Color Figure 16**
FAMILY: Chrysomelidae
SPECIES: *Disonycha conjugata* (Fabricius)
PHOTOGRAPHER: M. C. Thomas

slender with a few setae on each; apical tarsomere bearing a pair of large claws and a bisetose empodium between them.

Abdomen at base ventrally not tightly co-adapted to thorax, with five visible lightly punctate sterna (III-VII) and one visible densely punctate tergum (VII); segment VIII and genital segments completely invaginated and lightly sclerotized, sexually dimorphic; spiracles of segments I-VI functional, VII-VIII atrophied; terga increasingly well sclerotized from III-VII; terga IV-V and dorsal edges of sterna II-III with dense patches of microtrichia, tergum VI with apical palisade fringe. Male tergum VIII divided, IX-X entire, sternum VIII with long curved posteriorly directed lateral lobes, sternum IX present. Aedeagus small, subcylindrical; phallobase (basal piece) short, forming a nearly complete but asymmetrical ring; parameres long, nearly completely fused to one another and largely enclosing the slender median lobe. Female terga VIII-IX divided, X entire and situated between halves of IX, sternum VIII divided; ovipositor consisting of paired valvifers and heavily sclerotized coxites, the latter each with a scoop-like outer surface and a small stylus on inner side; spermatheca simple, bulbous.

General reviews and illustrations of adult morphology of *Sphaerites* can be found in Kryzhanovsky and Reichardt (1976) and especially Ôhara (1994), with further details in Hansen (1997; ovipositor), Stickney (1923; head capsule), Williams (1938; mouthparts), Forbes (1926; wing folding pattern), Kukalová-Peck and Lawrence (1993; wing venation with modern terminology), and Crowson (1974; internal anatomy).

**Description (larvae)**: Body elongate, slightly flattened, surfaces very lightly to moderately pigmented and sclerotized, with sparse vestiture. Head prognathous, protracted, strongly transverse, without distinct neck; epicranial sutures separate, without basal stem; stemmata apparently absent. Antenna less than half as long as head width, of three segments, pair of sensoria of second segment conical and situated on outer apex, lateral to insertion of apical segment. Labrum fused to head capsule to form a nasale bearing a single median tooth. Mandibles more or less symmetrical, each with a prominent curved acute apex, a single acute tooth medially, and an abruptly enlarged base which bears a penicillus or tuft of setae. Maxillary cardines distinct, transverse, separated by mentum; stipes elongate, without apical lobes; palp 4-segmented, basal segment with an articulated digitiform appendage on inner side. Labium consisting of prementum, mentum and submentum; ligula absent; palps 2-segmented, separated by less than the width of one of them. Gula present, elongate.

Thoracic terga and sterna each of 4 or more sclerotized plates. Legs short, of 5 segments including tarsungulus. Abdomen 10-segmented, largely membranous, tergal and sternal areas of segments I-VIII each with a transverse row of asperities and 22 or 9 small sclerites, respectively. Tergum IX much longer than preceding terga, of 8 small sclerites, and bearing pair of long, 4-segmented urogomphi at apex. Spiracles biforous, present ventrolaterally before mesothorax and dorsolaterally on abdominal segments I-VIII.

The above description is of first instar larvae of *S. glabratus*, reared and described by Nikitsky (1976), repeated in part by Newton (1991) and Hansen (1997). A different larva attributed to this species by Crowson (1974) was misidentified (Nikitsky 1976). Eggs large, white; pupae unknown.

**Habits and habitats.** These uncommon beetles are generally associated with decaying organic matter such as dung, carrion, fungi, fermenting fruit, and sap of dying or dead trees, but their ecology is poorly known. In Europe, *S. glabratus* has been associated especially with northern conifer forests; adults are attracted to sap flows on birch stumps, and have been observed to feed on the sap and mate; females laid eggs in nearby sap-impregnated soil, and larvae developed quickly with the next generation of adults emerging within a month (Nikitsky 1976). Larval feeding has not been reported, and that of adults requires confirmation; the mouthparts of larvae as well as adults are similar to those of histerids known to be predaceous on soft prey such as Diptera larvae. Like histerids, adults feign death when disturbed (P. Kovarik, pers. comm.).

**Status of the classification.** The relationship of *Sphaerites* to other beetles puzzled coleopterists for a long time. Horn (1880) and most other 19th century authors included the genus in the old broad concept of Silphidae, often (like Horn) even in the subgroup Silphini that includes modern silphids, while Ganglbauer (1899) among others recognized the distinctness of the genus by placing it in its own family but related it to clavicorn beetles such as Trogossitidae (as Ostomidae) and Nitidulidae. Sharp and Muir (1912) and Forbes (1926), based on male genitalia and wing venation and folding patterns, respectively, recognized a close relationship of *Sphaerites* to Histeridae and even more to *Syntelia*; Forbes placed these two genera together in his Synteliidae. More recent authors have consistently placed *Sphaerites* in its own family, while accumulating evidence including the discovery of larvae (Nikitsky 1976) has confirmed the close relationship of the family to Synteliidae and Histeridae. These three families are now invariably placed together in the polyphagan series Staphyliniformia, either in their own superfamily, Histeroidea (e.g., Crowson 1974), or in a monophyletic "histerid group" of families within Hydrophiloidea as a sister group to Hydrophilidae s.l. (e.g., Lawrence and Newton 1982). Recent phylogenetic analyses (Ôhara 1994, Hansen 1997, Slipinski and Mazur 1999) have supported Sphaeritidae as the most basal or primitive of the three families and the sister group of Synteliidae + Histeridae.

**Distribution.** The family includes a single genus with only four known species: *Sphaerites glabratus* (Fabricius, 1792), widespread across northern Europe and east to Mongolia, eastern Russia, and Japan; *S. dimidiatus* Jurecek, 1934, from Gansu and Sichuan provinces, China; *S. nitidus* Löbl, 1996, from Sichuan province, China; and *S. politus* Mannerheim, 1846, from western North America and (doubtfully) eastern Russia and Japan (Löbl 1996).

## Classification of the Nearctic Sphaeritidae

*Sphaerites* Duftschmid, 1805

One species, *S. politus* Mannerheim, 1846, southeastern Alaska south through British Columbia, Washington, Oregon and northern California, east to Alberta, Idaho and western Montana; locally common at carrion (Keen 1895), found also in bear dung and in or near compost (Hatch 1961), and in unbaited pitfall and flight traps in old-growth conifer forests (M. Ivie pers. comm.). This species was usually treated as a synonym of the widespread Palearctic species *S. glabratus* (Fabricius, 1792) (e.g., by Horn 1880, Schenkling 1931, McGrath and Hatch 1941) until Brown (1944) demonstrated differences between the two species and referred all North American records to *S. politus*, unique among *Sphaerites* species in the acutely projecting apex of the metatrochanter. *S. politus* was first reported from the Palearctic region by Adachi and Ohno (1962) and has now been recorded from various localities in Japan and eastern Russia (Kryzhanovsky and Reichardt 1976, Kryzhanovsky 1989, Ôhara 1994), but Löbl (1996) indicated that these records need confirmation and may actually refer to *S. glabratus*. Immature stages of *S. politus* are unknown but larvae should resemble those of the closely related *S. glabratus* described by Nikitsky (1976) and Newton (1991).

## Bibliography


ADACHI, T. and OHNO, M. 1962. Discovery of the family Sphaeritidae (Coleoptera) in Japan. Zoological Magazine [Dobutsugaku Zasshi], 71: 148-151 [in Japanese, English summary].

BROWN, W. J. 1944. Some new and poorly known species of Coleoptera, II. Canadian Entomologist, 76: 4-10.

CROWSON, R. A. 1974. Observations on Histeroidea, with descriptions of an apterous larviform male and of the internal anatomy of a male *Sphaerites*. Journal of Entomology (B), 42: 133-140.

FORBES, W. T. M. 1926. The wing folding patterns of the Coleoptera. Journal of the New York Entomological Society, 34: 42-115, pls. 7-18.

GANGLBAUER, L. 1899. Die Käfer von Mitteleuropa. Vol. 3, Familienreihe Staphylinoidea, II. Theil: Scydmaenidae, Silphidae, ... Carl Gerold's Sohn, Vienna. iii + 1046 pp.

HANSEN, M. 1997. Phylogeny and classification of the staphyliniform beetle families (Coleoptera). Biologiske Skrifter, Det Kongelige Danske Videnskabernes Selskab, 48: 1-339.

HATCH, M. H. 1961. The beetles of the Pacific Northwest. Part III: Pselaphidae and Diversicornia I. University of Washington Publications in Biology, 16 (3): ix + 503 pp.

HORN, G. H. 1880. Synopsis of the Silphidae of the United States with reference to the genera of other countries. Transactions of the American Entomological Society, 8: 219-319, pls. 5-7.

KEEN, J. H. 1895. List of Coleoptera collected at Massett, Queen Charlotte Islands, B. C. Canadian Entomologist, 27: 165-172, 217-220.

KRYZHANOVSKY, O. L. 1989. [Fams. Sphaeritidae, Synteliidae, Histeridae, pp. 294-310. In: Ler, P. A. (ed.), Keys to the Insects of the Far Eastern USSR in Six Volumes. Vol. 3, Coleoptera or Beetles, Part 1] [in Russian]. Nauka, Leningrad.

KRYZHANOVSKY, O. L. and REICHARDT, A. N. 1976. [Beetles of the superfamily Histeroidea (families Sphaeritidae, Histeridae, Synteliidae). In: Fauna USSR, Coleoptera, Vol. 5, No. 4] [in Russian]. Nauka, Leningrad. 434 pp.

KUKALOVÁ-PECK, J., and LAWRENCE, J. F. 1993. Evolution of the hind wing in Coleoptera. Canadian Entomologist, 125: 181-258.

LAWRENCE, J. F. and NEWTON, A. F., JR. 1982. Evolution and classification of beetles. Annual Review of Ecology and Systematics, 13: 261-290.

LÖBL, I. 1996. A new species of *Sphaerites* (Coleoptera: Sphaeritidae) from China. Mitteilungen der Schweizerischen Entomologischen Gesellschaft, 69: 195-200.

McGRATH, R. M. and HATCH, M. H. 1941. The Coleoptera of Washington: Sphaeritidae and Histeridae. University of Washington Publications in Biology, 10: 47-91.

NEWTON, A. F., JR. 1991. Sphaeritidae (Hydrophiloidea), pp. 359-360. In: Stehr, F. W. (ed), An introduction to immature insects of North America, Vol. 2. Kendall/Hunt Publishing Co., Dubuque, Iowa.

NIKITSKY, N. B. 1976. [On the morphology of the larva of *Sphaerites glabratus* and the phylogeny of Histeroidea] [in Russian]. Zoologicheskii Zhurnal, 55: 531-537.

ÔHARA, M. 1994. A revision of the superfamily Histeroidea of Japan (Coleoptera). Insecta Matsumurana (N.S.), 51: 1-283.

SCHENKLING, S. 1931. Fam. Sphaeritidae. In: Schenkling, S. (ed.), Coleopterorum Catalogus, Pars 117, pp. 1-2. W. Junk, Berlin.

SHARP, D. and MUIR, F. 1912. The comparative anatomy of the male genital tube in Coleoptera. Transactions of the Entomological Society of London, 1912: 477-642, pls. 42-78.

SLIPINSKI, S. A. and MAZUR, S. 1999. *Epuraeosoma*, a new genus of Histerinae and phylogeny of the family Histeridae (Coleoptera: Histeroidea). Annales Zoologici, 49: 209-230.

STICKNEY, F. S. 1923. The head-capsule of Coleoptera. Illinois Biological Monographs, 8: 1-104, pls. 1-26.

WILLIAMS, I. W. 1938. The comparative morphology of the mouthparts of the order Coleoptera treated from the standpoint of phylogeny. Journal of the New York Entomological Society, 46: 245-289.

# 15. HISTERIDAE Gyllenhal, 1808

by Peter W. Kovarik and Michael S. Caterino

**Family synonym:** Niponiidae Fowler, 1912

**Common name:** The clown beetles

The Histeridae is a distinctive group of mostly small, black, mainly predacious beetles. Most species are round or oval in dorsal view and stout. Some, however, are cylindrical, others are dorsoventrally flattened, and there are yet other manifestations which are completely bizarre. They are extremely compact in form, with retractile head and appendages. The elytra are truncate, exposing the last 2 abdominal terga. The posteriormost of these, the pygidium, is operculate. All species have geniculate antennae with a compact, usually 3-segmented club.

FIGURE 1.15 *Terapus* n. sp.

**Description:** Hister beetles range in length from 0.5-12 mm. Most species have adults that are black in color while others are rufescent, metallic blue or green, or marked with red. Many species have adults that are glabrous but some are finely to coarsely punctate while others still are hirsute.

The head is deflexed or occasionally porrect with supraorbital stria present or absent. The broad frons is smooth to punctate, convex to depressed, and with or without anterior and lateral margins. Frontal protuberances are occasionally present and a y-shaped endocarina is visible in some species. The indistinguishable clypeus is fused to the frons forming an epistoma which narrows abruptly anterior to the antennal insertions.

The often setose labrum is generally transverse and truncate or occasionally emarginate. The epipharynx is subdivided into a medial and lateral areas by 2 oblique rows of setae. The lateral and dorsal margins of the lateral areas are bordered with fringes of regularly spaced elongate setae that extend onto the medial area and are often peg-like. The lateral area is uniformly scaly. A dorsomedial cluster of short stiff setae generally borders the medial area dorsally. Several robust, often peg-like setae occur directly above this setal cluster. The medial area is generally covered with numerous sensillae. The mandibles are generally prominent, apically acute, and have a toothed inner margin in some species. The gular sutures are usually confluent posteriorly (in some species a small portion of the gula is visible posteriorly)and diverge anteriorly around the submentum.

**Acknowledgments.** We thank Dave Verity and Alexey Tishechkin for contributing to the keys and Rupert Wenzel, Alfred Newton, and Jeff Gruber for reading through and commenting on the draft.

The triangular to rectangular submentum is fused to the head capsule, and broadly articulated anteriorly with the mentum. The broad flat mentum can be carinate or furrowed. The apex of this structure can be truncate, projecting, or emarginate, and is laterally expanded in a few species. The base of the ligula underlies the mentum and only the apices of palpigers, paraglossae, and glossae (when present) are visible. The palpiger commonly bears a single minute seta and at least 1 pore on the ventral surface. The palpiger is laterally setose in some taxa. The labial palp is 3-segmented. The short basal palpal segment usually has a single minute lateral seta. Palpal segment II generally bears a pore and 4 short to medium-length setae apically and a variable number of lateral setae. The terminal palpal segment is variably setose. It is expanded, and dorsoventrally flattened in some tribaline taxa. Apical pores and sensilla digitiformia are often present. Palpiform sensilla in the distal membrane are either short and few in number or relatively long and numerous.

The galea of the maxilla in most taxa is sclerotized, relatively compact, and is densely crowned with simple, long, robust setae apically and mesally. In fungivorous taxa the galea is membranous, variably expanded, and often densly setose on the dorsal surface. The apex of the galea in some fungivores bears elongate spatulate setae. The lacinia is generally broadly attached to the mediostipes but is largely detached in some groups. The lacinial fringe mainly consists of slender tapered setae, but in a number of taxa distinct flattened and broad setae are present apically. The lacinial disc bears a robust tooth ventrally and usually several slender setae. The mediostipes is generally rugose and bears numerous small setae and usually a single pore. The often textured basistipes commonly has 2 pores and 2-3 elongate robust setae. The variably setose, and generally smooth cardo articulates proximally with the basistipes. This structure often bears a dorsal condyle. The typically strongly textured palpifer consistently bears an elongate, sometimes robust ventroapical seta and at least 1 ventral pore. The small dististipes occurs just above the ventral apex of the palpifer. The dorsal surface of the palpifer is variably setose and occasionally carinate while the dorsal apex commonly bears 1 to several elongate, robust setae.

The maxillary palp is 4-segmented. The proximal palpal segment consistently bears a pair of minute setae on the lateral margin and a cluster of pores in the apical membrane. Segment II generally has short lateral and lateroapical setae while segment III has 5 regularly spaced apical setae. The terminal palpal segment is variably setose and usually has a lateral cluster of sensilla digitiformia, several apical pores, and several isolated apical sensilla digitiformia. The apical membrane usually bears relatively few minute palpiform sensilla.

The antennal insertion is exposed and generally situated between the eye and mandible. However, antennae are occasionally inserted under frontal protuberances or on the frons. The antenna are elbowed with an elongate, curved scape. The funicle consists of 7 antennomeres and a compact, 3-segmented club which bears a variety of sensilla. Occasionally the sensilla are arranged in distinct plaques. The club sutures are usually complete, but are occasionally interrupted or absent.

The prothorax is excavate beneath for the reception of the prothoracic legs and sometimes the antennae. This structure is laterally margined, narrowed anteriorly (in most species), and anteriorly emarginate for the reception of the head. The pronotal disk is generally convex, smooth and glabrous but it can be punctate or striate, and is occasionally setose or costate. The procoxal cavities are open behind and the keeled prosternum is strongly produced between the procoxae and joins with the elevated mesosternum posteriorly. The keel can be flat, convex, or rarely carinate, and longitudinal striae are generally present. The prosternum is truncate to anteriorly lobate. In the Dendrophilinae and Abraeinae, the withdrawn antennae are shielded by the retracted protibiae and femora.

Antennal grooves are present alongside the prosternal keel. Saprinines similarly use retracted forelegs for antennal protection and nearly all taxa have distinct cavities for receiving the antennal club in front of the procoxae and next to the prosternal keel. In Onthophilinae, Tribalinae, Histerinae, and Hetaeriinae the anterior angles of the prothorax bear specialized cavities for reception of the antennae. The procoxae are transverse or rarely round. The protrochantin is hidden and the trochanter is fused to the femur. The femora are swollen and the protibiae are usually broad, flattened, and spinose to dentate along the outer margin. The protibia usually has a protarsal furrow on anterior surface, and has or lacks apical spurs. The meso- and metasternum are broad with the coxae well-separated and not projecting. The meso- and metatibiae are slender to broad, laterally spinose, (sometimes densely so) and usually have apical spurs. The tarsal formula is usually 5-5-5, rarely 5-5-4 and the generally unmodified claws are occasionally unequal or fused. The scutellum is visible in most species.

The elytra are generally rectangular (occasionally posteriorly tapered) and apically truncate and have a distinct, striate epipleuron laterally. The dorsal elytral striation is derived from a pattern of 8 longitudinal striae including an inner and outer subhumeral striae, 5 discal striae, and a sutural stria. The sutural stria is often continuous with a posterior marginal stria and an additional short oblique stria is often visible near the anterolateral corners.

Hind wing venation follows that of Kukalová-Peck and Lawrence (1993). The wing venation of *Dendrophilus* is rather complete and approximates that of *Sphaerites*. Veins present in *Dendrophilus* include ScA, ScP, R4, RA3, RA4, RP, RP1, RP2, RP3+4, MA1+2, MP3+4, CUA, CUA2+3+4, AA1+2, AP3+4, and J. In *Dendrophilus* and most other histerids, several veins including RP1, MP3+4 and CUA2+3+4 are reduced and represented solely by pigmented membrane. A reduction in wing length is common in histerids. For example, veins RP2 and RP3+4 both reach the wing margin in *Dendrophilus*, *Onthophilus*, and several histerine genera, whereas in many other genera, including *Epierus*, *Xestipyge*, *Hetaerius*, and *Geomysaprinus*, one or both of these veins end short of the wing margin or reach it only as pigmented membrane. A remarkable case of wing vein reduction occurs in *Plegaderus* where the only recognizable veins are the SC, MA1+2, RP2 and RP3+4 and CUA2+3+4. Of these RP2 is entirely pigmented membrane and covers a tiny fraction of its "normal" length. Vein RP3+4, while longer than RP2, is also reduced to pigmented membrane and stops well shy of the wing margin. Pigmentation patterns appear to be stable in some genera and quite variable in others.

The abdomen of histerids is usually short and broad. There are 7 tergites; I-IV are generally weakly sclerotized. Tergite IV bears central groove that is likely a fore wing locking structure. This apparently is a synapomorphy unique to the Histeridae (A. Newton pers. comm.). Tergite 6, the propygidium, is generally trapezoidal in shape. Tergite 7, the pygidium, is semicircular or triangular in shape. The propygidium and pygidium are significantly longer than the preceding terga and are both well-sclerotized and often punctate. The elytra conceal at least the 1st 5 tergites. The abdominal dorsum bears 4-5 pairs of spiracles all of which are hidden beneath the elytra. The 1st pair is much larger than the remainder. There are only 5 sternites present in histerids. The 1st visible sternite is considerably longer than those following it. This sternite may bear a pair of striae (termed lateral abdominal striae) which extend from the metacoxae to its posterior margin. The remaining visible sternites are short and may or may not be sculptured. Abdominal segments VIII-X generally remain concealed within the abdomen and in males are modified as copulatory structures whereas in females these segments form the ovipositor.

The terminal abdominal segments of male histerids essentially form a telescoping syringe while those of the female form an extensible tube for oviposition. While appearing quite different, both of these structures are products of differentially modified terminal abdominal segments which are normally concealed within the abdomen.

In males segment VIII is typically composed of a relatively large tergite and a pair of distally articulated processes or coxites. Coxites often bear minute setae laterally and longer setae apically. Coxal setae are extremely long in hololeptines, forming an elaborate comb. The coxites of male *Dendrophilus*, *Anapleus* and apparently some species of *Stictostix*, bear styli. Paired inflatable pubescent (velar) membranes are borne on the inner surface of coxites in *Onthophilus* and some saprinine genera including

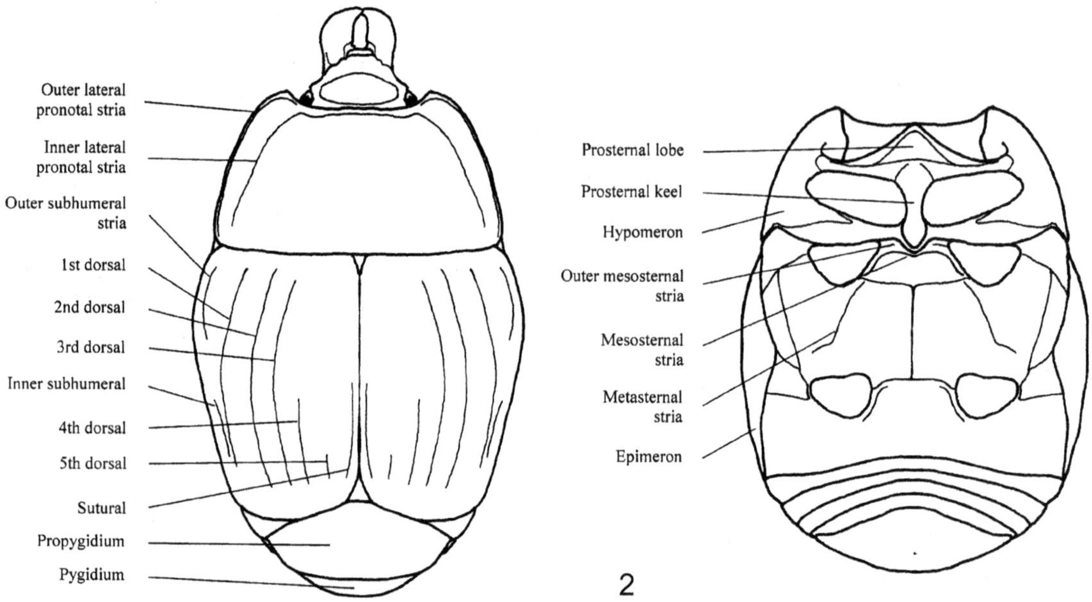


FIGURE 2.15. A generalized *Hister* showing external morphological terminology.

*Euspilotus* and *Hypocaccus*. When inflated, these structures apparently assist males in gaining purchase during copulation. The morphology of the male tergite is variable. The posterior margins are occasionally emarginate and in some cases the emargination is extreme, producing lateral tergal extensions or tails. The tergite often bears minute setae and cuticular cavities.

Segment IX of male histerids may or may not resemble the general structure of the preceding segment. A notable difference is an apparent fusion of the coxites forming a coxal carapace. This structure is longitudinally bisected by a suture in *Onthophilus* and is notably absent in *Stictostix*, *Peploglyptus*, and *Caerosternus*. The carapace bears prominent setae along its distal margin in many taxa and also a velar membrane in *Onthophilus*. The morphology of tergite IX is variable. In saprinines this structure is proximally emarginate and tailed. In most histerines and some tribalines, including *Stictostix*, *Peploglyptus*, and *Caerosternus*, this tergite is dorsomesally subdivided with the resulting plates variably reduced. Segment IX has a ventral sclerite, or spiculum gastrale, which is variable in structure. This structure is commonly long and slender and expanded at one or both ends.

The terminology of the aedeagus follows that of Lawrence and Britton (1991). There are essentially three parts to this structure: a basal phallobase, apical parameres, and a median lobe which is sandwiched between the parameres near their apex. The phallobase generally appears as an asymmetric short tube or globe-like structure and is usually joined to the parameres by a flexible membrane. However, in the Abraeinae, the phallobase is reduced and fused to the parameres. The phallobase is extremely long and tubular in some dendrophilines including *Carcinops*, *Paromalus*, and *Xestipyge* and is relatively long among members of the *Hister coenosus* group. The parameres are fused for most of their length in many taxa and tend to be relatively long and robust. In some genera such as *Onthophilus*, *Stictostix*, *Bacanius*, and *Gnathoncus*, the parameres are very long and slender with the bifid apex curving ventrad. In other genera including *Carcinops*, *Paromalus*, and *Xestipyge*, the parameres are short and separated for most of their length.

While the term coxite has not been applied to components of the male terminalia, morphological similarities between these structures in the two sexes suggest that both structures may be abdominal leg derivatives. The coxites of female histerids are articulated structures, with each bearing an articulated, apically setose structure termed a stylus. Styli are also present on the articulated coxites in some male histerids. The dorsal position of the coxal carapace has led specialists to conclude that this structure is the tergum of segment X. This is unlikely because this structure is located in the same relative position as the coxites of the previous segment. In addition, the carapace is divided by a longitudinal suture in some taxa indicating an apparent fusion of 2 separate structures. Finally, like the coxites of segment VIII, the distal margin of the carapace often bears relatively long setae and in some cases a velar membrane.

Eggs of histerids are usually ivory in color and elongate-oval (rarely round). In most histerids, the chorion is thin, smooth, and shiny. However, in species of *Epierus*, *Plagiogramma*, and *Platylomalus*, the chorion is pale brown and dull due to surface texturing. In these species the chorion is also leathery and partially retains its shape after larval eclosion.

Histerids are somewhat unusual in that their larvae have only 2 instars. Larvae are liquid feeders and digestion is extraoral (Kovarik 1995). The entire larval development for some species is completed in less than a week at optimal temperatures. Histerid larvae are generally elongate, parallel-sided, and range in size from a few millimeters to several centimeters. The body in cross section is generally rounded but can be dorsoventrally flattened as seen in some subcortical genera such as some *Platysoma* species and all *Hololepta*. Generally, the only conspicuously darkened and sclerotized structures are the head, pronotum, mesonotum, and urogomphi resulting in a body that appears to be mostly membranous. However, rather lightly pigmented, mainly small sclerites are present on all thoracic and abdominal segments. The thoracic legs of histerid larvae are extremely reduced and do not appear to play a significant role in locomotion. Movement results from anteriorly or posteriorly directed waves of muscular contractions of abdominal segments. Prolegs with retractor muscles are present in larvae of several taxa. The larval head and all body segments bear numerous setae and pores. A system for naming these structures was recently developed by Kovarik and Passoa (1993). Asperites appear to be universally present but vary in terms of their morphology and coverage. Enlarged asperites are generally found in association with ampullae and prolegs. The asperites on the prolegs of *Baconia* larvae are hook-like in appearance, resembling the crochets of caterpillars.

Terminology used in this description follows that of Kovarik and Passoa (1993) and Newton (1991). The head is prognathous, protracted and well-sclerotized. Epicranial sutures are present in the 1st instar only. The stem length is variable and the frontal arms are V-shaped or lyriform. The frontale has a maximum of 11 setae and 6 pores on each side. The parietal area of the head capsule generally has 19 setae and 13 pores on each side. Stemmata, if present, number 1 per side.

Antennae are 3-segmented. Segment I bears a total of 4 pores and the segment II bears a single pore. The pre-apical segment bears 1 or 2 conical or palpiform sensoria and 3 sensilla while segment III bears 6 apical sensilla. The sensilla of pre-apical and apical segments are short in most taxa but are quite long and setiform in a few taxa. The labroclypeus is completely fused to the head capsule forming a generally asymmetrical and variably toothed nasale. The mandibles are falcate and generally bear 1 or 2 teeth along the mesal margin. Two mandibular setae are present, 1 developed and apparent, the other vestigial and inconspicuous. The base of the mandible bears a penicillus.

The maxillary stipes is elongate with the mesal and occasionally lateral margins bearing a fringe of flexible setae. It also bears a total of 4 mechanorecepters and 3 pores. The maxillary palpi are generally 4- or rarely 5-segmented. The proximal palpomere represents a fusion of elements of the galea and palpi and bears a single pore and seta. The bisetose galea appears to articulate with the apex of the 1st palpomere. The terminal palpomere has numerous apical palpiform sensilla and a lateral sensillum digitformium.

The labial mentum is reduced and somewhat concealed with its distal portion membranous and proximal portion sclerotized. The prementum is laterally lobed or not and bears 3 setae, 2 pores and 2 peg-like sensilla in dorsoapical membrane. The dorsal surface of the prementum can be spinose or glabrous. The labial palpi are 2- or 3-segmented and rarely have a pore on the palpomere I. The terminal labial palpomere has numerous apical palpiform sensilla as well as a lateral sensillum digitorum. The tentorial fossa, when present, is centrally located on the ventral epicranium.

The pronotum covers the entire dorsal surface of the prothorax and generally bears 21or 22 setae and pores on each side. No setae are present on the pro-episternum whereas the subdivided pro-epimeron bears 2. The relatively large and subdivided presternum bears 9 setae on each side. The prosternite has 3 setae on each side. A pair of generally small lateral sternites is commonly found alongside the prosternite. The inconspicuous precoxites each bear 3 minute setae.

The large mesonotum bears 8 setae and 5 pores on each side. The long and narrow humeral tergites each have 5 minute setae and 3 pores. Intersegmental tergites or intertergites are present or lacking. A pair of tergal subdivisions termed lateral tergites each bear 7 setae and a single pore. The laterally located small anterior and large posterior pleurites each have 2 setae. The relatively small anterior sternites and precoxite bear 1 and 2 setae, respectively. The mesepimeron has 2 minute setae while the mesepisternum has only 1. The mesosternum has 4 or 5 setae on each side. Lateral sternites are present or absent.

The metathorax differs from the mesothorax in having the tergite further subdivided resulting in the formation of a small and unarmed dorsohumeral tergite and relatively large dorsolateral tergite which generally bears 3 or rarely 4 setae. The metanotum proper bears 4 setae and 2 pores. The anterior pleurite of the metathorax occasionally has an additional seta. The thoracic legs are 5-segmented and generally bear approximately 50 mechanoreceptors and several pores.

In abdominal segments I-VIII the tergum typically has 19 setae, 5 pores and up to 14 tergites on each side. A pair of egg bursters appear on the notum of abdominal segment I in the 1st instar larvae. The pleuron has 6 setae and up to 5 pleurites. The sternum has 17 setae and up to 13 sternites on each side. Generally 3 intertegites and 2 intersternites are present on these abdominal segments. In abdominal segment IX the tergum generally bears 9 setae, 2 pores and up to 10 tergites per side. The pleuron has 5 setae and up to 3 pleurites, whereas the sternum bears 5 setae and up to 5 sternites per side. Also commonly found on this segment are a pair of dorsal, 1- or 2-segmented urogomphi with a maximum of 13 setae and 5 pores. Occasionally a pair of lateral palpiform paragomphi are also present. Segment X or the pygopod bears 7 setae, a single pore and up to 5 sclerites per side.

The general form of histerid pupae is similar to that of the adult beetle. Functional spiracles are present only on abdominal segments I-IV. Histerid pupae develop in cells created by the larvae in or adjacent to the developmental substrate. These cells may or may not be reinforced with a proteinaceous cement. Those species that inhabit sand or relatively dry organic debris tend to be among the cement producers. The cement can be very tough which aids in preventing would-be predators ready access to the defenseless pupa. Pupal development generally takes at least a week at optimal temperatures.

Terminology used in this description follows that of Kovarik and Passoa (1993). Obvious features of the pupal head include a pair palpiliform processes, present on the dorsal surface of the labrum (labral papillae) in most non-histerine genera. The distal antennal scape of saprinines also bears papiliform processes or scape papillae. There are up to 3 groups of epicranial setae on the head of histerid pupae. They include the frontal, occular, and vertical epicranial groups. There are up to 2 groups of frontal setae including a medial and a ventral pair. Clypeal setae, if present, arise at or below the level of the anterior articulation of the mandible. Minute labral setae are occasionally present on the labrum.

Prothoracic setae are generally restricted to the notum. There are up to 4 groups of notal setae. They include the anterior, lateral, posterior, and interior pronotal groups. The pupal scutum and scutellum on the mesonotum may or may not bear scutal and scutellar setae. The metanotum generally lacks metanotal setae. The pupal elytra may or may not be costate. There are 2 distinct patterns of elytral setation. In one case setae are arranged in 3 discrete bands, each of which consists of numerous short setae. In the other, setation consists of generally less than 10 setae arising from various locations on the pupal elytron. The hind wing in some taxa bears a protuberance (wing tubercles) on the dorsal surface of the wing apex. Hind wing setae are known to occur in some taxa. In some cases pupal legs bear a single seta on the dorsodistal surface of the femora. The tergum of the pupal abdomen generally bears tergal and laterotergal setae. The setae arising on the sternites dorsolaterally are lateral setae and any arising on the sternites are termed sternal setae. In some taxa the pupal pygidium bears an external ridge down its midline. The pupal urogomphi that persist in most taxa have sclerotized and darkened apices.

**Habits and habitats.** Histerids are mainly predators of soft-bodied insect larvae and eggs, particularly those of cyclorraphan Diptera. Hence, substrates on which flies develop in numbers are among the best places to look for these beetles. Many muscoid fly larvae develop on the dung of large mammals and carrion. These are selectively preyed on by numerous species of *Hister*, *Margarinotus*, *Atholus*, *Phelister*, *Eupilotus*, *Saprinus*, *Xerosaprinus*, and *Xestipyge*. The volatile and odiferous byproducts of microbial degradation enable both flies and beetles to locate carrion and dung via olfaction. *Margarinotus guttifer* (Horn) is somewhat unusual among carrion attracted histerids. According to Wenzel (pers. comm.) this species is associated with carcasses in the dry decay phase. It probably feeds on larvae able to digest keratin.

A second guild of predatory histerids are specialists on dipteran larvae and eggs associated with rotting vegetation. Some of these beetles are attracted to rotting plants, especially palms and succulents in xeric areas. Included in this group are species of *Omalodes*, *Hololepta*, and *Iliotona* and *Carcinops*. Others selectively target dipteran larvae and eggs (including *Pseudotephritis* spp. [Diptera: Otitidae]) which feed on fermenting phloem and cambium (bast) of certain dead hardwoods. These taxa tend to be dorsoventrally flattened and include species of *Hololepta*, *Platysoma*, *Platylomalus*, *Carcinops*, and *Paromalus*. Many histerid species, including some *Spilodiscus*, *Hypocaccus*, *Monachister*, and *Philoxenus,* are partial to sand dunes and beaches where shifting sand covers and kills the grasses and forbs growing there. This decaying plant material is fed upon by fly and scarab larvae which in turn are prey for these histerids.

Yet another group of histerid beetles preys on diptera that develop on rotting fungi. These principally include species of *Hister* and *Margarinotus*. A fair number of histerid species inhabit hollow living trees. Some, including species of *Gnathoncus*, *Dendrophilus*, *Euspilotus rossi* (Wenzel), *Hister indistinctus* Say, *Paromalus seeversi* Wenzel, and *Chaetabraeus chandleri* Mazur apparently prefer relatively dry tree cavities in which a mammal or bird is nesting. Abundant prey items in this microhabitat include Diptera, Coleoptera, and Siphonaptera larvae. Species of *Bacanius* are known to inhabit relatively dry organic debris within tree cavities. *Merohister grandis* (Wenzel) is unusual in that it prefers wet tree cavities in mature forests. Larvae of some Diptera, including those of micropezid flies, inhabit this microhabitat in fair numbers. They serve as food for adults and larvae of *M. grandis*. Some histerids, including species of *Hololepta* and *Onthophilus*, are attracted to sap fluxes of trees. The tunnels and galleries of wood-boring beetles, especially bark beetle, serve as home for a number of cylindrical histerids. The larvae of these wood boring beetles serve as food for *Plegaderus*, *Teretrius*, *Teretriosoma*, and some *Platysoma*. Other histerids, including species of *Epierus*, *Bacanius*, and *Caerosternus americanus* are generally found in association with older dead trees and/or piles of sawdust. The adults of these species feed mainly on fungal spores.

There are many North American histerids which are obligate inhabitants of the burrows of reptiles, mammals, and even birds. Adults of most burrow-dwelling histerids feed on fly eggs and larvae which develop on dung deposits in the burrow. Adults of *Onthophilus* spp. will consume fly eggs (but not larvae) and also filter feed on the liquid coating of fresh dung. Certain morphological features such as elongate legs and antennae are common to many of these burrow-inhabiting taxa. *Chelyoxenus xerobatis* Hubbard is restricted to burrows of the gopher tortoise, *Gopherus polyphemus* (Daudin). Kangaroo rat burrows (*Dipodomys* spp.) are inhabited by species of *Eremosaprinus*. The burrows of ground squirrels (*Spermophilus* spp.) are inhabited by species of *Geomysaprinus* and *Aphelosternus*. Prairie dog burrow (*Cynomys* spp.) inhabitants include *Onthophilus cynomysi* Helava, *Saprinus discoidalis* LeConte, and an undescribed species of *Stictostix*. Woodchuck burrows (*Marmota monax* [Linnaeus]) are home to *Margarinotus egregius* (Casey) and *Geomysaprinus obsidianus* (Casey). *Geomysaprinus obscurus* was taken from the nest of a burrowing owl, *Athene cunicularia hypogea* (Bonaparte), in California (Lee and Ryckman 1954). While the majority of burrowing animals leave their burrow entrances open (including all of those mentioned above), pocket gophers (*Geomys* spp., *Thomomys* spp., etc.) typically plug their burrow systems during the daytime, making it difficult for diurnally active hister beetles to gain access. Accordingly, there appears to be little overlap in species of histerids inhabiting open vs. closed burrow systems. Obligate pocket gopher burrow-inhabiting histerids include species of *Geomysaprinus*, *Onthophilus*, *Spilodiscus*, *Margarinotus*, *Euspilotus*,

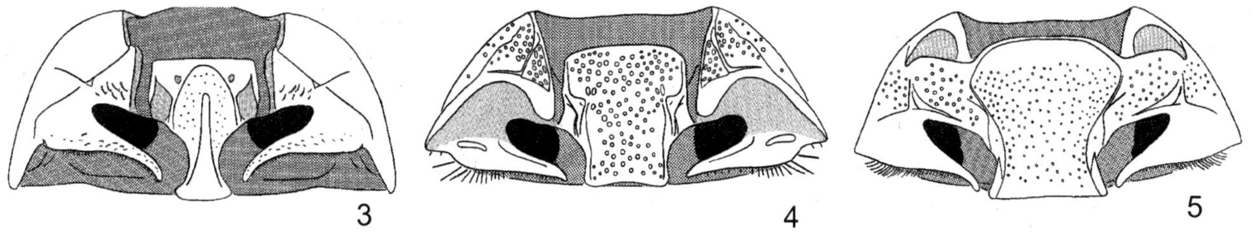


FIGURES 3.15 - 5.15. Fig. 3.15. Antennal grooves, cavities, and prosternum of *Hypocaccus* sp.; Fig. 4.15. Antennal grooves, cavities, and prosternum of *Anapleus marginatus* Lewis; Fig. 5.15. Antennal cavities and prosternum of *Stictostix* sp.

*Atholus*, and *Phelister*. Some of the pocket gopher burrow-dwellers give off a strong odor when handled. Pack rats (*Neotoma* spp.) nest above ground and several species of histerids are associated with their nests. These include *Margarinotus umbiculatus* (Casey), *Hister comes* Lewis, *H. criticus* Marseul, *H. humilis* (Fall), and *Onthophilus deflectus* Helava.

Some of the most interesting histerids from a morphological standpoint are those inhabiting ant and termite nests. Some of the myrmecophilus taxa have become so well-integrated into their host's society that they are actually fed by the ants. Others are treated as uninvited guests and mainly feed upon insect larvae in the refuse piles of the ants. Adult *Psiloscelis* actually capture and feed on adult ants (Carlton et al. 1996). The leafcutter ant *Atta texana* (Buckley) is the host of several histerids including *Reninus salvini* (Lewis), *Acritus attaphilus* Wenzel and an undescribed species of *Geomysaprinus*. Nests of *Pogonomyrmex* spp. are inhabited by species of *Hetaerius* and *Onthophilus* (MacKay 1983), while those of *Aphenogaster* are inhabited by *Tribalister* spp. *Formica* nests are inhabited by species of *Hetaerius* and *Psiloscelis*. *Lasius* nests are also home to *Hetaerius* spp. Species of *Terapus* and *Mroczkowskiella* inhabit nests of *Pheidole* spp. Finally *Ulkeus* spp., and possibly a species of *Chrysetaerius* from Arizona are associated with army ants (*Neivamyrmex* spp.).

A few histerids are coastal wrack inhabitants. They include *Neopachylopus sulcifrons* (Mannerheim) and species of *Baeckmanniolus* and *Halacritus*. Numerous dipteran taxa are known to breed in wrack and they likely serve as food for *N. sulcifrons* and *Baeckmanniolus* spp. One species, *Geocolus caecus* Wenzel, is blind and flightless and inhabits soil and leaf litter at cave entrances. *Stictostix californicus* (Horn), *Peploglyptus belfragei* LeConte and *Anapleus marginatus* Lewis are most frequently encountered in riparian situations and are apparently omnivorous. *Acritus* and *Aeletes* commonly occur in leaf litter or under tree bark. These tiny histers are thought to prey on mites.

**Economic importance**. As general predators, most histerids are considered to be economically neutral or moderately beneficial. However, a few species in typically pest-harboring habitats are of substantially greater importance. Numerous dung-inhabiting species have been found to be effective natural enemies of pest flies, both in cattle pastures (histerids in the genera *Hister, Phelister,* and *Euspilotus*) and in poultry houses (*Gnathoncus*, *Dendrophilus*, and *Carcinops*) and several species have been intentionally introduced outside of their native ranges for this purpose. Some species, particularly a *Teretriosoma*, have proven useful in the control of pest beetles of stored grains, and have likewise been widely introduced. Some of the subcortical predators (e.g., *Platysoma* and *Plegaderus*) are considered important natural enemies of bark beetles.

**Status of the classification**. While histerid classification has remained relatively stable since Bickhardt's (1916) treatment, there is still substantial work to be done. The monophyly of few of the higher taxa has been demonstrated but monophyly is clearly questionable for some of the subfamilies (Histerinae, Abraeinae, and Dendrophilinae) and tribes (Teretriini, Exosternini). Phylogenetic relationships among the higher taxa has likewise received little attention (but see Ohara 1994 and Slipinski and Mazur 1999). Several taxa have been revised for limited geographic areas but very few groups have been studied in their entirety. Most North American genera and species groups extend throughout the Holarctic or into the Neotropical region and need to be studied in this broader context. For identification of North American species, it is generally necessary to refer back to the works of J.E. LeConte (1845), Horn (1873) and Casey (1893, 1916). However, the family is receiving renewed attention, thanks in large part to the recent catalogues of Mazur (1984, 1997) and it is hoped that our knowledge of the American fauna will continue to improve in the near future.

**Distribution**. There are approximately 330 genera and 3,900 species found worldwide. Of these, 57 genera and 435 species are known from the United States.

## Key to the Nearctic Genera

(See Fig. 2.15 for external terminology)

| | | |
|---|---|---|
| 1 | Antennal cavities for receipt of retracted antennal club present alongside prosternal keel (Fig. 3.15); the funicle in repose curving inward (Saprininae) | 22 |
| — | Antennal cavities, if present, situated laterally (Figs. 4.15, 5.15); the funicle in repose curving outward | 2 |
| 2(1) | Labrum setose (Fig. 13.15) | 3 |
| — | Labrum without setae | 5 |
| 3(2) | Prosternal lobe with lateral extensions (alae) which conceal the retracted antennal funicle and partially conceal the antennal cavity from below (Fig. 5.15) | 4 |
| — | Prosternal lobe without lateral extensions, the retracted antennal funicle and club not concealed by alae (Fig. 4.15) (Dendrophilinae and Abraeinae) | 7 |
| 4(3) | Pronotum with at least 2 costae (Onthophilinae) | 33 |
| — | Pronotum lacking costae (Tribalinae) | 34 |

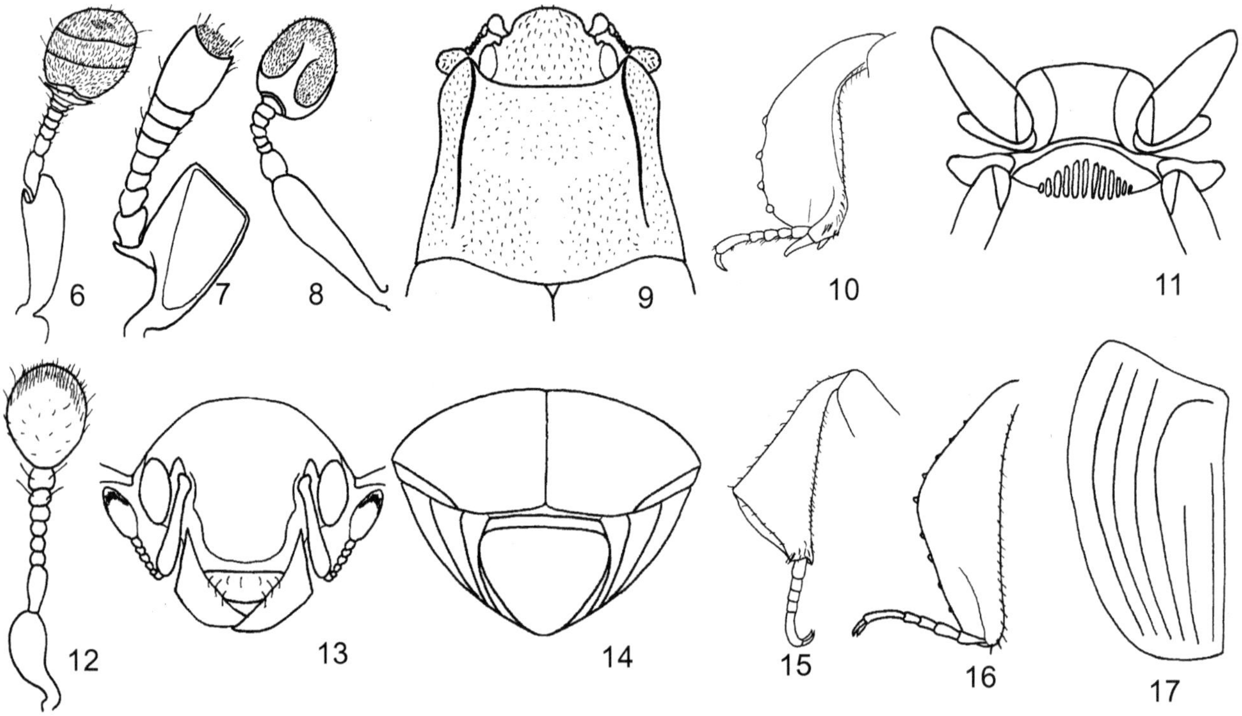


FIGURES 6.15 - 17.15. Fig. 6.15. Antenna of *Hister coenosus* Erichson; Fig. 7.15. Antenna of *Hetaerius tristriatus* Horn; Fig. 8.15. Antenna of *Teretrius* sp.; Fig. 9.15. Pronotum and head of *Plegaderus* sp.; Fig. 10.15. Protibia of *Platylomalus aequalis* (Erichson); Fig. 11.15. Pro- and mesosterna of *Aeletes* sp.; Fig. 12.15. Antenna of *Aeletes* sp.; Fig. 13.15. Face of *Aeletes* sp.; Fig. 14.15. Elytra and pygidia of *Bacanius* sp.; Fig. 15.15. Protibia of *Abraeus* sp.; Fig. 16.15. Protibia of *Chaetabraeus chandleri* Mazur; Fig. 17.15. Elytron of *Xestipyge* sp.

5(2) Antennal scape strongly expanded, triangular; antennal club completely setose, without annuli; mesosternum deeply emarginate, receiving prosternal projection; body form more or less cylindrical, approximately 5-6 mm in length; all femora and tibiae flat and expanded, nearly oval; meso- and metatibiae with apical tarsal furrows .......................................................... *Yarmister*

— Not matching this description; if antennal scape triangular, then club laterally sclerotized and mesosternum truncate or projecting anteriorly ... 6

6(5) Antennal scape cylindrical or slightly expanded apically (Fig. 6.15); protibiae with apical spurs; antennal club round or oval, laterally setose, annuli usually visible; labrum free (Histerinae) .. 38

— Antennal scape expanded apically, nearly triangular (Fig. 7.15); protibiae without apical spurs; antennal club variously sclerotized laterally, annuli obsolete; labrum fused to epistoma (though a suture may be visible) (Hetaeriinae) .................................................... 49

7(3) Body form cylindrical, parallel-sided; labrum plurisetose; mesosternum with acute anterior projection received by prosternal emargination; antennal club with medial longitudinal sclerotization (Fig. 8.15); mesepimeron produced dorsally between pronotum and elytron, visible from above; elytral striae obsolete (Teretriini) ......... 8

— Body form variable, generally round, oval or depressed; labrum bi- to multisetose; anterior mesosternal margin outwardly or inwardly arcuate, not acutely projecting; antennal club variable; mesepimeron not visible from above; elytral striae present or absent ................................. 9

8(7) Elytral length along midline approximately equal to that of pronotum; pygidium often with upper part convex or acute, lower part almost flat or concave; frons modified in several species, either strongly concave or with unusually deep antennal insertions ................................ *Teretriosoma*

— Elytral length along midline longer than that of pronotum; pygidium evenly convex; frons flat or evenly convex .................................... *Teretrius*

9(7) Pronotal disk with 2 longitudinal grooves, these frequently connected by a transverse groove (Fig. 9.15); prosternum broad, deeply excavate in a somewhat x-shaped pattern, densely setose within excavation; procoxae small, round, widely separated ........................................ *Plegaderus*

— Pronotal disk without longitudinal or transverse grooves; prosternum not excavate; procoxae transverse ...................................................... 10

10(9) Protibia with 2 apical spurs, the largest curved and nearly perpendicular to tibial axis (Fig. 10.15); elytra striate or not; generally greater than 2 mm in length ........................................................ 18

— Protibial spurs absent; elytra without striae; generally less than 2 mm in length ......................... 11

11(10) Antennal club with distinct, inwardly directed annuli; antennae inserted under frontal protuber-

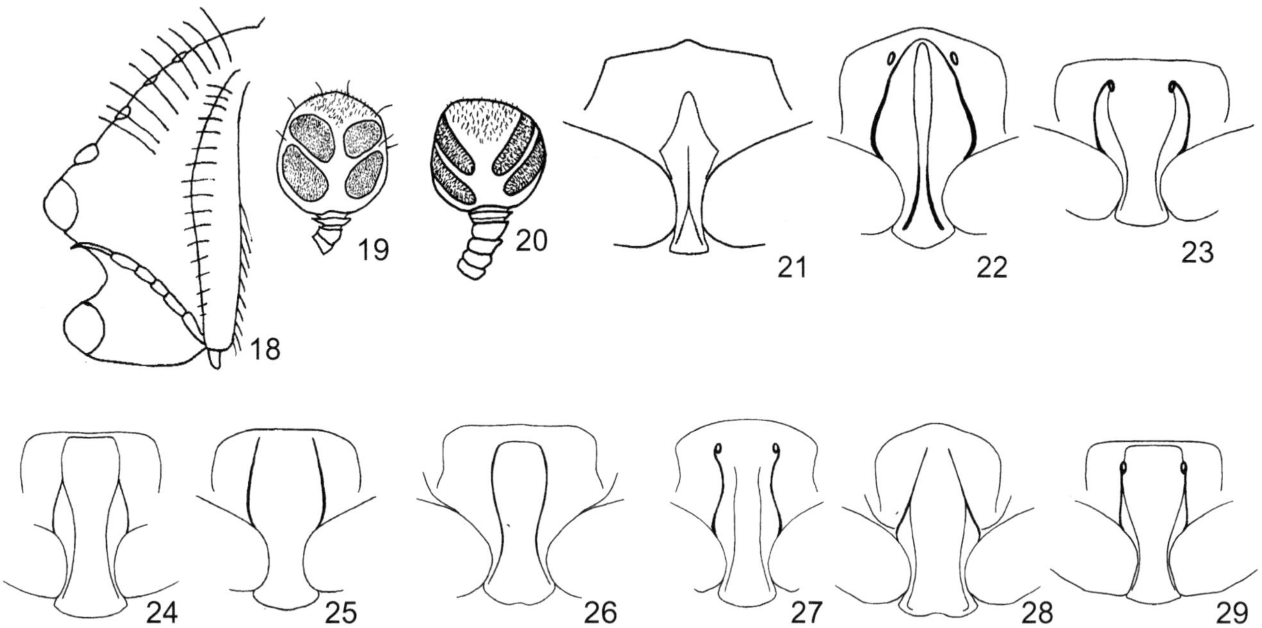


FIGURES 18.15 - 29.15. Fig. 18.15. Protibia of *Philoxenus desertorum* Mazur; Fig. 19.15. Antennal club of *Saprinus alienus* LeConte; Fig. 20.15. Antennal club of *Xerosaprinus* sp.; Fig. 21.15. Prosternum of *Neopachylopus sulcifrons* (Mannerheim); Fig. 22.15. Prosternum of *Hypocaccus* sp.; Fig. 23.15. Prosternum of *Xerosaprinus* sp.; Fig. 24.15. Prosternum of *Saprinus* sp.; Fig. 25.15. Prosternum of *Aphelosternus interstitialis* (LeConte); Fig. 26.15. Prosternum of *Eremosaprinus* sp.; Fig. 27.15. Prosternum of *Geomysaprinus* sp.; Fig. 28.15. Prosternum of *Gnathoncus* sp.; Fig. 29.15. Prosternum of *Euspilotus* sp.

ances; epipleuron separated from elytral dorsum by a distinct carina; protarsal furrow along outer margin of protibia; hind tarsi 5-segmented ........ ................................................................ *Anapleus*

— Antennal club sutures faint or obsolete, if visible, then outwardly arcuate; frons generally not protuberant over antennal insertions; elytron carinate or not; protarsal furrow either on apicolateral margin or anterior surface of tibia; hind tarsi 4 or 5-segmented .................................................... 12

12(11) Hind tarsi 4-segmented; upper and lower surfaces of antennal club with round, mostly glabrous sclerotized areas at base (Fig. 12.15) (Acritini) ..... 13

— Hind tarsi 5-segmented; antennal surfaces entirely setose ........................................................... 15

13(12) Scutellum hidden; epistoma with stria along lateral and frequently anterior margins (Fig. 13.15); mesosternal disk usually with row of parallel, longitudinal, sulciform punctures or sulci which extend anteriorly from mesometasternal suture (Fig. 11.15) ... *Aeletes*

— Scutellum visible, though sometimes minute; epistoma without marginal stria; mesosternum without sulciform, parallel, longitudinal punctures (in *Acritus* meso- and metasternal stria rarely may be very strongly and crenately punctate and superficially resemble sculpture found in *Aeletes)* .................... 14

14(13) Protibia distinctly expanded apically, approximately twice as long as wide, spinose along margin; mesosternum, especially mesosternal stria, anteriorly angulate; prosternal striae parallel at base, rather strongly divergent anteriorly . *Halacritus*

— Protibia slender, 3 or more times as long as wide, margin multisetose; anterior mesosternal margin outwardly arcuate, not angulate; prosternal striae variable .................................................... *Acritus*

15(12) Propygidium usually covered, or nearly covered by elytra (Fig. 14.15); pygidium inflexed and ventral or nearly so; body strongly convex, globular ............................................................ *Bacanius*

— Elytra not covering propygidium; pygidum and body shape variable ................................................. 16

16(15) Eyes absent; body form narrowly elongate oval, about twice as long as broad, moderately convex; wingless, soil-dwelling species ... *Geocolus*

— Eyes present; body form round, convex ........... 17

17(16) Protibia nearly triangular, widest just beyond middle (Fig. 15.15); frons protuberant above antennal insertions; margin of pronotum with 2 longitudinal carinae parallel; pronotum without row of basal punctures; propygidial length along midline approximately half that of pygidum; body rufescent ............................................................... *Abraeus*

— Protibia laterally rounded, widest nearer base than apex (Fig. 16.15); frons swollen, not protuberant above antennal insertions; marginal pronotal carinae divergent anteriorly; propygidial length along midline approximately equal to that of pygidum; row of punctures present along basal margin of pronotum; body brown or black .. *Chaetabraeus*

18(10) Sutures of antennal club inwardly arcuate; meso- and metatibiae expanded, similar in shape to protibia; elytron more or less margined by subcariniform outer subhumeral stria ................ ....................................................... *Dendrophilus*

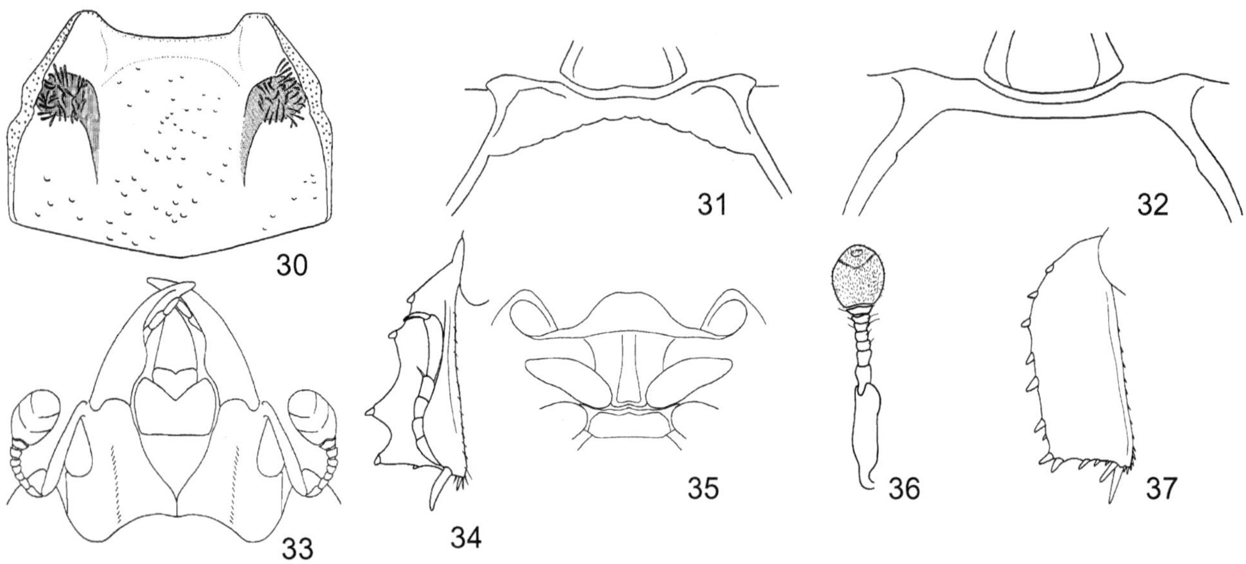


FIGURES 30.15 - 37.15. Fig. 30.15. Pronotum of *Peploglyptus belfragei* LeConte; Fig. 31.15. Meso- and metasterna of *Plagiogramma* sp.; Fig. 32.15. Meso- and metasterna of *Epierus* sp.; Fig. 33.15. Underside of head of *Hololepta* sp.; Fig. 34.15. Protibia of *Platysoma* sp.; Fig. 35.15. Prosternum of *Phelister* sp.; Fig. 36.15. Antenna of *Phelister* sp.; Fig. 37.15. Protibia of *Psiloscelis subopacus* LeConte.

— Sutures of antennal club outwardly arcuate; meso- and metatibiae slender; outer subhumeral stria abbreviated or absent, not subcariniform (Paromalini) ....................................................... 19

19(18) Elytra striate ....................................................... 20

— Elytra without complete striae, striae absent or only present as faint basal rudiments .................... 21

20(19) First tarsomere of hind tarsi about as long as next 2 combined and provided beneath with double row of stiff setae (about 5 or 6 on each side); 4th dorsal elytral stria arched to elytral suture at base, meeting sutural stria in some species (Fig. 17.15); pygidium occasionally transversely impressed or otherwise modified at apex ................*Xestipyge*

— Tarsomeres I-IV of hind tarsi subequal in length, 1st with only 1-2 pairs of ventral setae; 4th dorsal elytral stria not abruptly arched inward at base, not meeting base of sutural stria; pygidium unmodified ............................................. *Carcinops*

21(19) Body broad, very flattened; prosternal striae present ....................................................... *Platylomalus*

— Body subdepressed but convex above; prosternal striae absent ..................................... *Paromalus*

Saprininae

22(1) Frontal stria distinct, separating frons from epistoma ....................................................................... 23

— Frontal stria absent at middle, frons and epistoma confluent ....................................................... 26

23(22) Pronotal hypomeron setose ............................... 24

— Pronotal hypomeron bare ................................... 25

24(23) Protibiae strongly bidentate (Fig. 18.15); prosternal carinal striae visible along most of keel, anteriorly obsolete in some individuals; lateral prosternal striae variable, absent in some specimens .................................................. *Philoxenus*

— Protibiae with 2 strong apical marginal teeth, but with basal marginal teeth also moderately produced; prosternal keel with carinal striae absent, lateral striae distinct ....................... *Monachister*

25(23) Prosternum carinate, lateral striae weakly impressed, carinal striae visible at base only (Fig. 21.15); meso- and metatibiae densely spinose; Pacific coast only ................................. *Neopachylopus*

— Prosternum not carinate, lateral striae distinct, joined at prosternal apex; prosternal carinal striae more or less parallel, separate throughout length (Fig. 22.15); meso- and metatibiae variable, usually biseriately spinose; geographically widespread ..... ........................................................ *Hypocaccus*

26(22) Underside of antennal club with 4 distinct sensory plaques ('Reichardt's organ') (Figs. 19.15, 20.15) ....................................................................... 27

— Underside of antennal club with a single poorly defined sensory plaque or uniformly setose ..... 28

27(26) Preapical prosternal foveae present; prosternal keel with lateral prosternal striae terminating in foveae, carinal striae ascending to meet lateral striae in or slightly posterior to foveae (Fig. 23.15); sensory plaques elongate (Fig. 20.15); pronotal hypomeron setose ........................*Xerosaprinus*

— Preapical prosternal foveae absent; prosternal keel with carinal striae joining lateral striae, the single united striae continuing anteriorly around margin (Fig. 24.15); pronotal hypomeron setose or not ........................................................ *Saprinus*

28(26) Prosternal keel with only 1 pair of striae, either carinals or laterals (Figs. 25.15, 26.15) ........... 29
— Prosternal keel with 2 pairs of striae, these variously united or free (Figs. 27.15-29.15) ......... 30

29(28) Prosternal keel with lateral striae only, these descending slightly to anterior prosternal margin (Fig. 25.15); body elongate oval ........................ *Aphelosternus*
— Prosternal keel with carinal striae only; prosternal keel flat to weakly convex between them, these striae joining, ending freely, or terminating in preapical foveae (Fig. 26.15); body rounded ..... .................................................. *Eremosaprinus*

30(28) Tarsal claws of unequal length, outer claw 0.25 to 0.33 length of inner ........................ *Chelyoxenus*
— Tarsal claws of equal length .............................. 31

31(30) Carinal prosternal striae parallel or convergent, ending anteriorly without joining lateral striae (Fig. 27.15) ........................................ *Geomysaprinus*
— Carinal prosternal striae meeting lateral striae .. 32

32(31) Carinal prosternal striae widely separated at base, parallel to near middle of prosternum, then abruptly converging to apex; lateral prosternal striae descending to join carinals near base (Fig. 28.15); prosternal keel more or less flat between striae; preapical prosternal foveae absent ........ ......................................................... *Gnathoncus*
— Carinal prosternal striae closer at base, diverging in front of coxae to join lateral striae, the united striae ending anteriorly in preapical prosternal foveae (Fig. 29.15) .............................. *Euspilotus*

Onthophilinae

33(4) Pair of trichome-lined foveae anterolaterally on pronotal disc (Fig. 30.15) and prosternal keel; prosternal lobe produced; carinal prosternal striae present; umbilicately punctate ..... *Peploglyptus*
— Pronotum and prosternal keel lacking foveae; prosternal lobe not produced; carinal prosternal striae absent; sternum with coarse or fine non-umbilicate punctures ................................................... *Onthophilus*

Tribalinae

34(4) Antennae much longer than head capsule width; elytra feebly costate (western U.S.) .................. *Stictostix*
— Antennae less than head capsule width; elytra lacking costae ............................................................. 35

35(34) Elytra with discal striae ...................................... 36
— Elytra lacking discal striae; dorsal body outline nearly circular ................................ *Caerosternus*

36(35) Meso-metasternal stria present in addition to marginal metasternal stria (Fig. 31.15); marginal metasternal stria often evanescent medially .... .................................................. *Plagiogramma*
— Meso-metasternal stria absent (Fig. 32.15); marginal metasternal stria not evanescent medially .... 37

37(36) Pronotum with marginal stria only .............. *Epierus*
— Pronotum with lateral and marginal striae; elytra with sutural and 5th dorsal striae united anteriorly ... ....................................................... *Pseudepierus*

Histerinae

38(6) Antennal club annuli strongly inwardly directed, more or less V-shaped (Fig. 33.15); protarsal furrow sinuate (Fig. 34.15) .................................. 39
— Antennal annuli inwardly or outwardly arcuate, not V-shaped, obsolete in a few species; protarsal furrow straight ................................................ 42

39(38) Body strongly convex, round; elytral sutures finely impressed .......................................... *Omalodes*
— Body either more or less rectangular and depressed or cylindrical .................................................. 40

40(39) Mentum expanded laterally, concealing maxillary bases; head porrect (Fig. 33.15); mandibles strongly projecting; anterior corners of frons often acutely projecting; body large (greater than 6 mm in length), flat (Hololeptini) .................... 41
— Mentum not expanded, cardo and stipes visible; head deflexed; body smaller, depressed or cylindrical ................................................ *Platysoma*

41(40) Prosternum not carinate, rounded or truncate anteriorly; teeth on middle and hind tibiae unequally spaced, lower 2 arising from same process, more distant from upper than from each other ........... ............................................................ *Hololepta*
— Prosternum carinate, acutely produced anteriorly; teeth of hind and middle tibiae equally spaced .............................................................. *Iliotona*

42(38) Mesosternum sinuate anteriorly, projecting medially; prosternum emarginate at base (Fig. 35.15) ....................................................................... 43
— Mesosternum truncate or emarginate anteriorly; prosternum not emarginate ........................... 44

43(42) Body with metallic bluish iridescence; prosternal striae parallel to slightly divergent anteriorly; antennal club without visible annuli ......... *Baconia*
— Body rufescent to brown or black, without bluish tinge; prosternal striae convergent anteriorly in most specimens; antennal club with distinct inwardly arcuate subapical annulus (Fig. 36.15) ... ............................................................ *Phelister*

44(42) Elytron with epipleuron and dorsum separated by a cariniform stria; pronotum weakly explanate; antennal club without annuli ................. *Tribalister*
— Epipleuron and dorsum of elytron not separated by cariniform stria; pronotum convex; antennal club annulate (Histerini) ......................................... 45

45(44) Meso- and metatibiae broad and flat, more or less rectangular (Fig. 37.15); body surface, particularly frons, densely punctate, generally rufescent ......................................................... *Psiloscelis*
— Meso- and metatibiae slender or expanded only apically, not flat; body surface at most sparsely punctate, black or with red elytral maculations. ....................................................................... 46

46(45) Outer subhumeral stria complete (rarely with entire dorsal surface puncate with interspersed smooth bosses); protibiae generally shallowly excavate

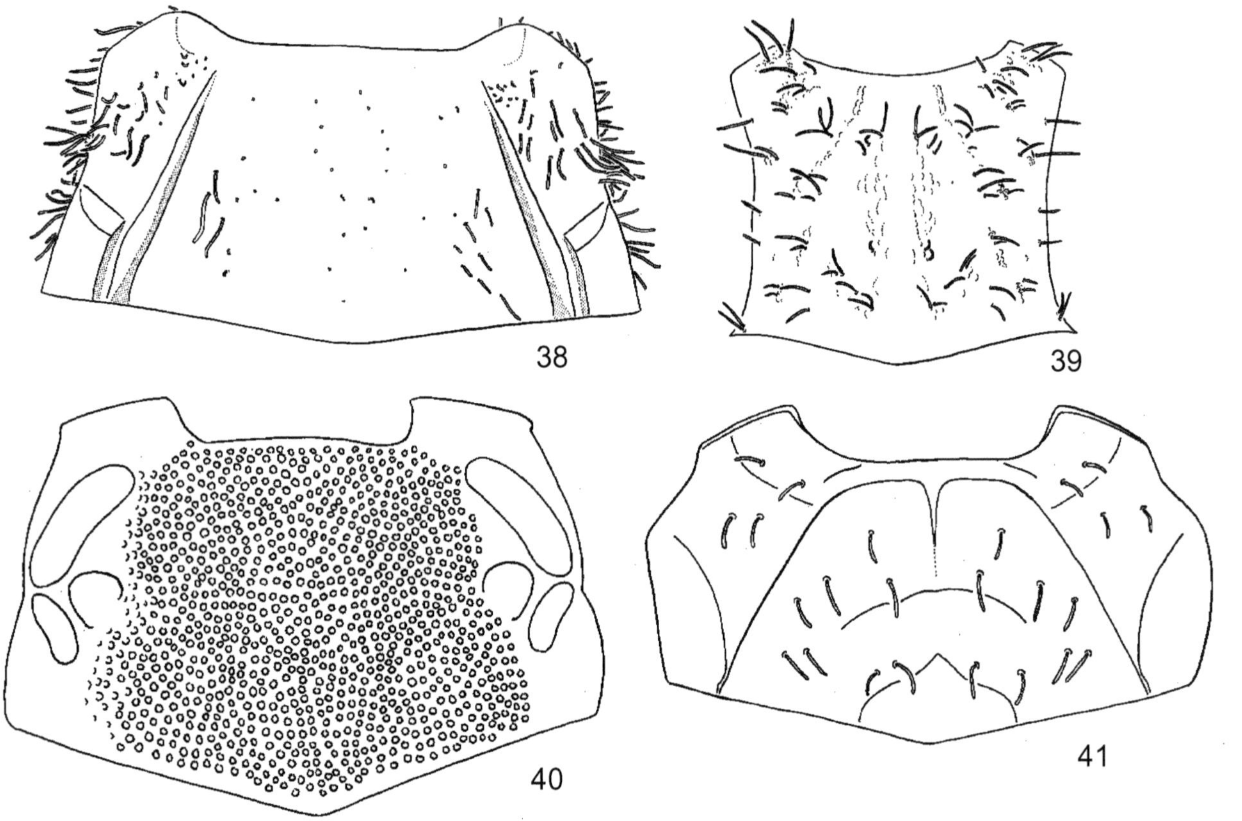


FIGURES 38.15 - 41.15. Fig. 38.15. Pronotum of *Hetaerius blanchardi* LeConte; Fig. 39.15. Pronotum of *Chrysetaerius* sp.; Fig. 40.15. Pronotum of *Terapus* sp.; Fig. 41.15. Pronotum of *Ulkeus* sp.

on anterior face between teeth; aedeagus with proximal apodemes short, gonopore tube elongate, sclerotized, gonopore distal; median lobe with dorsally articulating armature; spermathecae expanded apically, somewhat coiled at base .... .................................................... *Margarinotus*

— Outer subhumeral stria abbreviated or absent; anterior protibial excavations not present; aedeagus variable, gonopore usually not distal, armature, if present, may articulate laterally rather than dorsally; spermathecae elongate, sometimes weakly expanded apically, not coiled at base ................ 47

47(46) Pronotum with a single lateral stria, the pronotal disk punctate in anterior angles .............. *Merohister*

— Pronotum with 2 lateral striae, the outer of which may be abbreviated; pronotal disk generally smooth or very finely punctate in anterior corners ................................................................ 48

48(47) Mesosternum emarginate, sometimes weakly, receiving prosternal projection.................. *Hister*

— Mesosternum truncate, not receiving prosternal projection .................................................... *Atholus*

Hetaeriinae

49(6) Antennal club cylindrical (Fig. 7.15), sensilla restricted to truncated apex; pronotum with glandular lobe or lobes laterally which are demarcated from disk by an oblique furrow (Fig. 38.15); mesosternum usually with deep fossa on each side in antero-lateral angle .............................. *Hetaerius*

— Antennal club oval, sensilla present along apical and lateral margins (and occasionally on dorsoapical surface); mesosternum lacking fossae ............ 50

50(49) Pronotum subcylindrical (Fig. 39.15), without well-defined lateral margins (though a very fine subcariniform marginal stria is present), lateral angles inwardly arcuate, narrowest at middle; meso- and metafemora and tibiae cylindrical .... *Chrysetaerius*

— Pronotum with distinct lateral margins .................. 51

51(50) Pronotum densely punctate and bearing series of lateral protuberances (Fig. 40.15); elytra lacking discal striae; hind tibia expanded apically and with inner surface concave (Fig. 1.15)...................... *Terapus*

— Pronotum lacking lateral protuberances; elytra with discal striae; hind tibia not apically expanded and inner surface not concave .............................. 52

52(51) Pronotum, elytra, and pygidia lacking setae or trichomes ........................................................... 53

— Pronotum, elytra, and pygidium with setae or trichomes ........................................................................ 54

53(52) Elytron with 4 distinct outer dorsal striae .... *Reninus*

— Elytron with at most 3 poorly impressed outer dorsal striae ...................................................... *Euclasea*

54(52) Elytron with 4 dorsal striae and sutural stria entire; pronotum subdivided into broad discal and narrow lateral areas by a pair of fine striae (Fig. 41.15) ..... .................................................................. *Ulkeus*

— Elytron with at most a single distinct dorsal stria; pronotum not subdivided ........... *Mroczkowskiella*

### Classification of the Nearctic Genera

Histeridae

Abraeinae

Abraeini

*Chaetabraeus* Portevin 1929, 1 sp., *C. chandleri* Mazur 1991, central U.S. Inhabits moist accumulated organic debris in hollow oak trees.

*Abraeus* Leach 1817, 1 sp., *A. bolteri* LeConte 1880, California to Washington.

*Plegaderus* Erichson 1834, 12 spp., widely distributed. Some species are found in galleries of bark beetles, chiefly those attacking pine. They are predacious on bark beetle larvae.

Acritini

*Acritus* LeConte 1853, widely distributed. Found in association with rotting wood and in leaf litter. One species, *Acritus attaphilus*, inhabits refuse chambers of *Atta texana.*

subgenus *Acritus s. str.*, 10 sp.

subgenus *Pycnacritus* Casey 1916, 3 sp.

*Aeletes* Horn 1873, eastern United States to Texas and Arizona. Found beneath bark and in leaf litter.

subgenus *Aeletes s. str.*, 5 sp.

subgenus *Acritinus* Casey 1916, 2 sp.

*Halacritus* Schmidt 1793, 3 spp., West Virginia, Florida, and California. Includes species associated with wrack on coastal beaches.

*Paracritus* Brethes 1823.

Teretriini

*Teretrius* Erichson 1834, 8 spp., widely distributed. Predators of wood-boring Coleoptera. One species, *T. americanus* LeConte, has been found in association with wood of *Fagus* sp.

*Teretriosoma* Horn 1873, 5 spp., Florida and southwestern United States. Predators of wood-boring Coleoptera. Some species are found in association with wood of *Metopium* and *Leucaenea* spp.

Saprininae

*Hypocaccus* Thomson 1867, 37 spp., widely distributed; some species associated with lotic bars and beaches; others found on beaches of lakes and oceans.

subgenus *Hypocaccus s. str.*, 30 described and 4 undescribed spp., including 1 inhabiting the nests of *Pogonomyrmex* spp.

subgenus *Baeckmanniolus* Reichardt 1926, 3 spp., Atlantic and Pacific sandy coastal shores.

*Neopachylopus* Reichardt 1926, 1 sp., *N. sulcifrons* (Mannerheim 1843), a wrack associate of Pacific coastal shores.

*Monachister* Mazur 1991, 1 sp., *M. californicus* Mazur 1991, inhabits dune areas in southern California.

*Philoxenus* Mazur 1991, 1 sp., *P. desertorum* Mazur 1991, southern California. This species is a dune inhabitant.

*Geomysaprinus* Ross 1940, widely distributed.

subgenus *Geomysaprinus s. str.*, 4 described and 1 undescribed spp. Obligate inhabitants of *Geomys* spp. burrows.

subgenus *Priscosaprinus* Wenzel 1962, 23 described and 5 undescribed spp. Inhabiting burrows of *Thomomys*, *Geomys*, and *Cynomys* spp. One undescribed species inhabits nests of *Atta texana* (Moser 1963).

One undescribed sp. (incertae sedis) is an obligate burrow inabitant of *Thomomys* and *Spermophilus* spp.

*Xerosaprinus* Wenzel 1962, 18 spp., widely distributed. Partial to xeric habitats and attracted to dung and carrion.

subgenus *Auchmosaprinus* Wenzel 1962, 1 sp.

subgenus *Vastosaprinus* Wenzel 1962, 2 spp.

subgenus *Lophobregmus* Wenzel 1962, 1 sp.

subgenus *Xerosaprinus s. str.*, 14 spp.

*Saprinus* Erichson 1834, 9 spp., widely distributed; found in association with carrion and dung; some species inhabit rodent burrows, including those of *Cynomys* spp. One species, *S. pensylvanicus* (Paykull), is apparently restricted to sandy soils.

*Euspilotus* Lewis 1907, 28 spp., widely distributed. Attracted to carrion and dung; 1 species, *E. rossi*, known from flicker nests, another inhabits the burrows of *Geomys* sp.

subgenus *Hesperosaprinus* Wenzel 1962, 25 spp.

subgenus *Neosaprinus* Bickhardt 1909, 2 spp.

1 undescribed sp. (*incertae sedis*) from pocket gopher burrows

*Chelyoxenus* Hubbard 1894, 1 sp., *C. xerobatis* Hubbard 1894, Florida, Georgia, and South Carolina; an obligate burrow inhabitant of the gopher tortoise, *Gopherus polyphemus*.

*Aphelosternus* Wenzel 1962, 1 sp., *A. interstitialis* (LeConte 1851), California. Burrow inhabitants of *Spermophilus* spp. (D. Verity, pers. comm.).

*Gnathoncus* Duval 1858, 5 spp., generally distributed; found in bird nests and roosts and in hollow trees.

*Eremosaprinus* Ross 1939, 9 described and 4 undescribed spp., from California, Colorado, Arizona, and New Mexico. Obligate burrow inhabitants of *Dipodomys* spp.

Dendrophilinae

Dendrophilini

*Dendrophilus* Leach 1817, 5 spp., widely distributed. Found in relatively dry accumulated organic debris in hollow trees including *Quercus*, *Fagus*, *Magnolia*, and *Acer*. In some cases trees contained nests, including those of *Neotoma* spp. *Dendrophilus* spp. have been taken from gull nests and chicken dung.

Anapleini

*Anapleus* Horn 1873, 2 sp., 1 eastern, 1 southwestern United States. Eastern species, *A. marginatus*, usually inhabits riparian litter.

Bacaniini

*Bacanius* LeConte 1853, 7 spp., eastern United States and western coastal states. Commonly found in leaf litter or in association with decaying wood.

subgenus *Gomyister* Mazur 1984, 3 spp.

subgenus *Bacanius s. str.*, 4 spp.

*Geocolus* Wenzel 1944, 1 sp., *G. caecus* Wenzel 1944, Georgia and Alabama. A hypogeal species, collected in soil or in leaf litter at cave entrances.

Paromalini

*Platylomalus* Cooman 1948, 1 sp., *P. aequalis* (Erichson,1834), eastern United States to the Rocky Mountains. This species occurs beneath bark of decaying trees including *Liriodendron, Ulmus,* and *Populus.* They are predators of subcortical insects, especially dipteran larvae.

*Paromalus* auctorum, not Erichson 1834

*Paromalus* Erichson 1834, 8 spp., widely distributed, mostly under bark; 1 species, *P. seeversi* Wenzel, inhabits accumulated moist organic debris within hollow trees including *Quercus* and *Acer*. Predators of fly eggs and larvae.

*Microlomalus* Lewis 1907.

subgenus *Isolomalus* Lewis 1907, 2 sp.

subgenus *Paromalus s. str.*, 6 spp. They are predators of fly eggs and larvae.

*Carcinops* Marseul 1855, 14 spp., widely distributed, mostly southwestern United States.

subgenus *Carcinops s. str.*, 13 spp.

subgenus *Carcinopsida* Casey 1916, 1 sp., *C. opuntiae* (LeConte), found in association with rotting cacti including Saguaro.

*Xestipyge* Marseul 1862, 2 spp., eastern United States to Texas. Inhabiting rotting bales of hay (K. Stephan, pers. comm.) and leaf litter. They are attracted to dung and feed on fly eggs and larvae.

Onthophilinae

*Onthophilus* Leach 1817, 16 spp., widely distributed; 6 described and 4 undescribed species are obligate inhabitants of geomid rodent burrows; 1 species is restricted to burrows of *Cynomys* spp.; the remainder have been found associated with dung, carrion, tree wounds, nests of mice and *Neotoma* sp., rotting fungi, and plant material (Helava 1978). Adult activity and reproduction apparently coincides with cooler months of the year. Adults are filter feeders.

*Peploglyptus* LeConte 1880, 1 sp., *P. belfragei* LeConte 1880. Texas, Oklahoma, Arkansas, and Kansas. This species inhabits riparian areas. The trichome-lined fossae on the pronotum and prosternum suggest an association with ants (Kanaar 1981). One specimen was taken with ants from a rotting log near a creek (K. Stephan, pers. comm.)

Tribalinae

*Plagiogramma* Tarsia in Curia 1935, 1 sp., *P. subtropica* (Casey 1893), Mississippi, Florida. Adults are microphagous and fungal spore specialists; they inhabit riparian leaf litter.

*Epierus* Erichson 1834, 8 spp., 1, *E. pulicarius* Erichson widely distributed in eastern United States, the others chiefly southern and southwestern. Adult *Epierus* are both macrophagous predators and microphagous spore specialists (Wenzel in Moser *et al.* 1971). Adults of most species inhabit humus accumulating beneath the bark of decaying tree trunks. They actively tunnel through this material creating a network of stable runways. One species also has been collected from accumulated organic debris within tree cavities.

*Pseudepierus* Casey 1916, 1 sp., *P. gentilis* (Horn 1883), South Carolina, Indiana, Illinois, Kansas, Louisiana, and Arizona. Collected from leaf litter. Adults feed on fungal spores.

*Stictostix* Marseul 1870, 3 described and 1 undescribed spp. from western coastal states and Arizona, Idaho, Utah, and Oklahoma. One species frequents riparian habitats and has been found in association with ants; another was collected in the burrows of *Cynomys ludovicianus.*

*Caerosternus* LeConte 1852, 1 sp., *C. americanus* (LeConte 1845), eastern United States to east Texas. This species is nocturnally active and found in association with decaying hardwoods, pines, and lumber mill sawdust piles. Adults selectively feed on hyphomycete fungal spores.

[*Idolia* Lewis 1885, was previously reported from Texas. However, we have not confirmed any records north of the Mexican border.]

Histerinae

*Yarmister* Wenzel 1939, 1 sp., *Y. barberi* Wenzel 1939, Florida. The tribal affiliation of this monotypic genus has not been established. At the time of its description Wenzel raised the possibility that it is termitophilous.

Exosterinini

*Phelister* Marseul 1853, 13 described and 2 undescribed spp. from *Geomys* burrows. Eastern United States to Texas and Arizona. Attracted to dung and carrion.

*Baconia* Lewis 1885, 3 sp., 1 undescribed, eastern United States. Generally found in savannas beneath bark of decaying trees, especially *Quercus.*

*Tribalister* Horn 1873, 2 spp., Maryland, D. C., Indiana, Florida, Rhode Island, and southern California. Members of this genus are partial to sandy areas and are apparently nest inhabitants of *Aphenogaster* spp. (R. Wenzel, pers. comm.).

[*Pseudister* Bickhardt 1917, is not North American; *P. hospes* (Lewis 1902) reported from New York belongs to *Phelisteroides* and is not North American, but Brazilian.]

Omalodini

*Omalodes* Erichson 1834, 2 spp., Arizona, New Mexico and Florida. Found in association with moist rotting vegetation including gumbo limbo (M. Thomas, pers. comm.) and sotol.

Platysomatini

*Platysoma* Leach 1817, 14 spp., widely distributed. Found under tree bark. Predators of Diptera and Coleoptera larvae.
*Abbotia* Leach 1830.
subgenus *Platysoma s. str.*, 4 spp.
subgenus *Eurylister* Bickhardt 1920, 1 sp.
subgenus *Cylister* Cooman 1941, 5 spp.
subgenus *Cylistix* Marseul 1857, 4 spp.

Hololeptini

*Hololepta* Paykull 1811, widely distributed.
*Lionota* Marseul 1853.
*Lioderma* Marseul 1857.
subgenus *Hololepta s. str.*, 3 spp., found beneath bark of decaying trees including *Liriodendron*, *Populus*, *Ulmus*, and *Gleditsia*.
subgenus *Leionota* Dejean 1837, 7 spp., found in rotting vegetation including cactus, agave, and palms.

*Iliotona* Carnochan 1917, 1 sp., *I. cacti* LeConte 1851, Texas, New Mexico, Arizona, and southern California. Found in association with rotting cactus.

Histerini

*Atholus* Thomson 1862, 15 spp., widely distributed. One species inhabits *Geomys* burrows.
*Atholister* Reitter 1909.

*Margarinotus* Marseul 1853, 30 described spp., widely distributed, chiefly in deciduous forests or mixed forests containing oaks; primarily in mammal burrows and nests, including those of *Neotoma*, *Marmota*, and *Geomys*; others attracted to fungi; a few common species in decaying vegetation and carrion; adults of some species are predators on larvae of Noctuidae.
subgenus *Margarinotus s. str.*, 1 sp.
subgenus *Paralister* Bickhardt 1916, 5 spp.
subgenus *Stenister* Reichhardt 1920, 1 sp. (*M. obscurus* [Kugelann], introduced).
subgenus *Ptomister* Houlbert et Monnot 1922, 15 spp.
subgenus *Promethister* Kryzhanovskij 1966, 2 spp.
incertae sedis, 6 described and 1 undescribed spp.

*Hister* Linnaeus 1761, 33 spp., widely distributed, chiefly in forested areas; adults of a few species are found on carrion, others in fungi and mammal burrows, including nests of *Neotoma* sp.; a few species are restricted to the debris piles of leafcutter ants (*Atta* sp.); 1 species, *H. indistinctus* is commonly found in hollow trees; larvae and adults are predacious.
subgenus *Spilodiscus* Lewis 1906, 9 described and 1 undescribed spp., generally distributed in sandy areas, including sandy coasts of eastern and western United States. Several species are obligate inhabitants of pocket gopher burrows.
subgenus *Hister s. str.*, 25 spp.

*Merohister* Reitter 1909, 2 spp., *M. osculatus* (Blatchley) is generally found in association with fungi while *M. grandis* inhabits accumulated wet organic debris in hollow mature deciduous trees including *Fagus*, *Magnolia*, and *Celtis*. Both species occur in the central and eastern U.S.

*Psiloscelis* Marseul 1853, 6 spp., widely distributed in eastern North America, in the west restricted to higher elevation areas of the Rocky Mts. and Sierra Nevada. Though most species are associated with ants (primarily *Formica* spp.), *P. corrosa* Casey is known from *Spermophilus* burrows.

### Hetaeriinae

*Hetaerius* Erichson 1834, 23 described and 2 undescribed spp., widely distributed, chiefly western, with *Formica*, *Lasius*, and other ants.

*Mroczkowskiella* Mazur 1984, 3 spp., Florida, Alabama, Georgia, Texas, and Arizona; nest inhabitants of *Pheidole* spp.
*Echinodes* Zimmerman 1869.

*Reninus* Lewis 1889, 1 sp., *R. salvini* (Lewis 1888), Texas. Nest inhabitants of *Atta texana*.
*Renia* Lewis 1885, not Guenée 1854.

*Ulkeus* Horn 1885, 6 sp. (5 of which are undescribed), North Carolina, Tennessee, and Florida west to Texas and Arizona; guests of army ants (*Neivamyrmex* spp.)

*Terapus* Marseul 1862, 2 described spp. from Arizona and California and 3 undescribed spp. from Florida, Louisiana, Texas, and Arizona; guests of *Pheidole* spp.

*Chrysetaerius* Reichensperger 1923, 1 sp., undescribed, Arizona. Possibly a *Neivamyrmex* associate.

*Euclasea* Lewis 1888, 2 undescribed spp. from the southwest (Helava et al. 1985) and 1 from Florida.

### Bibliography


BICKHARDT, H. 1916. Histeridae, in P. Wytsman (ed.): Genera Insectorum, fasc. 166a.: pp. 1-112. La Haye.

BICKHARDT, H. 1917. Histeridae, in P. Wytsman (ed.): Genera Insectorum, fasc. 166b: pp 113-302. La Haye.

BØVING, A. G. and CRAIGHEAD, F. C. 1931. An Illustrated Synopsis of the Principal Larval Forms of the Order Coleoptera. p. 31, pls. 20, 21.

CARLTON, C. E., LESCHEN, R. A. B., and KOVARIK, P. W. 1996. Predation on adult blow flies by a Chilean hister beetle, *Euspilotus bisignatus* (Erichson). Coleopterists Bulletin, 50: 154.

CARNOCHAN, F. G. 1917. Hololeptinae of the United States. Annals of the Entomological Society of America, 10: 367-398.

CASEY, T. L. 1893. Coleopterological notices V. Annals of the New York Academy of Science, 7: 535-561.

CASEY, T. L. 1916. Memoirs on the Coleoptera, VII: 401-292. Lancaster.

CATERINO, M. S. 1998. A phylogenetic revision of *Spilodiscus* Lewis. Journal of Natural History, 32: 1129-1168.

CATERINO, M. S. 1999. The taxonomy and phylogenetics of the *coenosus* group of *Hister* Linnaeus. University of California Publications in Entomology, 119: 1-75 + plates.

CATERINO, M. S. 1999. Taxonomy and phylogeny of the *Hister servus* group: a Neotropical radiation. Systematic Entomology, 24: 351-376.

CROWSON, R. A. 1974. Observations on the Histeroidea, with descriptions of an apterous larviform male and of the internal anatomy of a male Sphaerites. Journal of Entomology (B), 42: 133-140.

DOWNIE, N. M. and ARNETT, R. H. 1995. Histeridae Clown Beetle Family, *in* The Beetles of Northeastern North America. pp. 604-628. Gainsville: The Sandhill Crane Press.

HATCH, M. H. 1926. New and noteworthy Histeridae from Alberta. Canadian Entomologist, 58: 272-276.

HATCH, M. H. 1929. Studies on Histeridae. Canadian Entomologist, 61: 76-95.

HATCH, M. H. 1938. Records of Histeridae from Iowa. Journal of the Kansas Entomological Society, 11: 19-20.

HATCH, M. H. 1961. Sphaeritidae and Histeridae pp. 253-276. *In* Hatch, M. H. 1961. Beetles of the Pacific Northwest. Seattle: University of Washington Press.

HELAVA, J.V.T., HOWDEN, H.F., and RITCHIE, A.J. 1985. A review of the New World genera of the myrmecophilous and termitophilous subfamily Hetaeriinae. Sociobiology, 10: 127-382.

HINTON, H. E. 1945. A key to the species of North American *Terapus*, with a description of a new species. Proceedings of the Royal Entomological Society of London, 14: 38-45.

HORN, G. H. 1873. Synopsis of the Histeridae of the United States. Proceedings of the American Philosophical Society, 13: 273-360.

HORN, G. H. 1880. Contributions to the coleopterology of the United States. No. 3. Transactions of the American Entomological Society, 8: 139-154.

JOHNSON, S. A., LUNDGREN, R. W., NEWTON, A. F. JR., THAYER, M. K., WENZEL, R. L., and WENZEL, M. R. 1992. Mazur's world catalogue of Histeridae: emendations, replacement names for homonyms, and an index. Polskie Pismo Entomologiczne, 61: 3-100.

KOVARIK, P. W. 1994. Pupal chaetotaxy of Histeridae with a description of the pupa of *Onthophilus kirni* Ross. Coleopterists Bulletin, 48: 254-260.

KOVARIK, P. W. 1995. Development of *Epierus divisus* Marseul. Coleopterists Bulletin, 49: 253-260.

KOVARIK, P. W. and PASSOA, S. 1993. Chaetotaxy of larval Histeridae based on a description of *Onthophilus nodatus* LeConte. Annals of the Entomological Society of America, 86: 560-576.

KOVARIK, P.W., D. S. VERITY, and J. C. MITCHELL. 1999. Two new Saprinine histerids from southwest North America. Coleopterists Bulletin, 53(2): 187-198.

KUKALOVÁ-PECK, J. and LAWRENCE, J. F. 1993. Evolution of the hind wing in Coleoptera. Canadian Entomologist, 125: 181-258.

LAWRENCE, J. F. and BRITTON, E. B. 1994. Australian Beetles. Melbourne: Melbourne University Press.

LECONTE, J. E. 1845, A monography of the North American *Histeroides*. Boston Journal of Natural History, 5: 32-86.

LEE, R. D. and RYCKMAN, R.E. 1954. Coleoptera and Diptera reared from owl nests. Bulletin of the Brooklyn Entomological Society, 49: 23-24.

LINDNER, W. 1967. Ökologie und Larvalbiologie einheimischer Histeriden. Zeitschrift für Morphologie und Ökologie der Tiere, 59: 341-380.

MACKAY, W. P. 1983. Beetles associated with the harvester ants *Pogonomyrmex montanus, P. subnitidus*, and *P. rugosus* (Hymenoptera: Formicidae). Coleopterists Bulletin, 37: 239-246.

MAZUR, S. 1984. A world catalogue of Histeridae. Polskie Pismo Entomologiczne, 54: 1-379.

MAZUR, S. 1991. New North American histerids. Annals of Warsaw Agricultural University SGGW. Forestry and Wood Technology, 42: 89-96.

MAZUR, S. 1997. A world catalogue of Histeridae. Genus (Suppl.): 1-373.

MOSER, J. C., THATCHER, R. C., and PICKARD, L. S. 1971. Relative abundance of the southern pine beetle associates in East Texas. Annals of the Entomological Society of America, 64: 72-77.

MOSER, J. C. 1963. Contents and structure of *Atta texana* nest in summer. Annals of the Entomological Society of America, 56: 286-291.

NEWTON, A. F., Jr. 1991. Histeridae. *in* Stehr, F. W. (ed.) Immature Insects. v. 2. Dubuque: Kendall/Hunt.

OHARA, M. 1994. A revision of the superfamily Histeroidea of Japan. Insecta Matsumurana, 51: 1-283.

ROSS, E. S. 1937. Studies in the genus *Hister*. The Pan-Pacific Entomologist, 13: 106-108.

ROSS, E. S. 1939. A new subgenus of North American *Saprinus*. Pan-Pacific Entomologist, 15: 39-43.

ROSS, E. S. 1940. A preliminary review of the North American species of *Dendrophilus*. Bulletin of the Brooklyn Entomological Society, 35: 103-108.

SHARP, D. and MUIR, F. 1912. The comparative anatomy of the male genital tube in Coleoptera. Transactions of the Entomological Society of London, 1912: 477-642. [pp. 512-513, pl. 55, figs. 79-82.]

SLIPINSKY, S. A., AND S. MAZUR. 1999. *Epuraeosoma*, a new genus of Histerinae and phylogeny of the family Histeridae. Annales Zoologici (Warszawa), 49: 209-230.

SUMMERLIN, J. W., BAY, D. E., HARRIS, R. L., and STAFFORD, K. C. I. 1982. Predation by four species of Histeridae on the horn fly. Annals of the Entomological Society of America, 75: 675-677.

SUMMERLIN, J. W., FINCHER, G. T., HUNTER, J. S., and BEERWINKLE, K. R., 1993. Seasonal distribution and diel flight activity of dung attracted histerids in open and wooded pastures in East-Central Texas. Southwestern Entomologist, 18: 251-261.

WENZEL, R. L., 1936. Short studies in the Histeridae. Canadian Entomologist, 68: 266-272.

WENZEL, R. L. and DYBAS, H., 1941. New and little known Neotropical Histeridae. Fieldiana, Zoology, 22: 433-472.

WENZEL, R. L., 1944. On the classification of histerid beetles. Fieldiana, Zoology, 28: 51-151.

WENZEL, R. L., 1955. The histerid beetles of New Caledonia. Fieldiana, Zoology, 37: 601-634.

WENZEL, R. L., 1960. Three new histerid beetles from the pacific northwest, with records and synonymies. Fieldiana, Zoology, 39: 447-463.

# 16. HYDRAENIDAE Mulsant, 1844

by Philip D. Perkins

**Family common name:** The minute moss beetles

**Family synonym:** Limnebiidae Mulsant, 1844

These minute aquatic or humicolous beetles superficially resemble small hydrophiloids, with which they can be found at the margins of aquatic habitats. The maxillary palpi of hydraenids are sometimes longer than the antennae as in some hydrophiloids, and the venter normally is clothed in hydrofuge pubescence. Hydraenids are readily distinguished by the abdominal structure, having six or seven visible abdominal sterna and a small intercoxal sternite between the metacoxae. Additionally, the antennal club in many hydraenids (and all Nearctic forms) comprises five pubescent antennomeres (Fig. 5.16), whereas hydrophiloids normally have only three.

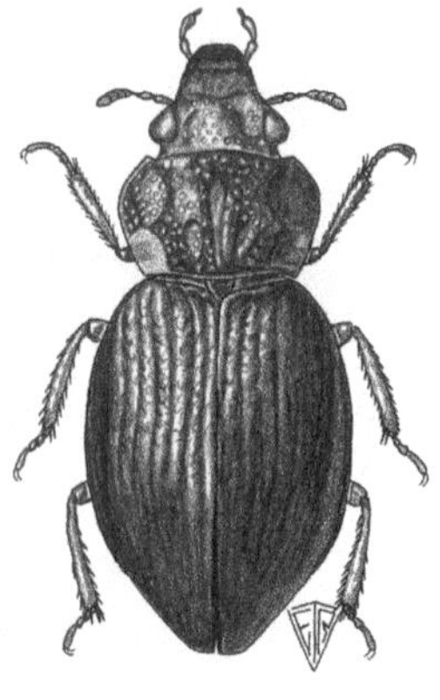

FIGURE 1.16. *Ochthebius puncticollis* LeConte.

**Description**: Shape elongate to oval, moderately convex; size ca. 0.5-3.0 mm in length; color black to testaceous, sometimes with metallic reflections; venter clothed in hydrofuge pubescence (rarely reduced); dorsum normally with sparse setae.

Head prominent, surface smooth or punctate and rugose. Eyes prominent; ocelli sometimes present near inner posterior borders of eyes. Antennae short, ca. 2/3 width of head, inserted near anterior margin of eyes, below clypeal margin; in repose held beneath eye, club positioned vertically behind eye in hypomeral antennal pocket, or more horizontally against hypomeron; antennomeres 7-11, primitively with 11, the last five forming a hydrofuge pubescent club, reductions in antennomeres involve both pre-club and club parts; Nearctic forms with 9 antennameres, 5 in club, pre-club antennomere cupuliform.

Clypeus distinct; labrum small to large, anterior border arcuate, emarginate or excavate. Mandibles concealed, except lateral surface, each with a small, chisel-shaped tooth and large grinding mola. Lacinia of maxilla small, with stout scraping processes or brushlike setae; galea variable in size, with simple setae to comblike rows of stout setae. Maxillary palpi with 4 palpomeres (first minute), proportions of palpomeres 2-4 variable, total length shorter than to much longer than antennae. Labial palpi with 3 palpomeres, small or very small, often indistinct and partly concealed by the mentum.

Thorax broader than the head, lateral borders sinuate or arcuate, rarely divergent behind; anterior and posterior borders evenly arcuate or slightly sinuate, often with narrow hyaline border. Pronotum usually with impressions or grooves. Hypomeron diverse in form and cuticular details that relate to exocrine secretion delivery systems, air capture function and protection of antennal club.

Elytra completely concealing abdomen, except sometimes apex exposed; normally with well-developed rows of punctures. Scutellum small, visible. Wings without m-cu loop, anal veins very reduced; wings sometimes reduced in size or absent.

Abdomen usually with 7 visible sterna, or with 6 visible when last ventrite (7th) is concealed by retraction. A small, normally triangular sclerite (intercoxal sternite) located between metacoxae.

Legs variable, rather short and stout to long and slender; femora and tibiae often with patches or rows of stout setae that are used in secretion grooming. Tarsi 5-5-5, the first three tarsomeres very small, last elongate, superficially tarsi therefore sometimes appearing 4-4-4 or 3-3-3.

Male genitalia of the trilobed type, without distinct basal piece, often very complex, sometimes with parameres reduced or absent. Females with more or less complex sclerotized spermatheca.

The larvae are campodeiform; mandible with basal mola; maxilla with well-developed normal galea and lacinia; palpiger normal, not appearing as a segment of the palpus. Thorax with well-developed legs. Abdomen 10-segmented, bearing urogomphi, the tenth segment with or without a pair of recurved hooks; spiracles of annular type.

**Habits and habitats.** Nearctic adults are aquatic, most species being found at the margins of streams and ponds. In western North America hydraenids are especially numerous where gravel/sand stream banks are not subject to frequent turnover, but are more or less sorted by particle size, resulting in interstitial living spaces. In these situations, adults of *Hydraena*, *Ochthebius* and *Limnebius* can be collected together and show some microhabitat partitioning (Perkins 1972, 1976). Some Nearctic species prefer brackish water or stagnant water with emergent vegetation and/or decaying leaves, a few are found associated with wet moss, and one species is intertidal (Perkins 1980). In other regions hydraenids are also found in these microhabitats as well as hygropetric (seeps and splash zones on rock faces), benthic, and humicolous microhabitats (Perkins and Balfour-Browne 1994). Larvae, with rare exceptions, are semiaquatic, being found in moist microhabitats near the adults.

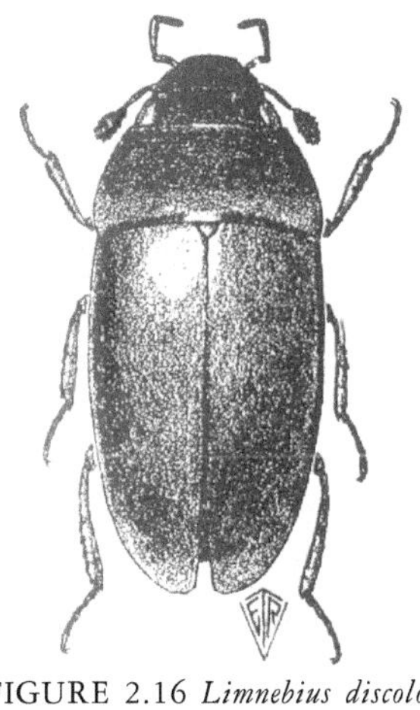

FIGURE 2.16 *Limnebius discolor* Casey

Adults and larvae of the generalist feeders graze on wet surfaces, largely those of stones, sand grains and plant matter, feeding on microscopic flora and fauna (algae, bacteria, protozoans, disintegrating particles, etc.). The mandibles of both stages have large molar grinding surfaces to process this heterogeneous material. The maxillae of adults are generally and primitively of the brush-like type, but the Ochthebiini have derived lacinia bearing enlarged, stout apical teeth. Feeding specializations undoubtedly occur within the general food list, particular to taxa and life stages, but this remains virtually unstudied.

Although 38 genera are currently known, about 85% of all described species are in three genera: *Hydraena*, *Limnebius* and *Ochthebius*. Adults of these three speciose genera each have a different type of "exocrine secretion delivery system" (ESDS) which increases the effectiveness of the respiratory bubble (e.g., Figs. 3.16, 4.16). The legs, which have specialized patches and rows of setae, are used to scrape and spread exocrine secretions along cuticular specializations that form the margins of

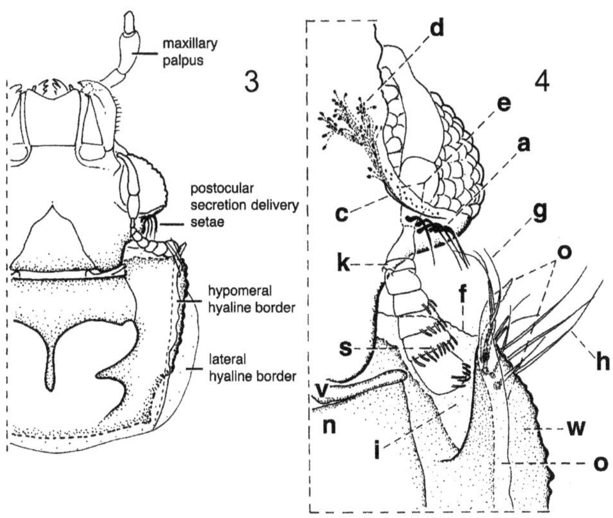


FIGURES 3.16 - 4.16. Fig. 3.16, *Ochthebius arenicolus* Perkins, ventral aspect of head and prothorax. Fig. 4.16, *Ochthebius*, ventral aspect of left side of head and adjacent area of prothorax, showing external cuticular features of antennal pocket and internal end-apparatus and ductules (d) of exocrine glands. Structures: (a) postocular secretion delivery setae, (c) postocular secretion delivery sulcus, (d) end-apparatus and ductules of exocrine glands, (e) periocular exocrine pores, (f) anterior hyaline border, (g) hypomeral antennal pocket setae, (h) anterolateral pronotal setae, (i) hypomeral antennal pocket, (k) cupule article of antenna, (n) prosternum, (o) hypomeral hyaline border, (s) sensilla of antennal club, (v) cervical sclerite, (w) wet-hypomeron (From Perkins, 1997).

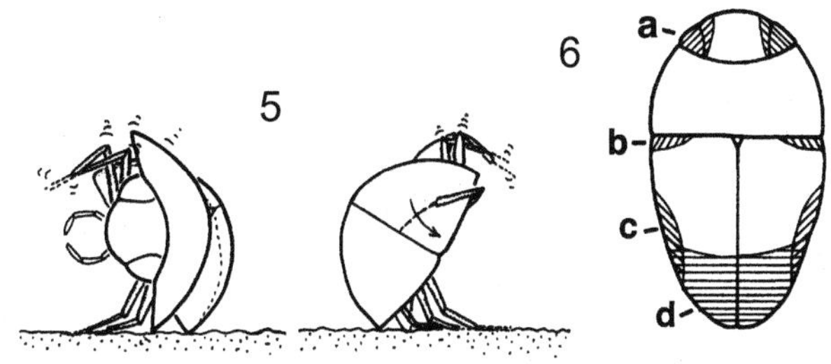


FIGURES 5.16 - 6.16, *Limnebius piceus*, schematic. Fig. 5.16, beetle balanced on edge, the tibiae of the left legs resting on the surface of a wet leaf, the right legs performing secretion-grooming, anterior (left) and posterior aspects. Fig. 6.16, dorsal areas groomed by legs: (a) protibia and protarsus, (b) protibia, (c) mesotibia, (d) metatibia and metatarsus. (From Perkins, 1997)

the respiratory bubble. This secretion grooming is performed out of the water, often with the beetle on edge (Figs. 5.16, 6.16). Streams and ponds are nearly worldwide, forming networks of microhabitats across a widely variable matrix of climatic, edaphic and biotic conditions. Among hydraenids, the three genera with specialized ESDS are overwhelmingly successful in these stream and pond microhabitats. Plesiomorphic, and less successful genera, lack specialized ESDS (Perkins 1997).

**Status of the classification.** The New World species were comprehensively revised by Perkins (1980); several new taxa collected since then are currently being described. A revised subfamily and tribal classification was recently presented, based upon a comparative morphological and behavioral study in which new character systems were described (Perkins 1997). The family is currently divided into four subfamilies and 12 tribes.

The phylogenetic position of the family has been the subject of much discussion. Hydraenids were classically grouped with the Hydrophilidae, due to similarities related to aquatic habits of adults. A pendulum-like swing between hydrophiloids and staphylinoids was started by Bøving and Craighead's (1931) monumental larval study. Crowson (most recently 1981) and Beutel (1994) argue for hydrophiloid relationships. Dybas (1976), Lawrence and Newton (1982), Newton and Thayer (1992), Hansen (1991, 1995, 1997) and Archangelsky (1998) place hydraenids as staphylinoids. It is worth noting that some of the oldest beetle fossils are hydraenids, from the lower Jurassic. The larvae retain several characteristics which were probably present in the common ancestor of staphylinoids and hydrophiloids, whereas the aquatic adaptations of adults superficially resemble those of hydrophiloids. Detailed study of hydraenid adult respiratory specializations for aquatic life (Perkins 1997) have not revealed any specialized structures or character systems that are shared with hydrophiloids. Likewise, study of the larvae of hydrophiloids has not revealed any derived characteristics which indicate a hydraenid relationship to basal hydrophiloids (Archangelsky 1998). Although subject to change, present evidence supports staphylinoid relationships.

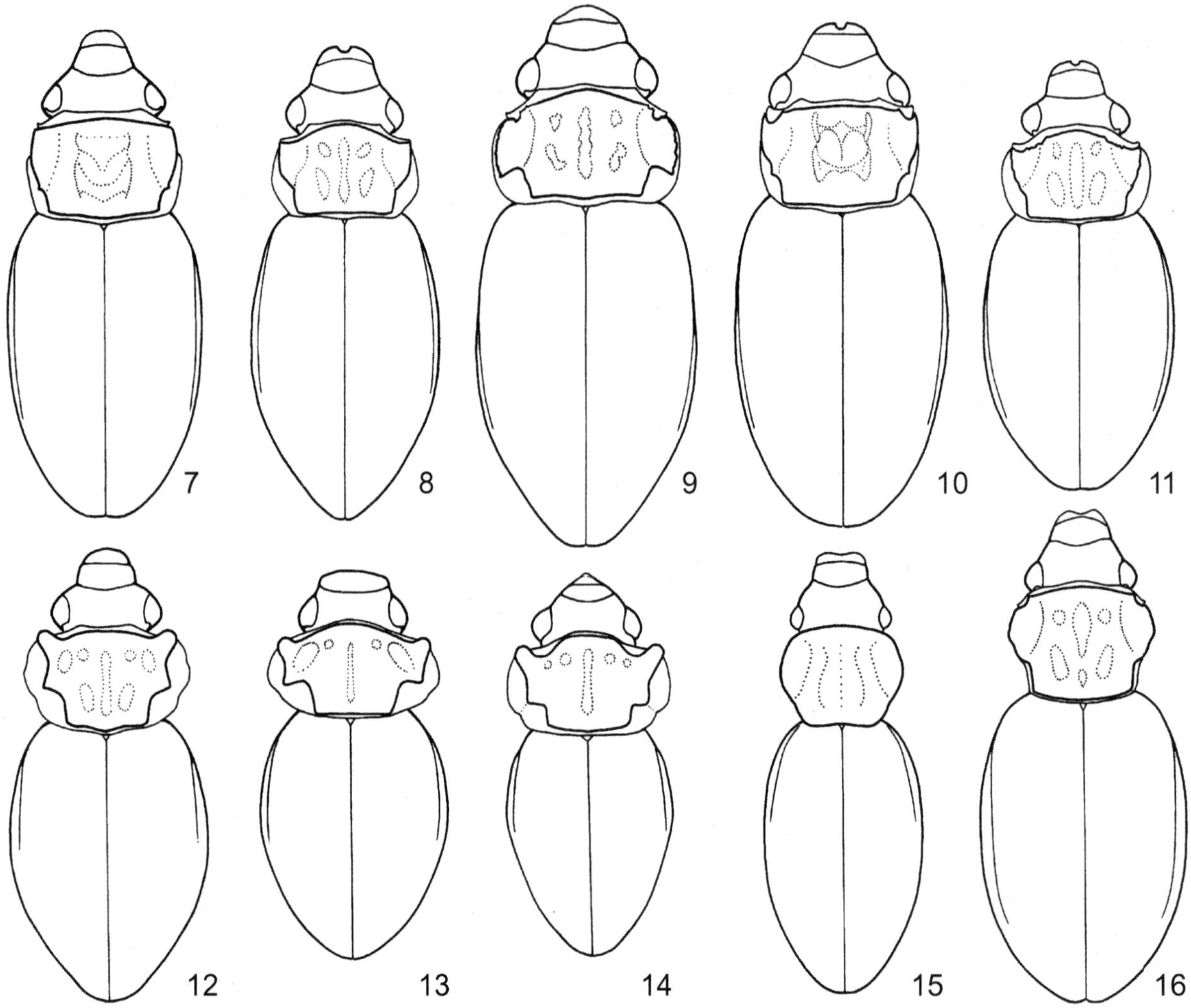


FIGURES 7.16 - 16.16, Habitus outlines. Fig. 7.16, *Ochthebius lineatus* LeConte; Fig. 8.16, *O. bisinuatus* Perkins; Fig. 9.16, *O. discretus* LeConte; Fig. 10.16, *O. gruwelli* Perkins; Fig. 11.16, *O. californicus* Perkins; Fig. 12.16, *Gymnochthebius nitidus* (LeConte); Fig. 13.16, *G. maureenae* Perkins; Fig. 14.16, *G. laevipennis* (LeConte); Fig. 15.16, *Neochthebius vandykei* (Knisch); Fig. 16.16, *Enicocerus benefossus* (LeConte). (From Perkins, 1980).

**Distribution.** This cosmopolitan family currently comprises 38 genera and about 1,200 described species. Many species (hundreds) remain to be described, especially from tropical and south temperate areas, and especially in the genus *Hydraena*. A total of 210 species are described from the Western Hemisphere. The 136 species found north of Guatemala comprise 9 shared with Mexico and areas south, 36 restricted to Mexico, 24 shared with Mexico and areas north, and 67 species which have their ranges north of Mexico.

### Key to the Nearctic Genera

1. Maxillary palpi very long, much longer than antennae, second palpomere very elongate and slender ..... ............................................................ *Hydraena*

— Maxillary palpi shorter, second palpomere not greatly elongate ............................................................ 2

2. Body contour evenly curved, pronotum not narrowed posteriorly and surface without impressions (Figs. 2.16, 5.16) ........................................... *Limnebius*

— Body contour interrupted between pronotum and elytra, pronotum with impressions ............................... 3

3. Pronotum without lateral hyaline borders; head without postocular secretion delivery setae (Fig. 15.16) ........................................................ *Neochthebius*

— Pronotum with lateral hyaline borders; head with postocular secretion delivery setae (Figs. 3.16, 4.16) ......................................................................... 4

4. Lateral hyaline border of prothorax within sinuation behind lateral depression; postocular area with secretion delivery setae blunt, secretion sulcus lacking (Fig. 16.16) ................................... *Enicocerus*

— Prothorax with lateral hyaline border extended laterally as far as or beyond lateral depression; postocular area with tapering secretion delivery setae and secretion sulcus (Figs. 3.16, 5.16) ............. 5

5. Pronotum lobate anterolaterally; aedeagus with mainpiece apically bifid (Figs. 12.16 - 14.16)....... ................................................ *Gymnochthebius*
— Pronotum not markedly lobate anterolaterally; aedeagus with preterminally inserted process, mainpiece not bifid at apex (Figs. 1.16, 7.16 - 11.16)................ ........................................................ *Ochthebius*

## Classification of the Nearctic Genera

### Hydraenidae

### Hydraeninae

**Diagnosis:** Members of this subfamily, redefined by Perkins (1997:195) are characterized by the following: the hypomeral antennal pocket is located in the anterior face of the hypomeron, generally anterior to most of the notosternal suture; the lateral portion of the pocket is formed by an extensive wet-hypomeron or specialized antennal pocket setae; the penultimate maxillary palpomere is neither longer than nor more markedly robust than the apical; and the specialized ESDS gland concentration is in the prothorax.

### Hydraenini

**Diagnosis:** This tribe, redefined by Perkins (1997:159) comprises two genera: *Hydraena* Kugelann 1794, a cosmopolitan genus with about 500 described species; and *Adelphydraena* Perkins 1989, a genus that retains several primitive characters and is represented by two species in Venezuela. The tribe is characterized by the following: the prosternal intercoxal process is expanded laterally behind the procoxae, closing the procoxal cavities (tip of each lateral process fitting into a small notch in corresponding postcoxal pronotal projection), the second article of the maxillary palpus is elongate and slender; and the last sternum in females has a pair of sensory clusters.

*Hydraena* Kugelann 1794, 29 spp., generally distributed, except Great Plains.

### Limnebiini

**Diagnosis:** This tribe, redefined by Perkins (1997:185) comprises two genera: *Limnebius* Leach 1915, a nearly cosmopolitan genus with about 110 described species; and *Laeliaena* Sahlberg 1900, a genus that retains several primitive characters and is represented by three species in Turkestan and northern India. The tribe is characterized by the following: the slightly concave hypomeral shape, lacking a hypomeral carina, and having a short row of antennal pocket setae, the smooth dorsal habitus, the proportions of the antennae, and the shape and chaetotaxy of the mentum.

*Limnebius* Leach 1915, 13 spp., montane western, eastern, and southern areas.

### Ochthebiinae

**Diagnosis:** Members of this subfamily, redefined by Perkins (1997:195) are characterized by the following: the hypomeral antennal pocket is in the ventral face of the hypomeron, bordered medially by the notosternal suture; the penultimate maxillary palpomere is more robust and longer than the ultimate; the specialized ESDS gland concentration is in the head; and the prothorax often has hypomeral and lateral hyaline borders.

### Ochthebiini

**Diagnosis:** This tribe, redefined by Perkins (1997:124), comprises 12 genera. The tribe is characterized by the following: the lacinia bears enlarged, stout apical teeth, the venter of the head has a large gular sclerite, the anterior wall of the tentorium is thickened, and the antenna has nine antennomeres (4 + 5).

*Ochthebius* Leach 1815, 43 spp., generally distributed, except Appalachian Mountains.

*Gymnochthebius* Orchymont 1943, 7 spp., generally southern areas, one species in northeastern United States and adjacent Canada.

*Neochthebius* Orchymont 1932, one species, *N. vandykei* (Knisch), Pacific Coast, in crevices of rocks in the intertidal zone (Fig. 15.16).

*Enicocerus* Stephens 1829, one species, *E. benefossus* (LeConte), from Quebec south to Virginia (Appalachian Mountains), and west to Indiana (Fig. 16.16).

## Bibliography

ARCHANGELSKY, M. 1998. Phylogeny of Hydrophiloidea (Coleoptera: Staphyliniformia) using characters from adult and preimaginal stages. Systematic Entomology, 23: 9-24.

BEUTEL, R. G. 1994. Phylogenetic analysis of Hydrophiloidea based on characters of the head of adults and larvae (Coleoptera: Staphyliniformia). Koleopterologische Rundschau, 64: 103-133.

BØVING, A. G. and CRAIGHEAD, F. C. 1931. An illustrated synopsis of the principal larval forms of the order Coleoptera. Brooklyn Entomological Society, Brooklyn, N.Y. viii + 351 pp.

CROWSON, R. A. 1981. The Biology of Coleoptera. Academic Press; Harcourt Brace Javanovich Publishers, New York.

DYBAS, H. S. 1976. The larval characters of featherwing and limulodid beetles and their family relationships in the Staphylinoidea (Coleoptera: Ptiliidae and Limulodidae). Fieldiana, Zoology, 70: 29-78.

HANSEN, M. 1991. A review of the genera of the Beetle family Hydraenidae (Coleoptera). Steenstrupia, 17: 1-52.

HANSEN, M. 1995. Evolution and classification of the Hydrophiloidea - a systematic review. pp. 321-353, *In:* Pakaluk, J. and Slipinski, S. A. (eds.). Biology, Phylogeny, and Classification of Coleoptera: Papers Celebrating the 80th Birthday of Roy A. Crowson. Muzeum I Instytut Zoologii PAN, Warsaw.

HANSEN, M. 1997. Phylogeny and classification of the staphyliniform beetle families (Coleoptera). Biologiske Skrifter, Det Kongelige Danske Videnskabernes Selskab 48: 1-339.

LAWRENCE, J. F. and A. F. NEWTON, JR. 1995. Families and subfamilies of Coleoptera (with selected genera, notes, references and data on family-group names). pp. 779-1006, *In:* Pakaluk, J. and Slipinski, S. A. (eds.). Biology, Phylogeny, and Classification of Coleoptera: Papers Celebrating the 80th Birthday of Roy A. Crowson. Muzeum I Instytut Zoologii PAN, Warsaw.

NEWTON, A. F., JR. and M. K. THAYER. 1992. Current classification and family-group names in Staphyliniformia (Coleoptera). Fieldiana: Zoology (N.S.), 67: 1-92.

PERKINS, P. D. 1972. A study of the Hydraenidae and Hydrophilidae (Coleoptera) of the San Gabriel River, with emphasis on larval taxonomy. Univ. Microfilms, Ann Arbor, Michigan. 257 pp.

PERKINS, P. D. 1976. Psammophilous aquatic beetles in southern California: A study of microhabitat preferences with notes on responses to stream alteration (Coleoptera: Hydraenidae and Hydrophilidae). Coleopterists Bulletin, 30: 309-324.

PERKINS, P. D. 1980. Aquatic beetles of the family Hydraenidae in the Western Hemisphere: classification, biogeography and inferred phylogeny (Insecta: Coleoptera). Quaestiones Entomologicae, 16 (1980): 3-554.

PERKINS, P. D. 1989. *Adelphydraena*, new genus, and two new species from Venezuela, and remarks on phylogenetic relationships within the subtribe Hydraenina (Coleoptera: Hydraenidae). Proceedings of the Biological Society of Washington, 102: 447-457.

PERKINS, P. D. 1997. Life on the effective bubble: exocrine secretion delivery systems (ESDS) and the evolution and classification of beetles in the family Hydraenidae (Insecta: Coleoptera). Annals of Carnegie Museum, 66(2): 89-207).

PERKINS, P. D., and J. BALFOUR-BROWNE. 1994. Contribution to the taxonomy of aquatic and humicolous beetles of the family Hydraenidae in southern Africa. Fieldiana: Zoology (New Series), 77: 1-159.

SPANGLER, P. J. 1991. Haliplidae, Noteridae, Hydraenidae, Hydrophilidae, Georyssidae, Chelonariidae, pp. 311-312, 314-322, 355-359, 394-395. *In:* Stehr, F. W. (ed.), Immature Insects. Vol. 2. Kendall/Hunt Publishing Co., Dubuque, Iowa.

# 17. PTILIIDAE Erichson, 1845

by W. Eugene Hall

**Family synonym:** Trichopterygidae Erichson, 1845

**Common name:** The featherwing beetles

The minute body size, slender whorls of setae on each antennomere and fringe of setae along the margin of the hindwings distinguish ptiliids from other beetles. Though the original family name Trichopterygidae refers to the 'hairy' appearance of the hindwings, some Ptiliidae exhibit vestigial or complete absence of hindwings.

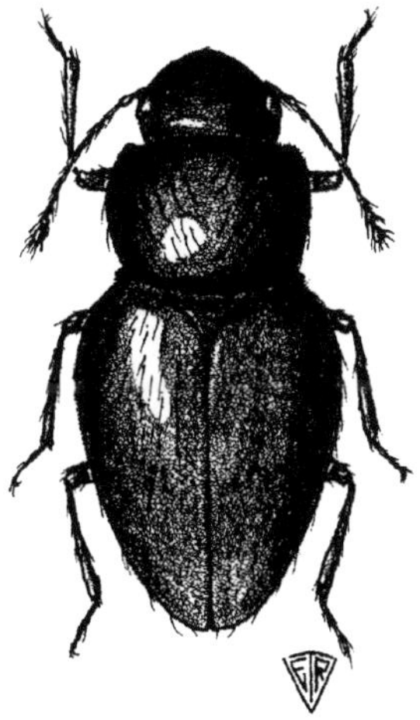

FIGURE 1.17. *Ptinidium evenescens* (Marsham).

**Description:** Form elongate, oval or limuloid, moderately to highly convex; size generally 0.40 mm to 1.2 mm in length with some tropical forms reaching near 4.0 mm, color ranging from yellow, brown, reddish-brown, black or gray, vestigial forms light brown or pale yellow; body usually densely pubescent with golden setae, at times pubescence sparse.

Head prognathous, normally round or subquadrate, rarely elongate and narrow; gular suture absent; tentorium simplified, lateral arms and corporotentorium moderately to greatly reduced. Antennae situated in front of eyes along anterior margin of head, consisting of 8 to 11 segments, each antennomere possessing a whorl of long setae at its apex, segments 1 and 2 moderately enlarged, the last 2 or 3 terminal segments usually forming a loose club. Eyes prominent, reduced, or absent. Mouthparts with mandibular molar area well-developed, apices of mandible elongate or reduced, lacinae variable, usually toothed and hooked at apex, galea 1- or 2-segmented with apex fringed, maxillary palpus segment III with 2 or 3 digitiform sensilla at base of segment IV, labial palps 1- to 3-segmented, mentum variable in shape.

Prothorax variable in form, moderately to strongly convex, usually wider than long, side margins nearly parallel to rounded, hind angles moderately to widely expanded, lateral margins rarely possessing glands or possible mycangia; prosternal process variable, absent, moderate or well-developed onto mesosternum; procoxal cavities opened, procoxae globular or cylindrical, prominent, generally contiguous. Mesosternum variable in form, discal region with wide or narrow median carina or elaborate mesosternal process; mesocoxae oval to subcylindrical, contiguous to moderately separated. Metasternum usually lacking modifications, occasionally possessing metasternal lines originating from mesocoxae or pleural region of sternum; metacoxae nearly contiguous to widely separated; hind femoral plates well-developed, posterior margin straight or concave. Metendosternite variable in form, depending on proximity of metacoxae. Tarsal formula 3-3-3, basal segment reduced, rarely modified, segment II vestigial, segment III long, slender, rarely modified; tarsal claws simple, equal or subequal in length; tarsal empodium with 1 or 2 setae. Mesoscutellum triangular, with or without sulci; metascutellum variable, short and broad to long and narrow.

Elytra either complete and rounded apically, covering to nearly covering abdominal segments, or shortened and exposing 3 to 5 abdominal tergites. Hindwings feather-like in appearance, possessing a fringe of setae along margins of membrane, membrane narrow or 'paddle'-shaped, basal stalk possessing one or two struts; venation reduced; folding pattern numerous to few transverse folds.

Abdomen generally with six visible sternites; number of spiracles 5-8; tergite VII of winged forms with palisade fringe on posterior margin; tergites II-VII (number variable) of winged forms possessing wingfolding strigulations/patches; femoral lines on sternite III abbreviated or complete, extended anterolaterally; posterior margin of pygidium variable, smooth, serrate or possessing a lobed, acute or bifid pygidial tooth; aedeagus of male lying symmetrically or asymmetrically in abdomen, variable in form, either complex and possessing lateral parameres, simple and tube-like or heavily sclerotized and short with lateral hook-like projections; spermatheca of female sclerotized, highly variable in form, possessing a well-defined 'spermathecal pump' .

**Larva:** Body shape linear, up to 2.0 mm in length; color white, yellow or light brown; setae sparse, simple. Head directed ventrally; epicranial sutures absent. Stemmata absent (except *Nossidium* from Panama); antennae 3-segmented, each segment possessing three to four setae; segment II possessing a sensory appendage of variable design. Mandibles symmetrical, prostheca present near middle, apex generally slender and possessing several teeth, molar region well-developed. Maxillary palpus 3-segmented; maxillary galea with or without fringe. Prothorax generally wider than abdominal segments; legs 4-segmented, possessing a single, simple claw. Spiracles annular. Urogomphi 1-segmented on abdominal segment IX, or absent. Anal hooks recurved, present ventrally on apex of abdominal segment X, rarely absent.

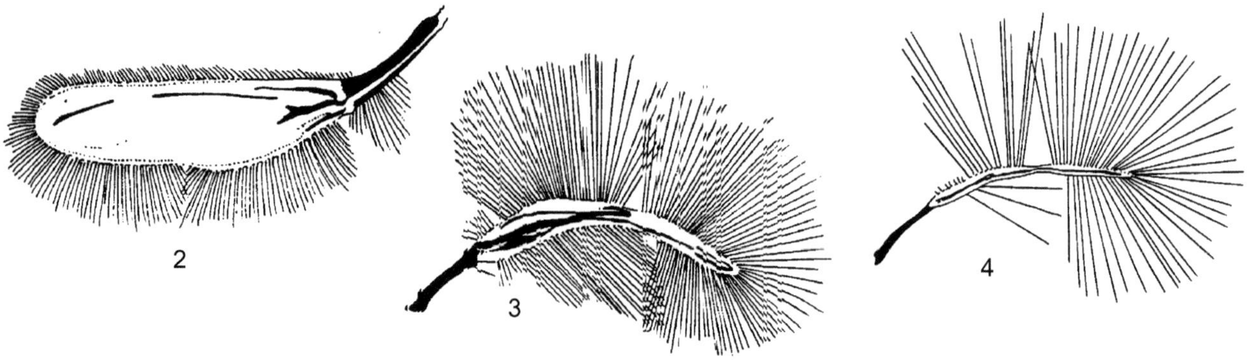


FIGURES 2.17 - 4.17. Ptiliid hindwings. Fig. 2.17, *Nossidium*; Fig. 3.17, *Acrotrichis*; Fig. 4.17, *Cylindrosella* (Fig. 4, Besuchet, 1971).

**Habits and habitats.** These smallest of all beetles live in a variety of moist habitats including rotten wood, mammal dung, fungus covered logs, tree holes, ant nests, banks of rivers and streams, forest litter and decaying organic detritus. Some groups are myrmecophiles or termitophiles. Ptiliids are microphagous as defined by Lawrence (1989), feeding mainly on spores and hyphae of fungi plus other organic particles. Leschen (1993) and Newton (1984) discuss the mycophagous feeding patterns of adults and larvae.

Featherwing beetles are associated with fungi on different levels. *Nossidium* and most genera of the tribe Nanosellini, with the exception of *Suterina*, generally inhabit a host fungus, usually of the Polyporaceae or Hydnaceae (Barber 1924; Dybas 1956, 1961, 1976; Graves 1960; Motschulsky 1868). Adults and larval nanosellines live together on the undersurface or within spore tubes of fungal fruiting bodies, feeding on spores or hyphae. Featherwing beetles also act as hosts to a variety of Laboulbeniales fungi (Tavares 1979; Hall, unpublished). These fungi can be observed attached to the exoskeleton of adult ptiliids.

Other ptiliid genera prefer moist habitats associated with bodies of water. *Actidium* inhabits sand or gravel banks along the edges of streams and rivers or on seashore algae. *Actinopteryx* is distributed along the east coast seashore and also inhabits mangrove swamps. *Motschulskium* adults and larvae live beneath seaweed piles on shores of the west coast.

Some genera of featherwing beetles (e.g., *Pteryx, Ptinella, Ptinellodes*) contain polymorphic species that occur in varying adult forms (Dybas 1978b; Matthews 1862; Taylor 1980, 1981). The 'normal morphs' possess normal morphological features, including well-developed eyes and wings, while 'vestigial morphs' exhibit reduced or absent eyes, hindwings and weak pigmentation. Ecological constraints play a role in the development of these polymorphic populations (Taylor 1980, 1981). The vestigial form in pterycine ptiliids is capable of normal reproductive functions. Pterycines normally occur in rotting logs or tree-holes and when conditions become unfavorable, winged normal morphs are produced, allowing for dispersal to a more favorable habitat.

Parthenogenetic species occur within some genera of Ptiliidae (e.g., *Acrotrichis, Bambara, Ptiliopycna, Ptinella*) with populations consisting entirely of females (Dybas 1966, 1978a; Taylor 1981).

Featherwing beetle adults and larvae generally inhabit the same niche, most likely feeding on identical food sources. Reproduction and life-cycle data have been noted for *Acrotrichis* (DeConinck and Coessens 1981; Hinton 1941) and *Ptinella* (Taylor 1980, 1981).

Species within a genus may appear identical superficially, yet spermatheca (sperm storage organ) design is variable and species specific. The diversity of this structure was recognized as far back as Flach (1889). The 'spermathecal pump' is unique to Ptiliidae within Coleoptera. Observations regarding spermathecal design and spermatozoan shape and length have been noted. Dybas and Dybas (1981, 1987) observed that sperm morphology is unique among species of *Bambara*. The length and morphology of these sperm apparently coincide with the proportions of the female spermatheca. Female ptiliids develop one egg at a time, the ovum being so large that at times it can occupy nearly half the length of the female body. Recent studies on male reproductive organs have shown diversity within these structures. De Marzo (1992) surveyed the internal male genitalia of seven genera within Ptiliidae, noting primitive and derived characteristics. Males of some ptiliid species are known to produce either extremely long sperm (Jamieson 1987; Taylor 1982) with the spermatozoa reaching twice the length of the adult male, or immobile, aflagellate sperm (Bacetti and DeConinck 1989). Male ptiliids occasionally exhibit secondary sexual characteristics. Numerous species within Acrotrichinae possess modified, bilobed basal protarsal segments, enlarged spurs on the hind trochanters, dense patches of setae on the metasternum or ventral abdominal apex. Males of *Micridium* (Ptiliinae) often possess a modified 'brush' or 'comb' between the metacoxae. On rare occasions males of the Nanosellini possess modified protarsal segments or enlarged spurs on the mesotarsi or metacoxae.

**Fossil history.** The ptiliid fossil record dates back to the Late Cretaceous and those described are representative of modern genera: *Ptilium tertiarium* from Germany (Statz and Horion 1937), *Ptinella oligocoenica* from Baltic amber (Parsons 1939) and *Microptilium geistautsi* from amber (Dybas 1961a).

**Status of the classification.** The present classification of Ptiliidae follows Lawrence and Newton (1995) with slight modifications, recognizing three subfamilies: Ptiliinae, Acrotrichinae and Cephaloplectinae. The family is in need of revision as the majority of North American ptiliid genera and species were described prior to 1900.

Morphology of ptiliid adults and larvae confirm their staphylinoid associations, possibly closest to the Hydraenidae. LeConte (1883) placed Ptiliidae within Clavicornia, between Staphylinidae and Hydroscaphidae. Matthews (1884) provided a small monograph on North American featherwing beetles and included the 'Hydroscaphina' as a tribe within Ptiliidae. Sharp and Muir (1912), using characters of the aedeagus, had difficulty placing Ptiliidae near any other group within Staphylinoidea. Based on hindwing-folding characters, Forbes (1924) suggested ptiliids shared "characters of the Liodid-Pselaphid series...and of the Nitidulidae." Boving and Craighead (1931) proposed a "leptinid association" which included Ptiliidae, Limnebiidae, Leptinidae and Anisotomidae, based on larval morphology. Crowson (1955, 1981) listed Ptiliidae as a basal group within Staphylinoidea. Dybas (1976), based on larval morphological evidence, proposed that ptiliids may be most closely associated with Hydraenidae. Lawrence and Newton (1982) support placement of Ptiliidae nearest to Hydraenidae and propose placement of Cephaloplectinae (=Limulodidae) within Ptiliidae. Newton and Thayer (1992) list a 'Pselaphidae + Ptiliidae + Scydmaenidae' classification. Lawrence and Newton (1995) propose a 'Hydraenidae + Ptiliidae + Agyrtidae' association. Hansen (1997) concluded that 'Agyrtidae + Leiodidae + Hydraenidae + Ptiliidae' form a monophyletic group within Staphyliniformia.

Ptiliid larvae are poorly known (Boving and Craighead 1931; Dybas 1976, 1991), yet possible evolutionary relationships to other staphylinoid groups have been proposed based on analysis of larval morphology. Dybas (1976) noted similarities of 'Ptiliidae + Limulodidae' to Hydraenidae, Leiodidae and Leptinidae, lending support to the "leptinid association" proposed by Boving and Craighead. Dybas (1976) notes both ptiliid and hydraenid larvae possess a mandibular mola and a single pair of ventral anal hooks on the abdominal apex. Immature cephaloplectines appear to lack ventral anal hooks, but Dybas states his description of larval Cephaloplectinae is tentative, the diagnosis being based on three specimens of a single North American species. Ptiliid larvae can be separated from hydraenids based on loss of epicranial sutures, one-segmented urogomphi and loss of stemmata. Immature stages of Leptinidae and Leiodidae, two families previously associated with Ptiliidae, lack a single pair of abdominal anal hooks (Newton 1991). Costa *et al.* (1988) erroneously illustrates epicranial sutures on the head capsule of *Acrotrichis*. Barber's (1924: Plate 7, Figs. 10.17-12.17) illustration of a nanoselline ptiliid larva is actually an immature Staphylinidae.

Keys and illustrations presented are based on a combination of Besuchet (1971, 1976), Dybas (1976, 1990), Hall (1999), Johnson (1985), and Sorensson (1997). Major sections of this key are based on modifications of Dybas (1990).

Due to their minute size, ptiliids must be slide-mounted to accurately observe morphological characters. Specimens can be cleared in cold KOH, rinsed in distilled water, then passed through progressions of 30-100% ETOH to prevent collapsing of the spermatheca, a key structure in species identification. Specimens should be mounted ventral side up, either in glycerin, Hoyers, or Euporal® (brief, short-term storage or permanent mounts, respectively).

**Distribution.** Featherwing beetles are widely distributed throughout the world. Presently (1999), more than 70 genera and 550 species have been described worldwide. Within North America, 27 genera and nearly 120 species are known but these numbers are conservative as numerous undescribed genera and species are present in museum and private collections.

### Key to Nearctic Ptiliidae Subfamilies

1. Antennae 8- to 10-segmented; form limuloid, convex; hindwings, eyes absent; prosternal process strongly developed, elongate and broad, extending over mesosternum; tibiae usually with longitudinal rows of spines; elytra truncate; associated with ants ... .............................................. Cephaloplectinae

— Not with above combination of characters ........... 2

2(1). Antennae 10 or 11-segmented; form elongate or oval; eyes present; hindwings present, prosternal process between procoxae absent; elytra complete, occasionaly ending in an apical fringe; mesosternal process generally elaborate, overlapping inner margins of mesocoxae; mesocoxae contiguous or nearly so; metasternal lines extending from mesocoxae; metacoxae nearly contiguous; pygidial tooth acute, bifid or lobed and toothed .............. .......................................... Ptiliinae (Nanosellini)

— Antennae 11-segmented; eyes normal or reduced; elytra complete or truncated; procoxae with or without prosternal process; hindwings normal or reduced; mesosternal process weakly to moderately developed; mesocoxae contiguous to moderately separated; metasternal lines, if present, extending from sides of metasternum; metacoxae nearly contiguous to widely separated; pygidium hind margin variable in form ........................................ 3

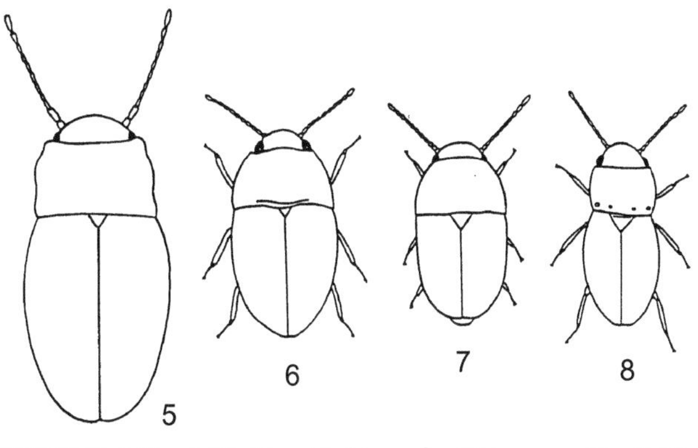


FIGURES 5.17 - 8.17. Dorsal habitus of ptiliid genera. Fig. 5.17, *Motschulskium*. Fig. 6.17, *Nossidum*. Fig. 7.17, *Bambara*. Fig. 8.17, *Ptenidium*. (Fig. 5, Dybas, 1990; 6, Besuchet, 1971).

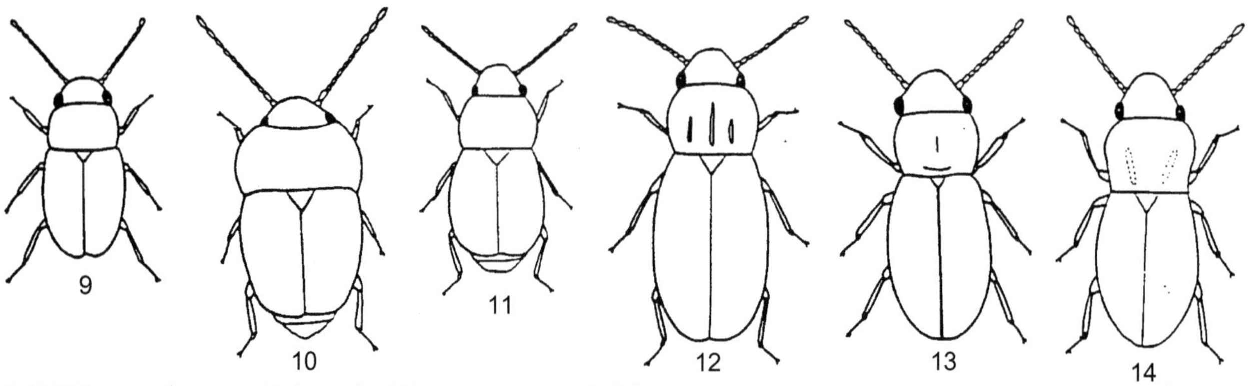


FIGURES 9.17 -14.17. Dorsal habitus of ptiliid genera. Fig. 9.17, *Ptiliola*. Fig. 10.17, *Actinopteryx*. Fig. 11.17, *Pteryx*. Fig. 12.17, *Ptilium*. Fig. 13.17, *Oligella*. Fig. 14.17, *Micridium* (Figs. 9-14, Besuchet, 1971).

3(2). Elytra complete, occasionally exposing last abdominal tergite, shortened in polymorphic forms; mesocoxal suture variable; eyes normal but reduced or absent in vestigial forms; procoxal cavity open or coxae moderately separated by a narrow prosternal process; posterior margin of pygidium variable in form; aedeagus tube-like, rarely possessing lateral lobes .................. Ptiliinae (Ptiliini)

— Elytra truncated or shortened; mesocoxal sutures straight; pygidium hindmargin bidentate or tridentate; aedeagus situated symmetrically in abdomen, lacking lateral lobes; metascutellum possessing a single, thick seta projecting off anterior margins ..................................................... Acrotrichiinae

### Key to Nearctic Genera of Ptiliini (Ptiliinae)

1. Hind coxae contiguous or nearly so; elytra complete, rounded at apices, not shortened or truncate, covering or nearly covering abdomen; hindwing membrane narrow or wide ....................................... 2

— Hind coxae separated by at least one-eighth metasternal width; elytra complete, shortened or truncate; hindwings, when present, consisting of a long, narrowed membrane ................................ 4

2(1). Mesocoxae slightly separated by a narrow, longitudinal mesosternal process, tapered posteriorly; eyes emarginate from behind with dorsal flange directed posteriorly; hindwing basal stalk possessing a single sclerotized strut, wing membrane narrow, marginal hairs approximately four times as long as width of membrane; pygidial apex lacking teeth (Figures: 7.17, 30.17, 67.17) ........ *Bambara*

— Mesocoxae separated by a median mesosternal process that is posteriorly subtruncate; eyes not emarginate posteriorly, lacking dorsal flange; hindwing basal stalk consisting of two sclerotized struts, wing membrane broad, paddle-shaped, marginal setae of membrane relatively short, not longer than width of membrane; pygidial apex with or without teeth ................................................................. 3

3(2). Venter of sternum possessing two sutures extending laterally from margins of mesocoxae; prothorax with sides evenly curved with a sharply defined dorsal groove along explanate margin; form strongly convex; dorsal body surface shining, covered in long, semierect golden setae; color reddish to light brown, prothorax at times daker than elytra (Figures: 2.17, 6.17, 28.17, 51.17, 63.17, 69.17) .......................................................... *Nossidium*

— Center of sternum possessing one suture extended laterally from margins of mesocoxae; prothorax with sides sinuate posteriorly, lacking dorsal marginal groove; body flattened, surface covered within pale, semirecumbent golden setae; color dark grey or black (Figures: 5.17, 29.17, 53.17, 68.17)....... .................................................... *Motschulskium*

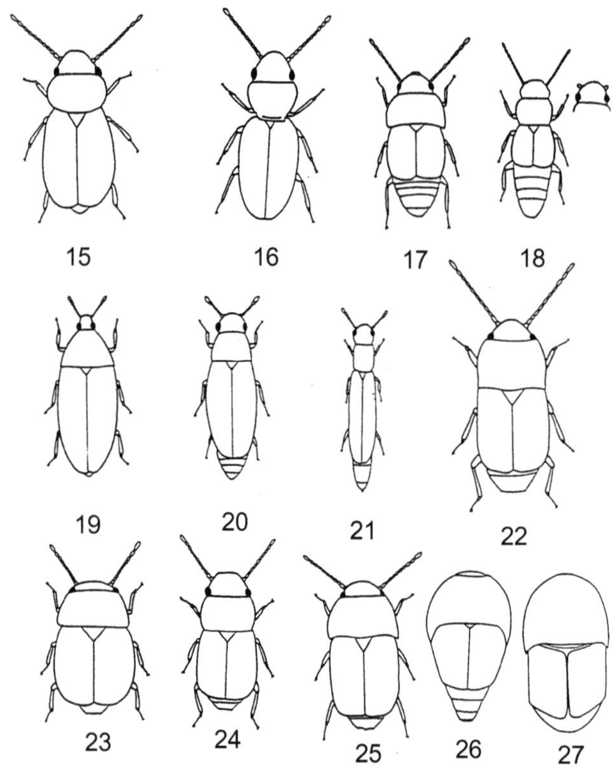


FIGURES 15.17 - 27.17. Dorsal habitus of ptiliid genera. Fig. 15.17, *Ptiliolum*. Fig. 16.17, *Actidium*. Fig. 17.17, *Ptinellodes*. Fig. 18.17, *Ptinella*. Fig. 19.17, *Suterina*. Fig. 20.17, *Nanosella*. Fig. 21.17, *Cylindrosella*. Fig. 22.17, *Smicrus*. Fig. 23.17, *Ptiliopycna*. Fig. 24.17, *Nephanes*. Fig. 25.17, *Acrotrichis*. Fig. 26.17, *Limulodes*. Fig. 27.17, *Paralimulodes* (Figs. 15-19, 26-27, Dybas, 1990; 20, 22, 23, Besuchet, 1971; 24, Seevers and Dybas, 1943; 25, Wilson *et al.*, 1954).

4(1). Eyes present or vestigial; elytra nearly complete, shortened or truncated; hindwings present or absent; polymorphic forms at times occurring together; pygidial hind margin smooth; body color brown to light yellow ........................................ 5
— Eyes present; elytra complete; hindwings fully developed; polymorphic forms unknown; pygidium hindmargin variable; mesocoxal pleural suture straight or curved anteriorly .............................. 7

5(4). Hind angles of prothorax moderately expanded posteriorly; females possessing claw-like structures, internally, near apex of abdomen (Figure: 17.17) ............................................................ *Ptinellodes*
— Hind angles of prothorax obtusely rounded or sharply angled, not expanded posteriorly ...................... 6

6(5). Prothorax narrower anteriorly than at base, widest behind middle, hind angles obtusely angulate, not sinuate or acute; side margins of prothorax narrowly explanate at middle then wider posteriorly (Figures: 11.17, 56.17, 74.17) .................. *Pteryx*
— Prothorax as wide or wider anteriorly than at base, widest anterior to middle, hind angles at times sinuate or broadly rounded; side margins of prothorax not explanate (Figures: 18.17, 37.17, 62.17)....... ................................................................. *Ptinella*

7(4). Prothoracic intercoxal process well developed, separating procoxae; foveae present along dorsal hind margin of prothorax; metasternal process subtruncate, extended anteriorly between mesocoxae; scutellum with basal groove; body surface shining; setae usually sparse; color reddish-brown to black (Figures: 8.17, 31.17, 52.17, 60.17, 64.17, 72.17) ....................................... *Ptenidium*
— Prothoracic intercoxal process weakly developed or absent; metasternum lacking such a process .... 8

8(7). Prothorax strongly constricted at base; mesocoxae separated by prolongation of mesosternal process; body surface granulate; color dark to light grey (Figures: 16.17, 35.17) ......................... *Actidium*
— Prothorax not strongly constricted basally, lacking above combination of characters ....................... 9

9(8). Elytral suture with 10 small tubercles or teeth, humerus with a small tooth or angulation; lateral mesocoxal suture straight, directed anteriorly at 45-degree angle (Figures: 9.17, 36.17, 61.17)..... ................................................................. *Ptiliola*
— Not with above combination of characters .......... 10

10(9). Mesosternal collar and median elevation well-developed, side arms of collar bent anteriorly; median carina ending between mesocoxae; lateral mesocoxal sutures straight, directed anteriorly; hind angles of prothorax produced posteriorly; pygidium of male composed of two fused abdominal tergites, tergites separate in female; body color dark (Figures: 10.17, 39.17) ........... *Actinopteryx*
— Mesosternum not as above ................................ 11

11(10). Pygidium posterior margin serrate, middle tooth somewhat large; body color brown (Figures: 15.17, 55.17, 70.17) .......................................... *Ptiliolum*
— Pygidium posterior margin dentate or not, never serrate ................................................................ 12

12(11). Pygidium with two apical teeth; mesosternum with a sharply defined median triangular elevation; prothorax with or without sulci ............................ 13
— Pygidium with or without single apical tooth; mesosternum not elevated or weakly elevated and demarcated at middle; longitudinal metasternal carinae present, partly or completely developed ... 14

13(12). Triangular mesosternal elevation with a median fovea possessing an internal extension laterally on each side; dorsal surface of prothorax with three well-developed longitudinal sulci (Figure 38.17) ......... ............................................................. *Millidium*
— Triangular mesosternal elevation lacking a fovea; dorsal surface of prothorax with or without longitudinal sulci; color brown, reddish-brown or grey (Figures 12.17, 34.17) ................................... *Ptilium*

14(12). Metacoxae broadly lamellate, separated by about one-third of the metasternal width; mesosternal shoulders sharply angulate; mesosternal process trifoliate; lateral meso/metasternal lines extended onto anterior half of metasternum; prothorax with or

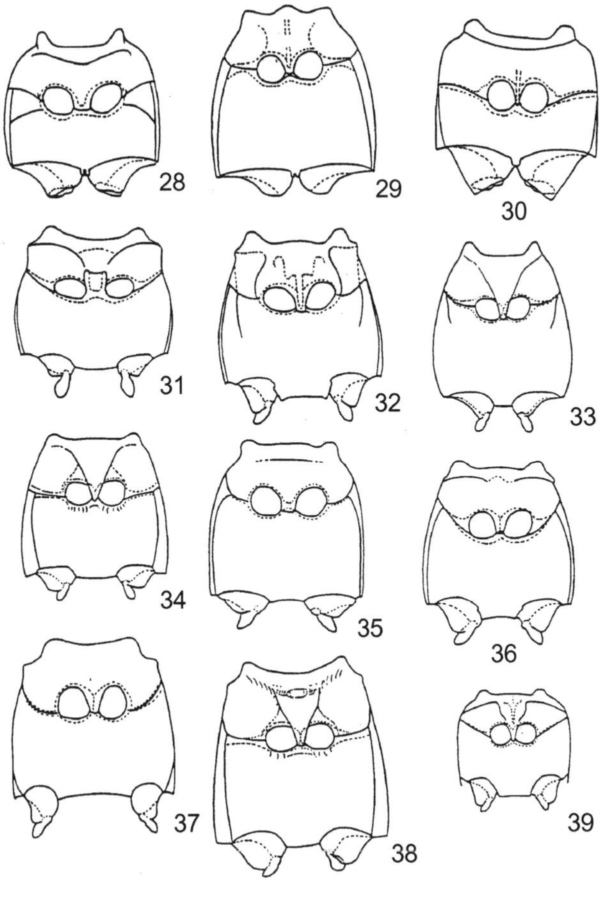


FIGURES 28.17 - 39.17. Ventral view, meso- and metathorax. Fig. 28.17, *Nossidium*. Fig. 29.17, *Mot-schulskium*. Fig. 30.17, *Bambara*. Fig. 31.17, *Ptenidium*. Fig. 32.17, *Micridium*. Fig. 33.17, *Oligella*. Fig. 34.17, *Ptilium*. Fig. 35.17, *Actidium*. Fig. 36.17, *Ptiliola*. Fig. 37.17, *Ptinella*. Fig. 38.17, *Millidium*. Fig. 39.17, *Actinopteryx* (Figs. 28-39, Dybas, 1990).

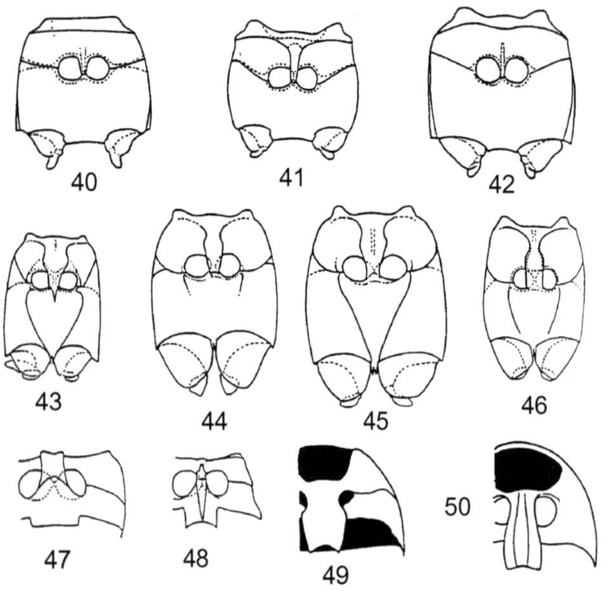


FIGURES 40.17 - 50.17. Figs. 40-48. Ventral view, meso- and metathorax. Fig. 40.17, *Nephanes*. Fig. 41.17, *Ptiliopycna*. Fig. 42.17, *Acrotrichis*. Fig. 43.17, *Nanosella*. Fig. 44.17, *Hydnosella*. Fig. 45.17, *Porophila*. Fig. 46.17, *Throscoptilium*. Fig. 47.17, *Limulodes*. Fig. 48.17, *Paralimulodes*. Figs. 49-50. Prothorax, ventral view. Fig. 49.17, *Limulodes*. Fig. 50.17, *Paralimulodes* (Figs. 40-44, Dybas, 1990; 45, 47, Seevers and Dybas, 1943; 48, Wilson, *et al.*, 1954).

without longitudinal sulci; body elongate-oval, convex; color yellow to brown (Figures 14.17, 32.17, 54.17) .................................................. *Micridium*

— Metacoxae narrowly lamellate, separated by about one-sixth of the metasternal width; mesosternal shoulders rounded; mesosternal process triangular, not elaborate; lateral metasternal lines reduced, extended onto anterior third of metasternum; prothorax with a shallow, median longitudinal depression; body flattened, linear; color yellow (Figures 13.17, 33.17) .......................................... *Oligella*

## Key to Nearctic Genera of Nanosellini (Ptiliinae)

1. Antennae 11-segmented; ectodermal glands present on lateral margins of prothorax; mesosternal process simple, flattened, nearly contiguous with metasternum between mesocoxae; metasternal lines long, narrowed posteriorly, faintly converging between metacoxae; elytral strudulatory file absent; posterior margin of pygidium possessing a median lobe flanked by a single tooth on each side (Figures 45.17, 74.17) ....................................... *Porophila*

— Antennae 10-segmented, clavate, last 3-4 segments forming a loose club; prothoracic ectodermal glands absent; mesosternal process elaborate; metasternal lines ending anterior to, on or between metacoxae; longitudinal striations of elytral venter present; posterior margin of pygidium smooth or possessing an acute or bifid apical tooth ...... 2

2(1). Head elongate, narrow, approximately one third as wide as prothorax; mesosternal process with lateral margins nearly parallel posteriorly; metasternal lines ending on metacoxae; pygidium rounded, apical tooth visible laterally; color yellow; occurs at base of trees or on bark of fungus covered logs (Figure 19.17) ........................................ *Suterina*

— Head normal, nearly two thirds as wide as prothorax; mesosternal process variable, ranging from 'arrow-head' shaped to lateral margins parallel; metasternal lines ending anterior to or between metacoxae; inhabits undersides or spore tubes of Polyporaceae and Hydnaceae fungi; color brown to reddish-brown ............................................. 3

3(2). Cylindrical pits present on hind angles of prothorax; mesosternal process ending posteriorly on metasternum as acute tooth, anterior median carina near mesosternal collar weakly developed or absent; metasternal lines present but abbreviated; pygidial tooth acute, weakly bifid; occurs on fungus of the Hydnaceae (Figures 44.17, 71.17).... *Hydnosella*

— Cylindrical pits of prothorax absent; mesosternal process ending between mesocoxae or on metasternum, anterior median carina near mesostrenal collar present; metasternal lines complete, extended onto anterior half of metasternum or converging between metacoxae; pockets present along anterior margin of procoxal acetabulum or pockets absent; mesopleural suture curved anteriorly or extended laterally from mesocoxae; inhabiting fungi of the Polyporaceae .................................. 4

4(3). Mesosternal process 'arrow-head' shaped, narrowed posteriorly; procoxal pockets present; pygidial tooth acute or bifid ........................................... 5

— Mesosternal process with sides margins nearly parallel; procoxal pockets absent; pygidial tooth acute ......................................................................... 7

5(4). Form elongate-oval, not more than three times as long as wide; mesosternal process extended onto metasternum as a narrow apical tooth; pygidial tooth acute or bifid (Figures 20.17, 43.17) .......... ............................................................ *Nanosella*

— Form elongate-cylindrical, narrow, lateral margins of prothorax nearly parallel; mesosternal process extended or not extended onto metasternum ....... 6

6(5). Mesosternal process not extending onto metasternum; elytral fringe absent at apices; pygidium bifid (Figures 4.17, 21.17) .................. *Cylindrosella*

— Mesoternal process extending onto metasternum; elytral fringe present at apices .. *Cylindroselloides*

7(4). Mesosternal process extended onto metasternum as a narrow keel; mesocoxal pleural suture curved anteriorly; metasternal lines converging between metacoxae (Figure 46.17) ............ *Throscoptilium*

— Mesosternal process not extended onto metasternum; mesocoxal pleural suture straight; metasternal lines abbreviated, not converging between metacoxae ................................. *Throscoptiloides*

## Key to Nearctic Genera of Acrotrichinae

1. Hind angles of prothorax moderately to widely expanded posteriorly; pygidium hind margin

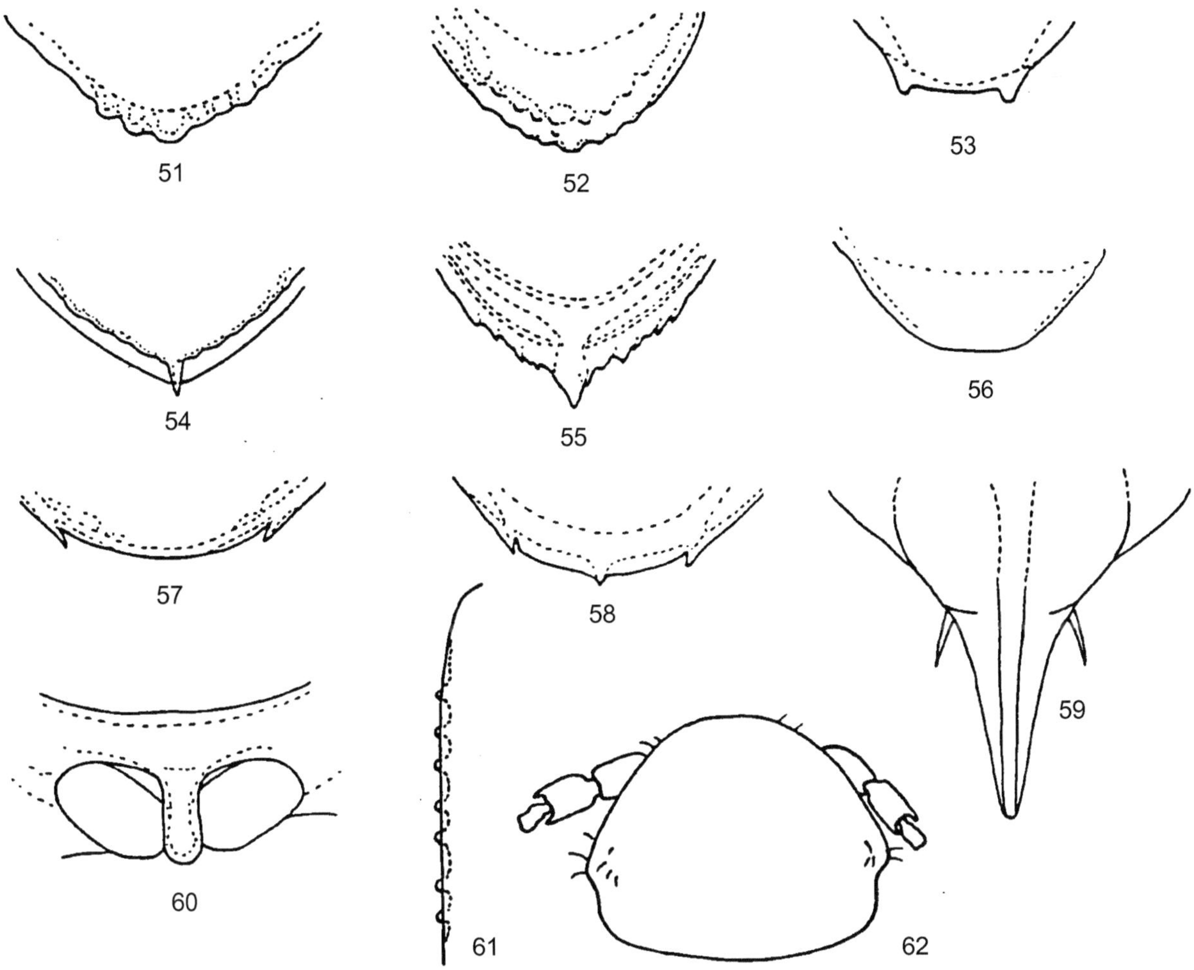


FIGURES 51.17 - 62.17. Figs. 51 - 58. Hind margin of pygidium. Fig. 51.17, *Nossidum*, female. Fig. 52.17, *Ptenidium*. Fig. 53.17, *Motschulskium*. Fig. 54.17, *Micridium*. Fig. 55.17, *Ptiliolum*. Fig. 56.17, *Pteryx*. Fig. 57.17, *Smicrus*. Fig. 58.17, *Acrotrichis*. Fig. 59.17, *Acrotrcihis*, metascutellum. Fig. 60.17, *Ptenidium*, prosternal process. Fig. 61.17, *Ptiliola*, denticules of elytral inner margin. Fig. 62.17, *Ptinella*, head of 'vestigial morph'.

bidentate or tridentate; spermatheca complex; color grey to black (Figures 3.17, 25.17, 42.17, 58.17, 59.17, 76.17) ..................................... *Acrotrichis*

— Hind angles of prothorax right-angled or obtusely rounded, not expanded posteriorly; median pygidial tooth present or absent ........................... 2

2(1). Prothoracic posterior angles obtusely rounded; mesosternal keel widened anteriorly, joined to mesosternal collar; form convex; pygidium with median and lateral teeth (Figures 23.17, 41.17) .. .......................................................... *Ptiliopycna*

— Prothoracic posterior angles obtusely or sharply angled; mesosternal keel not joined to mesosternal collar; form not highly convex; pygidium lacking median tooth .................................................... 3

3(2). Transverse line delimiting mesosternal collar indented anteriorly at middle, sinuate laterally; prothorax with sides straight or slightly sinuate at posterior third; body flattened; antennomeres 5-7 three to four times as long as broad, not abruptly constricted apically (Figures 22.17, 57.17) ................ *Smicrus*

— Transverse line of mesosternal collar straight or slightly sinuate, not sharply indented, its sides straight or slightly sinuate; antennomeres 5-8 abruptly constricted at apices (Figures 24.17, 40.17) .................................................. *Nephanes*

Key to Nearctic Genera of Cephaloplectinae

1. Antennae 8-segmented; prosternal process long, narrowed anteriorly, widened posteriorly, side margins nearly parallel, posterior margin curved anteriorly; metasternal process apex acute; mesocoxae contiguous; disc of metasternum with median carina ending acutely between metacoxae; color yellow (Figures 27.17, 48.17, 50.17) ....................... *Paralimulodes*

— Antennae 10-segmented; prosternal process broad, narrowed anteriorly, widened posteriorly, lateral margins rounded, posterior margin of process strongly curved anteriorly; metasternal process apex broad; mesocoxae nearly contiguous to slightly separated; disc of metasternum lacking median carina; color reddish-brown (Figures 26.17, 47.17, 49.17, 75.17) .. ............................................................ *Limulodes*

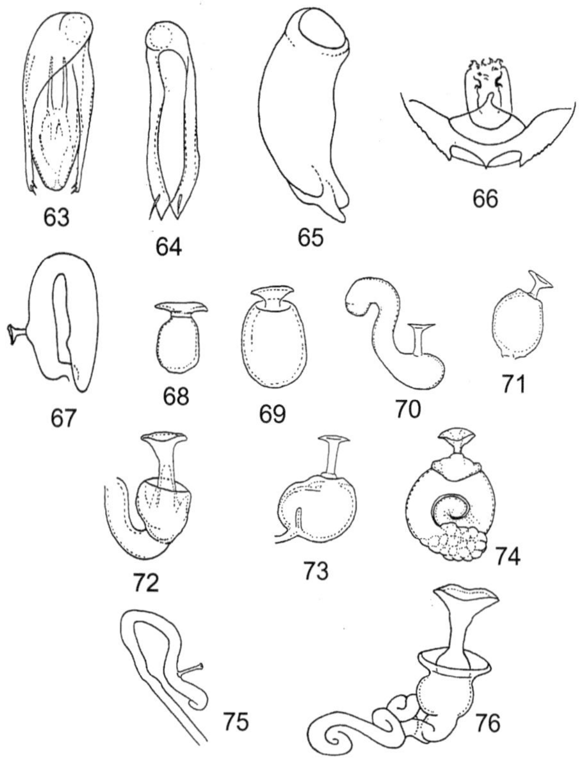


FIGURES 63.17 - 76.17. Figs. 63 - 66. Aedeagus. Fig. 63.17, *Nossidium*. Fig. 64.17, *Ptenidium*. Fig. 65.17, Ptiliine. Fig. 66.17, *Acrotrichis*. Figs. 67 - 76. Spermatheca. Fig. 67.17, *Bambara*. Fig. 68.17, *Motschulskium*. Fig. 69.17, *Nossidium*. Fig. 70.17, *Ptiliolum*. Fig. 71.17, *Hydnosella*. Fig. 72.17, *Ptenidium*. Fig. 73.17, *Porophila*. Fig. 74.17, *Pteryx*. Fig. 75.17, *Limulodes*. Fig. 76.17, *Acrotrichis*.

## Classification of the Nearctic Genera

### Ptiliidae Erichson 1845

Trichopterygidae Erichson 1845
Ptiliinae Erichson 1845
Cephaloplectinae Sharp 1883
Acrotrichinae Reitter 1909
Nanosellinae Barber 1924
Limulodidae Seevers and Dybas 1943

### Ptiliinae Erichson 1845

#### Ptilinini

Ptiliini is presently the largest, most diverse group within Ptiliidae. The tribe as currently defined is paraphyletic. The subfamily Ptiliinae is in dire need of revision.

The Nossidine-complex, including *Nossidium* and *Motschulskium*, possibly represents an undescribed subfamily within Ptiliidae (Dybas 1976; Hall, personal observ.). Nossidines are currently viewed as basal ptiliid stock. These genera and related forms possess primitive adult characters of the meso/metasternum, aedeagus, hindwings and wingfolding spicules. Nossidine larvae also exhibit primitive characteristics within Ptiliidae. De Marzo (1992) notes *Nossidium* possess primitive internal reproductive features compared to other ptiliid genera.

The Pterycine-complex, including *Pteryx*, *Ptinella*, and *Ptinellodes*, also forms a well-defined group within Ptiliini. Most pterycine genera possess glands located within the abdomen that are absent in other ptiliid genera. Pterycines usually inhabit rotten logs or tree holes and exhibit polymorphism in the adults (Matthews 1862; Park and Auerbach 1954; Taylor 1980, 1981). *Ptinella*-type pterycines have been collected in termite galleries (Hall, unpublished) from Arizona and Peru. Other pterycine genera not occurring in North America are known to be associated with termites: *Urotrianus* Silvestri 1946; *Pycnopteryx* Dybas 1955; *Dybasina* Lundgren 1983 [=*Termitopteryx* Dybas 1955]; and *Xenopteryx* Dybas 1961c.

Other genera of Ptiliini occurring outside of North America include *Achosia* Deane 1930 =?syn. *Ptinella*; *Africoptilium* Johnson 1967; *Astatopteryx* Perris 1862; *Championella* Matthews 1884 (=?syn. *Ptinella*); *Cissidium* Motschulsky 1855; *Cnemadoxia* Deane 1930; *Cochliarion* Deane 1930; *Dipentium* Johnson 1982; *Etronia* Deane 1931; *Euryptilium* Matthews 1872; *Gomyella* Johnson 1985; *Kimoda* Johnson 1985; *Kuschelidium* Johnson 1982; *Leaduadicus* Deane 1930; *Leptinla* Johnson 1985; *Malkinella* Dybas 1960; *Micridina* Johnson 1969a; *Microptilium* Matthews 1872; *Myrmicotrichis* Motschulsky 1855; *Neotrichopteryx* Deane 1931; *Notoptenidium* Johnson 1982; *Ptenidotonium* Johnson 1982; *Pterycodes* Matthews 1884; *Ptiliodes* Matthews 1882; *Rioneta* Johnson 1975; and *Skidmorella* Johnson 1971.

*Actidium* Matthews 1869
*Acteella* Motschulsky 1868
*Actella* Motschulsky 1868 (misspelling of *Acteella*)
*Ptenidula* Deane 1932

Seven species plus many undescribed, ranging from Illinois, Indiana, Wisconsin, North Carolina, Florida, Texas west to British Columbia, California, Oregon and southward to Guatemala. Prefers gravel and sand bars along streams and rivers. Specimens from San Carlos Bay, Sonora, Mexico were collected in cracks of rocks in a reef between tides.

*Actinopteryx* Matthews 1872

One species, *A. fucicola*, distributed along the east coast of the United States, possibly introduced (Dybas 1990). Occurs beneath kelp and organic seashore debris (Dybas 1976). It has been suggested (C. Johnson, pers. comm.; Hall, in prep.) that *Actinopteryx* may belong with Acrotrichinae, but until such formal placement has been published, the genus remains within Ptiliinae.

*Bambara* Vuillet 1911a
*Eurygene* Dybas 1966
*Trichopteryx* Nietner 1856

Seven species, ranging from Florida and the Gulf Coast west to Texas and Arizona. Some species parthenogenetic. Associated with plant debris and litter.

*Micridium* Motschulsky 1869

*Dilinium* Casey 1924

*Dilineum* Casey 1924 (misspelling in Dybas 1990)

One described North American species, *M. lineatum* LeConte 1863, plus several undescribed, widely distributed, occurring in leaf litter and tree-holes. A closely related genus, *Micridina* Johnson, occurs in Trinidad. Undescribed *Micridium*-type genera are associated with ants (*Atta*, *Solenopsis*) in Guatemala and Brazil (Hall, unpublished).

*Millidium* Motschulsky 1855

One species, *M. minutissimum* (Ljungh), Illinois, Washington, Wisconsin. Inhabits decayed vegetation, grass cuttings, horse manure.

*Motschulskium* Matthews 1872

One species, *M. sinuatocolle* Matthews 1872, known from the west coast of United States and Baja California, living beneath seashore seaweed piles and driftwood.

*Nossidium* Erichson 1845

Three species, plus many undescribed. Widely distributed, *Nossidium* is generally associated with fungi (Fogel and Peck 1975), forest leaf litter or decaying wood.

*Oligella* Motschulsky 1868

Undescribed species, possibly introduced, inhabiting leaf litter and stable debris (Dybas 1990).

*Ptenidium* Erichson 1845

*Anisarthria* Stephens 1830

*Epoptia* Deane 1930

subgenus *Gillmeisterium* Flach 1889

subgenus *Gressnerium* Ganglbauer 1899

subgenus *Matthewsium* Flach 1889

subgenus *Wankowizium* Flach 1889

Over a dozen described species within North America, widely distributed throughout the midwest, eastern and southwestern United States plus eastern Canada. Inhabit tree-holes, forest leaf litter, grass piles, wood rat nests, frass of bee nests. *Ptenidium* is generally distinguished from other ptiliids by its sparse pubescence and polished dorsal surface.

*Pteryx* Matthews 1859

*Aderces* Thomson 1858

Four species recorded from the United States, some polymorphic, plus many undescribed. Widely distributed, inhabiting decaying logs, tree-holes, forest leaf litter, sphagnum bogs.

*Ptiliola* Haldeman 1849

*Nanoptilium* Flach 1889

Three species recorded from North America and Canada. Widely distributed, inhabiting fungi, leaf litter, decayed organic material and mammal dung.

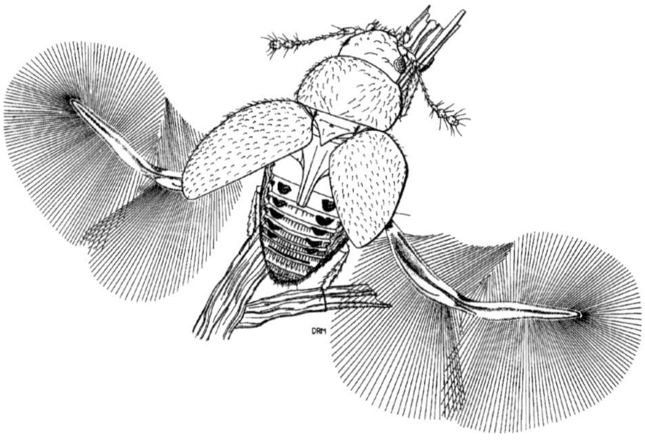

FIGURE 77.17. *Ptenidium* sp. (courtesy of D. Maddison).

*Ptiliolum* Flach 1888

*Euptilium* Flach 1889

*Trichoptilium* Flach 1889

*Typhloptilium* Flach 1889

One species plus numerous undescribed species occurring in northern and western United States and Canada. Found in forest leaf litter.

*Ptilium* Erichson 1845

*Epitomella* Motschulsky 1868

*Micrella* Motschulsky 1868

*Micrus* Motschulsky 1848

Thirteen species occurring in northwestern and southwestern United States, east to Pennsylvania and Rhode Island. Inhabits sawdust piles, dried or decayed vegetation, mammal nests, forest leaf litter.

*Ptinella* Motschulsky 1872

*Leaptiliodes* Deane 1932

*Neuglenes* Thomson 1859

*Plitium* Besuchet 1971

*Ptiliodina* Casey 1924

Three species recorded from North America, some polymorphic, plus many undescribed. Widely distributed, occurring under bark of trees, within tree-holes, sawdust piles and decaying logs.

*Ptinellodes* Matthews 1872

Dybas (1978) lists two species occurring in the southeastern United States, both of which exhibit polymorphic adult forms. Lives within tree holes, under bark on logs and forest leaf litter.

### PTILIINAE

#### NANOSELLINI

The Nanosellini form a well-defined group within Ptiliidae (Hall 1999; Sorensson 1997). Dybas (1976) questioned the subfamily status proposed by Barber (1924) and suggested possible tribal ranking within Ptiliinae, but offered no official nomenclatorial changes. Dybas (1990) later grouped the nanosellines as a tribe within Ptiliidae, but outside of Acrotrichinae, gave no subfamily or tribal definitions for other groups within Ptiliidae. Sorensson (1997) supports lowering Nanosellinae to tribal ranking within Ptiliinae. Recently, Hall (1999) conducted a phylogenetic analysis of the tribe, supporting the proposal that Nanosellini represents a monophyletic group within Ptiliidae. The sister group to Nanosellini is not clear but morphological analysis shows affinities to other genera currently placed within Ptiliinae.

With the exception of *Porophila*, all known North American genera of Nanosellini possess a well-defined patch of longitudinal striations located on the elytral venter (='stridulatory file' of Sorensson [1997]). This structure is absent within Ptiliidae outside of Nanosellini, though structures homologous to the nanoselline stridulatory file occur in other groups within Staphylinoidea (Hall, unpublished). Nanosellines also possess metasternal lines originating from the posterobasal margins of the mesocoxae and extended onto the metasternum, a feature absent in most other ptiliid genera.

Nanoselline genera outside of North America include *Baranowskiella* Sorensson 1997; *Fijisella* Hall 1999; *Fijiselloides* Hall 1999; *Garicaphila* Hall 1999; *Isolumpia* Deane 1931; *Limulosella* Hall 1999; *Mikado* Matthews 1889; *Nellosana* Johnson 1982; *Nellosanoides* Hall 1999; *Nepalumpia* Hall 1999; *Paratuposa* Deane 1931; *Phililumpia* Hall 1999; *Scydosella* Hall 1999; *Scydoselloides* Hall 1999; *Tasmangarica* Hall 1999; *Throscidium* Matthews 1872; *Throscosana* Hall 1999; *Vitusella* Hall 1999.

*Cylindrosella* Barber 1924

One species, *C. dampfi* Barber 1924 occurs in Mexico and southwestern United States. Inhabits the undersurface and spore tubes of polypore fungi.

*Cylindroselloides* Hall 1999

One species, *C. dybasi* Hall 1999, occurs in northeastern United States and southeastern Canada.

*Hydnosella* Dybas 1961

One species, *H. globitheca* Dybas 1961, occurs in Indiana and is immediately distinguished from other nanoselline genera by the presence of large circular pits near the prothoracic hind angles. Occurs on *Steccherinum*, a fungus of the Hydnaceae.

*Nanosella* Motschulsky 1868

*Mycophagus* Friedenreich 1883

*Ptilium fungi* LeConte 1863

Three species recorded from the United States (Barber 1924; Dury 1916; LeConte 1863; Motschulsky 1868). Based on examination of LeConte's type specimen of *P. fungi* and description of Motschulsky's *N. fungi* (Hall 1999), *P. fungi* is herein treated as a junior synonym of *N. fungi.* North American species listed under *Mycophagus* (*sensu* Barber 1924) are placed within *Nanosella* (Dybas 1961, 1976; Hall 1999). *Nanosella* is widely distributed in eastern and southern United States and parts of Mexico. Outside of North America the genus occurs in Central and South America, Africa, Japan and Australia (Hall, pers. observ.). Inhabits the undersurface of polypore fungi.

*Porophila* Dybas 1956

One species, *P. malkini* Dybas 1956, distributed in Oregon and northern California and distinguished from other North American nanosellines by the presence of 11 rather than 10 antennal segments and absence of an elytral stridulatory file. *Porophila*-type nanosellines are widespread throughout Mexico and south temperate regions of the world. Inhabits polypore fungi.

*Suterina* Dybas 1980

*Suterella* Dybas 1961

One species, *S. microcephala* (Dybas 1961), occurs in eastern United States. Undescribed species inhabit Florida, Costa Rica, Mexico, Panama, Venezuela and Pacific Islands (Hall, in prep.). Differs from other nanosellines by the proportionally reduced and narrow head. Unlike other nanosellines which generally inhabit polypore fungi, *Suterina* has been collected in leaf or forest litter or at the base of trees or stumps.

*Throscoptilium* Barber 1924

One species, *T. duryi* Barber 1924 occurs in Indiana, Ohio. Males of an undescribed species from North Carolina possess modified, recurved protarsal segments (Hall 1999) Inhabits the undersurface of polypore fungi.

*Throscoptiloides* Hall 1999

One species, *T. norcoensis* Hall 1999 occurs in eastern and southern regions of the United States.

### Acrotrichinae Reitter 1909

Within North America, Acrotrichinae is currently represented by four genera. *Acrotrichis* is abundant and widely distributed, being the most speciose genus of Ptiliidae with over 140 described species worldwide. All North American acrotrichines possess a single, curved spine anteriorly on each side of the metascutellum, a feature absent within other ptiliid subfamilies (Hall, pers. observ.).

Genera of Acrotrichinae occurring outside of North America include *Baeocrara* Thomson 1859, *Chirostirca* Johnson 1985, and *Storicricha* Johnson 1988.

*Acrotrichis* Motschulsky 1848

*Acratrichis* Motschulsky 1850 (misspelling of *Acrotrichis* )

*Cleopterium* Gistel 1856

*Cleopteryx* Gistel 1857

*Macdonaldium* Abdullah and Abdullah 1967

*Ptilopterium* Gistel 1850

*Trichopteryx* Kirby and Spence 1826 not Huebner 1825

subgenus *Capotrichis* Johnson 1969b

subgenus *Ctenopteryx* Flach 1889

subgenus *Flachiana* Sundt 1969

More than 35 described species of *Acrotrichis* are recorded from North and Central America, occurring in a variety of habitats ranging from leaf litter, mammal nests and dung to decaying fungi. Most species near or slightly exceeding 1.0 mm in length and are generally black with golden pubescence.

*Nephanes* Thomson 1859

*Elachys* Matthews 1860

*Titan* Matthews 1858

*Zamenhofia* Vuillet 1911b

Six species occurring in Arizona, Texas, Rhode Island, New York, Pennsylvania, Florida, Mississippi, Louisiana, and Iowa. Males of a species occurring in Arizona and Texas possess highly modified basal protarsal segments and thick, pegged setae on the metacoxae (Hall, pers. observ.).

*Ptiliopycna* Casey 1924

One species, *P. moerens* Matthews 1874 [new combination, Dybas 1978a], occurs in northeastern United States and adjacent region of Canada. Inhabits swamps, bogs and mosses.

*Smicrus* Matthews 1872

Two species occurring in Michigan, Indiana, Texas, Arizona and southern California. Inhabits damp environments, under plant debris near rivers and streams.

### Cephaloplectinae, Sharp 1883

Cephaloplectines are myrmecophiles and are distinguished from all other ptiliids by their limuloid shape, enlarged prosternal process and absence of eyes and hindwings. Neotropical forms (*Cephaloplectus*) represent the largest ptiliids, reaching near 4.0 mm in total length.

All species of Cephaloplectinae are associated with one or more species of ants and in North America occur with a variety of formicid genera, including *Aphaenogaster*, *Formica*, *Lasius*, *Neivamyrmex* and *Pheidole* (Bruch 1919; Matthews 1866; Seevers and Dybas 1943; Wilson *et al.* 1954). Adult cephaloplectines appear to feed on secretions of the ant exoskeleton and can be observed riding about on their hosts (Park 1933; Wilson *et al.* 1954). In the tropics, two or more genera of Cephaloplectinae are known to occur together in a single ant colony (Hall, pers. observ.).

The genus *Limulodes* was described by Matthews (1867) and placed within Ptiliidae. LeConte (1883) placed *Limulodes* within 'Trichopterygini'. Ganglbauer (1899) and Csiki (1911) list Limulodinae as a subfamily within Ptiliidae. Sharp (1883) described *Cephaloplectus* from Panama and erected the subfamily Cephaloplectinae within Staphylinidae. Seevers and Dybas (1943) combined Limulodinae and Cephaloplectinae, creating the family Limulodidae. Dybas (1976) presented evidence that Limulodidae was closely related to but removed from Ptiliidae. Lawrence and Newton (1982) proposed placing Cephaloplectinae (=Limulodidae) within Ptiliidae. Cephaloplectinae is retained in Newton and Thayer (1992) and Lawrence and Newton (1995). Cephaloplectines are herein retained as a subfamily. Within Ptiliidae, preliminary morphological analyses suggest cephaloplectines may be most closely related to pterycines (Hall, pers. observ.).

Genera of Cephaloplectinae occurring outside of North America include *Cephaloplectus*, *Eulimulodes* Mann 1926, and *Rodwayia* Lea 1907.

*Limulodes* Matthews 1866

*Ecitoxenus* Wasmann 1900

subgenus *Carinolimulodes* Seevers and Dybas 1943

subgenus *Cephaloplectodes* Seevers and Dybas 1943

subgenus *Ecitolimulodes* Seevers and Dybas 1943

subgenus *Idiolimulodes* Seevers and Dybas 1943

subgenus *Neolimulodes* Seevers and Dybas 1943

Three species occurring in North America, ranging from eastern Canada, Massachusetts, New York, Indiana, Illinois, Kentucky, D. C., Maryland, Florida, Iowa, Texas and Arizona.

*Paralimulodes* Bruch 1919

One species, *P. wasmanni,* is recorded from Argentina (Bruch 1919) and the United States (Wilson *et al.* 1954).

### Bibliography


ABDULLAH, M. and ABDULLAH, A. 1967. *Macdonaldium fungi*, a new genus and species of the feather-winged and smallest known beetles (Coleoptera: Ptiliidae) from East Como, Quebec. Entomological News, 78: 77-83.

BACETTI, B. and CONINCK, E. DE. 1989. Immotile, aflagellate spermatozoa in Ptiliidae coleopterans. Biology of the Cell, 67: 185-191.

BARBER, H.S. 1924. New Ptiliidae related to the smallest known beetle. Proceedings of the Entomological Society of Washington, 26(6): 167-168.

BESUCHET, C. 1971. Ptiliidae. *In*: H. Freude, K.W. Harde, G.A. Lohse (eds.), Die Käfer Mitteleuropus, 3. Krefeld, Goecke und Evers. pp. 311-334.

BESUCHET, C. 1976. Contribution à l'étude des Ptiliides paléarctiques (Coleoptera). Mitteilungen der Schweizerischen Entomologischen Gesellschaft, 49: 51-71.

BOVING, A. G., and CRAIGHEAD, F.C. 1931. An illustrated synopsis of the principal larval forms of the order Coleoptera. Entomologica Americana, (N.S.), 11: 1-351.

BRETHÉS, J. 1915. Description d'un nouveau genre et d'une nouvelle espèce de Ptiliidae (=Trichopterygidae) du Chili. Revista Chilena de Historia Natural, 19: 15-17.

BRUCH, C. 1919. Un nuevo coleóptero ecitófilo. Physis, 4: 579-582.

CASEY. T.L. 1924. Additions to the known Coleoptera of North America. Memoirs on the Coleoptera, Vol. 11. Lancaster Press, Lancaster, Pa, 347 pp.

COSTA, C., VANIN, S.A., and CASARI-CHEN, S.A. 1988. Larvas de Coleoptera do Brasil. Museu de Zoologia, Universidade de São Paulo, 282 pp., 165 pls.

CROWSON, R.A. 1960. The phylogeny of Coleoptera. Annual Review of Entomology, 5: 111-134.

CROWSON, R.A. 1981. The Biology of the Coleoptera. Academic Press, New York, xii + 802 pp.

CSIKI, E. 1911. Coleopterorum Catalogus. Volume VIII. Family Ptiliidae. W. Junk, Berlin: pp. 5-61.

DEANE, C. 1930. Trichopterygidae of Australia and Tasmania. Proceedings of the Linnean Society of New South Wales, 55:477-487.

DEANE, C. 1931. Trichopterygidae of Australia and adjacent islands. Proceedings of the Linnean Society of New South Wales, 56: 237-242.

DE CONICK, E. and COESSENS, R. 1981. Life cycle and reproductive pattern of *Acrotrichis intermedia* (Coleoptera: Ptiliidae) in experimental conditions. Journal of Natural History, 15: 1047-1055.

DE MARZO, L. 1992. Osservazioni anatomiche sui genitali interni maschili in alcuni Ptilidi (Coleoptera). Entomologica (Bari), 27: 107-115.

DYBAS, H.S. 1955. New feather-wing beetles from termite nests in the American tropics (Coleoptera: Ptiliidae). Fieldiana: Zoology, 37: 561-577.

DYBAS, H.S. 1956. A new genus of minute fungus-pore beetles from Oregon. Fieldiana, 34: 441-448.

DYBAS, H.S. 1960. A new genus of blind beetles from a cave in South Africa. Fieldiana, 39(36): 399-405.

DYBAS, H.S. 1961a. A new fossil featherwing beetle from baltic amber. Fieldiana, 44(1): 1-9.

DYBAS, H.S. 1961b. Two new genera of feather-wing beetles from the eastern United States. Fieldiana, 44(2): 11-18.

DYBAS, H.S. 1961c. A new genus of feather-wing beetles from termite nests in Bolivia. Fieldiana, 44(8): 57-62.

DYBAS, H.S. 1966. Evidence for parthenogenesis in the featherwing beetles, with a taxonomic review of a new genus and eight new species. Fieldiana, 51: 11-52.

DYBAS, H.S. 1976. The larval characters of featherwing and limulodid beetles and their family relationships in the Staphylinoidea. Fieldiana, 70(3): 29-78.

DYBAS, H.S. 1978a. The systematics geographical and ecological distribution of *Ptilopycna*, a nearctic genus of parthenogenetic featherwing beetles. American Midland Naturalist, 99(1): 83-100.

DYBAS, H.S. 1978b. Polymorphism in featherwing beetles, with a revision of the genus *Ptinellodes* (Coleoptera: Ptiliidae). Annals of the Entomological Society of America, 71(5): 695-714.

DYBAS, H.S. 1980. *Suterina* Dybas, replacement name for *Suterella* Dybas, preoccupied (Coleoptera: Ptiliidae). Coleopterists Bulletin, 34(3):261.

DYBAS, H.S. 1990. Ptiliidae, pp.1093-1112. *In*: D. Dindal (ed.), Soil Biology Guide. John Wiley and Sons, New York.

DYBAS, H.S. 1991. Ptiliidae, Limulodidae (Staphylinoidea), pp. 322-325. *In*: An introduction to immature insects of North America. Ed., F.W. Stehr. Vol. 2. Kendall and Hunt Publishing Co., Dubuque, Iowa. xvi + 975 pp.

DYBAS, L.K. and DYBAS, H.S. 1981. Coadaptation and taxonomic differentiation of sperm and spermathecae in featherwing beetles. Evolution, 34(1): 168- 174.

DYBAS, L.K. and DYBAS, H.S. 1987. Ultrastructure of mature spermatozoa of a minute featherwing beetle from Sri Lanka (Coleoptera: Ptiliidae: Bambara). Journal of Morphology, 191: 63-67.

ERICHSON, W.F. 1845. Naturgeschichte der Insecten Deuttschlands. Erste Abtheilung, Coleoptera, Vo. 3, pp. 1-320. Verlag der Nicolaischen Buchhandlung, Berlin.

FLACH, C. 1888. *In* G. Seidlitz, 1887-1891. Fauna Baltica. Die Kafer (Coleoptera) der Deutschen Ostseeprovinzen Russlands. Zweite neu bearbeitete Auflage. Hartungsche Verlagsduckerei, Konigsberg. vi + 192 + 818 pp.

FLACH, C. 1889. Bestimmungstabelle der Trichopterygidae des europäischen Faunengebietes. Verhandlungen der Zoologisch-botanischen Gesellschaft in Wien, 39: 481-532.

FOGEL, R., and PECK, S. 1975. Ecological studies of hypogeous fungi. I. Coleoptera associated with sporocarps. Mycologia, 67: 741-747.

FORBES, W. T. M. 1926. The wing folding patterns of the Coleoptera. Journal of the New York Entomological Society, 34: 42-115, pls. 7-18.

FRIEDENREICH, C.W. 1883. Pilzbewohnende Käfer in der Provinz Santa Catharina (Südbrasilien). Entomologische Zeitung (Stettin), 44: 375-380.

GANGLBAUER, L. 1899. Die Käfer von Mitteleuropa. Die Käfer der österreichisch-ungarischen Monarchie, Deutschlands, der Schweiz, sowie des französischen und italienischen Alpengebietes. Vol. 3, Familienreihe Staphylinoidea, II. Carl Gerold's Sohn, Vienna. iii + 1046 pp.

GISTEL, J. 1850. *In*: J. Gistel and T. Bromme, Handbuch der Naturgeschichte aller drei Reiche, für Lehrer und Lernende, für Schule und Haus. Hoffman'sche Verlags-Buchhandlung, Stuttgart. 1037 pp.

GISTEL, J. 1856. Die Mysterien der Europäischen Insectenwelt. T.Dannheimer, Kempten. xx + 530 pp.

GISTEL, J. 1857. Vacuna oder die Geheimnisse aus der organischen und Welt. Vol. 2. Schorner'schen Buchhandlung, Straubing, 1031 pp.

GRAVES, R.C. 1960. Ecological observations on the insects and other inhabitants of woody shelf fungi (Basidiomycetes: Polyporaceae) in the Chicago area. Annals of the Entomological Society of America, 53: 61-78.

GYLLENHAL, L. 1827. Insecta Svecica, Classis I. Coleoptera sive Eleuterata. Vol. 1, Part 4, x + 761 pp. F. Fleischer, Lipsiae (Leipzig).

HALDEMANN, S.S. 1848. Descriptions of North American Coleoptera, chiefly in the cabinet of J. L. Le Conte, M.D., with references to described species. Journal of the Academy of Natural Sciences of Philadelphia, (2)1: 95-110.

HALL, W. E. 1999. Generic revision of the tribe Nanosellini (Coleoptera: Ptiliidae: Ptiliinae). Transactions of the American Entomological Society, 125 (1-2): 29-126.

HAMMOND, P.M. 1979. Wing-folding mechanisms of beetles, with special refernce to investigations of Adephagan phylogeny (Coleoptera), pp. 113-180. *In*: Erwin, T.L., Ball, G.E., Whitehead, D.R. (eds.), Carabid Beetles: Their Evolution, Natural History and Classification. W. Junk, Dordrecht.

HANSEN, M. 1997. Phylogeny and classification of the staphyliniform beetle families (Coleoptera). The Royal Danish Academy of Sciences and Letters, Biologiske Skrifter, 48: 339pp.

HINTON, H.E. 1941. The immature stages of *Acrotrichis fascicularis* (Herbst) Col. Ptiliidae). Entomologists Monthly Magazine, 77: 245-250.

HLAVAC, T.F. 1975. The prothorax of Coleoptera (Except Bostrichiformia-Cucujiformia). Bulletin of the Museum of Comparative Zoology, 147: 137-183.

JAMIESON, B.G.M. 1987. The Ultrastructure and Phylogeny of Insect Spermatozoa. Cambridge University Press, 320 pp.

JOHNSON, C. 1967. Studies on Ethiopian Ptiliidae. 1, *Africoptilium* gen. n. from Central Africa. Entomologist, 100: 288-292.

JOHNSON, C. 1969a. A new genus and species of Ptiliidae (Col.) from Tamana Cave, Trinidad. Entomologist, 102: 145-148.

JOHNSON, C. 1969b. The genus *Acrotrichis* Motschulsky (Coleoptera: Ptiliidae) in the Ethiopian Region (Studies on Ethiopian Ptiliidae, No. 2). Revue de Zoologie et de Botanique Africaines, 79: 213-260, 7 pls.

JOHNSON, C. 1971. Some ptiliidae from the Philippine, Bismarck and Solomon Islands (Insecta, Coleoptera). Steenstrupia, 2 (4): 39-47.

JOHNSON, C. 1975. Ptiliidae. Mission entomologique du Musee Royal de l'Afrique Centrale aux Monts Uluguru, Tanzanie. Rev. Zool. Afr. 89 (3): 719-722.

JOHNSON C. 1982a. An introduction to the Ptiliidae (Coleoptera) of New Zealand. New Zealand Journal of Zoology, 9: 333-376.

JOHNSON, C. 1985. Revision of Ptiliidae (Coleoptera) occurring in the Mascarenes, Seychelles and neighbouring islands. Entomologica Basiliensia, 10: 159-237.

JOHNSON, C. 1988. Revision of Sri Lankan acrotrichines (Coleoptera: Ptiliidae). Revue Suisse de Zoologie, 95: 257-275.

KIRBY, W. and SPENCE, W. 1826. An introduction to entomology, or elements of the natural history of insects. Vol. 3. Longman, Rees, Orme, Brown and Green, London. viii + 732 pp , 20 pls.

LAWRENCE, J.L. 1989. Mycophagy in the Coleoptera: feeding strategies and morphological adaptations, pp. 1-23. *In*: N. Wilding, N.M. Collins, P.M. Hammond, J. F. Webber (eds.), Insect-Fungus Interactions. Academic Press, London, 344 pp.

LAWRENCE, J.L. and NEWTON, A.F., JR. 1982. The evolution and classification of beetles. Annual Review of Ecology and Systematics, 13: 261-290.

LAWRENCE, J.F. and NEWTON, , A.F., JR. 1995. Families and subfamilies of Coleoptera (with selected genera, notes, references and data on family-group names). pp. 779-1006. *In:* Pakaluk, J. and Slipinski, S.A. (ed.), Biology, Phylogfeny, and Classification of Coleoptera: Papers Celebrating the 80th Birthday of Roy A. Crowson.

LEA, A.M. 1907. On a new and remarkable genus of blind beetles from Australia and Tasmania of the family Trichopterygidae. Tasmanian Naturalist, 1: 14-16.

LECONTE, J.L. 1863. New species of North American Coleoptera. Part 1. Smithsonian Miscellaneous Collections, 167: 62-63.

LECONTE, J. and HORN, G.H. 1883. Classification of the Coleoptera of North America. Smithsonian Miscellaneous Collections, 507: xxxviii + 567 pp.

LESCHEN, R.A.B. 1993. Evolutionary patterns of feeding in selected Staphylinoidea (Coleoptera): shifts among food textures, pp. 59-104. *In*: Schaefer, C.W. and Leschen, R.A.B. (eds.), Functional Morphology of Insect Feeding. Thomas Say Publications in Entomology, Entomological Society of America, Lanham, Maryland.

LUNDGREN, R.W. 1983. *Dybasina*, a new name for *Termitopteryx* Dybas (Coleoptera: Ptiliidae). Coleopterists Bulletin, 37: 272.

MANN, W.M. 1926. New neotropical myrmecophiles. Journal of the Washington Academy of Sciences, 16: 448-455.

MATTHEWS, A. 1858. A synonymic list of the British Trichopterygidae. Zoologist, 16: 6104-6111.

MATTHEWS, A. 1860. Notes on the British Trichopterygidae, with descriptions of some new species. Zoologist, 18: 7063-7068.

MATTHEWS, A. 1862. A review of the genus *Ptinella*. Zoologist, 20: 8053-8060.

MATTHEWS, A. 1866. Description of a new genus of Trichopterygidae, lately discovered in the United States. Annals of the Lyceum of Natural History of New York, 8: 406-413, pl. 15.

MATTHEWS, A. 1872. Trichopterygia Illustrata et Descripta. E.W. Janson, London, 188 pp, 30 plates.

MATTHEWS, A. 1882. Descriptions of three new species of Trichopterygia, found by the Rev. T. Blackburn in the Sandwich Islands. Cistula Entomologica, 3 (26): 39-42, pl. 2.

MATTHEWS, A. 1884. Synopsis of North American Trichopterygida. Transactions of the American Entomological Society, 11: 113-156.

MATTHEWS, A. 1889. New genera and species of Trichopterygidae. Annals and Magazine of Natural History, (6) 3: 188-195.

MATTHEWS, A. 1900. Trichopterygia Illustrata et Descripta. Supplement. O.E. Janson and Son, London, 112 pp, 7 pls.

MOTSCHULSKY, V. 1844. Bemerkungen zu dem im Vten Bande der Zeitschrift für die Entomologie p. 192 von H.rn Maerkel gegebenen 'Beitraege zur Kenntniss der unter Ameisen lebenden Inseckten'. Bulletin de la Société Impériale des Naturalistes de Moscou, 17: 812-823.

MOTSCHULSKY, V. 1848. Kritische Beurtheilung von Dr. Erichson's Naturgeschichte der Insecten Deutschland's, und einiger anderen entomologischen Schriften in monographischer Zusammenstellung und besonderer Berucksichtigung der in Russland vorkommenden Arten. Bulletin de la Société Impériale des Naturalistes de Moscou, 21 (Première partie): 544-569, 1 pl.

MOTSCHULSKY, V. 1850. Kritische Beurtheilung von Dr. Erichson's Naturgeschichte der Insekten Deutschland's (Fortsetzung). Bulletin de la Société Impériale des Naturalistes de Moscou, 23 (Première partie): 195-257, 1 pl.

MOTSCHULSKY, V. 1855. Voyages. Lettre de M. de Motschulsky à M. Ménétriés. No. 2. A bord du bateau à vapeur United-States, 20 Mars 1854. Études Entomologiques, 4: 8-25.

MOTSCHULSKY, V. 1868. Énumération des nouvelles especes de Coléopterès, rapportés de ses voyages. Bulletin de la Société Impériale des Naturalistes de Moscou, 41(2): 170-192.

NEWTON, A.F., JR. 1984. Mycophagy in Satphylinoidea (Coleoptera), pp. 302-353. *In*: Q. Wheeler, M. Blackwell (eds.), Fungus-Insect Relationships: Perspectives in Ecology and Evolution. Columbia University Press, New York, 514 pp.

NEWTON, A.F., JR. 1991. Leiodidae, Leptinidae (Staphylinoidea), pp. 327-328, 330. *In*: F.W. Stehr (ed.), Immature Insects. Vol. 2. Kendall/Hunt Publishing Co., Dubuque, Iowa.

NEWTON, A.F., JR. and THAYER, M.K. 1992. Current classification and family-group names in Staphyliniformia (Coleoptera). Fieldiana: Zoology (N.S.), 67: 1-92.

PARK, O. 1933. Ecological study of the ptiliid myrmecocole, *Limulodes paradoxus* Matthews. Annals of the Entomological Society of America, 26: 255-261.

PARK, O., and AUERBACH, S. 1954. Further study of the tree-hole complex with emphasis on quantitative aspects of the fauna. Ecology, (35) 2: 208-222.

PARSONS, C.T. 1939. A ptiliid beetle from Baltic Amber in the Museum of Comparative Zoology. Psyche, 46: 62-64.

PAULIAN, R. 1941. Les premiers états des Staphylinoidea (Coleoptera). Étude de morphologie comparée. Mémoires du Muséum National d'Histoire Naturelle (N.S.), 15: 1-361.

PERRIS, E. 1862. Histoire des insectes du pin maritime. Supplément aux Coléoptères et rectifications. Annales de la Société Entomologique de France, (4) 2: 173-243, pls. 11-12.

SEEVERS, C.H. and DYBAS, H.S. 1943. A synopsis of the Limulodidae (Coleoptera): A new family proposed for myrmecophiles of the subfamilies Limulodinae (Ptiliidae) and Cephaloplectinae (Staphylinidae). Annals of the Entomological Society of America, 36: 546-586.

SHARP, D. 1883. Subfamily Cephaloplectinae, pp. 295-297. *In*: Biologia Centrali-Americana. Insecta. Coleoptera, (1) 2. Taylor and Francis, London.

SHARP, D. and MUIR, F. 1912. The comparative anatomy of the male genital tube in Coleoptera. Transactions of the Entomological Society of London, Part III: 477-642, pls. 42-78.

SILVESTRI, F. 1946. Primo contributo alla conoscenza dei Termitofili viventi con specie de *Syntermes*. Commentationes Pontificia Academia Scientiarum, 9: 515-559.

SORENSSON, M. 1997. Morphological and taxonomical novelties in the world's smallest beetles, and the first Old World record of Namosellini (Coleoptera: Ptiliidae) Systematic Entomology, 22: 257-283.

STATZ, G. and HORION, A. 1937. Ein fossiler Ptiliidenfund aus den mitteloligocanen Ablagerungen von Rott am Siebengebirge. Ent. Blatt., 38: 8-10.

SUNDT, E. 1969. Description of a new subgenus, *Flachiana*, and four new species of the genus *Acrotrichis* Motschulsky, 1848 (Col., Ptiliidae). Norsk Entomologisk Tidsskrift, 16: 49-53.

TAVARES, I. I. 1979. The Laboulbeniales and their arthropod hosts, pp. 229-258. *In*: L.H. Batra (ed.), Insect-Fungus Symbiosis. Nutrition, Mutualism and Commensalism. Allanheld, Osmun and Co., John Wiley and Sons, 276 pp.

TAYLOR, V. 1980. Coexistence of two species of *Ptinella* Motschulsky (Coleoptera: Ptiliidae) and the significance of their adaptation to different temperature ranges. Ecological Entomology, 5: 397-411.

TAYLOR, V. 1981. The adaptive and evolutionary significance of wing polymorphism and parthenogenesis in *Ptinella* Motschulsky (Coleoptera: Ptiliidae). Ecological Entomology, 6: 89-98.

TAYLOR, V.A. 1982. The giant sperm of a minute beetle. Tissue and Cell, (14) 1:13-123.

THOMSON, C.G. 1859. Skandinaviens Coleoptera, synoptiskt bearbetade. Vol. 1. Berlingska Boktryckeriet, Lund. 290 pp.

VUILLET, A. 1911a. Description d'un Trichopterygidae de l'Afrique occidentale française (Col.). Insecta (Rennes), 1: 159-161.

VUILLET, A. 1911b. Un nouveau Trichopterygidae du Soudan français (Col.). Insecta (Rennes), 1: 219-221.

WASMANN, E. 1900. Neue Dorylinengaste aus dem neotropischen und dem äthiopischen Faunengebiet. Zoologische Jahrbücher, Abtheilung für Systematik, Geographie und Biologie der Thiere, 14: 215-289, pls. 13-14.

WILSON, E.O., EISNER, T., and VALENTINE, B.D. 1954. The beetle genus *Paralimulodes* Bruch in North America, with notes on morphology and behaviour (Coleoptera: Limulodidae). Psyche: 154-161.

# 18. AGYRTIDAE C.G. Thomson, 1859

by Stewart B. Peck

**Common name:** The primitive carrion beetles

**Family synonyms:** Silphidae, part: inc. Lyrosominae, Pterolominae

Members of the Agyrtidae are small- to medium-sized, light to dark brownish beetles, 4-14 mm long, and are often associated with decaying organic material. They have traditionally been considered as part of the silphid carrion beetles, but they have lately been recognized as a separate family, and more closely related to leiodids than silphids (Newton and Thayer, 1992). Agyrtid larvae are the most generalized known in Staphylinoidea (Newton, 1997). The ancient origin of the family is indicated by the fact that one of the earliest known fossil staphylinoids, the early Jurassic genus *Mesecanus (= Mesagyrtes),* probably belongs here, although it was originally placed in Silphidae. These beetles are not frequently encountered, even by most collectors.

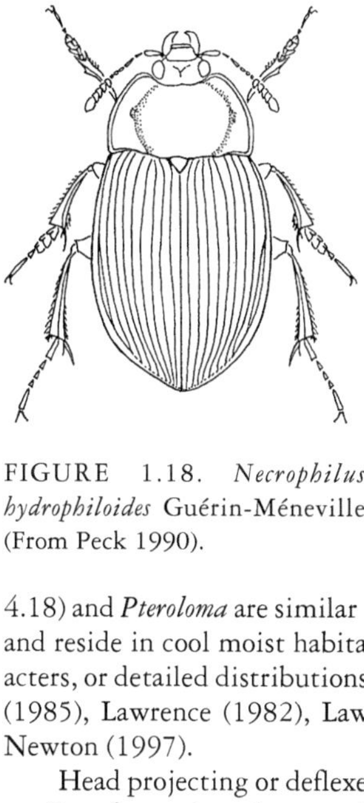

FIGURE 1.18. *Necrophilus hydrophiloides* Guérin-Méneville (From Peck 1990).

**Description:** Adult members of the family can be recognized by their oval to oblong-elongate, slightly flattened shape; generally small to medium (4-14 mm) size; brownish, usually shiny, nine- or ten-striate non-truncate elytra, which cover the entire abdomen; antennae filiform to gradually clavate, never with the eighth antennomere smaller than the seventh and ninth antennomeres; and there are five (rarely six) visible abdominal sterna (always six (rarely seven) in the Silphidae). Members of *Apteroloma* (Fig. 4.18) and *Pteroloma* are similar to carabids in general appearance and reside in cool moist habitats. Additional description, characters, or detailed distributions are given by Anderson and Peck (1985), Lawrence (1982), Lawrence and Newton (1982), and Newton (1997).

Head projecting or deflexed; eyes conspicuous, usually protruding; frontoclypeal suture distinct. Antennae usually ending in a distinct four- or five-segmented club, sometimes almost filiform or very weakly clubbed; antennal sensilla located in apical grooves (gutters) on the club segments in most genera; antennal insertions more or less exposed.

Thorax with pronotum larger than head. Pronotum with complete lateral edges, notosternal suture distinct. Prosternum short in front of the coxae and bearing a narrow but complete intercoxal process. Procoxae transverse, projecting, and subcontiguous, with large, exposed trochantins; procoxal cavities narrowly open posteriorly and open internally. Mesocoxae narrowly separated. Elytra completely covering the abdomen; punctation striate, as 9 or 10 longitudinal rows of punctures; epipleura well developed and complete, or extending to posterior four fifths. Metacoxae barely excavate mesally, contiguous, extending laterally to meet the elytra. Hindwing with a large, discrete jugal lobe, four anal veins, and simple, concave and convex, transverse folds; occasionally absent.

Abdomen with five (or rarely six) visible abdominal sterna; lateral portions of sternum 2 present beneath the metacoxae. Intersegmental areas lacking a pattern of microsclerites. Several basal terga membranous. Aedeagus may or may not have a separate basal piece; the parameres may be vestigial, fused, or absent.

Larvae (see Newton 1991) campodeiform; with moderately sclerotized terga and sterna; mandibles bearing several apical teeth; prostheca narrow and acute, associated with a brush of hairs; mandibles possessing a large and tuberculate molar part with dorsal and ventral rows of denticles; maxilla with distinct galea and lacinia; galea usually fringed; labium with strongly bilobed ligula; six ocelli on each side of head; urogomphi two-segmented, often with a multiannulate apical segment. Spiracles placed in posterolateral emarginations of the terga. Abdomen segment ten elongate and cylindrical, with eversible membranous lobes bearing numerous minute hooks.

**Habits and habitats.** What little is known of the biology of these beetles indicates that adults are scavengers of dead or decaying organic material. Habitat associations vary, but most species are found in moist or wet habitats, particularly along or under rocks at the margins of mountain streams or near high altitude snowfields (where they appear to scavenge on wind-blown insects), in leaf litter, and in association with some fungi in the soil or under bark. They seem to be cold-adapted and are primarily active in the cooler months of fall, winter, and spring. Larval habits are inadequately known, but are probably similar to those of the adults.

In contrast to silphids, agyrtids are rarely collected and are often difficult to find. Adults of some species, especially *Necrophilus,* will come to small carrion baits or carrion-baited traps in the cooler months or in cool cave entrances. However, most members of the family are rarely trapped, and are usually only hand-collected. Sifting fungi or material under rotting bark occasionally yields species of *Agyrtes* (Fig. 2.18) and *Ipelates* (Fig. 3.18). Species of *Apteroloma* (Fig. 4.18) and *Pteroloma* are found among gravel and moss on the banks of mountain streams, in washed-up river or beach debris,

**Acknowledgments.** I thank A.F. Newton, Jr., for his generous sharing of specimens and research results over many years, and all others who have helped my study of these beetles. Jarmila Peck prepared the illustrations and shared collecting trips for 28 years.

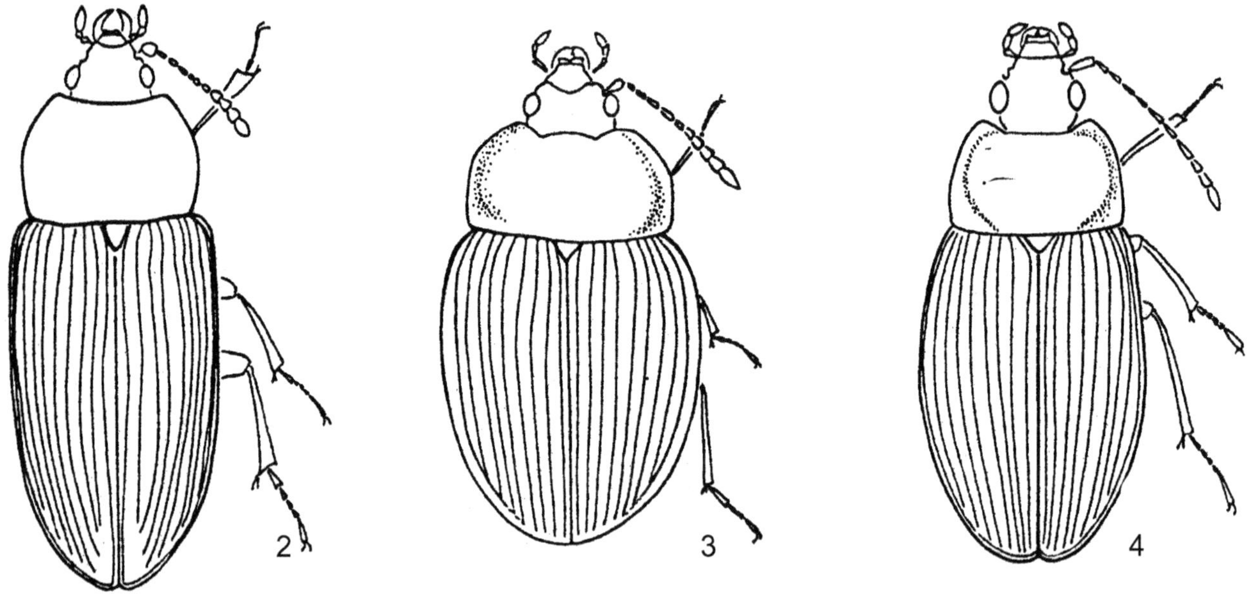


FIGURES 2.18 - 4.18. Fig. 2.18, *Agyrtes longulus* (LeConte), habitus. Fig. 3.18, *Ipelates latus* (Mannerheim), habitus. Fig. 4.18, *Apteroloma tenuicorne* (LeConte), habitus (From Peck 1990).

and, in the case of *Apteroloma,* on high altitude snowfields or under rocks at the edges of snowfields, where they scavenge on dead insects. *Lyrosoma* seemingly lives in Alaskan beach wrack. All species are night active. Many of the species are flightless. The group is of little economic importance.

**Status of the classification.** The North American species were first revised by Horn (1880) and subsequently reviewed by Hatch (1957), Miller and Peck (1979), Anderson and Peck (1985), and Peck (1990). The world subfamily ranking has been revised by Newton (1997).

**Distribution.** There are eight genera and 61 species worldwide (Newton 1997). They are almost all north temperate in distribution in North America, Europe, and Asia, but some extend southward to cool-temperate regions of high mountains on the Mexican Plateau, the Himalayan area, and Japan. Two disjunct relict species of *Zeanecrophilus* occur in New Zealand.

At present, six genera and 11 species are known in North America, north of Mexico; eight of these have distributions extending into Canada and Alaska. All are found in western North America except the one eastern *Necrophilus* species. Keys to species and details of biology and distribution are in Anderson and Peck (1985), Peck and Miller (1993), and Peck (1990).

## Key to the Nearctic Genera

1. Mandible without preapical teeth (Fig. 5.18); antenna with 2, or more, preapical antennomeres each with an apical groove containing a dense concentration of sensory setae (Fig. 9.18) ................................ 2

— Mandible with 1 or 2 large preapical teeth on inner margin (Fig. 6.18); antenna with all antennomeres lacking apical sensory grooves (Fig. 10.18) ............. 5

2(1). Elytron at middle with 10 rows of strongly punctate striae; dorsal ridge of elytral epipleuron depressed behind shoulder (Fig. 13.18); maxilla with last palpomere conspicuously swollen (Fig. 8.18) ...... ................................................................ *Agyrtes*

— Elytron at middle with 9 rows of weakly or strongly punctate striae; dorsal ridge of elytral epipleuron evenly rounded behind shoulder (Fig. 14.18); maxilla with last palpomere weakly or not swollen, subcylindrical or cylindrical (Fig. 7.18) .............. 3

3(2). Body form elongate; pronotum heart-shaped, widest in anterior one-half, much narrower than elytra at base (Fig. 11.18) .......................................... *Lyrosoma*

— Body form ovoid; pronotum not heart-shaped, almost as wide as elytra at base ................................... 4

4(3). Length greater than 8 mm; pronotum with lateral margins widely flattened (Fig. 1.18) ......... *Necrophilus*

— Length less than 8 mm; pronotum with lateral margins only narrowly flattened (Fig. 3.18) ........... *Ipelates*

5(1). Pronotum with distinct rounded depressions at middle of base and in posterior corners (Fig. 12.18) ....... ............................................................ *Pteroloma*

— Pronotum lacking such distinct basal depressions (Fig. 4.18) ................................................. *Apteroloma*

## Classification of the Nearctic Genera

### Pterolomatinae Thomson 1862

*Apteroloma* Hatch 1927, 4 spp., British Columbia to California and Arizona; other species mountains of Mexico; U.S. species key in Anderson and Peck (1985).

*Alloloma* Semenov-Tian-Shanskij 1932.
*Garytes* Mroczkowski 1966.
*Pterolorica* Hlisnikovsky 1968.

*Pteroloma* Gyllenhal 1827, 1 sp., *P. nebrioides* Brown 1933, Alberta, British Columbia, Montana.

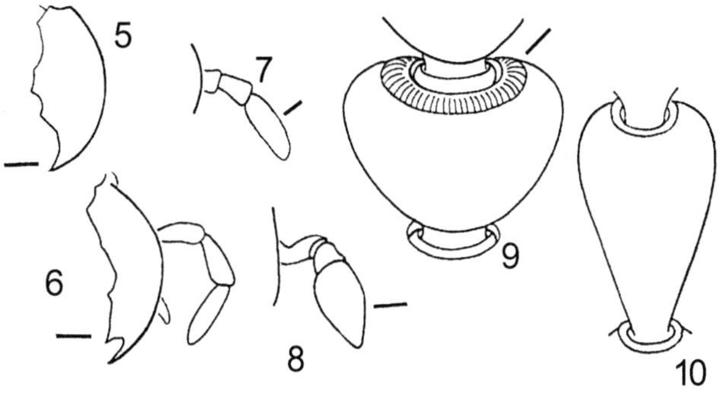


FIGURES 5.18 - 10.18. Fig. 5.18, *Necrophilus hydrophiloides* Guérin-Méneville, dorsal view of mandible. Fig. 6.18, *Pteroloma nebrioides* Brown, dorsal view of mandible. Fig. 7.18, *Ipelates latus* (Mannerheim), maxillary palpus, last segment normal. Fig. 8.18, *Agyrtes longulus*, maxillary palpus, last segment swollen. Fig. 9.18, *Necrophilus hydrophiloides*, antennomere 9 with groove containing sensory structures. Fig. 10.18, *Pteroloma nebrioides*, antennomere 9 without groove containing sensory structures.

*Adolus* Fischer 1828.
*Holocnemis* Schilling 1829.

Agyrtinae Thomson 1859

*Agyrtes* Frölich 1799, 2 spp., Pacific Coast; Alaska to Idaho to California; species key in Peck (1974), Anderson and Peck (1985).
*Lendomus* Casey 1924.
subgenus *Agyrtes* Frölich 1799.
subgenus *Agyrtecanus* Reitter 1901.

*Ipelates* Reitter 1884, 1 sp., *I. latus* (Mannerheim) 1852, Alaska to California.
*Pelates* Horn 1888, not Cuvier and Valenciennes 1829.
*Sphaeroloma* Portevin 1905.
*Pelatines* Cockerell 1906.
*Brachyloma* Portevin 1914, not Chambers 1878.
*Necrophilodes* Champion 1933; Hatch 1927 (subseq. missp. as *Necrophiloides*).

*Lyrosoma* Mannerheim 1853, 1 sp., *L. opacum* Mannerheim 1853, Alaska (Aleutian and Pribilof Islands).

Necrophilinae Newton 1997

*Necrophilus* Latreille 1829, 2 spp., *N. hydrophiloides* Guerin-Meneville 1835, Alaska to California, winged; *N. pettitii* Horn 1880, s. Ontario to n. Florida, west to Louisiana and Missouri, flightless.
*Necrobius* Gistel 1834.
*Paranecrophilus* Shibata 1969.
*Pseudosilpha* Schawaller 1978.

## Bibliography

ANDERSON, R.S. and PECK, S.B. 1985. The insects and arachnids of Canada and Alaska, part 13. The carrion beetles of Canada and Alaska. Coleoptera: Silphidae and Agyrtidae. Research Branch, Agriculture Canada, Ottawa, Publication 1778. 121 pp.

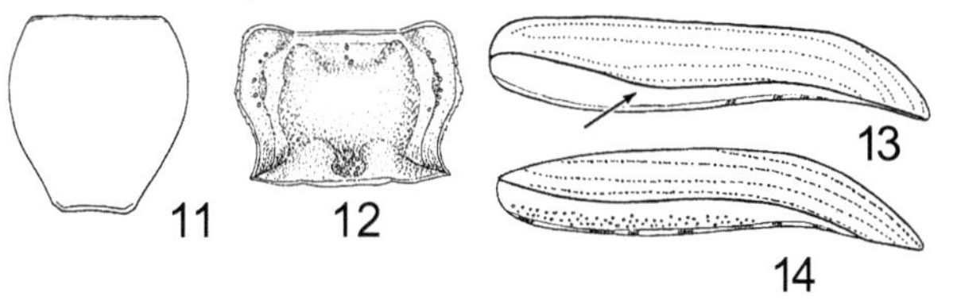


FIGURES 11.18 - 14.18. Fig. 11.18, *Lyrosoma opacum* Mannerheim, heart-shaped pronotum. Fig. 12.18, *Pteroloma nebroides*, pronotum, with basal depressions. Fig. 13.18, *Agyrtes longulus* (LeConte), outer edge of elytron, showing depression of epipleural ridge. Fig. 14.18, *Apteroloma tenuicorne* (LeConte), outer edge of elytron, showing evenly rounded epipleural ridge.

DINDAL, D. (Ed.) 1990. Soil Biology Guide. John Wiley & Sons, NY.

HATCH, M.H. 1957. The beetles of the Pacific northwest, part II. Staphyliniformia. Univ. Wash. Publs. Biol., Vol. 16. University of Washington Press, Seattle, WA. 386 pp. (Pacific Northwest spp.).

HORN, G.H. JR. 1880. Synopsis of the Silphidae of the United States with reference to the genera of other countries. Transactions of the American Entomological Society, 8: 219-319, pls. 5-7.

LAWRENCE, J.F. 1982. Coleoptera. pp. 482-553. *In* Parker, S.P., (ed). Synopsis and classification of living organisms. Vol. 2. McGraw Hill Co., New York.

LAWRENCE, J.F. and NEWTON, A.F. JR. 1982. Evolution and classification of beetles. Annual Review of Ecology and Systematics, 13: 261-290.

MILLER, S.E. and PECK, S.B. 1979. Fossil carrion beetles of Pleistocene California asphalt deposits, with a synopsis of Holocene California Silphidae (Insecta: Coleoptera: Silphidae). Transactions of the San Diego Society of Natural History, 19: 85-106.

NEWTON, A.F. 1991. Agyrtidae (Staphylinoidea). *In* Stehr, F.W. (ed.), Immature Insects, Vol. 2. Kendall/Hunt, Dubuque, Iowa.

NEWTON, A.F. 1997. Review of Agyrtidae (Coleoptera), with a new genus and species from New Zealand. Annals Zoologici (Warsaw), 47: 111-156.

NEWTON, A.F. and THAYER, M.K. 1992. Current classification and family-group names in Staphyliniformia (Coleoptera). Fieldiana: Zoology (N.S.), 67: 1-92.

PECK, S.B. 1974. A review of the *Agyrtes* (Silphidae) of North America. Psyche, 81: 501-506.

PECK, S.B. 1981. Distribution and biology of flightless carrion beetle *Necrophilus pettitii* in eastern North America (Coleoptera; Silphidae). Entomological News, 92: 181-185.

PECK, S.B. 1990. Insecta: Coleoptera: Silphidae and the associated families Agyrtidae and Leiodidae. pp. 1113-1136. *In* Dindal, D., (ed.). Soil Biology Guide. John Wiley and Sons, New York.

PECK, S.B. and MILLER, S.E. 1993. A catalogue of the Coleoptera of America north of Mexico. Family: Silphidae. Agricultural Research Service, U.S. Department of Agriculture, Washington, DC. Agriculture Handbook 529-28, 24 pp.

# 19. LEIODIDAE Fleming, 1821

by Stewart B. Peck

**Common name:** The round fungus beetles, including the small carrion beetles and mammal nest beetles

**Family synonyms:** Anisotomidae Stephens, 1829; Liodidae Reitter, 1884, incl. Camiaridae Jeannel, 1911; Catopidae Thomson, 1859; Cholevidae Kirby, 1837; Colonidae Horn, 1880; Leptinidae LeConte, 1866; Leptodiridae Hatch, 1933; Platypsyllidae Ritsema, 1869; Sogdiidae Lopatin, 1961

Most of these beetles are best distinguished by their eighth antennomere, which is usually notably smaller than the seventh and ninth. Many also have a rather thin integument (and break easily), and frequently have a granular surface, or transverse striations on the thorax and elytra, or elytra with punctate striae, or are smooth and glossy.

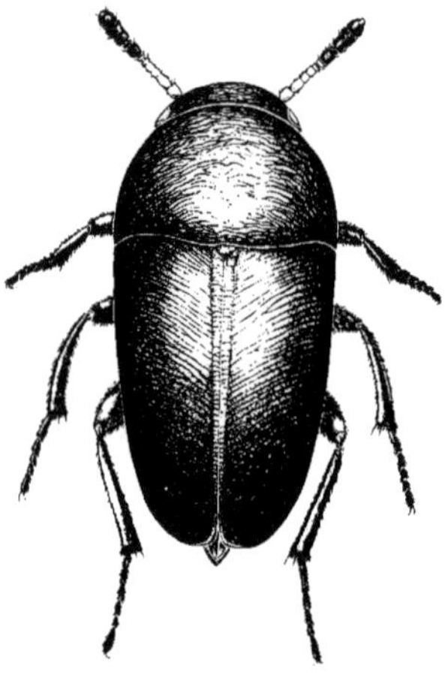

FIGURE 1.19. *Ptomaphagus consobrinus* (LeConte) (From Peck 1990).

**Description:** Shape broadly ovate to elongate, strongly convex to slightly flattened, and glabrous or clothed with decumbent or occasionally erect hairs; color pale brownish to brown to black, rarely with colored areas; body 1-8 mm (usually 1.5-4 mm) in length.

Head sometimes constricted posteriorly, forming a neck; frontoclypeal suture occasionally distinct. Antennae usually ending in a more or less distinct, interrupted five-segmented club, most often antennomere 8 is smaller than either 7 or 9; club occasionally with 3 or 4 antennomeres and not interrupted; periarticular gutters and usually one or more internal vesicles open on the distal surfaces of two or three apical segments; antennal insertions almost always exposed. Apical palpomere of maxilla often acute, occasionally much narrower than the preapical palpomere, and rarely enlarged.

Prothorax lateral edges almost always complete. Procoxae more or less contiguous, and trochantins either exposed or concealed, their cavities widely to narrowly open posteriorly and usually closed internally. Mesocoxae narrowly separated in most species. Elytra almost entire, concealing the abdomen completely, occasionally one or two abdominal terga exposed; epipleura usually well-developed but seldom complete. Metepisterna sometimes concealed by elytral epipleura. Metacoxae usually contiguous, either not excavate or with very slight plates mesally; sometimes the posterior face vertical, so that the metasternum and abdomen are on different planes; metacoxae usually extending laterally to meet elytra. Tarsi variable; usually 5-5-5; sometimes 4-4-4, 3-3-3, or heteromerous (5-5-4, 5-4-4). Basal tarsomeres of protarsi usually expanded in males, and narrow in females.

Abdomen usually with six visible sterna, occasionally five or four; first two rarely connate. Intersegmental areas without a pattern of microsclerites, several of the basal tergites membranous. Aedeagus sometimes with parameres attached to a separate basal piece.

Larvae (see Newton 1991a, 1991b) lightly to moderately sclerotized dorsally, occasionally short, broad, and heavily sclerotized; often clothed with modified setae. Ocelli often absent, but usually three on each side, occasionally five, two, or one. Mandibles usually with two or three apical teeth; prostheca usually slender, often well-developed and tuberculate or asperate, with dorsal and ventral tubercles or rows of denticles; mola occasionally reduced. Maxillae usually with apically distinct galea and lacinia; galea usually fringed. Urogomphi often with a multiannulate apical segment; in rare cases they may be absent.

**Habits and habitats.** Many leiodids, especially members of the Cholevinae (the small carrion beetles), are scavengers in leaf litter and moist decaying matter. Some occur in mammal or bird nests and burrows, tortoise burrows, in caves, and with harvester ants. Most can be trapped with dung or carrion baits. Several Leiodinae (the round fungus beetles) feed on spores or other tissue of various fungi, including Myxomycetes, Gasteromycetes, and Ascomycetes, while others, such as *Colon, Catopocerus,* and *Leiodes,* seemingly specialize in subterranean fungi (Newton 1984). The highly modified *Glacicavicola* has been found only in ice caves in northwestern North America. The Platypsyllinae (the mammal nest beetles) are scavengers or ectoparasites on some rodents and insectivores and may be found on the bodies or in the nests of their hosts (beavers, mountain beavers, shrews, moles, and various voles and mice). Various genera are flightless and eyeless or blind (most cave-inhabiting *Ptomaphagus,* all Platypsyllinae, all Catopocerinae). Others have reduced eyes and no flight wings and occur in deep soil and litter habitats, often at higher elevations. The winged species are best caught with baited traps or large area flight-intercept traps (Peck and Davies 1980), and these may measure seasonal activity and species diversity (Peck and

**Acknowledgments.** I thank A.F. Newton, Jr., for his generous sharing of specimens and research results over many years, and all others who have helped my study of these beetles. Jarmila Peck prepared the illustrations and shared collecting trips for 28 years.

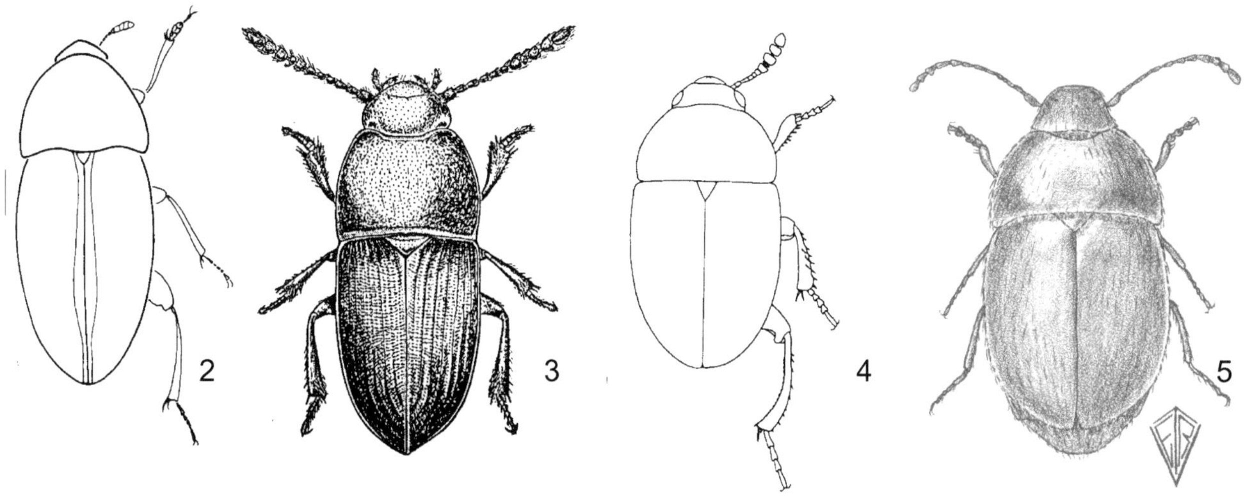


FIGURES 2.19-5.19 (not to same scale). Fig. 2.19, *Colon bidentatum* Sahlberg, habitus; Fig. 3.19, *Catopocerus appalachianus* Peck, habitus; Fig. 4.19, *Leiodes assimilis* (LeConte), habitus (after Baranowski, 1993); Fig. 5.19, *Leptinus americanus* (LeConte), habitus (Figs. 2.19-3.19 from Peck 1990).

Anderson 1985; Chandler and Peck 1992). The flightless ones are taken by Tulgren (Berlese) funnel extraction of sifted litter. The antennal periarticular gutters (also in Agyrtidae) and internal vesicles seemingly give increased sensory ability in food finding. The beetles are of little economic interest but of great ecological and evolutionary diversity.

**Status of classification.** This family is used here in the broad sense of Crowson (1981), Lawrence and Newton (1982), Newton (1998), and Peck (1990) to include groups often placed in up to six separate families, or included in the old and much broader concept of Silphidae. Many groups treated as tribes here are treated by other authors as subfamilies (see discussion and tribal classification in Newton and Thayer (1992)). A defining and shared derived character for the family (but difficult to see) is the presence of complex "vesicles" or sensilla-filled invaginations in the 2 or 3 preapical antennomeres (Peck 1977a). The distinct body forms and surface microsculpture of the subfamilies were the principal characters used to define these groups as families.

**Distribution.** This family (= Anistomidae; Camiaridae; Catopidae; Colonidae; Cholevidae; Leptinidae; Leptodiridae) contains about 250 genera and 3000 species worldwide (Newton 1998), which are usually placed in 6 to 8 subfamilies. All subfamilies occur in the Nearctic, except for Camiarinae, which occurs in south temperate countries. Catopocerinae are predominately a Nearctic subfamily. The very diverse litter- and cave-inhabiting tribe of the Palearctic, the Leptodirini, with over 725 species in 161 genera, contains only one genus and three species in the Nearctic. In America north of Mexico there are now 30 recognized genera and 324 species, with 128 species in Canada and Alaska (Peck 1991). Keys to genera and some species have been presented in Peck (1990), and in various recent revisions (see bibliography). Several North American genera are now being revised and generic limits and species composition will change somewhat as a consequence. The Neotropical fauna is summarized in Peck *et al.* (2000) and that of Mexico in Peck (2000).

## I. Key to the Nearctic Subfamilies

1. Head without occipital carina or crest ................ 2

— Head with occipital carina or elevated crest (Fig. 6.19, this may be hard to see if head is tightly retracted against pronotum; the crest is weak in *Platycholeus,* which has a tarsal formula of 5-5-5 in males and 4-5-5 in females)........................... 4

2(1). Antenna 10- or 11-segmented; club usually interrupted when 11-segmented (antennomere 8 usually smaller than 9 and 10 and without periarticular gutter or internal vesicles); cervical sclerites present; abdominal intersegmental membrane without minute sclerites; females with 5 or 6 visible abdominal sterna; body shape rounded or elongate .......................................................... 3

— Antenna 11-segmented; club gradual, of three to four uninterrupted antennomeres, 8 not markedly smaller than 9-10 (Fig. 7.19) (and with periarticular gutter and internal vesicles); cervical sclerites absent, intersegmental membranes between abdominal sterna with brick-wall pattern of minute sclerites; female with four visible abdominal sterna; body shape elongated (Fig. 2.19) Coloninae .................................................. *Colon*

3(2). Hind coxae separated by about a third their width; prosternum in front of coxae longer than coxal width; always without eyes; (Key II); body shape usually flattened (Fig. 3.19) ........ Catopocerinae

— Hind coxae not separated; prosternum in front of coxae much shorter than coxal width; usually with eyes; body generally round or oval (Fig. 4.19) (Key III) .............................................. Leiodinae

4(1). Occipital crest overlapping pronotum when head is in repose; cervical sclerites absent, procoxal cavities internally open behind; body dorsoventrally flattened (Fig. 5.19); eyes reduced; associated with small mammals (Key IV) .................... Platypsyllinae

— Occipital carina resting against front of pronotum when head is in repose (Fig. 6.19); cervical scler-

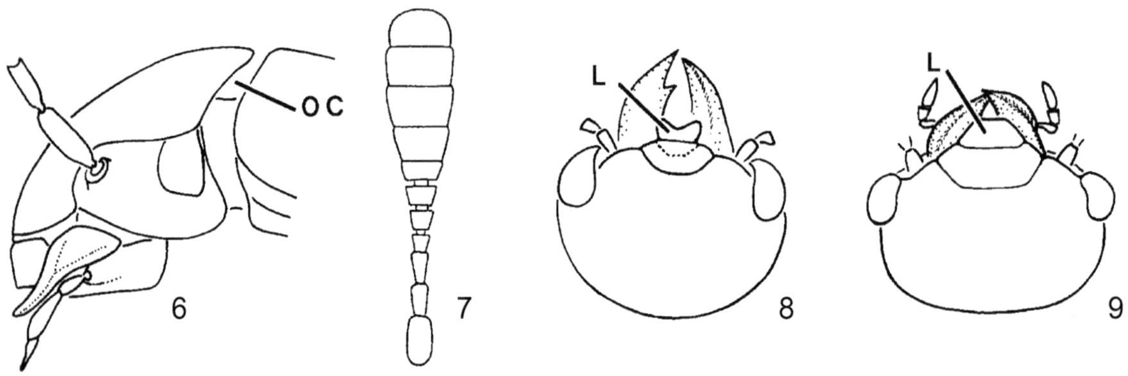


FIGURES 6.19-9.19. Fig. 6.19, Cholevinae, showing occipital carina (oc); Fig. 7.19, *Colon* sp., antenna with uninterrupted club; Fig. 8.19, *Hydnobius* sp., head with deeply emarginate labrum); Fig. 9.19, *Agathidium* sp., head without emarginate labrum (all from Peck 1990).

ites present; procoxal cavities internally closed behind; body and eyes usually normal; usually not associated with mammals (Key V) ................ ......................................................... Cholevinae

II. Key to the Nearctic Tribes and Genera of Catopocerinae

1. Appendages short, body ovoid and flattened, pronotum and elytra of similar widths, (Fig. 3.19); in soil habitats (Tribe Catopocerini) .................. ....................................................... *Catopocerus*

— Appendages conspicuously long, body elongate and rounded, pronotum much narrower than elytra; in northwestern caves (Tribe Glacicavicolini) ............................. *Glacicavicola*

III. Key to the Nearctic Tribes and Genera of Leiodinae

1. Labrum deeply emarginate apically (obscure in *Cyrtusa* and some *Hydnobius*) (Fig. 8.19) ......... 2

— Labrum shallowly or not at all emarginate apically (Fig. 9.19) ........................................................ 11

2(1). Tarsal formula 5-5-5 (Tribe Sogdini (= Hydnobiini)) ......................................................................... 3

— Tarsal formula 5-5-4 (Tribe Leiodini) ..................... 4

3(2). Antennal club with three antennomeres, antennomere 7 as small as 8 ............. *Triarthron*

— Antennal club with five antennomeres, antennomere 8 smaller than 7 ........... *Hydnobius*

4(2). Mesosternum vertical between the coxae (Fig. 10.19) (the Cyrtusa group) ............................... 5

— Mesosternum oblique between the coxae (Fig. 11.19) (the Leiodes group) ............................ 10

5(4). Antenna with 11 antennomeres, with interrupted 5-segmented club, antennomere 8 narrow and disc-shaped and sometimes hidden between 7 and 9 ......................................................................... 6

— Antenna with 10 antennomeres, with compact 3- or 4-segmented club .............................................. 8

6(5). Mesosternum with median longitudinal carina; both mandibles with a median tooth ............ *Anogdus*

— Mesosternum without median longitudinal carina; at most left mandible with a small tooth ............... 7

7(6). Underside of head without antennal grooves; anterior margin of clypeus not sinuous; mandible untoothed ......................................... *Liocyrtusa*

— Underside of head with distinct antennal grooves (Fig. 12.19); anterior margin of clypeus clearly sinuous; left mandible with a small tooth in anterior third .............................................. *Lionothus*

8(5). Tibia narrow; only the underside of the metatibia with spines, not on the outer surface; right mandible with a blunt tooth in anterior third ........... *Cyrtusa*

— Tibia widened; shovel-like; outer surface of metatibia strongly spined; left mandible with a large tooth at basal third .................................. 9

9(8). Antennal club with 4 antennomeres (Fig. 12.19) ... ........................................................ *Zeadolopus*

— Antennal club with 3 antennomeres ..... *Isoplastus*

10(4). Mesosternum with median longitudinal carina ...... ............................................................... *Leiodes*

— Mesosternum without median carina ..................... ................................................ *Ecarinosphaerula*

11(1). All tarsi with three tarsomeres; abdominal sternum 3 (first visible) with transverse carina (Fig. 13.19) (Tribe Scotocryptini) ......................... *Aglyptinus*

— All tarsi of at least four tarsomeres; abdominal sternum 3 (first visible) without transverse carina ... ....................................................................... 12

12(11). Tarsal formula 5-4-4, segmentation not sexually dimorphic, elytra usually transversely striolate; male with enlarged (tenent) setae on protarsi only; tibiae without longitudinal carinae (Tribe Pseudoliodini) ................................................. 13

— Tarsi sexually dimorphic, 5-5-4 in males, 5-4-4 or 4-4-4 in females; elytra not transversely striolate; male tenent setae usually on pro- and mesotarsi; tibiae with longitudinal carinae (Tribe Agathidiini) ....................................................................... 14

13(12). Elytra and pronotum with abundant fine transverse striolae; elytra without longitudinal rows of punctures; color paler (testaceous); mesosternal carina variable (Fig. 14.19) .......................... *Colenis*

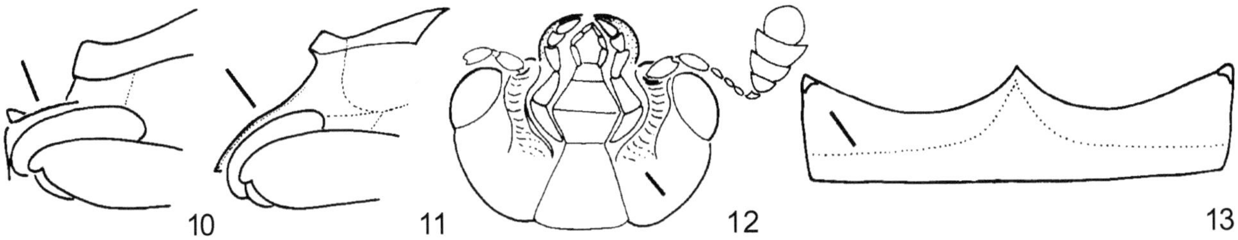


FIGURES 10.19-13.19. Fig. 10.19, *Cyrtusa* sp., mesosternum vertical between the coxae; Fig. 11.19, *Leiodes* sp., mesosternum oblique between the coxae; Fig. 12.19, *Zeadalophus* sp., with antennal grooves on underside of head, and 4 segmented antennal club; Fig. 13.19, *Aglyptinus laevis* (LeConte), abdominal sternum 3 (first visible) with transverse carina (all from Peck 1990).

— Elytra not as above, with longitudinal rows of punctures; color darker, mesosternal carina with transverse depression or notch (Fig. 15.19) ............. ...................................................... *Cainosternum*

14(12). Eighth antennomere distinctly smaller than seventh; antennal club with five antennomeres; head narrowed behind the eyes; body form convex to hemispherical, contractile ................ *Anisotoma*

— Eighth antennomere not (or slightly) smaller than seventh, antennal club abruptly formed, with three antennomeres; head often broad behind eyes (postocular tempora well developed); body form variable, sometimes oblong-eliptical and subdepressed, sometimes hemispherical and highly contractile .......................................... 15

15(14). Elytra with nine complete, punctate striae; head narrowed behind eyes; body form oblong-eliptical, subdepressed ................................ *Stetholiodes*

— Elytra with fewer than nine complete, punctate striae (or without striae); head often wide behind eyes; body form variable, often hemispherical and very contractile ...................................... *Agathidium*

## IV. Key to the Nearctic Genera of Platypsyllinae (= Leptininae, = Leptinidae)

1. Prosternum short and acute at apex, not extending between procoxae; length about 2 mm ............ ............................................................ *Leptinus*

— Prosternum produced posteriorly, extending beyond procoxae and appearing to separate them, or forming a broad flat plate, ending in a median lobe fringed with long setae ........................... 2

2(1). Prosternum produced posteriorly; separating the procoxae ........................................... *Leptinillus*

— Prosternum forming a broad, flat plate, ending in a median lobe fringed with long setae .................. ....................................................... *Platypsyllus*

## V. Key to the Nearctic Tribes and Genera of Cholevinae (= Leptodiridae, Catopidae, Cholevidae)

1. Posterior coxae contiguous, all tarsi with five tarsomeres ...................................................... 2

— Posterior coxae somewhat separated, tarsi all with five tarsomeres, except female protarsi with four tarsomeres (Tribe Leptodirini) ......... *Platycholeus*

2(1). Elytra with glossy surface, setal bases arranged in transverse or oblique strigae (Fig. 1.19) ......... 3

— Elytra with granular surface of irregularly arranged setal bases; strigae absent (Tribe Cholevini) ... 7

3(2). Hind margin of hind tibia with two inner long spines and an outer row or comb of short and equal spines only (Fig. 16.19) (Tribe Ptomaphagini), *Ptomaphagus* ................................................... 4

— Hind margin of hind tibia with two long inner spines (and perhaps a comb of short, equal spines), but more importantly, also with about four longer outer spines (Fig. 17.19) (Tribe Anemadini) ................ 6

4(3). Form oval or elongate oval; with long and erect hairs as well as short recumbent hairs; mesosternal carina high, low and effaced, or absent; compact and modified for life as guests in ant nests ....... ...................................... subgenus *Echinocoleus*

— Form elongate oval; with short recumbent hairs only, mesosternal carina present and usually well developed; may live with ants but not as highly modified myrmecophile ........................................... 5

5(4). Size medium (2.6 mm) to smaller; flight wings absent, eyes reduced to poorly defined collection of about 20 pigmented facets; aedeagal tip more elaborately sculptured, broader and blunter, southeastern United States, usually in litter or soil ............................... subgenus *Appadelopsis*

— Size medium (2.6 mm) or larger, flight wings usually present and eyes usually normal (wings absent, and eyes smaller or reduced to unpigmented spot in most cavernicolous species); aedeagal tip simple, elongate and pointed; widespread, in many habitats ....................... subgenus *Adelops*

6(3). Basal mesotarsomeres of male (with expanded protarsi) weakly dilated and spongy pubescent beneath; mesosternal carina more elevated and extending nearly to anterior margin of mesosternum ............................................................ *Nemadus*

— Basal mesotarsomeres of male (with expanded protarsi) not expanded; mesosternal carina feebly elevated, sometimes confined to region between mesocoxae ........................ *Dissochaetus*

7(2). Antennomeres 1-10 serrate (bipectinate); with anterior projections, giving sawtoothed appearance ..................................................... *Catoptrichus*

— Antennae normal, not serrate (bipectinate) ......... 8

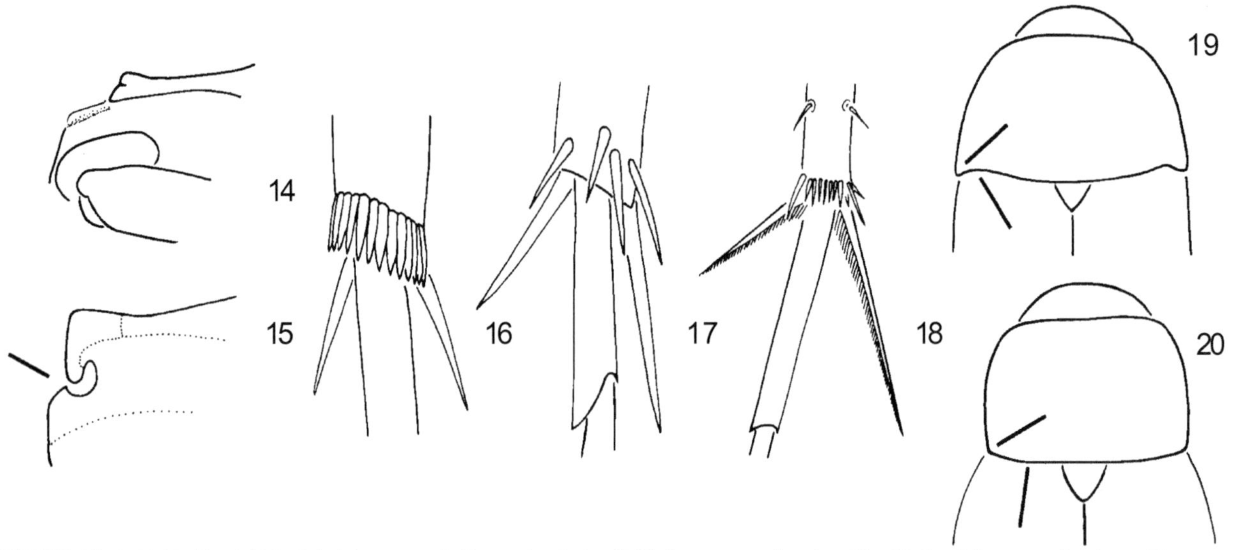


FIGURES 14.19-20.19. Fig. 14.19, *Colenis impunctata* LeConte, simple (undivided) mesosternal carina; Fig. 15.19, *Cainosternum imbricatum* Notman, divided (with notch) mesosternal carina; Fig. 16.19, *Ptomaphagus sp.,* apical metatibial comb of equal spines; Fig. 17.19, *Dissochaetus oblitus* (LeConte), apical metatibial spurs; Fig. 18.19, *Prionochaeta opaca* (Say), apical metatibial spurs; Fig. 19.19, *Sciodrepoides fumatus terminans* (LeConte), pronotum with rectangular posterior angles and undulating hind margin; Fig. 20.19, *Catops* sp., pronotum with rounded posterior angles and arcuate hind margin (all from Peck 1990).

8(7). Tibial spurs long serrate, longest metatibial spur as long as first metatarsal segment (Fig. 18.25); first mesotarsomere not dilated in males (with expanded protarsi) ............................ *Prionochaeta*

— Tibial spurs not long and serrate; much shorter than first tarsal segment, first mesotarsomere dilated in males (with expanded protarsi) ................... 9

9(8). Pronotum usually with base sinuate on either side just within the more or less distinctly rectangular posterior angles (Fig. 19.19); internal sac of aedeagus with a Y-shaped piece ...................... ..................................................... *Sciodrepoides*

— Pronotum usually with base arcuate, the hind angles obtuse or more or less evidently rounded (Fig. 20.19); internal sac of aedeagus without a "Y" shaped piece .......................................... *Catops*

### Classification of the Nearctic Genera

#### Subfamily Coloninae Horn 1880

*Colon* Herbst 1797, 42 species in 5 subgenera in litter habitats in North America (see Peck and Stephan 1996).

*Kolon* Herbst 1797
subgenus *Colon* Herbst 1797
subgenus *Myloechus* Latreille 1807
subgenus *Mesagyrtes* Brown 1895
subgenus *Eurycolon* Ganglbauer 1899
subgenus *Platycolon* Portevin 1907
subgenus *Chelicolon* Szymczakowski 1964
subgenus *Desmidocolon* Szymczakowski 1964
subgenus *Tricolon* Peck and Stephan 1996
subgenus *Striatocolon* Peck and Stephan 1996

#### Subfamily Catopocerinae Hatch 1927

##### Tribe Catopocerini Hatch 1927

*Catopocerus* Motschoulsky 1870, 14 spp., widespread North American (also in Siberia), in forest soil and deep litter (Peck 1974); all species are eyeless and wingless scavengers or feed on subterranean fungi (Fogel and Peck 1975).

*Homaeosoma* Austin 1880
*Pinodytes* Horn 1880
*Typholeiodes* Hatch 1935

##### Tribe Glacicavicolini Westcott 1968

*Glacicavicola* Westcott 1968, 1 sp., *G. bathyscioides* Westcott 1968, the ice-cave beetle, a highly evolved eyeless cave inhabitant, a scavenger in both cold lava tube and limestone caves in Idaho and Wyoming (Peck 1981; Westcott 1968).

#### Subfamily Leiodinae Fleming 1821
The round fungus beetles

##### Tribe Sogdini Lopatin 1961 (= Hydnobiini)

*Hydnobius* Schmidt 1847, 16 spp., mostly western, but several with a general northern distribution; subterranean fungi feeders (Hatch 1957).

*Triarthron* Märkel 1840, 2 spp., from Pennsylvania to Oregon (Hatch 1957).

*Triarthrum* Agassiz 1847

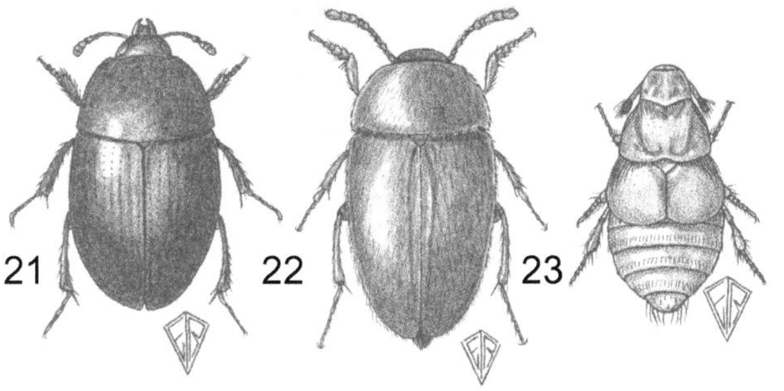


FIGURES 21.19 - 23.19. Dorsal habitus (not to same scale). Fig. 21.19. *Leiodes assimilis* (LeConte); Fig. 22.19. *Sciodrepoides terminans* (LeConte); Fig. 23.19. *Platypsyllus castoris* Ritsema.

Tribe Leiodini Fleming 1821

*Anogdus* LeConte 1866, 10 spp., in the eastern United States and British Columbia (Hatch 1957; Daffner 1988).
*Neocyrtusa* Brown 1937

*Cyrtusa* Erichson 1842, 2 spp., from Quebec and Michigan to British Columbia (Hatch 1957; Daffner 1988).
*Caenocyrta* Brown 1937

*Ecarinosphaerula* Hatch 1929, 1 sp., *E. ecarina* Hatch 1929, British Columbia and Nevada (Hatch 1957).

*Isoplastus* Hom 1880, 1 sp., *I. fossor* Horn 1880, from Quebec and Michigan to the District of Columbia, and Oklahoma.

*Leiodes* Latreille 1796, 72 spp., widely distributed, usually in forested habitats (Hatch 1957; see especially Baranowski 1993).
*Anisotoma* of Schmidt 1841: 143 (misidentification of *Anisotoma* Panzer)
*Liodes* Erichson 1845: 98 (unjustified emendation *of Leiodes* Latreille)
*Oosphaerula* Ganglbauer 1896
*Oreosphaerula* Ganglbauer 1899
*Pseudohydnobius* Ganglbauer 1899
*Pteromerula* Fleischer 1905
*Trichosphaerula* Fleischer 1905
*Parahydnobius* Fleischer 1908
*Eremosphaerula* Hlisnikovsky 1967

*Liocyrtusa* Daffner 1982, 3 spp., widely distributed (Daffner 1988).

*Lionothus* Brown 1937, 2 spp., widely distributed (Daffner 1988).
*Pseudocyrtusa* Portevin 1942.

*Zeadolopus* Broun 1903, 3 spp., Quebec to Michigan to Oklahoma and Florida.
*Apheloplastus* Brown 1937

Tribe Pseudoliodini Portevin 1926

*Cainosternum* Notman 1931, 1 sp., C. *imbricatum* Notman 1921, New York (Wheeler 1986).

*Colenis* Erichson 1845, 4 spp., in forests, from eastern United States and Canada to British Columbia (Peck 1998b).
*Carcharodus* Hlisnikovsky 1965 (preoccupied)
*Carcharodes* Hlisnikovsky 1965 (preoccupied)
*Mathewsionia* Hlisnikovsky 1965
*Colenodes* Peck 1998 (replacement name)

Tribe Scotocryptini Reitter 1884

*Aglyptinus* Cockerell 1906, 1 sp., *A. laevis* (LeConte) 1853, in the east from Ontario to Louisiana; usually associated with soft fungi.
*Aglyptus* LeConte 1866 (preoccupied)
*Agyptonotus* Champion 1913 (replacement name)

Tribe Agathidiini Westwood 1838

*Agathidium* Panzer 1797, 47 spp., widely distributed across the United States and Canada; usually in forested habitats and on slime molds (Hatch 1957; Russell 1979; Wheeler 1987, 1990). Taxonomy under study by Wheeler.
*Volvoxis* Kugelann 1794
*Agathidium* Illiger 1798
subgenus *Agathidium* Panzer 1797
subgenus *Cyphoceble* Thomson 1859
subgenus *Chaetoceble,* Sainte-Claire-Neville 1879
subgenus *Neoceble* Gozis 1886
subgenus *Euryceble* Hlisnikowsky 1964
subgenus *Rhabdoelytrum* Hlisnikowsky 1964
subgenus *Microceble* Angelini and De Marzo 1986
subgenus *Macroceble* Angelini 1993

*Anisotoma* Panzer 1797, 15 spp., generally distributed in forests over much of Canada and the United States; feeders on plasmodia and fruiting bodies of slime molds (Russell 1979; Wheeler 1979, 1987, 1990).
*Pentatoma* Schneider 1792
*Anisotoma* Illiger 1798
*Leiodes* of Schmidt 1841
*Eucyrta* Portevin 1927

*Stetholiodes* Fall 1910, 1 sp., *S. laticollis* Fall 1910, Indiana (Wheeler 1981).
*Agathodes* Portevin 1926
*Agathidiodes* Portevin 1944

Subfamily Cholevinae Kirby 1837

The small carrion beetles

Tribe Leptodirini Lacordaire 1854

*Platycholeus* Horn 1880, 3 spp., small-eyed winged *P. leptinoides* (Crotch) 1874, *P. opacellus* (Fall) 1909, and an undescribed species in the Pacific Northwest, the only North American representatives of this large and predominantly litter and cave dwelling tribe of Eurasia (see Newton 1998); in litter and under bark of logs and stumps, sometimes with ants and termites (Hatch 1957).

Tribe Anemadini Hatch 1928

*Nemadus* Thomson 1867, 10 spp., widespread North America, scavengers in forest litter and sometimes in caves and animal burrows; some species seem restricted to ant nests (Hatch 1933; Jeannel 1936; Fall 1937).

subgenus *Nemadus* Thomson 1867
subgenus *Eonargus* Iablokoff-Khnzorian 1959
subgenus *Laferius* Perkovsky 1994

*Dissochaetus* Reitter 1884, 3 spp., eastern, central, and southwestern United States, scavengers in forest litter and on carrion (Hatch 1933; Jeannel 1936; Peck 1999).

Tribe Cholevini Kirby 1837

*Catops* Paykull 1798, 10 spp. (and six undescribed), in forest litter and related habitats across North America south to Mexico only; carrion scavengers (Hatch 1933; Jeannel 1936; Peck, in prep.).

*Sciodrepa* Thomson 1862
*Lasiocatops* Reitter 1901

*Catoptrichus* Murray 1856, 1 sp., *C. frankenhauseri* (Mannerheim) 1852, in forest litter in the Pacific northwest, habits unknown, probably carrion scavenger (Hatch 1933).

*Prionochaeta* Horn 1880, 1 sp., *P. opaca* (Say) 1825, in forest litter and some caves in eastern North America; common on carrion (Peck 1977b).

*Sciodrepoides* Hatch 1933, 2 spp., subspecies of two Holarctic species, *S. fumatus terminans* (LeConte) 1850, and *S. watsoni hornianus* (Blanchard) 1915, commonly found as carrion scavengers in many field, forest and soil-related habitats across the northern half of the continent (Hatch 1933; Jeannel 1936).

Tribe Ptomaphagini Jeannel 1911

*Ptomaphagus* Illiger 1798, 52 spp., widespread in the United States and southern Canada, occupying a wide range of litter and soil habitats as scavengers, and extensions of these habitats such as ant nests and animal burrows; eighteen species are blind inhabitants of cave habitats in the southern United States (Peck 1973, 1984, 1998b); other such troglobites occur in Mexico and Central America.

*Adelops* Tellkampf 1884
subgenus *Ptomaphagus* Illiger 1798
subgenus *Adelops* Tellkarnpf 1844
subgenus *Echinocoleus* Horn 1885; Peck and Gnaspini 1997
subgenus *Merodiscus* Jeannel 1934
subgenus *Tupania* Szymczakowski 1961
subgenus *Appaladelopsis* Gnaspini 1996; Peck 1978 (for U.S. species, as *Adelopsis*)

Subfamily Platypsyllinae Ritsema 1869

The mammal nest beetles

*Leptinillus* Horn 1882, 2 spp.; *L. aplodontiae* Ferris 1918, the mountain beaver beetle, with mountain beaver, from California to Washington; and *L. validus* (Horn) 1872, the beaver nest beetle,with beaver throughout its range (Wood 1965).

*Leptinus* Müller 1817, 3 spp., the mouse nest beetles, distributed over much of eastern North America and the Pacific northwest; most frequently found in nests and fur of mice, shrews, and moles and occasionally in litter (Peck 1982). Their biology is not well-known, although it has been studied for *L. testaceus* Müller, the mouse nest beetle of Europe (Buckle 1976; Ising 1969).

*Playtpsyllus* Ritsema 1869, 1 sp., the highly modified and flea-appearing *P. castoris* Ritsema 1869, the beaver parasite beetle, a true ectoparasite of beavers (*Castor*) in North America and Eurasia (see Wood 1965).

*Platypsullus* Westwood 1869
*Playpsylla* LeConte 1872

## Bibliography


BARANOWSKI, R. 1993. Revision of the genus *Leiodes* Latreille of North and Central America (Coleoptera: Leiodidae). Entomologica Scandinavica, Supplement No. 42: 1-149.

BUCKLE, A.P. 1976. Studies on the biology and distribution of *Leptinus testaceus* Müller within a community of mixed small mammal species. Ecological Entomology, 1: 1-6.

CHANDLER, D.S. and PECK, S.B. 1992. An old-growth and a 40-year-old forest in New Hampshire: seasonality and diversity of their leiodid beetle faunas. Environmental Entomology, 21: 1283-1293.

CROWSON, R.A. 1981. The biology of the Coleoptera. Academic Press, New York. 802 pp.

DAFFNER, H. 1988. Revision der nordamerickanischen Arten der *Cyrtusa* – Verwandschaft (Coleoptera, Leiodidae, Leiodini). Annali dei Musei Civici, Rovereto (Italy), 4: 269-306.

DINDEL, D. (Ed.) 1990. Soil Biology Guide. John Wiley & Sons, NY.

FALL, H.C. 1937. The North American species of *Nemadus* Thom., with descriptions of new species (Coleoptera: Silphidae). Journal of the New York Entomological Society, 45: 335-340.

FOGEL, R. and PECK, S.B. 1975. Ecological studies of hypogeous fungi. I. Coleoptera associated with sporocarps. Mycologia, 67: 741-747.

GNASPINI, P. 1996. Phylogenetic analysis of the Tribe Ptomaphagini, with description of new Neotropical genera and species (Coleoptera, Leiodidae, Cholevinae, Ptomaphagini). Papéis Avulsos Zoologia (São Paulo), 39: 509-556.

HATCH, M.H. 1933. Studies on the Leptodiridae (Catopidae) with descriptions of new species. Journal of the New York Entomological Society, 41: 187-239.

HATCH, M.H. 1957. The beetles of the Pacific Northwest. Part II, Staphyliniformia. University of Washington Publications in Biology, vol. 16. University of Washington Press, Seattle, WA. 38 pp.

HORN, G.H. 1880. Synopsis of the Silphidae of the United States with reference to the genera of other countries. Transactions of the American Entomological Society, 8: 219-322.

ISING, E. 1969. Zur Biologie des *Leptinus testaceus* Müller 1817 (Insecta: Coleoptera). Zoologische Beiträge, 15: 393-456.

JANZEN, D.H. 1963. Observations on populations of adult beaver beetles, *Platypsyllus castoris* (Platypsyllidae: Coleoptera). The Pan-Pacific Entomologist, 34: 215-228.

JEANNEL, R. 1936. Monographie des Catopidae. Memoires de Museum National d'Histoire Naturelle, Nouvelle Serie, Tome 1. Paris. 433 pp.

LAWRENCE, J.F. and NEWTON, A.F. JR. 1980. Coleoptera associated with fruiting bodies of slime molds (Myxomycetes). Coleopterists Bulletin, 34: 129-143.

LAWRENCE, J.F. and NEWTON, A.F., JR. 1982. Evolution and classification of beetles. Annual Review of Ecology and Systematics, 13: 261-290.

NEWTON, A.F. JR. 1984. Mycophagy in Staphylinoidea (Coleoptera). pp. 302-353. *In* Wheeler, Q., and Blackwell, M., (eds.). Fungus-Insect Relationships: Perspectives in Ecology and Evolution. Columbia University Press, New York. 514 pp.

NEWTON, A.F. 1991a. Leiodidae (Staphylinoidea). pp. 327-329. *In* Stehr, F.W. (ed.), Immature Insects, Vol. 2.Kendal/Hunt, Dubuque, Iowa.

NEWTON, A.F. 1991b. Leptinidae (Staphylinoidea). p. 339. *In* Stehr, F.W. (ed.), Immature Insects, Vol. 2.Kendal/Hunt, Dubuque, Iowa.

NEWTON, A.F. JR. 1998. Phylogenetic problems, current classification and generic catalog of world Leiodidae (including Cholevidae). pp. 41-178. *In*: Phylogeny and evolution of subterranean and endogean Cholevidae (=Leiodidae Cholevinae). Proceedings of XX I.C.E. Firenze, 1996. Museo Regionale di Scienze Naturali, Torino (Italy).

NEWTON, A.F. JR. and THAYER, M.K. 1992. Current classification and family group names in Staphyliniformia (Coleoptera). Fieldiana: Zoology (n.s.), 67: 1-92.

PECK, S.B. 1973. A systematic revision and the evolutionary biology of the *Ptomaphagus (Adelops)* beetles of North America (Coleoptera: Leiodidae: Catopinae) with emphasis on cave-inhabiting species. Museum of Comparative Zoology (Harvard University) Bulletin, 145: 29-162.

PECK, S.B. 1974. The eyeless *Catopocerus* beetles (Leiodidae) of eastern North America. Psyche, 81: 377-397.

PECK, S.B. 1977a. An unusual sense receptor in internal antennal vesicles of *Ptomaphagus* (Coleoptera: Leiodidae). Canadian Entomologist, 109: 81-86.

PECK, S.B. 1977b. A review of the distribution and biology of the small carrion beetle *Prionochaeta opaca* of North America (Coleoptera: Leiodidae: Catopinae). Psyche, 83: 299-307.

PECK, S.B. 1978. Systematics and evolution of forest litter *Adelopsis* in the southern Appalachians (Coleoptera: Leiodidae: Catopinae). Psyche, 85: 355-382. (now *Ptomaphagus (Appadelopsis)*)

PECK, S.B. 1981. The Idaho cave beetle *Glacicavicola* also occurs in Wyoming. Coleopterists Bulletin, 35: 451-452.

PECK, S.B. 1982. A review of the ectoparasitic *Leptinus* beetles of North America (Coleoptera: Leptinidae). Canadian Journal of Zoology, 60: 1517-1527.

PECK, S.B. 1984. The distribution and evolution of cavernicolous *Ptomaphagus* beetles in the southeastern United States (Coleoptera: Leiodidae: Cholevinae) with new species and records. Canadian Journal of Zoology, 62: 730-740, illus.

PECK, S.B. 1990. Insecta: Coleoptera: Silphidae and the associated families Agyrtidae and Leiodidae. pp. 1113-1136. *In* Dindal, D., (ed.). Soil Biology Guide. John Wiley and Sons, New York.

PECK, S.B. 1991. Family Leiodidae. pp. 77-82. *In* Bousquet, Y., (ed.). Checklist of Beetles of Canada and Alaska. Research Branch, Agriculture Canada, Ottawa. Publication 1861/E.

PECK, S.B. 1998a. Revision of the *Colenis* of America north of Mexico (Coleoptera: Leiodidae: Leiodinae: Pseudoliodini). Canadian Entomologist, 130: 55-65.

PECK, S.B. 1998b. Cladistic biogeography of cavernicolous *Ptomaphagus* beetles (Leiodidae; Cholevinae; Ptomaphagini) in the United States, pp. 235-260. Proceedings of XX I.C.E. Firenze, 1966. Museo Regionale Scienze Naturali, Torino (Italy).

PECK, S.B. 1999. A review of the *Dissochaetus* (Coleoptera: Leiodidae; Cholevinae) of the United States and Canada. Canadian Entomologist, 131: 179-186.

PECK, S.B. 2000. Leiodidae (Coleoptera), the small scavenger beetles of Mexico. pp. 439-452. *In* Llorente, J., Soriano, F.G., Papavero, N. (eds.). Biodiversidad, Taxonomía y Biogeographía de Artrópodos de México. Vol. 2.

PECK, S.B. and ANDERSON, R.S. 1985. Seasonal activity and habitat associations of small carrion beetles (Coleoptera: Leiodidae: Cholevinae). Coleopterists Bulletin, 39: 347-353.

PECK, S.B. and DAVIES, A. 1980. Collecting small beetles with large-area "window" traps. Coleopterists Bulletin 34: 237-239.

PECK, S.B. and GNASPINI, P. 1997. Review of the myrmecophilous *Ptomaphagus*, subgenus *Echinocoleus* (new status) of North America (Coleoptera: Leiodidae; Ptomaphagini). Canadian Entomologist, 129: 93-104.

PECK, S.B., GNASPINI, P. and NEWTON, A.F. JR. 2000. Catalogue of the Leiodidae of Latin America (Mexico, West Indies, and Central and South America) (Insecta: Coleoptera). Giornale Italiano di Entomologia. (in press)

PECK, S.B. and STEPHAN, K. 1996. A revision of the genus *Colon* Herbst (Coleoptera: Leiodidae: Coloninae) of North America. Canadian Entomologist, 128: 667-741.

RUSSELL, L.K. 1979. Beetles associated with slime molds (Mycetozoa) in Oregon and California (Coleoptera: Leiodidae, Lathridiidae). Pan-Pacific Entomologist, 55: 1-9.

WESTCOTT, R.L. 1968. A new subfamily of blind beetle from Idaho ice caves with notes on its bionomics and evolution (Coleoptera: Leiodidae). Los Angeles County Museum Contributions to Science, 141: 1-14.

WHEELER, Q.D. 1979. Slime mold beetles of the genus *Anisotoma* (Leiodidae), evolution and classification. Systematic Entomology, 4: 251-309.

WHEELER, Q.D. 1981. Diagnosis and phylogenetic relationships of the monotypic genus *Stetholiodes* (Coleoptera: Leiodidae). Ohio Journal of Science, 81: 165-168.

WHEELER, Q.D. 1986. Rediscovery and cladistic placement of the genus *Cainosternum* (Coleoptera: Leiodidae). Annals of the Entomolological Society of America, 79: 377-383.

WHEELER, Q.D. 1987. A new species of *Agathidium* associated with an "epimycetic" slime mold plasmodium on *Pleurotus* fungi (Coleoptera: Leiodidae-Myxomycetes: Physarales-Basidiomycetes: Tricholomataceae). Coleopterists Bulletin, 41: 395-403.

WHEELER, Q.D. 1990. Morphology and ontogeny of postembryonic larval *Agathidium* and *Anisotoma* (Coleoptera: Leiodidae). American Museum Novitates, no. 2986. 46 pp.

WOOD, D.M. 1965. Studies on the beetles *Leptinillus validus* (Horn) and *Platypsyllus castoris* Ritsema (Coleoptera: Leptinidae) from beaver. Proceedings of the Entomological Society of Ontario, 95: 33-63.

# 20. SCYDMAENIDAE Leach, 1815

by Sean T. O'Keefe

**Family synonyms:** Scydmaenides Leach 1815; Scydmaenidae LeConte 1852; Anisosphaeridae; Tömösváry 1882

**Common name:** The ant-like stone beetles

The elongate elytra, five-segmented tarsi, six visible sternites, and clavate femora serve to distinguish these beetles from other minute staphylinoids (Fig. 1.20).

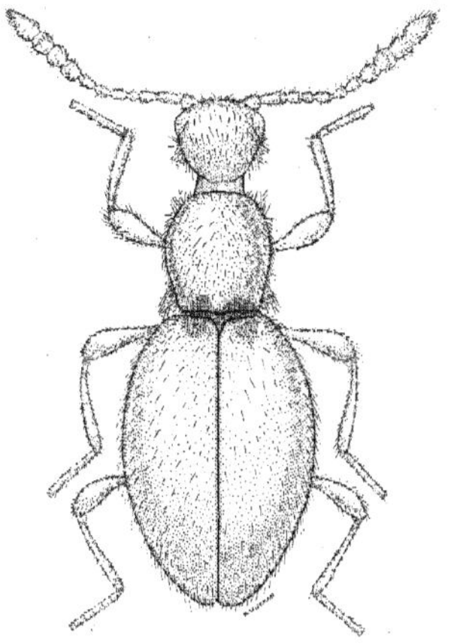

FIGURE 1.20. *Euconnus (Drastophus) laevicollis* (LeConte) (From O'Keefe 1998).

**Description:** Body elongate, slender to ovoid; slightly to very convex, 0.6-2.7 mm long; often distinctly constricted between head and pronotum and between pronotum and elytra, usually light to dark brown, sometimes black in color; often densely pubescent on head, pronotum, and elytra, pubescence usually long.

Head short and broad to ovoid; deflexed; distinctly constricted between vertex and occiput, except in Cephenniini; occiput formed into a "neck"; neck present but not abruptly constricted in *Chevrolatia* and *Lophioderus*; clypeofrontal region moderate in length and slightly narrowed anteriorly. Eyes anterior, median, or posterior; moderate in size, or absent, as in *Cephennium anophthalmicum*. Antennae inserted anterior to eyes, antennal insertions adjacent to widely separated; antennae composed of scape, pedicel and 9 flagellomeres; antennal club composed of distal 3 to 5 antennomeres, from indistinct to distinct in form. Vertex rounded, impressed (*Brachycepsis*, *Parascydmus*), excavated (*Taphroscydmus*), foveate (*Veraphis*), to highly sculptured (*Chevrolatia*). Mandibles usually planar, subtriangular with large basal area, narrowed to acute incisor, subapical teeth absent, although present in *Papusus*; prostheca present in most genera, quite varied in form; retinaculum present in many genera. Maxillary palpus composed of 4 palpomeres; palpomere III large, clavate; palpomere IV variable, although distinctly smaller than III (Figs. 26.2-8); apices of lacinea and galea densely covered with long setae. Labial palpus composed of 3 palpomeres; palpomeres II and III varied in form; mentum large.

Thorax with pronotum varied in shape, from rounded to quadrate in dorsal view, flat to distinctly convex, distinctly wider than head; most species with fovea or transverse furrow along posterior margin. Scutellum small, visible, although hidden in some genera, as in *Euconnus, Papusus, Leptoscydmus,* and *Microscydmus*. Elytra ovoid, convex as in Cyrtoscydmini and Scydmaenini, to rectangular, flat as in Chevrolatiini and Eutheiini; humeri varied in form with 0-2 basal foveae in broad basal impression; elytral apices entire in most genera, but also truncate (Chevrolatiini, Eutheiini), and in some forms fused (e.g., *Papusus*). Hind wings, if present, well-developed. Legs relatively short to moderate in length; procoxae projecting beyond prosternum, procoxal cavities open; mesocoxal cavities closed; trochanters relatively large and sublinear or rounded triangular; femora strongly clavate in distal half; protibiae and mesotibiae of most genera with dense patches of setae at distal end. Prosternum varied from subquadrate to very narrow before conical, contiguous procoxae, pubescence varied. Mesosternum raised (except flat in *Eutheia*), densely setose in many genera, bearing mid-ventral carina separating otherwise contiguous mesocoxae. Metasternum long, broad, occasionally carinate or medially excavate; metacoxae round to transverse, conical, widely separate to contiguous.

Abdomen with six visible sternites; intersegmental margins relatively straight, except in Scydmaenini, where they are strongly apically arcuate. Pygidium exposed, horizontal in Chevrolatiini, Eutheiini, *Leptoscydmus*, and *Taphroscydmus*; exposed, vertical in Scydmaenini; covered by elytra in Cephenniini and most Cyrtoscydmini. Male genitalia with median lobe bulbous, large (especially in Cyrtoscydmini), to elongate (Scydmaenini), to reduced, lightly sclerotized, distinctly curved (e.g., *Papusus*); parameres long, slender, articulated on dorsal surface of aedeagus, absent from some groups (e.g., *Lophioderus*). Ovipositor composed of paired, fused or unfused dorsolateral paraprocts, triangular dorsal proctiger, paired elongate ventral valvifers, paired elongate gonocoxae; gonostyli absent; spermatheca present, variable in shape from elongate, slender to spherical, accessory gland sometimes present.

Most scydmaenid larvae have been placed to genus by association with adults (Wheeler and Pakaluk 1983, Brown and Crowson 1980); although Schmid (1988a) had some success in rearing. Of the nearly 80 genera of Scydmaenidae, larvae have been associated only for *Cephennium*, *Coatesia, Eutheia*, *Euconnus*, *Leptomastax, Mastigus*, *Neuraphes*, *Scydmaenus*, *Scydmoraphes*, *Stenichnus,* and *Veraphis* (Brown and Crowson 1980, De Marzo 1984, Newton 1991, Schmid 1988a, Vit and DeMarzo 1989, Wheeler and Pakaluk 1983).

Larvae elongate, parallel-sided, straight and slightly flattened to ovate, ventrally curved and capable of rolling into a ball, to broadly ovate and strongly flattened; under 5 mm long. Head prognathous; 0, 1, or 3 stemmata on each side in close cluster. Antennae 2- or 3-segmented, antennomere II in most genera large and elongate with well-developed sensory appendage anterior to antennomere III, antennomere III small. Frontoclypeal suture absent. Labrum fused to head capsule. Mandible without molar area, retinaculum present in a few genera, apices acute. Maxillae without distinct articulating area; maxillary palpus 3-segmented in most genera, 2-segmented in a few genera, palpomere II in most genera large and elongate with prominent anterior sensorium. Labium with apex of prementum not forming a distinct ligula; labial palpus 2-segmented. Neck undifferentiated; thorax with well-sclerotized tergites; spiracles situated near antero-ventral margins of mesothorax. Legs 5-segmented, with bisetose claws. Abdomen with sclerotized tergites and sternites on segments I-IX, X forming a small ventral pygopod, XI with horny urogomphi in Eutheiini. Larvae have been reviewed by Böving and Craighead (1930), Brown and Crowson (1980), Newton (1991), Schmid (1988a), and Wheeler and Pakaluk (1983).

**Habits and habitats.** Scydmaenidae are moderately common in forest floor leaf litter, moss, rotting logs, tree hollows, and other moist habitats, although Wheeler and Pakaluk (1983) report collecting scydmaenids from dryer habitats. Members of *Papusus* and *Ceramphis* are known from the dry regions of the southwest deserts. Several species have been associated with ants (Lea 1910, 1912; Wasman 1894) or termites (Franz 1980). Adults and larvae are known to feed on oribatid mites (Schmid 1988b; Schuster 1966).

Schmid (1988b) discussed the morphological adaptations of scydmaenids for preying on oribatid mites. He studied the prey-capture techniques of 20 species of European Scydmaenidae and recognized two techniques of prey capture involving modifications of the mouthparts and tarsi for overcoming the hard-shelled mites. In the "hole scraping technique" members of the genus *Cephennium* grasp the mite from the dorsal side, hold it with a pair of suckers situated on the ventral side of their labium, and then pull the mite from the substrate. Using its legs, the beetle then orients the mite in order to clasp it easily. Then the beetle scrapes with its mandibles through the dorsal surface of the mite. In the "cutting technique" scydmaenids shear the mite cuticle along the weak areas of the ventral side, such as the hinge areas of the genital or anal plates. The forelegs are used to lift the mite off the ground and flip it onto its back. The beetle holds the mite with its legs during the feeding process and rotates the mite along its dorsoventral axis during the cutting process. The distal portion of the mandible is inserted into the hinge of the genital or anal plates and the plate is cut away. The retinaculum at the base of the mandible serves as support for the cutting operation. The body fluids of the mite are then lapped out of the opening.

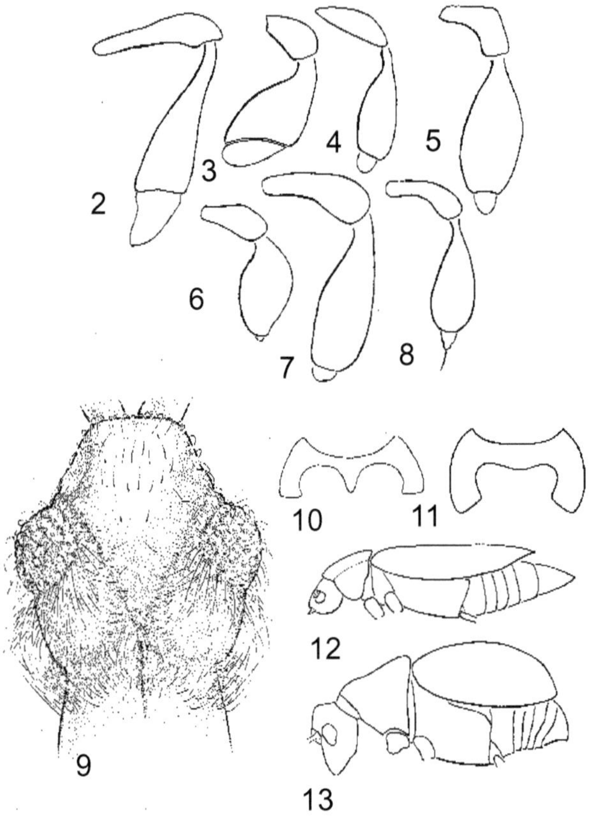


FIGURES 2.20 - 13.20. Fig. 2.20, *Papusus*, maxillary palp; Fig. 3.20, *Chevrolatia*, same; Fig. 4.20, *Scydmaenus*, same; Fig. 5.20, *Chelonoidum*, same; Fig. 6.20, *Veraphis*, same; Fig. 7.20, *Leptoscydmus*, same; Fig. 8.20, *Brachycepsis*, same; Fig. 9.20, *Chevrolatia*, head (From O'Keefe 1997); Fig. 10.20, *Chelonoidum*, Prosternum; Fig. 11.20, *Cephennium*, same; Fig. 12.20, *Eutheia*, lateral view; Fig. 13.20, *Ceramphis*, same (From O'Keefe 1997).

Adults can be readily collected using modified Berlese-Tulgren funnels or Winkler/Moczarski eclectors to process debris taken from potential habitats (Besuchet *et al.* 1987; Wheeler and McHugh 1987). They may be collected at times with flight intercept traps, pitfall traps, by searching underneath bark and stones, or at blacklights (Suter 1966).

**Status of classification:** The family Scydmaenidae has been recognized as a distinctive family of Staphylinoidea since 1815 (Leach 1815). Several competing hypotheses concerning the phylogenetic relationships of Scydmaenidae have been proposed. Brown and Crowson (1980) placed them near Leiodidae - Camiarinae or Silphidae. Bøving and Craighead (1930) placed them next to Pselaphidae, but Crowson (1955, 1960) believed that this pselaphid-scydmaenid similarity may represent parallel adaptation to feeding on mites. Naomi (1985) placed them next to Scaphidiidae and Lawrence and Newton (1982) placed them within their Staphylinine group. Their relationships remain unclear, pending ongoing phylogenetic analyses.

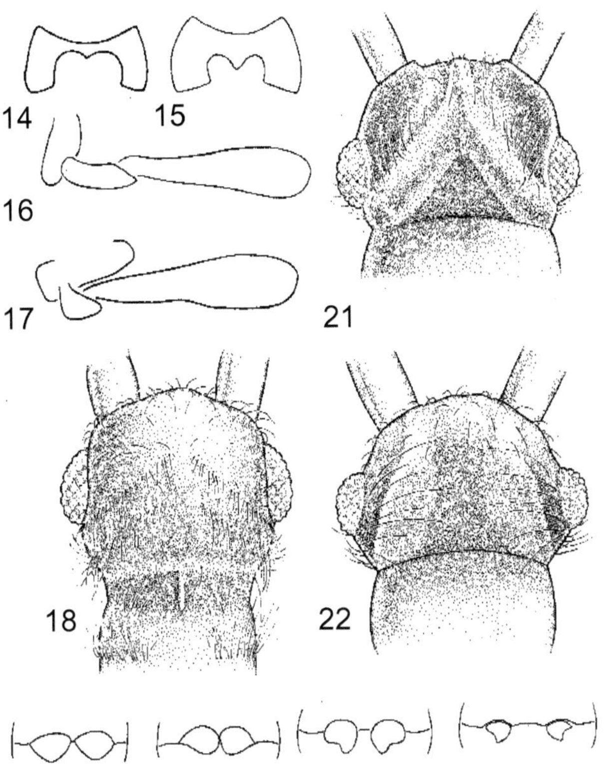


FIGURES 14.20-24.20. Fig. 14.20, prosternum, *Euthiconus*; Fig. 15.20, same, *Eutheia*; Fig. 16.20, coxa, trochanter, femur, *Scydmaenus*; Fig. 17.20, same, *Euconnus*; Fig. 18.20, head, *Lophioderus*; Fig. 19.20, metacoxae, *Microscydmus* (*s. str.*); Fig. 20.20, same, *Microscydmus* (*Delius*); Fig. 21.20, head, *Taphroscydmus;* Fig. 22.20, same, *Taphroscydmus;* Fig. 23.20, metacoxae, *Euconnus* (*Napochus*); Fig. 24.20, same, *Euconnus* (*s. str.*).

The subfamilial classification has not changed since Casey's (1897) proposal of splitting the family into two subfamilies (Newton and Thayer 1992). Based on larval characters, Brown and Crowson (1980: 58) suggested the possible elevation of Eutheiini and Cephenniini to subfamilies.

There has been confusion in the use of the name *Scydmaenus* between North American and European scydmaenid workers. In Casey's work (Casey 1897), the species covered under the name *Scydmaenus* were considered to belong to *Stenichnus* by Csiki (1919) and other workers. Although this was clarified in the world catalog (Csiki 1919), it was not corrected in the Leng checklist (Leng 1920) until the first supplement (Leng and Mutchler 1927).

Another publication of importance is that of Franz (1985), in which he designated type species for most of the genera and subgenera described by Casey (1897), placed the genera *Opresus*, *Delius*, and *Neladius* as either subgenera or synonyms of *Microscydmus*, placed *Acholerops* as a synonym of *Scydmaenus*, elevated *Brachycepsis* and *Parascydmus* to generic rank, placed *Drastophus* and *Noctophus* as synonyms of *Euconnus*, placed *Xestophus* as a synonym of *Psomophus*, and questioned the validity of the placement of *Connophron* as a synonym of *Napochus*. He later (Franz 1995) placed *Papusus* as a junior synonym of *Leptochromus*.

In the current classification of North American Scydmaenidae (O'Keefe, 1998) many of the changes proposed by Csiki (1919) and Franz (1985) are retained and the following alterations are included: 1) elevation of *Taphroscydmus* to generic rank; 2) removal of *Papusus* from synonymy with *Leptochromus*; 3) placement of *Ascydmus* and *Euthiodes* as junior synonyms of *Euthiconus*; 4) placement of *Neladius* as a subgenus of *Microscydmus*; 5) placement of *Noctophus*, *Drastophus*, and *Smicrophus* as subgenera of *Euconnus*; and 6) removal of *Xestophus* from synonymy with *Psomophus* and its placement as a subgenus of *Euconnus*.

The North American scydmaenid fauna has been treated in LeConte (1852), Casey (1897), Blatchley (1910), Marsh (1957), Downie and Arnett (1996), and O'Keefe (1996, 1997a, 1997b, 1997c, 1998).

**Distribution:** These beetles are found on all continents, except Antarctica, and major islands. There are over 4,500 described species placed in 80 genera, of which 217 described species (1997) in 18 genera are known from America north of Mexico.

### Key to the Nearctic Genera

1. Antennae elbowed between scape and pedicel (Fig. 25.20); scape as long as antennomeres II-IV combined; maxillary palpomere IV broadly triangular (Fig. 2.20) (Clidicini) .......... *Papusus* (Fig. 25.20)

— Antennae not elbowed between scape and pedicel; scape shorter than antennomeres II-IV combined; maxillary palpomere IV flat, or short, rounded, or acuminate (Figs. 3.20-8.20) .............................. 2

2(1). Base of pronotum with raised, medial, longitudinal carina (Fig. 20a.20); sides of neck with circlets of long setae; vertex of head longitudinally broadly impressed (Fig. 9.20); maxillary palpomere IV flat (Fig. 3.20) (Chevrolatiini) .......... *Chevrolatia* (Fig. 20.20)

— Base of pronotum without raised, medial, longitudinal carina; sides of neck with or without setae, setae never arranged in circlets; vertex of head, if impressed, not broadly, longitudinally; maxillary palpomere IV small, rounded or acuminate ....................................... 3

3(2). Posterior of head concealed beneath pronotum, not constricted (Figs. 27.20-28.20); pronotum nearly as wide as elytra; junction of pronotum and elytra not constricted (Cephenniini) ......................... 4

— Posterior of head not concealed beneath pronotum (except sometimes in Eutheiini, in which the elytra are subtruncate, Figs. 29.20-31.20); pronotum distinctly narrower than elytra; junction of pronotum and elytra constricted ..................................... 5

4(3). Eyes present; foveae present in basal pronotal angles (Fig. 27.20); procoxae separated by prosternal process (Fig. 10.20) .................................................. .................................... *Chelonoidum* (Fig. 27.20)

— Eyes absent; foveae absent in basal pronotal angles; procoxae not separated by prosternal process (Fig. 11.20) .................. *Cephennium* (Fig. 28.20)

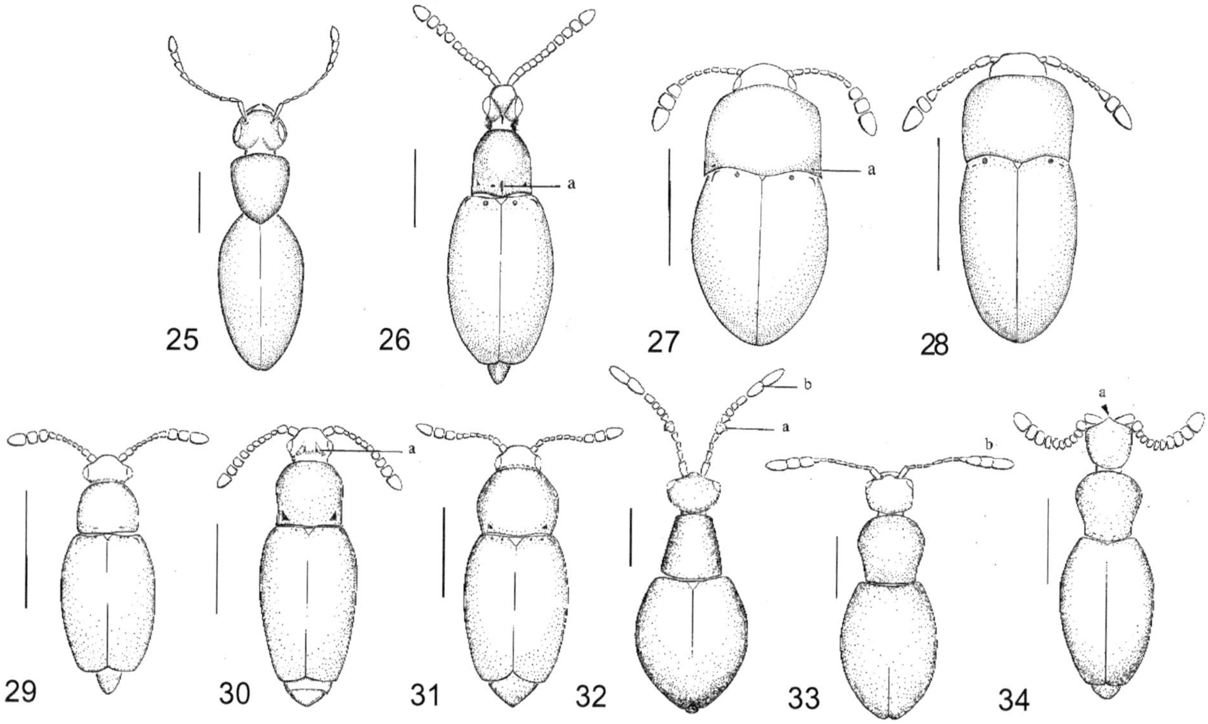


FIGURES 25.20 - 34.20. Fig. 25.20, *Papusus macer* Casey; Fig. 26.20, *Chevrolatia amoena* LeConte; Fig. 27.20, *Chelonoidum corporosum* (LeConte); Fig. 28.20, *Cephennium anophthalmicum* Brendel; Fig. 29.20, *Euthiconus lata* (Brendel); Fig. 30.20, *Eutheia* sp. from Texas; Fig. 31.20, *Veraphis cristata* (Brendel); Fig. 32.20, *Ceramphis deformata* (Horn); Fig. 33.20, *Scydmaenus motschulskii* (LeConte); Fig. 34.20, *Leptoscydmus caseyi* (Brendel).

5(3). Apex of elytra truncate or subtruncate (Figs. 29.20-31.20); pygidium exposed, horizontal (Fig. 12.20) .......... 6

— Apex of elytra entire; pygidium concealed (if exposed, then vertical (Fig. 13.20) [Scydmaenini] or vertex of head excavated [*Taphroscydmus*, Figs. 21.20-22.20]) .......... 9

6(5). Antennal insertions contiguous (Fig. 34.20) (Leptoscydmini) .......... *Leptoscydmus* (Fig. 34.20)

— Antennal insertions widely separated (Eutheiini) ...... 7

7(6). Probasisternum very short before coxae (Fig. 14.20) .......... *Euthiconus* (Fig. 29.20)

— Probasisternum at least 2/3 length of prothorax (Fig. 15.20) .......... 8

8(7). Two distinct foveae on vertex (Fig. 30a.20); mesosternum longitudinally carinate between coxae .......... *Veraphis* (Fig. 30.20)

— Vertex without foveae; mesosternum broad, flat between .......... *Eutheia* (Fig. 31.20)

9(5). Pygidium exposed, vertical (Fig. 13.20); metatrochanters elongate (Fig. 16.20); maxillary palpomere IV short, rounded (Fig. 7.20) (Scydmaenini) .......... 10

— Pygidium concealed (if exposed, then flat, as in Fig. 12.20, and vertex of head excavated [Figs. 21.20-22.20]); metatrochanters short (Figs. 17.20); maxillary palpomere IV long, acuminate (Fig. 8.20) (Cyrtoscydmini) .......... 11

10(9). Antennomere V expanded posteriorly (Fig. 32.20); sides of pronotum straight in dorsal view; antennal club composed of last 2 antennomeres (Fig. 32.20) .......... *Ceramphis* (Fig. 32.20)

— Antennomere V not modified; sides of pronotum rounded in dorsal view; antennal club composed of last 3 antennomeres (Fig. 33.20) .......... .......... *Scydmaenus* (Fig. 33.20)

11(10). Neck elongate, dorsally flattened, densely covered with long, erect setae (Fig. 18.20); males with antennomeres IV-VI modified into an arcuation .... .......... *Lophioderus* (Fig. 38.20)

— Neck distinct but not elongate, dorsally rounded, not densely covered with long, erect setae; antennomeres of males unmodified .......... 12

12(11). Antennal club composed of last 3 antennomeres; antennomere VII slightly wider and longer than VIII (Figs. 35.20, 36.20, 37.20); length 0.5-1.1 mm (*Microscydmus*) .......... 13

— Antennal club composed of last 3-5 antennomeres; antennomere VII smaller than VIII; length 1.0-2.7 mm .......... 15

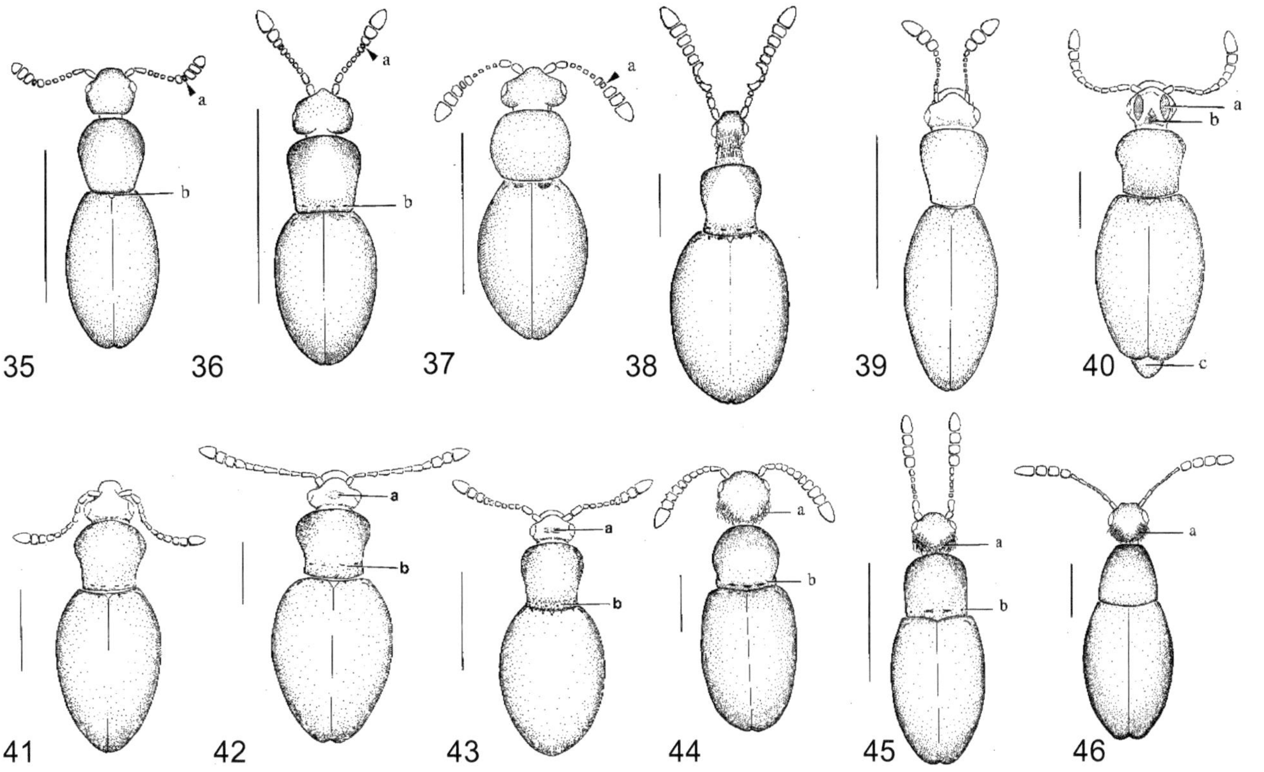


FIGURES 35.20 - 46.20. Fig. 35.20, *Microscydmus (Neladius) tenuis* (Casey); Fig. 36.20, *Microscydmus* (*s. str.*) *misellus* (LeConte); Fig. 37.20, *Microscydmus (Delius) robustulus* (Casey); Fig. 38.20, *Lophioderus arcifer* Casey; Fig. 39.20, *Catalinus angustus* (LeConte); Fig. 40.20, *Taphroscydmus californicus* (Motschulsky); Fig. 41.20, *Stenichnus perforatus* (Schaum); Fig. 42.20, *Brachycepsis subpunctatus* (LeConte); Fig. 43.20, *Parascydmus* sp.; Fig. 44.20, *Euconnus (Noctophus) schmitti* (Casey); Fig. 45.20, *Euconnus (Scopophus) gratus* Casey; Fig. 46.20, *Euconnus (Napochus) elongatum* (Casey).

13(12). In lateral perspective, eyes separate from base of head by twice their diameter; pronotum without basal foveae or dense lateral fringe; scutellum distinctly visible (Fig. 35.20) ................................................ ................. *Microscydmus* (*Neladius*) (Fig. 35.20)

— In lateral perspective, eyes separate from base of head about their diameter; pronotum with basal foveae, with or without dense lateral fringe of bristling setae; scutellum minute or not visible ................. 14

14(13). Metacoxae approximate, not contiguous (Fig. 19.20); scutellum hidden .................................... .................... *Microscydmus* (*s. str.*) (Fig. 36.20)

— Metacoxae contiguous (Fig. 20.20); scutellum barely visible ........... *Microscydmus* (*Delius*) (Fig. 37.20)

15(12). Eyes posterior, less than their diameter from base of neck; neck 2/3 as broad as head at its narrowest point; pronotum widest at anterior half; scutellum minute, but visible (Figs. 39.20-43.20) .... 16

— Eyes anterior or medial; neck half as broad as head at its widest point; pronotum widest at middle or at posterior half; scutellum hidden (Figs. 44.20-51.20) (*Euconnus*) ........................................... 20

16(15). Vertex rounded, not medially impressed ............ 17

— Vertex impressed or bi-impressed (Figs. 42.20, 43.20), or excavated posteriorly (Figs. 21.20-22.20) ... 18

17(16). Body elongate, slender; pronotum without foveae near posterior margin; antennal club distinct ............. .......................................... *Catalinus* (Fig. 39.20)

— Body ovoid, stouter; pronotum with 4 foveae near posterior margin; antennal club gradual ............ ....................................... *Stenichnus* (Fig. 41.20)

18(16). Vertex with deep triangular depression or head excavated (Figs. 21.20-22.20); pygidium exposed .............................. *Taphroscydmus* (Fig. 40.20)

— Vertex shallowly impressed or bi-impressed, never excavated (Figs. 42.20, 43.20); pygidium concealed ........................................................... 19

19(18). Pronotum with 4 basal foveae (Fig. 42.20); scutellum large; often light to dark brown in color ...... ................................. *Brachycepsis* (Fig. 42.20)

— Pronotum with 6 basal foveae (Fig. 43.20); scutellum minute; often black in color ......................... .................................. *Parascydmus* (Fig. 43.20)

20(15). Posterior of head behind eyes with dense patch of setae (Figs. 44.20-46.20); metacoxae nearly contiguous (Fig. 23.20); elytral pubescence dense ........................................................................ 21

— Posterior of head behind eyes without dense patch of setae (Figs. 47.20-51.20); metacoxae distinctly separated (Fig. 24.20); elytral pubescence absent or sparse ......................................................... 24

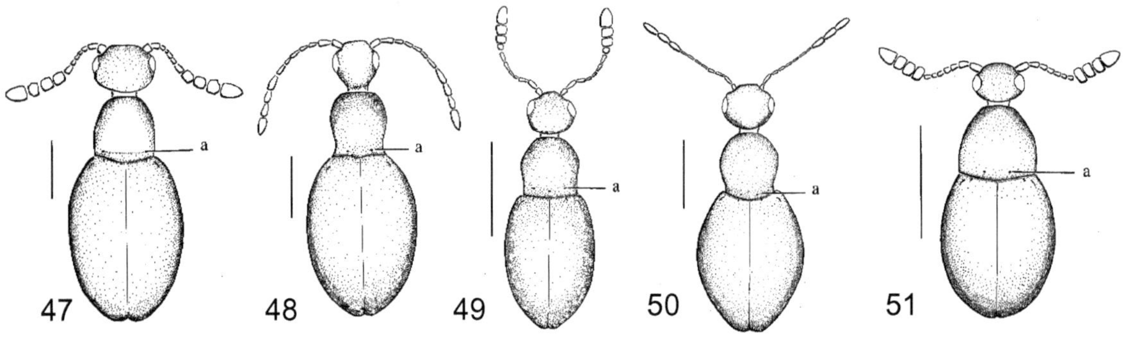


FIGURES 47.20 - 51.20. Fig. 47.20, *Euconnus (Pycnophus) rasus* (LeConte); Fig. 48.20, *Euconnus* (*s. str.*) *clavatus* (Say). Fig. 49.20, *Euconnus (Psomophus) fatuus* (LeConte); Fig. 50.20, *Euconnus (Xestophus) salinator* (LeConte); Fig. 51.20, *Euconnus (Smicrophus) laeviceps* (Casey).

21(20). Antennal club gradual, composed of last 5 antennomeres; 3 dorsal and 2 lateral foveae near posterior margin of pronotum (Fig. 44.20) .......... .................... *Euconnus* (*Noctophus*) (Fig. 44.20)
— Antennal club often distinct, composed of last 3-4 antennomeres; 0-4 foveae near posterior margin of pronotum .................................................... 22

22(21). Pronotum with 2 dorsal, or 2 dorsal and 2 lateral foveae along posterior margin of pronotum ....... ..................... *Euconnus* (*Scopophus*) (Fig. 45.20)
— Pronotum without foveae, but often with transverse groove .......................................................... 23

23(22). Antennal club composed of last 3 antennomeres. ..................................... *Euconnus* (*Napoconnus*)
— Antennal club composed of last 4 antennomeres. ...................... *Euconnus* (*Napochus*) (Fig. 46.20)

24(20). Vertex of head, pronotum, and elytra without setae; posterolateral of head with row of short setae ................. *Euconnus* (*Pycnophus*) (Fig. 47.20)
— At least pronotum pubescent; head and elytra either without setae or with long, sparsely scattered setae...................................................... 25

25(24). Foveae absent in posterolateral corners of pronotum; California ........ *Euconnus* (*Drastophus*) (Fig. 1.20)
— Foveae present in posterolateral corners of pronotum (Figs. 48.20-51.20); eastern and central U.S. ... 26

26(25). Head and elytra sparsely pubescent................. 27
— Head and elytra not pubescent or with a couple scattered setae ............................................. 28

27(26). Antennal club 4-segmented .................................. .............................. *Euconnus* (*s. str.*) (Fig. 48.20)
— Antennal club 3-segmented .................................. .................... *Euconnus* (*Psomophus*) (Fig. 49.20)

28(26). Antennal club composed of last 3 antennomeres; 2 dorsal and 2 lateral foveae near posterior margin of pronotum; pronotum widest at middle, antennomeres IX-XI longer than wide ................ ...................... *Euconnus* (*Xestophus*) (Fig.50.20)
— Antennal club composed of last 4 antennomeres; 2 dorsal foveae near posterior margin of pronotum; pronotum widest at posterior; antennomeres IX-XI wider than long ................................................ .................... *Euconnus* (*Smicrophus*) (Fig. 51.20)

## Classification of the Nearctic Genera

### Scydmaenidae

The tribal classification here follows that of Newton and Thayer (1992). Of the 12 tribes of Scydmaenidae in the world, seven occur in the Nearctic. Subfamilies are not used because the current classification of two subfamilies, Mastiginae and Scydmaeninae, does not represent accurate phylogenetic relationships. Numbers of genera in the world and species counts for genera are taken largely from the "World Catalog of the Genera of Scydmaenidae (Coleoptera)" by Alfred Newton and Herbert Franz (1998).

### Cephenniini

Members of Nearctic Cephenniini can easily be recognized from other Nearctic Scydmaenidae by the broad, oval body shape (Figs. 27.20-28.20), recessed head, and lack of a distinct constriction between the pronotum and elytra. Of the 11 described genera in the world, two occur in North America.

*Cephennium* Müller and Kunze, 1822

Of the over 100 described species of *Cephennium,* only one, *C. anopthalmicum* Brendel, is known from the Nearctic from central, coastal California.

*Chelonoidum* Strand, 1935

*Chelonoides* Croissandeau, 1894.

Of the dozen or so described species of *Chelonoidum,* four are known from the Nearctic. They are moderately common in the Pacific Northwest and also eastern U.S. See Casey (1897) for latest treatment.

### Chevrolatiini

Members of Chevrolatiini can be easily recognized from other Scydmaenidae by the raised, medial pronotal carina (Fig. 20.20) and sculpturing of the head (Fig. 9.20). A single genus is placed in the tribe.

*Chevrolatia* Jacquelin du Val, 1850,

Of the 11 species in the world, 4 are known from the Nearctic. A single rare species occurs in southeastern Arizona, two others from the Mississippi Valley region and southeastern U.S, and the fourth from Mexico. See O'Keefe (1997b) for latest treatment.

Clidicini

Members of the Nearctic Clidicini can be easily recognized from all other Nearctic Scydmaenidae by the elongate antennal scape and elbowed appearance between the scape and pedicel (Fig. 25.20). Of the three known genera in the world, only one occurs in the Nearctic.

*Papusus* Casey, 1897

This Nearctic genus contains a single described species, *P. macer* Casey, from southern California, three new species, one each from southern California, Arizona, and Utah, another new species from Sonora, Mexico, and four new species from Baja California. This genus is currently under revision (O'Keefe, in preparation).

Cyrtoscydmini

Members of Nearctic Cyrtoscydmini can be recognized from all other Nearctic Scydmaenidae by the narrow, acuminate maxillary palpomere IV (Fig. 8.20). Of the 44 genera in the tribe, 8 occur in the Nearctic.

*Brachycepsis* Brendel, 1889

This Nearctic genus is represented by 7 described species, from the western, midwestern, and northeastern U.S. There are several undescribed species (O'Keefe, unpublished). See Casey (1897) for the most recent treatment.

*Catalinus* Casey, 1897

This Nearctic genus is represented by a single described species, *C. angustus* (LeConte), from northern California. This group may be more widespread in the western U.S.

*Euconnus* Thomson, 1862

Members of this genus occur throughout the world. Of the over 2,500 described species, 113 are known from the Nearctic. Forty subgenera have been described, of which 10 occur in the Nearctic.

subgenus *Drastophus,* Casey 1897

*Drasterophus* Franz, 1985.

This Nearctic subgenus is represented by a single described species, *E. laevicollis* (Casey), from northern California. At least two new species are to be found in central coastal California (O'Keefe, unpublished).

subgenus *Euconnus* Thomson, 1862

Members of this group occur worldwide; 6 species have been described from the U.S. and perhaps as many as 20 more await description. They are more common in the midwestern and eastern U.S. See Casey (1897) for the most recent treatment.

subgenus *Napochus* Thomson, 1862

. *Connophron* Casey, 1897

*Glandularia* Schaufuss, 1889

Members of this subgenus occur worldwide; 84 species have been described from the U.S. and many still await description. They are commonly found in the midwestern and eastern U.S. See Casey (1897) for the most recent treatment.

subgenus *Napoconnus* Franz, 1957

Members of this subgenus are mostly tropical in distribution. The *Napochus* with a 3-segmented antennal club may belong to this group.

subgenus *Noctophus* Casey, 1897

This Nearctic subgenus is represented by a single species, *E. schmitti* (Casey), from the southeastern U.S. Other members of this group occur in the Antilles and northern South America.

subgenus *Psomophus* Casey, 1897

*Spanioconnus* Ganglbauer, 1899

*Diarthroconnus* Ganglbauer, 1900

*Psocophus* Franz, 1994

Members of this subgenus occur worldwide; 6 species have been described from the Nearctic. Two species have been described from Texas and the others are from the eastern U.S. See Casey (1897) for the most recent treatment.

subgenus *Pycnophus* Casey, 1897

*Nudatoconnus* Franz, 1980

Of the 70 species of this mostly tropical subgenus, only one, *E. rasus* (LeConte), is known from the Nearctic from eastern to northcentral U.S.

subgenus *Scopophus* Casey, 1897

This Nearctic subgenus is represented by 10 species occurring in the eastern and central U.S. See Casey (1897) for the most recent treatment.

subgenus *Smicrophus* Casey, 1897

This Nearctic subgenus is represented by 2 species from the eastern and central U.S. See Casey (1897) for the most recent treatment.

subgenus *Xestophus* Casey, 1897

This subgenus is represented by a single species, *E. salinator* (LeConte), from the northeastern U.S. Other members of this group occur in Central and South America.

*Euconnus* incertae sedis:

*Euconnus longiceps* Fall (O'Keefe 1997a) from the Pacific Northwest is the only described species of Nearctic *Euconnus* that is not placed into any existing subgenus. A few other undescribed species of *Euconnus,* mostly from the southeastern U.S., will not be able to be placed into existing subgenera. However, these are uncommonly encountered.

*Lophioderus* Casey, 1897

This Nearctic genus is represented by 36 species occurring in northern California and the Pacific Northwest and a single species from the southern Appalachians. See O'Keefe (1996) for the most recent treatment.

*Microscydmus* Saulcy and Croissandeau, 1893

Members of this genus occur worldwide. Of the nearly 170 described species, 5 are known from the Nearctic. See Casey (1897) for the most recent treatment.

subgenus *Delius* Casey, 1897

This Nearctic subgenus is represented by a single species, *M. robustulus* (Casey), from the eastern U.S.

subgenus *Microscydmus* (*s. str.*)

*Opresus* Casey, 1897

This worldwide subgenus is represented by 3 species from the eastern U.S.

subgenus *Neladius* Casey, 1897

This Nearctic subgenus is represented by a single species, *M. tenuis* (Casey), from the eastern U.S.

*Parascydmus* Casey, 1897

This Nearctic genus is represented by 4 described species from the eastern U.S. See Casey (1897) for the most recent treatment.

*Stenichnus* Thomson, 1859

*Scydmaenus* Schaum, 1844

subgenus (*Cyrtoscydmus*) Motschulsky, 1870

Members of this genus occur worldwide; 9 are described from the Nearctic. They occur in California, Iowa, and the eastern U.S. See Casey (1897) for the most recent treatment.

*Taphroscydmus* Casey, 1897

This Nearctic genus is represented by a single described species, *T. californicus* (Motchsulsky), from the Pacific Northwest and at least 4 undescribed species from the western U.S. (O'Keefe, unpublished).

### Eutheiini

Members of the Nearctic Eutheiini can be recognized from all other Nearctic Scydmaenidae by the subtruncate elytra and wide neck (Figs. 29.20-31.20). Of the 5 genera occurring in the world, 3 are known from the Nearctic.

*Eutheia* Stephens, 1830

*Euthia* Agassiz, 1847

*Euthiopsis* Müller, 1925

Of the 33 known species, 3 have been described from the United States and a fourth from Mexico. Members of this group are occasionally found in the Pacific Northwest and eastern U.S. Several undescribed species are known from the U.S. (O'Keefe, unpublished). See Casey (1897) and Marsh (1957) for current treatments.

*Euthiconus* Reitter, 1881

*Ascydmus* Casey, 1897

*Euthiodes* Brendel, 1893

Of the 6 described species, 2 are known from the Nearctic from the Midwest and northeastern U.S. See Casey (1897) for most current treatment.

*Veraphis* Casey 1897

Of the 11 described species, 8 are known from the Nearctic. These are occasionally found throughout the western and northeastern U.S. Several species await description (O'Keefe, unpublished). See Casey (1897) and Marsh (1957) for current treatments.

### Leptoscydmini

Members of Leptoscydmini can be recognized from all other Nearctic Scydmaenidae by the contiguous antennal insertions and shape of the head (Fig. 34.20). A single genus is in the tribe.

*Leptoscydmus* Casey, 1897

This Nearctic genus contains 2 described species from the northeastern U.S. See Casey (1897) for most current treatment.

### Scydmaenini

Members of the Nearctic Scydmaenini can be recognized from all other Nearctic Scydmaenidae by the exposed, vertical pygidium (Fig. 13.20), elongate metatrochanters, and compact antennal club. Of the 7 known genera in the world, 2 occur in the Nearctic.

*Ceramphis* Casey, 1897

This Nearctic genus is represented by a single species, *C. deformata* (Horn), from southern California and Arizona. See O'Keefe (1997c) for current treatment.

*Scydmaenus* Latreille, 1802

*Eumicrus* LaPorte, 1840

*Microstemma* Motschulsky, 1858

*Acholerops* Casey, 1897

Of the over 730 described species, 12 are known from the Nearctic. They are occasionally found in the southwestern, central, and eastern U.S. See Casey (1897) for most current treatment.

BIBLIOGRAPHY


BESUCHET, C., BURKHARDT, D. H., and LÖBL, I. 1987. The "Winkler/Moczarski" eclector as an efficient extractor for fungus and litter Coleoptera. Coleopterists Bulletin, 41: 392-394.

BLATCHLEY, W. S. 1910. Coleoptera or beetles (exclusive of the Rhynchophora) known to occur in Indiana with bibliography and descriptions of new species. The Nature Publishing Co., Indianapolis, IN. 1386 pp.

BÖVING, A. G. and CRAIGHEAD, F. C. 1931. An illustrated synopsis of the principal larval forms of the order Coleoptera. Entomologica Americana (N.S.), 11: 1-351.

BROWN, C. and CROWSON, R. A. 1980. Observations on scydmaenid (Col.) larvae with a tentative key to the main British genera. Entomologist's Monthly Magazine, 115: 49-59.

CASEY, T. L. 1897. Coleopterological notices, VII. Annals of the New York Academy of Science, 9: 285-684.

CROWSON, R. A. 1955. The Natural Classification of the Families of Coleoptera. Nathaniel Lloyd, London. 187 pp.

CROWSON, R. A. 1960. The phylogeny of Coleoptera. Annual Review of Entomology, 5: 111-134.

CSIKI, E. 1919. Scydmaenidae. *In:* Coleopterorum Catalogus, Pars 70 (S. Schenkling, editor). W. Junk, Berlin. 106 pp.

DE MARZO, L. 1984. Morfologia della larva e della pupa in *Mastigus pilifer* Kraatz (Coleoptera, Scydmaenidae). Entomologica [Bari], 19: 61-74.

DOWNIE, N. M. and ARNETT, R. H. 1996. The Beetles of Northeastern North America, Volume I. Introduction; Suborders Archostemata, Adephaga, and Polyphaga, thru Superfamily Cantharoidea. The Sandhill Crane Press, Gainesville, Florida.

FRANZ, H. 1980. Eine neue termitophile *Scydmaenus*-Art aus Südafrika. Entomologische Blätter, 76: 55-57.

FRANZ, H. 1985. Revision Caseyscher Scydmaenidentypen. Sitzungsberichte der Osterreichischen Akademie der Wissenshaften Mathematisch-Naturwissenschaftliche Klasse, Abt. 1, 194: 149-186.

LAWRENCE, J. F. and A. F. NEWTON, J. R. 1982. Evolution and classification of beetles. Annual Review of Ecology and Systematics, 13: 261-290.

LEA, A. M. 1910. Australian and Tasmanian Coleoptera inhabiting or resorting to the nests of ants, bees, and termites. Proceedings of the Royal Society of Victoria, Melbourne (N.S.), 23: 116-130.

LEA, A. M. 1912. Australian and Tasmanian Coleoptera inhabiting or resorting to the nests of ants, bees, and termites. Supplement. Proceedings of the Royal Society of Victoria, Melbourne (N.S.), 25: 31-78.

LEACH, W. E. 1815. Entomology [pp. 57-172]. *In:* Edinburgh Encyclopedia. Volume 9 (1) (Brewster, editor). Edinburgh.

LECONTE, J. L. 1852. Synopsis of the Scydmaenidae of the United States. Proceedings of the Academy of Natural Scences, Philiadelphia, 6: 149-157.

LENG, C. W. 1920. Catalog of the Coleoptera of America north of Mexico. Cosmos Press, Cambridge, MA. 470 pp.

LENG, C. W. and MUTCHLER, A. J. 1927. Supplement 1919 to 1924 (inclusive) to Catalog of the Coleoptera of America, north of Mexico. John D. Sherman, Mount Vernon, N.Y.

MARSH, G. A. 1957. Family Scydmaenidae [pp. 273-280]. *In:* The Beetles of the Pacific Northwest. Part II. Staphyliniformia (M. H. Hatch, editor). University of Washington Publications in Biology, 16.

NAOMI, S. -I. 1985. The phylogeny and higher classification of the Staphylinoidea and their allied groups (Coleoptera, Staphylinoidea). Esakia, 23: 1-27.

NEWTON, A. F. Jr. 1991. Scydmaenidae (Staphylinoidea) [pp. 330-334]. *In:* An introduction to immature insects of North America, Volume 2 (F.W. Stehr, editor). Kendall/ Hunt Publishing Co., Dubuque, Iowa.

NEWTON, A. F., Jr. and FRANZ, H. 1998. World catalog of the genera of Scydmaenidae (Coleoptera). Koleopterologische Rundschau, 68: 137-165.

NEWTON, A. F., Jr. and THAYER, M. K. 1992. Current classification and family-group names in Staphylinoidea (Coleoptera). Fieldiana: Zoology, (N.S.), 67: 1-92.

O'KEEFE, S. T. 1996. Revision of the Nearctic genus *Lophioderus* Casey (Coleoptera: Scydmaenidae). Thomas Say Publications in Entomology: Monographs; Entomological Society of America, Lanham, Maryland. 97 pp.

O'KEEFE, S. T. 1997a. *Euconnus longiceps* Fall, an odd ant-like stone beetle from the Pacific Northwest (Coleoptera: Scydmaenidae). Coleopterists Bulletin, 51: 277-283.

O'KEEFE, S. T. 1997b. Revision of the genus *Chevrolatia* Jacquelin du Val (Coleoptera: Scydmaenidae) for North America. Transactions of the American Entomological Society, 123: 163-185.

O'KEEFE, S. T. 1997c. Review of the Nearctic genus *Ceramphis* (Coleoptera: Scydmaenidae). Entomological News, 108: 335-344.

O'KEEFE, S. T. 1998. Notes on the classification of North American ant-like stone beetles (Coleoptera: Scydmaenidae). Coleopterists Bulletin, 52: 259-269.

SCHMID, R. 1988a. Die Larven der Ameisenkafer (Scydmaenidae, Staphylinoidea): Neu- und Nach-beschreibung mit einem vorlaufigen Bestim-mungsschlussel bis zur Gattung. Mitteilungen des Badischen Landesvereins fur Naturkunde und Natureschutz (N.F.), 14: 643-660.

SCHMID, R. 1988b. Morphologische Anpassungen in einem Rauber-Beute-System: Ameisenkafer (Scyd-maenidae, Staphylinoidea) und gepanzerte Milben (Acari). Zoologische Jahrbucher, Abteilung fur Systematik, Okologie und Geographie der Teire, 115: 207-228.

SCHUSTER, R. 1966. Uber den Beutefang des Ameisenkafers *Cephennium austriacus* Reitter. Naturwissenschaften 53: 113.

SUTER, W. 1966. Techniques for the collection of microcoleoptera of the families Pselaphidae, Ptiliidae, and Scydmaenidae. Coleopterists Bulletin, 20: 33038.

VIT, S. and DE MARZO, L. 1989. Description of the larva of *Leptomastax hypogeus* Pirazzoli (Coleoptera: Scydmaenidae). Archives des Sciences [Geneve], 42: 569-578.

WASMANN, E. 1894. Kritisches Verzeichniss der myrmekophilen und termitophilen Arthropoden. Mit angabe der Lebensweise und mit Beschreibung neuer Arten. Felix L. Dames, Berlin. xiii + 231 pp.

WHEELER, Q. D. and MCHUGH, J. V. 1987. A portable and convertable "Moczarski/Tullgren" extractor for fungus and litter Coleoptera. Coleopterists Bulletin, 41: 9-12.

WHEELER, Q. D. and PAKALUK, J. 1983. Descriptions of larval *Stenichnus (Cyrtoscydmus): S. turbatus* and *S. conjux*, with notes on their natural history (Coleoptera: Scydmaenidae). Proceedings of the Entomological Society of Washington, 85: 86-97.

# 21. SILPHIDAE Latreille, 1807

by Stewart B. Peck

**Common name:** The carrion beetles

**Family synonym:** Necrophoridae Kirby, 1837

Members of the family are large beetles, frequently found in association with decaying organic material. They are most commonly encountered at vertebrate carcasses and hence have the common name of carrion beetles. The habit of adult *Nicrophorus* of interring small vertebrate carcasses has also led to the use of the common names of burying beetles and sexton beetles. Adults can be easily recognized by their size; possession of clavate or capitate 11-segmented antennae; prominent fore coxae; and elytra which are truncate (Fig. 1.21), reticulate (Fig. 2.21), tricostate (Figs. 3.21, 4.21), or lacking costae, generally blackish, and usually with orange or red markings in *Nicrophorus.*

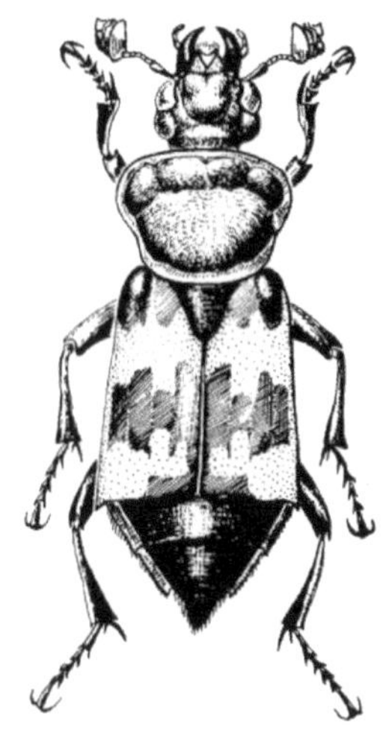

FIGURE 1.21. *Nicrophorus marginatus* Fabricius (From Peck 1990).

**Description:** Body 7-45 mm (usually 12-20 mm) long, ovate to moderately elongate and slightly to strongly flattened; usually glabrous dorsally, pronotum rarely pubescent. Males usually with broadly expanded protarsal segments, female protarsi resembling those on the other legs.

Head at least slightly constricted posteriorly; frontoclypeal suture occasionally distinct; gular sutures sometimes confluent. Antennae occasionally geniculate, with a long scape and a highly reduced pedicel; antennal insertions exposed; ending in a three-segmented, tomentose club, usually preceded by two or three enlarged but glabrous segments (Silphinae) or antennomeres 9-11 lamellate (Nicrophorinae).

Pronotum with complete lateral edges, sometimes explanate. Procoxae transverse, projecting, and contiguous, with large, exposed trochantins; their cavities widely open posteriorly and open internally. Mesocoxae usually moderately to very widely separated, rarely subcontiguous. Scutellum very large. Elytra sometimes truncate, exposing one or two abdominal terga (three to four in *Nicrophorus*); never striate; in Silphinae bearing none to three upraised costae (carinae) on each; a raised area (callus) may occur near the posterior end of the outermost costa; epipleura usually well-developed and complete almost to the apex. The elytra of *Nicrophorus* usually have broad colored bands (maculae) extending laterally to meet the epipleura. Tibial spurs occasionally enlarged, and tarsal claws sometimes apendiculate. Hindwing provided with a secondary hinge.

Abdomen with six (rarely seven) visible sterna; lateral portions of sternum 2 exposed; intersegmental areas marked with a pattern of microsclerites; the three basal terga membranous, tergum 5 with paired stridulatory files in *Nicrophorus.* Aedeagus basal piece reduced but distinct, parameres well-developed and symmetrical.

Larvae (see Newton 1991) campodeiform; with heavily sclerotized terga either bearing posterior spines or produced laterally beyond the edges of the sterna; sterna sclerotized or membranous; antennae long, with a very short sensorium; labrum composed of several sclerites; frontoclypeal suture laterally distinct, either six or only one ocelli on each side; mandibles without a mola or prostheca; mandibular apex more or less acute; mola divided apically; galea furnished with a large, dense brush of hairs; lacinial lobe spinose only on the lateral margins; ligula bilobed; urogomphi usually short, one- or two-segmented, and in most species segment 10 bears several eversible lobes.

**Habits and habitats.** Silphids are primarily scavengers and carrion feeders, but some species are phytophagous and may be garden pests *(Aclypea)* and others are predators of caterpillars or snails *(Dendroxena* in Europe). *Nicrophorus* spp. bury small mammal or bird carcasses and guard and feed their larvae. Their complex subsocial behavior has been extensively studied (reviewed in Anderson and Peck 1985; Scott 1996, 1997; see also Trumbo 1992). The largest North American species, *Nicrophorus americanus* Olivier, has vanished from most of its former extensive range, and is a federally listed endangered species (Lomolino *et al.* 1995; Backlund and Marrone 1997) with protected status.

**Status of the classification.** The family has historically included many other taxa currently not regarded as silphids. Most notable among these are members of the Agyrtidae, which, although still considered by some as silphids, are now interpreted as not being closely related to them. This once vaguely defined

**Acknowledgements.** I wish to thank A.F. Newton, Jr., for his generous sharing of specimens and research results over many years, and all others who have helped my studies of these beetles. Jarmila Peck prepared the illustrations and shared collecting trips for 28 years.

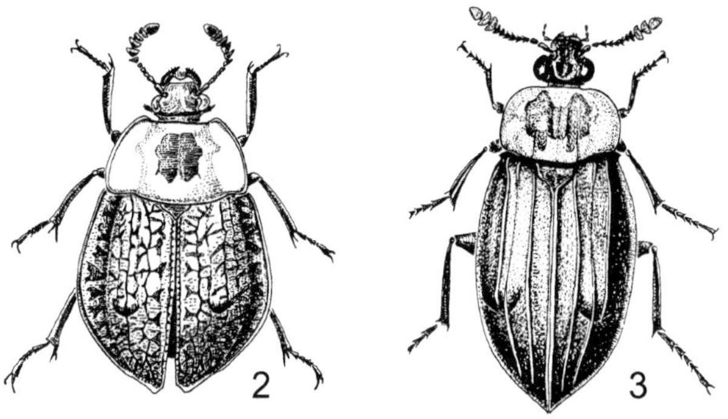


FIGURES 2.21-3.21. Fig. 2.21, *Necrophila americana* (Linnaeus), habitus. Fig. 3.21, *Oxelytrum discicolle* (Brullé), habitus (From Peck 1990).

group is now restricted to the larger carrion and burying beetles (e.g., Anderson and Peck 1985; Peck 1990) after the removal of some subfamilies or tribes that now form the separate family Agyrtidae (see above), and the removal of other groups to the families Leiodidae (e.g., Leptodirini and Estadiini) or Staphylinidae (e.g., Apateticinae, Trigonurinae, Microsilphinae) (Lawrence and Newton 1982). In this restricted sense, silphids are clearly monophyletic and closely allied to staphylinids (Lawrence and Newton 1982).

The North American silphids were first revised by LeConte (1853) and subsequently by Horn (1880). At that time, all species were placed into only two genera, *Nicrophorus* and *Silpha.* Revision of the world fauna by Portevin (1926) divided the genus *Silpha* into numerous genera. That revision incorporated genera proposed by Leach and also erected many new ones. This generic system has been refined and is now in widespread use in Europe and Asia, but the concepts were not generally applied consistently or correctly to the North American fauna. Only recently have these generic concepts been used for the Nearctic species in a way, as used here, which is consistent with that for the Palearctic species (Madge 1980). To complicate the matter, subgenera of *Silpha* and *Necrophila* are often used as genera by some European workers (but this does not affect the North American species).

Two species have been introduced into the Nearctic from the Palaearctic. *Dendroxena quadrimaculata* (Scopoli) was introduced intentionally into the northeastern U.S. for the control of gypsy moth larvae, but did not become established. *Silpha tristis* Illiger, a scavenger on dead insects, was seemingly accidentally introduced into southern California, and around Montreal, Quebec. Only the second population seems to be established (LaPlante 1997).

Much of the older literature is difficult to use in making determinations of North American silphids. Peck (1990) and Sikes and Peck (2000) give modern keys for the U.S. species. Anderson and Peck (1985) present keys and distribution maps for Canadian and Alaskan species. Distribution and bionomics of all or some of the U.S. species are summarized by Peck and Kaulbars (1987) and Ratcliffe (1996). Anderson (1982) and Anderson and Peck (1985) give data on all known larvae. The fauna of Latin America is reviewed by Peck and Anderson (1985) and phylogenetic relationships are proposed. A complete systematic catalogue of the Nearctic fauna is that of Peck and Miller (1989). Herman (1964) discusses the correct spelling of *Nicrophorus.* Many *Nicrophorus* species show variation in coloration and this led to the naming of many invalid species, subspecies, or varieties (Anderson and Peck 1986; Peck and Miller 1989).

**Distribution.** This family contains 15 genera and about 175 species worldwide (Newton 1995). These are contained in two subfamilies. Nicrophorinae contains the Asian genera *Eonecrophorus, Ptomascopus,* and the widely distributed genus *Nicrophorus,* with 74 species, which occurs throughout the Holarctic Region and extends into Southeast Asia to New Guinea and the Solomon Islands, northern Africa, the Caribbean island of Hispaniola, and Andean South America. The Silphinae, with 12 genera and 119 species, occur mainly in the Northern Hemisphere, but *Oxelytrum* extends throughout South America. Asian *Diamesus* extends into, and *Ptomaphilo* occurs widely in, Australia. *Silpha* and *Thanatophilus* extend to South Africa. At present we recognize 30 species in eight genera in North America north of Mexico. Only *Heterosilpha* is restricted to the Nearctic. Twenty-five species in six genera occur from Mexico southward (Anderson and Peck 1985).

### Key to the Nearctic Genera

1. Elytra shorter, truncate, exposing 3 or 4 abdominal terga, usually with red or orange irregular spots (Fig. 1.21); fifth abdominal tergum with stridulatory files; epistomal suture present (Fig. 5.21); second antennomere small, distinct, hidden in tip of first antennomere (Fig. 7.21) subfamily Nicrophorinae .................................. *Nicrophorus*

— Elytra longer, usually not truncate, at most exposing 1 or 2 abdominal terga, usually without large colored areas; fifth abdominal tergum lacking stridulatory files; epistomal suture absent (Fig. 6.21); second antennomere large, not hidden in tip of first antennomere (Fig. 8.21) subfamily Silphinae ...... 2

2(1). Pronotum with disc black, margins yellow (Figs. 2.21, 3.21) ................................................................ 3

— Pronotum entirely black, or, with disc black, margins orange red ....................................................... 4

3(2). Elytra with intervals between carinae smooth, and with apices drawn out to needlelike points (Fig. 3.21) ......................................................... *Oxelytrum*

— Elytra with intervals between carinae with coarse reticulate sculpturing, and with apices rounded, not drawn out to needlelike points (Fig. 2.21) ........... ......................................................... *Necrophila*

4(2). Eyes larger; pronotal postcoxal lobe low, rounded (Fig. 9.21); pronotum orbicular, widest toward middle; elytra with small red markings near tip (Fig. 4.21); males with hind femora greatly expanded ......... ........................................................... *Necrodes*

— Eyes smaller; pronotal postcoxal lobe well-developed, projecting, pointed (Fig. 10.21); pronotum not orbicular, widest toward base; elytra without small red markings near tip; males with hind femora not expanded ........................................................ 5

5(4). Head with short row of long erect hairs behind eyes (Fig. 6.21) ......................................................... 6

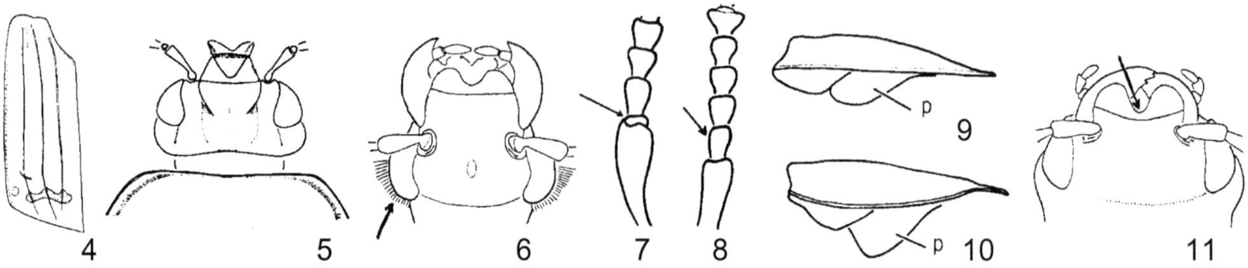


FIGURES 4.21-11.21. Fig. 4.21. *Necrodes surinamensis* (Fabricius), left elytron with three longitudinal costae. Fig. 5.21. *Nicrophorus defodiens* Mannerheim, head, with epistomal suture. Fig. 6.21, *Oiceoptoma inaequale* (Fabricius), head, epistomal suture absent. Fig. 7.21, *Nicrophorus orbicollis* Say, antenna, with small second segment. Fig. 8.21, *Necrodes surinamensis,* antenna, with normal sized second segment. Fig. 9.21, *Necrodes surinamensis,* lateral view of pronotum and rounded postcoxal lobe (p). Fig. 10.21, *Thanatophilus trituberculatus* (Kirby), lateral view of pronotum and pointed postcoxal lobe (p). Fig. 11.21, *Aclypea bituberosa* (LeConte), head with deeply emarginate labrum (Figs. 5, 9, 10 from Peck 1990).

— Head without row of long erect hairs behind eyes .... 7

6(5). Elytral shoulders with tooth; metafemur lacking carinae on inner face; elytra with intervals between carinae without reticulate sculpturing .............. ........................................................ *Oiceoptoma*

— Elytral shoulders rounded, not toothed; metafemur with 2 carinae on inner face; elytra with intervals between carinae with reticulate sculpturing .... ....................................................... *Heterosilpha*

7(5). Labrum deeply emarginate (Fig. 11.21); mesocoxae narrowly separated; pronotum with small smooth glossy areas (discal callosities) ............ *Aclypea*

— Labrum broadly, shallowly emarginate; mesocoxae widely separated; pronotum without smooth glossy discal callosities .................................. 8

8(7). Antennal club more robust, antennomere 2 shorter than 3, 8 shorter than 9; native to North America; widespread, common .................. *Thanatophilus*

— Antennal club more slender; antennomere 2 as long as 3, 8 as long as 9; introduced to North America, Quebec, rare ............................................ *Silpha*

### Classification of the Nearctic Genera

Silphinae Latreille 1807

*Aclypea* Reitter 1884; 2 spp., *A. bituberosa* (LeConte) 1859, and *A. opaca* (Linnaeus) 1758, western North America (see Anderson and Peck 1985).

*Blitophaga* Reitter 1884

*Heterosilpha* Portevin 1926; 2 spp., *H. aenescens* (Casey) 1880, and *H. ramosa* (Say) 1823, Oregon to Mexico (see Miller and Peck 1979; Peck and Kaulbars 1987).

*Necrodes* Leach 1815; 1 sp., *N. surinamensis* (Fabricius) 1775, widespread North America (see Peck and Kaulbars 1987; Ratcliffe 1972).

*Cyclophorus* Stephens 1829
*Asbolus* Bergroth 1884
*Protonecrodes* Portevin 1922

*Necrophila* Kirby and Spence 1828, 1 sp., *N. americana* (Linnaeus) 1758, widespread eastern North America (Peck and Kaulbars 1987).

subgenus *Necrophila* Kirby and Spence 1828
subgenus *Eosilpha* Semenov-Tian-Shanskij 1890
subgenus *Calosilpha* Portevin 1920
subgenus *Deutosilpha* Portevin 1920
subgenus *Chrysosilpha* Portevin 1921

*Oiceoptoma* Leach 1815; 3 spp., eastern North America (Peck 1990; Peck and Kaulbars 1987).

*Oeceoptoma* Agassiz 1847
*Isosilpha* Portevin 1920

*Oxelytrum* Gistel 1848; 1 sp., *O. discicolle* (Brullé) 1836, southern Texas southward (Anderson and Peck 1985).

*Hyponecrodes* Kraatz 1876
*Katanecrodes* Schouteden 1905
*Paranecrodes* Portevin 1921

*Silpha* Linnaeus 1758; *S. tristis* Illiger 1798, of Europe, established in southern Quebec (LaPlante 1997).

subgenus *Silpha* Linnaeus 1758
subgenus *Phosphuga* Leach 1817
subgenus *Ablattaria* Reitter 1884

*Thanatophilus* Leach 1815; 6 spp., widespread North America (Anderson and Peck 1985; Peck 1990; Peck and Kaulbars 1987)

*Pseudopelta* Bergroth 1884
*Philas* Portevin 1903
*Silphosoma* Portevin 1903
*Chalcosilpha* Portevin 1926

Nicrophorinae Kirby 1837
The burying beetles

*Nicrophorus* Fabricius 1775, 15 spp., generally distributed in the U.S. and Canada (Anderson and Peck 1985; Peck 1990; Peck and Kaulbars 1987).

*Necrophorus* Thunberg 1789 (unjust. emend. *Nicrophorus*)
*Necrophagas* Leach 1815
*Cyrtoscelis* Hope 1840
*Canthopsilus* Portevin 1914
*Necrocharis* Portevin 1923
*Necroxenus* Semenov-Tian-Shanskij 1926
*Eunecrophorus* Semenov-Tian-Shanskij 1933
*Necrocleptes* Semenov-Tian-Shanskij 1933
*Necrophorindus* Semenov-Tian-Shanskij 1933
*Necrophoniscus* Semenov-Tian-Shanskij 1933
*Necropter* Semenov-Tian-Shanskij 1933
*Nesonecrophorus* Semenov-Tian-Shanskij 1933
*Nesonecropter* Semenov-Tian-Shanskij 1933
*Stictonecropter* Semenov-Tian-Shanskij 1933
*Neonicrophorus* Hatch 1946

BIBLIOGRAPHY

ANDERSON, R.S. 1982. Resource partitioning in the carrion beetle (Coleoptera: Silphidae) fauna of southern Ontario: ecological and evolutionary considerations. Canadian Journal of Zoology, 60: 1314-1325.

ANDERSON, R.S. and PECK, S.B. 1984. Bionomics of Nearctic species of *Aclypea* Reitter: phytophagous "carrion" beetles (Coleoptera: Silphidae). Pan-Pacific Entomologist, 60 : 248-255.

ANDERSON, R.S. and PECK, S.B. 1985. The Insects and Arachnids of Canada and Alaska, part 13. The carrion beetles of Canada and Alaska (Coleoptera: Silphidae and Agyrtidae). Research Branch, Agriculture Canada, Ottawa, Publication 1778. 121 pp. (North American species, keys)

ANDERSON, R.S. and PECK, S.B. 1986. Geographic patterns of colour variation in North American *Nicrophorus* burying beetles (Coleoptera: Silphidae). Journal of Natural History, 20: 283-297.

BACKLUND, D.C. and MARRONE, G.M. 1997. New records of the endangered American burying beetle *Nicrophorus americanus* Olivier, in South Dakota. Coleopterists Bulletin, 51: 53-58.

BREWER, J.W. and BACON, T.R. 1975. Biology of the carrion beetle *Silpha ramosa* Say. Annals of the Entomological Society of America, 68 (5): 786-790.

HERMAN, L.H. JR. 1964. Nomenclatural consideration of *Nicrophorus* (Coleoptera: Silphidae). Coleopterists Bulletin, 18: 5-6.

HORN, G.H. 1880. Synopsis of the Silphidae of the United States with reference to the genera of other countries. Transactions of the American Entomological Society, 8: 219-319, pls. 5-7.

LAPLANTE, S. 1997. Premiéres évidences de l'etablissement d'une population de *Silpha tristis* Illiger (Coleoptera: Silphidae) en Amérique du Nord. Fabreries, 22: 85-93.

LAWRENCE, J.F. and NEWTON, A.F. 1982. Evolution and classification of beetles. Annual Review of Ecology and Systematics, 13: 261-290.

LECONTE, J.L. 1853. Synopsis of the Silphales of America, North of Mexico. Proceedings of the Academy of Natural Sciences of Philadelphia, 6: 274-287.

LOMOLINO, M.V., CREIGHTON, J.C., SCHNELL, G.D. and CERTAIN, D.L. 1995. Ecology and conservation of the endangered American burying beetle (*Nicrophorus americanus*). Conservation Biology, 9: 605-614.

MADGE, R.B. 1980. A catalogue of type-species in the family Silphidae (Coleoptera). Entomologica Scandinavica, 11: 353-362.

MILLER, S.E. and PECK, S.B. 1979. Fossil carrion beetles of Pleistocene California asphalt deposits with a synopsis of Holocene California Silphidae (Insecta: Coleoptera: Silphidae). Transactions of the San Diego Society of Natural History, 19: 85-106.

NEWTON, A.F. 1991. Silphidae (Staphylinoidea). pp. 339-341. *In* Stehr, F.W. (ed.), Immature Insects, Vol. 2. Kendall/Hunt, Dubuque, Iowa.

NEWTON, A.F. JR. 1995. unpublished World Catalogue of Silphidae.

PECK, S.B. 1990. Insecta: Coleoptera: Silphidae and the associated families Agyrtidae and Leiodidae. pp. 1113-1136. *In* Dindal, D., (ed.), Soil Biology Guide. John Wiley and Sons, New York. (keys to species)

PECK, S.B. and ANDERSON, R.S. 1982. The distribution and biology of the alpine-tundra carrion beetle *Thanatophilus coloradensis* (Wickham) in the Rocky Mountains of North America. Coleopterists Bulletin, 36: 112-115.

PECK, S.B. and ANDERSON, R.S. 1985. Taxonomy, phylogeny and biogeography of the carrion beetles of Latin America (Coleoptera: Silphidae). Quaestiones Entomologicae, 21: 247-317.

PECK, S.B. and KAULBARS, M.M. 1987. A synopsis of the distribution and bionomics of the carrion beetles (Coleoptera: Silphidae) of the conterminous United States. Proceedings of the Entomological Society of Ontario, 118: 47-81.

PECK, S.B. and MILLER, S.E. 1982. Type designations and synonymies for North American Silphidae (Coleoptera). Psyche, 89: 151-156.

PECK, S.B. and MILLER, S.E. 1993. A catalogue of the Coleoptera of America north of Mexico. Family: Silphidae. Agriculture Research Service, U.S. Department of Agriculture, Washington, D.C. Agriculture Handbook 529-28, 24 pp.

PORTEVIN, M.G. 1926. Les grands necrophages du globe. Encyclopedie Entomologique, 6: 1-270. Lechevalier, Paris.

RATCLIFFE, B.C. 1972. The natural history of *Necrodes surinamensis* (Fabr.) (Coleoptera: Silphidae). Transactions of the American Entomological Society, 98 (4): 359-410.

RATCLIFFE, B.C. 1996. The carrion beetles (Coleoptera: Silphidae) of Nebraska. University Nebraska State Museum, Bulletin 13, 100 pp.

SCOTT, M.P. 1996. Communal breeding in burying beetles. American Scientist, 84: 376-382.

SCOTT, M.P. 1997. The ecology and behavior of burying beetles. Annual Review of Entomology, 43: 595-618.

SIKES, D.S., and PECK, S.B. 2000. Description of *Nicrophorus hispaniola*, new species, from Hispaniola (Coleoptera: Silphidae) and a key to the species of *Nicrophorus* of the New World. Annals of the Entomological Society of America, 93: 391-397.

TRUMBO, S.T. 1992. Monogamy to communal breeding: exploitation of a broad resource base by burying beetles. Ecological Entomology, 17: 289-298.

Superfamily Staphylinoidea Latreille, 1802

Staphyliniformia Lameere, 1900; Brachelytra auctorum

# 22. STAPHYLINIDAE Latreille, 1802

by Alfred F. Newton, Margaret K. Thayer,
James S. Ashe, and Donald S. Chandler

**Common name:** The Rove Beetles

**Family synonyms:** Pselaphidae Latreille, 1802; Scaphidiidae Latreille, 1807; Micropeplidae Leach, 1815; Oxytelidae Fleming, 1821; Oxyporidae Fleming, 1821; Brathinidae LeConte, 1861; Dasyceridae Reitter, 1887

Staphylinidae is the largest or second-largest family of beetles, with over 46,200 known species placed in more than 3,200 genera; nearly 400 species are added each year, some from North America. It is a worldwide and ancient group, with a fossil history extending back some 200 million years. The family is abundantly represented in all but the driest parts of North America, and its 523 genera and over 4,100 named species make it the largest beetle family in this region.

Though staphylinids exhibit tremendous variation in form (e.g., Figs. 1-40.22), the vast majority can be distinguished from other beetles by their combination of short truncate elytra exposing more than half of the rather flexible abdomen, six or occasionally seven visible abdominal sterna, and (with few exceptions) contiguous procoxae. Pselaphinae are exceptional in having more rigid abdomens and often only five visible sterna, but are readily recognizable by their distinctive compact habitus (Figs. 3.22, 11-15.22) and usually distinctly clubbed antennae. Other atypical staphylinids that are easily recognizable by habitus alone include Micropeplinae (Fig. 9.22), Dasycerinae (Fig. 10.22), and Scaphidiinae (Fig. 4.22). A few other staphylinid genera have long elytra that cover all or nearly all of the abdomen, but these can be distinguished from most other beetles by their six visible sterna and lack of lobed fourth tarsomeres.

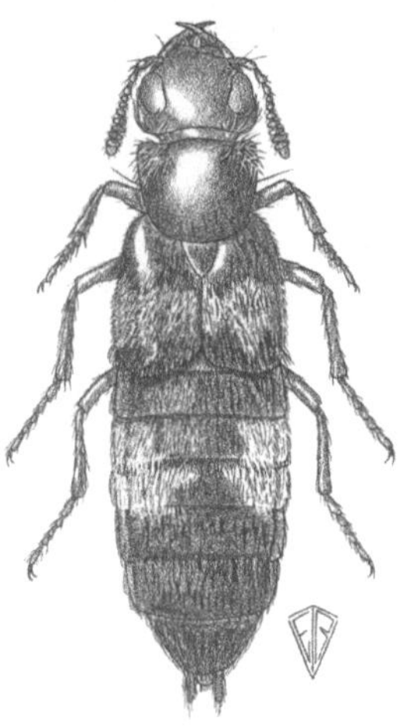

FIGURE 1.22. *Creophilus maxillosus villosus* (Gravenhorst) (After Arnett 1963).

**Description (Adults):** Length 1-35 mm (mostly 2-8 mm); form very elongate to ovoid (Figs. 1-40.22); color yellowish, reddish-brown, brown, or black, occasionally iridescent in part; generally well sclerotized; nearly glabrous to densely setose or pubescent; with or without surface microsculpture. Head variously shaped, prognathous to hypognathous, with or without distinct neck, sometimes with epistomal suture. Compound eyes usually present, pair of ocelli sometimes present. Antenna usually of 11 antennomeres, but with only 10, 9, or 3 in a few genera (also 4-8 in some non-North American ones), usually filiform but sometimes weakly or moderately clubbed. Labrum free, variable in shape, frequently emarginate anteriorly. Mandibles usually projecting and acute, sometimes apically blunt or not visible, often with one or more teeth on the mesal surface; with or without basal molar area. Maxilla with two distinct inner lobes (galea usually brush-like, lacinia usually comb-like), palp with four (rarely five) palpomeres. Labium variable in form, with a pair of palps usually composed of three palpomeres. Mentum present. Gular area well sclerotized, gular sutures separate, fused, or absent. One or two pair of cervical sclerites usually present.

Pronotum highly variable in shape, usually laterally margined. Scutellum usually visible, triangular. Elytra truncate, usu-

**Acknowledgments**: A. F. Newton and M. K. Thayer thank Amy Varsek, Jim Louderman, and Olga Helmy for testing various versions of the keys; Jan Klimaszewski for testing some of the Omaliinae keys; Ale Smetana for comments on the manuscript and for permission to cite newly discovered Chinese occurrences of a few genera; J. H. Frank for useful comments on the manuscript; and Lee Herman for comments on treatment of some paederine genera. J. S. Ashe thanks Sara Taliaferro for the wonderful habitus and structural illustrations for the keys to genera of the Aleocharinae; Terry Erwin and Dave Furth, U.S. National Museum of Natural History, for loan of the Casey Collection of Aleocharinae, without which production of the aleocharine keys would not have been possible; Rod Hanley for checking, proofing, and testing some of the keys; and Zack Falin for considerable help in organizing nomenclatural information. Development of the section on Aleocharinae was supported by NSF-PEET grant DEB-9521755 to the University of Kansas; a species database of Staphylinidae (excluding Pselaphinae) developed with the support of NSF grant BSR-8814449 to the Field Museum greatly facilitated compilation of the generic checklist and species counts; and work done under NSF grants BSR-8806625 and BSR-8906825 to the Field Museum contributed to preparation of the coverage of Omaliini and Staphylinina, respectively.

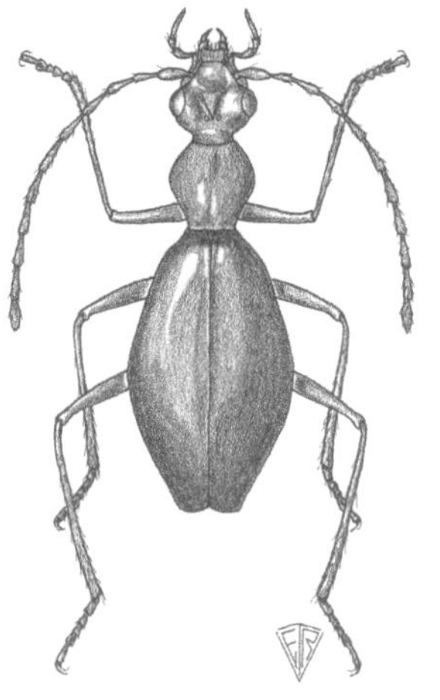

FIGURE 2.22. *Brathinus nitidus* LeConte (After Arnett 1963).

ally very short and exposing five to six abdominal segments dorsally (generally covering terga I-III), occasionally shorter (e.g., Figs. 27.22, 32.22) or longer, even covering entire abdomen (Figs. 2.22, 7.22, 10.22); sometimes striate; with or without epipleural fold. Hind wings usually present, compactly folded under elytra through action of a costal hinge proximal to the radial cell ("stigma"); wing venation moderately to extremely reduced, lacking cross-veins and cells. Pro- and metacoxae highly variable in size and shape, form often characteristic for a subfamily or other higher taxon; each pair of coxae usually contiguous, but sometimes moderately or widely separated; procoxal cavities usually open; and basal articulation of procoxa with trochantin variously exposed or concealed. Tarsi most often 5-5-5, but 3-3-3 in Pselaphinae and a few others, sometimes 4-4-4, 2-2-2, or heteromerous (in North America: 4-5-5 or 4-4-5 in many Aleocharinae; 5-4-4 in *Atanygnathus*); each tarsus normally with a pair of claws, one of them reduced or absent in some Pselaphinae; empodium between claws bearing 0-2 setae.

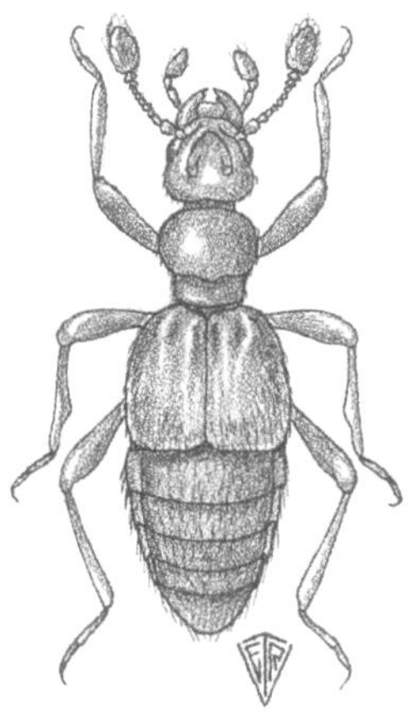

FIGURE 3.22. *Trimiomelba dubia* (LeConte) (After Arnett 1963).

Abdomen elongate, with six (morphological segments III-VIII) or seven (most Oxytelinae, II-VIII; Leptotyphlinae, III-IX) sterna visible ventrally and several terga usually visible dorsally, the exposed segments well sclerotized; in many Pselaphinae, many Scaphidiinae, and males of *Habrocerus*, segment VIII is reduced and largely or entirely hidden inside VII, so only five sterna are visible. Exposed segments usually with one or two pairs of paratergites laterally on each segment in addition to tergite and sternum, but some genera lacking paratergites or having most segments with tergum and sternum fused into a continuous ring (Fig. 55.22). Abdominal segments usually connected by relatively long intersegmental membranes (1/5 to 1/2 as long as each segment) that are often distinctly patterned and allow the abdomen to flex and telescope greatly. Spiracles usually present on segments I-VIII, placed in tergite or in membrane beside tergite, sometimes atrophied on segments III- or IV-VI, rarely atrophied on VIII or additional apical segments. In several subfamilies, one or more tergites with a pair of patches of wing-folding spicules (Hammond 1979). Several subfamilies with abdominal defensive glands associated (characteristically for each subfamily) with sternum IV, VII or VIII, tergite VII or IX, or elsewhere near abdominal apex (Araujo 1978; Dettner 1993).

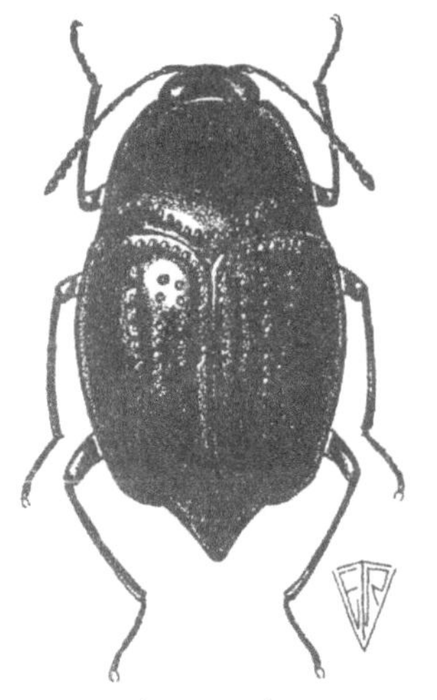

FIGURE 4.22. *Scaphidium piceum* Melsheimer (After Arnett 1963).

Sexual dimorphism usually restricted to internal genitalia, subtle or absent externally (adhesive setae on male protarsi common), but males of some genera with distinctive modifications of antennae, head, metasternum, various parts of the legs, or apical abdominal segments. Abdominal segments IX and X modified into a distinct genital segment in both sexes, often partly or largely exposed (especially in males). Male genitalia with aedeagus variable in shape; median lobe nearly always bulbous at base (to accommodate muscles used to evert the internal sac), with small foramen on one side (=dorsal, following Tikhomirova 1978 and Frank 1981a, contrary to usage of, e.g., Smetana and Campbell) for entry of ejaculatory duct; basal piece reduced or absent; parameres present (usually) or absent, basally articulated or not, sometimes elaborately developed and larger than the median lobe, or fused into a single paramere. Female genitalia usually not visible externally, ovipositor short and partly membranous, usu-

FIGURES 5.22-40.22. Selected genera of Staphylinidae. Fig. 5.22, *Omalium* (Omaliinae); Fig. 6.22, *Anthobium* (Omaliinae); Fig. 7.22, *Empelus* (Empelinae); Fig. 8.22, *Megarthrus* (Proteininae); Fig. 9.22, *Micropeplus* (Micropeplinae); Fig. 10.22, *Dasycerus* (Dasycerinae); Fig. 11.22, *Bibloporus* (Pselaphinae); Fig. 12.22, *Batrisodes* (Pselaphinae); Fig. 13.22, *Tychobythinus* (Pselaphinae); Fig. 14.22, *Pselaphus* (Pselaphinae); Fig. 15.22, *Adranes* (Pselaphinae); Fig. 16.22, *Tachinus* (Tachyporinae); Fig. 17.22, *Lordithon* (Tachyporinae); Fig. 18.22, *Habrocerus*, male (Habrocerinae); Fig. 19.22, *Tachyusa* (Aleocharinae); Fig. 20.22, *Myllaena* (Aleocharinae); Fig. 21.22, *Atheta* (*s. str.*) (Aleocharinae); Fig. 22.22, *Cordalia* (Aleocharinae); Fig. 23.22, *Xenodusa* (Aleocharinae); Fig. 24.22, *Trigonurus* (Trigonurinae); Fig. 25.22, *Renardia* (Osoriinae); Fig. 26.22, *Osorius* (Osoriinae); Fig. 27.22, *Carpelimus* (Oxytelinae); Fig. 28.22, *Oxytelus* (Oxytelinae); Fig. 29.22, *Oxyporus* (Oxyporinae); Fig. 30.22, *Stenus* (Steninae); Fig. 31.22, *Euaesthetus* (Euaesthetinae); Fig. 32.22, *Homeotyphlus* (Leptotyphlinae); Fig. 33.22, *Pseudopsis* (Pseudopsinae); Fig. 34.22, *Medon* (Paederinae); Fig. 35.22, *Astenus* (Paederinae); Fig. 36.22, *Homaeotarsus* (Paederinae); Fig. 37.22, *Diochus* (Staphylininae); Fig. 38.22, *Atrecus* (Staphylininae); Fig. 39.22, *Gauropterus* (Staphylininae); Fig. 40.22, *Quedius* (Staphylininae). Not all to same scale; setae not shown. (Modified from: 5, 6, 9, 10, 16-18, 27-31, 33-35, 38-40 — Lohse 1964; 11-14, 15 [*Claviger*] — Besuchet 1974; 19, 20, 22 [*Atemeles*], 23 — Lohse 1974; 7 — Hansen 1997a; 8, 32 — Newton 1990a; 21 — Benick and Lohse in Lohse 1974; 24 — Van Dyke 1934; 25 — Moore 1964d; 26 — Moore and Legner 1979; 36 — Hatch 1957; 37 — Smetana 1982.)

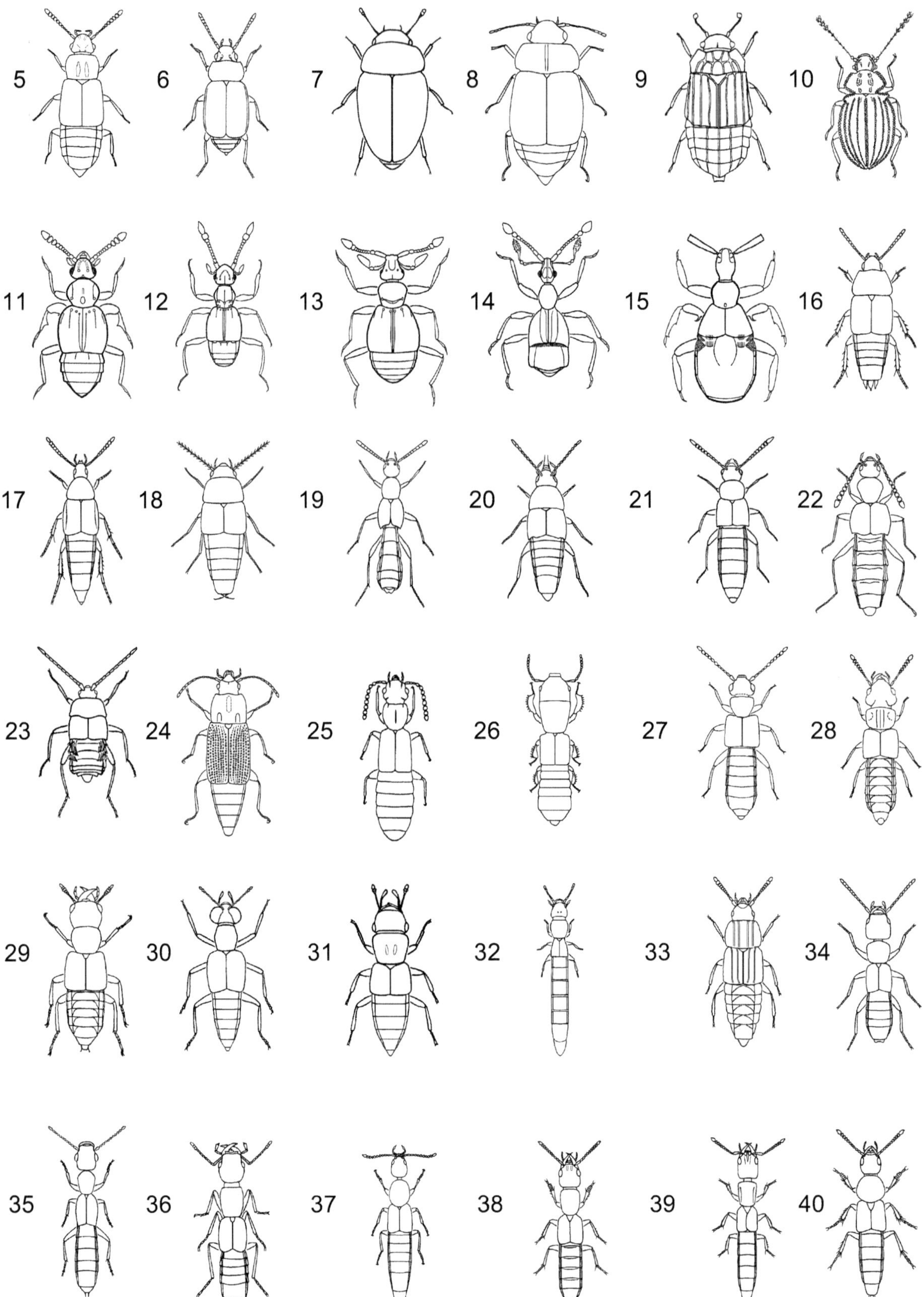
5 6 7 8 9 10
11 12 13 14 15 16
17 18 19 20 21 22
23 24 25 26 27 28
29 30 31 32 33 34
35 36 37 38 39 40

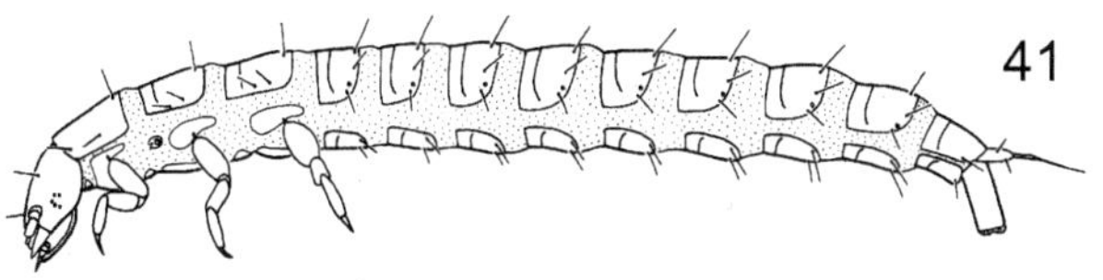

FIGURE 41.22. *Phloeocharis californica* Smetana and Campbell, larva (After Newton 1990a).

ally with a tergum X ("proctiger"), highly modified and usually divided tergum IX ("paraprocts"), two pairs of gonocoxites ("valvifer" and "coxite"), and a small setose stylus on each apical gonocoxite; these sclerites often greatly reduced or fused to one another, or largely membranous; sternum IX present but small, or absent.

Detailed and well-illustrated treatments of general staphylinid external morphology were given by Blackwelder (1936), Coiffait (1972), Tikhomirova (1973), and Naomi (1987-90); some more narrowly focused works are Evans (1965, feeding and digestive tract), Hammond (1979, wing folding, abdominal movement), Blum (1979, abdominal movement and wing folding), and Stork (1980, tarsal adhesive setae). Internal morphology is poorly known and has not generally been used in classification or identification, except for sclerotized parts of the genitalia (Sharp and Muir 1912, male genitalia; Tanner 1927, female genitalia), presence and structure of abdominal defensive glands (Araujo 1978; Dettner 1993, defensive glands and chemistry), and patterns of ovariole development (Welch 1993). Other useful morphological studies on Coleoptera in general are Forbes (1922, 1926, wing venation and folding), Stickney (1923, head capsule), Crowson (1938, 1944, metendosternite), Williams (1938, mouthparts), Hlavac (1975, prothorax), Caveney (1986, eye structure), and Kukalová-Peck and Lawrence (1993, wing venation and folding).

**Description (Larvae)**: Length about 0.5-25 mm; usually campodeiform and very elongate (Fig. 41.22), rarely broad and flat or C-shaped; color white to yellowish, sometimes with brown transverse bands or entirely brown above, or variegated; head, thoracic and abdominal sclerites, and appendages lightly to well sclerotized, with extensive exposed membrane between. Head usually prognathous, rarely hypognathous, with or without distinct neck, on each side with 1-6 stemmata or pigmented eyespots, or stemmata absent. Dorsum of head with Y-shaped ecdysial lines (epicranial sutures) which end in or near antennal insertions, stem of Y rarely absent. Antenna well developed, of 2-4 (usually 3) antennomeres; penultimate or (in 2-merous antennae) apical antennomere with large sensory appendage usually on anterior (inner) side but sometimes dorsal, ventral, or posterior in position. Frontal area without epistomal suture, but in most Pselaphinae with pair of eversible prey-capture structures. Labrum either articulated but less distinct than that of most beetle larvae, lacking internal apodemes (tormae) at hind margins and usually accompanied by 1-2 pairs of lateral sclerites or completely fused to head capsule to form a nasale which is usually toothed or spinose anteriorly. Mandibles never with contiguous molar lobes at base, their apices usually projecting and acute but sometimes multidentate or otherwise modified, rarely with a subapical pseudomola. Maxilla with a single distinct inner lobe (mala), which may be fixed or articulated at its base; maxillary palp large, of 2-4 (usually 3) palpomeres. Labium usually with a median ligula of variable shape (absent in Pselaphinae) and palp of 1-3 palpomeres. Mentum present or absent. Gular area variable, may be short and membranous with submental sclerite to long and completely sclerotized except for thin ecdysial line which may be Y-shaped. Posterior arms of tentorium attached directly to ventral surface of head rather than to transverse bridge.

Thoracic terga each consisting of a single sclerite with midlongitudinal ecdysial line; meso- and metaterga usually shorter than protergum and each with a transverse carina behind anterior margin; pleura and sterna variable, membranous or with one or more sclerites present on each segment. Pair of large annular spiracles present laterally between pro- and mesothorax. Legs well developed, usually of 5 segments (coxa, trochanter, femur, tibia, and clawlike tarsungulus; trochanter and femur fused in *Micropeplus)*; longest macroseta of leg located on trochanter; tarsungulus usually bisetose, rarely glabrous or with additional setae.

Abdomen elongate, 10-segmented. Segments I-VIII usually similar in appearance; tergum and sternum of each segment usually of a single sclerite or divided midlongitudinally by membranous area, with or without transverse carina behind anterior margin, tergum rarely with lateral projections; pair of annular spiracles present in or just outside lateral edges of tergum; one or two pairs of small pleural sclerites may be present. Segment IX consisting of tergum and sternum with or without transverse carina; tergum with pair of articulated appendages (urogomphi) at posterolateral edges; urogomphus of 1-3 segments (usually 2), rarely fixed or absent. Segment X cylindrical, transverse to elongate; apex with anus and eversible membranous lobes which usually bear small teeth (four large teeth in many Aleocharinae).

Larvae usually develop through three instars, rarely two, and possibly more than three in some species (unconfirmed). The instars are usually similar in structure and appearance, with minor differences in chaetotaxy and in shape and relative size of appendages, but differences between instars can be dramatic (e.g., in ectoparasitic *Aleochara* species). Later instars sometimes possess structures, such as cleaning combs on the protibia in most Staphylininae, that are lacking in the first instar. First instar larvae of some species possess one or more pairs of fine teeth on the dorsum of the head, thorax, or abdomen which probably function as egg-bursters. The embryological work of Tikhomirova and Melnikov (1970) demonstrated that the articulated mala of some staphylinid larvae is derived from an unarticulated mala as found in most staphylinids and in related Staphylinoidea, and is not homologous with the articulated galea of some other Coleoptera larvae.

Larvae of fewer than 3% of Nearctic staphylinid species have been described (Moore and Legner 1974d), but those of more than a fourth of the genera are known at least from exotic species (Newton 1990b), and at the subfamily level only that of Empelinae (among Nearctic taxa) remains undiscovered. Newton (1990a) included more than 70 Nearctic genera found commonly in litter and soil in a larval key, and Frank (1991) plus Newton (1991) reviewed immature stages more generally for North America. Keys to genera of staphylinid larvae of Europe by Kasule (1966, 1968, 1970), Pototskaya (1967), and Topp (1978) are also useful for the Nearctic fauna. Other general sources of relevant larval descriptions, keys, or morphological treatments include Beutel and Molenda (1997), Bøving and Craighead (1931), Costa *et al.* (1988), Hansen (1997a), Hayashi (1986), Paulian (1941), Peterson (1960), and Verhoeff (1919). Ashe and Watrous (1984) proposed a detailed system for larval chaetotaxy (setal naming) based on Aleocharinae, which Frania (1986b) modified for Paederini. Additional references on individual groups or genera are included in the "Classification of the genera" section later.

Eggs are usually spherical to oval, usually white or nearly so, sometimes with strong surface microsculpture; they are usually laid singly.

Pupae are exarate, unsclerotized, and with free appendages in most subfamilies, but are obtect and well sclerotized in Staphylininae. Pupae usually have long setiform processes, especially at the edges of the pronotum and sides of the abdomen; the spiracles on the basal 2-4 abdominal segments are elevated and functional, those on the remaining segments atrophied. Pupation is usually free in soil or other substrate, but at least some members of four subfamilies spin a silken cocoon (De Marzo 1988b, Frank and Thomas 1984c).

**Fossils.** The earliest known rove beetle, from upper Triassic deposits in Virginia, is more than 200 million years old (Fraser *et al.* 1996). More than two dozen extinct genera are known from mid-Jurassic to lower Cretaceous deposits in Eurasia (ca. 130-180 million years old); most of these resemble members of the modern subfamilies Omaliinae, Tachyporinae, Trigonurinae and Oxytelinae (Ryvkin 1985, 1990; Tikhomirova 1968, 1973). By the Cenozoic period (less than 70 million years ago), rich amber deposits from the Baltic Sea, Dominican Republic and elsewhere include hundreds of species, most of them referable to modern genera (e.g., Spahr 1981). Staphylinids are also abundant as Pleistocene or post-Pleistocene fossils and subfossils, which are usually referable to modern species (e.g., Elias 1994).

**Habits and habitats.** Rove beetles occur in almost every type of habitat, and eat almost everything except living tissues of higher plants (though *Apocellus* may do so on occasion: Chittenden 1915). A majority are predators of other insects and invertebrates, but all species of several subfamilies and some members of others feed on fungi or decaying organic matter. The casual observer frequently finds staphylinids on carrion or dung, but a majority of species live in forest leaf litter and mosses and in all sorts of decaying plant matter, where they are seldom found without special searching. They also abound under stones and plant debris, in marshes and bogs and along the shores of lakes, ponds and streams, where some species can "skim" over water; others are restricted to ocean shores, where a few species are routinely submerged by high tides (e.g., Moore 1964a, Moore and Legner 1976, Orth *et al.* 1978). Many species occur under bark of or in decaying logs, often in burrows of bark beetles, and some lurk or wander on trunks or foliage of living plants. Some species live in fungi where many are fungivores and others prey on flies and other fungus inhabitants, while others occur on flowers and eat pollen, and still others inhabit caves or occur only in high-altitude talus slopes. Many of the most unusual species are found only in ant and termite nests and may be highly modified to live and travel with their hosts (e.g., Hölldobler and Wilson 1990; Kistner 1979, 1982; Seevers 1957, 1965; Wilson 1971). Still others inhabit nests or burrows of birds (e.g., Hicks 1959, 1962, 1971), mammals (e.g., Israelson 1971), and even tortoises (reviewed by Jackson and Milstrey 1989). The Neotropical group Amblyopinina includes large so-called "parasites" of small mammals which are now thought to actually benefit their hosts by eating true parasites such as fleas (Ashe and Timm 1987). Some staphylinids are subsocial, showing parental care of larvae in special "nests" or burrows (Aleocharinae: *Eumicrota*, Ashe 1986c; Oxytelinae: *Platystethus*, Hinton 1944 and *Bledius*, Wyatt 1993; Oxyporinae: *Oxyporus*, Setsuda 1994). Staphylinid larvae, though less well-known than adults, are usually found in the same habitat and have similar feeding habits, being either predaceous on insects and other arthropods or feeding on fungi, decaying vegetable and animal matter, or microscopic flora and fauna. Larvae of certain species of Aleocharinae are ectoparasitic in fly puparia, with differentiation of instars into triungulin-like and more sedentary forms (e.g., Fuldner 1960), and many pselaphine larvae have special prey-capture structures on the head that have no analogue nor homologue in adults (e.g., De Marzo 1987). Large abdominal glands, commonly referred to as defensive glands, are found in adults of several subfamilies and larvae of most Aleocharinae; these have been shown in at least some cases to produce secretions effective in defense against predators (Dettner 1993; Klinger 1983; Steidle and Dettner 1995b). General references on staphylinid ecology and biology include Balduf (1935), Frank (1982), Hansen (1997b), Leschen (1993), Mank (1923), Newton (1984), Voris (1934), and Tikhomirova (1973). More specific sources are given in the "Classification of the genera" section later.

**Status of the classification.** As a result of much active work on the Staphylinidae and related families directed toward delimiting monophyletic groups, the scope and internal classification of the family have changed considerably in the nearly 40 years since the first edition of this book (Arnett 1963). However, phylogenetic studies at all levels are incomplete and often controversial, so that there is still not full consensus among all workers as to what taxa should be included, nor what their relationships are. The most significant of the recent changes accepted here and reflected in the following treatment are placing Pselaphidae and Scaphidiidae as subfamilies and Brathinidae as part of the omaliine tribe Anthophagini; adding *Empelus* (from Clambidae) and *Dasycerus* (from Lathridiidae) as subfamilies; splitting additional subfamilies out of Piestinae and Oxytelinae; and including Trichopseniinae and Hypocyphtinae in Aleocharinae.

These changes and other details of the higher-level classification used below are based on work presented in or reviewed by Lawrence and Newton (1982, 1995), Newton and Thayer (1988, 1992, 1995), and Ashe and Newton (1993), and are influenced by other work completed or in progress by all four authors of this chapter. Moore (1964d), followed by Moore and Legner (1974f, 1975a, 1979) and some other North American works, introduced an artificial system of subfamilies based on convenience in using his North American subfamily key; this system has not been adopted by other North American staphylinid systematists, nor used outside of North America. Coiffait (1972-84) elevated many staphylinid subfamilies to family rank with little explanation or justification; his system of ten families has been followed by some European workers but is not widely accepted there, nor used elsewhere. Other recent works using alternative higher-level classifications or proposing other groupings include Lohse (1964, 1974), Tikhomirova (1973), Naomi (1985), Beutel and Molenda (1997), and Hansen (1997a, 1997b). Additional sources for the classifications used at subfamily and lower ranks, as well as comments on alternative proposed rankings or concepts of some groups, are included in the discussion of each group in the "Classification of the genera" section later.

Blackwelder (1952) reviewed all generic names and implemented a great number of changes at this level that were reflected in the first edition of this book (Arnett 1963); some of these changes were later reversed (ICZN 1959, 1961a, 1961b, 1969, 1982, 1983, 1993). Newton and Thayer (1992) presented a number of changes in subfamily, tribal, and subtribal names, some relevant to the North American fauna, that were required by proper application of the International Code of Zoological Nomenclature then in effect (ICZN 1985).

The last worldwide species catalog for most Staphylinidae, the Coleopterorum Catalogus (Bernhauer and Schubert 1910-1916; Raffray 1911; Bernhauer and Scheerpeltz 1926; Scheerpeltz 1933, 1934), is very out of date. Modern catalogs are available for Scaphidiinae (Löbl 1997) and North American Pselaphinae (Chandler 1997). The most recent checklists for the rest of the family for North America present some problems: Moore and Legner (1975a) used the odd subfamily classification mentioned above; Poole and Gentili (1996) did not use a classification between family and genus and their list has a number of errors and omissions (see Smetana 1997 for many corrections for Staphylinidae). Older but still useful checklists include Leng (1920) and Blackwelder (1973a, 1973b, 1973c), and for Latin America Blackwelder (1944, including Mexico and the West Indies). Recent regional checklists are available for Canada and Alaska (Campbell and Davies 1991) and Florida (Frank 1986, Peck and Thomas 1998). Moore and Legner (1974e) presented a compact bibliography of North American literature on traditional staphylinids.

Great progress has been made in recent decades in modern revisionary studies of North American staphylinids, although much remains to be done and many genera (especially in the large subfamilies Aleocharinae and Paederinae) are still in need of such revision. References to detailed revisions or reviews of subfamilies or lower taxa are given in the "Classification of the genera" section later. More synthetic works on the family as a whole include keys to most genera (except Aleocharinae) in Arnett (1963), Moore and Legner (1974f, 1979), and for litter/soil genera Newton (1990a) plus Chandler (1990b). Regional identification guides to species include Blatchley (1910; Indiana), Downie and Arnett (1996; northeastern United States and southeastern Canada), Hatch (1957; Pacific Northwest states and provinces), and Blackwelder (1943; West Indies species, often found in southern Florida). Many adventive or synanthropic species are covered by Hinton (1945; stored products pests) and Coiffait (1972-84; Palearctic staphylinids, often adventive in North America).

**Distribution.** The family is worldwide, with 46,275 species described through 1998 (through Zoological Record, vol. 135); this represents an increase of 4,292 since 1987 (Newton 1990b). These species are currently placed in 3,210 genera and 31 subfamilies. In North America north of Mexico (the area covered by this chapter), some 4,153 described species are now known, an increase of more than 700 since the first edition of this book (Arnett 1963); these are placed in 521 genera in 25 subfamilies. Extralimital subfamilies are four from southern South America and the Australian region (Glypholomatinae Jeannel, 1962; Microsilphinae Crowson, 1950; Neophoninae Fauvel, 1905; and Solieriinae Newton and Thayer, 1992) and two from eastern Asia (Apateticinae Fauvel, 1895, and Protopselaphinae Newton and Thayer, 1995), each with only one or two genera. One subfamily (Empelinae) and 106 genera (**not** including Aleocharinae, for which the number is uncertain) are restricted to North America. Frank and Curtis (1979), using the former narrower family concept, estimated the actual fauna of Nearctic staphylinid species at 3,416, a number nearly achieved already for the same group. By updating their analysis and adding similar estimates for pselaphines, scaphidiines, and other recently incorporated genera, we estimate the final number of Nearctic staphylinid species will be between 4,800 and 5,000. In contrast, fewer than 1,000 staphylinid species have been recorded from Mexico but the actual fauna there is expected to exceed 5,000 (Navarrete-Heredia and Newton 1996). A separate guide to the staphylinid genera of Mexico is in preparation by the four authors of this chapter and José Luis Navarrete-Heredia, therefore we have made no attempt here to include exclusively Mexican genera and species.

## Keys to the Nearctic genera of Staphylinidae

Note: Identification of the smallest staphylinids, especially of the subfamily Aleocharinae, may require special study techniques, described at the beginning of Key VIII (Aleocharinae).

### I. Key to the Nearctic subfamilies of Staphylinidae

1. Antennae inserted posterior to a line drawn between anterior margins of eyes (i.e., antennae on surface of head between eyes) (Fig. 42.22) ............ 2
— Antennae inserted anterior to a line drawn between anterior margins of eyes (i.e., antennae at front or side margins of head; Figs. 1-3.22, 89-92.22) .. 4

2(1). Elytra long, exposing at most 2 complete abdominal terga; body very convex, broadly to elongately oval with smoothly curved lateral outline, surfaces usually polished and nearly or quite glabrous; hypognathous (Key IX) ................................ Scaphidiinae

— Elytra short, exposing 5-6 abdominal terga; body elongate, rarely oval, rarely with smoothly curved lateral outline, surfaces diverse but usually setose and not polished; usually prognathous, rarely hypognathous ............................................................. 3

3(2). Procoxa enclosed at base, trochantin not visible (Fig. 43.22); hind coxae separated, small; eyes bulbous (Fig. 30.22) (Key XIII) .............................. Steninae

— Procoxa not enclosed (Fig. 44.22), trochantin small but visible (arrow); hind coxae contiguous, large; eyes variable, but not huge and bulbous (Key VIII) ...................................... Aleocharinae (most)

4(1). Elytra covering whole abdomen, with lateral and 3 discal sharp longitudinal carinae (Fig. 10.22); antennomeres 3-7 extremely slender ............... .................................... (*Dasycerus*) Dasycerinae

— Elytra exposing some abdominal segments; if (rarely) covering all, other features not as above ........ 5

5(4). Body robust, relatively inflexible (Figs. 3.22, 11-15.22), usually reddish-brown; with deep conical foveae on the vertex and usually other parts of the body (in *Nisaxis*, vertexal foveae lacking, but many other foveae present); maxillary palps with 1-4 palpomeres plus a lightly sclerotized or unsclerotized apical pseudosegment (Fig. 45.22, arrow), therefore often appearing to have five palpomeres; tarsi 3-3-3, rarely apparently 2-2-2 (Bythinoplectitae only); abdominal sterna III and IV usually with very dense setal fringe (Fig. 104.22) covering a basal impression on sternum 4 (Key V) ............................................ Pselaphinae

— Body less strongly sclerotized, more flexible, usually darker brown or black; lacking deep vertexal foveae, though vertex sometimes with shallower tentorial pits; maxillary palps without apical pseudosegment, therefore appearing to have no more than (and usually exactly) four palpomeres; tarsi usually 5-5-5, sometimes 2-2-2, 3-3-3, or 4-4-4; abdominal sterna III and IV without dense setal fringe ................................................................. 6

6(5). Antennae of 9 antennomeres, the last one forming a large club (Fig. 9.22); posterior coxae well separated (Key IV) ................................ Micropeplinae

— Antennae of 10 or 11 antennomeres, not clubbed or with weak club of 2 or more antennomeres; posterior coxae contiguous ............................ 7

7(6). Last labial palpomere large, semilunar (Fig. 46.22); habitus as in Fig. 29.22 ........... (*Oxyporus*) Oxyporinae

— Last labial palpomere not semilunar ...................... 8

8(7). Head with a pair of ocelli (raised pale bumps) between the posterior margins of the eyes (Fig. 47.22, arrow) (Key II) ........................... Omaliinae (most)

— Head without ocelli ............................................... 9

9(8). Abdomen with complete sternum II (seven sterna can be counted) (Fig. 48.22) (Key XII) ................. .............................................. Oxytelinae (most)

— Abdominal sternum II absent or rudimentary, only six complete sterna visible (Fig. 49.22) (The minute, blind, wingless Leptotyphlinae often have seven exposed sterna, because the genital segment is permanently everted; however, segment VII is, as usual, the last segment with paratergites, allowing one to

42 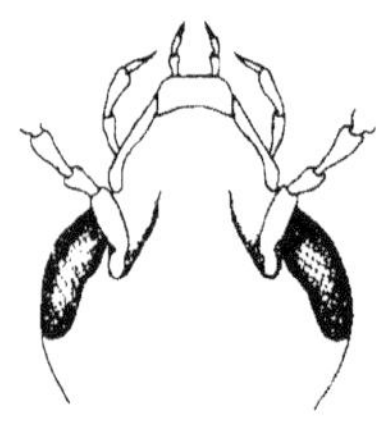
43 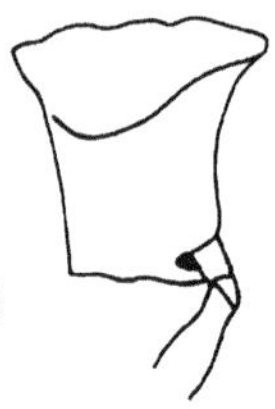
44 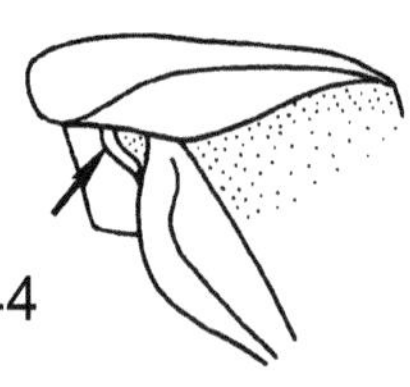
45 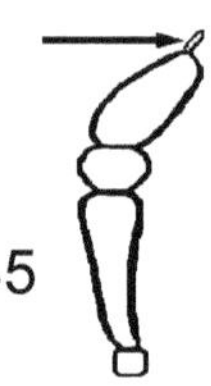
46 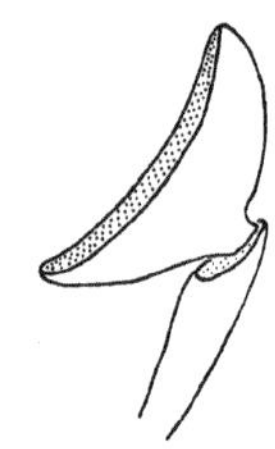
47 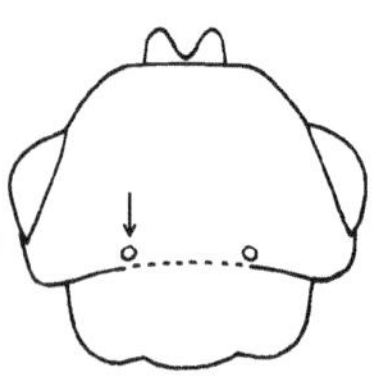
48 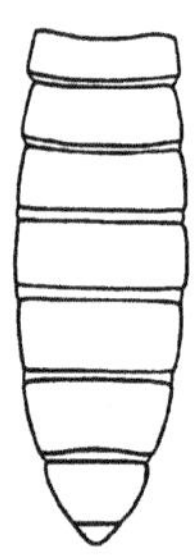
49 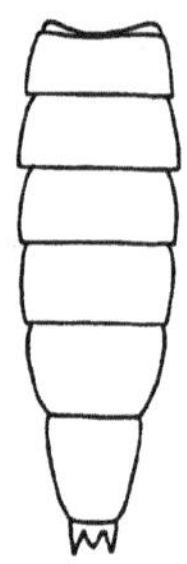
50 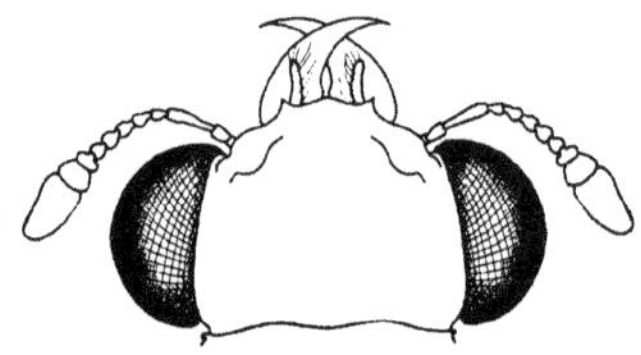
51 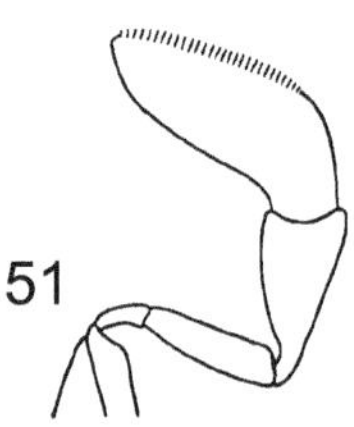

FIGURES 42.22-51.22. Fig. 42.22. *Aleochara lustrica* Say, head, dorsal (after Arnett 1963). Fig. 43.22. *Megalopinus caelatus* (Gravenhorst), prothorax, left lateral (after Newton 1990a). Fig. 44.22. *Atheta (s.l.)* sp., prothorax, left lateral, showing exposed trochantin (arrow) (after Newton 1990a). Fig. 45.22. Pselaphinae gen. sp., maxillary palp, showing digitiform apical pseudosegment (arrow). Fig. 46.22. *Oxyporus vittatus* Gravenhorst, apical labial palpomere (modified from Moore 1964d). Fig. 47.22. *Xylodromus concinnus* (Marsham), head, dorsal, showing ocellus (arrow) (modified from Newton 1990a). Fig. 48.22. *Oxytelus nimius* Casey, abdomen, ventral, showing seven visible sterna (after Newton 1990a). Fig. 49.22. *Erichsonius patella* (Horn), abdomen, ventral, showing six visible sterna (after Newton 1990a). Fig. 50.22. *Megalopinus* sp., head, dorsal (modified from Arnett 1963). Fig. 51.22. *Pinophilus densus* LeConte, maxillary palp (modified from Arnett 1963).

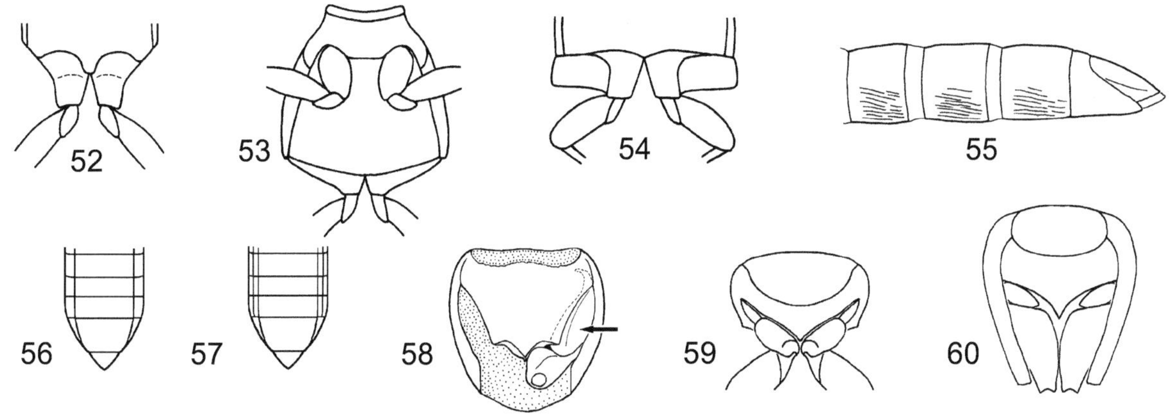


FIGURES 52.22-60.22. Fig. 52.22. *Erichsonius patella* (Horn), hind coxae, ventral (modified from Newton 1990a). Fig. 53.22. *Oxytelus nimius* Casey, meso- and metathorax, ventral (after Newton 1990a). Fig. 54.22. *Coproporus ventriculus* (Say), hind coxae, ventral (after Newton 1990a). Fig. 55.22. *Lispinus aequipunctatus* LeConte, abdominal segments V-VIII, left lateral. Fig. 56.22. Abdomen with 1 pair of paratergites on segments III-VII (schematic). Fig. 57.22. Abdomen with 2 pairs of paratergites on segments III-VII (schematic). Fig. 58.22. *Renardia nigrella* (LeConte), prothorax, ventral (right coxa removed), showing coxa and large trochantin (arrow). Fig. 59.22. *Siagonium punctatum* (LeConte), anterior coxae, anterolateral. Fig. 60.22. *Lathrobium amplipenne* Casey, anterior coxae, anterolateral.

determine that their first visible sternum is III, not II.) .................................................................... 10

10(9). Anterior margin of labrum with two long processes that are setose medially (Fig. 50.22); eyes huge, bulbous (Fig. 50.22) ........................................... ........................... (*Megalopinus*) Megalopsidiinae

— Labrum without such processes; eyes smaller . 11

11(10). Antennae of 10 antennomeres; size <2 mm; form broadly oval (Figs. 203-204.22) (Key VIII: D) ....... .............................. Aleocharinae: Hypocyphtini

— Antennae of 11 antennomeres; size and form variable ................................................................ 12

12(11). Last maxillary palpomere longer than penultimate, slightly arcuate, with apex obliquely truncate, apical edge narrowly concave with distinctive texture (Fig. 51.22) (Key XVII: I) .......................... ...................................... Paederinae: Pinophilini

— Last maxillary palpomere not so formed ........... 13

13(12). Metasternum with expanded plates covering part of femora (Fig. 165.22) (Key VIII: B) .................... ............................ Aleocharinae: Trichopseniini

— Metasternum without such plates ...................... 14

14(13). Hind coxae expanded posteriorly as plates partly covering femora, and elytra very long, covering all (or all but one) abdominal segments (Fig. 7.22) and antenna with apical 3 antennomeres forming abrupt club ......................... (*Empelus*) Empelinae

— Hind coxae usually without plates (e.g., Figs. 52-54.22); if with plates, then other characters not as above ....................................................... 15

15(14). Abdomen without lateral sutures separating most terga and sterna, each abdominal segment a solid ring (Fig. 55.22), abdomen cylindrical ............. 16

— Abdomen with lateral sutures separating terga and sterna, usually with paratergites forming prominent "margin" beside tergite (Figs. 56-57.22); if paratergites absent and sutures inconspicuous, abdomen tapering strongly from base to apex . 17

16(15). Labrum with toothed anterior margin; mandibles long and falciform (Key XIV) ........ Euaesthetinae (part)

— Labrum with anterior margin not toothed; mandibles short, not falciform (Key XI) ...... Osoriinae (most)

17(15). Minute, no more than 1.8 mm long; slender, blind, and wingless (Fig. 32.22); tarsi 2-2-2 or 3-3-3 (Key XV) ............................................. Leptotyphlinae

— Larger, body form variable, usually with eyes and wings; almost always with more than 3 tarsomeres ...................................................................... 18

18(17). Anterior coxae with only a small part exposed apically; anterior trochantins greatly enlarged (Fig. 58.22, arrow), larger than apical part of coxae and sometimes larger than pronotal hypomera; body extremely flattened, pronotum greatly narrowed basally, abdomen lacking paratergites (un-margined) (Fig. 25.22), head often larger than pronotum (Key XI) ...................................... Osoriinae: Eleusinini

— Anterior trochantins not greatly enlarged, smaller than exposed part of coxae or completely hidden under prosternum and pronotal hypomera; other features not as above .................................................. 19

19(18). Anterior coxae small and globular in anterolateral view (Fig. 59.22) ............................................ 20

— Anterior coxae larger, more elongate in anterolateral view (Fig. 60.22) ............................................ 23

20(19). Protrochantins concealed (Fig. 43.22) (Key XIV) ...... .......................................... Euaesthetinae (part)

— Protrochantins exposed (Fig. 61.22, arrow) ......... 21

21(20). Pronotum slightly wider at base than at apex; elytra longer, abdominal tergum 3 and at least part of 4 covered (Fig. 24.22) ..... (*Trigonurus*) Trigonurinae

— Pronotum distinctly narrower at base than at apex; elytra short, part or all of abdominal tergum 3 exposed ........................................................... 22

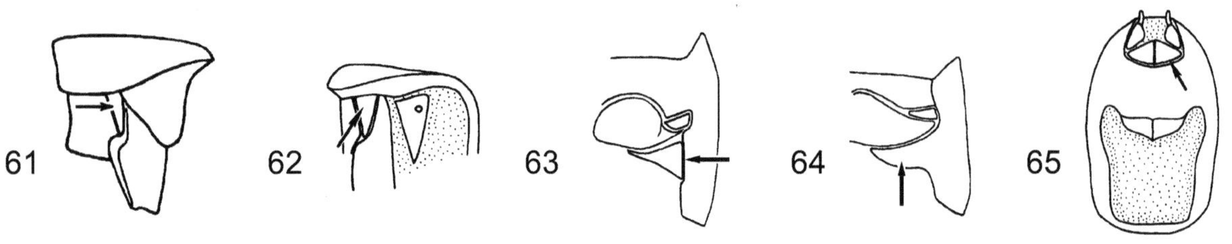


FIGURES 61.22-65.22. Fig. 61.22. *Derops divalis* (Sanderson), prothorax, left lateral, showing protrochantin (arrow) (modified from Newton 1990a). Fig. 62.22. *Trichophya* sp., prothorax, left lateral, showing protrochantin (arrow) and spiracular peritreme separated from hypomeron (modified from Newton 1990a). Fig. 63.22. *Olisthaerus substriatus* Gyllenhal, left side of prothorax, ventral, showing ridge (arrow) separating postcoxal sclerite and hypomeron (modified from Moore 1964d). Fig. 64.22. *Tachinus fimbriatus* Gravenhorst, left side of prothorax, ventral, showing postcoxal process of pronotal hypomeron (arrow) (modified from Moore 1964d). Fig. 65.22. *Nematolinus longicollis* (LeConte), prothorax, ventral, coxae removed, showing antesternal sclerites (arrow) (after Newton 1990a).

22(21). Protibia without spines externally; pronotal lateral margins serrate; body very flat (especially head); abdominal tergites IV and V with small median pair of cuticular combs (Key VI)...... Phloeocharinae (part)

— Protibia spinose externally; pronotal lateral margins smooth; body not extremely flat; abdominal tergites IV and V without such cuticular combs (Key X) ........................................................ Piestinae

23(19). Prothorax with triangular sclerite behind anterior coxa, either containing mesothoracic spiracle and separated from hypomeron by suture or membrane (Fig. 62.22) or (*Olisthaerus* only) without spiracle and separated from hypomeron by ridge (Fig. 63.22, arrow); lacking a solid postcoxal process like those in Figs. 61.22, 64.22.............. 24

— Prothorax without separate triangular sclerite, hypomeron may have a solidly attached triangular projection behind coxa (Fig. 64.22, arrow) or not (Fig. 65.22) ............................................... 26

24(23). Antennomeres 3-11 extremely slender, filamentous; anterior tarsi broadly dilated .............................
.............................. (*Trichophya*) Trichophyinae

— Antennae not filamentous; anterior tarsi not broadly dilated ............................................................ 25

25(24). Small, no more than 3 mm long (occasionally as large as 4-4.5mm), broad-bodied; if pronotum wider than elytra (together), then also broadly explanate laterally (Fig. 8.22); postcoxal sclerite of prothorax with spiracle (similar to Fig. 62.22) (Key III).........
......................................................... Proteininae

— Larger, no less than 6 mm long, slender-bodied; pronotum wider than elytra (together) and not explanate laterally; postcoxal sclerite of prothorax without spiracle (Fig. 63.22) .........................
................................ (*Olisthaerus*) Olisthaerinae

26(23). Pronotum and elytra (sometimes also head) longitudinally carinate or costate (Fig. 33.22); head and pronotum with coarse reticulate punctation; either abdominal tergite VIII with apical comb or elytra with a deep incision in apical margin (Key XVI) ................................................ Pseudopsinae

— Pronotum and elytra not carinate or costate, or if so then lacking other features above................. 27

27(26). Head constricted behind eyes to form a distinct (but not always sharp) neck clearly visible from above (e.g., Figs. 27.22, 34-40.22) ................. 28

— Sides of head parallel or converging uninterruptedly to base, not constricted to form a neck clearly visible from above (e.g., Figs. 16-18.22) .............. 30

28(27). Trochantin of front leg (Figs. 66.22, thin arrow; 60.22) flat, blade-like, often projecting anteriorly; exposed part of hind coxa (Fig. 52.22) narrow, more or less conical with transverse impression, apex strongly projecting posteriorly, anterior edge strongly convex, posterior face of coxa (hidden by metasternum in ventral view) vertical....... 29

— Trochantin of front leg (Figs. 61-62.22, arrows) with blunt, more or less straight external edge, not projecting or, rarely (*Vicelva*), concealed by prosternum and pronotal hypomera; hind coxa variable, usually transverse with lateral part visible in ventral view, apex not very prominent, anterior edge straight to slightly curved (Figs. 53-54.22) ....................................................... 34

29(28). Pronotum with large, opaque postcoxal process (Fig. 64.22, arrow; Fig. 66.22, thick arrow); intersegmental membranes of abdomen with "brick wall" pattern (Fig. 67.22) (Key XVII: A) ..............................
........................................ Paederinae: Paederini

— Pronotum without postcoxal process (Fig. 68.22) or with small translucent process (Fig. 69.22, long arrow); intersegmental membranes of abdomen usually with rounded or triangular pattern, or pattern indistinct (Fig. 70.22) (Key XVIII) .........................
.................................................... Staphylininae

30(27). Antennomeres 3-11 extremely slender, filamentous; habitus as in Fig. 18.22 (*Habrocerus*) .................
.................................................... Habrocerinae

— Antennae not filamentous................................. 31

31(30). Elytral epipleuron not delimited by a carina (Fig. 71.22) ............................................................ 32

— Elytral epipleuron delimited by a carina (Figs. 72-73.22, arrows) ................................................ 33

32(31). Flat, slender, pale body; apical maxillary palpomere with characteristic shape (Fig. 74.22); tarsal claws sharply curved; intertidal on Pacific coast (Key II).
.......... (*Giulianium*) Omaliinae: Aphaenostemmini

— Body otherwise; apical maxillary palpomere aciculate or conical; tarsal claws gradually arcuately curved; not coastal (*Phloeocharis, Ecbletus*) (Key VI) .............................. Phloeocharinae (part)

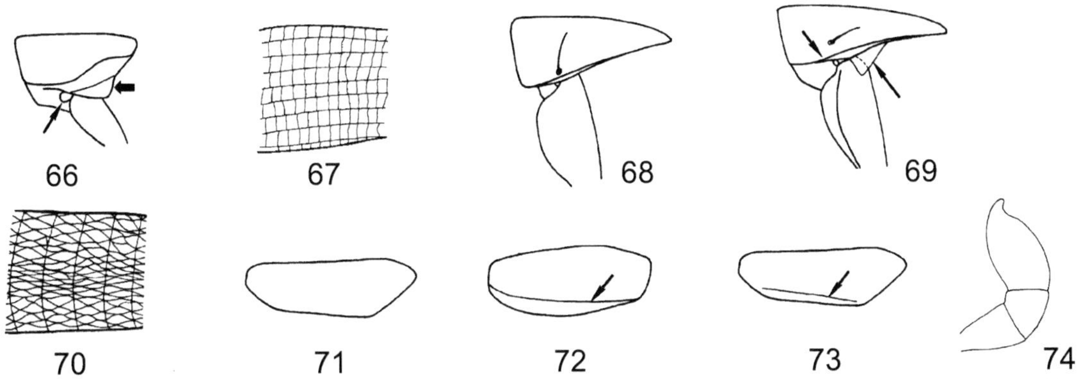


FIGURES 66.22-74.22. Fig. 66.22. *Homaeotarsus badius* (Gravenhorst), prothorax, left lateral, showing plate-like, projecting protrochantin (thin arrow) and postcoxal process of notum (thick arrow) (modified from Newton 1990a). Fig. 67.22. *Pinophilus densus* LeConte, intersegmental membrane between abdominal sterna with brick-wall-like pattern (after Arnett 1963). Fig. 68.22. *Philonthus caeruleipennis* Mannerheim, prothorax, left lateral (after Newton 1990a). Fig. 69.22. *Erichsonius patella* (Horn), prothorax, left lateral, showing small translucent postcoxal process (longer arrow) and confluence (shorter arrow) of superior and inferior marginal lines of pronotal hypomeron (after Newton 1990a). Fig. 70.22. *Philonthus alutaceus* Horn, intersegmental membrane between abdominal sterna without brick-wall-like pattern (after Arnett 1963). Fig. 71.22. Left elytron, lateral (schematic), without epipleural ridge (modified from Newton 1990a). Fig. 72.22. *Megarthrus pictus* Motschulsky, left elytron, lateral, showing complete epipleural ridge (arrow) (after Newton 1990a). Fig. 73.22. *Homaeotarsus badius* (Gravenhorst), left elytron, lateral, showing incomplete epipleural ridge (arrow) (after Newton 1990a). Fig. 74.22. *Giulianium campbelli* Moore, apical maxillary palpomeres (after Moore 1976).

33(31). Elytra each with several irregular impressed longitudinal striae (Key XII) ............................................ ...............(*Coprophilus*) Oxytelinae: Coprophilini

— Elytra each with at most an impressed sutural stria (may have rows of punctures [*Tachinus*], but not in impressed striae) (Key VII) ......... Tachyporinae

34(28). Abdomen with single pair of paratergites per segment (Fig. 56.22), paraterga planar, usually sloping dorsad toward sides (occasionally horizontal or sloping lateroventrad) ............................... 35

— Abdomen with two pairs of paratergites per segment (Fig. 57.22, lateral ones sometimes very slender, difficult to distinguish), so that paratergal area is somewhat convex upward ......................... 36

35(34). Abdomen with small defensive gland openings apically at each side of tergite IX; abdominal tergite VIII with apex serrate medially (*Deleaster*); or head, pronotum, and elytra very coarsely and densely punctate, with a pitted appearance (*Syntomium*); or small abdominal sternum II visible behind lateral parts of metacoxae (*Mitosynum*) (Key XII: A) .................. Oxytelinae: Deleasterini

— Abdomen without gland openings on tergite IX, but with defensive glands associated with internal anterior projection on base of sternum VIII; with none of other above features (Key II) ................ .................................................. Omaliinae (Part)

36(34). Tarsi 5-5-5 (Key VI) ................ Phloeocharinae (part)

— Tarsi 4-4-5 or other heteromerous combination (Key VIII) ........................................ Aleocharinae (part)

## II. Keys to the Nearctic tribes and genera of Omaliinae

### Key to Nearctic tribes

1. Tiny, flattened, parallel-sided, ca. 2.3 mm long and 0.33 mm wide; antennae moniliform; wingless, elytra short, covering only thorax; ocelli absent; eyes small, placed far forward on sides of head; apical maxillary palpomere with characteristic shape (Fig. 74.22); pronotum lacking lateral keel and postcoxal process; elytra without epipleural keel; posterior face of hind coxae oblique; intertidal on central and northern California coast ... .......................... (*Giulianium*) Aphaenostemmini

— Larger (particularly wider), variable in shape; antennae usually filiform or weakly clavate, seldom moniliform; usually winged, with elytra nearly always covering at least part of tergum III; ocelli usually present; eyes usually well-developed, not placed particularly far forward on head; apical maxillary palpomere otherwise; pronotum with postcoxal process and nearly always with lateral keel; elytra with epipleural keel; posterior face of hind coxae usually vertical, oblique in some Anthophagini; widespread ................................ 2

2(1). Tarsomeres 1-4 of all tarsi short, together usually no longer than tarsomere 5, broadened and with dense setae ventrally; apical 5-6 antennomeres wider than preceding; females of some species with elytra longer than those of males, sometimes prolonged into a rounded point near suture ..... ............................... (*Eusphalerum*) Eusphalerini

— Tarsomeres 1-4 of all tarsi not widened nor with dense ventral setae, often longer together (or even individually) than tarsomere 5; antennae of various forms; elytra only occasionally prolonged at sutural apex (females of some species of *Anthobium*) ....................................................... 3

3(2). Maxillary palpomere 3 strongly enlarged, pubescent, pear-shaped in some species, palpomere 4 much smaller, needle-like or short and conical (Figs. 75-76.22), often partly hidden in apex of palpomere 3 or by its apical setae; apices of elytra truncate (not rounded), elytra at least 1.5 times as long as pronotum; pair of small whitish or grayish patches of wing-folding spicules always present on abdominal tergite IV and sometimes also V (as in Fig.

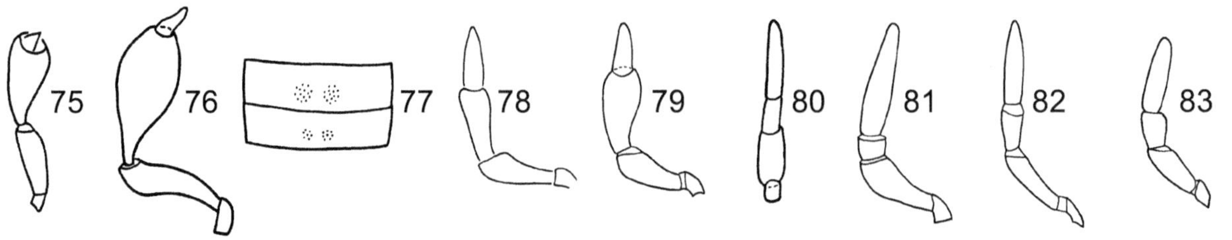


FIGURES 75.22-83.22. Fig. 75.22. *Ephelinus notatus* (LeConte), maxillary palp (after Campbell 1978c). Fig. 76.22. *Coryphium arizonense* (Bernhauer), maxillary palp. Fig. 77.22. *Coryphiomorphus hyperboreus* (Mäklin), wing-folding patches on abdominal tergites IV and V (after Campbell 1978c). Fig. 78.22. *Geodromicus brunneus* (Say), maxillary palp. Fig. 79.22. *Microedus austinianus* LeConte, maxillary palp. Fig. 80.22. *Artochia productifrons* Casey, maxillary palp. Fig. 81.22. *Lesteva pallipes* LeConte, maxillary palp. Fig. 82.22. *Olophrum obtectum* Erichson, maxillary palp. Fig. 83.22. *Anthobium* sp., maxillary palp.

77.22); mostly boreal, arctic, or alpine (Key C) ........................................................ Coryphiini

— Maxillary palp with different conformation (Figs. 78-83.22); wing-folding patches variable, not always present on tergite IV, and in some species entirely absent; if palp similar to that described above, then each elytron rounded at apex (*Orobanus*) or elytra no longer than pronotum and wing-folding patches completely lacking (*Omalorphanus*); in many habitats including boreal and alpine areas ........................................ 4

4(3). Hind tarsus with tarsomeres 1-4 together (measured along dorsal surface) nearly always longer than tarsomere 5; antennomeres 8-10 elongate to quadrate, never transverse (*Geodromicus* and *Microedus* have tarsomeres 1-4 not longer than 5, antennomeres 8-10 each distinctly longer than wide) (Key B) .................................. Anthophagini

— Hind tarsus with tarsomeres 1-4 together (measured along dorsal surface) shorter than or equal to tarsomere 5; antennomeres 8-10 nearly always transverse, occasionally quadrate, never elongate (a few exceptions having tarsomeres 1-4 longer than 5 have antennomeres 8-9, and nearly always 10, wider than long [a few scattered species] or labrum bilobed and temples in the form of rounded right angles [Fig. 47.22, one species of *Xylodromus*] or head with vertex elevated, delimited laterally by oblique ridges [as in Fig. 84.22; "*Omalium*" *flavidum* winged males]) (Key A) ...... ............................................................... Omaliini

A. Key to the Nearctic genera of Omaliini

1. Frons and vertex of head elevated above eyes, separated from eyes by ridges (parallel, Fig. 84.22, or slightly converging posteriorly), head sloping down laterally to eyes ........................................ 2

— Frons and vertex not set off by ridges from lateral parts of head (if ridges present anteriorly, middle part of head not distinctly on a higher plane than eyes) ............................................................... 4

2(1). Disk of pronotum evenly and strongly convex ........ ........................................................ *Pycnoglypta*

— Disk of pronotum flattened medially, usually with impressed areas beside midline and more laterally ..................................................................... 3

3(2). Elytra at least 1.5, usually more than 2, times as long as pronotum; eyes normal in size, if reduced still definitely multifaceted; ocelli present ................. ................................................ *Carcinocephalus*

— Elytra no more than 1.2 times as long as pronotum; eyes consisting of a single facet; ocelli absent ....................................................... *Omalonomus*

4(1). Apical maxillary palpomere distinctly narrower than the preceding one; body more or less flattened ........................................................................... 5

— Apical maxillary palpomere about the same width as the preceding one; body flattened or quite convex .................................................................. 7

5(4). Pronotum with elongate shallow impression on each side of midline in addition to impressed areas posterolaterally ............................ *Phloeonomus*

— Pronotum either flat or evenly convex in middle, with shallowly impressed area near each posterior corner ........................................................ 6

6(5). Pronotum flat in middle, often with a bump between the flat area and the impressed posterior corner; dark line (an internal ridge) sometimes visible along midline .................................... *Phloeostiba*

— Pronotum convex across middle, without median dark line ................................... *Paraphloeostiba*

7(4). Elytra very short, no longer than pronotum, exposing 1/2 or more of tergum III; body dark brown to black, legs sometimes paler; rocky ocean beaches of north Atlantic or coastal and river-edge sites in Alaska, Yukon, and Northwest Territory ................................................ *Micralymma*

— Elytra distinctly longer than pronotum (ratio ca. 1.5-2), covering all of tergum III (and usually at least part of tergum IV); body variously colored, yellowish- or reddish-brown to black; widely distributed ................................................................... 8

8(7). Mesosternum with fine median keel (only exceptions: two densely punctate Pacific Ocean beach species with very robust legs and numerous conspicuous heavy spines on the middle and hind tibiae); lateral margins of pronotum smooth; habitus resembling Fig. 5.22 ....................................... *Omalium*

— Mesosternum not carinate (if vaguely so, then pronotal margins finely crenulate) .................................. 9

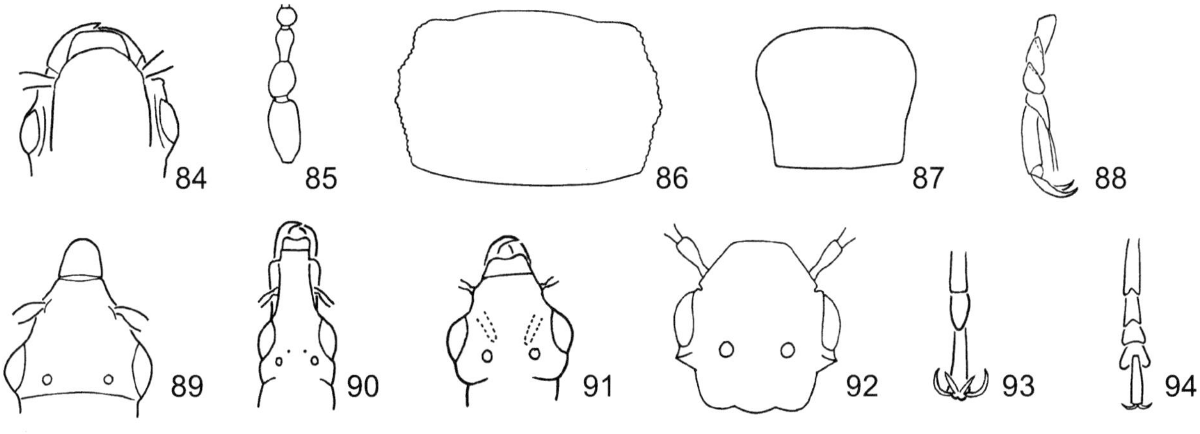


FIGURES 84.22-94.22. Fig. 84.22. *Pycnoglypta campbelli* Gusarov, head, dorsal, showing elevated vertex delimited by oblique ridges. Fig. 85.22. *Acrolocha leechi* Hatch, antennomeres 1-4. Fig. 86.22. *Acrulia tumidula* (Mäklin), pronotum, dorsal. Fig. 87.22. *Orobanus densus* Casey, pronotum, dorsal. Fig. 88.22. *Microedus austinianus* LeConte, front tarsus, showing lobed tarsomere 4. Fig. 89.22. *Artochia productifrons* Casey, head, dorsal. Fig. 90.22. *Tanyrhinus singularis* Mannerheim, head, dorsal. Fig. 91.22. *Trigonodemus striatus* LeConte, head, dorsal. Fig. 92.22. *Deinopteroloma pictum* (Fauvel), head, dorsal, showing notch between antenna and eye (after Newton 1990a). Fig. 93.22. *Anthophagus* sp., apex of tarsus, showing appendages on tarsomere 5. Fig. 94.22. *Pelecomalium flavescens* Casey, tarsus, showing bilobed tarsomere 4.

9(8). Hind tarsus no more than 1/2 as long as hind tibia; pronotum never evenly convex, always with at least faint impressions near middle or sides and/or lateral margins explanate; antennomere 3 extremely narrow at base (Fig. 85.22) such that at ca. 20X magnification it may look broken ............. 10

— Hind tarsus at least 3/5 as long as hind tibia; pronotum evenly convex or with impressions; antennomere 3 seldom so extremely thin at base, though usually thinner than antennomere 2 (if very thin [*Hapalaraea*], then pronotum evenly convex) ........................................................... 11

10(9). Pronotal lateral margin smooth; head and often pronotum with microsculpture between punctures; elytra with coarse punctures arranged in rough longitudinal striae (somewhat vague in one wingless species); widespread, including Pacific Northwest .......................................... *Acrolocha*

— Pronotal lateral margin finely crenulate and with two obtuse angles, one each in front and back of the midpoint (Fig. 86.22); head and pronotum without microsculpture between punctures; elytra densely confusedly punctate; California to Alaska ................................................................. *Acrulia*

11(9). Labrum strongly bilobed (Fig. 47.22); head transverse, with prominent rounded rectangular temples (Fig. 47.22) and very strong dorsal and lateral nuchal constrictions; head always with impressed tentorial pits in front of ocelli; body flattened and parallel-sided ............................................ *Xylodromus*

— Labrum at most faintly emarginate anteriorly; head triangular or quadrate, with or without tentorial pits; body shape various ................................ 12

12(11). Head with small tentorial pits in front of ocelli (sometimes each in a shallow groove) ..................... 13

— Head without tentorial pits................................. 14

13(12). Hind tarsi with tarsomeres 1-4 together much shorter than 5; head, pronotum, and elytra with fine and very dense mesh-like to granular microsculpture, except on frons ................................ *Omaliopsis*

— Hind tarsi with tarsomeres 1-4 together nearly as long as to longer than 5; head, pronotum, and elytra of most species without microsculpture, if microsculpture present, consisting of rather open meshes or transverse wavy lines.... *Phyllodrepa*

14(12). Convex, oval, small, less than 3 mm long; widespread ......................................................... *Hapalaraea*

— Flattened, parallel-sided, larger, over 3 mm long; Pacific and Rocky Mountain regions only ............... ....................................................... *Dropephylla*

### B. Key to the Nearctic genera of Anthophagini

1. All antennomeres usually densely and equally pubescent, first 3 occasionally less so; body pubescent dorsally; tibiae with dense pile-like pubescence all around ........................................ 2

— At least antennomeres 1-2 nearly glabrous, with only a few scattered long setae, dense pubescence usually beginning only on antennomere 4 or further apically; body sometimes pubescent (usually not densely), but tibiae lacking dense, even, pile-like pubescence ....................................... 9

2(1). Apical (fourth) maxillary palpomere shorter and distinctly narrower than third (no wider than in Figs. 78-79.22), fourth glabrous, third setose .......... 3

— Apical maxillary palpomere subequal in width to third (Figs. 80-83.22), both glabrous ......................... 5

3(2). Tarsomere 4 of all tarsi simple, not lobed beneath at apex; pronotum cordate, widest near front and distinctly narrower on basal 1/2 than at widest point (Fig. 87.22); apical maxillary palpomere less than 1/2 as wide and less than 2/3 as long as preceding; eyes densely setose, length of setae about equal to diameter of an eye facet......... *Orobanus*

— Tarsomere 4 of all tarsi with apex lobed beneath (Fig. 88.22), with apical fringe of setae on lobe; pronotum not much narrower at base than at widest point; apical maxillary palpomere 1/3 to 2/3 as wide and 0.4-0.8 times as long as preceding (Figs. 78-79.22); eyes glabrous ........................ 4

4(3). Head and pronotum with microsculpture; apical maxillary palpomere parallel-sided or slightly tapered toward apex (Fig. 79.22), no more than 0.6 times as wide as preceding......................... *Microedus*

— Head (except occasionally between ocelli) and pronotum without microsculpture; apical maxillary palpomere distinctly narrowed at base (Fig. 78.22), at its widest point 0.6 times as wide as preceding ..................................... *Geodromicus*

5(2). Pronotum cordate (Fig. 87.22), widest near front and much narrower at base than at widest point, not explanate laterally; eastern and western......... 6

— Pronotum more equally narrowed from widest point toward front and back, explanate at least posteriorly (exceptions: a few *Phlaeopterus* species with the pronotum somewhat cordate as above have ocelli and lateral impressions on the pronotum); western ............................................................ 7

6(5). Apical (fourth) maxillary palpomere about 4 times as long as third (Fig. 81.22); ocelli present; head with distinct (though not sharp) nuchal constriction and rounded temples; pronotum without distinct dimple-like impression at each side; eastern ..... .............................................................. *Lesteva*

— Apical maxillary palpomere about 2 times as long as third; without ocelli; head without nuchal constriction or temples; pronotum with distinct dimple-like impressions laterally just behind widest point; California to British Columbia ... *Vellica*

7(5). Head elongate and strongly narrowed anteriorly (Fig. 89.22); mouthparts prolonged, with labrum unusually narrow, about as long as wide, galea and lacinia of maxilla very long, stylet-like, extending beyond apex of labrum by about labral length; maxillary palp slender, fourth palpomere conical, slightly narrower than and about 3 times as long as third ................................................. *Artochia*

— Head not narrowed and prolonged anteriorly; labrum distinctly transverse, maxillae not so elongate; fourth maxillary palpomere about as wide as third, not extremely slender ..................................... 8

8(7). Head with very distinct nuchal line dorsally and laterally; elytral length seldom more than 2 times that of pronotum; head, pronotum, and elytra dorsally very finely densely punctate; mesosternum with no more than faint median carina .. *Unamis*

— Head with, at most, vague nuchal impressions; elytral length more than 2, usually more than 2.2, times that of pronotum; head, pronotum, and elytra dorsally without especially fine dense punctures; mesosternum with distinct median carina.......... ....................................................... *Phlaeopterus*

9(1). Elytra long (covering most of abdomen), with coarse punctures arranged in distinct longitudinal rows ...................................................................... 10

— Elytra long or short, but punctures not arranged in rows (exceptions: one species of *Acidota* has short elytra with punctures in rows; *Brathinus* species, Fig. 2.22, have long elytra with long erect setae in rows, but the punctures are not readily visible) ............................................................ 13

10(9). Pronotum widest at base, narrowed anteriorly, with apex narrower than base; head somewhat or greatly prolonged in front of eyes (Figs. 90-91.22) ........................................................................ 11

— Pronotum widest at middle or more anteriorly, head roughly quadrate to transverse, not prolonged in front of eyes .................................................. 12

11(10). Head (Fig. 90.22) extremely long and narrow, flat and prolonged as a beak in front of eyes; head behind eyes parallel-sided, with only faint temples; whole body reddish-brown........................... *Tanyrhinus*

— Head (Fig. 91.22) only slightly longer (to vague nuchal line) than wide across eyes, with distinctly rounded temples behind eyes; elytra brown to black, with yellow markings............................. *Trigonodemus*

12(10). Body broad, glabrous, oval in appearance; pronotum strongly transverse, widest at middle, broadly explanate with serrulate lateral margins; yellowish with dark brown markings on elytra and sometimes pronotum; head (Fig. 92.22) with sharp notch between eye and antennal base ........................ .................................................. *Deinopteroloma*

— Body narrowly elongate in appearance, with long erect setae; pronotum slightly transverse, widest at about anterior 1/4, with smooth narrowly explanate margins; yellowish brown without darker markings on elytra or pronotum; head without sharp notch between eye and antennal base ...................................................... *Anthobioides*

13(9). Abdomen (also pronotum, elytra) with dense coarse punctures, visible even at 8-10X magnification. .................................................................*Acidota*

— Abdomen with only fine or very fine punctures, only visible with higher magnification .................. 14

14(13). Head width (including eyes) greater than or about equal to that of pronotum at its widest point, pronotum much narrower at base than at its widest point (anterior to middle) .......................... 15

— Head (including eyes) narrower than pronotum at its widest point (usually the middle, occasionally anterior or posterior to that) ........................... 16

15(14). Elytra covering nearly entire abdomen; characteristic scydmaenid-like habitus (Fig. 2.22); dorsal surface of head, pronotum, and elytra impunctate; elytra with 4-5 sparse rows of long erect setae visible in anterior or posterior view; tarsi without a membranous lobe at base of each claw ........... ........................................................... *Brathinus*

— Elytra shorter, exposing most of abdomen; dorsal surface of head, pronotum, and elytra punctate; elytra with only short setae; tarsi with a small membranous lobe (Fig. 93.22) at base of each claw .... .......................................................*Anthophagus*

16(14). Gular sutures largely separate, but meeting for a short distance in the middle of the head ...... 17

— Gular sutures separate throughout their length... 20

17(16). Body glabrous; head and pronotum lacking microsculpture dorsally ................................ 18
— Body pubescent; microsculpture present dorsally on head and pronotum ................................ 19

18(17). Body flattened, only slightly convex, 2.4-4.4 mm long; head with vague nuchal impression dorsally and with small discrete tentorial pits in front of ocelli, sometimes in the form of small grooves; length of fourth (apical) maxillary palpomere only 1.2-1.3 times that of third .................... *Arpedium*
— Body moderately to strongly convex, 2.9-6.2 mm long; head without dorsal nuchal impression; tentorial pits usually represented by no more than a vague shallow impression; fourth maxillary palpomere 1.6-2.7 times as long as third (Fig. 82.22) ................................................ *Olophrum*

19(17). Elytra 1.1-2 times as long as pronotum; fourth (apical) maxillary palpomere basally about the same width as third, slightly tapered toward apex; ocelli present; head with distinct nuchal impression dorsally and laterally; far northern United States, Rocky Mountains, and across Canada ............... ........................................................ *Eucnecosum*
— Elytra shorter than or equal in length to pronotum; fourth maxillary palpomere conical, distinctly narrower than third; ocelli absent; head with a trace of nuchal impression laterally, none dorsally; known only from Oregon ........... *Omalorphanus*

20(16). Tarsomere 4 of all tarsi with lobe or extra setae apically beneath tarsomere 5 ........................ 21
— Tarsomere 4 without any apical elaboration ..... 22

21(20). Tarsomere 4 distinctly bilobed at apex (Fig. 94.22); antennomeres 7-10 (often 3-6 also) sharply truncate at apex; male with apical maxillary palpomere flattened and hatchet-shaped ...... *Pelecomalium*
— Tarsomere 4 with apical fringe or tuft of longer setae, but not distinctly bilobed; antennomeres 7-10 rounded or tapered at apex, not sharply truncate; male and female with apical maxillary palpomere more or less conical .. *Amphichroum*

22(20). Antennomere 4 smaller and less pubescent than 5; head, pronotum, and elytra glabrous and shiny dorsally; pronotum without microsculpture ... 23
— Antennomere 4 nearly identical to 5, occasionally slightly narrower but still equally densely pubescent; head, pronotum, and elytra pubescent or glabrous dorsally; pronotum sometimes with microsculpture ................................................ 24

23(22). Elytra with shallow longitudinal furrows, punctures somewhat in rows; ocelli prominent .................. ................................................ *Phyllodrepoidea*
— Elytra even in contour, without furrows, punctures not at all in rows; ocelli very reduced ....................... .................................................... *Mannerheimia*

24(22). Head, pronotum, and elytra glabrous dorsally... 25
— Head, pronotum, and elytra pubescent dorsally .. 26

25(24). Pronotum strongly transverse, width over 1.5 times length (Fig. 6.22), moderately to widely explanate laterally; narrow notch (Fig. 92.22) present between eye and antennal base ....................... *Anthobium*
— Pronotum quadrate to slightly transverse, width 1-1.25 times length, only narrowly explanate laterally; without notch between eye and antennal base ................................................ *Orochares*

26(24). Abdominal segments IV-VI lacking sutures between tergite and paratergites, the segments thus unmargined (but segment VII is margined); head without medial impression between dorsal tentorial pits; protibia and protarsus of male modified: tibia greatly thickened, tarsomeres 1-4 very short and wide, with large flattened setae ventrally, tarsomere 5 inflated, and anterior protarsal claw greatly enlarged; mountains of southern California ............................................. *Xenicopoda*
— Abdominal segments 3-7 normally "margined," with paratergites separated from each tergite by sutures; head with a medial impression between dorsal tentorial pits; male protibia slightly thicker than mesotibia, protarsi normal (claws equal, tarsomeres 1-4 slightly wider than those of female); northern United States, Rocky Mountains, and Canada ...................................... *Porrhodites*

C. Key to the Nearctic subtribes and genera of Coryphiini

1. Gular sutures nearly parallel or diverging anteriorly; nearly always brachypterous (subtribe Boreaphilina) .................................................... 2
— Gular sutures converging anteriorly; brachypterous or macropterous (subtribe Coryphiina) ............ 3

2(1). Pronotum transverse; anterior margin of labrum with long median tooth (longer than antennomere 2) and 2 shorter ones on each side of middle; mandibles extremely elongate, when closed, their tips reaching the middle of opposite eyes; Cascade Mountains of Washington ................... *Gnathoryphium*
— Pronotum longer than wide; anterior margin of labrum with row of numerous small tubercles; mandibles not so long, when closed reaching only to opposite antennal insertion; Alaska to Massachusetts and New Jersey ...................... *Boreaphilus*

3(1). Pronotum strongly transverse; head distinctly or slightly narrower than pronotum ..................... 4
— Pronotum very slightly wider than long to distinctly longer than wide; head usually at least as wide as pronotum, occasionally (*Eudectus*, southeastern United States) slightly narrower ...................... 6

4(3). Antennomeres 8-10 transverse; head with dorsal nuchal constriction weak or absent; disk of pronotum extremely convex, without deep impressions; body with long conspicuous setae.. ................................................................ *Haida*
— Antennomeres 8-10 quadrate; head with very strong dorsal nuchal constriction; disk of pronotum evenly slightly convex or, if more convex, widest before middle and with distinct impressions basally beside midline; body setae variable .... 5

5(4). Pronotum widest at middle, with lateral margins smooth; pronotal disk evenly slightly convex; head abruptly narrowed behind eyes, temples undeveloped; body with conspicuous long setae ................................................ *Pseudohaida*

— Pronotum widest in front of middle, with lateral margins at least slightly crenulate; pronotal disk with an impression basally on either side of midline; head with well-developed temples behind eyes; body with short inconspicuous setae ............ *Subhaida*

6(3). Pronotum elongate, surface with dense microsculpture between punctures; large, 3.6-4.8 mm long .............................. *Holoboreaphilus*

— Pronotum quadrate or slightly transverse, without microsculpture between punctures; usually smaller, nearly all under 3.5 mm long ............... 7

7(6). Abdominal tergite V as well as IV with pair of small whitish or grayish patches (Fig. 77.22) of wing-folding spicules near middle (patches on tergite 5 usually much smaller than those on 4, sometimes minute) ............................................................ 8

— Only abdominal tergite IV with pair of wing-folding patches ......................................................... 10

8(7). Lateral margins of pronotum smooth; body entirely dark; Alaska and Yukon Territory ........................ ............................................. *Coryphiomorphus*

— Lateral margins of pronotum faintly or distinctly crenulate; elytra sometimes with light markings, legs usually lighter than rest of body; more southern distribution ................................................ 9

9(8). Elytra dark with a discrete yellowish spot on disk or light with brown markings; mandibles with 2 subapical teeth; eastern United States and southeastern Canada ................................... *Ephelinus*

— Elytra uniformly dark, occasionally (*C. arizonense*) with a reddish-brown area extending backward from humerus; mostly western, 1 species northeastern United States and southeastern Canada ......................................................... *Coryphium*

10(7). Lateral margins of pronotum smooth; antennomeres 9-10 quadrate; Washington and Oregon ........... ..................................................... *Occiephelinus*

— Lateral margins of pronotum crenulate; antennomeres 9-10 transverse; southeastern United States ....................................... *Eudectus*

## III. Key to the Nearctic genera of Proteininae

1. Pronotum with median longitudinal impressed line, strongly explanate laterally, posterior corners notched; body flat and wide (Fig. 8.22) ............... ...................................................... *Megarthrus*

— Pronotum without impressed line, not explanate laterally, posterior corners entire; body moderately convex, more narrowly oval ............... *Proteinus*

## IV. Key to the Nearctic genera of Micropeplinae

1. Body with rough surface and elevated ridges (Fig. 9.22); abdominal segments IV-VII each with 1 pair paratergites ..................................... *Micropeplus*

— Body polished, without ridges; abdomen without paratergites ........................................... *Kalissus*

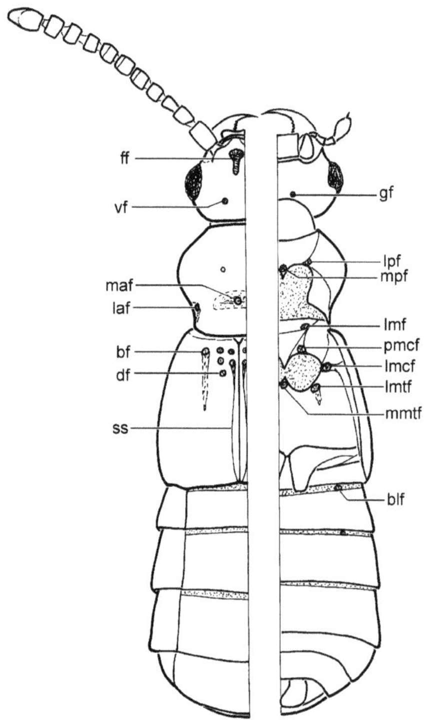


FIGURE 95.22. *Sonoma russelli* Chandler, dorsal (left) and ventral (right) halves, showing foveae: Head: vf, vertexal foveae; ff, frontal fovea; gf, gular foveae. Prothorax: maf, median antebasal fovea; laf, lateral antebasal foveae; mpf, median prosternal fovea; lpf, lateral prosternal foveae. Elytra: bf, basal foveae; df, discal foveae; ss, sutural stria. Pterothorax: lmf, lateral mesosternal foveae; pmcf, promesocoxal foveae; lmcf, lateral mesocoxal foveae; mmtf, median metasternal foveae; lmtf, lateral metasternal foveae. Abdomen: blf, basolateral foveae.

## V. Keys to the Nearctic supertribes, tribes, and genera of Pselaphinae (by D. S. Chandler)

### Key to Nearctic supertribes

1. Three antennomeres, first difficult to see, third much longer than other two combined (Fig. 96.22); first three visible tergites fused, forming a continuous plate (Fig. 15.22), so only three apparent tergites visible (Key N) .................... Clavigeritae

— Nine to eleven antennomeres, apical one at most as long as remaining ones combined; first three visible tergites free, five tergites visible (Figs. 11-14.22) .......................................................... 2

2(1). Elytra with foveae at base and also on disk (Fig. 95.22, df); prothorax with median prosternal fovea (Fig. 95.22, mpf); antenna clavate to moniliform, lacking distinct antennal club; basal two tarsomeres short, similar in length, third tarsomere much longer (Fig. 97.22) (Key A) ..................... Faronitae

— Elytra with foveae only at base (Fig. 11.22) or lacking; second and third tarsomeres much longer than basal tarsomere (Fig. 98.22) (except *Caccoplectus*, with single tarsal claw, Fig. 99.22); prothorax lacking median prosternal fovea; an-

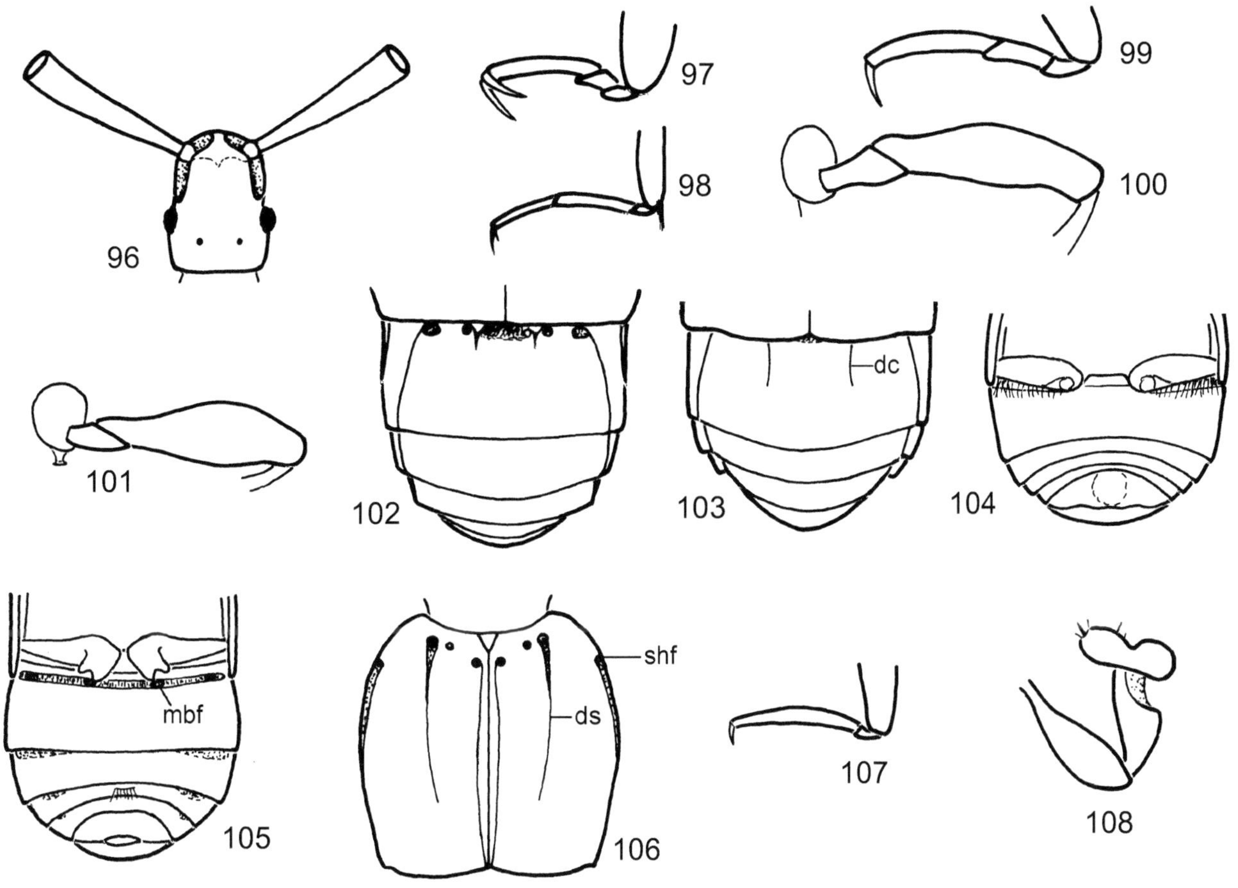


FIGURES 96.22-108.22. Fig. 96.22. *Fustiger knausii* Schaeffer, head, dorsal. Fig. 97.22. *Megarafonus ventralis* Casey, tarsus, lateral. Fig. 98.22. *Batrisodes lineaticollis* (Aubé), head, dorsal. Fig. 99.22. *Caccoplectus sentis* Chandler, tarsus, lateral. Fig. 100.22. *Hamotus elongatus* (Brendel), mesotrochanter and mesofemur, posterior. Fig. 101.22. *Reichenbachia tumidicornis* Casey, mesotrochanter and mesofemur, posterior. Fig. 102.22. *Batrisodes lineaticollis* (Aubé), abdomen, dorsal. Fig. 103.22. *Reichenbachia borealis* Casey, abdomen, dorsal; dc, discal carina. Fig. 104.22. *Reichenbachia borealis* Casey, abdomen, ventral. Fig. 105.22. *Oropus striatus* (LeConte), abdomen, ventral; mbf, mediobasal fovea. Fig. 106.22. *Saxet decora* (Casey), elytra, dorsal; ds, discal stria; shf, subhumeral fovea. Fig. 107.22. *Bythinoplectus gloydi* Park, tarsus, lateral. Fig. 108.22. *Bythinoplectus gloydi* Park, right maxillary palp, dorsal.

tenna typically with distinct club formed by apical 1-3 antennomeres .............................................. 3

3(2). Dorsal margin of mesotrochanter longer than apical trochanter width, mesotrochanter elongate (Fig. 100.22) (Key K) .................................. Pselaphitae

— Dorsal margin of mesotrochanter shorter than apical trochanter width, mesotrochanter nearly triangular (Fig. 101.22) ........................................ 4

4(3). Abdomen rounded laterally, lateral margins of paratergites usually smoothly fused with sterna, often present as short basal carinae on first tergite, mesal margins usually present as carinae on first four visible tergites (Figs. 12.22, 102.22) (Key F) .......................................................... Batrisitae

— Abdomen angulate laterally, lateral and mesal margins of paratergites clearly indicated on first three visible terga (Fig. 103.22), lateral margins of paratergites rarely smoothly fused with sterna ..................... 5

5(4). Metacoxae moderately to widely separated, usually flattened and not conically projecting at articulations with metatrochanters (Fig. 104.22) (exceptions have mesal margins of metacoxae narrowly separated and conically produced and elongate third maxillary palpomere [Speleobamini], or metacoxae contiguous and body nearly glabrous [*Eupsenius*]); elytron rarely (*Rybaxis*) with subhumeral fovea (Fig. 106.22, shf) (Key G) ........ ...................................................... Goniaceritae

— Metacoxae contiguous, conically projecting at articulations with metatrochanters (Fig. 105.22) (*Lemelba* and *Trichonyx* exceptions with metacoxae moderately separated, but have subhumeral foveae present); elytron usually with subhumeral fovea (Fig. 106.22, shf); third maxillary palpomere short and triangular, never elongate except in Bythinoplectitae (Fig. 108.22) ............................ 6

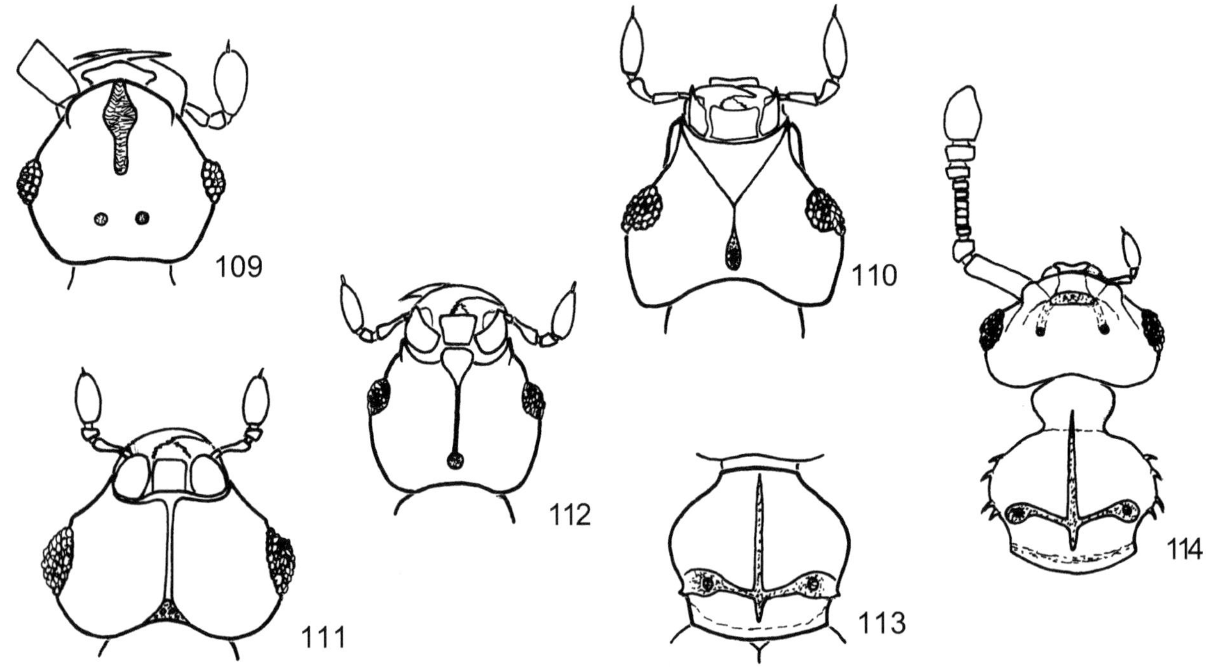


FIGURES 109.22-114.22. Fig. 109.22. *Megarafonus ventralis* Casey, head, dorsal. Fig. 110.22. *Sebaga notonoda* Park, head, ventral. Fig. 111.22. *Oropus striatus* (LeConte), head, ventral. Fig. 112.22. *Foveoscapha terracola* Park and Wagner, head, ventral. Fig. 113.22. *Oropus striatus* (LeConte), pronotum, dorsal. Fig. 114.22. *Rhexius substriatus* LeConte, head and pronotum, dorsal.

6(5). Tarsi apparently of two tarsomeres (Fig. 107.22); two tribes, one with head possessing large lateral excavations beneath antennal insertions that hold oddly formed maxillary palps at rest (Fig. 108.22) (Bythinoplectini), other tribe with members very small and narrowly elongate, abdomen clearly longer than rest of body (Mayetiini) (Key E) .............................................. Bythinoplectitae

— Tarsi of three tarsomeres (cf. Fig. 98.22); head never with large lateral excavations beneath antennae; abdomen shorter than rest of body (Key B) ....... ......................................................... Euplectitae

### A. Key to the Nearctic genera of Faronitae

1. Antennal tubercles separated by sulcus extending anteriorly from frontal fovea (Fig. 109.22); elytra not much longer than pronotum; metasternum about as long as mesocoxae; Pacific coast states ................................. *Megarafonus*

— Antennal tubercles connected by concave frontal bridge, clearly not divided by a sulcus extending from frontal fovea (Fig. 95.22); elytra half again as long as pronotum; metasternum longer than mesocoxae; Pacific coast states and Appalachian Mountains ............................................ *Sonoma*

### B. Key to the Nearctic tribes of Euplectitae

1. Head with Y-shaped gular carina; maxilla with cardo sharply angulate at lateroapical margin (Fig. 110.22); elytral bases with transverse carina; Oklahoma and Texas ......................... (*Sebaga*) Jubini

— Head ventrally with median longitudinal carina or sulcus (Figs. 111-112.22); maxilla with outer margin of cardo broadly to narrowly rounded; elytral bases denticulate or smooth, not carinate .................. 2

2(1). Head clearly transverse (Fig. 111.22); genae very broadly convex; prosternum concavely excavate to receive genae when head deflexed; pronotum with thin median sulcus from antebasal sulcus to near apex, antebasal sulcus straight or nearly so (Fig. 113.22) (Key C) ................................... Trogastrini

— Head triangular to elongate; genae smaller (Fig. 112.22); prosternum not excavate; pronotum usually lacking thin median discal sulcus, antebasal sulcus usually arcuate or biarcuate (Key D) ...... ........................................................... Euplectini

### C. Key to the Nearctic subtribes and genera of Trogastrini

1. Antenna geniculate, first antennomere at least 1/2 as long as remaining antennomeres combined (Fig. 114.22); pronotum with small apical lobe defined by a basal constriction (Fig. 114.22); eastern North America (subtribe Rhexiina) ................... *Rhexius*

— Antenna straight, first antennomere rarely longer than combined lengths of next two; pronotum lacking a distinct apical lobe (Fig. 113.22) (subtribe Trogastrina) ....................................................... 2

2(1). Lacking lateral metasternal foveae; males with antennomeres 5-6 laterally expanded, fourth visible tergite convex, lacking sulci . *Euboarhexius*

— Lateral metasternal foveae present (Fig. 95.22, lmtf); male antennomeres 5-6 normal, not expanded, presence of sulci on fourth visible tergite variable .. 3

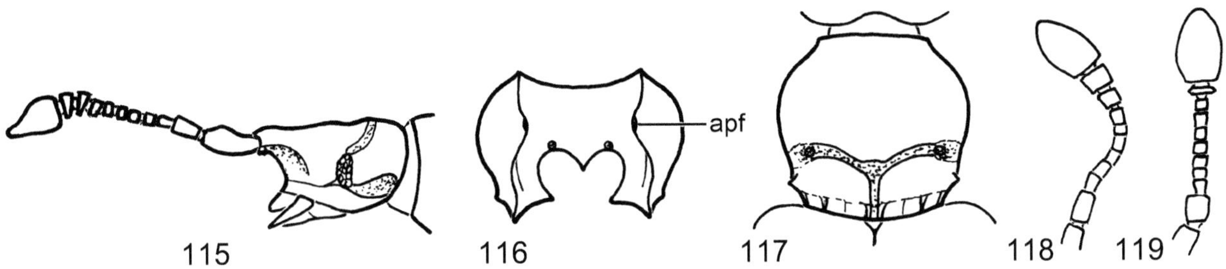


FIGURES 115.22-119.22. Fig. 115.22. *Rhinoscepsis bistriatus* LeConte, head, lateral. Fig. 116.22. *Euplectus confluens* LeConte, prosternum, ventral; apf, anteroprosternal fovea. Fig. 117.22. *Tetrascapha dentata* Park and Wagner, pronotum, dorsal. Fig. 118.22. *Trimioplectus australis* Chandler, antenna, lateral. Fig. 119.22. *Melba sulcatula* Casey, antenna, lateral.

3(2). Pronotum with tooth or blunt angulation on lateral margin adjacent to lateral fovea (Fig. 113.22); western North America .................................. *Oropus*

— Pronotum with lateral margin crenulate, lacking teeth or angulations .................................................. 4

4(3). First visible tergite with short discal carinae at base (cf Fig. 103.22, dc); males with anterior portion of vertex often modified, sulcate or tuberculate, fourth visible tergite convex, lacking sulci; eastern North America ........................... *Conoplectus*

— First visible tergite lacking discal carinae at base; males with vertex simple, fourth visible tergite with basolateral sulci (may be difficult to see); California ............................................ *Rhexidius*

D. Key to the Nearctic subtribes and genera of Euplectini

1. Head with frontal rostrum protruding as prominent antennal tubercle (Fig. 115.22), antennal insertions narrowly separated at apex (subtribe Rhinoscepsina) ................................................ 2

— Head with antennal insertions widely separated, frontal rostrum broad and low, lacking prominent antennal tubercle ............................................. 3

2(1). Pronotal disc with deep median longitudinal sulcus; head with curving glabrous sulci posteroventral and posterodorsal to eyes (Fig. 115.22); southeastern United States .............................. *Rhinoscepsis*

— Pronotal disc flattened, with at most a feeble longitudinal sulcus; lacking glabrous sulci beneath eyes; California .................................................. *Morius*

3(1). Prosternum with median longitudinal carina........ 4

— Prosternum flat or with middle smoothly raised (subtribes Trichonychina, Euplectina, Trimiina, and Bibloplectina) ............................................. 9

4(3). Mesocoxal cavities clearly separated by extensions of meso- and metasternum; promesocoxal foveae present (Fig. 95.22, pmcf); eastern North America (subtribe Panaphantina) .......................... *Thesium*

— Mesocoxal cavities contiguous; lacking promesocoxal foveae (subtribe Bibloporina) ............................ 5

5(4). Pronotum with antebasal foveae isolated, lacking antebasal sulcus (Fig. 11.22); northeastern United States ................................................ *Bibloporus*

— Pronotum with transverse sulcus connecting antebasal foveae ............................................. 6

6(5). Elytra lacking subhumeral foveae; Pacific Northwest ......................................................... *Abdiunguis*

— Elytra with subhumeral foveae (cf. Fig. 106.22, shf) ................................................................... 7

7(6). Anteroprosternal foveae present (Fig. 116.22, apf); pronotum with sharply defined median sulcus extending from base to antebasal sulcus; Pacific Northwest ............................................. *Hatchia*

— Anteroprosternal foveae lacking; pronotum with thin median carina extending from base to antebasal sulcus ............................................. 8

8(7). Head with vertexal foveae nude; Pacific coast states ............................................................. *Trisignis*

— Head with vertexal foveae densely setose; eastern North America .................................... *Eutyphlus*

9(3). Elytra lacking subhumeral foveae ...................... 10

— Elytra with subhumeral foveae (cf. Fig. 106.22, shf) ........................................................................ 18

10(9). Pronotum with teeth or bluntly angulate on lateral margins adjacent to lateral foveae; base with narrow constricted band, band with several short longitudinal carinae (Fig. 117.22); Pacific coast states ............................................ *Tetrascapha*

— Pronotum with lateral margins smooth adjacent to lateral foveae, lacking teeth or angulations; lacking basal band and carinae; eastern North America and Arizona ..................................................... 11

11(10). Lateral mesocoxal foveae present (Fig. 95.22, lmcf) ........................................................................ 12

— Lacking lateral mesocoxal foveae ........................ 14

12(11). Lateral metasternal foveae present (Fig. 95.22, lmtf); head with eyes barely visible in dorsal view; Arizona and Texas .............................. *Trimioarcus*

— Lacking lateral metasternal foveae; head with eyes easily visible in dorsal view; eastern North America ........................................................................ 13

13(12). Elytra with three basal foveae; lateral prosternal foveae present (Fig. 95.22, lpf); Pennsylvania...... .................................................. *Trigonoplectus*

— Elytra with two basal foveae; lacking lateral prosternal foveae; eastern North America ......... ......................................................... *Dalmosella*

14(11). Antennomere 11 at most as long as previous four combined (Fig. 118.22); with lateral prosternal foveae (Fig. 95.22, lpf); eastern North America .............. *Trimioplectus*

— Antennomere 11 at least as long as previous six combined (Fig. 119.22); lacking lateral prosternal foveae .......... 15

15(14). First visible tergite clearly half again as long as second; males with tubercle on vertex posterior to antennal insertions; eastern North America ....... *Trimiomelba*

— First two visible tergites subequal in length or first slightly longer; males with vertex smooth, lacking tubercle .......... 16

16(15). Body conspicuously granulate; first three visible tergites with short discal carinae at bases (cf. Fig. 103.22, dc); Arizona .......... *Zonaira*

— Body smooth, shining; discal carinae at base of first visible tergite only .......... 17

17(16). Lateral metasternal foveae large, easily visible (Fig. 95.22, lmtf); male mesofemur inflated; eastern North America and Arizona .......... *Melba*

— Lateral metasternal foveae small or lacking, difficult to see; male mesofemur only as wide as profemur; Arizona to California ........ *Allotrimium*

18(9). Pronotal disc with faint to distinct median longitudinal sulcus or distinct discal fovea (in *Oropodes*, *Euplecterga*, and *Foveoscapha* this is faint but distinct in almost all species, these genera only in Pacific coast states) .......... 19

— Pronotal disc convex to slightly flattened ........ 26

19(18). Anteroprosternal foveae present (Fig. 116.22, apf) .......... 20

— Anteroprosternal foveae lacking .......... 21

20(19). Head with vertexal foveae nude; lateral metasternal foveae absent; labral apex shallowly concave to convex; widespread, mostly eastern .. *Euplectus*

— Head with vertexal foveae densely setose; lateral metasternal foveae present; labral apex with U-shaped notch; eastern North America .......... *Leptoplectus*

21(19). Metacoxae separated by distance equal to 1/3 metasternal length .......... *Trichonyx*

— Metacoxae narrowly contiguous or very close at middle .......... 22

22(21). Head with vertexal foveae densely setose; eastern North America .......... 23

— Head with vertexal foveae nude; Pacific coast states .......... 24

23(22). Antennal tubercles present as low knobs, interantennal bridge between them level; vertexal foveae separated by less than one foveal width; first visible tergite lacking discal carinae .......... *Thesiastes*

— Antennal tubercles part of elongate ridges, interantennal bridge depressed at middle between them; vertexal foveae separated by one and a half or more foveal widths; first visible tergite with discal carinae .......... *Pycnoplectus*

24(22). Lateral metasternal foveae present (Fig. 95.22, lmtf) .......... 25

— Lateral metasternal foveae absent .......... 26

25(24). Elytra each with four basal foveae ... *Foveoscapha*

— Elytra each with two basal foveae ....... *Bontomtes*

26(24). Antennomere 10 asymmetrical in lateral view, triangular; promesocoxal foveae present (Fig. 95.22, pmcf) .......... *Euplecterga*

— Antennomere 10 symmetrical in lateral view, trapezoidal; promesocoxal foveae absent .......... *Oropodes*

27(18). Head with broad shallow sulci above and below eyes; Florida .......... *Acolonia*

— Head lacking sulci above and below eyes ........ 28

28(27). Metacoxae separated by 1/4 metasternal length; Florida .......... *Lemelba*

— Metacoxae contiguous to narrowly separated .... 29

29(28). Lateral prosternal foveae lacking, or very small and nude; Arizona .......... 30

— Lateral prosternal foveae present, densely setose (Fig. 95.22, lpf) .......... 31

30(29). Lateral metasternal foveae lacking ......... *Simplona*

— Lateral metasternal foveae combined into single median fovea between mesocoxae .......... *Allobrox*

31(29). Lateral metasternal foveae lacking .......... 32

— Lateral metasternal foveae present (Fig. 95.22, lmtf) .......... 33

32(31). Elytra each with two basal foveae; body surface smooth; eastern North America and southwestern United States .......... *Actiastes*

— Elytra each with three basal foveae; body surface granulate; eastern North America ......... *Ramecia*

33(29). Pronotum with median antebasal fovea bisected by thin longitudinal carina; eastern North America. .......... *Bibloplectus*

— Pronotum with median antebasal fovea lacking a bisecting carina, smoothly concave .......... 34

34(33). First visible tergite longest, third tergite about 2/3 length of first tergite (*Cupila clavicorne* with tergites subequal but pronotum transverse and with large setose lateral foveae, Fig. 120.22) .......... 35

— First through third visible tergites subequal in length .......... 37

35(34). First visible tergite with densely setose mediobasal impression between short discal carinae (cf. Fig. 103.22, dc) .......... *Pilactium*

— First visible tergite convex at base, area not densely setose .......... 36

36(35). Head clearly narrower than pronotum; pronotum transverse, with lateral margins slightly convex to straight adjacent to lateral foveae (Fig. 120.22); vertexal and lateral pronotal foveae densely setose; Pacific coast states .......... *Cupila*

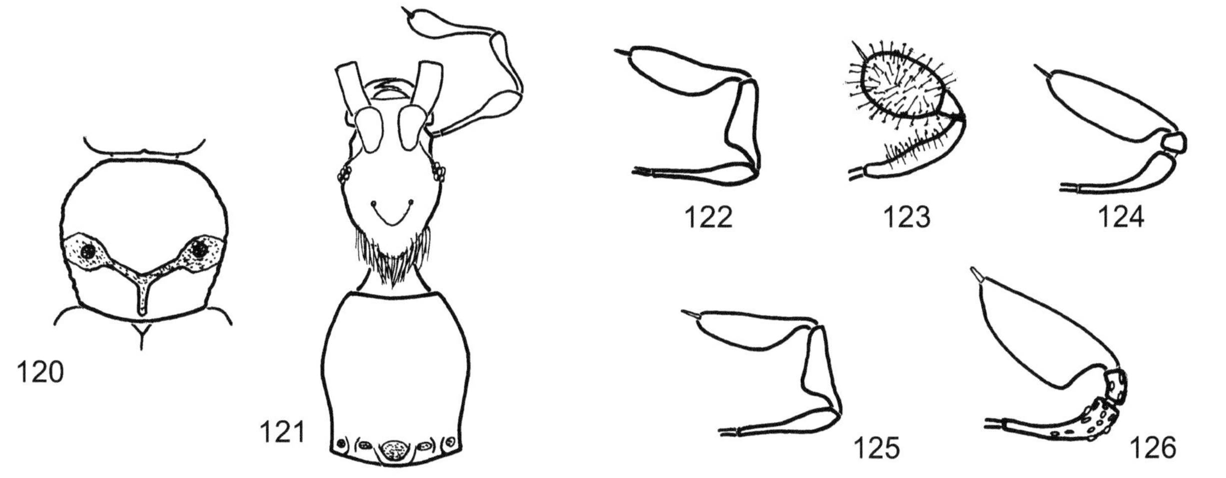


FIGURES 120.22-126.22. Fig. 120.22. *Cupila clavicornis* (Mäklin), pronotum, dorsal. Fig. 121.22. *Prespelea quirsfeldi* Park, head and pronotum, dorsal. Fig. 122.22. *Cylindrarctus comes* Casey, right maxillary palp, dorsal. Fig. 123.22. *Valda frontalis* Casey, right maxillary palp, dorsal. Fig. 124.22. *Machaerodes carinatus* (Brendel), right maxillary palp, dorsal. Fig. 125.22. *Custotychus daggyi* (Park), right maxillary palp, dorsal. Fig. 126.22. *Pselaptrichus magaliae* Chandler, right maxillary palp, dorsal.

— Head as wide as pronotum to slightly narrower; pronotum elongate, deeply constricted adjacent to lateral foveae; vertexal and lateral pronotal foveae nude; southeastern United States ......... ........................................................ *Dalmosanus*

37(34). Promesocoxal foveae present (Fig. 95.22, pmcf) ..... ............................................................ *Actizona*

— Lacking promesocoxal foveae ........................... 38

38(37). First through third visible tergites with distinct discal carinae; elytron with discal striae thin, clearly extending into apical ˚ of elytra (Fig. 106.22, ds); lacking median mesosternal fovea (best determined from cleared specimen); Alabama to Arizona ......................................................... *Saxet*

— Only first visible tergite with discal carinae, or carinae lacking; elytron lacking discal striae, or striae in narrow sulci that extend to elytral midpoint or slightly beyond it; median mesosternal fovea present (difficult to see unless procoxae rotated anteriorly or entire prothorax pulled slightly away from mesothorax) ............................................ 39

39(38). Head across eyes distinctly narrower than pronotum, head almost triangular ............ *Actium*

— Head across eyes as wide or nearly as wide as pronotum, head trapezoidal ........................... 40

40(39). Vertexal foveae small and nude; male mesofemur inflated; Arizona ............................. *Tomoplectus*

— Vertexal foveae large and densely setose; male mesofemur about as wide as profemur; eastern North America ................................ *Pseudactium*

### E. Key to the Nearctic tribes and genera of Bythinoplectitae

1\. Eyes present; oddly formed maxillary palps (Fig. 108.22) resting in large glabrous cavities below antennal insertions; head with frontal rostrum prominent and narrow; nine antennomeres; abdomen shorter than rest of body; Arizona (tribe Bythinoplectini) .............................................. 2

— Eyes absent; head lacking large cavities ventral to antennal insertions; frontal rostrum broad and indistinct; eleven antennomeres; body elongate and slender, less than 1 mm, abdomen longer than rest of body; Pacific coast states and southeastern United States ......... (tribe Mayetiini) *Mayetia*

2(1). Elytra each with two basal foveae (Fig. 95.22, bf); pronotum with lateral antebasal foveae distinct (cf. Fig. 95.22, laf) .............................. *Bythinoplectus*

— Elytra each with single basal fovea; pronotum lacking lateral antebasal foveae .. *Bythinoplectoides*

### F. Key to the Nearctic tribes and genera of Batrisitae

1\. Head with spine in place of eyes; eastern United States (tribe Amauropini) ....................... *Arianops*

— Head with eyes present, or if eyes lacking then ocular area flatly rounded or smoothly knobbed (tribe Batrisini) .................................................. 2

2(1). Pronotum smooth in basal 1/2, lacking paired basolateral tubercles; eastern North America and Arizona ................................................ *Arthmius*

— Pronotum with pair of basolateral tubercles adjacent to median antebasal fovea ....................... 3

3(4). Pronotal disc smooth or with pair of short lateral longitudinal carinae between faint to distinct median and lateral longitudinal sulci; metatibia with apical spur of coalesced setae (Fig. 98.22) .. ......................................................... *Batrisodes*

— Pronotal disc often with lateral longitudinal rows of short recurved spines on disc between faint median and lateral longitudinal sulci, or sulci lacking; metatibia with apical setae short and free, not forming a spur ............................................ 4

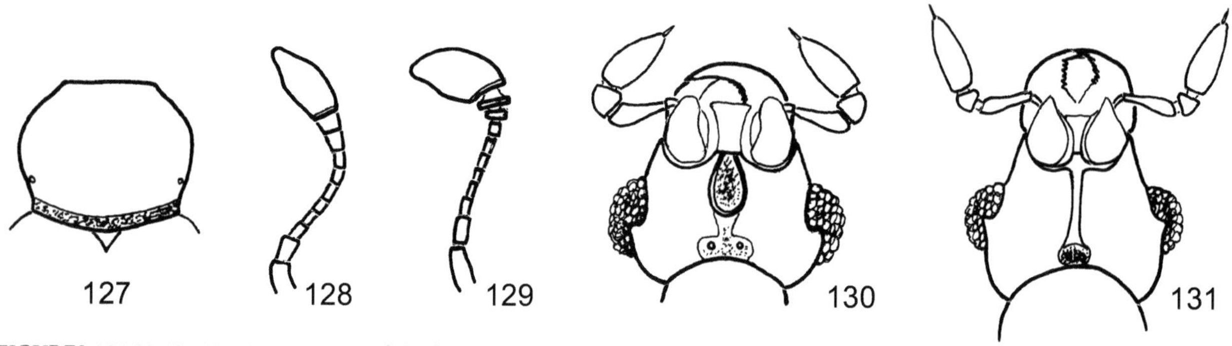


FIGURES 127.22-131.22. Fig. 127.22. *Anchylarthron cornutum* (Brendel), pronotum, dorsal. Fig. 128.22. *Pselaptus belfragei* LeConte, antenna, lateral. Fig. 129.22. *Scalenarthrus horni* LeConte, antenna, lateral. Fig. 130.22. *Decarthron abnorme* (LeConte), head, ventral. Fig. 131.22. *Reichenbachia tumidicornis* Casey, head, ventral.

4(3). Lacking eyes, rounded knob in place of each eye; pronotal disc smooth; troglobitic; Texas .......... ..................................................... *Texamaurops*

— Eyes distinct, or if reduced then area flatly rounded; pronotal disc often with paired rows of short recurved spines; eastern United States .............. ................................................ *Batriasymmodes*

## G. Key to the Nearctic tribes and some genera of Goniaceritae

1. Maxillary second, third, and fourth palpomeres elongate, third at least 1/2 length of fourth (Figs. 121-122.22) ........................................................... 2

— Maxillary third palpomere rarely much longer than wide, usually much less than 1/3 length of fourth (Fig. 130.22) ...................................................... 4

2(1). With long neck obscured by ruff of long setae (Fig. 121.22); metacoxae narrowly separated and conically produced at trochanteral articulations (cf. Fig. 105.22) (tribe Speleobamini) ............................. 3

— Neck short, lacking any long obscuring setae; metacoxae widely separated, not projecting at trochanteral articulations (cf. Fig. 104.22) (Key J) ................................................................ Tychini

3(2). Eyes distinct; foveae present on vertex, pronotum (Fig. 121.22), and elytra; southeastern United States .................................................. *Prespelea*

— Lacking eyes; vertex, pronotum and elytra lacking foveae; Alabama ............................... *Speleobama*

4(1). Fourth maxillary palpomere swollen, with erect capitate setae over surface (Fig. 123.22); metacoxa conically produced at trochanteral articulation (cf. Fig. 105.22); California (tribe Valdini) ............... *Valda*

— Maxillary palps of various shapes (Figs. 124-125.22), but lacking any obviously modified setae, all setae cylindrical; metacoxa barely produced at trochanteral articulation .................................................. 5

5(4). Frontal rostrum prominent (cf. Fig. 140.22); fourth maxillary palpomere enlarged, 2 or more times as wide as third (Figs. 124.22, 126.22); first antennomere elongate, at least as long as next three together (Key I) ...................................... Bythinini

— Lacking distinct frontal rostrum; fourth maxillary palpomere not so enlarged, no more than 2 times width of third (Fig. 130.22); first antennomere rarely as long as next three (Key H) .......... Brachyglutini

## H. Key to the Nearctic subtribes and genera of Brachyglutini

1. Body glabrous and polished; metacoxae narrowly separated; lacking discal striae on elytra; southeastern United States (subtribe Eupseniina)...... ........................................................... *Eupsenius*

— Body with distinct setae present over surface; metacoxae widely separated; discal striae on elytra present (cf. Fig. 106.22, ds) or absent ..... 2

2(1). Elytra lacking discal striae (subtribe Pselaptina) . 3

— Elytra with distinct discal striae extending posteriorly from a basal fovea (cf. Fig. 106.22, ds) ...... 6

3(2). Pronotum with distinct band of large coarse punctures at base (Fig. 127.22); eastern United States ..................................................... *Anchylarthron*

— Pronotum roughened to smooth in basal area, but not with large coarse punctures ............................... 4

4(3). Pronotum with median antebasal fovea distinct (Fig. 95.22, maf) ......................................... *Eutrichites*

— Pronotal disc smooth near base, lacking median antebasal fovea ................................................ 5

5(4). Joint of last two antennomeres straight (Fig. 128.22); mesosternum strongly carinate between mesocoxae; Texas to California .......... *Pselaptus*

— Last two antennomeres meeting at oblique angle (Fig. 129.22); mesosternum flat between mesocoxae; Arizona to California .......................... *Scalenarthrus*

6(2). Pronotum with lateral antebasal foveae indistinct and nude, or lacking; head ventrally with median oval fossa set off by carinoid edges (Fig. 130.22); ten antennomeres (subtribe Decarthrina) .......... ......................................................... *Decarthron*

— Pronotum with lateral antebasal foveae large and pubescent (Fig. 95.22, laf); head ventrally with median longitudinal carina (Fig. 131.22); eleven antennomeres (subtribe Brachyglutina) .......... 7

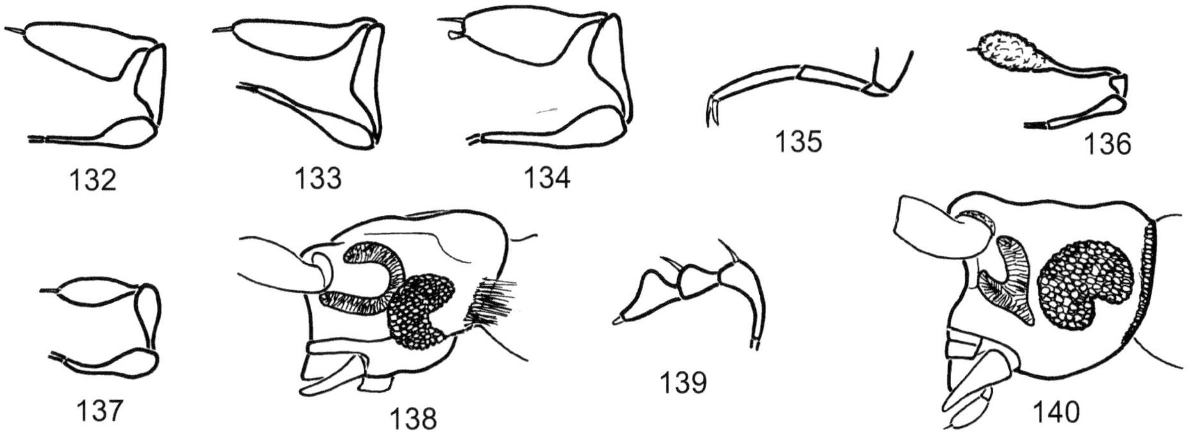


FIGURES 132.22-140.22. Fig. 132.22. *Nearctitychus sternalis* (Raffray), right maxillary palp, dorsal. Fig. 133.22. *Ouachitychus parvoculus* Chandler, right maxillary palp, dorsal. Fig. 134.22. *Lucifotychus testaceus* (Casey), right maxillary palp, dorsal. Fig. 135.22. *Tyrus corticinus* (Casey), tarsus, lateral. Fig. 136.22. *Pselaphus bellax* Casey, right maxillary palp, dorsal. Fig. 137.22. *Tyrus corticinus* (Casey), right maxillary palp, dorsal. Fig. 138.22. *Tmesiphorus costalis* LeConte, head, left lateral. Fig. 139.22. *Tmesiphorus costalis* LeConte, right maxillary palp, dorsal. Fig. 140.22. *Atinus monilicornis* (Brendel), head, left lateral.

7(6). Pronotum with sulcus connecting antebasal foveae; elytra with deep subhumeral sulcus on flanks, subhumeral fovea may be weakly indicated (cf. Fig. 106.22, shf) .......... *Rybaxis*

— Pronotum lacking sulcus between antebasal foveae; elytra smoothly rounded laterally, lacking subhumeral fovea or sulcus .......... 8

8(7). Head lacking vertexal foveae; pronotum with median antebasal fovea faint or lacking; eastern North America .......... *Nisaxis*

— Head with vertexal foveae present (Fig. 95.22, vf); pronotum with median antebasal fovea distinct (Fig. 95.22, maf) .......... 9

9(8). Pronotum with median and lateral antebasal foveae subequal, large and pubescent...... *Brachygluta*

— Pronotum with nude median antebasal fovea smaller than lateral foveae .......... 10

10(9). Pronotal disc convex; first visible tergite with discal carinae (Fig. 103.22, dc) short to elongate, usually easily seen at base .......... *Reichenbachia*

— Pronotal disc nearly flat, with shallow median longitudinal sulcus; discal carinae of first visible tergite very short and difficult to see; Florida islands .......... *Briaraxis*

I. Key to the Nearctic genera of Bythinini

1. Clypeus with a sharp median longitudinal carina; second and third maxillary palpomeres smooth (Fig. 124.22); eastern North America ....... *Machaerodes*

— Clypeus broadly convex to broadly concave at middle; second and third maxillary palpomeres studded with tubercles (Fig. 126.22), very rarely lacking them; eastern and western North America .......... 2

2(1). Eyes lacking; each elytron with at most one distinct basal fovea; second and fourth maxillary palpomeres each nearly as long as head; southeastern United States .......... 3

— Eyes present; each elytron with two distinct basal foveae; second and fourth maxillary palpomeres each no more than two-thirds head length; eastern and western North America .......... 4

3(2). Pronotum with antebasal sulcus lightly impressed or absent; sutural stria of elytron distinct at least basally (Fig. 95.22, ss) .......... *Speleochus*

— Pronotum with antebasal sulcus distinct; sutural stria of elytron lacking .......... *Subterrochus*

FIGURES 141.22-145.22. Fig. 141.22. *Hamotus elongatus* (Brendel), right maxillary palp, dorsal. Fig. 142.22. *Mipseltyrus mirus* Schuster, right maxillary palp, dorsal. Fig. 143.22. *Cedius ziegleri* LeConte, right maxillary palp, dorsal. Fig. 144.22. *Ctenisodes piceus* (LeConte), right maxillary palp, dorsal. Fig. 145.22. *Ctenisis raffrayi* Casey, right maxillary palp, dorsal.

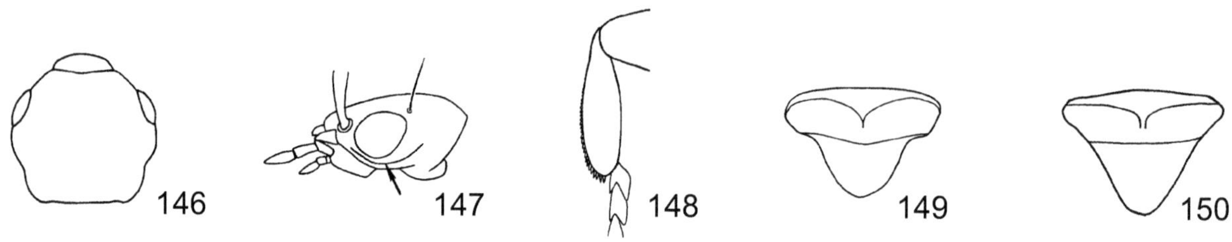


FIGURES 146.22-150.22. Fig. 146.22. *Derops divalis* (Sanderson), head, dorsal, with epistomal suture anteriorly (after Newton 1990a). Fig. 147.22. *Bolitobius kremeri* (Malkin), head, left lateral, showing postmandibular ridge (arrow) (modified from Newton 1990a). Fig. 148.22. *Sepedophilus basalis* (Erichson), right proleg, anterior (after Newton 1990a). Fig. 149.22. *Mycetoporus consors* LeConte, scutellum (after Campbell 1991). Fig. 150.22. *Bryophacis smetanai* Campbell, scutellum (after Campbell 1993b).

4(2). Pronotum with antebasal sulcus slightly and evenly arcuate between lateral foveae (Fig. 13.22); male genitalia asymmetrical; eastern North America .. ................................................ *Tychobythinus*

— Pronotum with antebasal sulcus smoothly angulate at middle; male genitalia symmetrical; western North America .............................. *Pselaptrichus*

J. Key to the Nearctic genera of Tychini

1. Vertexal foveae lateral, within one eye facet diameter of eye margin; third maxillary palpomere medially angulate in basal 1/2 (Fig. 122.22) ........... ................................................ *Cylindrarctus*

— Vertexal foveae on vertexal disc, distance from eyes about width of four to six eye facets; medial angulation of third maxillary palpomere variously placed ........................................................ 2

2(1). Third maxillary palpomere distinctly angulate basal to middle of mesal margin (Fig. 125.22); eastern North America ................................ *Custotychus*

— Third maxillary palpomere smoothly rounded on inner margin (Figs. 132-133.22) ........................... 3

3(2). Fourth maxillary palpomere broadest at basal 1/4 (Fig. 132.22); head with several distinct punctures between antennal tubercles and eyes; eastern United States ............................................ *Nearctitychus*

— Fourth maxillary palpomere broadest beyond basal 1/4 (Figs. 133-134.22); head lacking distinct punctures between antennal tubercles and eyes, appearing impunctate.......................................... 4

4(3). Third maxillary palpomere elongate and slender, 3/4 length of fourth (Fig. 133.22); lacking mediobasal foveae on second visible abdominal sternum; Arkansas .......................................... *Ouachitychus*

— Third maxillary palpomere widest near middle of mesal margin, about 2/3 length of fourth (Fig. 134.22); mediobasal foveae present on second visible sternum (Fig. 105.22, mbf) .................... 5

5(4). Fourth maxillary palpomere with small, apically broadened spine near apical pseudosegment (Fig. 134.22); male genitalia symmetrical ................ *Lucifotychus*

— Fourth maxillary palpomere with only apical pseudosegment; male genitalia asymmetrical; western United States ............... *Hesperotychus*

K. Key to the Nearctic tribes and some genera of Pselaphitae

1. Two basal tarsomeres short, only third tarsomere relatively long (Fig. 99.22); femur with ventral spinose tubercle or ridge near middle; Texas to Arizona (tribe Arhytodini) .................. *Caccoplectus*

— Basal tarsomere short, distal two tarsomeres relatively long (Fig. 135.22); femur lacking spinose tubercles or ridges .......................................... 2

2(1). Tarsi each with single claw; second and fourth maxillary palpomeres quite elongate, third very short (Fig. 136.22); first visible tergite longer than remaining tergites combined (Fig. 14.22) (tribe Pselaphini) ........................................................ 3

— Tarsi each with two subequal claws (Fig. 135.22); if second and fourth maxillary palpomeres elongate, then third at least 1/2 as long as fourth (Fig. 137.22); first visible tergite usually about as long as second ........................................................ 4

3(2). Center of vertex and pronotal base with flattened scale-like ("squamous") setae in transverse band,"sugary" in appearance; Texas ................ .................................................... *Neopselaphus*

— Head and pronotal base glabrous, shining, lacking transverse band of strongly flattened setae (Fig. 14.22); eastern North America to Pacific Northwest.................................................... *Pselaphus*

4(2). Head with semicircular setose sulcus partially enclosing antennal insertion (Fig. 138.22); second and third maxillary palpomeres each with lateral spine, fourth laterally lobed (Fig. 139.22) (tribe Tmesiphorini) ................................ *Tmesiphorus*

— Head lacking curved setose sulci around antennal insertions, a few genera with linear or Y-shaped (Fig. 140.22) bands of squamous setae; a few genera with spines on maxillary palps ......................... 5

5(4). Antennomeres all same width, antennal club lacking (tribe Ceophyllini) ....................... *Ceophyllus*

— Apical 1-4 antennomeres clearly longer or wider, always forming distinct antennal club ............ 6

6(5). Head with linear to Y-shaped (Fig. 140.22) band of flattened, scale-like ("squamous") setae below antennal insertion; at least head with squamous setae in foveae, sulci, and at head base (Key M)... ..........................................................Ctenistini

— Head lacking any bands of squamous setae beneath antennal insertions; body with only aciculate setae (Key L)..............................................Tyrini

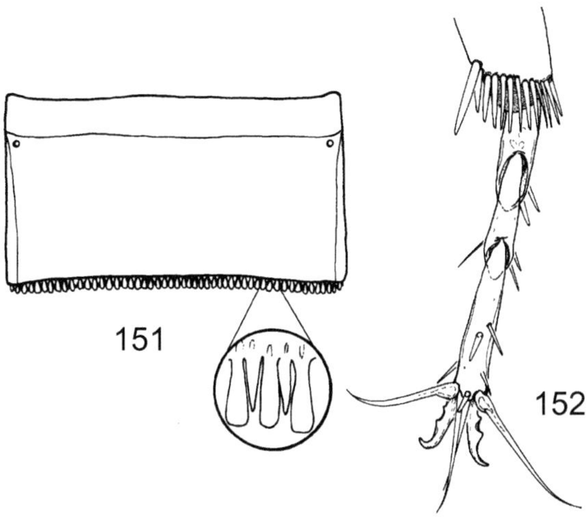


FIGURES 151.22-152.22. Fig. 151.22. *Deinopsis erosa* (Stephens), abdominal tergite IV with detail of apical ctenidium of cuticular projections. Fig. 152.22. *Deinopsis erosa* (Stephens), hind tarsus, with detail of tarsal claws.

### L. Key to the Nearctic subtribes and genera of Tyrini

1. Fourth maxillary palpomere enlarged and ovoid, with mesal longitudinal sulcus (Fig. 141.22) (subtribe Hamotina) .......... 2
— Fourth maxillary palpomere not much wider than third, mesal surface smooth, lacking sulcus (Fig. 142.22) (subtribe Tyrina) .......... 3

2(1). Antennal club formed by large eleventh antennomere; eastern North America .......... *Circocerus*
— Antennal club formed by last three antennomeres; southern United States .......... *Hamotus*

3(1). Pronotum with median tubercle near base, lacking antebasal foveae; Arizona .......... *Anitra*
— Pronotal base lacking tubercle, antebasal foveae present (Fig. 95.22, maf and laf) .......... 4

4(3). Last three maxillary palpomeres elongate (Fig. 142.22), together much longer than head; pronotum lacking sulcus between antebasal foveae .......... *Mipseltyrus*
— Maxillary palpomeres not so elongate (Figs. 137.22, 143.22), together much shorter than head; pronotum with sulcus connecting antebasal foveae .......... 5

5(4). Third maxillary palpomere strongly projecting medially (Fig. 143.22); eastern North America .......... *Cedius*
— Third maxillary palpomere elongate (Fig. 137.22) .......... *Tyrus*

### M. Key to the Nearctic genera of Ctenistini

1. Third maxillary palpomere with thick "pencil" of setae projecting from outer angle, palps easily seen (Fig. 144.22) .......... 2
— Maxillary palps lacking any tufts of laterally projecting setae, palps small and difficult to see (Fig. 140.22) .......... 3

2(1). Last two maxillary palpomeres strongly transverse, projecting laterally (Fig. 144.22); widespread .......... *Ctenisodes*
— Last two maxillary palpomeres only briefly angulate laterally, more elongate (Fig. 145.22); Arizona to California .......... *Ctenisis*

3(1). Metatibiae flattened, about same width throughout length; Texas to California .......... *Biotus*
— Metatibiae swollen in apical halves, circular in cross-section; southeastern United States .......... *Atinus*

### N. Key to the Nearctic genera of Clavigeritae

1. Eyes present; pronotum lacking lateral foveae, lateral margins rounded .......... *Fustiger*
— Eyes absent (Fig. 15.22); pronotum with short transverse grooves on lateral margins originating from lateral foveae .......... *Adranes*

### VI. Key to the Nearctic genera of Phloeocharinae

1. Clypeus strongly produced into tridentate beak that conceals most of labrum; vertex of head with pair of smooth round swellings (not ocelli, much larger in diameter), pronotum and elytra with longitudinal elevations or grooves; abdominal tergites III-VII with basolateral ridges (similar to Fig. 29.22); procoxal fissure nearly closed, trochantin concealed (similar to Fig. 43.22, but with short fissure anterolateral to coxa) .......... *Vicelva*
— Clypeus not produced, labrum fully exposed; head, pronotum and elytra without elevations or grooves; abdominal tergites without basolateral ridges; procoxal fissure open, trochantin exposed (similar to Figs. 61-62.22) .......... 2

2(1). Head not constricted immediately behind eye; if neck present, temple is at least as long as eye; epistomal suture present (similar to Fig. 146.22); elytra much longer than pronotum; mesosternum not carinate; at most with basal tarsomeres of male protarsi notably expanded and bearing tenent setae; body flat; widespread .......... 3
— Head constricted laterally immediately behind eye to form broad neck, so that temple is much shorter than eye or absent; epistomal suture absent; elytra much shorter than pronotum; mesosternum longitudinally carinate at least posteriorly; basal tarsomeres of pro- and mesotarsi noticeably expanded and bearing tenent setae in both sexes; body moderately convex; California and Oregon .......... 4

3(2). Apical maxillary palpomere nearly as wide at base as penultimate palpomere, short and subtriangular; at least antennomere 10 elongate; pronotal margins smooth; elytron without distinct epipleural keel (similar to Fig. 71.22); mesocoxae contiguous; California .......... *Ecbletus*
— Apical maxillary palpomere not more than 1/2 as wide as penultimate palpomere, long and slender; antennomeres 7-10 transverse; pronotal margins crenulate at least in part; elytron with epipleural keel (similar to Fig. 72.22, arrow); mesocoxae narrowly separated by mesosternal process; eastern North America, southwest to Arizona .......... *Charhyphus*

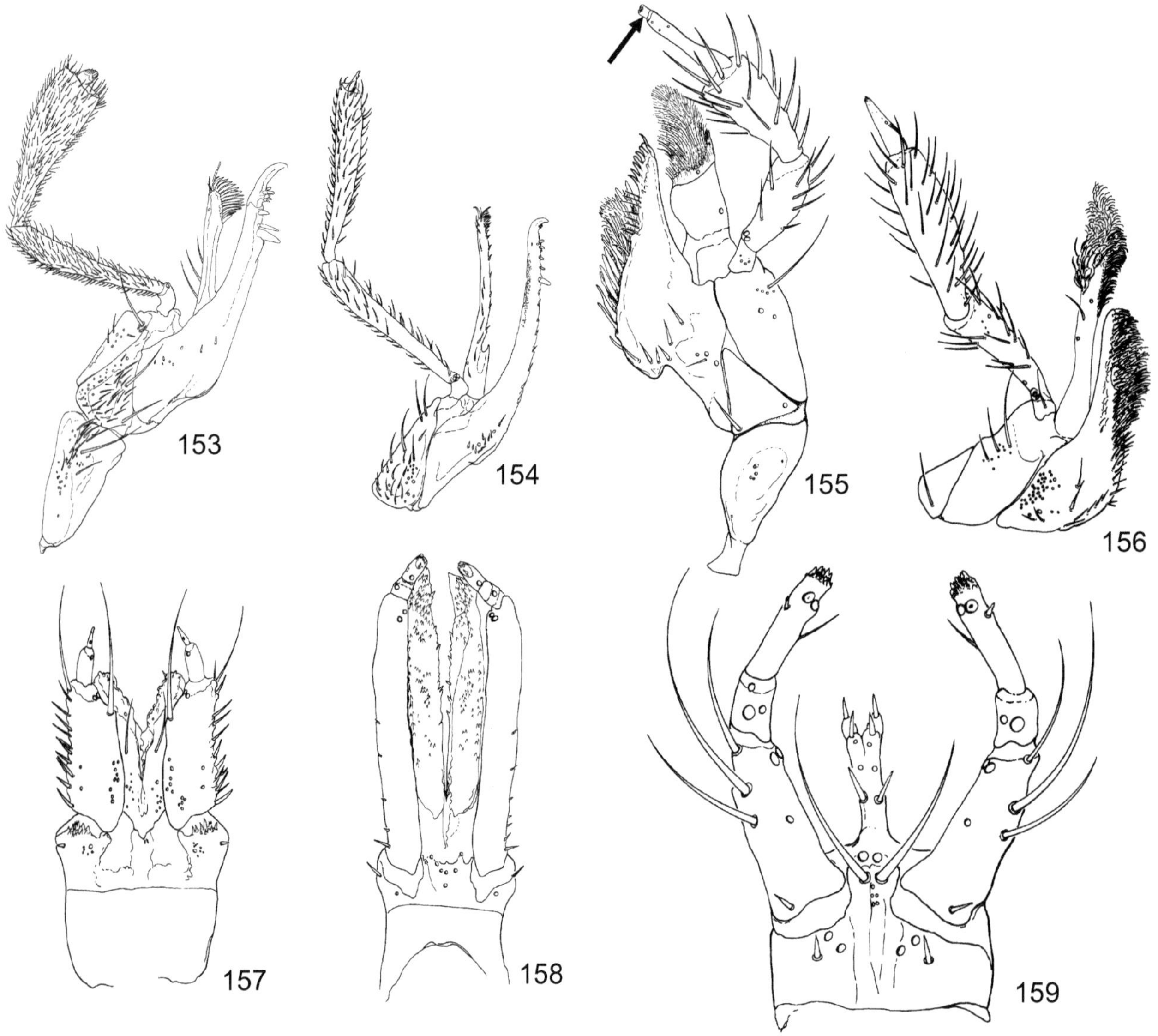


FIGURES 153.22-159.22. Fig. 153.22. *Deinopsis erosa* (Stephens), maxilla, ventral. Fig. 154.22. *Gymnusa variegata* Kiesenwetter, maxilla, ventral. Fig. 155.22. *Hoplandria lateralis* (Melsheimer), maxilla, ventral, showing apical pseudosegment (arrow). Fig. 156.22. *Drusilla canaliculata* (Fabricius), maxilla, ventral. Fig. 157.22. *Deinopsis erosa* (Stephens), labium, ventral. Fig. 158.22. *Gymnusa variegata* Kiesenwetter, labium, ventral. Fig. 159.22. *Hoplandria lateralis* (Melsheimer), labium, ventral.

4(2). Head with fine carina and groove above eye, delimiting small temple; elytron with epipleural keel, which is crenulate to toothed at humerus; protibia emarginate just before apex, outer edge bearing 4 short and strong, blunt spines; metatarsus very long, more than 2/3 as long as metatibia ........................................................................ *Dytoscotes*

— Head without carina or groove above eye; elytron without epipleural keel (similar to Fig. 71.22), although humerus may be crenulate or toothed; protibia not emarginate, without strong blunt spines; metatarsus shorter, less than 2/3 as long as metatibia .................................... *Phloeocharis*

## VII. Keys to the Nearctic tribes and genera of Tachyporinae

### Key to Nearctic tribes (Modified from Smetana 1983)

1. Pronotum strongly constricted basally, its base distinctly narrower than humeral width of elytra; sides of pronotum not strongly inflexed, fully visible in lateral view, including large well-sclerotized postcoxal process which forms a more or less equilateral triangle (Fig. 61.22) ....... (*Derops*) Deropini

— Pronotum not constricted basally, its base about as wide as or even wider than humeral width of elytra; sides of pronotum strongly inflexed so that hypomeron is scarcely or not at all visible in lateral view, postcoxal process absent or present as an elongate, lightly sclerotized lobe whose apex may be visible in lateral view ................................... 2

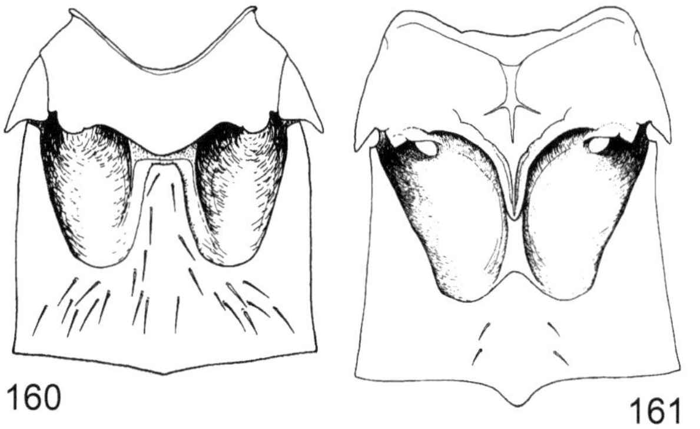


FIGURES 160.22-161.22. Fig. 160.22. *Zyras collaris* (Paykull), meso-metasternum. Fig. 161.22. *Stictalia* sp., meso-metasternum (mesocoxae narrowly separated).

2(1). Head without ridge below eye; elytron without distinct sutural stria (Key A) .................... Tachyporini

— Head with distinct ridge below eye (Fig. 147.22); elytron with impressed sutural stria (Key B) ....... ....................................................... Mycetoporini

### A. Key to the Nearctic Genera of Tachyporini

1. Protibia with ctenidium along external edge (Fig. 148.22); dorsum of head and pronotum moderately to densely, finely pubescent; procoxal cavities nearly or quite closed behind by meeting of postcoxal and prosternal processes in all but a few species .................................. *Sepedophilus*

— Protibia without external ctenidium; dorsum of head and pronotum glabrous or sparsely pubescent; procoxal cavities open behind or effectively closed by enlarged spiracular peritremes ....... 2

2(1). Mesosternum not carinate .................................. 3

— Mesosternum longitudinally carinate, at least posteriorly in front of or between mesocoxae ...... 5

3(2). Apical maxillary palpomere much narrower and shorter than penultimate palpomere ............. *Tachyporus*

— Apical maxillary palpomere as wide as, and as long or longer than, penultimate palpomere ............ 4

4(3). Antenna with fine recumbent pubescence starting from fourth or (usually) fifth antennomere; abdominal tergites IV-VI (first three fully visible) without long laterally projecting setae (but such setae may be present on the paratergites) and without spiracular openings ............................. *Tachinus*

— Antenna with fine recumbent pubescence starting from third antennomere; abdominal tergites IV-VI (first three fully visible) each with a long, laterally projecting seta at the posterolateral corner and with spiracular openings ............ *Nitidotachinus*

5(2). Elytral epipleuron inflexed but clearly visible in lateral view; body black; base of abdominal tergite III (normally largely covered by elytra) with transverse pruinose spot on each side of midline; size larger, about 4-7.5 mm long; Arizona to Texas ... ................................................ *Tachinomorphus*

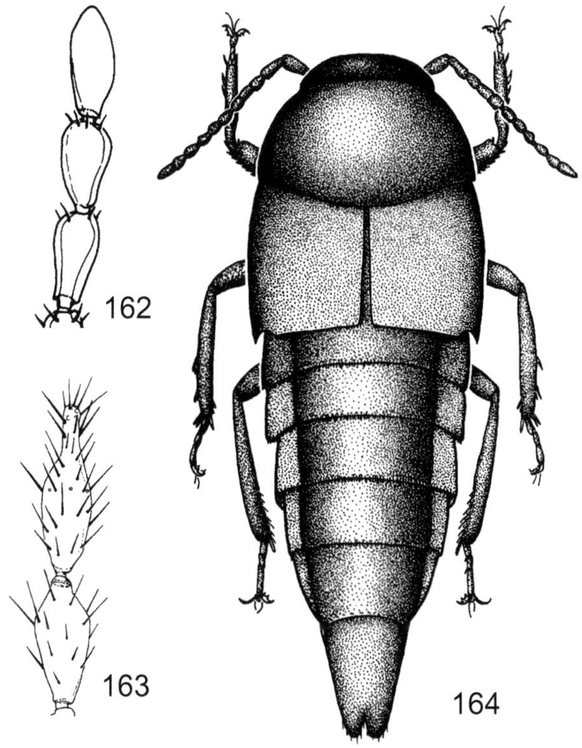


FIGURES 162.22-164.22. Fig. 162.22. *Deinopsis erosa* (Stephens), apical antennomeres. Fig. 163.22. *Adinopsis* sp., apical antennomeres. Fig. 164.22. *Deinopsis illinoisensis* Klimaszewski, length 4.1 mm.

— Elytral epipleuron strongly inflexed against underside of elytron, scarcely or not at all visible in lateral view (if clearly visible, then body bicolored); abdominal tergites without pruinose spots; size smaller, about 2-4.5 mm long; widespread .......... 6

6(5). Elytron unicolorous or bicolored with apical portion lighter in color; male sternum VII (penultimate visible) with apical margin truncate or slightly concave medially .................................. *Coproporus*

— Elytron bicolored, dark brown to black with apical portion yellow and with yellow sutural and humeral longitudinal bands; male sternum VII with apical margin deeply, broadly and concavely emarginate ................................................ *Cilea*

### B. Key to the Nearctic Genera of Mycetoporini (Modified from Campbell 1993b)

1. Apices of meso- and metatibiae each with 3 long spurs, with numerous smaller spines ctenidium-like, apices forming a straight edge ................ 2

— Apices of meso- and metatibiae each with 3 or 4 long spurs, apices of numerous smaller spines uneven, apices forming a jagged edge .......... 3

2(1). Elytron with a single row of discal punctures, occasionally also with 1-2 additional punctures; apical maxillary palpomere conical, not more than 1/2 as wide as densely pubescent penultimate palpomere; male sternum VIII (last visible) unmodified or with complex patches of modified setae .................... *Ischnosoma*

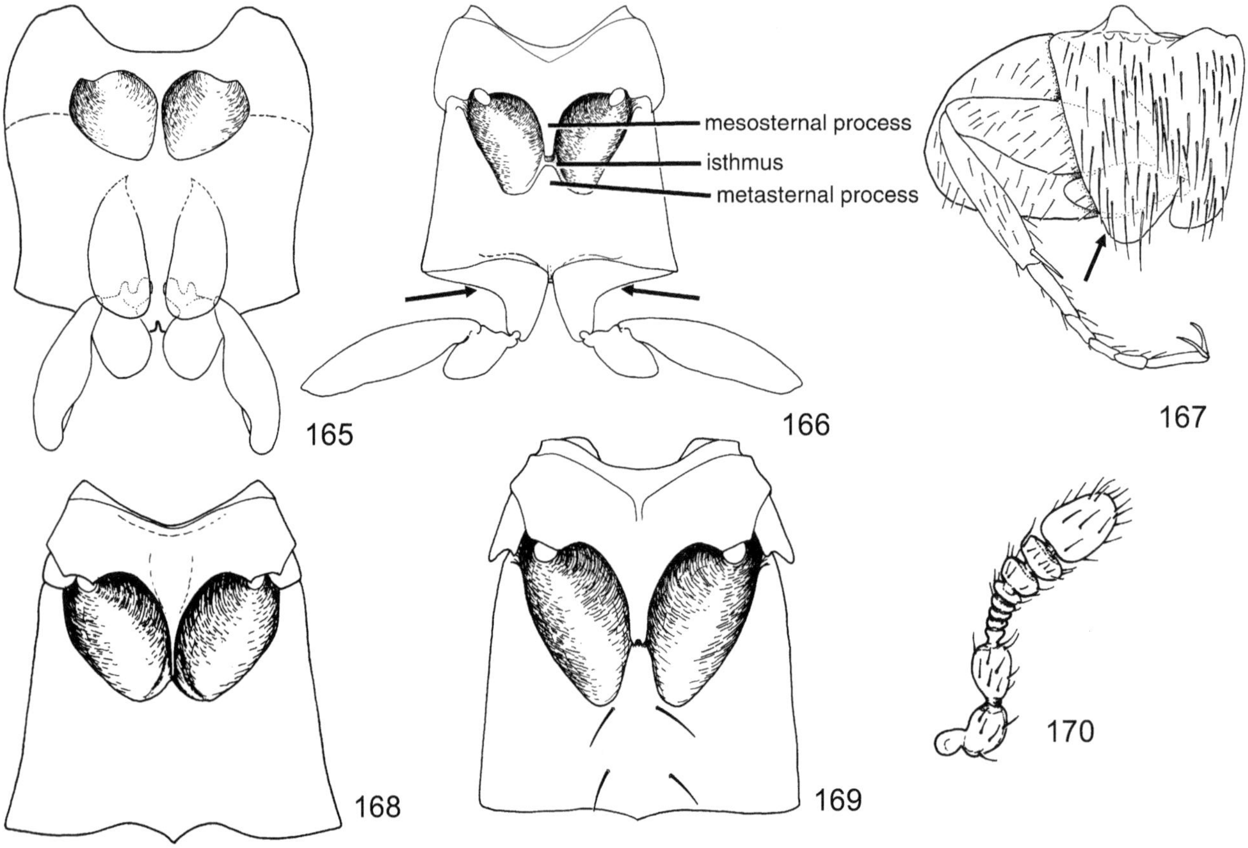


FIGURES 165.22-170.22. Fig. 165.22. *Trichopsenius frosti* Seevers, meso-metasternum, showing coxae indistinguishably fused to metasternum. Fig. 166.22. *Silusida marginella* Casey, meso-metasternum, showing large transverse coxae (arrows). Fig. 167.22. *Anacyptus testaceus* (LeConte), posterior coxa and leg, ventral, showing large ventral plate (arrow) covering base of femur. Fig. 168.22. *Moluciba grandipennis* Casey, meso-metasternum. Fig. 169.22. *Amarochara forticornis* (Boisduval and Lacordaire), meso-metasternum. Fig. 170.22. *Anacyptus testaceus* (LeConte), antenna.

— Elytron with 6 irregular rows of discal punctures; apical maxillary palpomere conical but only slightly narrower than coarsely, sparsely setose penultimate palpomere; male sternum VIII with patch of 2 or 3 pairs of oblique setae near middle of apical margin .................................. *Bryoporus*

3(1). Supraocular puncture and seta well developed, seta at least as long as eye (Fig. 147.22) .................. 4

— Supraocular puncture and seta reduced or obsolete: seta, if present, distinctly shorter than eye ......... 8

4(3). Elytron densely, evenly punctate, punctures not arranged in distinct rows .................. *Bolitopunctus*

— Elytron more sparsely punctate, with at least some punctures arranged in distinct rows ................. 5

5(4). Apical maxillary palpomere narrow, conical or aciculate, not more than 1/2 as wide as penultimate palpomere; apical labial palpomere distinctly narrower than penultimate palpomere ...................................... 6

— Apical maxillary palpomere conical, slightly narrower to subequal in width to penultimate palpomere; apical labial palpomere at least as wide as penultimate palpomere ......................................................... 7

6(5). Elytron without distinct microsculpture; basal carina of scutellum acutely angulate medially (Fig. 149.22); widespread ........................ *Mycetoporus*

— Elytron with dense, transverse waves of microsculpture; basal carina of scutellum narrowly divided medially (Fig. 150.22); British Columbia to Montana ...................... *Bryophacis* (in part; 1 sp.)

7(5). Pronotum and elytron with dense waves of microsculpture visible only under high magnification (above 100X); male sternum VII (penultimate visible) with distinct patterns of modified setae; larger, 5.6-10.8 mm long; widespread ..................... *Bolitobius*

— Pronotum and elytron with distinct waves of coarse, transverse microsculpture easily visible with low magnification (24X); male sternum VII not so ornamented; smaller, 4.2-5.6 mm long; Oregon to British Columbia .................................. *Neobolitobius*

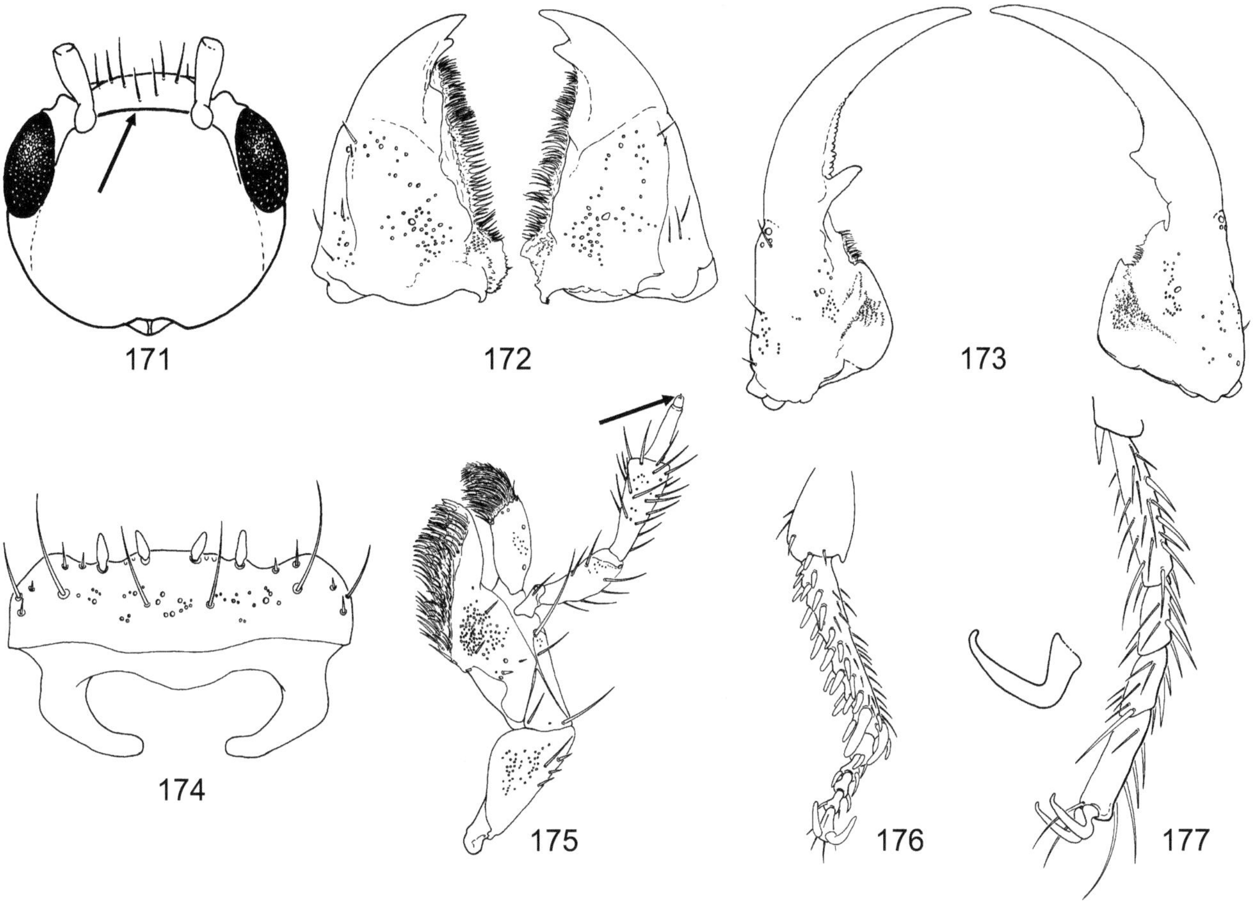


FIGURES 171.22-177.22. Fig. 171.22. *Oxypoda lividipennis* Mannerheim, head, showing frontal suture (arrow). Fig. 172.22. *Devia prospera* (Erichson), mandibles, dorsal. Fig. 173.22. *Gnathusa eva* Fenyes, mandibles, dorsal. Fig. 174.22. *Gnathusa eva* Fenyes, labrum, dorsal. Fig. 175.22. *Aleochara bimaculata* (Gravenhorst), maxilla, ventral, showing apical pseudosegment (arrow). Fig. 176.22. *Aleochara* sp., front tibia and tarsus. Fig. 177.22. *Bamona* sp., hind tarsus, with detail of tarsal claw.

8(3). Apical (fourth) maxillary palpomere distinctly shorter than third palpomere, second and third palpomeres moderately coarsely, sparsely pubescent; usually unicolored ........... *Bryophacis* (in part; most spp.)

— Apical (fourth) maxillary palpomere as long as or distinctly longer than third palpomere, second and third palpomeres with a few scattered setae only; most species distinctly bicolored ...................... 9

9(8). Apical maxillary palpomere conical; sides of elytra and of abdomen subparallel; preapical antennomeres broadly transverse; male sternum VIII (apical) with patches of modified setae ....... .......................................................... *Carphacis*

— Apical maxillary palpomere not distinctly tapered; sides of abdomen strongly converging from base to apex (Fig. 17.22); preapical antennomeres with length usually subequal to width, sometimes transverse; male sternum VIII only infrequently with patches of modified setae .................... *Lordithon*

## VIII. Keys to the tribes and genera of Nearctic Aleocharinae (by J. S. Ashe)

### Identifying Aleocharinae

Identifying genera of the subfamily Aleocharinae is distinctly challenging. This is primarily a result of the small size in most taxa (the average size is about 3 mm) and the consequently minute size of many distinctive features, coupled with the vast number of valid taxa, many of which are superficially similar. However, except possibly for some poorly characterized genera in the tribe Athetini, the following keys should make it possible to identify most genera in the North American fauna, and many are quite distinctive. Since one cannot make aleocharines larger, consistent aleocharine identification requires good optical equipment that is appropriate for the size of aleocharine specimens, well-prepared specimens, and patience and dedication. It is especially important to have well-prepared specimens; specimens which are contorted, dirty or embedded in large quantities of mounting medium so that the legs and ventral surface of the body are obscured usually cannot be identified until they are relaxed and remounted.

Correct counting of tarsomeres is central to identification using the keys to genera of Aleocharinae. This can be difficult for small to minute aleocharines, and the following techniques may be helpful: 1) view the tarsi under high magnification (70-100X magnification or more) against a bright background (i.e., back-lit); this frequently makes the divisions between the tarsomeres more apparent; 2) remove a leg, place it on a microscope slide as a dry mount, and view it with transmitted light using a compound microscope; you can glue the leg back onto the point with the specimen afterwards or 3) make a temporary or permanent microscope mount for transmitted light viewing.

Another important feature for identifying aleocharines that may be unfamiliar is the relative lengths of the mesosternal process, metasternal process, and the isthmus extending between the two. In most aleocharines, the meso- and metasternal processes are each defined by a fine bead or raised line that delimits the extent of the process between the mesocoxae. The "isthmus" is an extension of the metasternum anterior to the limit of the metasternal process proper and forms a bridge between the two processes. These are illustrated and labeled in Fig. 166.22. Proper examination of these processes often requires moving the anterior coxae or legs (see below).

In general, identification is greatly facilitated if one first mounts one or two cleared and dissected specimens of a series on microscope slides, either as temporary mounts in glycerine or Hoyer's medium, or as permanent slides in Canada balsam, Euparal, or another permanent medium. This is time consuming, but it makes it possible to reliably count tarsal segmentation and to clearly examine a number of minute, but distinctive, characteristics, especially in the mouthparts. Many of these same features can be seen with considerable effort and high magnification on well-prepared dried and pointed specimens, but they are near the limit of resolution of dissecting microscope optics and mistakes in interpretation are likely. Cleared and dissected specimens mounted on microscope slides, and examined using high quality compound microscope optics, are essential for serious taxonomic and phylogenetic study of aleocharines.

It is frequently necessary to adjust or move body parts, antennae or legs in order to examine important key characteristics. Moore *et al.* (1976) described a method of rapidly relaxing parts of dried and pinned specimens using Barber's solution that is very effective and useful.

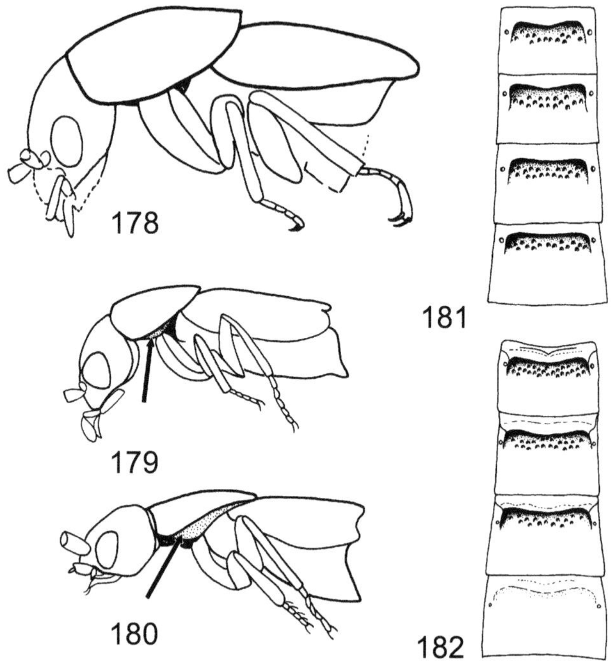


FIGURES 178.22-182.22. Fig. 178.22. *Oxypoda lividipennis* Mannerheim, lateral, showing pronotal hypomeron not exposed. Fig. 179.22. *Melanalia tabida* Casey, lateral, showing pronotal hypomeron (arrow) partially exposed. Fig. 180.22. *Calodera* sp., lateral, showing pronotal hypomeron (arrow) fully exposed. Fig. 181.22. *Calodera riparia* Erichson, abdominal tergites III-VI. Fig. 182.22. *Parocyusa rubicunda* (Erichson), abdominal tergites III-VI.

Key to Nearctic tribes and groups of tribes

1. Terga and sterna of abdominal segments III-VI each with apical ctenidium of short comb-like cuticular projections (Fig. 151.22) .............................. 2
— Terga and sterna of abdominal segments without apical ctenidia of comb-like cuticular projections ... 3

2(1). Tarsi 2-2-2 or 3-3-3; tarsal claws with a medial tooth (Fig. 152.22); maxilla (Fig. 153.22) and labium (Fig. 157.22) distinctive (Key A) ................ Deinopsini
— Tarsi 5-5-5; tarsal claws without medial tooth; maxilla (Fig. 154.22) and labium (Fig. 158.22) distinctive (*Gymnusa*) ................................. Gymnusini

3(1). Tarsi 5-5-5 ...................................................... Key B
— Tarsi other than 5-5-5 (at least front tarsi with only 4 tarsomeres) ...................................................... 4

4(3). Tarsi 4-4-5 or 4-4-4 (both front and middle tarsi with 4 tarsomeres) .................................................... 5
— Tarsi 4-5-5 ........................................................... 6

5(4). Tarsi 4-4-5 ...................................................... Key E
— Tarsi 4-4-4 ..................................................... Key D

6(4). Maxillary (Fig. 155.22, arrow) and (in most) labial (Fig. 159.22) palps with apical pseudosegment, appearing to have 5 and 4 palpomeres, respectively (Key C) .................................................... Hoplandriini
— Maxillary and labial palps without apical pseudosegments ................................................ 7

7(6). Abdomen distinctly petiolate, with basal segments narrowed to a distinct petiole and distal segments forming a more or less bulbous "gaster," body form distinctly myrmecoid (Fig. 264.22); mesocoxal cavities not margined behind; anterior margin of abdominal sternum IV with gland reservoirs (visible only in micro-slide preparations); associated with army ants of the genus *Neivamyrmex* (Key F)....... Crematoxenini
— Not exactly fitting the above description .............. 8

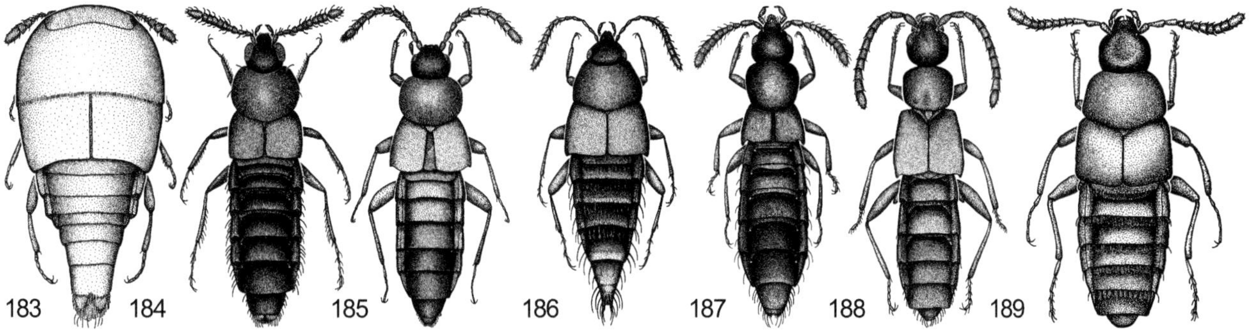


FIGURES 183.22-189.22. Fig. 183.22. *Anacyptus testaceus* (LeConte), length 1.1 mm. Fig. 184.22. *Aleochara lustrica* Say, length 5.9 mm. Fig. 185.22. *Devia prospera* (Erichson), length 4.5 mm. Fig. 186.22. *Oxypoda* sp., length 3.1 mm. Fig. 187.22. *Ocyustiba appalachiana* Lohse and Smetana, length 2.5 mm. Fig. 188.22. *Gyronycha* sp., length 3.8 mm. Fig. 189.22. *Acrimea fimbriata* Casey, length 2.1 mm.

8(7). Combination of apterous, with moderately to greatly shortened elytra, elytral length less than 2/3 pronotal length; head, pronotum, elytra, and abdominal sterna and at least some terga with numerous dark macrosetae; eyes small, 1/3 length of temples or less; pronotal disk deflexed onto prothoracic flanks, hypomeron not delimited from dorsum of pronotum by a marginal line, or marginal line present only in posterior 1/2; habitus as in Figs. 275-276.22; associated with ants of the genus *Liometopum* in the southwestern United States and adjacent Mexico (Key H) ........................................................ Sceptobiini

— Not exactly fitting all aspects of the above description (a few taxa of the Athetini, such as *Geostiba* and *Pontomalota*, have greatly shortened elytra and very small eyes, but the other features do not apply); free-living or myrmecophilous, but not associated with ants of the genus *Liometopum* ............................ 9

9(8). Pronotum broadest subapically, narrowed behind to a base not more than 3/4 greatest width of pronotum; pronotum with a slight to very deep medial longitudinal sulcus (Fig. 273.22); head with well-delimited narrow neck less than 1/2 head width (Fig. 274.22); sclerotized peritremes present around mesospiracular openings, peritreme size very small and only narrowly surrounding spiracles (visible only in micro-slide preparations) to very large and contiguous along midline; prosternum elongate behind level of procoxal insertions; abdominal tergite IV with distinctive medial gland opening; paramere of male copulatory organ with velum divided into 2 lobes (Key G) ........ Falagriini (except *Bryobiota*, which has 4-4-5 tarsi)

— Not exactly fitting above description................... 10

10(9). Metasternal process distinctly longer than very short mesosternal process (Fig. 160.22); galea and lacinia of maxilla usually moderately to greatly elongate (Fig. 156.22), length of galea equal to or greater than distance to base of galea from cardo; mesocoxae moderately to very broadly separated by broad meso- and metacoxal processes (Fig. 160.22) (Key I) ................................. Lomechusini

— Metasternal process not longer than mesosternal process and much shorter in most; galea and lacinia of maxilla not elongate, length of galea usually shorter than, or subequal to, distance to base of galea from cardo; mesocoxal cavities moderately (Fig. 291.22) to narrowly (Fig. 161.22) separated (in most) or contiguous (Key J) ............................ ..................... Athetini and Oxypodini: Tachyusina

A. Key to the Nearctic genera of Deinopsini

1. Tarsi 3-3-3, antennomere 11 without apical papilla (Fig. 162.22); habitus as in Fig. 164.22 ........... *Deinopsis*

— Tarsi 2-2-2; antennomere 11 with apical papilla (Fig. 163.22) ................................................ *Adinopsis*

B. Key to the Nearctic tribes and genera of Aleocharinae with 5-5-5 tarsi (Includes Trichopseniini, Mesoporini, Corotocini, Aleocharini, and some Oxypodini; Gymnusini treated above in first key)

1. Hind coxae apparently indistinguishably fused to metasternum, large trochanters appearing to articulate directly with the metasternum (Fig. 165.22); associated with termites of the genus *Reticulitermes* (tribe Trichopseniini) ......................................... 2

— Large transverse hind coxae clearly present (Fig. 166.22, arrows), trochanters clearly articulating with them; most not associated with termites . 3

2(1). Inner margin of each elytron with fine longitudinal raised line, margin of elytron mesal to raised line deflexed downward to produce a distinct V-shaped groove between the elytra; most specimens with abdomen distinctly swollen (physogastric) and showing areas of intersegmental membrane between the abdominal terga and sterna; mesosternal process very short and broadly pointed ................................. *Xenistusa*

— Inner margin of each elytron without fine longitudinal line or V-shaped groove between elytra; abdomen of most individuals parallel-sided or slightly convergent to apex, without areas of intersegmental membrane between abdominal segments (some individuals slightly physogastric, showing some abdominal inflation and intersegmental membrane); mesosternal process long and acutely pointed ........................... *Trichopsenius*

3(1). Hind coxa with lamella covering base of femur (Fig. 167.22, arrow); antennomeres 3-7 minute, 8-10 much larger, forming a distinct club (Fig. 170.22); minute (about 1.0 mm long), light brown to flavate, sublimuloid; habitus as in Fig. 183.22 (tribe Mesoporini) ......................................... *Anacyptus*

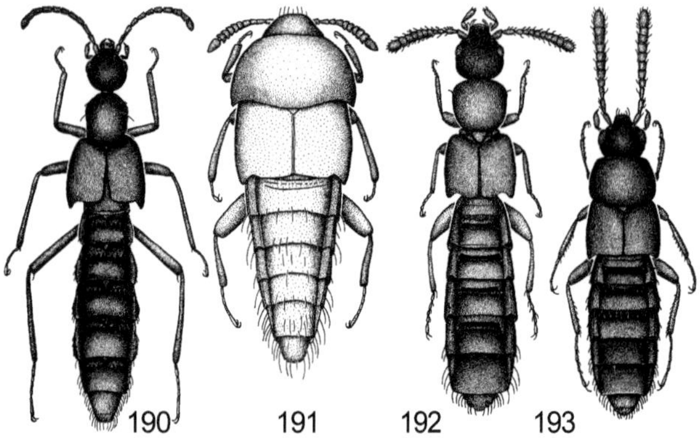


FIGURES 190.22-193.22. Fig. 190.22. *Blepharhymenus illectus* (Casey), length 4.6 mm. Fig. 191.22. *Decusa expansa* (LeConte), length 2.1 mm. Fig. 192.22. *Phloeopora* sp., length 2.6 mm. Fig. 193.22. *Dexiogyia* sp., length 2.4 mm.

— Hind coxa without lamella covering base of femur (Fig. 166.22); antennomeres of various forms, but 3-7 not minute and 8-10 not forming a distinct club; size various but most distinctly larger than 1.0 mm; sublimuloid, fusiform, or parallel-sided .............. 4

4(3). Maxillary (Fig. 175.22) and labial palps each with apical pseudosegment, thus appearing to have 5 and 4 palpomeres, respectively; front tibia with numerous short spines (Fig. 176.22); habitus as in Fig. 184.22 (tribe Aleocharini) ..................... *Aleochara*

— Maxillary and labial palps without apical pseudosegment, appearing to have 4 and 3 palpomeres, respectively; front tibia without spines in most .................................................. 5

5(4). Mentum and submentum fused; body moderately to strongly physogastric (abdomen swollen, with extensive areas of membrane exposed between sclerotized areas); associated with termites (tribe Corotocini: subtribe Eburniogastrina) ............. 6

— Mentum and submentum not fused, clearly separated by a suture; body not physogastric; usually not associated with termites (tribe Oxypodini, in part) .......................................................................... 7

6(5). Pronotum widest before middle (about 1/3 distance from apex to base), with apical and basal widths subequal; pronotal base straight or arcuate; basal labial palpomere moderate in length; body sparsely setose .......................... *Eburniogaster*

— Pronotum widest at or behind middle, with apex about 3/4 width of base; pronotal base shallowly lobed posteriorly; basal labial palpomere very long; body conspicuously setose, with long and short setae .................................... *Termitonidia*

7(5). Frontal suture present (Fig. 171.22) ..................... 8

— Frontal suture absent.......................................... 15

8(7). Pronotal hypomeron not (Fig. 178.22), or only narrowly (Fig. 179.22), visible in lateral view (when narrowly visible, a thin edge of the hypomeron is clearly visible, but most of it is deflexed under the pronotal margin and thus not visible in lateral view) ................................................................. 9

— Pronotal hypomeron clearly and broadly visible in lateral view (Fig. 180.22) .................................... 13

9(8). Pronotal hypomeron narrowly visible in lateral view (Fig. 179.22) .................................................. 10

— Pronotal hypomeron not at all visible in lateral view (Fig. 178.22) .................................................. 12

10(9). Mesosternal process very long and spiniform, at least 3 times as long as metasternal process (Fig. 168.22); meso- and metasternal processes contiguous ................................................ *Moluciba*

— Mesosternal and metasternal processes various but never as above .............................................. 11

11(10). Elytral length about 1.4-1.5 times length of pronotum; pronotum about 1.4 times as wide as long; basal abdominal sterna slightly, but distinctly, transversely impressed........... *Melanalia*

— Elytral length subequal to, or shorter than, length of pronotum; pronotum not more than 1.3 times as wide as long; basal abdominal sterna not transversely impressed (this couplet will only identify known North American species; European species have more broadly exposed pronotal hypomera, somewhat longer elytra and transversely impressed abdominal sterna) .... *Ocyusa*

12(9). Both mandibles with a subapical tooth (Fig. 172.22); body moderately large and robust (Fig. 185.22), 3.3 mm or more long; pronotum subelliptical, without hind angles; antennomere 3 distinctly longer than 2 ........................................................ *Devia*

— Mandibles without subapical tooth (right mandible may have more basal medial tooth); body of most smaller and more slender, most less than 2.5 mm long, habitus as in Fig. 186.22; pronotum with distinct hind angles; antennomere 3 as long as or shorter than 2 ..................................... *Oxypoda*

13(8). Eyes small, length less than length of postocular region; mesocoxae contiguous; mesocoxal cavities not delimited by a line; metasternal process not apparent; metasternum short and covered basally by the mesocoxae; elytra very short, about 1/2 as long as pronotum at midline; abdomen broadest at segments V-VI; abdominal tergites III-VII each increasing slightly in length; habitus as in Fig. 187.22; known only from high elevations in the Appalachian Mountains of North Carolina...... ......................................................... *Ocyustiba*

— Not exactly fitting above description ............... 14

14(13). Abdominal tergites III-VI with distinct and deep transverse impressions (Fig. 181.22); head with a distinct, but broad, neck about 2/3 head width; antennomeres 4-10 transverse to subquadrate; basal abdominal sterna with distinct transverse basal impressions ................................ *Calodera*

— Abdominal tergites III-V with distinct and deep transverse impressions (Fig. 182.22) (tergite VI with transverse basal cariniform line, but tergite not deeply impressed behind that line); head with only weakly delimited neck nearly as broad as base of head; antennomeres 4-7, in some species 4-10, distinctly elongate; basal abdominal sterna with, at most, weak and slightly developed transverse basal impressions ....................... *Parocyusa*

15(7). Mandibles extremely long and slender (Fig. 173.22); some species with anterior margin of labrum spinose (Fig. 174.22) ................................ *Gnathusa*

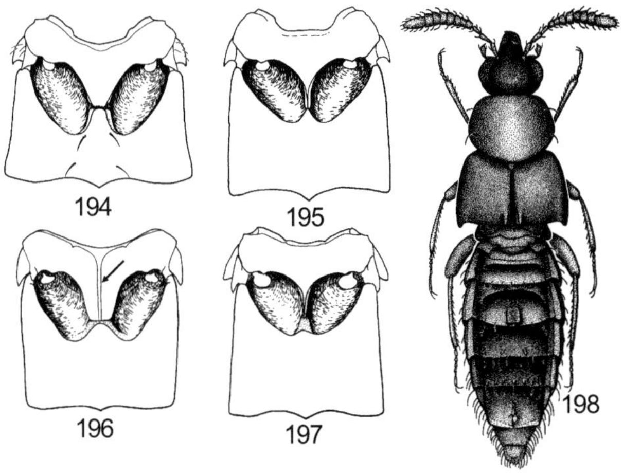


FIGURES 194.22-198.22. Fig. 194.22. *Hoplandria lateralis* (Melsheimer), meso-metasternum. Fig. 195.22. *Platandria* sp., meso-metasternum. Fig. 196.22. *Tinotus* sp., meso-metasternum, showing mesosternal carina (arrow). Fig. 197.22. *Tetrallus fenyesi* Bernhauer, meso-metasternum. Fig. 198.22. *Hoplandria lateralis* (Melsheimer), length 3.2 mm.

— Mandibles various, but not extremely long and slender; without spinose processes on anterior margin of labrum ..................................................... 16

16(15). Tarsal claws scythe-shaped (similar to Fig. 177.22) ..................................................................... 17

— Tarsal claws not scythe-shaped ........................... 18

17(16). Larger, 2.5-4.2 mm long; pronotum subquadrate; antennomeres 4-10 elongate (apical ones subquadrate in some species); habitus as in Fig. 188.22 ............................................. *Gyronycha*

— Smaller, 2.0 mm long or less, very slender; pronotum transverse; antennomeres 4-10 transverse ....... ............................................................ *Apimela*

18(16). Mesosternum carinate medially, at least at base (observing this character usually requires that the front coxae be moved) .................................. 19

— Mesosternum not at all carinate medially .......... 21

19(18). Mesosternum carinate from base to apex of mesosternal process; habitus as in Fig. 189.22 ... .............................................................. *Acrimea*

— Mesosternum carinate only at base or carina not attaining apex of mesosternal process .......... 20

20(19). Antennomere 4 subquadrate, 5-10 transverse and successively gradually broader toward antennal apex; pronotal hypomeron narrowly visible in lateral view; abdominal tergites III-VI with deep transverse basal impressions ......................................... *Pentanota*

— Antennomere 4 subquadrate to elongate, 5-10 elongate or 8-10 subquadrate to slightly transverse, antennomeres not or only slightly broader toward antennal apex; pronotal hypomeron broadly visible in lateral view; abdominal tergites III-V with deep transverse basal impressions ....... *Ocalea, Athetalia*

21(18). Form (Fig. 190.22) distinctive: head with narrow neck, head and pronotum subequal in width, elytra clearly broader than head, pronotum, or base of abdomen; abdominal tergites III-V with deep transverse impressions, the impressions with deep and coarse punctures and medial carina; abdomen widening and thickening from base to segment V where it is subequal in width to elytra ........ *Blepharhymenus*

— Basal impressions of abdominal segments III-V without medial carina; body form not as above .... 22

22(21). Pronotum without a lateral marginal bead, hypomeron broadly visible in lateral view, continuous with and not delimited from strongly convex dorsal surface of pronotum; myrmecophilous ......................................................... *Losiusa*

— Pronotum with a marginal bead, hypomeron clearly delimited from dorsal surface of pronotum, free-living or myrmecophilous .............................. 23

23(22). Body limuloid to sublimuloid in form (Fig. 191.22), with broad shield-like pronotum that covers base of moderately to strongly deflexed head (some *Myrmobiota* with weakly deflexed head), and strongly to moderately acuminate abdomen that tapers uniformly from broad base to more or less acutely pointed apex; myrmecophilous ........ 24

— Body not limuloid or sublimuloid, body form parallel-sided or subparallel-sided, abdomen not uniformly tapered from base to apex; free-living or myrmecophilous ........................................... 26

24(23). Antenna with 10 antennomeres; habitus as in Fig. 191.22 .................................................... *Decusa*

— Antenna with 11 antennomeres ......................... 25

25(24). Pronotum 1.7 to 2.0 times as wide as long; pronotal hypomeron not visible in lateral view; antenna long and slender, not or only slightly incrassate, antennomeres 4-10 elongate or subquadrate.... ........................................................... *Euthorax*

— Pronotum 1.3-1.5 times as wide as long; pronotal hypomeron visible in lateral view; antenna shorter and moderately to strongly incrassate, antennomeres 4-10 strongly transverse........... ....................................................... *Myrmobiota*

26(23). Abdominal tergites III-VI with deep transverse basal impressions (similar to Fig. 181.22)................. 27

— Abdominal tergites III-V with transverse basal impressions (similar to Fig. 182.22) .......................... 29

27(26). Abdominal sterna with (at most) very shallow and inconspicuous transverse basal impressions; smaller (3.0 mm long or less) and less robust (Fig. 192.22); elytra at most slightly broader than abdomen; somewhat flattened, usually found under bark of dead trees ................. *Phloeopora, Neodemosoma*

— Abdominal sterna IV-V with conspicuous transverse basal impressions; larger (usually 3.5 mm or longer, but one species of *Parocalea* 2.8-3.2 mm) and more robust; elytra conspicuously broader than abdomen; not noticeably flattened, not usually found under bark ........................................... 28

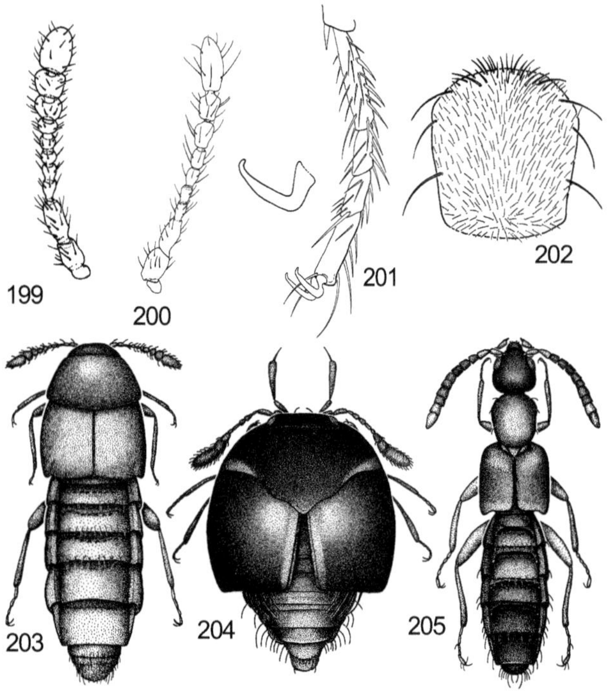


FIGURES 199.22-205.22. Fig. 199.22. *Oligota pusillima* (Gravenhorst), antenna. Fig. 200.22. *Cypha* sp., antenna. Fig. 201.22. *Bamona* sp., hind tarsus, with detail of tarsal claw. Fig. 202.22. *Bamona* sp., pronotum, showing pubescence pattern. Fig. 203.22. *Oligota* sp., length 1.4 mm. Fig. 204.22. *Holobus* sp., length 0.9 mm. Fig. 205.22. *Bamona* sp., length 2.7 mm.

28(27). Body sculpture and punctures fine and inconspicuous, body surface moderately to strongly shining; antennomere 4 transverse or quadrate, 5-10 strongly transverse, becoming broader to apex ........................................................... *Parocalea*

— Body sculpture and punctures very dense, rugose and prominent; body surface not strongly shining; antennomere 4 transverse to elongate, 5-10 quadrate to very elongate, not becoming noticeably broader to apex .......................... *Ilyobates*

29(26). Pronotum distinctly longer than wide ... *Longipeltina*

— Pronotum distinctly wider than long (at least 1.1 times as wide as long, usually much wider) ... 30

30(29). Antennomeres 4-6 elongate, 7-10 subquadrate, slightly elongate, or slightly transverse (not strongly transverse); myrmecophilous with ants of the genera *Formica*, *Camponotus*, and *Lasius* ....................................................... *Thyasophila*

— Antennomeres 5-10 strongly transverse ........... 31

31(30). Mesocoxae moderately broadly separated (Fig. 169.22); meso- and metasternal processes moderately broad and nearly equal in length ....... 32

— Mesocoxae narrowly separated; mesocoxal process slender, short and acute or long and acuminate ....... 33

32(31). Form robust and stout; about 3.5 mm long; surface of head and pronotum appearing dull due to dense covering of coarse, raised close-meshed microsculpture and small umbilicate punctures; head and pronotum covered with dense vestiture of short fine yellow pubescence; pronotum broadest subbasally; infraorbital (postgenal) carina present; myrmecophilous ................. *Pachycerota*

— Form more slender; smaller, 1.8-2.5 mm long; surface of head, pronotum and elytra smooth and shining, without microsculpture; head and pronotum without dense short fine yellow pubescence; pronotum broadest near or in front of middle; infraorbital (postgenal) carina absent; usually in leaf litter, moss or plant debris, some myrmecophilous ....................................... *Amarochara*

33(31). Light-colored (light reddish-brown), very slender, parallel-sided, minute, length about 1.8 mm; mesocoxal cavities not, or only very weakly, margined; mesosternal process short and acute; metasternal process very short and poorly (if at all) delimited .......................................... *Meotica*

— Not exactly fitting above description; mesocoxal cavities margined; mesosternal process long and slender; metasternal process clearly delimited and moderately long ...................................... 34

34(33). Ligula bifid; surface of head and pronotum densely asperately punctate and reticulate, appearing dull, not shining; labial palp subfiliform; base of abdomen distinctly narrower than elytra, habitus as in Fig. 193.22; apical margin of male tergite VIII with small denticles; apical margin of male sternum VIII not triangularly produced in the middle; usually found under bark of dead trees ............ ........................................................... *Dexiogyia*

— Ligula entire; surface of head and pronotum reticulate or not, but without dense covering of asperate punctures, surface shining or sub-shining; labial palp not subfiliform; base of abdomen subequal to width of elytra; apical margin of male tergite VIII without secondary sexual characteristics; apical margin of male sternum VIII produced into a rounded or triangular lobe; not usually found under bark of dead trees ............................... 35

35(34). Anterior and middle tibia with short spines as well as stiff hairs (similar to those of *Aleochara*, Fig. 176.22); usually found in mammal and bird nests ....................................................... *Haploglossa*

— Anterior and middle tibia with stiff hairs, but without short spines; usually found in plant and animal refuse in stables, granaries, and rarely in ant nests ................................................ *Crataraea*

### C. Key to the Nearctic genera of Hoplandriini (tarsi 4-5-5)

1. Mesocoxae broadly separated by broad meso- and metasternal processes (Fig. 194.22) .................. 2

— Mesocoxae narrowly separated by narrow meso- and metasternal processes (Fig. 195.22) .................. 3

2(1). Mesosternum with strong medial longitudinal carina; mesosternal process much longer than, and overlapping, shorter metasternal process (Fig. 196.22) .................................................. *Tinotus*

— Mesosternum without medial longitudinal carina; meso- and metasternal processes subequal in length, or metasternal process longer than mesosternal process (Fig. 194.22); habitus as in Fig. 198.22 .............................................. *Hoplandria*

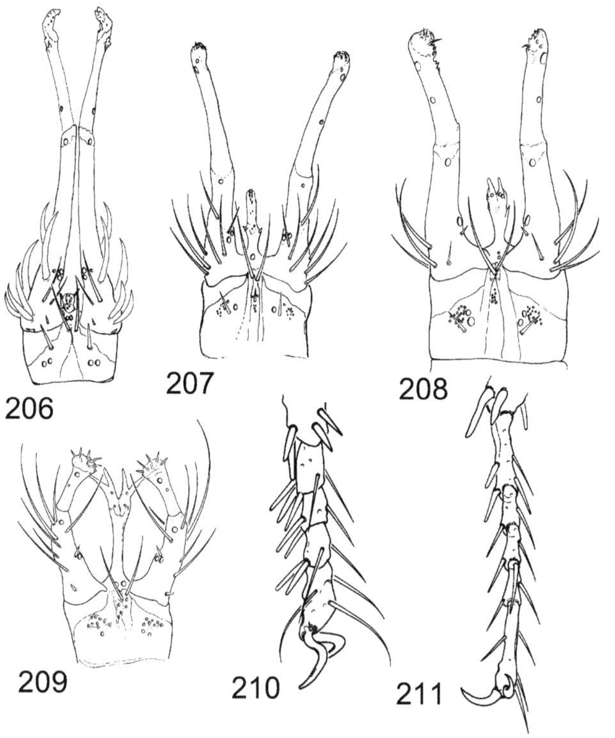


FIGURES 206.22-211.22. Fig. 206.22. *Myllaena gracilicornis* Fairmaire and Brisout, labium, ventral. Fig. 207.22. *Silusa rubra* Erichson, labium, ventral. Fig. 208.22. *Diestota rufipennis* (Casey), labium, ventral. Fig. 209.22. *Autalia puncticollis* Sharp, labium, ventral. Fig. 210.22. *Diglotta mersa* (Haliday), front tarsus. Fig. 211.22. *Aleochara sulcicollis* Mannerheim, hind tarsus, with "normal" tarsal claws.

3(1). Very small, 2 mm or less long; male tergite VIII without medial longitudinal carina or cariniform knob ................................................................ *Nosora*

— Larger, at least 3 mm long, male tergite VIII with medial longitudinal carina or cariniform knob ........ 4

4(3). Meso- and metasternal processes contiguous, or mesosternal process overlapping shorter metasternal process (Fig. 195.22); known from the eastern half of North America and the southwestern United States .............................. *Platandria*

— Meso- and metasternal processes separated by at least a short distance, isthmus present between processes (Fig. 197.22); known only from California ........................................................ *Tetrallus*

### D. Key to the Nearctic genera of Aleocharinae with 4-4-4 tarsi (Includes Hypocyphtini, Oxypodini: Meoticina in part, and some species of *Diaulota* [Liparocephalini])

1. Antenna with 10 antennomeres, apical antennomeres enlarged to form a loose to distinct club (Figs. 199-200.22); hind coxa with lamella covering base of femur (similar to Fig. 167.22) (tribe Hypocyphtini) .................................................. 2

— Antenna with 11 antennomeres, more or less filiform and elongate, apical antennomeres not forming a club; hind coxa without lamella covering base of femur ............................................................... 4

2(1). Head very broad and short, almost 2 times as wide as long; head strongly deflexed, hypognathous ................................................................ *Cypha*

— Head not more than 1.6 times as wide as long, not or only slightly deflexed ................................. 3

3(2). Body more or less parallel-sided (Fig. 203.22); pronotal hypomeron at least narrowly visible in lateral view ................................................. *Oligota*

— Body broadly ovoid, robust (Fig. 204.22); pronotal hypomeron not visible in lateral view... *Holobus*

4(1). Tarsal claws normal (long, slender, slightly curved); apex of mentum with broad and deep U-shaped emargination (Fig. 225.22); abdomen broadly oval, distinctly broader at segments V-VI than at base; elytra short, pronotum more than 1.3 times as long as elytra; in intertidal zones of rocky Pacific coast seashores (species with 4-4-5 tarsi will key out in Key E) ...................................... *Diaulota* (in part)

— Tarsal claws distinctly scythe-shaped (Fig. 201.22), not exactly fitting description above; not usually found in intertidal zones of Pacific coast seashores ............................................................ 5

5(4). Pronotal pubescence pattern with all setae directed anteriorly (Fig. 202.22); head and pronotum distinctly narrower than elytra (about 2/3 as wide); pronotum slightly wider than long; transverse impressions of abdominal tergites III-V without reticulation or distinct punctation; habitus as in Fig. 205.22 .................................................. *Bamona*

— Pronotal pubescence pattern with all setae directed laterally or posteriorly; head and pronotum only slightly narrower than elytra (about 9/10 as wide); pronotum wider than long; transverse impressions of abdominal tergites III-V reticulate and/or distinctly punctate ............................................... 6

6(5). Abdominal tergites III-VI with deep transverse depressions, impressions coarsely punctured...... ..................................................... *Leptobamona*

— Abdominal tergites III-V with transverse impressions, impressions reticulate but without coarse punctures ...................................................... *Alisalia*

### E. Key to the Nearctic genera of Aleocharinae with 4-4-5 tarsi (Includes Autaliini, Diglottini, Homalotini, Liparocephalini [except some species of *Diaulota*], Myllaenini, Philotermitini, Placusini, and intertidal genera of Athetini [*Thinusa*] and Falagriini [*Bryobiota*])

1. Labial palp long, slender, strongly and distinctly stylate (Figs. 206-208.22) ................................ 2

— Labial palp short or long, not distinctly stylate (Figs. 209.22, 241.22) .............................................. 6

2(1). Tarsal claws very slender, scythe-like (Fig. 210.22); body form distinctive (Fig. 250.22) with large, broadly rounded head, very small eyes, head as wide as or wider than pronotum, pronotum strongly narrowed behind; elytra much shorter

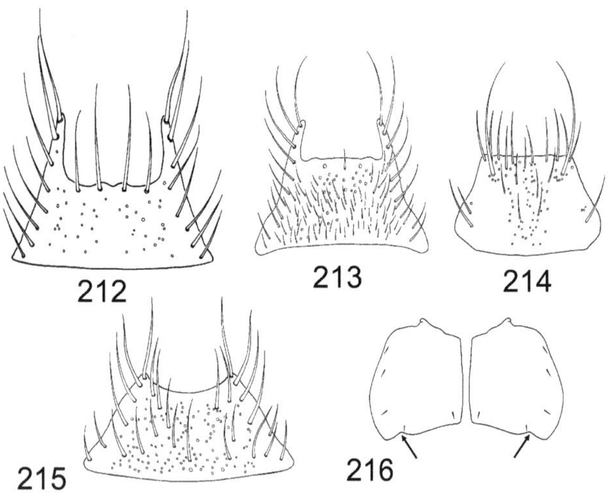


FIGURES 212.22-216.22. Fig. 212.22. *Myllaena* sp., mentum, ventral. Fig. 213.22. *Bryothinusa papuensis* Haghebaert, mentum, ventral. Fig. 214.22. *Pleurotobia trimaculata* (Erichson), mentum, ventral. Fig. 215.22. *Amblopusa alaskana* Ahn, mentum, ventral. Fig. 216.22. *Myllaena infuscata* Kraatz, elytra, showing sinuate posterior margin (arrows).

than pronotum (North American species only); metasternum greatly reduced; abdomen broadly oval in dorsal outline, at widest point wider than elytra; in intertidal zone of beaches (tribe Diglottini) .......... *Diglotta*

— Tarsal claws normally formed, not scythe-like (similar to Fig. 211.22); body form not exactly as above; most not normally found in intertidal zone of beaches (except *Bryothinusa*) .......... 3

3(2). Mentum with anterolateral margins produced into moderate to strong spiniform processes (Figs. 212-213.22) (tribe Myllaenini) .......... 4

— Mentum various, but anterolateral margins not produced as spiniform processes (similar to Figs. 214-215.22) (tribe Homalotini, in part) .......... 5

4(3). Body parallel-sided, strongly flattened (Fig. 251.22); head prognathous; pronotal hypomeron visible in lateral view; posterolateral margin of elytron not sinuate; in intertidal zone of beaches .......... *Bryothinusa*

— Body fusiform, with broad, dorsally broadly rounded pronotum and abdomen tapering from base to relatively acute apex (Fig. 20.22); head hypognathous; pronotal hypomeron not visible in lateral view; posterolateral margin of elytron sinuate (Fig. 216.22); usually riparian, along margins of freshwater streams, ponds, lakes and bogs .......... *Myllaena*

5(3). Mesocoxae narrowly separated by narrow meso- and metasternal processes (Fig. 217.22) (subtribe Silusina) .......... *Silusa*

— Mesocoxae broadly separated by broad meso- and metasternal processes (Fig. 218.22) (subtribe Diestotina) .......... *Diestota*

6(1). Head with narrow neck, less than 1/3 width of head; ligula distinctive, apex dichotomously branched into 4 divergent, acutely pointed processes (Fig. 209.22); mesocoxal cavities not margined behind; pronotum with 4 subbasal foveae; each elytron with 2 basal foveae; each of abdominal tergites III-V with deep transverse impression containing a large medial carina and smaller carinae laterally; habitus as in Fig. 252.22 (tribe Autaliini) .......... *Autalia*

— Without above combination of characters .......... 7

7(6). Combination of body, legs, and antennae black to piceous, abdomen brownish at base; body with coarse sculpture and inconspicuous pubescence; head hypognathous, narrower than elytra; pronotum very transverse, 2 times as wide as long, apex narrower than base and broadly emarginate around head, sides rounded, hind angles obtusely rounded, posterior margin slightly bisinuate (Fig. 253.22); pronotal hypomeron not visible in lateral view; mesocoxae narrowly separated; mesosternal process long, obtusely pointed, extending 3/4 length of mesocoxae and overlapping apex of much shorter and broader metasternal process (Fig. 219.22); hind tarsomere 1 subequal in length to 2+3 together; known only from Manitoba .......... *Cyphea*

— Not exactly fitting description above .......... 8

8(7). Associated with intertidal zones of Pacific seashores; elytra short, usually shorter than pronotum; hind wings very reduced or absent; metasternum very short; mesocoxae contiguous, mesocoxal cavities not margined posteriorly .......... 9

— Not associated with intertidal zones of Pacific seashores; other characters various .......... 14

9(8). Anterior and middle tibiae distinctly spinose (Fig. 220.22) (tribe Athetini, in part) .......... *Thinusa*

— Anterior and middle tibiae without spines .......... 10

10(9). Pronotum and base of head longitudinally impressed; head with a distinct neck; pronotum distinctly narrowed behind; ligula bifid apically; prementum with 2 medial setae (only visible in slide preparations); anterior margin of sternum IV with a distinct gland opening (only visible in slide preparations) (tribe Falagriini, in part) .......... *Bryobiota*

— Head and pronotum not longitudinally impressed; head without neck; pronotum not narrowed behind middle; ligula entire; prementum with 1 or no medial seta (only visible in slide preparations); anterior margin of sternum IV without gland opening (tribe Liparocephalini, major part) .......... 11

11(10). Abdominal tergites without deep transverse impressions; tarsi with large spatulate setae (Fig. 221.22); habitus as in Fig. 254.22 .......... *Liparocephalus*

— Abdominal tergites III-V or III-VI with deep transverse impressions; tarsi without spatulate setae .... 12

12(11). Lacinia distinctive: elongate, with a large apical spine, a row of four short spines more medially, and a large spine basal to these (Fig. 2-2-2.22); anterior margins of abdominal tergites III-VI with broad and deep V-shaped emarginations; habitus as in Fig. 255.22 .......... *Amblopusa*

— Lacinia with about 10 subequal spines in a row in apical third (Fig. 223.22); anterior margins of abdominal tergites III-VI straight, not emarginate .......... 13

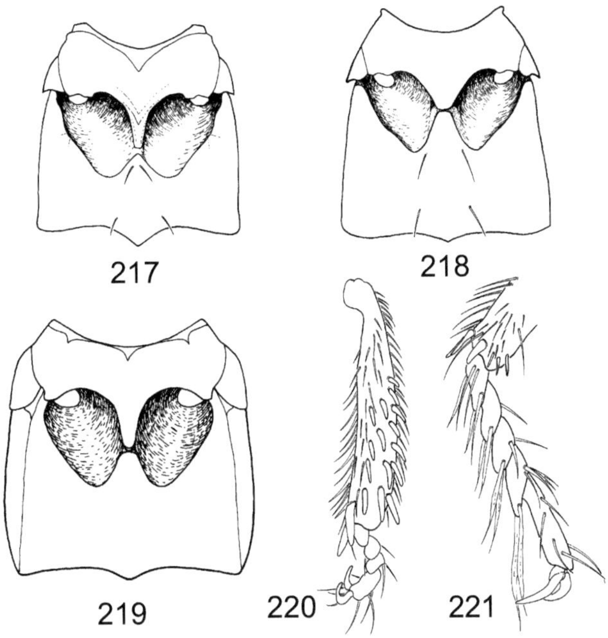


FIGURES 217.22-221.22. Fig. 217.22. *Silusa rubra* Erichson, meso-metasternum. Fig. 218.22. *Diestota rufipennis* (Casey), meso-metasternum. Fig. 219.22. *Cyphea wallisi* Fenyes, meso-metasternum. Fig. 220.22. *Thinusa maritima* (Casey), front tibia and tarsus. Fig. 221.22. *Liparocephalus cordicollis* LeConte, hind tarsus.

13(12). Mentum triangular, apex with a very deep and acute V-shaped emargination (Fig. 224.22); abdomen parallel-sided, not distinctly broader at segments V-VI than at base; pronotum less than 1.3 times elytra length .......................................... *Paramblopusa*

— Mentum trapezoidal, apex with broad and deep U-shaped emargination (Fig. 225.22); abdomen broadly oval, distinctly broader at segments V-VI than at base; pronotum more than 1.3 times as long as elytra (species with 4-4-4 tarsi will key out in Key D) ............................. *Diaulota* (in part)

14(8). Pronotum very broadly transverse and with a distinctive pubescence pattern: microsetae sparse or absent medially, increasing in number and length laterally, directed laterally and anterolaterally (Fig. 226.22); mandibles without a distinct patch of denticles in the ventral molar region; mesocoxae contiguous; mesocoxal cavities not margined behind (Fig. 232.22); habitus as in Fig. 256.22; associated with termites of the genus *Reticulitermes* (tribe Philotermitini) .............. ........................................................ *Philotermes*

— Not exactly fitting above description; pronotum transverse or not, but always with a different setal pattern; mandibles with or without patch of denticles in the ventral molar region; mesocoxal cavities margined behind; mesocoxae broadly separated, narrowly separated or contiguous; not found with termites ............................................................... 15

15(14). Ligula short and broadly rounded (Fig. 239.22); labial palp short, with 2 palpomeres; mandibles with "velvety patch" in the dorsal molar region modified into rows of large teeth (Fig. 227.22; visible only in slide preparations); mandibles without denticles in ventral molar region (visible only in slide preparations); mesocoxal cavities narrowly separated (tribe Placusini) ...................................... 16

— Ligula various, but not short and broadly rounded (Figs. 238.22, 240-241.22); labial palp with 2 or 3 palpomeres; "velvety patch" of dorsal molar region of mandibles made up of numerous microspinules, or microspinules reduced, or anterior row of microspinules modified into small denticles (Figs. 228-229.22; visible only in slide preparations), but not modified to rows of large teeth; mandibles with denticles in ventral molar region (Figs. 230-231.22; visible only in slide preparations); mesocoxal cavities broadly, moderately, or narrowly separated or contiguous (tribe Homalotini, in part) ......................................... 17

16(15). Head subquadrate, with distinct hind angles, with distinct neck about 1/2 head width; pronotum with anterolateral angles indistinct, broadly rounded, short sides almost straight behind them, posterior angles distinct, base strongly arcuate; abdominal tergite VII much longer than VI; habitus as in Fig. 257.22 ................................. *Euvira*

— Head transverse, without distinct hind angles or distinct neck; pronotum with anterolateral angles distinct, obtusely angulate, sides broadly rounded behind them; pronotal base evenly rounded or arcuate; abdominal tergite VII not much longer than VI ............................................................ *Placusa*

17(15). Mesocoxae very broadly separated by broad meso- and metasternal processes, separation almost or fully equal to width of mesocoxal cavity (Fig. 233.22); meso- and metasternal processes contiguous and meeting along a broad front between coxae (Fig. 233.22); 1 medial seta on prementum, or medial seta very reduced or absent (Fig. 240.22; visible only in slide preparations); lacinia distinctive, apex obliquely truncate and covered with numerous spinules, but setae and spines of inner surface reduced or absent (Fig. 236.22; visible only in slide preparations); associated with fresh fruiting bodies of basidiomycete fungi (subtribe Gyrophaenina)................................................. 18

— Mesocoxae moderately (Fig. 234.22), narrowly (Fig. 235.22) or not separated by meso-and metasternal processes, separation less than 1/2 width of mesocoxal cavity; meso- and metasternal processes contiguous or separated by an isthmus, but not meeting along a broad front between coxae; 2 medial setae on prementum, usually arranged one behind the other (Fig. 241.22; visible only in slide preparations); lacinia not as above (Fig. 237.22); associated with fruiting bodies of fungi or not ...... 22

18(17). Ligula split nearly to base into 2 lobes (Fig. 242.22); habitus as in Fig. 258.22 ............. *Agaricomorpha*

— Ligula entire (Fig. 240.22) .................................. 19

19(18). Body very compact and robust with very broad and short abdominal tergites, body broadly oval in dorsal outline, broadest at abdomen (Fig. 259.22); body with antero-posterior outline convex in lateral view; anterolateral margins of pronotum strongly deflexed; pronotal hypomeron not visible in lateral view; head deflexed and hypognathous, not fully visible from above .. *Encephalus*

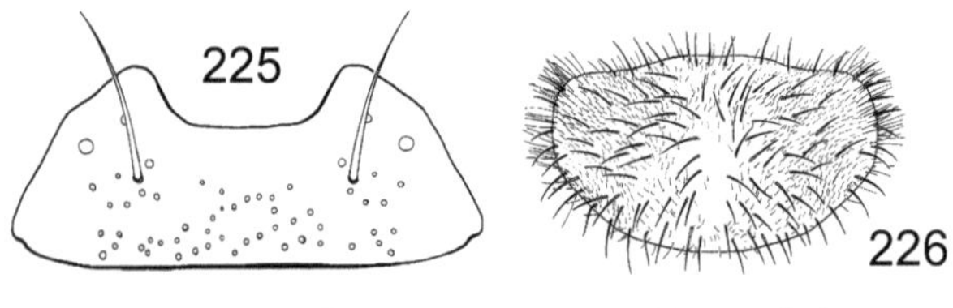


FIGURES 222.22-226.22. Fig. 222.22. *Amblopusa alaskana* Ahn, maxilla, ventral. Fig. 223.22. *Diaulota densissima* Casey, maxilla, ventral. Fig. 224.22. *Paramblopusa borealis* (Casey), mentum, ventral. Fig. 225.22. *Diaulota densissima* Casey, mentum, ventral. Fig. 226.22. *Philotermes pilosus* Kraatz, pronotum.

— Not exactly fitting above description .................. 20

20(19). Prothorax markedly transverse, 2 times or more as wide as long; setal patch on tergum X in form of a V-shaped row; pronotal hypomeron not visible in lateral view; antenna short, with antennomeres 4-10 markedly transverse, forming a loose parallel-sided club .................................... *Eumicrota*

— Prothorax less transverse, usually 1.2-1.7 times as wide as long; pronotal hypomeron nearly always visible in lateral view; setal patch of tergum X more or less quadrate; antenna various, but usually with 1 or more of antennomeres 4-10 quadrate or elongate ............................................ 21

21(20). Eyes extremely large, occupying most of lateral margins of head (Fig. 243.22); habitus as in Fig. 260.22 .......................................................... *Phanerota*

— Eyes moderate in size (Fig. 244.22)...... *Gyrophaena*

22(17). Ligula entire (undivided) at apex (Fig. 245.22) (subtribe Leptusina) ...................................... 23

— Ligula divided into 2 lobes at apex (Fig. 246.22).... 24

23(22). Pronotum with pubescence in midline directed anteriorly, and that on either side of midline directed laterally (similar to Fig. 308.22); ligula very long and slender, virtually as long as labial palpomeres 1 and 2 together (labial palpomeres 1 and 2 are partially fused and may appear as a single long palpomere); in rotting seaweed on beaches, in North America known only from Florida ........................ *Heterota*

— Pronotum with pubescence in midline directed posteriorly, and that on either side of midline directed lateroposteriorly or laterally (similar to Fig. 309.22); ligula variable, but seldom as long as labial palpomeres 1 and 2 together; not usually found in seaweed on beaches; widespread ................ .............................................. *Leptusa, Dianusa*

24(22). Body distinctly flattened; labial palp with 2 palpomeres (1 and 2 fused; Fig. 246.22) or with 3 indistinct palpomeres (1 and 2 partially fused); ligula about 1/2 as long as apparent first labial palpomere, apical 1/2 divided into 2 lobes (Fig. 246.22); basal tarsomere of hind tarsus subequal to second; patch of denticles in ventral molar region less dense and with fewer and smaller denticles (Fig. 231.22; visible only in slide preparations); under bark of dead trees (subtribe Homalotina) .................................................. 25

— Body generalized, not distinctly flattened; labial palp with 3 distinct palpomeres (Fig. 241.22); ligula elongate and slender, as long as or longer than labial palpomere 1, split into 2 lobes in apical 1/3-1/5 (Fig. 241.22); basal tarsomere of hind tarsus as long as tarsomeres 2 and 3 together; mandibles with very large patch of densely arranged denticles in ventral molar region (Fig. 230.22; visible only in slide preparations); associated with fresh and rotting mushrooms, especially leathery and fleshy polypores and persistent gilled mushrooms on logs (subtribe Bolitocharina) ................................... 29

25(24). Meso- and metasternal processes rounded or truncate apically; in North America known only from Florida ............................................... *Coenonica*

— Meso- and metasternal processes acutely pointed; widely distributed ......................................... 26

26(25). Tergite VIII produced in the middle as a long spine in both sexes ............................ *Anomognathus*

— Tergite VIII not produced in the middle as a single long spine (males may have other patterns of spines or processes as secondary sexual characters) ........................................................... 27

27(26). Head very large, elongate and flattened; temples at least 2 times as long as eyes; anterior pronotal angles distinctly angulate and acutely pointed, pronotal anterolateral angle forming an angle of less than 90 degrees; discal setae on tergum X arranged in a transverse row...... *Cephaloxynum*

— Head not unusually large, elongate or flattened, temples much less than 2 times as long as eyes; anterior pronotal angles less angulate and not acutely pointed, pronotal anterolateral angle forming an angle of 90 degrees or more; discal setae on tergum X arranged in a single transverse row or not............................................................. 28

28(27). Larger, more than 2.1 mm long; eyes large, distinctly longer than temples; pronotum appearing granulate because of uniform, coarse, close-meshed, raised reticulation; pronotum about 1.4 times as wide as long; discal setae on tergum X not arranged in a single transverse row; habitus as in Fig. 261.22 ......................................... *Homalota*

— Smaller, less than 1.2 mm long, much more slender; eyes small, distinctly shorter than temples; pronotum without raised reticulation; pronotum about 1.2 times as wide as long; discal setae on tergum X arranged in a transverse row; habitus as in Fig. 262.22 .................................... *Thecturota*

29(24). Pronotal hypomeron strongly inflexed throughout, not or only narrowly visible in lateral view .... 30

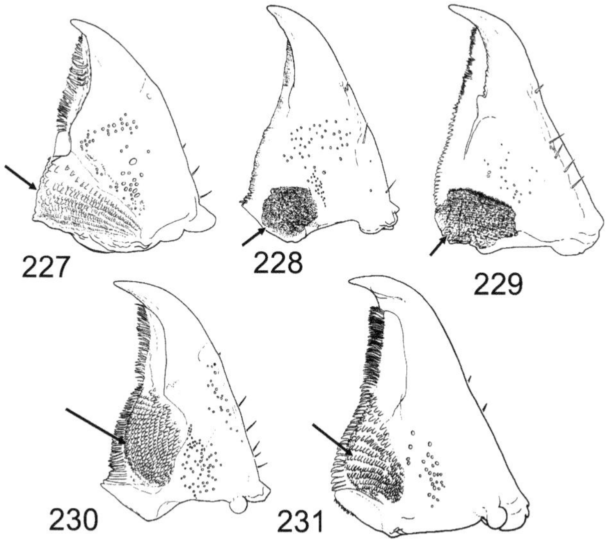


FIGURES 227.22-231.22. Fig. 227.22. *Placusa tachyporoides* (Waltl), mandible, dorsal, showing "velvety patch" (arrow) modified to transverse rows. Fig. 228.22. *Pleurotobia trimaculata* (Erichson), mandible, dorsal, showing "velvety patch" (arrow). Fig. 229.22. *Silusa rubra* Erichson, mandible, dorsal, showing "velvety patch" (arrow). Fig. 230.22. *Pleurotobia trimaculata* (Erichson), mandible, ventral, showing patch of spinules (arrow). Fig. 231.22. *Homalota* sp., mandible, ventral, showing patch of spinules (arrow).

— Pronotal hypomeron inflexed basally in some, but with at least anterior 1/2 broadly visible in lateral view ........................................................... 31

30(29). Antenna short and incrassate to apex, antennomeres 4-10 distinctly transverse; posterolateral angles of pronotum broadly rounded, not angulate; mesosternal process narrow, apex acutely pointed ............................................. *Silusida*

— Antenna longer and not noticeably incrassate to apex, antennomeres 4-10 subquadrate to slightly elongate; posterolateral angles of pronotum moderately to sharply angulate; mesosternal process broader, apex narrowly rounded, not pointed... ..................................................... *Phymatura*

31(29). Mesocoxae moderately widely separated by meso- and metasternal processes (Fig. 234.22), separation about 1/3 width of mesocoxal cavity; meso- and metasternal processes in contact, or virtually in contact, between coxae, isthmus between processes absent or very short (0.05 times length of mesocoxal cavity or less); both meso- and metacoxal processes broadly rounded or subtruncate; eyes large, about 1.5 times length of temples; habitus as in Fig. 263.22; known from the eastern half of North America ...... *Pleurotobia*

— Mesocoxae narrowly separated by meso- and metasternal processes (Fig. 235.22), separation 1/4 width of mesocoxal cavity or less; meso- and metasternal processes not in contact, isthmus between processes 0.1 or more times length of mesocoxal cavity; both meso- and metasternal processes narrowly rounded to acutely pointed; eyes smaller, less than 1.3 times as long as temples; known from northern and southwestern United States, Canada, and west coast of North America ...................................................... 32

32(31). Pronotum only slightly transverse, 1.3 times as wide as long or less; median lobe of aedeagus without distinct sclerotized internal plates; spermathecal duct sclerotized and complexly looped (Fig. 247.22); known only from the west coast of North America ............................................ *Stictalia*

— Pronotum moderately transverse, at least 1.4 times as wide as long; median lobe of aedeagus with distinct sclerotized internal plates; spermathecal duct not sclerotized (though it is very long and irregularly looped in *Hongophila*); known from northern and southwestern United States, Canada ................................................................. 33

33(32). At least some of antennomeres 5-10 distinctly elongate; male sternum VI without small projecting lobe medially on posterior margin; male sternum VII without clearly delimited medial patch of setose pores (though with broad band of pores basally); posterior triangular projection of male abdominal sternum VIII with distinct, but narrow, asetose margin (Fig. 248.22); known only from the northern United States and Canada east of the Rocky Mountains ......................................... *Neotobia*

— All of antennomeres 5-10 quadrate to transverse; male sternum VI with small projecting lobe medially on posterior margin (Fig. 249.22); male sternum VII with clearly delimited medial patch of setose pores (though with broad band of pores also present basally; Fig. 249.22); posterior triangular projection of male abdominal sternum VIII without distinct asetose margin; in North America known only from Arizona and New Mexico (also occurs in Mexico) ............................ *Hongophila*

F. Key to the Nearctic genera of Crematoxenini (4-5-5 tarsi) (Modified from Jacobson and Kistner 1992)

1. Eyes absent (Baja California, not North American). .................................................. *Pulicomorpha*

— Eyes present ..................................................... 2

2(1). Abdominal segment III petiolate and very long, nearly as long as rest of abdomen; head globose ...................................................... *Neobeyeria*

— Abdominal segment III petiolate but not nearly as long as remainder of abdomen; head not globose ................................................................. 3

3(2). Head distinctly longer than wide, about 1.3 times as long as wide; pronotum narrower in anterior 1/4; habitus as in Fig. 264.22 ............ *Probeyeria*

— Head quadrate or only slightly longer than wide, pronotum subequal in width throughout (except slightly wider around procoxal cavities) .. *Beyeria*

G. Key to the Nearctic genera of Falagriini (mostly 4-5-5 tarsi) (Modified from Hoebeke 1985, Ahn and Ashe 1995; *Bryobiota* will actually key out in Key E, but is also included here to provide a complete tribal key.)

1. Tarsi 4-4-5 ........................................... *Bryobiota*

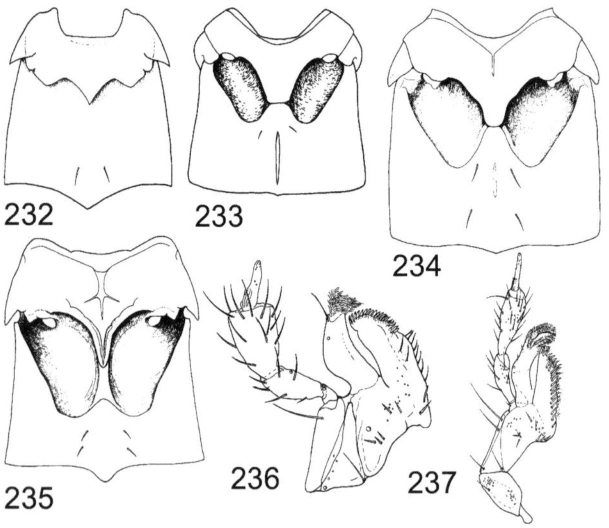


FIGURES 232.22-237.22. Fig. 232.22. *Philotermes pilosus* Kraatz, meso-metasternum. Fig. 233.22. *Gyrophaena simulans* Seevers, meso-metasternum. Fig. 234.22. *Pleurotobia trimaculata* (Erichson), meso-metasternum. Fig. 235.22. *Stictalia* sp., meso-metasternum. Fig. 236.22. *Gyrophaena affinis* Sahlberg, maxilla, ventral. Fig. 237.22. *Silusida marginella* Casey, maxilla, ventral.

— Tarsi 4-5-5 .......... 2

2(1). Abdominal tergite VIII with comb of minute denticles (Fig. 265.22) .......... 3
— Abdominal tergite VIII without comb of minute denticles .......... 9

3(2). Pronotal hypomeron not delimited from disk of pronotum by raised line; elytral punctation denser near scutellum (Fig. 266.22) .......... 4
— Pronotal hypomeron clearly delimited from disk of pronotum by raised line; elytral punctation denser near scutellum only in *Falagrioma* .......... 6

4(3). Mesosternum on a level ventral to metasternum; mesosternal process short, acute, not extending between mesocoxal cavities (Fig. 267.22); habitus as in Fig. 273.22 .......... *Aleodorus*
— Mesosternum on same level as metasternum; mesosternal process longer, distinctly extending between mesocoxal cavities .......... 5

5(4). Head large, subquadrate, usually with medial longitudinal impression on anterior 1/2 of vertex; habitus as in Fig. 274.22 .......... *Borboropora*
— Head smaller, rounded or transverse; vertex without medial longitudinal impression .......... *Lissagria*

6(3). Elytral punctation denser near scutellum (similar to Fig. 266.22) .......... *Falagrioma*
— Elytral punctation uniformly distributed, not noticeably denser near scutellum .......... 7

7(6). Scutellum bicarinate (Fig. 270.22) .......... *Falagria*
— Scutellum not bicarinate .......... 8

8(7). Scutellum coarsely granulose, with narrow smooth medial impression (Fig. 271.22) .......... *Leptagria*
— Scutellum without coarse granulations and without smooth medial impression .......... *Myrmecopora*

9(2). Mesospiracular peritremes small, ovoid to subtriangular, limited to small area around spiracle (similar to Fig. 269.22); pronotum faintly sulcate .......... *Falagriota*
— Mesospiracular peritremes large, quadrate or ovoid, occupying much of area behind prosternum (similar to Fig. 268.22); pronotum moderately to deeply sulcate .......... 10

10(9). Mesocoxal cavities margined posteriorly by fine bead; apical margin of abdominal tergite VIII densely fimbriate; elytron at base with small but distinct depression near humeral angle; scutellum without medial longitudinal carina; habitus as in Fig. 22.22 .......... *Cordalia*
— Mesocoxal cavities not margined posteriorly; apical margin of abdominal tergite VIII simple, not densely fimbriate; elytron without distinct impression near humeral angle; scutellum with medial longitudinal carina (Fig. 272.22), which may be complete or incomplete .......... *Myrmecocephalus*

## H. Key to the Nearctic genera of Sceptobiini (4-5-5 tarsi) (Modified from Danoff-Burg 1994)

1. Antenna short, not reaching beyond pronotum when extended posteriorly; antennomeres 8-10 strongly transverse; metafemur strongly compressed, blade-like; abdominal tergites IV-VI with 0-6 macrosetae on each lateral 1/2; greatest width of abdomen 0.9-1.3 times as wide as combined elytral width; habitus as in Fig. 275.22 .......... *Dinardilla*
— Antenna longer, reaching beyond pronotum when extended posteriorly; antennomeres 8-10 subquadrate or elongate; metafemur not, or only slightly, compressed, not blade-like; abdominal tergites IV-VI with 7-18 macrosetae in each lateral 1/2; greatest width of abdomen 1.2-2.0 times as wide as combined elytral width; habitus as in Fig. 276.22 .......... *Sceptobius*

## I. Key to the Nearctic genera of Lomechusini (4-5-5 tarsi) (Modified from Seevers 1978)

1. Head and prothorax with smooth intervals between large setigerous impressions; abdomen narrowed basally into weak petiole .......... *Ecitocala*
— Sculpture of head and prothorax different; abdomen not significantly narrowed basally .......... 2

2(1). Pronotum with a very deep medial longitudinal sulcus; habitus as in Fig. 277.22 .......... *Ecitonidia*
— Pronotum with, at most, a broad and shallow medial sulcus or groove, usually without pronotal sulcus .......... 3

3(2). Head and pronotum with prominent carinae; habitus as in Fig. 278.22 .......... *Ecitoxenidia*
— Head and pronotum without prominent carinae .... 4

4(3). Abdomen with prominent golden-yellow tufts of trichomes laterally on segments III-V; pronotum broadly transverse and strongly explanate and reflexed laterally (Fig. 23.22) .......... *Xenodusa*
— Abdomen without golden-yellow trichomes; pronotum not as above .......... 5

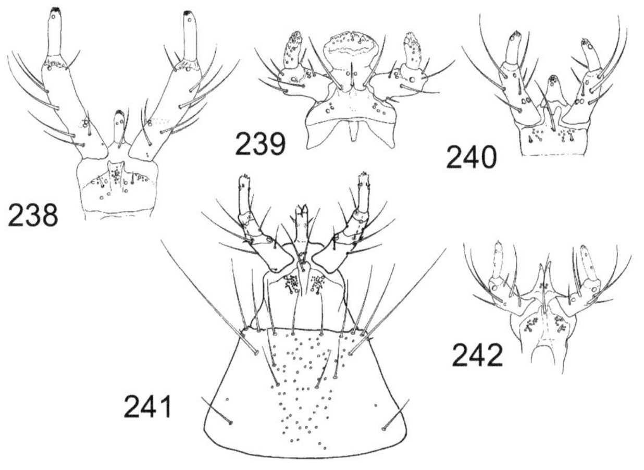


FIGURES 238.22-242.22. Fig. 238.22. *Liparocephalus cordicollis* LeConte, labium, ventral. Fig. 239.22. *Placusa tachyporoides* (Waltl), labium, ventral. Fig. 240.22. *Gyrophaena affinis* Sahlberg, labium, ventral. Fig. 241.22. *Pleurotobia trimaculata* (Erichson), labium and mentum, ventral. Fig. 242.22. *Agaricomorpha* sp., labium, ventral.

5(4). Eyes very large and coarsely faceted (Fig. 279.22), eyes occupying most of lateral margins of head .. ........................................................ *Tetradonia*
— Eyes normal sized and finely faceted ................ 6

6(5). Head with a distinct neck, neck 1/2 to 1/3 width of head ................................................................. 7
— Head without a neck, or with indistinct neck greater than 1/2 width of head .................................. 10

7(6). Antennomeres 1-4 elongate, 5-6 subquadrate and 7-10 transverse; very small, distinctive: no more than 1.7 mm long; body integument subglabrous and very shiny; body color light flavate or rufoflavate with darker markings on abdomen; pronotum slightly wider than long; pronotum without medial longitudinal furrow; abdominal terga and sterna with very long and very prominent black setae (setae longer than tergites); known species fully winged with elytra subequal to length of pronotum; habitus as in Fig. 280.22 ..... ............................................................. *Apalonia*
— All antennomeres elongate; not exactly fitting above description .......................................... 8

8(7). Disk of each elytron distinctly longitudinally impressed so that the dorsal surface of the elytra is flattened in comparison to the convex pronotum and forms a sharp angle with the slightly inflexed epipleural region (Fig. 281.22); abdominal tergite III with a broad smooth eminence, abdominal tergite VII with a small posteriorly-projecting medial spine; known only from California ........ *Trachyota*
— Not exactly fitting above description..................... 9

9(8). Larger, 2.5-5 mm long; head and pronotum moderately pubescent and at most slightly reticulate, shining; body color more or less uniformly dark brownish or reddish-brown; pronotum quadrate or slightly longer than wide; pronotum with broad and very shallow to prominent longitudinal medial furrow (indistinctly present only basally in some species); known species apterous with very short elytra (1/2 to 1/3 length of pronotum); habitus as in Fig. 282.22 ....................................... *Drusilla*
— Smaller, 1.8-2.2 mm long; head and pronotum with dense umbilicate punctation, surface not shining; body color rufo-flavate or slightly reddish-brown, with darker head in some and darker markings on abdomen in most; pronotum without any trace of medial longitudinal furrow; known species fully winged with elytra as long as, or slightly longer than, pronotum; habitus as in Fig. 283.22 ............................................................ *Meronera*

10(6). Body form distinctive (Fig. 284.22); 2.0 mm long; integument with very sparse pubescence and almost without microsculpture, strongly shining; pronotum about 1.25 times as wide as long; elytra about 1.3 times as long as pronotum; antennomeres 1-4 very slender, 5-6 slightly transverse, and 7-10 very transverse, antenna incrassate toward apex (antennomere 11 about 2 times as wide as 5), and antennomere 11 as long as 8, 9, and 10 combined ...... *Xesturida*
— Not exactly fitting combination of characters above ....................................................................... 11

11(10). Antenna compact and robust, with cylindrical antennomeres arranged so that pedicels of antennomeres not, or at most slightly, visible; sterna with vestiture of long bristle-like setae; habitus as in Fig. 285.22 .................. *Dinocoryna*
— Antenna not compact and robust, pedicels of antennomeres clearly visible; sterna without vestiture of long bristle-like setae ................ 12

12(11). Small, slender, more or less flattened, 2.5-3.0 mm long; head transverse, widest behind eyes; pronotum slightly to moderately transverse, lateral areas of disc of pronotum with a broad and shallow to very distinct submarginal longitudinal depression on each side; habitus as in Fig. 286.22; myrmecophilous, in nests of *Neivamyrmex* army ants ................................................ *Microdonia*
— Larger, more robust, 2.8-6 mm long (most greater than 3 mm); habitus as in Fig. 287.22; other features not exactly as above; free-living or myrmecophilous, but not usually found with *Neivamyrmex* ........................................ *Zyras*

### J. Key to the Nearctic genera of Aleocharinae with 4-5-5 tarsi but lacking other characteristics of Hoplandriini, Crematoxenini, Falagriini, Sceptobiini, and Lomechusini (Includes Oxypodini, major part; Athetini, major part)

Cautionary note: The tribe Athetini is by far the most difficult tribe in the Aleocharinae. It is clear that the current classification of the North American taxa is completely inadequate, and many currently recognized subtribes, genera, and subgenera cannot be clearly delimited or diagnosed. Furthermore, many genera are heterogeneous, currently including species that do not all appear to be closely related, which makes them even more difficult to diagnose. The tribe and its genera in North America require detailed study and complete revision in the context of current classification of European athetines. The key provided

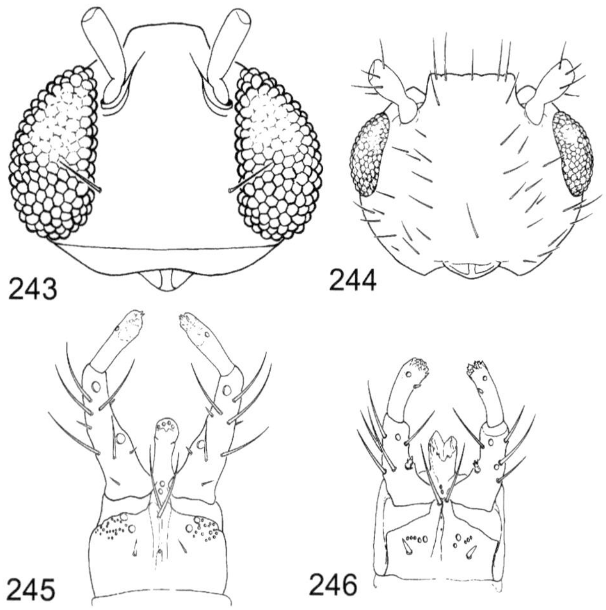


FIGURES 243.22-246.22. Fig. 243.22. *Phanerota carinata* (Say), head. Fig. 244.22. *Gyrophaena* sp., head. Fig. 245.22. *Leptusa eximia* Kraatz, labium, ventral. Fig. 246.22. *Homalota* sp., labium, ventral.

here is the best that can be done in the absence of such a revision. It should effectively identify taxa included in couplets 1-23. After that, many of the characters are more difficult to interpret and the genera are less clearly delimited. Some couplet leads end at groups of genera (or parts of genera) for whose separation unambiguous and uniform characters are not yet known. This does not imply that the clustered genera are regarded as synonyms, but only that they cannot be reliably separated at present. Genera that are better-defined, but variable, are keyed out in more than one couplet if necessary.

1. Male abdominal sternum VIII distinctive, apex broadly and deeply emarginate, lateral margins elongated into slender processes, each of which has a short row of spines, and sternum VIII with very long black setae near apical margin (Fig. 288.22); small, 1.1-1.2 mm long; pronotum broad, 1.5 times as wide as long; head, pronotum, and elytra densely asperate ................. *Strophogastra*

— Male abdominal sternum VIII not as above; not exactly fitting above description ........................ 2

2(1). Body form distinctive (Fig. 318.22), length 1.5 mm, uniformly light rufo-flavate; pronotum strongly transverse, 1.6 times as wide as long; pronotal hypomeron not visible in lateral view; antenna short, antennomeres 5-10 strongly transverse, 10 2 times as wide as long, antenna distinctly incrassate; abdominal tergite III slightly transversely impressed, tergites IV-V not impressed; male with abdominal tergite VIII bi-emarginate (tricuspid), and sternum VIII produced apically into a broad triangular lobe; female with apical margin of abdominal tergite VIII broadly emarginate medially; labial palp with 2 palpomeres (1 and 2 partially fused); ligula slightly elongate (about 1.5 times as long as basal width) and bifid at apex; known only from Florida ................. *Schistacme*

— Not exactly fitting above combination of characters ................................................................ 3

3(2). Head with well-defined narrow neck, neck not more than 1/3 as wide as head; head subquadrate; pronotum broadest in apical third, narrower basally; antenna short, antennomeres 5-10 transverse; abdominal tergites III-V deeply transversely impressed (tergite VI distinctly, but less deeply impressed), each with numerous coarse punctures; known only from California .... *Gnypetella*

— Neck, if present, greater than 1/2 head width; not exactly fitting the above description ............. 4

4(3). Abdominal tergites III-V with deep basal transverse impressions having series of parallel ridges and coarse punctures, often with medial longitudinal carina in impressions; elongate, slender, with slightly to moderately clavate abdomen (Fig. 19.22), abdominal segments V-VII broader and thicker than basal segments ..................................... *Tachyusa*

— Abdominal tergites III-V without parallel ridges or medial longitudinal carina; not exactly fitting above description ........................................................ 5

5(4). Postclypeus abruptly deflexed (Fig. 299.22); body covered with pile of fine light-colored pubescence ................................................ *Brachyusa*

— Postclypeus not abruptly deflexed; body pubescence various ............................................................. 6

6(5). Pronotal pubescence pattern distinctive: setae in midline directed posteriorly in apical 1/2 and anteriorly in basal 1/2 (Fig. 304.22); all antennomeres elongate; pronotum subquadrate, broadest near apical margins, narrower at base (Fig. 304.22). Body and legs more or less covered with dense vestiture of short light-colored pubescence; described species light reddish-yellow with or without darker head, elytral markings, and apical abdominal segments; known only from Texas ................ *Teliusa*

— Pronotal pubescence pattern different; not exactly fitting above description ...................................... 7

7(6). Mesocoxae moderately broadly separated (Figs. 290-291.22) ..................................................... 8

— Mesocoxae narrowly separated (similar to Fig. 235.22) or contiguous ................................................. 13

8(7). Maxillary palpomere 3 very large and strongly incrassate (Fig. 289.22), palpomere 4 small and cone-shaped; head and pronotum distinctly narrower than elytra, head narrowed behind eyes to neck about 1/2 head width; anterolateral margins of pronotum rounded and deeply depressed; pronotal hypomeron fully visible in lateral view; male antennomere 10 very large, at least 2 times as long as 9; habitus as in Fig. 319.22 ......................................... *Callicerus*

— Not exactly fitting the above description ............. 9

9(8). Antenna elongate, antennomeres 1-8 longer than broad, 9-10 subquadrate to slightly transverse. ........................................................... *Gnypeta*

— Antenna short to moderately long, antennomeres 5-10 transverse ........................................... 10

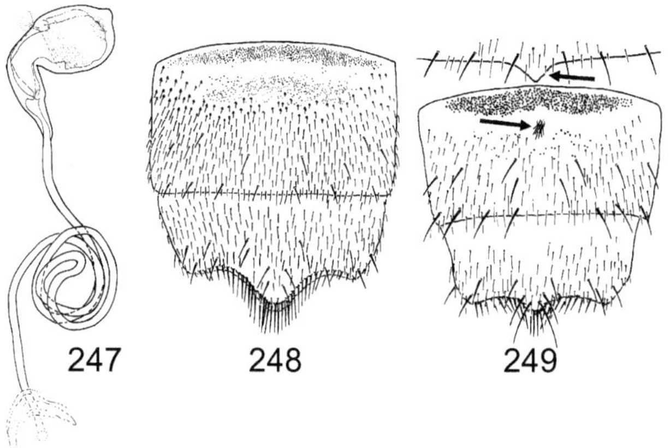


FIGURES 247.22-249.22. Fig. 247.22. *Stictalia californica* Casey, spermatheca. Fig. 248.22. *Neotobia alberta* Ashe, male abdominal sterna VII-VIII. Fig. 249.22. *Hongophila arizonica* Ashe, male abdominal sterna VI-VIII, showing lobe on posterior margin of tergite VI (arrow) and medial patch of setose pores on tergite VII (arrow).

10(9). Head and pronotum distinctly narrower than elytra (elytra at least 1.2 times as wide as pronotum); abdomen oval in outline, broadest at segment V and broader than elytra at widest point; body usually with long dense pubescence; habitus as in Fig. 320.22 .................................................. *Trichiusa*

— Head, pronotum, elytra and abdomen not as above; without long dense pubescence .................. 11

11(10). Labial palp with 2 distinct or 3 indistinct palpomeres (1 and 2 partly or completely fused), last labial palpomere distinctive: dilated along mesal margin and with numerous sensory structures on mesal margin (Fig. 293.22); metasternal process subequal in length to mesosternal process and distinctly shorter than isthmus; males with concentration of setigerous pores on frons of head ........................................................ *Thamiaraea*

— Labial palp with 3 palpomeres, unmodified; metasternal process distinctly shorter than mesosternal process, moderately broad isthmus present or absent; males without concentration of setigerous pores on frons of head ........... 12

12(11). Meso- and metasternal processes fully or very nearly contiguous, isthmus absent; mesosternal process extending 2/3-3/4 length of mesocoxae; apical 1/3 of ligula divided into 2 lobes; pronotum strongly convex in cross section; pronotal hypomeron narrowly and incompletely visible in lateral view ............................................. *Alaobia*

— Meso- and metasternal processes not contiguous, isthmus broad and subequal in length to metasternal process; mesosternal process extending to near middle of mesocoxae; ligula divided into 2 lobes nearly to the base; prothorax flatter, only slightly curved in cross section; pronotal hypomeron fully visible in lateral view ................................ *Earota*

13(7). Body form distinctive (Fig. 321.22): pronotum 1.2-1.3 times as wide as long, rather robust, widest in apical 1/2 and narrowest at base (base about 3/4 as wide as greatest width), pronotal hypomeron fully visible in lateral view; elytra shorter than pronotum, outer apical angles deeply sinuate; anterior and middle legs with rows of small but robust spines; head, pronotum and elytra finely granulose owing to raised fine-meshed reticulation; on sandy beaches of the Pacific coast ..... ...................................................... *Pontomalota*

— Body not as above; not usually found on beaches ...................................................................... 14

14(13). Body very elongate, parallel-sided and subcylindrical (Fig. 322.22); antennomeres 4-10 strongly transverse, antenna very strongly incrassate to the apex; front and middle tibia with rows of spines; in the receptacles of ripe native figs; in North America known only from Florida.. ........................................................... *Charoxus*

— Not exactly fitting above description .............. 15

15(14). Pronotum exceptionally broad and shield-like, at least 1.6 times as wide as long ....................... 16

— Pronotum not more than 1.5 times as wide as long ....................................................................... 17

16(15). Pronotum with all setae directed posteriorly or posterolaterally (similar to Fig. 309.22); pronotum about 1.6 times as wide as long, basal angles distinct and angulate, male with pronotal disk strongly longitudinally impressed, impression about 0.6 times as wide as pronotum and more deeply impressed near base (Fig. 305.22); some females with shallow basal impression on disk of pronotum; medium-sized, 3.1-3.5 mm long; inquilines with ants .................................. *Goniusa*

— Pronotum with setae in midline directed anteriorly (Fig. 306.22); pronotum about 1.7-1.8 times as wide as long, basal angles broadly rounded to slightly angulate, sides broadly and evenly arcuate (Fig. 306.22); pronotum without impressions in either sex; larger, 3.3-5.0 mm long; usually in leaf litter, moss, log/leaf litter and other forest floor litter habitats, not with ants ....................... *Lypoglossa*

17(15). Pronotal hypomeron not at all visible in lateral view (Fig. 295.22) ............. *Acrotona, Strigota, Mocyta*

— Pronotal hypomeron narrowly (Figs. 296-297.22) to broadly (similar to Fig. 298.22) visible in lateral view ................................................................ 18

18(17). Pronotal hypomeron fully and completely visible in lateral view (Fig. 298.22) ................................ 19

— Pronotal hypomeron narrowly to moderately visible in lateral view (at least part of hypomeron inflexed and not visible in lateral view) (Figs. 296-297.22) ...................................................................... 31

19(18). Eyes obviously and distinctly shorter than (Fig. 301.22) or much shorter than (Fig. 300.22) length of temples behind eyes ................................ 20

— Eyes subequal to (Fig. 302.22) or distinctly longer than (Fig. 303.22) length of temples behind eyes ...................................................................... 39

20(19). Pronotal setal pattern with all microsetae directed from lateral margins toward the middle (Fig. 307.22); male tergite III with lateroapical angles prolonged as moderate to long spines; male tergite VII with a weak to moderately strong medial carina ............................................... *Seeversiella*

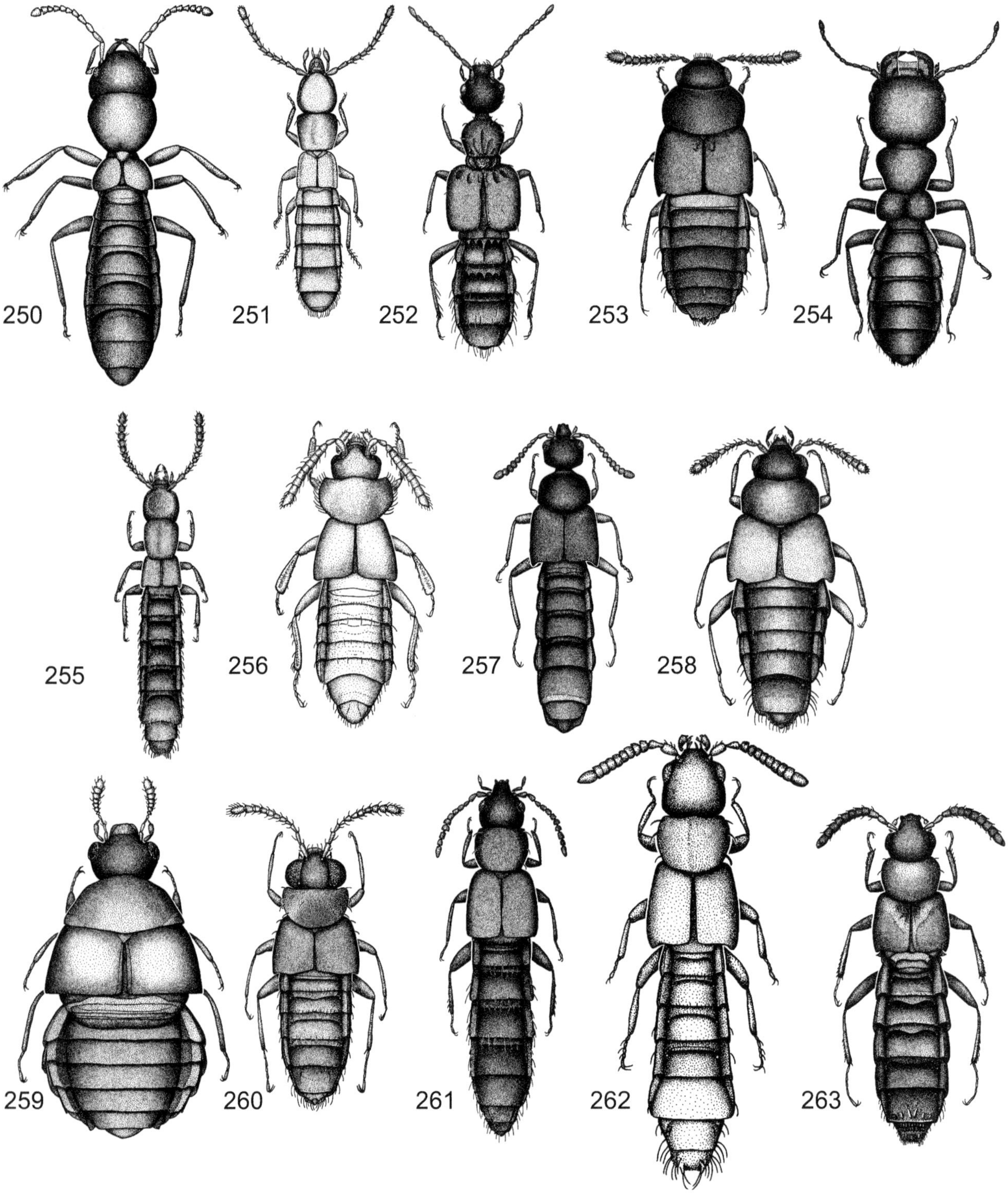


FIGURES 250.22-263.22. Fig. 250.22. *Diglotta pacifica* Fenyes, length 2.1 mm. Fig. 251.22. *Bryothinusa catalinae* Casey, length 2.2 mm. Fig. 252.22. *Autalia rivularis* (Gravenhorst), length 2.0 mm. Fig. 253.22. *Cyphea wallisi* Fenyes, length 1.5 mm. Fig. 254.22. *Liparocephalus cordicollis* LeConte, length 4.0 mm. Fig. 255.22. *Amblopusa alaskana* Ahn, length 3.3 mm. Fig. 256.22. *Philotermes pilosus* Kraatz, length 3.2 mm. Fig. 257.22. *Euvira diazbatresae* Ashe and Kistner, length 2.7 mm. Fig. 258.22. *Agaricomorpha apacheana* Seevers, length 1.9 mm. Fig. 259.22. *Encephalus complicans* Stephens, length 2.2 mm. Fig. 260.22. *Phanerota dissimilis* (Say), length 2.3 mm. Fig. 261.22. *Homalota plana* (Gyllenhal), length 3.0 mm. Fig. 262.22. *Thecturota histrio* Casey, length 1.5 mm. Fig. 263.22. *Pleurotobia trimaculata* (Erichson), length 3.8 mm.

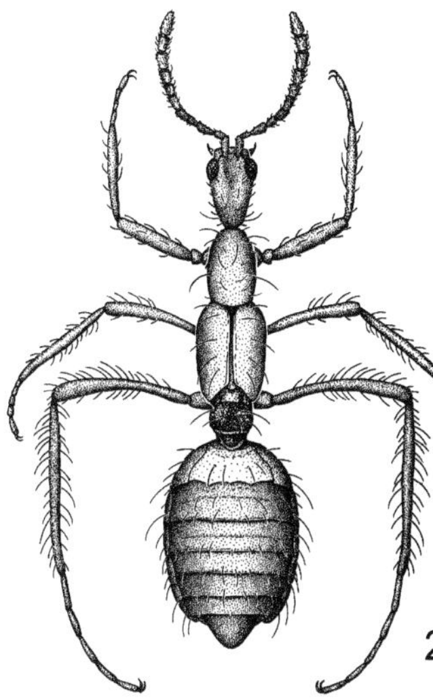
264

FIGURE. 264.22. *Probeyeria pulex* (Sanderson), length 3.2 mm.

— Pronotal setal pattern different; male characters various, but never with lateroapical margins of tergite III produced as spinose processes ......... 21

21(20). Pronotal pubescence pattern with setae in midline directed anteriorly in at least apical 1/2 to 3/4 ........................................................................ 22

— Pronotal pubescence pattern with setae in midline directed posteriorly or posterolaterally ......... 26

22(21). Antenna short with antennomeres 5-10 distinctly transverse; eyes about 1/3 length of temples behind eyes; pronotum slightly wider than long, widest in apical third and slightly narrowed basally, disc with a distinct but shallow subbasal impression; mesosternal process acute, extending to near middle of mesocoxae, metasternal process long and very spiniform, extending to near middle of mesocoxae, attaining or nearly attaining mesocoxal process; abdominal tergite VII much shorter than VI (only slightly more than 1/2 as long); known only from Florida ........ *Asthenesita*

— Antenna relatively long, antennomeres 5-10 slightly elongate to quadrate, or 8-10 moderately to slightly transverse (the genus *Valenusa*, the type of which is missing antennomeres 7-11, is keyed out here) ........................................................ 23

23(22). Long and slender, 3.2 mm long, uniformly rufotestaceous in color; pronotum narrow, 1.2 times as wide as long; elytra short, less than 1.1 times as long as pronotum; mesosternal process extending 0.6 times length of mesocoxae, metasternal process very short, extending less than 0.07 times length of mesocoxae, isthmus very long (0.8 times as long as metasternal process) and subcarinate; known only from California *Valenusa*

— Not exactly fitting description above ................... 24

24(23). Small, flattened, and parallel-sided, with large prognathous head (as large as or larger than pronotum; Fig. 323.22); eyes about 0.7 times as long as temples; antennomere 3 shorter than 2, antennomere 3 distinctly club-shaped, with narrow base and bulbous apical 1/2; abdominal segment VII much longer than VI (about 1.4 times as long) ............................................ *Hydrosmectina*

— Not exactly fitting above description ................... 25

25(24). Ligula divided to base into 2 divergent lobes; mesocoxae contiguous; apical process of median lobe of aedeagus not acute or prominent; condylite of paramere not enlarged ................... *Boreophilia*

— Ligula divided into 2 lobes only in apical 1/2; mesocoxae narrowly separated; apical process of median lobe of aedeagus acute and prominent; condylite of paramere greatly enlarged in most (see Lohse *et al.* 1990, figures 193, 198), as large as paramerite in some ........................ *Boreostiba*

26(21). Tergite VII very long, at least 1.6 times as long as VI; eyes very small, less than 0.3 times as long as temples; intercoxal processes in different planes (apex of mesosternal process distinctly ventral to unmargined metasternal process); mesocoxal cavities not margined behind; body very light rufoflavate, uniformly covered with moderately dense pile of microsetae; small, slender, and parallel-sided, about 1.5 mm long; known from Rhode Island ..................................................... *Sipaliella*

— Not exactly fitting the above description ......... 27

27(26). Elytra distinctly longer than pronotum .............. 28

— Elytra subequal to, or shorter than, pronotum ... 29

28(27). Intercoxal processes subequal in length, isthmus extremely short or nearly absent; color light rufoflavate; pronotum with strong, distinct reticulate microsculpture, pronotum only slightly wider than long (about 1.05 times as wide as long), widest in apical third, with bisinuate sides converging slightly to obsolete basal angles and arcuate basal margin; abdominal tergite VI longer than VII; known only from California .................. *Gaenima*

— Intercoxal processes not subequal in length, isthmus present; darkly colored, dark brown to piceous; pronotum without strong, distinct microsculpture; pronotum distinctly transverse, at least 1.3 times as wide as long, widest in apical third, sides rounded; abdominal tergite VII longer than VI; known from California (and Europe) ....... ......................................................... *Parameotica*

29(27). Eyes moderately large, only slightly shorter than temples (0.7-0.9 times as long); infraorbital carina present in basal 1/3 but absent below eyes; elytra subequal to length of pronotum; pronotum broadly oval, 1.2-1.3 times as wide as long ....... .......................................... *Crephalia, Ousipalia*

— Eyes small to very small, no more than 1/2 as long as temples; infraorbital carina completely absent in most (faintly present in basal 1/5 of head in some); elytra distinctly shorter than pronotum; pronotum subquadrate to quadrate, not more than 1.1 times as wide as long ................................................ 30

30(29). Body black or very dark brown; median lobe of aedeagus with apex divided into 2 elongate processes; known only from arctic and subarctic zones of North America .............. *Pseudousipalia*

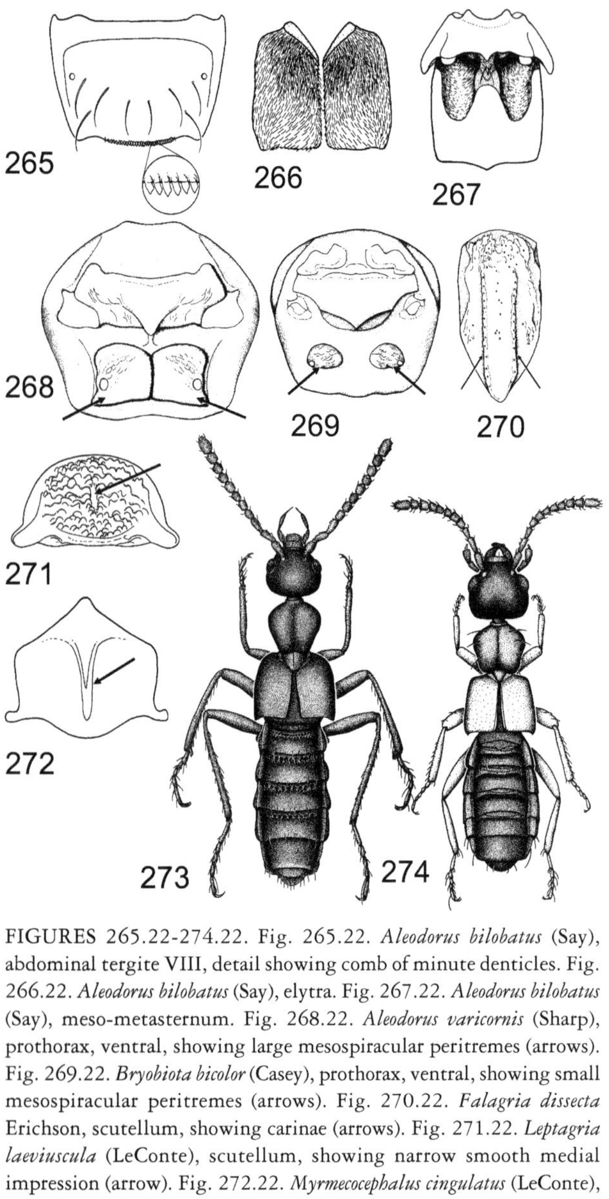


FIGURES 265.22-274.22. Fig. 265.22. *Aleodorus bilobatus* (Say), abdominal tergite VIII, detail showing comb of minute denticles. Fig. 266.22. *Aleodorus bilobatus* (Say), elytra. Fig. 267.22. *Aleodorus bilobatus* (Say), meso-metasternum. Fig. 268.22. *Aleodorus varicornis* (Sharp), prothorax, ventral, showing large mesospiracular peritremes (arrows). Fig. 269.22. *Bryobiota bicolor* (Casey), prothorax, ventral, showing small mesospiracular peritremes (arrows). Fig. 270.22. *Falagria dissecta* Erichson, scutellum, showing carinae (arrows). Fig. 271.22. *Leptagria laeviuscula* (LeConte), scutellum, showing narrow smooth medial impression (arrow). Fig. 272.22. *Myrmecocephalus cingulatus* (LeConte), scutellum, showing medial carina (arrow). Fig. 273.22. *Aleodorus bilobatus* (Say), length 3.6 mm. Fig. 274.22. *Borboropora quadriceps* (LeConte), length 3.3 mm.

— Body usually medium brown, light brown or flavate; median lobe of aedeagus entire, apex not divided into 2 elongate processes; widely distributed, but not known from arctic and subarctic zones of North America ........................................ *Geostiba*

31(18). Antenna with at least some antennomeres distinctly elongate or quadrate; pronotal pubescence with setae in midline directed anteriorly (Fig. 308.22); lateral margins of pronotum and elytra, and middle and hind tibiae with distinct macrosetae (bristles); pronotum usually with moderately dense asperate punctation (a few species will not key out here because they have fully exposed pronotal hypomera) ........................... *Dimetrota*

— Antenna with antennomeres 4-10 distinctly transverse (antennomere 5 or 10 may be very weakly transverse or subquadrate in *Synaptina* and *Iotota*); other characters various ........................................... 32

32(31). Ligula short, broad, and entire (Fig. 294.22); eyes very small, length of eyes less than length of temples; head shape distinctive, either broadly oval or widened basally so that greatest width of head is behind eyes; mesocoxal cavities not margined (subgenus *Amischa*) or margined (subgenus *Colposura*) ............................................ *Amischa*

— Ligula divided into 2 lobes; other features various, but not occurring in same combination as above ...................................................................... 33

33(32). Pronotum with setae in midline directed anteriorly (similar to Fig. 308.22); pronotum usually with asperate punctation .......................................... ............................ *Datomicra, Pseudota* (in part)

— Pronotum with all setae directed posteriorly or posterolaterally; pronotum usually without asperate punctation ...................................... 34

34(33). Infraorbital carina incomplete (weakly present in basal 1/2 at most); eyes small, length less than or subequal to that of temples; variously colored, usually medium to dark brown ....................... 35

— Infraorbital carina complete to maxillary insertion; eyes distinctly longer than temples; mostly light-colored with light brown or flavate bodies (except for some *Pseudota* and *Fusalia* that are primarily medium to dark brown), some with darker head and elytral and abdominal markings ...... 36

35(34). Eyes small, length subequal to or slightly less than temples (eyes of known species not less than 0.8 times length of temples) (Seevers 1978 mentioned elongate carinules on tergite VIII as a distinctive feature of this genus, but it is not clear what he meant) .................................. *Synaptina*

— Eyes very small, length 0.5-0.7 times that of temples .................................................................. *Iotota*

36(34). Pronotum with all setae directed posteriorly; light brown or flavate, some with darker head, elytral, and abdominal markings ................................ 37

— Pronotum with setae directed posterolaterally (Fig. 309.22); body color various .............. *Anatheta* (Type species, *A. planulicollis* Casey, only; *A. curata* Casey is not congeneric and does not seem to fit any described genus), *Canasota* (in part), *Pseudota* (in part)

37(36). Relatively large, 3.5 mm long; body ground color medium brown with slightly darker abdomen and elytral apices .......................................... *Fusalia*

— Smaller, less than 3.0 mm long; body ground color light brown to flavate; with or without darker markings on body ................................................. 38

38(37). Metasternal process distinctly longer than mesosternal process; body color pattern distinctive: ground color pale flavate, with apical 1/3 of elytra and abdominal tergites VI and VIII dark piceous or black; male secondary sexual characters distinctive: tergite VIII with strong spine on each side, emarginate between spines, emargination with broad lobe medially, apex of lobe

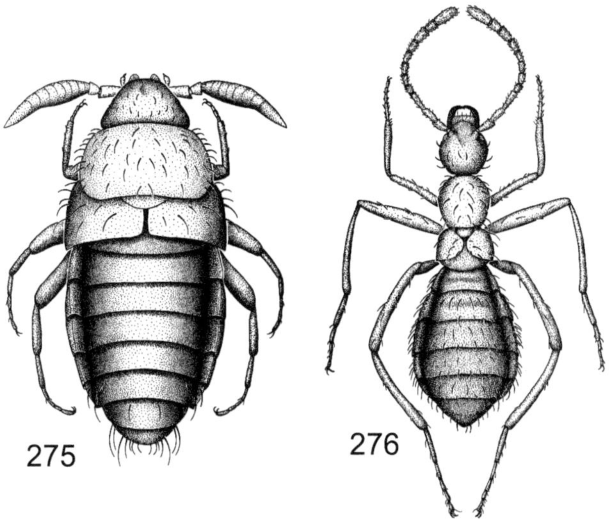

FIGURES 275.22-276.22. Fig. 275.22. *Dinardilla liometopi* Wasmann, length 3.0 mm. Fig. 276.22. *Sceptobius schmidti* (Wasmann), length 2.7 mm.

subtruncate and emarginate medially (Fig. 312.22) ................................................................ *Sableta*

— Metasternal process not longer than mesosternal process; body color and secondary sexual characteristics not as above ........ *Pancota, Canasota* (in part)

39(19). Mesocoxal cavities not margined behind; anterior and middle tibia with rows of small spines; head, pronotum and elytra with prominent, very dense, reticulate microsculpture, appearing granulose; on seashores of Pacific coast ........... *Tarphiota*

— Mesocoxal cavities margined; anterior and middle tibia without spines; with or without dense granulose microsculpture (usually without) .. 40

40(39). Antenna distinctive: antennomeres 1 and 2 slender at base and swollen distally, 1 moderately compressed, 3 subquadrate, and 4-10 transverse and incrassate (Fig. 313.22); abdomen enlarged apically, widest and thickest at segment VII (about 1.5 times as wide there as at base); abdominal segment VII about 2 times as long as segment VI; habitus as in Fig. 324.22 ............................................ *Clusiota*

— Not exactly fitting above description ................... 41

41(40). Empodial bristles much longer than tarsal claws (Fig. 315.22); pronotum narrow, not more than 1.1 times as wide as long .................................. *Aloconota*

— Empodial bristles not longer than tarsal claws; pronotum various, but more than 1.1 times as wide as long in most ................................................ 42

42(41). Abdominal tergites III-VI each with deep transverse basal impression ............................................ 43

— Abdominal tergite VI without deep transverse basal impression, tergites III-V or III-IV with at most moderately deep transverse basal impressions (sometimes none) .................................................... 46

43(42). Body reddish-castaneous, strongly shining, without reticulation; pubescence long and rather silky; pronotum slightly narrower at base than elytra; all setae in midline of pronotum directed anteriorly; body not subparallel-sided ........... *Euromota*

— Not exactly fitting above description; body color different; integuments at least slightly reticulated; setae in midline of pronotum various, directed anteriorly or posteriorly; base of pronotum and elytra subequal in width; body subparallel-sided ........................................................................ 44

44(43). Infraorbital carina complete to maxillary insertion; all setae in midline of pronotum directed anteriorly ....................................................... *Anopleta*

— Infraorbital carina incomplete or absent; at least setae in basal 1/2 to 1/4 of midline of pronotum directed posteriorly ....................................... 45

45(44). Infraorbital carina present on basal part of head; pubescence on midline of pronotum directed anteriorly in apical 1/2 to 3/4 and posteriorly in basal 1/2 to 1/4; in North America known only from Arctic regions of Quebec ............ *Bessobia*

— Infraorbital carina absent; pubescence on midline of pronotum directed posteriorly in apical 2/3 and anteriorly in basal 1/3; in North America known only from Alaska, New Hampshire, and arctic regions of Quebec ........................ *Dinaraea*

46(42). Antennomeres 4-10 distinctly elongate, or 4-9 distinctly elongate and 10 subquadrate; pubescence in midline of pronotum directed anteriorly and pubescence on sides of pronotum directed laterally (similar to Fig. 308.22) ....................... 47

— At least some, sometimes all, of antennomeres 4-9 distinctly subquadrate or transverse; pubescence in midline of pronotum various, directed anteriorly or posteriorly, lateral setae directed posteriorly, posterolaterally or laterally .................... 52

47(46). Infraorbital carina, complete to maxillary insertion ........................................................................ 48

— Infraorbital carina incomplete, absent near maxillary insertion (usually missing from apical 1/4 to 1/2 of head) ................................................... 50

48(47). Slender, parallel-sided, 2.5 mm long; head prognathous, with a slight but distinct medial depression on frons; eyes large, slightly bulging outside line of head, with coarse setae between facets; pronotum very slightly transverse, 1.3 times as wide as long, slightly narrowed basally; mesocoxae contiguous; meso- and metasternal processes very short, extending less than 1/5 length of coxal cavities, isthmus very long (about 1.5 times as long as mesosternal process; Fig. 292.22); abdominal segments VI and VII much longer than segments IV and V, segment VII slightly longer than VI; habitus as in Fig. 325.22; known only from California ............................ *Doliponta*

— Not exactly fitting above description ................... 49

49(48). Antennomere 11 with a subbasal "spongy" sensory patch on each side (Fig. 314.22) .......................... ................................................ *Stethusa* (in part)

— Antennomere 11 without subbasal "spongy" sensory patches ......................... *Philhygra* (in part)

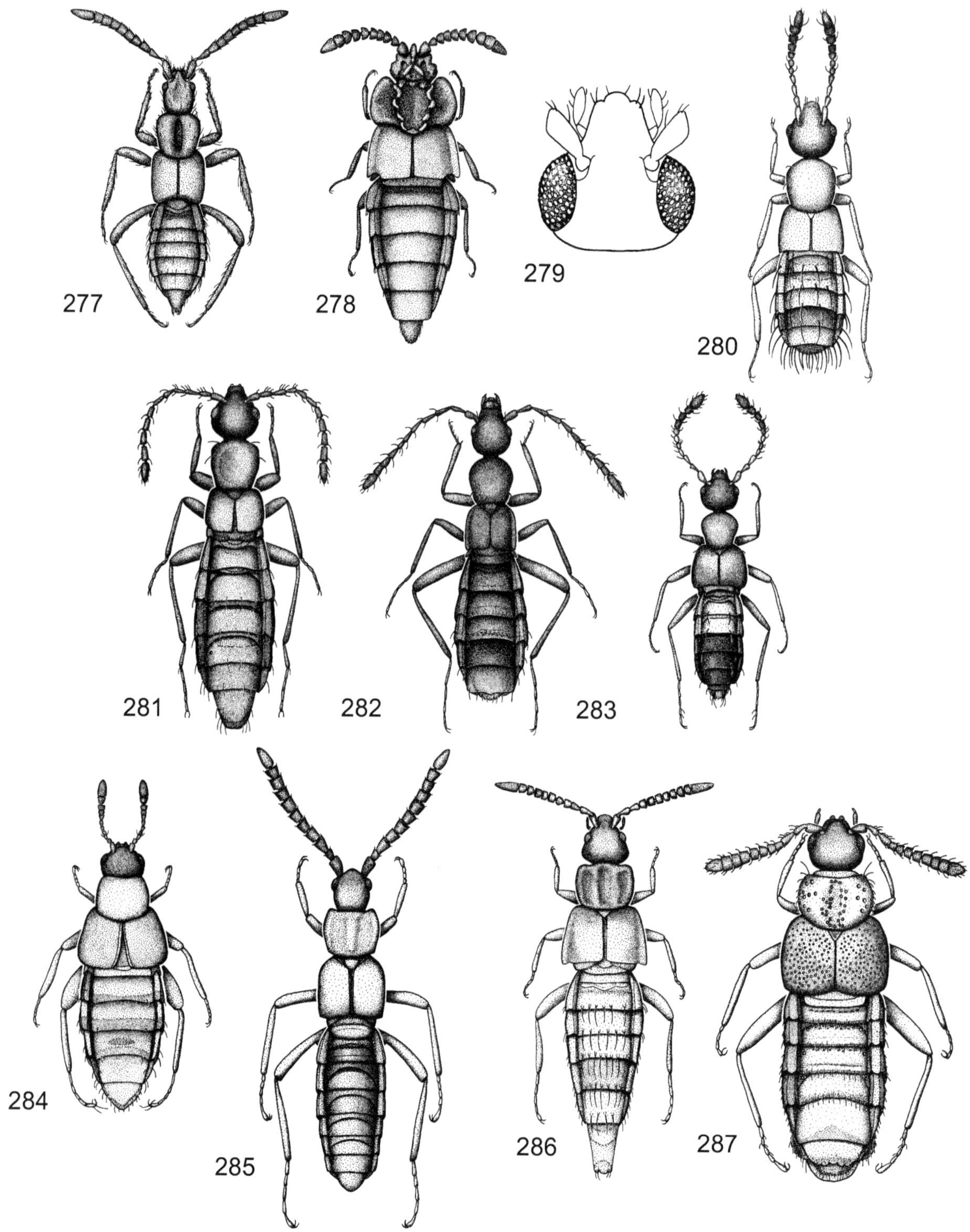


FIGURES 277.22-287.22. Fig. 277.22. *Ecitonidia wheeleri* Wasmann, length 2.9 mm. Fig. 278.22. *Ecitoxenidia brevicornis* Seevers, length 2.3 mm. Fig. 279.22. *Tetradonia megalops* (Casey), head. Fig. 280.22. *Apalonia seticornis* Casey, length 1.3 mm. Fig. 281.22. *Trachyota lativentris* Casey, length 3.0 mm. Fig. 282.22. *Drusilla* sp., length 3.0 mm. Fig. 283.22. *Meronera venustula* (Erichson), length 2.2 mm. Fig. 284.22. *Xesturida laevis* Casey, length 2.0 mm. Fig. 285.22. *Dinocoryna carolinensis* Seevers, length 3.8 mm. Fig. 286.22. *Microdonia nitidiventris* (Brues), length 3.0 mm. Fig. 287.22. *Zyras rudis* (LeConte), length 4.4 mm.

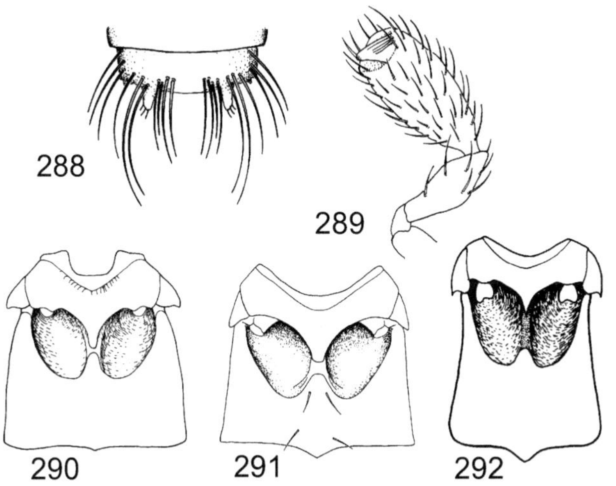


FIGURES 288.22-292.22. Fig. 288.22. *Strophogastra pencillata* Fenyes, male abdominal sternum VIII. Fig. 289.22. *Callicerus* sp., maxillary palp. Fig. 290.22. *Gnypeta* sp., meso-metasternum. Fig. 291.22. *Thamiaraea hospita* (Märkel), meso-metasternum. Fig. 292.22. *Doliponta veris* (Fenyes), meso-metasternum.

50(47). Larger, 3.5-4.4 mm long; head, pronotum, and occasionally abdomen with weak microsculpture, surface shining to strongly shining; elytra with granulose sculpture ................................ *Atheta*
— Smaller, not over 3.1 mm long, most smaller; other combination of features not exactly as above .. ........................................................................ 51

51(50). Smaller, 1.4-2.0 mm long, sub-depressed, slender, elongate; abdominal tergite VII distinctly longer than VI .......................................... *Hydrosmecta*
— Larger, mostly more than 2.0 mm long; not noticeably slender or elongate; abdominal tergite VII not noticeably longer than VI ............. *Philhygra* (in part)

52(46). Antennomeres 5-10, and usually 4, distinctly transverse (4 sometimes subquadrate or slightly elongate) ............................................................... 53
— At least some of antennomeres 5-10 elongate or quadrate, antennomere 4 distinctly elongate to slightly elongate ............................................ 59

53(52). Pronotal pubescence directed anteriorly in midline (similar to Fig. 308.22) ................................... 54
— Pronotal pubescence directed posteriorly in midline (similar to Fig. 309.22) ................................... 56

54(53). Metasternal process very short, hardly, or not at all, extending between the mesocoxae; isthmus as long as the mesosternal process, mesosternal process on a plane ventral to the metasternal process .................................... *Noverota* (in part)
— Meso- and metasternal processes different (the following couplet may not separate all species correctly) ............................................................. 55

55(54). Small, 1.5 mm long or less; pronotum with finely asperate punctation .......................... *Microdota*
— Larger, most larger than 2.0 mm long; pronotum rarely with asperate punctation .......... *Xenota* (in part)

56(53). Pronotal pubescence pattern unique: all setae in midline and sides subparallel and directed posteriorly except for several basal rows that are moderately to strongly transverse (Fig. 310.22) ................. 57
— All pronotal pubescence directed posteriorly or posterolaterally ............................................. 58

57(56). Metasternal process very short and hardly extending between coxae; isthmus and mesosternal processes subequal in length; 1.8 mm long; dark-colored, piceous to black, without obvious macrosetae on body ........................... *Omegalia*
— Mesosternal and metasternal processes subequal in length (mesosternal process slightly longer) and contiguous near middle of mesocoxal cavities, isthmus absent; 1.5 mm long; shining, reddish-brown, with small macrosetae on abdomen and sides of pronotum ...................... *Micratheta*

58(56). Small, 1.5 mm long, pale rufoflavate color; body subparallel, moderately densely hairy in appearance and moderately reticulate; head with dense erect pubescence; antenna distinctly incrassate apically; known only from Mississippi ................ *Phasmota*
— Not exactly matching above combination of characters .................. *Micrearota, Noverota* (in part); *Xenota* (in part), *Halobrecta*

59(52). Infraorbital carina present, complete to maxillary insertion ......................................................... 60
— Infraorbital carina incomplete or absent (not present near maxillary insertion) ................................ 61

60(59). Antennomere 11 with a subbasal "spongy" sensory patch on each side (Fig. 314.22) .......................... ................................................ *Stethusa* (in part)
— Antennomere 11 without a subbasal "spongy" sensory patch .... *Xenota* (in part), *Philhygra* (in part)

61(59). Pubescence in midline of pronotum directed posteriorly in apical 1/2 to 2/3 and anteriorly in posterior 1/2 to 1/3, pronotal pubescence on each side directed lateroposteriorly and more or less transversely at base (Fig. 311.22); body without macrosetae or bristles ..................... *Paradilacra*
— Pubescence at midline and on sides not exactly as above; body with macrosetae or bristles ....... 62

62(61). Pubescence in midline of pronotum directed posteriorly, pubescence on sides directed posteriorly or posterolaterally (similar to Fig. 309.22).... .............................................................. *Liogluta*
— Pubescence in midline of pronotum directed anteriorly, pubescence on sides directed laterally..... 63

63(62). Tarsal setae exceptionally long (Fig. 316.22); punctation and microsculpture of head, pronotum, and elytra unusually fine and close, surface appearing dull; infraorbital carina absent ............ *Adota*
— Tarsal setae not exceptionally long (Fig. 317.22); punctation and microsculpture of head, pronotum, and elytra not unusually fine and close, surface not appearing dull; infraorbital carina incomplete, present only at base of head ................ *Philhygra* (in part)

### IX. Key to the Nearctic tribes and genera of Scaphidiinae

1. Antenna very slender, most or all of apical 5 antennomeres strongly asymmetrical and flattened (Fig. 326.22); scutellum scarcely if at all visible; smaller, generally 1-2.5 mm long (tribe Scaphisomatini) .............................................. 4
— Antenna more robust (Fig. 4.22), apical 5 antennomeres symmetrical, moderately or not at all flattened; scutellum clearly visible; larger, generally 3-5 mm long ............................................ 2

2(1). Eye strongly emarginate near antennal insertion (tribe Scaphidiini) ............................ *Scaphidium*
— Eye not emarginate ............................................ 3

3(2). Tibiae (especially mesotibia) with distinct external spines; pronotum widest at base, sides gradually converging anteriorly (tribe Cypariini) .. *Cyparium*
— Tibiae without external spines; pronotum widest near middle, sides sinuately constricted toward base (tribe Scaphiini) ......................... *Scaphium*

4(1). Third antennomere very short, usually 1/2 or less as long as fourth, more or less triangular and flattened (Fig. 326.22); first visible abdominal sternum usually with slightly elevated plate behind each metacoxa .......................................................... 5
— Third antennomere cylindrical and elongate, usually more than 1/2 length of fourth; first visible abdominal sternum without elevated post-coxal plates . 6

5(4). Apical maxillary palpomere long and slender, sides gradually tapering to apex ............ *Scaphisoma*
— Apical maxillary palpomere enlarged, flattened, triangular, with striate external edge ..... *Caryoscapha*

6(4). Body of normal shape, broader than high; metacoxae separated widely, by more than width of metafemur ........................................... *Baeocera*
— Body strongly laterally compressed, higher than wide; metacoxae separated narrowly, by less than width of metafemur ............................. *Toxidium*

### X. Key to the Nearctic genera of Piestinae

1. Head with curved V- or U-shaped groove extending between antennal insertions; pronotum distinctly impressed along midline, sides subparallel for median third or more but abruptly constricted in basal third; mesosternum carinate; Arizona ..... *Piestus*
— Head without V- or U-shaped groove between antennal insertions; pronotum not impressed along midline, sides less parallel and gradually constricted from about middle toward base; meso-sternum not carinate; widespread ........................................ 2

2(1). Head between antennal insertions with pair of impressions (female) or pair of conspicuous horns near antennae (male); mandibles each with a dorsal tooth which is longer than mandibular apex in males and most females; disk of elytron with multiple, often irregular, rows of punctures or striae; larger, about 4-6 mm long; widespread in western and northeastern North America, south to Georgia ..................................................... *Siagonium*

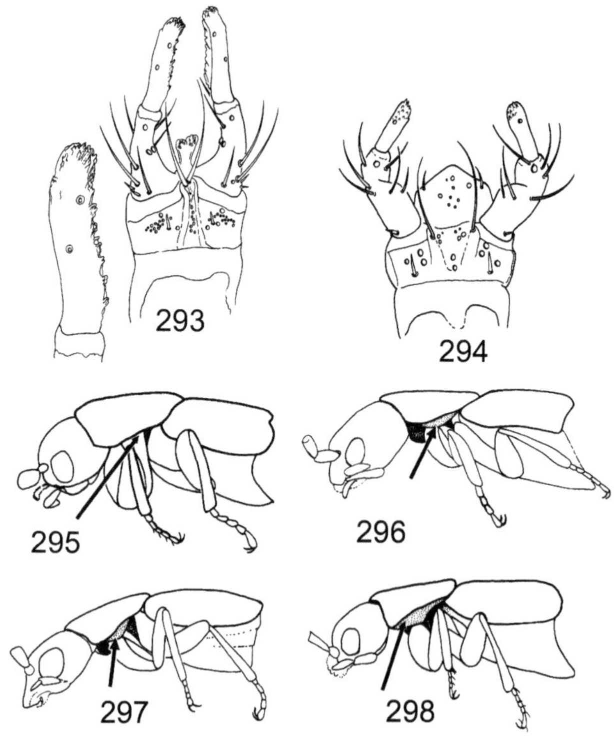


FIGURES 293.22-298.22. Fig. 293.22. *Thamiaraea hospita* (Märkel), labium, ventral, with detail of apical palpomere. Fig. 294.22. *Amischa* sp., labium, ventral. Fig. 295.22. *Acrotona prudens* Casey, lateral, showing pronotal hypomeron concealed (arrow). Fig. 296.22. *Amischa* sp., lateral, showing pronotal hypomeron (arrow) partially exposed. Fig. 297.22. *Dimetrota opinata* Casey, lateral, showing pronotal hypomeron (arrow) partially exposed. Fig. 298.22. *Atheta granulata* (Mannerheim), lateral, showing pronotal hypomeron (arrow) fully exposed.

— Head between antennal insertions neither impressed nor with horns; mandibles without dorsal teeth; elytron with impressed sutural stria, but disk with scattered punctures only; smaller, about 2-3 mm long; Florida ......................................... *Hypotelus*

### XI. Key to the Nearctic tribes, subtribes, and genera of Osoriinae

1. Membranous suture present between each abdominal tergum and sternum; pronotum gradually but strongly narrowed toward base for basal 1/3 or more; body very flat (tribe Eleusinini) ............. 2
— Abdominal tergum and sternum of each of segments III-VII fused into complete ring (Fig. 55.22); pronotal shape diverse, rarely gradually and strongly narrowed toward base; body flat to convex or cylindrical ................................................................. 3

2(1). Head without groove near dorsal edge of eye; pronotum longer than wide (rarely only as long as wide), with sides subparallel for the anterior 2/3 (Fig. 25.22); elytron with fine sutural stria.......... ............................................................ *Renardia*

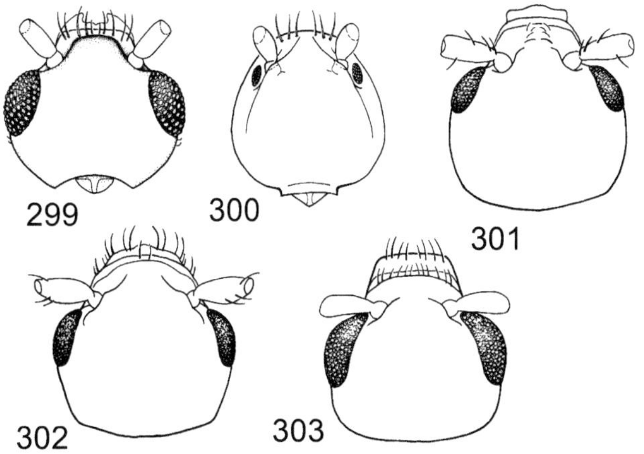


FIGURES 299.22-303.22. Fig. 299.22. *Brachyusa* sp., head. Fig. 300.22. *Geostiba alticola* Lohse, head. Fig. 301.22. *Crephalia prolongata* (Casey), head, showing eyes somewhat shorter than temples. Fig. 302.22. *Atheta granulata* (Mannerheim), head, showing eyes subequal to temples. Fig. 303.22. *Adota definita* Casey, head, showing eyes larger than temples.

— Head with groove extending posteriorly from dorsal edge of eye; pronotum distinctly wider than long, with sides subparallel for at most the anterior 1/2; elytron without sutural stria ...................... *Eleusis*

3(1). Procoxa with deep groove and carina on mesal surface (Fig. 327.22, long arrow); protibia with inner edge concave and bearing ctenidium (Fig. 327.22, short arrow); abdominal segment VIII (last visible) with "extra" tergite at apex (Fig. 328.22, =dorsum of permanently everted genital segment); body convex, often nearly cylindrical (tribe Osoriini) ......................... 4

— Procoxa with (Fig. 329.22, arrow) or without carina on mesal surface; protibia with inner edge straight, without ctenidium; abdominal segment VIII without "extra" tergite at apex (but part of eversible genital segment may be visible); body generally more or less flat (tribe Thoracophorini) ............. 6

4(3). Small, < 2 mm long; eyes absent; pronotum without lateral margin; elytron without epipleural keel (similar to Fig. 71.22); Texas ....................................... .......................... Unnamed Genus near *Rhadopsis*

— Larger, > 2.5 mm long; eyes present; pronotum with lateral margin; elytron with epipleural keel (similar to Fig. 72.22, arrow); widespread in eastern and southwestern United States ............................... 5

5(4). Antenna not geniculate, scape about as long as next 2 antennomeres combined; external edge of protibia straight to slightly convex, at most with a few slender spines ............................ *Holotrochus*

— Antenna geniculate, scape about as long as or longer than next 3 antennomeres combined; external edge of protibia strongly convex, strongly spinose .................................................... *Osorius*

6(3). Procoxae separated ventrally by a flat or convex process of prosternum (subtribe Lispinina) ............. 7

— Procoxae contiguous ............................................. 8

7(6). Abdominal segments with coarse diagonal to nearly longitudinal strigae, at least ventrally (Fig. 55.22) ............................................................... *Lispinus*

— Abdominal segments without coarse strigae ........... ............................................................ *Nacaeus*

8(6). Procoxal fissure open, trochantin exposed (similar to Fig. 64.22) (subtribe Clavilispinina) ................. ...................................................... *Clavilispinus*

— Procoxal fissure closed, trochantin concealed (similar to Fig. 43.22) ............................................... 9

9(8). Elytra (sometimes also head or pronotum) costate; body surfaces generally roughly sculptured and microsculptured, dull; abdominal sternum VIII extended dorsally in front of tergum VIII (Fig. 330.22) (subtribe Thoracophorina) .......... *Thoracophorus*

— Elytra not costate; body coarsely punctate or not, with or without depressions along midline of pronotum but otherwise without rough sculpture, more or less shining; abdominal sternum VIII normal, not visible from above (subtribe Glyptomina) ........................................................................ 10

10(9). Apex of mentum with a comb of spiniform setae; first visible abdominal segment gradually narrowed toward base; tarsi 4-4-4 .......... *Lispinodes*

— Apex of mentum with a single median spine; first visible abdominal segment with abruptly narrowed basal 1/2; tarsi 3-3-3 .............................. *Espeson*

## XII. Keys to the Nearctic tribes and genera of Oxytelinae

### Key to Nearctic tribes

1. Abdomen normal, sternum II not visible or very short and fused to front of sternum III, hence only 6 complete sterna can be counted (Fig. 49.22) ..... 2

— Abdomen with complete sternum II which is separated by membrane from sternum III, hence 7 complete sterna can be counted (Fig. 48.22) ............ 3

2(1). Abdominal segments III-VI each with 2 slender paratergites per side; elytron with about 6 distinct punctate striae; abdomen long, slender, more or less parallel-sided ........ (*Coprophilus*) Coprophilini

— Abdominal segments III-VI each with a single broad paratergite per side; elytron at most with a sutural stria; abdomen broad, sides convex and usually forming greatest body width (Key A) ........... ......................................................... Deleasterini

3(1). Mesocoxae contiguous or slightly separated by slender mesosternal process; tarsi 2-2-2, 3-3-3, 4-4-4 or 5-5-5 (Key B) ........................... Thinobiini

— Mesocoxae moderately to widely separated by metasternal process (separation about 1/2 of mesocoxal width or more); tarsi 3-3-3 (Key C) .... ............................................................ Oxytelini

### A. Key to the Nearctic genera of Deleasterini

1. Larger, about 5-8 mm long; form not compact, antennae and legs very long, length of metatibia subequal to maximum body width; apical maxillary palpomere elongate and stout ..... *Deleaster*

— Smaller, about 1.5-3 mm long; form compact, antennae and legs of moderate length, length of metatibia much less than maximum body width; apical maxillary palpomere broad at base but subulate .................................................................. 2

2(1). Labrum with anterior margin emarginate; elytra, from apex of scutellum, at least as long as pronotum; widespread ...................................... *Syntomium*

— Labrum with anterior margin broadly convex; elytra, from apex of scutellum, distinctly shorter than pronotum; southeastern Canada ...... *Mitosynum*

B. Key to the Nearctic genera of Thinobiini

1. Apical maxillary palpomere nearly as wide as, and much longer than, penultimate palpomere; southeastern United States ............................. *Manda*

— Apical maxillary palpomere much narrower than, and usually shorter than, penultimate palpomere, subulate or aciculate; widespread .......................... 2

2(1). Abdominal tergites III-VII each with curved basolateral ridge (similar to Fig. 28.22); tarsi 2-2-2 ........................................................................ 3

— Abdominal tergites III-VII without curved basolateral ridges; tarsi 3-3-3, 4-4-4, or 5-5-5 ..................... 4

3(2). Procoxal fissure present, trochantin well exposed in lateral view (Fig. 331.22); elytron without epipleural ridge; widespread ............... *Thinobius*

— Procoxal fissure absent, trochantin concealed in lateral view (similar to Fig. 332.22); elytron with epipleural ridge; Texas .......................... *Neoxus*

4(2). External edge of protibia nearly straight to strongly convex, with longitudinal row of several to many slender or stout spines as well as fine setae .. 5

— External edge of protibia more or less straight, with fine setae but no distinct spines .................... 6

5(4). Antenna not or weakly geniculate; body moderately flat; tarsi 3-3-3; western states and provinces only ........................................................... *Aploderus*

— Antenna strongly geniculate; body convex, often subcylindrical; tarsi usually 4-4-4 (rarely 3-3-3); widespread ............................................ *Bledius*

6(4). Apical maxillary palpomere small, subulate, at base more than 2/3 as wide as preceding one; procoxal fissure long, open and exposing trochantin (similar to Fig. 331.22) or closed (similar to Fig. 3-3-3.22) ............................................ *Ochthephilus*

— Apical maxillary palpomere minute, subulate or acicular, at base less than 1/2 as wide as preceding one; procoxal fissure short, closed, trochantin concealed ....................................................... 7

7(6). Tarsi 3-3-3 (Fig. 334.22), basal two tarsomeres very short; pronotal impressions diverse but generally not including transverse curved impression across middle of base ................................. *Carpelimus*

— Tarsi 5-5-5, but basal 3 tarsomeres very short, compressed, and closely associated (Fig. 335.22); pronotal impressions (in most species) including transverse curved impression across middle of base .................................................................. 8

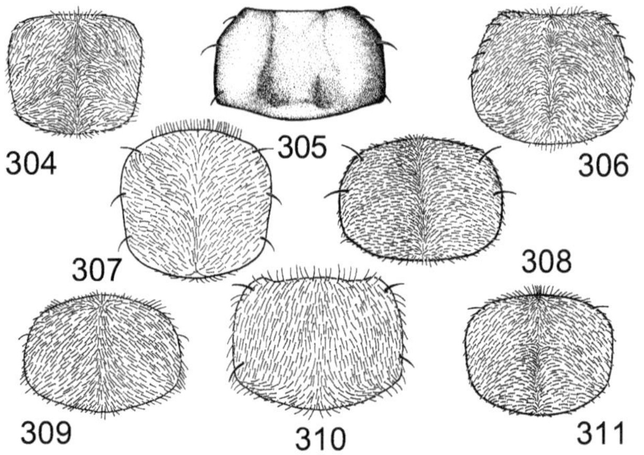


FIGURES 304.22-311.22. Fig. 304.22. *Teliusa alutacea* Casey, pronotum, showing pubescence pattern. Fig. 305.22. *Goniusa obtusa* (LeConte), male pronotum. Fig. 306.22. *Lypoglossa franclemonti* Hoebeke, pronotum, showing pubescence pattern. Fig. 307.22. *Seeversiella bispinosa* Ashe, pronotum, showing pubescence pattern. Fig. 308.22. *Dimetrota opinata* Casey, pronotum, showing pubescence pattern. Fig. 309.22. *Canastota flaveola* (Melsheimer), pronotum, showing pubescence pattern. Fig. 310.22. *Micratheta caudex* (Casey), pronotum, showing pubescence pattern. Fig. 311.22. *Paradilacra persola* Casey, pronotum, showing pubescence pattern.

8(7). Abdominal tergite VIII (last large tergite) with posterior margin broadly rounded; pronotum without transverse curved impression across middle of base; Pacific beaches from California to Washington .................................................... *Teropalpus*

— Abdominal tergite VIII (last large tergite) with posterior margin truncate to emarginate; pronotum (in most species) with transverse curved impression across middle of base; widespread in North America ......................................... *Thinodromus*

C. Key to the Nearctic genera of Oxytelini

1. Abdominal tergites III-VII without curved basolateral ridges; pronotal hypomeron narrow; procoxal fissure widely open and trochantin exposed (similar to Fig. 44.22) .............................. *Platystethus*

— Abdominal tergites III-VII each with curved basolateral ridge (Fig. 28.22); pronotal hypomeron broad, procoxal fissure absent and trochantin concealed ............................................................. 2

2(1). Neck, at narrowest point, 1/2 or less as wide as head behind eyes; pronotum globose or quadrate, not transverse, disc evenly and strongly convex ..................................................... *Apocellus*

— Neck, at narrowest point, more than 1/2 as wide as head behind eyes; pronotum transverse, shape diverse, disc usually of low and uneven convexity with longitudinal impressions .................... 3

3(2). Scutellum with diamond-shaped impression (Fig. 336.22); abdominal tergite II also with curved basolateral ridge ................................ *Oxytelus*

— Scutellum with tripartite impression, usually of a fleur-de-lis shape (Fig. 337.22); abdominal tergite II without curved basolateral ridge ...... *Anotylus*

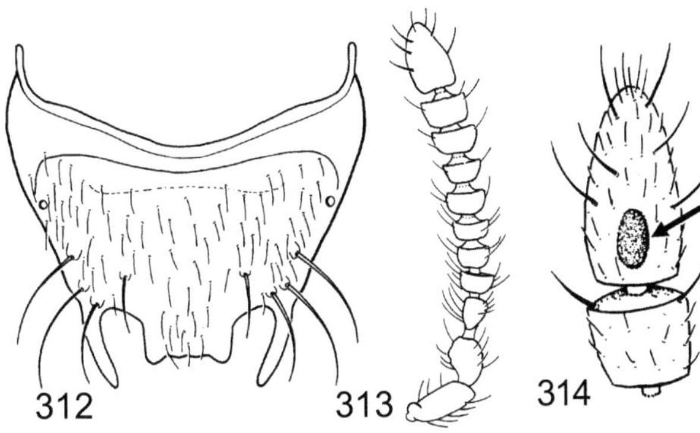


FIGURES 312.22-314.22. Fig. 312.22. *Sableta infulata* Casey, male abdominal tergite VIII. Fig. 313.22. *Clusiota claviventris* Casey, antenna. Fig. 314.22. *Stethusa clarescans* Casey, apical antennomeres, showing "spongy" basal sensory patch (arrow).

### XIII. Key to the Nearctic genera of Steninae

1. Eyes large but occupying only about 2/3 of side of head, the remaining 1/3 with broadly rounded temples; labium normal, not protrusible; head, pronotum and elytra with metallic blue reflection, elytron with large reddish to yellow spot near middle .................................................. *Dianous*

— Eyes very large and occupying entire side of head (Fig. 30.22), temples virtually absent; labium highly modified, protrusible; head, pronotum and elytra rarely with metallic blue reflection, usually black, elytra occasionally with red spots ............. *Stenus*

### XIV. Key to the Nearctic tribes and genera of Euaesthetinae

1. Abdominal segments with 1 pair of paratergites, hence margined (Fig. 31.22); tarsi 5-5-5 or 4-4-4 .......................................................................... 2

— At least abdominal segments IV-VI (visible segments 2-4) without paratergites, tergum and sternum of each segment fused into a complete ring (similar to Fig. 55.22); tarsi 5-5-5 (tribe Fenderiini) ........ 4

2(1). Tarsi 5-5-5; rare, known from British Columbia (tribe Nordenskioldiini) ......................... *Nordenskioldia*

— Tarsi 4-4-4; common, widespread (tribe Euaesthetini) .......................................................................... 3

3(2). Head without deep impressions between eyes; body more or less densely punctate, may be shining but not polished; labrum with row of comb-like teeth in front; first visible abdominal tergite simple ............................................ *Euaesthetus*

— Head with deep transverse impression or pair of impressions (dorsal tentorial pits) between eyes; body sparsely punctate, polished and strongly shining; labrum without teeth but may be crenulate; first visible abdominal tergite with fine median carina .................................................. *Edaphus*

4(1). Antennal insertions separated by more than 3 times length of first antennomere; without distinct impressions between eyes; eyes large, at least 3 times as long as temples; pronotum with short median sulcus before middle and pair of short paramedian sulci behind middle; elytra subequal in length to, and together much wider than, pronotum; widespread .................. *Stictocranius*

— Antennal insertions separated by about 2 times length of first antennomere; with pair of deep impressions (tentorial pits) between eyes; eyes minute, at most 1/3 as long as temples; pronotum without median sulcus but with pair of long paramedian sulci extending most of its length; elytra much shorter than, and together subequal in width to, pronotum; Washington to California .... .............................................................. *Fenderia*

### XV. Key to the Nearctic genera of Leptotyphlinae (Modified in part from Coiffait 1962)

1. Tarsi 2-2-2; elytron with humeral tooth; gular sutures separate throughout; Alaska .................... .................................................... *Chionotyphlus*

— Tarsi 3-3-3; elytron without humeral tooth; gular sutures separate or fused in part; western and southeastern United States ...................................... 2

2(1). Labrum with strong median tooth, without or with 1-2 pairs of lateral teeth; gular sutures separate posterior to tentorial pits but fused anterior to pits into single narrow suture for distance at least 1/2 as long as sutures posterior to pits; Florida .................................................. *Cubanotyphlus*

— Labrum without distinct median tooth, without or with 1-2 pairs of paramedian teeth; gular sutures completely separate, or fused at or just anterior to tentorial pits, if fused the fused portion anterior to pits is short and thick, much less than 1/2 as long as sutures posterior to pits; western United States ............................................................... 3

3(2). Gular sutures mostly separate posterior to tentorial pits, but contiguous at or immediately posterior to pits .............................................................. 4

— Gular sutures completely separate posterior to tentorial pits (Fig. 338.22) .................................... 6

4(3). Side of head more or less straight for no more than anterior 1/3, strongly convex in posterior 2/3 before neck; mandible with two large teeth on mesal edge, serrate between these ............ *Neotyphlus*

— Side of head more or less straight for about anterior 2/3, rounded in posterior 1/3 before neck; mandible with a single large tooth on mesal edge, not serrate ............................................................ 5

5(4). Second tarsomere at least 2 times as long as the first, these two together about as long as third tarsomere; labrum with anterior edge distinctly emarginate, edentate, with 3 pairs of long setae along edge but otherwise glabrous; aedeagus with pair of large, apically setose parameres .............. *Heterotyphlus*

— Second tarsomere subequal in length to the first, these two together much shorter than third tarsomere; labrum with anterior edge convex, with pair of small paramedian teeth and about 7 pairs of setae of various lengths not confined to edge; aedeagus with pair of small, glabrous parameres ..... *Prototyphlus*

6(3). Ridge on front of head between antennae obsolete medially; mandible with a single large tooth on mesal edge, not serrate; labrum with 3 pairs of setae near anterior edge and 2 pairs on disk .... .................................................... *Homeotyphlus*

— Ridge on front of head between antennae complete; mandible, in addition to large preapical tooth, with more basal second large tooth, or serrations, or both; labrum with 2 pairs of setae near anterior edge and one pair on disk ............................... 7

7(6). Aedeagus completely lacking parameres, at repose in abdomen with median foramen and usual points of paramere attachment facing left side of beetle ...................................................... *Xenotyphlus*

— Aedeagus with pair of minute, glabrous parameres, at repose in abdomen with median foramen and adjacent parameres facing right side of beetle ........ 8

8(7). Median lobe of aedeagus slender and tubular, internal sac with long slender copulatory pieces that are slightly exposed in repose; gular sutures completely separate ...................... *Cainotyphlus*

— Median lobe of aedeagus with broad base and very long, slender, curved apical shaft, internal sac with short broad copulatory pieces that are largely exposed in repose; gular sutures confluent immediately anterior to tentorial pits ......................................................... *Telotyphlus*

### XVI. Key to the Nearctic genera of Pseudopsinae (Modified from Newton 1982a)

1. Apical maxillary palpomere less than 1/3 as wide as penultimate palpomere; procoxal fissure widely open, trochantin well exposed; abdominal tergites with basolateral ridges (Fig. 33.22), apical tergite with comb on posterior margin .............. ......................................................... *Pseudopsis*

— Apical maxillary palpomere subequal in width to penultimate palpomere; procoxal fissure narrowly open or closed, trochantin barely or not visible; abdominal tergites without basolateral ridges, apical tergite without comb on posterior margin.. 2

2(1). Antenna densely pubescent from seventh antennomere to apex; procoxal fissure closed, trochantin concealed; mesocoxae narrowly separated by meso-and metasternal processes; dorsum of head with pairs of occipital and supraocular carinae ............................ *Nanobius*

— Antenna densely pubescent from eighth antennomere to apex; procoxal fissure narrowly open, trochantin not concealed; mesocoxae contiguous; dorsum of head without carinae .................................... 3

3(2). Subocular carina present; lateral margin of pronotum coarsely serrate; abdominal tergites without basal longitudinal carinae .... *Asemobius*

— Subocular carina absent; lateral margin of pronotum with prominent spine; first four abdominal tergites each with basal midlongitudinal carina ............. *Zalobius*

### XVII. Keys to the Nearctic tribes, subtribes, and genera of Paederinae

#### Key to Nearctic tribes

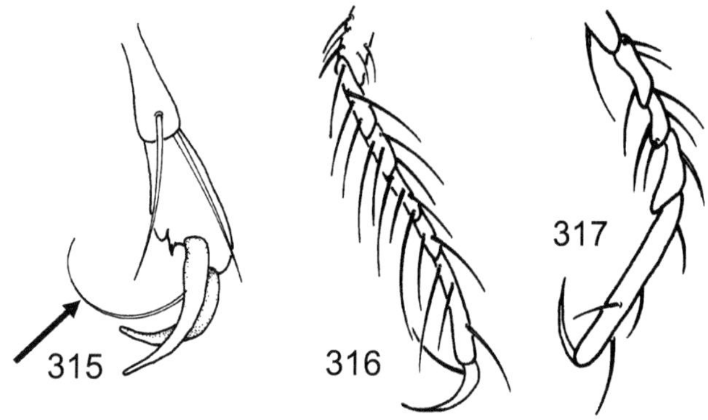


FIGURES 315.22-317.22. Fig. 315.22. *Aloconota insecta* (Thomson), tarsal claws and empodial seta (arrow). Fig. 316.22. *Adota setositarsus* (Casey), hind tarsus, showing unusually long setae. Fig. 317.22. *Philhygra bellula* (Casey), hind tarsus, showing "normal" setae.

1. Apical (fourth) maxillary palpomere narrower and much shorter than third (Figs. 339-342.22) (Key A) ............................................................ Paederini

— Apical (fourth) maxillary palpomere at least as long as third, flattened and with broad oblique apex having distinctive fuzzy or spongy texture (Fig. 51.22) (Key I) ...................................... Pinophilini

#### A. Key to the Nearctic subtribes of Paederini

1. Apical maxillary palpomere broad, compressed, truncate and pubescent (Fig. 339.22); usually with head, elytra and abdominal apex black or dark blue, pronotum and basal part of abdomen yellow to orange (a few species all dark)............................ .......................................... (*Paederus*) Paederina

— Apical maxillary palpomere minute or conical, cylindrical or subulate (Figs. 340-342.22), neither broad nor pubescent; body sometimes with red, orange, or yellow parts but not as above ........ 2

2(1). Antenna strongly geniculate at first joint (Fig. 36.22), scape very elongate, as long as following 3 to 4.5 antennomeres together; maxillary palp with apical palpomere conical to subulate and usually at least 1/2 as wide at base as the maximum width of the preapical palpomere (Fig. 341.22) (Key H) ........... ....................................................... Cryptobiina

— Antenna not geniculate, scape nearly always as long as 2 following antennomeres (3 in some *Scopaeus*, which have an extremely narrow neck); apical maxillary palpomere usually narrower...................... 3

3(2). Width of neck at narrowest point less than 1/5 maximum head width; gular sutures fused, at least near base if not completely (Key E).............. Stilicina

— Neck more than 1/5 as wide as head (narrower in *Scopaeus*, which has gular sutures narrowly to very narrowly separate); gular sutures separate or fused........................................................... 4

4(3). Apex of hind tibia with well-developed and complete ctenidium on both posterior and anterior faces, setae of posterior ctenidium longer (Fig. 343.22); gular sutures always separate; pronotum without microsculpture (Key B) ....................... Lathrobiina

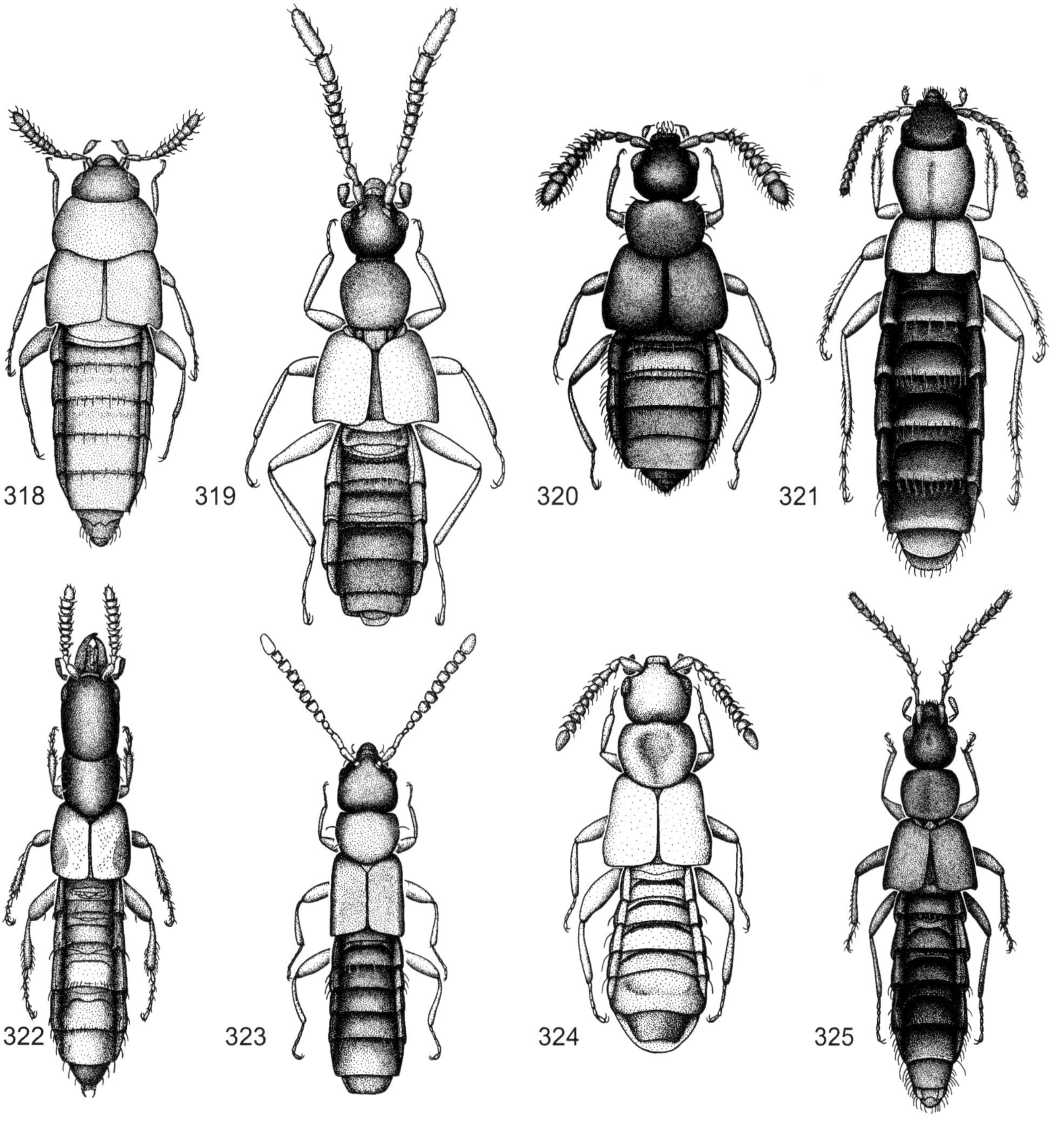


FIGURES 318.22-325.22. Fig. 318.22. *Schistacme obtusa* Notman, length 1.6 mm. Fig. 319.22. *Callicerus* sp., length 2.7 mm. Fig. 320.22. *Trichiusa robustula* Casey, length 1.5 mm. Fig. 321.22. *Pontomalota terminalia* Ahn and Ashe, length 5.0 mm. Fig. 322.22. *Charoxus major* Kistner, length 4.2 mm. Fig. 323.22. *Hydrosmectina subtilissima* Kraatz, length 1.3 mm. Fig. 324.22. *Clusiota claviventris* Casey, length 1.5 mm. Fig. 325.22. *Doliponta veris* (Fenyes), length 3.0 mm.

— Apex of hind tibia without well-developed and complete ctenidium on anterior side, though usually present on posterior side (Fig. 344.22; if present anteriorly, then gular sutures fused); gular sutures fused or separate; pronotum with or without microsculpture .................................................. 5

5(4). Tarsomere 4 of all tarsi lobed ventrally (similar to Fig. 345.22); gular sutures fused .............................. 6

— Tarsomere 4 of all tarsi simple; gular sutures fused or separate ....................................................... 7

6(5). Labrum short, transverse, with two sharp teeth near midline (Fig. 346.22), neither labrum nor its teeth extending beyond closed mandibles (*Astenus*) ............................................................ Astenina

— Labrum large, protruding medially, covering most of closed mandibles and lacking apical teeth (Key F) ...................................................... Stilicopsina

7(5). Dorsal body surface finely rugose or tuberculate; gular sutures fused; prosternum with median carina (Key G) ...................................... Echiasterina

— Dorsum densely punctate or not, but not tuberculate and seldom rugose (if so, then gular sutures separate); gular sutures usually separate, occasionally fused; prosternum with or without median carina ........................................................ 8

8(7). Pronotum wider than long or occasionally quadrate (ratio of length to width <1.1, usually <1); neck more than 1/3 as wide as head, usually more than 2/5 as wide; gular sutures usually separate, though sometimes very close, occasionally fused (in three genera known from Arizona and New Mexico, two of which have the neck only 1/3-2/5 as wide as head) (Key C) ................................................. Medonina

— Pronotum longer than wide (ratio of length to width >1, usually >1.1); neck no more than about 1/3 as wide as head (in *Scopaeus* <1/5); gular sutures always separate, though sometimes very close together (Key D) ................................ Scopaeina

### B. Key to the Nearctic genera of Lathrobiina

1. Pronotum longer than wide; head and pronotum sparsely but obviously punctate, pronotal midline usually impunctate .................................... 2

— Pronotum distinctly transverse; head and pronotum lacking or with very few punctures on disk, though with conspicuous large dark setae more marginally ..................................................................... 3

2(1). Elytron with epipleural carina above lateral margin (similar to Figs. 72-73.22, arrows) .... *Lobrathium*

— Elytron without epipleural carina (similar to Fig. 71.22) ............................................... *Lathrobium*

3(1). Labrum with pair of teeth, or at least acute projections, at sides of median notch ......................... ................................. *Acalophaena, Paederopsis*

— Labrum without teeth, sides of median notch rounded ........................................... *Dacnochilus*

### C. Key to the Nearctic genera of Medonina

1. Antenna verticillate (with a few long prominent erect setae on each antennomere); antennomere 2 about as wide as antennomere 1, antennomeres 3-11 narrower, slender and all about the same width ................................................ *Thinocharis*

— Antenna normal, not verticillate, antennomeres 3-11 increasing at least slightly in maximum width toward antennal apex ...................................... 2

2(1). Gular sutures fused for at least 1/3 of their length; known from Arizona and New Mexico ............. 3

— Gular sutures completely separate; widespread . 5

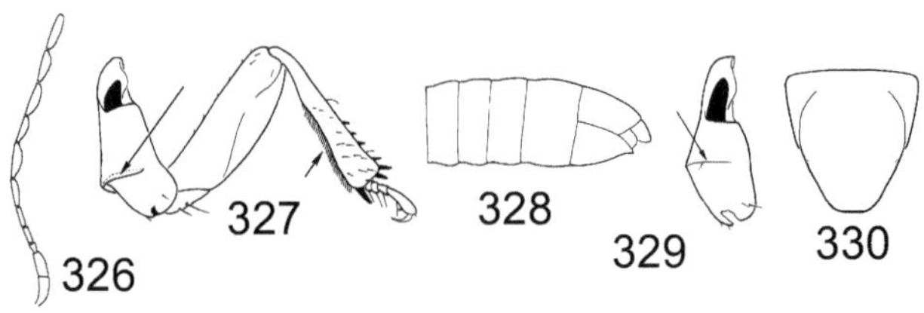


FIGURES 326.22-330.22. Fig. 326.22. *Scaphisoma commune* Löbl, antenna. Fig. 327.22. *Holotrochus australicus* Oke, left proleg, anterior, showing procoxal groove (long arrow) and protibial comb (short arrow) (after Newton 1990c). Fig. 328.22. *Osorius latipes* (Gravenhorst), abdominal segments IV-VIII, showing "extra" tergite (of genital segment). Fig. 329.22. *Myrmelibia kingi* (Lea), left procoxa, anterior, showing coxal carina (arrow) (after Newton 1990c). Fig. 330.22. *Thoracophorus sculptus* (Blackburn), abdominal segment VIII, dorsal (after Newton 1990c).

3(2). Gular sutures fused only near base of head; neck narrow, no more than 1/3 as wide as head ............... ........................................................... *Stilocharis*

— Gular sutures fused over most of their length; neck wider, at least 2/5 as wide as head .................. 4

4(3). Labrum with a small median tooth and 2 pairs of more lateral teeth; head and pronotum densely contiguously punctate, without ground sculpture between punctures ............................ *Deroderus*

— Labrum notched medially, without a median tooth; head and pronotum with non-contiguous umbilicate punctures and distinct fine ground sculpture between them ............................ *Neomedon*

5(2). Large, over 6 mm long; pronotum with pronounced mid-longitudinal impunctate line, otherwise densely covered with very small but distinct punctures; labrum with only a single blunt median tooth set in from margin (Fig. 347.22) ............ *Achenomorphus*

— Smaller, under 6 mm (usually under 5 mm) long; pronotal midline nearly always punctate, pronotal surface with very fine micropunctures and/or distinctly visible punctures; labrum without single median tooth (except in some *Lithocharis*), but usually with paired teeth .................................. 6

6(5). Head and pronotum very finely and densely micropunctate, appearing uniformly finely granulate, without obvious separate punctures; labrum without paired teeth, with or without a blunt median one ........................................... *Lithocharis*

— Head and pronotum with sparse to dense punctures (if punctures contiguous, then large enough to give a rugose appearance); labrum without median tooth, usually with paired teeth ............... 7

7(6). Head sparsely finely punctate dorsally (barely visible with 20X magnification), ventrally sparsely but much more coarsely punctate (clearly visible with 20X magnification); head and pronotum shiny, without microsculpture between punctures; tiny, ca. 2.5 mm long; known only from Florida ......... ........................................................... *Ophiooma*

— Head sparsely to densely punctate, but size and spacing of punctures not drastically different dorsally and ventrally; usually distinctly larger, if as small as 2.5 mm then head and pronotum

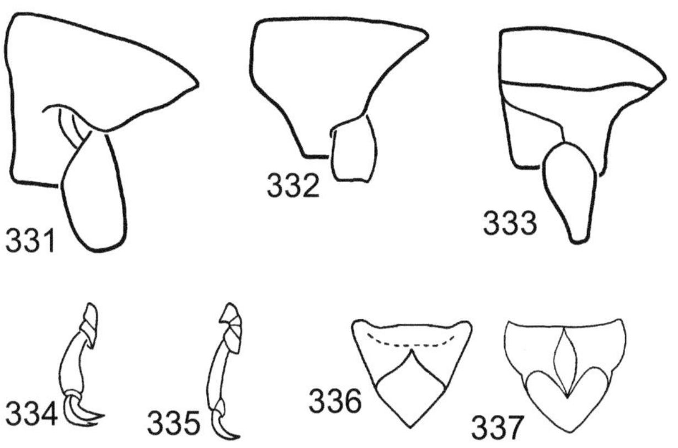


FIGURES 331.22-337.22. Fig. 331.22. *Thinobius* sp., prothorax, left lateral (after Herman 1970b). Fig. 332.22. *Pareiobledius pruinosus* (Bernhauer), prothorax, left lateral (after Herman 1970b). Fig. 333.22. *Bledius coulteri* Hatch, prothorax, left lateral (after Herman 1970b). Fig. 334.22. *Carpelimus* sp., tarsus (after Herman 1970b). Fig. 335.22. *Thinodromus* sp., tarsus (after Herman 1970b). Fig. 336.22. *Oxytelus nimius* Casey, scutellum (after Newton 1990a). Fig. 337.22. *Anotylus exasperatus* (Kraatz), scutellum (after Herman 1970b).

densely microsculptured; widespread (including Florida) ........ 8

8(7). Gular sutures closest before middle of head, diverging posteriorly from there (Fig. 340.22) .............. ........ *Sunius, Xenomedon*

— Gular sutures closest at middle or more posteriorly ........ *Medon*

### D. Key to the Nearctic genera of Scopaeina

1\. Neck no more than 1/5 as wide as head .... *Scopaeus*

— Neck at least 1/4, usually about 1/3, as wide as head ........ *Orus*

### E. Key to the Nearctic genera of Stilicina (Modified from Blackwelder 1939)

1\. Head coarsely umbilicately punctured or with coarse and deep elongate punctures; without dense ground sculpture throughout ........ 2

— Head not or indistinctly umbilicately punctate; with dense ground sculpture ........ 4

2(1). Punctures of head not very dense; labrum with median tooth ........ *Acrostilicus*

— Punctures of head very dense; labrum with teeth in pairs, but without median tooth ........ 3

3(2). Head emarginate at base; labral teeth separated by 2 times their average width, notch between them rounded; punctation of pronotum and head very fine and dense; body stout ........ *Pachystilicus*

— Head not emarginate at base; labral teeth (more medial pair if four) less widely separated, notch between them not rounded; punctation of pronotum and head coarse, sometimes elongate; body slender ........ *Rugilus*

4(1). Head suborbicular (occasionally parallel-sided and truncate posteriorly with rounded temples, similar to Fig. 34.22); labrum with prominent teeth, not shallowly emarginate; body with only normal setae ........ *Eustilicus*

— Head widest just behind eyes, strongly obliquely narrowed to neck; labrum with two small teeth set in shallow emargination occupying median 2/5 of labral apex; setae of head, pronotum, and elytra modified into short stiff erect dark bristles ........ *Megastilicus*

### F. Key to the Nearctic genera of Stilicopsina

1\. Pronotum strongly produced anteriorly like a neck (Fig. 348.22), about anterior third of lateral margin of pronotum concave in dorsal or ventral view; pronotum slightly longer than wide (length at least 1.1 times width) ........ *Stamnoderus*

— Pronotum not produced anteriorly, anterolateral margin of pronotum not concave in dorsal or ventral view; pronotum as long as wide, rounded octagonal in shape ........ *Stilicopsis*

### G. Key to the Nearctic genera of Echiasterina

1\. Head, pronotum, and elytra with longitudinal carinae, surface between them tuberculate; temples forming approximately a right angle in dorsal view; pronotum parallel-sided in basal 2/3; anterior margin of labrum coarsely serrate, with 4 obtuse teeth; larger, about 3.7 mm long ........ *Myrmecosaurus*

— Body without longitudinal carinae, surface very densely punctate (finely granulate in appearance) but not tuberculate; temples rounded; pronotum widest at about apical 1/4, narrowed toward base; labrum with four acute teeth, the median pair larger than the lateral; smaller, under 2.5 mm long ........ *Echiaster*

### H. Key to the Nearctic genera of Cryptobiina

1\. Body with distinctive color pattern: thorax dorsally and ventrally, abdominal tergites III, VIII, most of VII, and front part of head black, remainder orangish; head strongly obliquely narrowed behind eyes ........ *Lissobiops*

— Body usually unicolored, reddish-brown to black, or occasionally with some parts red or orange, but not as above; head more parallel-sided with rounded temples, rarely strongly obliquely narrowed behind eyes ........ 2

2(1). Elytron with slightly oblique epipleural carina above lateral margin (Fig. 73.22, arrow) .... *Homaeotarsus*

— Elytron without epipleural carina (similar to Fig. 71.22) ........ 3

3(2). Labrum with pair of small acute or obtusely rounded teeth near middle, notched between them; small, 4-6.5 mm long (except one Arizona species reported as 8 mm) ........ *Ochthephilum*

— Labrum without teeth, whole front margin emarginate in very shallow V-shape; larger, 7-9 mm long ........ *Biocrypta*

### I. Key to the Nearctic subtribes and genera of Pinophilini

1. Abdominal segments III-VI margined, each with 1-2 pairs of lateral sclerites separating tergite and sternum (similar to Figs. 56-57.22), surface without checkerboard-like pattern of punctures (subtribe Pinophilina) ...................................... 2

— Abdominal segments III-VI unmargined, each consisting of a complete ring, without lateral sclerites (similar to Fig.55.22), and with checkerboard-looking pattern of very coarse punctures (subtribe Procirrina) ........................... *Palaminus*

2(1). Elytra and abdominal apex red, rest of body dark . ......................................................... *Araeocerus*

— Elytra darker, reddish brown to black like rest of body (one species with elytral apex narrowly pale) ......................................................................... 3

3(2). Tarsomere 4 of middle and hind tarsi conspicuously lobed ventrally (Fig. 345.22), tarsomere 4, including lobe, at least as long as tarsomere 3; internal edge of mandibles with a well-developed slightly bifid tooth near middle ....................... *Pinophilus*

— Tarsomere 4 of middle and hind tarsi only slightly lobed ventrally, tarsomere 4, including lobe (but not setae), shorter than tarsomere 3; internal edge of mandibles with only a small tooth near base . ....................................................... *Lathropinus*

### XVIII. Keys to the Nearctic tribes, subtribes, and genera of Staphylininae

#### Key to Nearctic tribes

1. With strongly sclerotized plate (solid or divided into two contiguous sclerites) in front of prosternum (Fig. 65.22, arrow), which has a broadly concave anterior edge to receive it ............................. 2

— Without distinct sclerotized plate in front of prosternum, which has a straight, concave or convex anterior edge ........................................... 3

2 (1). Elytra with slightly convex sutural edges that overlap when elytra are closed; neck at narrowest point rarely more than 1/3 as wide as head (Key B) ..................................................... Xantholinini

— Elytra with straight sutural edges that meet in straight line when elytra are closed; neck at narrowest point nearly 1/2 as wide as head (Key A) ..................... ................................................................ Othiini

3(1). Apical maxillary palpomere minute, aciculate, less than 1/4 as long as penultimate palpomere (Fig. 349.22); neck at narrowest point less than 1/3 as wide as head ......................... (*Diochus*) Diochini

— Apical maxillary palpomere larger, at least 1/3 as long as penultimate palpomere; neck, if present, more than 1/3 as wide as head ....................... 4

4(3). Antennae inserted at anterior margin of front, with insertions closer to each other than to eyes; southern Texas ............. (*Platyprosopus*) Platyprosopini

— Antennae inserted at anterolateral margin of front, or on front, with insertions closer to eyes than to each other; widespread (Key C) ...... Staphylinini

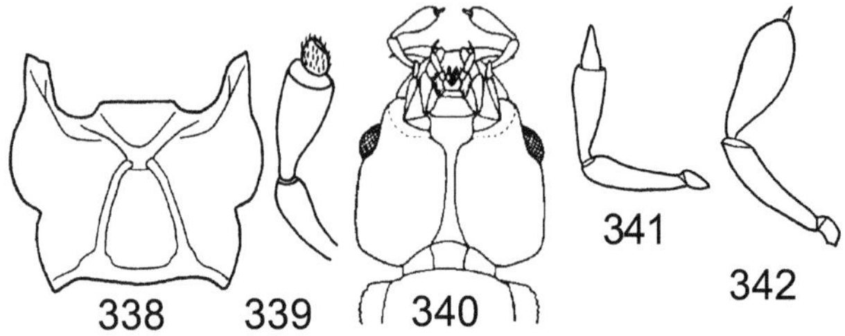


FIGURES 338.22-342.22. Fig. 338.22. *Homeotyphlus* sp., head capsule, ventral. Fig. 339.22. *Paederus obliteratus* LeConte, maxillary palpomeres 2-4 (after Newton 1990a). Fig. 340.22. *Sunius confluentus* (Say), head, ventral (after Arnett 1963). Fig. 341.22. *Homaeotarsus bicolor* (Gravenhorst), maxillary palp. Fig. 342.22. *Astenus* sp., maxillary palp.

### A. Key to the Nearctic genera of Othiini (Modified from Smetana 1982)

1. Disk of pronotum with a single pair of punctures near middle; apical maxillary palpomere fusiform, distinctly longer than penultimate palpomere; widespread ........................................... *Atrecus*

— Disk of pronotum with at least 4 pairs of punctures forming paramedian rows; apical maxillary palpomere conical, not longer than penultimate palpomere; California to British Columbia .......... ............................................................ *Parothius*

### B. Key to the Nearctic genera of Xantholinini (Modified from Smetana 1982)

1. Superior marginal line of pronotal hypomeron turning downward well behind middle and joining or almost joining inferior line in front of procoxae (Fig. 350.22) ........................................................... 2

— Superior marginal line of pronotal hypomeron turning downward only close to anterior angle and not joining inferior line, occasionally very fine and more or less fading out anteriorly (Fig. 351.22) ......................................................................... 5

2(1). Apical maxillary palpomere very small, conical to almost aciculate, distinctly narrower and considerably shorter than penultimate palpomere; ocular grooves on head very long, linear; neck extremely narrow, at narrowest point 1/5 or less as wide as head ........................................... *Neoxantholinus*

— Apical maxillary palpomere moderately large to large, more or less fusiform, at least as long as penultimate palpomere; ocular grooves on head obsolete to absent; neck wider, at narrowest point at least 1/4 as wide as head ............................ 3

3(2). Pronotal disc with at least 5 punctures (=dorsal row) on either side of midline; generally smaller, 5-8 mm long .............................................. *Nudobius*

— Pronotal disc impunctate except near margins; larger, 8-16 mm long ........................................ 4

4(3). Punctation of head even in spacing, punctures not forming coalescent grooves; pronotum with lateral rows of punctures situated on evenly convex surface .............................. *Thyreocephalus*

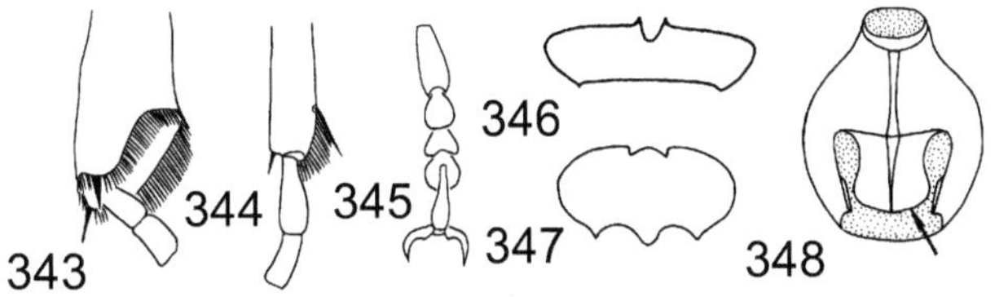


FIGURES 343.22-348.22. Fig. 343.22. *Lathrobium armatum* Say, apex of left metatibia, anterior, showing apical ctenidia on anterior and posterior faces. Fig. 344.22. *Achenomorphus corticinus* (Gravenhorst), apex of left metatibia, anterior, showing apical ctenidium only on posterior face. Fig. 345.22. *Pinophilus opacus* LeConte, tarsus, showing lobed tarsomere 4. Fig. 346.22. *Astenus* sp., labrum (after Newton 1990a). Fig. 347.22. *Achenomorphus corticinus* (Gravenhorst), labrum (after Newton 1990a). Fig. 348.22. *Stamnoderus monstrosus* (LeConte), prothorax, ventral, coxae removed, showing broadly expanded prosternum (arrow) (after Newton 1990a).

— Punctation of head uneven in spacing, punctures forming coalescent grooves running posteromediad from each eye; pronotum with lateral rows of punctures situated in impressed arcuate grooves *Gauropterus*

5(1). Apical maxillary palpomere more or less fusiform, its greatest width more than 2/3 width of apex of preceding one .................................................. 6

— Apical maxillary palpomere conical to subulate or cylindrical, its greatest width less than 2/3 width of apex of preceding one .............................. 13

6(5). Metatibia with apical and subapical ctenidia, latter more or less interrupted medially but extended proximally along outer margin of metatibia; aedeagus with parameres in form of characteristic, laterally compressed flaps ....... *Gyrohypnus*

— Metatibia with apical ctenidium only; aedeagus with parameres dorsoventrally compressed, not flap-shaped, or parameres almost absent ................ 7

7(6). Ocular puncture (bearing long seta) situated far from inner margin of eye, distance separating ocular punctures from each other not more than 2.5 times as long as distance separating each ocular puncture from inner margin of eye .................... 8

— Ocular puncture (bearing long seta) situated close to inner margin of eye, distance separating ocular punctures from each other at least 3 times as long as distance separating each ocular puncture from inner margin of eye ......................... 10

8(7). Punctation of head very fine and sparse; scutellum with some extremely fine, hardly noticeable punctures ..................................... *Xestolinus* (in part)

— Punctation of head, at least laterally, coarse and dense; scutellum with at least 2 distinct punctures on apical portion ..................................... 9

9(8). Abdomen bicolored, nearly black at base with apical 2 segments bright yellowish-red .................. ........................................................ *Lissohypnus*

— Abdomen unicolored, usually dark ..... *Xantholinus*

10(7). Pronotum, except for smooth median strip, densely punctate, so that distinct dorsal and lateral rows of punctures are not apparent .......................... ............................................ *Stictolinus* (in part)

— Pronotum with distinct dorsal and lateral rows of punctures ...................................................... 11

11(10). Deflexed portion of temples distinctly flattened and separated from dorsal side of head by impunctate, occasionally slightly elevated strip ... *Neohypnus*

— Deflexed portion of temples continuously convex with dorsal side of head and not appreciably separated from it ..................................................... 12

12(11). Gula moderately long, gular sutures contiguous posteriorly; pronotum with at least 8 punctures in each dorsal row; abdominal tergite VIII of male with broadly emarginate apex bearing dense row of black pegs perpendicular to edge ................ ......................................................... *Hypnogyra*

— Gula very long, gular sutures at least narrowly separated; pronotum with at most 6 punctures in each dorsal row; apical abdominal tergite of male not emarginate, without black pegs ...... *Oxybleptes*

13(5). Protarsus with basal 4 tarsomeres in both sexes strongly expanded laterally (less so in female), with pad of dense pale hairs ventrally ........... 14

— Protarsomeres 1-4 not appreciably expanded in either sex and without pad of dense pale hairs ventrally ............................................................... 15

14(13). Surface of head dull and finely rugose due to very dense sculpture with punctures longitudinally confluent and interspaces forming narrow longitudinal ridges; third antennomere elongate, almost as long as second antennomere .............. .................................................... *Stenistoderus*

— Surface of head shiny, punctation simple, very sparse and fine; third antennomere very short, much shorter than second antennomere ......... ........................................................... *Microlinus*

15 (13). Wingless; penultimate abdominal tergite without whitish apical margin ........... *Habrolinus* (in part)

— Winged; penultimate abdominal tergite with fine whitish apical margin ............................................ 16

16 (15). Prosternum obtusely transversely elevated, posterior margin truncate, without intercoxal process; western North America ................. *Hesperolinus*

— Prosternum more or less longitudinally carinate medially, or at least obtusely longitudinally elevated posteromedially, with small angulate or triangular intercoxal process; widespread ..................... 17

17(16). Punctation of elytron unevenly spaced, with tendency to form more or less irregular longitudinal rows at least laterally ...................................... 18

— Punctation of elytra more or less evenly spaced, without tendency to form longitudinal rows . 21

18(17). Punctation of head very fine and sparse; scutellum with only extremely fine, hardly noticeable punctures ..................................... *Xestolinus* (in part)

— Punctation on lateral portions of head coarse to very coarse, and at least moderately dense; scutellum with fine but distinct punctures ... 19

19(18). At least 1 of dorsal rows of punctures on pronotal disc with no more than 5 punctures; ocular grooves on head about as long and nearly as deep and distinct as frontal grooves; aedeagus without parameres ........................ *Phacophallus*

— Dorsal rows of punctures on pronotal disc with 6 or more punctures each; ocular grooves on head much shorter, shallower, and less distinct than frontal grooves .............................................. 20

20(19). Penultimate maxillary palpomere more than 1.5 times as long as wide, moderately widened toward apex; aedeagus with moderately developed, symmetrical parameres .................................... *Leptacinus*

— Penultimate maxillary palpomere less than 1.3 times as long as wide, not widened toward apex; aedeagus with strong, very asymmetrical parameres ........................................ *Lepitacnus*

21(17). Apical maxillary palpomere slightly shorter to feebly longer than penultimate palpomere ......... 22

— Apical maxillary palpomere distinctly shorter than penultimate palpomere .................................. 23

22(21). Elytron moderately long, at side about as long as pronotum at midline ............. *Stictolinus* (in part)

— Elytron very long, at side distinctly longer than pronotum at midline ............ *Habrolinus* (in part)

23(21). Posterior face of meso- and metatibia with at least 1 subapical as well as apical ctenidium; if subapical ctenidium indistinct, then head at least anteromedially with very fine and dense granulose sculpture ....... *Lithocharodes* (in part)

— Posterior face of meso- and metatibia with apical ctenidium only ............................................... 24

24(23). Prosternum with nearly complete longitudinal carina at midline ................................................ 25

— Prosternum obtusely longitudinally elevated posteromedially but without longitudinal carina ........................................................................ 26

25(24). Head very finely punctate, completely rounded posteriorly; abdominal tergites with conspicuously numerous and long semierect setae..... *Crinolinus*

— Head coarsely punctate, of rounded quadrangular shape; abdominal tergites with at most a few long setae along with short pubescence ................. ...................................................... *Linohesperus*

26(24). Head densely punctate, the punctures separated by little more than their diameters .................... ....................................... *Lithocharodes* (in part)

— Head moderately densely to sparsely punctate, the punctures separated by distinctly to much more than their diameters ........................................ 27

27(26). Antenna with second and third antennomeres subequal in length; mesotibia with numerous strong spines on outer margin; elytron at side slightly shorter than pronotum at midline ......... ................................................................. *Zenon*

— Antenna with third antennomere shorter than second; mesotibia with only a few fine spines on outer margin; elytron at side distinctly longer than pronotum at midline ............ *Habrolinus* (in part)

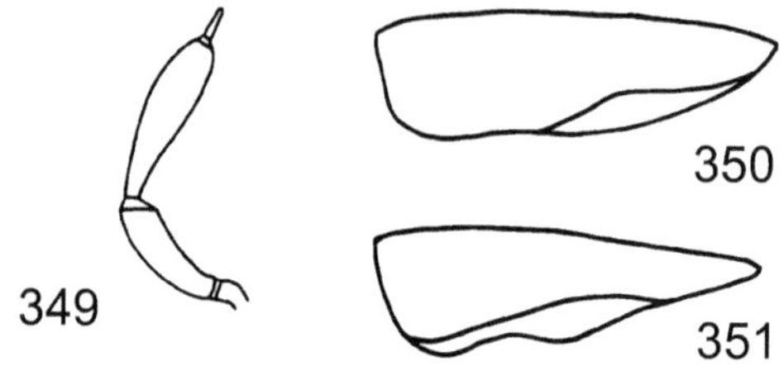


FIGURES 349.22-351.22. Fig. 349.22. *Diochus schaumi* Kraatz, maxillary palp (after Newton 1990a). Fig. 350.22. *Nudobius arizonicus* Casey, pronotum, left lateral. Fig. 351.22. *Neohypnus picipennis* (LeConte), pronotum, left lateral.

C. Key to the Nearctic subtribes of Staphylinini

1. Superior marginal line of pronotal hypomeron (lateral margin of pronotum) completely visible in dorsal view, including at anterior pronotal angle, not connected to inferior marginal line of hypomeron (Fig. 352.22) .................................. 2

— Superior marginal line of pronotal hypomeron deflexed ventrad behind anterior pronotal angle and therefore not visible from above at anterior pronotal angle, often fused anteriorly (Figs. 69.22, short arrow; 68.22) to inferior marginal line of pronotal hypomeron ........................................ 4

2(1). Disk of pronotum (in North American genera) multipunctate to densely punctate, with impunctate median line; pronotal hypomeron less strongly inflexed, meeting prosternum at low or no angle and fused to it without distinct suture; tarsi always 5-5-5 (Key F)...................................... Xanthopyginia (in part: *Tympanophorus*)

— Disk of pronotum (in North American genera) with at most a few (usually 1-2) pairs of setae; pronotal hypomeron nearly always very strongly inflexed anteriorly, meeting prosternum at distinct angle (up to 90°) in both ventral and anterior view, the junction indicated by distinct notosternal suture; tarsi 5-4-4 in one genus .................................... 3

3(2). Tarsi 5-4-4; maxillary and labial palps extremely elongate, former almost as long as exposed part of head capsule; each apical tarsomere with conspicuous arched long seta (longer than claws) dorsally at apex .................................................. .........................(*Atanygnathus*) Tanygnathinina

— Tarsi 5-5-5; maxillary and labial palps not extremely elongate, former much shorter than exposed part of head capsule; apical tarsomeres without long apico-dorsal setae (Key D) .................. Quediina

4(1). Apical maxillary palpomere small, aciculate; pronotal hypomeron with 3 marginal lines (usual superior and inferior lines, plus an oblique line between them), all uniting behind anterior angle; body very flat; Arizona ........................ (*Holisus*) Hyptiomina

— Apical maxillary palpomere subequal to or longer than third, or if smaller then not aciculate; pronotal hypomeron with usual superior and inferior marginal lines only; body shape diverse; widespread ......................................................................... 5

352

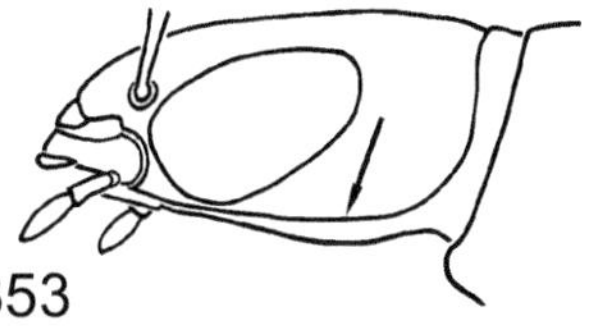

353

354

FIGURES 352.22-354.22. Fig. 352.22. *Xanthopygus xanthopygus* (Nordmann), pronotum, left lateral slightly ventral. Fig. 353.22. *Anaquedius vernix* (LeConte), head, left lateral, showing infraorbital ridge (arrow). Fig. 354.22. *Platydracus viridanus* (Horn), labium, ventral, showing notched ligula (arrow) (after Newton 1990a).

5(4). Disc of pronotum either more or less densely punctate and setose with punctures usually separated by less than their diameters, or disk impunctate and glabrous; ligula in most species distinctly notched (Fig. 354.22, arrow), emarginate or bilobed at apex; tarsal empodium with 2 or more setae; always large, 12-30 mm long (Key E) .................. ........................................................ Staphylinina

— Pronotal punctation diverse, central part of disc often with only two rows of punctures, rarely densely punctate with punctures separated by less than their diameters, never impunctate (in North American species); ligula entire, rounded or slightly sinuate apically; tarsal empodium glabrous (except *Xanthopygus*); usually smaller, 3-20 mm long ........................................................ 6

6(5). Pronotal hypomeron without postcoxal process (Fig. 68.22) (Key G) ................. Philonthina (major part)

— Pronotal hypomeron with distinct translucent postcoxal process (Fig. 69.22, long arrow) ........ 7

7(6). Superior lateral line of pronotal hypomeron united with inferior line near front angle of pronotum (Fig. 69.22, short arrow); abdominal tergites without basolateral ridges; small, 3-7 mm long; widespread ............ Philonthina (in part: *Erichsonius*)

— Superior and inferior lateral marginal lines of pronotal hypomeron widely separate throughout (Fig. 352.22); first three visible abdominal tergites with basolateral ridges (similar to Figs. 28-29.22); large, 15-20 mm long; Arizona to Texas .... Xanthopygina (in part: *Xanthopygus*)

### D. Key to the Nearctic genera of Quediina

1. Antenna strongly geniculate, first antennomere elongate, usually as long as next four combined (similar to Fig. 36.22); protarsal claws longer than others ............................................ *Acylophorus*

— Antenna not geniculate, first antennomere not longer than next two combined; protarsal claws not longer than others ...................................... 2

2(1). Dorsal side of all tarsomeres setose; very small to very large species, 1.7-21 mm long; widespread ......................................................................... 3

— Dorsal side of apical four tarsomeres of middle and hind legs smooth and without setae except for a few at apex of each tarsomere; large species, 9-14 mm long; eastern North America ................ 5

3(2). Apical maxillary palpomere subulate, thin and sharp, shorter than and at base only about 1/2 as wide as apex of penultimate palpomere . *Heterothops*

— Apical maxillary palpomere more or less fusiform, not shorter than and about as wide as apex of penultimate palpomere ..................................... 4

4(3). Head without infraorbital ridge; pronotum elongate, subhexagonal, with distinct hind angles, and with 3 pairs of paramedian setae on disk; Oregon to British Columbia ....................................... *Beeria*

— Head usually with infraorbital ridge (Fig. 353.22, arrow), at least posteriorly; pronotum about as wide as long or transverse, sides and base broadly and continuously rounded and therefore without distinct hind angles, with at most 2 pairs of paramedian setae on disk; widespread . *Quedius*

5(2). Protibia spinose along outer edge and with long spur at inner front angle; pronotum transverse, sides broadly rounded; head between eyes smooth, with only a few macrosetae near eyes .......................................................... *Anaquedius*

— Protibia setose, without spines along outer edge and without spur at inner front angle; pronotum about as wide as long, sides subparallel; head between eyes smooth at middle but becoming coarsely and moderately densely punctate near eyes .............................................. *Hemiquedius*

### E. Key to the Nearctic genera of Staphylinina

1. Central part of pronotal disc apparently impunctate (may have nonseta-bearing micropunctures visible at high magnifications); superior and inferior marginal lines of pronotal hypomeron well separated throughout, superior line fading out behind anterior pronotal angle ..................................... 2

— Central part of pronotal disc more or less densely punctate and setose with punctures usually separated by less than their diameters; superior and inferior marginal lines of pronotal hypomeron fused behind anterior pronotal angle, or at least closely approximate anteriorly and both continued beyond anterior pronotal angle ............... 4

2(1). General color deep black, with distinctive white pubescence especially on anterior angles of pronotum, across middle of elytra and on abdominal terga IV-V (Fig. 1.22); middle of inner edge of right mandible with 3 small teeth, 1 dorsal and 2 ventral; winged species, widespread .................. *Creophilus*

— General color dark brown or variegated yellow and brown, without white pubescence; middle of inner edge of right mandible with two prominent teeth in same plane; wingless species, confined to Pacific beaches .......................................... 3

3(2). Protibia laterally expanded, posterior face concave; elytra with convex sutural edges which overlap slightly when closed; translucent postcoxal process of pronotum a low flange; color variegated yellow and brown ............................. *Thinopinus*

— Protibia not laterally expanded, posterior face convex; elytra with straight sutural edges which meet in straight line when closed; translucent postcoxal process of pronotum large, triangular; color uniform dark brown .................... *Hadrotes*

4(1). Anterior angle of pronotum prominent, acute; mesosternum with complete midlongitudinal carina ........................................................ *Ontholestes*

— Anterior angle of pronotum not prominent or acute; mesosternum not carinate ................................ 5

5(4). Postocular macrosetal puncture (Fig. 355.22, arrow) nearer to base of head than to eye; middle of inner edge of each mandible with at least 2 teeth, usually with 3 teeth in different planes (2 dorsal/1 ventral on left mandible, 1 dorsal/2 ventral on right); paramere of aedeagus not closely appressed to median lobe and somewhat dorsoventrally flexile, with several long normal setae near apex but never with peg setae .................................. *Platydracus*

— Postocular macrosetal puncture (Fig. 356.22, arrow) nearer to eye than to base of head; middle of inner edge of each mandible with at most 2 teeth in same plane, left mandible rarely with third more distal and slightly ventral tooth; paramere of aedeagus with large base that is closely and immovably appressed to median lobe, sometimes with black peg setae on inner face near apex ........................... 6

6(5). Pronotal hypomeron with distinct translucent postcoxal process which may be large and triangular (Fig. 69.22, long arrow) or reduced to a low flange; superior marginal line of hypomeron usually visible from above in less than basal 1/2 of pronotum, deflexed anterior to that ................. 7

— Pronotal hypomeron without any postcoxal process; superior marginal line of hypomeron usually visible from above in more than basal 1/2 of pronotum, deflexed anterior to that ................. 8

7(6). Right mandible with 1, left mandible with 2, distinct teeth near middle; apical labial and maxillary palpomeres glabrous, fusiform; color in most species not totally black; paramere of aedeagus with black tubercles (peg setae) on inner face near apex ............................................. *Dinothenarus*

— Right and left mandibles each with a single distinct tooth near middle, or edentate; apical labial and maxillary palpomeres setose, more or less enlarged and flattened (more so in males), with obliquely truncate apices; color (including appendages and setae) totally black; paramere of aedeagus without peg setae .................. *Tasgius*

8(6). Left mandible with 3 large teeth, right mandible with large securiform (hatchet-like) lobe which is as wide as rest of mandible at same point; apical labial palpomere subfusiform, narrower than penultimate palpomere and at most with a few inconspicuous setae; superior marginal line of pronotal hypomeron fused to inferior line well behind front angle of pronotum; mesocoxal cavity not margined posteromedially; color black with red elytra and legs, with gold tomentum on neck, scutellum, and in patches on some abdominal tergites ............................................ *Staphylinus*

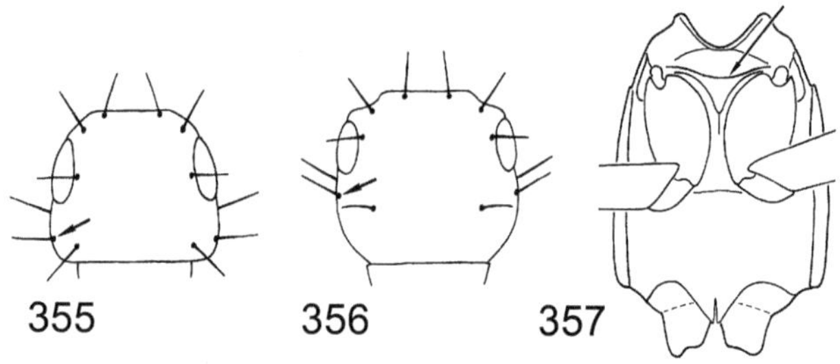


FIGURES 355.22-357.22. Fig. 355.22. *Platydracus viridanus* (Horn), head, dorsal, showing postocular seta (arrow) (after Newton 1990a). Fig. 356.22. *Dinothenarus badipes* (LeConte), head, dorsal, showing postocular seta (arrow) (after Newton 1990a). Fig. 357.22. *Neobisnius ludicrus fauveli* Smetana, meso- and metathorax, ventral, showing transverse mesosternal carina (arrow).

— Left and right mandibles each with 1-2 large teeth; apical labial palpomere more or less enlarged and flattened, at least as wide as penultimate palpomere and sparsely setose; superior marginal line of pronotal hypomeron approaching inferior line anterior to procoxae but not fused to it at all or only at front angle; mesocoxal cavity margined posteromedially; color black, with or without metallic bronze reflection or reddish legs or some light-colored setae on abdomen ........... *Ocypus*

### F. Key to the Nearctic genera of Xanthopygina

1. Head small, not more than 2/3 as wide as pronotum; pronotum orbicular except for truncate anterior edge, lateral marginal line completely visible from above; apical labial palpomere greatly enlarged, nearly 2 times as wide as penultimate palpomere, with obliquely truncate apex of distinctive texture; widespread ....................... *Tympanophorus*

— Head larger, at least 5/6 as wide as pronotum; pronotum not orbicular, sides subparallel or slightly angulate before middle, lateral marginal line anteriorly more or less deflexed ventrad and not clearly visible from above behind anterior angle; apical labial palpomere not greatly expanded, not or little wider than penultimate palpomere; known from Arizona to Texas .......... ...................................................... *Xanthopygus*

### G. Key to the Nearctic genera of Philonthina (Modified from Smetana 1995)

1. Apical labial palpomere expanded, much wider than apical maxillary palpomere; body very elongate, cylindrical, abdomen when fully extended at least half again as long as length of head, pronotum and elytra combined; known from Texas ........... ................................................................ *Flohria*

— Apical labial palpomere not expanded, not or scarcely wider than apical maxillary palpomere; body generally less elongate and not cylindrical, abdomen usually little longer than length of head through elytra; widespread .............................. 2

2(1). Largest lateral macrosetal puncture of pronotum separated from lateral margin by at most little more than the width of the puncture (Fig. 68.22) ............... 3
— Largest lateral macrosetal puncture of pronotum separated from lateral margin by about 3 times the width of the puncture or more (Fig. 69.22) .................... 7

3(2). Protarsus in both sexes with basal 4 tarsomeres more or less dilated (usually less so in female), each tarsomere bearing modified pale setae on ventral surface in addition to regular unmodified marginal setae .................................................. 4
— Protarsus in both sexes with basal 4 tarsomeres not dilated, each bearing only regular unmodified marginal setae on ventral surface .......................... 6

4(3). Apical maxillary palpomere more or less fusiform, never subulate, usually more than 1.3 times as long as penultimate palpomere; labial palp moderately long to long, apical palpomere at least 1.5 times as long as penultimate palpomere ........... ......................................................... *Philonthus*
— Apical maxillary palpomere subulate, not more than 1.3 times as long as penultimate palpomere; labial palp short, apical palpomere about 1.3 times as long as penultimate palpomere .................... 5

5(4). Apical maxillary palpomere about 1.3 times as long as penultimate palpomere, which is distinctly narrower than slightly swollen second palpomere; apical labial palpomere distinctly narrower than slightly swollen penultimate palpomere ........... ....................................................... *Gabronthus*
— Apical maxillary palpomere about as long as penultimate palpomere, which is about as wide as second palpomere; apical labial palpomere about as wide as penultimate palpomere ........... ............................................................. *Rabigus*

6(3). Apical labial palpomere slender, distinctly narrower than penultimate palpomere ................. *Gabrius*
— Apical labial palpomere not distinctly narrower than penultimate palpomere ........................... *Bisnius*

7(2). Protarsus in both sexes with basal 4 tarsomeres each bearing modified pale setae on ventral face, in addition to regular unmodified marginal setae, and each almost always variably dilated (usually less so in female) ............................................. 8
— Protarsus in both sexes with basal 4 tarsomeres each bearing only regular unmodified marginal setae on ventral face, and each not dilated ...... ......................................................... *Belonuchus*

8(7). Hind tarsus short, basal tarsomere at most as long as apical tarsomere; apical maxillary palpomere large but distinctly subulate ............................ 9
— Hind tarsus moderately long, basal tarsomere at least slightly longer than apical tarsomere; apical maxillary palpomere more or less fusiform or cylindrical, not subulate ............................................ 10

9(8). Second antennomere swollen, much wider than third antennomere and about as wide as basal antennomere; pronotal hypomeron with distinct translucent postcoxal process (Fig. 69.22, long arrow); mesosternum without transverse carina ..... ......................................................... *Erichsonius*
— Second antennomere not swollen, not distinctly wider than third antennomere and distinctly narrower than basal antennomere; pronotal hypomeron without postcoxal process (Fig. 68.22); mesosternum with transverse carina (Fig. 357.22, arrow) ............................................... *Neobisnius*

10(8). Outer edge of protibia without spines ................. ..................................................... *Laetulonthus*
— Outer edge of protibia with spines .................... 11

11(10). Maxillary and labial palps elongate, penultimate maxillary palpomere about 2 times as long as wide, penultimate labial palpomere more than 2 times as long as wide; mesosternum with rounded intercoxal process; nearly always in forest habitats .................................................... *Hesperus*
— Maxillary and labial palps short, penultimate maxillary palpomere about 1.3 times as long as wide, penultimate labial palpomere about as long as wide; mesosternum with acute intercoxal process; ocean beaches .............................. *Cafius*

### Classification of the Nearctic Genera of Staphylinidae

The current classification of staphylinids overall and in North America was reviewed in the Introduction. In the following detailed classification and generic list, taxa above the generic level (i.e., subfamily through subtribe) are listed in a presumed "natural" sequence suggesting phylogenetic relationships as far as these are known or can be estimated, but genera are listed alphabetically within these higher taxa. Synonymies are not necessarily complete but include those synonyms that have been used in North America. Subgenera, even when in general use elsewhere in the world, have not been widely used in North America and recently have been deliberately abandoned by several authors because they have seldom been adequately defined and applied to all species. Thus, subgenera are also listed in synonymy but with indication of their status. The list of genera and number of species recorded from North America below is believed to be complete through literature published in 1999. Except for Aleocharinae, where the overall distribution and size of most genera are vague or unknown, notes on the extralimital distribution of each genus (if any) are also given. "North America" and "Nearctic" as used here refer to Canada and the continental United States (including Alaska), but not Mexico.

### Summary of Nearctic Subfamilies, Supertribes, Tribes and Subtribes

I. Subfamily OMALIINAE
- 1. Tribe Omaliini
- 2. Tribe Eusphalerini
- 3. Tribe Anthophagini
- 4. Tribe Coryphiini
  - 4a. Subtribe Boreaphilina
  - 4b. Subtribe Coryphiina
- 5. Tribe Aphaenostemmini

II. Subfamily EMPELINAE

III. Subfamily PROTEININAE
- 6. Tribe Proteinini

IV. Subfamily MICROPEPLINAE

V. Subfamily DASYCERINAE

VI. Subfamily PSELAPHINAE
- VIa. Supertribe FARONITAE
- VIb. Supertribe EUPLECTITAE
  - 7. Tribe Trogastrini
    - 7a. Subtribe Trogastrina
    - 7b. Subtribe Rhexiina
  - 8. Tribe Euplectini
    - 8a. Subtribe Rhinoscepsina
    - 8b. Subtribe Panaphantina
    - 8c. Subtribe Trisignina
    - 8d. Subtribe Bibloporina
    - 8e. Subtribe Trichonychina
    - 8f. Subtribe Euplectina
    - 8g. Subtribe Bibloplectina
    - 8h. Subtribe Trimiina
  - 9. Tribe Jubini
- VIc. Supertribe BYTHINOPLECTITAE
  - 10. Tribe Bythinoplectini
    - 10a. Subtribe Bythinoplectina
  - 11. Tribe Mayetiini
- VId. Supertribe BATRISITAE
  - 12. Tribe Batrisini
    - 12a. Subtribe Batrisina
  - 13. Tribe Amauropini
- VIe. Supertribe GONIACERITAE
  - 14. Tribe Brachyglutini
    - 14a. Subtribe Brachyglutina
    - 14b. Subtribe Pselaptina
    - 14c. Subtribe Decarthrina
    - 14d. Subtribe Eupseniina
  - 15. Tribe Bythinini
  - 16. Tribe Valdini
  - 17. Tribe Tychini
  - 18. Tribe Speleobamini
- VIf. Supertribe PSELAPHITAE
  - 19. Tribe Tyrini
    - 19a. Subtribe Tyrina
    - 19b. Subtribe Hamotina
  - 20. Tribe Ceophyllini
  - 21. Tribe Tmesiphorini
  - 22. Tribe Ctenistini
  - 23. Tribe Arhytodini
  - 24. Tribe Pselaphini
- VIg. Supertribe CLAVIGERITAE
  - 25. Tribe Clavigerini

VII. Subfamily PHLOEOCHARINAE

VIII. Subfamily OLISTHAERINAE

IX. Subfamily TACHYPORINAE
- 26. Tribe Deropini
- 27. Tribe Tachyporini
- 28. Tribe Mycetoporini

X. Subfamily TRICHOPHYINAE

XI. Subfamily HABROCERINAE

XII. Subfamily ALEOCHARINAE
- 29. Tribe Gymnusini
- 30. Tribe Deinopsini
- 31. Tribe Mesoporini
- 32. Tribe Trichopseniini
- 33. Tribe Aleocharini
  - 33a. Subtribe Aleocharina
- 34. Tribe Hoplandriini
- 35. Tribe Oxypodini
  - 35a. Subtribe Oxypodina
  - 35b. Subtribe Dinardina
  - 35c. Subtribe Meoticina
  - 35d. Subtribe Blepharhymenina
  - 35e. Subtribe Tachyusina
- 36. Tribe Corotocini
  - 36a. Subtribe Eburniogastrina
- 37. Tribe Hypocyphtini
- 38. Tribe Myllaenini
- 39. Tribe Diglottini
- 40. Tribe Liparocephalini
- 41. Tribe Autaliini
- 42. Tribe Homalotini
  - 42a. Subtribe Bolitocharina
  - 42b. Subtribe Gyrophaenina
  - 42c. Subtribe Homalotina
  - 42d. Subtribe Leptusina
  - 42e. Subtribe Silusina
  - 42f. Subtribe Diestotina
- 43. Tribe Placusini
- 44. Tribe Philotermitini
- 45. Tribe Athetini
  - 45a. Subtribe Acrotonina
  - 45b. Subtribe Athetina
  - 45c. Subtribe Dimetrotina
  - 45d. Subtribe Geostibina
- 46. Tribe Crematoxenini
- 47. Tribe Falagriini
- 48. Tribe Sceptobiini
- 49. Tribe Lomechusini
  - 49a. Subtribe Lomechusina
  - 49b. Subtribe Myrmedoniina

XIII. Subfamily TRIGONURINAE

XIV. Subfamily SCAPHIDIINAE
- 50. Tribe Cypariini
- 51. Tribe Scaphiini
- 52. Tribe Scaphidiini
- 53. Tribe Scaphisomatini

XV. Subfamily PIESTINAE

XVI. Subfamily OSORIINAE
- 54. Tribe Eleusinini
- 55. Tribe Thoracophorini
  - 55a. Subtribe Clavilispinina
  - 55b. Subtribe Lispinina
  - 55c. Subtribe Thoracophorina
  - 55d. Subtribe Glyptomina
- 56. Tribe Osoriini

XVII. Subfamily OXYTELINAE
- 57. Tribe Deleasterini
- 58. Tribe Coprophilini
- 59. Tribe Thinobiini
- 60. Tribe Oxytelini

XVIII. Subfamily OXYPORINAE

XIX. Subfamily MEGALOPSIDIINAE

XX. Subfamily STENINAE

XXI. Subfamily EUAESTHETINAE
- 61. Tribe Nordenskioldiini
- 62. Tribe Fenderiini
- 63. Tribe Euaesthetini

XXII. Subfamily LEPTOTYPHLINAE
- 64. Tribe Neotyphlini

XXIII. Subfamily PSEUDOPSINAE

XXIV. Subfamily PAEDERINAE
- 65. Tribe Paederini
  - 65a. Subtribe Lathrobiina
  - 65b. Subtribe Medonina
  - 65c. Subtribe Scopaeina
  - 65d. Subtribe Stilicina
  - 65e. Subtribe Stilicopsina
  - 65f. Subtribe Astenina
  - 65g. Subtribe Echiasterina
  - 65h. Subtribe Cryptobiina
  - 65i. Subtribe Paederina
- 66. Tribe Pinophilini
  - 66a. Subtribe Pinophilina
  - 66b. Subtribe Procirrina

XXV. Subfamily STAPHYLININAE
- 67. Tribe Platyprosopini
- 68. Tribe Diochini
- 69. Tribe Othiini
- 70. Tribe Xantholinini
- 71. Tribe Staphylinini
  - 71a. Subtribe Quediina
  - 71b. Subtribe Tanygnathinina
  - 71c. Subtribe Staphylinina
  - 71d. Subtribe Xanthopygina
  - 71e. Subtribe Philonthina
  - 71f. Subtribe Hyptiomina

I. Subfamily OMALIINAE MacLeay 1825

Usually recognizable by possession of a pair of ocelli near the hind margin of the head, though some genera and species lack or appear to lack them. North American Omaliinae have open front coxal cavities; well-developed pronotal postcoxal process (except *Giulianium*); front coxae conical and prominent; a conspicuous blunt trochantin usually exposed by the open procoxal fissure (most Coryphiini with closed fissure); tarsi 5-5-5; apical maxillary palpomere about as wide as preapical or narrower, sometimes subulate; antennae inserted under the lateral margins of the frons, filiform, clavate, or moniliform; posterior face of metacoxae vertical or occasionally oblique; and abdomen with six visible sterna, a single pair of paratergites on segments 3 or 4 to 7 (7 only in *Xenicopoda*), intersegmental membranes with brick-wall-like pattern of sclerites, all spiracles well-developed and functional, and a well-developed defensive gland complex associated with an anterior projection of abdominal sternum 8 (Klinger 1980; visible only in cleared or dissected specimens). Omaliinae are generally broader in body form than "typical" staphylinids, with a shorter and less flexible abdomen; members of several genera have relatively long elytra (e.g., Figs. 2.22, 6.22), sometimes covering the whole abdomen, a feature also found in Scaphidiinae (Fig. 4.22), Empelinae (Fig. 7.22), and Dasycerinae (Fig. 10.22). Newton and Thayer (1995) divided the subfamily into seven tribes, of which five are Holarctic, one Palearctic (Hadrognathini Portevin, 1929, possibly = Omaliini), and one (Corneolabiini Steel, 1950) southern temperate, occurring in Australia, New Zealand, and southern South America. About 117 genera are placed in the subfamily worldwide; 55 of these are presently known from North America (five first recorded here), represented by just over 200 species. Dettner and Reissenweber (1991) discussed the defensive secretions of Omaliinae and Proteininae, their effectiveness, and their possible use in phylogenetic studies. Larvae of twenty genera occurring in North America have been described (Steel 1970, 19 based on British species, nearly all reared; Thayer 1985a, larva and pupa of *Brathinus*); most species appear to overwinter as larvae. Newton (1990a) included twelve genera of Omaliinae in his key to larvae of soil Staphylinidae.

1. Tribe Omaliini MacLeay 1825
Recognizable by the combination of: hind tarsi with tarsomeres 1-4 together shorter than 5 (except in *Xylodromus*, some

*Phyllodrepa* species, and a few scattered others); lacking dense setae ventrally on all tarsi of both sexes (as found in Eusphalerini); antennae usually with club of 5-6 (occasionally 3 or 7) antennomeres, antennomeres 8-10 nearly always transverse (quadrate in a few species); maxillary palpomere 4 usually as wide as 3 (if narrower, not minute and needle-like as in Coryphiini); body either flattish and parallel-sided or oval in dorsal view and convex; posterior margin of abdominal tergite III with a narrow transverse groove having tiny setae in front of it that project across it (lost in wingless and a few other species). Females have a sclerotized spermatheca and usually a characteristic internal genital sclerite, though these are only visible in cleared specimens. There are 43 described genera worldwide, 15 of which occur in North America, including a tentative first record of *Carcinocephalus*. The species occur primarily in wooded areas, though a few are at least partly synanthropic. Many are associated with rotting plant or animal material, as either predators or saprophages, and a few are pollen-feeders. Despite favoring humid microhabitats, they are not usually associated with water (e.g., streams, ponds) except for *Pycnoglypta*, found in marshy habitats, and some *Omalium* and *Micralymma* species that are littoral or riparian. The tribe is worldwide in distribution and is the only tribe of Omaliinae that occurs in the tropics. Relatively few are montane or alpine in distribution, but some are fall- or winter-active in the lowlands.

*Acrolocha* Thomson 1858

*Elonium* Leach 1819 (rejected name; ICZN 1993)

Three described species, *A. diffusa* (Fauvel 1878), Alberta and Wisconsin to New York and Newfoundland, *A. helferi* Steel 1957 and *A. leechi* Hatch 1957, California to British Columbia, Idaho, and Utah, and two undescribed species in California and the Rocky Mountains (into Mexico); in forest litter, fungi, and carrion. One of the undescribed species and *A. leechi* are flightless. Larvae of at least some species will eat insect food. Most of the nine species included in *Acrolocha* by Hatch (1957) actually belong in *Omalium*, *Hapalaraea*, *Acrulia*, and a probable new genus. Revision and key: Steel (1957) (all but *A. leechi*); Hatch (1957) (Pacific Northwest species except *A. helferi*). Biology and larva: Steel (1970). Seven additional species occur from Europe to Japan.

*Acrulia* Thomson 1858

One species, *A. tumidula* (Mäklin 1853), Alaska to Oregon; forest litter and moss, under bark of logs or trees, sometimes associated with Scolytidae; wings reduced. *Acrolocha crenulata* Hatch 1957 appears to be conspecific with this, but there also is an undescribed species occurring from British Columbia to California. Larvae will eat freshly killed insects, but adults might be scavengers or mycophages. Biology: Deyrup and Gara (1978), Steel (1970). Larva: Steel (1970). Four additional species occur in the western Palearctic, one of which is recorded from fungusy leaves, twigs, and logs (Koch 1989).

*Carcinocephalus* Bernhauer 1903 (placement of species somewhat uncertain)

Three species?, western and eastern; on fungi, under bark, in leaf and log litter. *Omalium flavidum* Hamilton 1895 (northeastern United States, southeastern Canada; wing dimorphic, in forest litter) and *Omalium exsculptum* Mäklin 1852 (=*Acrolocha rugosa* Hatch 1957) are provisionally keyed out here, along with a species recorded by Keen (1895) as "*Homalium arpedinum* Fauvel, n. sp.," but never described. The latter two species occur from British Columbia and Manitoba south to California. Redescription of *O. flavidum*: Thayer (1992). Four species are known from central and southern Europe as far east as the Balkans; the genus has not been recorded previously from North America.

*Dropephylla* Mulsant and Rey 1880

Two species, *D. cacti* (Schwarz 1899), southern California to New Mexico, and *D. longula* (Mäklin 1852), Alaska to Alberta and Montana; in flowers, forest litter, under bark, sometimes associated with and feeding on Scolytidae. *Dropephylla* has sometimes been treated as a subgenus of *Hapalaraea* or *Phyllodrepa*, but it appears to be a distinct group based on both larval (Steel 1970) and adult (Thayer, unpublished) characters. Both adults and larvae are predaceous, though some adults of *D. cacti* can feed on pollen. Larva: Steel (1970). Biology: Steel (1970), Furniss (1995; *D. longula*, as *Hapalaraea longula*). About 30 species are known from the Palearctic region and northern India, and one has been seen from Mexico.

*Hapalaraea* Thomson 1858

Two described species, *H. hamata* (Fauvel 1878), along United States-Canada boundary, south to Illinois, Arkansas, and Georgia, and *H. megarthroides* (Fauvel 1878), Pacific coast and Rocky Mountain states and provinces, and an undescribed species from British Columbia to Idaho and California; in flowers and leaf litter, possibly mammal and bird nests. Hatch (1957) placed these two species in *Acrolocha*; *A. barri* Hatch, 1957, is the same as *H. hamata*, as pointed out by Steel (1959). As used here, the genus does not include the former subgenus *Phyllodrepa*, which is treated as a separate genus (q.v.). Both adults and larvae in culture feed on freshly killed insects. Biology and larva: Steel (1970). Three additional species occur in Europe.

*Micralymma* Westwood 1838

Two species, *M. marinum* (Ström 1783), coastal from Massachusetts to Newfoundland, and *M. brevilingue* Schiødte 1845, Alaska, Yukon, and Northwest Territories; both flightless, found on ocean beaches and river shores. Revision and key to subspecies: Steel (1958, 1962). Adults and larvae of *M. marinum* are predaceous, possibly (though not exclusively) on the intertidal collembolan *Anurida maritima* (Guérin). In Britain, *M. marinum* appears to overwinter in the adult stage but may do so in the larval stage in North America. Larva: Steel (1970). Biology and distribution: Steel (1970), King *et al.* (1979), Elliott *et al.* (1983), Thayer (1985b). An additional species, also flightless, occurs in the Caucasus, *M. marinum* also occurs in the coast of Greenland and northern Europe, and *M. brevilingue* occurs in both Siberia and Greenland.

*Omaliopsis* Jeannel 1940 [NEW RECORD FOR NORTH AMERICA]

One undescribed species, from southern California, Arizona, and New Mexico, extending into Mexico (Nuevo León and Oaxaca), at middle to high elevations (1500 m in California, 1,800 to 3,600 m elsewhere), taken in small numbers in *Ceanothus* flowers, carrion trap, and pitfall trap. Little is known of the biology of *Omaliopsis* species; in other parts of the world they have been collected in decaying plant leaves and stems (Jeannel 1940, eastern Africa; Bolivia and Ecuador), in flowers (southern Africa), under bark of logs (Chile, Ecuador, eastern Africa), in fungi or rotting fruit (southern and eastern Africa, eastern Australia), in leaf litter (Chile, Peru, Australia, Madagascar, eastern and southern Africa) and in carrion and dung traps (Australia, Chile, southern Africa). There are presently only 7 described species placed in the genus, but another 20 or so need to be moved here, and numerous undescribed species have been seen (including from Taiwan, southern China, and eastern India in addition to the southern temperate and montane tropical areas listed above). Larvae are undescribed, although those of *O. russata* have been collected in large numbers under bark of an *Araucaria* log in Chile.

*Omalium* Gravenhorst 1802 (Fig. 5.22)

*Homalium* Ljungh 1804 (unjustified emendation)

Fourteen described species, generally distributed, and several new ones; in decomposing material of all kinds, fungi, some on ocean beaches, in birds' nests, or in synanthropic situations. Some species are flightless. Both adults and larvae in culture feed on freshly killed insects; some British species overwinter as adults. Larva: Steel (1970). Biology: Hicks (1959, 1962, 1971), Steel (1970). Over 70 species occur in the Palearctic region, India, and Central America (not yet described), a few of these being adventive in other regions (including at least two in North America). Aside from adventive species, "*Omalium*" species so far recorded from Central America and the southern hemisphere actually belong to other genera.

*Omalonomus* Campbell and Peck 1990

One species, *O. relictus* Campbell and Peck 1990, southern Alberta and Saskatchewan, Utah, Washington; flightless, in soil and forest litter. Recognition, distribution: Campbell and Peck (1990). The genus is restricted to North America.

*Paraphloeostiba* Steel 1960 [NEW RECORD FOR NORTH AMERICA]

One adventive species, *P. gayndahensis* (MacLeay 1871), collected on flowers of cultivated *Annona cherimola* (cherimoya) in southern California This species was originally described from Australia (where it is widespread), and was transferred to *Paraphloeostiba* by Steel (1960). It has since been recorded from New Zealand (as early as 1944: Kuschel 1990), and from France, Italy, and the Canary Islands (Tronquet 1998). It has been collected in fermenting fruit, other decaying plant material, badger dung, and flowers of Araceae (Steel 1960; Kuschel 1990; Tronquet 1998, 2000; Thayer unpublished). Tronquet (2000) reported it apparently excluding native species of *Proteinus* from decaying plant material in France. There are 30 species currently included in the genus, distributed from eastern India to Taiwan, through Indonesia to New Guinea, the New Hebrides, and Australia.

*Phloeonomus* Heer 1839

Two species, *P. suffusus* Casey 1893, Alaska to California, and *P. laesicollis* Mäklin 1852, widely distributed (including Mexico); under bark of logs and trees, sometimes associated with Scolytidae. Key: Casey (1893). Larva: Steel (1970). Biology: Deyrup and Gara (1978), Steel (1970). North American records of *P. pusillus* (Gravenhorst, 1806) appear to be misidentifications of *P. laesicollis*; *P. pusillus* is Palearctic only, as recognized by Casey (1893). The generic usage here corresponds to *Phloeonomus* (*s. str.*) of past North American usage, with *Phloeostiba* treated as a separate genus (as in Steel 1970; Zanetti 1987; Watanabe 1990). Adults and larvae of the two British species fed on banana but not insects in culture (Steel 1970), but adults and larvae of *P. laesicollis* will feed on insects (Thayer, unpublished). There are about 40 described species worldwide; members of *Phloeonomus* and closely related genera are the only truly tropical species of Omaliinae. The species keyed out as "*Phloeonomus* (*sensu lato*)" is an apparently adventive tropical species of the *Phloeonomus* group collected on flowers of cultivated *Annona cherimola* (cherimoya) in southern California. Its identity and proper generic placement are not entirely certain.

*Phloeostiba* Thomson 1858

*Distemmus* LeConte 1861

One species, *P. lapponicus* (Zetterstedt 1838), widely distributed except southeastern United States, also Palearctic; under bark of logs and trees. This genus has often been treated as a subgenus of *Phloeonomus*, but more recently separated from it (e.g., Steel 1970; Zanetti 1987; Watanabe 1990). Adults of the two British species fed on banana but not insects in culture, but larvae ate either; adults might normally be sap-feeders. The British species appear to overwinter as adults. Biology and larva: Steel (1970). There are six described and a few new species from the Palearctic, southeast Asia, and Africa.

*Phyllodrepa* Thomson 1859

Eleven described and several new species, widely distributed but with more species in the West; in flowers, litter, bird and mammal nests and middens. One species, *P. floralis* (Paykull 1789), is apparently an adventive European species. This genus has often been treated as a subgenus of *Hapalaraea*, but appears to be a distinct group, as discussed by Steel (1970). Adults and larvae are predaceous, though some adults may also be pollen-feeders. Larva: Steel (1970). Biology: Hicks (1959, 1971), Steel (1970). There are about 20 Palearctic species; a few species recorded from South America are variously adventive or misplaced.

*Pycnoglypta* Thomson 1858
Three species, *P. aptera* Campbell 1983, Manitoba and Michigan to Nova Scotia; *P. campbelli* Gusarov 1995, Alaska through Illinois to Nova Scotia; and *P. heydeni* Eppelsheim 1886, Alaska, Yukon, and Northwest Territories; flightless (two species) or wing-dimorphic (*P. campbelli*) in damp litter and around bases of plants in swamps, bogs, and wet prairies, and one record from a vole nest. Old records from North Carolina and Georgia are probably misidentifications. Campbell (1983a) and earlier authors have recorded *P. lurida* (Gyllenhal, 1813) from North America, but Gusarov (1995) showed that North American specimens are not *P. lurida* and described them as *P. campbelli*. Key: Campbell (1983a; two species, *P. campbelli* as *P. lurida*). Recognition notes, distribution: Gusarov (1995). There are nominally six additional species in the Palearctic, but placement of three of them here is doubtful. An additional species, *"Omalium" fractum* Fauvel, 1878, (eastern United States and Ontario; in forest litter and under bark) will also key out here, but probably belongs in a separate genus rather than *Pycnoglypta*.

*Xylodromus* Heer 1839
Two described species, *X. concinnus* (Marsham 1802), Pacific coast and across Canada and northern United States (also Palearctic), and *X. capito* Casey, 1893, Wisconsin to Maryland, and undescribed species from eastern and western North America; in mammal and bird nests, forest litter, one (*X. concinnus*) partly synanthropic, occasionally in stored food products or in barns. It is not clear whether *X. concinnus* is indigenous or adventive in North America. Old North American records of *X. depressus* (Gravenhorst, 1802) are doubtful; these may refer to *Dropephylla longula* and/or *X. concinnus*. At least the larvae are predaceous. Larva: Steel (1970). Biology: Hicks (1959, 1962, 1971), Steel (1970). There are 11 Palearctic species in addition to *X. concinnus*, which is also adventive in New Zealand, Australia, and southern South America.

2. Tribe Eusphalerini Hatch 1957
Anthobiini *auctorum*
Eusphalerini, containing the single Holarctic and northern Oriental genus *Eusphalerum*, can be recognized by having tarsomeres 1-4 broadened and with dense setae ventrally on all tarsi of both sexes, tarsomeres 1-4 short, together usually no longer than 5, and the apical 5-6 antennomeres wider than preceding ones. The females of some species have elytra longer than the males, sometimes covering the entire abdomen and sometimes prolonged into a narrowly rounded point near the suture; in some species the sexes differ in color.

*Eusphalerum* Kraatz 1858
*Anthobium* auctorum (misidentified, not Leach 1819)
*Onibathum* Tottenham 1939 (subgenus) (syn. of *Eusphalerum* (*s. str.*) in Zanetti 1987)
Twenty-seven species, widely distributed, mostly western, few central; pollen-feeding in flowers of shrubs, herbs, and trees, often collected with *Amphichroum* and/or *Pelecomalium* in the West. One species recorded from Newfoundland, *E. torquatum* (Marsham 1802), is a European species; the validity of this record is questionable. Most North American species have at least the head, pronotum, and elytra pale yellowish-brown. Though adults seem to be universally pollen-feeders, the larvae are not and their food remains uncertain. Keys: Casey (1893; 8 Californian species, as *Anthobium*), Hatch (1957; Pacific Northwest species). Larva: Steel (1970). Biology and defensive glands: Steel (1970), Klinger and Maschwitz (1977), Klinger (1978, 1983), Drugmand (1992). The genus is divided into four subgenera, at least two of which, including the typical subgenus, occur in North America, but most of the North American species have not been assigned to subgenera and Hatch's (1957) usage of subgenera may not be correct. Around 200 species occur in the Palearctic and northern Oriental regions and one has been seen from Baja California.

3. Tribe Anthophagini Thomson 1859
This group has been something of a taxonomic dumping-ground. It is difficult characterize, and as presently delimited is unlikely to be a natural, or monophyletic, group. As circumscribed here, Anthophagini normally have the hind tarsi relatively long, with tarsomeres 1-4 together longer than 5; the antennae filiform, with antennomeres 8-10 quadrate to elongate; and the maxillary palpomere 4 usually as wide and as long as (or longer than) 3 (if narrower, not minute and needle-like as in Coryphiini); many have relatively elongate "loose-jointed" bodies and long legs, and several genera have long elytra that cover most of the abdomen. There are about 40 genera in the Holarctic and Oriental regions, of which 27 occur in North America. Members of many genera are predators associated with streams, other water sources (waterfalls, bogs, marshes), or snowfields, although at least three genera have turned to pollen-feeding as adults, and a few may be mycophagous or saprophagous. Many are montane or arctic/alpine in distribution, and several are cool-season-active. The genus *Brathinus* is included here for the first time in a North American faunal treatment, following Hammond (1971) and Thayer (1985a), and two formerly Palearctic genera, *Mannerheimia* and *Phyllodrepoidea*, are recorded from North America for the first time.

*Acidota* Stephens 1829
Three species, *A. crenata* (Fabricius 1792), northern transcontinental, southward to New Mexico and Virginia (also Palearctic), *A. quadrata* (Zetterstedt 1838), flightless, Alaska and across Canada (also Palearctic), and *A. subcarinata* Erichson 1839, eastern North America; in litter and moss in forests and bogs. Some species apparently do not eat insect food (Steel 1970), but at least adults of some do (Thayer unpublished). Revision and key: Campbell (1982b). Biology and larva: Steel (1970). There are ten additional species from the Palearctic and Taiwan; North American records of the European *A. cruentata* Mannerheim 1830 appear to be based on misidentifications.

*Amphichroum* Kraatz 1858
*Stachygraphis* Horn 1883
Three species, *A. floribundum* LeConte 1863, *A. maculatum* Horn 1882, and *A. maculicolle* (Mannerheim 1843), all from Pacific coast states and provinces (the first also in Baja California); pollen-feeding in flowers of shrubs, herbs, and trees, including *Alnus* and *Salix*,

often collected with *Pelecomalium* and/or *Eusphalerum*. Key: Casey (1886b; most species transferred to *Pelecomalium* by Casey 1893). Eleven species are known from the Palearctic region and India.

*Anthobioides* Campbell 1987

One species, *A. pubescens* Campbell 1987, western Washington; in wet moss and litter along mountain creeks. Recognition: Campbell (1987). The genus is restricted to North America.

*Anthobium* Leach 1819 (Fig. 6.22)

*Deliphrum* Erichson 1839

*Lathrimaeum* Erichson 1839

Twelve species, Pacific coast and Rocky Mountain states and provinces, South Dakota, and Virginia. Adults and larvae are predators as far as known, found on carrion, dung, mushrooms, in conifer and hardwood litter, in mammal and bird nests. *Anthobium* and *Deliphrum* have been separated in some European and North American literature, but were synonymized by Moore (1966). There is a question regarding the placement of some North American species, and the complex of genera related to *Anthobium* needs worldwide revision. Keys to some species: Casey (1893; as *Lathrimaeum*), Hatch (1957; Pacific Northwest species). Biology and larva: Steel (1970). There are about 40 species widespread in the Palearctic and in India, and one has been seen from Mexico.

*Anthophagus* Gravenhorst 1802

One apparently undescribed species, northern California (Moore 1966); Palearctic species mostly montane or alpine, adults found in flowers, on foliage, occasionally in marmot nests, and larvae usually near streams. Both adults and larvae are active predators, the adults of some species foraging for aphids and other prey on the foliage of shrubs (often near water). Larva: Steel (1970). Biology: Horion (1963), Claridge and Murphy (1967), Steel (1970). There are about 40 species distributed throughout the Palearctic region.

*Arpedium* Erichson 1839

Three described species, *A. angulare* Fauvel 1878, northeastern United States and southeastern Canada, *A. cribratum* Fauvel 1878, Iowa to Ohio, and *A. schwarzi* Fauvel 1878, eastern United States and Ontario; usually in flood debris, litter, or moss near water, occasionally in vole nests or beaver houses. One or two undescribed species have been seen from central and eastern North America. Campbell (1984b) separated *Eucnecosum* from *Arpedium* for the first time in North American literature. He treated *A. angulare* as a synonym of *A. cribratum* (which he also recorded from British Columbia), but they appear to be distinct species: the former is flightless, the latter fully winged, and their male characters differ. Revision and key to most spp: Campbell (1984b). Larvae are undescribed; adults and larvae of *A. angulare* will feed on insect food (Thayer, unpublished). An additional species, *A. quadrum* (Gravenhorst 1806), occurs in Europe, but records of its occurrence in North America appear to be based on misidentifications.

*Artochia* Casey 1893

Two species, *A. productifrons* Casey 1893 and *A. californica* Bernhauer 1912, Alaska to California, Utah, Idaho; in stream edge moss and litter, probably predaceous.The genus is restricted to North America, and the two specific names may be synonyms.

*Brathinus* LeConte 1852 (Fig. 2.22)

Three species, *B. nitidus* LeConte 1852, eastern United States and southeastern Canada, *B. varicornis* LeConte 1852, northeastern United States and southeastern Canada, and *B. californicus* Hubbard 1894, California to Washington and Idaho; adults and larvae predaceous, living in riparian or bog moss and litter, *B. nitidus* sometimes in caves. The genus was originally put in Scydmaenidae, later placed in Silphidae or its own family Brathinidae, before settling in the Omaliinae (initially placed there by Casey 1897, supported by Hammond 1971 and Thayer 1985a). Keys: Hammond (1971), Frank *et al.* (1987). Biology and larva: Thayer (1985a). Distribution and ecology: Peck (1975). Two additional species are known from Japan.

*Deinopteroloma* Jansson 1946

*Mathrilaeum* Moore 1966

Two species, *D. pictum* (Fauvel 1878), California to British Columbia, and *D. subcostatum* (Mäklin 1852), California to Alaska; in forest litter, riparian litter, mushrooms. Both North American species were originally placed in *Anthobium* (as *Lathrimaeum*). Revision and key: Smetana (1985). Nine additional species are known from southern and southeastern Asia (India and Nepal to Taiwan).

*Eucnecosum* Reitter 1909

*Revelstokea* Hatch 1957

Three Holarctic species in Alaska and across Canada, *E. brachypterum* (Gravenhorst 1802) (also Maine), *E. brunnescens* (Sahlberg 1871) (also Colorado and northern New York), and *E. tenue* (LeConte 1863) (also Rocky Mountain states and northern Michigan); boreal, in forest and tundra litter, usually near water, *E. brunnescens* occasionally in bird nests. Most individuals are flightless, but occasional specimens of *E. tenue* and *E. brachypterum* are fully winged. Adults and larvae are predaceous, and at least some species may overwinter as adults. There may be an additional species from Siberia, described in "*Microlymma*," but that may not be distinct from the three listed. Revision and key: Campbell (1984b). Biology and larva: Steel (1970; as *Arpedium*).

*Geodromicus* Redtenbacher 1856

*Psephidonus* Gistel 1856

Twelve species, generally distributed, mostly montane and western (as far north as Alaska); adults and larvae predaceous, riparian, often in moss or wet litter. They may overwinter as adults. The name *Psephidonus* has been used for this genus following Blackwelder (1952), but Zerche (1987) dated *Geodromicus* as having been published earlier in 1856. Keys: Casey (1893; nine species), Hatch (1957; Pacific Northwest species). Biology and larva: Steel (1970). Over 100 species have been described from Europe to Japan, including the Himalayas.

*Lesteva* Latreille 1796
*Pseudolesteva* Casey 1893
*Tevales* Casey 1893
*Paralesteva* Casey 1905
*Lesta* Blackwelder 1952
Two species, the flightless *L. cribratula* (Casey 1893), Kentucky, Tennessee, Virginia, and North Carolina and fully winged *L. pallipes* LeConte 1863, widespread in central and eastern United States and Canada; both species riparian, in litter and moss, sometimes in caves. Adults and larvae are active predators. Key: Casey (1893; as *Pseudolesteva pallipes* and *Tevales cribratulus*). Biology and larva: Steel (1970). About 75 species are known from Europe to Japan, including the Himalayas.

*Mannerheimia* Mäklin 1880 [NEW RECORD FOR NORTH AMERICA]
One apparently new species, far northern California; on mushrooms. Eighteen species are known from Europe to the Himalayas and Japan, mostly alpine or subalpine, in litter or dung and under stones (Koch 1989).

*Microedus* LeConte 1874
Seven described and several new species, one (*M. austinianus* LeConte 1874) northern transcontinental, the rest western from Alaska to California and Alberta to Arizona; larvae and probably adults predaceous, mostly montane, riparian, on banks or in moss and litter. Keys: Hatch (1957; six species), Moore and Legner (1972b; seven species). Larvae are undescribed. Species have been described only from North America, but an undescribed species has been seen from China.

*Olophrum* Erichson 1839
*Lathrium* LeConte 1850
Seven species, six boreal and alpine, Alaska south to Arizona and across the Canada -United States border region to Newfoundland, and one (*O. obtectum* Erichson 1840), eastern United States and southeastern Canada; in litter, moss, and other vegetation at stream, pond, or bog edges, occasionally in rodent nests or middens. One species is flightless, and one wing dimorphic. At least among the British species, some appear to overwinter as larvae and some as adults. Revision and key: Campbell (1983b). Biology and larva: Steel (1970). Four of the boreal North American species are Holarctic, and there are at least 30 additional Palearctic species known.

*Omalorphanus* Campbell and Chandler 1987
One species, *O. aenigma* Campbell and Chandler 1987, Oregon; flightless, found in litter in conifer forest, and active on snow in spring. Campbell and Chandler (1987). The genus is restricted to North America.

*Orobanus* LeConte 1878
Seven species, some flightless, Pacific coast and Rocky Mountain states and provinces, east to Nebraska; in flood debris and moss, on and under stones along streams. Key: Mank (1934; six species), Hatch (1957; four species). The genus is restricted to North America.

*Orochares* Kraatz 1858
*Paradeliphrum* Hatch 1957 (one of originally included species transferred to *Porrhodites*)
Two species, *O. suteri* Campbell 1984, Wisconsin and Illinois, on logs, and *O. tumidus* (Hatch 1957), British Columbia and Oregon; both fall-active. Fauvel's (1878) record of the European *O. angustatus* (Erichson 1839) was regarded as doubtful by Campbell (1984c). Revision and key: Campbell (1984c). World checklist: Thayer (1993). Four species are known from Europe, Iran, and Japan, one of which is recorded from dung and rotting plant material (Koch 1989).

*Pelecomalium* Casey 1886
*Heterops* Mannerheim 1843 (not Blanchard 1842)
Twelve species, Alaska to California, Idaho, and Nevada, and northeastern U.S.-southeastern Canada; pollen-feeders in flowers of shrubs, herbs, and trees, including *Ceanothus*, *Alnus*, and *Salix*, often collected with *Amphichroum* and *Eusphalerum* in the West. Keys: Casey (1893); Hatch (1957; 3 Pacific Northwest species). The genus is restricted to North America.

*Phlaeopterus* Motschulsky 1853
*Tilea* Fauvel 1878
Fourteen species, Pacific coast and Rocky Mountain states and provinces; montane to alpine, along streams, seeps, and snowfields. Key: Casey (1893; seven species), Hatch (1957; 11 species). Larvae are undescribed. The genus is restricted to North America.

*Phyllodrepoidea* Ganglbauer 1895 [NEW RECORD FOR NORTH AMERICA]
One apparently new species, California; in forest pitfall trap. Larva: Steel (1970). Biology: Steel (1970), Crowson (1982). The single described European species, *P. crenata* (Gravenhorst 1802), is found under tight or moldy bark of logs (mostly hardwood), in lichens and moss, or on slime molds, often active fall or winter and overwintering as both adults and larvae (Steel 1970, Crowson 1982, Koch 1989). They may feed on fungal spores or conidia (Crowson 1982).

*Porrhodites* Kraatz 1858
The genus contains only two species, *P. fenestralis* (Zetterstedt 1828), Alaska and British Columbia to Labrador, also Wyoming and Palearctic, found in mouse nests, under dung, in carrion and rotting plant debris; and *P. inflatus* (Hatch 1957), a wing-dimorphic species from northeastern US, southeastern Canada, and Pacific coastal and Rocky Mountain states and provinces, in litter, moss, and on or under snow in fall and early winter. Campbell (1984c). Biology: Koch (1989; Europe).

*Tanyrhinus* Mannerheim 1852
One species, *T. singularis* Mäklin 1852, Alaska to California and perhaps Arizona; in litter, moss, under logs and on fungi. The genus is restricted to North America.

*Trigonodemus* LeConte 1863
Two species, *T. fasciatus* Leech 1939, British Columbia and Oregon, and *T. striatus* LeConte 1863, northeastern United States

and southeastern Canada; in fungi and rotting logs. Review and key: Smetana (1996). Six additional species are known from China, Taiwan, and Japan.

*Unamis* Casey 1893
Five species, British Columbia to California, Idaho to Arizona; montane and alpine, in moss, wet debris, or on banks along streams. Keys: Hatch (1957), Moore and Legner (1972a). Species have been described only from North America, but an undescribed species has been seen from China.

*Vellica* Casey 1885
One species, *V. longipennis* Casey 1885, British Columbia to California; montane and alpine, in wet debris and moss along streams. The genus is restricted to North America, but doubtfully distinct from *Phlaeopterus*.

*Xenicopoda* Moore and Legner 1971
One species, *X. helenae* Moore and Legner 1971, coastal mountains of southern California; pollen-feeding in flowers of *Ceanothus crassifolius*. Redescription: Thayer (1978). The genus is restricted to North America.

4. Tribe Coryphiini Jacobson 1908
Coryphiini can be separated from nearly all other Omaliinae in North America by the form of the maxillary palps with palpomere 3 enlarged, pear-shaped in some species; palpomere 4 small, narrow, needle-shaped or short and conical, often partly hidden in end of 3 or among the setae there. *Orobanus* and *Omalorphanus* (both Anthophagini) have the fourth palpomere nearly as small as Coryphiini, but the former has the apices of the elytra broadly arcuate rather than truncate and the latter lacks a distinct nuchal constriction and has the elytra shorter than the pronotum. The tribe occurs only in the Northern Hemisphere, primarily at high latitudes and/or altitudes. There are 24 genera world wide, 10 of which occur in North America. With a few exceptions, Coryphiini are cold-loving and generally occur in arctic or alpine areas, often near cold streams or snowfields. In a revision of the North American genera and species, Campbell (1978c) questioned inclusion of *Ephelinus* in Coryphiini and excluded *Haida*, *Pseudohaida*, and *Eudectoides* from it. Zerche (1990, 1993; revision and phylogenetic analysis of Palearctic genera and species) maintained that *Ephelinus* definitely belongs within Coryphiini and suggested that *Haida*, *Pseudohaida*, and *Eudectoides* were certainly allied with the group and might be the sister group of the rest of the tribe. Zerche (1990), on the basis of the Palearctic fauna, divided Coryphiini into two subtribes, both of which occur in North America. He did not, however, assign all North American genera to subtribes, and placement here of some North American genera in Coryphiina is provisional.

4a. Subtribe Boreaphilina Zerche 1990
Zerche (1990) characterized this tribe as having slightly raised oblique ridges at the back of the head, the surface between them sloping gradually to the neck; gular sutures close together, nearly parallel; ocelli (absent in some) placed relatively far forward and more or less close together; mandibles elongate; and pronotum more strongly convex than in Coryphiina. All of these characters are somewhat variable within the subtribe. Coryphiini can be separated from nearly all other Omaliinae in North America by the form of the maxillary palps with palpomere 3 enlarged, pear-shaped in some species; palpomere 4 small, narrow, needle-shaped or short and conical, often partly hidden in end of 3 or among the setae there. Zerche (1990,1993) placed six genera in this subtribe, including the following two found in North America, although *Gnathoryphium* does not have the characteristic pronotal features. The parallel gular sutures appear to be the most consistent and readily observable feature of the subtribe.

*Boreaphilus* Sahlberg 1832
One species, *B. henningianus* Sahlberg 1832, Alaska, Yukon Territory, Manitoba to New Brunswick, and New Jersey (also northern Palearctic); nearly always flightless, found in damp or wet places, active late fall and early spring in the eastern part of its range. Description, North American distribution: Campbell (1978c). There are about 20 additional Palearctic species of the genus distributed from Europe to Japan.

*Gnathoryphium* Campbell 1978
One species, *G. mandibulare* Campbell 1978, Washington; flightless, found near snow field and in mountain stream debris. Description, distribution: Campbell (1978c). The genus is restricted to North America.

4b. Subtribe Coryphiina Jacobson 1908
Zerche (1990) characterized this tribe as having vertex of head strongly convex posteriorly, sloping steeply toward neck; body (except mandibles and ocelli) uniformly black; anteriorly converging gular sutures; and ocelli usually placed far back on the head and distant from each other. As in Boreaphilina, these features are not present in all members of the subtribe. Zerche (1990, 1993) placed fifteen genera here, not including the strictly North American ones, which are provisionally placed here on the basis of his subtribal characters.

*Coryphiomorphus* Zerche 1988
One species, *C. hyperboreus* (Mäklin 1880), Alaska and Yukon Territory (also northern Scandinavia and northwest Russia); under stones near stream, near snowfields. Biology: Campbell (1978c; as *Coryphium*), Zerche (1990). One other species is known from the High Tatra Mountains of eastern Europe.

*Coryphium* Kirby 1834
Three species, *C. arizonense* (Bernhauer 1912), British Columbia and Idaho to southern California, *C. brunneum* (Hatch 1957), flightless, southeastern British Columbia and northern Idaho, and *C. nigrum* Campbell 1978, northern New York and Quebec to Massachusetts and Nova Scotia; winter active or in cold habitats, variously in squirrel middens, peat, or stream-edge moss and litter. Key: Campbell (1978c; including *C. hyperboreum*, and *C. vandykei*,

which were moved to *Coryphiomorphus* and *Occiephelinus*, respectively, by Zerche 1990). According to Zerche (1990), none of the North American species placed in *Coryphium* are congeneric with Palearctic *Coryphium* species, but this situation has not yet been resolved. Larvae are undescribed. Sixteen species occur from western Europe to the Himalayas and Japan.

*Ephelinus* Cockerell 1906
*Ephelis* Fauvel 1878 (not Lederer 1863)
Three species, *E. guttatus* (LeConte 1863), southeastern United States, *E. notatus* (LeConte 1863), eastern United States and Ontario, and *E. pallidus* (LeConte 1863), Kansas and Illinois; in litter near water, mostly fall and spring. Key: Campbell (1978c). Campbell (1978c) expressed reservations about including *Ephelinus* in Coryphiini, but Zerche (1990) argued that it did indeed belong here, although he did not place it to subtribe. The genus is restricted to North America.

*Eudectus* Redtenbacher 1857
*Eudectoides* Campbell 1978 (synonymized by Zerche 1990)
One species, *E. crassicornis* LeConte 1885, Florida and Louisiana; one series known from palmetto and oak litter. Campbell (1978c) moved this species from *Eudectus* to *Eudectoides*, which he excluded from Coryphiini, but Zerche (1990, 1993) regarded *Eudectoides* as a synonym of *Eudectus*, which he placed in the subtribe Coryphiina. *Eudectus* also includes seven northern Palearctic species ranging from Scotland to Japan.

*Haida* Keen 1897
Three species, Alaska to Idaho and California; forest litter, sometimes near streams. Key: Campbell (1978c). A fourth species, *Omalium callosum* Mäklin 1852, from Alaska also belongs here. Campbell (1978c) excluded the genus from Coryphiini, but Zerche (1990) included it; it may be closely related to *Eudectus*. Species have been described only from North America, but a species from China has been discovered recently.

*Holoboreaphilus* Campbell 1978
One Holarctic species, *H. nordenskioldi* (Mäklin 1878), Alaska and northern Canada, also northern Russia; flightless, in arctic tundra, usually near streams or pools. Larvae are undescribed. Distribution, biology: Campbell (1978c), Zerche (1990).

*Occiephelinus* Hatch 1957
One species, *O. vandykei* Hatch 1957, Washington and Oregon; wet moss or debris along cold streams, under stones near snowfields. Campbell (1978c) synonymized the genus with *Coryphium*, but Zerche (1990) resurrected it (see *Coryphium*). The genus is restricted to North America.

*Pseudohaida* Hatch 1957
One species, *P. rothi* Hatch 1957, Oregon, British Columbia; old growth forest. Review: Campbell (1978c). Distribution, ecology: Campbell and Winchester (1993). Campbell (1978c) moved the second of Hatch's original species to *Subhaida*. He also excluded the genus from Coryphiini, but Zerche (1990) regarded it as belonging there. The genus is restricted to North America.

*Subhaida* Hatch 1957
Seven species, Rocky Mountains and westward to Alaska, British Columbia, and California; in litter and moss in forest, sometimes near streams, some species fall-active; four of the species are flightless. Key: Campbell (1978c). Larva: Newton (1990a; in key). The genus is restricted to North America.

5. Tribe Aphaenostemmini Peyerimhoff 1914
Body extremely elongate, flattened, parallel-sided, lacking ocelli, neck without lateral constriction, antenna without club, and posterior face of metacoxa oblique. This group was recently redefined by Newton and Thayer (1995) to include two genera, *Aphaenostemmus* Peyerimhoff 1914 (known from northern Africa to northern India, most recently placed in its own subfamily) and *Giulianium* (from California, originally placed in Phloeocharinae). Nothing is known of their biology beyond the intertidal habitat of *Giulianium*.

*Giulianium* Moore 1976
Three species, *G. campbelli* Moore 1976, and *G. newtoni* Ahn and Ashe 1999, northern and central coastal California, and *G. alaskanum* Ahn and Ashe 1999, coastal Alaska, under boards in intertidal zone. Key: Ahn and Ashe (1999). The genus is known from western North America and Japan.

II. Subfamily EMPELINAE Newton and Thayer 1992
Recognizable as small, convex, clambid-like beetles with hypognathous head, elytra nearly covering abdomen (see Fig. 7.22, no more than three abdominal segments exposed), antenna with loose trimerous club, mesothoracic spiracles connected by a sclerotized band behind procoxae, each femur with ventral tibial groove, metacoxae posteriorly excavate to receive retractile femora, spiracles on abdominal tergites I-VIII functional, and without ocelli, epistomal suture, elytral striae, or pronotal postcoxal process. They also have antennal insertions at front margin of head, concealed by a small shelf-like projection; protrochantin exposed; tarsi 5-5-5; and abdomen with six visible sterna and one pair of paratergites on segments III-VII. The single included genus was originally placed in Clambini (then part of Silphidae) and later Clambidae (e.g., Hatch 1957; Arnett 1963), but moved to Leiodidae by Crowson (1955). Crowson (1960) suggested it was a very primitive staphylinoid, and Hammond (1971) proposed that it had strong affinities with Proteininae, noting the presence of the eighth sternum gland projection found in that subfamily, Omaliinae, and related subfamilies. The relationships of the subfamily within the Omaliine Group were discussed by Newton and Thayer (1992, 1995) but remain uncertain; Hansen (1997a) treated it as a distinct family. This is the only staphylinid subfamily restricted to North America and the only one in that area whose larva remains unknown.

*Empelus* LeConte 1861 (Fig. 7.22)
One species, *E. brunnipennis* (Mannerheim 1852), southern Alaska to California; in forest leaf litter, fungi, and moss at stream edge. Review: Newton and Thayer (1992, 1995). The genus is restricted to North America.

III. Subfamily PROTEININAE Erichson 1839
Pteroniinae Moore 1964

Broad-bodied staphylinids, with postcoxal process of pronotum absent but with large triangular, well-sclerotized spiracular peritreme behind front coxa; abdomen with six visible sterna, one pair of paratergites on segments III-VII, intersegmental membranes without brick-wall-like pattern of sclerites, spiracles on segments 4-6 nonfunctional, reduced or absent; antennae inserted under the sides of the frons; anterior coxae transverse, subconical, and somewhat prominent, protrochantins broadly exposed; hind coxae transverse; tarsi 5-5-5 (4-4-4 in some southern hemisphere taxa). Like the Omaliinae and other related groups, they have a well-developed omaliine-type defensive gland associated with an anterior projection of abdominal sternum 8 (Klinger 1980), but this can only be seen by clearing or dissecting specimens. Proteininae are small beetles, under 3 mm. long, living in fungi, under bark, and in leaf litter and decaying vegetation, and the North American species are probably saprophagous or mycophagous. Five tribes with a total of 11 genera are currently recognized (Newton and Thayer 1995); four of these (Anepiini Steel 1966, Austrorhysini Newton and Thayer 1995, Nesoneini Steel 1966, and Silphotelini Newton and Thayer 1995, with a total of eight genera) are restricted to temperate areas of the southern hemisphere, while the fifth (Proteinini) occurs in the northern hemisphere, tropics, and some southern temperate areas. Two genera and 22 species are known from North America. Larvae: Steel (1966).

6. Tribe Proteinini Erichson 1839
Relatively broad-bodied, with long elytra for Staphylinidae, aedeagus with parameres fused or absent, female genitalia characteristically modified. The tribe includes three genera, widespread but predominantly Holarctic, two of which occur in North America; the third, *Metopsia* Wollaston 1854, includes eight species in the western Palearctic. Proteinini are usually associated with decaying animal or plant material or fungi. Biology: Newton (1984; feeding habits), Dettner and Reissenweber (1991; defensive secretions).

*Megarthrus* Stephens 1829 (Fig. 8.22)
Twelve species, widespread but mostly western and eastern; saprophagous or mycophagous, found in leaf litter and rotten wood, fungi, dung, carrion, and flood debris. Revision and key: Cuccodoro and Löbl (1996). Larva: Kemner (1925), Steel (1966). Biology: Newton (1984), Cuccodoro (1995). Over 100 additional species are known, from all regions except Australia, New Zealand, and temperate South America.

*Proteinus* Latreille 1796
*Pteronius* Blackwelder 1952
Ten species, two generally distributed, the rest western; believed to be saprophagous or mycophagous, sometimes predaceous, found in decaying fungi, carrion, and plant debris. Key, identification notes for some species: Hatch (1957), Frank (1979b). Conservation of the name *Proteinus*: ICZN (1969). Larva: Frank (1991), Steel (1966). Biology: Newton (1984), Cuccodoro (1995). Two species recorded from North America are also Palearctic, and 23 additional species are described from the Neotropical, Palearctic, and Oriental regions.

IV. Subfamily MICROPEPLINAE Leach 1815

Recognizable by antennae with nine antennomeres, the ninth enlarged into an oval club; pronotum strongly transverse, explanate at sides; elytra with longitudinal ridges (weak in *Kalissus*); prothorax with deep excavations underneath for reception of the antennae; procoxae separated by a prosternal process; all three pairs of coxae widely separated; tarsi 4-4-4 (appearing 3-3-3, first tarsomere very small); spiracles on abdominal segments 3-6 reduced and nonfunctional. They also have antennal insertion concealed under frontal shelf; protrochantin exposed; and abdomen with six visible sterna and 0-1 pairs of paratergites on segments IV-VII. There are five described genera, widespread in northern temperate and tropical areas of the world; two of these occur in North America, represented by 15 species. Campbell (1968 and supplements cited below) revised the subfamily for the New World, and Campbell (1984a) catalogued the North American species. Matthews (1970) presented Pliocene records of *Micropeplus* and *Kalissus* from Alaska. The group has been treated as either a subfamily of Staphylinidae (e.g., Campbell 1968, 1973a; Newton and Thayer 1992, 1995) or as a separate family (e.g., Hinton and Stephens 1941; Campbell 1978b,1984a, 1989; Newton 1984; Löbl and Burckhardt 1988).

*Kalissus* LeConte 1874
One species, *K. nitidus* LeConte 1874, coastal British Columbia and Washington; on mud flat and pebbly lake margin. Campbell (1968, 1973a). The genus is restricted to North America.

*Micropeplus* Latreille 1809 (Fig. 9.22)
Fourteen species, generally distributed, one of them Holarctic; probable saprophages or mold feeders in a variety of habitats, including fungi, leaf litter, lake shores or marshy areas, and mammal and bird nests. Revision and keys, etc.: Campbell (1968, 1973a, 1978b, 1989). Biology: Hinton and Stephens (1941), Newton (1984). Larva: Newton (1991). Pupa: Hinton and Stephens (1941). Nearly 50 additional species occur across Mexico, Europe, and Asia.

V. Subfamily DASYCERINAE Reitter 1887

Easily recognizable among staphylinids by the distinctive, latridiid-like habitus alone (Fig. 10.22). Other features include very slender, verticillate antennae with apical antennomeres wider, antennae inserted under a large knob on the side of the frons; roughly sculptured and tuberculate head and pronotum and tricostate elytra, all usually with encrusted surfaces; presence of pselaphine-like foveae on the thorax and abdomen; 3-3-3 tarsi; and a small complex glandular structure of the omaliine type on the eighth sternum. They have the procoxal fissure narrowly open, the protrochantin partly exposed; and abdomen with six visible sterna and one pair of paratergites on segments V-VI. The single included genus was originally placed in Latridiidae, but removed to a family of its own in Staphylinoidea by Crowson (1955), and later placed as a subfamily of Staphylinidae most closely related to Pselaphinae and Protopselaphinae (Thayer 1987, Newton and Thayer 1995). Four species occur in North America.

*Dasycerus* Brongniart 1800 (Fig. 10.22)
Four species, three in the southern Appalachian Mountains from western Virginia to northern Alabama, and one in the coastal ranges and Sierra Nevada of central California; flightless forest inhabitants, associated with fungusy logs and debris on the forest floor where adults and larvae are evidently mycophagous on a variety of fungi. Revision and key: Löbl and Calame (1996). Biology: Löbl and Calame (1996), Newton (1991), Wheeler and McHugh (1994). Larva: Newton (1991). The genus also includes 13 species, a few of them fully winged, that are widespread through the Palearctic and Oriental regions.

VI. Subfamily PSELAPHINAE Latreille 1802 (by D. S. Chandler)

Pselaphinae were maintained as a separate family until Newton and Thayer (1995) thoroughly documented their placement in the "Omaliine group" of subfamilies in the Staphylinidae. Pselaphines are readily recognized by their distinctive appearance: typically more compact than most other staphylinids; small (0.5-5.5 mm long, average 1.5); head and pronotum narrower than the elytra; antennae apically clubbed in all but Faronitae; and possession of a unique pattern of foveation (Fig. 95.22; somewhat reduced in the more derived groups). They also have antennae inserted under shelf-like frontal projections, usually with 11 antennomeres, but with only 10, 9, or 3 in a few genera (also 4-8 in some non-North American ones); maxillary palp with 1-4 (usually 4) palpomeres and a small lightly sclerotized or unsclerotized apical pseudosegment (appearing to be a fifth palpomere in most); procoxal fissure closed, trochantin concealed; tarsi 3-3-3 (apparently 2-2-2 in Bythinoplectitae); abdomen with first three or (usually) four segments each with one pair of paratergites, which are sometimes mostly or entirely fused to tergite and/or sternum (Clavigeritae with paratergites of first three visible segments fused to each other); and five or six visible abdominal sterna.

Pselaphines are predators, and do not feed on molds as may be implied by the older common name, "ant-like mold beetles" (which presumably referred to their frequent occurrence in leaf mold, i.e., humus). They are potentially important as indicators of old-growth habitats (Vit 1985, Chandler 1987, Balleto and Casale 1991). The few described larvae of this large group were recently characterized by De Marzo (1985, 1987, 1988a), Newton (1991), and Kaupp (1997), as were the eggs (De Marzo 1986b). Ecology and behavior of some species have been treated for larvae (De Marzo and Vit 1982, De Marzo 1986a, 1988a), pupae (De Marzo 1988b), free-living adults (Engelmann 1956; Park 1932b, 1947, 1956; Reichle 1967; De Marzo and Vit 1982), myrmecophilous and termitophilous adults (Park 1932a, 1933, 1935a, 1935b; Akre and Hill 1973; Kistner 1982; Leschen 1991), and attraction to ultraviolet light (Wolda and Chandler 1996). Important morphological features were recently described by De Marzo (1989, ovarioles), De Marzo and Vovlas (1989, setose foveae and defensive glands), and Ohishi (1986, abdominal structure). The most recent catalog for North America (Chandler 1997) summarized unpublished and published biological information for all species, as well as taxonomic information. The classification here follows the world classification of Newton and Chandler (1989) as modified by Newton and Thayer (1995), with some exceptions noted in the Euplectitae; it thus differs slightly from older classifications used in the most recent catalog (Chandler 1997) and identification keys to genera (Chandler 1990b). Of the 1,136 genera worldwide, 100 are known from North America, 46 of them restricted to that area; extralimital occurrence of others is noted briefly in the generic treatments. There are 710 species currently known from North America.

VIa. Supertribe Faronitae Reitter 1882
The Faronitae are considered to be the basal group within the Pselaphinae, based on the presence of the maximum number of foveae, and similarity of form to other staphylinids: antenna clavate to moniliform and lacking a distinct club; tarsi with basal two tarsomeres short and subequal in length, third tarsomere much longer and with two equal claws; and metacoxae contiguous, each conically produced at articulation with trochanter. This group is not divided into tribes, and holds 19 genera primarily restricted to temperate regions in the northern and southern hemispheres.

*Megarafonus* Casey 1897
*Nanorafonus* Schuster and Marsh 1958 (subgenus)
Four species, Pacific Northwest and California; most frequent in conifer leaf litter, also in hardwood leaf litter. Revision and key: Schuster and Marsh (1958b). An additional species, *M. (Nafonus) fundus* Park 1943, is found in cloud forests of central Mexico.

*Sonoma* Casey 1886 (Fig. 95.22)
*Rafonus* Casey 1893
Twenty-five species, Pacific coast states and Appalachian Mountains; most in conifer leaf litter, a few in oak or *Rhododendron* leaf litter, one species only in rotten wood. Revision and key: Marsh and Schuster (1962). Larva: Newton (1991).

VIb. Supertribe Euplectitae LeConte 1861
The Euplectitae are difficult to consistently separate from the Goniaceritae, which is reflected in the movement of the Trichonychini to the Goniaceritae, and recently back again (Besuchet 1999). The Euplectitae are considered more primitive in consis-

tently possessing more foveae; contiguous metacoxae each of which is conically projecting at the junction with the trochanter; mesofemur with dorsal margin extending to near the mesocoxa; and abdomen generally extended with all segments freely moveable. This diverse group is apparently poorly represented in tropical areas, though this may be in part a collecting artifact due to their comparatively small size. There are six tribes and 15 subtribes in the Euplectitae, many of these proposed by Jeannel (see Newton and Chandler 1989) and defined on rather questionable characters involving form of the male genitalia and distributions. Four of the six tribes occur in North America, while the Pteracmini Jeannel 1962, and Raffrayiini Jeannel 1949, are found in southern temperate regions. The validity of all tribes and subtribes in this complex needs to be reevaluated. Generic revision and keys to genera of North and Central America: Grigarick and Schuster (1980).

7. Tribe Trogastrini Jeannel 1949

Trogastrini are most diverse in the Neotropical and Australian regions, with a few genera occurring in North America and western Europe. They are characterized by head wide and transverse, mouthparts usually small in comparison; genae broadly swollen; and pronotum often with a distinct thin longitudinal discal sulcus crossing the straight transverse antebasal sulcus, and often somewhat angularly produced at the anterior margin. Five subtribes with 24 genera have been placed here. Two of these subtribes, Mitrametopina Park 1952 (*Mitrametopus*, Neotropical) and Trisignina (*Trisignis*, Nearctic), have been misplaced in this tribe previously, but are here placed in Euplectini, with *Mitrametopus* near *Euplectus* (Euplectina) and *Trisignis* near *Abdiunguis* (Bibloporina), although the two respective subtribes are not synonymized here. Two of the remaining three subtribes occur in North America, while the third (Phtegnomina Park 1951) includes one South American genus. Three genera previously placed in the subtribe Trogastrina (see below) are here placed in the subtribe Euplectina of Euplectini.

7a. Subtribe Trogastrina Jeannel 1949

The pronotum of members of the Trogastrina is angularly rounded at the anterior margin, and the antennal scape is no longer than the combined length of the next two antennomeres. Three genera (*Bontomtes*, *Foveoscapha*, and *Tetrascapha*) have been placed in the Trogastrina (Newton and Chandler 1989, Chandler 1997) on the basis of the only moderately reduced posterior claw of each tarsus; the posterior claw is variably distinct in a number of genera in the Euplectitae. These three genera are here treated in the Euplectini, subtribe Euplectina. The Trogastrina still include 17 genera found in the New World, Australia, and western Europe.

*Conoplectus* Brendel 1888
*Prorhexius* Raffray 1890
*Hexirhexius* Grigarick and Schuster 1980

Five species, eastern North America; in conifer and hardwood leaf litter, one species taken occasionally with ants. Revision and key: Carlton (1983). Other species are found in Mexico and Central America.

*Euboarhexius* Grigarick and Schuster 1966

Four species, two each in the Appalachian Mountains and Pacific Northwest; the eastern species occur in *Rhododendron* leaf litter, the western species in mixed conifer and hardwood leaf litter. Revision and key: Carlton and Allen (1986).

*Oropus* Casey 1886

Thirty-two species, California and Pacific Northwest; in various conifer and hardwood leaf litters, occasionally in mosses. Revision and key: Schuster and Grigarick (1960); synonymy: Chandler (1999).

*Rhexidius* Casey 1887

Nine species, California; in redwood and some hardwood leaf litters. Revision and key: Schuster and Grigarick (1962). More species are present in Mexico.

7b. Subtribe Rhexiina Park 1951

This subtribe is based on the single genus *Rhexius*, a large Neotropical genus that has penetrated into eastern North America. Members of this group are readily recognized by the anterior pronotal apex forming a basally constricted lobe, and the geniculate antenna with the antennal scape being at least as long as the next 3-4 antennomeres.

*Rhexius* LeConte 1849

Six species, eastern North America, placed in the typical subgenus; in leaf litter, rotten wood, and grass roots. Revision and key: Chandler (1990a). This large Neotropical genus, which reaches its northern limit in eastern North America, includes one other subgenus.

8. Tribe Euplectini LeConte 1861

The Euplectini include 10 subtribes in the most recent catalog by Newton and Chandler (1989). The subtribes Trisignina (Nearctic) and Mitrametopina (Neotropical) are added here from Trogastrini and Trichonychina from Goniaceritae (following the recent synonymy of Trichonychini with Euplectini by Besuchet 1999), and really the tribes Pteracmini and Raffrayiini (neither North American) belong in this group also. Jeannel established most of these groups in a series of papers (see Newton and Chandler 1989 for complete list), basing them on a few variable morphological characters, aedeagal structure, and distribution patterns. Very few of these groups can be supported based on the characters used to define them, and the relevant subtribes are only cursorily characterized below. Members of eight subtribes are found in North America.

8a. Subtribe Rhinoscepsina Bowman 1934

The Rhinoscepsina are characterized by the narrow, prominent frontal rostrum, with the antennal insertions being quite close. This is a strictly New World group.

*Morius* Casey 1893

One species, *M. occidens* Casey 1893, California; in redwood and hardwood leaf litter. Redescription: Schuster (1959); Grigarick and Schuster (1980).

*Rhinoscepsis* LeConte 1878
One species, *R. (Rhinoscepsis) bistriatus* LeConte 1878, southeastern United States; in pine and oak leaf litter. Redescription: Grigarick and Schuster (1980). This is the northernmost representative of a large Neotropical genus, which includes one other subgenus.

8b. Subtribe Panaphantina Jeannel 1950
The Panaphantina are characterized by the mesocoxae being separated by extensions of the meso- and metasternum. Five genera are included, with all but the European genus *Panaphantus* being basically Neotropical.

*Thesium* Casey 1884
*Apothinus* Sharp 1887
One species, *T. cavifrons* (LeConte 1863), eastern North America, also two undescribed species known from Arizona; from tree holes and rotten wood, less common in leaf litter. Revision: Chandler (1990a). This is the northernmost member of a large Neotropical genus.

8c. Subtribe Trisignina Park and Schuster 1955
The original description of this group (Park and Schuster 1955) does not clearly separate it from any of the other subtribes. The posterior tarsal claws are well developed for Euplectini, and are more typical of Trogastrini, but not remarkably so. The carinate prosternum places it near the Bibloporina as mentioned by Park and Schuster (1955), and it is indeed close to genera such as *Abdiunguis* and *Eutyphlus* of that subtribe.

*Trisignis* Park and Schuster 1955
Two species, *T. helferi* Park and Schuster 1955 and *T. marshi* Park and Schuster 1955, Pacific Northwest and California; usually from redwood leaf litter intermixed with leaves from other plants. Redescription: Grigarick and Schuster (1980).

8d. Subtribe Bibloporina Park 1951
The Bibloporina are readily recognized by the presence of the median longitudinal carina of the prosternum. However, studies on the Australian fauna have revealed that this feature can be variably present within some genera, though it does appear to be invariant in North American groups. Many of the bibloporine genera are actually closest to some groups in the Pteracmini, while *Bibloporus* is not close at all to the other genera of the North American fauna.

*Abdiunguis* Park and Wagner 1962
One species, *A. fenderi* Park and Wagner 1962, Pacific Northwest; from mosses in the vicinity of springs or in flood debris. Redescription: Grigarick and Schuster (1980).

*Bibloporus* Thomson 1859 (Fig. 11.22)
*Faliscus* Casey 1884
One species, *B. bicanalis* (Casey 1884), northeastern North America; in white pine leaf litter. Redescription: Grigarick and Schuster (1980). Biology: Chandler (1987). This rare species is the only North American member of this small Holarctic and African genus.

*Eutyphlus* LeConte 1880
*Nicotheus* Casey 1884
*Planityphlus* Park 1956 (subgenus)
Five species, eastern North America centered on the Appalachian Mountains; primarily from *Rhododendron* and other hardwood leaf litter. Revision and key: Park (1956).

*Hatchia* Park and Wagner 1962
One species, *H. carinata* Park and Wagner 1962, Washington. Description: Grigarick and Schuster (1980).

8e. Subtribe Trichonychina Reitter 1882
The Trichonychini are most clearly characterized by the separated metacoxae, which are separated by a distance of one-third the metasternal length inTrichonyx. Besuchet (1999) recently transferred this group to Euplectini from its former position as a tribe of Goniaceritae. It is a western Palearctic group of four genera, one species of which is adventive in North America.

*Trichonyx* Chaudoir 1845
One adventive European species, *T. sulcicollis* (Reichenbach 1816), now established at two old nursery inspection sites in New York state; found under bark with the ants *Ponera pennsylvanica* Buckley 1866, and *Lasius brunneus* (Latreille 1798). Redescription: Grigarick and Schuster (1980). Distribution and biology: Cooper (1961). Larva: Besuchet (1956).

8f. Subtribe Euplectina LeConte 1861
The Euplectina consist of the Euplectini left after all the more distinctive genera are removed to other subtribes. Included genera lack a prosternal carina and a prominent frontal rostrum, the mesocoxae are contiguous, and the antennal club is formed by the apical 3-4 antennomeres. In Jeannel's (1959) most recent definition, this group also has the fourth visible tergite longer than the third, and the antebasal transverse sulcus on the pronotum must be present. However, several genera near the type genus, *Euplectus*, lack this sulcus.

*Acolonia* Casey 1893
One species, *A. cavicollis* (LeConte 1878), central Florida; under pine bark. Revision: Wagner (1975).

*Bontomtes* Grigarick and Schuster 1980
One species, *B. ripariae* Grigarick and Schuster 1980, Oregon; from moss on log. Description: Grigarick and Schuster (1980).

*Euplecterga* Park and Wagner 1962
*Euplecturga*; Grigarick and Schuster 1976 (misspelling)
Three species, California and Pacific Northwest; in Douglas-fir leaf litter. Revision and key: Grigarick and Schuster (1976).

*Euplectus* Leach 1817
*Diplectellus* Reitter 1909 (subgenus)
Thirteen species, widespread, mostly eastern; 11 native species are strongly associated with rotten wood, tree holes, or under bark,

but two adventive European species are found in compost, grass clippings, or spilled grain. Revision and key: Wagner (1975). Larva: Böving and Craighead (1931), Wagner (1975), De Marzo (1987, European species). Biology: Park *et al.* (1950). This is a large, worldwide genus; only the adventive species have been assigned to subgenera.

*Foveoscapha* Park and Wagner 1962
One species, *F. terracola* Park and Wagner 1962, Pacific Northwest; in hardwood and fern leaf litter. Redescription: Grigarick and Schuster (1980).

*Leptoplectus* Casey 1908
One species, *L. pertenuis* (Casey 1884), eastern North America; primarily in rotten wood and tree holes, also in leaf litter. Revision: Wagner (1975). Additional species occur in Central America.

*Oropodes* Casey 1893
Six species, California and Pacific Northwest; in conifer and hardwood leaf litter. Revision and key: Grigarick and Schuster (1976).

*Pseudactium* Casey 1908
*Ramecia auctorum* (misidentified, not Casey 1893)
*Racemia* Newton and Chandler 1989
Eleven species, eastern North America; in hardwood leaf litter, possibly associated with old-growth forests (Carlton and Chandler 1994). Revision and key: Carlton and Chandler (1994).

*Pycnoplectus* Casey 1897
Thirteen species, eastern North America; strongly associated with rotten wood, tree holes, or under bark. Revision and key: Wagner (1975).

*Ramecia* Casey 1893
*Liniolis* Grigarick and Schuster 1980
Two species, *R. capitula* (Casey 1884) and *R. crinita* (Brendel 1865), eastern North America; under bark of maple and oak logs. Generic description (as *Liniolis*): Grigarick and Schuster (1980). Taxonomy and species separation: Carlton and Chandler (1994).

*Saxet* Grigarick and Schuster 1980
Three species, Alabama to Arizona; one species in oak leaf litter. Generic description: Grigarick and Schuster (1980). Key (Arizona species): Chandler (1985).

*Tetrascapha* Schuster and Marsh 1957
Three species, California and Pacific Northwest; often in alder litter, or in litter or mosses around springs. Redescription: Grigarick and Schuster (1980).

*Thesiastes* Casey 1893
Four species, eastern North America; primarily from tree holes and rotten wood, also in grass clumps. Revision and key: Chandler (1990a). Two other species occur in northern Mexico.

*Trimioplectus* Brendel 1890
Three species, eastern North America; strongly associated with hardwood tree holes. Revision and key: Chandler (1990a).

8g. Subtribe Bibloplectina Jeannel 1959
Jeannel (1959) based this subtribe on those euplectines with a transverse antebasal sulcus on the pronotum, the fourth visible tergite not longer than the third, and the aedeagus not constricted near the middle.

*Bibloplectus* Reitter 1881
*Bibloplectodes* Jeannel 1949 (subgenus)
Ten species, eastern North America; from hardwood and conifer leaf litter, and mosses. Revision and key: Chandler (1990a). In addition to one species from northern Mexico, this genus includes more than 30 Old World species; subgenera have not been used recently in North America.

*Trigonoplectus* Bowman 1934
Two species, *T. minutus* Bowman 1934 and *T. rostratus* Bowman 1934, western Pennsylvania. Revision and key: Bowman (1934). Generic redescription: Grigarick and Schuster (1980).

8h. Subtribe Trimiina Bowman 1934
This subtribe includes those genera that have antennomere 11 basically forming the antennal club, antennomere 10 being very short and appressed to 11, and antennomere 9 being about the same size as 8. The antennal club is as long as or longer than the combined length of the previous 4 antennomeres.

*Actiastes* Casey 1897
Eight species, mainly eastern North America, with one southwestern species (*A. desertorum* Grigarick and Schuster 1971) extending south into northern Mexico; from hardwood leaf litter (often *Rhododendron* in Appalachian Mountains), under pine bark, and in tree holes. Revision and key: Grigarick and Schuster (1971).

*Actium* Casey 1886
*Proplectus* Raffray 1890
*Basolum* Casey 1897
Thirty-five species, widely distributed but mainly western, a majority known only from California; primarily from conifer leaf litter, also under bark, in mosses and in various hardwood leaf litters, particularly when mixed with pine needles. Revision and key: Grigarick and Schuster (1971). Other species occur in Mexico and the West Indies.

*Actizona* Chandler 1985
Two species, *A. trifoveata* (Park 1963) and *A. chuskae* Chandler 1985, boreal North America, south in Rocky Mountains to Arizona; from beech tree holes, or rotten birch and beech logs in coniferous forests. Description: Chandler (1985). Biology: Chandler (1993).

*Allobrox* Fletcher 1928
*Cutrimia* Park 1945
One species, *Allobrox stephani* Grigarick and Schuster 1977, Arizona; in rotten conifer logs, also oak and sycamore leaf litter. Redescription: Grigarick and Schuster (1980). Other species occur in Mexico.

*Allotrimium* Park 1943
Two species, *A. excavatum* Chandler 1985 and *A. tuberculatum* Chandler 1985, Arizona and California; in rotten conifer logs, or oak and cottonwood leaf litter. Generic redescription: Grigarick and Schuster (1980). Key: Chandler (1985). Other species occur in Mexico.

*Cupila* Casey 1897
Three species, Pacific Northwest from southern Alaska to northern California; in conifer and hardwood leaf litter, also mosses. Revision and key: Grigarick and Schuster (1968).

*Dalmosanus* Park 1952
*Pygmactium* Schuster and Grigarick 1968
Three species, southeastern United States; strongly associated with rotten wood and tree holes, series from conifer or hardwood leaf litter usually entirely females. Revision and key: Chandler (1990a); generic synonymy: Chandler (1999). Many undescribed species occur in Mexico and Central America.

*Dalmosella* Casey 1897
One species, *D. tenuis* Casey 1897, eastern North America; in rotten wood and tree holes. Redescription: Grigarick and Schuster (1980).

*Lemelba* Park 1953
One species, *L. davisi* Park 1953, southern Florida and some Caribbean Islands; from hardwood leaf litter and rotten wood. Redescription: Grigarick and Schuster (1980).

*Melba* Casey 1897
Six species in the typical subgenus, eastern North America and Arizona; primarily in hardwood leaf litter, also in rotten hardwood logs. Revision and key: Chandler (1990a). Other species and subgenera occur in Mexico and Central America.

*Pilactium* Grigarick and Schuster 1970
Two species, *P. summersi* Grigarick and Schuster 1970 and *P. benedictae* Grigarick and Schuster 1978, Oregon and California; from pine and oak leaf litter. Generic redescription: Grigarick and Schuster (1980).

*Simplona* Casey 1897
Two species, *S. arizonica* Casey 1897 and *S. dybasi* Chandler 1985, Arizona; in rotting conifer logs and under bark of conifers. Generic redescription: Grigarick and Schuster (1980). Key: Chandler (1985). Other species occur in Mexico.

*Tomoplectus* Raffray 1898
One species, *T. apache* Chandler 1985, Arizona. Description: Chandler (1985). Other species occur in Mexico.

*Trimioarcus* Park 1952
Two species, *T. pajarito* Chandler 1985 and *T. musamator* Chandler 1992, Arizona and Texas, respectively; in leaf litter. Generic redescription: Grigarick and Schuster (1980). Other species occur in Mexico.

*Trimiomelba* Casey 1897 (Fig. 3.22)
One species, *T. dubia* (LeConte 1849), eastern North America; primarily in leaf litter. Revision: Chandler (1999). Generic redescription: Grigarick and Schuster (1980).

*Trimium* Aubé 1833
One adventive European species, *T. (Trimium) brevicorne* (Reichenbach 1816), described by LeConte in 1878 as *T. discolor* from Louisiana (Chandler 1990a). This species has been repeatedly intercepted at ports-of-entry, but it has not been determined that it is truly established in North America. Larva of European species: De Marzo (1987).

*Zonaira* Grigarick and Schuster 1980
Two species, *Z. puncticollis* (LeConte 1878) and *Z. trilinea* Grigarick and Schuster 1980, Arizona, the latter species also in northern Mexico; the first species is associated with ants, the second with rotting saguaro cacti or sotol. Generic description: Grigarick and Schuster (1980). Key: Chandler (1985).

9. Tribe Jubini Raffray 1904
Jubinini Raffray 1908 (misspelling)
Auxenocerini Jeannel 1962
This tribe is characterized by the presence of the sharply projecting cardo of the maxilla. Also the elytron is usually carinate along the anterior margins, the head usually has a prominent Y-shaped gular carina, and the pronotum is almost always strongly constricted at the base. The Jubini are a Neotropical group, with 14 genera described. Revision and keys to genera and species: Park (1952).

*Sebaga* Raffray 1891
One species, *S. ocampi* Park 1945, Oklahoma and Texas, also in eastern Mexico; from leaf litter or leaf litter at cave entrances. Revision and key: Park (1952). This species is the northernmost representative of a large Neotropical genus.

VIc. Supertribe Bythinoplectitae Schaufuss 1890
The Bythinoplectitae include three disparate tribes, and are primarily characterized by the apparent reduction of three tarsomeres to two. This group clearly is near the Euplectitae, and may represent separate derived lineages from that group. The Bythinoplectitae have dorsal margin of mesofemur extending to near mesocoxa; metacoxae contiguous, each conically produced at articulation with trochanter; tarsi each with a single claw, the

posterior claw lost; second tarsomere greatly reduced, indicated by small tab fused to dorsal surface of first tarsomere (Coulon 1989). The Bythinoplectini (75 genera) and Dimerini Raffray 1908 (7 genera; not North American) are basically tropical in distribution, while the Mayetiini (3 genera) are mainly Holarctic but extend into the tropics. The very long, slender, staphylinid-like appearance of species of the latter two tribes resulted in their frequent placement within Staphylinidae long before the staphylinid affinities of all pselaphines became clear.

10. Tribe Bythinoplectini Schaufuss 1890
Pyxidicerini Raffray 1904

Members of this tribe are characterized by the large lateral subantennal excavations that hold the maxillary palps at rest, and the prominent narrow frontal rostrum. The maxillary palps are oddly formed and asymmetrical, with the third and fourth segments often transverse or globular. Only one of the two included subtribes is found in North America, but the other mainly Old World subtribe (Pyxidicerina Raffray 1904) includes one genus from western Mexico.

10a. Subtribe Bythinoplectina Schaufuss 1890
Zethopsina Jeannel 1952

This subtribe has the subantennal excavations open laterally, with the posterior margins of the excavations extending laterally to the anterior margins of the eyes. This group holds 65 genera, two of which occur in Arizona. Revision and key: Comellini (1985); key to world genera: Coulon (1989).

*Bythinoplectoides* Comellini 1985

One species, *B. gigas* Comellini 1985, Arizona and western Mexico; found with the ant *Solenopsis aurea* Wheeler 1906. Description: Comellini (1985). The single included species was originally described from Jalisco, Mexico.

*Bythinoplectus* Reitter 1882

One species, *B. gloydi* Park 1949, Arizona and northwestern Mexico; associated with the ant *Paratrechina terricola* (Buckley 1866) (Coulon 1989, as *P. melanderi*). Revision and key: Comellini (1985). A large Neotropical genus, barely extending into the United States in southern Arizona.

11. Tribe Mayetiini Winkler 1925

The Mayetiini are characterized by loss of all foveae except vertexal and gular foveae; body elongate and narrow, less than 1 mm long, abdomen longer than rest of body; and antennae with compact club of two antennomeres. Two of the three included genera are confined to Africa.

*Mayetia* Mulsant and Rey 1875

Eighteen species, Pacific coast states and the southeastern United States; associated with sods or soils, and sometimes in leaf litter; in California often taken in soils with Protura. Revisions: Schuster *et al.* (1960, California species); Carlton and Robison (1996; eastern U.S. species, with key). All of the 130+ known Holarctic species are eyeless and wingless, but recently two species with fully developed eyes and metathoracic wings were described from Peru and Nepal.

VId. Supertribe Batrisitae Reitter 1882

The Batrisitae are most distinct by the trend in fusion of the paratergites to the abdominal tergites and sterna, and the abdominal foveae projecting internally to meet and fuse, forming solid internal struts. This is the basis for the rigid cylindrical abdomen typical of most batrisites. Other features are head lacking median gular ridge; mesofemur with dorsal margin near mesotrochanter; tarsal claws unequal, with posterior claw setiform; first tarsomere short, second and third tarsomeres elongate; and aedeagus asymmetrical. Three tribes are recognized, two represented in North America and one (Metopiasini Raffray 1904) restricted to the Neotropical region. The Batrisitae include 223 genera, which are primarily found in tropical regions with only a few genera in northern or southern temperate areas.

12. Tribe Batrisini Reitter 1882

This group is defined by antennal scapes normal in length, emarginate at dorsal and ventral apices; ocular-mandibular carinae present; eyes rounded to flat when reduced in troglobitic species; pronotum often with both inner and outer pairs of basolateral foveae. North American genera have the lateral margins of the paratergites present only on the first visible tergite, where they are indicated by short basal carinae. The mesal paratergal margins are distinct as carinae in most groups. The Batrisini include about 195 genera, and are almost entirely tropical in distribution. All North American genera are members of the largest subtribe, Batrisina. Three other subtribes, currently restricted to Africa, are based upon several characters that have proven to be variably developed within several batrisine genera. The subtribes are not further characterized here.

12a. Subtribe Batrisina Reitter 1882

*Arthmius* LeConte 1849

Five species, eastern United States and Arizona; from pine and hardwood leaf litter. Revision and key: Casey (1893). These are the northernmost representatives of an enormous Neotropical genus.

*Batriasymmodes* Park 1951
*Extollodes* Park 1965 (subgenus)
*Speleodes* Park 1965 (subgenus)

Fifteen species, eastern North America, centered on the Appalachian Mountains; some species are troglobites or troglophiles, others are found in leaf litter and rotten wood, while *B. (Batriasymmodes) monstrosus* (LeConte 1849) is associated with ants in the genera *Lasius* and *Amblyopone* (Park 1935b). Revision and key: Park (1965).

*Batrisodes* Reitter 1882 (Fig. 12.22)
*Babnormodes* Park 1951 (subgenus)
*Declivodes* Park 1951 (subgenus)
*Elytrodes* Park 1951 (subgenus)
*Excavodes* Park 1951 (subgenus)
*Pubimodes* Park 1951 (subgenus)
*Empinodes* Park 1953 (subgenus)
*Spifemodes* Park 1953 (subgenus)
Seventy-two species, widespread; some species are found in rotten wood, leaf litter or mosses, others with ants of the genera *Lasius*, *Formica*, *Aphaenogaster*, and *Camponotus*, and still others are troglobites or troglophiles. Revisions and keys: Park (1947, eastern spp.; 1956, *Babnormodes*; 1958, cavernicolous species; 1960, cavernicolous *Babnormodes*); Park and Wagner (1962, Pacific Northwest species); Grigarick and Schuster (1962b, *Empinodes*); Chandler (1992b, Texas cave species; 1999, new synonymies). Ecological studies on *B. (Excavodes) lineaticollis* (Aubé 1833) (=*B. globosus* (LeConte 1849), synonymized by Chandler 1999) and other species: Park (1932b, 1935a, 1935b, 1947, 1956); Park *et al.* (1953); Reichle (1966); Engelmann (1956); Chandler (1987). Larva: Bøving and Craighead (1931); larva and larval behavior of European species, De Marzo (1985, 1986a, 1987). One troglobitic Texas species, *B. (Excavodes) texanus* Chandler 1992, is on the federal list of threatened and endangered species as the "Coffin Cave mold beetle." This genus also includes a small number of Palearctic species.

*Texamaurops* Barr and Steeves 1963
One species, *T. reddelli* Barr and Steeves 1963, Texas; a troglobite. Redescription: Chandler (1992b). It is on the federal list of threatened and endangered species as the "Kretschmarr Cave mold beetle."

13. Tribe Amauropini Jeannel 1948
Amauropsini Jeannel 1948 (incorrect original spelling)
The separation of this tribe from Batrisini is somewhat questionable, but amauropines may be recognized by being eyeless, with the area normally occupied by the eyes protruding as an acute spine. The tribe includes 12 genera, one endemic to eastern North America and the other 11 restricted to the western Palearctic region. All members are found in deep leaf litter, under stones, or are troglobites.

*Arianops* Brendel 1893
*Anops* Brendel 1891 (not Oken 1815, etc.)
*Eusanops* Brendel 1893
*Arispeleops* Park 1951
Thirty-three species, southeastern United States north to Pennsylvania and west to Oklahoma, mainly in the Appalachian and Ouachita/Ozark Mountains; most frequently found on the underside of large stones, occasionally in deep leaf litter, but three species occur with ants of the genus *Amblyopone* (Barr 1974). Revision and key: Barr (1974).

VIe. Supertribe Goniaceritae Reitter 1882
Bryaxinae LeConte 1861 (based on rejected generic name)
This group is characterized by antennal scape lacking a ventral emargination at its apex, apicodorsal margin usually straight; mesotrochanter short, with dorsal extension of femur close to coxal articulation; metacoxae widely separated; tarsal claws unequal, the posterior claw setiform or lacking; first tarsomere short, second and third tarsomeres elongate, second tarsomere as long as or longer than third tarsomere; aedeagus symmetrical or nearly so, and parameres usually present and symmetrical. The Goniaceritae are found in all biogeographic regions and include 14 tribes, 5 of which occur in North America. Besuchet (1999) removed Trichonychini to synonymy with Euplectitae: Euplectini.

14. Tribe Brachyglutini Raffray 1904
This tribe is defined by maxillary palps with third segments short and triangular, fourth segments ovoid to elongate-conical, usually obliquely truncate at bases; ocular-mandibular carinae present; first visible sternum narrowly visible at base between metacoxae and laterally, second visible sternum much longer than first at midline; and posterior tarsal claws reduced. Four of the six subtribes included here are represented in North America.

14a. Subtribe Brachyglutina Raffray 1904
The Brachyglutina are characterized by median gular ridge complete from mouthparts to gular fovea; and elytron with discal stria at least in basal half. The Pselaptina are separated from the Brachyglutina by the lack of discal striae or impressions on the elytra. In the New World this separation holds up well, but is completely invalidated in Australia and the Oriental region by intermediate genera. However, the two subtribes are kept separate here. The Brachyglutina include 62 genera, and are represented in all biogeographic regions.

*Brachygluta* Thomson 1859
*Nisa* Casey 1886
Twenty-three species, widespread; found under leaf litter or stones, or in mosses along streams, ponds, or marshes. Revisions and keys: Bowman (1934), Brendel (1890). Larva of European species: De Marzo (1987). A few species are found in northern Mexico, but most of the species in this large genus are Palearctic; North American species have not been assigned to subgenera used in the Palearctic region.

*Briaraxis* Brendel 1894
One species, *B. depressa* Brendel 1894, southern Florida islands, also circum-Caribbean; under wrack and stones on beaches. Redescription: Chandler (1992a).

*Nisaxis* Casey 1886
Four species, coastal eastern United States to southern Texas; found under debris or grass litter in marine or saline marshes. Revision and key: Bowman (1934). This genus is also represented in Mexico and Central America.

*Reichenbachia* Leach 1826

*Reichenbachius* Casey 1906

Sixty species, widespread; frequently taken from leaf litter or mosses in wet areas, occasionally under stones, logs, or beach debris. Revision and key to western species: Grigarick and Schuster (1967); taxonomy of eastern species: Chandler (1989). Larva of European species: Kaupp (1997). Biology: Reichle (1966, 1967, 1969). This enormous genus is found in all biogeographic regions, though it is poorly represented in southern temperate areas.

*Rybaxis* Saulcy 1876

Thirteen species, eastern North America to the Pacific Northwest; from various leaf litters and mosses on bog and swamp margins. Revision and key: Fall (1927). Biology: Reichle (1966, 1967, 1969). This large genus is found in all biogeographic regions except the Neotropics.

14b. Subtribe Pselaptina Park *et al.* 1976

The validity of the Pselaptina is questioned above in the discussion of the characteristics of the Brachyglutina. This subtribe was characterized by Park *et al.* (1976) as having a median gular carina, and lacking discal striae or impressions on the elytra. The former character is present in all Brachyglutina, and the latter character cannot be used successfully at the world level to define a distinct group. The two groups are maintained as separate entities here, with the Pselaptina holding 28 genera primarily from the southern biogeographic regions. This group has been basically formed from genera that share the derived loss of many foveae, sulci, and the elytral striae.

*Anchylarthron* Brendel 1887

*Verticinotus* Brendel 1888

Three species, eastern United States; in leaf litter. Revision and key: Bowman (1934). Species from Mexico are myrmecophilous, and the North American species may be also.

*Eutrichites* LeConte 1880

Two species, *E. zonatus* (Brendel 1865) and *E. arizonensis* Carlton 1989, eastern North America and Arizona; found in grass debris, sawdust, and old river drift. Revision and key: Carlton (1989). This small genus is widespread in the Neotropics.

*Pselaptus* LeConte 1880

One species, *P. belfragei* LeConte 1880, Texas to California. Redescription: Bowman (1934). This is the northernmost member of a poorly known Neotropical genus.

*Scalenarthrus* LeConte 1880

*Cylindrembolus* Schaufuss 1887

*Abryxis* Raffray 1890

One species, *S. horni* LeConte 1880, Arizona and California, also in Mexico; found in oak and sycamore leaf litter and with the ant *Solenopsis aurea* Wheeler 1906. Redescription: Bowman (1934). This is the northernmost representative of a large Neotropical genus.

14c. Subtribe Decarthrina Park 1951

Decarthronina Park 1951 (incorrect original spelling)

Members of the Decarthrina are definitively characterized by the median, oval, carinate impression of the gula just posterior to the mouthparts. Three New World genera are placed in this subtribe; only one is found in North America.

*Decarthron* Brendel 1865

Twenty-four species, throughout North America, most species in the eastern United States; in a wide variety of hardwood leaf litters and mosses, particularly near water, but one species (*D. formiceti* (LeConte 1849)) is obligately associated with ants of the genus *Formica*. Revision and key to eastern species: Park (1958). Biology: Reichle (1966, 1969). This large New World genus is most diverse in the Neotropics.

14d. Subtribe Eupseniina Park 1951

Members of this subtribe lack a median longitudinal gular carina, and appear to be glabrous over the most of the body. The subtribe includes two Neotropical genera, one of them found as far north as the southeastern United States.

*Eupsenius* LeConte 1849

Two species, southeastern United States; in hardwood leaf litter. Key: Park *et al.* (1976). This small genus has many undescribed species throughout the circum-Caribbean region.

15. Tribe Bythinini Raffray 1890

The Bythinini have the third maxillary palpomere short, second and fourth palpomeres elongate and the fourth swollen; often with the frontal rostrum prominent and first antennomere elongate; ocular-mandibular carina present; first visible sternum about as long as second at midline and clearly visible; and posterior claw of each tarsus reduced. The Bythinini are a Holarctic group, currently holding 22 genera. Three subtribes have been recognized, all represented in North America (see Newton and Chandler 1989, Chandler 1997), but Löbl and Kurbatov (1995) discussed many problems in defining and differentiating these and formally eliminated the use of subtribes, which we follow here.

*Machaerodes* Brendel 1890

One species, *M. carinatus* (Brendel 1865), eastern United States centered on the Appalachian Mountains; taken from various leaf litters. Redescription: Bowman (1934).

*Pselaptrichus* Brendel 1889

*Vestitrichus* Park 1953 (subgenus)

Thirty-seven species, California and Pacific Northwest; found in various leaf litters or mosses, often near streams or springs. Revisions and keys: Schuster and Marsh (1956); Park and Wagner (1962, Pacific Northwest species).

*Speleochus* Park 1951
Three species, northern Alabama; troglobites. Revision and key: Park (1960).

*Subterrochus* Park 1960
Three species, northern Alabama; troglobites. Revision and key: Park (1960).

*Tychobythinus* Ganglbauer 1896 (Fig. 13.22)
Five species, eastern North America centered on the Appalachian Mountains; in mosses or swamp litter, *Rhododendron* leaf litter, and (three species) associated with caves. Revision and key: Besuchet (1982a). Larva of European species: De Marzo (1987). Biology: Reichle (1966, 1967, 1969). This genus has many species in the Palearctic region.

16. Tribe Valdini Park 1953
Defined by fourth maxillary palpomere enlarged and swollen, with short, apically clubbed setae over the surface, third palpomere short and second elongate; frontal rostrum distinct, ocular-mandibular carina present; metacoxa conically produced at trochanteral articulation; first visible sternum easily seen, about as long as second at middle; and claws of each tarsus similar in size. The tribe includes only a single genus and species from California.

*Valda* Casey 1893
One species, *V. frontalis* Casey 1893, northern California; under bark of oaks, incense-cedar, and pines. Description: Casey (1893).

17. Tribe Tychini Raffray 1904
Members of the Tychini have: second, third, and fourth maxillary palpomeres elongate, third clearly longer than wide; antennae closely inserted on broad to narrow frontal rostrum; metacoxae widely separated, each not produced at trochanteral articulation; first visible sternum shorter than second; and posterior claw of each tarsus reduced. The Tychini form a largely Holarctic group, with *Atychodea* Reitter, occurring in the Oriental region (Borneo). Generic revision and key: Chandler (1988b).

*Custotychus* Park and Wagner 1962
Five species, eastern United States; in various leaf litters and moss. Generic redescription, composition: Chandler (1988b).

*Cylindrarctus* Schaufuss 1887
Nine species, eastern United States; in various leaf litters near water. Revision and key: Chandler (1988a).

*Hesperotychus* Schuster and Marsh 1958
Fourteen species, western United States; from various leaf litters. Revision: Schuster and Marsh (1958a); generic redescription: Chandler (1988b).

*Lucifotychus* Park and Wagner 1962
*Hylotychus* Grigarick and Schuster 1962
Twenty species, Canada and United States except for Southwest and Great Plains states; from various leaf litters, often in wet areas with mosses. Revisions and keys: Park and Wagner (1962, Pacific Northwest species), Grigarick and Schuster (1962a, some western species), Chandler (1991, eastern species). Biology: Reichle (1966, 1969), Chandler (1987).

*Nearctitychus* Chandler 1988
One species, *N. sternalis* (Raffray 1904), eastern United States, primarily east of the Appalachian Mountains; in various leaf litters. Generic description: Chandler (1988b).

*Ouachitychus* Chandler 1988
One species, *O. parvoculus* Chandler 1988, Arkansas; in deep leaf litter. Description: Chandler (1988b).

18. Tribe Speleobamini Park 1951
The two genera of Speleobamini have second, third, and fourth maxillary palpomeres elongate; antennae closely inserted on prominent frontal rostrum; neck elongate, with cervical constriction covered by ruff of setae; metathoracic coxae contiguous; first visible sternum short, second much longer at middle; posterior claws of each tarsus absent. This group is endemic to the southern Appalachian Mountain region of the United States.

*Prespelea* Park 1953
*Fusjuguma* Park 1956 (subgenus)
Two species, southern Appalachian Mountains; in deep leaf litter. Revision and key: Park (1956).

*Speleobama* Park 1951
One species, *S. vana* Park 1951, northern Alabama; a troglobite. Description: Park (1951).

VIf. Supertribe Pselaphitae Latreille 1802
This group is characterized by: mesotrochanter comparatively long, dorsal margin of mesofemur distant from coxal articulation; first tarsomere short, second and third tarsomeres comparatively elongate; metacoxae at least narrowly separated, usually widely separated; abdominal tergites free. The form of the maxillary palps frequently varies between genera, often at least the fourth palpomere is enlarged or modified. Members of this supertribe are found in all biogeographic regions. Currently 15 tribes are included, with 6 of these occurring in North America.

19. Tribe Tyrini Reitter 1882
The Tyrini are characterized by elongate head with the clypeal apex broadly rounded; ocular-mandibular carinae lacking; maxillary palps lacking lateral spines or large setae; tarsi with two distinct claws. Form of the maxillary palps is useful for discrimination of most of the genera. There are 81 genera in this tribe, and members are found in all biogeographic regions, being most diverse in Australia, the Oriental region, and the Neotropics. There are four recognized subtribes, two of these being widespread and

found in North America while two are confined to the Old World tropics.

19a. Subtribe Tyrina Reitter 1882
The Tyrina are characterized by third and fourth maxillary palpomeres elongate, typically the third palpomere 1/2 as long as the fourth or longer; metacoxae clearly distant. The group currently includes 61 genera and is found on all continents.

*Anitra* Casey 1893
One species, *A. glaberula* Casey 1893, Arizona; associated with the ant *Paratrechina terricola* (Buckley 1866) (as *P. melanderi*). Redescription: Chandler (1974b).

*Cedius* LeConte 1849
*Sinistrocedius* Newton and Chandler 1989 (subgenus)
Three species, eastern North America; associated with ants (*Aphaenogaster*, *Lasius*, *Formica*, *Camponotus*) in rotting logs or under bark. Revision and key: Park (1949); synonymy: Chandler (1999).

*Mipseltyrus* Park 1953
Four species, northern California and southern Appalachian Mountains; in various leaf litters.

*Tyrus* Aubé 1833
*Pytna* Casey 1887.
Three species, North America except Great Plains; typically found under bark of coniferous trees, but also under bark of deciduous trees or in rotten wood. Revision and key: Casey (1897); synonymy: Chandler (1999).

19b. Subtribe Hamotina Park 1951
Members of this subtribe have third maxillary palpomere small, fourth palpomere enlarged and ovoid; and metacoxae narrowly separated. There are 18 genera in the Hamotina, 13 of them restricted to the New World, two of which occur in the Nearctic.

*Circocerus* Motschulsky 1855
*Cercocerus*; LeConte 1861 (misspelling)
*Upoluna* Schaufuss 1886
One species, *C. batrisioides* Motschulsky 1856, southeastern United States; associated with ants (*Aphaenogaster*) in rotten wood, also found in tree holes and deciduous leaf litter. Redescription: Bowman (1934). Biology: Park (1935b). Another species occurs in Panama.

*Hamotus* Aubé 1844
*Hamotoides* Schaufuss 1888 (subgenus)
Five species in the subgenus *Hamotoides*, southeastern United States, Texas, New Mexico and Arizona, three of these species also found in Mexico; under bark of cottonwood and pine, in rotting saguaro and cardon cacti, in leaf litter, and in caves. Key: Chandler (1974a, western U.S. species). Larva of Brazilian species: Costa *et al.* (1988). This large genus is mainly Neotropical, where the typical subgenus occurs.

20. Tribe Ceophyllini Park 1951
This tribe is characterized by the lack of an antennal club and the expanded and leaf-like third maxillary palpomere. This group, based on a single genus and species, should probably be placed in the Tyrina, as the maxillary palps are not at all unusual for that group, and the antennal form is a common modification found in myrmecophiles.

*Ceophyllus* LeConte 1849
One species, *C. monilis* LeConte 1849, eastern North America; associated with several species of ants in the genus *Lasius*. Redescription: Bowman (1934). Biology: Park (1932a, 1935b).

21. Tribe Tmesiphorini Jeannel 1949
Members of the Tmesiphorini are characterized by head usually with setose semicircular sulcus partially encircling insertion of antennal scape; maxillary palps with third and/or fourth palpomeres angulate, often bearing lateral spines; ocular-mandibular carinae absent; protibia medially thickened and curved; and each tarsus with two claws. There are 23 genera currently included, with most of these being in the Afrotropical, Oriental, and Australian regions.

*Tmesiphorus* LeConte 1849
Two species, *T. carinatus* (Say 1824) and *T. costalis* LeConte 1849, throughout the eastern United States north to Ontario and west to Texas; in leaf litter, rotten wood, under bark, and in caves, casually to strongly associated with termites (*Reticulitermes*) and ants (*Aphaenogaster*, *Formica*). Redescription: Bowman (1934). Biology: Engelmann (1956), Park (1933, 1935b). This large genus is found in all major biogeographic regions except the Neotropics and New Zealand.

22. Tribe Ctenistini Blanchard 1845
The Ctenistini are characterized by head with straight or Y-shaped impression filled with squamous (thickened and flattened) setae below antennal insertions; ocular-mandibular carina absent; third and/or fourth maxillary palpomeres bearing lateral spines or pencils of setae and often strongly angulate laterally; and each tarsus with two equal claws. Often there are squamous setae over much of the body, but these may be restricted to foveae, sulci, and ventral articulations of the major body parts. Thirty-two genera are currently placed in the Ctenistini, and the tribe is represented in all biogeographic regions.

*Atinus* Horn 1868
Two species, *A. monilicornis* (Brendel 1866), southeastern United States, and *A. brevicornis* Casey 1893, Arizona and Texas; associated with ants of the genera *Prenolepis* and *Pheidole*. Revision and key: Casey (1893). This small genus is also known from northern Mexico.

*Biotus* Casey 1887
One species, *B. formicarius* Casey 1887, Texas to California; collected with ants. Redescription: Casey (1893). This small genus also occurs in northern Mexico.

*Ctenisis* Raffray 1890
One species, *C. raffrayi* Casey 1893, southwestern United States, also in Mexico and Central America; associated with the ant *Messor julianus* (Pergande 1894) (Chandler 1976, as *Veromessor*). Redescription: Casey (1893). This small genus includes about eight additional Neotropical species.

*Ctenisodes* Raffray 1897
*Pilopius* Casey 1897
Fourteen species in North America; several species are known only from leaf litter and rotten wood, but probably all are at least loosely associated with ants, including the known host genera *Cyphomyrmex*, *Formica*, *Ischnomyrmex*, and *Novomessor*. Revision and key: Casey (1897). Biology: Park *et al.* (1949, 1953), Mickey and Park (1956), Park (1964). One species each is also known from northern Mexico and Japan.

23. Tribe Arhytodini Raffray 1890
Holozodini Raffray 1900
The Arhytodini are characterized by ocular-mandibular carina absent; maxillary palp reduced and quite small; first two visible abdominal tergites subequal in length; first two tarsomeres short, about equal in length dorsally; and each tarsus with a single claw. There are often flattened ("squamous") setae in sulci, foveae, and articulations of the major body parts. Biological information is lacking for most of the arhytodine genera, although a few species of the Neotropical genus *Rhytus* have been associated with ants. Ten genera are known, of which half are confined to Madagascar, one is known from Zaire, and four are Neotropical with one extending into the southwestern United States. One undescribed genus has been seen from Australia and New Guinea.

*Caccoplectus* Sharp 1887
Four species, Arizona and Texas, two of these species also found in Mexico or Central America; all specimens collected at ultraviolet light. Revisions and keys: Chandler (1976), Chandler and Wolda (1986). This large Neotropical genus includes more than 30 additional species.

24. Tribe Pselaphini Latreille 1802
Members of the Pselaphini are characterized by ocular-mandibular carina absent; first maxillary palpomere long and narrow, often 1/2 as long as second palpomere, second and fourth palpomeres frequently very long; gula slightly to greatly swollen; first visible abdominal tergite longest, typically longer than remaining tergites combined; and each tarsus with a single claw. Twenty-four genera are known, found in all biogeographic regions, with the greatest number of genera recorded from Africa and Australia.

*Neopselaphus* Jeannel 1951
One species, *N. mexicanus* (Park 1945), Texas, also in Mexico and Central America; collected at ultraviolet light. Revision and key: Besuchet (1982b). This Neotropical genus includes about eight species.

*Pselaphus* Herbst 1792 (Fig. 14.22)
Four species, eastern North America to Pacific Northwest; in leaf litter from swamps, or in mosses. Revision and key: Casey (1894); synonymy: Chandler (1999). Larva and biology of European species: De Marzo (1987, 1988a, 1988b). This is a mainly Holarctic genus of three dozen species.

VIg. Supertribe Clavigeritae Leach 1815
All members of this group are obligate myrmecophiles. They are characterized by short antennae with only 3-6 antennomeres (3 in North American genera); ocular-mandibular carina absent; maxillary palp small, often of only one palpomere; mesotrochanter elongate, dorsal extension of femur distant from coxal articulation; each tarsus with a single claw; and abdomen with first three visible tergites fused into a single tergal plate. Besuchet (1991) reduced the number of tribes from 14 to three, one of these diverse and widespread and two (Tiracerini Besuchet 1986; Colilodionini Besuchet 1991) with a single genus each confined to the Australian and Oriental regions, respectively.

25. Tribe Clavigerini Leach 1815
Adranini Chenu and Desmarest 1857
Clavigerodini Schaufuss 1882
Fustigerini Jeannel 1949
Members of the Clavigerini have first antennomere very short, often difficult to see; clypeus broadly arcuate, with lateral cavities partially enclosing the reduced maxillary palps; head cylindrical, frontal rostrum strongly projecting; and aedeagus with phallobase lacking a dorsal diaphragm. This tribe includes 96 genera, most of these found in the Afrotropical and Oriental regions.

*Adranes* LeConte 1849 (Fig. 15.22)
Six species, eastern United States to Pacific Northwest and California; obligately associated with ants in the genera *Aphaenogaster* and *Lasius*. Revision and key: Wickham (1901). Biology: Akre and Hill (1973), Hill *et al.* (1976), Kistner (1982), Park (1932a, 1935b, 1964).

*Fustiger* LeConte 1866
Two species, *F. fuchsii* Brendel 1866 and *F. knausii* Schaeffer 1906, eastern and southwestern United States to California; obligately associated with ants in the genus *Crematogaster*. Revision and key: Bowman (1934). Biology: Park (1935b), Leschen (1991). This large genus includes about 50 additional species, mainly Neotropical but some from the Old World tropics.

VII. Subfamily PHLOEOCHARINAE Erichson 1839
This subfamily has not been well defined and is difficult to characterize; historically it has been a dumping ground for relatively primi-

tive staphylinids that do not fit well elsewhere. Herman (1972a) suggested three characteristics that might form a useful definition: procoxa without mesal groove, abdominal tergites IV and V each with a pair of distinctive cuticular combs, and hypopharynx distinctive; Smetana (1983) added another: antennal insertion more or less concealed from above. Genera remaining here (each with at least three of these characteristics) are generally elongate slender staphylinids with apparently predatory mouthparts (this includes known larvae as well as adults); they lack defensive glands of the omaliine or oxyteline type, the sublimuloid shape and dorsally exposed antennal insertion of tachyporines and related families, or the special characters of other subfamilies. They have: procoxal fissure open (with protrochantin broadly exposed) or nearly closed (with protrochantin concealed); tarsi 5-5-5; and abdomen with six visible sterna and one or two pairs of paratergites per segment. Genera removed include *Pseudopsis* (placed here by some authors) to Pseudopsinae (Herman 1975, Newton 1982a); *Rimulincola* to Tachyporinae: Deropini as a synonym of *Derops* (Smetana 1983); and *Giulianium* to Omaliinae: Aphaenostemmini (Newton and Thayer 1995). *Olisthaerus*, placed variously here or in its own subfamily Olisthaerinae, is probably closely related to phloeocharines (Ashe and Newton 1993), but further study of these subfamilies and a few other odd genera is needed. Of the seven genera placed in this subfamily at present, five occur in North America (represented by six species) and two are endemic to this area; the remaining two genera are endemic to Australia and New Zealand, and a related undescribed genus occurs in Chile.

*Charhyphus* Sharp 1887
*Triga* Fauvel 1878 (not Gray 1867)
*Trigites* Handlirsch 1907
*Pseudeleusis* Bernhauer 1923

Two species, *C. picipennis* (LeConte 1863), widespread in eastern North America from Wisconsin and Ontario to Nova Scotia, south in the Appalachian Mountains to northern Georgia, and *C. arizonensis* Herman 1972, Arizona; adults and larvae found together under bark, especially of hardwoods. Key: Herman (1972a). Larva: Frank (1991). This genus also includes one species in eastern Siberia and two (one undescribed) in Mexico and Guatemala.

*Dytoscotes* Smetana and Campbell 1980
One species, *D. pacificus* Smetana and Campbell 1980, western Oregon and northwestern California; a flightless species, rare in forest leaf litter. Review: Smetana and Campbell (1980). This genus, closely allied to *Phloeocharis*, is endemic to North America.

*Echletus* Sharp 1887
One species, *E. leechi* Moore 1965, California, known only from two collections at low elevations of the Sierra Nevada (Calaveras and Tuolumne Counties); nothing is known about its habitat or biology. Descriptions: Moore (1965), Moore and Legner (1973). This very rare genus includes one other species, known only from its original collection on the Pearl Islands of Panama (Sharp 1887: 708).

*Phloeocharis* Mannerheim 1830
One species, *P. californica* Smetana and Campbell 1980, coastal mountains and Sierra Nevada of central California; a flightless species, found with its larvae in forest leaf litter. Description: Smetana and Campbell (1980). Larvae: Kasule (1968), Newton (1990a; in key). This genus also includes more than 30 species in the western Palearctic region, most of them (including many flightless species) with restricted distributions in the mountains of circum-Mediterranean countries of southern Europe and northwest Africa.

*Vicelva* Moore and Legner 1973
One species, *V. vandykei* (Hatch 1957) (= *V. paradoxica* Moore and Legner 1973), western Oregon north to southern Alaska; rare, probably in leaf litter or wet debris along streams. Description and synonymy: Moore and Legner (1973), Moore (1974). This odd, isolated genus also occurs in northeastern Russia (Ryabukhin 1999).

## VIII. Subfamily OLISTHAERINAE Thomson 1858

Very similar to Phloeocharinae and sometimes included in that subfamily (see above), but spicule patches of the abdominal ter-gites are absent, and the postcoxal process of the pronotum is of unique structure (separated by a suture from the rest of the pronotum and slightly moveable). They also have antennae inserted at the sides of the frons, insertions exposed in dorsal view; protrochantin exposed; tarsi 5-5-5; and abdomen with six visible sterna and two pairs of paratergites per segment. Only the genus *Olisthaerus*, with two species, is currently included here. The odd species *Habrocerus magnus* LeConte 1878, recently excluded from Habrocerinae but not placed elsewhere (Assing and Wunderle 1995), may belong here as suggested by larval structure (Paulian 1941: 273, compare to *Olisthaerus* larva described by Saalas 1917: pl. I), but this species lacks the near-articulation of the postcoxal process of the pronotum.

*Olisthaerus* Dejean 1833
Two species, *O. megacephalus* (Zetterstedt 1828) and *O. substriatus* (Gyllenhal 1827), Alaska and across Canada to Quebec, the latter species also in New England and New York; both also widespread in the northern Palearctic region; under bark of dead conifers. Key: Fauvel (1878: 188). Biology and larva: Saalas (1917: 296-305, pl. I). The genus includes only these two Holarctic species.

## IX. Subfamily TACHYPORINAE MacLeay 1825

Body (except Deropini) more or less sublimuloid (Figs. 16-17.22), small head more or less retractile to level of eyes and lacking distinct neck, broad pronotum and elytra, and relatively short tapered abdomen; antennal insertion more or less visible from above, anterior to eye; antenna not verticillate; elytron with epipleural keel; procoxa large, prominent, with widely exposed trochantin; tarsi 5-5-5; abdomen with six visible sterna and 0-2 pairs of paratergites per segment. At least five tribes with a total of 40 genera are included here, one (Megarthropsini Cameron

1919, three genera) confined to the Oriental region, one (Vatesini Seevers 1958, one genus) to the Neotropical region, and the remaining three with representatives in North America (two other nominal tribes, Symmixini Bernhauer 1915, from the Himalayas and Cordobanini Bernhauer 1910, from Mexico, each with one genus, are of doubtful validity). The known North America fauna includes 195 species placed in 17 genera. Key to tribes: Smetana (1983).

26. Tribe Deropini Smetana 1983

Body not at all limuloid, very elongate and narrow, antennae and legs very long; head not retractile, with slight but distinct neck well behind eyes; pronotal hypomeron not strongly inflexed, completely visible in lateral view. The single odd genus included here has been placed in various subfamilies including Phloeocharinae, but has sexually modified apical abdominal segments and other features typical of members of the tribe Tachyporini in spite of a very un-tachyporine habitus.

*Derops* Sharp 1889

*Rimulincola* Sanderson 1947

One species, *D. divalis* (Sanderson 1947), Ozarks and associated mountains of eastern Oklahoma, Arkansas, Missouri, and southern Illinois; found in damp to wet leaf litter and decayed vegetation in deep rock crevices or along streams, and occasionally in caves. Review: Smetana (1983). The genus also includes about eight species in the Oriental and eastern Palearctic regions.

27. Tribe Tachyporini MacLeay 1825

Head without ridge below the eye; elytron without distinct sutural stria; pronotal hypomeron strongly inflexed and scarcely or not at all visible in lateral view except for apex of elongate postcoxal process (if present); procoxal cavities often closed behind by postcoxal process of pronotum (most *Sepedophilus*) or by enlarged spiracular peritremes (most other genera). Twenty-four genera are included in this tribe worldwide; seven mostly widespread genera occur in North America, and none are endemic to this area.

*Cilea* Jacquelin du Val 1856

*Leucoparyphus* Roger 1856

One species, *C. silphoides* (Linnaeus 1767), widely distributed in the United States and southern Canada, also nearly cosmopolitan; usually synanthropic, found in various decaying plant materials including compost and grass piles, haystacks, rotting fruit and dung. Review: Campbell (1975b). Larva: Kasule (1968), Topp (1978). This genus also includes one species from Guatemala and about 20 species from Asia and Africa.

*Coproporus* Kraatz 1857

*Erchomus* Motschulsky 1857

Nine species, eastern and southern United States and the Pacific Northwest, under bark of dead trees and logs, in rotting cacti, on fungi of various kinds, and in plant debris. Revision and key: Campbell (1975b). Larva: Frank (1991), Newton (1990a). The genus includes more than 200 species worldwide, mainly in the Old and New World tropics.

*Nitidotachinus* Campbell 1993

Five species, widespread in hilly or mountainous areas, absent from Great Plains and southeastern coastal plain states; found especially in wet debris and moss along streams and in other periaquatic habitats. Revision and key: Campbell (1973c, revised 1988, 1993a). The genus, recently removed from *Tachinus*, includes six additional species from Japan, China, and Taiwan (Campbell 1993a).

*Sepedophilus* Gistel 1856

*Conurus* auctorum (misidentification)

*Conosoma* auctorum (misidentification)

*Conosomus* auctorum (misidentification)

Thirty species, including three adventive from Europe, widespread but mainly eastern; many species found mainly in forest leaf litter and debris or under bark of old logs, but some species are associated with polypore tree fungi or fleshy fungi, and at least some adults and larvae are mycophagous. Revision and key: Campbell (1976). Larvae: Newton (1984, 1990a); exotic species, as *Conosoma*: Kasule (1968), Topp (1978); species misidentified as *Scaphisoma*: Böving and Craighead (1931), etc. (see Kasule 1966, Newton 1984). Biology: Newton (1984). This genus is worldwide with more than 300 named species.

*Tachinomorphus* Kraatz 1859

*Physetoporus* Horn 1877

Four species, three in Arizona and one in western Texas, all four also in Mexico or further south; found in plant material in a fermentation stage of decay, especially large cacti or fruits. Revision and key: Campbell (1973b). In addition to three further species in Central and South America, this genus includes more than a dozen species in tropical areas of Africa and Asia to Australia.

*Tachinus* Gravenhorst 1802 (Fig. 16.22)

*Tachinoderus* Motschulsky 1857 (subgenus)

Forty-two species, including three adventive from Europe, generally distributed; usually associated with decaying organic matter including rotting mushrooms, dung and carrion, some species in leaf litter, wet debris near streams, or in mammal burrows. Revision and key: Campbell (1973c); supplements and revised keys: Campbell (1975a, 1988). Larvae: Böving and Craighead (1931), Kasule (1968), Newton (1990a). Four additional species are found in Mexico and Central America, and about 120 additional species in Eurasia.

*Tachyporus* Gravenhorst 1802

*Palporus* Campbell 1979 (subgenus)

Thirty species, at least two adventive from Europe, generally distributed; some species found on vegetation in meadows or fields, others in leaf litter and other ground debris in forests or near water. Revision and key: Campbell (1979b). Larvae: Frank (1991), Lipkow (1966), Newton (1990a). Biology of exotic species: Lipkow (1966). Five additional species are found in Mexico and Central America, and at least 70 more in the Old World, mainly in Eurasia.

28. Tribe Mycetoporini Thomson 1859
Bolitobiini Horn 1877

Body elongate, usually strongly tapered to narrow head and abdominal apex (Fig. 17.22); head with distinct ridge below eye (postmandibular ridge, according to Smetana and Davies 2000); elytron with impressed sutural stria; pronotal hypomeron strongly inflexed and not visible in lateral view, postcoxal process absent. Most species are largely glabrous and some have contrasting yellow or orange and brown or black color patterns. The generic classification of this group has been in rapid flux due to changes in generic concepts as well as in application of several familiar generic names; the tribal name has also been changed. All nine genera currently recognized are found in North America, but many tropical and southern temperate species placed in old "catch-all" genera like *Bryoporus* and *Bolitobius* need to be restudied. Key to genera: Campbell (1993b).

*Bolitobius* Leach 1819
*Megacronus* Stephens 1829
*Bryocharis* Boisduval and Lacordaire 1835

Three species, *B. cingulatus* Mannerheim 1830, widespread in both Nearctic and Palearctic regions, and *B. fenderi* (Hatch 1957) and *B. kremeri* Malkin 1944, Oregon to British Columbia; found in forest leaf litter or associated with fungi of various kinds. This genus, formerly known as *Bryocharis*, also includes many species in the Palearctic region, but is in need of revision for all areas, and is known to include a few new species in the United States and Mexico. Larva (as *Bryocharis*): Kasule (1968), Pototskaya (1967).

*Bolitopunctus* Campbell 1993

Two rare species, *B. muricatulus* (Hatch 1957), northern California to British Columbia and Idaho, in forest leaf litter and wood debris, and *B. punctatissimus* Campbell 1993, southern California, found in a wood rat (*Neotoma* sp.) nest. Revision and key: Campbell (1993b). This genus is endemic to North America.

*Bryophacis* Reitter 1909

Seven species, widespread in western North America with two species extending to eastern Canada and northeastern United States; found mainly in forest leaf litter or associated with decaying logs, also in periaquatic situations near streams or lakes. Revision and key: Campbell (1993b). Recently separated from *Bryoporus*, this genus includes at least nine additional species from the Palearctic region.

*Bryoporus* Kraatz 1857

Three species, *B. rufescens* LeConte 1863 and *B. testaceus* LeConte 1863, both widely distributed in the United States and southern Canada, and *B. niger* Campbell 1993, Arizona and New Mexico to Mexico; found mainly in forest leaf litter and damp debris, especially near water. Revision and key: Campbell (1993b). The genus includes about 50 additional species throughout the New and Old World tropical and warm temperate regions.

*Carphacis* Gozis 1886

Three species, *C. dimidiatus* (Erichson 1839) and *C. intrusus* (Horn 1877), eastern United States and southeastern Canada, and *C. nepigonensis* (Bernhauer 1912), widespread in northeastern and western North America, also in western Mexico; common in rotting mushrooms and other soft fungi, occasionally in other decaying organic matter or leaf litter. Revision and key: Campbell (1980). The genus, recently separated from *Lordithon*, includes eight additional Palearctic species.

*Ischnosoma* Stephens 1829

Thirteen species, widespread in North America, one of these (*I. splendidum* (Gravenhorst 1806)) also in Europe; found mainly in forest leaf litter, some species especially in wet debris or moss along streams or in other periaquatic situations. Revision and key: Campbell (1991). Larva of European species (as *Mycetoporus*): Topp (1978). Recently separated from *Mycetoporus* by Campbell (1991), this genus includes five additional species from Mexico and Central America and an indeterminate number from throughout much of the Old World.

*Lordithon* Thomson 1859 (Fig. 17.22)
*Bolitobius* auctorum (misidentified, not Leach 1819)
*Bobitobus* Tottenham 1939 (subgenus)

Twenty-three species, generally distributed; found commonly on mushrooms and sometimes on tree fungi, where adults are predators of fly larvae, but the biology is poorly known. Revision and key: Campbell (1982c). This genus, long known as *Bolitobius*, includes at least 100 additional species, mainly in Eurasia but with some in Mexico and Central America, Peru, southern South America, and Australia.

*Mycetoporus* Mannerheim 1830
*Schinomosa* Tottenham 1939

Eighteen species, generally distributed, one of these (*M. segregatus* Campbell 1991) also in Mexico; found mainly in forest leaf litter, humus and moss, but some species usually at edges of streams and lakes, in squirrel middens, or in nests of *Neotoma* and other small ground-dwelling mammals. Revision and key: Campbell (1991). Larva (identified to genus only, could also be *Ischnosoma*): Kasule (1968), Newton (1990a). This genus in its former broad sense included more than 150 species and was nearly worldwide except for South America; with the removal of *Ischnosoma*, the remaining number of species and distribution are uncertain, but the genus apparently does not occur south of Mexico in this hemisphere.

*Neobolitobius* Campbell 1993

One species, *N. varians* (Hatch 1957), Oregon, Washington, and British Columbia; a winter-active species of unknown habits. Campbell (1993b). This genus is endemic to North America.

X. Subfamily TRICHOPHYINAE Thomson 1858

Antennomeres 3-11 extremely slender, verticillate, antennal insertions on front margin of head, exposed in dorsal view; body compact, tapered anteriorly and posteriorly, moderately to densely punctate and setose; head with distinctive neck constriction that allows molding of head to prothorax; five maxillary palpomeres, fourth large and spindle-shaped, fifth minute and hyaline; pronotum without post-coxal lobes, with spiracle in large triangular well-sclerotized peritreme; protrochantin exposed; elytron without epipleural keel; metacoxa transverse, flat, not excavate; tarsi 5-5-5; and abdomen with six visible sterna and two pairs of paratergites per segment. Only one genus is included here; it is probably the sister group to the small subfamily Habrocerinae, and these two groups together may be the sister group to Aleocharinae (Ashe and Newton 1993). There are four North American species.

*Trichophya* Mannerheim 1830
*Eumitocerus* Casey 1886
Four species, three indigenous to southwestern United States from Oregon to California and east to Colorado and western Texas, and the adventive European species *T. pilicornis* (Gyllenhal 1810), now widespread from British Columbia to Oregon and east to Newfoundland and Virginia; found mainly in forest leaf litter and ground squirrel middens, *T. pilicornis* also in caves. Review, key, world checklist and larval description: Ashe and Newton (1993). The genus also includes 11 species indigenous to the Palearctic and Oriental regions, two new species in Arizona and New Mexico, and at least a half dozen new species from Mexico through Nicaragua.

XI. Subfamily HABROCERINAE Mulsant and Rey 1877

Antennomeres 3-11 extremely slender, verticillate; body very compact, sublimuloid (Fig. 18.22), with a few large setae only, shining; labrum cleft; apical (fourth) maxillary palpomere subulate; pronotum without postcoxal lobes, without enlarged and well-sclerotized spiracular peritremes; elytron with epipleural keel; metacoxa excavate, concealing most of femur in repose; tarsi 5-5-5; male genitalia extremely modified, incorporating abdominal segment VIII (apical) as well as genital segment, so females have six visible abdominal sterna and males only five (but with long appendages of segment VIII usually showing as in Fig. 18.22). Antennal insertions on sides of frons, slightly concealed from dorsal view; protrochantins exposed; abdomen with one pair of paratergites per segment. There are only two genera included here, one widely distributed (including three species in North America) and one (*Nomimocerus* Coiffait and Sáiz 1965, with five species) confined to southern South America. World review and key to genera: Assing and Wunderle (1995).

*Habrocerus* Erichson 1839 (Fig. 18.22)
Three species, one native (*H. schwarzi* Horn 1877, eastern Canada and northeastern United States); one adventive from Europe (*H. capillaricornis* (Gravenhorst 1806), widespread); and one that has been excluded from this genus but not placed elsewhere yet (*H. magnus* LeConte 1878, northeastern United States and eastern Canada - see Olisthaerinae); in litter, wood debris, and fungi. Key: Assing and Wunderle (1995). Larva: Kasule (1968), Paulian (1941), Topp (1978). *Habrocerus* includes 12 other species, 11 (including the widely adventive *H. capillaricornis*) widely distributed in Eurasia and one each in Costa Rica and Brazil.

XII. Subfamily ALEOCHARINAE Fleming 1821 (by J. S. Ashe)

North American Aleocharinae can be recognized by the combination of antennae usually inserted into vertex between the eyes; posterior coxae expanded laterally under the femur; and the distinctive aedeagus, with large and usually multiarticulated parameres. They also have 11 antennomeres (10 in Hypocyphtini and *Decusa*); procoxal fissure open, trochantin exposed; tarsi usually 4-5-5, 5-5-5, or 4-4-5, in a few genera 4-4-4, 3-3-3 or 2-2-2; and abdomen with six visible sterna, usually two pairs of paratergites per segment (absent in Trichopseniini). The form of the parameres is unique to, and uniform among, the Aleocharinae and is a shared derived feature that provides strong evidence that the subfamily is a monophyletic group (Hammond 1975, Ashe 1994). All but a few basal groups of Aleocharinae also have unique defensive glands on tergite VII of adults and tergum VIII of larvae (Steidle and Dettner 1993, Frank and Thomas 1984c).

The subfamily is in great need of comprehensive revision and study at all taxonomic levels. As currently delimited the Aleocharinae include 51 tribes and numerous subtribes, over 1000 described and probably valid genera and over 12,000 described species in the world fauna. However, the true diversity is much higher, with many thousands of species and numerous higher taxa remaining to be described from throughout the world, especially from tropical areas. In North America, 21 tribes, 183 genera, and about 1385 described species are known. The current classification is very unsatisfactory, with mostly inadequately characterized genera assigned to poorly defined tribes or subtribes. Work published in the last two decades has provided a somewhat firmer foundation for study of North American Aleocharinae. The most important of these is Seevers (1978). This work provided keys to genera, diagnoses and reclassification of most genera, and a revised tribal arrangement. However, Seevers' research, on which the published work was based, was not yet complete at the time of his death. Consequently, there are many omissions, inconsistencies and incomplete sections in his work. Nonetheless, the tribal and subtribal classification that Seevers provided is the best available (though still very inadequate, especially for the tribes Oxypodini and Athetini), and the classification used here is based primarily on that proposed by Seevers, with addition of emendations and revisions that have been published since. Other important recent general works on the North American Aleocharinae include Lohse *et al.* (1990, supplemented by Lohse 1991) (Revision and keys to the Arctic Aleocharinae) and Downie and Arnett (1996) (Keys to the aleocharines of the Northeastern United States). However, the keys in all of these works lack adequate illustrations and are very difficult to use (even for the specialist). In addition to these more general guides there have been some important revisions of tribes or subtribes. Those of particular importance include Klimaszewski (1984) (Aleocharini), Hoebeke (1985) and Ahn and Ashe (1995) (Falagriini); Danoff-Burg (1994) (Sceptobiini), Ahn and Ashe (1996a) (Liparocephalini), Ashe (1984a) (Gyrophaenina), Ashe (1992) (Bolitocharina), Jacobson and Kistner (1992) (Crematoxenini), Klimaszewski and Peck (1986) (cavernicolous species), Pasteels and Kistner (1971) (Trichopseniini), works by Seevers on the myrmecophilous (1965) and termitophilous (1957) Staphylinidae, and others mentioned below. Because of the generally limited knowledge and numerous taxonomic complexities of this group, the indications of extralimital distributions of higher taxa should not be regarded as comprehensive or complete. Information on extralimital distributions of genera is largely limited to occurrence in Mexico; it is not clear how many genera are restricted to North America.

29. Tribe Gymnusini Heer 1839
Fusiform body shape with deflexed head; without a tergal gland on the anterior margin of abdominal tergum VII; ctenidium of comb-like cuticular projections on the posterior edges of abdominal segments III-VI; 6 setigerous pores on dorsal surface of head; distinctive mouthparts (see Klimaszewski 1979 for details); and 5-5-5 tarsi. All are associated with riparian areas along the margins of marshes, bogs, ponds, and streams. World revision of the tribe: Klimaszewski (1979). The tribe contains two genera (Holarctic Region and Auckland Islands), one of which occurs in North America.

*Gymnusa* Gravenhorst 1806
Seven species, widely distributed in the northeastern and northwestern United States, Canada, and Alaska. Key: Klimaszewski (1979).

30. Tribe Deinopsini Sharp 1883
Rather flattened, tear-drop body shape with strongly deflexed head; without a tergal gland on the anterior margin of abdominal tergum VII; with ctenidium of comb-like cuticular projections on the posterior edges of abdominal segments III-V; distinctive mouthparts (see Klimaszewski 1979 for details); and 2-2-2 or 3-3-3 tarsi (North American taxa; other deinopsine taxa with 5-5-5 tarsi do not occur in North America). All are inhabitants of marshes, bogs, stream and lake margins, and similar riparian habitats. World revision of the tribe: Klimaszewski (1979). This worldwide tribe contains three genera, two of which occur in North America.

*Adinopsis* Cameron 1919
Three species, southeastern United States, Maryland, and Michigan. Key: Klimaszewski (1979, 1982a).

*Deinopsis* Matthews 1838 (Fig. 164.22)
    *Dinopsis* Agassiz 1846
Eleven species, widely distributed in eastern United States and Canada. Key to most species: Klimaszewski (1979, supplemented by 1980, 1982a). Klimaszewski (1979) did not treat *D. americana* Kraatz because he was unable to locate type material; however, this species is here included in the count of North American species.

31. Tribe Mesoporini Cameron 1959
Mostly small to minute, tear-drop shaped to sublimuloid aleocharines with 5-5-5 tarsi; without a tergal gland on the anterior margin of abdominal tergum VII; with antennomeres 3-7 minute, 8-10 much larger, forming a distinct "club;" and with a large lamella on the posterior coxa that covers the base of the femur. This tribe contains seven genera worldwide, one of them in North America.

*Anacyptus* Horn 1877 (Fig. 183.22)
    *Microcyptus* Horn 1883
One species, *A. testaceus* (LeConte 1863), Arizona, Florida, Georgia, Mississippi, North Carolina, "Lake Superior", also known from Mexico; usually under bark, especially pine bark, often among termites (reported with *Prorhinotermes*, *Neotermes*, and *Reticulitermes*) and occasionally reported with ants (*Neivamyrmex*) (Seevers 1957).

32. Tribe Trichopseniini LeConte and Horn 1883
Tarsi 5-5-5; paratergites absent; posterior coxae absent (possibly indistinguishably fused to the metasternum) so that the trochanters appear to articulate directly with the metasternum; and presence of a large lamella on the metasternum covering at least the base of the hind femur. All species are obligate inquilines in the nests of termites. World revision of the tribe: Seevers (1957), Kistner (1969), Pasteels and Kistner (1971). Status as subfamily and fossils: Kistner (1998). This tribe contains fifteen genera worldwide; two of these occur in North America.

*Trichopsenius* Horn 1877
Five species, eastern United States and California; obligate associates of termites of the genus *Reticulitermes*. Key: Seevers (1957). Larvae and pupae: Kistner and Howard (1980). Biology, host integrating mechanisms, behavior of selected species: Howard (1976, 1978, 1979, 1980), Howard and Kistner (1978), Howard *et al.* (1982).

*Xenistusa* LeConte 1880
Four species, Florida, Mississippi, and Texas; obligate associates of termites of the genus *Reticulitermes*. Seevers (1957) synonymized all three of LeConte's original species of *Xenistusa* under the name *X. cavernosa* LeConte 1880, and described one additional new species. Seevers proposed that the three LeConte species were based on individuals from the same termite colony that showed different amounts of distortion due to drying. Pasteels and Kistner (1971) questioned this interpretation based on new alcohol-preserved material and, by implication, recognized all of LeConte's species as valid pending further study.

Larva and pupa: Kistner and Howard (1980). Biology and behavior: Howard (1978, 1980), Howard *et al.* (1982).

33. Tribe Aleocharini Fleming 1821
Tarsi 5-5-5; maxillary and labial palps with a pseudosegment on the last palpomere (so that they appear to have 5 and 4 palpomeres, respectively) (Lohse 1974, Seevers 1978). Most North American taxa have the pronotal hypomera not, or only narrowly, visible in lateral aspect, and at least some have a reticulated velum on the parameres of the aedeagus (Seevers 1978). Revision of North American representatives of tribe, Klimaszewski (1984). This tribe contains at least fifteen genera worldwide, placed in three subtribes; only one genus occurs in North America.

33a. Subtribe Aleocharina Fleming 1821
Aleocharina are distinguished from the other (non-North American) subtribes of Aleocharini by lacking their distinctive shape of male abdominal segment IX, according to Kistner (1970a, 1970b); the members of both extralimital subtribes are termitophilous.

*Aleochara* Gravenhorst 1802 (Fig. 184.22)
*Coprochara* Mulsant and Rey 1874 (subgenus)
*Eucharina* Casey 1906 (not Agassiz 1860)
*Funda* Blackwelder 1952 (new name for *Eucharina* Casey 1906)
*Xenochara* Mulsant and Rey 1874 (subgenus)
*Polychara* Mulsant and Rey 1874
*Isochara* Bernhauer 1901
*Maseochara* Sharp 1883 (subgenus)
*Pinalochara* Casey 1906
*Emplenota* Casey 1884 (subgenus)
*Aidochara* Casey 1906 (subgenus?)
*Calochara* Casey 1906 (subgenus)
*Oreochara* Casey 1906
*Echochara* Casey 1906 (subgenus)
*Rheocharella* Casey 1906

Fifty-three species, generally distributed; adults primarily predators of Diptera eggs, larvae, and pupae at dung, carrion, or rotting fungi, seaweed, or cacti; larvae ectoparasitoids of cyclorrhaphous Diptera pupae near such habitats. Revision and keys to subgenera and species: Klimaszewski (1984). Larvae: Frank (1991), Fuldner (1960). Biology: Fuldner (1960), Klimaszewski (1984; North American host records), Klimaszewski and Jansen (1993; review of biology), Maus *et al.* (1998; host records), Peschke and Fuldner (1977; host records, biology), Peschke and Metzler (1982; defensive and pheromonal secretions). This very large, worldwide genus has been divided into sixteen subgenera, of which the above six plus the typical subgenus occur in North America (Klimaszewski 1984). The status of *Aidochara* Casey is uncertain. Klimaszewski (1984) did not mention *Aidochara* in his revision of Aleochara. The type specimen in the Casey Collection (from California) is missing from its point, and is presumably lost. No other specimens from the type locality, or from the type series are known; other specimens labeled as *Aidochara planiventris* in the Casey Collection are from Rhode Island and are not likely to be conspecific. Though *Aidochara* is tentatively treated as a subgenus of *Aleochara* here, resolution of the status of this name will probably require designation of a neotype.

34. Tribe Hoplandriini Casey 1910
Tarsi 4-5-5; pseudosegment present on the last palpomere of the maxillary and (on most) labial palps (so that they appear to have 5 and 4 palpomeres, respectively); velum of paramere with reticulated pattern of sclerotized supports. This tribe contains twenty-one described genera worldwide, five of them occurring in North America.

*Hoplandria* Kraatz 1857 (Fig. 198.22)
*Arrhenandria* Génier 1989 (subgenus)
*Genosema* Notman 1920 (subgenus)
*Lophomucter* Notman 1920 (subgenus)

Twelve species, *H. (Arrhenandria.) laeviventris* Casey 1910, widely distributed in east, *H. (Genosema) pulchra* Kraatz 1857, Florida to Texas, *H. (s. str.)*, three species, eastern North America and Texas to Arizona, *H. (Lophomucter)*, seven eastern species in three species groups; frequently found in carrion, dung and other rotting organic material. Revision and key: Génier (1989). One of the *H. (s. str.)* described species and numerous unnamed species occur in Mexico.

*Nosora* Casey 1911
One species, *N. meticola* Casey 1911, Arizona and Texas and an additional species in Mexico; a large series of one species was found in *Datura* flowers. Key: Casey (1911).

*Platandria* Casey 1893
Four species, eastern half of United States, southwestern United States and Oregon; frequently found on flowering shrubs. Revision and key: Génier and Klimaszewski (1986).

*Tetrallus* Bernhauer 1905
Five species, California; frequently found in flowers of various shrubs.

*Tinotus* Sharp 1883
*Exaleochara* Keys 1907

Seventeen species, widely distributed. Key: Casey (1893).

35. Tribe Oxypodini Thomson 1859
The tribe Oxypodini as recognized by Seevers (1978) is heterogeneous and very poorly characterized. Seevers (1978) gave the following combination of characteristics: tarsi 5-5-5, 4-5-5 or 4-4-4; head without a distinct neck in most, neck present in some; frontal suture present or absent; terminal antennomere with or without coeloconic sensilla; mouthparts generalized; mesocoxae narrowly separated in most, moderately separated in a few; intercoxal processes slender in most; abdominal tergum IX narrowly subdivided at base (see Seevers 1978 for description and discussion); parameres of aedeagus with striated velums; and median lobe of aedeagus with an elongate compressor plate and without an athetine bridge. This large and poorly defined tribe contains about 146 genera worldwide, 40 of which occur in North America.

35a. Subtribe Oxypodina Thomson 1859

Seevers (1978) did not provide a characterization of this subtribe, in which he appeared to include all those Oxypodini that cannot be included in one of the other subtribes. Features shared by members of this subtribe include tarsi 5-5-5; frontal suture present or absent; antennomere 11 with or without coeloconic sensilla; pronotal hypomeron visible or not in lateral aspect. This is a heterogeneous assemblage that cannot be satisfactorily characterized; it includes 24 genera in North America.

*Acrimea* Casey 1911 (Fig. 189.22)
Three species, Washington, Oregon and Idaho.

*Amarochara* Thomson 1858
  *Nasirema* Casey 1893
  *Lasiochara* Ganglbauer 1895
Four species, Indiana, Iowa, Pennsylvania, and Rhode Island; usually found in leaf litter, litter associated with fallen logs or near water, and occasionally with ants (Seevers 1978).

*Athetalia* Casey 1910
Five species, California to British Columbia. Lohse and Smetana (1985) moved *Homalota vasta* Mäklin 1853, from *Athetalia*, where it was placed by Seevers (1978), to *Liogluta*. They also pointed out that two of the remaining five species in *Athetalia*, *A. metlakatlana* (Bernhauer 1909) and *A. oregonensis* (Bernhauer 1909), belong "near *Atheta*". However, they did not formally assign these two species to genera in the Athetini, so they are still included in *Athetalia* in this list pending further study. Seevers (1978) followed Casey (1910a) in placing *Athetalia* in the tribe Athetini; however, Lohse and Smetana (1985: 298) noted that *Athetalia* is an oxypodine (tarsi 5-5-5). Examination of the type and other described species of *Athetalia* shows this to be true. No consistent way to separate *Athetalia* and *Ocalea* is evident, and they are therefore keyed out at the same couplet, but they are not placed in synonymy pending detailed studies of the mouthparts and genitalia.

*Calodera* Mannerheim 1830
One species, *C. infuscata* Blatchley 1910, Indiana, and at least two undescribed species from the New England states, California, and Alberta.

*Crataraea* Thomson 1858
One species, *C. suturalis* (Mannerheim 1830), scattered localities throughout North America; frequently found under plant and animal refuse in stables, granaries and similar situations, and occasionally with ants of the genera *Lasius* and *Formica* according to Seevers (1978). This is a European species probably adventive in North America.

*Devia* Blackwelder 1952 (Fig. 185.22)
  *Dasyglossa* Kraatz 1856 (not Illiger 1807)
One species, *D. congruens* (Casey 1893), widespread in northern United States and Canada; most frequently found in leaf litter and similar forest floor habitats.

*Dexiogyia* Thomson 1858 (Fig. 193.22)
Six species, widespread in eastern North America and California; under bark of dead trees, especially pine.

*Gnathusa* Fenyes 1909
Four species, Alaska, Alberta, British Columbia, California, and Yukon Territory; usually found in moss and leaf litter.

*Haploglossa* Kraatz 1856
  *Microglotta* Kraatz,1862
  *Microglossa* Stein 1868
Two species, Virginia and Oklahoma; associated with mammal and bird nests, including one from bank swallow nests. Revision and key: Klimaszewski and Ashe (1991). Larva and biology of European species: Drugmand (1990), Paulian (1941).

*Ilyobates* Kraatz 1856
  *Gennadota* Casey 1906
Two species, eastern Canada and New York.

*Longipeltina* Bernhauer 1912
One species, *L. bakeri* Bernhauer 1912, California.

*Melanalia* Casey 1911
One species, *M. tabida* Casey 1911, California.

*Moluciba* Casey 1911
One species, *M. grandipennis* Casey 1911, British Columbia.

*Neodemosoma* Pace 1989
Two species, North Carolina and Ohio. Pace (1989: 23) described *Neodemosoma* to contain *Leptusa semirufa* Casey 1906 and *Leptusa exposita* Casey 1911, and pointed out that it belonged to the Oxypodini. Examination of the types of both species does not show any way to distinguish *Neodemosoma* from *Phloeopora* (except by the slightly broader pronotum of *Neodemosoma*; however, a few *Phloeopora* have similarly broad pronota). *Neodemosoma* is not placed in formal synonymy with *Phloeopora* in this list pending detailed study of the mouthparts and genitalia.

*Ocalea* Erichson 1837
  *Isoglossa* Casey 1893
  *Rheobioma* Casey 1906
Seven species, western, Colorado and Arizona to California and British Columbia. Larva of European species: Paulian (1941), Topp (1978).

*Ocyusa* Kraatz 1856
  [subgenus *Cousya* Mulsant and Rey 1875, probably not North American]
  [subgenus *Mniusa* Mulsant and Rey 1875, probably not North American]
Five species, Alaska, California, Iowa, North Carolina, Rhode Island, and Yukon Territory. Key to some species: Seevers (1978). Larva of European species: Topp (1978).

*Ocyustiba* Lohse and Smetana 1988 (Fig. 187.22)
One species, *O. appalachiana* Lohse and Smetana 1988, North Carolina; in forest floor litter, fallen leaves, and moss in high elevation coniferous and deciduous forests in the Appalachian Mountains (Lohse and Smetana 1988a).

*Oxypoda* Mannerheim 1830 (Fig. 186.22)
*Hylota* Casey 1906
*Sphenoma* Mannerheim 1830 (subgenus)
*Disochara* Thomson 1858 (subgenus)
*Bessopora* Thomson 1859 (subgenus)
*Demosoma* Thomson 1859 (subgenus)
*Baeoglena* Thomson 1867 (subgenus)
*Podoxya* Mulsant and Rey 1874 (subgenus)
*Paroxypoda* Ganglbauer 1895 (subgenus)
One hundred two species, widely distributed. Three additional described species are known from Mexico. Larva of European species: Topp (1978). This genus is divided into at least 15 subgenera, but few of the North American species are assigned to subgenus; the above seven subgenera as well as the typical subgenus have been reported in North America.

*Pachycerota* Casey 1906
One species, *P. duryi* Casey 1906, Massachusetts to Iowa. Found in ant nests.

*Parocalea* Bernhauer 1901
Two described species and at least two undescribed species, northwestern North America, Alaska, Yukon Territory, and Northwest Territories. Key to arctic species: Lohse *et al.* (1990).

*Parocyusa* Bernhauer 1902
*Chilopora* Kraatz 1856 (not Haime 1854)
*Tetralaucopora* Bernhauer 1928
*Chiloporata* Strand 1935
Two species, eastern North America, New York and North Carolina. Key (as *Tetralaucopora*): Seevers (1978).

*Pentanota* Bernhauer 1905
One species, *P. meuseli* Bernhauer 1905, described from Russia (Lake Baikal), but reported by Seevers (1978) from Alaska based on a specimen in the Bernhauer collection. Herman, in Seevers (1978), questioned whether the genus *Pentanota* is distinct from *Ilyobates*.

*Phloeopora* Erichson 1837 (Fig. 192.22)
Eight species, widespread in North America; under the bark of dead trees. One described and several undescribed species are known from Mexico.

[*Stichoglossa* Fairmaire and Laboulbène 1856, not North American; species placed in *Dexiogyia*]

*Thyasophila* Fairmaire and Laboulbène 1856
*Thiasophila* Kraatz 1856
*Myrmecodelus* Motschulsky 1857
Four species, Iowa, Massachusetts, New York; myrmecophilous with ants of the genera *Formica*, *Camponotus*, and *Lasius*.

35b. Subtribe Dinardina Mulsant and Rey 1873
The subtribe Dinardina is not characterized by any distinctive feature. Most members of this tribe can usually be recognized by a combination of tarsi 5-5-5; body form limuloid to sublimuloid; frontal suture absent; 10 or 11 antennomeres; pronotum rather broad and shield-like, pronotal width at least 1/3 broader than long (except *Losiusa*, in which it is only 1/10 broader than long); anterior margin of pronotum covering base of more-or-less deflexed head; abdomen acuminate (moderately to strongly narrowed from base to apex); and association with ants (typically *Formica*, *Aphaenogaster* and *Lasius*) (adapted from Seevers 1978). Key to North American genera: Seevers (1978). Seevers (1978) assigned ten genera to this subtribe; four of them occur in North America.

*Decusa* Casey 1900 (Fig. 191.22)
One species, *D. expansa* (LeConte 1866), eastern United States from District of Columbia to Indiana.

*Euthorax* Solier 1849
*Myrmecochara* Kraatz 1857
*Eurynotida* Casey 1906
Five species, Arizona, Colorado, District of Columbia, Louisiana, Texas; often in association with ants of the genus *Solenopsis*. One described species and at least two undescribed species are known from Mexico.

[*Homoeusa* Kraatz 1856, not North American?]

*Losiusa* Seevers 1978
One species, *L. angusticollis* Seevers 1978, Massachusetts; from an ant nest.

*Myrmobiota* Casey 1893
*Soliusa* Casey 1900
Three species, Iowa, Massachusetts, New York.

35c. Subtribe Meoticina Seevers 1978
Tarsi 4-4-4 or 5-5-5; tarsal claws scythe-shaped (called "falcate" by Seevers 1978) (except *Meotica*); body form slender, more or less dorsoventrally compressed; sides subparallel in most (except *Bamona*); head with neck about 1/3 head width; head without frontal suture; terminal antennomere with pair of coeloconic sensilla; middle coxae narrowly separated; mesocoxal process slender and very acutely pointed, not extended beyond middle of coxae; metacoxal process very short to almost absent in most; mesocoxal acetabula not or only weakly margined in most (moderately margined in *Gyronycha*); pronotal hypomera fully visible in lateral aspect; abdominal tergites III-V or III-IV with transverse impressions (modified from Seevers 1978). Most members of this subtribe are relatively easy to recognize by presence of the scythe-

shaped tarsal claws and distinctive body form. The genus *Meotica*, which does not have scythe-shaped tarsal claws, seems misplaced with the other genera. Seevers (1978) assigned six genera to this subtribe, all of which occur in North America.

*Alisalia* Casey 1911
Eight species, widely scattered from California to New England. One unnamed species is known from Mexico.

*Apimela* Mulsant and Rey 1874
*Gyronychina* Casey 1911
*Gampsonycha* Bernhauer 1912
Four species, California and Nevada.

*Bamona* Sharp 1883 (Fig. 205.22)
Three species, North Carolina and California; adults usually along margins of lakes and streams. At least four unnamed species are known from Mexico.

*Gyronycha* Casey 1893 (Fig. 188.22)
Six species, California, North Carolina, New York, and Texas; usually along the margins of lakes and streams. At least two unnamed species are known from Mexico.

*Leptobamona* Casey 1911
One species, *L. pertenuis* (Casey 1911), New Jersey.

*Meotica* Mulsant and Rey 1873
One species, *M. exilis* (Erichson 1837), Maine; usually in leaf litter. This is a European species probably adventive in the United States (Seevers 1978).

35d. Subtribe Blepharhymenina Klimaszewski and Peck 1986
This subtribe is based on the distinctive body form of *Blepharhymenus*. Tarsi 5-5-5; head broadly rounded behind eyes to a distinct, narrow neck (less than 1/2 head width); head without frontal suture; pronotum subquadrate to slightly elongate; head and pronotum subequal in width and each much narrower than elytra; elytra much broader than abdomen at base; abdomen widened and thickened from base to become as broad as elytra beyond segment V; abdominal tergites III-V with deep transverse impressions, impressions with coarse punctures and a strong medial carina. Only a single North America genus is assigned to this subtribe.

*Blepharhymenus* Solier 1849 (Fig. 190.22)
*Echidnoglossa* Wollaston 1864
*Colusa* Casey 1885
Twenty-five species, Iowa, Michigan, Colorado, Utah, and Oregon, with most described species occurring in California. Keys: Casey (1885b: 291; 1893: 315).

35e. Subtribe Tachyusina Thomson 1859
Tarsi 4-5-5; frontal suture absent; terminal antennomere without coeloconic sensilla; mesocoxae separated by moderately broad intercoxal processes; pronotal pubescence directed posteriorly (except in *Teliusa*); median lobe of aedeagus with a distinctive, prominent triangular crista (see Seevers 1978 for details and illustrations). Seevers (1978) made the case that there is no justification for classifying members of the Tachyusina in the tribe Falagriini, as has been frequently done in the past. The Tachyusina lack the divided paramere velum, mesospiracular peritremes, longitudinally sulcate pronotum, and other features that characterize the Falagriini. Seevers (1978) assigned five North American genera to this subtribe.

*Brachyusa* Mulsant and Rey 1874
*Tetralina* Casey 1911
Four species, California, British Columbia, and Montana. Key: Seevers (1978).

*Gnypeta* Thomson 1858
*Euliusa* Casey 1906
*Gnypetoma* Casey 1906
Forty species, widely distributed; frequently found in riparian habitats and debris along the margins of marshes, ponds, lakes, and streams. Two additional described species are known from Mexico.

*Gnypetella* Casey 1906
Two species, California.

[*Hygropora* Kraatz 1856, not North American (species transferred to *Brachyusa*)]

*Tachyusa* Erichson 1837 (Fig. 19)
Twenty-one species, widespread throughout North America; characteristic inhabitants of riparian habitats along the margins of marshes, ponds, lakes and streams. Three additional described species are known from Mexico. Larva of European species: Topp (1978). *Tachyusa* is frequently treated as one of six subgenera of the worldwide genus *Ischnopoda* Stephens 1835; all described North American species are in *Tachyusa*. Following Seevers (1978), *Tachyusa* is here treated at the generic level and recognized as the valid generic name for the North American species.

*Teliusa* Casey 1906
Two species, Texas.

36. Tribe Corotocini Fenyes 1918
Tarsi 5-5-5 (some taxa not found in North America have 4-4-4); mentum and submentum fused (suture absent); antennomere 11 with pair of coeloconic sensilla; mesocoxal cavities not margined; hind coxae triangular; abdomen moderately to strongly physogastric in most. All species are inquilines in the nests of termites. World revision of tribe: Jacobson *et al.* (1986). This tribe contains 62 described genera in ten subtribes; only one subtribe and two genera occur in North America.

36a. Subtribe Eburniogastrina Jacobson *et al.* 1986
Tarsi clearly 5-5-5 with distinct and well-developed joints between tarsomeres 4 and 5; and abdominal tergites II-VII with a

longitudinal slit between each spiracle and the lateral margin of the tergite (see Jacobson *et al.* 1986 for details). The subtribe contains two genera, both occurring in North America.

*Eburniogaster* Seevers 1938
Two species, Arizona and Texas, one additional species known from Mexico; all species with known hosts occur with termites of the genus *Tenuirostritermes* (Jacobson *et al.* 1986).

*Termitonidia* Seevers 1938
One species, *T. lunata* Seevers 1938, Arizona, three additional species known from Mexico; all species with known hosts occur with termites of the genus *Tenuirostritermes* (Jacobson *et al.* 1986). Key: Seevers (1957).

37. Tribe Hypocyphtini Laporte 1835
Oligotini Thomson 1859
Mostly very small to minute aleocharines; tarsi 4-4-4; 10 antennomeres, 8-10 slightly to moderately enlarged to form a loose club; hind coxa with a large ventral lamella that covers the base of the femora. This tribe is most frequently discussed in the literature as Oligotini; it contains five described genera, three of which occur in North America.

*Cypha* Leach 1819
*Hypocyphtus* Gyllenhal 1827
Three species, British Columbia, Colorado, and Pennsylvania.

*Holobus* Solier 1849 (Fig. 204.22)
*Somatium* Wollaston 1854
Six species, widely distributed in eastern North America and California; at least some species are found on vegetation and are predators of phytophagous mites. Biology and larvae (as *Oligota*): Badgley and Fleschner (1956), Moore *et al.* (1975), Paulian (1941).

*Oligota* Mannerheim 1830 (Fig. 203.22)
*Microcera* Mannerheim 1830
*Goliota* Mulsant and Rey 1873
*Logiota* Mulsant and Rey 1873
Ten species, widely distributed.

38. Tribe Myllaenini Ganglbauer 1895
Tarsi 4-4-5; labial palps very long and stylate; maxilla very long and stylate, galea very slender with setae only at the apex, lacinia long and stylate with widely scattered teeth internally; mentum with anterolateral margins produced into moderate to strong spinose processes; tarsal claws not scythe-like. The tribe contains five described genera, two of which occur in North America.

*Bryothinusa* Casey 1904 (Fig. 251.22)
One species, *B. catalinae* Casey 1904, California; intertidal on Pacific beaches. Key: Moore and Legner (1975b). Larva: Moore and Orth (1979a). One additional described species is known from the west coast of Mexico.

*Myllaena* Erichson 1837 (Fig. 20.22)
*Centroglossa* Matthews 1838
Twenty-two species, widespread in North America; associated with riparian habitats along the margins of marshes, ponds, lakes and streams. Revision and key: Klimaszewski (1982b). Four additional described species are known from Mexico.

39. Tribe Diglottini Jacobson 1909
Recognized by the combination of: tarsi 4-4-5 (4-4-4 in European species); tarsal claws scythe-like; labial palps very long, thin and stylate; maxillary lobes very long and slender with widely scattered teeth on lacinia; body form distinctive with large, broadly rounded head, small eyes, head as broad as or broader than pronotum, pronotum strongly narrowed posteriorly; elytra shorter than pronotum (North American species only), abdomen broadly oval in dorsal outline, broader at widest point than elytra. Found in intertidal zone of beaches. This tribe contains two genera, one of which occurs in North America.

*Diglotta* Champion 1887 (Fig. 250.22)
*Diglossa* Haliday 1837 (not Wagler 1832)
Three species, two on the Pacific coast from Baja California to Oregon, and one on the Atlantic coast, New Jersey; intertidal. Note that the European species have 4-4-4 tarsi rather than the 4-4-5 tarsi characteristic of North American taxa. Revision: Haghebaert (1991), Moore and Orth (1979b). Larva of European species: Kemner (1925).

40. Tribe Liparocephalini Fenyes 1918
Distinguished by the combination of: tarsi 4-4-5 (4-4-4 in some *Diaulota*); body densely covered with very fine, short setae; elytra very short, flight wings absent; labial palps with 2 palpomeres (palpomeres 1 and 2 fused); ligula of prementum long, slender and entire at apex; one medial seta present on prementum; mandibles without denticles in ventral molar area; metasternum greatly shortened; middle coxae contiguous. World revision of tribe: Ahn (1996a), Ahn and Ashe (1995, 1996a, 1996b). This tribe contains four genera distributed around the Pacific rim from Baja California to Japan; all four occur in North America. All species are inhabitants of intertidal regions.

*Amblopusa* Casey 1893 (Fig. 255.22)
*Boreorhadinus* Sawada 1991
Two species, Pacific coast from Alaska to California. Revision, key, and larval description: Ahn and Ashe (1996b).

*Diaulota* Casey 1893
*Genoplectes* Sawada 1955
Six species, Pacific coast from Alaska to Baja California. Revision, key, and larval description: Ahn (1996a).

*Liparocephalus* Mäklin 1853 (Fig. 254.22)
Two species, Pacific coast from Alaska to southern California. Review, key, and larval description: Chamberlin and Ferris (1929), Ahn (1997a).

*Paramblopusa* Ahn and Ashe 1996
One species, *P. borealis* (Casey 1906), Pacific coast from Alaska to Oregon. Revision: Ahn and Ashe (1996b).

41. Tribe Autaliini Thomson 1859
The single North American genus can be distinguished by tarsi 4-4-5; head with very narrow neck, less than 1/3 head width; ligula of prementum elongate with apex dichotomously branched into four divergent, acutely pointed processes; mesocoxal cavities not margined behind; pronotum with four subbasal foveae; each elytron with two basal foveae; each of abdominal tergites III-V with deep transverse impression containing a large medial carina and smaller lateral carinae. These characteristics do not all apply to the other (non-North American) genera assigned to the Autaliini, and the tribe is very difficult to characterize on a worldwide basis. This tribe contains ten genera; only one occurs in North America.

*Autalia* Leach 1819 (Fig. 252.22)
Four species, widely distributed; usually in leaf litter, compost piles, and similar habitats of decomposing organic matter. Key: Hoebeke (1988a).

42. Tribe Homalotini Heer 1839
Bolitocharini Thomson 1859
Homalotini are a large and structurally diverse group that is difficult to characterize. Most members can be recognized by a combination of tarsi 4-4-5; mandible with a patch or rows of denticles in the ventral molar region; bases of medial setae of prementum very close together, setal insertions in contact in many, setae displaced one behind the other in some; medial pseudopore field of prementum very narrow. This taxon is most frequently referred to in the literature as the tribe Bolitocharini. However, Newton and Thayer (1992) pointed out that the name Homalotini has priority as the correct name for this tribe. This large tribe includes about 123 genera; in North America there are six subtribes and 22 genera, two of them not assigned to subtribe.

42a. Subtribe Bolitocharina Thomson 1859
Characterization: Labial palps of 3 palpomeres, not styliform; ligula elongate and slender, as long as or longer than first labial palpomere, bifid in apical 1/3-1/5; mandibles with very large patch of densely arranged denticles in ventral molar regions; medial setae of prementum arranged one behind the other; medial pseudopore field of prementum very narrow and without pseudopores; many with male secondary sexual characteristics consisting of medial or lateral carina or knob on tergites VII and VIII, and broadly emarginate margin of tergite VIII with 3-5 denticles in each half of emargination (adapted from Ashe 1992). Larvae and adults of all members of the Bolitocharina occur in association with macroscopic fruiting bodies of fungi, especially the Polyporaceae and lignicolous Agaricales. World revision and phylogeny of genera of the subtribe: Ashe (1992). Feeding biology and behavior: Ashe (1993). This subtribe includes nine genera worldwide, of which six occur in North America.

[*Ditropalia* Casey 1906 (=*Bolitochara* Mannerheim 1830), not North American]

*Hongophila* Ashe 1992
One species, *H. arizonica* Ashe 1992, Arizona and New Mexico. At least two undescribed species occur in Mexico.

*Neotobia* Ashe 1992
One species, *N. alberta* Ashe 1992, northern United States and Canada from the Rocky Mountains to the east coast.

*Phymatura* Sahlberg 1876
*Venusa* Casey 1906
One species, *P. blanchardi* (Casey 1893), widely distributed in the eastern half of North America.

*Pleurotobia* Casey 1906 (Fig. 263.22)
*Phymaturosilusa* Roubal 1932
Three species, widely distributed in eastern North America. At least one undescribed species is present in Arkansas. Ashe (1992) suggested that all three described species may be synonyms of *P. trimaculata* (Erichson 1839), but this has not been confirmed. Larval description and biology: Ashe (1990).

*Silusida* Casey 1906
One species, *S. marginella* Casey 1906, widely distributed in the eastern half of North America.

*Stictalia* Casey 1906
Sixteen described species and an undetermined number of undescribed species, western North America from Alaska to southern California, one species as far east as the Canadian Rocky Mountains. Key: Casey (1906).

42b. Subtribe Gyrophaenina Kraatz 1856
Labial palps of 2 palpomeres; 1 medial seta on prementum; ligula elongate and entire at apex, or broad and rounded, or divided to near base into 2 lobes; lacinia of maxilla distinctive, apex obliquely truncate and densely covered with small spines, and inner face of lacinia without spines or setae; mesocoxae very broadly separated, meso- and metacoxal processes meeting and broadly contiguous between coxae, isthmus between processes absent. All species are inhabitants of fresh mushrooms (Polyporaceae and Agaricales) as both larvae and adults (Ashe 1984a, 1984b). World revision, phylogeny of genera of subtribe, biology, and keys to genera: Ashe (1984a). Larvae of many genera: Ashe (1986b). Feeding behavior and biology: Ashe (1993). This subtribe includes 18 genera worldwide, five of them occurring in North America.

[*Agaricochara* Kraatz 1856, not North American (Ashe 1984a: 264)]

*Agaricomorpha* Ashe 1984 (Fig. 258.22)
One described species, *A. apacheana* (Seevers 1951), and at least two undescribed species in Arizona and New Mexico; adults and larvae

most commonly on woody or leathery polypore mushrooms. Generic description and biology: Ashe (1984a, 1984b). Larva: Ashe (1986b). At least three unnamed species are known from Mexico.

*Encephalus* Kirby 1832 (Fig. 259.22)
One species, *E. americanus* Seevers 1951, Montana, Illinois, New Mexico, and Alberta. Generic redescription: Ashe (1984a).

*Eumicrota* Casey 1906
Seven species, eastern half of North America, Arizona, and New Mexico; adults and larvae most commonly on fleshy polypore mushrooms and persistent gilled mushrooms on logs, less commonly on woody polypore mushrooms. Revision and key: Seevers (1951); generic redescription: Ashe (1984a). Larva: Ashe (1986b). Biology: Ashe (1984a, 1984b, 1986c, 1987). Several undescribed species are known from Mexico.

*Gyrophaena* Mannerheim 1830
Sixty-two species, distributed throughout North America; adults and larvae most commonly on fleshy gilled mushrooms, some species abundant on fleshy polypores or persistent gilled mushrooms on logs. At least 14 undescribed species are known, mostly from the mountain systems of the southwestern United States. Revision and key: Seevers (1951); generic redescription: Ashe (1984a). Larva: Ashe (1986b). Biology: Ashe (1984a, 1984b, 1987). At least 13 described species and an undetermined number of undescribed species occur in Mexico.

*Phanerota* Casey 1906 (Fig. 260.22)
Five species, eastern half of United States; adults and larvae most commonly on fleshy gilled mushrooms, some species abundant on fleshy polypores or persistent gilled mushrooms on logs. Revision and keys: Seevers (1951), Ashe (1986d); generic redescription: Ashe (1984a). Larva: Ashe (1981, 1986b). Biology: Ashe (1981, 1982, 1984b, 1987). Two described species and an undetermined number of undescribed species occur in Mexico.

42c. Subtribe Homalotina Heer 1839
Tarsi 4-4-5; body moderately to strongly dorsoventrally flattened, more-or-less parallel-sided; labial palps not styliform, of 2 or 3 palpomeres (mostly 2); neck broad; middle coxae very narrowly separated or contiguous. All species occur under bark of dead trees. This subtribe includes four genera in North America.

*Anomognathus* Solier 1849
*Theetura* Thomson 1858
*Thectura* Thomson 1859
One species, *A. americanus* (Casey 1893), New York. Larva (genus, in key): Topp (1978).

*Cephaloxynum* Bernhauer 1907
Two species, Texas and Florida. Review of genus and tribal placement: Newton (1988).

*Homalota* Mannerheim 1830 (Fig. 261.22)
*Epipeda* Mulsant and Rey 1871
*Lampromalota* Cameron 1920
*Mimomalota* Cameron 1920
Nine species, widely distributed; presumed to be microphagous on surface fungal hyphae and spores. Feeding biology: Ashe (1993).

*Thecturota* Casey 1893 (Fig. 262.22)
*Oligurota* Casey 1893
*Hemithecta* Casey 1911
Eleven species, New York, Virginia, Iowa, Texas, Arizona, Nevada, and California.

42d. Subtribe Leptusina Fenyes 1918
This subtribe includes many very generalized homalotines and is difficult to characterize. Most can be recognized by a combination of tarsi 4-4-5; labial palp with 2 palpomeres (palpomeres 1 and 2 partially or fully fused), substylate in some; ligula elongate and entire apically (in most); mesocoxal cavities narrowly divided by slender meso- and metacoxal processes, or mesocoxae virtually contiguous. This subtribe includes three genera in North America.

*Dianusa* Casey 1906
Two species, California and New York. Seevers (1978) placed *Dianusa* as a synonym of *Leptusa (Eucryptusa)*, but Pace (1989: 23) specifically assigned two species to *Dianusa* as a valid genus. Casey (1911) also named a species, *Eucryptusa (Dianusa) bakeri*, that Pace (1989) did not mention; by default, it remains in *Leptusa (Eucryptusa)* where Seevers (1978) placed it. Characters that will allow unambiguous separation of *Dianusa* and *Leptusa* have not been found, and the two genera appear together in the key.

[*Euryusa* Erichson 1837, not North American. The athetine *Goniusa obtusa* (LeConte 1866) was originally described as a species of *Euryusa*. *Euryusa* is a European genus, and there is currently no evidence that any species of *Euryusa* occur in North America.]

*Heterota* Mulsant and Rey 1873
One adventive species, *H. plumbea* (Waterhouse 1858), Florida; in seaweed on beaches. First report of species in North America, description, and distribution: Frank and Thomas (1984b). Frank and Thomas (1984b) discussed the similarity between *Heterota* and *Leptusa* (especially *Leptusa* subgenus *Halmaeusa*). Following their discussion, *Heterota* is here placed in the Leptusina rather than in its own subtribe as done by Fenyes (1918-21).

*Leptusa* Kraatz 1856
*Eucryptusa* Casey 1906 (subgenus)
*Ulitusa* Casey 1906 (subgenus)
*Dysleptusa* Pace 1982 (subgenus)
*Adoxopisalia* Pace 1989 (subgenus)
*Boreoleptusa* Pace 1989 (subgenus)

*Heteroleptusa* Pace 1989 (subgenus)
Fifteen species, widely distributed throughout North America. Pace (1989) provided a complete world revision of *Leptusa* including keys to the subgenera, species, and subspecies, as well as illustrations of habitus and important diagnostic features of most taxa. Larvae of exotic species: Pace (1989), Topp (1978). Seventy-two subgenera are described from throughout the world; the above six subgenera occur in North America (Pace 1989).

42e. Subtribe Silusina Fenyes 1918
Tarsi 4-4-5, labial palps unusually long and filiform or stylate, sutures between palpomeres indistinct in many; ligula elongate and entire apically; mesocoxal cavities relatively narrowly separated by narrow meso- and metacoxal processes. As considered here, the Silusina do not include those taxa usually assigned to the subtribe Diestotina (or tribe Diestotini). This subtribe includes one genus in North America.

*Silusa* Erichson 1837
*Stenusa* Kraatz 1856 (subgenus)
Eight species, Georgia, Iowa, New York, North Carolina, and California. At least two undescribed species are known from Mexico. Two subgenera have been recognized including the typical subgenus, both occurring in North America.

42f. Subtribe Diestotina Mulsant and Rey 1871
Tarsi 4-4-5, labial palps of 2 palpomeres, unusually long and filiform or stylate, sutures between palpomeres indistinct in many; ligula divided into 2 lobes apically; mesocoxal cavities broadly separated by broad meso- and metacoxal processes; metasternal process as long as, or longer than, mesosternal process. This subtribe includes one genus in North America. Pace (1986) reassigned *Aphelogiossa* to subgeneric status under *Diestota* and placed several genera in synonymy with it.

*Diestota* Mulsant and Rey 1870
*Aphelogiossa* Casey 1893 (subgenus)
*Amenusa* Casey 1906
*Pectusa* Casey 1911
*Orthodiatelus* Notman 1920
*Elachistarthron* Notman 1920
Six species, Mississippi, California, Arizona, North Carolina, and Florida. Five described species are also known from Mexico. Five subgenera are recognized (key to subgenera, Pace 1986), only one of which occurs in North America.

Homalotini not assigned to subtribe:

*Coenonica* Kraatz 1857
One species, *C. puncticollis* Kraatz 1857, Florida; under bark of dead logs. First report of this genus and species in North America, description, habitus illustration, and distribution: Frank and Thomas (1984b).

*Cyphea* Fauvel 1863 (Fig. 253.22)
One species, *C. wallisi* Fenyes 1921, Manitoba. Larva of European species: Paulian (1941), Topp (1978).

43. Tribe Placusini Mulsant and Rey 1871
Tarsi 4-4-5; labrum rounded medially with small a-sensillum; mandible with dorsobasal "velvety patch" modified to transverse rows of large teeth; mandibles without rows of denticles in ventral molar region; ligula short, broad, and apically rounded; two medial setae of prementum with bases distant from each other; labial palps short, of 2 palpomeres; mesocoxae narrowly separated by slender intercoxal processes, or mesocoxae virtually contiguous (modified from Ashe 1991b). Redefinition and characterization of tribe based on larval and adult characteristics: Ashe (1991b). This tribe contains two genera worldwide; both occur in North America.

*Euvira* Sharp 1883 (Fig. 257.22)
*Crimalia* Casey 1911
One species, *E. quadriceps* (Casey 1911), Mississippi, and at least one undescribed species in Michigan, Kansas, North Carolina, Ohio, and Florida. The biology of *Euvira* is poorly understood; however, larvae and adults of one Mexican species were abundant in the larval web nest of a communal pierid butterfly (Ashe and Kistner 1989), and adults have been found in rolled leaf nests of various lepidopteran caterpillars in Michigan and Costa Rica (Ashe, unpublished data). Larva: Ashe and Kistner (1989). Four described and at least three additional unnamed species are known from Mexico.

*Placusa* Erichson 1837
*Calpusa* Mulsant and Rey 1871 (subgenus)
Eight species, widely distributed; under bark of dead trees, presumed microphagous on subcortical fungal hyphae and spores. Larva: Ashe (1991a 1991b). Feeding biology: Ashe (1993). Three additional described species occur in Mexico. Two subgenera are recognized including the typical subgenus; both occur in North America but the North American species have not been assigned to subgenera.

44. Tribe Philotermitini Seevers 1957
Tarsi 4-4-5; pronotum very broadly transverse with a distinctive pubescence pattern, microsetae sparse or absent medially, macrosetae increasing in number and distinctly longer laterally, microsetae directed laterally and anterolaterally; mandibles without a distinct patch of denticles in ventral molar region; mesocoxae contiguous; mesocoxal cavities not margined behind. All species are obligate inquilines in the nests of termites. This tribe contains two genera; one occurs in North America.

*Philotermes* Kraatz 1857 (Fig. 256.22)
Seven species, widespread in eastern United States from Massachusetts to Florida and west to Illinois and Texas; associated with termites of the genus *Reticulitermes*. Key to most species: Seevers (1957). Redescription of genus: Kistner and Gut (1977). Biology and behavior: Howard (1978), Howard *et al.* (1982).

45. Tribe Athetini Casey 1910

This tribe is very difficult to characterize; Seevers (1978) proposed the following combination of characters: tarsi 4-5-5; maxillary palps of 4 palpomeres; labial palps of 3 palpomeres in most (basal 2 palpomeres fused in a few); head without a prominent neck in most; mouthparts generalized; medial lobe of aedeagus with compressor plate oval (not elongate), and with a transverse sclerotized strip called the "athetine bridge" by Seevers (1978) (see Seevers 1978 for a description and discussion of this feature). Seevers (1978) suggested that the "athetine bridge" may be a synapomorphy for the Athetini. The Athetini, with thousands of species worldwide, are one of the most successful groups of small beetles. They are the dominant micropredators in many habitats including carcasses, dung, decaying fruit, rotting vegetation. They also occur in bird and mammal nests, riparian areas of lakes and streams, ant nests and occasionally on flowers (Seevers 1978).

The tribe Athetini is by far the most difficult tribe in the Aleocharinae. The current classification of North American athetine taxa is completely inadequate, and many of the genus-level taxa currently recognized, and the subtribes proposed by Seevers (1978), cannot be clearly delimited or diagnosed. Furthermore, many recognized genera are heterogeneous and include species that do not appear to be congeneric. The classification of the tribe, and the genus-level taxa in North America, requires detailed study and complete revision. This very large and poorly characterized tribe contains about 173 genera worldwide. Following Seevers (1978) and modifications proposed by Lohse *et al.* (1990) and Lohse and Smetana (1985), 64 genera are recognized here as occurring in North America. Not all genera are assigned to subtribes. Larvae of "*Atheta*" in a broad sense spanning several genera and subtribes here were reviewed by Topp (1975), and a detailed scheme for naming setae based on *Atheta coriaria* (Kraatz 1856) proposed by Ashe and Watrous (1984).

45a. Subtribe Acrotonina Seevers 1978

This poorly defined subtribe includes those athetines that have a relatively convex pronotum and pronotal hypomera that are not visible in lateral aspect (Seevers 1978). Three genera are assigned to this subtribe in North America.

*Acrotona* Thomson 1859
- *Coprothassa* Thomson 1859 (subgenus)
- *Colpodota* Mulsant and Rey 1873 (subgenus)
- *Achromota* Casey 1893 (subgenus)
- *Eurypronota* Casey 1893 (subgenus)
- *Ancillota*, Casey 1910 (subgenus)
- *Aremia* Casey 1910 (subgenus)
- *Arisota* Casey 1910 (subgenus)
- *Dolosota* Casey 1910 (subgenus)
- *Microlia* Casey 1910 (subgenus)
- *Neada* Casey 1910 (subgenus)
- *Reania* Casey 1910 (subgenus)

Forty-nine species (not including *Acrotona* (*sensu stricto*) of Seevers 1978, see comment under *Mocyta* below), widely distributed.

*Mocyta* Musant and Rey 1874
- *Acrotona (sensu stricto)* of Seevers 1978

Nineteen species (including those treated as *Acrotona* (*sensu stricto*) by Seevers 1978), widely distributed. Lohse *et al.* (1990) placed Seevers' [North American] *Acrotona* (*sensu stricto*) in synonymy with *Mocyta*. Because of the diverse array of species included in *Acrotona* by Seevers (1978), it is not possible to unambiguously separate *Mocyta*, *Strigota*, and *Acrotona* (*sensu* Seevers). This group of genera badly needs revision.

*Strigota* Casey 1910
- *Eustrigota* Casey 1911

Fifteen species, widely distributed. The character used by Seevers (1978) to separate *Strigota* from *Acrotona* (fine white pilose pubescence) is very difficult to distinguish, and is not used in the key for separation of these taxa.

45b. Subtribe Athetina Casey 1910
- Xenotae Seevers 1978 (*nomen nudum*)

No unique features that characterize this subtribe as delimited by Seevers (1978) are evident, and Seevers did not provide a characterization. Seevers (1978) likewise used a subtribal heading "Xenotae," which would be properly spelled Xenotina, but did not define this group or separate it from Athetina. Together, these groups include those athetines that have medium-sized to large eyes and fully exposed pronotal hypomera. Seevers (1978) used Athetina in a different sense than the tribal usage of Lohse *et al.* (1990) for the Arctic species. The Athetina include 27 North America genera (not including *Schistoglossa* Kraatz — see below).

*Adota* Casey 1910
- *Panalota* Casey 1910

Ten species, Alaska and California; littoral or intertidal. Lohse and Smetana (1985: 282) noted that *Panalota* is a junior synonym of *Adota*. Habitat: Lohse and Smetana (1985).

*Alaobia* Thomson 1858

One species, *A. sparreschneideri* (Munster 1922), Alaska and Yukon Territory.

*Aloconota* Thomson 1858
- *Aloconota* Thomson 1858 (subgenus)
- *Taphrodota* Casey 1906 (subgenus)
- *Terasota* Casey 1906 (subgenus)

Nine species, New York, Virginia, North Carolina, Iowa, and California. Larva of European species: Topp (1975).

*Anopleta* Mulsant and Rey 1874

One undescribed species, Washington and Alberta (Seevers 1978).

*Atheta* Thomson 1858 (Fig. 21.22)
- *Megista* Mulsant and Rey 1874
- *Elytrusa* Casey 1906
- [subgenus *Dochmonota* Thomson 1859, not North American?]

[subgenus *Peliolurga* Tottenham 1939, not North American]

Seven species, Alaska and throughout the Canadian arctic. Key to arctic species: Lohse *et al.* (1990). Larvae of European species: Ashe and Watrous (1984), Topp (1975). The name *Atheta* was conserved in this sense by ICZN (1961b).

[*Badura* Mulsant and Rey 1873, probably not North American]

*Bessobia* Thomson 1858
One species, *B. cryptica* Lohse 1990, Quebec.

*Boreostiba* Lohse 1990
Five species, throughout arctic North America. Key: Lohse *et al.* (1990).

*Clusiota* Casey 1910 (Fig. 324.22)
One species, *C. claviventris* Casey 1910, British Columbia.

*Dinaraea* Thomson 1858
Three species, Alaska, Quebec, and New Hampshire (Lohse *et al.* 1990). Larvae of European species: Topp (1975).

*Euromota* Casey 1906
One species, *E. lucida* Casey 1906, Virginia to New York.

*Halobrecta* Thomson 1858
One species, *H. algophila* (Fenyes 1909), California.

*Hydrosmecta* Thomson 1858
Thirteen species, New York, Wisconsin, and California. Seevers (1978) noted that all species assigned to this genus may not be congeneric.

*Hydrosmectina* Ganglbauer 1895 (Fig. 323.22)
One species, *H. macra* Fenyes 1921, California; in sand on the shores of rivers and creeks. Habitat: Fenyes (1918-21: 238; 1921).

*Iotota* Casey 1910
Three species, California.

*Lamiota* Casey 1910
Three species, British Columbia.

*Liogluta* Thomson 1858
*Athetota* Casey 1906
*Pseudomegista* Bernhauer 1907
*Anepsiota* Casey 1893 (subgenus)
Fifteen species, widely distributed. Key to arctic species: Lohse *et al.* (1990). Lohse and Smetana (1985: 286) noted that *Pseudomegista nigropolita* (Bernhauer 1907) is a species of *Liogluta*, and that *Pseudomegista* is a junior synonym of *Liogluta*. In addition, Lohse *et al.* (1990) placed *Athetota* as a synonym of *Liogluta* and treated *Anepsiota* as a subgenus of *Liogluta*.

*Micratheta* Casey 1910
One species, *M. caudex* (Casey 1910), Virginia.

*Micrearota* Casey 1910
Nine species, northeastern North America and Texas.

*Microdota* Mulsant and Rey 1873
*Hilara* Mulsant and Rey 1873
Twenty-seven species, widespread in North America. *Microdota* is difficult to distinguish from *Datomicra*, except that the pronotal hypomera are fully exposed in lateral aspect in *Microdota* and partially exposed in *Datomicra* (Seevers 1978).

*Noverota* Casey 1910
Seven species, widespread in eastern North America.

*Omegalia* Casey 1910
Two species, California.

*Paradilacra* Bernhauer 1909
*Dilacra* Thomson 1858 (*sensu* Casey 1910a, 1911)
Thirteen species, widespread in western North America.

*Phasmota* Casey 1910
One species, *P. ingratula* (Casey 1910), Mississippi.

*Philhygra* Mulsant and Rey 1873
*Hygroecia* Mulsant and Rey 1873 (*sensu* Brundin 1944)
*Metaxya* Mulsant and Rey 1873 (*sensu* Casey 1910a, 1911; Bernhauer and Scheerpeltz 1926)
*Homalotusa* Casey 1906
*Amphibitherion* Notman 1921
Fifty-five species, generally distributed. Lohse *et al.* (1990) placed *Metaxya* (*sensu* Casey 1910a) and *Homalotusa* as synonyms of *Philhygra*. The generic assignment of species included in this genus very badly needs study; they form a heterogeneous group, and all may not be congeneric.

*Schistoglossa* Kraatz 1856
Seevers (1978) proposed that *Atheta reticula* Casey 1910 (as *reticulata*) from Virginia should be placed in *Schistoglossa*, along with two North American species described by Brundin (1943). Herman (in Seevers 1978) noted that he was unable to locate the species cited by Seevers in Brundin's paper, and could not find them listed in the Zoological Record. Examination of *A. reticula* shows that it does not have the apically bifid mandibles that characterize *Schistoglossa viduata* (Erichson 1837) (the type species of the genus). Therefore, unless additional information is discovered, there is no clear evidence that *Schistoglossa* occurs in North America.

*Stethusa* Casey 1910
Thirteen species, widespread in the eastern half of North America. Seevers (1978) transferred many of the species included in *Hypatheta* Fenyes 1918 to *Stethusa*, and proposed that *Hypatheta* may be "in part" a synonym of *Stethusa*. However, Seevers states

that he did not examine the European representatives of *Hypatheta* which include the type species *H. castanoptera* (Mannerheim 1830). Consequently, *Hypatheta* is not treated as a synonym of *Stethusa* in this work pending comparative examination of appropriate type specimens.

*Valenusa* Casey 1906
One species, *V. parallela* Casey 1906, California.

*Xenota* Mulsant and Rey 1874
  *Atheta* (*sensu stricto*) Thomson 1858 (*sensu* Ganglbauer 1895, Casey 1910a, 1911, Bernhauer and Scheerpeltz 1926, Fenyes 1918-21)
  *Philhygra* Mulsant and Rey 1873 (*sensu* Casey 1910a, not Brundin 1944)
  *Mycota* Mulsant and Rey 1874
  *Tetropla* Mulsant and Rey 1874
  *Ceritaxa* Mulsant and Rey 1873 (*sensu* Bernhauer 1909) (subgenus)
  *Delphota* Casey 1910 (subgenus)
  *Donesia* Casey 1910 (subgenus)
  *Dralica* Mulsant and Rey 1874 (*sensu* Bernhauer 1907) (subgenus)
  *Halobrecthina* Bernhauer 1909 (subgenus)
    *Rovalida* Casey 1910
  *Nemota* Casey 1910 (subgenus)
  *Traumoecia* Mulsant and Rey 1874 (*sensu* Bernhauer 1909) (subgenus)
One hundred sixty-seven species, generally distributed, were included in this genus by Seevers (1978). However, Lohse and Smetana (1985) noted that Seevers (1978) included species representing several genera in his *Xenota*. Consequently, *Xenota* (*sensu* Seevers 1978) is a heterogeneous group and is not the same as *Xenota* Mulsant and Rey. Lohse and Smetana (1985) transferred several species from *Xenota* (*sensu* Seevers 1978) to other athetine genera, but most of the species included here in *Xenota* have not been critically studied. This fact helps explain the difficulty of providing a key that will reliably identify specimens of *Xenota* (*sensu* Seevers 1978).

45c. Subtribe Dimetrotina Seevers 1978
This very poorly characterized subtribe includes those athetines that have a less convex pronotum than in the Acrotonina, so that the pronotal hypomera are partially visible in lateral aspect (Seevers 1978). Ten genera are assigned to this subtribe in North America.

*Amischa* Thomson 1858
  *Colposura* Casey 1893
Twelve species, widely distributed. Casey (1893) proposed *Colposura* as a subgenus of *Amischa*; Seevers (1978) questioned the usefulness of this taxon as a subgenus.

*Anatheta* Casey 1910
One species, *A. planulicollis* (Casey 1910), Kansas. *Anatheta curata* (Casey 1910) does not seem to be congeneric, and its placement needs additional study (Seevers 1978).

*Canastota* Casey 1910
Seven species, widespread in the eastern half of North America. Partial key: Seevers (1978).

*Datomicra* Mulsant and Rey 1874
  *Hilarina* Casey 1910
  *Micromota* Casey 1910
  *Monadia* Casey 1910
  *Oligomia* Casey 1910
  *Taxicerella* Casey 1910
Twenty-four species, widely distributed. It is difficult to separate specimens of *Datomicra* from those of *Dimetrota*, except by the smaller size of *Datomicra* (Seevers 1978).

[*Daya* Fauvel 1878, not Bleeker 1877 (replaced by *Homia* Blackwelder 1952), not North American]

*Dimetrota* Mulsant and Rey 1873
  *Coproceramius* Gistel 1857
  *Dalotia* Casey 1910
  *Engamota* Casey 1910
  *Dimetrotina* Casey 1911
Forty-three species, widely distributed. Description of genus and key to arctic species: Lohse *et al.* (1990). All of the North American species assigned to this genus by Casey do not appear to be congeneric (Seevers 1978).

*Fusalia* Casey 1911
One species, *F. brittoni* (Casey 1911), Connecticut.

*Pancota* Casey 1906
  *Dolosota* Casey 1910 (in part)
Thirteen species, widespread in eastern North America.

*Pseudota* Casey 1910
Fifteen species, northeastern and northwestern North America and Arizona. *Pseudota* is very difficult to distinguish from *Dimetrota* (Seevers 1978).

*Sableta* Casey 1910
One species, *S. inflata* Casey 1910, Mississippi; frequently on fungusy logs and on polypore mushrooms on logs.

*Synaptina* Casey 1910
  *Rhodeota* Casey 1911
Five species, Rhode Island, Iowa, and California.

45d. Subtribe Geostibina Seevers 1978
Seevers (1978) placed in this subtribe genera that are characterized by the combination of small eyes; short elytra (probably indicating lack, or reduction, of hind wings); fully exposed pronotal hypomera; and incrassate antennae with antennomeres 4-10 transverse. He included seven genera in this subtribe. Lohse and Smetana (1988b) pointed out that Seevers' concept of the

Geostibina unites a heterogeneous assemblage of genera that are not closely related to each other.

*Anaduosternum* Notman 1922
One species, *A. brevipennis* Notman 1922, New Jersey.

*Asthenesita* Casey 1893
One species, *A. pallens* Casey 1893, Florida.

*Crephalia* Casey 1910
Three species, Missouri, New York, and Pennsylvania.

*Gaenima* Casey 1911
One species, *G. impedita* Casey 1911, California.

*Geostiba* Thomson 1858
*Evanystes* Gistel 1856
*Sipalia* Mulsant and Rey 1873 (*sensu* Casey 1910a, 1911, Bernhauer and Scheerpeltz 1926)
*Glossola* Fowler 1888
*Typhlusida* Casey 1906
*Sibiota* Casey 1906 (subgenus)
*Sonomota* Casey 1911 (subgenus)
*Ditroposipalia* Scheerpeltz 1951 (subgenus)
*Lioglutosipalia* Scheerpeltz 1951 (subgenus)
Nine species, Oregon and the Appalachian Mountains of North Carolina and Tennessee. Review and key: Lohse and Smetana (1988b), augmented by Pace (1997).

*Ousipalia* Gozis 1886
Four species, California, Colorado, and Alberta. Lohse and Smetana (1988b) pointed out that Seevers (1978) was incorrect to place *Ousipalia* as a junior synonym of *Geostiba*, and that the species that Casey placed in *Ousipalia* correctly belong in that genus. Characters that will reliably separate *Ousipalia* from *Crephalia* have not been found and they are keyed out in the same couplet.

*Sipaliella* Casey 1911
One species, *S. filaria* (Casey 1911), Rhode Island.

Athetini not assigned to subtribe:

*Boreophilia* Benick 1973
Thirteen species, throughout Arctic North America. Key: Lohse *et al.* (1990).

*Callicerus* Gravenhorst 1802 (Fig. 319.22)
One undescribed species, Ontario; in groundhog (*Marmota monax*) burrow. Two subgenera have been described; the only known North America species belongs in the typical subgenus.

*Charoxus* Sharp 1883 (Fig. 322.22)
One species, *C. spinifer* Frank 1996, Florida; adults and larvae inside ripe syconia of Figs. Revision and keys: Kistner (1981), Frank and Thomas (1997). Biology: Frank and Thomas (1997), Ashe (unpublished data). Kistner (1981) placed *Charoxus* in the tribe Aleocharini; Frank and Thomas (1997) transferred it to the Athetini because of the presence of a "athetine bridge" on the median lobe of the aedeagus of *Charoxus*. Two additional described species occur in Mexico.

*Doliponta* Blackwelder 1952 (Fig. 325.22)
*Lipodonta* Fenyes 1921 (not Nitzsch 1820)
One species, *D. veris* (Fenyes 1921), California; flood debris. Description, habitat: Fenyes (1921).

*Earota* Mulsant and Rey 1874
*Macroterma* Casey 1906
Five species, New York, New Hampshire, Iowa, Indiana, and California.

*Goniusa* Casey 1906
Two species, Massachusetts, Pennsylvania, District of Columbia, and Washington state; associated with ants of the genus *Formica*. Seevers (1978) stated that one species occurs as far west as Texas. Key, overview of biology and larval description: Kistner (1976). Kistner (1976) transferred *Goniusa* to the Lomechusini (as Zyrasini), but Seevers (1978) did not, so it is here left in the Athetini pending further study.

*Lypoglossa* Fenyes 1918
Two species, widely distributed throughout Rocky Mountains, west coast and northern half of North America; in leaf litter, leaf/log litter, and other forest floor debris. Revision and key: Hoebeke (1992).

*Parameotica* Ganglbauer 1895
Two species, California.

*Pontomalota* Casey 1885 (Fig. 321.22)
Two species, widespread on Pacific coast of North America from Baja California to Alaska; on sandy beaches among and under beach wrack and seaweed, adults occasionally present on sand in large numbers near maximum extent of waves. Revision and key: Ahn and Ashe (1992).

*Pseudousipalia* Lohse 1990
One species, *P. microptera* Lohse 1990, Alaska and Yukon Territory.

*Schistacme* Notman 1920 (Fig. 318.22)
One species, *S. obtusa* Notman 1920, Florida. *Schistacme* was originally placed in the "group Silusae" of the Homalotini (previously Bolitocharini) by Notman (1920) and was treated as the "*Schistacme* Group" of the Homalotini by Seevers (1978). However, examination of the type shows that the tarsi are 4-5-5 rather than 4-4-5, and *Schistacme* should be placed in the Athetini.

*Seeversiella* Ashe 1986
One species, *S. bispinosa* Ashe 1986, Arizona, New Mexico, Colorado, and Michigan; in mixed conifer and aspen litter or wood

chips and other woody debris at the base of ponderosa pine trees. Review: Ashe (1986a).

*Strophogastra* Fenyes 1921
One species, *S. penicillata* Fenyes 1921, Manitoba and Alberta.

*Tarphiota* Casey 1893
Two species, Pacific coast from California to Alaska; among beach wrack and seaweed on sandy beaches. Revision and key: Ahn (1996b).

*Thamiaraea* Thomson 1858
Two species, Kansas, North Carolina, Louisiana, and Pennsylvania; usually at fermenting sap flows on trees. Revision and key: Hoebeke (1988b).

*Thinusa* Casey 1893
Two species, Pacific coast from California to Alaska; among beach wrack and seaweed on sandy beaches. Revision and key: Ahn (1997b). *Thinusa* has previously been placed in the tribe Phytosini because it has 4-4-5 tarsi; however, Ahn and Ashe (1992) indicated it was closely related to *Pontomalota* and the two genera should be included in a monophyletic group that probably included *Tarphiota* as well. Ahn (1997b) formally moved *Thinusa* to the Athetini.

*Trichiusa* Casey 1893 (Fig. 320.22)
Nineteen species, widely distributed.

46. Tribe Crematoxenini Mann 1921
The characters used by Jacobson and Kistner (1992) to define the Crematoxenini are found in different combinations among some New World Lomechusini. Consequently, this tribe is difficult to separate from the Lomechusini. Jacobson and Kistner (1992) characterized the tribe Crematoxenini with the following combination of features: tarsi 4-5-5 segmented; anterior margin of abdominal sternum IV with well-developed gland reservoirs; base of abdomen greatly narrowed to form a distinct ant-like "petiole" in most taxa; and pronotum without a distinct median sulcus. Seevers (1978) also noted that the mesocoxal cavities are not margined. Most also share the following characteristics with the Lomechusini: mesosternal process very short, hardly extended between the coxal cavities; metasternal process very long, extended between coxae to near middle of coxae or more. All species for which hosts are known are inquilines with army ants of the genus *Neivamyrmex*. World revision of tribe: Jacobson and Kistner (1992). This exclusively New World tribe includes ten genera; three or four occur in North America.

*Beyeria* Fenyes 1910
Two species, Arizona. Key: Jacobson *et al.* (1987). One additional species, *B. sauvis* Mann, occurs in Mexico.

*Neobeyeria* Jacobson *et al.* 1987
One species, *N. arizonensis* Jacobson *et al.* 1987, Arizona.

*Probeyeria* Seevers 1965 (Fig. 264.22)
One species, *P. pulex* (Sanderson 1943), Arizona, Arkansas, and Kansas. Summary of behavior: Akre and Rettenmeyer (1966).

[*Pulicomorpha* Mann 1924, two species in Baja California Sur; inquilines with the army ant *Neivamyrmex leonardi* (Wheeler 1915). Seevers (1978) suggested that the genus may occur in the southwestern United States. Key: Jacobson and Kistner (1992).]

47. Tribe Falagriini Mulsant and Rey 1873
Characterization: Tarsi 4-5-5 (4-4-5 in *Bryobiota*); head with well-defined neck, neck narrow, less than 1/2 head width; pronotum narrowed basally, with a slight to very strong medial longitudinal sulcus; prosternum usually elongate behind procoxae, often extending laterally to pronotal hypomera; peritremes present around mesosternal spiracles, peritreme size ranging from very small and only narrowly surrounding spiracles, to very large and conspicuous but not contiguous along midline, to very large and contiguous along midline; male copulatory organ with velum of paramerite and condylite clearly separated into 2 lobes; and anterior margin of abdominal sternum IV with distinctive gland opening (see Ahn and Ashe 1995). Revision and phylogeny of North American genera and species: Hoebeke (1985), revised by Ahn and Ashe (1995). This tribe contains about 30 genera worldwide; 11 genera occur in North America.

*Aleodorus* Say 1830 (Fig. 273.22)
*Chitalia* Sharp 1883
Four species, widely distributed. Revision and key: Hoebeke (1985). One additional described species and an undetermined number of undescribed species are known from Mexico.

*Borboropora* Kraatz 1862 (Fig. 274.22)
*Pseudoscopaeus* Weise 1877
*Aneurota* Casey 1893
*Orthagria* Casey 1906
Two species, eastern United States. Revision and key: Hoebeke (1985).

*Bryobiota* Casey 1893
Two species, Pacific coast from British Columbia to Baja California; intertidal zone of seashores. Revision and key: Ahn and Ashe (1995).

*Cordalia* Jacobs 1925 (Fig. 22.22)
*Cardiola* Mulsant and Rey 1874 (not Broderip 1834)
*Strandiodes* Bernhauer 1930
*Cardiolita* Strand 1933
One species, *C. obscura* (Gravenhorst 1802), northeastern North America, Colorado, and Washington, also in Europe. Review: Hoebeke (1985). Larva: Topp (1978).

*Falagria* Leach 1819
Two species, *F. dissecta* Erichson 1839, widely distributed, and the Palearctic *F. sulcata* (Paykull 1789), northern United States

and southern Canada. Revision and key: Hoebeke (1985). Three additional described species are known from Mexico.

*Falagrioma* Casey 1906
*Anaulacaspis* Ganglbauer 1895 (*sensu* Jarrige 1946)
One species, *F. socorroensis* Hoebeke 1985, New Mexico. Review: Hoebeke (1985).

*Falagriota* Casey 1906
One species, *F. occidua* (Casey 1906), California and Oregon. Review: Hoebeke (1985).

*Leptagria* Casey 1906
One species, *L. perexilis* Casey 1906, Texas, Florida and New York, in horse and cow dung. Review: Hoebeke (1985), Frank *et al.* (1989).

*Lissagria* Casey 1906
*Omoschema* Notman 1920
Two species, southeastern United States and California. Revision and key: Hoebeke (1985).

*Myrmecocephalus* MacLeay 1873
*Stilicioides* Broun 1880
*Stenagria* Sharp 1883
*Lorinota* Casey 1906
Eight species, widely distributed, one cosmopolitan. Revision and key: Hoebeke (1985). An additional four described and at least 12 undescribed species are known from Mexico.

*Myrmecopora* Saulcy 1864
One species, *M. vaga* (LeConte 1866), Lake Superior region; usually in riparian areas of lakes and streams, though at least some European species occur in the intertidal zone of seashores. Three subgenera have been recognized but only the typical subgenus is represented in North America. This genus was excluded from Falagriini by Hoebeke (1985) but returned by Ahn and Ashe (1995).

48. Tribe Sceptobiini Seevers 1978
Characterization: Members of this tribe are very similar to those of the Falagriini in most tribal characteristics, and Seevers (1978) placed the tribe Sceptobiini together with the Falagriini into the supertribe Falagriinea. These two tribes share 4-5-5 tarsi; male copulatory organ with velum of paramerite and condylite clearly separated into 2 lobes; and anterior margin of abdominal sternum IV with distinctive gland opening. They differ in that sceptobiines have macrosetae on pronotum, elytra, most abdominal tergites, and all abdominal sterna; sceptobiines are all apterous with moderately to greatly shortened elytra; sceptobiines lack the sulcate pronotum and well-defined narrow neck that characterize most Falagriini; and sceptobiines are all inquilines with ants of the genus *Liometopum* (Danoff-Burg 1994). Revision and phylogeny of tribe: Danoff-Burg (1994). Behavior: Danoff-Burg (1996). This exclusively New World tribe includes only two genera known from the southwestern United States and Mexico. All species are inquilines in the nests and raiding columns of ants of the genus *Liometopum*.

*Dinardilla* Wasmann 1901 (Fig. 275.22)
*Dinardella* Mann 1914 (error for *Dinardilla*)
Two species, Arizona, Colorado, New Mexico, and Texas, also in Mexico. Revision and key: Danoff-Burg (1994). Behavior: Danoff-Burg (1996).

*Sceptobius* Sharp 1883 (Fig. 276.22)
*Apteronina* Wasmann 1901
*Symbiochara* Fenyes 1909
Three species, Arizona, California, Colorado, New Mexico, and Texas, two of these also in Mexico. Revision and key: Danoff-Burg (1994). Behavior: Danoff-Burg (1996).

49. Tribe Lomechusini Fleming 1821
Myrmedoniini Thomson 1867
Zyrasini Bradley 1930
Tarsi 4-5-5; metasternal process distinctly longer than very short mesosternal process; galea and lacinia of maxilla of most moderately to greatly elongated, length of galea equal to or greater than distance to base of galea from cardo; mesocoxae moderately to very broadly separated by broad meso- and metasternal processes. This tribe is most frequently referred to in the literature as Myrmedoniini or Zyrasini. However, Newton and Thayer (1992) pointed out that the name Lomechusini has priority as the correct name for this tribe. Many genera (though not all) and species are associated with ants. This poorly defined tribe includes about 152 genera worldwide; two recognized subtribes and 13 genera (not all assigned to subtribe) occur in North America.

49a. Subtribe Lomechusina Fleming 1821
Lomechusini with broad, explanate pronota and prominent tufts of golden-yellow trichomes on the abdomen. Seevers (1978) called this taxon the "Xenodusae group." One genus occurs in North America.

*Xenodusa* Wasmann 1894 (Fig. 23.22)
*Pseudolomechusa* Mann 1914 (proposed as subgenus of *Xenodusa*)
Four species, northeastern United States, western North America, southwestern United States, and California; myrmecophilous. Adults spend summer in nests of *Formica* ants, where they breed, and winter in nests of *Camponotus*. Revision and key: Hoebeke (1976). Larva and biology: Wheeler (1911). One additional described species occurs in Mexico.

49b. Subtribe Myrmedoniina Thomson 1867
Zyrasina Bradley 1930
This subtribe cannot be adequately characterized. It appears to be made up of those relatively generalized lomechusines that lack prominent specializations for myrmecophily. Seevers (1978)

called this taxon the "Drusillae group." Six North American genera are assigned to this subtribe.

*Apalonia* Casey 1906 (Fig. 280.22)
Two species, Florida and Kansas.

*Drusilla* Leach 1819 (Fig. 282.22)
*Astilbus* Dillwyn 1829
*Myrmedonia* Erichson 1837
Two described species, *D. cavicollis* Casey 1906, Alaska, and *D. canaliculata* (Fabricius 1787), a European species adventive in eastern North America; also at least three undescribed species in North Carolina, Arkansas, and Oklahoma. Larva of European species: Paulian (1941), Topp (1978).

*Meronera* Sharp 1887 (Fig. 283.22)
*Merona* Sharp 1883 (not Norman 1865)
One species, *M. venustula* (Erichson 1839), widespread in eastern North America. Biology and larva: Ashe (1985). Ashe (1993) reported that larvae and adults feed on fungal mycelia and hyphae and described presumptive correlated mouthpart features. Seevers (1978) placed *Meronera* in the Oxypodini subtribe Tachyusina. However, *Meronera* has the broadly separated mesocoxae and the elongate metasternal process (much longer than the short mesosternal process) that is characteristic of the Lomechusini. One additional described species and at least two unnamed species are known from Mexico.

*Tetradonia* Wasmann 1894
*Chlorotusa* Casey 1906
One species, *T. megalops* (Casey 1906), Texas. The host of *T. megalops* is not known, but other *Tetradonia* species from Central and South America are known to be associated with army ants, especially various species of *Eciton*. World revision, key, and biology: Jacobson and Kistner (1998).

*Trachyota* Casey 1906 (Fig. 281.22)
Two species, California. Seevers (1978) placed *Trachyota* in the Oxypodini subtribe Tachyusina. However, *Trachyota* has the broadly separated mesocoxae and the elongate metasternal process (much longer than the short mesosternal process) that is characteristic of the Lomechusini. The species of *Trachyota* are surprisingly similar to some North American species of *Drusilla*. The *Drusilla* species have shorter elytra and do not have the distinctive longitudinal furrows on the elytra and the abdominal modifications that characterize specimens of *Trachyota* (see key or Seevers 1978 for a description of these).

*Zyras* Stephens 1835 (Fig. 287.22)
*Myrmoecia* Mulsant and Rey 1873
*Platyusa* Casey 1885
*Nototaphra* Casey 1893
Sixteen species, widely distributed in eastern North America. Key: Seevers (1978). Larva of European species: Topp (1978). At least six additional described species occur in Mexico. The name *Zyras* was conserved by ICZN (1961a). *Zyras* is a large and complex genus with over 500 species worldwide assigned to at least 35 subgenera (Seevers 1965, 1978). The North American species have not been assigned to subgenera.

Lomechusini not assigned to tribe:

*Dinocoryna* Casey 1893 (Fig. 285.22)
*Ecitonusa* Wasmann 1897
Six species, Florida to North Carolina and west to Kansas and Arizona; all inquilines with *Neivamyrmex* army ants. Key to most species, Seevers (1959). Summary of behavior: Akre and Rettenmeyer (1966).

*Ecitocala* Frank 1981
One species, *E. rugosa* Frank 1981, Florida. Description and habitus drawing: Frank and Thomas (1981a).

*Ecitonidia* Wasmann 1900 (Fig. 277.22)
One species, *E. wheeleri* Wasmann 1900. Arkansas, Colorado, Kansas, and Arizona; inquilines with *Neivamyrmex* army ants.

*Ecitoxenidia* Wasmann 1909 (Fig. 278.22)
Four species, Alabama, Arizona, Florida, North Carolina, and Texas; inquilines with army ants of the genus *Neivamyrmex* as far as known. Revision and key: Kistner *et al.* (1996). One additional species occurs in Mexico.

*Microdonia* Casey 1893 (Fig. 286.22)
Four species, Arizona, Florida, Kansas, and Texas, two of these and two additional described species in Mexico; all inquilines with various species of army ants of the genus *Neivamyrmex*. Revision and key: Kistner *et al.* (1996). Summary of behavior: Akre and Rettenmeyer (1966).

*Xesturida* Casey 1906 (Fig. 284.22)
One species, *X. laevis* Casey 1906, Mississippi. One undescribed species is reported from Mexico (Seevers 1978).

XIII. Subfamily TRIGONURINAE Reiche 1865

Body compact (Fig. 24.22), almost nitidulid-like; pronotum wider at base than at apex; elytra together much longer than wide, extending well beyond metasternum and leaving only five abdominal terga exposed; each elytron with five or more rows of coarse punctures; tarsi 5-5-5; abdomen with six visible sterna, one pair of paratergites per segment. Antennal insertions at sides of frons, concealed under shelf-like projection; protrochantins exposed. The single included Holarctic genus (primarily North American, with seven species here) has usually been placed in a loosely defined Piestinae, but also has been placed in Silphidae or, after discovery of its distinctive and primitive larva (Pototskaya 1976), in its own staphylinid subfamily (e.g., Newton and Thayer 1992). Certain fossils from the Upper Jurassic or lower Cretaceous have been compared to it (Ryvkin 1990).

*Trigonurus* Mulsant 1847 (Fig. 24.22)
Seven species, British Columbia to southern California; adults and larvae found under bark, especially of conifer logs. Key: Blackwelder (1941), augmented by Hatch (1957: 241). Larva of European species: Pototskaya (1976). The genus also includes one species each from southwestern Europe and the Caucasus region.

XIV. Subfamily SCAPHIDIINAE Latreille 1807

Body form characteristic (Fig. 4.22): broadly oval, compact, very convex and shiny, with long slender legs and long but truncate elytra that leave only 1-2 abdominal segments fully exposed; head small, somewhat retracted and hypognathous, antennae inserted on frons between eyes, neck absent; procoxal fissure very short and narrow to absent, protrochantin always concealed; abdomen more or less conical, with six visible sterna and 0-2 pairs of paratergites per segment; tarsi 5-5-5. This distinctive subfamily has usually been treated as a family of its own, but on the basis of larval structure, male genitalia and other characters belongs in Staphylinidae (e.g., Kasule 1966, Lawrence and Newton 1982, Leschen and Löbl 1995). Currently (Löbl 1997) it is divided into four tribes, all represented in North America. As far as known the entire group, with some 1360 species in 45 genera worldwide but most diverse in the tropics, is mycophagous on a wide variety of fungi and slime molds (Newton 1984). Seven genera containing 70 species are known from North America. Larvae were reviewed most recently by Kompantsev and Pototskaya (1987) and Newton (1991).

50. Tribe Cypariini Achard 1924
Eye not emarginate near antennal insertion; antennal club compact with only slightly flattened antennomeres; pronotum widest at base, sides steadily converging anteriorly; tibiae (especially mesotibia) distinctly spinose. This tribe includes a single genus.

*Cyparium* Erichson 1845
Two species, *C. ater* Casey 1900, southern Texas, and *C. concolor* (Fabricius 1801) (=*C. flavipes* LeConte 1860), southeastern United States; on mushrooms and coral fungi in which the larvae feed. Key: Casey (1900: 56). Biology: Newton (1984). Larvae of exotic species: Hayashi (1986), Kompantsev and Pototskaya (1987), Newton (1991). The genus is pantropical extending into eastern Asia and North America, with nearly 50 known species.

51. Tribe Scaphiini Achard 1924
Eye not emarginate near antennal insertion; antennal club compact with only slightly flattened antennomeres; pronotum widest near middle, sides sinuately constricted toward base; tibiae not spinose. Except for one Nearctic species, this group of three genera and 15 species is confined to Eurasia and possibly South Africa.

*Scaphium* Kirby 1837
One species, *S. castanipes* Kirby 1837, Alaska and British Columbia east to Quebec and New Hampshire; associated with mushrooms on which the larvae feed. Biology and larva: Ashe (1984c). The genus also includes two Palearctic species and one doubtfully recorded from South Africa.

52. Tribe Scaphidiini Latreille 1807
Eye strongly emarginate near antennal insertion; antennal club loose (Fig. 326.22) with distinctly flattened antennomeres; pronotum widest at base, sides steadily converging anteriorly; tibiae not spinose. As redefined by Leschen and Löbl (1995) this tribe includes the groups Cerambyciscaphini Pic 1915, and Diateliini Achard 1924, but excludes Scaphiini; their revision of generic limits leaves only four genera, three endemic to the Oriental region with one species each and one nearly worldwide genus.

*Scaphidium* Olivier 1790 (Fig. 4.22)
Three to five species (in need of revision), Ontario and Quebec and widespread in the eastern United States, west to Iowa and Texas and possibly Colorado; associated with old logs and particularly with certain polypore fungi (including resupinate types) on these. Key to most species: Casey (1900: 55). Biology: Leschen (1994), Newton (1984). Larvae of exotic species: Hayashi (1986), Kompantsev and Pototskaya (1987), Newton (1991). This genus is nearly worldwide but mainly tropical, with more than 260 species.

53. Tribe Scaphisomatini Casey 1893
Scaphisomini auctorum
Antenna very slender, antennomeres 3-11 filiform and/or flattened and asymmetrical, forming a very loose antennal club; scutellum scarcely if at all visible. As recently redefined by Leschen and Löbl (1995) this group included the former tribes Toxidiini Achard 1924, and Heteroscaphini Achard 1914, as subtribes, in addition to the subtribes Baeocerina Achard 1924, and Baeoceridiina Achard 1924, but Löbl (1997) discarded all these subtribes as not monophyletic. This tribe includes 37 genera and is best represented in the Old World tropics.

*Baeocera* Erichson 1845
*Sciatrophes* Blackburn 1903
*Eubaeocera* Cornell 1967
Thirty-nine species, widely distributed in eastern North America southwest to Arizona, also one species extending to the Pacific Northwest, but none in California or Oregon. Species are commonly collected in forest leaf litter and around rotting logs, often associated with various fungi, but the only reared species breed in slime molds. The name *Baeocera* was conserved in Opinion 1221 (ICZN 1982). Key: Löbl and Stephan (1993). Biology: Lawrence and Newton (1980), Newton (1984), Stephenson *et al.* (1994). Larva: Newton (1991). This genus is worldwide with about 215 species.

*Caryoscapha* Ganglbauer 1899
One species, *C. americanum* Löbl 1987, Arkansas, Florida, Oklahoma and Illinois; found on various soft fungi and reared several times from *Hericium*. Description: Löbl (1987). Larva and biology: Ashe (1984c, as *Scaphisoma terminata*), Leschen (1988). This

genus also includes three Palearctic species, and is closely related to *Scaphisoma* in which it has often been placed as a subgenus.

*Scaphisoma* Leach 1815
*Scaphiomicrus* Casey 1900
Twenty-two species, widely distributed in southern Canada and throughout the United States; found especially on logs, often associated with polypore fungi, less commonly on mushrooms or various other fungi and slime molds. Partial keys: Casey (1893: 523-524; 1900: 59-60), Leschen *et al.* (1990). Biology: Leschen (1988), Leschen *et al.* (1990), Newton (1984), Stephenson *et al.* (1994). Larvae: Hanley (1996), Leschen (1988), Newton (1991); earlier misidentifications of *Sepedophilus* larvae as *Scaphisoma* were corrected by Kasule (1966) and Newton (1984). This nearly worldwide genus includes more than 540 known species.

*Toxidium* LeConte 1860
Two species, *T. compressum* Zimmermann 1869 and *T. gammaroides* LeConte 1860, Ontario and Quebec south to Florida and Texas; associated with various polypore fungi on old logs. Key: Casey (1893: 522). Biology: Leschen (1994), Newton (1984). About 40 species occur throughout the tropical regions of the world.

XV. Subfamily PIESTINAE Erichson 1839

Body relatively elongate and depressed; antennae inserted under shelflike corners of frons; procoxae small, globose; protrochantin exposed; abdomen long and parallel-sided, with six visible sterna and one or two pairs of paratergites per segment; tarsi 5-5-5. Various tribes, subtribes and genera once included here have been transferred to the subfamilies Osoriinae and Micropeplinae by Blackwelder (1942), Pseudopsinae by Newton (1982a), Trigonurinae and Apateticinae by Newton and Thayer (1992), etc. The seven genera remaining here are similar to Osoriinae except for having abdominal paratergites; they are not divided into tribes. Four genera are Old World only (two in eastern Asia, one in Australia and one in New Zealand), one Holarctic, and two mainly Neotropical but extend into the southern United States. Most species occur under bark of decaying trees where they are saprophages or mycophages; mandibles of several genera have possible mycangia (Crowson and Ellis 1969). There are five species in three genera in North America.

*Hypotelus* Erichson 1839
One species, *H. hostilis* Fauvel 1865, Florida, also Mexico and northern Central America; probably under bark of decaying trees as are some other Neotropical species. *H. capito* (LeConte 1880) from Texas was moved to the genus *Cephaloxynum* of Aleocharinae by Newton (1988). *Hypotelus* includes about seven additional species, widespread in the Neotropical region including the West Indies.

*Piestus* Gravenhorst 1806
One species, *P. (Piestus) extimus* Sharp 1887, southern Arizona, also in Mexico; on decaying cacti and sotol (*Dasylirion* spp.). Larvae of Neotropical species: Böving and Craighead (1931), Frank (1991), Paulian (1941). This genus includes about 50 species throughout most of the Neotropical region, placed in seven subgenera by Scheerpeltz (1952); most species are found under bark or in forest litter.

*Siagonium* Kirby and Spence 1815
*Prognathus* Berthold 1827
Three species, widespread in western and northeastern North America but not known from California or Florida; found commonly with larvae under the bark of decaying trees, especially conifers. Key: Moore (1975a). Larvae of Palearctic species: Kasule (1968), Paulian (1941). The most widespread species, *S. punctatum* (LeConte 1866), extends into central Mexico; the genus also includes more than a dozen species from Eurasia.

XVI. Subfamily OSORIINAE Erichson 1839

Abdomen not "margined," i.e., entirely without paratergites, and (except Eleusinini) on most segments with tergum and sternum completely fused into a solid ring; antennae inserted under shelflike corner of frons; mandibles with well-developed molar lobes at base, their apices often with multiple teeth or scooplike; procoxal fissure open or closed, protrochantin exposed or concealed; tarsi 5-5-5, 4-4-4, or 3-3-3; abdomen with six visible sterna. The few other staphylinids with unmargined abdomens (certain genera of Paederinae and Euaesthetinae, and some *Stenus* species) are predators with slender, more or less falcate mandibles lacking molar lobes, while Osoriinae, as far as known, are saprophages or mycophages. This subfamily as currently defined includes 103 genera in four tribes, three of them represented in North America (by 31 species), while the fourth (Leptochirini Sharp 1887) is widespread in the Old and New World tropics but does not extend north of Mexico (Blackwelder 1942; Moore and Legner 1974b, 1974c; Newton and Thayer 1992).

54. Tribe Eleusinini Sharp 1887
Body flat; pronotum strongly narrowed at base (Fig. 25.22); prosternum exceptionally long before coxae; procoxae small and globose, contiguous; procoxal fissure large and exposing exceptionally enlarged trochantin; each abdominal segment with membranous suture between tergum and sternum. This tribe has been placed variously as a subtribe of Thoracophorini (Blackwelder 1942) or a separate subfamily, but treated as a tribe in Steel's (1950) generic review. It includes only three genera, two of which occur in North America while the third is widespread through the Oriental and Pacific regions.

*Eleusis* Laporte 1835
*Isomalus* Erichson 1839 (subgenus)
Two species, *E. pallida* (LeConte 1863), widespread in eastern United States, and *E. humilis* (Erichson 1840) (=*E. fasciata* (LeConte 1863)), southwestern United States but also widespread in the New and Old World tropics; under bark of dead trees, especially *Populus* and other hardwoods. Key: Horn (1871). Bi-

ology: Hamilton (1888). This genus of nearly 200 species is pantropical with a few species in warm temperate areas of Asia and Australia as well as North America.

*Renardia* Motschulsky 1865 (Fig. 25.22)
*Eumalus* Sharp 1887

Two species, *R. nigrella* (LeConte 1863) (? =*jubilaea* Motschulsky 1865) and *R. canadensis* (Horn 1871), in need of revision, widespread in North America from British Columbia south to California and Arizona in the west and Ontario and New York south to Louisiana in the east; under bark of decaying trees, sometimes in galleries of Scolytidae where adults and larvae feed on ambrosia fungi. Generic synonymy: Hammond (1970). Biology: Deyrup and Gara (1978), Newton (1984). The genus includes three more named and several undescribed species from Mexico to Panama.

55. Tribe Thoracophorini Reitter 1909
Lispinini Bernhauer and Schubert 1910

Body flat to convex; procoxae contiguous or separated by prosternal process, usually small and globose, with or without a transverse sulcus on the anterior face; procoxal fissure open and exposing small trochantin, or closed; protibia without ctenidium on inner edge; genital segment eversible, normally not or only slightly exposed. The tribe is further divided into four subtribes, all of which occur in North America.

55a. Subtribe Clavilispinina Newton and Thayer 1992
Paralispinina Blackwelder 1942 (based on preoccupied generic name)

Body relatively convex; procoxae contiguous, procoxal fissure open and exposing trochantin; abdominal segment VIII (last) with sternum normal, or extended dorsally in front of tergum. About eight genera are included in this inadequately defined group (Newton 1990c).

*Clavilispinus* Bernhauer 1926
*Ancaeus* Fauvel 1865 (not Adams 1861)
*Paralispinus* Bernhauer 1921 (not Eichelbaum 1913)
*Neolispinodes* Bernhauer 1937 (subgenus)

Six species, widely distributed. Two species, *C. exiguus* (Erichson 1840) and *C. rufescens* (LeConte 1863), are found under bark in the southeastern United States; the former also occurs widely in the New and Old World tropics. The remaining four species, *C. californicus* (LeConte 1863), *C. laevicauda* (LeConte 1866), *C. prolixus* (LeConte 1877) and *C. rufescens* (Hatch 1957), are endemic to northern and southwestern North America and probably do not belong in this genus; some species are found under bark, others in ant nests of *Formica* and *Camponotus* in rotting logs. One species originally placed here (*C. mariae* (Hatch 1957)) was transferred to the genus *Stictocranius* of Euaesthetinae by Newton (1989). Key to Neotropical species including the first listed above: Irmler (1991). Larva of *C. californicus*: Newton (1990a). Biology: Schwarz (1894, as *Lispinus*). The genus includes about 20 species total and is pantropical.

55b. Subtribe Lispinina Bernhauer and Schubert 1910

Body relatively flat; procoxae separated by a flat process of the prosternum; procoxal fissure open, trochantin exposed; abdominal segment VIII (last) with sternum extended dorsally in front of tergum. Three widespread mainly tropical genera are included here, two of which occur in the United States.

*Lispinus* Erichson 1839
*Pseudolispinodes* Bernhauer 1926
*Spinilus* Blackwelder 1942

Two species, *L. aequipunctatus* LeConte 1868 and *L. obscurus* LeConte 1863, southwestern United States, the latter also Ohio and Alabama; under bark of dead conifers and in rotting cacti. Key: LeConte (1877: 250). Larva of exotic species: Paulian (1941; as *Lispinus grandiceps* Fauvel). About 220 additional species are known from tropical regions and Japan.

*Nacaeus* Blackwelder 1942
*Pseudolispinodes* of Blackwelder 1942 (misidentified, not Bernhauer 1926)
*Tannea* Blackwelder 1952 (subgenus)

Two species, *N. tenuis* (LeConte 1863) (=*N. tenella* auctorum, not Erichson 1840), southern United States from Arizona to Florida, north to Pennsylvania?, and an unidentified species from Connecticut; under bark, especially of conifers. Larvae: Frank (1991), Paulian (1941; as *Lispinus aethiops* Eppelsheim 1895). About 150 species occur mainly in tropical regions.

55c. Subtribe Thoracophorina Reitter 1909

Body relatively convex, in Nearctic species with carinae and often rough sculpturing on head, pronotum and elytra; gular sutures fused at middle; procoxae small, contiguous; procoxal fissure closed, trochantin concealed; abdominal segment VIII (last) with sternum extended dorsally in front of tergum and fused to tergum near midline. Eleven genera are included here, ten confined to the Neotropical or Oriental regions and one found in North America (Newton 1990c).

*Thoracophorus* Motschulsky 1837
*Glyptoma* of Duponchel 1845 (misidentified, not Erichson 1839)

Three or four species, eastern United States and southeastern Canada, at least two of these species also in the West Indies and Mexico; under bark and in leaf litter, occasionally in woody fungi. Key to Neotropical species: Irmler (1985). Larvae: Bohác (1978), Burakowski and Newton (1992). Biology: Burakowski and Newton (1992), Newton (1984). This worldwide genus includes more than 50 species and is most diverse in Latin America.

55d. Subtribe Glyptomina Newton and Thayer 1992
Calocerina Blackwelder 1942 (based on preoccupied generic name)

Body relatively convex, without carinae; gular sutures present only as a semicircle near base of head, absent from middle; procoxae very small, contiguous (in Nearctic species); procoxal fissure closed, trochantin concealed; abdominal segment VIII (last) with normal sternum that is not extended dorsally in front of tergum or fused with tergum. This largely tropical group of about six genera is currently under review; three undescribed species of two widespread genera occur in North America.

*Espeson* Schaufuss 1882
Two undescribed species, Florida and North Carolina; under bark, in rotting logs and in forest litter. First reported from United States by Newton (1990a; in key). This genus occurs throughout the Old and New World tropics, with about 20 species remaining after removal of many species to *Lispinodes* by Lecoq (1994, as *Pseudespeson*).

*Lispinodes* Sharp 1880
*Pseudespeson* Lecoq 1994
One undescribed species, Illinois and Michigan; in forest leaf litter. This genus, recently augmented by species removed from *Espeson* by Lecoq (1994, as *Pseudespeson*), now contains nearly two dozen rare species scattered through the Old and New World tropics, with a radiation of ten endemic species on the Hawaiian Islands.

56. Tribe Osoriini Erichson 1839
Body convex to cylindrical; procoxae contiguous, conical and prominent, with a transverse sulcus on the anterior face; procoxal fissure open, trochantin exposed; protibia with a well-developed ctenidium along much of inner face; genital segment permanently exposed, well sclerotized, fixed to truncate apex of abdominal segment VIII. This tribe is divided into two subtribes, with about 67 genera, mainly tropical. Two genera belonging to the typical subtribe are nearly worldwide, and occur in the United States; most other genera are from tropical or southern temperate regions and are more localized. An undescribed genus and species of blind flightless osoriine, possibly belonging near the African genus *Rhabdopsis* Fagel 1955, has been found in Texas and is included in the key.

*Holotrochus* Erichson 1839
*Neotrochus* Blackwelder 1943
Three species, southeastern and southwestern United States; in forest leaf litter. Blackwelder (1943) removed most New World species including these to his new genus *Neotrochus*, but this genus has since been synonymized with *Holotrochus* (Irmler 1982). Key: Chapin (1928). Larvae: Kasule (1966), Newton (1990a; genus in key). One species often included here, *H. laevicauda* (LeConte 1866) belongs in the genus *Clavilispinus*. The genus is nearly worldwide with about 100 species, but is not well defined; species from some regions such as New Zealand have been split off recently into endemic genera.

*Osorius* Latreille 1829 (Fig. 26.22)
*Molosoma* Say 1830
Nine species, eastern and southwestern United States; in soil and forest leaf litter, sometimes in rotting logs or under objects such as old corn stalks in damp fields; one species may be a lawn pest. Key: Notman (1925). Larvae: Costa *et al.* (1988), Frank (1991), Newton (1990a; genus in key). Biology: Smith *et al.* (1978). This genus is nearly worldwide with more than 300 species.

XVII. Subfamily OXYTELINAE Fleming 1821

The defining characteristic of this subfamily is a unique pair of defensive glands opening onto the ninth tergum (Herman 1970b, Czarniaski and Staniec 1997). The second abdominal sternum, usually very short and fused to the front of the third sternum in Coleoptera, is large and free in all but the small basal groups Deleasterini and Coprophilini, so that seven complete abdominal sterna are visible, uniquely separating most members of the subfamily from other staphylinids. Other characteristics include antennal insertion at side of frons, concealed under tiny to conspicuous shelf-like projection; procoxal fissure broadly or narrowly open or closed, protrochantin exposed or concealed; tarsi most often 3-3-3 or 5-5-5, but 4-4-4 or 2-2-2 in a few genera; abdomen with one (Deleasterini) or two pairs of paratergites per segment.

Herman (1970b) redefined the subfamily and revised the generic and tribal concepts worldwide, recognizing only two tribes, Coprophilini and Oxytelini. Newton (1982b) modified Herman's (1970b) phylogenetic analysis and demonstrated that Herman's Coprophilini was not monophyletic, but did not make formal changes at that time (in part because data on the proper names to apply to parts of "Coprophilini" were lacking). Hansen (1996) broke up Herman's "Coprophilini" into three tribes and added additional coprophiline genera to Herman's Oxytelini for the European fauna. Here, Oxytelini *sensu* Herman (1970b) is retained but his Coprophilini is divided into three tribes in a manner consistent with Newton's (1982b) phylogenetic analysis, and also close to earlier classifications including Hatch (1957). All four tribes are represented in North America, with a total of 17 of the 45 known genera; about 262 species are known from this area.

57. Tribe Deleasterini Reitter 1909
Syntomiini Böving and Craighead 1931
Mesocoxae contiguous or very narrowly separated by mesosternal process; tarsi 5-5-5; abdomen with one pair of broad paratergites per segment, and with the normal six complete abdominal sterna visible. This basal group, shown to be monophyletic and the sister group to all remaining oxytelines by Newton (1982b), also includes the genera *Euphanias* Fairmaire and Laboulbène 1856 (Eurasia) and *Oxypius* Newton 1982 (Australia).

*Deleaster* Erichson 1839
Three species, *D. concolor* LeConte 1866, California to British Columbia, *D. trimaculatus* Fall 1910, Colorado, New Mexico, and Arizona (also northwestern Mexico), and the European *D. dichrous* (Gravenhorst 1802), adventive in northeastern North America from Manitoba to Newfoundland and south to Ohio. Found rarely in forest litter and in wet debris or under rocks near streams, where adults may be predators (Herman 1970b:

364). Larva: Topp (1978). The genus includes another half dozen species from Eurasia and Ethiopia.

*Mitosynum* Campbell 1982
One species, *M. vockerothi* Campbell 1982, New Brunswick; known from a single collection in pan traps at the edge of a *Sphagnum* bog. The genus includes only this flightless species, and is very closely related to *Syntomium* (Campbell 1982a).

*Syntomium* Curtis 1828
*Eunonia* Casey 1904
Four species, British Columbia to Oregon, two of these also extending east to Quebec, Newfoundland and northeastern United States; in forest leaf litter or associated with rotting logs. Key to most species: Hatch (1957: 86). Larvae: Kasule (1968), Newton (1990a), Paulian (1941), Schmid (1988, as "Typ 3 Larve").
This genus also includes several Palearctic species.

58. Tribe Coprophilini Heer 1839 (*sensu stricto*)
Mesocoxae narrowly separated by mesosternal process, or contiguous; tarsi 5-5-5; abdomen with two pairs of slender paratergites per segment, and with the normal six complete abdominal sterna visible. This tribe in the restricted sense used here (and by Hansen 1996 as well as Hatch 1957 and some earlier workers) is equivalent to the "*Coprophilus* group" of three very similar genera discussed by Herman (1970b) and Newton (1982b), and not the much larger but paraphyletic tribe "Coprophilini" proposed by Herman (1970b) and followed by many subsequent workers. Even in this narrow sense the group is difficult to define as adults, but known larvae have strongly falcate urogomphi and lack stemmata, both uniquely derived features within Oxytelinae (Newton 1982b). The other two included genera are *Homalotrichus* Solier 1849 (southern South America and Australia) and *Coprostygnus* Sharp 1886 (New Zealand).

*Coprophilus* Latreille 1829
*Elonium* Leach 1819
Three species, *C. castoris* Campbell 1979, Quebec, New Brunswick and New York, in beaver lodges; *C. sexualis* Leech 1939, British Columbia to Oregon; and the European *C. striatulus* (Fabricius 1792) recently discovered in Ontario, Quebec and New York. Another species, *C. vandykei* Hatch 1957, was transferred to the genus *Vicelva* of Phloeocharinae by Moore (1974). Key: Campbell (1979a); adventive species: Hoebeke (1995a). Larva: Kasule (1968), Newton (1990a; genus in key), Paulian (1941). The genus, whose name was recently conserved (ICZN 1993), also includes nearly 20 Palearctic species.

59. Tribe Thinobiini Sahlberg 1876
Carpelimini Hatch 1957
Mesocoxae contiguous or very narrowly separated by mesosternal process; tarsi 2-2-2, 3-3-3, 4-4-4 or 5-5-5; abdomen with two pairs of slender paratergites per segment, and with an "extra" segment so that seven complete sterna are visible (also in Oxytelini). This tribe includes all those "Coprophilini" of Herman (1970b) and Newton (1982b) that have the "extra" abdominal sternum, thus excluding the genera here placed in Deleasterini and Coprophilini *sensu stricto*. This is the largest oxyteline tribe, with about 20 genera worldwide, most of them including species found in wet biotopes where adults and larvae may be alga-feeders. Most of the genera form part of a monophyletic group, but a few genera, including *Manda* in the Nearctic region, are less definitely included here and may make this group paraphyletic with respect to the closely related Oxytelini (see Newton 1982b).

*Aploderus* Stephens 1833
*Haploderus* Erichson 1839 (unjustified emendation)
Seven species, western North America from California to Alaska and Idaho; found in forest leaf litter, in damp debris near streams, and associated with dung and carrion. Partial keys: Casey (1889b: 73); Hatch (1957: 94). Larva: Böving and Craighead (1931; genus in key); Staniec (1997; larva, biology of European species). The genus also includes several species from Eurasia.

*Bledius* Leach 1819
*Microbledius* Herman 1972
*Psamathobledius* Herman 1972
Ninety species, generally distributed in the United States and southern Canada; found especially in damp sandy or muddy areas along streams, rivers and ponds, in brackish or saltwater flats near the ocean or inland salt lakes, and even in the intertidal zone of ocean beaches. Adults of at least some species harvest algae from the wet surface and store it in living chambers excavated in the mud or sand; females lay eggs and rear larvae in these special tunnels, which can be sealed during high tide or periods of high water. This genus has been divided into as many as 11 subgenera, but Herman (1986) demonstrated that most of these are not monophyletic groups and proposed a system of 34 species groups instead, ten of which occur in North America. Review with partial revision, phylogenetic analyses and world catalog: Herman (1972b, 1976, 1983, 1986). Defensive glands: Steidle and Dettner (1995a, 1995b). Larvae: Griffiths and Griffiths (1983), Moore (1964a), Paulian (1941), Pototskaya (1967), Wyatt (1993). Biology: Griffiths and Griffiths (1983), Herman (1986), Wyatt (1986, 1993). The genus includes about 450 species from all regions, but is very poorly represented on oceanic islands or areas with cool or cold temperate climates.

*Carpelimus* Leach 1819 (Fig. 27.22)
*Trogophloeus* Mannerheim 1830
*Troginus* Mulsant and Rey 1878 (subgenus)
*Paratrogophloeus* Hatch 1957 (subgenus)
About 79 species, badly in need of revision, generally distributed throughout North America; found especially in periaquatic situations such as wet debris or moss along streams and ponds, but some species mainly in forest leaf litter. Partial keys: Casey (1889a), Hatch (1957: 91). Redefinition and reassignment of many species to other genera: Herman (1970b). Larvae: Moore (1964a), Moore and Legner (1974a); as *Trogophloeus*: Kasule (1968), Topp (1978). Even after removal of several former subgenera and many species to the genera *Thinodromus*, *Teropalpus* and others by Herman (1970b), this remains a very large and

complex genus with several hundred additional species worldwide. Many species (in North America and elsewhere) have not been assigned to the remaining, poorly defined subgenera.

*Manda* Blackwelder 1952
*Acrognathus* Erichson 1839 (not Agassiz 1836)
One species, *M. nearctica* Moore 1964, Florida and Georgia, usually collected at UV light and perhaps living on damp swampy ground as does one exotic species. Description: Moore (1964b). The genus also includes two western Palearctic species.

*Neoxus* Herman 1970
One species, *N. crassicornis* (Casey 1889), Texas, transferred from *Thinobius* by Herman (1970b). The genus includes one additional rare species from Panama.

*Ochthephilus* Mulsant and Rey 1856
*Ancyrophorus* Kraatz 1858
Three species, widespread in the west from Alaska to California and Alberta to New Mexico, in the east from Ontario and Newfoundland south through the Appalachian Mountains to Alabama; found in damp debris and moss along streams. Key: Hatch (1957: 90, as *Ancyrophorus*). Larva and biology of European species: Bourne (1975). This genus includes about forty additional species in the Palearctic and northern Oriental regions.

*Teropalpus* Solier 1849
One species, *T. lithocharinus* (LeConte 1877), California to Washington, found on marine beaches. Habitat, illustration: Moore (1979b). This genus was elevated from a subgenus of *Carpelimus* by Herman (1970b); it includes eight other species, all from temperate coasts of the southern hemisphere, with one of those species adventive in England.

*Thinobius* Kiesenwetter 1844
About 24 species, widely distributed but mostly western; found mainly in periaquatic situations, near streams and ponds as well as on intertidal mud flats. Partial key: Casey (1889b: 79). Biology and larva: Kincaid (1961). The genus includes nearly one hundred additional species and is nearly worldwide, with at least a half dozen inadequately defined subgenera to which many species have not been assigned.

*Thinodromus* Kraatz 1858
Six species, widespread in the United States and southern Canada; usually found in wet debris or moss near streams and ponds. This genus was elevated from a subgenus of *Carpelimus* by Herman (1970b); it includes about 80 additional species worldwide.

60. Tribe Oxytelini Fleming 1821
Mesocoxae separated by a broad metasternal process; tarsi 3-3-3; abdomen with two pairs of slender paratergites per segment, and with an "extra" sternum (also in Thinobiini). This tribe of about 14 genera worldwide is more homogenous morphologically than Thinobiini and may be monophyletic (Herman 1970b, Newton 1982b).

*Anotylus* Thomson 1859
Eighteen species, at least five of them adventive, generally distributed; many species are found mainly on dung, carrion and other decomposing organic matter, and others in forest leaf litter, with a few species reported from mammal or ant nests. Partial keys (as *Oxytelus* in part): Casey (1893: 382), Hatch (1957: 95); additional adventive species: Campbell and Tomlin (1984). World overview and biology: Hammond (1976). Larvae, as *Oxytelus* in part: Kasule (1968), Paulian (1941), Pototskaya (1967). This genus was treated as a subgenus of *Oxytelus* until Herman (1970b) and Hammond (1976) removed it with many other subgenera to a newly defined, very large genus of more than 350 species worldwide. Subgenera were not recognized by Hammond (1976).

*Apocellus* Erichson 1839
*Ocaleomorpha* Fleischer 1921
Nine species, widely distributed in North America; found variously in forest leaf litter, on dung, or associated with ants including *Solenopsis* and *Atta* in Mexico and probably the southern United States; one species reported as an occasional greenhouse pest. Key to most species: Casey (1885a: 153). Biology: Chittenden (1915), Kistner (1982: 74). This genus is restricted to the New World, with about 30 additional species found throughout the Neotropical region.

*Oxytelus* Gravenhorst 1802 (Fig. 28.22)
*Epomotylus* Thomson 1859 (subgenus)
*Tanycraerus* Thomson 1859 (subgenus)
Nine species, three of them adventive, generally distributed; most species associated with dung or other decaying matter. Partial keys: Frank and Thomas (1981b), also (including *Anotylus* spp.) Casey (1893: 382) and Hatch (1957: 95). Larvae: Frank (1991), Paulian (1941; *Oxytelus* in part). This genus was redefined and greatly restricted by Herman (1970b) and Hammond (1976) who removed many subgenera and a majority of species formerly included here to *Anotylus*; it is still a large genus with about 200 species worldwide.

*Platystethus* Mannerheim 1830
Three species: *P. americanus* Erichson 1840, a widespread Nearctic species especially common on cattle dung, also found in Mexico; *P. spiculus* Erichson 1840, a widespread Neotropical species found across southernmost United States from California to Florida, also mainly in dung; and the Palearctic species *P. degener* Mulsant and Rey 1879 (=*P. cornutus* auctorum, not Gravenhorst 1802; Moore 1979a), first recorded from North America in 1971 and now found from Nebraska and Iowa east to Ontario, New York and Georgia, associated with streamedge habitats. At least some dung-associated species feed on dung and are subsocial, with parental care of larvae. Review and key: Moore and Legner (1971); supplemental notes and distribution: Frank (1976), Hu and Frank (1995b), Moore (1979a). Biology and larvae: Hinton (1944), Legner and Moore (1978), Pierre

(1944). The genus includes about 45 additional species throughout much of the world, except the Australian region.

XVIII. Subfamily OXYPORINAE Fleming 1821

Antennae inserted at sides of the head near the eyes, filiform or clavate; mandibles prominent; apical labial palpomere very large, strongly securiform; procoxae large, conical, prominent, the protrochantin broadly exposed; middle coxae widely separated; posterior coxae transverse; tarsi 5-5-5; abdomen with six visible sterna and two pairs of paratergites per segment. The subfamily includes a single isolated genus, with 14 North American species.

*Oxyporus* Fabricius 1775 (Fig. 29.22)
*Pseudoxyporus* Nakane and Sawada 1956 (subgenus)
Fourteen species, widespread in the United States and southern Canada but mainly eastern; adults and larvae are found in and feed on mushrooms and other fleshy fungi, where females may construct and protect special egg-hatching chambers. Revision and keys to subgenera, species and subspecies: Campbell (1969, revised 1978a). Biology and larvae: Hanley and Goodrich (1994, 1995a, 1995b), Goodrich and Hanley (1995), Leschen and Allen (1988), Setsuda (1994), Frank (1991). Another 75 species are known from the Holarctic and Oriental regions and Central America.

XIX. Subfamily MEGALOPSIDIINAE Leng 1920

Eyes very large and prominent, occupying most of side of head; antennae very short (much shorter than head width), apical 2-3 antennomeres forming distinct club; labrum with two very elongate processes (Fig. 50.22); body extremely shiny; procoxae small, conical, procoxal fissure closed and trochantin concealed; pronotum with transverse or oblique sulci; tarsi 5-5-5; abdomen with six visible sterna and two pairs of paratergites per segment. The subfamily includes a single isolated genus, two species of which occur in North America.

*Megalopinus* Eichelbaum 1915
*Megalops* Erichson 1839 (not Lacepède 1803, etc.)
*Megalopsidia* Leng 1918
Two species, *M. caelatus* (Gravenhorst 1802) and *M. rufipes* (LeConte 1863), both widespread in the southeastern United States from Florida north to Indiana and west to Oklahoma and Louisiana, the latter species also Texas, Arizona, Mexico, and Panama; found rarely, usually in association with decaying trees and fungusy logs, but the biology and immature stages of this genus are virtually unknown. Key: Downie and Arnett (1996: 459). Larva: Newton (1990a; genus in key). The genus includes well over 100 species from throughout the Neotropical region as well as much of the Old World tropics, Japan, and Australia. Five subgenera were recognized in Benick's (1951) world overview, with the two North American species placed in the typical subgenus, but recent workers have not used subgenera.

XX. Subfamily STENINAE MacLeay 1825

Antennae inserted on the vertex between the eyes (as in Aleocharinae, but unlike that subfamily the posterior coxae are conical, not expanded laterally, and the anterior coxae are small, the procoxal fissure closed and trochantin concealed); antennae slender but with apical 2-3 antennomeres forming a weak club; eyes very large and prominent (Fig. 30.22) as in *Megalopinus* (but in contrast to that genus the body is more densely punctate or rugose, and pronotum not sulcate); tarsi 5-5-5; abdomen with six visible sterna and 0-1 pairs of paratergites per segment, in the former case the terga and sterna sometimes fused into solid rings. This subfamily, closely related to Megalopsidiinae and especially Euaesthetinae, includes only two genera, both present in North America (totaling 169 species).

*Dianous* Leach 1819
Two species, *D. chalybaeus* LeConte 1863 and *D. nitidulus* LeConte 1874, widespread across Canada and in the northeastern United States south to New York and Indiana, the latter species also widespread throughout the mountains of the western United States; in damp debris and on rocks along mountain streams. Key: Downie and Arnett (1996). Larvae of European species: Kasule (1966), Topp (1978). Distribution map: Puthz (1981). The genus as redefined in Puthz's (1981) world overview includes more than 90 additional species in Eurasia.

*Stenus* Latreille 1796 (Fig. 30.22)
*Hemistenus* Motschulsky 1860 (subgenus)
*Hypostenus* Rey 1884 (subgenus)
*Areus* Casey 1884
*Nestus* Rey 1884 (subgenus)
*Tesnus* Rey 1884 (subgenus)
*Mutinus* Casey 1884
*Parastenus* Heyden 1905 (subgenus)
One hundred and sixty-seven species, generally distributed; found in diverse habitats, especially on rocks or plants near streams or other aquatic situations, on vegetation in general, and on or in forest leaf litter and debris. *Stenus* species are specialized predators of Collembola and other arthropods; adults have a unique protrusible labium used in prey capture, and at least some have special pygidial glands that allow them to skim across water surfaces. Key to most subgenera: Hatch (1957: 248); partial keys: Casey (1884), Sanderson (1946, 1957), Hatch (1957: 248), Puthz (1971b, 1972, 1988), Downie and Arnett (1996: 484). Morphology, world overview: Puthz (1971a). Biology, prey capture: Betz (1996, 1998, 1999), Jenkins (1960), Weinreich (1968). Larvae (genus only, or exotic species): Frank (1991), Hayashi (1986), Kasule (1966), Weinreich (1968). This is one of the largest beetle genera, with about 1800 species widely distributed throughout the world. Six subgenera have been widely used, all of them in North America, but the subgenera are understood to be rather artificial and not all species have been assigned to them (Puthz 1971a).

XXI. Subfamily EUAESTHETINAE Thomson 1859

Antennae slender but apical 2-3 antennomeres forming more or less distinct club, inserted in front of eyes but distinctly in from side of head and mandibular articulation; mandibles slender, falcate, without molar lobes; labrum usually with denticulate or crenulate anterior margin; anterior coxae small, conical, more or less prominent; procoxal fissure closed, trochantin concealed; abdomen with six visible sterna, and with one pair of paratergites per segment or terga and sterna fused into solid rings; tarsi 5-5-5 or 4-4-4 (5-5-4 in some exotic genera). The subfamily was divided into six tribes by Scheerpeltz (1974), three of which occur in North America; two others (Alzadaesthetini Scheerpeltz 1974; Austroesthetini Cameron 1944) are restricted to temperate areas of the southern hemisphere and one (Stenaesthetini Bernhauer and Schubert 1911, with heteromerous tarsi) is widespread in the Old and New World tropics. Five of the 26 genera, totaling 22 known species, occur in North America.

61. Tribe Nordenskioldiini Bernhauer and Schubert 1911
Tarsi 5-5-5; abdominal segments margined (i.e., most with paratergites). This tribe contains two genera, one (*Edaphosoma* Scheerpeltz 1976) restricted to the Himalayas and the other Holarctic.

*Nordenskioldia* Sahlberg 1880
One species, *N. columbiana* Puthz 1974, British Columbia; found in *Salix* litter along a mountain stream. This genus includes only one other rare species, from Siberia (Puthz 1974).

62. Tribe Fenderiini Scheerpeltz 1974
Tarsi 5-5-5; abdominal segments not margined laterally. This tribe contains two genera, one Holarctic and the other endemic to North America.

*Fenderia* Hatch 1957
One species, *F. capizzii* Hatch 1957, southwestern Washington, western Oregon, and northern California (but probably including several undescribed species); flightless and nearly blind, found in forest leaf litter. Originally placed in Osoriinae because of its unmargined abdomen (Hatch 1957), this genus was moved to Euaesthetinae by Puthz (1974) and Scheerpeltz (1974); it includes only this species.

*Stictocranius* LeConte 1866
Two species, *S. puncticeps* LeConte 1866, widespread in eastern North America from New Hampshire and Georgia west to Ontario, Illinois, Arkansas, and Mississippi, and *S. mariae* (Hatch 1957), rare in Oregon, California and possibly Utah; found in forest leaf litter. Distribution and distinguishing characters: Newton (1989), Puthz (1974, 1989). This genus was erroneously placed in Austroesthetini by Scheerpeltz (1974), who thought the tarsi were 4-4-4; it includes one other species from Sichuan, China.

63. Tribe Euaesthetini Thomson 1859
Tarsi 4-4-4; abdominal segments margined (i.e., most with paratergites). Twelve genera, mainly tropical, are placed in this tribe; two widespread genera occur in North America.

*Edaphus* Motschulsky 1857
Four species, widespread in the eastern United States from Pennsylvania to Florida and west to Kansas and Texas; found in forest leaf litter. Key: Puthz (1974). The genus includes about 100 additional species, mainly in the Old and New World tropics but extending into warm temperate areas.

*Euaesthetus* Gravenhorst 1806 (Fig. 31.22)
Twenty species, throughout eastern North America, west to Alaska, Washington and New Mexico; mainly in forest leaf litter or in damp periaquatic habitats. Key: Bernhauer (1928); additions: Puthz (1998). Larvae: Frank (1991), Kasule (1966), Newton (1990a). This genus includes 14 Eurasian species as well as seven from Mexico to Costa Rica and Cuba.

XXII. Subfamily LEPTOTYPHLINAE Fauvel 1874

Tiny (no more than 1.8 mm long), form extremely long and slender (Fig. 32.22); without body foveae of pselaphine type; elytra short and fused to body, immovable; eyes and wings absent; antennal insertions concealed under swellings on middle of frons; mandible often with membranous prostheca but never with molar lobe; procoxal fissure open or closed, trochantin usually concealed in either case; abdomen with two pairs of paratergites on most segments, of 6 normal segments, but genital segment often large and everted giving appearance of 7 segments; tarsi 3-3-3 or 2-2-2. The system of five tribes established by Coiffait (1963, 1972) is still followed, although the characterizations of some tribes have changed (Pace 1996). All five tribes are found in the Mediterranean region; two (Cephalotyphlini Coiffait 1963; Entomoculiini Coiffait 1957) are confined to that area plus the Canary Islands, Leptotyphlini is also known from central Africa and Indonesia, and Metrotyphlini Coiffait 1963, is also known from two genera in South America (Chile and Venezuela) and undescribed taxa from Belize and Mexico. The tribe Neotyphlini, the most primitive and widespread, is the only one known from North America, where it is represented by nine genera and 13 species. Worldwide, 41 genera of Leptotyphlinae have been described, but undoubtedly many genera and species remain to be found. Little is known about the biology or microhabitat of most genera and species; they are usually collected by Berlese extraction of leaf litter and soil, or washing of deep soil, and are thought to be predators of mites, Collembola and other small soil or litter arthropods. Key to world tribes: Coiffait (1963, 1972). Overview, morphology and biology: Pace (1996). Larvae: Newton (1990a; subfamily in key); Pace (1996; European genus).

64. Tribe Neotyphlini Coiffait 1963
Procoxal fissure present at least as small notch; maxillary palpomeres 2-3 about equally broad, palpomere 4 much nar-

rower; abdominal sterna without deep transverse sulci; tarsi 3-3-3 or 2-2-2. The tribe Metrotyphlini, known from undescribed species in Mexico and perhaps to be found in North America, is similar but has no trace of a procoxal fissure. All but one of the nine North American genera of this tribe are endemic to North America; *Cubanotyphlus* also includes two species from Cuba. Undescribed species possibly representing additional genera have been seen from California, Oregon, Washington, Idaho, and New Mexico, and through much of Mexico and Central America. This tribe is also known from Chile (five genera), southern Europe (two genera) and South Africa (one genus), plus two undescribed genera from Australia. Checklist of most New World species: Frank and Thomas (1984a).

*Cainotyphlus* Coiffait 1962
Two species, *C. pacificum* Coiffait 1962 and *C. walkeri* Coiffait 1962, California (Marin and Madera Counties). Key: Coiffait (1962).

*Chionotyphlus* Smetana 1986
One species, *C. alaskensis* Smetana 1986, Alaska (Fairbanks area); found in forests of aspen, birch and other hardwoods with spruce, by Berlese extraction of leaf litter and peaty soil immediately below. Smetana (1986) considered this species to be a Tertiary relict surviving in the unglaciated area of central Alaska.

*Cubanotyphlus* Coiffait and Decou 1972
One species, *C. largo* Frank 1984, Florida (Key Largo); found in wood rat nests (*Neotoma floridana* Ord) in tropical hardwood forest. According to Frank and Thomas (1984a) these nests provide one of the few moist soil-like habitats on the low limestone islands. The two other known species of this genus, from elevations of 420 and 1000 m in Cuba's Oriente Province, were found by soil-washing forest litter.

*Heterotyphlus* Coiffait 1962
   *Heteroleptotyphlus* Moore and Legner 1975 (misspelling)
Two species, *H. helferi* Coiffait 1962 and *H. vicinus* Coiffait 1962, California (Sonoma and Marin Counties). Key: Coiffait (1962).

*Homeotyphlus* Coiffait 1962 (Fig. 32.22)
One species, *H. maritimus* Coiffait 1962, California (Mendocino County). Description: Coiffait (1962).

*Neotyphlus* Coiffait 1959
   *Neoleptotyphlus* Moore and Legner 1975 (misspelling)
One species, *N. californicus* Coiffait 1959, California (?Santa Clara County). Review: Coiffait (1962).

*Prototyphlus* Coiffait 1962
One species, *P. gigas* Coiffait 1962, California (Mendocino County). Description: Coiffait (1962).

*Telotyphlus* Coiffait 1962
Two species, *T. sonomensis* Coiffait 1962 and *T. tuolumnensis* Coiffait 1962, California (Sonoma and Tuolumne Counties). Key: Coiffait (1962).

*Xenotyphlus* Coiffait 1962
Two species, *X. californicus* Sáiz 1975 and *X. taylorensis* Coiffait 1962, California (San Diego and Marin Counties); the former species was found in humus under *Quercus agrifolia* or *Ceanothus leucodermis* trees at 1300 m elevation. Distinguishing characters: Sáiz (1975).

## XXIII. Subfamily PSEUDOPSINAE Ganglbauer 1895

The definitive characteristic of this subfamily is the unique presence of a fine stridulatory file on either side of the genital segment, which also has the lateral sclerites fused dorsally in front of the tergum in both sexes. Other diagnostic features include coarse punctation of head and pronotum; longitudinal carinae or costae of the pronotum, elytra and sometimes of the head; antennae not or weakly clubbed; and either a comb at the apex of the apical abdominal tergite or a deep incision of the posterior margin of each elytron. Pseudopsinae also have: antennal insertions on sides of frons, concealed under shelf-like projections; procoxal fissure broadly or narrowly open or closed, leaving the protrochantin exposed, barely visible, or concealed; tarsi 5-5-5 (one species 3-3-3); abdomen with six visible sterna and one or two pairs of paratergites per segment. This subfamily was redefined by Newton (1982a) to include three genera endemic to North America, formerly placed in Oxytelinae or Piestinae, as well as the widely distributed genus *Pseudopsis*. Phylogenetic analysis of and key to genera: Newton (1982a); world revision of species: Herman (1975, 1977). There are 12 species in North America.

*Asemobius* Horn 1895
One species, *A. caelatus* Horn 1895, California, Washington, and southern British Columbia; rare, found once in flood debris. Review: Herman (1977).

*Nanobius* Herman 1977
One species, *N. serricollis* (LeConte 1875), southern California to southern British Columbia; in forest leaf litter. Review: Herman (1977). Larva (genus in key): Newton (1990a).

*Pseudopsis* Newman 1834 (Fig. 33.22)
Eight species, widespread through western North America from California to Alaska, east to Alberta, Wyoming and New Mexico, and in eastern North America from Manitoba to Newfoundland south to southern Illinois and North Carolina; found mainly in forest leaf litter, sometimes in dung or fungi of various kinds and in or near nests of ground-dwelling mammals. Revision, phylogenetic analysis and key: Herman (1975). Larva (genus in key): Newton (1990a). This genus also includes 20 species in the Palearctic region and China, 14 species from Mexico to Ecuador, and one species each in New Zealand and southern Chile plus Argentina.

*Zalobius* LeConte 1874
Two species, *Z. nancyae* Herman 1977 and *Z. spinicollis* (LeConte 1874), central California to southern British Columbia, east to Alberta and Idaho; in wet debris and moss along mountain streams or at seeps. Revision and key: Herman (1977).

XXIV. Subfamily PAEDERINAE Fleming 1821

North American Paederinae can be recognized by having hind coxa narrow, triangular, its apex strongly projecting posteriorly, posterior face vertical, and anterior edge strongly convex; pronotum with a large, opaque postcoxal process of the pronotal hypomeron; protrochantin exposed, thin and plate-like, roughly equal in length and width though sometimes quite small; antennae inserted along the anterior margin of the head, insertions concealed in dorsal view; abdomen with six visible sterna, two pairs of paratergites per segment (terga and sterna fused into rings in one genus), intersegmental membranes having a brick-wall-like pattern of minute sclerites; and apical maxillary palpomere either much smaller than previous one or larger and flattened, with apex oblique. All adults have 5-5-5 tarsi, but a few with short and lobed fourth tarsomeres may look as if they are 4-4-4. Larvae of Paederinae are unique among Staphylinidae in having paired trichobothria (very long slender setae) on the stipes of the maxilla, the sides of the head, and the sides of the pronotum; first instars of at least some have egg-bursters made of sharp spicules at the back of the vertex. As far as known, Paederinae have only two, instead of the usual three, larval instars (Frania 1986b). The subfamily is divided into two tribes, both of which are widely distributed worldwide and in North America. A much-needed phylogenetic analysis and worldwide revision of the genera and higher taxa of the subfamily is in progress by L. H. Herman and will likely result in numerous changes to what is shown below. There are over 220 genera in the group worldwide, of which 36 (with 436 described species) occur in North America. Because of the many differences among classifications, it is difficult to give exact figures for the number of genera in each subtribe worldwide. Casey (1905, 1910b) provided the last taxonomic treatment of the North American fauna. Relatively little has been published on the habits of the North American species; larvae of 14 of the genera occurring here have been diagnosed or described (Kasule 1970, Watrous 1981, Frania 1986b, Newton 1990a). Newton (1990a) included 17 genera in his key to larvae of soil Staphylinidae, but some of them were not separated individually. Kellner and Dettner (1992) described the ventral abdominal glands, first mentioned by Herman (1981), that occur in some members of the subfamily and discussed their potential usefulness as phylogenetic characters.

65. Tribe Paederini Fleming 1821
Nearctic Paederini can be recognized by their having last maxillary palpomere smaller than the preceding one, conical or acicular except in the subtribe Paederina, where it is short, compressed, and pubescent. They are found especially in damp places, but also under bark of logs, some in caves, and some in ant nests; many can be collected at ultraviolet (UV) or other lights. There are about 200 genera in the tribe worldwide, of which 32 occur in North America, placed below in nine subtribes. There has been considerable disagreement among classifications at the subtribal and generic levels, with some authors (notably Casey 1905) splitting several of the genera treated below into multiple genera and/or recognizing different numbers of subtribes. The classification presented below represents a compromise among various existing classifications and convenience of identification and should be regarded as provisional. The following arrangement of subtribes largely follows that of Casey (1905), except for combining his Lithochares with Medonina; many of his genera have been synonymized with others, so using his keys can be somewhat confusing. Keys to subtribes and genera: Casey (1905), Blackwelder (1939; genera only). Kasule (1970) described and keyed larvae of six genera, all of which occur in North America, and Newton (1990a) partially keyed out 16 North American genera. Three subtribes do not occur in North America: Acanthoglossina Coiffait 1982 (Old World and Galapagos Islands), Cylindroxystina Bierig 1943 (Neotropical), and Dolicaonina Casey 1905 (Old World and Neotropical). A character given fundamental importance in previous keys to genera of Paederini is the posterior and lateral expansion of the prosternum toward the edges of the pronotal hypomeron, and in some cases fusion of the two; use of this character was avoided in the keys here because it is very difficult to observe in typical specimens, but its occurrence is noted below.

65a. Subtribe Lathrobiina Laporte 1835
Recognizable by the combination of: metatibiae at apex having a ctenidium (closely placed series of even-length setae) on both sides; protarsi strongly dilated; head and pronotum shining, without microsculpture. The subtribe as delimited here is worldwide in distribution and includes about ten genera, four or five of them found in North America. The species of many genera are riparian, others endogean, and some (at least in Europe) cavernicolous. Keys to genera and species: Casey (1905: 69-76 ff.; recognizing far more genera than here).

*Acalophaena* Sharp 1886
*Calophaena* Lynch 1884 (not Klug 1821)
Two species, *A. compacta* Casey 1905, Arizona, and an unidentified species, Texas; biology unknown, many Neotropical records of UV or other light collecting. Casey's species may be a senior synonym of *Paederopsis gloydi* (see below). Twelve additional species are described from Mexico through Argentina.

*Dacnochilus* LeConte 1861
Two species, *D. laetus* LeConte 1863, Florida through Texas, and *D. fresnoensis* Leech 1939, southern California and Arizona; possibly associated with ants or termites, some UV or other light records from Mexico. Distinguishing features: Leech (1939). Biology: Moore (1973). One additional species is described from Venezuela.

*Lathrobium* Gravenhorst 1802
*Lathrobium*, *sensu stricto* (subgenus)
*Litolathra* Casey 1905
*Tetartopeus* Czwalina 1888 (subgenus)
*Deratopeus* Casey 1905 (synonymized by Watrous 1980)
*Abletobium* Casey 1905 (subgenus)
*Apteralium* Casey 1905 (subgenus)
*Lathrobioma* Casey 1905 (subgenus)
*Lathrobiopsis* Casey 1905 (subgenus)
*Lathrolepta* Casey 1905 (subgenus)

Nearly 70 species (one adventive and one Holarctic), mostly eastern; riparian in moss or damp litter in forests, marshes, bogs; at least some come to UV light, but some species are flightless. Keys to (sub)genera and species: Casey (1905); Watrous (1980; subgenus *Tetartopeus*). Biology: Watrous (1980). Larva: Kasule (1970), Watrous (1981), Newton (1990a; in key, including *Lobrathium*). The genus as delimited here has been divided into seven subgenera (some authors include *Lobrathium* also), all of which occur in North America; some of these have been treated as genera by some authors. Over 300 species are described from all other parts of the world; some European species have been found in mammal or bird nests (Drugmand 1989), but may simply be overwintering there (A. Smetana, pers. comm.).

*Lobrathium* Mulsant and Rey 1878
*Lathrobiella* Casey 1905
*Lathrotaxis* Casey 1905
*Eulathrobium* Casey 1905 (subgenus)
*Lathrotropis* Casey 1905
*Pseudolathra* Casey 1905 (subgenus)
*Linolathra* Casey 1905
*Microlathra* Casey 1905
*Paralathra* Casey 1905

Nearly 70 species (two adventive), generally distributed; riparian litter, moss, and debris; at least some come to UV light, but some species are flightless. Keys to (sub)genera and species: Casey (1905); Watrous (1981; subgenus *Eulathrobium*). Biology, larva: Moore (1964e), Watrous (1981; subgenus *Eulathrobium*). Larva: Newton (1990a; in key with *Lathrobium*). The genus is divided into four subgenera, three of which including the typical subgenus occur in North America; some of these are treated as genera by some authors. About 90 species are known from outside North America, worldwide except Australian and Pacific regions.

*Paederopsis* Wasmann 1912
One species, *P. gloydi* Kistner 1955, Arizona; biology unknown, possibly myrmecophilous. Kistner(1955) correctly moved *Paederopsis* from Staphylinini to Paederini: Lathrobiina (though without using the subtribal name). This species is probably the same as *Acalophaena compacta* (see above). Three additional species of *Paederopsis* are known from Brazil.

65b. Subtribe Medonina Casey 1905
Recognizable by the combination of: pronotal length and width subequal; neck more than 1/3, usually more than 2/5, as wide as head; tarsomeres without ventral lobes; metatibia nearly always with apical ctenidium only on posterior surface. The subtribe is worldwide in distribution, and may include a few other genera in addition to the ten found in North America. Found in leaf litter, flood debris, and other decaying material, occasionally ant nests; some attracted to UV light. Keys to genera and species: Casey (1905; as Lithochares and Medones). Frania (1986a, 1986b) questioned the allocation of genera to this subtribe and Stilicina (particularly that *Eustilicus* may belong in Medonina), but did not make a definitive recommendation for changes.

*Achenomorphus* Motschulsky 1858
One species, *A. corticinus* (Gravenhorst 1802), Quebec to Manitoba to Florida; in litter and compost, occasionally in mushrooms or nests of *Formica obscuripes* Forel 1886, comes to UV light. Larva: Newton (1990a; in key, with three other genera). Biology: Weber (1935). Thirteen additional species occur in the Neotropics, tropical Asia, and Japan.

*Deroderus* Sharp 1886
*Polymedon* Casey 1905 (not Osten-Sacken 1877)
*Lypomedon* Blackwelder 1952 (new name for *Polymedon* Casey)

One species, *D. tabacinus* (Casey 1905), Arizona; in decaying cacti. Larva: Frania (1986b), Newton (1990a; in key, with three other genera). Frania (1986a) synonymized *Lypomedon* with *Deroderus*. Three additional species occur from Mexico through Panama.

*Lithocharis* Dejean 1833
*Metaxyodonta* Casey 1886
*Pseudomedon* Mulsant and Rey 1878 (subgenus)
*Ramona* Casey 1886

Nine species (at least two adventive), widely distributed, some flightless; in forest leaf litter, flood debris, moss in bogs, compost, some at UV light. Keys: Casey (1886a; Californian species, including *Sunius* also), Sanderson (1945). Larva: Frank (1991), Newton (1990a; in key, with three other genera), Paulian (1941). Over 100 additional species are known worldwide.

*Medon* Stephens 1833 (Fig. 34.22)
*Oxymedon* Casey 1905
*Platymedon* Casey 1889 (subgenus)
*Paramedon* Casey 1905
*Medonodonta* Casey 1905 (subgenus)
*Tetramedon* Casey 1905 (subgenus)
*Nitimedon* Hatch 1957 (subgenus)

Fifty-one species, nearly all western; in litter or soil, a few reported to be myrmecophilous, in nests of *Formica* spp. Keys to (sub)genera and species: Casey (1905: 150-155 ff.). Larva: Kasule (1970; identity uncertain), Frania (1986b), Newton (1990a; in, key with three other genera). Biology: Kistner (1982),Weber (1935). Close to 300 species worldwide outside North America, some of them found in caves or mammal or wasp nests (Drugmand 1989).

*Neomedon* Sharp 1886
Two species, *N. arizonensis* Casey 1905, Arizona and *N. piciventris* Fall 1907, New Mexico; under conifer bark. Ten additional species known from Mexico and Guatemala, some from forest litter.

*Ophiooma* Notman 1920
One flightless species, *O. rufa* Notman 1920, Florida; debris on beach. *Ophiooma* was originally placed in Xantholinini by Notman (1920), and was transferred to Paederinae by Smetana (1982) on the advice of L. H. Herman. This genus has not been closely compared with other Medonina, and it is unclear whether it should remain a separate genus. The genus is restricted to North America.

*Stilocharis* Sharp 1886
One or more unnamed species, Arizona; in leaf litter. Recorded by Frania (1986a). Larva: Frania (1986b), Newton (1990a; in key, with three other genera). One species is described from Guatemala, but several undescribed species are known from Mexico.

[*Stilomedon* Sharp 1886, as redelimited by Frania (1986a), is not North American]

*Sunius* Stephens 1829
*Hemimedon* Casey 1905
*Lena* Casey 1886
*Sunius*, *sensu stricto* (subgenus)
*Oligopterus* Casey 1886
*Medonella* Casey 1905
*Micromedon* Casey 1905 (not Luze 1911)
*Hypomedon* Mulsant and Rey 1878 (subgenus)
*Caloderma* Casey 1886 (subgenus)
*Trachysectus* Casey 1886 (subgenus)
Twenty-six species (two adventive), widely distributed but mostly western; in damp litter, under bark, at UV light. Keys to (sub)genera and species: Casey (1905). Larva: Newton (1990a; in key with three other genera). This genus is divided into four subgenera, all of which occur in North America. Over 100 additional species are known, worldwide in distribution.

*Thinocharis* Kraatz 1859
*Sciocharis* Lynch 1884 (subgenus)
*Sciocharella* Casey 1905 (subgenus)
Five species, Indiana and Texas eastward; in forest litter (including litter under bark of logs), on polypore fungi, at UV light. The genus is divided into three subgenera, two of which occur in North America. There are over 120 additional species, distributed worldwide.

*Xenomedon* Fall 1912
One species, *X. formicaria* Fall 1912, southern California; found in nests of the ant *Formica francoeuri* Bolton 1995 (as *Formica pilicornis*). Fall (1912) said that the genus falls between *Sunius* and *Medon* in Casey's (1905) key, but did not indicate clearly how it differs from those; its validity is uncertain.

65c. Subtribe Scopaeina Mulsant and Rey 1878
Recognizable by the combination of fourth maxillary palpomere small, aciculate; neck narrow, not more than 1/3 as wide as the head (not more than 1/5 in *Scopaeus*); gular sutures separate though sometimes very close; pronotum longer than wide; prosternum not laterally expanded dorsal to procoxae; tarsomeres without ventral lobes; and metatibia with well-developed apical ctenidium only on posterior surface. Both included genera occur in North America. Mostly riparian in litter. Keys to (sub)genera and species: Casey (1905: 191-192 ff.).

*Orus* Casey 1884
*Leucorus* Casey 1905 (subgenus)
*Pycnorus* Casey 1905 (subgenus)
*Nivorus* Herman 1965 (subgenus)
Fifteen named and one undescribed species, widespread but mostly western; mainly riparian in forest and marsh habitats, occasionally on fungus and at UV light. Revision and keys: Herman (1965a, 1965b). The genus is restricted to the Western Hemisphere, and all four subgenera including the typical subgenus occur in North America (*Nivorus* represented only by an undescribed species). Four additional named species are known from Guatemala, the West Indies, and Surinam.

*Scopaeus* Erichson 1839
*Polyodontus* Solier 1849 (not Eysenhardt 1818)
*Leptorus* Casey 1886
*Pseudorus* Casey 1910
*Scopaeodera* Casey 1886 (subgenus)
*Scopaeoma* Casey 1905 (subgenus)
*Scopaeopsis* Casey 1905 (subgenus)
Forty-two species, widely distributed; in litter, primarily riparian, often attracted to UV light. Keys to subgenera (as genera) and many species: Casey (1905). Larva: Newton (1990a; in key). This worldwide genus is divided into four subgenera, all of which including the typical subgenus occur in North America; over 360 additional species occur worldwide.

65d. Subtribe Stilicina Casey 1905
Recognizable by the combination of apical maxillary palpomere slender, aciculate; gular sutures fused near base or over nearly whole length; and neck less than 1/5 as wide as head. North American genera also have pronotum quadrate to slightly elongate; tarsomeres without ventral lobes; and metatibia with well-developed apical ctenidium only on posterior surface. Found in leaf litter, flood debris, and other decaying material, a few in ant nests or tortoise burrows; some attracted to UV light. The subtribe is worldwide in distribution, and five of the roughly six genera occur in North America. Keys to genera and species: Casey (1905: 219). Frania (1986a, 1986b) questioned the boundaries between this subtribe and Medonina (particularly placement of *Eustilicus*), but did not make a definitive recommendation for changes; the grouping used here follows Casey (1905) and Blackwelder (1939).

*Acrostilicus* Hubbard 1896

One species, *A. hospes* Hubbard 1896, Florida; in burrows of the gopher tortoise, *Gopherus polyphemus*. Hubbard gave only a sketchy diagnosis of the genus and species (comparing it only to *Stilicopsis*), and never published the fuller description he planned. Blackwelder (1939) keyed it out near the genera here included in Stilicina. Its status needs to be reevaluated.

*Eustilicus* Sharp 1886

*Eustilicus*, *sensu stricto* (subgenus)

*Omostilicus* Casey 1905

*Trochoderus* Sharp 1886 (subgenus)

*Stilicolina* Casey 1905

Six species, Arizona to New Jersey and South Carolina; riparian litter, deep forest litter, caves, possibly mammal nests. Revision and key to most species (as *Stilicolina*): Herman (1970a); see also Frania (1994). Larva: Frania (1986b), Newton (1990a; in key, with three other genera). Ecology: Herman (1970a), Frania (1991, 1994). Frania (1986a, 1986b) provided some evidence that this genus may belong in Medonina rather than Stilicina. Seven additional species have been described from Mexico and Guatemala, and numerous new species are known to occur as far south as Brazil, especially in Mexico.

*Megastilicus* Casey 1889

One species, *M. formicarius* Casey 1889, Massachusetts and Ontario to Iowa; myrmecophilous, a nest predator found with the ants *Formica ulkei* Emery 1893, *F. subsericea* Say 1836, and *F. exsectoides* Forel 1886. Biology: Park (1935a), Wickham (1900). Observations by Park (1935a) of host ants in the laboratory attacking and perhaps being killed by the beetles suggest that this species may have defensive chemicals.

*Pachystilicus* Casey 1905

Two species, *P. quadriceps* (LeConte 1880), California, Missouri, Massachusetts, biology unknown, and *P. hanhami* (Wickham 1898), British Columbia to Manitoba, New Brunswick; taken with ants. Key: Casey (1905). Biology: Wickham (1898). The genus is apparently restricted to North America, though Frania (1986a) suggested that it may be a synonym of *Rugilus*.

*Rugilus* Leach 1819

*Stilicus* Berthold 1827

*Stilicosoma* Casey 1905

Eleven species (three adventive), widely distributed; mainly in damp litter and other plant debris, also fungi, dung, or carrion, attracted to UV light. Key: Casey (1905; as *Stilicus*); see also Hoebeke (1995b) concerning adventive species and synonymies. Larva: Kasule (1970), Frania (1986b), Newton (1990a; in key with three other genera), Topp (1978). Frania (1986a) suggested that perhaps Casey's *Stilicosoma* (*S. rufipes* only) should be reseparated from *Rugilus*. Kellner and Dettner (1992) agreed on the basis of abdominal gland structure. Over 200 additional species occur worldwide.

65e. Subtribe Stilicopsina Casey 1905

Recognizable by the combination of apical maxillary palpomere minute, barely visible at 50X magnification, ventrally lobed fourth tarsomeres, and very large edentate labrum. North American genera also have the prosternum expanded laterally and caudally, only very narrowly separated from the pronotal hypomera; the gular sutures fused; pronotum quadrate to elongate; tarsomeres without ventral lobes; and metatibia with small apical ctenidium only on posterior surface. The subtribe contains up to five mostly tropical genera, two of which occur in North America. As far as known, the North American species live in more or less damp litter. Keys to genera and species: Casey (1905: 231 ff.).

*Stamnoderus* Sharp 1886

Three species, eastern United States, and one new one, Arizona; in litter, at UV light. Key: Casey (1905: 233). There are about 14 species in the Neotropical region.

*Stilicopsis* Sachse 1852

Two species, *S. paradoxa* Sachse 1852 and *S. subtropica* Casey 1910, southeastern United States; flightless, in damp forest litter. Distinguishing characters: Casey (1910b: 191). About forty more species are known from tropical areas worldwide.

65f. Subtribe Astenina Hatch 1957

Suniina Sharp 1886 (based on misidentified type genus)

Recognizable by the combination of apical maxillary palpomere minute, ventrally lobed fourth tarsomeres and bidentate labrum. North American genera also, like Echiasterina, have the prosternum expanded laterally and caudally, fused with the pronotal hypomera, and the gular sutures fused. The two tribes are sometimes combined, and the boundary between them is not clear on a world basis. One genus occurs in North America, and there are up to three other Palearctic and Old World tropical genera. Little appears to be known of their biology; many are found in litter.

*Astenus* Dejean 1833 (Fig. 35.22)

*Sunius* of Erichson 1839 (misidentified, not Stephens 1829)

Twenty-four species, generally distributed; in forest, prairie, and lake shore litter, compost, and at UV light; some species are flightless. Key to most species (as *Sunius*): Casey (1905: 236-244). Larva: Kasule (1970), Newton (1990a; in key), Topp (1978). Nearly 400 additional species are known worldwide, especially from warm temperate or tropical areas; some European species are found in bird, mammal, or ant (*Formica*) nests (Drugmand 1989).

65g. Subtribe Echiasterina Casey 1905

Recognizable by the combination of apical maxillary palpomere minute or aciculate, prosternum expanded laterally and caudally, connate with the pronotal hypomera, and fourth tarsomeres simple, not ventrally lobed. Members of the subtribe share with Astenina this prosternal structure and fusion of the gular sutures. The two tribes are sometimes combined, and the boundary between them is not clear on a world basis. In addition to the two genera occurring in North America (only one of which is native), there are a few other Neotropical genera. Little is known of their biology; most seem to be found in litter, and some are myrmecophilous.

*Echiaster* Erichson 1839
*Leptogenius* Casey 1886 (subgenus)
Two species, *E. brevicornis* (Casey 1886) and *E. ludovicianus* Casey 1905, Indiana, Texas, and southeastern United States; in forest litter, prairie sod, and moss. Keys to subgenera (as genera) and species: Casey (1905: 245-248). Larva: Newton (1990a; in key). The genus is divided into three subgenera, two of which (including the typical subgenus) occur in the United States.

*Myrmecosaurus* Wasmann 1909
One species, *M. ferrugineus* Bruch 1932, Florida, Alabama and Louisiana; apparently flightless, in nests of the red imported fire ant, *Solenopsis invicta* Buren. Review and key to all known species: Frank (1977). Larva (other species): Paulian (1941), Silvestri (1946). Biology: Tschinkel (1992) found that they follow the raiding trails of *S. invicta*. Six species in addition to *M. ferrugineus* are known from South America,and it is assumed that *M. ferrugineus* was accidentally introduced to North America from Argentina with the ants.

65h. Subtribe Cryptobiina Casey 1905
Readily recognizable by the geniculate antennae, bent anteriorly from the very elongate basal antennomere (as long as the following 3-4.5 antennomeres combined; Fig. 36.22); they also have the fourth maxillary palpomere conical; neck wide, more than 1/3 (often more than 3/5) as wide as head; and protarsi not dilated. There are about eleven genera worldwide, four of which occur in North America. Most are in riparian or other damp litter, and many are attracted to UV light. Keys to genera and species: Casey (1905: 22-27 ff.; much narrower generic concepts than currently). The generic limits have changed considerably in the last century, and outside North American many species are currently misallocated among *Homaeotarsus*, *Biocrypta*, and *Ochthephilum*.

*Biocrypta* Casey 1905
Two species, *B. magnolia* Blatchley 1917, Florida, and *B. prospiciens* (LeConte 1878), Texas, Oklahoma, and Arizona; in leaf litter and rotten logs. Distinguishing features: Blatchley (1917). Larva: Newton (1990a; in key, with *Homaeotarsus*). Another ten or more species are described from the Neotropical region.

*Homaeotarsus* Hochhuth 1851 (Fig. 36.22)
*Spirosoma* Motschulsky 1857
*Hesperobium* Casey 1886 (subgenus)
*Gastrolobium* Casey 1905 (subgenus)
Thirty-six species, generally distributed; in forest, marsh, and prairie litter, often riparian, sometimes under bark of logs. A few records of *H. bicolor* (Gravenhorst 1802) from nests of *Formica ulkei* Emery 1893 (Park 1935a) are probably accidental. Keys to some species: Horn (1885; as *Cryptobium*), Casey (1905: 32-49; as *Gastrolobium* and *Hesperobium*). Larva: Newton (1990a; in key with *Biocrypta*). Two of six recognized subgenera occur in North America; the genus is worldwide (except Europe and the Australian region), with 30 or so additional species described and more awaiting description, especially in the Neotropics.

*Lissobiops* Casey 1905
One rare species, *L. serpentinus* (LeConte 1863), North Carolina, Alabama, and Indiana; biology unknown. Separation of *Lissobiops* from *Homaeotarsus* is somewhat questionable, as it appears to be very similar to two species of that genus; it is likely that either the two genera should be combined or those two species should be transferred from *Homaeotarsus* to *Lissobiops*.

*Ochthephilum* Stephens 1829
*Cryptobium* Mannerheim 1830
*Ababactus* Sharp 1885 (subgenus)
*Cryptobiella* Casey 1905 (subgenus)
*Neobactus* Blackwelder 1939 (subgenus)
Six species, mostly southern, some flightless; in litter and moss in boggy or damp areas. Keys to some species: Horn (1885; as *Cryptobium*). It is not clear whether the European *O. fracticorne* (Paykull 1800) is indigenous or adventive here; it being partly synanthropic (Drugmand 1989) argues for the latter, though Campbell and Davies (1991) marked it as Holarctic. Larva (as *Cryptobium*): Kasule (1970), Paulian (1941), Topp (1978). The genus is divided into five subgenera, three of which occur in North America. Close to 300 species are described worldwide.

65i. Subtribe Paederina Fleming 1821
Recognizable by the form of the last maxillary palpomere: narrower than the preceding palpomere, wider than long, compressed, pubescent, truncate at apex (Fig. 339.22); protarsi strongly dilated. The subtribe is variously treated as containing a single genus or up to fourteen genera, although the additional generic names have been applied only to regional faunas, and a comprehensive worldwide revisionary study is badly needed. None of the additional generic names has been applied to North American species, although Casey (1905) erected two genera that were later synonymized with *Paederus* (*sensu stricto*). Frank (1988) reviewed the taxonomic history of the generic names and provided the most recent placement of all species group names.

*Paederus* Fabricius 1775
*Leucopaederus* Casey 1905
*Paederillus* Casey 1905
Fifteen species, generally distributed; in litter or among leaves in riparian zones or marshes, some species abundant in wet agricultural fields in some parts of the world; some species are flightless. Key: Casey (1905; treated as three genera; some species have been synonymized). Larva: Ahmed (1957), Frank (1991), Kasule (1970), Newton (1990a; in key), Paulian (1941). Biology: Ahmed (1957), Frank and Kanamitsu (1987). The latter work is a detailed review of the literature on natural history and the medical and agricultural significance of *Paederus* species; pederin and related hemolymph compounds in a number of species can cause dramatic dermatitis in humans, and may prove useful in some kinds

of medical therapy. At least some species can walk rapidly on the surface of water (Heberdey 1943). Frank (1988) reviewed the use of subgeneric names in *Paederus* (*sensu lato*) and provided the most recent placement of all species group names. The genus in the narrowest sense is divided into eleven subgenera. Of the species found in North America, only the Holarctic *P.* (*s. str.*) *riparius* (Linnaeus 1758) has been placed to subgenus. Over 600 species of *Paederus* (*sensu lato*) occur worldwide, about 3/4 of them either placed in *Paederus* (*sensu stricto*) or not placed to subgenus.

66. Tribe Pinophilini Nordmann 1837

Recognizable by having the fourth maxillary palpomere as long as the third, hatchet-shaped, and with its apex oblique and spongy-looking (Fig. 51.22); the basal four tarsomeres of the protarsi are extremely widened in both sexes. Both recognized subtribes occur in North America. There are twenty-six genera worldwide, primarily in subtropical and tropical areas, four of which occur in North America. Little is known of pinophiline biology; at least some are litter inhabitants, while others occur on foliage. Key to three genera: Casey (1910b: 192; uses *Araeocerus* for *Lathropinus* below).

66a. Subtribe Pinophilina Nordmann 1837

Separable from Procirrina by having the abdomen distinctly margined laterally, two pairs of paratergites per segment. Nineteen genera worldwide, of which three occur in North America.

*Araeocerus* Nordmann 1837

One species, *A. elegans* Fall 1932, Texas; biology unknown. Abarbanell and Ashe (1989) did not mention this member of the North American fauna. Several additional species from the Neotropical region and Australia are placed here.

*Lathropinus* Sharp 1886

Two species, *L. picipes* (Erichson 1840), southeastern United States, and *L. obsidianus* (Casey 1910), Texas; biology unknown. Key: Casey (1910b). Both species were originally described in *Araeocerus*, but were transferred to *Lathropinus* by Abarbanell and Ashe (1989). Eighteen additional species are known from Mexico through South America.

*Pinophilus* Gravenhorst 1802

Six species, throughout the eastern and southwestern United States; at least some in litter, also attracted to UV light. Revision and key: Abarbanell and Ashe (1989). Behavior: Holcomb (1978). There are over 200 additional species, occurring nearly worldwide, except Europe and the Australian region. Some Central American species are reported to occur on stream banks (Sharp 1886).

66b. Subtribe Procirrina Bernhauer and Schubert 1912

Separable from Pinophilina by having the abdominal segments cylindrical, tergum and sternum of each fused into a ring, lacking paratergites. Seven genera are placed here, only one of which occurs in North America. Many members of this group are found on foliage, and some tropical genera appear to be canopy-dwellers.

*Palaminus* Erichson 1839

Fourteen species, mostly in the eastern United States, also Arizona and California; found on foliage, also in leaf litter, and attracted to UV light; some species are flightless. Key: Notman (1929; world key). Larva: Newton (1990a; in key), Silvestri (1946; misidentified as ?*Somoleptus laevis* larva). The genus is divided into two subgenera, but the North American species are unplaced. About 300 species are described worldwide, mostly Neotropical and Afrotropical, in the Palearctic region occurring only in Japan.

XXV. Subfamily STAPHYLININAE Latreille 1802

Antennae inserted along or very close to the anterior margin of the head, insertions visible from above or only partly concealed; head usually with distinct neck; pronotal hypomeron generally narrow and without large well-sclerotized postcoxal process, at most with thin and usually small translucent process (Fig. 69.22, arrow), or process absent (Fig. 68.22); procoxal fissure widely open, trochantin fully exposed, flat and bladelike; procoxae large, exserted; tarsi 5-5-5 (5-4-4 in *Atanygnathus*); elytron without epipleural keel; abdomen with six visible sterna, two pairs of paratergites per segment; abdominal intersegmental membranes with pattern of regular hexagonal sclerites or small and very irregular angulate or rounded sclerites. Paederinae are similar but have abdominal intersegmental membranes with a brick-wall pattern, antennal insertions usually more distinctly concealed under frontal elevations, and pronotal hypomeron usually with a well-developed postcoxal process. This subfamily has been divided into as many as six subfamilies (e.g., Moore and Legner 1975a), and more commonly into two, Xantholininae and Staphylininae (e.g., Coiffait 1972, Smetana 1982). However, the subfamily in the current broad sense is clearly monophyletic, being unique among staphylinids in possessing obtect pupae and with other synapomorphies of larvae and adults (e.g., Kasule 1970). Subdividing it presents difficulties in placing odd groups such as Platyprosopini, and may result in some paraphyletic groups (Smetana 1995: 47). Of the 292 genera worldwide, 58 are known from North America, with 546 described species. Keys to tribes and genera (except Staphylinini): Smetana (1982). Key to tribes and many genera of larvae: Kasule (1970), Pototskaya (1967). Key to tribes and many genera of larvae: Kasule (1970), Pototskaya (1967).

67. Tribe Platyprosopini Lynch 1884

Antennal insertions closer to each other than either is to eye; neck distinct but very broad, only slightly narrower than head; pronotal hypomeron strongly inflexed and scarcely visible from the side, without postcoxal process; prosternum with straight anterior margin, without antesternal sclerites; abdominal intersegmental membranes apparently with "brick-wall" pattern but actually of hexagonal sclerites; aedeagus trilobed with pair of long setose parameres. Only one isolated genus is included here.

*Platyprosopus* Mannerheim 1830
One species, *P. mexicanus* Sharp 1887 (=*P. texanus* Moore 1963), southern Texas; also widespread in Mexico; probably found on river floodplains and similar wet habitats like other species of the genus. Synonymy, distribution: Smetana (1982). Larvae (of exotic species): Irmler (1977), Paulian (1941), Pototskaya (1967). This genus is pantropical with about fifty species total.

68. Tribe Diochini Casey 1906
Antennae not geniculate, insertions closer to eyes than to each other; apical maxillary palpomere minute, aciculate, penultimate two palpomeres large and finely pubescent; neck narrow, less than a third as wide as head; pronotal hypomeron strongly inflexed and only partly visible from the side, without postcoxal process; prosternum with emarginate anterior margin, with slight sclerotization of membrane in front of sternum but without distinct antesternal sclerites; abdominal intersegmental membranes with regular pattern of hexagonal sclerites; aedeagus trilobed with pair of long setose parameres. In addition to the nearly worldwide genus *Diochus*, this tribe includes *Antarctothius* from extreme southern South America and *Coomania* from the Oriental region.

*Diochus* Erichson 1839 (Fig. 37.22)
One species, *D. schaumi* Kraatz 1860 (often erroneously identified as *D. nanus* Erichson 1839); throughout the eastern United States, also southern Ontario and Quebec; found in forest litter especially near water, and in various periaquatic situations. Synonymy and distribution: Smetana (1982). Larva: Newton (1990a; in key). The genus includes more than forty species throughout the tropics and in many warm temperate areas.

69. Tribe Othiini Thomson 1859
Antennae not geniculate; neck about 1/2 as wide as head or slightly wider; prothorax with a large sclerite or pair of sclerites in front of emarginate sternum; elytral suture straight, margins of elytra not overlapping; abdominal intersegmental membranes with pattern of small irregular angular (often triangular) sclerites; aedeagus trilobed with pair of long setose parameres. Revision and keys to genera and species: Smetana (1982). Three additional genera occur in Eurasia, the East Indies, and New Zealand.

*Atrecus* Jacquelin du Val 1856 (Fig. 38.22)
*Baptolinus* Kraatz 1857
*Gyrohypnus* of Thomson 1859 (misidentified, not Leach 1819)
Five species, western and northeastern United States and southern Canada; usually under loose bark of dead trees, especially conifers, sometimes in old galleries of Scolytidae. Key: Smetana (1982). Larvae (of exotic species): Kasule (1970), Paulian (1941), Pototskaya (1967), Saalas (1917). Biology: Deyrup and Gara (1978). This genus is strictly Holarctic, with five additional Palearctic species.

*Parothius* Casey 1906
Two species, *P. californicus* (Mannerheim 1843) and *P. punctatus* Smetana 1982, British Columbia south to central California; under bark of conifers. Key: Smetana (1982). This genus is endemic to western North America.

70. Tribe Xantholinini Erichson 1839
Antennae more or less geniculate, with approximate bases; neck narrow, usually a third or less as wide as head, rarely (Fig. 39.22) nearly 1/2 as wide; prothorax with a large sclerite or pair of sclerites in front of emarginate sternum (Fig. 65.22, arrow); elytral edges convex and overlapping; abdominal intersegmental membranes with pattern of small irregular angular (often triangular) sclerites; aedeagus strongly modified from simple trilobed, with basal bulb greatly enlarged, apex of median lobe reduced or absent, and paired parameres reduced and glabrous; body always very elongate and slender (Fig. 39.22). This worldwide tribe includes more than 75 genera, of which 23 are known from North America (four represented only by adventive species); seven genera are endemic to this area. Revision and keys to all North American genera and species: Smetana (1982), supplemented by Smetana (1988, 1990). Keys to larvae of some genera: Kasule (1970), Paulian (1941), Pototskaya (1967).

*Crinolinus* Smetana 1982
One very rare species, *C. intonsus* Smetana 1982, Washington and Nevada, of unknown habits. The genus is endemic to western North America.

*Gauropterus* Thomson 1860 (Fig. 39.22)
One European species, *G. fulgidus* (Fabricius 1787), widespread across the northern United States, probably adventive here in the nineteenth century and now nearly cosmopolitan; in compost piles and other decaying matter, synanthropic. The genus includes about 30 additional species from most geographic regions.

*Gyrohypnus* Leach 1819
*Hyponygrus* Tottenham 1940
Three species, two (*G. fracticornis* (Müller 1776) and *G. angustatus* Stephens 1832) adventive from Europe and now widely distributed, and *G. campbelli* Smetana 1982, southeastern Canada and northeastern United States; in compost, dung and other decaying organic matter, mammal and bird nests, and in litter or moss near water. Key: Smetana (1982). Larvae: Pototskaya (1967), Topp (1978). The genus includes about a dozen additional Palearctic species. The application of the name *Gyrohypnus* was fixed by ICZN (1983).

*Habrolinus* Casey 1906
*Timagenes* Smetana 1982 (subgenus)
Eight species, five flightless ones in the subgenus *Habrolinus* and three strongly dimorphic species (winged males, flightless females) in the subgenus *Timagenes*, central and northern California and southwestern Oregon, in forest litter. Revised key: Smetana (1988). This genus is endemic to western North America.

*Hesperolinus* Casey 1906
*Leiolinus* Casey 1906

One species, *H. parcus* (LeConte 1863), California, Nevada, Utah and Alberta; found in open habitats near water and possibly parthenogenetic (Smetana 1988). The genus is apparently endemic to western North America.

*Hypnogyra* Casey 1906
*Phalacrolinus* Coiffait 1972
Two species, *H. gularis* (LeConte 1880) and *H. micans* Casey 1906, Ontario and northeastern United States west to Iowa and south to Mississippi and Texas; found especially in treeholes or in decaying logs and stumps, sometimes with ants or birds nesting there. Key: Smetana (1982). The genus includes at least five additional species from Eurasia.

*Lepitacnus* Smetana 1982
One species, *L. pallidulus* (LeConte 1880), California, Arizona and southeastern United States as far north as Indiana, also known from Sonora, Mexico; in litter, old sawdust piles or decaying palm tissue. The genus includes at least one additional undescribed species from Mexico.

*Leptacinus* Erichson 1839
*Xanthophius* Motschulsky 1860
*Leptacinodes* Casey 1906
Three European species, *L. batychrus* (Gyllenhal 1827), *L. intermedius* Donisthorpe 1936, and *L. pusillus* (Stephens 1833), all apparently adventive in the nineteenth century and now widely distributed in North America; in decaying organic matter including compost, especially in synanthropic situations. Key: Smetana (1982). Larvae (of exotic species): Kasule (1970), Paulian (1941). The genus includes about 80 species from all areas.

*Linohesperus* Smetana 1982
*Hesperolinus* auctorum (misidentified, not Casey 1906)
Twenty-five species, western United States and Canada, east to Nebraska; mainly in forest leaf litter, also in debris near water and in *Neotoma* nests. Revised key: Smetana (1988). This genus is apparently endemic to western North America.

*Lissohypnus* Casey 1906
One species, *L. texanus* Casey 1906, Texas and Louisiana, also reported from the Galapagos Islands; in forest litter especially near water. The genus includes only this species at present, but some Neotropical "*Xantholinus*" species probably belong here.

*Lithocharodes* Sharp 1876
*Nematolinus* Casey 1906
*Oligolinus* Casey 1906
Seven species, eastern and southwestern United States and southeastern Canada; mainly in forest leaf and log litter and in damp debris along streams, some species mainly in periaquatic situations in swamps. Key: Smetana (1982, partially revised 1988). The genus also includes about 20 species in the Neotropical region.

[*Megalinus* Mulsant and Rey 1877, Palearctic only]

*Microlinus* Casey 1906
One species, *M. pusio* (LeConte 1880), South Carolina and Florida, also in the West Indies; found in treeholes, under bark and in buttress litter especially in swampy areas. At least one additional species of this genus occurs in Mexico and Central America.

*Neohypnus* Coiffait and Sáiz 1964
Twenty-eight species, widely distributed; most species found in forest litter or in debris near water, some species in decaying organic matter including dung and rotting cacti, or in treeholes or under bark. Key: Smetana (1982, partially revised 1988). Larva and biology: Hu and Frank (1995a). The genus occurs throughout the Neotropical region with an undetermined number of species, including most of the former "*Xantholinus*" from that region (Smetana 1982).

*Neoxantholinus* Cameron 1944
One rare species, *N. cristatus* Smetana 1982, southeastern United States (Florida and Arkansas); probably under bark of dying or dead trees like other species of the genus. About 20 additional species are known from the Neotropical and Australian regions.

*Nudobius* Thomson 1860
Five species, widely distributed; found with their larvae under bark of dead trees, especially of *Pinus* and other conifers, sometimes in galleries of Scolytidae. Key: Smetana (1982). Larvae: Kasule (1970), Paulian (1941), Struble (1930). Biology: Deyrup and Gara (1978), Struble (1930). The genus includes about thirty additional species from Africa, Eurasia, Mexico and Central America.

*Oxybleptes* Smetana 1982
Five species, eastern United States and southeastern Canada, plus Oregon; apparently active in fall, sometimes found in flight or swarming, but habits unknown and possibly different for each sex. Key: Smetana (1982, partially revised 1988). This genus is endemic to North America.

*Phacophallus* Coiffait 1956
Two species, *P. parumpunctatus* (Gyllenhal 1827) and *P. tricolor* (Kraatz 1859), adventive from the Palearctic and Oriental regions, respectively, and now widely distributed in North America (and elsewhere); in decaying organic matter including compost, especially in synanthropic situations. Key: Smetana (1982). This genus includes at least four additional Old World species, but may include further species still placed in *Leptacinus* (Smetana 1982).

*Stenistoderus* Jacquelin du Val 1857
*Leptolinus* Kraatz 1857
One species, *S. rubripennis* (LeConte 1880), Ontario and eastern United States west to Nebraska and Texas; in leaf litter and debris, especially near water. This genus includes about 15 additional species from the Palearctic and Afrotropical regions.

*Stictolinus* Casey 1906
Four species, California and Oregon, and northeastern United States and southeastern Canada; found in forest litter, or in mosses and wet debris near water, occasionally in mammal burrows or nests. Key: Smetana (1982). This genus is endemic to North America.

*Thyreocephalus* Guérin-Méneville 1844
*Saurohypnus* Sharp 1885
Two rare species, *T. arizonicus* Smetana 1982, southern Arizona, and *T. nigerrimus* (Sharp 1887), southern Texas; in leaf litter or associated with decaying logs. Key: Smetana (1982). Larvae (of exotic species): Marucci and Clancy (1952), Paulian (1941). The genus includes dozens of species throughout the New and Old World tropics.

*Xantholinus* Dejean 1821
Two European species, *X. linearis* (Olivier 1794) and *X. longiventris* Heer 1839, adventive in the west (British Columbia to California and Utah) and northeast (Nova Scotia through Pennsylvania); in decaying organic matter, often in synanthropic situations. Key: Smetana (1982). Larvae (mostly of exotic species): Costa *et al.* (1988), Frank (1991), Kasule (1970), Paulian (1941), Pototskaya (1967). This genus in the strict sense includes about 150 species confined to temperate areas of the Old World (Smetana 1982), but many species from other regions originally placed here have not yet been reassigned; all Neotropical "*Xantholinus*" species probably belong in *Neohypnus* or *Lissohypnus*. This genus was treated as *Gyrohypnus* in Arnett (1963) and elsewhere, but the name *Xantholinus* was fixed in the present sense by ICZN (1983).

*Xestolinus* Casey 1906
One species, *X. abdominalis* Casey 1906, widespread but rare in western and northeastern North America. The genus also includes one species from Japan.

*Zenon* Smetana 1982
One species, *Z. armifer* Smetana 1982, known from a single specimen from Denver, Colorado. The genus is endemic to North America.

71. Tribe Staphylinini Latreille 1802
Antennal bases generally well separated and closer to eye than to each other, antennae seldom geniculate; neck usually distinct but broad, seldom less than 1/2 as wide as head; pronotal hypomeron variably inflexed, with or without a thin and usually small translucent post-coxal process; anterior margin of prosternum usually straight, sometimes convex or concave, never with antesternal sclerite; elytral suture usually straight, elytral margins rarely overlapping (one Pacific coast flightless species in North America); tarsi nearly always 5-5-5 (5-4-4 in *Atanygnathus*); abdominal intersegmental membranes with pattern of small irregular rounded sclerites; aedeagus with parameres fused at least at base, usually fused into a single lobe which often bears black tubercles (peg setae) on side facing median lobe, paramere sometimes greatly reduced or absent. Most if not all Staphylinini also have paired eversible defensive glands (Dettner 1993, Huth and Dettner 1990, Jefson *et al.* 1983).

This very large tribe is divided into eight subtribes, six of which occur in North America. Those not occurring here are Amblyopinina Seevers 1944, including six genera of mammal "ectoparasites" (now thought to be commensals, see Ashe and Timm 1987) from the Neotropical region and possibly Australia and Anisolinina Hayashi 1993, from the Neotropical and Oriental regions. Many of these subtribes are not adequately defined, however, and some genera (e.g., *Flohria* in North America) are difficult to place in them; see Smetana (1977, 1984) and Hayashi (1993, 1997). Keys to most subtribes: Smetana (1995), Smetana and Davies (2000). Of the 200 genera worldwide, 31 occur in North America, six of them endemic.

71a. Subtribe Quediina Kraatz 1857
Head usually with infraorbital ridge which is an extension of nuchal carina; pronotal hypomeron usually strongly inflexed and scarcely visible in lateral view (except in *Beeria*), joining the prosternum at a distinct angle and separated from it by a distinct suture; superior marginal line of hypomeron not deflexed, fully visible from above; tarsi 5-5-5. This worldwide subtribe includes at least 40 genera, of which six occur in North America and three are endemic to this area. Revision and keys to genera and species: Smetana (1971, supplemented 1973, 1976, 1978, 1981b).

*Acylophorus* Nordmann 1837
*Amacylophorus* Smetana 1971 (subgenus)
*Palpacylophorus* Smetana 1971 (subgenus)
Eighteen species, widely distributed; found in wet moss and debris near water, especially in swamps, marshes and bogs. Revision and key: Smetana (1971, partly revised 1978). A nearly worldwide genus, with more than 100 additional species, especially in the Old World tropics.

*Anaquedius* Casey 1915
One species, *A. vernix* (LeConte 1878), eastern United States and southeastern Canada; found in wet moss and debris near water, especially in swamps, marshes and bogs. Review: Smetana (1971, revised map 1978). Larva: LeSage (1984). This genus is endemic to North America.

*Beeria* Hatch 1957
One odd *Philonthus*-like species, *B. nematocera* (Casey 1915) (=*B. punctata* Hatch 1957), western Oregon to southern Alaska; found in wet talus (heaped stone fragments) along cold mountain streams. Review, ecology and placement: Smetana (1977, supplemented 1978, 1981b). This genus is endemic to North America.

*Hemiquedius* Casey 1915
One species, *H. ferox* (LeConte 1878), eastern United States and southeastern Canada; found in wet moss and debris near water, especially in swamps, marshes and bogs. Review: Smetana (1971, revised map 1978). This genus is endemic to North America.

*Heterothops* Stephens 1829
Eighteen species, widely distributed; found mainly in forest leaf litter or associated with decaying logs and treeholes, some species in galleries and nests of various mammals or in wet biotopes. Revision and key: Smetana (1971, revised 1978). Larvae (genus in key or exotic species): Hinton (1945), Kasule (1970), Newton (1990a), Paulian (1941), Topp (1978). This genus is nearly worldwide with about 120 additional species.

*Quedius* Stephens 1829 (Fig. 40.22)
*Raphirus* Stephens 1829 (subgenus)
*Sauridus* Mulsant and Rey 1876
*Quediellus* Casey 1915
*Microsaurus* Dejean 1833 (subgenus)
*Quediochrus* Casey 1915
*Anastictodera* Casey 1915
*Quedionuchus* Sharp 1884 (subgenus)
*Distichalius* Casey 1915 (subgenus)
*Megaquedius* Casey 1915 (subgenus)
*Paraquedius* Casey 1915 (subgenus)
Ninety-one species, generally distributed; four species are probably naturally Holarctic, and four others are adventive from Europe. Most species are found in forest leaf and log litter and mosses, but some are found under bark of decaying logs, in galleries and nests of various mammals, in caves, or in wet habitats. Most of the seven subgenera recognized and keyed by Smetana (1971) are also widespread. Revision and key: Smetana (1971, partly revised 1973, 1976, 1981b). Larvae: Voris (1939b); exotic species or genus in key: Kasule (1970), Newton (1990a), Paulian (1941), Pototskaya (1967). This large and morphologically diverse genus includes more than 800 species worldwide, although many of these (especially from southern temperate areas) probably do not belong in it.

71b. Subtribe Tanygnathinina Reitter 1909
Similar to Quediina, but with reduced tarsal formula (5-4-4), infraorbital ridge originating near mandibular articulation and not connected to nuchal carina, and maxillary and labial palps extremely elongated. This group, recently separated from Quediina by Smetana (1984), includes only one widespread genus.

*Atanygnathus* Jacobson 1909
*Tanygnathus* Erichson 1839 (not Wagler 1832)
*Tanygnathinus* Reitter 1909
Three species, southeastern United States, north to New Hampshire and Michigan, and west to Arizona and California; found in wet debris and moss in swamps, marshes, bogs and other very wet biotopes. Key: Smetana (1971). This genus, with about 40 additional species, is mainly pantropical extending into warm temperate regions.

71c. Subtribe Staphylinina Latreille 1802
Head usually without infraorbital ridge, rarely with postmandibular ridge originating near base of mandible and not connected to nuchal ridge; pronotal hypomeron not strongly inflexed, more or less fully visible in lateral view; superior marginal line of hypomeron deflexed to ventral surface well behind anterior pronotal angle and often approaching or joining the inferior line, or fading out without approaching it, in any case not visible from above anteriorly; ligula usually distinctly notched or emarginate; tarsi 5-5-5; size always large, about 12-30 mm long.

About 33 genera are included here worldwide, a majority of them known only from Eurasia and Africa. Seven of the nine genera found in North America are also more speciose in the Old World, but two genera with one species each are endemic to the Pacific coast of North America; a tenth genus, *Leistotrophus*, is endemic to the Neotropics. The limits of this group are not settled; *Creophilus*, *Hadrotes* and *Thinopinus* have often been placed in Xanthopygina (e.g., Arnett 1963, Hatch 1957, Hayashi 1997) but probably belong here based on genitalic, larval and adult characters (e.g., Coiffait 1956, Kasule 1970, Smetana and Davies 2000). Adding to confusion have been nomenclatural problems with the names *Staphylinus* and *Creophilus* which also affect the group name (resolved by ICZN 1959) and widely different generic concepts used in Europe and elsewhere (especially affecting *Staphylinus*); the generic classification here follows the recent study of Smetana and Davies (2000). The last revision of North American species was by Horn (1879). Larvae: Bohác (1982), Frank (1979a), Kasule (1970), Voris (1939a).

*Creophilus* Leach 1819 (Fig. 1.22)
One common species, *C. maxillosus* (Linnaeus 1758), very widely distributed, also in Mexico and northern Central America, the West Indies, and throughout the Palearctic region, and on many islands elsewhere; found on carrion of all kinds where adults feed on maggots, in natural habitats of many kinds as well as in synanthropic situations. The exceptionally wide distribution and synanthropy of this species suggest human influence at least in part, but there is some geographic differentiation suggesting that a broad pre-human distribution is natural. Most North American specimens can be distinguished easily from Palearctic specimens by having a more extensive pattern of white setae on the body, and are often treated as the subspecies *C. m. villosus* (Gravenhorst 1802), but the typical (Palearctic) subspecies occurs in western Alaska (Hatch 1938) and occasional individuals have been found in northeastern North America (A. Newton, unpublished observations). Morphology: Blackwelder (1936), Dajoz and Caussanel (1969). Biology and larva: Abbott (1938), Dajoz and Caussanel (1969), Greene (1996), Kramer (1955); chemical defense: Huth and Dettner (1990), Jefson *et al.* (1983). The genus includes a dozen species in all and is nearly worldwide, but apparently absent from mainland Africa south of the Sahara.

*Dinothenarus* Thomson 1858
*Staphylinus* auctorum (in part)
*Trichoderma* auctorum (misidentification)
*Parabemus* Reitter 1909 (subgenus)
Six species, five in the subgenus *Parabemus*: *D. badipes* (LeConte 1863), *D. luteipes* (LeConte 1861), *D. nigrellus* (Horn 1879), *D. pleuralis* (LeConte 1861), and *D. saphyrinus* (LeConte 1861),

widely distributed across Canada and the northern United States, south to West Virginia in the east and southern California in the west; found under rocks and debris or in leaf litter, especially near water. The sixth species, *D. (Dinothenarus) capitatus* (Bland 1864), is rare in northern North America from Newfoundland to British Columbia, south to West Virginia; found at dung and carrion of various kinds in northern forests. This brightly colored species has usually been placed in *Ontholestes* in the North American literature, but was correctly transferred to this subgenus by Smetana (e.g., 1981a). Key to subgenera: Smetana and Davies (2000); keys to most species: Hatch (1957: 176, as *Abemus* and *Parabemus*); Horn (1879: 186, part A of key). Larvae (including European species): Bohác (1982), Frank (1979a, 1991), Paulian (1941). Chemical defense: Huth and Dettner (1990). This genus as redefined by Smetana and Davies (2000) includes about 20 additional species from Eurasia, especially the mountains of central and southern Asia.

*Hadrotes* Mannerheim 1852
One species, *H. crassus* (Mannerheim 1846), Alaska to southern California (also Baja California); flightless, restricted to ocean beaches where it is found under decaying seaweed and other debris. Larva: Moore (1964c). Biology: Orth *et al.* (1978). This genus is endemic to North America.

*Ocypus* Leach 1819
*Staphylinus* auctorum (in part)
*Goerius* Westwood 1827
*Matidus* Motschulsky 1845 (subgenus)
*Pseudocypus* Mulsant and Rey 1876 (subgenus)
Four adventive European species: *O. (Pseudocypus) aeneocephalus* (DeGeer 1774), first found at Vancouver in 1932 and still known only from southwestern British Columbia; *O. (Ocypus) olens* (Müller 1764), first found in southern California in 1931 and now known from California to Washington and southwestern Arizona; and *O. (Matidus) brunnipes* (Fabricius 1781) and *O. (Matidus) nitens* (Schrank 1781) (= *O. similis* (Fabricius 1792)), first found in North America in the 1960s and 1940s, respectively, and still known only from eastern Massachusetts and southeastern New Hampshire (Newton 1987). All four species are found mainly in disturbed or synanthropic habitats, where *O. olens* may actually contribute to control of a less welcome introduced species, the brown garden snail (*Helix aspersa* Müller). Key to subgenera: Smetana and Davies (2000); key to species: Coiffait (1974: 423). Biology (*O. olens*): Moore (1975b), Orth *et al.* (1975, 1976), Nield (1976). Larvae: Boháс (1982), Frank (1979a), Kasule (1970), Paulian (1941), Pototskaya (1967). Chemical defense: Huth and Dettner (1990). This genus as redefined by Smetana and Davies (2000) includes more than 140 species in Eurasia and Ethiopia, placed in four subgenera.

*Ontholestes* Ganglbauer 1895
Two species, the indigenous *O. cingulatus* (Gravenhorst 1802), widespread and common in eastern and northwestern North America, and the adventive European species *O. murinus* (Linnaeus 1758) which has so far been found only in the Avalon Peninsula of Newfoundland (Smetana 1981a). The species are predators of flies and other insects attracted to dung of all kinds, also frequently found on carrion, decaying soft fungi or other rotting organic matter. The species once called *O. capitatus* (Bland 1864) belongs in the genus *Dinothenarus* (Smetana and Davies 2000). Key: Smetana (1981a). Larva: Kasule (1970), Peterson (1960), Voris (1939a). Biology: Alcock (1991), Huth and Dettner (1990). The genus includes about 30 species total, mostly in Eurasia but with three species in Africa and one in South America.

*Platydracus* Thomson 1858
*Staphylinus* auctorum (in part)
Twenty-seven species, including six undescribed species, widely distributed but mainly eastern United States and Arizona; many species are found on dung, carrion and rotting fungi where they prey on other insects attracted to these substances, but some are found only in ground litter, under bark, in wet areas including edges of lakes, or on ocean beaches. This former subgenus of *Staphylinus* includes most North American (and virtually all Neotropical) species once placed in *Staphylinus* (see under that genus). Checklist with new combinations and synonymies: Smetana and Davies (2000). Partial keys: Hatch (1957: 178, as subgenus *Platydracus*); Horn (1879: 187, section B of *Staphylinus* key); no adequate full key available. Larvae: LeSage (1977, as *Staphylinus*), Schmidt (1994a), Voris (1939a, as *Staphylinus*). This genus includes about 200 species in Eurasia and Africa, and 70 additional species in the Neotropical region. The New World fauna is under revision by A. F. Newton and more than 100 new species will be added to this fauna, including six new Nearctic species.

*Staphylinus* Linnaeus 1758
One species, *S. ornaticauda* LeConte 1863, very rare in northeastern North America, from Manitoba to Nova Scotia south to Minnesota and Michigan; found mainly in or near bogs. This species has usually been misidentified in the North American literature as *S. erythropterus* Linnaeus 1758, or *S. caesareus* Cederhjelm 1798 (e.g., Horn 1879), but no authentic Nearctic records of those Palearctic species are known. This genus as restricted by Smetana and Davies (2000) includes only six additional species in the Palearctic region. Larvae of European species: Bohác (1982), Frank (1979a), Pototskaya (1967). Chemical defense: Huth and Dettner (1990).

*Staphylinus* was for a long time a large genus of uncertain limits, which has been divided recently into more than 20 genera and subgenera in the Old World (e.g., Coiffait 1956, 1974). The separation of the former subgenus *Platydracus* as a distinct genus has been widely accepted and is strongly supported by characters of the mandibles, chaetotaxy, aedeagus, and larval legs; this removed nearly 300 species from *Staphylinus*, including virtually all tropical species. The remaining 220+ species of *Staphylinus sensu lato*, mainly from the Palearctic and northern Oriental regions, have been reviewed by Smetana and Davies (2000) who removed most of these species to their revised concepts of the genera *Dinothenarus*, *Ocypus*, and *Tasgius* (q.v.).

*Tasgius* Stephens 1829
*Staphylinus auctorum* (in part)
*Paratasgius* Jarrige 1952
*Rayacheila* Motschulsky 1845 (subgenus)
*Alapsodus* Tottenham 1939
*Allocypus* Coiffait 1964

Three adventive European species: *T. (Tasgius) ater* (Gravenhorst 1802), widespread in North America (from which it was originally described, before being recognized as European); and *T. (Rayacheila) melanarius* (Heer 1839) and *T. (Rayacheila) winkleri* (Bernhauer 1906), both in northeastern and northwestern North America where they were first found in the 1930s (east) or 1940s (west); found especially under debris near water including marine beaches, but also common away from water in synanthropic situations including urban areas. Prior to Newton (1987), North American specimens of the last two species were erroneously reported as *T. globulifer* (Fourcroy 1785), which evidently does not occur in North America. Key to subgenera: Smetana and Davies (2000); key to species: Coiffait (1974: 532, as *Tasgius* and *Alapsodus (Allocypus)*). Larvae: Boháс (1982), Paulian (1941). Chemical defense: Huth and Dettner (1990). This genus as redefined by Smetana and Davies (2000) includes about 44 species from the western and central Palearctic region and the Himalayas.

*Thinopinus* LeConte 1852
One species, *T. pictus* LeConte 1852, Alaska to southern California (also Baja California); flightless, restricted to the intertidal zone of sandy ocean beaches, where adults prey on amphipods at low tide at night. Larva: Böving and Craighead (1931), Craig (1970). Biology: Craig (1970), Malkin (1958), Richards (1983). This genus is endemic to North America.

71d. Subtribe Xanthopygina Sharp 1884
Head with or without postmandibular ridge; or with carina originating near base of mandible and not connected to nuchal carina; pronotal hypomeron moderately inflexed, more or less visible in lateral view; superior marginal line of hypomeron continued to anterior pronotal angle without deflection or with slight deflection that conceals it in dorsal view, never approaching or joining the inferior line which continues independently to the front of the pronotum or fades out before then; ligula not distinctly notched or emarginate; tarsi 5-5-5; size fairly large, about 10-20 mm long.

The subtribal limits are uncertain; *Creophilus*, *Thinopinus* and *Hadrotes*, among other genera, have been included here by many twentieth century authors (e.g., Hayashi 1997), but were excluded by Sharp (1884) when he first briefly defined the group and have recently been returned to Staphylinina (q. v.). The single Nearctic species now included in *Flohria* was long placed in one or another xanthopygine genus, but *Flohria* was also excluded from Xanthopygina by Sharp (1884), and perhaps is best placed in Philonthina as defined by Smetana and Davies (2000). Two dozen Neotropical genera definitely belong here, but several genera restricted to the Old World tropics as well as *Tympanophorus* have been placed here by some authors; clearly, this group needs better definition. Larvae have been described for only two Central American species (Irmler 1979, Quezada *et al.* 1969).

[*Gastrisus* Sharp 1876, not North American (see under *Flohria*)]

[*Trigonopselaphus* Gemminger and Harold 1868 (=*Trigonophorus* Nordmann 1837, not Stephens 1829), not North American (see under *Flohria*)]

*Tympanophorus* Nordmann 1837
Two species, *T. puncticollis* Erichson 1840 (=*T. borealis* (Hatch 1957)), northeastern North America from New Brunswick west to Minnesota and south to Alabama, also British Columbia; and *T. concolor* Sharp 1884, Arizona and New Mexico (also south at least to Costa Rica); habits unknown, found occasionally in flight or pitfall traps or in litter, and once at the mouth of a fox burrow. This genus includes only about a dozen species but these are widely scattered through the New and Old World tropics, extending north to Japan as well as south to Australia and New Caledonia.

*Xanthopygus* Kraatz 1857
*Lampropygus* Sharp 1884
*Heteropygus* Bernhauer 1906

One species, *X. xanthopygus* Nordmann 1837 (=*X. cacti* Horn 1868), Arizona and Texas (also south to Guatemala); found in rotting cacti. *Xanthopygus borealis* Hatch 1957 belongs in the genus *Tympanophorus* as a synonym of *T. puncticollis* (see Moore and Legner 1975a: 46). Larva and biology of a related Central American species: Quezada *et al.* (1969). This genus includes nearly 40 species from the Neotropical region.

71e. Subtribe Philonthina Kirby 1837
Head without infraorbital or postmandibular ridge; pronotal hypomeron variably inflexed, more or less visible to concealed in lateral view; superior lateral line of pronotal hypomeron deflexed behind anterior pronotal angle and thus concealed anteriorly in dorsal view, often approaching or joining the inferior line; hypomeron fused to prosternum without angular junction or distinct suture; ligula not distinctly notched or emarginate; tarsi 5-5-5, empodium glabrous; 3-18 mm long.

This subtribe, often combined with Staphylinina, includes more than three dozen genera worldwide although some additional tropical or southern temperate genera may belong here; 12 genera occur in North America. Revision of most North American genera and species: Smetana (1995); see also Frank (1983).

*Belonuchus* Nordmann 1837
Nine species, eastern United States north to Ontario and Quebec and southwest to Arizona and southern California; most species are found especially on rotting fruits, cacti, logs or other organic matter in a fermentation stage of decay, some also on dung and carrion, but two southwestern species are apparently obligate cavernicoles. Revision and key: Smetana (1995). Larva and biology: Mank (1923), Silvestri (1945). This pantropical

genus includes about 200 species, most from the Neotropical region.

*Bisnius* Stephens 1829
*Gefyrobius* Thomson 1859
Thirty-four species, including four adventive from the Palearctic region and one Holarctic species; found in a wide variety of habitats including forest litter or debris and mosses near water, on dung, carrion and decaying fungi, and often in nests of various mammals and birds. Revision and key: Smetana (1995). Larvae (as *Philonthus cephalotes* (Gravenhorst 1802), *P. fimetarius* (Gravenhorst 1802), *P. puella* (Nordmann 1837) and/or *P. sordidus* (Gravenhorst 1802)): Boller (1983), Kasule (1970), Paulian (1941), Pototskaya (1967). This genus has only recently been separated from *Philonthus* (Smetana 1995) and includes at least three dozen additional species from the Palearctic region.

*Cafius* Stephens 1829
*Remus* Holme 1837 (subgenus)
*Bryonomus* Casey 1885 (subgenus)
*Euremus* Bierig 1934 (subgenus)
*Pseudoremus* Koch 1936 (subgenus)
Thirteen species, ten on the Pacific coast from Alaska to California, and five on the Atlantic and Gulf coasts from Newfoundland to Texas (two species on both coasts); common in wrack and other debris on marine beaches. Review and key: Orth and Moore (1980); revised key to southeastern species: Frank *et al.* (1986). Larvae and biology: James *et al.* (1971), Orth *et al.* (1978). This worldwide genus of 50 species restricted to seacoasts has been divided into six subgenera, but recent reviews of North American species have deliberately avoided using subgenera.

*Erichsonius* Fauvel 1874
*Actobius* Fauvel 1876
*Sectophilonthus* Tottenham 1949 (subgenus)
Nineteen species, widely distributed; found in leaf litter, mosses and damp debris near streams, springs, lakes, marshes or other damp situations. Revision and key: Frank (1975); key modification and subgeneric assignments: Frank (1981a). Larva: Schmidt (1996). The genus includes more than 120 additional species from Eurasia and Africa, but is absent from the Australian and Neotropical regions. The former genus *Actobius* was generally used in a broader sense to include *Neobisnius* as well as *Erichsonius* species.

*Flohria* Sharp 1884
One rare species, *F. subcoerulea* (LeConte 1863), Texas (also Mexico); habits unknown, found once in a forest litter sample. This species has generally been placed in the genus *Trigonopselaphus* by North American authors, but was moved to *Gastrisus* by Scheerpeltz (1972). It belongs in neither of these typical Neotropical xanthopygine genera, but rather in the isolated Mexican genus *Flohria*, and is probably conspecific with the only species hitherto placed in that genus, *F. laticornis* Sharp 1884. *Flohria* has not been placed to subtribe, but the glabrous tarsal empodium suggests placement in Philonthina as defined by Smetana and Davies (2000).

*Gabrius* Stephens 1829
Forty-two species including six adventive from Europe, widely distributed, especially western; found mainly in forest leaf litter, squirrel middens or associated with dead trees, or in wet debris or moss near water, the adventive species often in compost or other synanthropic habitats. Revision and key: Smetana (1995). Larvae: Kasule (1970), Pototskaya (1967). The genus includes more than three hundred species, mostly in Eurasia and Africa, with a few in the Neotropical and none in the Australian regions (Schillhammer 1997).

*Gabronthus* Tottenham 1955
Two adventive species, the African *G. mgogoricus* Tottenham 1955, Florida, Mississippi, and North Carolina, and the European *G. thermarum* (Aubé 1850), Michigan and Quebec south to Oklahoma and North Carolina; found mainly in compost, farm animal droppings and various other mostly synanthropic situations. Review and key: Smetana (1995). The genus includes more than 30 Old World species, mainly tropical.

*Hesperus* Fauvel 1874
Six species, throughout the eastern United States, north to Ontario and Quebec, plus Montana and Arizona; at least the two common species are associated with old trees (especially treeholes) or decaying logs. Revision and key: Smetana (1995). The genus is nearly worldwide with more than 200 species, mainly in the Old World tropics.

*Laetulonthus* Moore and Legner 1972
Two species, *L. laetulus* (Say 1834), eastern North America to Texas, and *L. nobilis* (Bernhauer 1910), Arizona and New Mexico; associated with trees, found in tree hollows and in debris at the base of live trees and under bark and in rotting wood of decaying logs. Review and key: Smetana (1995). This genus is apparently endemic to North America.

*Neobisnius* Ganglbauer 1895
Fourteen species, including the adventive *N. villosulus* (Stephens 1832), widely distributed; found mainly in moist habitats including the margins of rivers, marshes and lakes. Revision and key: Frank (1981b). Larva: Schmidt (1994b), Topp (1978). This genus, which together with *Erichsonius* formed the old genus *Actobius*, is worldwide with about 100 species.

*Philonthus* Stephens 1829
*Spatulonthus* Tottenham 1955
*Paragabrius* Coiffait 1963
*Palaeophilonthus* Coiffait 1972
One hundred and twelve species, including 17 species adventive from Europe or elsewhere, very widespread; found in a wide variety of habitats including forest leaf and log litter, wet debris and mosses near water, on dung, carrion and decaying fungi, and

in nests or middens of various mammals and birds as well as the gopher tortoise (*Gopherus polyphemus*), the adventive species often in compost or other synanthropic situations. Revision and key: Smetana (1995). Larvae: Boller (1983), Kasule (1970), Moore (1977), Paulian (1941), Pototskaya (1967). Eggs and biology: Hu and Frank (1995c, 1997). Subgenera for this very large worldwide genus have been used by some authors, but not by Smetana (1995). Although well over 1000 species are included here at present, this number may drop as poorly known tropical faunas are studied and the narrower generic concepts now applied to the Holarctic fauna (e.g., Smetana 1995) are applied to those.

*Rabigus* Mulsant and Rey 1876

One species, *R. laxellus* (Casey 1915), across southern Canada and northern United States, south in the Rocky Mountains to New Mexico and Arizona; found in open habitats such as fields, meadows and forest edges. Review: Smetana (1995). This genus, recently separated from *Philonthus*, also includes more than a dozen Palearctic species.

71f. Subtribe Hyptiomina Casey 1906

In addition to the "extra" (third) marginal line of the pronotal hypomeron (unique in New World Staphylininae) and very flat body, this group is characterized by aciculate apical maxillary and labial palpomeres; fused gular sutures; protibia with a crude "antennal cleaner" of about three oblique rows of stout setae; and short 5-5-5 tarsi with tarsomeres 1-4 transverse and empodium glabrous. A single genus belongs here after the removal of several genera or species to other subfamilies (Newton 1988).

*Holisus* Erichson 1839

*Hyptioma* Casey 1906

One unidentified species, Arizona (Chiricahua Mountains, Rustler Park). This genus includes 30 species throughout most of the Neotropical region, and one species in central Africa; most species are found under the bark of decaying logs.

## Bibliography


ABARBANELL, N. R. and ASHE, J. S. 1989. Revision of the species of *Pinophilus* Gravenhorst (Coleoptera: Staphylinidae) of America north of Mexico. Fieldiana: Zoology (N.S.), 54: iii + 32 pp.

ABBOTT, C. E. 1938. The development and general biology of *Creophilus villosus* Gray. Journal of the New York Entomological Society, 46: 49-53.

AHMED, M. A. 1957. Life-history and feeding habits of *Paederus alfierii* Koch (Coleoptera: Staphylinidae). Bulletin de la Société Entomologique d'Egypte, 41: 129-143.

AHN, K.J. 1996a. A review of *Diaulota* Casey (Coleoptera: Staphylinidae: Aleocharinae), with description of a new species and known larvae. Coleopterists Bulletin, 50: 270-290.

AHN, K.J. 1996b. Revision of the intertidal aleocharine genus *Tarphiota* (Coleoptera: Staphylinidae). Entomological News, 107: 177-185.

AHN, K.J. 1997a. A review of *Liparocephalus* Mäklin (Coleoptera: Staphylinidae: Aleocharinae) with descriptions of larvae. PanPacific Entomologist, 73: 79-92.

AHN, K.J. 1997b. Revision and systematic position of the intertidal genus *Thinusa* Casey (Coleoptera: Staphylinidae: Aleocharinae). Entomologica Scandinavica, 28: 75-81.

AHN, K.J. and ASHE, J. S. 1992. Revision of the intertidal aleocharine genus *Pontomalota* Casey (Coleoptera: Staphylinidae) with a discussion of its phylogenetic relationships. Entomologica Scandinavica, 23: 347-359.

AHN, K.J. and ASHE, J. S. 1995. Systematic position of the intertidal genus *Bryobiota* Casey and a revised phylogeny of the falagriine genera of America north of Mexico (Coleoptera: Staphylinidae: Aleocharinae). Annals of the Entomological Society of America, 88: 143-154.

AHN, K.J. and ASHE, J. S. 1996a. Phylogeny of the intertidal aleocharine tribe Liparocephalini (Coleoptera: Staphylinidae). Systematic Entomology, 21: 99-114.

AHN, K.J. and ASHE, J. S. 1996b. Revision of the intertidal aleocharine genus *Amblopusa* Casey and description of the new genus *Paramblopusa* (Coleoptera: Staphylinidae). Journal of the New York Entomological Society, 103: 138-154 [1995].

AHN, K.-J. and ASHE, J. S. 1999. Two new species of *Giulianium* Moore from the Pacific coasts of Alaska and California (Coleoptera: Staphylinidae: Omaliinae). Pan-Pacific Entomologist, 75: 159-164.

AKRE, R. D. and HILL, W. B. 1973. Behavior of *Adranes taylori*, a myrmecophilous beetle associated with *Lasius sitkaensis* in the Pacific Northwest (Coleoptera: Pselaphidae; Hymenoptera: Formicidae). Journal of the Kansas Entomological Society, 46: 526-536.

AKRE, R. D. and RETTENMEYER, C. W. 1966. Behavior of Staphylinidae associated with army ants (Formicidae: Ecitonini). Journal of the Kansas Entomological Society, 39: 745-782.

ALCOCK, J. 1991. Adaptive mate-guarding by males of *Ontholestes cingulatus* (Coleoptera: Staphylinidae). Journal of Insect Behavior, 4: 763-771.

ARAUJO, J. 1978. Anatomie comparée des systèmes glandulaires de défense chimique des Staphylinidae. Archives de Biologie, 89: 217-249.

ARNETT, R. H., JR. 1963. The beetles of the United States (a manual for identification). Catholic University of America Press, Washington, D.C. xi + 1112 pp.

ASHE, J. S. 1981. Studies of the life history and habits of *Phanerota fasciata* Say (Coleoptera: Staphylinidae: Aleocharinae) with notes on the mushroom as a habitat and descriptions of the immature stages. Coleopterists Bulletin, 35: 83-96.

ASHE, J. S. 1982. Evidence about species status of *Phanerota fasciata* (Say) and *Phanerota dissimilis* (Erichson) (Coleoptera: Staphylinidae: Aleocharinae) from host mushroom relationships. Coleopterists Bulletin, 36: 155-161.

ASHE, J. S. 1984a. Generic revision of the subtribe Gyrophaenina (Coleoptera: Staphylinidae: Aleocharinae) with a review of the described subgenera and major features of evolution. Quaestiones Entomologicae, 20: 129-349.

ASHE, J. S. 1984b. Major features of the evolution of relationships between gyrophaenine staphylinid beetles (Coleoptera: Staphylinidae: Aleocharinae) and fresh mushrooms, pp. 227-255. In: Wheeler, Q. and Blackwell, M. (eds.), Fungus/Insect Relationships: Perspectives in Ecology and Evolution. Columbia University Press, New York.

ASHE, J. S. 1984c. Description of the larva and pupa of *Scaphisoma terminata* Melsh. and the larva of *Scaphium castanipes* Kirby with notes on their natural history (Coleoptera: Scaphidiidae). Coleopterists Bulletin, 38: 361-373.

ASHE, J. S. 1985. Fecundity, development and natural history of *Meronera venustula* (Erichson) (Coleoptera: Staphylinidae: Aleocharinae). Psyche, 92: 181-204.

ASHE, J. S. 1986a. *Seeversiella bispinosa*, a new genus and species of athetine Aleocharinae (Coleoptera: Staphylinidae) from North America. Journal of the New York Entomological Society, 94: 500-511.

ASHE, J. S. 1986b. Structural features and phylogenetic relationships among larvae of genera of gyrophaenine staphylinids (Celeoptera: Staphylinidae: Aleocharinae). Fieldiana: Zoology (N.S.), 30: 1-60.

ASHE, J. S. 1986c. Subsocial behavior among gyrophaenine staphylinids (Coleoptera: Staphylinidae, Aleocharinae). Sociobiology, 12: 315-320.

ASHE, J. S. 1986d. *Phanerota cubensis* and *Phanerota brunnessa* n. sp., with a key to the species of *Phanerota* occurring in Florida (Coleoptera: Staphylinidae). Florida Entomologist, 69: 236-245.

ASHE, J. S. 1987. Egg chamber production, egg protection and clutch size among fungivorous beetles of the genus *Eumicrota* (Coleoptera: Staphylinidae) and their evolutionary implications. Zoological Journal of the Linnean Society, 90: 255-273.

ASHE, J. S. 1990. Natural history, development and immatures of *Pleurotobia tristigmata* (Erichson) (Coleoptera: Staphylinidae: Aleocharinae). Coleopterists Bulletin, 44: 4-4-5-460.

ASHE, J. S. 1991a. The larvae of *Placusa* Mannerheim (Coleoptera: Staphylinidae), with notes on their feeding habits. Entomologica Scandinavica, 21: 477-485 [1990].

ASHE, J. S. 1991b. The systematic position of *Placusa* Erichson and *Euvira* Sharp: the tribe Placusini described (Coleoptera: Staphylinidae: Aleocharinae). Systematic Entomology, 16: 383-400.

ASHE, J. S. 1992. Phylogeny and revision of genera of the subtribe Bolitocharina (Coleoptera: Staphylinidae: Aleocharinae). University of Kansas Science Bulletin, 54: 335-406.

ASHE, J. S. 1993. Mouthpart modifications correlated with fungivory among aleocharine staphylinids (Coleoptera: Staphylinidae: Aleocharinae), pp. 105-130. In: Schaefer, C. W. and Leschen, R. A. B. (eds.), Functional morphology of insect feeding. Thomas Say Publications in Entomology, Entomological Society of America, Lanham, Maryland.

ASHE, J. S. 1994. Evolution of aedeagal parameres of aleocharine staphylinids (Coleoptera: Staphylinidae: Aleocharinae). Canadian Entomologist, 126: 475-491.

ASHE, J. S. and KISTNER, D. H. 1989. Larvae and adults of a new species of *Euvira* (Coleoptera: Staphylinidae: Aleocharinae) from the nests of the communal pierid butterfly *Eucheira socialis* with a redescription of the genus *Euvira*. Sociobiology, 15: 85-106.

ASHE, J. S. and NEWTON, A. F., JR. 1993. Larvae of *Trichophya* and phylogeny of the tachyporine group of subfamilies (Coleoptera: Staphylinidae) with a review, new species and characterization of the Trichophyinae. Systematic Entomology, 18: 267-286.

ASHE, J. S. and TIMM, R. M. 1987. Predation by and activity patterns of "parasitic" beetles of the genus *Amblyopinus* (Coleoptera: Staphylinidae). Journal of Zoology (London), 212: 429-437.

ASHE, J. S. and WATROUS, L. E. 1984. Larval chaetotaxy of Aleocharinae (Staphylinidae) based on a description of *Atheta coriaria* Kraatz. Coleopterists Bulletin, 38: 165-179.

ASSING, V. and WUNDERLE, P. 1995. A revision of the species of the subfamily Habrocerinae (Coleoptera: Staphylinidae) of the world. Revue Suisse de Zoologie, 102: 307-359.

BADGLEY, M. E. and FLESCHNER, C. A. 1956. Biology of *Oligota oviformis* Casey (Coleoptera: Staphylinidae). Annals of the Entomological Society of America, 49: 501-502.

BALDUF, W. V. 1935. The bionomics of entomophagous Coleoptera. J. S. Swift Co., St. Louis, etc. 220 pp.

BALLETO, E. and CASALE, A. 1991. Mediterranean insect conservation, pp. 121-142. In: Collins, N. M. and Thomas, J. A. (eds.), The conservation of insects and their habitats. Academic Press, London.

BARR, T. C. JR. 1974. The eyeless beetles of the genus *Arianops* Brendel (Coleoptera, Pselaphidae). Bulletin of the American Museum of Natural History, 154 (1): 1-51.

BENICK, L. 1951. Spezielles und Allgemeines über die Subfam. Megalopsidiinae (Col. Staph.). Entomologische Blätter für Biologie und Systematik der Käfer, 47: 58-87.

BERNHAUER, M. 1928. Übersicht über die nordamerikanischen *Euaesthetus*-Arten (Col. Staph.). Deutsche Entomologische Zeitschrift, 1928: 38-40.

BERNHAUER, M. and SCHEERPELTZ, O. 1926. Staphylinidae VI. In: Coleopterorum Catalogus, Pars 82: 499-988. W. Junk, Berlin.

BERNHAUER, M. and SCHUBERT, K. 1910. Staphylinidae I. In: Coleopterorum Catalogus, Pars 19: 1-86. W. Junk, Berlin.

BERNHAUER, M. and SCHUBERT, K. 1911. Staphylinidae II. In: Coleopterorum Catalogus, Pars 29: 87-190. W. Junk, Berlin.

BERNHAUER, M. and SCHUBERT, K. 1912. Staphylinidae III. In: Coleopterorum Catalogus, Pars 40: 191-288. W. Junk, Berlin.

BERNHAUER, M. and SCHUBERT, K. 1914. Staphylinidae IV. In: Coleopterorum Catalogus, Pars 57: 289-408. W. Junk, Berlin.

BERNHAUER, M. and SCHUBERT, K. 1916. Staphylinidae V. In: Coleopterorum Catalogus, Pars 67: 409-498. W. Junk, Berlin.

BESUCHET, C. 1956. Larves et nymphes de Psélaphides (Coléoptères). Revue Suisse de Zoologie, 63: 697-705.

BESUCHET, C. 1974. 24. Familie: Pselaphidae, pp. 305-362. In: Freude, H., Harde, K. W. and Lohse, G. A. (eds.), Die Käfer Mitteleuropas, Vol. 5, Staphylinidae II (Hypocyphtinae und Aleocharinae), Pselaphidae. Goecke and Evers, Krefeld.

BESUCHET, C. 1982a. Contribution à l'étude des Bythinini cavernicoles néarctiques (Coleoptera: Pselaphidae). Revue Suisse de Zoologie, 89: 49-53.

BESUCHET, C. 1982b. Le genre *Neopselaphus* Jeann. (Coleoptera: Pselaphidae). Revue Suisse de Zoologie, 89: 797-807.

BESUCHET, C. 1991. Révolution chez les Clavigerinae (Coleoptera, Pselaphidae). Revue Suisse de Zoologie, 98: 499-515.

BESUCHET, C. 1999. Psélaphides paléarctiques nouveaux ou méconnus (Coleoptera: Staphylinidae: Pselaphinae). Revue Suisse de Zoologie, 106: 789-811.

BETZ, O. 1996. Function and evolution of the adhesion-capture apparatus of *Stenus* species (Coleoptera, Staphylinidae). Zoomorphology, 116: 15-34.

BETZ, O. 1998. Comparative studies on the predatory behaviour of *Stenus* spp. (Coleoptera: Staphylinidae): the significance of its specialized labial apparatus. Journal of Zoology (London), 244: 527-544.

BETZ, O. 1999. A behavioural inventory of adult *Stenus* species (Coleoptera: Staphylinidae). Journal of Natural History, 33: 1691-1712.

BEUTEL, R. G. and MOLENDA, R. 1997. Comparative morphology of selected larvae of Staphylinoidea (Coleoptera, Polyphaga) with phylogenetic implications. Zoologischer Anzeiger, 236: 37-67.

BLACKWELDER, R. E. 1936. Morphology of the coleopterous family Staphylinidae. Smithsonian Miscellaneous Collections, 94 (13): 102 pp.

BLACKWELDER, R. E. 1939. A generic revision of the staphylinid beetles of the tribe Paederini. Proceedings of the United States National Museum, 87: 93-125.

BLACKWELDER, R. E. 1941. A monograph of the genus *Trigonurus* (Coleoptera: Staphylinidae). American Museum Novitates, 1124: 1-13.

BLACKWELDER, R. E. 1942. Notes on the classification of the staphylinid beetles of the groups Lispini and Osoriinae. Proceedings of the United States National Museum, 92: 75-90.

BLACKWELDER, R. E. 1943. Monograph of the West Indian beetles of the family Staphylinidae. Bulletin of the United States National Museum, 182: viii + 658 pp.

BLACKWELDER, R. E. 1944. Checklist of the Coleopterous insects of Mexico, Central America, the West Indies, and South America. Part 1. Bulletin of the United States National Museum, 185, pp. ixii + 1188.

BLACKWELDER, R. E. 1952. The generic names of the beetle family Staphylinidae, with an essay on genotypy. Bulletin of the United States National Museum, 200: iv + 483 pp.

BLACKWELDER, R. E. 1973a. Checklist of the Staphylinidae of Canada, United States, Mexico, Central America and the West Indies. North American Beetle Fauna Project, Family No. 15 (Yellow Version). Biological Research Institute of America, Inc., Siena College, Loudonville, New York. 165 pp.

BLACKWELDER, R. E. 1973b. Checklist of the Pselaphidae of Canada, United States, Mexico, Central America and the West Indies. North American Beetle Fauna Project, Family No. 16 (Red Version). Biological Research Institute of America, Inc., Latham, New York. 33 pp.

BLACKWELDER, R. E. 1973c. Checklist of the Scaphidiidae of Canada, United States, Mexico, Central America and the West Indies. North American Beetle Fauna Project, Family No. 24 (Red Version). Biological Research Institute of America, Inc., Latham, New York. 3 pp.

BLATCHLEY, W. S. 1910. An illustrated descriptive catalog of the Coleoptera or beetles (exclusive of the Rhynchophora) known to occur in Indiana. The Nature Publishing Co., Indianapolis, Indiana. 1386 pp.

BLATCHLEY, W. S. 1917. On some new or noteworthy Coleoptera from the west coast of Florida, II. Canadian Entomologist, 49: 236-240.

BLUM, P. 1979. Zur Phylogenie und ökologischen Bedeutung der Elytrenreduktion und Abdomenbeweglichkeit der Staphylinidae (Coleoptera): Vergleichend und funktionsmorphologische Untersuchungen. Zoologische Jahrbücher, Abteilung für Anatomie, 102: 533-582.

BOHÁC, J. 1978. Description of the larva and pupa of *Thoracophorus brevicristatus* (Coleoptera, Staphylinidae). Acta Entomologica Bohemoslovaca, 75: 394-399.

BOHÁC, J. 1982. The larval characters of Czechoslovak species of the genera *Abemus* Muls. et Rey, *Staphylinus* L. and *Ocypus* Sam. (Staphylinidae, Coleoptera). Studie Ceskoslovenská Akademie Ved, 4: 1-96, pls. 1-27.

BOLLER, F. 1983. Zur Larvalmorphologie der Gattung *Philonthus* Curtis (Coleoptera, Staphylinidae). Spixiana, 6: 113-131.

BOURNE, J. D. 1975. *Ochthephilus aureus* (*Ancyrophorus aureus*) Fauv. (Coleoptera Staphylinidae: Oxytelinae). Notes écologiques et description morphologique de la larve. Mitteilungen der Schweizerischen Entomologischen Gesellschaft, 48: 233-236.

BÖVING, A. G. and CRAIGHEAD, F. C. 1931. An illustrated synopsis of the principal larval forms of the order Coleoptera. Entomologica Americana (N.S.), 11: 13-51 [1930].

BOWMAN, J. R. 1934. The Pselaphidae of North America. Privately published, Pittsburgh, Pennsylvania. 149 pp.

BRENDEL, E. 1890. In: Brendel, E. and Wickham, H. F. The Pselaphidae of North America: A monograph. Bulletin of the Laboratory of Natural History, State University of Iowa, 1: 216-304, pls. 6-9.

BRUNDIN, L. 1943. Zur Kenntnis einiger in die *Atheta* Untergattung *Metaxya* M. and R. gestellten Arten (Col. Staphylinidae). Lunds Universitets Årsskrift (N.F.), (2) 39 (4): 1-37, pls. 1-7.

BURAKOWSKI, B. and NEWTON, A. F., JR. 1992. The immature stages and bionomics of the myrmecophile *Thoracophorus corticinus* Motschulsky, and placement of the genus (Coleoptera, Staphylinidae, Osoriinae). Annali del Museo Civico di Storia Naturale "G. Doria," 89: 17-42.

CAMPBELL, J. M. 1968. A revision of the New World Micropeplinae (Coleoptera: Staphylinidae) with a rearrangement of the world species. Canadian Entomologist, 100: 225-267.

CAMPBELL, J. M. 1969. A revision of the New World Oxyporinae (Coleoptera: Staphylinidae). Canadian Entomologist, 101: 225-268.

CAMPBELL, J. M. 1973a. New species and records of New World Micropeplinae (Coleoptera: Staphylinidae). Canadian Entomologist, 105: 569-576.

CAMPBELL, J. M. 1973b. A revision of the genus *Tachinomorphus* (Coleoptera: Staphylinidae) of North and Central America. Canadian Entomologist, 105: 1015-1034.

CAMPBELL, J. M. 1973c. A revision of the genus *Tachinus* (Coleoptera: Staphylinidae) of North and Central America. Memoirs of the Entomological Society of Canada, 90: 11-37.

CAMPBELL, J. M. 1975a. New species and records of *Tachinus* (Coleoptera: Staphylinidae) from North America. Canadian Entomologist, 107: 87-94.

CAMPBELL, J. M. 1975b. A revision of the genera *Coproporus* and *Cilea* (Coleoptera: Staphylinidae) of America north of Mexico. Canadian Entomologist, 107: 175-216.

CAMPBELL, J. M. 1976. A revision of the genus *Sepedophilus* Gistel (Coleoptera: Staphylinidae) of America north of Mexico. Memoirs of the Entomological Society of Canada, 99: 1-89.

CAMPBELL, J. M. 1978a. New species of *Oxyporus* (Coleoptera: Staphylinidae) from North America. Canadian Entomologist, 110: 805-813.

CAMPBELL, J. M. 1978b. New species and records of New World Micropeplidae (Coleoptera). II. Canadian Entomologist, 110: 1247-1258.

CAMPBELL, J. M. 1978c. A revision of the North American Omaliinae (Coleoptera: Staphylinidae). 1. The genera *Haida* Keen, *Pseudohaida* Hatch, and *Eudectoides* new genus. 2. The tribe Coryphiini. Memoirs of the Entomological Society of Canada, 106: 1-87.

CAMPBELL, J. M. 1979a. *Coprophilus castoris*, a new species of Staphylinidae (Coleoptera) from beaver lodges in eastern Canada. Coleopterists Bulletin, 33: 223-228.

CAMPBELL, J. M. 1979b. A revision of the genus *Tachyporus* Gravenhorst (Coleoptera: Staphylinidae) of North and Central America. Memoirs of the Entomological Society of Canada, 109: 1-95.

CAMPBELL, J. M. 1980. A revision of the genus *Carphacis* des Gozis (Coleoptera: Staphylinidae) of North America. Canadian Entomologist, 112: 935-953.

CAMPBELL, J. M. 1982a. *Mitosynum vockerothi*, a new genus and new species of Coleoptera (Staphylinidae: Oxytelinae) from eastern Canada. Canadian Entomologist, 114: 687-691.

CAMPBELL, J. M. 1982b. A revision of the North American Omaliinae (Coleoptera: Staphylinidae). 3. The genus *Acidota* Stephens. Canadian Entomologist, 114: 1003-1029.

CAMPBELL, J. M. 1982c. A revision of the genus *Lordithon* Thomson of North and Central America (Coleoptera: Staphylinidae). Memoirs of the Entomological Society of Canada, 119: 11-16.

CAMPBELL, J. M. 1983a. A new species of *Pycnoglypta* Thomson (Coleoptera: Staphylinidae) from eastern Canada. Canadian Entomologist, 115: 361-370.

CAMPBELL, J. M. 1983b. A revision of the North American Omaliinae (Coleoptera: Staphylinidae). The genus *Olophrum* Erichson. Canadian Entomologist, 115: 577-622.

CAMPBELL, J. M. 1984a. A catalog of the Coleoptera of America north of Mexico. Family: Micropeplidae. United States Department of Agriculture, Agriculture Handbook 529-24: x + 5 pp.

CAMPBELL, J. M. 1984b. A revision of the North American Omaliinae (Coleoptera: Staphylinidae). The genera *Arpedium* Erichson and *Eucnecosum* Reitter. Canadian Entomologist, 116: 487-527.

CAMPBELL, J. M. 1984c. A review of the North American species of the omaliine genera *Porrhodites* Kraatz and *Orochares* Kraatz (Coleoptera: Staphylinidae). Canadian Entomologist, 116: 1227-1249.

CAMPBELL, J. M. 1987. *Anthobioides pubescens*, an unusual new genus and species of Omaliinae (Coleoptera: Staphylinidae) from Washington. Canadian Entomologist, 119: 1027-1042.

CAMPBELL, J. M. 1988. New species and records of North American *Tachinus* Gravenhorst (Coleoptera: Staphylinidae). Canadian Entomologist, 120: 231-295.

CAMPBELL, J. M. 1989. *Micropeplus nelsoni*, a new species from the Cascade Range of Washington (Coleoptera: Micropeplidae). Coleopterists Bulletin, 43: 305-310.

CAMPBELL, J. M. 1991. A revision of the genera *Mycetoporus* Mannerheim and *Ischnosoma* Stephens (Coleoptera: Staphylinidae: Tachyporinae) of North and Central America. Memoirs of the Entomological Society of Canada, 156: 1-169.

CAMPBELL, J. M. 1993a. A review of the species of *Nitidotachinus* new genus (Coleoptera: Staphylinidae: Tachyporinae). Canadian Entomologist, 125: 521-548.

CAMPBELL, J. M. 1993b. A revision of the genera *Bryoporus* Kraatz and *Bryophacis* Reitter and two new related genera from America North of Mexico (Coleoptera: Staphylinidae: Tachyporinae). Memoirs of the Entomological Society of Canada, 166: 1-85.

CAMPBELL, J. M. and CHANDLER, D. S. 1987. *Omalorphanus aenigma*, an unusual new genus and species of Omaliinae (Coleoptera: Staphylinidae) from Oregon. Canadian Entomologist, 119: 315-327.

CAMPBELL, J. M. and DAVIES, A. 1991. Families Micropeplidae, Staphylinidae, Scaphidiidae, Pselaphidae, pp. 84, 86-129. In: Bousquet, Y. (ed.), Checklist of Beetles of Canada and Alaska. Publication 1861/E, Research Branch, Agriculture Canada, Ottawa.

CAMPBELL, J. M. and PECK, S. B. 1990. *Omalonomus relictus*, an unusual new genus and new species (Coleoptera: Staphylinidae, Omaliinae) of blind rove beetle; a preglacial (Tertiary?) relict in the Cypress Hills, Alberta-Saskatchewan, Canada. Canadian Entomologist, 122: 949-961.

CAMPBELL, J. M. and TOMLIN, A. D. 1984. The first record of the Palearctic species *Anotylus insecatus* (Gravenhorst) (Coleoptera: Staphylinidae) from North America. Coleopterists Bulletin, 37: 309-313 [1983].

CAMPBELL, J. M. and WINCHESTER, N. N. 1993. First record of *Pseudohaida rothi* Hatch (Coleoptera: Staphylinidae: Omaliinae) from Canada. Journal of the Entomological Society of British Columbia, 90: 83.

CARLTON, C. E. 1983. Revision of the genus *Conoplectus* Brendel (Coleoptera: Pselaphidae). Coleopterists Bulletin, 37: 55-80.

CARLTON, C. E. 1989. Revision of the genus *Eutrichites* LeConte (Coleoptera: Pselaphidae). Coleopterists Bulletin, 43: 105-119.

CARLTON, C. E. and ALLEN, R. T. 1986. Revision of the genus *Euboarhexius* Grigarick and Schuster (Coleoptera: Pselaphidae). Coleopterists Bulletin, 40: 285-296.

CARLTON, C. E. and CHANDLER, D. S. 1994. Revision of the Nearctic genus *Pseudactium* Casey (Coleoptera: Pselaphidae: Euplectinae). Coleopterists Bulletin, 48: 171-190.

CARLTON, C. E. and ROBISON, H. W. 1996. Notes on *Mayetia pearsei* Schuster, Marsh and Park, with a revised key to eastern *Mayetia* species (Staphylinidae: Pselaphinae). Coleopterists Bulletin, 50: 244-250.

CASEY, T. L. 1884. Revision of the Stenini of America north of Mexico. Collins Printing House, Philadelphia. 206 pp., 1 pl.

CASEY, T. L. 1885a. 1885a. Contributions to the descriptive and systematic Coleopterology of North America. Part II. Collins Printing House, Philadelphia. pp. 61-124 [1884], 125-198 [1885].

CASEY, T. L. 1885b. New genera and species of Californian Coleoptera. Bulletin of the California Academy of Sciences, 1: 283-337.

CASEY, T. L. 1886a. Revision of the California species of *Lithocharis* and allied genera. Bulletin of the California Academy of Sciences, 2 (5): 1-40.

CASEY, T. L. 1886b. Descriptive notices of North American Coleoptera I. Bulletin of the California Academy of Sciences, 2 (6): 157-264, pl. 7.

CASEY, T. L. 1889a. A preliminary monograph of the North American species of *Trogophloeus*. Annals of the New York Academy of Sciences, 4: 322-383.

CASEY, T. L. 1889b. Coleopterological notices, I. Annals of the New York Academy of Sciences, 5: 39-198.

CASEY, T. L. 1893. Coleopterological notices, V. Annals of the New York Academy of Science, 7: 281-606, pl. 1.

CASEY, T. L. 1897. Coleopterological notices, VII. Annals of the New York Academy of Sciences, 9: 285-684.

CASEY, T. L. 1900. Review of the American Corylophidae, Cryptophagidae, Tritomidae and Dermestidae, with other studies. Journal of the New York Entomological Society, 8: 51-172.

CASEY, T. L. 1905. A revision of the American Paederini. Transactions of the Academy of Science of St. Louis, 15: 17-248.

CASEY, T. L. 1906. Observations on the staphylinid groups Aleocharinae and Xantholinini, chiefly of America. Transactions of the Academy of Science of St. Louis, 16: 125-434.

CASEY, T. L. 1910a. New species of the staphylinid tribe Myrmedoniini, pp. 1-183. In: Memoirs on the Coleoptera, Vol. 1. New Era, Lancaster, Pennsylvania.

CASEY, T. L. 1910b. Synonymic and descriptive notes on the Paederini and Pinophilini, pp. 184-201. In: Memoirs on the Coleoptera, Vol. 1. New Era, Lancaster, Pennsylvania.

CASEY, T. L. 1911. New American species of Aleocharinae and Myllaeninae, pp. 1-245. In: Memoirs on the Coleoptera, Vol. 2. New Era, Lancaster, Pennsylvania.

CAVENEY, S. 1986. The phylogenetic significance of ommatidium structure in the compound eyes of polyphagan beetles. Canadian Journal of Zoology, 64: 1787-1819.

CHAMBERLIN, J. C. and FERRIS, G. F. 1929. On *Liparocephalus* and allied genera (Coleoptera; Staphylinidae). PanPacific Entomologist, 5: 137-143, 153-162.

CHANDLER, D. S. 1974a. The *Hamotus* of Arizona (Coleoptera: Pselaphidae). PanPacific Entomologist, 49: 378-382 [1973].

CHANDLER, D. S. 1974b. A redefinition of the Tyrini with the addition of *Anitra* Casey (Coleoptera: Pselaphidae). PanPacific Entomologist, 50: 162-164.

CHANDLER, D. S. 1976. A revision of the genus *Caccoplectus* (Coleoptera: Pselaphidae). Coleopterists Bulletin, 29: 301-316 [1975].

CHANDLER, D. S. 1985. The Euplectini of Arizona (Coleoptera: Pselaphidae). Entomography, 3: 107-126.

CHANDLER, D. S. 1987. Species richness and abundance of Pselaphidae (Coleoptera) in oldgrowth and 40-year-old forests in New Hampshire. Canadian Journal of Zoology, 65: 608-615.

CHANDLER, D. S. 1988a. A revision of the Nearctic genus *Cylindrarctus* (Coleoptera: Pselaphidae). Transactions of the American Entomological Society, 114: 129-146.

CHANDLER, D. S. 1988b. A cladistic analysis of the world genera of Tychini (Coleoptera: Pselaphidae). Transactions of the American Entomological Society, 114: 147-165.

CHANDLER, D. S. 1989. Synonymies and notes on the *Reichenbachia* of eastern North America (Coleoptera: Pselaphidae). Coleopterists Bulletin, 43: 379-389.

CHANDLER, D. S. 1990a. The Pselaphidae (Coleoptera) of Latimer County, Oklahoma, with revisions of four genera from eastern North America. Part I. Faroninae and Euplectinae. Transactions of the American Entomological Society, 115: 503-529 [1989].

CHANDLER, D. S. 1990b. Insecta: Coleoptera: Pselaphidae, pp. 1175-1190. In: Dindal, D. L. (ed.), Soil Biology Guide. John Wiley and Sons, New York.

CHANDLER, D. S. 1991. The *Lucifotychus* of eastern North America (Coleoptera: Pselaphidae). Psyche, 98: 47-56.

CHANDLER, D. S. 1992a. Notes on *Briaraxis depressa* (Coleoptera: Pselaphidae). Entomological News, 103: 15-18.

CHANDLER, D. S. 1992b. The Pselaphidae (Coleoptera) of Texas caves. Texas Memorial Museum, Speleological Monographs, 3: 241-253.

CHANDLER, D. S. 1993. Identity and biological notes on *Actizona trifoveata* (Park) (Coleoptera: Pselaphidae). Coleopterists Bulletin, 47: 289-290.

CHANDLER, D. S. 1997. A catalog of the Coleoptera of America north of Mexico. Family: Pselaphidae. United States Department of Agriculture, Agriculture Handbook 52931: ix + 118 pp.

CHANDLER, D. S. and WOLDA, H. 1986. Seasonality and diversity of *Caccoplectus*, with a review of the genus and description of a new genus, *Caccoplectinus* (Coleoptera: Pselaphidae). Zoologische Jahrbücher, Abteilung für Systematik, Ökologie und Geographie der Tiere, 113: 469-524.

CHAPIN, E. A. 1928. The North American species of *Holotrochus* Erichson (Coleoptera, Staphylinidae), with descriptions of two new species. Proceedings of the Entomological Society of Washington, 30: 65-67.

CHITTENDEN, F. H. 1915. The violet rove beetle. United States Department of Agriculture Bulletin, No. 264, 4 pp.

CLARIDGE, M. F. and MURPHY, D. M. 1967. Ecological notes on some *Anthophagus* species in Britain and northern Norway (Coleoptera: Staphylinidae). Entomologist, 100: 184-188.

COIFFAIT, H. 1956. Les "Staphylinus" et genres voisins de France et des régions voisines; essai de paléobiogéographie. Mémoires du Muséum National d'Histoire Naturelle (Série A: Zoologie), 8: 177-224, 22 pls.

COIFFAIT, H. 1962. Les Leptotyphlitae (Col. Staphylinidae) de Californie. Revue Française d'Entomologie, 29: 154-166.

COIFFAIT, H. 1963. Les Leptotyphlitae (Col. Staphylinidae) du Chili, systématique et biogéographie de la sousfamille, pp. 371-383. In: Delamare Deboutteville, C. and Rapoport, E. (eds.), Biologie de l'Amérique Australe, Vol. 2, Études sur la Faune du Sol. Centre National de la Recherche Scientifique, Paris.

COIFFAIT, H. 1972-84. Coléoptères Staphylinidae de la région paléarctique occidentale. Vols. 15. Nouvelle Revue d'Entomologie, Supplements 2 (2), ix + 651 pp., 6 pls.; 4 (4), 593 pp.; 8 (4), 364 pp.; 12 (4), 440 pp.; 13 (4), 424 pp.

COMELLINI, A. 1985. Notes sur les Psélaphides néotropicaux (Coleoptera). 5 La tribu des Pyxidicerini. Revue Suisse de Zoologie, 92: 707-759.

COOPER, K. W. 1961. Occurrence of the European pselaphid beetle *Trichonyx sulcicollis* (Reichenbach) in New York state. Entomological News, 72: 90-92.

COSTA, C., VANIN, S. A. and CASARICHEN, S. A. 1988. Larvas de Coleoptera do Brasil. Museu de Zoologia, Universidade de Sao Paulo, Sao Paulo. 282 pp., 165 pls.

COULON, G. 1989. Révision générique des Bythinoplectini Schaufuss, 1890 (=Pyxidicerini Raffray, 1903, syn. nov.) (Coleoptera, Pselaphidae, Faroninae). Mémoires de la Société Royale Belge d'Entomologie, 34: 12-82.

CRAIG, P. C. 1970. The behavior and distribution of the intertidal sand beetle, *Thinopinus pictus* (Coleoptera: Staphylinidae). Ecology, 51: 1012-1017.

CROWSON, R. A. 1938. The metendosternite in Coleoptera: a comparative study. Transactions of the Royal Entomological Society of London, 87: 397-415, pls. 113.

CROWSON, R. A. 1944. Further studies on the metendosternite in Coleoptera. Transactions of the Royal Entomological Society of London, 94: 273-310, pls. 110.

CROWSON, R. A. 1955. The natural classification of the families of Coleoptera. Nathaniel Lloyd, London. (seen as 1967 reprint, E. W. Classey, Hampton. 187 pp.)

CROWSON, R. A. 1960. The phylogeny of Coleoptera. Annual Review of Entomology, 5: 111-134.

CROWSON, R. A. 1982. Observations on *Phyllodrepoidea crenata* (Gravenhorst) (Col., Staphylinidae). Entomologist's Monthly Magazine, 118: 125-126.

CROWSON, R. A. and ELLIS, I. 1969. Observations on *Dendrophagus crenatus* (Cucujidae) and some comparisons with piestine Staphylinidae. Entomologist's Monthly Magazine, 104: 161-169 [1968].

CUCCODORO, G. 1995. Two new species of *Megarthrus* (Coleoptera, Staphylinidae, Proteininae) and a note on "water loading" behaviour. Journal of Zoology (London), 236: 253-264.

CUCCODORO, G. and LÖBL, I. 1996. Revision of the rovebeetles of the genus *Megarthrus* of America north of Mexico (Coleoptera, Staphylinidae, Proteininae). Mitteilungen der Münchner Entomologischen Gesellschaft, 86: 145-188.

CZARNIAWSKI, W. and STANIEC, B. 1997. Notes on the structure of defensive organ openings of some Oxytelinae (Coleoptera: Staphylinidae). Polskie Pismo Entomologiczne, 66: 33-43.

DAJOZ, R. and CAUSSANEL, C. 1969. Morphologie et biologie d'un coléoptère prédateur: *Creophilus maxillosus* (L.) (Staphylinidae). Cahiers des Naturalistes, Bulletin des Naturalistes Parisiens (N.S.), 24: 65-102 [1968].

DANOFFBURG, J. A. 1994. Evolving under myrmecophily: a cladistic revision of the symphilic beetle tribe Sceptobiini (Coleoptera: Staphylinidae: Aleocharinae). Systematic Entomology, 19: 25-45.

DANOFFBURG, J. A. 1996. An ethogram of the antguest beetle tribe Sceptobiini (Coleoptera: Staphylinidae; Formicidae). Sociobiology, 27: 287-328.

DE MARZO, L. 1985. Organi erettili e ghiandole tegumentali specializzate nelle larve di *Batrisodes oculatus* Aubé: studio morfoistologico (Coleoptera, Pselaphidae). Entomologica [Bari], 20: 125-145.

DE MARZO, L. 1986a. Osservazioni etologiche sulle larve de *Batrisodes oculatus* Aubè (Coleoptera Pselaphidae). Frustula Entomologica (N.S.), 78: 501-506.

DE MARZO, L. 1986b. Morfologia delle uova in alcuni Pselafidi (Coleoptera). Entomologica [Bari], 21: 155-163.

DE MARZO, L. 1987. Morfologia delle larva matura in alcuni Pselafidi (Coleoptera). Entomologica [Bari], 22: 97-135.

DE MARZO, L. 1988a. Comportamento predatorio nelle larve di *Pselaphus heisei* Herbst (Coleoptera, Pselaphidae). Atti del XV Congresso Nazionale Italiano di Entomologia, 1988: 817-824.

DE MARZO, L. 1988b. Costruzione della loggia pupale e del bozzolo in alcuni Pselafidi (Coleoptera). Entomologica [Bari], 23: 161-169.

DE MARZO, L. 1989. Note di anatomia sui genitali interni in alcuni Pselafidi (Coleoptera). Entomologica [Bari], 24: 99-105.

DE MARZO, L. and VIT, S. 1982. Note sulla presenza di *Batrisodes oculatus* Aubè (Coleoptera, Pselaphidae) in una grotta di Puglia. Entomologica [Bari], 17: 149-162.

DE MARZO, L. and VOVLAS, N. 1989. Strutture ed organi esoscheletrici in *Batrisodes oculatus* (Aubé) (Coleoptera, Pselaphidae). Entomologica [Bari], 24: 113-125.

DETTNER, K. 1993. Defensive secretions and exocrine glands in freeliving staphylinid beetles their bearing on phylogeny (Coleoptera: Staphylinidae). Biochemical Systematics and Ecology, 21: 143-162.

DETTNER, K. and REISSENWEBER, F. 1991. The defensive secretion of Omaliinae and Proteininae (Coleoptera: Staphylinidae): its chemistry, biological and taxonomic significance. Biochemical Systematics and Ecology, 19: 291-303.

DEYRUP, M. A. and GARA, R. I. 1978. Insects associated with Scolytidae (Coleoptera) in western Washington. PanPacific Entomologist, 54: 270-282.

DOWNIE, N. M. and ARNETT, R. H., JR. 1996. The Beetles of Northeastern North America. Vol. 1. Sandhill Crane Press, Gainesville, Florida. 880 pp.

DRUGMAND, D. 1989. Distribution et phénologie des Paederinae de Belgique (Coleoptera, Staphylinidae). Documents de Travail de l'Institut Royal des Sciences Naturelles de Belgique, 55: 1-52.

DRUGMAND, D. 1990. Le genre *Haploglossa* Kraatz, 1856 en Europe occidentale. Approches systématique, écologique et ontogénique (Coleoptera: Staphylinidae: Aleocharinae). Annales de la Société Entomologique de France (N.S.), 26: 83-91.

DRUGMAND, D. 1992. Pariade de *Eusphalerum luteum* (Marsham, 1802) (Coleoptera, Staphylinidae, Omaliinae). Bulletin et Annales de la Société Royale Belge d'Entomologie, 128: 35-36.

ELIAS, S. A. 1994. Quaternary Insects and their Environments. Smithsonian Institution Press, Washington and London. xiii + 284 pp.

ELLIOTT, P., KING, P. E. and FORDY, M. R. 1983. Observations on staphylinid beetles living on rocky shores. Journal of Natural History, 17: 575-581.

ENGELMANN, M. D. 1956. Observations on the feeding behavior of several pselaphid beetles. Entomological News, 67: 19-24.

EVANS, M. E. G. 1965. A comparative account of the feeding methods of the beetles *Nebria brevicollis* (F.) (Carabidae) and *Philonthus decorus* (Grav.) (Staphylinidae). Transactions of the Royal Society of Edinburgh, 66: 91-109 [1963-64].

FALL, H. C. 1912. Four new myrmecophilous Coleoptera. Psyche, 19: 9-12.

FALL, H. C. 1927. The North American species of *Rybaxis*. Psyche, 34: 218-226.

FAUVEL, A. 1878. Les staphylinides de l'Amérique du Nord. Bulletin de la Société Linnéenne de Normandie, (3) 2: 167-266.

FENYES, A. 191821. Coleoptera. Fam. Staphylinidae, subfam. Aleocharinae. In: Wytsman, P. (ed.), Genera Insectorum, Fasc. 173ac. M. Nijhoff, The Hague and L. Desmet-Verteneuil, Brussels. 453 pp., 7 pls.

FENYES, A. 1921. New genera and species of Aleocharinae with a polytomic synopsis of the tribes. Bulletin of the Museum of Comparative Zoology, 65: 17-36.

FORBES, W. T. M. 1922. The wing-venation of the Coleoptera. Annals of the Entomological Society of America, 15: 328-345, pls. 29-35.

FORBES, W. T. M. 1926. The wing folding patterns of the Coleoptera. Journal of the New York Entomological Society, 34: 42-115, pls. 7-18.

FRANIA, H. 1986a. Status of *Eustilicus* Sharp, *Trochoderus* Sharp, *Deroderus* Sharp, and *Stilocharis* Sharp (Coleoptera: Staphylinidae: Paederinae: Paederini) with implications for classification of the Medonina and Stilicina. Canadian Journal of Zoology, 64: 467-480.

FRANIA, H. 1986b. Larvae of *Eustilicus* Sharp, *Rugilus* Leach, *Deroderus* Sharp, *Stilocharis* Sharp, and *Medon* Stephens (Coleoptera: Staphylinidae: Paederinae: Paederini), and their phylogenetic significance. Canadian Journal of Zoology, 64: 2543-2557.

FRANIA, H. 1991. Displacement of one taxon by another as the cause of certain ecological shifts in *Eustilicus* Sharp (Coleoptera: Staphylinidae): a test of the evidence. Proceedings of the Entomological Society of Washington, 93: 437-448.

FRANIA, H. 1994. Phylogeny and biogeography of *Eustilicus* Sharp (Coleoptera: Staphylinidae: Paederinae): reevaluation based upon a new and relict species from the Edwards Plateau of Texas. Canadian Entomologist, 126: 493-501.

FRANK, J. H. 1975. A revision of the New World species of the genus *Erichsonius* Fauvel (Coleoptera: Staphylinidae). Coleopterists Bulletin, 29: 177-203.

FRANK, J. H. 1976. *Platystethus spiculus* Er. (Staphylinidae) in Florida. Coleopterists Bulletin, 30: 157-158.

FRANK, J. H. 1977. *Myrmecosaurus ferrugineus*, an Argentinian beetle from fire ant nests in the United States. Florida Entomologist, 60: 31-36.

FRANK, J. H. 1979a. Larval morphology and the classification of *Staphylinus (sensu lato)* (Col., Staphylinidae). Entomologist's Monthly Magazine, 114: 235-238.

FRANK, J. H. 1979b. A new species of *Proteinus* Latreille (Coleoptera: Staphylinidae) from Florida. Florida Entomologist, 62: 329-340.

FRANK, J. H. 1981a. A new *Erichsonius* species from Arizona with discussion on phylogeny within the genus (Coleoptera: Staphylinidae). Coleopterists Bulletin, 35: 97-106.

FRANK, J. H. 1981b. A revision of the New World species of the genus *Neobisnius* Ganglbauer (Coleoptera: Staphylinidae: Staphylininae). Occasional Papers of the Florida State Collection of Arthropods, 1: pp. ivii + 160.

FRANK, J. H. 1982. The parasites of the Staphylinidae (Coleoptera). Agricultural Experiment Stations, Institute of Food and Agricultural Sciences, University of Florida, Bulletin 824 (technical), vii + 118 pp.

FRANK, J. H. 1983. New records of Philonthini from the circumCaribbean region (Coleoptera: Staphylinidae). Florida Entomologist, 66: 473-481.

FRANK, J. H. 1986. A preliminary checklist of the Staphylinidae (Coleoptera) of Florida. Florida Entomologist, 69: 363-382.

FRANK, J. H. 1988. *Paederus*, sensu lato (Coleoptera: Staphylinidae): an index and review of the taxa. Insecta Mundi, 2: 97-159.

FRANK, J. H. 1991. Staphylinidae (Staphylinoidea), pp. 341-352. In: Stehr, F. W. (ed.), Immature Insects, Vol. 2. Kendall/Hunt Publishing Co., Dubuque, Iowa.

FRANK, J. H., CARLYSLE, T. C. and REY, J. R. 1986. Biogeography of the seashore Staphylinidae *Cafius bistriatus* and *C. rufifrons* (Insecta: Coleoptera). Florida Scientist, 49: 148-161.

FRANK, J. H. and CURTIS, G. A. 1979. Trend lines and the number of species of Staphylinidae. Coleopterists Bulletin, 33: 133-149.

FRANK, J. H., HABECK, D. H. and PECK, S. B. 1987. The distribution of *Brathinus nitidus* (Coleoptera: Staphylinidae) and a new key to the North American species. Coleopterists Bulletin, 41: 137-140.

FRANK, J. H. and KANAMITSU, K. 1987. *Paederus, sensu lato* (Coleoptera: Staphylinidae): Natural history and medical importance. Journal of Medical Entomology, 24: 155-191.

FRANK, J. H., KLIMASZEWSKI, J. and PECK, S. B. 1989. *Leptagria perexilis* and *Myllaena audax* (Coleoptera: Staphylinidae) in Florida. Florida Entomologist, 72: 717-718.

FRANK, J. H. and THOMAS, M. C. 1981a. Myrmedoniini (Coleoptera, Staphylinidae, Aleocharinae) associated with army ants (Hymenoptera, Formicidae, Ecitoninae) in Florida. Florida Entomologist, 64: 138-146.

FRANK, J. H. and THOMAS, M. C. 1981b. *Oxytelus incisus* Motschulsky and *O. pennsylvanicus* Erichson (Coleoptera, Staphylinidae, Oxytelinae) in Florida. Florida Entomologist, 64: 399-405.

FRANK, J. H. and THOMAS, M. C. 1984a. *Cubanotyphlus largo*, a new species of Leptotyphlinae (Coleoptera: Staphylinidae) from Florida. Canadian Entomologist, 116: 1411-1417.

FRANK, J. H. and THOMAS, M. C. 1984b. *Heterota plumbea* and *Coenonica puncticollis* in Florida (Coleoptera: Staphylinidae). Florida Entomologist, 67: 409-417.

FRANK, J. H. and THOMAS, M. C. 1984c. Cocoonspinning and the defensive function of the median gland in larvae of Aleocharinae (Coleoptera, Staphylinidae): a review. Quaestiones Entomologicae, 20: 7-23.

FRANK, J. H. and THOMAS, M. C. 1997. A new species of *Charoxus* (Coleoptera: Staphylinidae) from native figs (*Ficus* spp.) in Florida. Journal of the New York Entomological Society, 104: 70-78 [1996].

FRASER, N. C., GRIMALDI, D. A., OLSEN, P. E. and AXSMITH, B. 1996. A Triassic Lagerstätte from eastern North America. Nature, 380: 615-619.

FULDNER, D. 1960. Beiträge zur Morphologie und Biologie von *Aleochara bilineata* Gyll. und *A. bipustulata* L. (Coleoptera: Staphylinidae). Zeitschrift für Morphologie und Ökologie der Tiere, 49: 312-386.

FURNISS, M. M. 1995. Biology of *Dendroctonus punctatus* (Coleoptera: Scolytidae). Annals of the Entomological Society of America, 88: 173-182.

GÉNIER, F. 1989. A revision of the genus *Hoplandria* Kraatz of America North of Mexico (Coleoptera: Staphylinidae, Aleocharinae). Memoirs of the Entomological Society of Canada, 150: 1-59.

GÉNIER, F. and KLIMASZEWSKI, J. 1986. Review of the types of the genus *Platandria* Casey with a key to the species (Coleoptera: Staphylinidae: Aleocharinae). Coleopterists Bulletin, 40: 201-216.

GOODRICH, M. A. and HANLEY, R. S. 1995. Biology, development and larval characters of *Oxyporus major* (Coleoptera: Staphylinidae). Entomological News, 106: 161-168.

GREENE, G. L. 1996. Rearing techniques for *Creophilus maxillosus* (Coleoptera: Staphylinidae), a predator of fly larvae in cattle feedlots. Journal of Economic Entomology, 89: 848-851.

GRIFFITHS, C. L. and GRIFFITHS, R. J. 1983. Biology and distribution of the littoral rove beetle *Psamathobledius punctatissimus* (LeConte) (Coleoptera: Staphylinidae). Hydrobiologia, 101: 203-214.

GRIGARICK, A. A. and SCHUSTER, R. O. 1962a. Notes on Tychini from western North America (Coleoptera: Pselaphidae). PanPacific Entomologist, 38: 169-177.

GRIGARICK, A. A. and SCHUSTER, R. O. 1962b. Species of the genus *Batrisodes* from the Pacific slope of western North America (Coleoptera: Pselaphidae). PanPacific Entomologist, 38: 199-213.

GRIGARICK, A. A. and SCHUSTER, R. O. 1967. *Reichenbachia* found in the United States west of the continental divide (Coleoptera: Pselaphidae). University of California Publications in Entomology, 47: 1-45.

GRIGARICK, A. A. and SCHUSTER, R. O. 1968. A revision of the genus *Cupila* Casey (Coleoptera: Pselaphidae). PanPacific Entomologist, 44: 38-44.

GRIGARICK, A. A. and SCHUSTER, R. O. 1971. A revision of *Actium* Casey and *Actiastes* Casey (Coleoptera: Pselaphidae). University of California Publications in Entomology, 67: 1-56.

GRIGARICK, A. A. and SCHUSTER, R. O. 1976. A revision of the genus *Oropodes* Casey (Coleoptera: Pselaphidae). PanPacific Entomologist, 52: 97-109.

GRIGARICK, A. A. and SCHUSTER, R. O. 1980. Discrimination of genera of Euplectini of North and Central America (Coleoptera: Pselaphidae). University of California Publications in Entomology, 87: vi + 56 pp., 79 pls.

GUSAROV, V. I. 1995. Two new species of *Pycnoglypta* Thomson (Col., Staphylinidae) from North America and from the Far East of Russia. Entomologist's Monthly Magazine, 131: 229-242.

HAGHEBAERT, G. 1991. A review of the *Diglotta* of the world (Coleoptera, Staphylinidae, Aleocharinae). Bulletin et Annales de la Société Royale Belge d'Entomologie, 127: 223-234.

HAMILTON, J. 1888. Catalogue of the myrmophilous Coleoptera with bibliography and notes. Canadian Entomologist, 20: 161-166.

HAMMOND, P. M. 1970. Some problematic Motschulsky species of Staphylinidae. Entomologist's Monthly Magazine, 106: 67-70.

HAMMOND, P. M. 1971. The systematic position of *Brathinus* LeConte and *Camioleum* Lewis (Coleoptera: Staphylinidae). Journal of Entomology (B), 40: 63-70.

HAMMOND, P. M. 1975. The phylogeny of a remarkable new genus and species of gymnusine staphylinid (Coleoptera) from the Auckland Islands. Journal of Entomology (B), 44: 153-173.

HAMMOND, P. M. 1976. A review of the genus *Anotylus* C. G. Thomson (Coleoptera: Staphylinidae). Bulletin of the British Museum (Natural History), Entomology, 33: 139-187, pls. 13.

HAMMOND, P. M. 1979. Wingfolding mechanisms of beetles, with special reference to investigations of Adephagan phylogeny (Coleoptera), pp. 113-180. In: Erwin, T. L., Ball, G. E. and Whitehead, D. R. (eds.), Carabid Beetles: Their Evolution, Natural History, and Classification. W. Junk, Dordrecht.

HANLEY, R. S. 1996. Immature stages of *Scaphisoma castaneum* Motschulsky (Coleoptera: Staphylinidae: Scaphidiinae), with observations on natural history, fungal hosts and development. Proceedings of the Entomological Society of Washington, 98: 36-43.

HANLEY, R. S. and GOODRICH, M. A. 1994. Natural history, development and immature stages of *Oxyporus stygicus* Say (Coleoptera: Staphylinidae: Oxyporinae). Coleopterists Bulletin, 48: 213-225.

HANLEY, R. S. and GOODRICH, M. A. 1995a. The Oxyporinae (Coleoptera: Staphylinidae) of Illinois. Journal of the Kansas Entomological Society, 67: 394-414 [1994].

HANLEY, R. S. and GOODRICH, M. A. 1995b. Review of mycophagy, host relationships and behavior in the New World Oxyporinae (Coleoptera: Staphylinidae). Coleopterists Bulletin, 49: 267-280.

HANSEN, M. 1996. Katalog over Danmarks biller; Catalogue of the Coleoptera of Denmark. Entomologiske Meddelelser, 64: 12-31 [in Danish and English].

HANSEN, M. 1997a. Phylogeny and classification of the staphyliniform beetle families (Coleoptera). Biologiske Skrifter, Det Kongelige Danske Videnskabernes Selskab, 48: 13-39.

HANSEN, M. 1997b. Evolutionary trends in "staphyliniform" beetles (Coleoptera). Steenstrupia, 23: 43-86.

HATCH, M. H. 1938. Report on the Coleoptera collected by Dr. Victor B. Scheffer on the Aleutian Islands in 1937. PanPacific Entomologist, 14: 145-149.

HATCH, M. H. 1957. The beetles of the Pacific Northwest. Part II. Staphyliniformia. University of Washington Publications in Biology 16: ix + 384 pp.

HAYASHI, N. 1986. [Larvae, pp. 202-218, pls. 11-13] [in Japanese]. In: Morimoto, K. and Hayashi, N. (eds.), The Coleoptera of Japan in Color, Vol. 1. Hoikusha Publishing Co., Osaka, Japan.

HAYASHI, Y. 1993. Studies on the Asian Staphylininae, I (Coleoptera, Staphylinidae). Elytra, 21: 281-301.

HAYASHI, Y. 1997. Studies on the Asian Staphylininae (Coleoptera, Staphylinidae), III. The characteristics of the Xanthopygini. Elytra, 25: 475-402.

HEBERDEY, R. F. 1943. Ein Wasserläufer unter den Käfern (*Paederus rubrothoracicus* Gze.). Zeitschrift für Morphologie und Ökologie der Tiere, 40: 361-376.

HERMAN, L. H., JR. 1965a. A revision of *Orus* Casey. I. Subgenus *Leucorus* Casey and a new subgenus (Coleoptera: Staphylinidae). Coleopterists Bulletin, 18: 112-121 [1964].

HERMAN, L. H., JR. 1965b. Revision of *Orus*. II. Subgenera *Orus*, *Pycnorus* and *Nivorus* (Coleoptera: Staphylinidae). Coleopterists Bulletin, 19: 73-90.

HERMAN, L. H., JR. 1970a. The ecology, phylogeny, and taxonomy of *Stilicolina* (Coleoptera, Staphylinidae, Paederinae). American Museum Novitates, 2412: 1-26.

HERMAN, L. H., JR. 1970b. Phylogeny and reclassification of the genera of the rovebeetle subfamily Oxytelinae of the world (Coleoptera, Staphylinidae). Bulletin of the American Museum of Natural History, 142: 343-454.

HERMAN, L. H., JR. 1972a. A revision of the rovebeetle genus *Charhyphus* (Coleoptera, Staphylinidae, Phloeocharinae). American Museum Novitates, 2496: 1-16.

HERMAN, L. H., JR. 1972b. Revision of *Bledius* and related genera. Part I. The *aequatorialis, mandibularis,* and *semiferrugineus* groups and two new genera (Coleoptera, Staphylinidae, Oxytelinae). Bulletin of the American Museum of Natural History, 149: 111-254.

HERMAN, L. H., JR. 1975. Revision and phylogeny of the monogeneric subfamily Pseudopsinae for the world (Staphylinidae, Coleoptera). Bulletin of the American Museum of Natural History, 155: 241-317.

HERMAN, L. H., JR. 1976. Revision of *Bledius* and related genera. Part II. The *armatus, basalis,* and *melanocephalus* groups (Coleoptera, Staphylinidae, Oxytelinae). Bulletin of the American Museum of Natural History, 157: 71-172.

HERMAN, L. H., JR. 1977. Revision and phylogeny of *Zalobius*, *Asemobius*, and *Nanobius*, new genus (Coleoptera, Staphylinidae, Piestinae). Bulletin of the American Museum of Natural History, 159: 45-86.

HERMAN, L. H., JR. 1981. Revision of the subtribe Dolicaonina of the New World, with discussions of phylogeny and the Old World genera (Staphylinidae, Paederinae). Bulletin of the American Museum of Natural History, 167: 327-520.

HERMAN, L. H., JR. 1983. Revision of *Bledius*. Part III. The *annularis* and *emarginatus* groups (Coleoptera, Staphylinidae, Oxytelinae). Bulletin of the American Museum of Natural History, 175: 11-45.

HERMAN, L. H., JR. 1986. Revision of *Bledius*. Part IV. Classification of species groups, phylogeny, natural history, and catalogue (Coleoptera, Staphylinidae, Oxytelinae). Bulletin of the American Museum of Natural History, 184: 13-67.

HICKS, E. A. 1959. Checklist and bibliography on the occurrence of insects in birds' nests. Iowa State College Press, Ames, Iowa. 681 pp.

HICKS, E. A. 1962. Checklist and bibliography on the occurrence of insects in birds' nests. Supplement I. Iowa State Journal of Science, 36: 233-348.

HICKS, E. A. 1971. Checklist and bibliography on the occurrence of insects in birds' nests. Supplement II. Iowa State Journal of Science, 46: 123-338.

HILL, W. B., AKRE, R. B. and HUBER, J. D. 1976. Structure of some epidermal glands in the myrmecophilous beetle *Adranes taylori* (Coleoptera: Pselaphidae). Journal of the Kansas Entomological Society, 49: 367-384.

HINTON, H. E. 1944. Some general remarks on subsocial beetles, with notes on the biology of the staphylinid, *Platystethus arenarius* (Fourcroy). Proceedings of the Royal Entomological Society of London (A), 19: 115-128.

HINTON, H. E. 1945. A Monograph of the Beetles associated with Stored Products. Vol. 1. British Museum (Natural History), London. viii + 443 pp.

HINTON, H. E. and STEPHENS, F. L. 1941. Notes on the food of *Micropeplus*, with a description of the pupa of *M. fulvus* Erichson (Coleoptera, Micropeplidae). Proceedings of the Royal Entomological Society of London (A), 16: 29-32.

HLAVAC, T. F. 1975. The prothorax of Coleoptera: (Except Bostrichiformia-Cucujiformia). Bulletin of the Museum of Comparative Zoology, 147: 137-183.

HOEBEKE, E. R. 1976. A revision of the genus *Xenodusa* (Staphylinidae, Aleocharinae) for North America. Sociobiology, 2: 108-143.

HOEBEKE, E. R. 1985. A revision of the rove beetle tribe Falagriini of America north of Mexico (Coleoptera: Staphylinidae: Aleocharinae). Journal of the New York Entomological Society, 93: 913-1018.

HOEBEKE, E. R. 1988a. A new species of rove beetle, *Autalia phricotrichosa* (Coleoptera: Staphylinidae: Aleocharinae), from Mexico, with a key to the New World species of *Autalia*. Coleopterists Bulletin, 42: 87-93.

HOEBEKE, E. R. 1988b. Review of the genus *Thamiaraea* Thomson in North America (Coleoptera: Staphylinidae: Aleocharinae) with description of a new species. Journal of the New York Entomological Society, 96: 16-25.

HOEBEKE, E. R. 1992. Taxonomy and distribution of the athetine genus *Lypoglossa* Fenyes (Coleoptera: Staphylinidae: Aleocharinae) in North America, with description of a new species. Journal of the New York Entomological Society, 100: 381-398.

HOEBEKE, E. R. 1995a. *Coprophilus striatulus* (Coleoptera: Staphlinidae): confirmation of establishment of a Palearctic oxyteline rove beetle in North America. Entomological News, 106: 15.

HOEBEKE, E. R. 1995b. Three Palearctic species of *Rugilus* Leach in North America (Coleotpera: Staphylinidae, Paederinae): redescriptions, new synonymy, and new records. Insecta Mundi, 9: 69-80.

HOLCOMB, M. 1978. Observations on the morphology and behavior of *Pinophilus parcus* LeConte (Coleoptera, Staphylinidae, Pinophilinae). Folia Entomológica Mexicana, 3940: 209-210.

HÖLLDOBLER, B. and WILSON, E. O. 1990. The ants. Belknap Press of Harvard University Press, Cambridge, Massachusetts. xii + 732 pp.

HORION, A. 1963. Faunistik der mitteleuropäischen Käfer, Vol. 9: Staphylinidae, 1. Teil, Micropeplinae bis Euaesthetinae. Aug. Feyel, ÜberlingenBodensee. xii + 412 pp.

HORN, G. H. 1871. Remarks on the species of the genus *Isomalus* Er. of the United States. Transactions of the American Entomological Society, 3: 297-299.

HORN, G. H. 1879. Synopsis of the species of *Staphylinus* and the more closely allied genera inhabiting the United States. Transactions of the American Entomological Society, 7: 185-200.

HORN, G. H. 1885. A study of the species of *Cryptobium* of North America. Transactions of the American Entomological Society, 12: 85-106, pls. 1-2.

HOWARD, R. W. 1976. Observations on behavioral interactions between *Trichopsenius frosti* Seevers (Coleoptera: Staphylinida) and *Recticulitermes flavipes* (Kollar) (Isoptera: Rhinotermitidae). Sociobiology, 2: 189-192.

HOWARD, R. W. 1978. Proctodeal feeding by termitophilous Staphylinidae associated with *Reticulitermes virginicus* (Banks). Science, 201: 541-543.

HOWARD, R. W. 1979. Mating behavior of *Trichopsenius frosti*: physogastric *Reticulitermes flavipes* queens serve as sexual aggregation centers. Annals of the Entomological Society of America, 72: 127-129.

HOWARD, R. W. 1980. Trailfollowing by termitophiles. Annals of the Entomological Society of America, 73: 36-38.

HOWARD, R. W. and KISTNER, D. H. 1978. The eggs of *Trichopsenius depressus* and *T. frosti* (Coleoptera: Staphylinidae, Trichopseniinae) with a comparison to those of their host termites *Reticulitermes virginicus* and *R. flavipes* (Isoptera: Rhinotermitidae, Heterotermitinae). Sociobiology, 3: 99-106.

HOWARD, R. W., MCDANIEL, C. A. and BLOMQUIST, G. J. 1982. Chemical mimicry as an integrating mechanism for three termitophiles associated with *Reticulitermes virginicus* (Banks). Psyche, 89: 157-167.

HU, G. Y. and FRANK, J. H. 1995a. Biology of *Neohypnus pusillus* (Sachse) (Coleoptera: Staphylinidae) and its predation on immature horn flies in the laboratory. Coleopterists Bulletin, 49: 43-52.

HU, G. Y. and FRANK, J. H. 1995b. New distributional records for *Platystethus* (Coleoptera: Staphylinidae: Oxytelinae) with notes on the biology of *P. americanus*. Florida Entomologist, 78: 137-144.

HU, G. Y. and FRANK, J. H. 1995c. Structural comparison of the chorion surface of five *Philonthus* species (Coleoptera: Staphylinidae). Proceedings of the Entomological Society of Washington, 97: 582-589.

HU, G. Y. and FRANK, J. H. 1997. Predation on the Horn Fly (Diptera: Muscidae) by five species of *Philonthus* (Coleoptera: Staphylinidae). Environmental Entomology, 26: 1240-1246.

HUTH, A. and DETTNER, K. 1990. Defense chemicals from abdominal glands of 13 rove beetle species of subtribe Staphylinina (oleoptera: Staphylinidae, Staphylininae). Journal of Chemical Ecology, 16: 2691-2711.

ICZN. 1959. Opinion 546. Designation under the plenary powers of a type species in harmony with accustomed usage for the genus "*Staphylinus*" Linnaeus, 1758 (Class Insecta, Order Coleoptera). Opinions and Declarations Rendered by the International Commission on Zoological Nomenclature, 20: 141-151.

ICZN. 1961a. Opinion 599. *Bolitochara* Mannerheim, 1831 (Insecta, Coleoptera): designation of a typespecies under the plenary powers. Bulletin of Zoological Nomenclature, 18: 238-240.

ICZN. 1961b. Opinion 600. *Ischnopoda* Stephens, 1835, and *Tachyusa* Erichson, 1837 (Insecta, Coleoptera); designations of typespecies under the plenary powers. Bulletin of Zoological Nomenclature, 18: 241-243.

ICZN. 1969. Opinion 876. *Proteinus* Latreille, 1796 (Insecta, Coleoptera): designation of a typespecies under the plenary powers. Bulletin of Zoological Nomenclature, 26: 14-15.

ICZN. 1982. Opinion 1221. *Baeocera* Erichson, 1845 (Insecta, Coleoptera): designation of type species. Bulletin of Zoological Nomenclature, 39: 175-177.

ICZN. 1983. Opinion 1250. *Gyrohypnus* Samouelle, 1819, ex Leach MS, *Xantholinus* Dejean, 1821, ex Dahl, and *Othius* Stephens, 1829, ex Leach MS (Insecta, Coleoptera): type species designated for these genera. Bulletin of Zoological Nomenclature, 40: 85-87.

ICZN. 1985. International Code of Zoological Nomenclature, Third Edition, adopted by the XX General Assembly of the International Union of Biological Sciences. International Trust for Zoological Nomenclature, London. xx + 338 pp.

ICZN. 1993. Opinion 1722. *Acrolocha* Thomson, 1858 (Insecta, Coleoptera): conserved, and *Coprophilus* Latreille, 1829: *Staphylinus striatulus* Fabricius, 1792 designated as the type species. Bulletin of Zoological Nomenclature, 50: 164-165.

IRMLER, U. 1977. Revision der neotropischen *Platyprosopus* Arten (Coleoptera Staphylinidae) und Beschreibung der Larve von *Platyprosopus minor* Sharp. Studies on Neotropical Fauna and Environment, 12: 57-70.

IRMLER, U. 1979. Taxonomie, Verbreitung und Biologie der neotropischen Staphylinidengattung *Xenopygus* Bernh. (Coleoptera, Staphylinidae). Entomologische Blätter, 75: 30-36.

IRMLER, U. 1982. Descriptions of new Neotropical *Holotrochus* and a key to the species of the genus (Coleoptera: Staphylinidae). Coleopterists Bulletin, 35: 379-397 [1981].

IRMLER, U. 1985. Neue Arten der Gattungen *Aneucamptus* und *Thoracophorus* (Col., Staphylinidae) aus der Neotropis. Entomologische Blätter, 81: 41-58.

IRMLER, U. 1991. Neue Arten der Gattung *Clavilispinus* Blackwelder (Col., Staphylinidae) aus der Neotropis. Entomologische Blätter, 87: 85-91.

ISRAELSON, G. 1971. On the coleopterous fauna of the subterranean tunnel systems of small mammals, with particular reference to burrows of voles in Finland. Notulae Entomologicae, 51: 113-123.

JACKSON, D. R. and MILSTREY, E. G. 1989. The fauna of gopher tortoise burrows, pp. 86-98. In: Diemer, J. E. *et al.* (eds.), Gopher Tortoise Relocation Symposium proceedings. State of Florida Game and Fresh Water Fish Commission, Nongame Wildlife Program Technical Report No. 5, 109 pp.

JACOBSON, H. R. and KISTNER, D. H. 1992. Cladistic study, taxonomic restructuring, and revision of the myrmecophilous tribe Crematoxenini with comments on its evolution and host relationships (Coleoptera: Staphylinidae; Hymenoptera: Formicidae). Sociobiology, 20: 91-201.

JACOBSON, H. R. and KISTNER, D. H. 1998. A redescription of the myrmecophilous genus *Tetradonia* and a description of a new, closely related, freeliving genus, *Tetradonella* (Coleoptera: Staphylinidae). Sociobiology, 31: 151-279.

JACOBSON, H. R., KISTNER, D. H. and ABDELGALIL, F. A. 1987. A redescription of the myrmecophilous genera *Probeyeria*, *Beyeria* and the description of a closely related new genus from Arizona (Coleoptera: Staphylinidae). Sociobiology, 13: 307-338.

JACOBSON, H. R., KISTNER, D. H. and PASTEELS, J. M. 1986. Generic revision, phylogenetic classification, and phylogeny of the termitophilous tribe Corotocini (Coleoptera: Staphylinidae). Sociobiology, 12: 12-45.

JAMES, G. J., MOORE, I. LEGNER, E. F. 1971. The larval and pupal stages of four species of *Cafius* (Coleoptera: Staphylinidae) with notes on their biology and ecology. Transactions of the San Diego Society of Natural History, 16: 279-289.

JEANNEL, R. 1959. Révision des Psélaphides de l'Afrique intertropicale. Annales du Musée Royal du Congo Belge, Tervuren (Série 81: Sciences Zoologiques), 75: 17-42.

JEANNEL, R. 1962. Les Psélaphides de la Paléantarctide occidentale, pp. 295-479. In: Delamare Deboutteville, C. and Rapoport, E. (eds.), Biologie de l'Amérique Australe, Vol. 1, Études sur la Faune du Sol. Centre National de la Recherche Scientifique, Paris.

JEANNEL, R. 1964. Révision des Psélaphides de l'Afrique Australe, pp. 23-217. In: The Humicolous Fauna of South Africa: Pselaphidae and Catopidae (Coleoptera) (N. Leleup Expedition 19601961). Transvaal Museum Memoir No. 15, Transvaal Museum, Pretoria.

JEFSON, M., MEINWALD, J., NOWICKI, S., HICKS, K. and EISNER, T. 1983. Chemical defense of a rove beetle (*Creophilus maxillosus*). Journal of Chemical Ecology, 9: 159-180.

JENKINS, M. F. 1960. On the method by which *Stenus* and *Dianous* (Coleoptera: Staphylinidae) return to the banks of a pool. Transactions of the Royal Entomological Society of London, 112: 1-14.

KASULE, F. K. 1966. The subfamilies of the larvae of Staphylinidae (Coleoptera) with keys to the larvae of the British genera of Steninae and Proteininae. Transactions of the Royal Entomological Society of London, 118: 261-283.

KASULE, F. K. 1968. The larval characters of some subfamilies of British Staphylinidae (Coleoptera) with keys to the known genera. Transactions of the Royal Entomological Society of London, 120: 115-138.

KASULE, F. K. 1970. The larvae of Paederinae and Stapylininae (Coleoptera: Staphylinidae) with keys to the known British genera. Transactions of the Royal Entomological Society of London, 122: 49-80.

KAUPP, A. 1997. Beitrag zur Larvalmorphologie der Palpenkäfer (Coleoptera, Pselaphidae). Entomologische Blätter, 93: 57-68.

KEEN, J. H. 1895. List of Coleoptera collected at Massett, Queen Charlotte Islands, B. C. Canadian Entomologist, 27: 165-172, 217-220.

KELLNER, R. L. L. and DETTNER, K. 1992. Comparative morphology of abdominal glands in Paederinae (Coleoptera: Staphylinidae). International Journal of Insect Morphology and Embryology, 21: 117-135.

KEMNER, N. A. 1925. Zur Kenntnis der Staphylinidenlarven, I. Die Larven der Tribus Proteinini und Diglossini. Entomologisk Tidskrift, 46: 61-77, pls. 1-2.

KINCAID, T. 1961. The ecology and morphology of *Thinobius frizzelli* Hatch, an intertidal beetle. The Calliostoma Co., Seattle, Washington. 15 pp. 6 pls.

KING, P. E., FORDY, M. and ALKHALIFA, M. S. 1979. Observations on the intertidal *Micralymma marinum* (Col., Staphylinidae). Entomologist's Monthly Magazine, 115: 133-136.

KISTNER, D. H. 1955. A new species of *Paederopsis* from Arizona (Coleoptera; Staphylinidae). Chicago Academy of Sciences, Natural History Miscellanea, No. 142, 4 pp.

KISTNER, D. H. 1969. Revision of the termitophilous subfamily Trichopseniinae (Coleoptera, Staphylinidae), I. The genus *Schizelythron* Kemner. Entomological News, 80: 44-53.

KISTNER, D. H. 1970a. New termitophilous Staphylinidae (Coleoptera) from Hodotermitidae (Isoptera) nests. Journal of the New York Entomological Society, 78: 2-16.

KISTNER, D. H. 1970b. New termitophiles associated with *Longipeditermes longipes* (Haviland) II. The genera *Compactopedia*, *Emersonilla*, *Hirsitilla*, and *Limulodilla*. Journal of the New York Entomological Society, 78: 17-32.

KISTNER, D. H. 1976. Revision and reclassification of the genus *Goniusa* Casey with a larval description and ant host records (Coleoptera: Staphylinidae). Sociobiology, 2: 83-95.

KISTNER, D. H. 1979. Social and evolutionary significance of social insect symbionts, pp. 339-413. In: Hermann, H. R. (ed.), Social Insects, Vol. 1. Academic Press, New York.

KISTNER, D. H. 1981. The reclassification of the genus *Charoxus* Sharp with the description of new species (Coleoptera: Staphylinidae). Journal of the Kansas Entomological Society, 54: 587-598.

KISTNER, D. H. 1982. The social insects' bestiary, pp. 12-44. In: Hermann, H. R. (ed.), Social Insects, Vol. 3. Academic Press, New York.

KISTNER, D. H. 1998. New species of termitophilous Trichopseniinae (Coleoptera: Staphylinidae) found with *Mastotermes darwiniensis* in Australia and in Dominican Amber. Sociobiology, 31: 51-64, 71-76.

KISTNER, D. H., ASHE, J. S. and JACOBSON, H. R. 1996. New species of the myrmecophilous genera *Microdonia* and *Ecitoxenidia* from Mexico (Coleoptera, Staphylinidae) with a review of previously described species. Sociobiology, 27: 47-78.

KISTNER, D. H. and GUT, L. J. 1977. The genus *Philotermes* Kraatz with a description of a new species (Coleoptera: Staphylinidae). Sociobiology, 2: 273-282.

KISTNER, D. H. and HOWARD, R. W. 1980. First report of larval and pupal Trichopseniinae: external morphology, taxonomy and behavior (Coleoptera: Staphylinidae). Sociobiology 5: 3-20.

KLIMASZEWSKI, J. 1979. A revision of the Gymnusini and Deinopsini of the world (Coleoptera: Staphylinidae, Aleocharinae). Agriculture Canada Monograph 25: 11-69.

KLIMASZEWSKI, J. 1980. Two new species of Deinopsini from the Afrotropical and Nearctic regions, with notes on two other species of this tribe (Coleoptera, Staphylinidae). Polskie Pismo Entomologiczne 50: 109-120.

KLIMASZEWSKI, J. 1982a. A revision of the Gymnusini and Deinopsini of the world (Coleoptera: Staphylinidae). Supplementum 2. Canadian Entomologist 114: 317-335.

KLIMASZEWSKI, J. 1982b. Studies of Myllaenini (Coleoptera: Staphylinidae, Aleocharinae) 1. Systematics, phylogeny, and zoogeography of Nearctic *Myllaena* Erichson. Canadian Entomologist 114: 181-242.

KLIMASZEWSKI, J. 1984. A revision of the genus *Aleochara* Gravenhorst of America north of Mexico (Coleoptera: Staphylinidae, Aleocharinae). Memoirs of the Entomological Society of Canada 129: 1-211.

KLIMASZEWSKI, J. and ASHE, J. S. 1991. The oxypodine genus *Haploglossa* Kraatz in North America (Coleoptera: Staphylinidae: Aleocharinae). Giornale Italiano di Entomologia 5: 409-416.

KLIMASZEWSKI, J. and JANSEN, R. 1993. Systematics, biology and distribution of *Aleochara* Gravenhorst from southern Africa. Part I: subgenus *Xenochara* Mulsant and Rey (Coleoptera: Staphylinidae). Annals of the Transvaal Museum 36: 53-107.

KLIMASZEWSKI, J. and PECK, S. B. 1986. A review of the cavernicolous Staphylinidae (Coleoptera) of eastern North America: Part I. Aleocharinae. Quaestiones Entomologicae 22: 51-113.

KLINGER, R. 1978. Ein artspezifisches Sexualpheromon auf der Cuticula der Weibchen von *Eusphalerum minutum* L. Naturwissenschaften, 65: 5-97.

KLINGER, R. 1980. The defensive gland of Omaliinae (Coleoptera: Staphylinidae), II. Comparative gross morphology and revision of the classification within the genus *Eusphalerum*. Entomologica Scandinavica, 11: 454-457.

KLINGER, R. 1983. Eusphaleren, blütenbesuchende Staphyliniden. 1) Zur Biologie der Käfer (Col., Staphylinidae). Deutsche Entomologische Zeitschrift (N.F.), 30: 37-44.

KLINGER, R. and MASCHWITZ, U. 1977. The defensive gland of Omaliinae (Coleoptera: Staphylinidae), I. Gross morphology of the gland and identification of the scent of *Eusphalerum longipenne* Erichson. Journal of Chemical Ecology, 3: 401-410.

KOCH, K. 1989. Die Käfer Mitteleuropas. E1. Ökologie, Band 1. Goecke and Evers, Krefeld. 440 pp.

KOMPANTSEV, A. V. and POTOTSKAYA, V. A. 1987. [New data on larvae of scaphidiid beetles (Coleoptera, Scaphidiidae), pp. 87-100. In: Ecology and morphology of insects living in fungus substrates] [in Russian]. Nauka, Moscow.

KRAMER, S. 1955. Notes and observations on the biology and rearing of *Creophilus maxillosus* (L.) (Coleoptera, Staphylinidae). Annals of the Entomological Society of America, 48: 375-380.

KUKALOVÁ-PECK, J. and LAWRENCE, J. F. 1993. Evolution of the hind wing in Coleoptera. Canadian Entomologist, 125: 181-258.

LAWRENCE, J. F. and NEWTON, A. F., JR. 1980. Coleoptera associated with the fruiting bodies of slime molds (Myxomycetes). Coleopterists Bulletin, 34: 129-143.

LAWRENCE, J. F. and NEWTON, A. F., JR. 1982. Evolution and classification of beetles. Annual Review of Ecology and Systematics, 13: 261-290.

LAWRENCE, J. F. and NEWTON, A. F., JR. 1995. Families and subfamilies of Coleoptera (with selected genera, notes, references and data on family-group names), pp. 779-1006. In: Pakaluk, J. and Slipinski, S. A. (eds.), Biology, phylogeny and classification of Coleoptera: Papers celebrating the 80th birthday of Roy A. Crowson. Muzeum i Instytut Zoologii PAN, Warszawa.

LECONTE, J. L. 1877. On certain genera of Staphylinidae Oxytelini, Piestidae, and Micropeplidae, as represented in the fauna of the United States. Transactions of the American Entomological Society, 6: 213-252.

LECOQ, J.C. 1994. Un nouveau genre et une nouvelle espèce d'Osoriinae de Sierra Leone: *Pseudespeson rossii* (Coleoptera, Staphylinidae). In: Ricerche biologiche in Sierra Leone (Parte IV). Accademia Nazionale dei Lincei, 267: 299-306.

LEECH, H. B. 1939. Three new species of Nearctic rove beetles from the Pacific Coast (Coleoptera, Staphylinidae). Canadian Entomologist, 71: 258-261.

LEGNER, E. F. and MOORE, I. 1978. The larva of *Platystethus spiculus* Erichson (Coleoptera: Staphylinidae) and its occurrence in bovine feces in irrigated pastures. Psyche, 84: 158-164 [1977].

LENG, C. W. 1920. Catalogue of the Coleoptera of America, North of Mexico. Cosmos Press, Cambridge, Massachusetts. x + 470 pp.

LESAGE, L. 1977. Stades immatures de staphylins, I. La larve et la nymphe de *Staphylinus cinnamopterus* Gravenhorst (Coleoptera: Staphylinidae). Naturaliste Canadien, 104: 235-238.

LESAGE, L. 1984. The larva of *Anaquedius vernix* (Coleoptera: Staphylinidae). Canadian Entomologist, 116: 189-196.

LESCHEN, R. A. B. 1988. The natural history and immatures of *Scaphisoma impunctatum* (Coleoptera: Scaphidiidae). Entomological News, 99: 225-232.

LESCHEN, R. A. B. 1991. Behavioral observations on the myrmecophile *Fustiger knausii* (Coleoptera: Pselaphidae: Clavigerinae) with a discussion of grasping notches in myrmecophiles. Entomological News, 102: 215-2-2-2.

LESCHEN, R. A. B. 1993. Evolutionary patterns of feeding in selected Staphylinoidea (Coleoptera): shifts among food textures, pp. 59-104. In: Schaefer, C. W. and Leschen, R. A. B. (eds.), Functional Morphology of Insect Feeding. Thomas Say Publications in Entomology, Entomological Society of America, Lanham, Maryland.

LESCHEN, R. A. B. 1994. Retreat-building by larval Scaphidiinae (Staphylinidae). Mola, 4: 3-5.

LESCHEN, R. A. B. and ALLEN, R. T. 1988. Immature stages, life histories and feeding mechanisms of three *Oxyporus* spp. (Coleoptera: Staphylinidae: Oxyporinae). Coleopterists Bulletin, 42: 321-3-3-3.

LESCHEN, R. A. B. and LÖBL, I. 1995. Phylogeny of Scaphidiinae with redefinition of tribal and generic limits (Coleoptera: Staphylinidae). Revue Suisse de Zoologie, 102: 425-474.

LESCHEN, R. A. B., LÖBL, I. and STEPHAN, K. 1990. Review of the Ozark Highland *Scaphisoma* (Coleoptera: Scaphidiidae). Coleopterists Bulletin, 44: 274-294.

LIPKOW, E. 1966. Biologischökologische Untersuchungen über *Tachyporus* Arten und *Tachinus rufipes* (Col., Staphyl.). Pedobiologia, 6: 140-177.

LÖBL, I. 1987. Contribution to the knowledge of the genus *Caryoscapha* Ganglbauer (Coleoptera: Scaphidiidae). Coleopterists Bulletin, 41: 385-391.

LÖBL, I. 1997. Catalogue of the Scaphidiinae (Coleoptera: Staphylinidae). Instrumenta Biodiversitatis I, Muséum d'Histoire Naturelle, Genève. xii + 190 pp.

LÖBL, I. and BURCKHARDT, D. 1988. *Cerapeplus* gen. n. and the classification of micropeplids (Coleoptera: Micropeplidae). Systematic Entomology, 13: 57-66.

LÖBL, I. and CALAME, F. G. 1996. Taxonomy and phylogeny of the Dasycerinae (Coleoptera: Staphylinidae). Journal of Natural History, 30: 247-291.

LÖBL, I. and KURBATOV, S. A. 1995. New *Tychobythinus* (Coleoptera, Staphylinidae, Pselaphinae) from east and southeast Asia. Mitteilungen der Schweizerischen Entomologischen Gesellschaft, 68: 297-304.

LÖBL, I. and STEPHAN, K. 1993. A review of the species of *Baeocera* Erichson (Coleoptera, Staphylinidae, Scaphidiinae) of America north of Mexico. Revue Suisse de Zoologie, 100: 675-733.

LOHSE, G. A. 1964. 23. Familie: Staphylinidae. In: Freude, H., Harde, K. W. and Lohse, G. A. (eds.), Die Käfer Mitteleuropas, Vol. 4, Staphylinidae I (Micropeplinae bis Tachyporinae), 264 pp. Goecke and Evers, Krefeld.

LOHSE, G. A. 1974. 23. Familie: Staphylinidae, pp. 1-304. In: Freude, H., Harde, K. W. and Lohse, G. A. (eds.), Die Käfer Mitteleuropas, Vol. 5, Staphylinidae II (Hypocyphtinae und Aleocharinae), Pselaphidae, 381 pp. Goecke and Evers, Krefeld.

LOHSE, G. A. 1991. Corrections to revision of Arctic Aleocharinae of North America (Coleoptera: Staphylinidae). Coleopterists Bulletin, 45: 20.

LOHSE, G. A., KLIMASZEWSKI, J. and SMETANA, A. 1990. Revision of arctic Aleocharinae of North America (Coleoptera: Staphylinidae). Coleopterists Bulletin, 44: 121-202.

LOHSE, G. A. and SMETANA, A. 1985. Revision of the types of species of Oxypodini and Athetini (sensu Seevers) described by Mannerheim and Mäklin from North America (Coleoptera: Staphylinidae). Coleopterists Bulletin, 39: 281-300.

LOHSE, G. A. and SMETANA, A. 1988a. A new genus and a new species of Aleocharinae from the Appalachian Mountains of North Carolina (Coleoptera: Staphylinidae). Coleopterists Bulletin, 42: 265268.

LOHSE, G. A. and SMETANA, A. 1988b. Four new species of *Geostiba* Thomson from the Appalachian Mountains of North Carolina, with a key to North American species and synonymic notes (Coleoptera: Staphylinidae: Aleocharinae). Coleopterists Bulletin, 42: 269278.

MALKIN, B. 1958. Protective coloration in *Thinopinus pictus*. Coleopterists Bulletin, 12: 20.

MANK, E. W. 1934. New species of *Orobanus*. PanPacific Entomologist, 10: 121-124.

MANK, H. G. 1923. The biology of the Staphylinidae. Annals of the Entomological Society of America, 16: 220-237.

MARSH, G. A. and SCHUSTER, R. O. 1962. A revision of the genus *Sonoma* Casey (Coleoptera: Pselaphidae). Coleopterists Bulletin, 16: 33-56.

MARUCCI, P. E. and CLANCY, D. W. 1952. The biology and laboratory culture of *Thyreocephalus albertisi* (Fauvel) in Hawaii. Proceedings of the Hawaiian Entomological Society, 14: 525-532.

MATTHEWS, J. V., JR. 1970. Two new species of *Micropeplus* from the Pliocene of western Alaska with remarks on the evolution of Micropeplinae (Coleoptera: Staphylinidae). Canadian Journal of Zoology, 48: 779-788.

MICKEY, G. H. and PARK, O. 1956. Cytological observations on Pselaphid beetles. Ohio Journal of Science, 56: 155-164.

MOORE, I. 1964a. The Staphylinidae of the marine mud flats of southern California and northwestern Baja California (Coleoptera). Transactions of the San Diego Society of Natural History 13: 269-284.

MOORE, I. 1964b. *Manda*, a genus new to the Nearctic region (Coleoptera: Staphylinidae). Coleopterists Bulletin, 18: 57-58.

MOORE, I. 1964c. The larva of *Hadrotes crassus* (Mannerheim) (Coleoptera: Staphylinidae). Transactions of the San Diego Society of Natural History, 13: 309-312.

MOORE, I. 1964d. A new key to the subfamilies of the Nearctic Staphylinidae and notes on their classification. Coleopterists Bulletin, 18: 83-91.

MOORE, I. 1964e. Notes on *Lobrathium subseriatum* Leconte with a description of the larva (Coleoptera: Staphylinidae). Bulletin of the National Speleological Society, 26: 119-120.

MOORE, I. 1965. *Echletus*, a genus of staphylinid beetle new to the Nearctic region (Coleoptera: Staphylinidae). PanPacific Entomologist, 41: 44-46.

MOORE, I. 1966. Notes on the Nearctic Anthophagini with a key to the genera (Coleoptera: Staphylinidae). Coleopterists Bulletin, 20: 47-56.

MOORE, I. 1973. A note on *Dacnochilus fresnoensis* (Coleoptera: Staphylinidae). PanPacific Entomologist, 49: 42.

MOORE, I. 1974. Notes on *Vicelva* Moore and Legner with new synonymy (Coleoptera: Staphylinidae). Coleopterists Bulletin, 28: 214.

MOORE, I. 1975a. The distribution of *Siagonium* (Coleoptera: Staphylinidae) in North America. Journal of the Kansas Entomological Society, 48: 96-100.

MOORE, I. 1975b. The Devil's Coach Horse and the escargots. Newsletter of the Michigan Entomological Society, 20 (4): 1, 8.

MOORE, I. 1976. *Giulianium campbelli*, a new genus and species of marine beetle from California (Coleoptera: Staphylinidae). PanPacific Entomologist, 52: 56-59.

MOORE, I. 1977. The larva of *Philonthus nudus* (Sharp), a seashore species from Washington (Coleoptera: Staphylinidae). Proceedings of the Entomological Society of Washington, 79: 405-408.

MOORE, I., HLAVAC, J. and FROMMER, S. 1976. Instant relaxing of insects. Coleopterists Bulletin, 30: 99-100.

MOORE, I. and LEGNER, E. F. 1971. A review of the Nearctic species of *Platystethus* (Coleoptera: Staphylinidae). PanPacific Entomologist, 47: 260-264.

MOORE, I. and LEGNER, E. F. 1972a. A new alpine species of *Unamis* from California (Coleoptera: Staphylinidae). Coleopterists Bulletin, 26: 21-22.

MOORE, I. and LEGNER, E. F. 1972b. A new species of *Microedus* from the Sierra Nevada Mountains (Coleoptera: Staphylinidae). Coleopterists Bulletin, 26: 75-77.

MOORE, I. and LEGNER, E. F. 1973. The genera of the subfamilies Phloeocharinae and Olisthaerinae of America north of Mexico with description of a new genus and new species from Washington (Coleoptera: Staphylinidae). Canadian Entomologist, 105: 35-41.

MOORE, I. and LEGNER, E. F. 1974a. The larva and pupa of *Carpelimus debilis* Casey (Coleoptera: Staphylinidae). Psyche, 80: 289-294 [1973].

MOORE, I. and LEGNER, E. F. 1974b. The genera of the Lispininae of America north of Mexico (Coleoptera: Staphylinidae). Coleopterists Bulletin, 28: 77-84.

MOORE, I. and LEGNER, E. F. 1974c. The genera of the Osoriinae of America north of Mexico (Coleoptera: Staphylinidae). Coleopterists Bulletin, 28: 115-119.

MOORE, I. and LEGNER, E. F. 1974d. A catalogue of the taxonomy, biology and ecology of the developmental stages of the Staphylinidae (Coleoptera) of America north of Mexico. Journal of the Kansas Entomological Society, 47: 469-478.

MOORE, I. and LEGNER, E. F. 1974e. Bibliography (1758 to 1972) to the Staphylinidae of America North of Mexico (Coleoptera). Hilgardia, 42: 511-547.

MOORE, I. and LEGNER, E. F. 1974f. Keys to the genera of the Staphylinidae of America North of Mexico exclusive of the Aleocharinae (Coleoptera: Staphylinidae). Hilgardia, 42: 548-563.

MOORE, I. and LEGNER, E. F. 1975a. A catalogue of the Staphylinidae of America North of Mexico (Coleoptera). University of California Division of Agricultural Sciences Special Publication, 3015: 1-514.

MOORE, I. and LEGNER, E. F. 1975b. A study of *Bryothinusa* (Coleoptera: Staphylinidae), comparing a tabular and a dichotomous key to the species. Bulletin of the Southern California Academy of Sciences, 74: 109-112.

MOORE, I. and LEGNER, E. F. 1976. Intertidal rove beetles (Coleoptera: Staphylinidae), pp. 521-551. In: Cheng, L. (ed.), Marine insects. North-Holland Publishing Co., Amsterdam.

MOORE, I. and LEGNER, E. F. 1979. An illustrated guide to the genera of the Staphylinidae of America North of Mexico exclusive of the Aleocharinae (Coleoptera). University of California Division of Agricultural Sciences Priced Publication, 4093: 1-332.

MOORE, I., LEGNER, E. F. and BADGLEY, M. E. 1975. Description of the developmental stages of the mite predator, *Oligota oviformis* Casey, with notes on the osmeterium and its glands (Coleoptera: Staphylinidae). Psyche, 82: 181-188.

MOORE, I. and ORTH, R. E. 1979a. Notes on *Bryothinusa* with description of the larva of *B. catalinae* Casey (Coleoptera: Staphylinidae). Psyche, 85: 183-189 [1978].

MOORE, I. and ORTH, R. E. 1979b. *Diglotta legneri*, a new seashore beetle from California (Coleoptera: Staphylinidae). Coleopterists Bulletin, 33: 337-340.

NAOMI, S.I. 1985. The phylogeny and higher classification of the Staphylinidae and their allied groups (Coleoptera, Staphylinoidea). Esakia, 23: 1-27.

NAOMI, S.I. 1987-90. Comparative morphology of the Staphylinidae and the allied groups (Coleoptera, Staphylinoidea), Parts I-XI. Kontyû [1987-88]/Japanese Journal of Entomology [1989-90], 55: 450-458, 666-675; 56: 67-77, 241-250, 506-513, 727-738; 57: 82-90, 269-277, 517-526, 720-733; 58: 16-23.

NAVARRETEHEREDIA, J. L. and NEWTON, A.F., JR. 1996. Staphylinidae (Coleoptera), pp. 369-380. In: Llorente, J. E., García, A. N. and González, E. (eds.), Biodiversidad, taxonomía y biogeografía de artrópodos de México: hacia una síntesis de su conocimiento. Instituto de Biología, Universidad Autónoma de México, México D.F., México.

NEWTON, A. F., JR. 1982a. Redefinition, revised phylogeny, and relationships of Pseudopsinae (Coleoptera, Staphylinidae). American Museum Novitates, 2743: 1-13.

NEWTON, A. F., JR. 1982b. A new genus and species of Oxytelinae from Australia, with a description of its larva, systematic position, and phylogenetic relationships (Coleoptera, Staphylinidae). American Museum Novitates, 2744: 1-24.

NEWTON, A. F., JR. 1984. Mycophagy in Staphylinoidea (Coleoptera), pp. 302-353. In: Wheeler, Q. and Blackwell, M. (eds.), Fungus Insect Relationships: Perspectives in Ecology and Evolution. Columbia University Press, New York.

NEWTON, A. F., JR. 1987. Four *Staphylinus (sensu lato)* species new to North America, with notes on other introduced species (Coleoptera: Staphylinidae). Coleopterists Bulletin, 41: 381-384.

NEWTON, A. F., JR. 1988. Fooled by flatness: subfamily shifts in subcortical Staphylinidae (Coleoptera). Coleopterists Bulletin, 42: 255-262.

NEWTON, A. F., JR. 1989. *Paralispinus mariae* Hatch (Coleoptera: Staphylinidae: Osoriinae) transferred to genus *Stictocranius* (Euaesthetinae). Coleopterists Bulletin, 43: 145-146.

NEWTON, A. F., JR. 1990a. Insecta: Coleoptera: Staphylinidae adults and larvae, pp. 1137-1174. In: Dindal, D. L. (ed.), Soil Biology Guide. John Wiley and Sons, New York.

NEWTON, A. F., JR. 1990b. Larvae of Staphyliniformia (Coleoptera): where do we stand? Coleopterists Bulletin, 44: 205-210.

NEWTON, A. F., JR. 1990c. *Myrmelibia*, a new genus of myrmecophile from Australia, with a generic review of Australian Osoriinae (Coleoptera: Staphylinidae). Invertebrate Taxonomy, 4: 81-94.

NEWTON, A. F., JR. 1991. Micropeplidae, Dasyceridae, Scaphidiidae, Pselaphidae (Staphylinoidea), pp. 334-339, 353-355. In: Stehr, F. W. (ed.), Immature Insects, Vol. 2. Kendall/Hunt Publishing Co., Dubuque, Iowa.

NEWTON, A. F., JR. and CHANDLER, D. S. 1989. World catalog of the genera of Pselaphidae (Coleoptera). Fieldiana: Zoology (N.S.), 53: 1-93.

NEWTON, A. F., JR. and THAYER, M. K. 1988. A critique on Naomi's phylogeny and higher classification of Staphylinidae and allies (Coleoptera). Entomologia Generalis, 14: 63-72.

NEWTON, A. F., JR. and THAYER, M. K. 1992. Current classification and family-group names in Staphyliniformia (Coleoptera). Fieldiana: Zoology (N.S.), 67: 1-92.

NEWTON, A. F., JR. and THAYER, M. K. 1995. Protopselaphinae new subfamily for *Protopselaphus* new genus from Malaysia, with a phylogenetic analysis and review of the Omaliine Group of Staphylinidae including Pselaphidae (Coleoptera), pp. 219-320. In: Pakaluk, J. and Slipinski, S. A. (eds.), Biology, phylogeny and classification of Coleoptera: Papers celebrating the 80th birthday of Roy A. Crowson. Muzeum i Instytut Zoologii PAN, Warszawa.

NIELD, C. E. 1976. Aspects of the biology of *Staphylinus olens* (Müller), Britain's largest staphylinid beetle. Ecological Entomology, 1: 117-126.

NOTMAN, H. 1920. Staphylinidae from Florida in the collection of the American Museum of Natural History, with descriptions of new genera and species. Bulletin of the American Museum of Natural History, 42: 693-732.

NOTMAN, H. 1925. A synoptic review of the beetles of the tribe Osoriini from the Western Hemisphere. Proceedings of the United States National Museum, 67 (11): 1-26.

NOTMAN, H. 1929. New species of *Palaminus* from the West Indies, together with a synoptic review of the genus. American Museum Novitates, 386: 1-17.

OHISHI, H. 1986. Consideration of internal morphology for the taxonomy of Pselaphidae, pp. 111-130. In: Papers on entomology presented to Prof. Takehiko Nakane in commemoration of his retirement. Japanese Society of Coleopterology, Tokyo.

ORTH, R. E. and MOORE, I. 1980. A revision of the species of *Cafius* Curtis from the west coast of North America with notes of the east coast species (Coleoptera: Staphylinidae). Transactions of the San Diego Society of Natural History, 19: 181-211.

ORTH, R. E., MOORE, I., FISHER, T. W. and LEGNER, E. F. 1975. A rove beetle, *Ocypus olens*, with potential for biological control of the brown garden snail, *Helix aspersa*, in California, including a key to the Nearctic species of *Ocypus*. Canadian Entomologist, 107: 1111-1116.

ORTH, R. E., MOORE, I., FISHER, T. W. and LEGNER, E. F. 1976. Biological notes on *Ocypus olens*, a predator of Brown Garden Snail, with descriptions of the larva and pupa (Coleoptera: Staphylinidae). Psyche, 82: 292-298 [1975].

ORTH, R. E., MOORE, I., FISHER, T. W. and LEGNER, E. F. 1978. Yearround survey of Staphylinidae of a sandy beach in southern California (Coleoptera). Wasmann Journal of Biology, 35: 169-195.

OTOTSKAYA, V. A. 1967. [Classification Key of the Larvae of Staphylinidae in the European Part of the USSR] [in Russian]. Nauka, Moscow. 120 pp.

OTOTSKAYA, V. A. 1976. [Morphology of the larva of *Trigonurus asiaticus* Reiche and interconnection of the tribe Trigonurini with other tribes of the subfamily Piestinae (Coleoptera, Staphylinidae), pp. 13-21. In: Mamaev, B. M. (ed.), Evolutionary Morphology of Insect Larvae.] [in Russian]. Nauka, Moscow.

PACE, R. 1986. Aleocharinae del Perù (Coleoptera, Staphylinidae). Redia, 69: 417-467.

PACE, R. 1989. Monografia del genere *Leptusa* Kraatz (Coleoptera, Staphylinidae). Memorie del Museo Civico di Storia Naturale di Verona (2) (A: Biologie), 8: 1-307.

PACE, R. 1996. Fauna d'Italia, XXXIV. Coleoptera: Staphylinidae: Leptotyphlinae. Edizioni Calderini, Bologna. viii + 328 pp.

PACE, R. 1997. Aleocharinae attere del North Carolina, Tennessee e Oregon (Coleoptera, Staphylinidae). Bollettino del Museo Regionale di Scienze Naturali, Torino, 15: 101-110.

PARK, O. 1932a. The myrmecocoles of *Lasius umbratus mixtus aphidicola* Walsh. Annals of the Entomological Society of America, 25: 77-88.

PARK, O. 1932b. The food of *Batrisodes globosus* (LeC.) (Coleop.: Pselaphidae). Journal of the New York Entomological Society, 40: 377-378.

PARK, O. 1933. The food and habits of *Tmesiphorus costalis* Lec. (Coleop.: Pselaphidae). Entomological News, 44: 149-151.

PARK, O. 1935a. Beetles associated with the moundbuilding ant *Formica ulkei* Emery. Psyche, 42: 216-231.

PARK, O. 1935b. Further records of beetles associated with ants (Coleop., Hymen.). Entomological News, 46: 212-215.

PARK, O. 1947. Observations on *Batrisodes* (Coleoptera: Pselaphidae), with particular reference to the American species east of the Rocky Mountains. Bulletin of the Chicago Academy of Sciences, 8: 45-132, pls. 1-11.

PARK, O. 1949. New species of Nearctic pselaphid beetles and a revision of the genus *Cedius*. Bulletin of the Chicago Academy of Sciences, 8: 315-343, pls. 1-8.

PARK, O. 1951. Cavernicolous pselaphid beetles of Alabama and Tennessee, with observations on the taxonomy of the family. Geological Survey of Alabama, Museum Paper 31, 107 pp.

PARK, O. 1952. A revisional study of Neotropical pselaphid beetles. Part One. Tribes Faronini, Pyxidicerini and Jubini. Chicago Academy of Sciences, Special Publication No. 9 (1): 1-49.

PARK, O. 1956. New or little known species of pselaphid beetles from southeastern United States. Journal of the Tennessee Academy of Science, 31: 54-100.

PARK, O. 1958. New or little known species of pselaphid beetles chiefly from southeastern United States. Journal of the Tennessee Academy of Science, 33: 39-74.

PARK, O. 1960. Cavernicolous pselaphid beetles of the United States. American Midland Naturalist, 64: 66-104.

PARK, O. 1964. Observations upon the behavior of myrmecophilous pselaphid beetles. Pedobiologia, 4: 129-137.

PARK, O. 1965. Revision of the genus *Batriasymmodes* (Coleoptera: Pselaphidae). Transactions of the American Microscopical Society, 84: 184-201.

PARK, O., AUERBACH, S. and CORLEY, G. 1950. The treehole habitat with emphasis on the pselaphid beetle fauna. Bulletin of the Chicago Academy of Sciences, 9: 19-57.

PARK, O., AUERBACH, S. and WILSON, M. 1949. Pselaphid beetles of an Illinois prairie: the fauna, and its relation to the prairie peninsula hypothesis. Bulletin of the Chicago Academy of Sciences, 8: 267-276, pl. 2.

PARK, O., AUERBACH, S. and WILSON, M. 1953. Pselaphid beetles of an Illinois prairie: the population. Ecological Monographs, 23: 1-15.

PARK, O. and SCHUSTER, R. O. 1955. A new subtribe of pselaphid beetles from California. Chicago Academy of Sciences, Natural History Miscellanea, No. 148, 6 pp.

PARK, O. and WAGNER, J. A. 1962. Family Pselaphidae, pp. 4-31, pls. 1-10. In: Hatch, M. H. (ed.), The Beetles of the Pacific Northwest, Part III: Pselaphidae and Diversicornia I. University of Washington Publications in Biology, 16: ix + 503 pp [1961].

PARK, O., WAGNER, J. A. and SANDERSON, M. W. 1976. Review of the pselaphid beetles of the West Indies (Coleoptera: Pselaphidae). Fieldiana: Zoology, 68: xi + 90 pp.

PASTEELS, J. M. and KISTNER, D. H. 1971. Revision of the termitophilous subfamily Trichopseniinae (Coleoptera: Staphylinidae). II. The remainder of the genera with a representational study of the gland systems and a discussion of their relationships. Miscellaneous Publications of the Entomological Society of America, 7: 351-399.

PAULIAN, R. 1941. Les premiers états des Staphylinoidea (Coleoptera). Étude de morphologie comparée. Mémoires du Muséum National d'Histoire Naturelle (N.S.), 15: 13-61, pls. 1-3.

PECK, S. B. 1975. A review of the distribution and habitats of North American *Brathinus* (Coleoptera: Staphylinidae: Omaliinae). Psyche, 82: 59-66.

PECK, S. B. and THOMAS, M. C. 1998. A distributional checklist of the beetles (Coleoptera) of Florida. Arthropods of Florida and Neighboring Land Areas, 16: i-viii + 1-180.

PESCHKE, K. and FULDNER, D. 1977. Review and new investigations of the life history of parasitoid Aleocharinae (Coleoptera; Staphylinidae). Zoologische Jahrbücher, Abteilung für Systematik, Ökologie und Geographie der Tiere, 104: 242-262 [in German].

PESCHKE, K. and METZLER, M. 1982. Defensive and pheromonal secretion of the tergal gland of *Aleochara curtula*, I. The chemical composition. Journal of Chemical Ecology, 8: 774-783.

PETERSON, A. 1960. Larvae of Insects: An Introduction to Nearctic Species. Part II: Coleoptera, Diptera, Neuroptera, Siphonaptera, Mecoptera, Trichoptera. Privately published, Columbus, Ohio. v + 416 pp.

PIERRE, F. 1944. Description de la larve de *Platystethus cornutus* Grav. et aperçu de sa biologie (Col. Staphylinidae). Revue Française d'Entomologie, 10: 170-174.

POOLE, R. W. and GENTILI, P. 1996. Nomina Insecta Nearctica: a check list of the insects of North America. Vol. 1: Coleoptera, Strepsiptera. EIS, Entomological Information Services, Rockville, MD. 827 pp.

POTOTSKAYA, V. A. 1967. [Classification Key of the Larvae of Staphylinidae in the European Part of the USSR] [in Russian]. Nauka, Moscow. 120 pp.

POTOTSKAYA, V. A. 1976. [Morphology of the larva of *Trigonurus asiaticus* Reiche and interconnection of the tribe Trigonurini with other tribes of the subfamily Piestinae (Coleoptera, Staphylinidae), pp. 13-21. In: Mamaev, B. M. (ed.), Evolutionary Morphology of Insect Larvae.] [in Russian]. Nauka, Moscow.

PUTHZ, V. 1971a. Revision der afrikanischen Steninenfauna und allgemeines über die Gattung *Stenus* Latreille (Coleoptera Staphylinidae). Annales du Musée Royal de l'Afrique Centrale, Tervuren (Série 81: Sciences Zoologiques), 187: 13-76.

PUTHZ, V. 1971b. Über die Gruppe des *Stenus cautus* Erichson (Coleoptera, Staphylinidae). Entomologisk Tidskrift, 92: 242-254.

PUTHZ, V. 1972. Das Subgenus "*Hemistenus*" (Col., Staphylinidae). Annales Entomologici Fennici, 38: 75-92.

PUTHZ, V. 1974. A new revision of the Nearctic *Edaphus* species and remarks on other North American Euaesthetinae (Coleoptera, Staphylinidae). Revue Suisse de Zoologie, 81: 911-932.

PUTHZ, V. 1981. Was ist *Dianous* Leach, 1819, was ist *Stenus* Latreille, 1796? Die Aporie des Stenologen und ihre taxonomischen Konsequenzen (Coleoptera, Staphylinidae). Entomologische Abhandlungen (Dresden), 44: 87-132.

PUTHZ, V. 1988. Revision der nearktischen Steninenfauna 3 (Coleoptera, Staphylinidae). Neue Arten und Unterarten aus Nordamerika. Entomologische Blätter, 84: 132-164.

PUTHZ, V. 1989. The male of *Stictocranius mariae* (Hatch) (Coleoptera, Staphylinidae). Entomologische Blätter, 85: 165-166.

QUEZADA, J. R., AMAYA, C. A. and HERMAN, L. H., JR. 1969. *Xanthopygus cognatus* Sharp (Coleoptera: Staphylinidae), an enemy of the Coconut Weevil, *Rhynchophorus palmarum* L. (Coleoptera: Curculionidae) in El Salvador. Journal of the New York Entomological Society, 77: 264-269.

RAFFRAY, A. 1911. Pselaphidae. In: Coleopterorum Catalogus, Pars 27. W. Junk, Berlin. 2-2-2 pp.

REICHLE, D. E. 1966. Some pselaphid beetles with boreal affinities and their distribution along the postglacial fringe. Systematic Zoology, 15: 330-344.

REICHLE, D. E. 1967. The temperature and humidity relations of some bog pselaphid beetles. Ecology, 48: 208-215.

REICHLE, D. E. 1969. Distribution and abundance of bog-inhabiting pselaphid beetles. Transactions of the Illinois State Academy of Science, 62: 233-264.

RICHARDS, L. J. 1983. Feeding and activity patterns of an intertidal beetle. Journal of Experimental Marine Biology and Ecology, 73: 213-224.

RYVKIN, A. B. 1985. [Beetles of the family Staphylinidae from the Jurassic of Transbaikalia, pp. 88-91. In: Rasnitsyn, A. P. (ed.), Jurassic insects of Siberia and Mongolia.] [in Russian]. Trudy Paleontologicheskogo Instituta Akademiya Nauk SSSR, 211: 1-192, 18 pls.

RYVKIN, A. B. 1990. [Family Staphylinidae Latreille, 1802, pp. 52-66. In: Rasnitsyn, A. P. (ed.), Late Mesozoic insects of Eastern Transbaikalia.] [in Russian]. Trudy Paleontologicheskogo Instituta Akademiya Nauk SSSR, 239: 1-2-2-2, 16 pls.

SAALAS, U. 1917. Die Fichtenkäfer Finnlands, studien über die Entwicklungsstadien, Lebenweise und geographische Verbreitung der an *Picea excelsa* Link lebenden Coleopteren nebst einer LarvenBestimmungstabelle. Annales Academiae Scientiarum Fennicae, (A) 8: 15-47.

SÁIZ, F. 1975. Une nouvelle espèce de Leptotyphlinae de Californie (U.S.A.) (Coleopt. Staphylinidae). Nouvelle Revue d'Entomologie, 5: 43-45.

SANDERSON, M. W. 1945. A new North American species of *Lithocharis* (Coleoptera: Staphylinidae). Proceedings of the Entomological Society of Washington, 47: 94-97.

SANDERSON, M. W. 1946. Nearctic *Stenus* of the *croceatus* group (Coleoptera, Staphylinidae). Annals of the Entomological Society of America, 39: 425-430.

SANDERSON, M. W. 1957. North American *Stenus* of the *advenus* complex including a new species from Illinois (Coleoptera: Staphylinidae). Transactions of the Illinois State Academy of Science, 50: 281-286.

SCHEERPELTZ, O. 1933. Staphylinidae VII: Supplementum I. In: Coleopterorum Catalogus, Pars 129: 990-1500. W. Junk, Berlin.

SCHEERPELTZ, O. 1934. Staphylinidae VIII: Supplementum II. In: Coleopterorum Catalogus, Pars 130: 1501-1881. W. Junk, Berlin.

SCHEERPELTZ, O. 1952. Revision der Gattung *Piestus* Gravh. (Coleoptera Staphylinidae). Revista Chilena de Entomología, 2: 281-305.

SCHEERPELTZ, O. 1972. Eine neue Art der Gattung *Trigonopselaphus* GemmingerHarold, nebst einer Dichotomik der jetzt zu dieser Gattung gehörigen Arten, Bemerkungen über die aus dieser Gattung auszuscheidenden Arten und neue, zum Teil auf diesen ausgeschiedenen Arten gegründete Gattungen (Col. Staphylinidae, Subfam. Staphylininae, Tribus Xanthopygini). Mitteilungen der Münchner Entomologischen Gesellschaft, 62: 31-48.

SCHEERPELTZ, O. 1974. Coleoptera: Staphylinidae, pp. 43-394. In: Hanström, B., Brinck, P. and Rudebeck, G. (eds.), South African Animal Life, Vol 15. Swedish Natural Science Research Council, Stockholm.

SCHILLHAMMER, H. 1997. Taxonomic revision of the Oriental species of *Gabrius* Stephens (Coleoptera: Staphylinidae). Monographs on Coleoptera, 1: 1-139.

SCHMID, R. 1988. Die Larven der Ameisenkäfer (Scydmaenidae, Staphylinoidea): Neu und Nachbeschreibung mit einem vorläufigen Bestimmungsschlüssel bis zur Gattung. Mitteilungen des Badischen Landesvereins für Naturkunde und Naturschutz (N.F.), 14: 643-660.

SCHMIDT, D. A. 1994a. Notes on the biology and a description of the egg, third instar larva and pupa of *Platydracus tomentosus* (Gravenhorst) (Coleoptera: Staphylinidae). Coleopterists Bulletin, 48: 310-318.

SCHMIDT, D. A. 1994b. Notes on the biology and a description of the egg, third instar larva and pupa of *Neobisnius sobrinus* (Coleoptera: Staphylinidae). Transactions of the Nebraska Academy of Sciences, 21: 55-61.

SCHMIDT, D. A. 1996. Description of the immatures of *Erichsonius alumnus* and *E. pusio* (Horn) (Coleoptera: Staphylinidae). Coleopterists Bulletin, 50: 205-215.

SCHUSTER, R. O. 1959. Notes on *Morius occidens* Casey with a description of the male (Coleoptera: Pselaphidae). PanPacific Entomologist, 37: 95-97.

SCHUSTER, R. O. and GRIGARICK, A. A. 1960. A revision of the genus *Oropus* Casey (Coleoptera: Pselaphidae). Pacific Insects, 2: 269-299.

SCHUSTER, R. O. and GRIGARICK, A. A. 1962. A revision of the genus *Rhexidius* Casey (Coleoptera: Pselaphidae). PanPacific Entomologist, 38: 1-14.

SCHUSTER, R. O. and MARSH, G. A. 1956. A revision of the genus *Pselaptrichus* Brendel (Coleoptera: Pselaphidae). University of California Publications in Entomology, 11: 117-158.

SCHUSTER, R. O. and MARSH, G. A. 1958a. A new genus of Tychini from California (Coleoptera: Pselaphidae). PanPacific Entomologist, 34: 125-137.

SCHUSTER, R. O. and MARSH, G. A. 1958b. A study of the North American genus *Megarafonus* Casey (Coleoptera: Pselaphidae). PanPacific Entomologist, 34: 187-194.

SCHUSTER, R. O. and MARSH, G. A. 1994. Construction of the egg chamber and protection of the eggs by female *Oxyporus japonicus* Sharp (Coleoptera, Staphylinidae, Oxyporinae). Japanese Journal of Entomology, 62: 803-809.

SHARP, D. 1883-1887. Fam. Staphylinidae, pp. 145-747, 775-802, pls. 5-19. In: Biologia CentraliAmericana, Insecta, Coleoptera, Vol. 1 (2). Taylor and Francis, London.

SHARP, D. and MUIR, F. 1912. The comparative anatomy of the male genital tube in Coleoptera. Transactions of the Entomological Society of London, 1912: 477-642, pls. 42-78.

SILVESTRI, F. 1945. Descrizione e biologia del coleottero stafilinide *Belonuchus formosus* Grav. introdotto in Italia per la lotta contro Ditteri Tripaneidi. Bollettino del Laboratorio di Entomologia Agraria di Portici, 5: 312-328.

SILVESTRI, F. 1946. Contribuzione alla conoscenza dei mirmecofili IIIIV. Bollettino del Laboratorio di Entomologia Agraria di Portici, 6: 52-69.

SMETANA, A. 1971. Revision of the tribe Quediini of America north of Mexico Coleoptera: Staphylinidae). Memoirs of the Entomological Society of Canada, 79: vi + 303 pp.

SMETANA, A. 1973. Revision of the tribe Quediini of America north of Mexico (Coleoptera: Staphylinidae). Supplementum 2. Canadian Entomologist, 105: 1421-1434.

SMETANA, A. 1976. Revision of the tribe Quediini of America north of Mexico (Coleoptera: Staphylinidae). Supplementum 3. Canadian Entomologist, 108: 169-184.

SMETANA, A. 1977. The nearctic genus *Beeria* Hatch. Taxonomy, distribution and ecology (Coleoptera: Staphylinidae). Entomologica Scandinavica, 8: 177-190.

SMETANA, A. 1978. Revision of the tribe Quediini of America north of Mexico (Coleoptera: Staphylinidae). Supplementum 4. Canadian Entomologist, 110: 815-840.

SMETANA, A. 1981a. *Ontholestes murinus* (Linné 1758) in North America (Coleoptera: Staphylinidae). Coleopterists Bulletin, 35: 125-126.

SMETANA, A. 1981b. Revision of the tribe Quediini of America north of Mexico (Coleoptera: Staphylinidae). Supplementum 5. Canadian Entomologist, 113: 631-644.

SMETANA, A. 1982. Revision of the subfamily Xantholininae of America north of Mexico (Coleoptera: Staphylinidae). Memoirs of the Entomological Society of Canada, 120: iv + 389 pp.

SMETANA, A. 1983. The status of the staphylinid genera *Derops* Sharp and *Rimulincola* Sanderson (Coleoptera). Entomologica Scandinavica, 14: 269-279.

SMETANA, A. 1984. Le "culte de l'édéage": Réflexions additionnelles, suivies d'une discussion sur le concept de la soustribu Heterothopsi Coiffait 1978 (Coleoptera, Staphylinidae). Nouvelle Revue d'Entomologie (N.S.), 1: 277-282.

SMETANA, A. 1985. Systematic position and review of *Deinopteroloma* Jansson, 1946, with descriptions of four new species (Coleoptera, Silphidae and Staphylinidae (Omaliinae)). Systematic Entomology, 10: 471-499.

SMETANA, A. 1986. *Chionotyphlus alaskenis* n.g., n. sp., a Tertiary relict from unglaciated interior Alaska (Coleoptera, Staphylinidae). Nouvelle Revue d'Entomologie (N.S.), 3: 171-187.

SMETANA, A. 1988. Revision of the subfamily Xantholininae of America north of Mexico (Coleoptera: Staphylinidae). Supplementum 1. Canadian Entomologist, 120: 525-558.

SMETANA, A. 1990. Revision of the subfamily Xantholininae of America north of Mexico (Coleoptera: Staphylinidae). Supplementum 2. Coleopterists Bulletin, 44: 83-87.

SMETANA, A. 1995. Rove beetles of the subtribe Philonthina of America north of Mexico (Coleoptera: Staphylinidae). Classification, phylogeny and taxonomic revision. Memoirs on Entomology, International, 3: x + 946 pp.

SMETANA, A. 1996. A review of the genus *Trigonodemus* LeConte, 1863, with descriptions of two new species from Asia (Coleoptera: Staphylinidae: Omaliinae). Coleoptera, Schwanfelder Coleopterologische Mitteilungen, 19: 1-18.

SMETANA, A. 1997. Book review: Poole, R. W. and Gentili, P. (editors) 1996, Nomina Insecta Nearctica; A check list of the insects of North America; Volume 1, Coleoptera, Strepsiptera. Koleopterologische Rundschau, 67: 265-269.

SMETANA, A. and CAMPBELL, J. M. 1980. A new genus and two new phloeocharinae species from the Pacific coast of North America (Coleoptera: Staphylinidae). Canadian Entomologist, 112: 1061-1069.

SMITH, R. L., LANZARO, G. C., WHEELER, J. E. and SNYDER, A. 1978. Ecology and behavior of *Osorius planifrons*. Annals of the Entomological Society of America, 71: 752-755.

SPAHR, U. 1981. Systematischer Katalog der Bernstein und Kopal-Käfer (Coleoptera). Stuttgarter Beiträge zur Naturkunde (B: Geologie und Paläontologie), 80: 107 pp.

STEEL, W. O. 1950. Notes on Staphylinidae, chiefly from New Zealand. (2) A new genus and three new species of Eleusiini. Transactions of the Royal Society of New Zealand, 78: 213-235.

STEEL, W. O. 1957. Notes on the Omaliinae. (8). The genus *Acroocha* Thomson. Entomologist's Monthly Magazine, 93: 157-164.

STEEL, W. O. 1958. Notes on the Omaliinae (Col., Staphylinidae). (9). The genus *Micralymma* Westwood. Entomologist's Monthly Magazine, 94: 138-142.

STEEL, W. O. 1959. [Review] The beetles of the Pacific Northwest. Part 2, Staphyliniformia, by M. H. Hatch. Entomologist's Gazette, 10: 82.

STEEL, W. O. 1962. Notes on the Omaliinae (Col., Staphylinidae). (11). The genus *Micromalium* Melichar with further notes on *Micralymma* Westwood. Entomologist's Monthly Magazine, 97: 237-239 [1961].

STEEL, W. O. 1966. A revision of the staphylinid subfamily Proteininae (Coleoptera) I. Transactions of the Royal Entomological Society of London, 118: 285-311.

STEEL, W. O. 1970. The larvae of the genera of Omaliinae with particular reference to the British fauna. Transactions of the Royal Entomological Society of London, 122: 1-47.

STEIDLE, J. L. M. and DETTNER, K. 1993. Chemistry and morphology of the tergal gland of freeliving adult Aleocharinae (Coleoptera: Staphylinidae) and its phylogenetic significance. Systematic Entomology, 18: 149-168.

STEIDLE, J. L. M. and DETTNER, K. 1995a. The chemistry of the abdominal gland secretion of six species of the rove beetle genus *Bledius*. Biochemical Systematics and Ecology, 23: 757-765.

STEIDLE, J. L. M. and DETTNER, K. 1995b. Abdominal gland secretion of *Bledius* rove beetles as an effective defence against predators. Entomologia Experimentalis et Applicata, 76: 211-216.

STEPHENSON, S. L., WHEELER, Q. D., MCHUGH, J. V. and FRAISSINET, P. R. 1994. New North American associations of Coleoptera with Myxomycetes. Journal of Natural History, 28: 921-936.

STICKNEY, F. S. 1923. The head capsule of Coleoptera. Illinois Biological Monographs, 8: 1-104, pls. 1-26.

STORK, N. E. 1980. A scanning electron microscope study of tarsal adhesive setae in the Coleoptera. Zoological Journal of the Linnean Society, 68: 173-306.

STRUBLE, G. H. 1930. The biology of certain Coleoptera associated with bark beetles in western yellow pine. University of California Publications in Entomology, 5: 105-134.

TANNER, V. M. 1927. A preliminary study of the genitalia of female Coleoptera. Transactions of the American Entomological Society, 53: 5-49, pls. 2-15.

THAYER, M. K. 1978. Redescription of *Xenicopoda* Moore and Legner (Coleoptera: Staphylinidae, Omaliinae) with supplementary notes. Psyche, 84: 142-149 [1977].

THAYER, M. K. 1985a. The larva of *Brathinus nitidus* LeConte and the systematic position of the genus (Coleoptera: Staphylinidae). Coleopterists Bulletin, 39: 174-184.

THAYER, M. K. 1985b. *Micralymma marinum* (Stroem) in North America: biological notes and new distributional records (Coleoptera: Staphylinidae). Psyche, 92: 49-55.

THAYER, M. K. 1987. Biology and phylogenetic relationships of *Neophonus bruchi*, an anomalous south Andean staphylinid (Coleoptera). Systematic Entomology, 12: 389-404.

THAYER, M. K. 1992. Discovery of sexual wing dimorphism in Staphylinidae (Coleoptera): *"Omalium" flavidum*, and a discussion of wing dimorphism in insects. Journal of the New York Entomological Society, 100: 540-573.

THAYER, M. K. 1993. *Osellia* Zanetti a junior synonym of *Orochares* Kraatz, with a checklist of *Orochares* species (Coleoptera: Staphylinidae: Omaliinae). Coleopterists Bulletin, 47: 285-287.

TIKHOMIROVA, A. L. 1968. [Jurassic staphylinid beetles of the Karatau (Coleoptera, Staphylinidae), pp. 139-154. In: Rohdendorf, B. B. (ed.), Jurassic Insects of the Karatau.] [in Russian]. Akademiya Nauk SSSR, Moscow.

TIKHOMIROVA, A. L. 1973. [Morphoecological features and phylogeny of the Staphylinidae (with a catalog of the fauna of the USSR)] [in Russian]. Akademiya Nauk SSSR, Moscow. 190 pp.

TIKHOMIROVA, A. L. and MELNIKOV, O. A. 1970. The late embryogenesis of Staphylinidae and nature of aleocharo and staphylinomorphoüs larvae. Zoologischer Anzeiger, 184: 76-87.

TOPP, W. 1975. Zur Larvalmorphologie der Athetae (Col., Staphylinidae). Stuttgarter Beiträge zur Naturkunde (A: Biologie) No. 268, 23 pp.

TOPP, W. 1978. Bestimmungstabelle für die Larven der Staphylinidae, pp. 304-334. In: Klausnitzer, B. (ed.), Ordnung Coleoptera (Larven). W. Junk, The Hague.

TSCHINKEL, W. R. 1992. Brood raiding in the fire ant, *Solenopsis invicta* (Hymenoptera: Formicidae): laboratory and field observations. Annals of the Entomological Society of America, 85: 638-646.

VAN DYKE, E. C. 1934. The North American species of *Trigonurus* Muls. et Rey (Coleoptera: Staphylinidae). Bulletin of the Brooklyn Entomological Society, 29: 177-183.

VERHOEFF, K. W. 1919. Studien über die Organisation und Biologie der Staphylinoidea. IV. Zur Kenntnis der Staphyliniden-Larven. V. Zur Kenntnis der Oxyteliden-Larven. Archiv für Naturgeschichte, 85 (A) (6): 1-111, pls. 1-4.

VIT, S. 1985. Quelque éléments de la faune Coléoptérologique résistant à la destruction de l'ancienne forêt de pantano de Policoro (Basilicata). Annai del Museo Civico di Storia Naturale di Genova, 85: 307-331.

VORIS, R. 1934. Biologic investigations on the Staphylinidae (Coleoptera). Transactions of the Academy of Science of St. Louis, 28: 231-261.

VORIS, R. 1939a. The immature stages of the genera *Ontholestes, Creophilus* and *Staphylinus*; Staphylinidae (Coleoptera). Annals of the Entomological Society of America, 32: 288-300, pls. 1-3.

VORIS, R. 1939b. Immature staphylinids of the genus *Quedius* (Coleoptera: Staphylinidae). Entomological News, 50: 151-155, 188-190, pl. 1.

WAGNER, J. A. 1975. Review of the genera *Euplectus*, *Pycnoplectus*, *Leptoplectus*, and *Acolonia* (Coleoptera: Pselaphidae) including Nearctic species north of Mexico. Entomologica Americana, 49: 125-207.

WATANABE, Y. 1990. A taxonomic study on the subfamily Omaliinae from Japan. Memoirs of the Tokyo University of Agriculture, 31: 55-391.

WATROUS, L. E. 1980. *Lathrobium (Tetartopeus)*: Natural history, phylogeny and revision of the Nearctic species (Coleoptera, Staphylinidae). Systematic Entomology, 5: 303-338.

WATROUS, L. E. 1981. Studies of *Lathrobium (Lobrathium)*: revision of the *grande* species group (Coleoptera: Staphylinidae). Annals of the Entomological Society of America, 74: 144-150.

WEBER, N. A. 1935. The biology of the thatching ant, *Formica rufa obscuripes* Forel, in North Dakota. Ecological Monographs, 5: 165-206.

WEINREICH, E. 1968. Über den Klebfangapparat der Imagines von *Stenus* Latr. (Coleopt., Staphylinidae) mit einem Beitrag zur Kenntnis der Jugendstadien dieser Gattung. Zeitschrift für Morphologie der Tiere, 62: 162-210.

WELCH, R. C. 1993. Ovariole development in Staphylinidae (Coleoptera). Invertebrate Reproduction and Development, 23: 225-234.

WHEELER, Q. D. and MCHUGH, J. V. 1994. A new southern Appalachian species, *Dasycerus bicolor* (Coleoptera: Staphylinidae: Dasycerinae), from declining endemic fir forests. Coleopterists Bulletin, 48: 265-271.

WHEELER, W. M. 1911. Notes on the myrmecophilous beetles of the genus *Xenodusa*, with a description of the larva of *X. cava* LeConte. Journal of the New York Entomological Society, 19: 163-169.

WICKHAM, H. F. 1898. On Coleoptera found with ants (fourth paper). Psyche, 8: 219-221, pl. 6.

WICKHAM, H. F. 1900. On Coleoptera found with ants. Fifth paper. Psyche, 9: 3-5.

WICKHAM, H. F. 1901. Two new blind beetles of the genus *Adranes*, from the Pacific Coast. Canadian Entomologist, 33: 25-28.

WILLIAMS, I. W. 1938. The comparative morphology of the mouthparts of the order Coleoptera treated from the standpoint of phylogeny. Journal of the New York Entomological Society, 46: 245-289.

WILSON, E. O. 1971. The insect societies. Belknap Press of Harvard University Press, Cambridge, Massachusetts. x + 548 pp.

WOLDA, H. and CHANDLER, D. S. 1996. Diversity and seasonality of tropical Pselaphidae and Anthicidae (Coleoptera). Proceedings of the Koninklijke Nederlandse Akademie van Wetenschappen, 99: 313-3-3-3.

WYATT, T. D. 1986. How a subsocial intertidal beetle, *Bledius spectabilis*, prevents flooding and anoxia in its burrow. Behavioral Ecology and Sociobiology 19: 323-331.

WYATT, T. D. 1993. Submarine beetles: when the tide comes in, they batten down the hatches. Natural History, 102 (7): 6, 89.

ZANETTI, A. 1987. Fauna d'Italia, XXV. Coleoptera: Staphylinidae, Omaliinae. Edizioni Calderini, Bologna. xii + 472 pp.

ZERCHE, L. 1987. Zur Synonymie von *Geodromicus* Redtenbacher (Coleoptera, Staphylinidae, Omaliinae). Beiträge zur Entomologie, 37: 137-138.

ZERCHE, L. 1990. Monographie der Paläarktischen Coryphiini (Coleoptera, Staphylinidae, Omaliinae). Akademie der Landwirtschaftswissenschaften der Deutschen Demokratischen Republik, Berlin. 413 pp.

ZERCHE, L. 1993. Monographie der paläarktischen Coryphiini(Coleoptera, Staphylinidae, Omaliinae). Supplementum 1. Beiträge zur Entomologie 43: 319-374.

# Taxonomic Index

(Note: **Taxon names of genus rank and above, both valid and synonyms, are indexed in the keys and classification sections.)**

## A

## C

## D

## E

## H

## M

## N

## O

## P

## Q

## R



## S

## U



## V



## X

## Y



## Z